edition 4

INTERMEDIATE ALGEBRA

James Streeter
Late Professor of Mathematics
Clackamas Community College
Donald Hutchison
Clackamas Community College
Barry Bergman
Clackamas Community College
Louis Hoelzle
Bucks County Community College

Boston Burr Ridge, IL Dubuque, IA Madison, WI New York
San Francisco St. Louis Bangkok Bogotá Caracas Kuala Lumpur
Lisbon London Madrid Mexico City Milan Montreal New Delhi
Santiago Seoul Singapore Sydney Taipei Toronto

McGraw-Hill Higher Education

A Division of The ***McGraw-Hill*** *Companies*

INTERMEDIATE ALGEBRA, FOURTH EDITION

Published by McGraw-Hill, a business unit of The McGraw-Hill Companies, Inc., 1221 Avenue of the Americas, New York, NY 10020.

This book is printed on acid-free paper.

1 2 3 4 5 6 7 8 9 0 VNH/VNH 0 9 8 7 6 5 4 3 2 1

ISBN 0–07–231691–8
ISBN 0–07–237671–6 (AIE)

Publisher: *William K. Barter*
Senior sponsoring editor: *David Dietz*
Developmental editor: *Burrston House*
Editorial coordinator: *Beatrice Wikander*
Marketing manager: *Mary K. Kittell*
Senior project manager: *Susan J. Brusch*
Media technology lead producer: *Steve Metz*
Production supervisor: *Enboge Chong*
Designer: *K. Wayne Harms*
Interior designer: *Sheilah Barrett*
Cover designer: *Jamie O'Neal*
Cover photo (jet): *Stone/Chad Slattery*
Senior photo research coordinator: *Carrie K. Burger*
Supplement producer: *Brenda A. Ernzen*
Compositor: *Interactive Composition Corporation*
Typeface: *10/12 Times Roman*
Printer: *Von Hoffmann Press, Inc.*

Chapter Openers: 1: © R. Lord/The Image Works; 2: © Hank Morgan/Photo Researchers, Inc.; 4: © Michelle Bridwell/PhotoEdit; 6: © Lawrence Migdale/Tony Stone Images; 7: © Susan Van Etten/Index Stock; 10: © Kathy McLaughlin/The Image Works; 11: © Chris Jones/The Stock Market

Interior: page 34: © PhotoDisc/Vol. 41; p. 37: © PhotoDisc, Vol. 54; p. 70: © PhotoDisc/Vol. 28; p. 89: © PhotoDisc/Vol. 25; p. 172: © PhotoDisc/Vol. 49; p. 186: © PhotoDisc/Vol. 55; p. 319: © PhotoDisc/Vol. 48; p. 327: © Corbis CD/Vol. 30; p. 394: © Corbis CD/Vol. 20; p. 470: © Corbis CD/Vol. 166; p. 518: © PhotoDisc/Vol.10; p. 550: © Corbis CD/Vol. 133; p. 620: © Corbis CD/Vol. 26; p. 634: © Corbis CD/Vol. 165; p. 737: © PhotoDisc/Vol. 41; p. 743: © PhotoDisc/Vol. 42; p. 755: © Corbis CD/Vol. 102; p. 809: © Corbis CD/Vol.174; p. 907: © Corbis CD/Vol. 122; p. 917: © Corbis CD/Vol. 126

www.mhhe.com

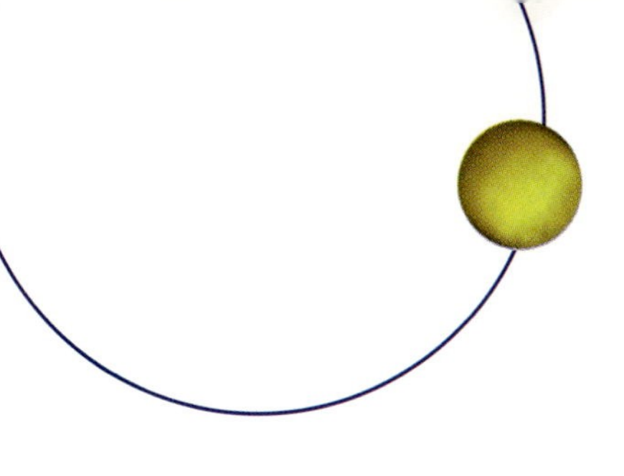

About the Authors

Donald Hutchison spent his first 10 years of teaching working with disadvantaged students. He taught in an inner-city elementary school and an inner-city high school. Don also worked with physically and mentally challenged children in state agencies in New York and Oregon.

In 1982, Don completed his graduate work in mathematics. He was then hired by Jim Streeter to teach at Clackamas Community College. Through Jim's tutelage, Don developed a fascination with the relationship between teaching and writing mathematics. He has come to believe that his best writing is a result of his classroom experience, and his best teaching is a result of the thinking involved in manuscript preparation.

Don is active in several professional organizations. He was a member of the ACM committee that undertook the writing of computer curriculum for the 2-year college. From 1989 to 1994 he was chair of the AMATYC Technology in Mathematics Education committee. He was president of ORMATYC from 1996 to 1998.

Barry Bergman has enjoyed teaching mathematics to a wide variety of students over the years. He began in the field of adult basic education, and moved into the teaching of high school mathematics in 1977. He taught at that level for 11 years, at which point he served for a year as a K-12 mathematics specialist for his county. This work allowed him to help promote the emerging NCTM standards in his region.

In 1990 Barry began the present portion of his career, having been hired to teach at Clackamas Community College. He maintains a strong interest in the appropriate use of technology in the learning of mathematics.

Throughout the past 23 years, Barry has played an active role in professional organizations. As a member of OCTM, he contributed several articles and activities to the group's journal. Recently, he served as an officer of ORMATYC for 4 years, and participated on an AMATYC committee to provide feedback and reactions to NCTM's revision of the standards.

Louis Hoelzle has been teaching at Bucks County Community College for 30 years. In 1989, Lou became chair of the Mathematics Department at Bucks County Community College. He has taught the entire range of courses from arithmetic to calculus, giving him an excellent view of the current and future developmental needs of students.

Over the past 36 years, Lou has also taught physics courses at 4-year colleges, which has enabled him to have the perspective of the practical applications of mathematics. Lou has always focused his writing on the student.

Lou is also active in several professional organizations. He has served on the Placement and Assessment committee and the Grants committee for AMATYC. He was president of PSMATYC from 1997 to 1999.

Dedication

This book is dedicated to the family I've lived with and loved for the past 30 years. Claudia, my beautiful and loving partner, and Trinh, Micol, Jake, and Christine. They are the essence of my being.

Donald Hutchison

This book is dedicated to my wife Marcia, who encouraged me to enter into this work, and who, along with our wonderful boys Joel and Adam, gave me support through the duration of the project.

Barry Bergman

This book is dedicated to my wife and children, who have shared my joys, sorrows, successes, and failures over 36 years. Rose, my friend, my inspiration, and part of my very being, and Beth, Ray, Amy, Oscar, Meg, Johanna, and Patrick, the joys of my life.

Louis Hoelzle

This series is dedicated to the memory of James Arthur Streeter, an artisan with words, a genius with numbers, and a virtuoso with pictures from 1940 until 1989.

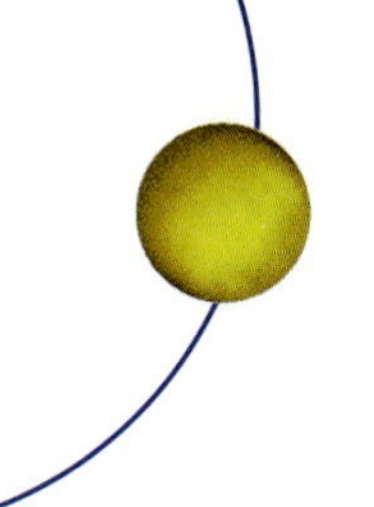

Table of Contents

Preface

STATEMENT OF PHILOSOPHY

We believe that the key to learning mathematics, at any level, is active participation. When students are active participants in the learning process, they have the opportunity to construct their own mathematical ideas and make connections to previously studied material. Such participation leads to understanding, success, and confidence. We developed this text with that philosophy in mind and integrated many features throughout the book to reflect that philosophy. The *Check Yourself* exercises are designed to keep the students active and involved with every page of exposition. The calculator references involve students actively in the development of mathematical ideas. Although we fully develop algebraic solutions to linear and quadratic equations, we also demonstrate solutions graphically. We have found that many students better understand the nature of a solution if they can visualize it. Almost every exercise set has application problems, challenging exercises, writing exercises, and/or collaborative exercises. Each exercise is designed to awaken interest and insight within students. Not all of the exercises will be appropriate for every student, but each one provides another opportunity for both the instructor and the student. Our hope is that every student who uses this text will be a better mathematical thinker as a result.

FEATURES OF THE FOURTH EDITION

In the previous three editions of *Intermediate Algebra,* we were fortunate to have great instructors to work with at Clackamas Community College and Bucks County Community College. Coupled with the excellent reviews provided by McGraw-Hill, we were able to put together a text that reflected the ideas of good teachers and the needs of a variety of students.

In this edition, McGraw-Hill, in conjunction with Burrston House, gave us even more. We were given the opportunity to participate in discussion groups with faculty in seven different states. These groups, a combination of users of this text and at least nine other texts, inspired us with tough questions, creative solutions, and helpful suggestions. It was a testament to the value of group work. Many of those discussion participants are mentioned in the acknowledgements that follow this preface. All of them deserve thanks.

For those who have used previous editions of the text, we will describe the specific changes from the third edition.

1. We have added Chapter One, a review of information related to the real number system.
2. Separate sections have been developed discussing the process of solving applications of intermediate algebra topics.
3. The chapter on systems of equations now follows the chapter on graphing linear equations.
4. The *ac* method of factoring and the trial-and-error method are treated in separate sections.
5. The material related to radical expressions has been expanded.
6. Graphical solutions to linear equations are now presented in a separate section that appears after algebraic methods have been developed.

Each of the features of this edition was scrutinized by the discussion groups. Almost all were modified in some way. Every supplement to this text was thoroughly discussed. More than ever, we are confident that the entire learning package is of value to students of intermediate algebra. We will describe each of the features of that package.

PEDAGOGICAL FEATURES

Application Areas

Each chapter opens with a real-world vignette that showcases an example of how mathematics is used in a wide variety of jobs and professions. Exercise sets for each section then feature one or more modeling/word problems that relate to the chapter-opening vignette. The application areas and the chapter each area appears in are:

Application Area	Chapter
Meteorology	1
Quality Science	2
Social Science	3
Medicine	4
Business	5
Civil Engineering	6
Political Science	7
Economics	8
Engineering	9
Pyrotechnics	10
Pharmacology	11

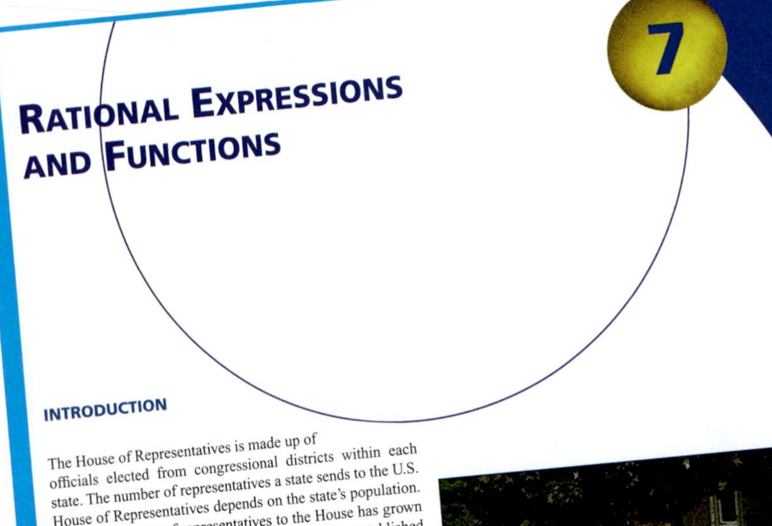
7

RATIONAL EXPRESSIONS AND FUNCTIONS

INTRODUCTION

The House of Representatives is made up of officials elected from congressional districts within each state. The number of representatives a state sends to the U.S. House of Representatives depends on the state's population. The total number of representatives to the House has grown from 106 in 1790 to 435, the maximum number established in 1930. These 435 representatives are apportioned to the 50 states on the basis of population. This apportionment is revised after every decennial (10-year) census.

If a particular state has a population, A, and its number of representatives is equal to a, then $\frac{A}{a}$ represents the ratio of people in the state to their total number of representatives in the U.S. House. It follows that the total population of the country, P, and its total number of representatives, r, is represented by the ratio $\frac{P}{r}$. If another state, with population, E, has e number of representatives, then $\frac{E}{e}$ should also be equal to $\frac{P}{r}$ and to $\frac{A}{a}$, if the apportionment is fair.

A comparison of these ratios for states in 1990 finds Pennsylvania with 546,880 people per representative and Arizona with 610,800—Arizona was above the national average of 571,750 people per representative for 1990, and Pennsylvania below. This is not so much a result of political backroom bargaining as it is a result of ratios that do not divide out evenly—there are remainders when the proportions are solved. Should the numbers be rounded up or down? If they are all rounded down, the total is too small, if rounded up, the total number of representatives would be more than the 435 seats in the House. So, because all the states cannot be treated equally, the question of what is fair and how to decide who gets the additional representatives has been debated in Congress since its inception.

Section Objectives

Objectives for each section are clearly identified.

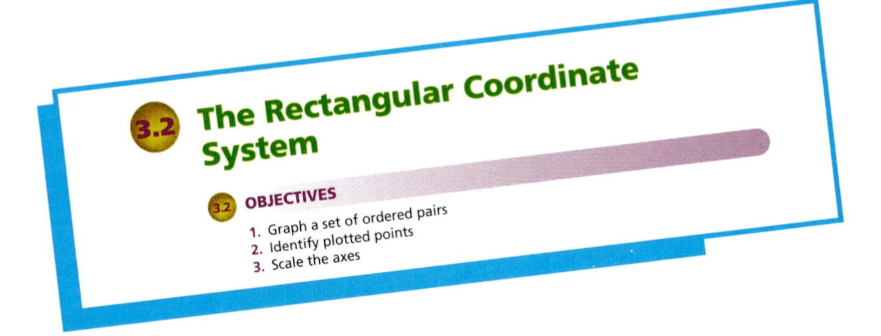

3.2 The Rectangular Coordinate System

3.2 OBJECTIVES

1. Graph a set of ordered pairs
2. Identify plotted points
3. Scale the axes

NOTE We find the common factors 4, x, and y in the numerator and denominator. We divide the numerator and denominator by the common factor $4xy$. Note that

$$\frac{4xy}{4xy} = 1$$

NOTE We have *divided* the numerator and denominator by the common factor $x - 2$. Again note that

$$\frac{x - 2}{x - 2} = 1$$

CAUTION

Pick any value other than 0 for the variable x, and substitute. You will quickly see that

$$\frac{x + 2}{x + 3} \neq \frac{2}{3}$$

Marginal Notes and Caution Icons

Marginal notes are provided throughout and are designed to help the students focus on important topics and techniques. Caution icons point out potential trouble spots.

Check Yourself Exercises

These exercises have been the hallmark of the text; they are designed to actively involve students throughout the learning process. Each example is followed by an exercise that encourages students to solve a problem similar to the one just presented. Answers are provided at the end of the section for immediate feedback.

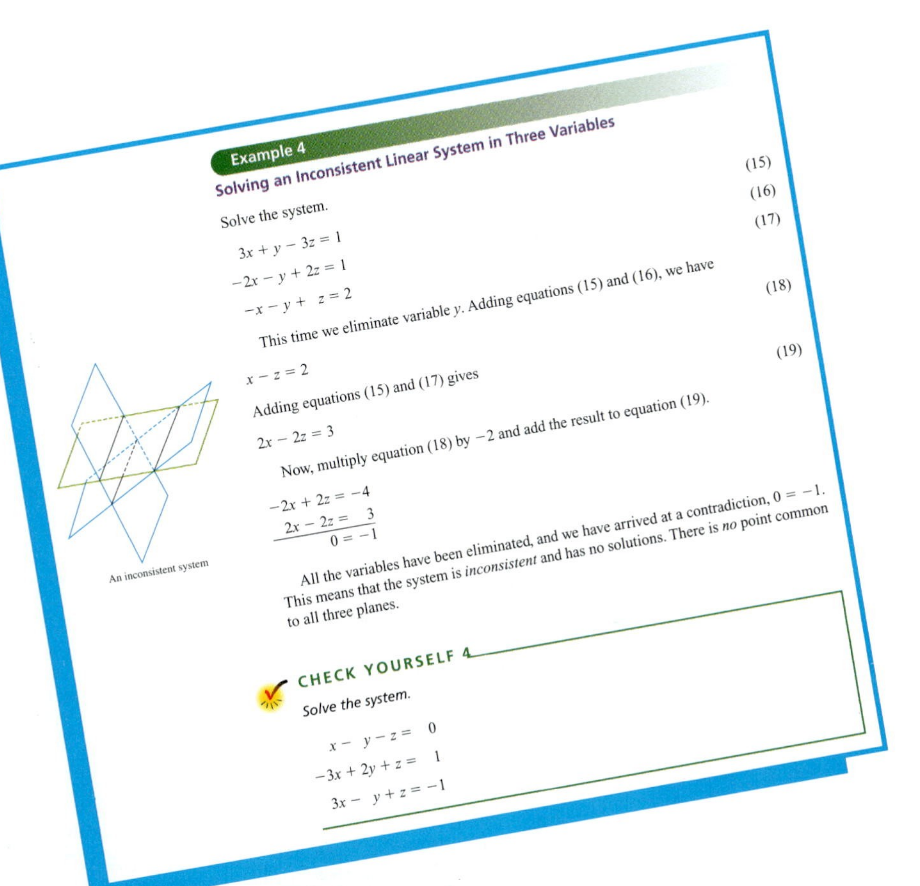

Example 4

Solving an Inconsistent Linear System in Three Variables

Solve the system.

$$3x + y - 3z = 1 \quad (15)$$
$$-2x - y + 2z = 1 \quad (16)$$
$$-x - y + z = 2 \quad (17)$$

This time we eliminate variable y. Adding equations (15) and (16), we have

$$x - z = 2 \quad (18)$$

Adding equations (15) and (17) gives

$$2x - 2z = 3 \quad (19)$$

Now, multiply equation (18) by -2 and add the result to equation (19).

$$\begin{aligned} -2x + 2z &= -4 \\ 2x - 2z &= 3 \\ \hline 0 &= -1 \end{aligned}$$

All the variables have been eliminated, and we have arrived at a contradiction, $0 = -1$. This means that the system is *inconsistent* and has no solutions. There is *no* point common to all three planes.

An inconsistent system

CHECK YOURSELF 4

Solve the system.

$$x - y - z = 0$$
$$-3x + 2y + z = 1$$
$$3x - y + z = -1$$

Comprehensive Exercise Sets and Challenge Exercises

Complete exercise sets are at the end of each section and after the summary at the end of each chapter. These exercises were designed to reinforce basic skills and develop critical thinking and communication abilities. Exercise sets include writing and word problems, collaborative and group exercises, and challenge exercises, all denoted by distinctive icons. The calculator icon points out examples and exercises that illustrate when the calculator can best be used for further understanding of the concept at hand. Answers to odd-numbered exercises are provided at the end of each exercise set.

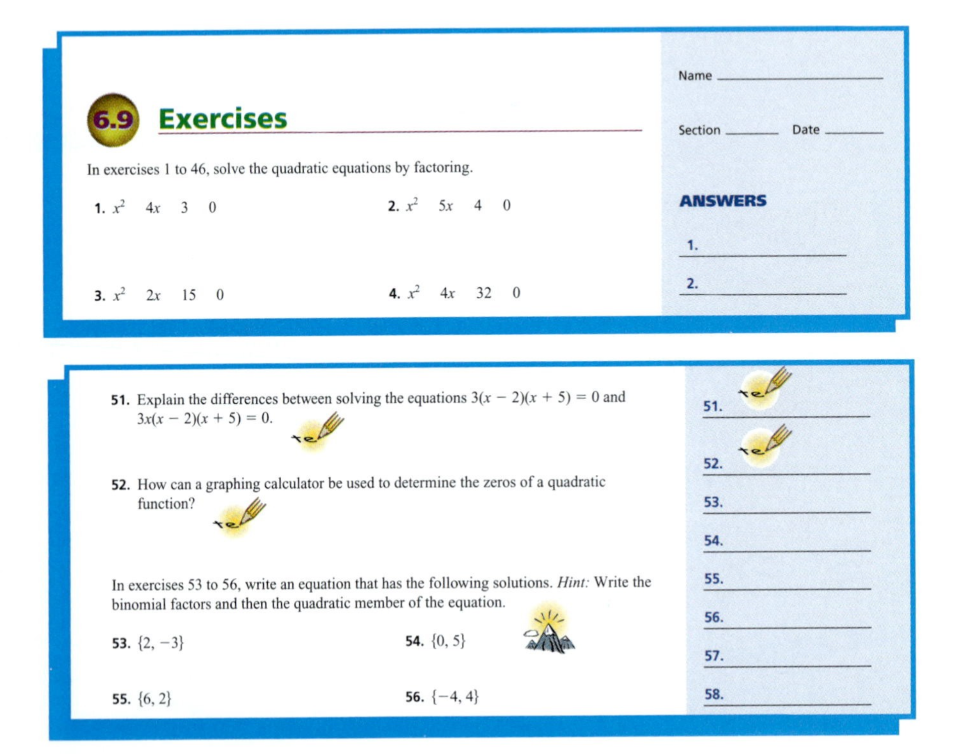

6.9 Exercises

Name

Section Date

In exercises 1 to 46, solve the quadratic equations by factoring.

1. $x^2 \quad 4x \quad 3 \quad 0$ **2.** $x^2 \quad 5x \quad 4 \quad 0$

3. $x^2 \quad 2x \quad 15 \quad 0$ **4.** $x^2 \quad 4x \quad 32 \quad 0$

ANSWERS

1.

2.

51. Explain the differences between solving the equations $3(x - 2)(x + 5) = 0$ and $3x(x - 2)(x + 5) = 0$.

52. How can a graphing calculator be used to determine the zeros of a quadratic function?

In exercises 53 to 56, write an equation that has the following solutions. *Hint:* Write the binomial factors and then the quadratic member of the equation.

53. $\{2, -3\}$ **54.** $\{0, 5\}$

55. $\{6, 2\}$ **56.** $\{-4, 4\}$

51.

52.

53.

54.

55.

56.

57.

58.

Summary and Summary Exercises

These comprehensive sections give students an opportunity to practice and review important concepts at the end of each chapter. Answers can be found in the *Instructor's Resource Manual.* Section references aid in summarizing the material effectively.

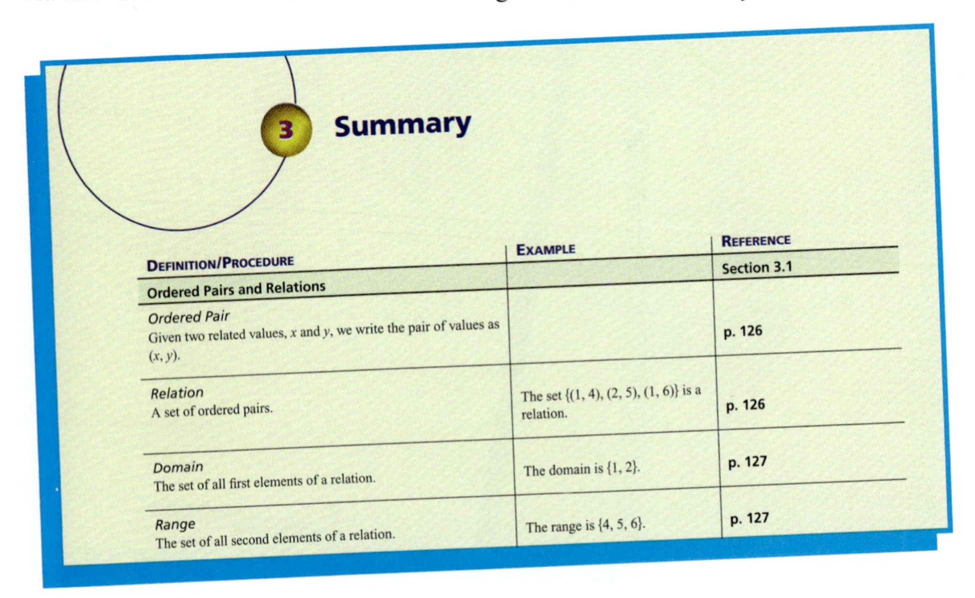

3 Summary

DEFINITION/PROCEDURE	EXAMPLE	REFERENCE
Ordered Pairs and Relations		Section 3.1
Ordered Pair Given two related values, x and y, we write the pair of values as (x, y).		p. 126
Relation A set of ordered pairs.	The set $\{(1, 4), (2, 5), (1, 6)\}$ is a relation.	p. 126
Domain The set of all first elements of a relation.	The domain is $\{1, 2\}$.	p. 127
Range The set of all second elements of a relation.	The range is $\{4, 5, 6\}$.	p. 127

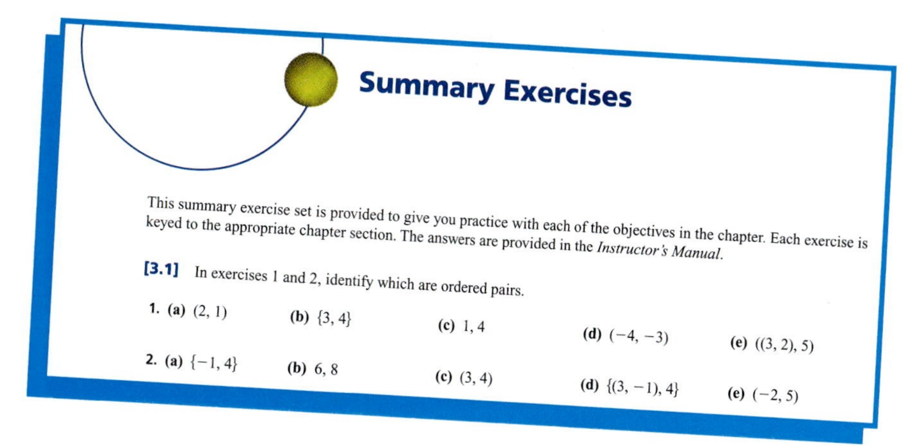

Summary Exercises

This summary exercise set is provided to give you practice with each of the objectives in the chapter. Each exercise is keyed to the appropriate chapter section. The answers are provided in the *Instructor's Manual*.

[3.1] In exercises 1 and 2, identify which are ordered pairs.

1. (a) $(2, 1)$ (b) $\{3, 4\}$ (c) $1, 4$ (d) $(-4, -3)$ (e) $((3, 2), 5)$
2. (a) $\{-1, 4\}$ (b) $6, 8$ (c) $(3, 4)$ (d) $\{(3, -1), 4\}$ (e) $(-2, 5)$

Cumulative Tests

These tests help students build on what was previously covered and give them more opportunity to build skills necessary in preparing for midterm and final exams. Answers are at the back of the book.

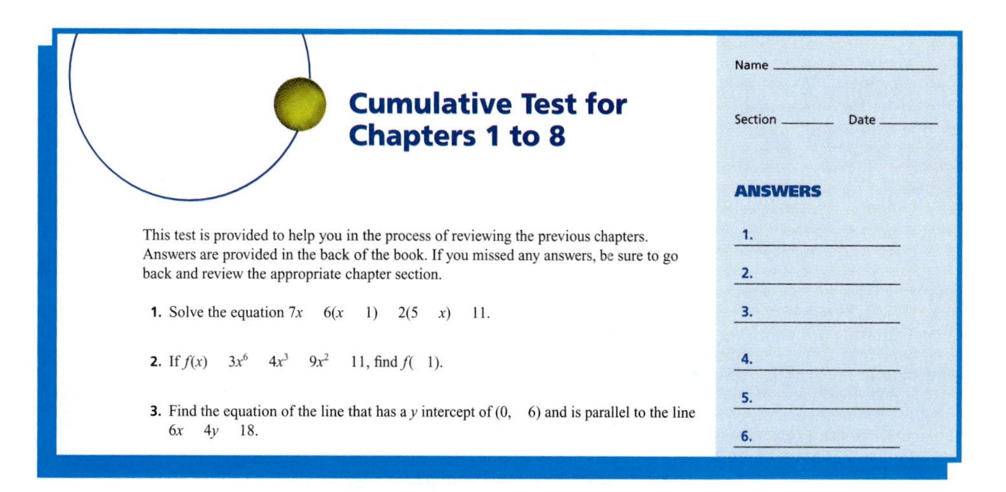

Cumulative Test for Chapters 1 to 8

Name ______

Section ______ Date ______

ANSWERS

1. ______
2. ______
3. ______
4. ______
5. ______
6. ______

This test is provided to help you in the process of reviewing the previous chapters. Answers are provided in the back of the book. If you missed any answers, be sure to go back and review the appropriate chapter section.

1. Solve the equation $7x \quad 6(x \quad 1) \quad 2(5 \quad x) \quad 11$.
2. If $f(x) \quad 3x^6 \quad 4x^3 \quad 9x^2 \quad 11$, find $f(\quad 1)$.
3. Find the equation of the line that has a y intercept of $(0, \quad 6)$ and is parallel to the line $6x \quad 4y \quad 18$.

Self-Tests

Each chapter ends with a self-test to give students confidence and guidance in preparing for in-class tests. Answers are at the back of the book.

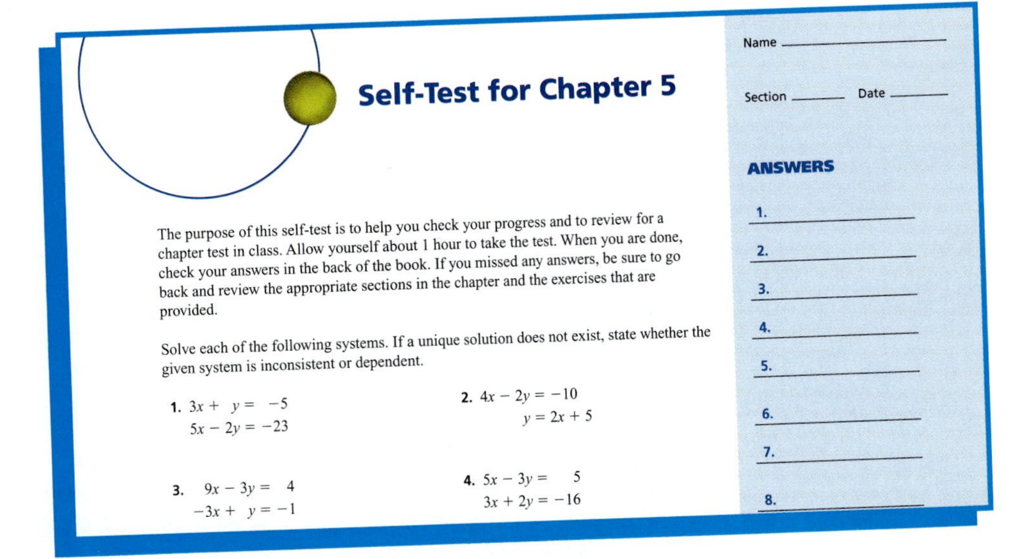

Self-Test for Chapter 5

Name ______

Section ______ Date ______

ANSWERS

1. ______
2. ______
3. ______
4. ______
5. ______
6. ______
7. ______
8. ______

The purpose of this self-test is to help you check your progress and to review for a chapter test in class. Allow yourself about 1 hour to take the test. When you are done, check your answers in the back of the book. If you missed any answers, be sure to go back and review the appropriate sections in the chapter and the exercises that are provided.

Solve each of the following systems. If a unique solution does not exist, state whether the given system is inconsistent or dependent.

1. $3x + y = -5$
   $5x - 2y = -23$
2. $4x - 2y = -10$
   $y = 2x + 5$
3. $9x - 3y = 4$
   $-3x + y = -1$
4. $5x - 3y = 5$
   $3x + 2y = -16$

SUPPLEMENTS

A comprehensive set of ancillary materials for both the student and the instructor is available with this text.

Annotated Instructor's Edition
This ancillary includes answers to all exercises and tests. These answers are printed in a second color for ease of use by the instructor and are located on the appropriate pages throughout the text.

Instructor's Solutions Manual
The manual provides worked-out solutions to the pre-tests, section exercises, summary exercises, self-tests, and cumulative tests.

Print Test Bank
The print test bank contains (1) a diagnostic pre-test for each chapter; (2) five forms of chapter tests; (3) two forms of cumulative tests; and (4) four forms of final tests. Answers to all these tests are provided.

Print and Computerized Testing
The testing materials provide an array of formats that allow the instructor to create tests using both algorithmically generated test questions and those from a standard test bank. This testing system enables the instructor to choose questions either manually or randomly by section, question type, difficulty level, and other criteria. Testing is available for Windows, and Macintosh computers.

Student's Solutions Manual
The manual provides worked-out solutions to the odd-numbered exercises in the text.

Streeter Video Series
The video series gives students additional reinforcement of the topics presented in the book. The videos were developed especially for the Streeter pedagogy, and features are tied directly to the main text's individual chapters and section objectives. The videos feature an effective combination of learning techniques, including personal instruction, state-of-the-art graphics, and real-world applications. Students are encouraged to work examples on their own and check their results with those provided.

Streeter Smart Tutorial CD-ROM
This interactive CD-ROM is a self-paced tutorial specifically linked to the text and reinforces topics through unlimited opportunities to review concepts and practice problem solving. The CD-ROM contains chapter- and section-specific tutorials, multiple-choice questions with feedback, and algorithmically generated questions. It requires virtually no computer training on the part of students and supports Windows and Macintosh computers.

In addition, a number of other technology and web-based ancillaries are under development; they will support the ever-changing technology needs in developmental mathematics. For further information about these or any supplements, please contact your local McGraw-Hill sales representative.

Website and Online Learning Center
Web-based, interactive learning is provided for your students. Student and intructor resources include learning objectives, PowerPoint slides, online quizzing, audio/visual tutorial and practice, and web links.

NetTutor
NetTutor is a web-based tutorial system available through the Online Learning Center. Using a "white board" technology with built-in math notation, students may interact with live skilled tutors who are familiar with this textbook and its methods.

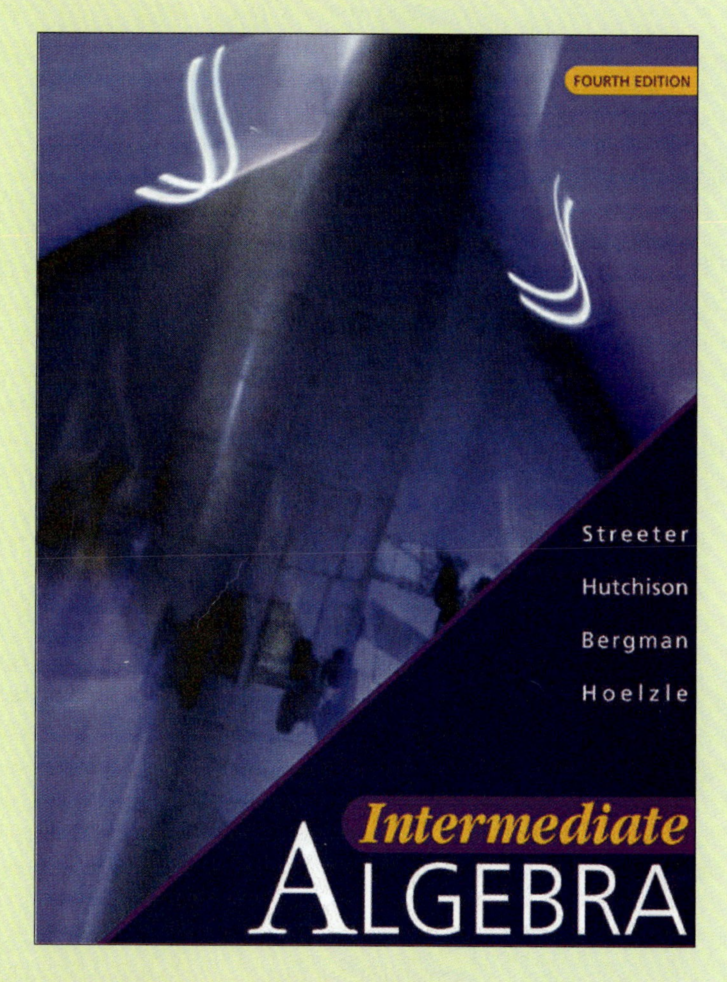

MCGRAW-HILL IS PROUD TO OFFER AN EXCITING NEW SUITE OF MULTIMEDIA PRODUCTS AND SERVICES CALLED COURSE SOLUTIONS.

Designed specifically to help you with your individual course needs, Course Solutions will assist you in integrating your syllabus with our premier titles and state-of-the-art new media tools that support them.

AT THE HEART OF COURSE SOLUTIONS YOU'LL FIND:

- Fully integrated multimedia
- A full-scale Online Learning Center
- A Course Integration Guide

AS WELL AS THESE UNPARALLELED SERVICES:

- McGraw-Hill Learning Architecture
- McGraw-Hill Course Consultant Service
- Visual Resource Library (VRL) Image Licensing
- McGraw-Hill Student Tutorial Service
- McGraw-Hill Instructor Syllabus Service
- PageOut Lite
- PageOut: The Course Website Development Center
- Other Delivery Options

COURSE SOLUTIONS truly has the solutions to your every teaching need. Read on to learn how we can specifically help you with your classroom challenges.

SPECIAL ATTENTION
to your specific needs.

These "perks" are all part of the extra service delivered through McGraw-Hill's Course Solutions:

MCGRAW-HILL LEARNING ARCHITECTURE

Each McGraw-Hill *Online Learning Center* is ready to be ported into our *McGraw-Hill Learning Architecture*—a full course management software system for Local Area Networks and Distance Learning Classes. Developed in conjunction with Top Class software, *McGraw-Hill Learning Architecture* is a powerful course management system available upon special request.

MCGRAW-HILL COURSE CONSULTANT SERVICE

In addition to the *Course Integration Guide,* instructors using Course Solutions textbooks can access a special curriculum-based *Course Consultant Service* via a Web-based threaded discussion list within each *Online Learning Center.* A McGraw-Hill Course Solutions Consultant will personally help you—as a text adopter—integrate this text and media into your course to fit your specific needs. This content-based service is offered in addition to our usual software support services.

VISUAL RESOURCE LIBRARY (VRL) IMAGE LICENSING

Most of our Course Solutions titles are accompanied by a *Visual Resource Library (VRL) CD-ROM,* which features text figures in electronic format. Previously, use of these images was restricted to in-class presentation only. Now, McGraw-Hill will license adopters the right to use appropriate VRL image files—FREE OF CHARGE—for placement on their local website! Some restrictions apply. Consult your McGraw-Hill sales representative for more details.

MCGRAW-HILL INSTRUCTOR SYLLABUS SERVICE

For *new* adopters of Course Solutions textbooks, McGraw-Hill will help correlate all text, supplement, and appropriate materials and services to your course syllabus. Simply call your McGraw-Hill sales representative for assistance.

PAGEOUT LITE

Free to Course Solutions textbook adopters, *PageOut Lite* is perfect for instructors who want to create their own website. In just a few minutes, even novices can turn their syllabus into a website using *PageOut Lite.*

PAGEOUT: THE COURSE WEBSITE DEVELOPMENT CENTER

For those that want the benefits of *PageOut Lite's* no-hassle approach to site development, but with even more features, we offer *PageOut: The Course Website Development Center.*

PageOut shares many of *PageOut Lite's* features, but also enables you to create links that will take your students to your original material, other website addresses, and to *McGraw-Hill Online Learning Center* content. This means you can assign *Online Learning Center* content within your syllabus-based website. *PageOut's* gradebook function will tell you when each student has taken a quiz or worked through an exercise, automatically recording the scores for you. *PageOut* also features a discussion board list where you and your students can exchange questions and post announcements, and an area for students to build personal web pages.

OTHER DELIVERY OPTIONS

Online Learning Centers are also compatible with a number of full-service online course delivery systems or outside educational service providers. For a current list of compatible delivery systems, contact your McGraw-Hill sales representative.

And for your students...

MCGRAW-HILL STUDENT TUTORIAL SERVICE

Within each *Online Learning Center* resides a FREE *Student Tutorial Service.* This web-based "homework hotline"—available via a threaded discussion list—features guaranteed, 24-hour response time on weekdays.

www.mhhe.com/streeter

Acknowledgments

Those familiar with the publishing process will attest that change is inevitable. The same is true at McGraw-Hill. The difference there is that change invariably seems to lead to something positive. We have been most fortunate to work with our editors Bill Barter, Erin Brown, and Bea Wikander. All three manage that fine line by being both demanding managers and supportive coworkers. We have appreciated their talents, energy, and time. As always, we encourage prospective authors to talk with the staff at McGraw-Hill. It will be a valuable use of your time.

We would like to thank the many reviewers who reviewed and improved this text:

Alisher Abdullayev, National University (CA)
Linda Beattie, Western New Mexico University (NM)
Andrew Beiderman, Community College of Baltimore City-Essex (MD)
Ernie Berman, City Colleges of Chicago-Truman (IL)
Lucille Croom, Hunter College (NY)
Dennis Donohue, Community College of Southern Nevada (NV)
Judith Downey, University of Nebraska-Omaha (NE)
William Dunn, Las Positas College (CA)
Doris Holland, Tarrant County College (TX)
Mark Litrell, Rio Hondo College (CA)
Rodolfo Maglio, Oakton Community College (IL)
Mike McAfee, Mt. Hood Community College (OR)
Kathy C. Nickell, College of Du Page (IL)
Michael Polley, Southeastern Community College (IA)
Larry Pontaski, Pueblo Community College (CO)
Virginia Puckett, Miami-Dade Community College-North (FL)
Mary Rice, University of Illinois-Chicago (IL)
Eugenia Shipe, Penn State University-McKeesport (PA)
Carolyn Thomas, San Diego City College (CA)
Carla Thompson, Tulsa Community College (OK)

We would also like to thank those who contributed to the development of the third edition:

Jackie Cohen, Augusta College (GA)
Maura Corley, Henderson Community College (KY)
Jon Davidson, Southern State Community College (OH)
Kenneth Hufham, Cape Fear Community College (NC)
Robert Farinelli, Community College of Allegheny County (PA)
James Warren Fightmaster, Virginia Western Community College (VA)
Roseann Foglio, Gloucester Community College (NJ)
Brenda Foster, Central Virginia Community College (VA)
David Freeman, Marietta College (OH)
Dorothy Gotway, University of Missouri at St. Louis (MO)
Pauline Graveline, State University of New York-College of Technology at Canton (NY)
John Jacobs, Massachusetts Bay Community College (MA)
Billie L. James, University of South Dakota (SD)
Norma James, New Mexico State University (NM)
Conrad Johnson, Central Florida Community College (FL)
Diane L. Johnson, Humboldt State University (CA)
Robert Malena, Community College of Allegheny County (PA)
Laurie McManus, St. Louis University (MO)
Valerie Melvin, Cape Fear Community College (NC)
Fred Monaco, State University of New York-College of Technology at Canton (NY)

Cheryl Raboin, Eastern Wyoming College (WY)
Janice Rech, University of Nebraska at Omaha (NE)
Thomas Roe, South Dakota State University (SD)
Rakesh Rustagi, Northeastern Illinois University (IL)
Debra Sample, Arkansas State University (AR)
Hugh Sanders, Georgia College (GA)
Richard Semmler, Northern Virginia Community College (VA)
Randy Taylor, Las Positas College (CA)
Carol Walker, Hinds Community College (MS)
Raymond Whaley, Waubonsee College (IL)

We are especially grateful to the following focus group participants, who offered many valuable suggestions for the fourth edition:

Mary Kay Abbey, Montgomery College (MD)
Cynthia Albee, Tarrant County College (TX)
Sabah Alquaddoomi, Pasadena City College
Dimos Arsenidis, California State University Long Beach
Deidre Baker, College of Alameda (CA)
Jerry Bartolomeo, Nova Southeastern University (FL)
Palma Benko, Passaic County Community College (NJ)
Connie Bish, Las Positas College (CA)
Barbara Britton, Malcolm X (IL)
Brenda D. Brown, Prince George's Community College (MD)
Eleanor Browne, Richland College (TX)
Linda Burton, Miami Dade Community College-Homestead
Bob Caldwell, El Camino College (CA)
Donna Carlson, College of Lake County (IL)
Jim Castro, California State University Northridge
Yong S. Colen, Monroe College (NY)
Jorge Cossio, Miami-Dade Community College-Kendall
Charles Dietz, Charles County Community College (MD)
C. Toland Draper, College of Alameda (CA)
C. Wayne Ehler, Anne Arundel Community College (MD)
Gerald Floyd, Malcolm X (IL)
Dorothy Fujimura, California State University Hayward
Judy Godwin, Collin County Community College (TX)
David E. Gustafson, Tarrant County College (TX)
Garry Hart, California State University Dominguez Hills
Laxman Hegde, Frostburg State University (MD)
Celeste Hernandez, Richland College (TX)
Nancy Johnson, Broward Community College (FL)
Rosamma Joseph, Oakton Community College (IL)
Rosemary M. Karr, Collin County Community College (TX)
Joseph Kazimir, East Los Angeles College
Surinder K. Khurana, Fullerton College (CA)
Serge Kuznetsov, Kennedy-King College (IL)
Shirley Lathrop, Truman College (IL)
William Lepowsky, Laney College (CA)
Greg Liano, Brookdale Community College (NJ)
Sandy Lynn, Foothill College (CA)
Alice Madson, Kankakee Community College (IL)
Dorothy S. Marshall, Edison Community College
Bob Martin, Tarrant County College, NE Campus (TX)
Marilyn Massey-Moss, Collin County Community College (TX)
Victor Mastrovincenzo, Hudson County Community College (NJ)

William M. Mays, Gloucester County College (NJ)
Shyla McGill, Columbia College (IL)
Janet McLaughlin, S.C, Montclair State University (NJ)
Constance McNair, Broward Community College (FL)
Michael J. Morse, East Los Angeles College
Ann C. Mugavero, College of Staten Island
Shoeleh Mutameni, Morton College (IL)
Pat Newell, Edison Community College
Louise Olshan, County College of Morris (NJ)
Michael N. Payne, College of Alameda (CA)
Gary Piercy, Moraine Valley Community College (IL)
Virginia Puckett, Miami-Dade Community College-North
Linda Retterath, Mission College (CA)
Beth Rinehart, Anne Arundel Community College (MD)
David Ross, College of Alameda (CA)
Radha Sankaran, Passaic County Community College (NJ)
David P. Schaefer, College of Lake County (IL)
Richard Semmlen, Northern Virginia Community College
Luz V. Shin, Los Angeles Valley College
Alexis Thurman, County College of Morris (NJ)
Anthony Valenti, Nova University (FL)
Frissell Walker, College of Alameda (CA)
Anne Walsh, Monroe College (NJ)
Martin Weissman, Essex County College (NJ)
Pam Zener, Governors State University (IL)

We thank all of the students whom we have taught, talked to, questioned, and tested. This text was created for them. We also thank our community college compatriots. Professionals such as Betsy Farber, Alice Hayden, Susan Hopkirk, and Mark Yannotta are constantly providing us with both intentional and inadvertent guidance in our writing projects.

Donald Hutchison
Barry Bergman
Louis Hoelzle

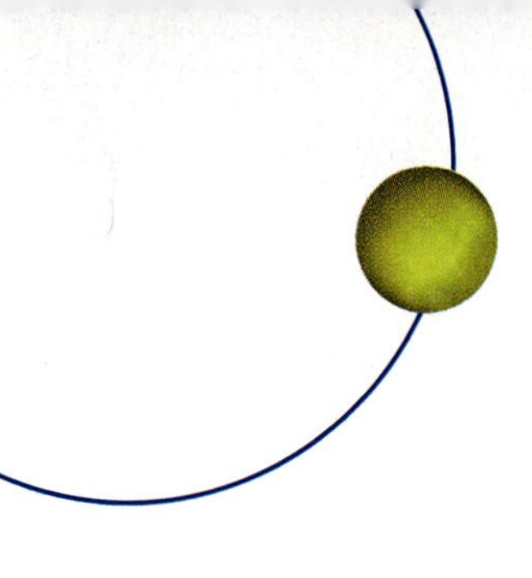

To the Student

You are about to begin a course in algebra. We made every attempt to provide a text that will help you understand what algebra is about and how to effectively use it. Our goal has been to build a bridge between beginning algebra and college algebra, a course in which the function concept is central. Your progress through the course will depend on the amount of time and effort you devote to the course and your previous background in math. There are some specific features in this book that will aid you in your studies. Here are some suggestions about how to use this book. (Keep in mind that a review of *all* the chapter and summary material will further enhance your ability to grasp later topics and to move more effectively through the text.)

1. If you are in a lecture class, make sure that you take the time to read the appropriate text section *before* your instructor's lecture on the subject. Then take careful notes on the examples that your instructor presents during class.
2. After class, work through similar examples in the text, making sure that you understand each of the steps shown. Examples are followed in the text by *Check Yourself* exercises. Algebra is best learned by being involved in the process, and that is the purpose of these exercises. Always have a pencil and paper at hand, and work out the problems presented and check your results immediately. If you have difficulty, go back and carefully review the previous exercises. Make sure you understand what you are doing and why. The best test of whether you understand a concept lies in your ability to explain that concept to one of your classmates. Try working together.
3. At the end of each chapter section you will find a set of exercises. Work these carefully to check your progress on the section you have just finished. You will find the solutions for the odd-numbered exercises following the exercise set. If you have difficulties with any of the exercises, review the appropriate parts of the chapter section. If your questions are not completely cleared up, by all means do not become discouraged. Ask your instructor or an available tutor for further assistance. A word of caution: Work the exercises on a regular (preferably daily) basis. Again, learning algebra requires becoming involved. As is the case with learning any skill, the main ingredient is practice.
4. When you complete a chapter, review by using the *Summary.* You will find all the important terms and definitions in this section, along with examples illustrating all the techniques developed in the chapter. Following the summary are *Summary Exercises* for further practice. The exercises are keyed to chapter sections, so you will know where to turn if you are still having problems.
5. When you finish with the *Summary Exercises,* try the *Self-Test* that appears at the end of each chapter. This test will give you an actual practice test to work as you review for in-class testing. Again, answers are provided.
6. Finally, an important element of success in studying algebra is the process of regular review. We provide a series of *Cumulative Tests* throughout the textbook, beginning at the end of Chapter 2. These tests will help you review not only the concepts of the chapter that you have just completed but those of previous chapters. Use these tests in preparation for any mid-term or final exams. If it appears that you have forgotten some concepts that are being tested, don't worry. Go back and review the sections where the idea was initially explained, or the appropriate chapter summary. That is the purpose of the cumulative tests.

We hope that you will find our suggestions helpful as you work through this material, and we wish you the best of luck in the course.

Donald Hutchison
Barry Bergman
Louis Hoelzle

THE REAL NUMBERS

1

INTRODUCTION

Listen to any winter weather forecast in the northern United States, and you will probably hear something about the wind-chill factor. Just what is the wind-chill factor?

The speed of the wind does not actually effect the temperature, but it does effect how cold it feels. Scientists have conducted many studies to determine the relationship between the temperature, the wind speed, and how cold it actually feels outside. As a result of these studies, a formula has been developed that is used to compute the wind-chill factor. The table below uses that formula to calculate the effect.

The formula used to create this table has two inputs (temperature and wind speed) and one output (wind-chill factor). Equations of this type will be quite accessible for you by the time you finish this text.

Wind-Chill Factor in Degrees Fahrenheit

Wind Speed (mph)	Actual Temperature (Fahrenheit)								
	50	40	30	20	10	0	−10	−20	−30
5	48	37	27	16	6	−5	−15	−26	−36
10	40	28	16	4	−9	−21	−33	−46	−58
15	36	22	9	−5	−18	−36	−45	−58	−72
20	32	18	4	−10	−25	−39	−53	−67	−82
25	30	16	0	−15	−29	−44	−59	−74	−88
30	28	13	−2	−18	−33	−48	−53	−79	−94
35	27	11	−4	−20	−35	−49	−67	−82	−98
40	26	10	−6	−21	−37	−53	−69	−85	−100
45	25	10	−7	−22	−38	−54	−70	−86	−100

1.1 Sets

1.1 OBJECTIVES

1. Use set notation
2. Recognize the basic subsets of the real numbers
3. Plot numbers on a number line

In the fall of 2000, there were approximately 3,200,000 people studying mathematics in 2-year and 4-year colleges and universities in the United States. One-half of these were studying in 2-year institutions. The following table, compiled by the National Science Foundation (http://www.nsf.gov), shows the mathematics enrollment in community colleges over the last 30 years.

Fall Enrollments in Math Programs; 2-Year Institutions (thousands)

Course level	1970	1980	1990	1995	2000 (projected)
All math courses	555	900	1241	1384	1600
Remedial	191	482	724	790	950
Precalculus	134	188	245	295	330
Calculus	59	97	128	139	150
Other	171	133	144	160	170

Note that, in the fall of 2000, there were approximately 950,000 students in 2-year colleges taking a remedial course similar to the one in which you are enrolled. Why are all of these people taking these classes? (Did I hear you say, "Because it's required!"?) This is a required class because both curriculum designers and employers have concluded that applied mathematics (courses for which this is a prerequisite) is important preparation for many, if not most, careers.

In 1998, the Bureau of Labor Statistics (http://stats.bls.gov), surveyed employers about the importance of mathematics. Following are just a few of the careers in which applied mathematics was considered important:

Accountants and auditors
Aircraft pilots and flight engineers
Computer programmers
Cost estimators
Dentists
Drafters
Electrical and electronic technicians
Engineering technicians
Financial managers
Insurance sales workers
Landscape architects
Management analysts and consultants
Optometrists
Pharmacists
Physicians
Psychologists
Real estate agents and brokers
Science technicians
Securities and financial services sales workers
Sociologists
Surveyors
Tool programmers, numerical control
Underwriters
Urban and regional planners
Veterinarians

Among the many mathematical topics that have applications in most of these fields is the idea of a set.

A **set** is a collection of objects, symbols, or numbers. We often describe a set by enclosing its contents, or a description of its contents, in braces.

NOTE Describing a set by listing its elements is called the *roster method* of defining a set.

$\{1, 2, 3\}$ is a set containing the first three counting numbers.

$\{z, x, c, v, b, n, m\}$ is a set containing the letters found on the bottom row of a computer keyboard.

NOTE This is read, "the set of all x such that x is a day of the week." This way of describing a set is called *set-builder notation.*

$\{x \mid x \text{ is a day of the week}\}$ is a set containing seven elements.

Frequently, we represent a set with a capital letter so that we may more easily refer to it. For example, we might say

$D = \{x \mid x \text{ is a day of the week}\}$

Any object or symbol that is contained in a set is called an **element,** or a **member,** of the set. The symbol $\in$ is used to indicate that an object is an element of a set.

$3 \in \{1, 2, 3, 4\}$ indicates that 3 is contained in the set $\{1, 2, 3, 4\}$.

Saturday $\in D$ indicates that Saturday is an element of the set named above.

NOTE The symbol $\notin$ is used to indicate that an element is *not* a member of a set.

January $\notin D$ indicates that January is *not* an element of D.

CHECK YOURSELF 1

Complete each statement with the symbol $\in$ or $\notin$.

If $V = \{\text{a, e, i, o, u}\}$

(a) a ____ V **(b)** t ____ V

Suppose that we define a set A as

$A = \{\text{a, b, c, d}\}$

NOTE Notice that the order in which we list the elements of a set is immaterial.

and then a set B as

$B = \{\text{d, c, b, a}\}$

Here the two sets consists of the same elements—the first four letters of the alphabet. We can write

NOTE Of course, this is read, "set A equals set B."

$A = B$

In general, we say that two sets A and B are equal whenever the two sets have exactly the same elements.

If the elements of a set can be ordered and we wish to indicate that a set continues as described, we use an ellipsis, three dots that mean "and so on."

$\{\text{a, b, c, . . . , z}\}$	describes the entire alphabet
$\{1, 2, 3, . . . , 100\}$	describes the first 100 counting numbers
$\{1, 3, 5, . . .\}$	describes the positive odd numbers

You will notice that the last set described above *ends* with an **ellipsis** (three dots). This indicates that the elements continue without end. A set that has no end is said to be **infinite.**

A set that has some specific number of elements is said to be **finite.**

The set {a, b, c, d} is a finite set.

The set {2, 4, 6, . . . , 50} is a finite set.

NOTE This is the set of positive even numbers. They continue without end.

But the set {2, 4, 6, . . .} is an infinite set.

CHECK YOURSELF 2

Describe each set as finite or infinite.

(a) {1, 3, 5, . . .} infinite **(b)** {1, 2, 3, . . . , 99} finite

(c) {all prime numbers} infinite

NOTE Recall that a prime number has *only* itself and 1 as factors.

Several numeric sets are used so commonly in mathematics that they have readily identifiable names. The rich history (or prehistory, in this case) of mathematics began with tallying or counting. The set of numbers used for counting is called the set of **natural numbers** and is designated by the capital letter N. In set notation, we write

NOTE Including zero in this set, we have the set of whole numbers.

$W = \{0, 1, 2, 3, \ldots\}$

$N = \{1, 2, 3, \ldots\}$

An equally important mathematical set is the set of **integers.** This is how we describe the set of integers:

NOTE Notice that every natural number is *also* an element of the set of integers.

$Z = \{\ldots, -2, -1, 0, 1, 2, \ldots\}$

We see that the set continues without end in *both* the positive and negative directions. Also note the choice of the letter Z to designate this particular set. This comes from the word *Zahl,* the German word for "number."

NOTE A rational number has the form

$$\frac{p}{q}$$

in which p and q are integers and q cannot be 0.

We also will be referring to the set of **rational numbers.** A rational number is one that can be written as a fraction that is the ratio of two integers. Because ratios can be considered as **quotients** (and also because R will be used to designate a different set of numbers) we denote the set of rational numbers with the letter Q.

Because we cannot list the rational numbers in any meaningful fashion, we define that set by *describing* the elements of that set instead:

NOTE Because any integer can be written as the ratio of two integers, namely itself and 1, every integer is also a rational number.

$$Q = \left\{ \frac{p}{q} \,\middle|\, p, q \in Z, q \neq 0 \right\}$$

This is read "the set of elements of the form p over q, such that p and q are integers and q is not equal to zero."

There is another important characterization of the set of rational numbers. The decimal representation of any rational number is either a terminating decimal or a repeating decimal.

So 0.45 and 0.825 name rational numbers.

Also 0.3333. . . and 0.272727. . . name rational numbers

NOTE This is no surprise because

$$0.45 = \frac{45}{100},$$

the *ratio* of two integers.

$\frac{1}{3} = 0.\overline{3}$ (the bar denotes the repeating pattern)

$\frac{3}{11} = 0.\overline{27}$

CHECK YOURSELF 3

Complete each statement with the symbols ∈ or ∉.

(a) 3 ______ Z (b) -3 ______ N

(c) $\frac{2}{3}$ ______ Z (d) 0.25 ______ Q

(e) $\frac{2}{3}$ ______ N (f) -5 ______ Z

(g) $-\frac{3}{4}$ ______ Q (h) $0.\overline{35}$ ______ Q

NOTE The disturbing fact that $\sqrt{2}$ cannot be represented as the ratio of two integers was known to the Pythagoreans (400 B.C.), and it contradicted their belief that all lengths could be represented by ratios of integers.

NOTE Q' is read "Q prime."

Not every number can be expressed as the ratio of two integers. For example, it can be shown that the square root of 2 (denoted $\sqrt{2}$) cannot be written as the ratio of integers and is therefore *not* a rational number.

Numbers such as $\sqrt{2}$, $\sqrt[3]{7}$, and π are called **irrational numbers.** None of their decimal representations will ever repeat or terminate. The set of irrational numbers can be designated Q'.

Now, if we combine the set of rational numbers with the set of irrational numbers, we call this new set the set of **real numbers** and use the letter R to designate the set. This is the set to which we refer most often in algebra.

CHECK YOURSELF 4

For the set $\left\{-4, 2.3, \sqrt{6}, -\pi, 0, \frac{3}{4}, 0.\overline{36}, 7\right\}$, *which of the elements are*

(a) Irrational numbers (b) Real numbers (c) Natural numbers
(d) Rational numbers (e) Integers

A convenient way to "picture" the set of real numbers is with a number line. This number line is constructed by drawing a straight line and then choosing a point to correspond to 0. This point is called the **origin** of the number line.

The standard convention is to allow positive numbers to increase to the right and negative numbers to decrease to the left. This is represented in the number line shown below.

NOTE Zero is neither positive nor negative.

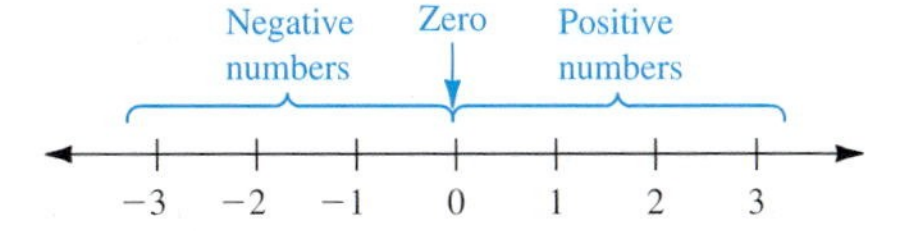

Each point on the line corresponds to a number called the **coordinate** of that point. The set of all numbers that correspond to a point on the number line makes up the set of real numbers.

NOTE Later in the text, we will encounter numbers that cannot be represented on the real number line.

Every real number will correspond to exactly one point on the line, and every point corresponds to exactly one real number.

Locate the point corresponding to each element of the set

$$\left\{-1, \sqrt{2}, \frac{5}{2}, -2.5, \pi\right\}$$

on the real number line.

NOTE An approximation for $\sqrt{2}$ is 1.414, and for π, 3.14.

$-2.5 \quad -1 \quad \sqrt{2} \quad \frac{5}{2} \quad \pi$

$-3 \quad -2 \quad -1 \quad 0 \quad 1 \quad 2 \quad 3$

CHECK YOURSELF 5

Locate each element of set A on the number line, when

$$A = \left\{-\pi, -2, \sqrt{3}, -\frac{3}{4}, 3.5\right\}$$

$-4 \quad -3 \quad -2 \quad -1 \quad 0 \quad 1 \quad 2 \quad 3 \quad 4$

The following diagram summarizes the relationships between the various numeric sets that have been introduced in this section.

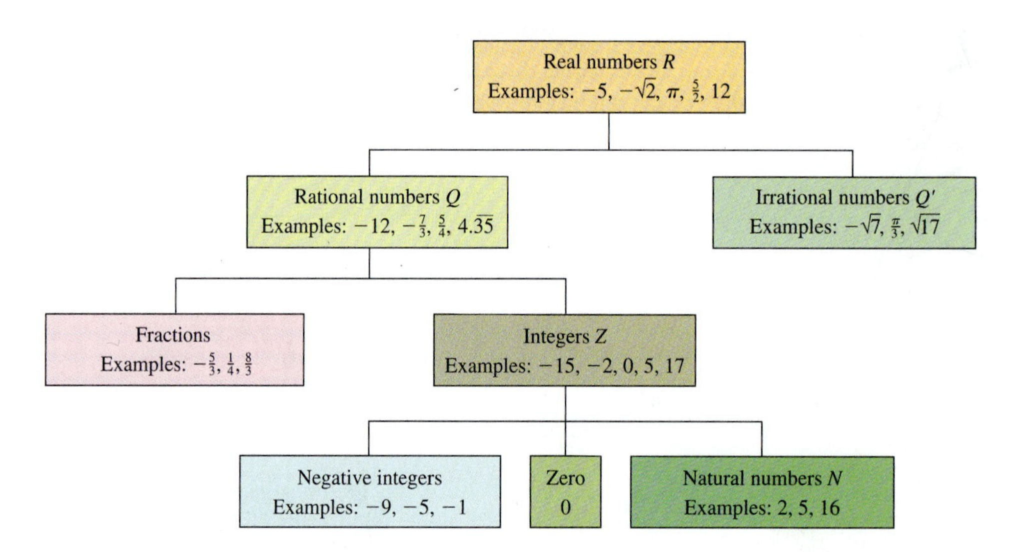

CHECK YOURSELF ANSWERS

1. **(a)** $a \in V$; **(b)** $t \notin V$ **2.** **(a)** Infinite; **(b)** finite; **(c)** infinite

3. **(a)** $3 \in Z$; **(b)** $-3 \notin N$; **(c)** $\frac{2}{3} \notin Z$; **(d)** $0.25 \in Q$; **(e)** $\frac{2}{3} \notin N$; **(f)** $-5 \in Z$; **(g)** $-\frac{3}{4} \in Q$; **(h)** $0.\overline{35} \in Q$

4. **(a)** $\sqrt{6}, -\pi$; **(b)** all; **(c)** 7; **(d)** $-4, 2.3, 0, \frac{3}{4}, 0.\overline{36}$, and 7; **(e)** $-4, 0, 7$

5. $-\pi \quad -2 \quad -\frac{3}{4} \quad \sqrt{3} \quad 3.5$

$-4 \; -3 \; -2 \; -1 \; 0 \; 1 \; 2 \; 3 \; 4 \; 5$

Name ____________________

Section ________ Date ________

1.1 Exercises

Name all the numeric sets to which each of the following numbers belong. The numeric sets consist of the natural numbers N, integers Z, rational numbers Q, irrational numbers Q', and real numbers R.

1. 5

2. -3.4

3. -3

4. 3.2

5. $\sqrt{5}$

6. 0

7. -3.1416

8. $-\sqrt{7}$

9. $\sqrt{4}$

10. $\frac{3}{4}$

11. π

12. $\sqrt{144}$

13. $-\frac{2}{5}$

14. $0.\overline{36}$

15. 2π

16. 2.35

On the number lines provided, graph the points named by each element of the set.

17. $\{3, 6, 9\}$

0

18. $\{-2, 0, 2\}$

0

19. $\left\{-\frac{1}{3}, 3, \frac{5}{2}\right\}$

0

20. $\left\{\frac{1}{2}, \frac{5}{4}, -\frac{7}{3}\right\}$

0

21. $\{-\sqrt{3}, \sqrt{2}\}$

0

22. $\{-\sqrt{5}, -\sqrt{3}, \sqrt{2}\}$

0

Describe each set as finite or infinite.

23. $\{3, 6, 9, \ldots\}$

24. $\{4, 8, 12, 16, \ldots\}$

25. $\{4, 8, 12, 16, 20, \ldots 48\}$

26. $\{5, 10, 15, 25, 30, \ldots\}$

27. {all even numbers}

28. $\{1, 2, 3, 5, 8, 13, \ldots 55\}$

ANSWERS

1. ____
2. ____
3. ____
4. ____
5. ____
6. ____
7. ____
8. ____
9. ____
10. ____
11. ____
12. ____
13. ____
14. ____
15. ____
16. ____
17. ____
18. ____
19. ____
20. ____
21. ____
22. ____
23. ____ 24. ____
25. ____ 26. ____
27. ____ 28. ____

ANSWERS

29. ________
30. ________
31. ________
32. ________
33. ________
34. ________
35. ________
36. ________
37. ________
38. ________
39. ________
40. ________
41. ________
42. ________
43. ________
44. ________
45. ________
46. ________
47. ________
48. ________
49. ________
50. ________
51. ________
52. ________
53. ________
54. ________
55. ________

Insert $\in$ or $\notin$ so that the statement is true.

29. 2 _____ Z **30.** 3.4 _____ Z **31.** -3 _____ N

32. $\sqrt{3}$ _____ Z **33.** $\frac{5}{3}$ _____ Z **34.** -4 _____ Z

35. $-\frac{2}{7}$ _____ Q **36.** $\frac{23}{6}$ _____ Q **37.** $\sqrt{3}$ _____ R

38. $\sqrt{3}$ _____ Q' **39.** $\frac{8}{0}$ _____ R **40.** $\sqrt{25}$ _____ R

41. 0 _____ N **42.** $0.\overline{24}$ _____ Q'

From the set $\left\{-4, -\frac{5}{3}, 0, \frac{4}{3}, \sqrt{3}, 3\pi, 4.5, 0.\overline{21}, 8\right\}$, list the elements that are

43. Natural numbers **44.** Rational numbers **45.** Irrational numbers

46. Real numbers **47.** Integers

Determine whether the given statements are true or false.

48. All odd integers are real numbers.

49. All even integers are rational numbers.

50. Every negative real number is a natural number.

51. Every integer is a real number.

52. All natural numbers are integers.

53. Every real number is an integer.

54. All integers are natural numbers.

55. Zero is not a rational number.

Answers

1. N, Z, Q, R **3.** Z, Q, R **5.** Q', R **7.** Q, R **9.** N, Z, Q, R
11. Q', R **13.** Q, R **15.** Q', R **17.**

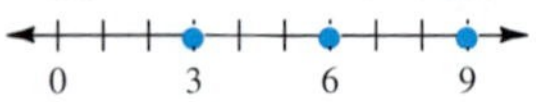

19.

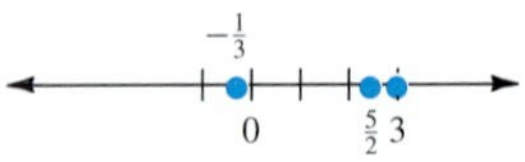

21.

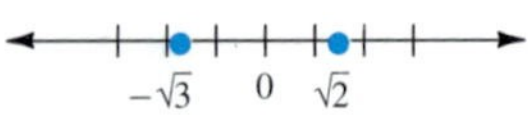

23. Infinite

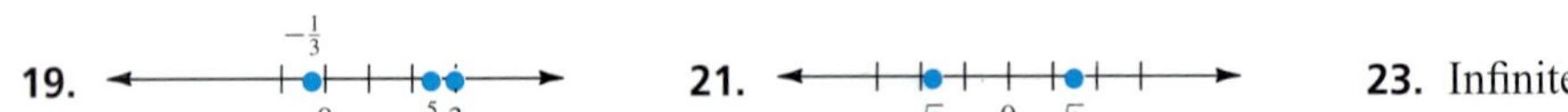

25. Finite **27.** Infinite **29.** $\in$ **31.** $\notin$ **33.** $\notin$ **35.** $\in$
37. $\in$ **39.** $\notin$ **41.** $\notin$ **43.** 8 **45.** $\sqrt{3}, 3\pi$ **47.** $-4, 0, 8$
49. True **51.** True **53.** False **55.** False

Operations and Properties

OBJECTIVES

1. Represent the four arithmetic operations using variables
2. Evaluate expressions using the order of operations
3. Recognize and apply the properties of addition
4. Recognize and apply the properties of multiplication
5. Recognize and apply the distributive property

NOTE Francois Viete (1540–1603), a French mathematician, first introduced the practice of using letters to represent known and unknown quantities.

The process of combining two elements of a set to produce a third element is called a **binary operation.** There are four basic binary operations: addition, subtraction, multiplication, and division. In algebra, we write these operations as follows:

$x + y$ is called the sum of x and y, or x plus y.

$x - y$ is called the difference of x and y, or x minus y.

NOTE We do not use a multiplication sign × because of the possible confusion with the letter *x*.

xy (or $x \cdot y$) is the product of x and y, or x times y.

$\frac{x}{y}$ (or $x \div y$) is the quotient of x and y, or x divided by y.

NOTE The symbols + and – first appeared in print in a book by Johann Widman (1489). The symbol × dates to a text by William Oughtred (1631).

Each of the above is an example of an **expression.** An expression is a meaningful collection of numbers, variables, and operations.

Algebraic expressions frequently involve more than one of the operation symbols that we have seen thus far in this section. For instance, when we are given an expression to evaluate such as

$$2 + 3 \cdot 4$$

we must agree on the order in which the indicated operations are to be performed. If we don't, we can end up with different results after the evaluation. For instance, if we were to add first, in this case we would have

$$\underbrace{2 + 3}_{\text{Add first}} \cdot 4 = \underbrace{5 \cdot 4}_{\text{Then multiply}} = 20$$

However, if we multiply first, we have

$$2 + \underbrace{3 \cdot 4}_{\text{Multiply}} = \underbrace{2 + 12}_{\text{Then add}} = 14$$

Because we get different answers depending on the order in which we do the operations, the language of algebra would not be clear unless we agreed on which of the methods of evaluation shown above is correct.

NOTE This means that $2 + 3 \cdot 4 = 14$ and the second approach shown above is the correct one.

To avoid this difficulty, we will agree that the multiplication in an expression such as

$$2 + 3 \cdot 4$$

should always be done *before* the addition.

NOTE An algorithm is a step-by-step process for solving a problem.

We refer to our procedure as the **order of operations.** The following algorithm gives us a set of rules, defining the order in which the operations should be performed.

NOTE The most common grouping symbols are parentheses, brackets, fraction bars, absolute value signs, and radicals.

Rules and Properties: Order of Operations

1. Simplify within the innermost grouping symbol, and work outward until all grouping symbols are removed.
2. Evaluate any expressions involving exponents.
3. Perform any multiplication and division, working from left to right.
4. Then do any addition and subtraction, again working left to right.

Example 1

Evaluating Expressions

Evaluate each expression.

(a) $2 \cdot 4 + 3 = 8 + 3$ Multiply first.

$= 11$ Then do the addition.

(b) $2(4 + 3) = 2 \cdot 7$ Simplify within the grouping symbol.

$= 14$ Then multiply.

NOTE Remember:

$7^2 = 7 \cdot 7 = 49$

Two factors

$2^3 = 2 \cdot 2 \cdot 2 = 8$

Three factors

(c) $2(4 + 3)^2 = 2(7)^2$ Add inside the parentheses.

$= 2 \cdot 49$ Evaluate the power.

$= 98$ Multiply.

(d) $3 + 5 \cdot 2^3 - 3 = 3 + 5 \cdot 8 - 3$ Evaluate the power.

$= 3 + 40 - 3$ Multiply.

$= 43 - 3$ Add and then subtract—from left to right.

$= 40$

CHECK YOURSELF 1

Evaluate each expression.

(a) $50 - 6 \cdot 8$ 2 **(b)** $3(25 - 20)$ 15

(c) $3(25 - 20)^2$ 75 **(d)** $17 + 2 \cdot 3^3$ 71

There are several properties of the two primary operations, addition and multiplication, that are very important in the study of algebra. The following table describes several of those properties for real numbers a, b, and c.

Property	Addition	Multiplication
Closure	$a + b \in R$	$a \cdot b \in R$
Associative	$(a + b) + c = a + (b + c)$	$(a \cdot b) \cdot c = a \cdot (b \cdot c)$
Commutative	$a + b = b + a$	$a \cdot b = b \cdot a$
Identity	$a + 0 = a$	$a \cdot 1 = a$
Inverse	$a + (-a) = 0$	$a \cdot \frac{1}{a} = 1$

NOTE The multiplicative inverse of a is also called the *reciprocal* of a. This is the property that allows us to *define division* by any nonzero number.

Example 2 illustrates the use of the properties introduced above.

Example 2

Identifying Properties of Multiplication

State the property used to justify each statement.

NOTE The *grouping* has been changed.

(a) $2 + (3 + b) = (2 + 3) + b$

Associative property of addition

NOTE Because $\frac{3}{2}$ is the reciprocal of $\frac{2}{3}$.

(b) $\left(\frac{2}{3}\right)\left(\frac{3}{2}\right) = 1$

Multiplicative inverse

(c) $(-3)(-4)$ is a real number

Closure property of multiplication

(d) $1(5) = 5$

Multiplicative identity

NOTE Only the *order* has been changed.

(e) $2 + (x + y) = 2 + (y + x)$

Commutative property of addition

CHECK YOURSELF 2

State the property used to justify each statement.

(a) $(9)(-7)$ is a real number

(b) $2 + x + y = 2 + y + x$

(c) $\left(\frac{1}{3} \cdot 3\right)xy = 1 \cdot xy$

(d) $0 + x + y = x + y$

(e) $12(3ab) = (12 \cdot 3)ab$

In addition to the specific properties for addition and multiplication, we have one property that involves both operations.

Rules and Properties: Distributive Property

For any real numbers *a*, *b*, and *c*,

$a(b + c) = ab + ac$

In words, multiplication distributes *over* addition.

The following example illustrates the use of the distributive property.

Example 3

Using the Distributive Property

Use the distributive property to simplify each expression.

NOTE "Distribute" the multiplication by 4 over $3x$ and 7.

(a) $4(3x + 7) = 4(3x) + 4(7)$

$= 12x + 28$

Simplify.

(b) $7(3x + 2y + 5) = 7(3x) + 7(2y) + 7(5)$

$= 21x + 14y + 35$

CHECK YOURSELF 3

Use the distributive property to simplify each expression.

(a) $5(4a + 5)$ (b) $4(2x^2 + 5x)$

(c) $6(4a + 3b + 7c)$ (d) $5(p + 5q)$

One of the most important uses of the distributive property relates to the combining of like terms. Example 4 illustrates.

Example 4

Combining Like Terms

Combine all like terms.

(a) $2x + 5x = (2 + 5)x = 7x$

(b) $3a + 4b + 7a - 3b = (3 + 7)a + (4 - 3)b = 10a + b$

CHECK YOURSELF 4

Combine all like terms.

(a) $3x + 12x$ (b) $2a + b + 7a - b$

(c) $2x^2 + 9x + 2 + x^2 - 3x$

CHECK YOURSELF ANSWERS

1. **(a)** 2; **(b)** 15; **(c)** 75; **(d)** 71
2. **(a)** Closure property of multiplication; **(b)** commutative property of addition; **(c)** multiplicative inverse; **(d)** additive identity; **(e)** associative property of multiplication
3. **(a)** $20a + 25$; **(b)** $8x^2 + 20x$; **(c)** $24a + 18b + 42c$; **(d)** $5p + 25q$
4. **(a)** $15x$; **(b)** $9a$; **(c)** $3x^2 + 6x + 2$

1.2 Exercises

Name ____________

Section ______ Date ______

Translate each of the following statements, using symbols.

1. The sum of 10 and x
2. x plus 5
3. 12 more than p
4. The sum of m and 25
5. n increased by 1
6. s increased by 3
7. m minus 14
8. 5 less than b
9. Subtract 1 from x.
10. 25 minus a
11. The product of m and n
12. The quotient of b and 2
13. s divided by 4
14. 7 times b
15. 2 times the difference of c and d
16. Twice the sum of a and b
17. 4 less than the product of r and s
18. 11 more than 2 times w
19. The sum of c and 4, divided by d
20. The difference of m and 2, divided by n

Apply the order-of-operations algorithm to evaluate the following expressions.

21. $5 + 4 \cdot 6$
22. $7 + 5 \cdot 3$
23. $7(8 - 2)$
24. $3(12 - 7)$
25. $6(8 - 4)^2$
26. $4(12 - 6)^2$
27. $(4 + 3)(5 + 3)$
28. $(5 + 6)(3 + 1)$
29. $4 + 3 \cdot 5 + 2$
30. $5 + 6 \cdot 2 + 1$
31. $(7 + 5)(7 - 5)$
32. $(12 + 3)(12 - 3)$
33. $7 + 5 \cdot 7 - 5$
34. $11 + 2 \cdot 11 - 3$
35. $9^2 - 5^2$
36. $12^2 - 3^2$
37. $(9 - 5)^2$
38. $(11 - 2)^2$

ANSWERS

1. ____________
2. ____________
3. ____________
4. ____________
5. ____________
6. ____________
7. ____________
8. ____________
9. ____________
10. ____________

11. ______	12. ______
13. ______	14. ______
15. ______	16. ______
17. ______	18. ______
19. ______	20. ______
21. ______	22. ______
23. ______	24. ______
25. ______	26. ______
27. ______	28. ______
29. ______	30. ______
31. ______	32. ______
33. ______	34. ______
35. ______	36. ______
37. ______	38. ______

ANSWERS

39. ____ 40. ____
41. ____ 42. ____
43. ____ 44. ____
45. ____ 46. ____
47. ____ 48. ____
49. ____ 50. ____
51. ____ 52. ____
53. ____ 54. ____
55. ____ 56. ____
57. ____ 58. ____
59. ____ 60. ____
61. ____ 62. ____
63. ____ 64. ____
65. ____ 66. ____
67. ____ 68. ____
69. ____
70. ____
71. ____
72. ____
73. ____
74. ____
75. ____
76. ____
77. ____
78. ____

39. $16 \div 2^3 \cdot 2 - 3 + 11$

40. $10 - 3 \cdot 8 \div 4 + 3$

41. $-12 - 8 \div 4$

42. $48 \div 8 - 14 \div 2$

43. $(2^3 + 3)^2 + 12 \div 3 \cdot 2$

44. $(4 \cdot 3 + 13) \div 5 \cdot 3^2$

45. $3[35 - 3(6 - 2)^2]$

46. $3[14 - 2(5 - 3)^3]$

47. $\dfrac{5 - 15}{2 + 3}$

48. $\dfrac{4 - (-8)}{2 - 5}$

49. $\dfrac{-6 + 18}{-2 - 4}$

50. $\dfrac{-4 - 21}{3 - 8}$

51. $\dfrac{(5)(-12)}{(-3)(5)}$

52. $\dfrac{(-8)(-3)}{(2)(-4)}$

In each exercise, apply the commutative and associative properties to rewrite the expression. Then simplify the result.

53. $(b + 5) + 3$

54. $(x + 2) + 8$

55. $8 + (6 + a)$

56. $10 + (2 + y)$

57. $(2x + 5) + 12$

58. $(2w + 2) + 10$

59. $8 + (p + 6)$

60. $6 + (2m + 12)$

61. $(8 + a) + (-8)$

62. $-2 + (p + 2)$

63. $2(8x)$

64. $6(2b)$

65. $\dfrac{1}{4}(4w)$

66. $6p\left(\dfrac{1}{6}\right)$

67. $\left(\dfrac{2}{7}\right)\left(\dfrac{7}{2}\right)\left(\dfrac{1}{m}\right)m$

68. $\left(\dfrac{3}{4}\right)(b)\left(\dfrac{4}{3}\right)$

In each exercise apply the distributive property to rewrite the expression. Then simplify the result when possible.

69. $5(2m + 3)$

70. $2(4p + 5)$

71. $4a(a + 4)$

72. $6b(b + 5)$

73. $\dfrac{1}{2}(4a + 10)$

74. $\dfrac{1}{3}(6y + 15)$

75. $5(3a + 2b + 4)$

76. $6(3m + 6n + 7)$

77. $\dfrac{1}{2}(4a + 6b + 2c)$

78. $\dfrac{2}{3}(3x + 6y + 9z)$

In each exercise, apply the distributive property to simplify the expression.

79. $8b + 2b$

80. $10a + 2a$

81. $3m + 4m + m$

82. $b + 11b + 3b$

83. $\frac{2}{3}a + \frac{4}{3}a$

84. $\frac{2}{5}b + \frac{8}{5}b$

85. $\frac{1}{2}a + \frac{1}{3}a$

86. $\frac{3}{4}m + \frac{5}{6}m$

In each exercise, apply the appropriate properties to rewrite the expression. Then simplify the result.

87. $6x + (2 + 3x)$

88. $5p + (3 + 9p)$

89. $8y + (2y + 5)$

90. $8m + (3 + 4m)$

91. $2x + 9 + 4x + 6$

92. $2a + 1 + 9a + 7$

93. $3b + 2b + 5 + 4b$

94. $6x + 7 + 8x + 10$

95. $3 + 7y + (-3) + y$

96. $w + (-7) + 2w + 7$

97. $2 + 3(2y + 1) + 3y$

98. $5 + 2(3b + 3) + 4b$

99. $2y^2 + 3y(2 + y) + 3y$

100. $5n + 3n(n + 2) + 2n^2$

State the property used to justify the following statements.

101. $2 + 8 = 8 + 2$

102. $3 \cdot 6$ is a real number

103. $2(y + 5) = 2y + 10$

104. $6(2x) = (6 \cdot 2)x$

105. $4 + (5 + 6) = (4 + 5) + 6$

106. $4 + (5 + 6) = (5 + 6) + 4$

107. $18b + 6 + 12b = 18b + 12b + 6$

108. $18b + (12b + 6) = (18b + 12b) + 6$

109. $(18b + 12b) + 6 = (18 + 12)b + 6$

110. $\left(\frac{2}{7}\right)\left(\frac{7}{2}\right) = 1$

111. $\frac{3}{5} + \left(-\frac{3}{5}\right) = 0$

112. $(y + 3)(y + 2) = y(y + 2) + 3(y + 2)$

ANSWERS

79. ______ 80. ______
81. ______ 82. ______
83. ______ 84. ______
85. ______ 86. ______
87. ______ 88. ______
89. ______ 90. ______
91. ______ 92. ______
93. ______ 94. ______
95. ______ 96. ______
97. ______
98. ______
99. ______
100. ______
101. ______
102. ______
103. ______
104. ______
105. ______
106. ______
107. ______
108. ______
109. ______
110. ______
111. ______
112. ______

ANSWERS

113. ______
114. ______
115. ______
116. ______
117. ______
118. ______
119. ______
120. ______
121. ______
122. ______
123. ______
124. ______
125. ______
126. ______
127. ______
128. ______

Determine whether each statement is true or false. If it is false, rewrite the right side of the equation to make it a true statement.

113. $3 + 5(y + 4) = 3 + 5y + 4$

114. $10 + 5x + 5 = 5(2 + x + 5)$

115. $7b + 8b = 15b$

116. $3a + (10 + 2a) = (3a + 2a) + 10$

117. $4(3w + 3) = 12w + 7$

118. $\frac{1}{5}y + \frac{2}{5}y = \frac{3}{10}y$

119. $3m + 4m + 1 = 7m + 1$

120. $3b + 2b + 5 = 10b$

121. $6y + (-6)y = 0$

122. $4b + (-4b) = b$

123. $2n + 6n = 8n^2$

124. $3a + a = 3a^2$

125. A local baker observed that the sales in her store in May were twice the sales in April. She also observed that the sales in June were three-fourths the sales in April. Use variables to describe the sales of the bakery in each of the 3 months.

126. Computer Corner noted that the sales of computers in August were three-fourths of the sales of computers in July. The sales of computers in September were five-sixths of the sales of computers in July. Use variables to describe the number of computers sold in each of the 3 months.

127. Create an example to show that subtraction of signed numbers is *not* commutative.

128. Create an example to show that division of signed numbers is *not* associative.

Answers

1. $10 + x$ **3.** $p + 12$ **5.** $n + 1$ **7.** $m - 14$ **9.** $x - 1$ **11.** mn
13. $\frac{s}{4}$ **15.** $2(c - d)$ **17.** $rs - 4$ **19.** $\frac{c + 4}{d}$ **21.** 29 **23.** 42
25. 96 **27.** 56 **29.** 21 **31.** 24 **33.** 37 **35.** 56 **37.** 16
39. 12 **41.** −14 **43.** 129 **45.** −39 **47.** −2 **49.** −2
51. 4 **53.** $b + 8$ **55.** $14 + a$ **57.** $2x + 17$ **59.** $p + 14$ **61.** a
63. $16x$ **65.** w **67.** 1 **69.** $10m + 15$ **71.** $4a^2 + 16a$
73. $2a + 5$ **75.** $15a + 10b + 20$ **77.** $2a + 3b + c$ **79.** $10b$ **81.** $8m$
83. $2a$ **85.** $\frac{5}{6}a$ **87.** $9x + 2$ **89.** $10y + 5$ **91.** $6x + 15$
93. $9b + 5$ **95.** $8y$ **97.** $9y + 5$ **99.** $5y^2 + 9y$
101. Commutative property of addition **103.** Distributive property
105. Associative property of addition **107.** Commutative property of addition
109. Distributive property **111.** Additive inverse **113.** False, $3 + 5y + 20$
115. True **117.** False, $12w + 12$ **119.** True **121.** True
123. False, $8n$ **125.** April: x; May: $2x$; June: $\frac{3}{4}x$ **127.**

1.3 Inequalities and Absolute Value

1.3 OBJECTIVES

1. Use the notation of inequalities
2. Graph inequalities
3. Use the absolute value notation

Let's now consider two relations on the set of real numbers. These are the relations of order or inequality known as **less than** or **greater than.**

The set of real numbers is an ordered set. Given any two numbers, we can determine whether one number is less than, equal to, or greater than the other. Let's see how this is expressed symbolically.

We use the **inequality symbol** < to represent "less than," and we write

$a < b$ This is read "*a* is less than *b*"

to indicate that *a* is less than *b*. The number line gives us a clear picture of the meaning of this statement. The point corresponding to *a* must lie *to the left* of the point corresponding to *b*.

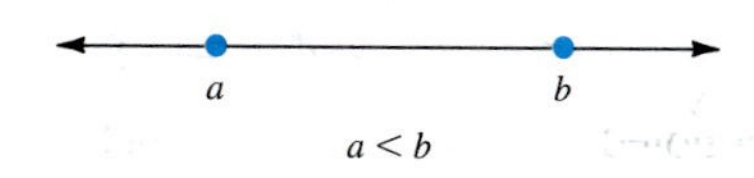

$a < b$

NOTE Notice that

$a > b$

and

$b < a$

are equivalent statements. The symbol "points to" the smaller quantity.

Similarly, the inequality symbol > represents "greater than," and the statement

$a > b$ This is read "*a* is greater than *b*"

indicates that *a* is greater than *b* and means that the point corresponding to *a* on the number line lies *to the right* of the point corresponding to *b*.

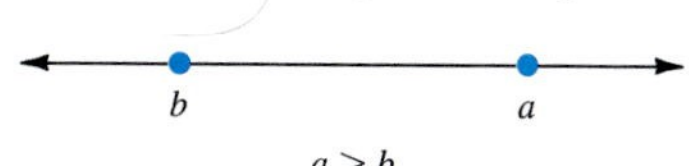

$a > b$

The following example illustrates the use of the inequality symbols.

Example 1

Establishing the Direction of Inequalities

Complete each statement by inserting the symbol < or > between the given numbers.

(a) 2 ______ 8

$2 < 8$

(b) 2.786 ______ 2.78

$2.786 > 2.78$

(c) −23 ______ −5

$-23 < -5$

(d) $\sqrt{2}$ ______ 1.4

$\sqrt{2} > 1.4$ Recall that 1.414 is an approximation for $\sqrt{2}$.

CHECK YOURSELF 1

Insert an inequality symbol that makes each of the following a true statement.

(a) 5 ______ −2

(b) 3.14 ______ π

(c) −10 ______ −15

(d) $\sqrt{15}$ ______ 4

(e) 9.78 ______ 9.87

(f) −1.3 ______ $-\frac{4}{3}$

Suppose we are given an inequality of the form

$x > -1$

The **solution set** for an inequality (as it is for an equation) is the set of all values for the variable that make the inequality a true statement. A convenient way to picture that solution set is by a graph on a number line. The following example illustrates.

Example 2

Graphing Inequalities

Graph the following set.

$\{x \mid x < 4\}$.

NOTE This set is read "the set of all x such that x is less than 4."

We want to include all real numbers less than 4, that is, *to the left* of 4 on the number line.

NOTE The parenthesis at 4 means that the point corresponding to 4 is *not included* in the graph. Such a graph is called an *open half line.*

CHECK YOURSELF 2

Graph the following sets.

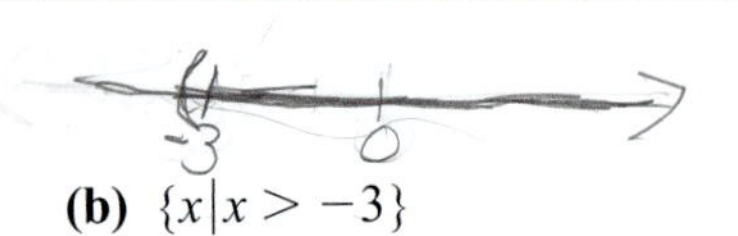

(a) $\{x \mid x < 5\}$ **(b)** $\{x \mid x > -3\}$

Two other symbols, $\leq$ and $\geq$, are also used in writing inequalities. In each case they combine the inequality symbols for less than or greater than with the symbol for equality. The following shows the use of these new symbols. The statement

$a \leq b$

NOTE This combines the symbols < and = and means that either $a < b$ or $a = b$.

is read "a is less than or equal to b." Similarly,

$a \geq b$

NOTE Here either $a > b$ or $a = b$.

is read "a is greater than or equal to b." We consider the graph of inequalities involving these symbols in our next example.

Example 3

Graphing Inequalities

Graph of the following set.

$\left\{x \mid x \geq \frac{7}{2}\right\}$

Here we want all numbers to the right of $\frac{7}{2}$ *and including* $\frac{7}{2}$.

NOTE Here the bracket at $\frac{7}{2}$ means that the point corresponding to $\frac{7}{2}$ *is included* in the graph. Such a graph is called a *closed half line.*

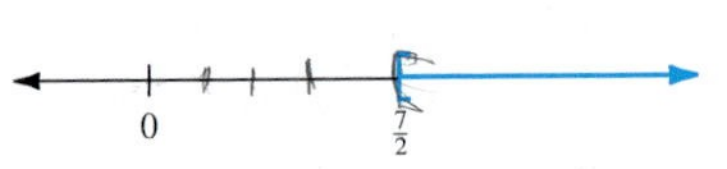

CHECK YOURSELF 3

Graph each of the following sets.

(a) $\{x \mid x \leq 7\}$ **(b)** $\left\{x \mid x \geq -\frac{4}{3}\right\}$

Note: You may very well encounter a different notation for indicating the graphs of inequalities. This involves the use of circles to represent open and closed half lines. For example, the graph of $\{x \mid x > 3\}$ can be drawn as

NOTE The open circle is used to indicate the *open* half line, extending to the right, and *not including* 3.

and the graph of $\{x \mid x \leq -2\}$ as

NOTE The closed circle is used to indicate the *closed* half line, extending to the left, and *including* −2.

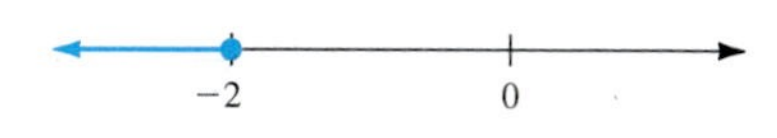

Our subsequent work with inequalities in Chapters 2, 4, 5 and 9 involves the use of a **double-inequality** statement such as

$$-3 < x < 4$$

This statement combines the two inequalities

NOTE The word "and" is implied in any double-inequality statement.

$x > -3$ *and* $x < 4$

or,

$$\{x \mid x > -3 \text{ and } x < 4\}$$

That is why it is sometimes called a **compound inequality.**

In our next example we look at the graphs of inequalities that have this form.

Example 4

Graphing Compound Inequalities

Graph of the following set.

$$\{x \mid -3 < x < 4\}$$

For the solution set of this double inequality, we want all points that lie to the right of -3 $(x > -3)$ and to the left of 4 $(x < 4)$. This means that we should include all points that lie *between* -3 and 4.

NOTE The parentheses indicate that the endpoints, −3 and 4, are *not included* in the graph. This is called an *open interval.*

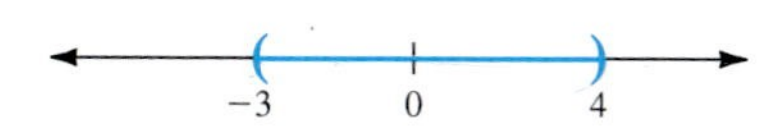

CHECK YOURSELF 4

Graph each of the following sets.

(a) $\{x \mid -1 < x < 6\}$ **(b)** $\{x \mid -2 \leq x < 8\}$

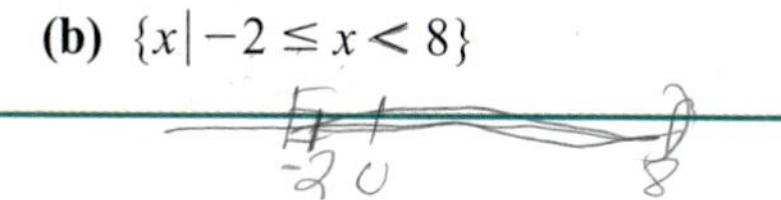

Once again, we refer to the number line to introduce our final topic of this section. If we locate the number 4 and its additive inverse, −4, on the number line, we see that both numbers correspond to points that are the same distance (4 units) from the origin.

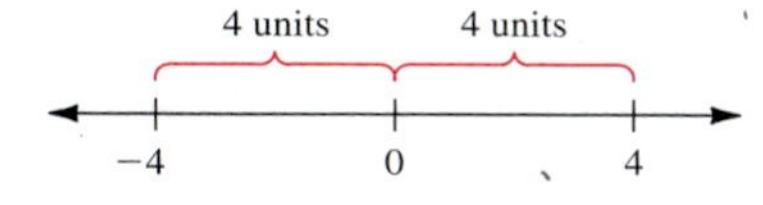

When we are concerned not with the direction (left or right) of a number from the origin, but only with the distance from the origin, we refer to that number's absolute value.

An **absolute value** is the distance (on the number line) between the point named by that real number and the origin. We indicate the absolute value of a number with vertical bars.

In general, we can define the absolute value of any real number a as

NOTE If a is *negative*, then its additive inverse, $-a$, must be *positive* and we want a *positive* absolute value. Say $a = -2$, then $|-2| = -(-2) = 2$.

$$|a| = \begin{cases} a & \text{if } a \text{ is positive} \\ 0 & \text{if } a \text{ is zero} \\ -a & \text{if } a \text{ is negative} \end{cases}$$

The use of the absolute value notation is illustrated in our final example.

Example 5

Evaluating Absolute Value Expressions

Evaluate each of the following expressions.

(a) $|32| = 32$ (b) $|-2.5| = 2.5$

(c) $|\sqrt{2}| = \sqrt{2}$ (d) $|-\sqrt{2}| = \sqrt{2}$

(e) $-|-5| = -5$ (f) $|-3| + |-7| = 3 + 7 = 10$

NOTE $|-5|$ is 5, so $-|-5|$ must be -5.

CHECK YOURSELF 5

Evaluate each of the following expressions.

(a) $|121|$ (b) $|-3.4|$ (c) $|\sqrt{3}|$

(d) $|-\sqrt{5}|$ (e) $-|-8|$ (f) $|-9| + |-2|$

CHECK YOURSELF ANSWERS

1. **(a)** $5 > -2$; **(b)** $3.14 < \pi$; **(c)** $-10 > -15$; **(d)** $\sqrt{15} < 4$; **(e)** $9.78 < 9.87$; **(f)** $-1.3 > -\frac{4}{3}$

2. **(a)** $\{x | x < 5\}$ **(b)** $\{x | x > -3\}$

3. **(a)** $\{x | x \le 7\}$ **(b)** $\left\{x \middle| x \ge -\frac{4}{3}\right\}$

4. **(a)** $\{x | -1 < x < 6\}$ **(b)** $\{x | -2 \le x < 8\}$

(This is a *half open* interval.)

5. **(a)** 121; **(b)** 3.4; **(c)** $\sqrt{3}$; **(d)** $\sqrt{5}$; **(e)** -8; **(f)** 11

Name ______________________

Section ________ Date ________

1.3 Exercises

Insert an inequality symbol or an equal sign to make each of the following a true statement.

1. 8 __>__ 3
2. -3 ______ 7
3. -6 __≤__ -4
4. -2 ______ -3
5. -2.6 __>__ -3.8
6. -7.40 ______ -7.4
7. $-\frac{5}{3}$ __<__ $-\frac{5}{4}$
8. -1.2 ______ $-\frac{4}{3}$
9. $\sqrt{2}$ __>__ 1.4
10. $-\sqrt{3}$ ______ -1.4
11. 1.75 __≥__ $\frac{7}{4}$
12. $\frac{8}{3}$ ______ 2.33
13. $|-3|$ __>__ -3
14. $|-3|$ ______ $|-1|$
15. $-|5|$ __≤__ $|-5|$
16. $|4|$ ______ $-|-4|$

Write each of the following inequalities in words.

17. $y \geq 2$ y is greater than or equal to 2
18. $x > 3$
19. $m < -3$ m is less than -3
20. $n \leq 5$
21. $a \geq b$ a is greater than or equal to b
22. $r < s$
23. $b > 0$ b is greaters than 0
24. $y \leq 0$
25. $-3 < p < 7$ -3 less than p is less than 7
26. $-4 \leq y \leq -2$

Graph each of the following sets. Assume x represents a real number.

27. $\{x \mid x < 4\}$
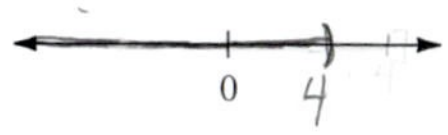
28. $\{x \mid x \geq 5\}$
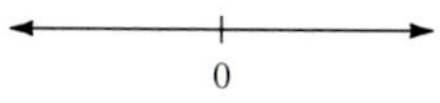
29. $\{x \mid x > -4\}$
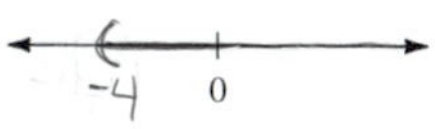
30. $\{x \mid x \leq -2\}$
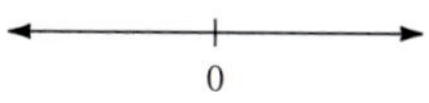
31. $\{x \mid -4 \geq x\}$
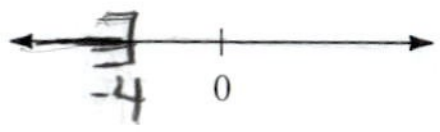
32. $\{x \mid 4 < x\}$
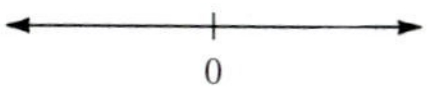
33. $\{x \mid x \geq 4\}$
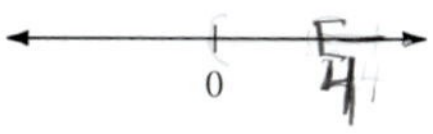
34. $\{x \mid 0 \geq x\}$
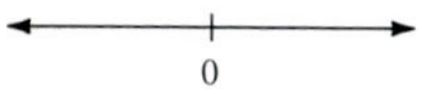

ANSWERS

1.	2.
3.	4.
5.	6.
7.	8.
9.	10.
11.	12.
13.	14.
15.	16.

17. ______
18. ______
19. ______
20. ______
21. ______
22. ______
23. ______
24. ______
25. ______
26. ______
27. ______
28. ______
29. ______
30. ______
31. ______
32. ______
33. ______
34. ______

ANSWERS

35. ______
36. ______
37. ______
38. ______
39. ______
40. ______
41. ______
42. ______
43. ______
44. ______
45. ______
46. ______
47. ______
48. ______
49. ______ 50. ______
51. ______ 52. ______
53. ______ 54. ______
55. ______ 56. ______
57. ______ 58. ______
59. ______ 60. ______
61. ______ 62. ______
63. ______ 64. ______

35. $\{x \mid 1 < x < 2\}$

36. $\{x \mid -3 \le x \le 5\}$

37. $\{x \mid -4 < x \le -1\}$

38. $\{x \mid 2 \le x < 5\}$

Rewrite each of the following statements, using inequality symbols. Then graph the solution set for each inequality. Assume that x represents a real number.

39. x is less than 3

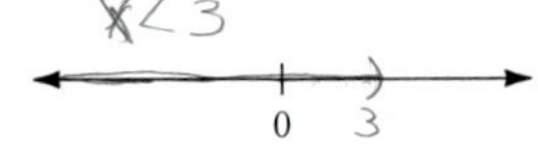

40. x is more than -2

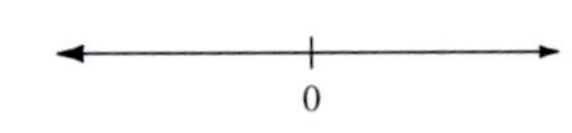

41. x is at least -1

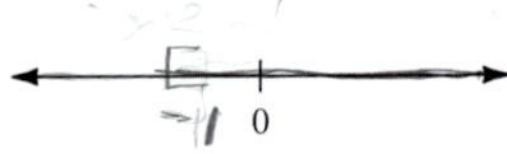

42. x is no more than 5

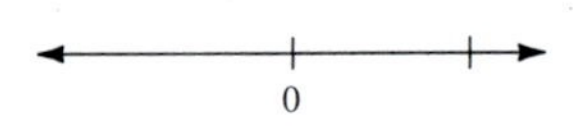

43. x is greater than 4

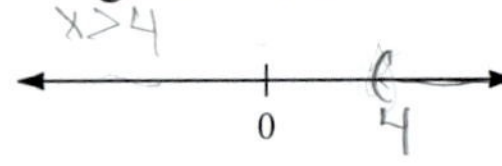

44. x is at least 2

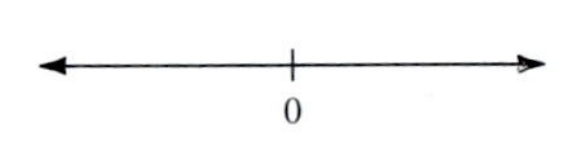

45. x is no more than -2

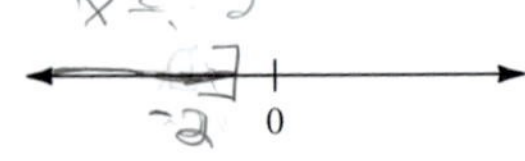

46. x is not less than -2

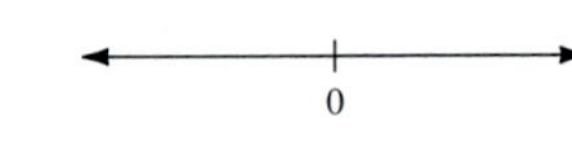

47. 2 is less than x, and x is less than 4

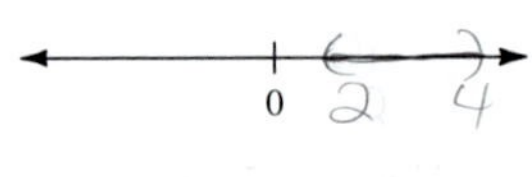

48. -4 is less than or equal to x, and x is less than -2

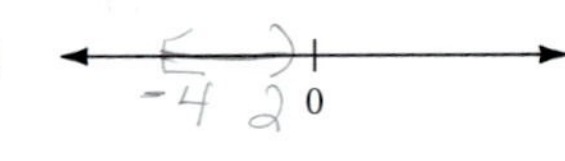

Write each of the following expressions without the absolute value symbol.

49. $|5|$

50. $|-5|$

51. $|-2.5|$

52. $|4.5|$

53. $\left|\frac{5}{6}\right|$

54. $-\left|\frac{7}{8}\right|$

55. $-|1.2|$

56. $|-4.5|$

57. $-|-3|$

58. $-|-6|$

59. $-\left|\frac{2}{3}\right|$

60. $\left|-\frac{7}{2}\right|$

61. $|-3| + |-5|$

62. $|-2| + |-6|$

63. $-(|3| + |-4|)$

64. $-(|-4| + |-8|)$

Label each statement as true or false. If it is false, explain.

65. $|-2| \geq 2$

66. $|0| = 0$

67. $|-12| = |12|$

68. $|2 - 3| > |0 - 1|$

69. $|-x| = x$ *Hint:* For this statement to be true, it must be true for all values of the variable.

70. $-|a| = -a$

71. The absolute value of any real number is positive or zero.

72. Some real numbers have no absolute value.

73. The absolute value of any real number is equal to the absolute value of its additive inverse.

74. There is only one real number that is equal to its own absolute value.

75. Do you think that the following statement is true?

$|a + b| = |a| + |b|$ for all numbers a and b

When we don't know whether such a statement is true, we refer to the statement as a **conjecture.** We may "test" the conjecture by substituting specific numbers for the variables.

Test the conjecture using two positive numbers for a and b.

Test again using a positive number for a and 0 for b.

Test again using two negative numbers.

Now try using one positive number and one negative number.

Summarize your results in a rule that you think is true.

In exercises 76 to 80, test the given conjecture, as you did in exercise 75.

76. Do you think that the following statement is true?

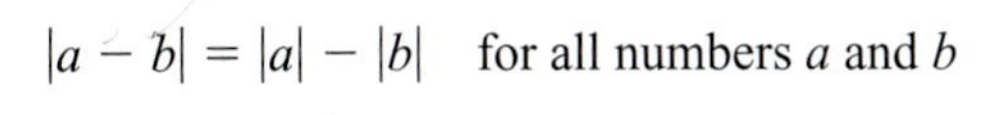

$|a - b| = |a| - |b|$ for all numbers a and b

77. Do you think that the following statement is true?

$|a \cdot b| = |a| \cdot |b|$ for all numbers a and b

78. Do you think that the following statement is true?

$|a - b| = |b - a|$ for all numbers a and b

ANSWERS

65. ______

66. ______

67. ______

68. ______

69. ______

70. ______

71. ______

72. ______

73. ______

74. ______

75. ______

76. ______

77. ______

78. ______

ANSWERS

79. 

80. 

81.

79. Do you think that the following statement is true?

$$\left|\frac{a}{b}\right| = \frac{|a|}{|b|} \quad \text{for all numbers } a \text{ and } b$$

80. Do you think that the following statement is true?

$$|a| = |-a| \quad \text{for any number } a$$

81. If a represents a positive number and b represents a negative number, determine whether the given expression is positive or negative.

(a) $|b| + a$ **(b)** $b + (-a)$ **(c)** $(-b) + a$ **(d)** $-b + |-a|$

Answers

1. $>$ **3.** $<$ **5.** $>$ **7.** $<$ **9.** $>$ **11.** $\geq$ or $\leq$ **13.** $>$
15. $<$ **17.** y is greater than or equal to 2 **19.** m is less than -3
21. a is greater than or equal to b **23.** b is greater than 0
25. -3 is less than p and p is less than 7

27.  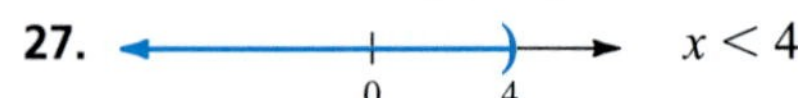 $x < 4$ **29.** $x > -4$

31.  $x \leq -4$ **33.** $x \geq 4$

35. $1 < x < 2$

37. $-4 < x \leq -1$

39.  $x < 3$ **41.** $x \geq -1$

43. $x > 4$ **45.** $x \leq -2$

47. $2 < x < 4$ **49.** 5 **51.** 2.5 **53.** $\frac{5}{6}$

55. -1.2 **57.** -3 **59.** $-\frac{2}{3}$ **61.** 8 **63.** -7 **65.** True

67. True

69. False, it would not be true for negative values of x, for example, $|-(-1)| \neq -1$.

71. True **73.** True **75.** **77.**  **79.** 

81. **(a)** Positive; **(b)** negative; **(c)** positive; **(d)** positive

1.4 Positive Integer Exponents and Scientific Notation

OBJECTIVES

1. Use the properties of exponents
2. Use scientific notation

Exponents are used as a short-hand form for repeated multiplication. Instead of writing

$a \cdot a \cdot a \cdot a \cdot a$

we write

NOTE We call *a* the base of the expression and 5 the exponent or the power.

a^5

which we read as "*a* to the fifth power."

Definitions: Exponential Form

In general, for any real number *a* and any natural number *n*,

$$a^n = \underbrace{a \cdot a \cdot \cdots \cdot a}_{n \text{ factors}}$$

An expression of this type is said to be in **exponential form.** We call *a* the **base** of the expression and *n* the **exponent,** or the **power.**

Let's consider what happens when we multiply two expressions in exponential form with the same base.

NOTE We expand the expressions and apply the associative property to regroup.

$$a^4 \cdot a^5 = \underbrace{(a \cdot a \cdot a \cdot a)}_{4 \text{ factors}}\underbrace{(a \cdot a \cdot a \cdot a \cdot a)}_{5 \text{ factors}}$$

$$= \underbrace{a \cdot a \cdot a \cdot a \cdot a \cdot a \cdot a \cdot a \cdot a}_{9 \text{ factors}}$$

$$= a^9$$

Notice that the product is simply the base taken to the power that is the sum of the two original exponents.

In fact, in general, the following holds:

Rules and Properties: First Property of Exponents

NOTE This is our *first property of exponents*
$a^m \cdot a^n = a^{m+n}$

For any real number *a* and natural numbers *m* and *n*,

$$a^m \cdot a^n = \underbrace{(a \cdot a \cdot \cdots \cdot a)}_{m \text{ factors}}\underbrace{(a \cdot a \cdot \cdots \cdot a)}_{n \text{ factors}}$$

$$= \underbrace{a \cdot a \cdot \cdots \cdot a}_{m + n \text{ factors}}$$

$$= a^{m+n}$$

Example 1

Simplifying Expressions

Simplify each expression.

(a) $b^4 \cdot b^6 = b^{10}$

(b) $(2a)^3 \cdot (2a)^4 = (2a)^7$

(c) $(-2)^5(-2)^4 = (-2)^9$

(d) $10^7 \cdot 10^{11} = 10^{18}$

CHECK YOURSELF 1

Simplify each product.

(a) $(5b)^6(5b)^5$ $5b^{11}$

(b) $(-3)^4(-3)^3$ -3^7

(c) $10^8 \cdot 10^{12}$ 10^{20}

(d) $(xy)^2(xy)^3$ xy^5

Applying the commutative and associative properties of multiplication, we know that a product such as

$2x^3 \cdot 3x^2$

can be rewritten as

$(2 \cdot 3)(x^3 \cdot x^2)$

or as

$6x^5$

We expand on the ideas illustrated above in our next example.

Example 2

Simplifying Expressions

Using the first property of exponents together with the commutative and associative properties, simplify each product.

NOTE Multiply the coefficients and *add* the exponents by Property 1. With practice you will *not need to write* the regrouping step.

(a) $(5x^4)(3x^2) = (5 \cdot 3)x^4x^2 = 15x^6$

(b) $(x^2y^3)(x^2y^4) = (x^2 \cdot x^2)(y^3 \cdot y^4) = x^4y^7$

(c) $(4c^5d^3)(3c^2d^2) = (4 \cdot 3)(c^5c^2)(d^3d^2) = 12c^7d^5$

CHECK YOURSELF 2

Simplify each expression.

(a) $(4a^2b)(2a^3b^4)$ $8a^5b^5$

(b) $(3x^4)(2x^3y)$ $6x^7y$

We now consider a second property of exponents that can be used to simplify quotients of expressions in exponential form that have the same base.

Consider the quotient

$$\frac{a^6}{a^4}$$

If we write this in expanded form, we have

$$\frac{\overbrace{a \cdot a \cdot a \cdot a \cdot a \cdot a}^{\text{6 factors}}}{\underbrace{a \cdot a \cdot a \cdot a}_{\text{4 factors}}}$$

This can be reduced to

NOTE Divide the numerator and denominator by the four common factors of a.

$$\frac{a \cdot a \cdot a \cdot a \cdot a \cdot a}{a \cdot a \cdot a \cdot a} \quad \text{or} \quad a^2$$

NOTE Notice that $\frac{a}{a} = 1$, when $a \neq 0$.

This means that

$$\frac{a^6}{a^4} = a^2$$

Rules and Properties: Second Property of Exponents

NOTE This is our *second property of exponents.* We write $a \neq 0$ to avoid division by zero.

In general, for any real number a ($a \neq 0$) and natural numbers m and n, $m > n$,

$$\frac{a^m}{a^n} = a^{m-n}$$

Example 3

Simplifying Expressions

Simplify each expression.

(a) $\frac{x^{10}}{x^4}$ x6 (b) $\frac{a^8}{a^7}$ a (c) $\frac{63w^8}{7w^5}$ 9w3

(d) $\frac{-32a^4b^5}{8a^2b}$ -4a2b4 (e) $\frac{10^{16}}{10^6}$ 10 10

NOTE Subtract the exponents, applying the second property.

(a) $\frac{x^{10}}{x^4} = x^{10-4} = x^6$

NOTE Notice that $a^1 = a$; there is no need to write the exponent of 1 because it is understood.

(b) $\frac{a^8}{a^7} = a^{8-7} = a$

NOTE We *divide* the coefficients and subtract the exponents.

(c) $\frac{63w^8}{7w^5} = 9w^{8-5} = 9w^3$

NOTE Divide the coefficients and subtract the exponents for *each* variable.

(d) $\frac{-32a^4b^5}{8a^2b} = -4a^{4-2}b^{5-1} = -4a^2b^4$

(e) $\frac{10^{16}}{10^6} = 10^{16-6} = 10^{10}$

CHECK YOURSELF 3

Simplify each expression.

(a) $\dfrac{y^{12}}{y^5}$ **(b)** $\dfrac{x^9}{x^8}$ **(c)** $\dfrac{45r^8}{-9r^6}$

(d) $\dfrac{49a^6b^7}{7ab^3}$ **(e)** $\dfrac{10^{13}}{10^5}$

Suppose that we have an expression of the form

$(a^2)^4$

This can be written as

$\underbrace{(a \cdot a)(a \cdot a)(a \cdot a)(a \cdot a)}_{2 \cdot 4 \text{ or } 8 \text{ factors}}$ or a^8

This suggests in general, the following:

Rules and Properties: Third Property of Exponents

For any real number a and natural numbers m and n,

$(a^m)^n = a^{mn}$

NOTE This is our *third property of exponents.*

Our next example illustrates the use of this third property of exponents.

Example 4

Simplifying Expressions

Using Property 3 of exponents, simplify each expression.

NOTE We *multiply* the exponents.

(a) $(x^3)^5 = x^{3 \cdot 5} = x^{15}$

(b) $(a^2)^8 = a^{2 \cdot 8} = a^{16}$

(c) $(10^2)^3 = 10^{2 \cdot 3} = 10^6$

Be Careful! Students sometimes confuse $(x^3)^5$ or x^{15} with $x^3 \cdot x^5$ or x^8. In the first case we *multiply* the exponents, in the second we *add!*

CHECK YOURSELF 4

Simplify each expression.

(a) $(b^4)^7$ **(b)** $(10^3)^3$ **(c)** b^4b^7

Let's develop another property for exponents. An expression such as

$(2x)^5$

can be written in expanded form as

$$\underbrace{(2x)(2x)(2x)(2x)(2x)}_{\text{5 factors}}$$

We could use the commutative and associative properties to write this product as

$(2 \cdot 2 \cdot 2 \cdot 2 \cdot 2)(x \cdot x \cdot x \cdot x \cdot x)$

or

2^5x^5

NOTE Notice that each factor of the base has been raised to the fifth power.

This suggests our fourth property of exponents.

Rules and Properties: Fourth Property of Exponents

For any real numbers a and b and any natural number m,

$(ab)^m = a^m b^m$

NOTE This is our *fourth property of exponents.*

The use of this fourth property is illustrated in our next example.

Example 5

Simplifying Expressions

Simplify each expression.

(a) $(xy)^5 = x^5y^5$

(b) $(10a)^4 = 10^4 \cdot a^4 = 10{,}000a^4$

(c) $(2p^2q^3)^3 = 2^3(p^2)^3(q^3)^3$

$= 8p^6q^9$

NOTE Notice that we also apply the third property in simplifying this expression.

CHECK YOURSELF 5

Simplify each expression.

(a) $(ab)^7$ **(b)** $(4p)^3$ **(c)** $(3m^4n^2)^2$

Our fifth (and final) property of exponents can be established in a similar fashion to the fourth property. It deals with the power of a quotient rather than the power of a product.

Rules and Properties: Fifth Property of Exponents

For any real numbers a and b ($b \neq 0$) and natural number m,

$$\left(\frac{a}{b}\right)^m = \frac{a^m}{b^m}$$

NOTE This is our *fifth property of exponents.*

Our next example shows the application of this property.

Example 6

Simplifying Expressions

Simplify each expression.

(a) $\left(\frac{x^2}{y}\right)^3 = \frac{(x^2)^3}{y^3} = \frac{x^6}{y^3}$ — Property 5, Property 3

(b) $\left(\frac{2a}{b^3}\right)^4 = \frac{(2a)^4}{(b^3)^4}$ — Property 5

$= \frac{2^4a^4}{b^{12}} = \frac{16a^4}{b^{12}}$ — Property 4, Property 3

CHECK YOURSELF 6

Simplify each expression.

(a) $\left(\frac{m^3}{n}\right)^4$ (b) $\left(\frac{3t^2}{s^3}\right)^3$

As we have seen, more complicated expressions require the use of more than one of our properties, for simplification. The next example illustrates other such cases.

Example 7

Simplifying Expressions

Use the properties of exponents to simplify the following expressions.

(a) $(2x)^2(2x)^4 = (2x)^{2+4} = (2x)^6$ — Property 1

$= 2^6x^6 = 64x^6$ — Property 4

(b) $\frac{(x^4)^3}{(x^3)^2} = \frac{x^{12}}{x^6} = x^{12-6} = x^6$ — Property 3, Property 2

(c) $\frac{6a^4b^5}{3a^2b} = 2a^{4-2}b^{5-1} = 2a^2b^4$ — Property 2, Division

(d) $\frac{7.5 \times 10^{14}}{2.5 \times 10^3} = \frac{7.5}{2.5} \times 10^{14-3}$ — Property 2

$= 3 \times 10^{11}$ — Divide.

CHECK YOURSELF 7

Simplify each expression.

(a) $(3y)^2(3y)^3$ **(b)** $\dfrac{(3a^2)^3}{9a^3}$ **(c)** $\dfrac{25x^3y^4}{5x^2y}$

The following table summarizes the five properties of exponents introduced in this section.

General Form	Example
I. $a^m a^n = a^{m+n}$	$x^2 \cdot x^3 = x^5$
II. $\dfrac{a^m}{a^n} = a^{m-n}$ $a \neq 0, m > n$	$\dfrac{5^7}{5^3} = 5^4$
III. $(a^m)^n = a^{mn}$	$(z^5)^4 = z^{20}$
IV. $(ab)^m = a^m b^m$	$(4x)^3 = 4^3x^3 = 64x^3$
V. $\left(\dfrac{a}{b}\right)^m = \dfrac{a^m}{b^m}$ $b \neq 0$	$\left(\dfrac{2}{3}\right)^6 = \dfrac{2^6}{3^6} = \dfrac{64}{729}$

Before leaving the properties of exponents, we would like to make an important extension of one of the properties. In Property 2, suppose that we now allow *m to equal n.* We then have

$$\frac{a^m}{a^m} = a^{m-m} = a^0 \tag{1}$$

But we know that it is also true that

$$\frac{a^m}{a^m} = 1 \tag{2}$$

Comparing (1) and (2), it then seems reasonable to make the following definition.

NOTE With this definition $\dfrac{a^m}{a^n} = a^{m-n}$ when $m \geq n$.

NOTE We must have $a \neq 0$ because the form 0^0 is called *indeterminate.* It is considered in later mathematics classes.

Definitions: The Zero Exponent

For any real number a, $a \neq 0$,

$a^0 = 1$

Example 8

Using Zero as an Exponent

Use the above definition to simplify each expression.

(a) $10^0 = 1$ **(b)** $(a^3b^2)^0 = 1$

(c) $6x^0 = 6 \cdot 1 = 6$

NOTE Notice that in $6x^0$ the zero exponent applies *only* to x.

CHECK YOURSELF 8

Simplify each expression.

(a) 25^0 **(b)** $(m^4n^2)^0$ **(c)** $8s^0$

You may have noticed that throughout this section we have frequently used 10 as a base in our examples. You will find that experience useful as we discuss scientific notation.

We begin the discussion with a calculator exercise. On most (scientific) calculators, if you find 2.3 times 1000, the display will read

2300.

Multiply by 1000 a second time. Now you will see

2300000.

Multiplying by 1000 a third time will result in the display

NOTE This must equal 2,300,000,000.

2.3 09

And multiplying by 1000 again yields

2.3 12

NOTE Consider the following table

$2.3 = 2.3 \times 10^0$
$23 = 2.3 \times 10^1$
$230 = 2.3 \times 10^2$
$2300 = 2.3 \times 10^3$
$23{,}000 = 2.3 \times 10^4$
$230{,}000 = 2.3 \times 10^5$

Can you see what is happening? This is the way calculators display very large numbers. The number on the left is always between 1 and 10, and the number on the right indicates the number of places the decimal point must be moved to the right to put the answer in standard (or decimal) form.

This notation is used frequently in science. It is not uncommon, in scientific applications of algebra, to find yourself working with very large or very small numbers. Even in the time of Archimedes (287–212 B.C.), the study of such numbers was not unusual. Archimedes estimated that the universe was 23,000,000,000,000,000 meters in diameter.

In scientific notation, his estimate for the diameter of the universe would be

$$2.3 \times 10^{16} \text{ m}$$

In general, we can define scientific notation as follows:

Definitions: Scientific Notation

Any number written in the form

$a \times 10^n$

in which $1 \leq a < 10$ and n is an integer, is written in scientific notation.

Example 9

Writing Numbers in Scientific Notation

Write each of the following numbers in scientific notation.

(a) 120,000 **(b)** 88,000,000

(c) 520,000,000 **(d)** 4,000,000,000

NOTE Notice the pattern for writing a number in scientific notation.

NOTE The exponent on the 10 shows the *number of places* we must move the decimal point so that the multiplier will be a number between 1 and 10.

(a) $120000. = 1.2 \times 10^5$

Five places — The power is 5.

(b) $88{,}000{,}000. = 8.8 \times 10^7$

Seven places — The power is 7.

(c) $520{,}000{,}000. = 5.2 \times 10^8$

Eight places

NOTE To convert back to standard or decimal form, the process is simply reversed.

(d) $4{,}000{,}000{,}000. = 4 \times 10^9$

Nine places

CHECK YOURSELF 9

Write in scientific notation.

(a) 212,000,000,000,000,000   **(b)** 5,600,000

2.12×10^{17} 5.6×10^6

Example 10

Applying Scientific Notation

Light travels at a speed of 3.05×10^8 meters per second (m/s). There are approximately 3.15×10^7 s in a year. How far does light travel in a year?

We multiply the distance traveled in 1 s by the number of seconds in a year. This yields

NOTE Multiply the coefficients, add the exponents.

$$(3.05 \times 10^8)(3.15 \times 10^7) = (3.05 \cdot 3.15)(10^8 \cdot 10^7)$$
$$= 9.6075 \times 10^{15}$$

NOTE Notice that $9.6075 \times 10^{15} \approx 10 \times 10^{15} = 10^{16}$

For our purposes we round the distance light travels in a year to 10^{16} m. This unit is called a **light-year,** and it is used to measure astronomical distances.

Example 11

Applying Scientific Notation

The distance from Earth to the star Spica (in Virgo) is 2.2×10^{18} m. How many light-years is the star Spica away from Earth?

NOTE We divide the distance (in meters) by the number of meters in 1 light-year.

$$\frac{2.2 \times 10^{18}}{10^{16}} = 2.2 \times 10^{18-16}$$
$$= 2.2 \times 10^2 = 220 \text{ light-years}$$

CHECK YOURSELF 10

The farthest object that can be seen with the unaided eye is the Andromeda galaxy. This galaxy is 2.3×10^{22} m from Earth. What is this distance in light-years?

CHECK YOURSELF ANSWERS

1. **(a)** $(5b)^{11}$; **(b)** $(-3)^7$; **(c)** 10^{20}; **(d)** $(xy)^5$ **2.** **(a)** $8a^5b^5$; **(b)** $6x^7y$
3. **(a)** y^7; **(b)** x; **(c)** $-5r^2$; **(d)** $7a^5b^4$; **(e)** 10^8 **4.** **(a)** b^{28}; **(b)** 10^9; **(c)** b^{11}
5. **(a)** a^7b^7; **(b)** $64p^3$; **(c)** $9m^8n^4$ **6.** **(a)** $\frac{m^{12}}{n^4}$; **(b)** $\frac{27t^6}{s^9}$
7. **(a)** $243y^5$; **(b)** $3a^3$; **(c)** $5xy^3$ **8.** **(a)** 1; **(b)** 1; **(c)** 8
9. **(a)** 2.12×10^{17}; **(b)** 5.6×10^6 **10.** 2.3×10^6 or 2,300,000 light-years

1.4 Exercises

Write each product in exponential form. Identify the base and the exponent.

1. $a \cdot a \cdot a \cdot a$

2. $x \cdot x \cdot x \cdot x \cdot x \cdot x$

3. $2 \cdot 2 \cdot 2$

4. $10 \cdot 10 \cdot 10 \cdot 10 \cdot 10$

5. $(3x)(3x)(3x)(3x)$

6. $(5p)(5p)(5p)(5p)(5p)$

Simplify each of the following products.

7. $y^3 \cdot y^4$

8. $x^4 \cdot x^6$

9. $p^6 \cdot p^5 \cdot p^4$

10. $a^2 \cdot a^3 \cdot a^4 \cdot a^5$

11. $2^5 \cdot 2^0$

12. $(-2)^6(-2)^4$

13. $(-3)^3(-3)^4(-3)^6$

14. $7^5 \cdot 7^4 \cdot 7^2 \cdot 7^0$

15. $2 \cdot a^5 \cdot a^3 \cdot a^5$

16. $4 \cdot b^5 \cdot b^2 \cdot b^4$

Use the first property of exponents together with the commutative and associative properties to simplify the following products.

17. $(x^2y^3)(x^3y^2)$

18. $(a^5b)(a^2b^4)$

19. $(m^3n^2)(m^5n^3)(m^2n^3)$

20. $(p^2q^3)(p^4q)(p^5q^0)$

21. $(2a^5)(4a^3)(-3a^3)$

22. $(2b^4)(-4b)(-3b^5)$

23. $(6s^2)(3s^4)(s^0)(-2s^2)$

24. $(2a^2)(4a)(a^3)(2a^2)$

25. $(5xy^3)(3x^3y)(2xy)$

26. $(-2ab)(6a^3b)(-3a^2b^0)$

27. $(rs^2t)(r^2s^4t)(r^5st)$

28. $(xyz)(x^7y^4z^6)(x^3yz)(xyz^4)$

Use the second property of exponents to simplify each expression.

29. $\dfrac{a^{10}}{a^6}$

30. $\dfrac{x^{24}}{x^{17}}$

31. $\dfrac{a^6b^{12}}{a^4b^4}$

32. $\dfrac{x^6y^9}{xy^4}$

33. $\dfrac{x^6y^5z^2}{x^2yz^0}$

34. $\dfrac{a^8b^5c^3}{a^2bc^0}$

Name ____________

Section ______ Date ______

ANSWERS

1. ____
2. ____
3. ____
4. ____
5. ____
6. ____

7. ____ 8. ____
9. ____ 10. ____
11. ____ 12. ____
13. ____ 14. ____
15. ____ 16. ____
17. ____ 18. ____
19. ____ 20. ____
21. ____ 22. ____
23. ____ 24. ____
25. ____ 26. ____
27. ____ 28. ____
29. ____ 30. ____
31. ____ 32. ____
33. ____ 34. ____

ANSWERS

35. ______
36. ______
37. ______
38. ______
39. ______ 40. ______
41. ______ 42. ______
43. ______ 44. ______
45. ______ 46. ______
47. ______ 48. ______
49. ______ 50. ______
51. ______
52. ______
53. ______
54. ______
55. ______
56. ______
57. ______
58. ______
59. ______
60. ______
61. ______
62. ______
63. ______
64. ______
65. ______
66. ______

35. $\dfrac{28m^2n^6}{7mn^2}$

36. $\dfrac{36r^6s^6}{12r^2s}$

37. $\dfrac{(2y - 1)^8}{(2y - 1)^6}$

38. $\dfrac{(2x + 3)^5}{(2x + 3)^4}$

Use the properties of exponents to simplify each expression.

39. $(a^6)^3$

40. $(p^6)^4$

41. $(3x^2)(x^3)^4$

42. $(b^2)(2b^2)^3$

43. $(2w^3)(w^4)(w^2)$

44. $(5x^2)(3x^5)(x)$

45. $(m^2n)(m^4)^2(m^4)^0$

46. $(a^3)^2(a^4b)(b^2)^4$

47. $(rs^3t)(r^3)^3(s^3)^2(t^4)^2$

48. $(a^3bc^2)(a^3)(b^2)^3(c^3)^0$

Simplify each expression.

49. $(3b^4)^3(b^2)^3$

50. $(2a^3)^2(a^2)^3(a^4)$

51. $(3w^2)^2(2w^2)^3$

52. $(1 \times 10^3)^2 \cdot (2 \times 10^2)^4$

53. $(x^2y^4)^3(xy^2)^0$

54. $(m^3n^4)^4(m^2n)^3$

55. $(ab^4c)^4(abc^3)^8(a^6bc)^5$

56. $(x^3y^3z^3)^0(xy^3z)^2(x^4yz^3)$

57. $(3a^3)(2b^4)^3$

58. $(2m^3)(5m^3)^2$

59. $(3b^3)^2(b^0)^6$

60. $(4a^0)^2(a^4)^3$

61. $(2x^3)^4(3x^4)^2$

62. $(3a^2)^3(2a^5)^3$

63. $\left(\dfrac{2x^5}{y^3}\right)^2$

64. $\left(\dfrac{2a^5}{3b^8}\right)^3$

65. $\left(\dfrac{m^8}{n^4}\right)\left(\dfrac{3n^7}{m^3}\right)$

66. $\left(\dfrac{x^9}{y^2}\right)\left(\dfrac{3y^5}{2x^3}\right)$

67. $(-8m^2n)(-3m^4n^5)^4$

68. $(5a^5b)(-3a^3b^4)^3$

69. $-w^3(3w^3)^2(2w^4)^3$

70. $-x^3(4x^4)(3x^0)^3$

71. $\left(\dfrac{3a^8b^3}{2x^2y^7}\right)\left(\dfrac{x^6y^5}{a^3b^0}\right)^2$

72. $\left(\dfrac{6x^5y^4}{5cd}\right)\left(\dfrac{c^3d}{xy^3}\right)$

Express each number in scientific notation.

73. The distance from Mars to the sun: 141,000,000 mi

74. The diameter of Earth's orbit: 186,000,000 mi

75. The diameter of Jupiter: 88,000 mi

76. The amount of free oxygen on Earth: 1,500,000,000,000,000,000,000 g

77. The mass of the moon is approximately 7.37×10^{22} kg. If this were written in standard or decimal form, how many zeros would follow the second digit 7?

78. Scientists estimate the mass of our sun to be 1.98×10^{24} kg. If this number were written in standard or decimal form, how many zeros would follow the digit 8?

79. The distance from Pluto to the sun is 5.91×10^{12} mi. If this number were written in standard or decimal form, how many zeros would follow the digit 1?

80. The distance light travels in 100 years is 5.8×10^{14} mi. If this number were written in standard or decimal form, how many zeros would follow the digit 8?

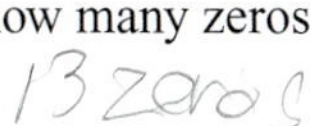

ANSWERS

67. ______
68. ______
69. ______
70. ______
71. ______
72. ______
73. ______
74. ______
75. ______
76. ______
77. ______
78 ______
79. ______
80. ______

ANSWERS

81. ______
82. ______
83. ______
84. ______
85. ______
86. ______
87. ______
88. ______
89. ______
90. ______
91. ______
92. ______
93. ______
94. ______
95. ______
96. ______
97. ______
98. ______
99. ______
100. ______
101. ______
102. ______
103. ______
104. ______

In the expressions below, perform the indicated calculations. Write your result in scientific notation.

81. $(2 \times 10^5)(3 \times 10^3)$

82. $(3.3 \times 10^7)(2 \times 10^4)$

83. $\dfrac{9 \times 10^9}{3 \times 10^6}$

84. $\dfrac{7.5 \times 10^{11}}{1.5 \times 10^7}$

85. $\dfrac{(3.3 \times 10^{15})(9 \times 10^{10})}{(1.1 \times 10^8)(3 \times 10^6)}$

86. $\dfrac{(6 \times 10^{12})(4.8 \times 10^6)}{(1.6 \times 10^7)(3 \times 10^2)}$

87. Alkaid, the most distant star in the Big Dipper, is 2.1×10^{18} m from Earth. Approximately how long does it take light, traveling at 10^{16} m/year, to travel from Alkaid to Earth?

88. Megrez, the nearest of the Big Dipper stars, is 6.6×10^{17} m from Earth. Approximately how long does it take light, traveling at 10^{16} m/year, to travel from Megrez to Earth?

Do each of the following problems.

89. Write 8^3 as a power of 2. (Remember that $8 = 2^3$.)

90. Write 16^6 as a power of 2.

91. Write 9^{12} as a power of 3.

92. Write 81^8 as a power of 3.

93. Write 3^8 as a power of 9.

94. Write 3^{18} as a power of 27.

95. Write 2^9 as a power of 8.

96. Write 10^{20} as a power of 100.

Assume that n is an integer such that all exponents are positive numbers. Then simplify each of the following expressions.

97. $a^{2n} \cdot a^{4n}$

98. $x^{n+1} \cdot x^{2n}$

99. $\dfrac{r^{n+4}}{r^{n+1}}$

100. $(w^n)^{4n}$

101. $(a^{n+2})^n$

102. $\dfrac{(x^{3n})(x^{n+3})}{x^{4n}}$

103. $\dfrac{(w^n)(w^{4n+5})}{w^{5n}}$

104. Do some research to discover the meaning of the word "googol." What are the origins of this term? What connection does it have with this section?

Answers

1. a^4, base a, exponent 4 **3.** 2^3, base 2, exponent 3 **5.** $(3x)^4$, base $3x$, exponent 4 **7.** y^7 **9.** p^{15} **11.** 2^5 **13.** $(-3)^{13}$ **15.** $2a^{13}$ **17.** x^5y^5 **19.** $m^{10}n^8$ **21.** $-24a^{11}$ **23.** $-36s^8$ **25.** $30x^5y^5$ **27.** $r^8s^7t^3$ **29.** a^4 **31.** a^2b^8 **33.** $x^4y^4z^2$ **35.** $4mn^4$ **37.** $(2y-1)^2$ **39.** a^{18} **41.** $3x^{14}$ **43.** $2w^9$ **45.** $m^{10}n$ **47.** $r^{10}s^9t^9$ **49.** $27b^{18}$ **51.** $72w^{10}$ **53.** x^6y^{12} **55.** $a^{42}b^{29}c^{33}$ **57.** $24a^3b^{12}$ **59.** $9b^6$ **61.** $144x^{20}$ **63.** $\dfrac{4x^{10}}{y^6}$ **65.** $3m^5n^3$ **67.** $-648m^{18}n^{21}$ **69.** $-72w^{21}$ **71.** $\dfrac{3a^2b^3x^{10}y^3}{2}$ **73.** 1.41×10^8 **75.** 8.8×10^4 **77.** 20 **79.** 10 **81.** 6×10^8 **83.** 3×10^3 **85.** 9×10^{11} **87.** 210 years **89.** 2^9 **91.** 3^{24} **93.** 9^4 **95.** 8^3 **97.** a^{6n} **99.** r^3 **101.** a^{n^2+2n} **103.** w^5

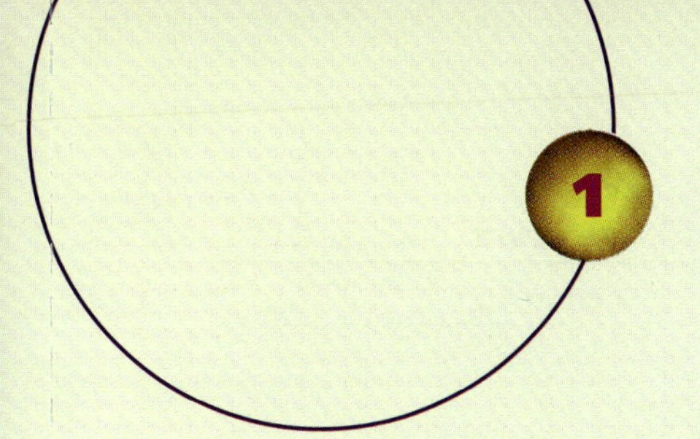

Summary

Definition/Procedure	Example	Reference
Sets		**Section 1.1**
A *set* is a collection of objects, symbols, or numbers. We can describe a set by listing its contents in braces. Sets are often named by capital letters.	$A = \{1, 2, 3, 4\}$ is a set containing the first four counting numbers.	
Any object contained in a set is called an *element,* or a *member,* of that set. We write $$a \in A$$ to indicate that a is an element of set A. We can also write $$a \notin A$$ to indicate that a is *not* an element of set A.	For the set A above $3 \in A$ $5 \notin A$	
Two sets are *equal* if they have exactly the same elements. We write $$A = B$$ to indicate that set A is equal to set B.	If $B = \{4, 3, 2, 1\}$, $A = B$	
A set whose elements continue indefinitely is said to be an *infinite* set, and ellipses (. . .) are often used in describing infinite sets.	$E = \{2, 4, 6, \ldots\}$, the set of positive even numbers, is an infinite set.	
A set that has some specific number of elements is called a *finite* set.	$A = \{1, 2, 3, 4\}$ is a finite set. It has four elements.	**p. 3**
Numeric Sets The set of *natural numbers* is N, and $$N = \{1, 2, 3, \ldots\}$$	$5 \in N \quad -3 \notin N \quad \frac{2}{3} \notin N$	
The set of *integers* is denoted Z, and $$Z = \{\ldots, -2, -1, 0, 1, 2, \ldots\}$$	$7 \in Z \quad -8 \in Z \quad \frac{3}{4} \notin Z$	
The set of *rational numbers* is denoted Q, and $$Q = \left\{\frac{p}{q} \middle\| p, q \in Z, q \neq 0\right\}$$	$-\frac{3}{4} \in Q \quad -5 \in Q \quad \sqrt{2} \notin Q$	
The set of *irrational numbers* is denoted Q', and Q' consists of all numbers that *cannot* be expressed as the ratio of two integers—that is, they are *not* rational.	$\sqrt{2} \in Q' \quad \sqrt[3]{5} \in Q'$ $\pi \in Q' \quad \frac{4}{5} \notin Q'$	
The set of *real numbers* is denoted R, and R combines the set of rational numbers with the set of irrational numbers.	$\frac{2}{3} \in R \quad -8 \in R$ $\sqrt{3} \in R \quad \sqrt{-2} \notin R$	
The set of real numbers can be pictured on the *real number line.* Zero (the origin) Negative numbers Positive numbers −3 −2 −1 0 1 2 3 Every real number corresponds to exactly one point on the number line, and every point on that line corresponds to exactly one real number.	Zero is neither positive nor negative.	**p. 6**

Continued

DEFINITION/PROCEDURE	EXAMPLE	REFERENCE
Operations and Properties		**Section 1.2**
Order of Operations **1.** Simplify within the innermost grouping symbol, and then work outward until all grouping symbols are removed. **2.** Evaluate any expressions involving exponents. **3.** Perform any multiplication and division, working from left to right. **4.** Then do any addition and subtraction, again working from left to right.	$4(5 - 2)^2 + 7 = 4(3)^2 + 7$ $= 4 \cdot 9 + 7$ $= 36 + 7$ $= 43$	**p. 10**
For any real numbers a, b, and c: *Closure Properties* $a + b$ is a real number $a \cdot b$ is a real number	$3 + 4$ is a real number. $3 \cdot 4$ is a real number.	**p. 10**
Commutative Properties $a + b = b + a$ $a \cdot b = b \cdot a$	$7 + 5 = 5 + 7$ $7 \cdot 5 = 5 \cdot 7$	**p. 10**
Associative Properties $(a + b) + c = a + (b + c)$ $(a \cdot b) \cdot c = a \cdot (b \cdot c)$	$(2 + 7) + 4 = 2 + (7 + 4)$ $(2 \cdot 7) \cdot 4 = 2 \cdot (7 \cdot 4)$	**p. 10**
Identities There exists a unique real number 0 such that $a + 0 = 0 + a = a$ The number 0 is called the *additive identity.* There exists a unique real number 1 such that $a \cdot 1 = 1 \cdot a = a$ The number 1 is called the *multiplicative identity.*	$7 + 0 = 7$ $-8 + 0 = -8$ $3 \cdot 1 = 3$ $(-5)(1) = -5$	**p. 10**
Inverse Properties For any real number a, there exists a unique real number $-a$ such that $a + (-a) = (-a) + a = 0$ and $-a$ is called the *additive inverse,* or the *opposite,* of a. For any real number a ($a \neq 0$), there exists a unique number $\frac{1}{a}$ such that $a \cdot \frac{1}{a} = \frac{1}{a} \cdot a = 1$ $\frac{1}{a}$ is called the *multiplicative inverse,* or the *reciprocal,* of a.	$3 + (-3) = 0$ $(3)\left(\frac{1}{3}\right) = 1$	**p. 10**
Distributive Property $a(b + c) = ab + ac$	$5(4 + 3) = 5 \cdot 4 + 5 \cdot 3$	**p. 11**
Combining Like Terms $ax + bx = (a + b)x$	$4a + 9a = 13a$	**p. 12**

Continued

DEFINITION/PROCEDURE	EXAMPLE	REFERENCE
Inequalities and Absolute Value		**Section 1.3**
Inequalities The inequality relations are as follows: **1.** *Less than,* which we denote with the symbol $<$. We write $a < b$ to indicate that a is less than b. Graphically this means that the point corresponding to a must lie *to the left* of the point corresponding to b.  **2.** *Greater than,* which we denote with the symbol $>$. We write $a > b$ to indicate that a is greater than b. Graphically this means that the point corresponding to a must lie *to the right* of the point corresponding to b.  Two other symbols, $\le$ and $\ge$, are also used in writing inequalities. The statement $a \le b$ is read "a is less than or equal to b." The statement $a \ge b$ is read "a is greater than or equal to b."	Graph $\{x\|x < 3\}$.  The parenthesis means that 3 is *not* included in the graph. Graph $\{x\|x > -4\}$.  Graph $\{x\|x \le -1\}$. 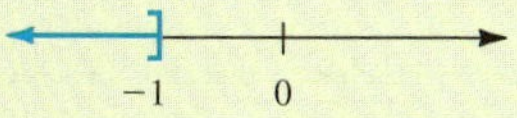 The bracket means that -1 is included in the graph.	p. 17
Double Inequalities An inequality of the form $-4 < x < 3$ is called a *double inequality* and combines the two inequality statements $x > -4$ and $x < 3$	Graph $\{x\|-4 < x < 3\}$. 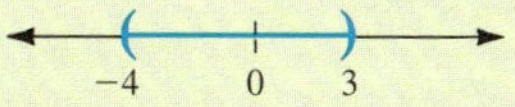	p. 19
Absolute Value The *absolute value* of a real number is the distance (on the number line) between the point named by that real number and the origin. The absolute value of a number a is denoted $\|a\|$. In general, we can define the absolute value of any real number a as $\|a\| = \begin{cases} a & \text{if } a \text{ is positive} \\ 0 & \text{if } a \text{ is zero} \\ -a & \text{if } a \text{ is negative} \end{cases}$	$\|7\| = 7$ $\|0\| = 0$ $\|-7\| = -(-7) = 7$	p. 20
Integer Exponents and Scientific Notation		**Section 1.4**
Exponents The notation: $a^n = \underbrace{a \cdot a \cdot a \cdot \cdots \cdot a}_{n \text{ factors}}$ (Exponent: n; Base: a) The number or letter used as a factor, here a, is called the *base* of the expression. The *exponent,* or *power,* here n, tells us how many times the base is used as a factor. An expression such as 5^3 is said to be in *exponential form.* The expression written as $5 \cdot 5 \cdot 5$ is in *expanded form.*	$3^4 = 3 \cdot 3 \cdot 3 \cdot 3 = 81$ $m^2n^3 = m \cdot m \cdot n \cdot n \cdot n$	p. 25

Continued

DEFINITION/PROCEDURE	EXAMPLE	REFERENCE
Integer Exponents and Scientific Notation		**Section 1.4**
Properties of Exponents For any real numbers a and b and natural numbers m and n: Property 1: $a^m \cdot a^n = a^{m+n}$ Property 2: $\frac{a^m}{a^n} = a^{m-n} \quad a \neq 0, m \geq n$ Property 3: $(a^m)^n = a^{m \cdot n}$ Property 4: $(ab)^m = a^m b^m$ Property 5: $\left(\frac{a}{b}\right)^m = \frac{a^m}{b^m} \quad b \neq 0$	$x^5 \cdot x^7 = x^{5+7} = x^{12}$ $\frac{x^7}{x^5} = x^{7-5} = x^2$ $(x^5)^3 = x^{5 \cdot 3} = x^{15}$ $(2xy)^3 = 2^3x^3y^3 = 8x^3y^3$ $\left(\frac{x^2}{3}\right)^2 = \frac{(x^2)^2}{3^2} = \frac{x^4}{9}$	**pp. 25–30**
The Zero Exponent $a^0 = 1 \quad a \neq 0$	$5^0 = 1$	**p. 31**
Scientific Notation Scientific notation is a useful way of expressing very large or very small numbers through the use of powers of 10. Any number written in the form $a \times 10^n$ in which $1 \leq a < 10$ and n is an integer, is said to be written in scientific notation.	$38{,}000{,}000. = 3.8 \times 10^7$ 7 places	**p. 32**

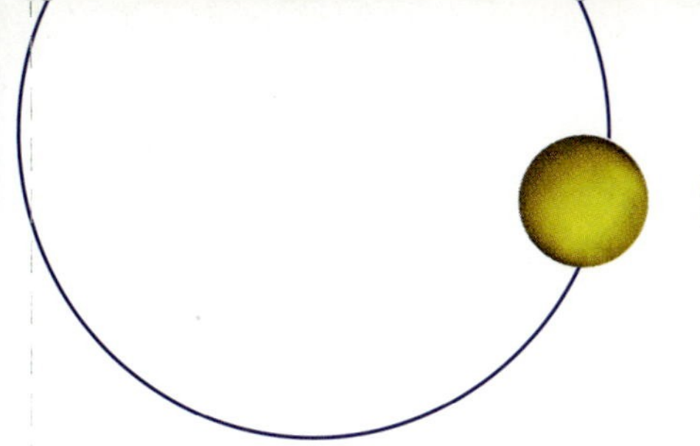

Summary Exercises

This summary exercise set is provided to give you practice with each of the objectives of the chapter. Each exercise is keyed to the appropriate chapter section. The answers are provided in the instructor's manual that accompanies this text. Your instructor will provide guidelines on how to best use these exercises in your instructional program.

[1.1] Name all the numeric sets to which each of the following numbers belong. The numeric sets consist of the natural numbers N, integers Z, rational numbers Q, irrational numbers Q', and real numbers R.

1. 0

2. -9

3. $-\sqrt{3}$

4. $\frac{4}{3}$

5. 3π

6. $-\frac{5}{8}$

7. $1.\overline{74}$

8. 3.25

[1.1] On the number line provided, graph the points corresponding to the elements of each of the given sets.

9. $\{1, 4, 7\}$

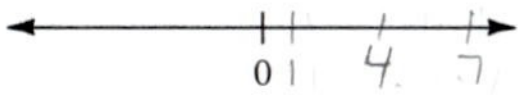

10. $\{-4, 0, 4\}$

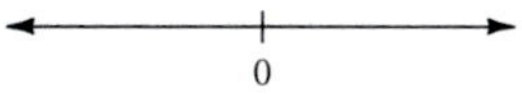

11. $\left\{-\frac{2}{3}, \frac{5}{3}, \frac{7}{3}\right\}$

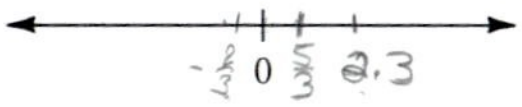

12. $\{-\sqrt{5}, -\sqrt{2}, \sqrt{3}\}$

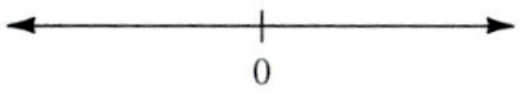

[1.2] Translate the following statements to symbols.

13. a increased by 4

14. 6 times y

15. 8 less than w

16. b decreased by 7

17. Twice the difference of n and 2

18. The product of 3 and a, divided by b

19. The quotient of the sum of x and y, divided by 3

20. 4 less than the product of s and t

[1.2] Evaluate each of the following expressions.

21. $2 + 5 \cdot 3$

22. $5(4 - 3)$

23. $5 + 3 \cdot 5 - 3$

24. $(5 + 3)(5 - 3)$

25. $6^2 - 2^2$

26. $2 + 4 \cdot 3^2$

27. $(3 + 4)4^2$

28. $2 + (3 \cdot 5)^2$

[1.2] Write each of the following, using symbols. Then simplify your results.

29. The sum of 6 and 92

30. The product of 5 and 7

31. 8 added to the difference of 12 and 7

32. The difference of 4 cubed and 3 cubed

[1.2] Apply the appropriate properties to rewrite the given expressions. Then simplify your results.

33. $5 + (a + 2)$

34. $4(3xy)$

35. $-8 + (m + 8)$

36. $5w\left(\frac{1}{5}\right)$

37. $\left(\frac{3}{7}\right)(7a)\left(\frac{5}{3}\right)$

38. $3(x + 2y + 3z)$

39. $5y + 2y + y$

40. $2(3a + 4b + 6a)$

41. $4p + 7p + 5 + p$

42. $2b + 7 + 3b + 2$

43. $3 + 4(2c + 1) + 3c$

44. $4 + 2w(w + 1) + 2w$

[1.2] State the property that is used to justify each of the following statements.

45. $25x + (-25x) = 0$

46. $(25x)\left(\frac{1}{25x}\right) = 1$

47. $a + (3a + 4) = (a + 3a) + 4$

48. $2x + 4 + 6x = 2x + 6x + 4$

49. $3(2x) = (3 \cdot 2)x$

50. $4(2a + 1) = 4 \cdot 2a + 4 \cdot 1$

51. $5y + 0 = 5y$

52. $(1)(4w) = 4w$

[1.3] Graph each of the following sets.

53. $\{x | x > 3\}$

0

54. $\{x | x \le 5\}$

0

55. $\left\{x \middle| x < \frac{7}{2}\right\}$

0

56. $\{x | x \ge -6\}$

0

57. $\{x | -3 < x < 6\}$

0

58. $\{x | -5 \le x < -1\}$

0

[1.3] Write each of the following expressions without the absolute value symbol.

59. $|9|$

60. $|-4.3|$

61. $-|3|$

62. $-|-3.5|$

63. $|4| + |-4|$

64. $|-3| + |-5|$

65. $-(|3| + |-4|)$

66. $-(|-7| + |-9|)$

[1.4] Write each product in exponential form. Identify the base and the exponent.

67. $b \cdot b \cdot b \cdot b \cdot b$

68. $(6w)(6w)(6w)(6w)$

[1.4] Simplify each expression, using the properties of exponents.

69. $w^3 \cdot w^8$

70. $(-2)^2(-2)^4$

71. $(x^3y^2)(x^7y^4)$

72. $(5a^0b^3)(-2a^2b^4)$

73. $\frac{a^{14}}{a^8}$

74. $\frac{(3x + 1)^7}{(3x + 1)^3}$

75. $(m^2n)^4$

76. $(3a^2b^5)^3$

77. $(3b^2)^0(-2b^5)^2$

78. $\left(\frac{c}{d^3}\right)^2$

79. $\left(\frac{2a^2b^4}{c^3}\right)^3$

80. $\left(\frac{x^8}{y^5}\right)\left(\frac{y^3}{3x^2}\right)^2$

[1.4] Write each of the following numbers in scientific notation.

81. 6,500,000

82. 34,000,000,000

[1.4] Write each of the following numbers in decimal form.

83. 3×10^5

84. 8.2×10^8

Perform the indicated operations, and express the result in scientific form.

85. $(3.2 \times 10^5)(2 \times 10^7)$

86. $\dfrac{7.5 \times 10^6}{1.5 \times 10^2}$

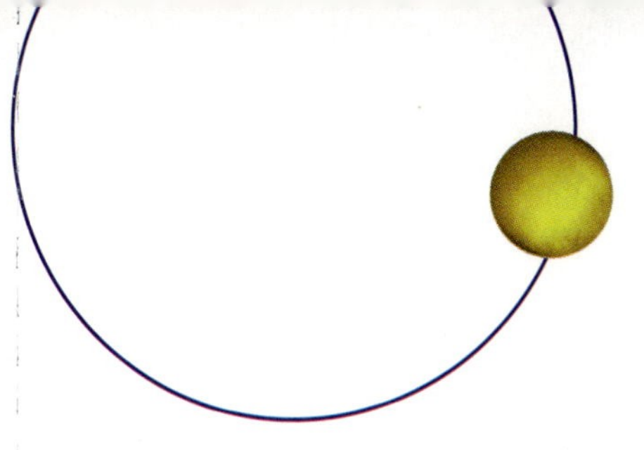

Self-Test for Chapter 1

Name ____________________

Section ________ Date ________

ANSWERS

1. (a)
(b)
(c)
(d)
(e)
2.
3.
4.
5.
6.
7.
8.
9.
10.
11.
12.
13.

The purpose of this self-test is to help you check your progress and to review for a chapter test in class. Allow yourself about an hour to take the test. When you are done, check your answers in the back of the book. If you missed any problems, be sure to go back and review the appropriate sections in the chapter and the exercises that are provided.

For the set of numbers

$$\left\{-6, -\frac{2}{3}, 4.1, -\sqrt{3}, 0, \pi, 9, \frac{5}{4}, 0.\overline{78}\right\}$$

which numbers belong to the following numeric sets?

1. (a) Natural numbers N **(b)** Integers Z
(c) Rational numbers Q **(d)** Irrational numbers Q'
(e) Real numbers R

Translate each of the following statements, using symbols.

2. 2 times the sum of x and y

3. 8 less than p, divided by t

Evaluate each of the following expressions.

4. $2 + 5 \cdot 3^2$

5. $(2 - 5)4^2$

State the property that is used to justify each of the following statements.

6. $y + (2y + 5) = (y + 2y) + 5$

7. $4(2c + 7) = 4 \cdot 2c + 4 \cdot 7$

Graph each of the following sets.

8. $\{x | x > -5\}$

0

9. $\{x | -1 \le x \le 4\}$

0

Write the following expressions without absolute value symbols.

10. $|-3| - |-6|$

11. $-|5| + |-7|$

Simplify each expression. Assume all variables represent nonzero real numbers.

12. $(x^4y^5)(3x^0y^2)$

13. $(2x^2y^3)^2$

ANSWERS

14. ______

15. ______

16. ______

17. ______

18. ______

19. ______

20. ______

14. $(4b^5c^3)^0(-3b^3c^4)^2$

15. $\left(\dfrac{x}{y^3}\right)^2$

16. $\left(\dfrac{x^{13}}{y^2}\right)\left(\dfrac{y^3}{x^2}\right)^4$

17. $\dfrac{(-x^2y^3)^2}{(2xy)^3}$

Simplify each expression by combining like terms.

18. $-2c + 7b - 3c - 10b$

19. $6a + 5(3a + 4) - 7$

Write the following number in scientific notation.

20. 2,530,000

LINEAR EQUATIONS AND INEQUALITIES

2

INTRODUCTION

Quality control is exercised in nearly all manufacturing processes. Samples of the product are taken at each stage, and various measurements are made to see if the product fits within given specifications.

In the pharmaceutical-making process, great caution must be exercised to ensure that the medicines and drugs are pure and contain precisely what is indicated on the label. Guaranteeing such purity is a task the quality control division of the pharmaceutical company assumes.

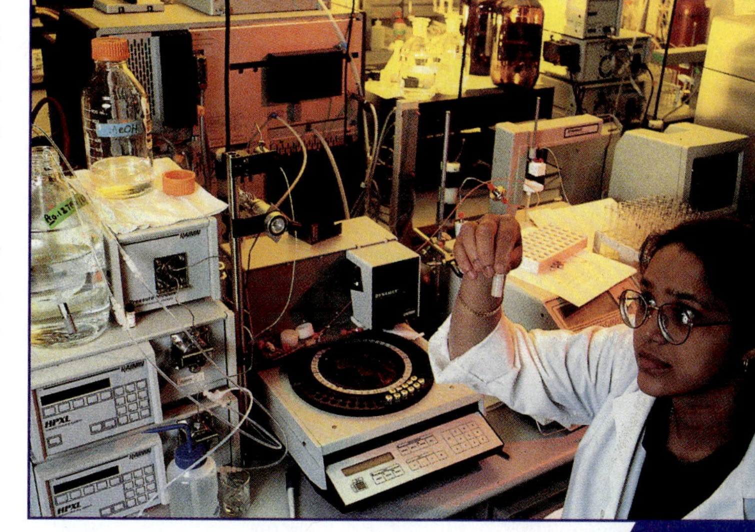

A lab technician working in quality control must run a series of tests on samples of every ingredient, even simple ingredients such as salt (NaCl). One such test is a measure of how much weight is lost as a sample is dried. The technician must set up a 3-hour procedure that involves cleaning and drying bottles and stoppers and then weighing them while they are empty and again when they contain samples of the substance to be heated and dried. At the end of the procedure, to compute the percentage of weight loss from drying, the technician uses the formula

$$L = \frac{W_g - W_f}{W_g - T} \cdot 100$$

in which L = percentage loss in drying

W_g = weight of container and sample

W_f = weight of container and sample after drying process completed

T = weight of empty container

This equation is more useful when solved for one of the variables, here W_f or T. In this chapter, you will learn how to solve such an equation.

2.1 Solutions of Linear Equations in One Variable

2.1 OBJECTIVES

1. Identify a linear equation
2. Combine like terms to solve an equation

We begin this chapter by considering one of the most important tools of mathematics—the equation. The ability to recognize and solve various types of equations and inequalities is probably the most useful algebraic skill you will learn, and we will continue to build on the methods developed here throughout the remainder of the text. To start, let's describe what we mean by an equation.

An **equation** is a mathematical statement in which two expressions represent the same quantity. An equation has three parts:

$$\underbrace{5x + 6}_{\text{Left side}} \quad \underset{\text{Equals sign}}{=} \quad \underbrace{2x - 3}_{\text{Right side}}$$

The equation simply says that the expression on the left and the expression on the right represent the same quantity.

In this chapter, we will work with a particular kind of equation.

NOTE Linear equations are also called **first-degree equations** because the highest power of the variable is the first power, or first degree.

Definitions: Linear Equation in One Variable

A **linear equation in one variable** is any equation that can be written in the form

$ax + b = 0$

in which a and b are any real numbers

$a \neq 0$

NOTE We also say the solution *satisfies* the equation.

The **solution** of an equation in one variable is any number that will make the equation a true statement. The **solution set** for such an equation is simply the set consisting of all solutions.

Example 1

Checking Solutions

Verify that -3 is a solution for the equation.

$5x + 6 = 2x - 3$

Replacing x with -3 gives

NOTE We use the question mark over the symbol of equality when we are checking to see if the statement is true.

$$5(-3) + 6 \stackrel{?}{=} 2(-3) - 3$$

$$-15 + 6 \stackrel{?}{=} -6 - 3$$

$$-9 = -9 \qquad \text{A true statement.}$$

CHECK YOURSELF 1

Verify that 7 is a solution for this equation.

$5x - 15 = 2x + 6$

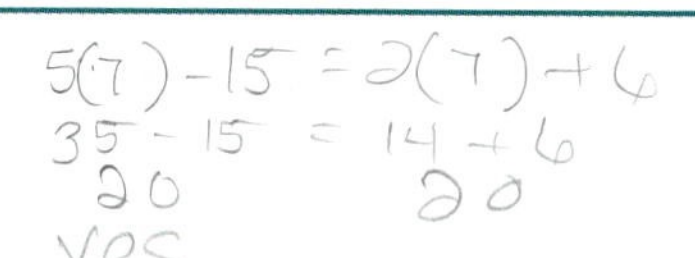

Solving linear equations in one variable will require using **equivalent equations.**

Definitions: Equivalent Equations

Two equations are *equivalent* if they have the same solution set.

For example, the three equations

$5x + 5 = 2x - 4 \qquad 3x = -9 \qquad x = -3$

NOTE You can easily verify this by replacing x with -3 in each equation.

are all equivalent because they all have the same solution set, $\{-3\}$. Note that replacing x with -3 will give a true statement in the third equation, but it is not as clear that -3 is a solution for the other two equations. This leads us to an equation-solving strategy of *isolating* the variable, as is the case in the equation $x = -3$.

To form equivalent equations that will lead to the solution of a linear equation, we need two properties of equations: addition and multiplication. The addition property is defined here.

NOTE Adding the same quantity to both sides of an equation gives an equivalent equation, which holds true whether c is positive or negative.

Rules and Properties: Addition Property of Equations

If $\quad a = b$

then $\quad a + c = b + c$

Recall that subtraction can always be defined in terms of addition, so

$a - c = a + (-c)$

The addition property also allows us to *subtract* the same quantity from both sides of an equation.

The multiplication property is defined here.

NOTE Multiplying both sides of an equation by the same nonzero quantity gives an equivalent equation.

Rules and Properties: Multiplication Property of Equations

If $\quad a = b$

then $\quad ac = bc \quad$ when $c \neq 0$

It is also the case that division can be defined in terms of multiplication, so

$$\frac{a}{c} = a \cdot \frac{1}{c} \qquad c \neq 0$$

The multiplication property allows us to *divide* both sides of an equation by the same nonzero quantity.

Example 2

Applying the Properties of Equations

Solve for x.

$$3x - 5 = 4 \qquad (1)$$

NOTE Why did we add 5? We added 5 because it is the *opposite* of -5, and the resulting equation will have the variable term on the left and the constant term on the right.

We start by using the addition property to add 5 to both sides of the equation.

$$3x - 5 + 5 = 4 + 5$$

$$3x = 9 \qquad (2)$$

Now we want to get the x term alone on the left with a coefficient of 1 (we call this *isolating* the x). To do this, we use the multiplication property and multiply both sides by $\frac{1}{3}$.

NOTE We choose $\frac{1}{3}$ because $\frac{1}{3}$ is the *reciprocal* of 3 and $\frac{1}{3} \cdot 3 = 1$

$$\frac{1}{3}(3x) = \frac{1}{3}(9)$$

$$\left(\frac{1}{3} \cdot 3\right)(x) = 3$$

$$\text{So, } x = 3 \qquad (3)$$

In set notation, we write $\{3\}$, which represents the set of all solutions. No other value of x makes the original equation true. We could also use set-builder notation. We write $\{x \mid x = 3\}$, which is read, "Every x such that x equals three." We will use both notations throughout the text.

Because any application of the addition or multiplication properties leads to an equivalent equation, equations (1), (2), and (3) in Example 2 all have the same solution, 3.

To check this result, we can replace x with 3 in the original equation:

$$3(3) - 5 \stackrel{?}{=} 4$$

$$9 - 5 \stackrel{?}{=} 4$$

$$4 = 4 \quad \text{A true statement.}$$

You may prefer a slightly different approach in the last step of the solution above. From equation (2),

$$3x = 9$$

The multiplication property can be used to *divide* both sides of the equation by 3. Then,

$$\frac{3x}{3} = \frac{9}{3}$$

$$x = 3$$

Of course, the result is the same.

CHECK YOURSELF 2

Solve for x.

$$4x - 7 = 17$$

The steps involved in using the addition and multiplication properties to solve an equation are the same if more terms are involved in an equation.

Example 3

Applying the Properties of Equations

Solve for x.

$5x - 11 = 2x - 7$

Our objective is to use the properties of equations to isolate x on one side of an equivalent equation. We begin by adding 11 to both sides.

NOTE Adding 11 puts the constant term on the right.

$$5x - 11 + 11 = 2x - 7 + 11$$

$$5x = 2x + 4$$

We continue by adding $-2x$ to (or subtracting $2x$ from) both sides. We can do this because of our addition property of equations.

NOTE If you prefer, write

$5x - 2x = 2x - 2x + 4$

Again:

$3x = 4$

$$5x + (-2x) = 2x + (-2x) + 4$$

We have now isolated the variable term on the left side of the equation.

$3x = 4$

In the final step, we multiply both sides by $\frac{1}{3}$.

NOTE This is the same as dividing both sides by 3. So

$\frac{3x}{3} = \frac{4}{3}$

$x = \frac{4}{3}$

$$\frac{1}{3}(3x) = \frac{1}{3}(4)$$

$$x = \frac{4}{3}$$

In set notation, we write $\left\{\frac{4}{3}\right\}$. We leave it to you to check this result by substitution.

CHECK YOURSELF 3

Solve for x.

$7x - 12 = 2x - 9$

Both sides of an equation should be simplified as much as possible *before* the addition and multiplication properties are applied. If like terms are involved on one side (or on both sides) of an equation, they should be combined before an attempt is made to isolate the variable. Example 4 illustrates this approach.

Example 4

Applying the Properties of Equations with Like Terms

Solve for x.

NOTE Notice the like terms on the left and right sides of the equation.

$8x + 2 - 3x = 8 + 3x + 2$

Here we combine the like terms $8x$ and $-3x$ on the left and the like terms 8 and 2 on the right as our first step. We then have

$5x + 2 = 3x + 10$

We can now solve as before.

$5x + 2 - 2 = 3x + 10 - 2$ Subtract 2 from both sides.

$5x = 3x + 8$

Then,

$5x - 3x = 3x - 3x + 8$ Subtract $3x$ from both sides.

$2x = 8$

$\frac{2x}{2} = \frac{8}{2}$ Divide both sides by 2.

$x = 4$ or $\{4\}$

The solution is 4, which can be checked by returning to the *original equation.*

CHECK YOURSELF 4

Solve for x.

$7x - 3 - 5x = 10 + 4x + 3$

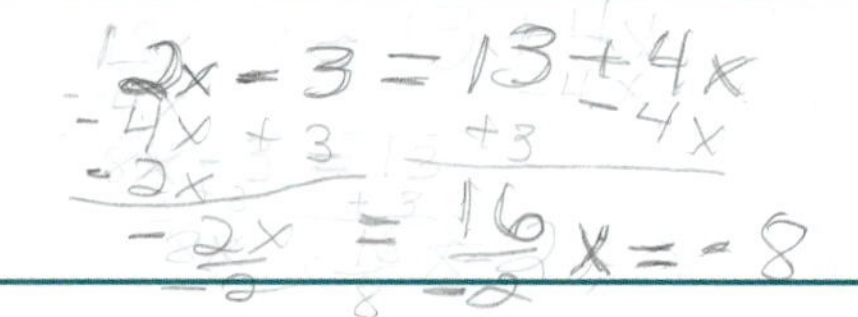

If parentheses are involved on one or both sides of an equation, the parentheses should be removed by applying the distributive property as the first step. Like terms should then be combined before an attempt is made to isolate the variable. Consider Example 5.

Example 5

Applying the Properties of Equations with Parentheses

Solve for x.

$x + 3(3x - 1) = 4(x + 2) + 4$

First, apply the distributive property to remove the parentheses on the left and right sides.

$x + 9x - 3 = 4x + 8 + 4$

Combine like terms on each side of the equation.

$10x - 3 = 4x + 12$

NOTE Recall that to isolate the x, we must get x alone on the left side with a coefficient of 1.

Now, isolate variable x on the left side.

$10x - 3 + 3 = 4x + 12 + 3$ Add 3 to both sides.

$10x = 4x + 15$

$10x - 4x = 4x - 4x + 15$ Subtract $4x$ from both sides.

$6x = 15$

$\frac{6x}{6} = \frac{15}{6}$ Divide both sides by 6.

$x = \frac{5}{2}$ or $\left\{\frac{5}{2}\right\}$

The solution is $\frac{5}{2}$. Again, this can be checked by returning to the original equation.

CHECK YOURSELF 5

Solve for x.

$$x + 5(x + 2) = 3(3x - 2) + 18$$

NOTE The LCM of a set of denominators is also called the **lowest common denominator (LCD).**

To solve an equation involving fractions, the first step is to multiply both sides of the equation by the **least common multiple (LCM)** of all denominators in the equation. This will clear the equation of fractions, and we can proceed as before.

Example 6

Applying the Properties of Equations with Fractions

Solve for x.

$$\frac{x}{2} - \frac{2}{3} = \frac{5}{6}$$

First, multiply each side by 6, the least common multiple of 2, 3, and 6.

$$6\left(\frac{x}{2} - \frac{2}{3}\right) = 6\left(\frac{5}{6}\right)$$

$$6\left(\frac{x}{2}\right) - 6\left(\frac{2}{3}\right) = 6\left(\frac{5}{6}\right) \quad \text{Apply the distributive property.}$$

$$\overset{3}{\cancel{6}}\left(\frac{x}{\underset{1}{\cancel{2}}}\right) - \overset{2}{\cancel{6}}\left(\frac{2}{\underset{1}{\cancel{3}}}\right) = \overset{1}{\cancel{6}}\left(\frac{5}{\underset{1}{\cancel{6}}}\right) \quad \text{Simplify.}$$

Next, isolate the variable x on the left side.

NOTE The equation is now cleared of fractions.

$$3x - 4 = 5$$

$$3x = 9$$

$$x = 3 \qquad \text{or} \qquad \{3\}$$

The solution, 3, can be checked as before by returning to the original equation.

CHECK YOURSELF 6

Solve for x.

$$\frac{x}{4} - \frac{4}{5} = \frac{19}{20}$$

Be sure that the distributive property is applied properly so that *every term* of the equation is multiplied by the LCM.

Example 7

Applying the Properties of Equations with Fractions

Solve for x.

$$\frac{2x - 1}{5} + 1 = \frac{x}{2}$$

First, multiply each side by 10, the LCM of 5 and 2.

$$10\left(\frac{2x - 1}{5} + 1\right) = 10\left(\frac{x}{2}\right)$$

$$\overset{2}{\cancel{10}}\left(\frac{2x - 1}{\underset{1}{\cancel{5}}}\right) + 10(1) = \overset{5}{\cancel{10}}\left(\frac{x}{\underset{1}{\cancel{2}}}\right)$$ Apply the distributive property on the left. Reduce.

$$2(2x - 1) + 10 = 5x$$

$$4x - 2 + 10 = 5x$$

$$4x + 8 = 5x$$ Next, isolate x. Here we isolate x on the right side.

$$8 = x \quad \text{or} \quad \{8\}$$

The solution for the original equation is 8.

CHECK YOURSELF 7

Solve for x.

$$\frac{3x + 1}{4} - 2 = \frac{x + 1}{3}$$

Thus far, we have considered only equations of the form $ax + b = 0$, in which $a \neq 0$. If we allow the possibility that $a = 0$, two additional equation forms arise. The resulting equations can be classified into three types depending on the nature of their solutions.

1. An equation that is true for only particular values of the variable is called a **conditional equation.** Here the equation can be written in the form

   $ax + b = 0$

   in which $a \neq 0$. This case was illustrated in all our previous examples and exercises.
2. An equation that is true for all possible values of the variable is called an **identity.** In this case, *both a* and b are 0, so we get the equation $0 = 0$. This will be the case if both sides of the equation reduce to the same expression (a true statement).
3. An equation that is never true, no matter what the value of the variable, is called a **contradiction.** For example, if a is 0 but b is nonzero, we end up with something like $4 = 0$. This will be the case if both sides of the equation reduce to a false statement.

Example 8 illustrates the second and third cases.

Example 8

Identities and Contradictions

(a) Solve for x.

$$2(x - 3) - 2x = -6$$

Apply the distributive property to remove the parentheses.

$$2x - 6 - 2x = -6$$

$$-6 = -6 \quad \text{A \textit{true} statement.}$$

NOTE See the definition of an identity, above. By adding 6 to both sides of this equation, we have $0 = 0$.

Because the two sides reduce to the true statement $-6 = -6$, the original equation is an *identity,* and the solution set is the set of all real numbers.

(b) Solve for x.

$$3(x + 1) - 2x = x + 4$$

Again, apply the distributive property.

$$3x + 3 - 2x = x + 4$$

$$x + 3 = x + 4$$

$$3 = 4 \quad \text{A \textit{false} statement.}$$

NOTE See the earlier definition of a contradiction. Subtracting 3 from both sides, we have $0 = 1$.

Because the two sides reduce to the false statement $3 = 4$, the original equation is a contradiction. There are no values of the variable that can satisfy the equation. The solution set has nothing in it. We call this the **empty set** and write $\{\}$ or $\varnothing$.

CHECK YOURSELF 8

Determine whether each of the following equations is a conditional equation, an identity, or a contradiction.

(a) $2(x + 1) - 3 = x$ **(b)** $2(x + 1) - 3 = 2x + 1$ **(c)** $2(x + 1) - 3 = 2x - 1$

NOTE An **algorithm** is a step-by-step process for problem solving.

An organized step-by-step procedure is the key to an effective equation-solving strategy. The following algorithm summarizes our work in this section and gives you guidance in approaching the problems that follow.

Step by Step: Solving Linear Equations in One Variable

Step 1 Remove any grouping symbols by applying the distributive property.

Step 2 Multiply both sides of the equation by the LCM of any denominators, to clear the equation of fractions.

Step 3 Combine any like terms that appear on either side of the equation.

Step 4 Apply the addition property of equations to write an equivalent equation with the variable term on *one side* of the equation and the constant term on the *other side.*

Step 5 Apply the multiplication property of equations to write an equivalent equation with the variable isolated on one side of the equation.

Step 6 Check the solution in the *original* equation.

NOTE If the equation derived in step 5 is always true, the original equation was an identity. If the equation is always false, the original equation was a contradiction.

When you are solving an equation for which a calculator is recommended, it is often easiest to do all calculations as the last step.

Example 9

Evaluating Expressions Using a Calculator

Solve the following equation for x.

$$\frac{185(x - 3.25) + 1650}{500} = 159.44$$

Following the steps of the algorithm, we get

$$\frac{185x - 185 \cdot 3.25 + 1650}{500} = 159.44$$ Remove parentheses.

$$185x - 185 \cdot 3.25 + 1650 = 159.44 \cdot 500$$ Multiply by the LCM.

$$185x = 159.44 \cdot 500 + 185 \cdot 3.25 - 1650$$ Apply the addition property.

$$x = \frac{159.44 \cdot 500 + 185 \cdot 3.25 - 1650}{185}$$ Isolate the variable.

Now, remembering to insert parentheses around the numerator, we use a calculator to simplify the expression on the right.

$x = 425.25$ or $\{425.25\}$

CHECK YOURSELF 9

Solve the following equation for x.

$$\frac{2200(x + 17.5) - 1550}{75} = 2326$$

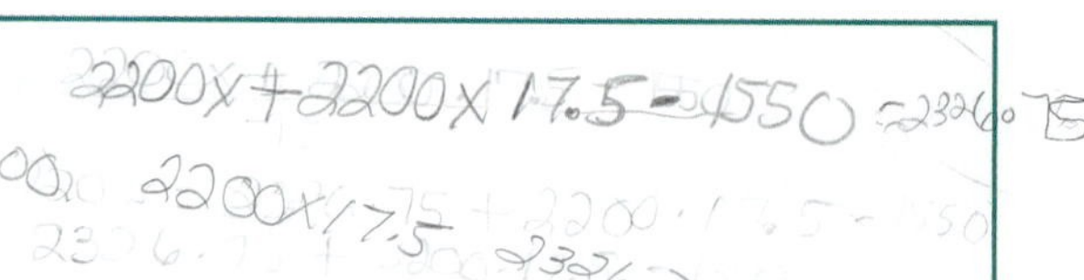

CHECK YOURSELF ANSWERS

1. $5(7) - 15 \stackrel{?}{=} 2(7) + 6$
$35 - 15 \stackrel{?}{=} 14 + 6$
$20 = 20$
A true statement.
2. $\{6\}$ **3.** $\left\{\frac{3}{5}\right\}$ **4.** $\{-8\}$ **5.** $\left\{-\frac{2}{3}\right\}$

6. $\{7\}$ **7.** $\{5\}$ **8.** **(a)** Conditional; **(b)** contradiction; **(c)** identity **9.** $\{62.5\}$

2.1 Exercises

In exercises 1 to 14, solve each equation, and check your results. Express each answer in set notation.

1. $5x - 8 = 17$

2. $4x + 9 = -11$

3. $8 - 7x = -41$

4. $-7 - 4x = 21$

5. $7x - 5 = 6x + 6$

6. $9x + 4 = 8x - 3$

7. $8x - 4 = 3x - 24$

8. $5x + 2 = 2x - 5$

9. $7x - 4 = 2x + 26$

10. $11x - 3 = 4x - 31$

11. $4x - 3 = 1 - 2x$

12. $8x + 5 = -19 - 4x$

13. $2x + 8 = 7x - 37$

14. $3x - 5 = 9x + 22$

In exercises 15 to 32, simplify and then solve each equation. Express your answer in set notation.

15. $5x - 2 + x = 9 + 3x + 10$

16. $5x + 5 - x = -7 + x - 2$

17. $7x - 3 - 4x = 5 + 5x - 13$

18. $8x - 3 - 6x = 7 + 5x + 17$

19. $5x = 3(x - 6)$

20. $2(x - 15) = 7x$

21. $5(8 - x) = 3x$

22. $7x = 7(6 - x)$

23. $2(2x - 1) = 3(x + 1)$

24. $3(3x - 1) = 4(3x + 1)$

25. $8x - 3(2x - 4) = 17$

26. $7x - 4(3x + 4) = 9$

27. $7(3x + 4) = 8(2x + 5) + 13$

28. $-4(2x - 1) + 3(3x + 1) = 9$

Name ____________________

Section ________ Date ________

ANSWERS

1. ________ 2. ________
3. ________ 4. ________
5. ________ 6. ________
7. ________ 8. ________
9. ________ 10. ________
11. ________ 12. ________
13. ________
14. ________
15. ________
16. ________
17. ________
18. ________
19. ________
20. ________
21. ________
22. ________
23. ________
24. ________
25. ________
26. ________
27. ________
28. ________

ANSWERS

29. ______
30. ______
31. ______
32. ______
33. ______
34. ______
35. ______
36. ______
37. ______
38. ______
39. ______
40. ______
41. ______
42. ______
43. ______
44. ______
45. ______
46. ______
47. ______
48. ______
49. ______
50. ______

29. $9 - 4(3x + 1) = 3(6 - 3x) - 9$

30. $13 - 4(5x + 1) = 3(7 - 5x) - 15$

31. $5 - 2[x - 2(x + 1)] = 55 - 4[x - 3(x + 2)]$

32. $7 - 5[x - 3(x + 2)] = 25 - 2[x - 2(x - 3)]$

In exercises 33 to 46, clear fractions and then solve each equation. Express your answer in set notation.

33. $\frac{2x}{3} - \frac{5}{3} = 3$

34. $\frac{3x}{4} + \frac{1}{4} = 4$

35. $\frac{x}{6} + \frac{x}{5} = 11$

36. $\frac{x}{6} - \frac{x}{8} = 1$

37. $\frac{2x}{3} - \frac{x}{4} = \frac{5}{2}$

38. $\frac{5x}{6} + \frac{2x}{3} = \frac{5}{6}$

39. $\frac{x}{5} - \frac{x - 7}{3} = \frac{1}{3}$

40. $\frac{x}{6} + \frac{3}{4} = \frac{x - 1}{4}$

41. $\frac{5x - 3}{4} - 2 = \frac{x}{3}$

42. $\frac{6x - 1}{5} - \frac{2x}{3} = 3$

43. $\frac{2x + 3}{5} - \frac{2x - 1}{3} = \frac{8}{15}$

44. $\frac{3x}{5} - \frac{3x - 1}{2} = \frac{11}{10}$

45. $0.5x - 6 = 0.2x$

46. $0.7x - 7 = 0.3x - 5$

In exercises 47 to 56, classify each equation as a conditional equation, an identity, or a contradiction.

47. $3(x - 1) = 2x + 3$

48. $2(x + 3) = 2x + 6$

49. $3(x - 1) = 3x + 3$

50. $2(x + 3) = x + 5$

51. $3(x - 1) = 3x - 3$

52. $2(x + 3) = 3x + 5$

53. $3x - (x - 3) = 2(x + 1) + 2$

54. $5x - (x + 4) = 4(x - 2) + 4$

55. $\frac{x}{2} - \frac{x}{3} = \frac{x}{6}$

56. $\frac{3x}{4} - \frac{2x}{3} = \frac{x}{6}$

In exercises 57 to 60, use a calculator to solve the given equations for x. Round your answer to two decimal places and use set notation.

57. $\frac{63(x - 2.45) + 325}{200} = 3$

58. $\frac{47(x + 3.15) - 263}{315} = 11$

59. $\frac{-23x - 14(x - 9.75)}{23.46} = 15.75$

60. $\frac{-15.25x + 12(2x - 11.23)}{-15.6} = 8.4$

61. What is the common characteristic of equivalent equations?

62. What is meant by a *solution* to a linear equation?

63. Define **(a)** identity and **(b)** contradiction.

64. Why does the multiplication property of equation not include multiplying both sides of the equation by 0?

Label exercises 65 to 70 true or false.

65. Adding the same value to both sides of an equation creates an equivalent equation.

66. Multiplying both sides of an equation by 0 creates an equivalent equation.

67. To clear an equation of fractions, we multiply both sides by the GCF of the denominator.

68. The multiplication property of equations allows us to divide both sides by the same nonzero quantity.

ANSWERS

51. ________

52. ________

53. ________

54. ________

55. ________

56. ________

57. ________

58. ________

59. ________

60. ________

61. ________

62. ________

63. ________

64. ________

65. ________

66. ________

67. ________

68. ________

ANSWERS

69. ____________

70. ____________

69. Some equations have more than one solution.

70. No matter what value is substituted for x, the expressions on either side of the equals sign have the same value.

Answers

1. $\{5\}$ **3.** $\{7\}$ **5.** $\{11\}$ **7.** $\{-4\}$ **9.** $\{6\}$ **11.** $\left\{\frac{2}{3}\right\}$ **13.** $\{9\}$ **15.** $\{7\}$ **17.** $\left\{\frac{5}{2}\right\}$ **19.** $\{-9\}$ **21.** $\{5\}$ **23.** $\{5\}$ **25.** $\left\{\frac{5}{2}\right\}$ **27.** $\{5\}$ **29.** $\left\{-\frac{4}{3}\right\}$ **31.** $\{-13\}$ **33.** $\{7\}$ **35.** $\{30\}$ **37.** $\{6\}$ **39.** $\{15\}$ **41.** $\{3\}$ **43.** $\left\{\frac{3}{2}\right\}$ **45.** $\{20\}$ **47.** Conditional **49.** Contradiction **51.** Identity **53.** Contradiction **55.** Identity **57.** $\{6.82\}$ **59.** $\{-6.30\}$ **61.**  **63.** **65.** True **67.** False **69.** True

2.2 Literal Equations and Formulas

2.2 OBJECTIVE

1. Solve a literal equation for a specified variable

Many problems in algebra require the use of **formulas** for their solution. Formulas are simply equations that express a relationship between more than one variable or letter. You are already familiar with a number of examples. For instance,

$$P = R \cdot B \qquad A = \frac{1}{2}h \cdot b \qquad P = 2L + 2W$$

are formulas for percentage, the area of a triangle, and the perimeter of a rectangle, respectively.

One useful application of the equation-solving skills we considered in Section 2.1 is in rewriting these formulas, also called **literal equations,** in more convenient equivalent forms.

NOTE A *literal equation* is any equation that involves more than one variable or letter.

Generally, that more convenient form is one in which the original formula or equation is solved for a particular variable or letter. This is called **solving the equation for a variable,** and the steps used in the process are very similar to those you saw earlier in solving linear equations.

Consider the following example.

Example 1

Solving a Literal Equation

Solve the formula

$d = r \cdot t \qquad$ for t

This formula gives distance d in terms of a rate r and time t.

To solve for t means to isolate t on one side of the equation. This can be done by dividing both sides by r. Given

$$d = r \cdot t$$

we use the multiplication property of equations to divide by r, the coefficient of t.

$$\frac{d}{r} = \frac{r \cdot t}{r}$$

$$\frac{d}{r} = t$$

We usually write the equation in the equivalent form with the desired variable on the left. So

$$t = \frac{d}{r}$$

We now have t in terms of d and r, as required.

CHECK YOURSELF 1

Solve the formula $C = 2\pi r$ for r.

Solving a formula for a particular variable may require the use of both properties of equations, as the following example illustrates.

Example 2

Solving a Literal Equation

Solve the formula

$P = 2L + 2W$ for L

This formula gives the perimeter of a rectangle P in terms of its width W and its length L.

To solve for L, start by using the addition property of equations to subtract $2W$ from both sides.

NOTE We want to isolate the term with the variable we are solving for—here L.

$$P = 2L + 2W$$
$$P - 2W = 2L + 2W - 2W$$
$$P - 2W = 2L$$

We now use the multiplication property to divide both sides by 2:

$$\frac{P - 2W}{2} = \frac{2L}{2}$$
$$\frac{P - 2W}{2} = L$$
$$L = \frac{P - 2W}{2}$$

NOTE This result can also be written as

$$L = \frac{P}{2} - W$$

This gives L in terms of P and W, as desired.

CHECK YOURSELF 2

Solve the formula $ax + by = c$ for y.

You may also have to apply the distributive property in solving for a variable. Consider the following example.

Example 3

Solving a Literal Equation

Solve the formula

$A = P(1 + rt)$ for r

This formula gives the amount A in an account earning simple interest, with principal P, interest rate r, and time t.

First, we use the distributive property to remove the parentheses on the right.

$$A = P(1 + rt) = P + Prt$$

We now subtract P from both sides.

$$A - P = P - P + Prt$$

$$A - P = Prt$$

Finally, to isolate r, we divide by Pt, the coefficients of r on the right.

$$\frac{A - P}{Pt} = \frac{Prt}{Pt}$$

$$\frac{A - P}{Pt} = r$$

$$r = \frac{A - P}{Pt}$$

CHECK YOURSELF 3

Solve the equation for n.

$$S = 180(n - 2)$$

Often it is necessary to apply the multiplication property, to clear the literal equation of fractions, as the first step of the solution process. This is illustrated in Example 4.

Example 4

Solving a Literal Equation

Solve the formula for C.

$$D = \frac{C - S}{n}$$

This formula gives the yearly depreciation D for an item in terms of its cost C, its salvage value S, and the number of years n.

As our first step, we multiply both sides of the given equation by n to clear of fractions.

$$D = \frac{C - S}{n}$$

$$nD = n\left(\frac{C - S}{n}\right)$$

$$nD = C - S$$

NOTE On the *right* note that $\frac{n}{n} = 1$ and multiplying by 1 leaves $C - S$.

We now add S to both sides.

$$nD + S = C - S + S$$

$$nD + S = C$$

$$C = nD + S$$

and the cost C is now represented in terms of n, D, and S.

CHECK YOURSELF 4

Solve the formula $V = \frac{1}{3}\pi r^2 h$ for h.

CHECK YOURSELF ANSWERS

1. $r = \frac{C}{2\pi}$ **2.** $y = \frac{c - ax}{b}$ **3.** $n = \frac{S + 360}{180}$ **4.** $h = \frac{3V}{\pi r^2}$

Name ______

Section ______ Date ______

Exercises

In exercises 1 to 24, solve each of the formulas for the indicated variable.

1. $V = Bh$ for h
2. $P = RB$ for B
3. $C = 2\pi r$ for r
4. $e = mc^2$ for m
5. $V = LWH$ for H
6. $I = Prt$ for r
7. $V = \pi r^2 h$ for h
8. $S = 2\pi rh$ for r
9. $V = \frac{1}{3}Bh$ for B
10. $V = \frac{1}{3}\pi r^2 h$ for h
11. $I = \frac{E}{R}$ for R
12. $V = \frac{KT}{P}$ for T
13. $ax + b = 0$ for x
14. $y = mx + b$ for x
15. $P = 2L + 2W$ for W
16. $ax + by = c$ for y
17. $D = \frac{C - S}{n}$ for S
18. $D = \frac{R(100 - x)}{100}$ for R
19. $R = C(1 + r)$ for r
20. $A = P(1 + rt)$ for t
21. $A = \frac{1}{2}h(B + b)$ for b
22. $L = a + (n - 1)d$ for n
23. $F = \frac{9}{5}C + 32$ for C
24. $C = \frac{5}{9}(F - 32)$ for F

ANSWERS

1. ______ 2. ______
3. ______ 4. ______
5. ______ 6. ______
7. ______ 8. ______
9. ______
10. ______
11. ______
12. ______
13. ______
14. ______
15. ______
16. ______
17. ______
18. ______
19. ______
20. ______
21. ______
22. ______
23. ______
24. ______

ANSWERS

25. ______

26. ______

27. ______

28. ______

29. ______

30. ______

31. ______

32. ______

33. ______

34. ______

Solve each of the following exercises using the indicated formula from exercises 1 to 24.

25. A rectangular solid has a base with length 6 cm and width 4 cm. If the volume of the solid is 72 cubic centimeters (cm^3), find the height of the solid. See exercise 5.

26. A cylinder has a radius of 4 inches (in.). If its volume is 144π cubic inches ($in.^3$), what is the height of the cylinder? See exercise 7.

27. A principal of $2000 was invested in a savings account for 4 years. If the interest earned for that period was $480, what was the interest rate? See exercise 6.

28. The retail selling price of an item, R, was $20.70. If its cost, C, to the store was $18, what was the markup rate, r? See exercise 19.

29. The radius of the base of a cone is 3 cm. If the volume of the cone is 24π cm^3, find the height of the cone. See exercise 10.

30. The volume of a pyramid is 30 $in.^3$. If the height of the pyramid is 6 in., find the area of its base, B. See exercise 9.

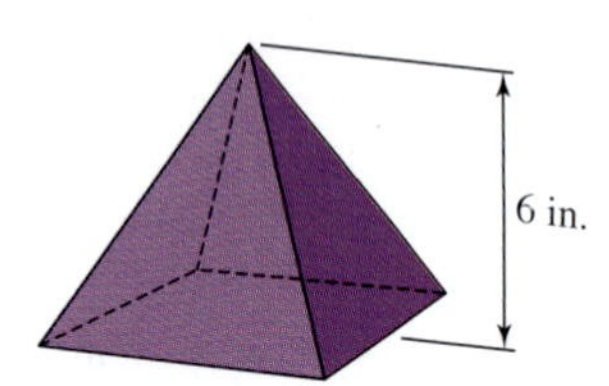

31. If the perimeter of a rectangle is 60 ft and its length is 18ft, find its width. See exercise 15.

32. The yearly depreciation, D, for a piece of machinery was $1500 over 8 years. If the cost of the machinery was $15,000, what was its salvage value, S? See exercise 17.

33. A principal of $5000 was invested in a time-deposit account paying 9% annual interest. If the amount in the account at the end of a certain period was $7250, for how long was the money invested? See exercise 20.

34. The area of a trapezoid is 36 $in.^2$. If its height is 4 in. and the length of one of the bases is 11 in., find the length of the other base. See exercise 21.

Answers

1. $h = \frac{V}{B}$ **3.** $r = \frac{C}{2\pi}$ **5.** $H = \frac{V}{LW}$ **7.** $h = \frac{V}{\pi r^2}$ **9.** $B = \frac{3V}{h}$

11. $R = \frac{E}{I}$ **13.** $x = -\frac{b}{a}$ **15.** $W = \frac{P - 2L}{2}$ **17.** $S = C - nD$

19. $r = \frac{R - C}{C}$ **21.** $b = \frac{2A - hB}{h}$ **23.** $C = \frac{5}{9}(F - 32)$ or $C = \frac{5F - 160}{9}$

25. 3 cm **27.** 6% **29.** 8 cm **31.** 12 ft **33.** 5 years

2.3 Problem Solving and Applications of Linear Equations

2.3 OBJECTIVES

1. Apply the five-step strategy for solving an application
2. Solve motion problems
3. Find a break-even point
4. Solve a literal equation

"I have a problem!"

How often have you, or a friend, started a conversation with this statement? And, more important, what do you do to solve the problem? George Polya, in his book *How to Solve It,* contends that there are four parts to solving any problem.

1. Understand the problem
2. Devise a plan
3. Carry out the plan
4. Look back

This approach is useful for solving any problem. One of the reasons that mathematics is a required course in most programs is that, in a math class, you are constantly practicing the art of problem solving. To help you remember to use Polya's approach on problems that you encounter outside of the classroom, we will consistently use a series of steps similar to those prescribed above.

Recall that an **algorithm** is a series of steps that, when followed, solves a problem. The following five-step algorithm is essentially the same as Polya's approach, but for one thing. We have divided the "devise a plan" step into two steps that are more directly relevant to solving mathematical problems. Here is our five-step approach:

Step by Step: Solving Applications

Step 1 Read the problem carefully to determine the unknown quantities.
Step 2 Choose a variable to represent the unknown. Express all other unknowns in terms of this variable.
Step 3 Translate the problem to the language of algebra to form an equation.
Step 4 Solve the equation, and answer the question of the original problem.
Step 5 Verify your solution by returning to the original problem.

Our first two applications fall into the category of **uniform-motion problems.** Uniform motion means that the speed of an object does not change over a certain distance or time. To solve these problems, we will need a relationship between the distance traveled, represented by d, the rate (or speed) of travel, r, and the time of that travel, t. In general, the relationship for the distance traveled d, rate r, and time t, is expressed as

$$d = r \cdot t$$

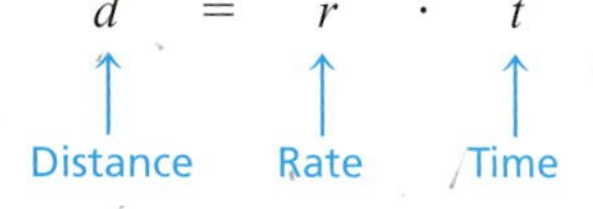

This is the key relationship, and it will be used in all motion problems. Let's see how it is applied in Example 1.

Example 1

Solving a Motion Problem

On Friday morning Ricardo drove from his house to the beach in 4 h. In coming back on Sunday afternoon, heavy traffic slowed his speed by 10 mi/h, and the trip took 5 h. What was his average speed (rate) in each direction?

Step 1 We want the speed or rate in each direction.

Step 2 Let x be Ricardo's speed to the beach. Then $x - 10$ is his return speed.

It is always a good idea to sketch the given information in a motion problem. Here we would have

Going → $x \frac{\text{mi}}{\text{h}}$ for 4 h

Returning ← $x - 10 \frac{\text{mi}}{\text{h}}$ for 5 h

Step 3 Because we know that the distance is the same each way, we can write an equation, using the fact that the product of the rate and the time each way must be the same.

So

Distance (going) = distance (returning)

Time · rate (going) = time · rate (returning)

$$4x = 5(x - 10)$$

Time · rate (going) → $4x$; Time · rate (returning) → $5(x - 10)$

A chart can help summarize the given information. We begin by filling in the information given in the problem.

	Rate	Time	Distance
Going	x	4	
Returning	$x - 10$	5	

Now we fill in the missing information. Here we use the fact that $d = rt$ to complete the chart.

	Rate	Time	Distance
Going	x	4	$4x$
Returning	$x - 10$	5	$5(x - 10)$

From here we set the two distances equal to each other and solve as before.

Step 4 Solve.

$$4x = 5(x - 10)$$

$$4x = 5x - 50$$

$$-x = -50$$

$$x = 50 \text{ mi/h}$$

NOTE x was his rate going; $x - 10$, his rate returning.

So Ricardo's rate going to the beach was 50 mi/h, and his rate returning was 40 mi/h.

Step 5 To check, you should verify that the product of the time and the rate is the same in each direction.

CHECK YOURSELF 1

A plane made a flight (with the wind) between two towns in 2 h. Returning against the wind, the plane's speed was 60 mi/h slower, and the flight took 3 h. What was the plane's speed in each direction?

Example 2 illustrates another way of using the distance relationship.

Example 2

Solving a Motion Problem

Katy leaves Las Vegas for Los Angeles at 10 A.M., driving at 50 mi/h. At 11 A.M. Jensen leaves Los Angeles for Las Vegas, driving at 55 mi/h along the same route. If the cities are 260 mi apart, at what time will they meet?

Step 1 Let's find the time that Katy travels until they meet.

Step 2 Let x be Katy's time.

Then $x - 1$ is Jensen's time. Jensen left 1 h later!

Again, you should draw a sketch of the given information.

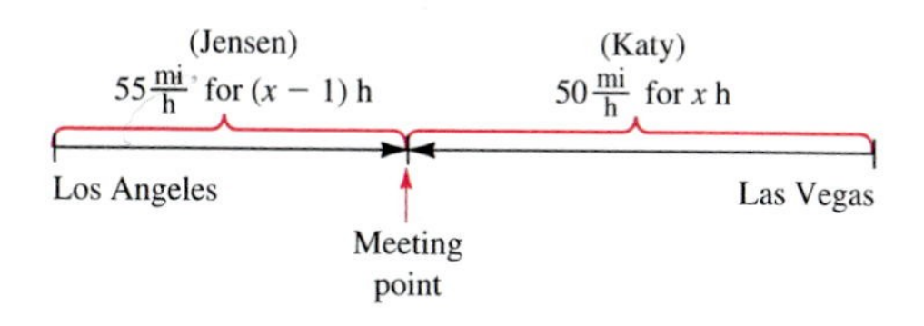

Step 3 To write an equation, we will again need the relationship $d = rt$. From this equation, we can write

$$\text{Katy's distance} = 50x$$

$$\text{Jensen's distance} = 55(x - 1)$$

As before, we can use a table to solve.

	Rate	Time	Distance
Katy	50	x	$50x$
Jensen	55	$x - 1$	$55(x - 1)$

From the original problem, the sum of those distances is 260 mi, so

$$50x + 55(x - 1) = 260$$

Step 4

$$50x + 55(x - 1) = 260$$
$$50x + 55x - 55 = 260$$
$$105x - 55 = 260$$
$$105x = 315$$
$$x = 3 \text{ h}$$

NOTE Be sure to answer the question asked in the problem.

Finally, because Katy left at 10 A.M., the two will meet at 1 P.M. We leave the check of this result to you.

CHECK YOURSELF 2

At noon a jogger leaves one point, running at 8 mi/h. One hour later a bicyclist leaves the same point, traveling at 20 mi/h in the opposite direction. At what time will they be 36 mi apart?

Next, we consider an application from business. But we will need some new terminology. The total cost of manufacturing an item consists of two types of costs. The **fixed cost,** sometimes called the **overhead,** includes costs such as product design, rent, and utilities. In general, this cost is constant and does not change with the number of items produced. The **variable cost,** which is a cost per item, includes costs such as material, labor, and shipping. The variable cost depends on the number of items being produced.

A typical cost equation might be

$$C = 3.30x + 5000$$

(3.30x: Variable cost; 5000: Fixed cost)

in which total cost, C, equals variable cost times the number of items produced, x, plus the fixed cost.

The total **revenue** is the income the company makes. It is calculated as the product of the selling price of the item and the number of items sold. A typical revenue equation might be

$$R = 7.50x$$

(7.50: Selling price per item; x: Number of items sold)

and total revenue equals an item's selling price times the number sold.

The **break-even point** is that point at which the revenue equals the cost (the company would exactly break even without a profit or a loss).

Let's apply these concepts in Example 3.

Example 3

Finding the Break-Even Point

A firm producing DVDs finds that its fixed cost is \$5000 per month and that its variable cost is \$3.50 per DVD. The cost of producing x DVDs is then given by

$$C = 3.50x + 5000$$

The firm can sell the units at \$7.50 each, so the revenue from selling x units is

$$R = 7.50x$$

Find the break-even point.

Because the break-even point is that point at which the revenue equals the cost, or $R = C$, from our given equations we have

$$\underbrace{7.50x}_{\text{Revenue}} = \underbrace{3.50x + 5000}_{\text{Cost}}$$

Solving as before gives

$$4x = 5000$$

$$x = 1250$$

The firm will break even (no profit or loss) by producing and selling exactly 1250 DVDs each month.

CHECK YOURSELF 3

A firm producing lawn chairs has fixed costs of \$525 per week. The variable cost is \$8.50 per chair, and the revenue per chair is \$15.50. This means that the cost equation is

$$C = 8.50x + 525$$

and the revenue equation is

$$R = 15.50x$$

Find the break-even point.

Let's turn now to an application of our work in solving literal equations for a specified variable.

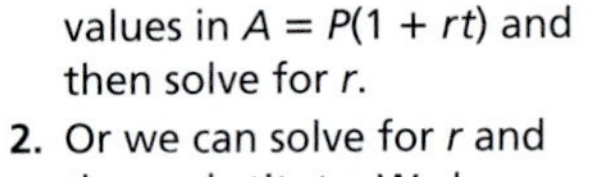

Example 4

Solving an Interest Rate Application

Suppose a principal of \$1000 is invested in a mutual fund account for 5 years. For the amount in the account to be \$1400 at the end of that period, what must the interest rate be?

NOTE From the given problem, we know values for A, P, and t. Two strategies are possible.

1. We can substitute the known values in $A = P(1 + rt)$ and then solve for r.
2. Or we can solve for r and then substitute. We have illustrated this approach.

In Example 3 of Section 2.2, we solved the formula

$$A = P(1 + rt)$$

for r with the result

$$r = \frac{A - P}{Pt}$$

In our problem, $A = \$1400$, $P = \$1000$, and $t = 5$ years. Substituting, we have

$$r = \frac{1400 - 1000}{(1000)(5)} = \frac{400}{5000} = 0.08$$

The necessary interest rate is 8%.

CHECK YOURSELF 4

Suppose that a principal of \$5000 is invested in a time deposit fund for 3 years. If the amount in the account at the end of that period is \$6050, what was the annual interest rate?

Note, in the previous example and exercise, that once we solved for r in the original equation, we were able to easily use the result with different sets of data. You can compare this to first substituting the known values and then having to solve for r in each case separately.

Solving for a specified variable also has significance in computer programming. In using a formula for the computation of the value of a variable, you must solve the formula for that variable to use the formula in a program.

CHECK YOURSELF ANSWERS

1. $180 \frac{\text{mi}}{\text{h}}$ with the wind and $120 \frac{\text{mi}}{\text{h}}$ against **2.** At 2 P.M. **3.** 75 chairs **4.** 7%

2.3 Exercises

Name

Section Date

Solve the following applications.

ANSWERS

1.

2.

3.

4.

5.

6.

7.

8.

1. **Speed.** On her way to a business meeting, Kim took the freeway, and the trip took 3 h. Returning, she decided to take a side road, and her speed along that route averaged 9 mi/h slower than on the freeway. If Kim's return trip took $3\frac{1}{2}$ h and the distance driven was the same each way, find her average speed in each direction.

2. **Speed.** Beth was required to make a cross-country flight in training for her pilot's license. When she flew from her home airport, a steady 30-mi/h wind was behind her, and the first leg of the trip took 5 h. When she returned against the same wind, the flight took 7 h. Find the plane's speed in still air and the distance traveled on each leg of the flight.

3. **Speed.** Craig was driving on a 220-mi trip. For the first 3 h he traveled at a steady speed. At that point, realizing that he would be late to his destination, he increased his speed by 10 mi/h for the remaining 2 h of the trip. What was his driving speed for each portion of the trip?

4. **Distance.** Robert can drive to work in 45 min, whereas if he decides to take the bus, the same trip takes 1 h 15 min. If the average rate of the bus is 16 mi/h slower than his driving rate, how far does he travel to work?

5. **Time.** At 9 A.M., Tom left Boston for Baltimore, traveling at 45 mi/h. One hour later, Andrea left Baltimore for Boston, traveling at 50 mi/h along the same route. If the cities are 425 mi apart, at what time did Tom and Andrea meet?

6. **Time.** A passenger bus left a station at 1 P.M., traveling north at an average rate of 50 mi/h. One hour later, a second bus left the same station, traveling south at a rate of 55 mi/h. At what time will the two buses be 260 mi apart?

7. **Distance.** On Tuesday, Malia drove to a conference and averaged 54 mi/h for the trip. When she returned on Thursday, road construction slowed her average speed by 9 mi/h. If her total driving time was 11 h, what was her driving time each way, and how far away from her home was the conference?

8. **Time.** At 8:00 A.M., Robert left on a trip, traveling at 45 mi/h. One-half hour later, Laura discovered that Robert forgot his luggage and left along the same route, traveling at 54 mi/h, to catch up with him. When did Laura catch up with Robert?

ANSWERS

9. ____________

10. ____________

11. ____________

12. ____________

13. ____________

9. **Business.** A firm producing gloves finds that its fixed cost is \$4000 per week and its variable cost is \$8.50 per pair. The revenue is \$13.50 per pair of gloves, so that cost and revenue equations are, respectively,

$C = 8.50x + 4000$ and $R = 13.50x$

Find the break-even point for the firm.

10. **Business.** A company that produces calculators determines that its fixed cost is \$8820 per month. The variable cost is \$70 per calculator: the revenue is \$105 per calculator. The cost and revenue equations, respectively, are given by

$C = 70x + 8820$ and $R = 105x$

Find the number of calculators the company must produce and sell to break even.

11. **Business.** A firm that produces scientific calculators has a fixed cost of \$1260 per week and variable cost of \$6.50 per calculator. If the company can sell the calculators for \$13.50, find the break-even point.

12. **Business.** A publisher finds that the fixed cost associated with a new paperback is \$18,000. Each book costs \$2 to produce and will sell for \$6.50. Find the publisher's break-even point.

13. **Business.** A firm producing flashlights finds that its fixed cost is \$2400 per week, and its variable cost is \$4.50 per flashlight. The revenue is \$7.50 per flashlight, so the cost and revenue equations are, respectively,

$C = 4.50x + 2400$ and $R = 7.50x$

Find the break-even point for the firm (the point at which the revenue equals the cost).

ANSWERS

14. ____

15. ____

16. ____

17. ____

14. Business. A company that produces portable television sets determines that its fixed cost is $8750 per month. The variable cost is $70 per set, and the revenue is $105 per set. The cost and revenue equations, respectively, are given by

$C = 70x + 8750$ and $R = 105x$

Find the number of sets the company must produce and sell to break even.

15. Business. An important economic application involves supply and demand. The number of units of a commodity that manufacturers are willing to **supply,** S, is related to the market price, p. A typical supply equation is

$$S = 40p - 285 \qquad (1)$$

(Generally the supply increases as the price increases.)

The number of units that consumers are willing to buy, D, is called the **demand,** and it is also related to the market price. A typical demand equation is

$$D = -45p + 1500 \qquad (2)$$

(Generally the demand decreases as the price increases.)

The price at which the supply and demand are equal (or $S = D$) is called the **equilibrium price** for the commodity. The supply and demand equations for a certain model portable radio are given in equations (1) and (2). Find the equilibrium price for the radio.

16. Business. The supply and demand equations for a certain type of computer modem are

$S = 25p - 2500$ and $D = -40p + 5300$

Find the equilibrium price for the modem.

17. Business. You find a new bicycle that you like, and you plan to ride it for exercise several days a week. You are also happy to find that this very model is on sale for 22% off. You speak to the salesclerk, who begins writing up the sale by first adding on your state's 7.8% sales tax. He then takes off the 22%. "No!", you object, "You should take the 22% off first and then add the sales tax." The salesclerk says he is sorry, but he has been instructed to first calculate the amount of tax. Who is correct? Defend your position using algebra.

ANSWERS

18. ______

18. Density and Body Fat. You have probably heard the question, "Which weighs more, a ton of feathers or a ton of bricks?" The usual response of "bricks" is, of course, incorrect. A ton of anything weighs 2000 pounds. Why do people miss this question so often? We know from experience that bricks are heavier than feathers. However, what we mean is that bricks have a much higher *density* than feathers (they weigh more per unit volume). A cubic meter of bricks weighs far more than a cubic meter of feathers.

We often compare the density of a substance to the density of water. Water has a density of 1 g/cm^3. Things denser than water (such as body muscle and bone) sink, whereas things less dense (such as body fat) float.

The concept of density is important in analyzing the significance of the percentage of body fat you have. There are many ways to measure, or estimate, your percentage of body fat. Most health clubs and many college health classes offer the opportunity to find this percentage. By multiplying percentage of body fat times your weight, you can calculate the actual pounds of fat in your body.

However, we are looking for a more useful measure. Body density, D, can be calculated using the following formula.

$$D = \frac{1}{\frac{A}{a} + \frac{B}{b}}$$

in which A = proportion of body fat

B = proportion of lean body tissue $(1 - A)$

a = density of fat body tissue in grams/cubic cm (approximately 0.9 g/cm^3)

b = density of lean body tissue in grams/cubic cm (approximately 1.1 g/cm^3)

Use this formula to compute the following.

1. Substituting $1 - A$ for B, 0.9 for a, and 1.1 for b, solve the formula so that D is expressed in terms of A.
2. Find the values associated with body fat proportions of 0.1, 0.15, 0.2, 0.25, and 0.3.
3. Use other resources (health professionals, health clubs, the Internet, and the library) to find another measure of body density. How do the two methods compare?

Answers

1. 63 mi/h going; 54 mi/h returning **3.** 40 mi/h for 3 h; 50 mi/h for 2 h
5. 2:00 P.M. **7.** 5 h going; 6 h returning; 270 mi **9.** 800 pairs of gloves
11. 180 calculators **13.** 800 flashlights **15.** $21 **17.**

2.4 Linear Inequalities

OBJECTIVES

1. Solve and graph the solution set for a linear inequality
2. Solve and graph the solution set for a compound inequality

In Section 2.1 we defined a linear equation in one variable as an equation that could be written in the form

$$ax + b = 0$$

in which a and b are real numbers and $a \neq 0$.

A linear inequality in one variable is defined in a similar fashion.

Definitions: Linear Inequality

A *linear inequality* can be written in the form

$$ax + b < 0$$

in which a and b are real numbers and $a \neq 0$.

The inequality symbol $<$ can be replaced with any of the other inequality symbols $>$, $\leq$, or $\geq$, so that

$$ax + b > 0 \qquad ax + b \leq 0 \qquad \text{and} \qquad ax + b \geq 0$$

are also linear inequalities.

Fortunately your experience with linear equations in Section 2.1 provides the groundwork for solving linear inequalities. You will see many similarities.

NOTE We can also say that a solution *satisfies* the inequality.

A **solution** for a linear inequality in one variable is any real number that will make the inequality a true statement when the variable is replaced by that number. The **solution set** for a linear inequality is the set of all solutions.

Our strategy for solving linear inequalities is, for the most part, identical to that used for solving linear equations. We write a sequence of equivalent inequalities to isolate the variable on one side of the inequality symbol.

Writing equivalent inequalities for this purpose requires two properties. First:

NOTE Adding the same quantity to both sides of an inequality gives an equivalent inequality.

Rules and Properties: Addition Property of Inequalities

If $a < b$, then $a + c < b + c$.

This addition property is similar to that seen earlier in solving equations. As before, because subtraction is defined in terms of addition, the property also allows us to subtract the same quantity from both sides of an inequality without changing the solutions.

Our second property, dealing with multiplication, has an important difference. We begin by writing the true inequality

$$2 < 3$$

Multiplying both sides of that inequality by the same *positive* number, say 3, gives

$3(2) < 3(3)$

$6 < 9$ Another true statement!

Note that the new inequality has the same sense (points in the same direction) as the original inequality.

However, if we now multiply both sides of the inequality by a *negative* number, say -3, we have

$-3(2) < -3(3)$

$-6 < -9$ A *false* statement!

To make this a true statement, we must *reverse the sense* of the inequality to write

$-6 > -9$

This suggests that if we multiply both sides of an inequality by a negative number, we must reverse the sense of the inequality to form an equivalent inequality. From this discussion we can now state our second property.

Rules and Properties: Multiplication Property of Inequalities

If $a < b$

then $ac < bc$ when c is a *positive* number $(c > 0)$

and

$ac > bc$ when c is a *negative* number $(c < 0)$

NOTE Multiplying both sides of an inequality by a positive number gives an equivalent inequality.

NOTE Multiplying both sides of an inequality by a negative number and reversing the sense give an equivalent inequality.

Again, because division is defined in terms of multiplication, this property also allows us to divide both sides of an inequality by the same nonzero number, reversing the sense of the inequality if that number is negative.

We will use these properties in solving inequalities in much the same way as we did in solving equations, with the one significant difference pointed out above.

The following examples illustrate the solution process for linear inequalities.

Example 1

Solving a Linear Inequality

Solve and graph the inequality

$$4x - 3 < 5 \qquad (1)$$

First, we add 3 to both sides.

NOTE As in solving equations, we apply the addition property and *then* the multiplication property, to isolate the variable.

$$4x - 3 + 3 < 5 + 3$$

$$4x < 8 \qquad (2)$$

We now divide both sides of the inequality by 4, to isolate the variable on the left.

$$\frac{4x}{4} < \frac{8}{4}$$

$$x < 2 \qquad (3)$$

Because inequalities (1), (2), and (3) are all equivalent, the solution set for the original inequality consists of all numbers that are less than 2. That set can be written

$\{x|x < 2\}$

The graph of the solution set is

NOTE Recall that the parenthesis means 2 is *not* included in the solution set.

CHECK YOURSELF 1

Solve and graph the inequality.

$5x + 7 > 22$

Example 2

Solving a Linear Inequality

Solve and graph the inequality

$3x - 5 \geq 5x + 3$

Add 5 to both sides.

$$3x - 5 + 5 \geq 5x + 3 + 5$$
$$3x \geq 5x + 8$$

Subtract $5x$ from both sides.

$$3x - 5x \geq 5x - 5x + 8$$
$$-2x \geq 8$$

We must now divide both sides by -2. Because the divisor is negative, we reverse the sense of the inequality.

$$\frac{-2x}{-2} \leq \frac{8}{-2}$$
$$x \leq -4$$

The solution set consists of all numbers that are less than or equal to -4 and is graphed below.

$\{x|x \leq -4\}$

NOTE Here we use the bracket to indicate that -4 is included in the solution set.

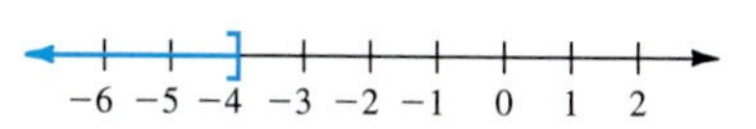

CHECK YOURSELF 2

Solve and graph.

$4x + 3 \leq 7x - 12$

In working with more complicated inequalities, as was the case with equations, any signs of grouping must be removed, and like terms combined, before the properties of inequalities are applied to isolate the variable.

Example 3

Solving a Linear Inequality

Solve and graph

$5 - 3(x - 2) \le 1 - x$

First, remove the parentheses on the left and combine like terms.

$5 - 3x + 6 \le 1 - x$

$-3x + 11 \le 1 - x$

We now proceed as before.

$-3x + 11 - 11 \le 1 - 11 - x$ Subtract 11.

$-3x \le -10 - x$

$-3x + x \le -10 - x + x$ Add x.

$-2x \le -10$

$\frac{-2x}{-2} \ge \frac{-10}{-2}$ Divide by -2, reversing the sense of the inequality.

$x \ge 5$

The solution set $\{x \mid x \ge 5\}$ is graphed below.

CHECK YOURSELF 3

Solve and graph.

$4 - 2(x + 5) \ge -9 - 4x$

If fractions are involved in an inequality, you should apply the multiplication property to clear the inequality of fractions as your first step.

Example 4

Solving a Linear Inequality

Solve and graph the inequality.

$$\frac{3x + 2}{6} - 1 < \frac{x}{3}$$

$$6\left(\frac{3x + 2}{6} - 1\right) < 6\left(\frac{x}{3}\right)$$

NOTE Multiply both sides by 6, the LCM of 6 and 3.

$$6\left(\frac{3x + 2}{6}\right) - 6(1) < 6\left(\frac{x}{3}\right)$$

NOTE Apply the distributive property on the left.

$$3x + 2 - 6 < 2x$$

The inequality is now cleared of fractions, and we proceed as before:

$$3x - 4 < 2x$$

$$3x < 2x + 4$$

$$x < 4$$

The solution set is graphed below.

$\{x | x < 4\}$

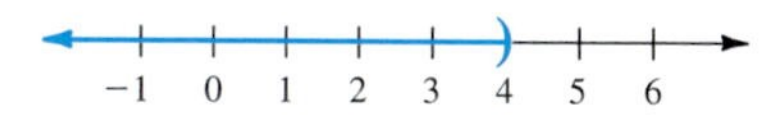

CHECK YOURSELF 4

Solve and graph the inequality.

$$\frac{5x - 1}{4} - 2 > \frac{x}{2}$$

The following algorithm summarizes our work thus far in this section in solving linear inequalities.

Step by Step: Solving Linear Inequalities in one Variable

Step 1 Clear the inequality statement of any fractions by using the multiplication property.

Step 2 Remove any grouping symbols, and combine like terms on either side of the inequality.

Step 3 Apply the addition property to write an equivalent inequality with the variable term on one side of the inequality and the constant term on the other.

Step 4 Apply the multiplication property to write an equivalent inequality with the variable isolated on one side of the inequality. Be sure to reverse the sense of the inequality if you multiply or divide by a negative number.

Step 5 Graph the solution set of the original inequality.

Let's now consider two types of inequality statements that arise frequently in mathematics. Consider a statement such as

$$-2 < x < 5$$

NOTE In the compound inequality the word "and" is understood.

It is called a **double** or **compound inequality** because it combines the two inequalities

$$-2 < x \quad \text{and} \quad x < 5$$

To solve a compound inequality means to isolate the variable in the middle term, as the following example illustrates.

Example 5

Solving a Compound Inequality

Solve and graph the compound inequality

$$-3 \leq 2x + 1 \leq 7$$

First, we subtract 1 from each of the three members of the compound inequality.

NOTE We are really applying the addition property to each of the *two* inequalities that make up the compound inequality statement.

$$-3 - 1 \le 2x + 1 - 1 \le 7 - 1$$

or

$$-4 \le 2x \le 6$$

We now divide by 2 to isolate the variable x.

$$\frac{-4}{2} \le \frac{2x}{2} \le \frac{6}{2}$$

$$-2 \le x \le 3$$

The solution set consists of all numbers between -2 and 3, including -2 and 3, and is written

$$\{x \mid -2 \le x \le 3\}$$

That set is graphed below.

NOTE Because the points at both ends of the interval are included, we sometimes call this a *closed interval*.

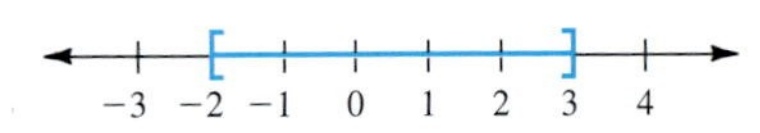

Note: Our solution set is equivalent to

$$\{x \mid x \ge -2 \text{ and } x \le 3\}$$

Look at the individual graphs.

$\{x \mid x \ge -2\}$

$\{x \mid x \le 3\}$

$\{x \mid x \ge -2 \text{ and } x \le 3\}$

Because the connecting word is "and," we want the *intersection* of the sets, that is, those numbers common to both sets.

NOTE Using set notation, we could write this as

$\{x \mid x \ge -2\} \cap \{x \mid x \le 3\}$

The $\cap$ represents the intersection of the two sets.

CHECK YOURSELF 5

Solve and graph the inequality.

$$-5 \le 2x - 3 \le 3$$

When the coefficient of the variable is a negative number, care must be taken in isolating the variable.

Example 6

Solving a Compound Inequality

Solve and graph the compound inequality

$$-3 < 4 - 3x < 13$$

Subtract 4 from each member of the inequality.

$$-7 < -3x < 9$$

Now we must divide by -3. The sense of the inequality is reversed whenever we divide by a negative number.

$$\frac{-7}{-3} > \frac{-3x}{-3} > \frac{9}{-3}$$

$$\frac{7}{3} > x > -3$$

In the standard smallest-to-largest format, we have

$$-3 < x < \frac{7}{3}$$

The solution consists of all numbers between -3 and $\frac{7}{3}$ and is written

$$\left\{x \middle| -3 < x < \frac{7}{3}\right\}$$

That set is graphed below.

NOTE This is sometimes called an *open interval.*

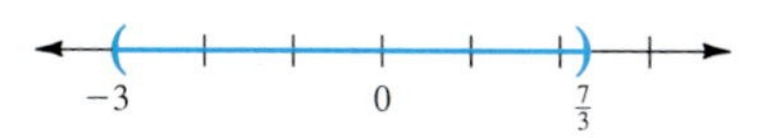

CHECK YOURSELF 6

Solve and graph the double inequality.

$$-5 < 3 - 2x < 5$$

A compound inequality may also consist of two inequality statements connected by the word "or." The following example illustrates the solution of that type of compound inequality.

Example 7

Solving a Compound Inequality

Solve and graph the inequality

$$2x - 3 < -5 \quad \text{or} \quad 2x - 3 > 5$$

In this case we must work with each of the inequalities *separately.*

$$2x - 3 < -5 \quad \text{or} \quad 2x - 3 > 5$$

NOTE Add 3.

$$2x < -2 \qquad 2x > 8$$

NOTE Divide by 2.

$$x < -1 \qquad x > 4$$

The graph of the solution set is shown.

$$\{x|x < -1 \text{ or } x > 4\}$$

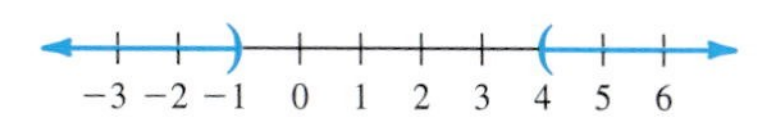

NOTE In set notation we write the union as

$\{x|x < -1\} \cup \{x|x > 4\}$

Note that because the connecting word is "or" in this case, the solution set of the original inequality is the *union* of the two sets, that is, those numbers that belong to either or both of the sets.

CHECK YOURSELF 7

Solve and graph the inequality.

$3x - 4 \le -7 \quad \text{or} \quad 3x - 4 \ge 7$

The following chart summarizes our discussion of solving linear inequalities and the nature of the solution sets of the types of inequalities we have considered in this section.

Type of Inequality	Graph of Solution Set
$ax + b < c$	If $a > 0$: 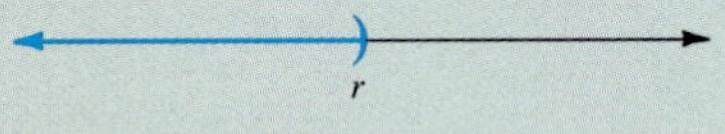 If $a < 0$: 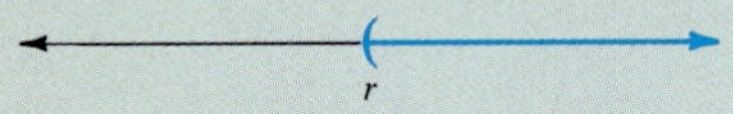
$-c < ax + b < c$	
$ax + b < -c \quad \text{or} \quad ax + b > c$	

CHECK YOURSELF ANSWERS

1. $\{x \mid x > 3\}$ 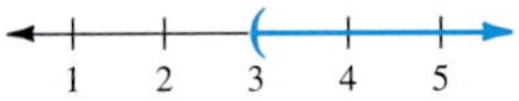
2. $\{x \mid x \ge 5\}$ 
3. $\left\{x \mid x \ge -\dfrac{3}{2}\right\}$ 
4. $\{x \mid x > 3\}$ 
5. $\{x \mid -1 \le x \le 3\}$ 
6. $\{x \mid -1 < x < 4\}$ 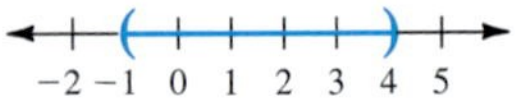
7. $\left\{x \mid x \le -1 \text{ or } x \ge \dfrac{11}{3}\right\}$ 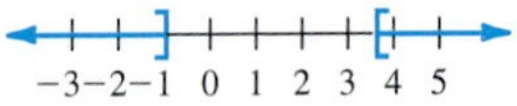

Name ________________

Section ________ Date ________

2.4 Exercises

Solve each of the following inequalities. Then graph the solution set.

1. $x - 2 < 5$

2. $x + 3 > -4$

3. $x + 5 \geq 3$

4. $x - 4 \leq -2$

5. $5x > 25$

6. $4x < -12$

7. $-3x \leq -15$

8. $-7x > 21$

9. $2x + 3 < 10$

10. $5x - 3 \leq 17$

11. $-2x - 7 \geq 5$

12. $-3x + 4 < -4$

13. $5 - 3x < 14$

14. $2 - 5x \geq 22$

15. $3x - 4 > 2x + 5$

16. $4x + 3 \leq 3x + 11$

17. $8x + 2 \leq 2x + 10$

18. $5x - 1 > x + 9$

19. $7x - 3 > 2x - 13$

20. $9x + 2 \geq 2x - 19$

21. $4x - 3 \leq 6x + 5$

22. $7x - 1 \leq 10x - 6$

23. $5 - 3x > 2x + 3$

24. $7 - 5x > 3x - 9$

Simplify and then solve each of the following inequalities.

25. $5(2x - 1) \leq 25$

26. $3(3x + 1) > -15$

27. $4(5x + 1) > 3(3x + 5)$

28. $3(2x + 4) \leq 5(3x - 3)$

29. $3(x - 1) - 4 < 2(3x + 1)$

30. $3(3x - 1) - 4(x + 3) \leq 15$

31. $3(x + 7) - 11 > 2(3x - 5) + x$

32. $3(2x + 7) - 5 \leq 4(x + 1) - x$

ANSWERS

1. ________________
2. ________________
3. ________________
4. ________________
5. ________________
6. ________________
7. ________________
8. ________________
9. ________________
10. ________________
11. ________________
12. ________ 13. ________
14. ________________
15. ________ 16. ________
17. ________ 18. ________
19. ________________
20. ________________
21. ________________
22. ________ 23. ________
24. ________ 25. ________
26. ________ 27. ________
28. ________ 29. ________
30. ________ 31. ________
32. ________________

ANSWERS

33. ______
34. ______
35. ______
36. ______
37. ______
38. ______
39. ______
40. ______
41. ______
42. ______
43. ______
44. ______
45. ______
46. ______
47. ______
48. ______
49. ______
50. ______
51. ______
52. ______
53. ______
54. ______
55. ______
56. ______
57. ______
58. ______

Clear of fractions and then solve each of the following inequalities.

33. $\frac{x-4}{3} < 5$

34. $\frac{x+5}{2} \geq -3$

35. $\frac{x+2}{-3} \leq 3$

36. $\frac{x-2}{-4} > -6$

37. $\frac{x}{2} - \frac{x}{3} \geq 2$

38. $\frac{x}{4} - 2 < \frac{x}{5}$

39. $\frac{x}{5} - \frac{x-7}{3} < \frac{1}{3}$

40. $\frac{x}{4} - \frac{4x+3}{20} < \frac{1}{5}$

41. $\frac{x-3}{2} - \frac{x+5}{5} \leq \frac{1}{2}$

42. $\frac{x+5}{4} - \frac{x+1}{3} \leq \frac{2}{3}$

Solve each of the following compound inequalities. Then graph the solution set.

43. $3 \leq x + 1 \leq 5$

44. $-2 < x - 3 < 3$

45. $-8 < 2x < 4$

46. $-6 \leq 3x \leq 9$

47. $1 \leq 2x - 3 \leq 6$

48. $-2 < 3x - 5 < 4$

49. $-1 < 5 - 3x < 8$

50. $-7 \leq 3 - 2x \leq 8$

Solve each of the following compound inequalities. Then graph the solution set.

51. $x - 1 < -3$ or $x - 1 > 3$

52. $x + 2 < -5$ or $x + 2 > 5$

53. $2x - 1 < -7$ or $2x - 1 > 7$

54. $2x + 3 < -3$ or $2x + 3 > 3$

55. $3x - 1 < -7$ or $3x - 1 > 7$

56. $4x + 3 < -5$ or $4x + 3 > 5$

Suppose that the revenue a company will receive from producing and selling x items is given by R and the cost of those items by C. The company will make a profit only if the revenue is greater than the cost, that is, when $R > C$. Use this information to find the number of items that must be produced and sold for the company to make a profit.

57. $R = 50x, \quad C = 1000 + 30x$

58. $R = 800x, \quad C = 24{,}000 + 500x$

Recall that the *average* of a group of test scores is the sum of those test scores divided by the number of scores. Use this information to solve the following problems.

59. Suppose that Kim has scores of 83, 94, and 91 on three 100-point tests in her chemistry class thus far. Describe the set of scores on the 100-point final test that will give her an average of 90 or above, so that she will receive an A for the course.

Hint: If x represents her final score, then

$83 + 94 + 91 + x$

will give her total score for the four tests.

60. Robert has scores of 78, 85, 70, and 83 on four tests. Describe the set of scores he must have on the 100-point final to average 80 or above for the course.

Solve each of the following problems.

61. A college must decide how many sections of intermediate algebra to offer during the fall quarter. Each section should contain a maximum of 35 students, and the college anticipates that a total of 400 students will enroll for the sections. How many sections should be offered?

Hint: If x represents the number of sections, then $\frac{400}{x}$ will give the number of students per section. Establish an inequality from the given information. Note that you can clear the inequality of fractions by multiplying by x because x is the number of sections and must be a positive number. Also keep in mind that x must be a *whole* number.

62. A student-activities director must order buses for a football game. He anticipates that 300 students will sign up, and the capacity of each bus is 40 people. How many buses should he have available?

63. The mileage markers on a freeway begin at marker 0 at the southern border of a state and continue to increase toward the northern border. The legal maximum speed on the freeway is 65 mi/h, and the legal minimum speed is 45 mi/h. If you enter the freeway at marker 100 and travel north for 2 h, what is the possible range of values for the nearest marker you could legally reach?

Hint: Because distance = rate · time, the minimum distance can be calculated as (45)(2) and the maximum distance as (65)(2). Let m be the marker you could legally reach, and establish a double-inequality statement for the solution.

64. You enter the freeway at marker 240 and now travel south for 3 h. What is the possible range of values for the nearest marker you could legally reach?

65. A new landfill must last at least 30 years for it to receive an operating permit from the local community. The proposed site is capable of receiving 570×10^6 metric tons (t) of refuse over its lifespan. How much refuse can the landfill accept each year and still meet the conditions of its permit?

ANSWERS

59. ______

60. ______

61. ______

62. ______

63. ______

64. ______

65. ______

66. ______________

66. A garbage burner must receive at least 1350 tons of trash per day to be economical enough for a community to build it. Local laws restrict truck weight to a 15-ton limit. How many truck deliveries per day will be necessary to supply the burner with its daily requirement of trash?

Answers

1. $\{x \mid x < 7\}$

3. $\{x \mid x \geq -2\}$

5. $\{x \mid x > 5\}$

7. $\{x \mid x \geq 5\}$

9. $\left\{x \mid x < \frac{7}{2}\right\}$

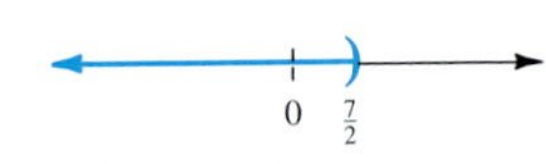

11. $\{x \mid x \leq -6\}$

13. $\{x \mid x > -3\}$

15. $\{x \mid x > 9\}$

17. $\left\{x \mid x \leq \frac{4}{3}\right\}$

19. $\{x \mid x > -2\}$

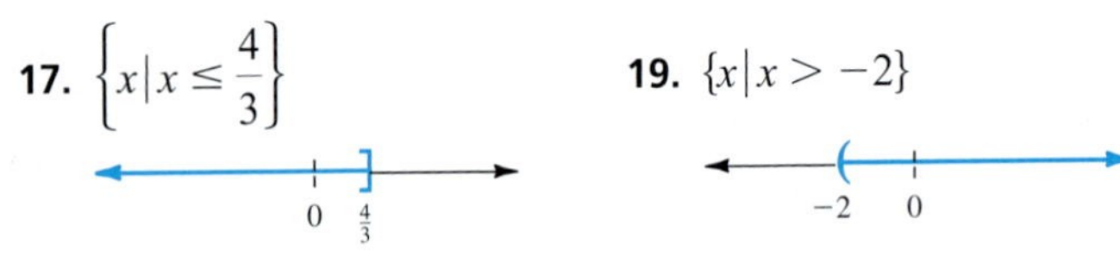

21. $\{x \mid x \geq -4\}$

23. $\left\{x \mid x < \frac{2}{5}\right\}$

25. $\{x \mid x \leq 3\}$

27. $\{x \mid x > 1\}$ **29.** $\{x \mid x > -3\}$ **31.** $\{x \mid x < 5\}$ **33.** $\{x \mid x < 19\}$
35. $\{x \mid x \geq -11\}$ **37.** $\{x \mid x \geq 12\}$ **39.** $\{x \mid x > 15\}$ **41.** $\{x \mid x \leq 10\}$

43. $\{x \mid 2 \leq x \leq 4\}$

45. $\{x \mid -4 < x < 2\}$

47. $\left\{x \mid 2 \leq x \leq \frac{9}{2}\right\}$

49. $\{x \mid -1 < x < 2\}$

51. $\{x \mid x < -2 \quad \text{or} \quad x > 4\}$

53. $\{x \mid x < -3 \quad \text{or} \quad x > 4\}$

55. $\left\{x \mid x < -2 \quad \text{or} \quad x > \frac{8}{3}\right\}$

57. $x > 50$ **59.** $x \geq 92$ **61.** $x \geq 12$ **63.** $190 \leq m \leq 230$
65. $x \leq 19 \times 10^6$ t

2.5 Absolute Value Equations and Inequalities

2.5 OBJECTIVES

1. Solve an absolute value equation in one variable
2. Solve an absolute value inequality in one variable

Equations and inequalities may involve the absolute value notation in their statements. In this section we build on the tools developed in Sections 2.1 and 2.4 and on our earlier work with absolute value for the necessary solution techniques.

NOTE Technically we mean the distance between the *point corresponding* to x and the *point corresponding* to 0, the origin.

Recall from Section 1.3 that the absolute value of x, written $|x|$, is the distance between x and 0 on the number line. Consider, for example, the absolute value equation

$$|x| = 4$$

This means that the distance between x and 0 is 4, as is pictured below.

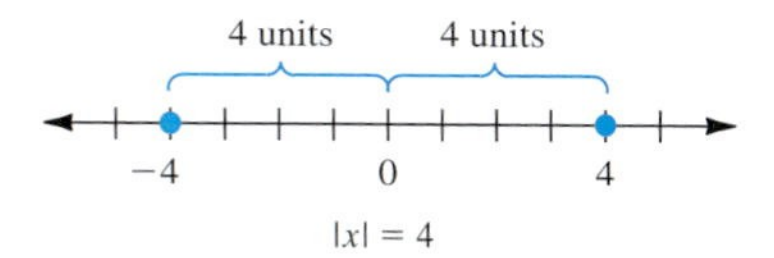

As the sketch illustrates, $x = 4$ and $x = -4$ are the two solutions for the equation.

This observation suggests the more general statement.

CAUTION

p must be positive because an equation such as $|x - 2| = -3$ has no solution. The absolute value of a quantity must always be equal to a nonnegative number.

Rules and Properties: Absolute Value Equations—Property 1

For any positive number p, if

$$|x| = p$$

then

$$x = p \quad \text{or} \quad x = -p$$

This property allows us to "translate" an equation involving absolute value to two linear equations that we can then solve separately. The following example illustrates.

Example 1

Solving an Absolute Value Equation

Solve for x:

$$|3x - 2| = 4$$

From Property 1 we know that $|3x - 2| = 4$ is equivalent to the equations

$$3x - 2 = 4 \quad \text{or} \quad 3x - 2 = -4$$

NOTE Add 2.

$$3x = 6 \qquad\qquad 3x = -2$$

NOTE Divide by 3.

$$x = 2 \qquad\qquad x = -\frac{2}{3}$$

CAUTION

Be Careful! A common mistake is to solve *only* the equation $3x - 2 = 4$. You must solve *both* of the equivalent equations to find the two required solutions.

The solutions are $-\frac{2}{3}$ and 2. These solutions are easily checked by replacing x with $-\frac{2}{3}$ and 2 in the original absolute value equation.

CHECK YOURSELF 1

Solve for x.

$|4x + 1| = 9$

An equation involving absolute value may have to be rewritten before you can apply Property 1. Consider the following example.

Example 2

Solving an Absolute Value Equation

Solve for x:

$|2 - 3x| + 5 = 10$

To use Property 1, we must first isolate the absolute value on the left side of the equation. This is easily done by subtracting 5 from both sides for the result:

$|2 - 3x| = 5$

We can now proceed as before by using Property 1.

$2 - 3x = 5 \quad$ or $\quad 2 - 3x = -5$

NOTE Subtract 2.

$-3x = 3 \qquad -3x = -7$

NOTE Divide by -3.

$x = -1 \qquad x = \frac{7}{3}$

The solution set is $\left\{-1, \frac{7}{3}\right\}$.

CHECK YOURSELF 2

Solve for x.

$|5 - 2x| - 4 = 7$

In some applications more than one absolute value is involved in an equation. Consider an equation of the form

$|x| = |y|$

Because the absolute values of x and y are equal, x and y are the same distance from 0. This means they are either *equal* or *opposite in sign.* This leads to a second general property of absolute value equations.

Rules and Properties: Absolute Value Equations—Property 2

If

$|x| = |y|$

then

$x = y$ or $x = -y$

Let's look at an application of this second property in our next example.

Example 3

Solving an Absolute Value Equation

Solve for x:

$$|3x - 4| = |x + 2|$$

By Property 2, we can write

$3x - 4 = x + 2$ or $3x - 4 = -(x + 2)$

$3x - 4 = -x - 2$

$3x = x + 6$ $\qquad$ $3x = -x + 2$

$2x = 6$ $\qquad$ $4x = 2$

$x = 3$ $\qquad$ $x = \frac{1}{2}$

The solution set is $\left\{\frac{1}{2}, 3\right\}$.

CHECK YOURSELF 3

Solve for x.

$$|4x - 1| = |x + 5|$$

We started this section by noting that the solution set for the equation

$$|x| = 4$$

consists of those numbers whose distance from the origin is equal to 4. Similarly, the solution set for the absolute value inequality

$$|x| < 4$$

consists of those numbers whose distance from the origin is *less than* 4, that is, all numbers between -4 and 4. The solution set is pictured below.

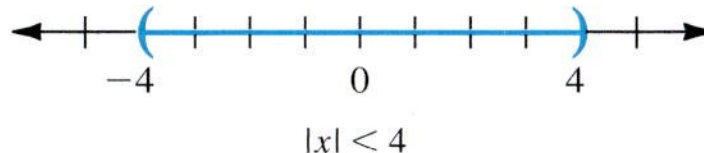

NOTE The solution set would be

$\{x \mid -4 < x < 4\}$

The solution set can be described by the compound inequality

$$-4 < x < 4$$

and this suggests the following general statement.

Rules and Properties: Absolute Value Inequalities—Property 1

For any positive number p, if

$$|x| < p$$

then

$$-p < x < p$$

Let's look at an application of Property 1 in solving an absolute value inequality.

Example 4

Solving an Absolute Value Inequality

Solve and graph the solution set of

$$|2x - 3| < 5$$

NOTE With Property 1 we can *translate* an absolute value inequality to an inequality *not* involving absolute value that can be solved by our earlier methods.

From Property 1, we know that the given absolute value inequality is equivalent to the compound inequality

$$-5 < 2x - 3 < 5$$

Solving as before, we isolate the variable in the center term.

$-2 < 2x < 8$ Add 3 to all three parts.

$-1 < x < 4$ Divide by 2.

The solution set is

$$\{x \mid -1 < x < 4\}$$

The graph is shown below.

NOTE Notice that the solution is an open interval on the number line.

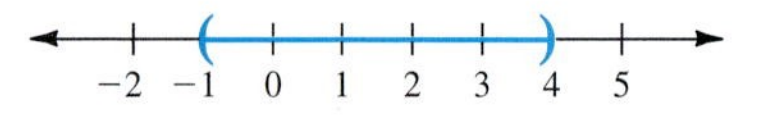

CHECK YOURSELF 4

Solve and graph the solution set.

$$|3x - 4| \le 8$$

We know that the solution set for the absolute value inequality

$$|x| < 4$$

consists of those numbers whose distance from the origin is *less than* 4. Now consider the solution set for

$$|x| > 4$$

It must consist of those numbers whose distance from the origin is *greater than* 4. The solution set is pictured below.

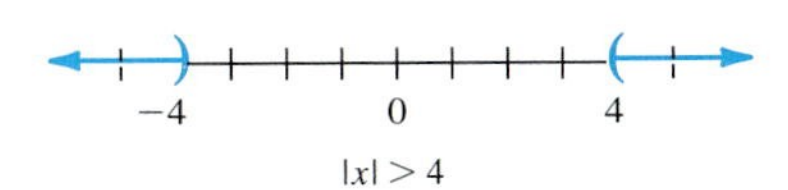

The solution set can be described by the compound inequality

$x < -4$ or $x > 4$

and this suggests the following general statement.

Rules and Properties: Absolute Value Inequalities—Property 2

For any positive number p, if

$|x| > p$

then

$x < -p$ or $x > p$

Let's apply Property 2 to the solution of an absolute value inequality.

Example 5

Solving an Absolute Value Inequality

NOTE Again we *translate* the absolute value inequality to the compound inequality *not* involving absolute value.

Solve and graph the solution set of

$|5x - 2| > 8$

From Property 2, we know that the given absolute value inequality is equivalent to the compound inequality

$5x - 2 < -8$ or $5x - 2 > 8$

Solving as before, we have

NOTE Add 2.

$5x < -6$ or $5x > 10$

NOTE Divide by 5.

$x < -\frac{6}{5}$ $\quad$ $x > 2$

NOTE You could describe the solution set as

$\left\{x \middle| x < -\frac{6}{5}\right\} \cup \{x|x > 2\}$

The solution set is $\left\{x \middle| x < -\frac{6}{5} \text{ or } x > 2\right\}$ and the graph is shown below.

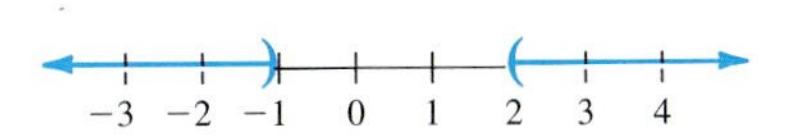

CHECK YOURSELF 5

Solve and graph the solution set.

$|3 - 2x| \geq 9$

The following chart summarizes our discussion of absolute value inequalities.

NOTE As before, p must be a positive number.

Type of Inequality	Equivalent Inequality	Graph of Solution Set
$\lvert ax + b \rvert < p$	$-p < ax + b < p$	
$\lvert ax + b \rvert > p$	$ax + b < -p$ or $ax + b > p$	

CHECK YOURSELF ANSWERS

1. $\left\{-\frac{5}{2}, 2\right\}$ **2.** $\{-3, 8\}$ **3.** $\left\{2, -\frac{4}{5}\right\}$

4. $\left\{x \,\middle|\, -\frac{4}{3} \le x \le 4\right\}$

5. $\{x \mid x \le -3 \text{ or } x \ge 6\}$

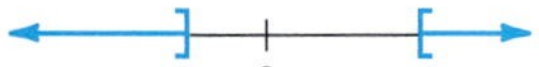

Name ____________

Section ________ Date ________

2.5 Exercises

Solve each of the following absolute value equations.

1. $|x| = 5$

2. $|x| = 7$

3. $|x - 2| = 3$

4. $|x + 5| = 6$

5. $|x + 6| = 0$

6. $|x - 3| = 0$

7. $|3 - x| = 7$

8. $|5 - x| = 4$

9. $|2x - 3| = 9$

10. $|3x + 5| = 11$

11. $|5 - 4x| = 1$

12. $|3 - 6x| = 9$

13. $\left|\frac{1}{2}x + 5\right| = 7$

14. $\left|\frac{2}{3}x - 4\right| = 6$

15. $\left|4 - \frac{3}{4}x\right| = 8$

16. $\left|3 - \frac{2}{5}x\right| = 9$

17. $|3x + 1| = -2$

18. $|5x - 2| = -3$

Rewrite each of the following absolute value equations, and then solve the equations.

19. $|x| - 3 = 2$

20. $|x| + 4 = 6$

21. $|x - 2| + 3 = 5$

22. $|x + 5| - 2 = 5$

23. $|2x - 3| - 1 = 6$

24. $|3x + 5| + 2 = 4$

ANSWERS

1. ____________
2. ____________
3. ____________
4. ____________
5. ____________
6. ____________
7. ____________
8. ____________
9. ____________
10. ____________
11. ____________
12. ____________
13. ____________
14. ____________
15. ____________
16. ____________
17. ____________
18. ____________
19. ____________
20. ____________
21. ____________
22. ____________
23. ____________
24. ____________

ANSWERS

25. ______
26. ______
27. ______
28. ______
29. ______
30. ______
31. ______
32. ______
33. ______
34. ______
35. ______
36. ______
37. ______
38. ______
39. ______
40. ______
41. ______
42. ______
43. ______
44. ______
45. ______
46. ______
47. ______
48. ______

25. $\left|\frac{1}{2}x + 2\right| - 3 = 5$

26. $\left|\frac{1}{3}x - 4\right| + 3 = 9$

27. $8 - |x - 4| = 5$

28. $10 - |2x + 1| = 3$

29. $|3x - 2| + 4 = 3$

30. $|5x - 3| + 5 = 3$

Solve each of the following absolute value equations.

31. $|2x - 1| = |x + 3|$

32. $|3x + 1| = |2x - 3|$

33. $|5x - 2| = |2x + 4|$

34. $|7x - 3| = |2x + 7|$

35. $|x - 2| = |x + 1|$

36. $|x + 3| = |x - 2|$

37. $|2x - 5| = |2x - 3|$

38. $|3x + 1| = |3x - 1|$

39. $|x - 2| = |2 - x|$

40. $|x - 4| = |4 - x|$

Find and graph the solution set for each of the following absolute value inequalities.

41. $|x| < 5$

42. $|x| > 3$

43. $|x| \geq 7$

44. $|x| \leq 4$

45. $|x - 4| > 2$

46. $|x + 5| < 3$

47. $|x + 6| \leq 4$

48. $|x - 7| \geq 5$

49. $|3 - x| > 5$

50. $|5 - x| < 3$

51. $|x - 7| < 0$

52. $|x + 5| \geq 0$

53. $|2x - 5| < 3$

54. $|3x - 1| > 8$

55. $|3x + 4| \geq 5$

56. $|2x + 3| \leq 9$

57. $|5x - 3| > 7$

58. $|6x - 5| < 13$

59. $|2 - 3x| < 11$

60. $|3 - 2x| \geq 11$

61. $|3 - 5x| \geq 7$

62. $|7 - 3x| < 13$

63. $\left|\frac{3}{4}x - 5\right| < 7$

64. $\left|\frac{2}{3}x + 5\right| \geq 3$

On some popular calculators there is a special absolute value function key. It is usually labeled "abs." To register an absolute value, you press this key and then put the desired expression in parentheses. For the expression $|x + 3|$, enter abs$(x + 3)$. Rewrite each expression in calculator form.

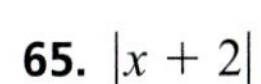

65. $|x + 2|$

66. $|x - 2|$

67. $|2x - 3|$

68. $|5x + 7|$

69. $|3x + 2| - 4$

70. $|4x - 7| + 2$

71. $2|3x - 1|$

72. $-3|2x + 8|$

ANSWERS

49. ______
50. ______
51. ______
52. ______
53. ______
54. ______
55. ______
56. ______
57. ______
58. ______
59. ______
60. ______
61. ______
62. ______
63. ______
64. ______
65. ______
66. ______
67. ______
68. ______
69. ______
70. ______
71. ______
72. ______

Answers

1. $\{-5, 5\}$ **3.** $\{-1, 5\}$ **5.** $\{-6\}$ **7.** $\{-4, 10\}$ **9.** $\{-3, 6\}$

11. $\left\{1, \frac{3}{2}\right\}$ **13.** $\{-24, 4\}$ **15.** $\left\{-\frac{16}{3}, 16\right\}$ **17.** No solution

19. $\{-5, 5\}$ **21.** $\{0, 4\}$ **23.** $\{-2, 5\}$ **25.** $\{-20, 12\}$ **27.** $\{1, 7\}$

29. No solution **31.** $\left\{-\frac{2}{3}, 4\right\}$ **33.** $\left\{-\frac{2}{7}, 2\right\}$ **35.** $\left\{\frac{1}{2}\right\}$ **37.** $\{2\}$

39. All real numbers

41. $\{x \mid -5 < x < 5\}$

−5 0 5

43. $\{x \mid x \le -7 \text{ or } x \ge 7\}$

−7 0 7

45. $\{x \mid x < 2 \text{ or } x > 6\}$

0 2 6

47. $\{x \mid -10 \le x \le -2\}$

−10 −2 0

49. $\{x \mid x < -2 \text{ or } x > 8\}$

−2 0 8

51. No solution

53. $\{x \mid 1 < x < 4\}$

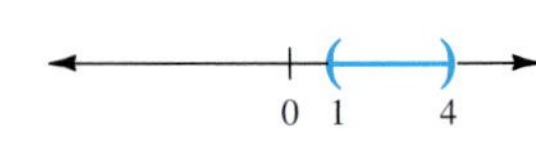

55. $\left\{x \mid x \le -3 \text{ or } x \ge \frac{1}{3}\right\}$

−3 0 $\frac{1}{3}$

57. $\left\{x \mid x < -\frac{4}{5} \text{ or } x > 2\right\}$

$-\frac{4}{5}$ 0 2

59. $\left\{x \mid -3 < x < \frac{13}{3}\right\}$

−3 0 $\frac{13}{3}$

61. $\left\{x \mid x \le -\frac{4}{5} \text{ or } x \ge 2\right\}$

$-\frac{4}{5}$ 0 2

63. $\left\{x \mid -\frac{8}{3} < x < 16\right\}$

$-\frac{8}{3}$ 0 16

65. abs(x + 2) **67.** abs(2 * x − 3) **69.** abs(3 * x + 2) − 4

71. 2 * abs(3 * x − 1)

2.6 Variation

2.6 OBJECTIVES

1. Solve problems involving direct variation
2. Solve problems involving inverse variation
3. Solve problems involving direct, inverse, and joint variation

NOTE Considering these special functions is called the study of *variation* because the quantity varies.

NOTE The idea that the circumference is a constant times the diameter was known to the Babylonians as early as 2000 B.C.

We have seen that equations can describe a relationship between two quantities. One particular type of equation is so common that there exists a precise terminology to describe the relationship. That type of equation is our focus in this section. In each case we will see that a quantity is *varying* with respect to one or more other quantities in a particular fashion.

Suppose that our quantities are related in a manner such that one is a constant multiple of the other. There are many real-world applications.

The circumference of a circle is a constant multiple of the length of its diameter:

$$C = \pi d$$

Circumference — The constant pi — Diameter

If a rate or speed is constant, the distance traveled is a constant multiple of time.

$$d = rt$$

Distance — Rate—a constant — Time

If you earn a fixed hourly pay, your total pay is a constant multiple of the number of hours that you worked.

$$T = ph$$

Total pay — Hourly pay—a constant — Hours worked

In all the cases above, the changes in one variable are proportional to the changes in the other. For instance, if the diameter of a circle is *doubled,* its circumference is *doubled.* This leads to our first definition.

Definitions: Direct Variation

If y is a constant multiple of x, we write

$y = kx$ $\quad$ k is a constant

and say that y *varies directly* as x, or that y is *directly proportional* to x. The constant k is called the *constant of variation*.

Typically, in a variation problem, you will be given the type of variation involved and related values for the variables. From this information you can determine the constant of variation and, therefore, the equation relating the quantities. The following examples illustrate.

Example 1

Solving a Direct Variation Problem

If y varies directly as x and $y = 40$ when $x = 8$, find the equation relating x and y. Also find the value of y when $x = 10$.

Because y varies directly with x, from our definition we have

NOTE As you will see, in direct variation, as the absolute value of one variable *increases*, the absolute value of the other also *increases*.

$$y = kx$$

We need to determine k, and this is easily done by letting $x = 8$ and $y = 40$:

$$40 = k(8)$$

$$k = 5$$

The desired equation relating x and y is then

$$y = 5x$$

To complete the example, if $x = 10$,

$$y = 5(10) = 50$$

CHECK YOURSELF 1

If y varies directly as x and y = 72 when x = 9, find the value of y when x = 12.

As we said, problems of variation occur frequently in applications of mathematics to many fields. The following is a typical example.

Example 2

Solving a Direct Variation Application

NOTE Example 2 is an application of Hooke's law.

In physics, it is known that the force F needed to stretch a spring x units varies directly with x. If a force of 18 lb stretches a spring 3 in., how far will the same spring be stretched by a force of 30 lb?

From the problem we know that $F = kx$, and letting $F = 18$ and $x = 3$, we have

$$18 = k(3)$$

$$k = 6$$

Therefore, $F = 6x$ relates the two variables. Now, to find x when $F = 30$, we write

$$30 = 6x$$

$$x = 5$$

So the force of 30 lb would stretch the spring 5 in.

CHECK YOURSELF 2

The pressure at a point under water is directly proportional to the depth. If the pressure at a depth of 2 ft is 125 lb/ft^2, find the pressure at a depth of 10 ft.

Many applications require that one variable be directly proportional to some power of a second variable. For instance, we might say that y is directly proportional to the square of x and write

NOTE We also say that "y varies directly with the square of x."

$$y = kx^2$$

Consider the following example.

Example 3

Solving a Direct Variation Problem

The distance s that an object will fall from rest (neglecting air resistance) varies directly with the square of the time t. If an object falls 64 ft in 2 s, how far will it fall in 5 s?

The relating equation in this example is

$$s = kt^2$$

By letting $s = 64$ and $t = 2$, we can determine k:

$$64 = k(2)^2$$

$$64 = 4k$$

so

$$k = 16$$

We now know that the desired equation is $s = 16t^2$, and substituting 5 for t, we have

$$s = 16(5)^2 = 400 \text{ ft}$$

CHECK YOURSELF 3

The distance that an object falls from rest varies directly with the square of time t. If an object falls 144 ft in 3 s, how far does it fall in 6 s?

NOTE Perhaps the most common example is the relationship between rate and time. The faster something travels (rate *increases*), the sooner it arrives (time *decreases*).

If two quantities are related so that an *increase* in the absolute value of the first gives a proportional *decrease* in the absolute value of the second, we say that the variables *vary inversely* with each other. This leads to our second definition.

Definitions: Inverse Variation

If y *varies inversely* as x, we write

$$y = \frac{k}{x} \quad k \text{ is a constant}$$

We can also say that y is *inversely proportional* to x.

Example 4

Solving an Inverse Variation Problem

If y varies inversely as x and $y = 18$ when $x = \frac{1}{2}$, find the equation relating x and y. Also find the value of y when $x = 3$.

Because y varies inversely as x, we can write

$$y = \frac{k}{x}$$

Now with $y = 18$ and $x = \frac{1}{2}$, we have

$$18 = \frac{k}{\frac{1}{2}}$$

$$9 = k$$

We now have the desired equation relating x and y:

$$y = \frac{9}{x}$$

and when $x = 3$,

$$y = \frac{9}{3} = 3$$

CHECK YOURSELF 4

If w is inversely proportional to z and w = 15 when z = 5, find the equation relating z and w. Also find the value of w when z = 3.

Let's consider an application that involves the idea of inverse variation.

Example 5

Solving an Inverse Variation Application

The intensity of illumination I of a light source varies inversely as the square of the distance d from that source. If the illumination 4 ft from the source is 9 footcandles (fc), find the illumination 6 ft from the source.

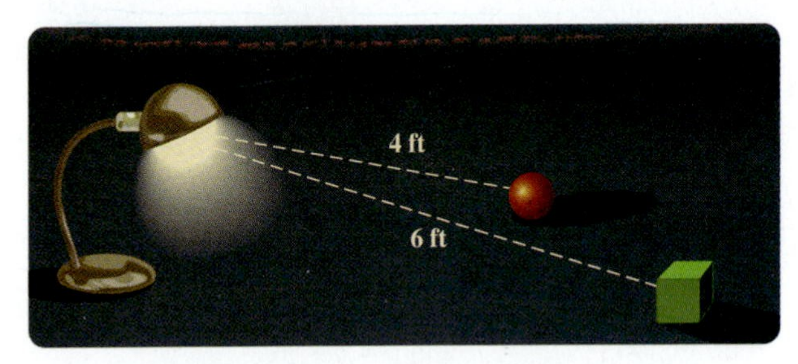

From the given information we know that

$$I = \frac{k}{d^2}$$

Letting $d = 4$ and $I = 9$, we first find the constant of variation k:

$$9 = \frac{k}{4^2} \quad \text{or} \quad 9 = \frac{k}{16}$$

and

$$k = 144$$

The equation relating I and d is then

$$I = \frac{144}{d^2}$$

and when $d = 6$,

$$I = \frac{144}{6^2} = \frac{144}{36} = 4 \text{ fc}$$

CHECK YOURSELF 5

At a constant temperature, the volume of a gas varies inversely as the pressure. If a gas has volume 200 ft^3 under a pressure of 40 pounds per square inch (lb/in.^{2}), what will be its volume under a pressure of 50 lb/in.^{2}?

It is also common for one quantity to depend on *several* others. We can find a familiar example from geometry.

The volume of a cylinder depends on its height and the square of its radius.

This is an example of *joint variation*. We say that the volume V varies jointly with the height h and the square of the radius r. We can write

NOTE You probably recognize that k, the constant of variation, is π in this case.

$$V = khr^2$$

In general:

Definitions: Joint Variation

If z *varies jointly* as x and y, we write

$z = kxy$ $\quad$ k is a constant

The solution techniques for problems involving joint variation are similar to those used earlier, as Example 6 illustrates.

Example 6

Solving a Joint Variation Problem

Assume that z varies jointly as x and y. If $z = 100$ when $x = 2$ and $y = 20$, find the value of z if $x = 4$ and $y = 30$.

From the given information we have

$$z = kxy$$

Letting $z = 100$, $x = 2$, and $y = 20$ gives

$$100 = k(2)(20) \qquad \text{or} \qquad k = \frac{5}{2}$$

The equation relating z with x and y is then

$$z = \frac{5}{2}xy$$

and when $x = 4$ and $y = 30$, by substitution

$$z = \frac{5}{2}(4)(30) = 300$$

CHECK YOURSELF 6

Assume that r varies jointly as s and t. If r = 64 when s = 3 and t = 8, find the value of r when s = 16 and t = 12.

Once again there are many physical applications of the concept of joint variation. The following is a typical example.

Example 7

Solving a Joint Variation Application

The "safe load" for a wooden rectangular beam varies jointly as its width and the square of its depth.

If the safe load of a beam 2 in. wide and 8 in. deep is 640 lb, what is the safe load of a beam 4 in. wide and 6 in. deep?

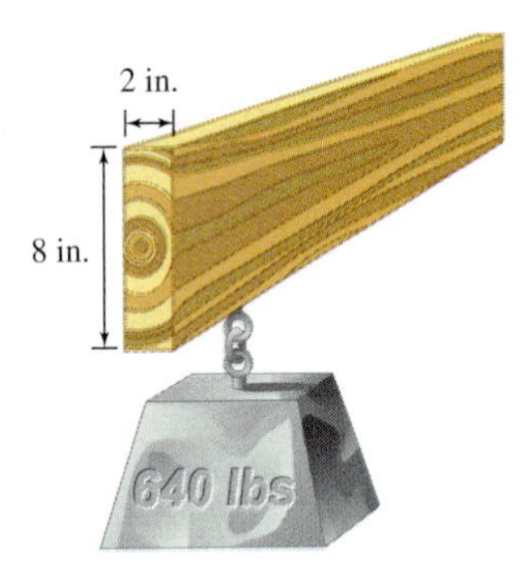

From the given information,

NOTE S is the safe load, w the width, and d the depth.

$$S = kwd^2$$

Substituting the given values yields

$$640 = k(2)(8)^2$$

$$k = 5$$

We then have the equation

$$S = 5wd^2$$

and, for the 4 by 6 in. beam,

$$S = 5(4)(6)^2 = 720 \text{ lb}$$

CHECK YOURSELF 7

The force of a wind F blowing on a vertical wall varies jointly as the surface area of the wall A and the square of the wind velocity v.

If a wind of 20 mi/h has a force of 100 lb on a wall with area 50 ft², what force will a wind of 40 mi/h produce on the same wall?

There is one final category of variation problems. This category involves applications in which inverse variation is combined with direct or joint variation in stating the equation relating the variables.

These are called *combined variation* problems. In general, a typical statement form is as follows:

Definitions: Combined Variation

If z varies directly as x and inversely as y, we write

$z = \frac{kx}{y}$ in which k is a constant

Example 8

Solving a Combined Variation Problem

Assume that w varies directly as x and inversely as the square of y. When $x = 8$ and $y = 4$, $w = 18$. Find w if $x = 4$ and $y = 6$.

From the given information we can write

$$w = \frac{kx}{y^2}$$

Substituting the known values, we have

$$18 = \frac{k \cdot 8}{4^2} \quad \text{or} \quad k = 36$$

We now have the equation

$$w = \frac{36x}{y^2}$$

and letting $x = 4$ and $y = 6$, we get

$$w = \frac{36 \cdot 4}{6^2} = 4$$

CHECK YOURSELF 8

Ohm's law for an electric circuit states that the current I varies directly as the electromotive force E and inversely as the resistance R.

If the current is 10 amperes (A), the electromotive force is 110 volts (V) and the resistance is 11 ohms (Ω). Find the current for an electromotive force of 220 V and a resistance of 5 Ω.

Translating verbal problems to algebraic equations is the basis for all variation applications. The following table gives some typical examples from our work in this section.

Variation Statement	Algebraic Equation
y varies directly as x.	$y = kx$
y varies inversely as x.	$y = \frac{k}{x}$
z varies jointly as x and as y.	$z = kxy$
z varies directly as x and inversely as y.	$z = \frac{kx}{y}$

All four basic types of variation problems involve essentially the same solution technique. The following algorithm summarizes the steps involved in all the variation problems that we have considered.

Step by Step: Solving Problems Involving Variation

Step 1 Translate the given problem to an algebraic equation involving the constant of variation k.

Step 2 Use the given values to find that constant.

Step 3 Replace k with the value found in step 2 to form the general equation relating the variables.

Step 4 Substitute the appropriate values of the variables to solve for the corresponding value of the desired unknown quantity.

CHECK YOURSELF ANSWERS

1. $y = 96$ **2.** 625 lb/ft^2 **3.** 576 ft **4.** $w = \frac{75}{z}$, $w = 25$

5. 160 ft^3 **6.** 512 **7.** 400 lb **8.** $I = \frac{kE}{R}$ and $k = 1$, so $I = 44$ A

Name ____________

Section ________ Date ________

2.6 Exercises

Translate each of the following statements of variation to an algebraic equation, using k as the constant of variation.

1. s varies directly as the square of x.

2. z is directly proportional to the square root of w.

3. r is inversely proportional to s.

4. m varies inversely as the cube of n.

5. V varies directly as T and inversely as P.

6. A varies jointly as x and y.

7. V varies jointly as h and the square of r.

8. t is directly proportional to d and inversely proportional to r.

9. w varies jointly as x and y and inversely as the square of z.

10. p varies jointly as r and the square of s and inversely as the cube of t.

Find k, the constant of variation, given each of the following sets of conditions.

11. y varies directly with x; $y = 54$ when $x = 6$.

12. m varies inversely with p; $m = 5$ when $p = 3$.

13. r is inversely proportional to the square of s; $r = 5$ when $s = 4$.

14. u varies directly with the square of w; $u = 75$ when $w = 5$.

ANSWERS

1. ____________
2. ____________
3. ____________
4. ____________
5. ____________
6. ____________
7. ____________
8. ____________
9. ____________
10. ____________
11. ____________
12. ____________
13. ____________
14. ____________

ANSWERS

15. ______
16. ______
17. ______
18. ______
19. ______
20. ______
21. ______
22. ______
23. ______
24. ______
25. ______
26. ______
27. ______

15. V varies jointly as x and y; $V = 100$ when $x = 5$ and $y = 4$.

16. w is directly proportional to u and inversely proportional to v; $w = 20$ when $u = 10$ and $v = 3$.

17. z varies directly as the square of x and inversely as y; $z = 20$ when $x = 2$ and $y = 4$.

18. p varies jointly as r and the square of q; $p = 144$ when $q = 6$ and $r = 2$.

19. m varies jointly as n and the square of p and inversely as r; $m = 40$ when $n = 5$, $p = 2$, and $r = 4$.

20. x varies directly as the square of y and inversely as w and z; $x = 8$ when $y = 4$, $w = 3$, and $z = 2$.

Solve each of the following variation problems.

21. Let y vary directly with x. If $y = 60$ when $x = 5$, find the value of y when $x = 8$.

22. Suppose that z varies inversely as the square of w and that $z = 3$ when $w = 4$. Find the value of z when $w = 6$.

23. Variable A varies jointly with x and y, and $A = 120$ when $x = 6$ and $y = 5$. Find the value of A when $x = 8$ and $y = 3$.

24. Let p be directly proportional to q and inversely proportional to the square of r. If $p = 3$ when $q = 8$ and $r = 4$, find p when $q = 9$ and $r = 6$.

25. Suppose that s varies directly with r and inversely with the square of t. If $s = 4$ when $r = 12$ and $t = 6$, find the value of s when $r = 8$ and $t = 4$.

26. Variable p varies jointly with the square root of r and the square of q. If $p = 72$ when $r = 16$ and $q = 3$, find the value of p when $r = 25$ and $q = 2$.

Solve each of the following variation problems.

27. The length that a spring will stretch varies directly as the force applied to the spring. If a force of 10 lb will stretch a spring 2 in., what force will stretch the same spring 3 in.?

ANSWERS

28. ____________

29. ____________

30. ____________

31. ____________

32. ____________

33. ____________

34. ____________

35. ____________

36. ____________

37. ____________

28. If the temperature of a gas is held constant, the volume occupied by that gas varies inversely as the pressure to which the gas is subjected. If the volume of a gas is 8 ft^3 when the pressure is 12 lb/in.^{2}, find the volume of the gas if the pressure is 16 lb/in.^{2}.

29. If the current, in amperes, in an electric circuit is inversely proportional to the resistance, the current is 55 A when the resistance is 2 Ω. Find the current when the resistance is 5 Ω.

30. The distance that a ball rolls down an inclined plane varies directly as the square of the time. If the ball rolls 36 ft in 3 s, how far will it roll in 5 s?

31. The volume of a right circular cone varies jointly as the height and the square of the radius. If the volume of the cone is 15π cm^3 when the height is 5 cm and the radius is 3 cm, find the volume when the height is 6 cm and the radius is 4 cm.

32. The safe load of a rectangular beam varies jointly as its width and the square of its depth. If the safe load of a beam is 1000 lb when the width is 2 in. and the depth is 10 in., find its safe load when the width is 4 in. and the depth is 8 in.

33. The stopping distance (in feet) of an automobile varies directly as the square of its speed (in miles per hour). If a car can stop in a distance of 80 ft from 40 mi/h, how much distance will it take to stop from a speed of 60 mi/h?

34. The period (the time required for one complete swing) of a simple pendulum is directly proportional to the square root of its length. If a pendulum with length 9 cm has a period of 3.3 s, find the period of a pendulum with length 16 cm.

35. The distance (in miles) that a person can see to the horizon from a point above Earth's surface is directly proportional to the square root of the height (in feet) of that point. If a person 100 ft above Earth's surface can see 12.5 mi, how far can an observer in a light airplane at 3600 ft see to the horizon?

36. The illumination produced by a light source on a surface varies inversely as the square of the distance of that surface from the source. If a light source produces an illumination of 48 fc on a wall 4 ft from the source, what will be the illumination (in footcandles) of a wall 8 ft from the source?

37. The electrical resistance of a wire varies directly as its length and inversely as the square of its diameter. If a wire with length 200 ft and diameter 0.1 in. has a resistance of 2 Ω, what will be the resistance in a wire of length 400 ft with diameter 0.2 in.?

ANSWERS

38. ______________________

39. ______________________

40. ______________________

38. The frequency of a guitar string varies directly as the square root of the tension on the string and inversely as the length of the string. If a frequency of 440 cycles per second, or *hertz* (Hz), is produced by a tension of 36 lb on a string of length 60 cm, what frequency (in hertz) will be produced by a tension of 64 lb on a string of length 40 cm?

39. The temperature of the steam from a geothermal source is inversely proportional to the distance it is transported. Write an algebraic equation relating temperature to distance.

40. Power available from a wind generator varies jointly as the square of the diameter of the rotor and the cube of the wind velocity. Write an algebraic equation relating power, rotor diameter, and wind speed.

Answers

1. $s = kx^2$ **3.** $r = \frac{k}{s}$ **5.** $V = \frac{kT}{P}$ **7.** $V = khr^2$ **9.** $w = \frac{kxy}{z^2}$

11. 9 **13.** 80 **15.** 5 **17.** 20 **19.** 8 **21.** 96 **23.** 96

25. 6 **27.** 15 lb **29.** 22 A **31.** 32π cm^3 **33.** 180 ft

35. 75 mi **37.** 1 Ω **39.** $t = \frac{k}{d}$

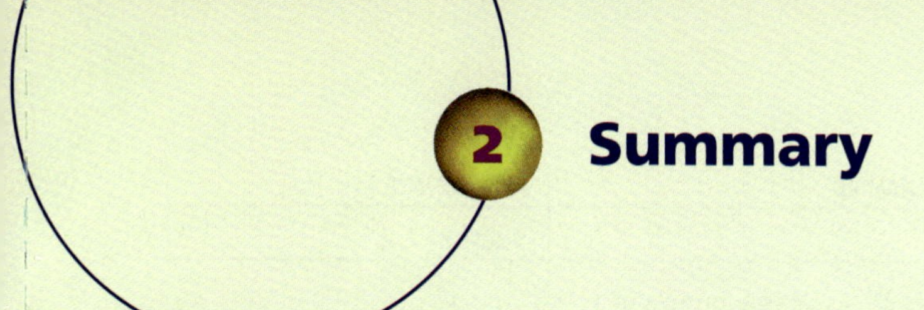

2 Summary

DEFINITION/PROCEDURE	EXAMPLE	REFERENCE
Solutions to Linear Equations in One Variable		**Section 2.1**
A **linear equation in one variable** is any equation that can be written in the form $$ax + b = 0$$ in which a and b are any real numbers and $a \neq 0$. To solve a linear equation means to find its solution. A **solution** of an equation in one variable is any number that will make the equation a true statement. The **solution set** consists of all solutions. Two equations are **equivalent** if they have the same solution set. Forming a sequence of equivalent equations that will lead to the solution of a linear equation involves two properties of equations.	$5x - 6 = 3x + 2$ is a linear equation. The variable appears only to the first power. The solution for the equation above is 4 because $5 \cdot 4 - 6 \stackrel{?}{=} 3 \cdot 4 + 2$ $14 = 14$ is a true statement. $5x - 6 = 3x + 2$ and $2x = 8$ are equivalent equations. Both have 4 as the solution.	**pp. 50–51**
Addition Property of Equations If $a = b$, then $a + c = b + c$. In words, adding the same quantity to both sides of an equation gives an equivalent equation.	If $x - 3 = 7$, then $x - 3 + 3 = 7 + 3$.	**p. 51**
Multiplication Property of Equations If $a = b$, then $ac = bc$, $c \neq 0$. In words, multiplying both sides of an equation by the same nonzero quantity gives an equivalent equation.	If $2x = 8$, then $\frac{1}{2}(2x) = \frac{1}{2}(8)$.	**p. 51**
Solving Linear Equations in One Variable **Step 1** Multiply both sides of the equation by the LCM of all denominators, to clear the equation of fractions. **Step 2** Remove any grouping symbols by applying the distributive property. **Step 3** Combine like terms that appear on either side of the equation. **Step 4** Apply the addition property of equations to write an equivalent equation with the variable term on *one side* of the equation and the constant term on the *other side*. **Step 5** Apply the multiplication property of equations to write an equivalent equation with the variable isolated on one side of the equation. **Step 6** Check the solution in the *original* equation.	Solve $4(x + 1) - 5x = 1$ Remove grouping symbols. $4x + 4 - 5x = 1$ Combine like terms. $-x + 4 = 1$ Subtract 4. $-x = -3$ Divide by -1. $x = 3$ To check: $\frac{3+1}{5} - \frac{3}{4} \stackrel{?}{=} \frac{1}{20}$	**p. 57**

Continued

DEFINITION/PROCEDURE	**EXAMPLE**	**REFERENCE**
Literal Equations and Formulas		**Section 2.2**
Formulas and **literal equations** express a relationship between more than one variable or letter. Solving a formula or literal equation for a variable means isolating that specified variable on one side of the equation. The steps used in the process are very similar to those used in solving linear equations.	$P = 2L + 2W$ is a formula or a literal equation. To solve for L: $P - 2W = 2L$ $L = \dfrac{P - 2W}{2}$	**p. 63**
Problem Solving and Applications of Linear Equations		**Section 2.3**
Solving Applications	One number is 3 less than twice another. If the sum of the numbers is 27, find the two numbers.	
Step 1 Read the problem carefully to determine the unknown quantities.	**1.** The unknowns are the two numbers.	
Step 2 Choose a variable to represent the unknown or unknowns.	**2.** Let x be the first number. Then $2x - 3$ is the second.	
Step 3 Translate the problem to the language of algebra to form an equation.	**3.** $\underbrace{x + 2x - 3}_{\text{Sum of the numbers}} = 27$	
Step 4 Solve the equation and answer the question of the original problem.	**4.** Solving as before gives $x = 10$ and $2x - 3 = 17$	
Step 5 Verify your solution by returning to the original problem.	**5.** The sum of the numbers is 27, and 17 is 3 less than twice 10.	**p. 69**
Linear Inequalities		**Section 2.4**
A linear inequality in one variable is any inequality that can be written in the form $ax + b < 0$ in which a and b are real numbers and $a \neq 0$.	$5x - 7 < 2x + 4$ is a linear inequality. The inequalities $ax + b > 0 \quad ax + b \le 0$ and $\quad ax + b \ge 0$ are also linear inequalities.	**p. 79**
Addition Property of Inequalities If $a < b$, then $a + c < b + c$.	If $x - 3 < 7$ $x - 3 + 3 < 7 + 3$ $x < 10$	**p. 79**
Multiplication Property of Inequalities If $a < b$ then $\quad ac < bc \quad$ when c is a *positive* number and $ac > bc \quad$ when c is a *negative* number. Solving linear inequalities involves essentially the same procedures as solving linear equations. The following algorithm is applied.	If $\frac{1}{3}x < 5$ $3\left(\frac{1}{3}\right)x < 3 \cdot 5$ $x < 15$ If $-x < 4$ $x > -4$	**p. 80**

Continued

DEFINITION/PROCEDURE	EXAMPLE	REFERENCE
Linear Inequalities		**Section 2.4**
Solving Linear Inequalities in One Variable **Step 1** Clear the inequality statement of any fractions by using the multiplication property. **Step 2** Remove any grouping symbols and combine like terms. **Step 3** Apply the addition property to write an equivalent inequality with the variable term on *one side* of the inequality and the constant term on the *other side.* **Step 4** Apply the multiplication property to write an equivalent inequality with the variable isolated on one side of the inequality. Be sure to reverse the sense of the inequality if you multiply or divide by a negative number. **Step 5** Graph the solution set of the original inequality.	$2x - 5 \geq 7x + 15$ Add 5 to both sides. $2x \geq 7x + 20$ Subtract $7x$ from both sides. $-5x \geq 20$ Divide both sides by -5. $x \leq -4$ Note that we *reverse the sense* of the inequality. The graph of the solution set $\{x \mid x \leq -4\}$ is -4 0	**p. 83**
The **compound inequalities** considered here fall into two categories: **1.** $ax + b > -c$ and $ax + b < c$ which can be written in the double-inequality form as $-c < ax + b < c$ This type of inequality statement is solved by isolating the variable in the middle term. **2.** $ax + b < -c$ or $ax + b > c$ This type of inequality statement is solved by considering each inequality separately.	$-5 < 3x < 7$ means that $-5 < 3x$ and $3x > 7$ So $-\frac{5}{3} < x$ and $x < \frac{7}{3}$	**p. 84**
Absolute Value Equations and Inequalities		**Section 2.5**
To solve absolute value equations, the following property is applied. *Absolute Value Equations—Property 1* For any positive number p, if $\lvert x \rvert = p$ then $x = -p$ or $x = p$ To solve an equation involving absolute value, translate the equation to the two equivalent linear equations. Those equations can then be solved separately.	$\lvert 2x - 5 \rvert = 7$ is equivalent to $2x - 5 = -7$ or $2x - 5 = 7$ so $x = -1$ or $x = 6$	**p. 91**

Continued

DEFINITION/PROCEDURE	EXAMPLE	REFERENCE
Absolute Value Equations and Inequalities		**Section 2.5**
Absolute Value Inequalities—Property 1 and Property 2 **1.** For any positive number p, if $|x| < p$ then $-p < x < p$ To solve this form of inequality, translate to the equivalent compound inequality and then solve as before.	$|3x - 5| < 7$ is equivalent to $-7 < 3x - 5 < 7$ This yields $-2 < 3x < 12$ $-\frac{2}{3} < x < 4$	
2. For any positive number p, if $|x| > p$ then $x < -p$ or $x > p$ To solve this form of inequality, translate to the equivalent compound inequality and then solve as before.	$|2 - 5x| \geq 12$ is equivalent to $2 - 5x \leq -12$ or $2 - 5x \geq 12$ This yields $x \geq \frac{14}{5}$ or $x \leq -2$	p. 94
Variation		**Section 2.6**
Direct Variation If y *varies directly* as x (or y is *directly proportional* to x), we write $y = kx$ in which k is the *constant of variation.*	If y varies directly as x and $y = 64$ when $x = 4$, find the equation relating x and y. $y = kx$ $64 = k \cdot 4$ so $k = 16$ so $y = 16x$	p. 101
Inverse Variation If y *varies inversely* as x (or y is *inversely proportional* to x), we write $y = \frac{k}{x}$	If r varies inversely as the square of s, the relating equation is $r = \frac{k}{s^2}$	p. 103
Joint Variation If z *varies jointly* as x and y, we write $z = kxy$	If m varies jointly as n and the square root of p, the relating equation is $m = kn\sqrt{p}$	p. 105
Combined Variation If z varies directly as x and inversely as y, we write $z = \frac{kx}{y}$	If V varies directly as T and inversely as P, the relating equation is $V = \frac{kT}{P}$	p. 107

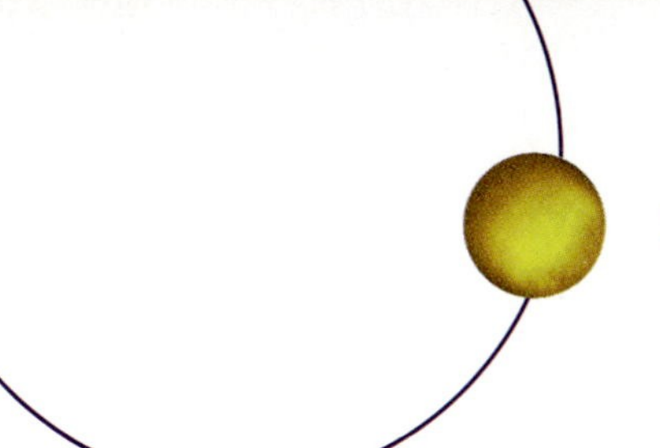

Summary Exercises

This summary exercise set is provided to give you practice with each of the objectives of the chapter. Each exercise is keyed to the appropriate chapter section. The answers are provided in the instructor's manual that accompanies this text. Your instructor will provide guidelines on how to best use these exercises in your instructional program.

[2.1] Solve the following equations. Express your answer in set notation.

1. $4x - 5 = 23$

2. $7 - 3x = -8$

3. $5x + 2 = 6 - 3x$

4. $7x - 3 = 2x + 12$

5. $2x - 7 = 9x - 35$

6. $5 - 3x = 2 - 6x$

7. $7x - 3 + 2x = 5 + 6x + 4$

8. $2x + 5 - 4x = 3 - 6x + 10$

9. $3(x - 5) = x + 1$

10. $4(2x - 1) = 6x + 5$

11. $7x - 3(x - 2) = 30$

12. $8x - 5(x + 3) = -10$

13. $7(3x + 1) - 13 = 8(2x + 3)$

14. $3(2x - 5) - 2(x - 3) = 11$

15. $\frac{2x}{3} - \frac{x}{4} = 5$

16. $\frac{3x}{4} - \frac{2x}{5} = 7$

17. $\frac{x}{2} - \frac{x + 1}{3} = \frac{1}{6}$

18. $\frac{x + 1}{5} - \frac{x - 6}{3} = \frac{1}{3}$

[2.2] Solve for the indicated variable.

19. $P = RB$ for R

20. $I = Prt$ for t

21. $S = 2\pi rh$ for h

22. $S = \frac{1}{2}gt^2$ for g

23. $y = mx + b$ for m

24. $A = P(1 + rt)$ for r

[2.3] Solve each of the following problems.

25. A principal of \$5000 was invested in a savings account paying 6% annual interest. If the interest earned over a certain period was \$1200, for how long was the money invested? See exercise 20.

26. A cylinder has lateral surface area 96π square inches (in.^{2}). If the radius of the cylinder is 6 in., find the height of the cylinder. See exercise 21.

27. A principal of \$3000 was invested in a money market fund. If the amount in the account was \$3720 at the end of 3 years, what was the annual interest rate? See exercise 24.

[2.3] Solve each of the following applications.

28. Lisa left Friday morning, driving on the freeway to visit friends for the weekend. Her trip took 4 h. When she returned on Sunday, heavier traffic slowed her average speed by 6 mi/h, and the trip took $4\frac{1}{2}$ h. What was her average speed in each direction, and how far did she travel each way?

29. A bicyclist started on a 132-mi trip and rode at a steady rate for 3 h. He began to tire at that point and slowed his speed by 4 mi/h for the remaining 2 h of the trip. What was his average speed for each part of the journey?

30. At noon, Jan left her house, jogging at an average rate of 8 mi/h. Two hours later, Stanley left on his bicycle along the same route, averaging 20 mi/h. At what time will Stanley catch up with Jan?

31. At 9 A.M., David left New Orleans for Tallahassee, averaging 47 mi/h. Two hours later, Gloria left Tallahassee for New Orleans along the same route, driving 5 mi/h faster than David. If the two cities are 391 mi apart, at what time will David and Gloria meet?

32. A firm producing running shoes finds that its fixed costs are $3900 per week, and its variable cost is $21 per pair of shoes. If the firm can sell the shoes for $47 per pair, how many pairs of shoes must be produced and sold each week for the company to break even?

[2.4] Solve each of the following inequalities. Then graph the solution sets.

33. $3x - 2 > 10$

34. $5x - 3 \le -18$

35. $5 - 3x \le 3$

36. $7 - 4x \ge 15$

37. $9x - 3 < 7x - 13$

38. $5 - 3x > 2 - 6x$

39. $2x - 5 \ge 7x - 10$

40. $4 - 3x < 14 + 2x$

41. $4(5x - 4) \ge 3(3x + 2)$

42. $3(2x - 1) > 2(x - 4) - 11$

43. $\frac{x}{2} - \frac{x+8}{5} < \frac{1}{2}$

44. $\frac{x+3}{4} - \frac{x-1}{3} > \frac{2}{3}$

[2.4] Solve each of the following compound inequalities. Then graph the solution sets.

45. $3 < x + 5 < 7$

46. $-2 \le 3x + 4 \le 10$

47. $-5 \le 3 - 2x \le 5$

48. $-4 < 5 - 3x < 4$

49. $3x - 1 < -7$ or $3x - 1 > 7$

50. $2x + 5 \le -9$ or $2x + 5 \ge 9$

[2.5] Solve each of the following absolute value equations.

51. $|x + 3| = 5$

52. $|3x - 2| = 7$

53. $|7 - x| = 3$

54. $|5 - 3x| = 14$

55. $|2x + 1| - 3 = 6$

56. $7 - |x - 3| = 5$

57. $|3x - 1| = |x + 5|$

58. $|x - 5| = |x + 3|$

[2.5] Solve each of the following absolute value inequalities. Then graph the solution sets.

59. $|x| \le 3$

60. $|x + 3| > 5$

61. $|x - 7| > 4$

62. $|3 - x| < 6$

63. $|2x + 7| > 5$

64. $|3x - 1| < 14$

65. $|3x + 4| < 11$

66. $|5x + 2| \ge 12$

67. $|3 - 2x| \ge 15$

68. $|5 - 3x| < 11$

69. $\left|\frac{2x - 1}{3}\right| < 5$

70. $\left|\frac{2x + 1}{3}\right| \ge 5$

[2.6] Translate each of the following statements of variation to an algebraic equation, using k as the constant of variation.

71. d varies directly as the square of t.

72. r varies inversely as the square root of s.

73. y is directly proportional to x and inversely proportional to w.

74. z varies jointly as the cube of x and the square root of y.

[2.6] Find k, the constant of variation, given each of the following sets of conditions.

75. y varies directly as the cube root of x; $y = 12$ when $x = 8$.

76. p is inversely proportional to the square of q; $p = 3$ when $q = 4$.

77. r varies jointly as s and the square of t; $r = 150$ when $s = 2$ and $t = 5$.

78. t varies directly as the square of u and inversely as v; $t = 36$ when $u = 3$ and $v = 2$.

[2.6] Solve each of the following variation problems.

79. Let z vary inversely as the square of w. If $z = 3$ when $w = 4$, find the value of z when $w = 2$.

80. Suppose that s varies directly as the square of t and that $s = 90$ when $t = 3$. Find the value of s when $t = 5$.

81. The variable m varies jointly as p and the square of n. If $m = 144$ when $n = 2$ and $p = 3$, find the value of m when $n = 3$ and $p = 4$.

82. Let p be directly proportional to the square of q and inversely proportional to r. If $p = 2$ when $q = 2$ and $r = 12$, find the value of p when $q = 4$ and $r = 24$.

83. The distance that a ball will fall (neglecting air resistance) is directly proportional to the square of time. If the ball falls 64 ft in 2 s, how far will it fall in 5 s?

84. If the temperature of a gas is held constant, the volume occupied by that gas varies inversely as the pressure. A gas has volume 200 ft^3 when it is subjected to a pressure of 20 $lb/in.^2$. What will its volume be under a pressure of 25 $lb/in.^2$?

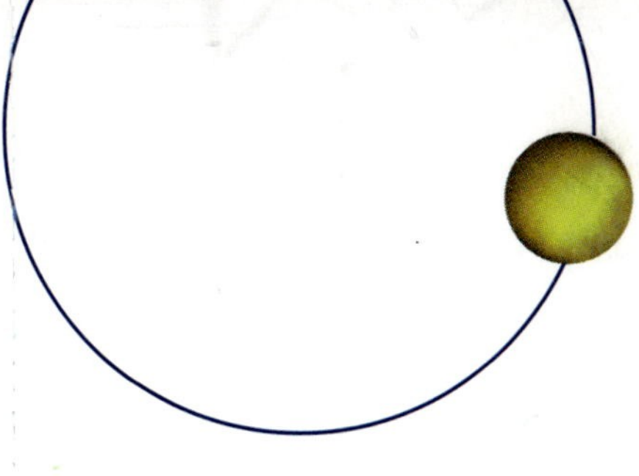

Self-Test for Chapter 2

Name ____________

Section ______ Date ______

ANSWERS

1. ____
2. ____
3. ____
4. ____
5. ____
6. ____
7. ____
8. ____
9. ____
10. ____
11. ____
12. ____
13. ____
14. ____
15. ____
16. ____

The purpose of this self-test is to help you check your progress and to review for a chapter test in class. Allow yourself about an hour to take the test. When you are done, check your answers in the back of the book. If you missed any questions, be sure to go back and review the appropriate sections in the chapter and the exercises that are provided.

Solve each of the following equations.

1. $7 - 5x = 3$

2. $7x + 8 = 30 - 4x$

3. $5x - 3(x - 5) = 19$

4. $\dfrac{x + 3}{4} - \dfrac{x}{2} = \dfrac{3}{8}$

Solve each of the following literal equations for the indicated variables.

5. $A = P(1 + rt)$ for r

6. $A = \dfrac{1}{2}h(B + b)$ for h

Solve the following applications.

7. At 10 A.M., Sandra left her house on a business trip and drove at an average rate of 45 mi/h. One hour later, Adam discovered that Sandra had left her briefcase behind, and he began driving at 55 mi/h along the same route. When will Adam catch up with Sandra?

8. A firm producing flashlights finds that its fixed cost is \$2400 per week and its variable cost is \$4.50 per flashlight. The revenue is \$7.50 per flashlight, so the cost and revenue equations are, respectively,

$C = 4.50x + 2400$ and $R = 7.50x$

Find the break-even point for the firm.

Solve each of the following inequalities, then graph the solution set.

9. $5x - 3 \le 17$

10. $3x + 7 < 5(x - 2)$

11. $\dfrac{x + 1}{2} - \dfrac{1}{3} > \dfrac{x + 3}{6}$

12. $-5 \le 3 - 2x \le 7$

Solve each of the following equations and inequalities.

13. $|3x - 5| = 7$

14. $|2x - 3| = |x + 1|$

15. $|4x - 3| < 9$

16. $|5 - 4x| \ge 13$

ANSWERS

17. ______

18. ______

19. ______

20. ______

17. Let s vary inversely as the cube of t. If $s = 16$ when $t = 2$, find the value of s when $t = 4$.

18. Variable p is jointly proportional to r and the square root of s. If $p = 80$ when $r = 4$ and $s = 16$, find the value of p when $r = 3$ and $s = 25$.

19. Suppose that z is directly proportional to x and inversely proportional to the square of y. If $z = 32$ when $x = 4$ and $y = 2$, find the value of z when $x = 6$ and $y = 4$.

20. The pressure at a point under water varies directly as the depth. If the pressure at a point 4 ft below the surface of the water is 250 pounds per square foot (lb/ft^2), find the pressure at a depth of 8 ft.

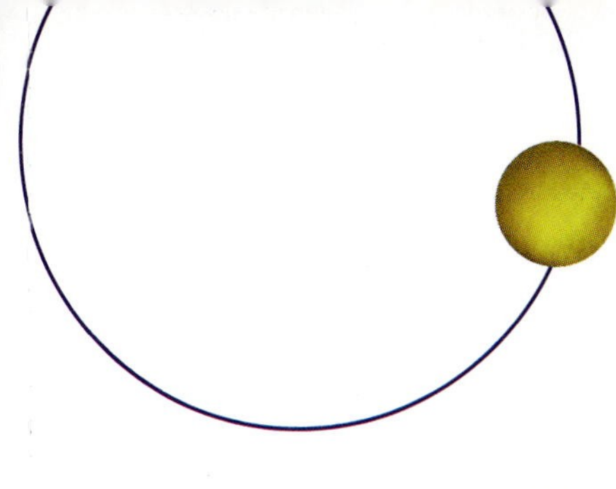

Cumulative Test for Chapters 1 and 2

Name ____________

Section ______ Date ______

ANSWERS

1. ____________
2. ____________
3. ____________
4. ____________
5. ____________
6. ____________
7. ____________
8. ____________
9. ____________
10. ____________
11. ____________
12. ____________
13. ____________
14. ____________
15. ____________
16. ____________

This test is provided to help you in the process of reviewing the previous chapters. Answers are provided in the back of the book. If you missed any answers, be sure to go back and review the appropriate chapter sections.

1. For the set of numbers $\{-5, 3, 4.1, 5, 0, \sqrt{3}, -7\}$, which belong to **(a)** the natural numbers, **(b)** the integers?

2. Translate the statement "5 less than p, divided by s" into symbols.

3. Simplify the expressions **(a)** $(x^3y^5)^2$ **(b)** $\dfrac{(x^4y^7)^2}{xy^6}$

4. Write $-|7| - |-7|$ without using absolute value symbols.

5. Graph the set $\{x|-2 < x \leq 5\}$.

6. Write 4,370,000,000 in scientific notation.

7. Name the property that justifies the statement $x + (3w + 5) = (x + 3w) + 5$.

8. Evaluate the expression $34 + 24 \div 6 \cdot 4 - 5^2$.

9. Combine like terms in the expression $7x - 2(2x - 5) - 7$.

Solve each of the following equations and check the results.

10. $5c - 7 = 3c - 2$

11. $3t + 11 = 5(t - 2)$

12. $\frac{1}{3}y - 5 = 8 - \frac{5}{6}y$

Solve each of the following equations for the indicated variable.

13. $A = \frac{1}{2}h(B + b)$ for B

14. $7p - 4y + 12 = 0$ for p

Solve the following applications.

15. Carla and Jake leave town at the same time, traveling in opposite directions. If Carla travels at 55 mi/h and Jake travels at 35 mi/h, how long will it take for them to be 330 mi apart?

16. Randolph can drive to work in 45 minutes (min) whereas if he decides to take the bus, the same trip takes 1 h 15 min. If the average rate of the bus is 16 mi/h slower than his driving rate, how far does he travel to work?

ANSWERS

17. ______

18. ______

19. ______

20. ______

21. ______

22. ______

23. ______

24. ______

25. ______

Solve each inequality, and graph its solution set.

17. $3x - 5 > 5x + 9$

18. $-11 \le -3x + 4 \le -2$

Solve each of the following equations or inequalities.

19. $8 - |x - 4| = 5$

20. $|x - 4| = |4 - x|$

21. $|5 - 2x| > 5$

22. $|x + 6| \le 4$

23. Find the constant of variation k, given that m varies inversely with p, and that $m = 5$ when $p = 3$.

24. Let y vary directly with x. If $y = 60$ when $x = 5$, find the value of y when $x = 8$.

25. The current, in amperes, in an electric circuit is inversely proportional to the resistance. If the current is 55 A when the resistance is 2Ω, find the current when the resistance is 5Ω.

THE COORDINATE PLANE AND FUNCTIONS

INTRODUCTION

Data are usually presented in one of three forms: a table, a formula, or a graph. Graphs often show trends that may not be easy to see when data are read from a table, or when a rule is given in a formula.

Economists are among the many professionals who use graphs to show connections between two sets of data. For example, an economist may use a graph to look for a connection between two different measures for the quality of life in various countries.

One way of measuring the quality of life in a country is the per capita gross domestic product (GDP). The GDP is the total value of all goods and services produced by all businesses and individuals over the course of 1 year. To find the per capita GDP, we divide that total value by the population of the country.

Other economists, including some who wrote an article in *Scientific American* in May 1993, question this method. Rather than comparing GDP among countries, they use survival rate (life expectancy) to measure the quality of life. The following graph compares life expectancy and GDP.

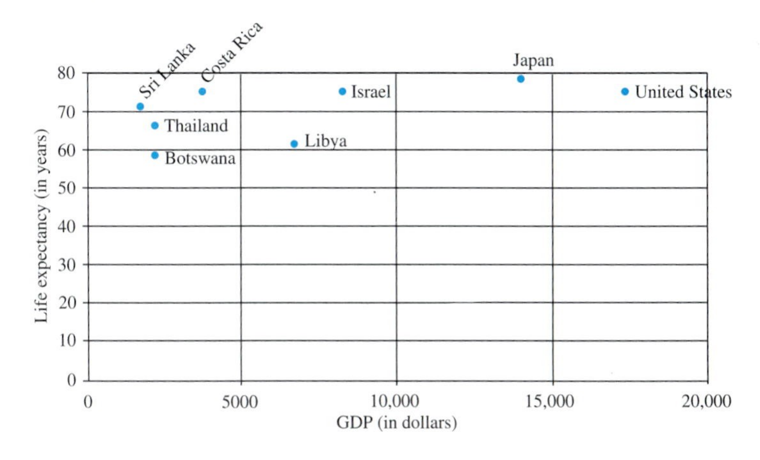

A graph such as this one helps us to visualize complex information. This kind of graph is created using ordered pairs. We will introduce ordered pairs in this chapter and then use them throughout the book.

3.1 Ordered Pairs and Relations

3.1 OBJECTIVES

1. Identify ordered pairs
2. Identify a relation
3. Identify the domain and range of a function

In Chapter 1, we introduced the idea of a set of numbers. In this chapter, we will look at pairs of numbers.

Given the two related values x and y, we write the pair as (x, y). The set of all possible ordered pairs of real numbers is written as $\{(x, y) \mid x \in R, y \in R\}$.

NOTE We read this as, "the set of ordered pairs, (x, y) in which x is a real number and y is a real number. Note that sets always use braces to enclose their contents.

The ordered pair $(2, -3)$ is different from the ordered pair $(-3, 2)$. By contrast, the set $\{2, -3\}$ is identical to the set $\{-3, 2\}$.

Example 1

Identifying Ordered Pairs

Which of the following are ordered pairs?

(a) $(2, -\pi)$ **(b)** $\{2, -4\}$ **(c)** $(1, 3, -1)$

(d) $\{(1, -5), (9, 0)\}$ **(e)** $2, 5$

Only (a) is an ordered pair. (b) is a set (it uses braces instead of parentheses), (c) has three numbers instead of two, (d) is a set of ordered pairs, and (e) is simply a list of two numbers.

CHECK YOURSELF 1

Which of the following are ordered pairs?

(a) $\left\{\frac{1}{2}, -3\right\}$ **(b)** $\left(-3, \frac{1}{3}\right)$ **(c)** $\{(5, 0)\}$

(d) $(1, -5)$ **(e)** $-3, 6$

Ordered pairs are not necessarily made up of two numbers. Given something like (John Doe, 123-45-6789), we have an ordered pair. In this case, it is a name paired with that person's Social Security number (SSN).

Definitions: Relation

A set of ordered pairs is called a **relation.**

We usually denote a relation with a capital letter.

Given

R = {(John Doe, 123-45-6789), (Jacob Smith, 987-65-4321), (Julia Jones, 111-22-3333)}

we have a relation, which we call R. In this case, there are three ordered pairs in the relation R.

Within this relation, there are two interesting sets. The first is the set of names, which happens to be the set of first elements. The second is the set of SSNs, which is the set of second elements. Each of these sets has a name.

Definitions: Domain

The set of first elements in a relation is called the **domain** of the relation.

Example 2

Finding the Domain of a Relation

Find the domain of each relation.

(a) $A = \{(\text{Ben Bender}, 58), (\text{Carol Clairol}, 32), (\text{David Duval}, 29)\}$

The domain of A is {Ben Bender, Carol Clairol, David Duval}.

(b) $B = \left\{\left(5, \frac{1}{2}\right), (-4, -5), (-12, 10), (-16, \pi)\right\}$

The domain of B is $\{5, -4, -12, -16\}$.

CHECK YOURSELF 2

Find the domain of each relation.

(a) $R = \{(\text{Secretariat}, 10), (\text{Seattle Slew}, 8), (\text{Charismatic}, 5), (\text{Gallant Man}, 7)\}$

(b) $S = \left\{\left(-\frac{1}{2}, \frac{3}{4}\right), (0, 0), (1, 5), (\pi, \pi)\right\}$

NOTE $X \rightarrow$ Domain
$Y \rightarrow$ Range
Many students find it helpful to remember that domain and range occur in alphabetical order.

Definitions: Range

The set of second elements in a relation is called the **range** of the relation.

Example 3

Finding the Range of a Relation

Find the range for each relation.

(a) $A = \{(\text{Ben Bender}, 58), (\text{Carol Clairol}, 32), (\text{David Duval } 29)\}$

The range of A is $\{58, 32, 29\}$

(b) $B = \left\{\left(5, \frac{1}{2}\right), (-4, -5), (-12, 10), (-16, \pi)\right\}$

The range of B is $\left\{\frac{1}{2}, -5, 10, \pi\right\}$

CHECK YOURSELF 3

Find the range of each relation.

(a) $R = \{(\text{Secretariat}, 10), (\text{Seattle Slew}, 8), (\text{Charismatic}, 5), (\text{Gallant Man}, 7)\}$

(b) $S = \left\{\left(-\frac{1}{2}, \frac{3}{4}\right), (0, 0), (1, 5), (\pi, \pi)\right\}$

CHECK YOURSELF ANSWERS

1. (b) and (d) are ordered pairs
2. **(a)** The domain of R is {Secretariat, Seattle Slew, Charismatic, Gallant Man};
   (b) The domain of S is $\left\{-\frac{1}{2}, 0, 1, \pi\right\}$
3. **(a)** The range of R is $\{10, 8, 5, 7\}$; **(b)** The range of S is $\left\{\frac{3}{4}, 0, 5, \pi\right\}$

Name ______________

Section ________ Date ________

3.1 Exercises

In exercises 1 to 4, identify the ordered pairs.

1. **(a)** $(3, -5)$ **(b)** $\{7, 9\}$ **(c)** $(2, 5)$ **(d)** $5, 2$ **(e)** $((3, 1), 4)$

2. **(a)** $\{7, 23\}$ **(b)** $(1, 0, (5, 6))$ **(c)** $\left(\frac{1}{2}, -1\right)$ **(d)** $[5, 6]$ **(e)** $(23, 7)$

3. **(a)** $18, 67$ **(b)** $(-3, -9)$ **(c)** $\{3, 9\}$ **(d)** $(3, 7, -3)$ **(e)** $[12, 56]$

4. **(a)** $\{45, 67]$ **(b)** $(9, 3)$ **(c)** $5, 8$ **(d)** $(11, -3, 9)$ **(e)** $[5, 2]$

In exercises 5 to 20, find the domain and range of each relation.

5. A = {(Colorado, 21), (Edmonton, 5), (Calgary, 18), (Vancouver, 17)}

6. B = {(Eric Lindros, 88), (Mark Recchi, 8), (John LeClair, 10), (Keith Primeau, 25)}

7. C = {(John Adams, −16), (John Kennedy, −23), (Richard Nixon, −5), (Harry Truman, −11)}

8. E = {(Utah, 27), (San Antonio, 28), (Minnesota, 24), (Denver, 19)}

9. $F = \left\{\left(\text{St. Louis}, \frac{1}{2}\right), \left(\text{Denver}, -\frac{3}{4}\right), \left(\text{Green Bay}, \frac{7}{8}\right), \left(\text{Dallas}, -\frac{4}{5}\right)\right\}$

10. $G = \left\{(\text{Chamber}, \pi), (\text{Testament}, 2\pi), \left(\text{Rainmaker}, \frac{1}{2}\right), (\text{Street Lawyer}, 6)\right\}$

11. $\{(1, 2), (3, 4), (5, 6), (7, 8), (9, 10)\}$

12. $\{(2, 3), (3, 5), (4, 7), (5, 9), (6, 11)\}$

13. $\{(1, 2), (4, 6), (3, 3), (5, 4), (6, 1)\}$

14. $\{(3, 4), (5, 7), (6, 1), (2, 2), (4, 3)\}$

15. $\{(1, 2), (1, 3), (1, 4), (1, 5), (1, 6)\}$

16. $\{(3, 4), (3, 6), (3, 8), (3, 9), (3, 10)\}$

17. $\{(1, 5), (2, 5), (3, 6), (2, 4), (4, 5)\}$

18. $\{(2, 8), (3, 9), (2, 9), (3, 8), (4, 7)\}$

19. $\{(-1, 3), (-2, 4), (-3, 5), (4, 4), (5, 6)\}$

20. $\{(-2, 4), (1, 4), (-3, 4), (5, 4), (7, 4)\}$

ANSWERS

1. ________ 2. ________

3. ________ 4. ________

5. ________

6. ________

7. ________

8. ________

9. ________

10. ________

11. ________

12. ________

13. ________

14. ________

15. ________

16. ________

17. ________

18. ________

19. ________

20. ________

21. ____________

22. ____________

23. ____________

24. ____________

25. ____________

26. ____________

21. The stock prices for a given stock over a week's time are displayed in a table. List this information as a set of ordered pairs using the day of the week as the domain.

Day	1	2	3	4	5
Price	$9\frac{1}{8}$	8	$8\frac{7}{8}$	$9\frac{1}{4}$	9

22. Food Purchases. In the snack department of the local supermarket, candy costs \$1.58 per pound. For 1 to 5 pounds, write the cost of candy as a set of ordered pairs.

In exercises 23 to 26, write a set of ordered pairs that describes each situation. Give the domain and range of each relation.

23. The first element is an integer between -3 and 3. The second coordinate is the cube of the first coordinate.

24. The first element is a positive integer less than 6. The second coordinate is the sum of the first coordinate and -2.

25. The first element is the number of hours worked 10, 20, 30, 40; the second coordinate is the salary at \$6 per hour.

26. The first coordinate is the number of toppings on a pizza (up to 4); the second coordinate is the price of the pizza, which is \$9 plus \$1 per topping.

Answers

1. (a) and (c) **3.** (b)
5. D: {Colorado, Edmonton, Calgary, Vancouver}; R: {21, 5, 18, 17}
7. D: {John Adams, John Kennedy, Richard Nixon, Harry Truman}; R: $\{-16, -23, -5, -11\}$
9. D: {St. Louis, Denver, Green Bay, Dallas}; R: $\left\{\frac{1}{2}, -\frac{3}{4}, \frac{7}{8}, -\frac{4}{5}\right\}$
11. D: {1, 3, 5, 7, 9}; R: {2, 4, 6, 8, 10} **13.** D: {1, 3, 4, 5, 6}; R: {1, 2, 3, 4, 6}
15. D: {1}; R: {2, 3, 4, 5, 6} **17.** D: {1, 2, 3, 4}; R: {4, 5, 6}
19. D: $\{-1, -2, -3, 4, 5\}$; R: {3, 4, 5, 6}
21. $\left\{\left(1, 9\frac{1}{8}\right), (2, 8), \left(3, 8\frac{7}{8}\right), \left(4, 9\frac{1}{4}\right), (5, 9)\right\}$
23. $\{(-2, -8), (-1, -1), (0, 0), (1, 1), (2, 8)\}$; D: $\{-2, -1, 0, 1, 2\}$; R: $\{-8, -1, 0, 1, 8\}$
25. {(10, 60), (20, 120), (30, 180), (40, 240)}; D: {10, 20, 30, 40}; R: {60, 120, 180, 240}

3.2 The Rectangular Coordinate System

OBJECTIVES

1. Graph a set of ordered pairs
2. Identify plotted points
3. Scale the axes

In Chapter 1, we used a number line to locate and visualize real numbers. Such a line has been used by mathematicians for so many years that we do not know who gets credit for creating the number line. Locating and visualizing ordered pairs is a different story.

NOTE In the eighteenth century, René Descartes, a French philosopher and mathematician, created a way of graphing ordered pairs.

A rectangular coordinate system consists of two perpendicular number lines, called **axes,** with the positive directions defined by "up" and "right." If the two lines have the same scale, we sometimes refer to the system as a Cartesian coordinate system in honor of René Descartes.

NOTE The development of the **coordinate system** was part of an effort to combine the knowledge of geometry with that of algebra.

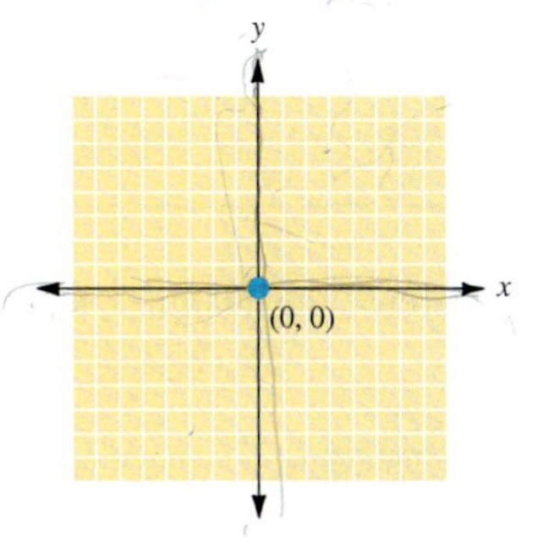

Definitions: Axes

The horizontal line is called the ***x* axis.** The vertical line is the ***y* axis.** Together they are called the ***x* and *y* axes** (pronounced "axees").

Definitions: Coordinate Plane

The plane (a flat surface that continues forever in every direction) containing the x and y axes is called the **coordinate plane.**

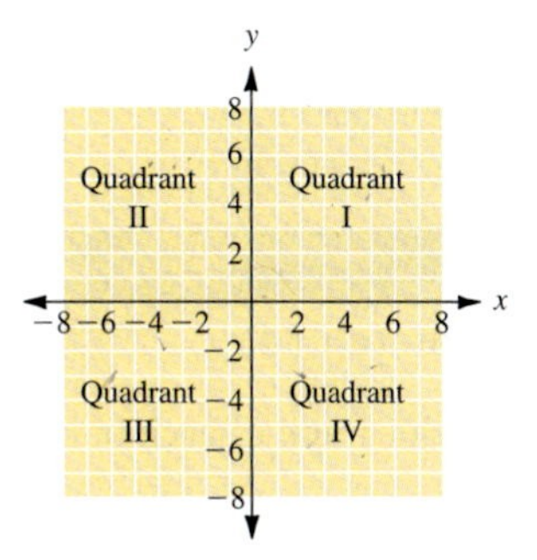

The x and y axes divide the coordinate plane into four parts called quadrants.

Every point in the coordinate plane can be described using an ordered pair of real numbers. And, conversely, every ordered pair can be plotted as a point in the plane.

Given an ordered pair (2, 4), 2 is called the ***x* coordinate** and 4 is called the ***y* coordinate.** If either coordinate of an ordered pair is equal to 0, the associated point lies on one of the axes. Every point that does not lie on the x or y axis can be plotted in one of the four quadrants.

Example 1

Graphing Ordered Pairs

Graph the ordered pair $(3, -5)$.

Beginning at the origin, we move 3 units in the x direction (positive is to the right) and -5 units in the y direction (negative is down).

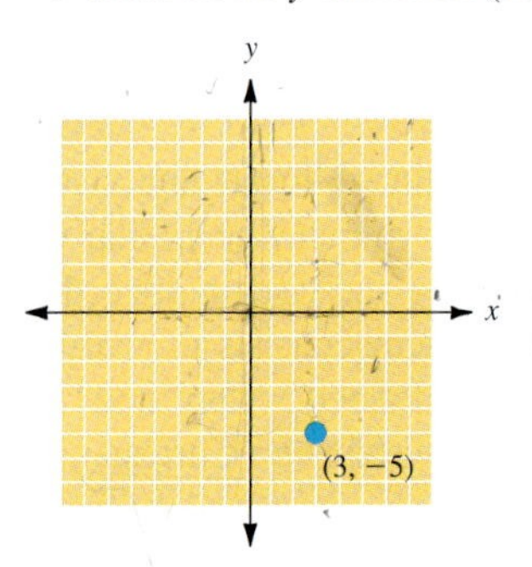

CHECK YOURSELF 1

Graph the ordered pairs $\{(0, 3), (-2, -4), (3, -1)\}$.

In general, the following pattern shows the quadrant in which a given point is located.

Quadrant	Sign Pattern
I	(+, +)
II	(−, +)
III	(−, −)
IV	(+, −)

If a given point lies on an axis, it fits one of the following patterns:

Axis	Pattern
x	$(x, 0)$
y	$(0, y)$

We will use these patterns in Example 2.

Example 2

Locating the Graph of an Ordered Pair

Identify the quadrant or axis for the graph of each point.

(a) $\left(3, -\frac{1}{4}\right)$

The pattern indicates quadrant IV.

(b) $(0, 6)$

The pattern indicates the y axis.

CHECK YOURSELF 2

Identify the quadrant or axis for the graph of each point.

(a) $(-3, 0)$ **(b)** $\left(-\frac{7}{2}, 6\right)$ **(c)** $(\pi, 3)$

To find the ordered pair associated with a plotted point, we move vertically from the point to find the x coordinate and horizontally from the point to find the y coordinate.

Example 3

Identifying Plotted Points

Find the ordered pair associated with each point.

NOTE When no scale appears on the grid, we assume that each division on each axis is one unit.

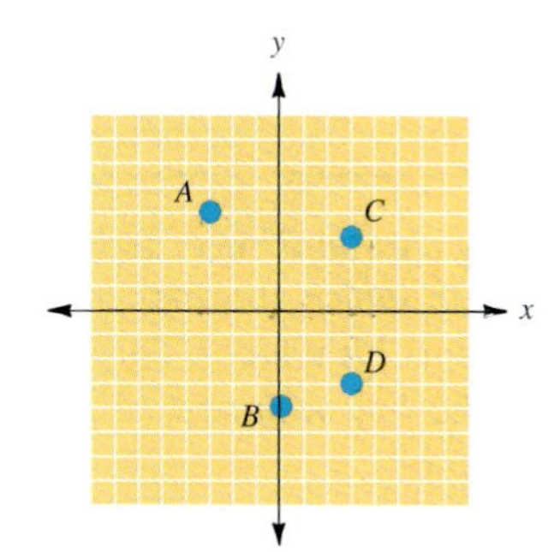

From point A, a vertical line meets the x axis at -3. A horizontal line meets the y axis at 4. The ordered pair is $(-3, 4)$. B is associated with $(0, -4)$, C with $(3, 3)$, and D with $(3, -3)$.

CHECK YOURSELF 3

Find the ordered pair associated with each point.

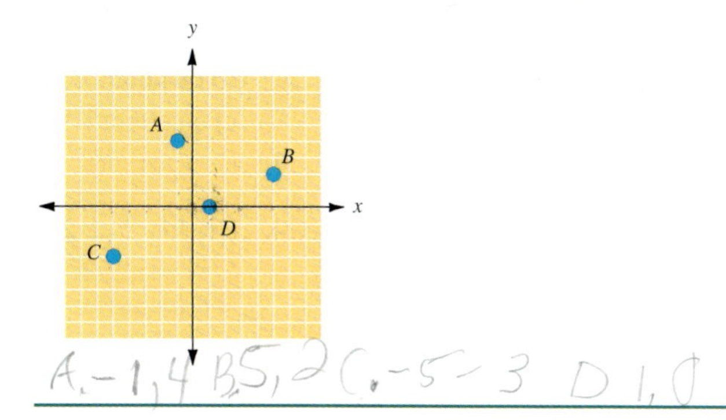

NOTE The same decisions must be made when you are using a graphing calculator. When graphing this kind of relation on a calculator, you must decide what the appropriate **viewing window** should be.

Example 4

Reading Ordered Pairs Given Scaled Axes

A survey of residents in a large apartment building was recently taken. The following points represent ordered pairs in which the first number is the number of years of education a person has had, and the second number is his or her income (in thousands of dollars). Estimate, and interpret, each ordered pair represented.

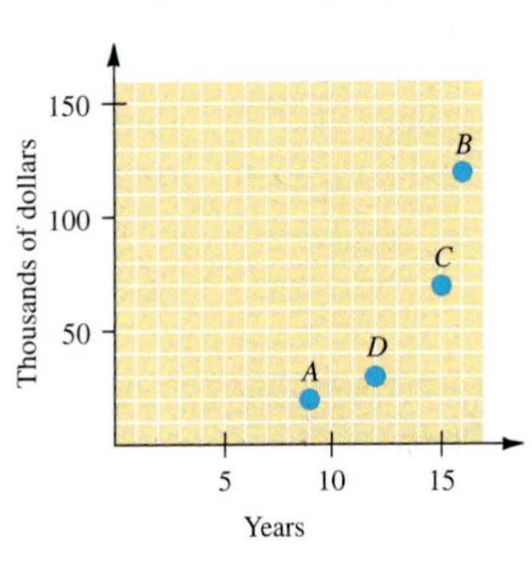

A is (9, 20), B is (16, 120), C is (15, 70), and D is (12, 30). Person A completed 9 years of education and made $20,000 in 2001. Person B completed 16 years of education and made $120,000 in 2001. Person C had 15 years education and made $70,000. Person D had 12 years and made $30,000.

It is not obvious from this graph how to predict income from years of education, but you might suspect that in most cases, more education results in more income.

CHECK YOURSELF 4

Each year on his son's birthday, Armand records his son's weight. The following points represent ordered pairs in which the first number represents his son's age and the second number represents his weight. For example, point A indicates that when his son was 1 year old, the boy weighed 14 pounds. Estimate each ordered pair represented.

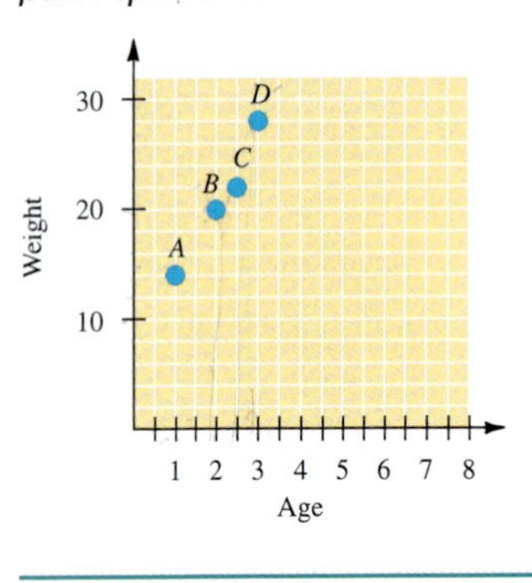

CHECK YOURSELF ANSWERS

1.

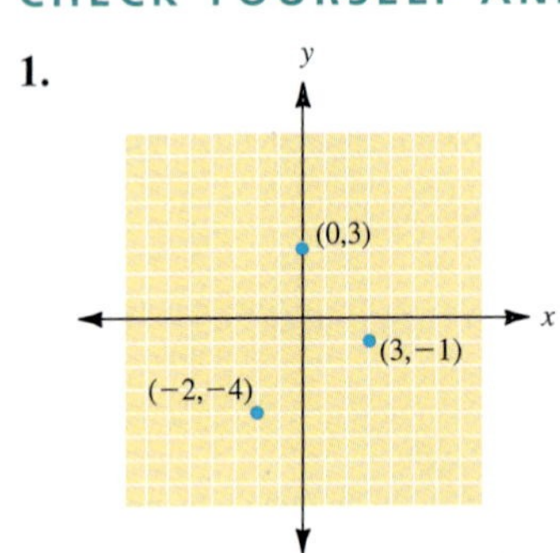

2. **(a)** x axis; **(b)** quadrant II; **(c)** quadrant I
3. $A(-1, 4)$, $B(5, 2)$, $C(-5, -3)$, and $D(1, 0)$
4. $A(1, 14)$, $B(2, 20)$, $C\left(2\frac{1}{2}, 22\right)$, and $D(3, 28)$

Name ______________________

Section ________ Date ________

3.2 Exercises

In exercises 1 to 6, graph each set of ordered pairs.

1. $\{(3, 5), (-4, 6), (-2, 6), (5, -6)\}$

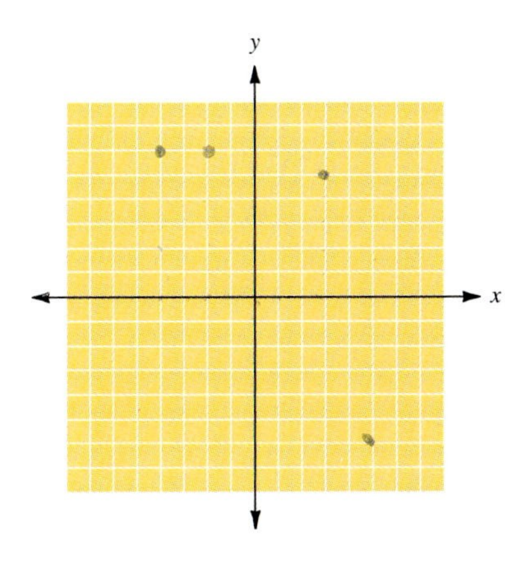

2. $\left\{\left(-5, -\frac{1}{2}\right), (6, 0), (-3, -4), (-1, 4)\right\}$

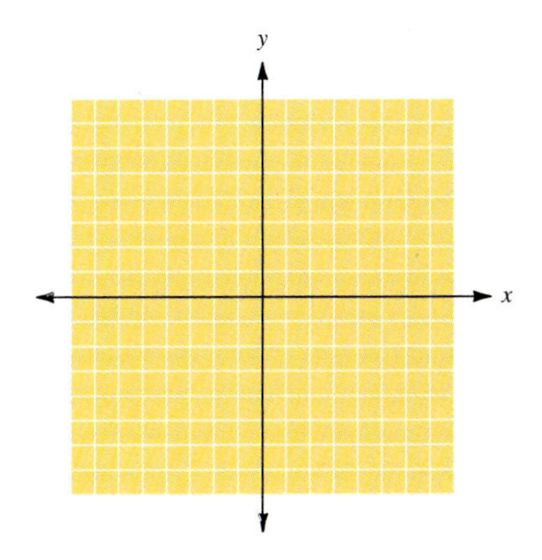

3. $\left\{(0, -5), (2, -3), \left(-1, \frac{5}{2}\right), \left(-5, \frac{3}{4}\right)\right\}$

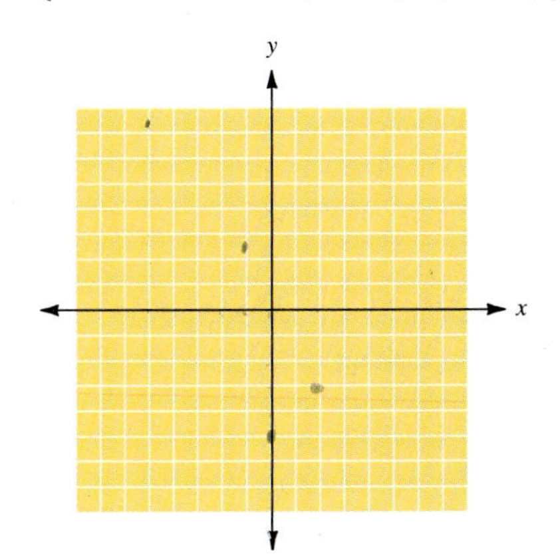

ANSWERS

1. ______________________

2. ______________________

3. ______________________

ANSWERS

4. ______

5. ______

6. ______

4. $\left\{(-1, -2), (5, -2), (-3, 4), \left(\frac{1}{2}, 2\right)\right\}$

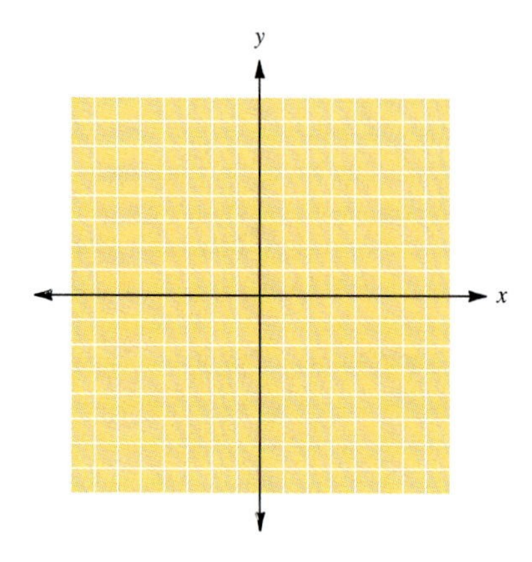

5. $\{(-1, 6), (3, -5), (-2, -5), (1, 4)\}$

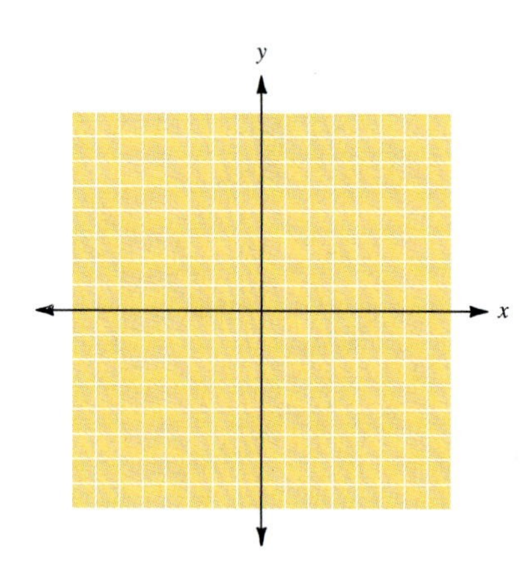

6. $\{(3, 5), (-2, 0), (0, -2), (-1, -5)\}$

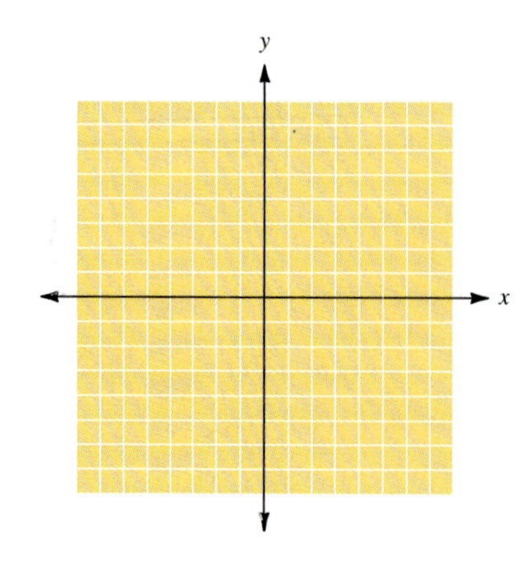

In exercises 7 to 18, give the quadrant in which each of the following points is located or the axis on which the point lies.

7. (4, 5)

8. $(-3, 2)$

9. $(-4, -3)$

10. $(2, -4)$

11. (5, 0)

12. $(-5, 7)$

13. $(-4, 7)$

14. $(-3, -7)$

15. $(0, -7)$

16. $(-3, 0)$

17. $\left(5\frac{3}{4}, -3\right)$

18. $\left(-2, 4\frac{5}{6}\right)$

In exercises 19 to 28, give the coordinates (ordered pairs) associated with the points indicated in the figure.

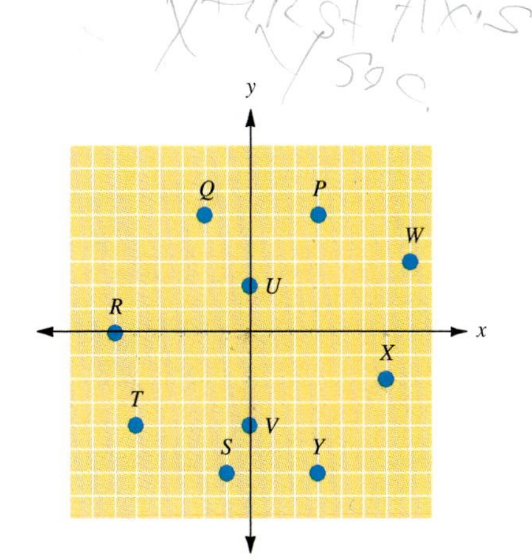

19. P

20. Q

21. R

22. S

23. T

24. U

25. V

26. W

27. X

28. Y

ANSWERS

7. ______
8. ______
9. ______
10. ______
11. ______
12. ______
13. ______
14. ______
15. ______
16. ______
17. ______
18. ______
19. ______
20. ______
21. ______
22. ______
23. ______
24. ______
25. ______
26. ______
27. ______
28. ______

ANSWERS

29. ______________________

30. ______________________

29. A company has kept a record of the number of items produced by an employee as the number of days on the job increases. In the following figure, points correspond to an ordered-pair relationship in which the first number represents days on the job and the second number represents the number of items produced. Estimate each ordered pair produced. In your own words, state the meaning of the graph.

As the # of day increases the # of items produce increased

30. In the following figure, points correspond to an ordered-pair relationship between height and age in which the first number represents age and the second number represents height. Estimate each ordered pair represented.

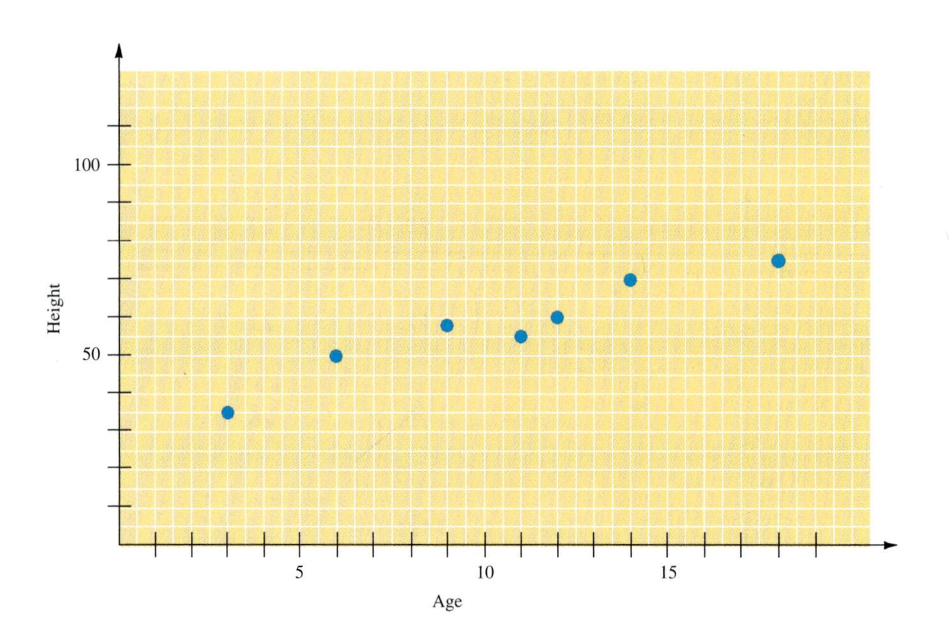

In exercises 31 and 32, plot the points whose coordinates are given in the table. Scale the axes appropriately.

31. Grades. The table gives the time, x, in hours invested in studying for four different algebra exams and the resulting grade, y.

x	4	3	2	8	1
y	85	80	70	95	60

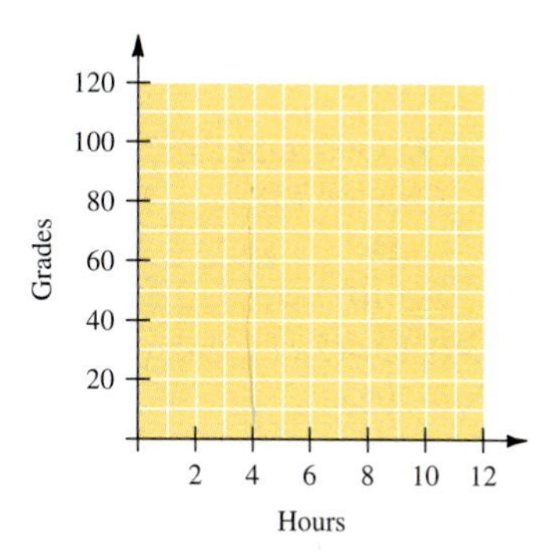

32. Fuel efficiency. The table gives the speed, x, of a car in miles per hour and the approximate fuel efficiency, y, in miles per gallon.

x	45	55	60	65	70
y	30	25	20	20	15

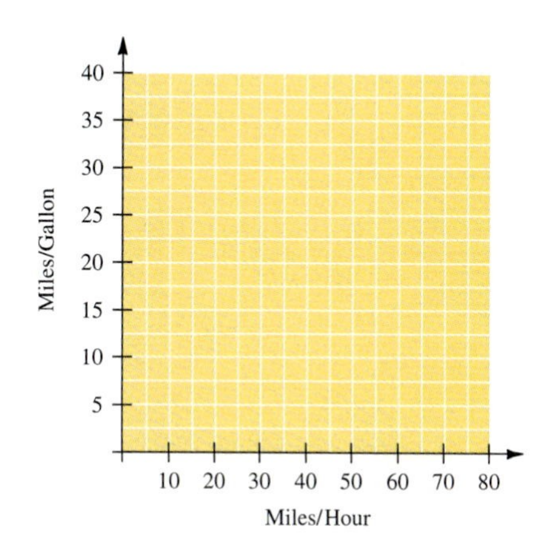

ANSWERS

31. ____________________

32. ____________________

ANSWERS

33. ______________________

34. ______________________

33. Graph the points with coordinates (1, 2), (2, 3), and (3, 4). What do you observe? Give the coordinates of another point with the same property.

34. Graph points with coordinates $(-1, 3)$, $(0, 0)$, and $(1, -3)$. What do you observe? Give the coordinates of another point with the same property.

Answers

1.

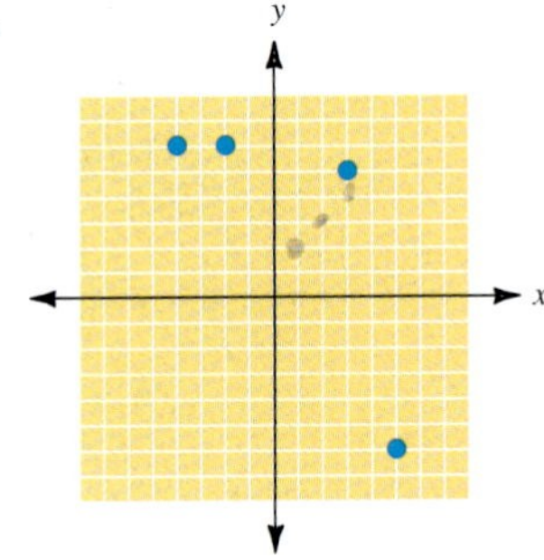

3.

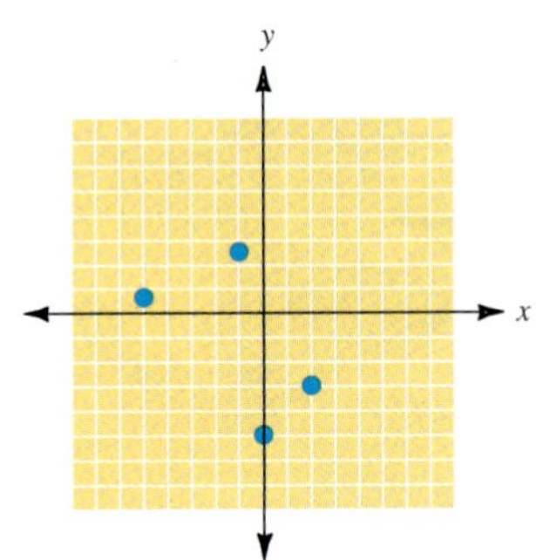

5.

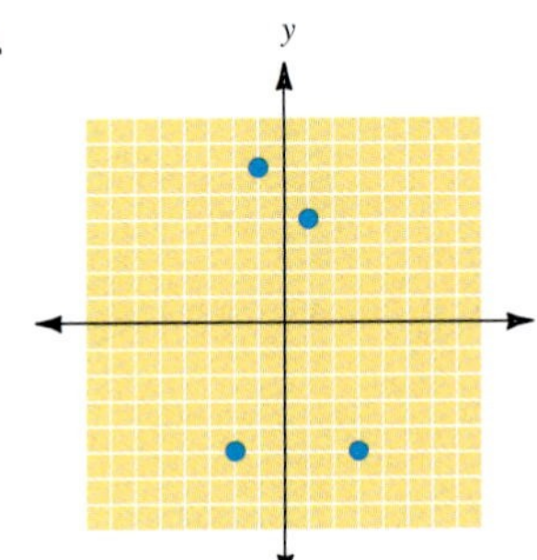

7. Quadrant I **9.** Quadrant III
11. x axis **13.** Quadrant II **15.** y axis
17. Quadrant IV **19.** (3, 5) **21.** $(-6, 0)$
23. $(-5, -4)$ **25.** $(0, -4)$ **27.** $(6, -2)$
29. Points are (1, 30), (2, 45), (3, 60), (4, 60), (5, 75), (6, 90), (7, 95)

31.

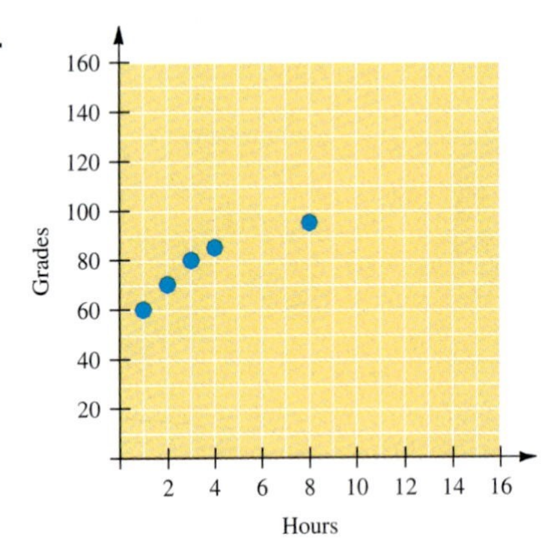

33. The y coordinate is 1 more than the x coordinate; (4, 5)

3.3 An Introduction to Functions

3.3 OBJECTIVES

1. Evaluate an expression
2. Determine whether a relation is a function
3. Use the vertical line test
4. Identify the domain and range of a function

Variables can be used to represent numbers whose value is unknown. Using addition, subtraction, multiplication, division, and exponentiation, these numbers and variables form expressions such as:

$3 + 5 \qquad 7x - 4 \qquad x^2 - 3x - 4 \qquad x^4 - x^2 + 2$

If a specific value is given for the variable, we **evaluate the expression.**

Example 1

Evaluating Expressions

Evaluate the expression $x^4 - 2x^2 + 3x + 4$ for the indicated value of x.

(a) $x = 0$

Substituting 0 for x in the expression yields:

$$(0)^4 - 2(0)^2 + 3(0) + 4 = 0 - 0 + 0 + 4$$
$$= 4$$

(b) $x = 2$

Substituting 2 for x in the expression yields:

$$(2)^4 - 2(2)^2 + 3(2) + 4 = 16 - 8 + 6 + 4$$
$$= 18$$

(c) $x = -1$

Substituting -1 for x in the expression yields:

$$(-1)^4 - 2(-1)^2 + 3(-1) + 4 = 1 - 2 - 3 + 4$$
$$= 0$$

CHECK YOURSELF 1

Evaluate the expression $2x^3 - 3x^2 + 3x + 1$ for the indicated value of x.

(a) $x = 0$ **(b)** $x = 1$ **(c)** $x = -2$

We could design a machine whose function would be to crank out the value of an expression for each given value of x. We could call this machine something simple such as f, our **function machine.** Our machine might look like this.

For example, when we put -1 into the machine, the machine would substitute -1 for x in the expression, and 5 would come out the other end because

$$2(-1)^3 + 3(-1)^2 - 5(-1) - 1 = -2 + 3 + 5 - 1 = 5$$

Note that, with this function machine, an input of -1 will always result in an output of 5. One of the most important aspects of a function machine is that each input has a unique output.

In fact, the idea of the function machine is very useful in mathematics. Your graphing calculator can be used as a function machine. You can enter the expression into the calculator as Y_1 and then evaluate Y_1 for different values of x.

Generally, in mathematics, we do not write $Y_1 = 2x^3 + 3x^2 - 5x - 1$. Instead, we write $f(x) = 2x^3 + 3x^2 - 5x - 1$, which is read "$f$ of x is equal to . . . ." Instead of calling f a function machine, we say that f is a function of x. The greatest benefit of this notation is that it lets us easily note the input value of x along with the output of the function. Instead of "the value of Y_1 is 155 when $x = 4$," we can write $f(4) = 155$.

Example 2

Evaluating a Function

Given $f(x) = x^3 + 3x^2 - x + 5$, find the following.

(a) $f(0)$

Substituting 0 for x in the expression on the right, we get

$$(0)^3 + 3(0)^2 - (0) + 5 = 5$$

(b) $f(-3)$

Substituting -3 for x in the expression on the right, we get

$$\begin{aligned}(-3)^3 + 3(-3)^2 - (-3) + 5 &= -27 + 27 + 3 + 5 \\ &= 8\end{aligned}$$

(c) $f\left(\frac{1}{2}\right)$

Substituting $\frac{1}{2}$ for x in the expression on the right, we get

$$\begin{aligned}\left(\frac{1}{2}\right)^3 + 3\left(\frac{1}{2}\right)^2 - \left(\frac{1}{2}\right) + 5 &= \frac{1}{8} + 3\left(\frac{1}{4}\right) - \frac{1}{2} + 5 \\ &= \frac{1}{8} + \frac{3}{4} - \frac{1}{2} + 5 \\ &= \frac{1}{8} + \frac{6}{8} - \frac{4}{8} + 5 \\ &= \frac{3}{8} + 5 \\ &= 5\frac{3}{8} \text{ or } \frac{43}{8}\end{aligned}$$

CHECK YOURSELF 2

Given $f(x) = 2x^3 - x^2 + 3x - 2$, find the following.

(a) $f(0)$ **(b)** $f(3)$ **(c)** $f\left(-\frac{1}{2}\right)$

We can rewrite the relationship between x and $f(x)$ in Example 2 as a series of ordered pairs.

$f(x) = x^3 + 3x^2 - x + 5$

From this we found that

$$f(0) = 5, \qquad f(-3) = 8, \qquad \text{and} \qquad f\left(\frac{1}{2}\right) = \frac{43}{8}$$

NOTE Because $y = f(x)$, $(x, f(x))$ is another way of writing (x, y).

There is an ordered pair, which we could write as $(x, f(x))$, associated with each of these. Those three ordered pairs are

$$(0, 5), \qquad (-3, 8), \qquad \text{and} \qquad \left(\frac{1}{2}, \frac{43}{8}\right)$$

Example 3

Finding Ordered Pairs

Given the function $f(x) = 2x^2 - 3x + 5$, find the ordered pair $(x, f(x))$ associated with each given value for x.

(a) $x = 0$

$f(0) = 2(0)^2 - 3(0) + 5 = 5$, so the ordered pair is $(0, 5)$.

(b) $x = -1$

$f(-1) = 2(-1)^2 - 3(-1) + 5 = 10$. The ordered pair is $(-1, 10)$.

(c) $x = \frac{1}{4}$

$f\left(\frac{1}{4}\right) = 2\left(\frac{1}{4}\right)^2 - 3\left(\frac{1}{4}\right) + 5 = \frac{35}{8}$. The ordered pair is $\left(\frac{1}{4}, \frac{35}{8}\right)$.

CHECK YOURSELF 3

Given $f(x) = 2x^3 - x^2 + 3x - 2$, find the ordered pair associated with each given value of x.

(a) $x = 0$ **(b)** $x = 3$ **(c)** $x = -\frac{1}{2}$

In Section 3.1, we defined a relation as a set of ordered pairs. In the following example, we will determine which relations can be modeled by a function machine.

Example 4

Modeling with a Function Machine

Determine which relations can be modeled by a function machine.

(a) The set of all possible ordered pairs in which the first element is a U.S. state and the second element is a U.S. senator from that state.

We can not model this relation with a function machine. Because there are two senators from each state, each input does not have a unique output. In the picture, New Jersey is the input, but New Jersey has two different senators.

(b) The set of all ordered pairs in which the input is the year and the output is the U.S. Open golf champion of that year.

This relation can be modeled with the function machine. Each input has a unique output. In the picture, an input of 2000 gives an output of Tiger Woods. Every time the input is 2000, the output will be Tiger Woods.

(c) The set of all ordered pair in the relation R, when

$R = \{(1, 3), (2, 5), (2, 7), (3\ -4)\}$

This relation cannot be modeled with a function machine. An input of 2 can result in two different outputs, either 5 or 7.

(d) The set of all ordered pairs in the relation S, when

$S = \{(-1, 3), (0, 3), (3, 5), (5, -2)\}$

This relation can be modeled with a function machine. Each input has a unique output.

CHECK YOURSELF 4

Determine which relations can be modeled by a function machine.

(a) The set of all ordered pairs in which the first element is a U.S. city and the second element is the mayor of that city.
(b) The set of all ordered pairs in which the first element is a street name and the second element is a U.S. city in which a street of that name is found.
(c) The relation $R = \{(-2, 3), (-4, 9), (9, -4)\}$.
(d) The relation $S = \{(1, 2), (3, 4), (3, 5)\}$.

The idea of a function machine leads us to a more formal definition of a function.

Definitions: Function

A function is a set of distinct ordered pairs (a relation) in which no two first coordinates are equal.

In our next example, the set of ordered pairs will be represented by a table.

Example 5

Identifying a Function

For each table of values below, decide whether the relation is a function.

(a)

x	y
-2	1
-1	1
1	3
2	3

(b)

x	y
-5	-2
-1	3
-1	6
2	9

(c)

x	y
-3	1
-1	0
0	2
2	4

Part (a) represents a function. No two first coordinates are equal. Part (b) is not a function because -1 appears as a first coordinate with two different second coordinates. Part (c) is a function.

CHECK YOURSELF 5

For each table of values below, decide whether the relation is a function.

(a)

x	y
-3	0
-1	1
1	2
3	3

(b)

x	y
-2	-2
-1	-2
1	3
2	3

(c)

x	y
-2	0
-1	1
0	2
0	3

We defined a function in terms of ordered pairs. A set of ordered pairs can be specified in several ways; here are the most common.

Rules and Properties: Ordered Pairs

1. We can present ordered pairs in a list or table, as in Example 5.
2. We can give a rule or equation that will generate ordered pairs.
3. We can use a graph to indicate ordered pairs. The graph can show distinct ordered pairs, or it can show all the ordered pairs on a line or curve.

Let's look at a graph of the ordered pairs from Example 5 to introduce the **vertical line test,** which is a graphic test for identifying a function.

(a)

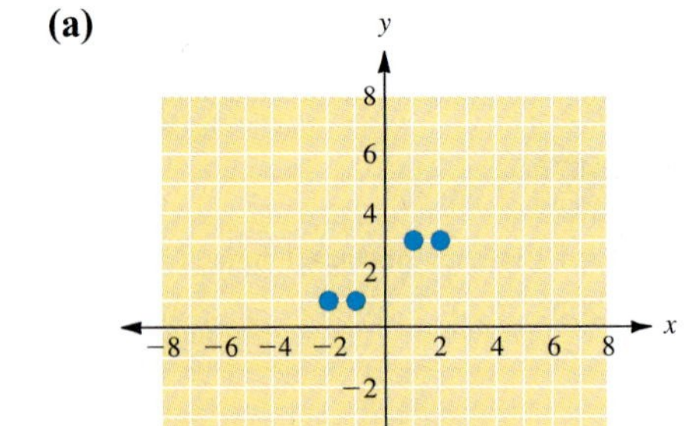

(b)

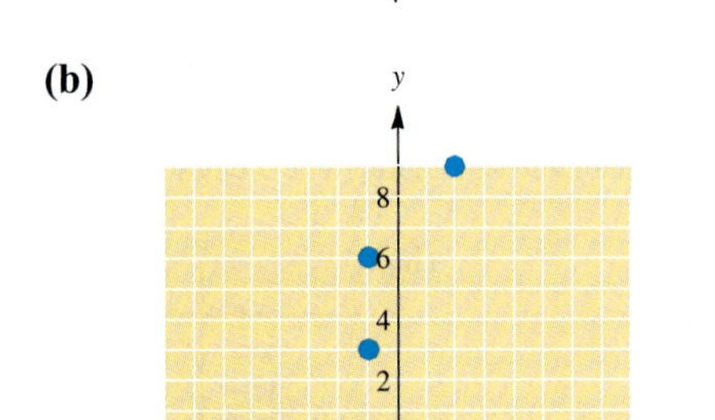

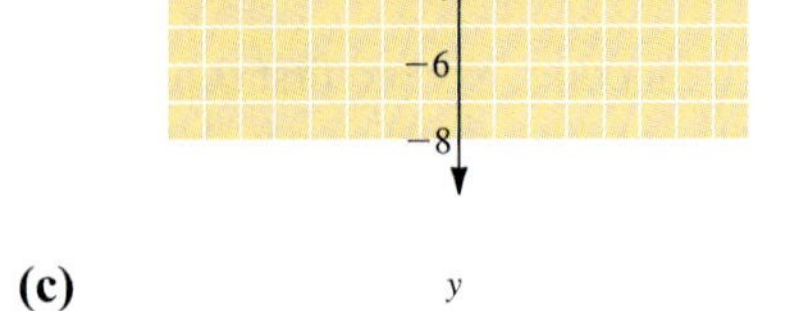

(c)

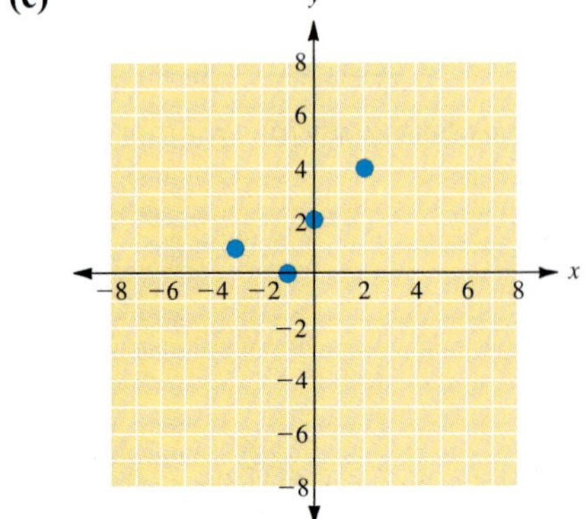

Notice that in the graphs of relations (a) and (c), there is no vertical line that can pass through two different points of the graph. In relation (b), a vertical line can pass through the two points that represent the ordered pairs $(-1, 3)$ and $(-1, 6)$. This leads to the following definition.

Definitions: Vertical Line Test

A relation is a function if no vertical line can pass through two or more points on its graph.

Example 6

Identifying a Function

For each set of ordered pairs, plot the related points on the provided axes. Then use the vertical line test to determine which of the sets is a function.

(a) $\{(0, -1), (2, 3), (2, 6), (4, 2), (6, 3)\}$

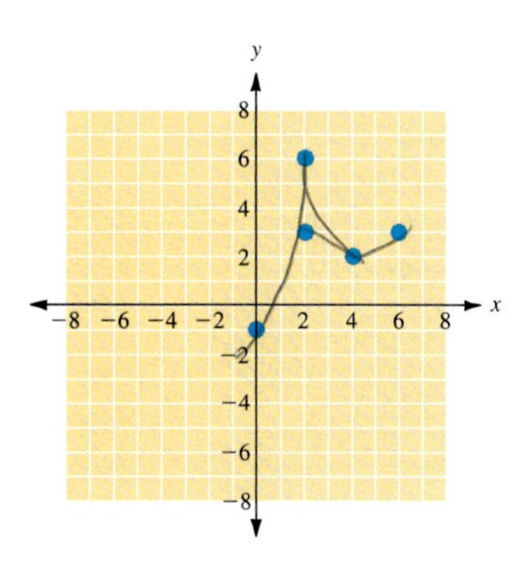

Because a vertical line can be drawn through the points (2, 3) and (2, 6), the relation does not pass the vertical line test. This is not a function.

(b) $\{(1, 1), (2, 0), (3, 3), (4, 3), (5, 3)\}$

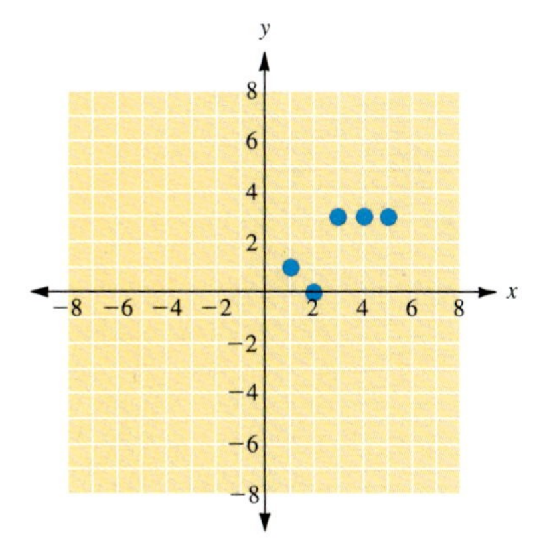

This is a function. Although a horizontal line can be drawn through several points, no vertical line passes through more than one point.

CHECK YOURSELF 6

For each set of ordered pairs, plot the related points. Then use the vertical line test to determine which of the sets is a function.

(a) $\{(-2, 4), (-1, 4), (0, 4), (1, 3), (5, 5)\}$

(b) $\{(-3, -1), (-1, -3), (1, -3), (1, 3)\}$

The vertical line test can be used to determine whether a graph is the graph of a function.

Example 7

Identifying a Function

Which of the following graphs represents the graph of a function?

(a)

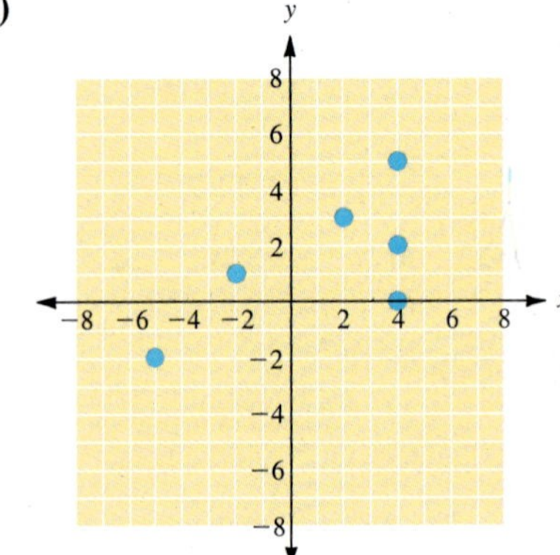

(b)

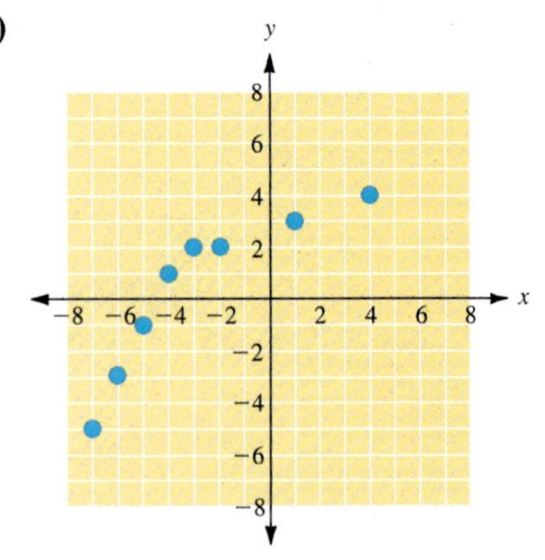

(c)

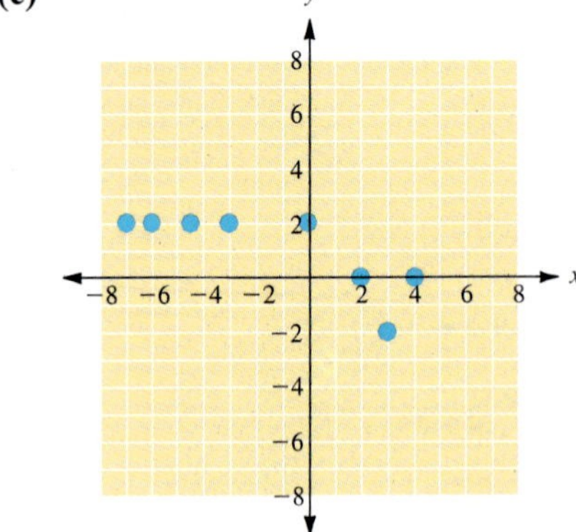

Part (a) is not a function, part (b) is a function, and part (c) is a function.

CHECK YOURSELF 7

Which of the following graphs represents the graph of a function?

(a)

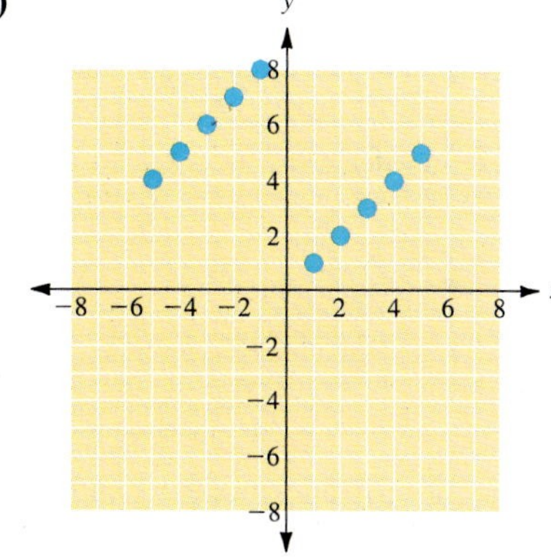

(b)

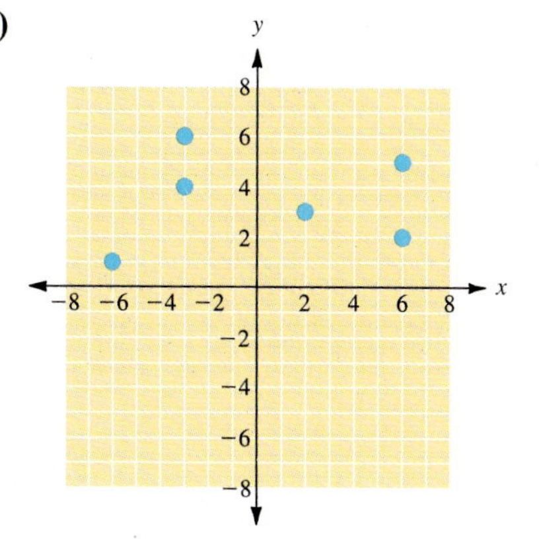

(c)

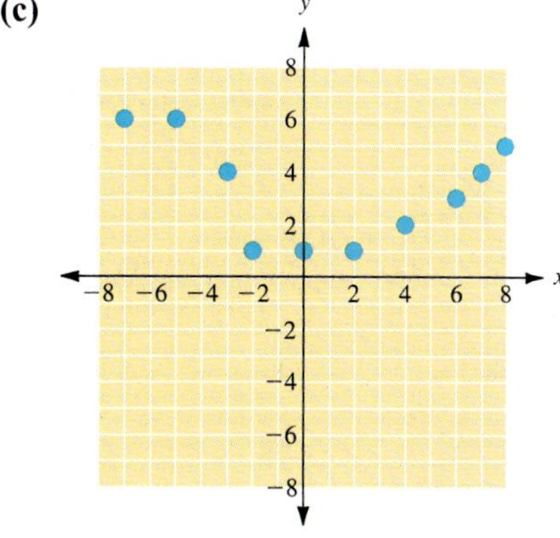

Example 8

Identifying a Function

Which of the following graphs represents the graph of a function?

(a)

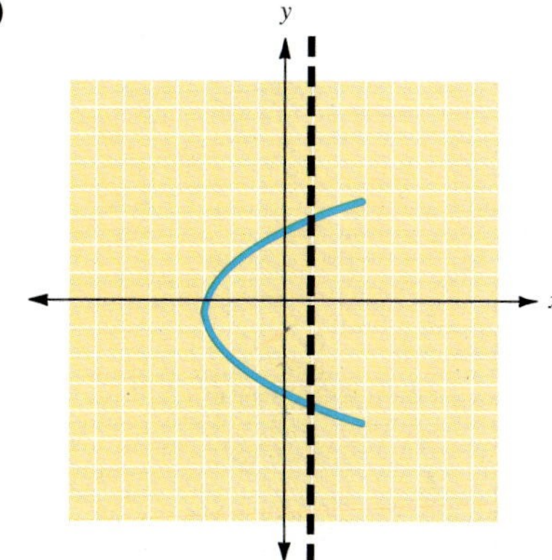

(b)

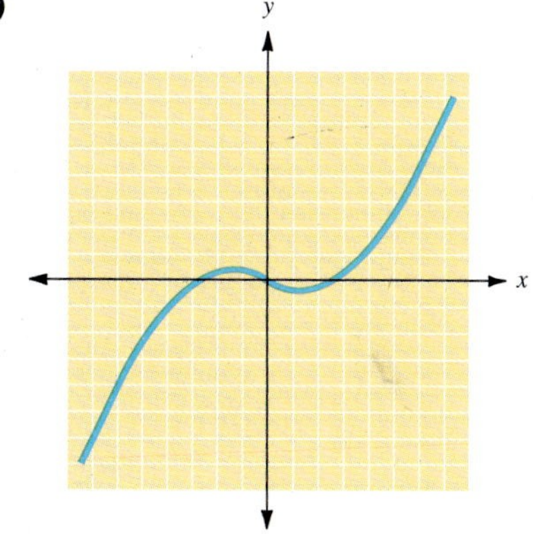

NOTE Curves, like the number line, are made up of a continuous set of points.

(c)

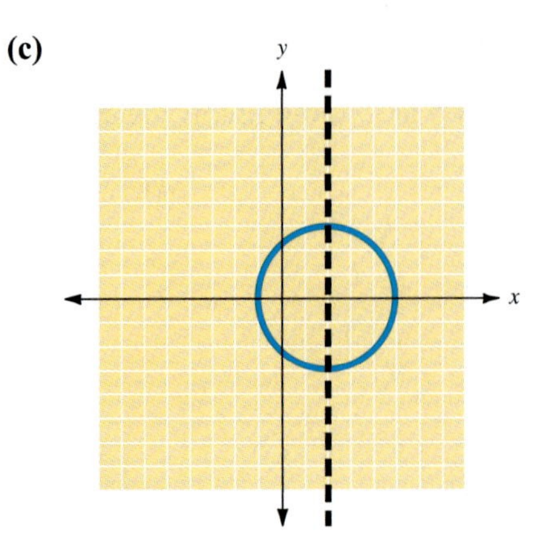

Part (a) is not a function; it does not pass the vertical line test. Part (b) is a function because it passes the vertical line test. Part (c) is not a function.

CHECK YOURSELF 8

Which of the following graphs represents the graph of a function?

(a)

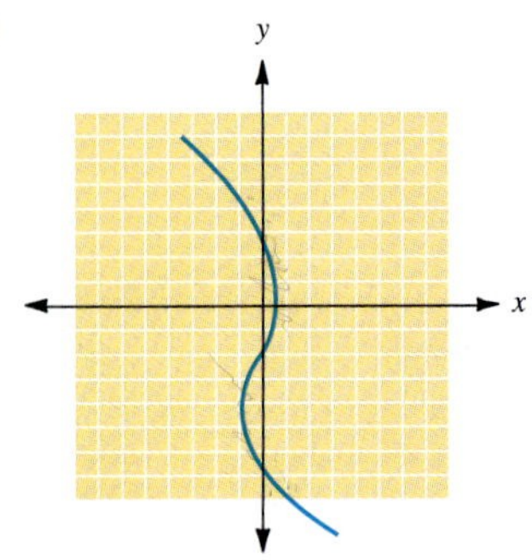

(b)

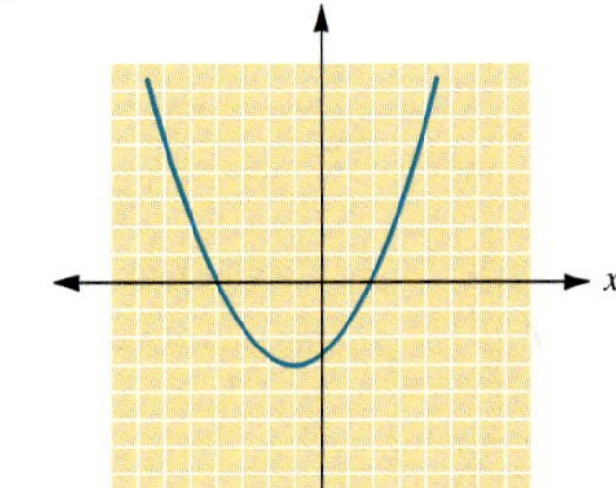

(c)

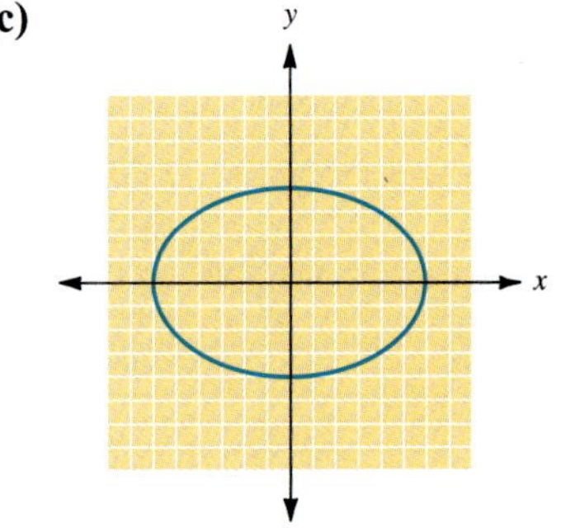

The graph of a relation can be used to determine the domain and range of the relation.

Example 9

Identifying Functions, Domain, and Range

For each of the following graphs, determine whether the relation is a function, find the domain of the relation, and find the range of the relation.

(a)

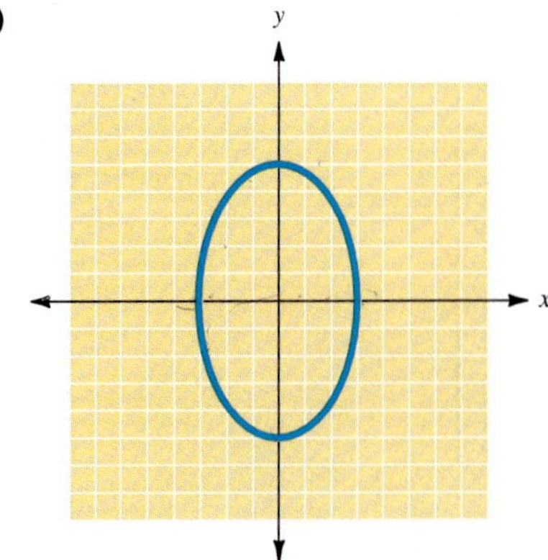

NOTE You might ask yourself "What values do we see used as x coordinates in the graph?" Answer: "All values between -3 and 3, inclusive."

This is not a function.

The domain is the set of x values from -3 to 3, inclusive. We write $D = \{x \mid -3 \leq x \leq 3\}$.

The range is the set of y values from -5 to 5, inclusive. We write $R = \{y \mid -5 \leq y \leq 5\}$.

(b)

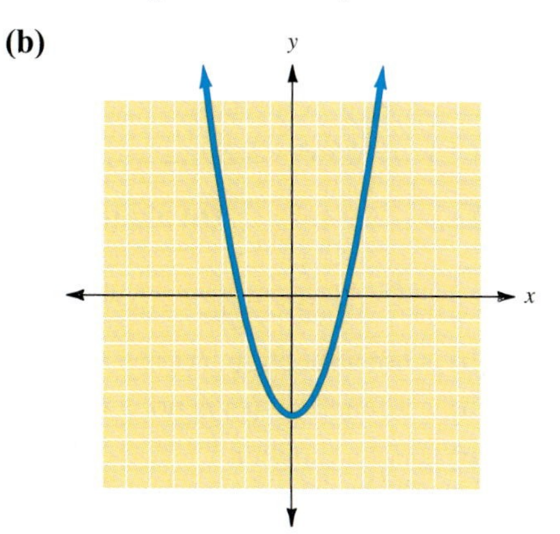

This is a function.

The graph continues forever in both the positive x direction and the negative x direction. Because there is a point on this graph related to every x value, the domain is the set of all real x values. We write

$D = \{x \mid x \in R\}$

The range is the set of y values greater than or equal to -5. We write $R = \{y \mid y \geq -5\}$.

CHECK YOURSELF 9

For each of the following graphs, determine whether the relation is a function, find the domain of the relation, and find the range of the relation.

(a) **(b)**

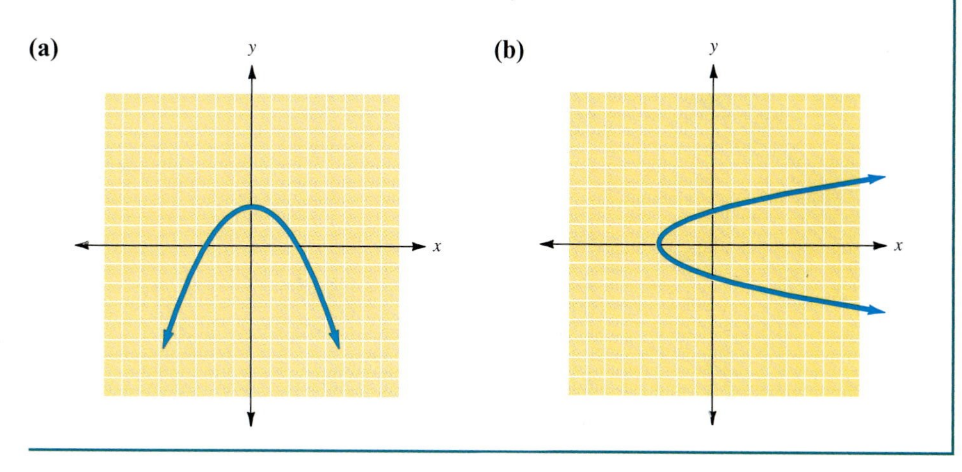

At this point, you may be wondering how the concept of function relates to anything outside the study of mathematics. A function is a relation that yields a single output (y value) each time a specific input (x value) is given. Any field in which predictions are made is building on the idea of functions. Here are a few examples:

- The physicist looks for the relationship that uses a planet's mass to predict its gravitational pull.
- The economist looks for the relationship that uses the tax rate to predict the employment rate.
- The business marketer looks for the relationship that uses an item's price to predict the number that will be sold.

- The college board looks for the relationship between tuition costs and the number of students enrolled at the college.
- The biologist looks for the relationship that uses temperature to predict a body of water's nutrient level.

CHECK YOURSELF ANSWERS

1. **(a)** 1; **(b)** 3; **(c)** -33 2. **(a)** -2; **(b)** 52; **(c)** -4
3. **(a)** $(0, -2)$; **(b)** $(3, 52)$; **(c)** $\left(-\frac{1}{2}, -4\right)$
4. **(a)** Is a function; **(b)** is not a function; **(c)** is a function; **(d)** is not a function
5. **(a)** Is a function; **(b)** is a function; **(c)** is not a function
6. **(a)** Is a function; **(b)** is not a function
7. **(a)** Is a function; **(b)** is not a function; **(c)** is a function
8. **(a)** Is not a function; **(b)** is a function; **(c)** is not a function
9. **(a)** This is a function; $D = \{x \mid x \in R\}$; $R = \{y \mid y \leq 2\}$
   (b) This is not a function; $D = \{x \mid x \geq -3\}$; $R = \{y \mid y \in R\}$

Name ______________________

Section ________ Date ________

Exercises

In exercises 1 to 10, evaluate each function for the value specified.

1. $f(x) = x^2 - x - 2$; find **(a)** $f(0)$, **(b)** $f(-2)$, and **(c)** $f(1)$.

2. $f(x) = x^2 - 7x + 10$; find **(a)** $f(0)$, **(b)** $f(5)$, and **(c)** $f(-2)$.

3. $f(x) = 3x^2 + x - 1$; find **(a)** $f(-2)$, **(b)** $f(0)$, and **(c)** $f(1)$.

4. $f(x) = -x^2 - x - 2$; find **(a)** $f(-1)$, **(b)** $f(0)$, and **(c)** $f(2)$.

5. $f(x) = x^3 - 2x^2 + 5x - 2$; find **(a)** $f(-3)$, **(b)** $f(0)$, and **(c)** $f(1)$.

6. $f(x) = -2x^3 + 5x^2 - x - 1$; find **(a)** $f(-1)$, **(b)** $f(0)$, and **(c)** $f(2)$.

7. $f(x) = -3x^3 + 2x^2 - 5x + 3$; find **(a)** $f(-2)$, **(b)** $f(0)$, and **(c)** $f(3)$.

8. $f(x) = -x^3 + 5x^2 - 7x - 8$; find **(a)** $f(-3)$, **(b)** $f(0)$, and **(c)** $f(2)$.

9. $f(x) = 2x^3 + 4x^2 + 5x + 2$; find **(a)** $f(-1)$, **(b)** $f(0)$, and **(c)** $f(1)$.

10. $f(x) = -x^3 + 2x^2 - 7x + 9$; find **(a)** $f(-2)$, **(b)** $f(0)$, and **(c)** $f(2)$.

ANSWERS

1. ____________
2. ____________
3. ____________
4. ____________
5. ____________
6. ____________
7. ____________
8. ____________
9. ____________
10. ____________

ANSWERS

11. ______
12. ______
13. ______
14. ______
15. ______
16. ______
17. ______
18. ______
19. ______
20. ______
21. ______
22. ______
23. ______
24. ______
25. ______
26. ______

In exercises 11 to 18, determine which of the relations are also functions.

11. $\{(1, 6), (2, 8), (3, 9)\}$

12. $\{(2, 3), (3, 4), (5, 9)\}$

13. $\{(-1, 4), (-2, 5), (-3, 7)\}$

14. $\{(-2, 1), (-3, 4), (-4, 6)\}$

15. $\{(1, 3), (1, 2), (1, 1)\}$

16. $\{(2, 4), (2, 5), (3, 6)\}$

17. $\{(-1, 1), (2, 1), (2, 3)\}$

18. $\{(2, -1), (3, 4), (3, -1)\}$

In exercises 19 to 24, decide whether the relation, shown as a table of values, is a function.

19.

x	y
3	1
−2	4
5	3
−7	4

20.

x	y
−2	3
1	4
5	6
2	−1

21.

x	y
2	3
4	2
2	−5
−6	−3

22.

x	y
1	5
3	−6
1	−5
−2	−9

23.

x	y
−1	2
3	6
6	2
−9	4

24.

x	y
4	−6
2	3
−7	1
−3	−6

In exercises 25 to 30, for each set of ordered pairs, plot the related points on the graph. Then use the vertical line test to determine which sets are functions.

25. $\{(-3, 1), (-1, 2), (-2, 3), (1, 4)\}$

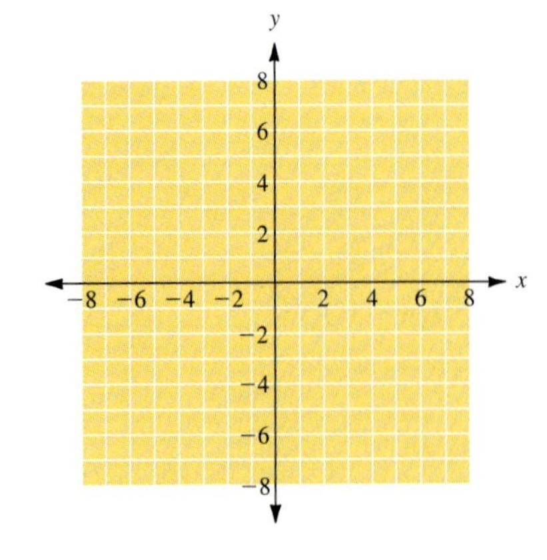

26. $\{(2, 2), (1, 1), (3, 3), (4, 5)\}$

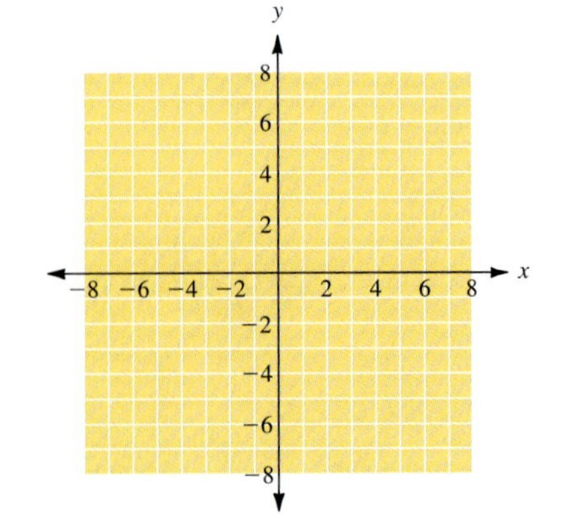

27. $\{(-1, 1), (2, 2), (3, 4), (5, 6)\}$

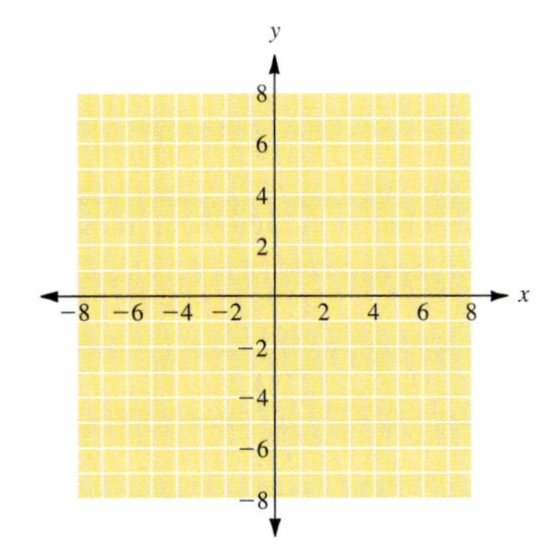

28. $\{(1, 4), (-1, 5), (0, 2), (2, 3)\}$

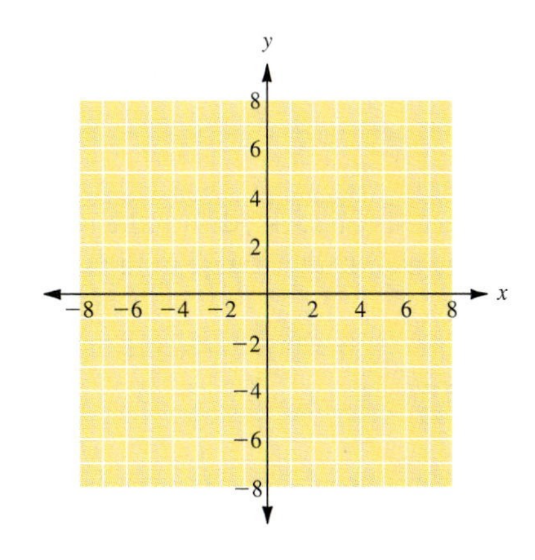

29. $\{(1, 2), (1, 3), (2, 1), (3, 1)\}$

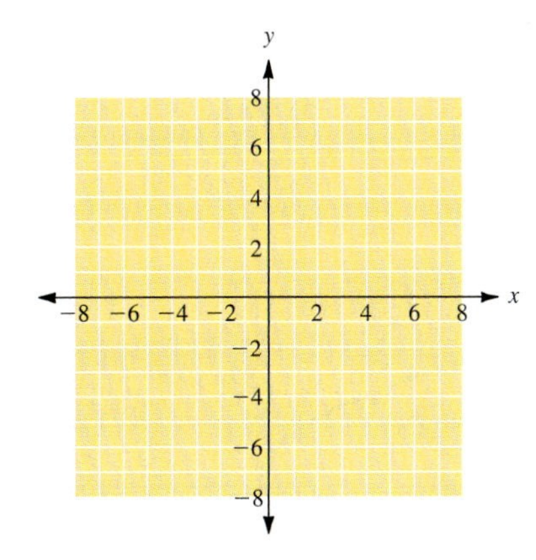

30. $\{(-1, 1), (3, 4), (-1, 2), (5, 3)\}$

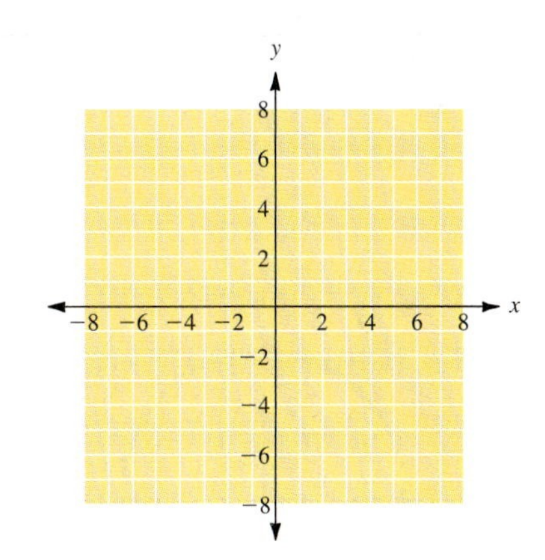

Your graphing calculator can be used to evaluate a function for a specific value of x. If $f(x) = 3x^2 - 7$, and you wish to find $f(-3)$,

1. Use the [Y =] key to enter $Y_1 = 3x^2 - 7$.
2. Select [TABLE] ([2nd] [GRAPH]), and choose -3 for x.
3. The table will give you a value of 20 for Y_1.

Use that technique to evaluate the functions in exercises 31 to 34.

31. $f(x) = 3x^2 - 5x + 7$; find **(a)** $f(-5)$, **(b)** $f(5)$, and **(c)** $f(12)$.

32. $f(x) = 4x^3 - 7x^2 + 9$; find **(a)** $f(-6)$, **(b)** $f(6)$, and **(c)** $f(10)$.

ANSWERS

27. ______

28. ______

29. ______

30. ______

31. ______

32. ______

33. ______

34. ______

35. ______

36. ______

37. ______

38. ______

39. ______

40. ______

33. $f(x) = 3x^4 - 6x^3 + 2x^2 - 17$; find **(a)** $f(-3)$, **(b)** $f(4)$, and **(c)** $f(7)$.

34. $f(x) = 5x^7 + 8x^4 - 9x^2 - 13$; find **(a)** $f(-4)$, **(b)** $f(-3)$, and **(c)** $f(2)$.

For exercises 35 to 44, determine whether the relation is a function, find the domain of the relation, and find the range of the relation.

35.

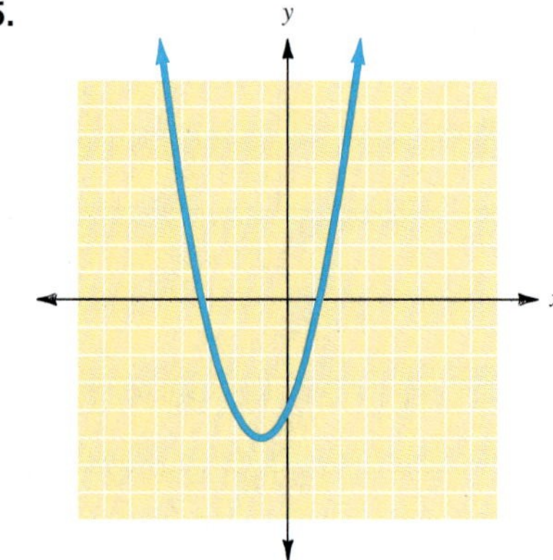

36.

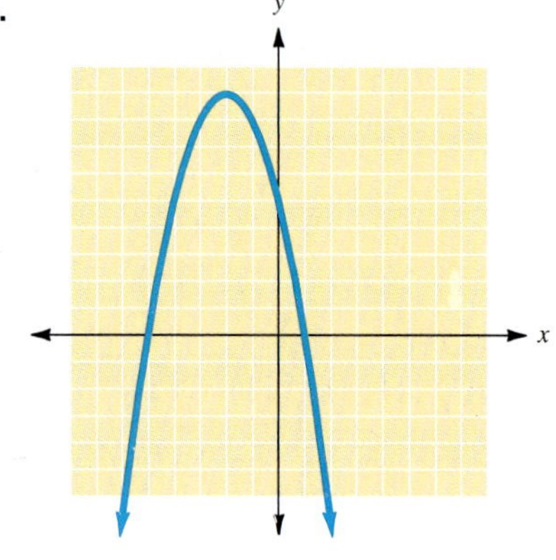

37.

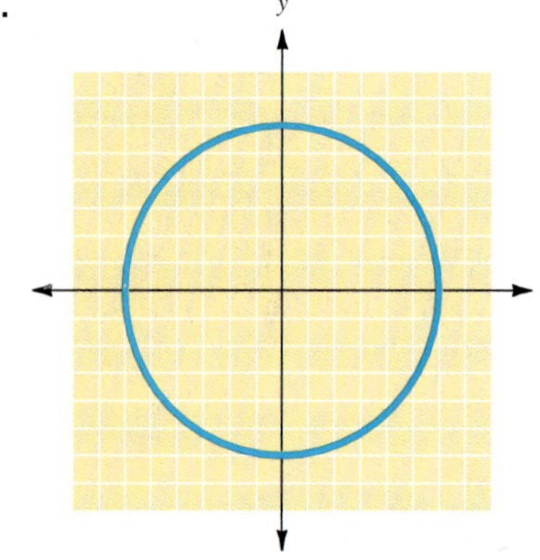

38.

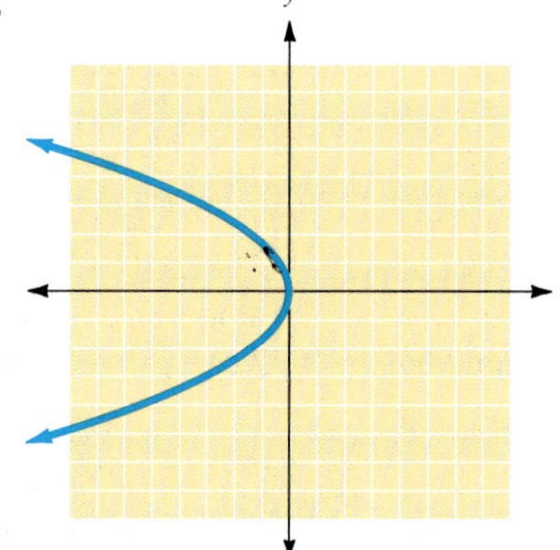

39.

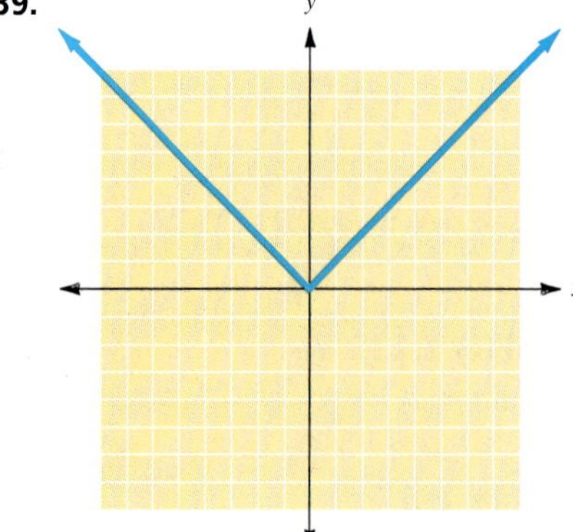

40.

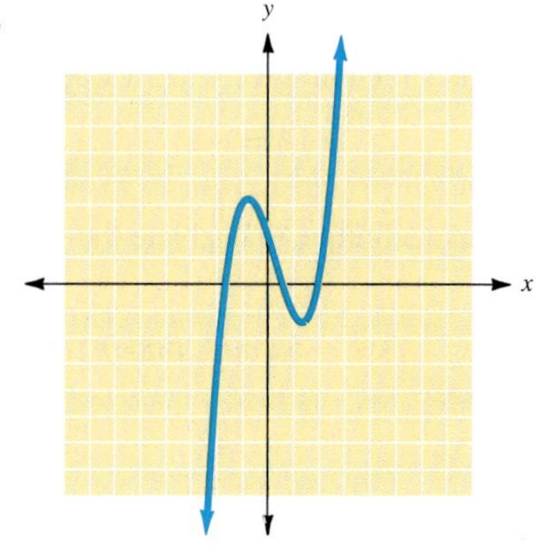

41.

42.

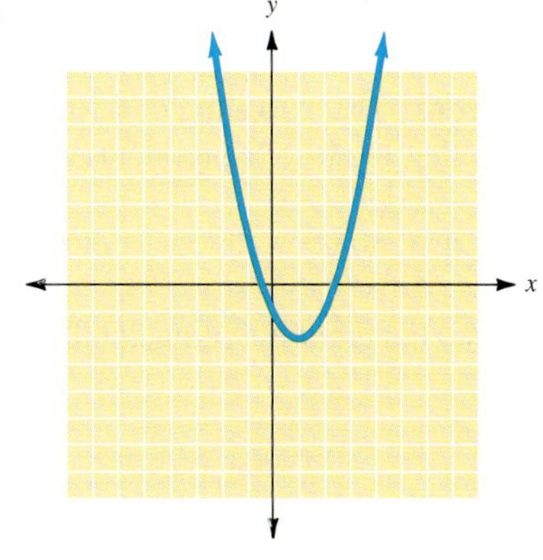

43.

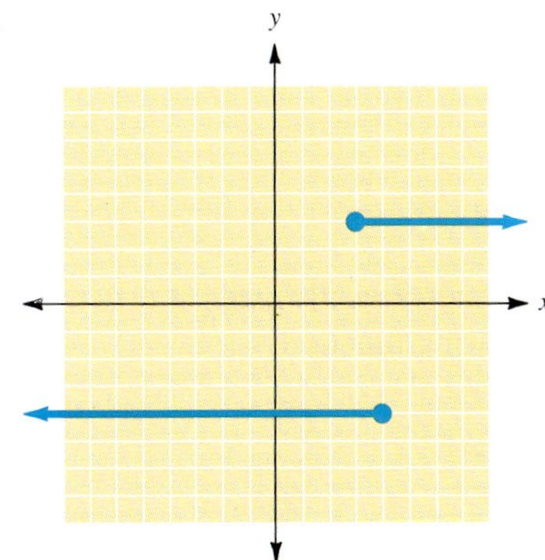

44.

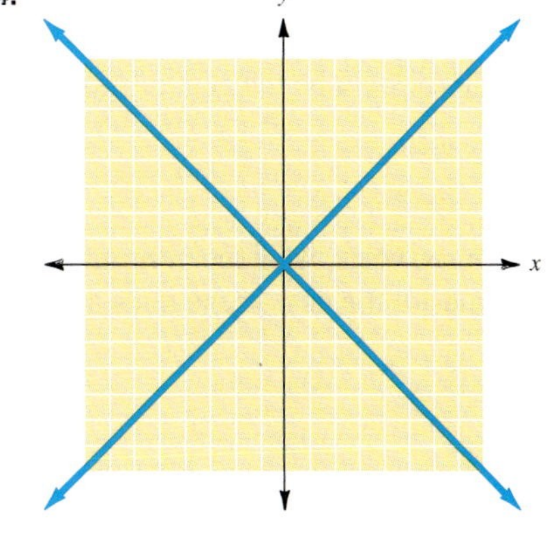

45. The following table shows the average hourly earnings for blue collar workers from 1947 to 1993. These figures are given in "real" wages, which means that the *purchasing power* of the money is given rather than the actual dollar amount. In other words, the amount earned for 1947 is not the actual amount listed here; in fact, it was much lower. The amount you see here is the amount in dollars that 1947 earnings could buy in 1947 compared to what 1993 wages could buy in 1993.

Year	Average Hourly Earnings (in 1993 dollars)
1947	$ 6.75
1967	10.67
1973	12.06
1979	12.03
1982	11.61
1989	11.26
1991	10.95
1993	10.83

Make a Cartesian coordinate graph of this data, using the year as the domain and the hourly earnings as the range. You will have to decide how to set up the axes so that the data all fit on the graph nicely. (*Hint:* Do not start the year at 0!) In complete sentences, answer the following questions: What are the trends that you notice from reading the table? What additional information does the graph show? Is this relation a function? Why or why not?

ANSWERS

41. ____________

42. ____________

43. ____________

44. ____________

45. ____________

46. ______

47. ______

48. ______

49. ______

Solve the following application problems.

46. Profit. The marketing department of a company has determined that the profit for selling x units of a product is approximated by the function

$$f(x) = 50\sqrt{x} - 0.25x - 600$$

Find the profit in selling 2500 units.

47. Cost. The inventor of a new product believes that the cost of producing the product is given by the function

$$C(x) = 1.75x + 7000$$

What would be the cost of producing 2000 units of the product?

48. Phone cost. A phone company has two different rates for calls made at different times of the day. These rates are given by the following function

$$C(x) = \begin{cases} 24x + 33 & \text{between 5 P.M. and 11 P.M.} \\ 36x + 52 & \text{between 8 A.M. and 5 P.M.} \end{cases}$$

when x is the number of minutes of a call and C is the cost of a call in cents.

(a) What is the cost of a 10-minute call at 10:00 A.M.?

(b) What is the cost of a 10-minute call at 10:00 P.M.?

49. Accidents. The number of accidents in 1 month involving drivers x years of age can be approximated by the function

$$f(x) = 2x^2 - 125x + 3000$$

Find the number of accidents in 1 month that involved **(a)** 17-year-olds and **(b)** 25-year-olds.

50. ________________

51. ________________

50. Stopping distance. The distance x, (in feet) that a car will skid on a certain road surface after the brakes are applied is a function of the car's velocity v (in miles per hour). The function can be approximated by

$$x = f(v) = 0.017v^2$$

How far will the car skid if the brakes are applied at **(a)** 55 mph? **(b)** 70 mph?

51. Science. An object is thrown upward with an initial velocity of 128 ft/s. Its height h after t seconds is given by the function

$$h(t) = -16t^2 + 128t$$

What is the height of the object at **(a)** 2 s? **(b)** 4 s? **(c)** 6 s?

Answers

1. **(a)** -2; **(b)** 4; **(c)** -2 **3.** **(a)** 9; **(b)** -1; **(c)** 3 **5.** **(a)** -62; **(b)** -2; **(c)** 2
7. **(a)** 45; **(b)** 3; **(c)** -75 **9.** **(a)** -1; **(b)** 2; **(c)** 13 **11.** Function
13. Function **15.** Not a function **17.** Not a function
19. Function **21.** Not a function **23.** Function
25. Function **27.** Function

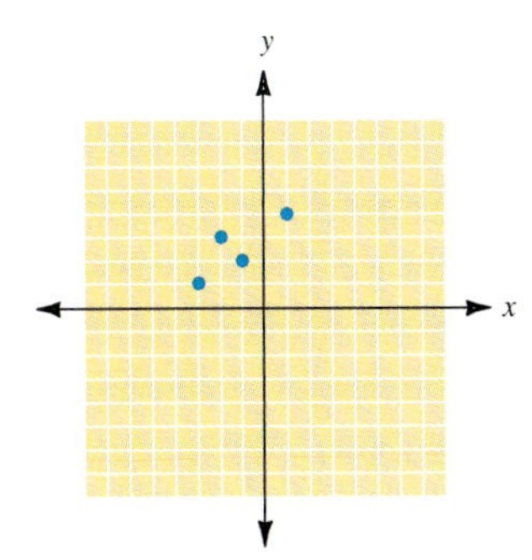

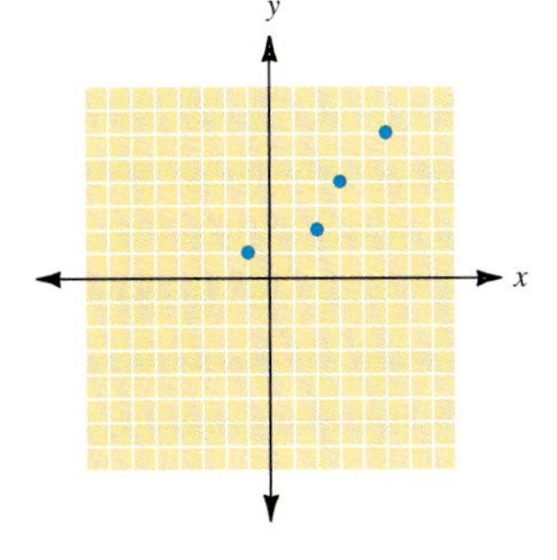

29. Not a function

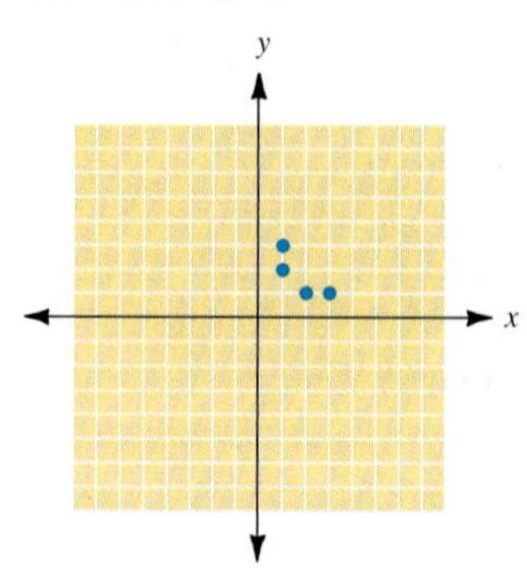

31. 107, 57, 379

33. 406, 399, 5226

35. Function; $D = \{x \mid x \in R\}$
$R = \{y \mid y \geq -5\}$

37. Not a function; $D = \{x \mid -6 \leq x \leq 6\}$
$R = \{y \mid -6 \leq y \leq 6\}$

39. Function; $D = \{x \mid x \in R\}$
$R = \{y \mid y \geq 0\}$

41. Function; $D = \{x \mid x \in R\}$
$R = \{y \mid y \geq 3\}$

43. Not a function; $D = \{x \mid x \in R\}$
$R = \{y \mid y = -4 \text{ or } 3\}$

45.

47. \$10,500

49. **(a)** 1453; **(b)** 1125

51. **(a)** 192 ft; **(b)** 256 ft; **(c)** 192 ft

3.4 Reading Values from a Graph

OBJECTIVES

1. Given x, find the function value on the graph
2. Given a function value, find the related x value
3. Find the x and y intercepts from a graph

In Section 3.2, we learned to read the coordinates of a point by drawing a vertical line from the point to the x axis to find the x coordinate and then drawing a horizontal line from the point to the y axis to find the y coordinate. A graph of a curve (including a graph of a straight line) is actually the graph of an infinite number of connected points. Finding the coordinates of any point on a curve is exactly the same as finding the coordinates of an isolated point.

Keep in mind that although we usually say something like, "Find the coordinates of the point . . . ," every time we read a graph we are able to only *estimate* the coordinates.

Example 1

Reading Values from a Graph

Find the coordinates of the labeled points.

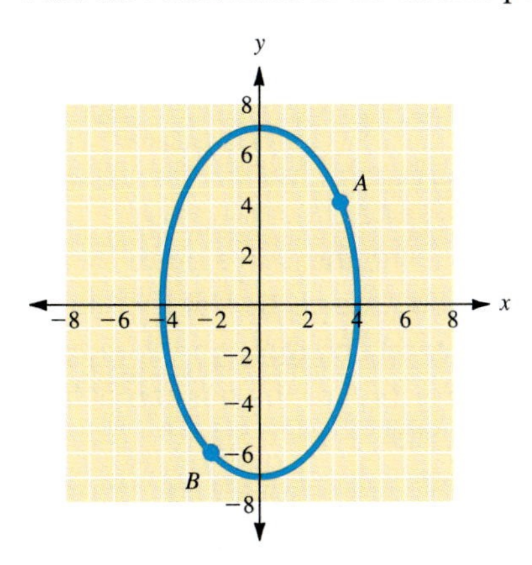

Point A has an x coordinate of 3 and a y coordinate of 4. Point A represents the ordered pair (3, 4). Point B represents the ordered pair $(-2, -6)$.

CHECK YOURSELF 1

Find the coordinates of the labeled points.

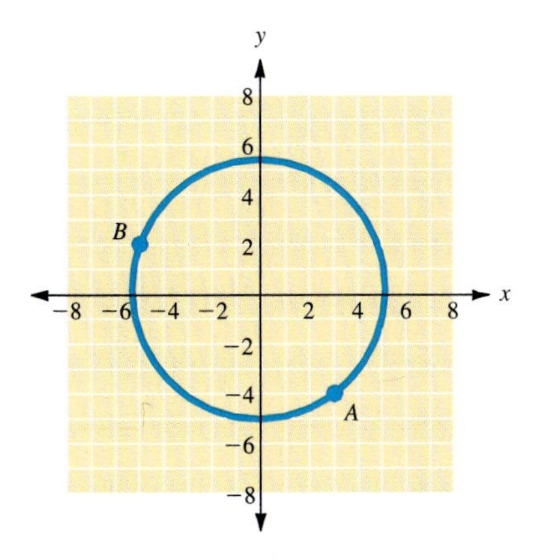

If a graph is the graph of a function, then every ordered pair (x, y) can be thought of as $(x, f(x))$.

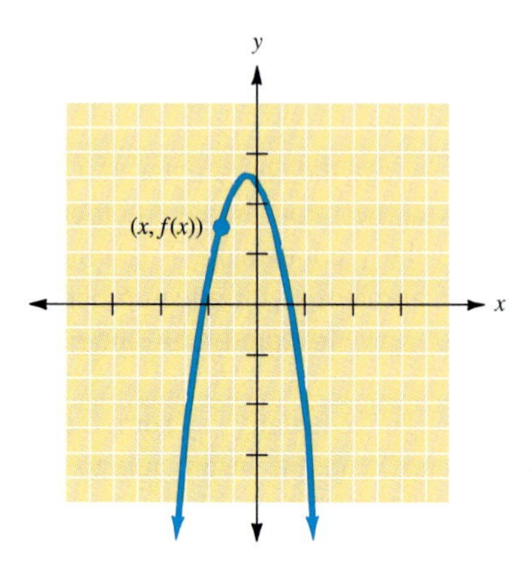

For a specific value of x, let's call it a, we can find $f(a)$ with the following algorithm.

Step by Step: Reading Function Values from Graphs

Step 1 Draw a vertical line through a on the x axis.
Step 2 Find the point of intersection of that line with the graph.
Step 3 Draw a horizontal line through the graph at that point.
Step 4 Find the intersection of the horizontal line with the y axis.
Step 5 $f(a)$ is that y value.

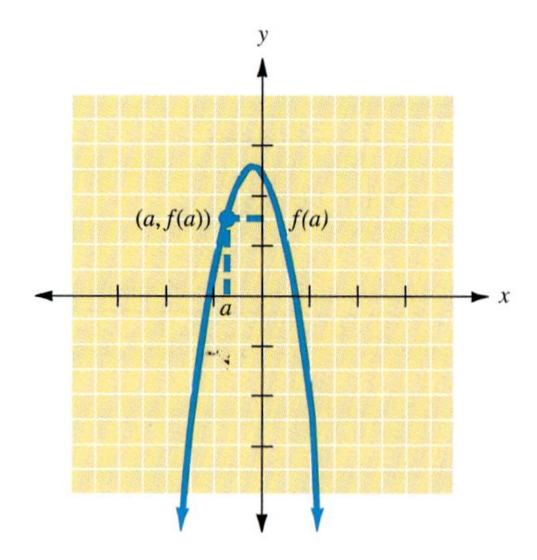

Example 2 illustrates this algorithm.

Example 2

Finding the Function Value on a Graph Given x

Consider the following graph of the function f. Use the graph to estimate $f(2)$.

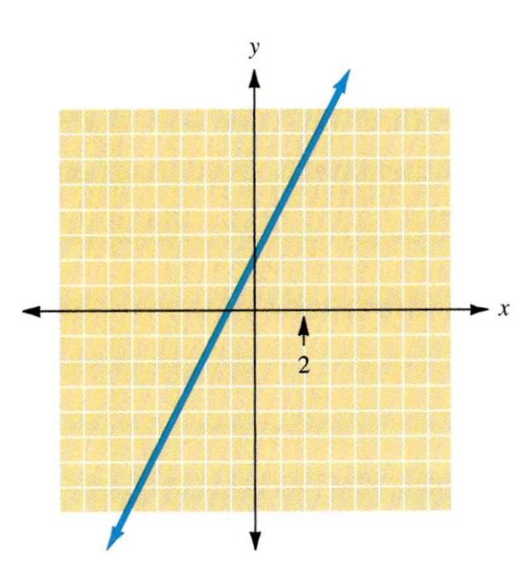

$f(2)$ is a y value. It is the y value that is paired with an x value of 2. Locate the number 2 on the x axis, draw a vertical line to the graph of the function, and then draw a horizontal line to the y axis, as shown below.

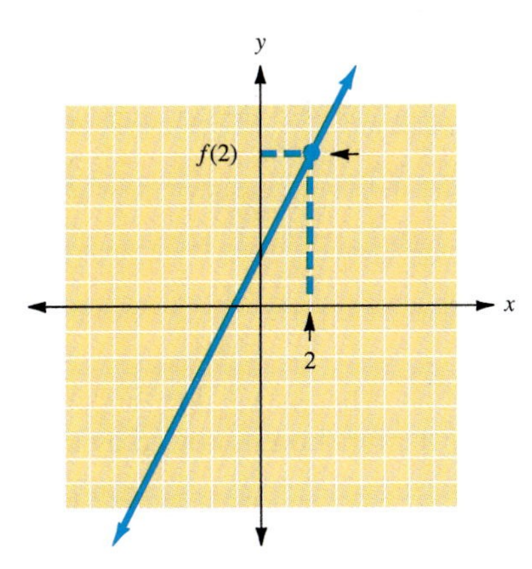

The coordinates of the point are $(2, f(2))$. The y value of the point is $f(2)$. Read the y value of this point on the y axis. It appears that $f(2)$, the y value of the point, is approximately 6. Therefore, $f(2) = 6$.

CHECK YOURSELF 2

Using the graph of the function f in Example 2, estimate each of the following.

(a) $f(1)$ **(b)** $f(-1)$ **(c)** $f(-3)$

In the preceding problem, you were given the x value and were asked to find the corresponding function value or y value. Now you will do the opposite operation. You will be given the function value and then you will need to find the corresponding x value(s).

If given the function value, we can find the associated x value by using the following algorithm.

Step by Step: Finding x Values from Function Values

Step 1 Find the given function value on the y axis.

Step 2 Draw a horizontal line through that point.

Step 3 Find every point on the graph that intersects the horizontal line.

Step 4 Draw a vertical line through each of those points of intersection.

Step 5 Each point of intersection of the vertical lines and the x axis gives an x value.

Example 3

Finding the x Value from a Graph Given the Function Value

Use the following graph of the function f to find all values of x such that $f(x) = -5$.

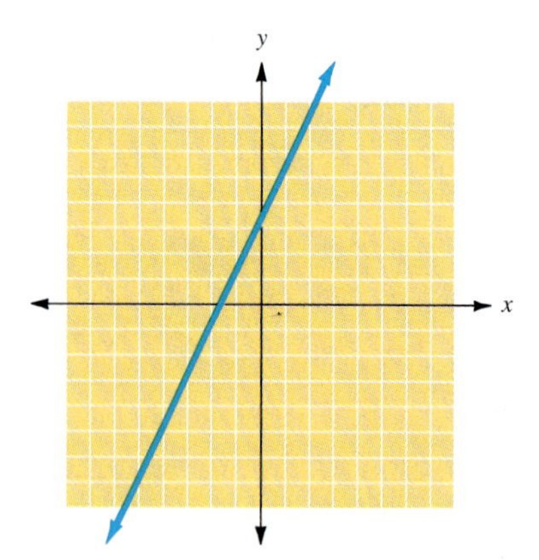

This time -5 is a function value, or y value. Locate -5 on the y axis, and draw a horizontal line to the graph of the function, followed by a vertical line to the x axis, as shown below.

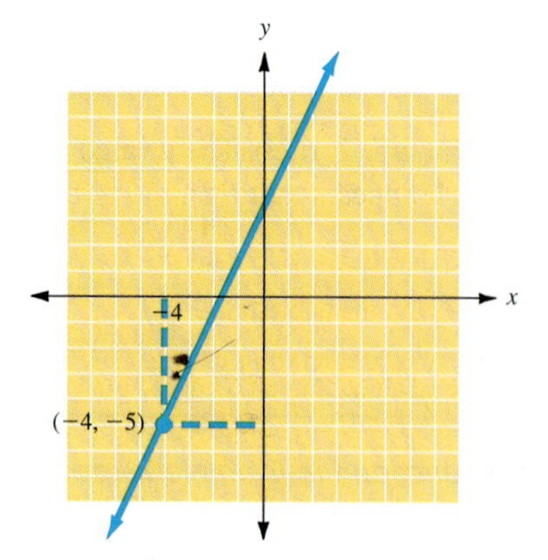

The solution of $f(x) = -5$ is $x = -4$. In particular, $f(-4) = -5$.

CHECK YOURSELF 3

Use the following graph to find all values of x such that

(a) $f(x) = 1$ **(b)** $f(x) = 5$ **(c)** $f(x) = -1$

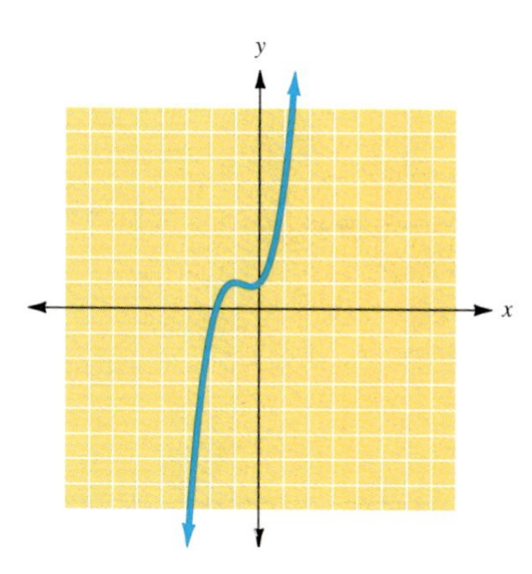

Among the most important values that can be read from graphs are the values of the x and y intercepts.

Example 4

Finding x and y Intercepts

Find the x and y intercepts from the graph.

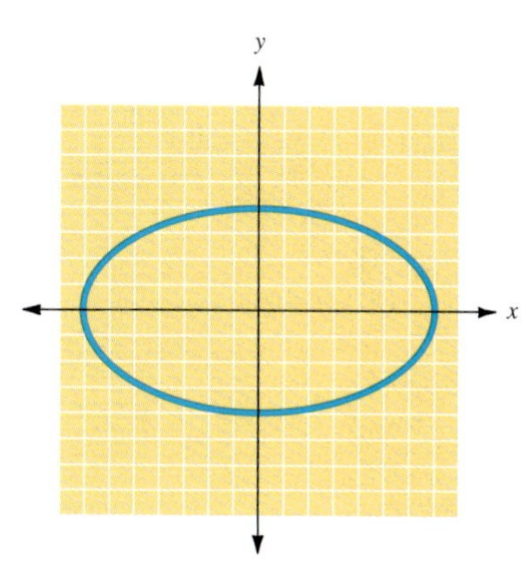

The x coordinate of the x intercept is the value of x for which $y = 0$. It is the x value of any point on the graph that touches the x axis. This graph touches the x axis at (7, 0) and also at (−7, 0). The x coordinates of the x intercepts are (7, 0) and (−7, 0).

The y coordinate of the y intercept is the value of y when $x = 0$. It is the y value of any point that touches the y axis. This graph touches the y axis at (0, 4) and (0, −4). The y intercepts are (0, 4) and (0, −4).

CHECK YOURSELF 4

Find the x and y intercepts from the graph.

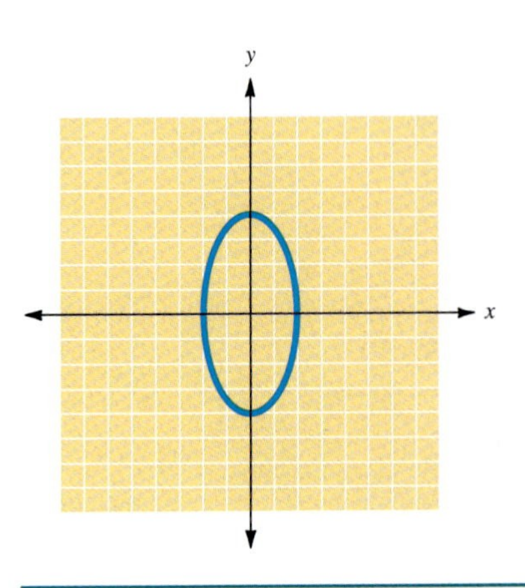

CHECK YOURSELF ANSWERS

1. **(a)** $(3, -4)$; **(b)** $(-5, 2)$
2. **(a)** $f(1) = 4$; **(b)** $f(-1) = 0$; **(c)** $f(-3) = -4$
3. **(a)** $x = -1, 0$; **(b)** $x = 1$; **(c)** $x = -2$
4. x int: $(-2, 0), (2, 0)$; y int: $(0, -4), (0, 4)$

Name ______________________

Section ________ Date ________

3.4 Exercises

In exercises 1 to 12, find the coordinates of the labeled points. Assume that each small square is a 1-unit square.

ANSWERS

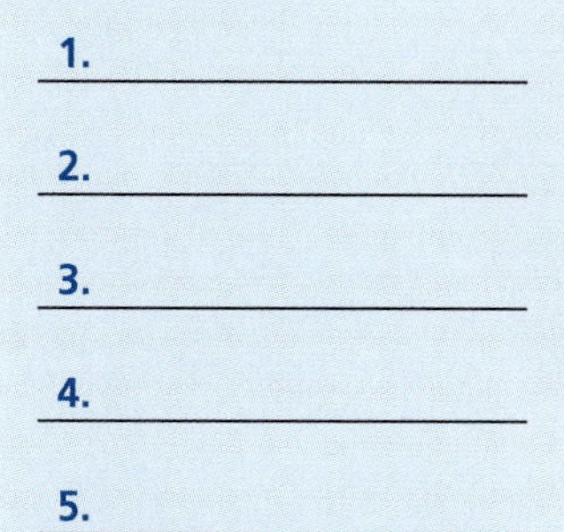

1. ______________

2. ______________

3. ______________

4. ______________

5. ______________

6. ______________

1.

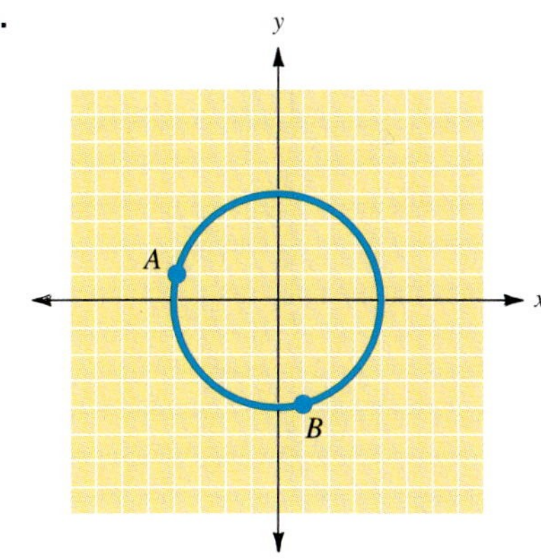

2.

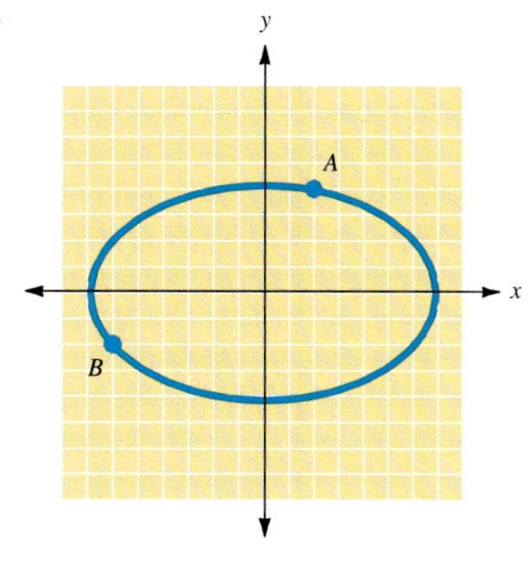

3.

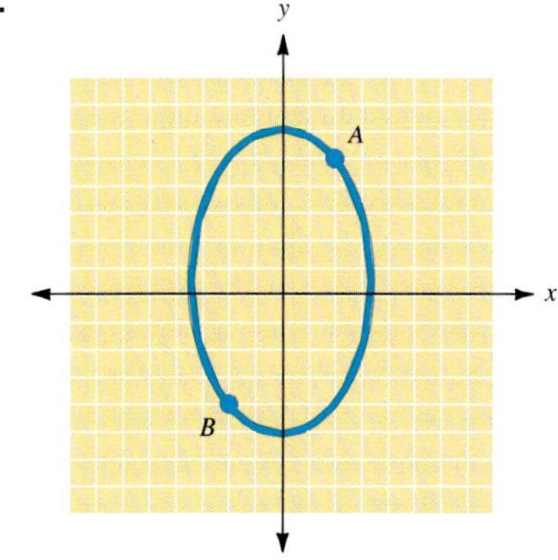

4.

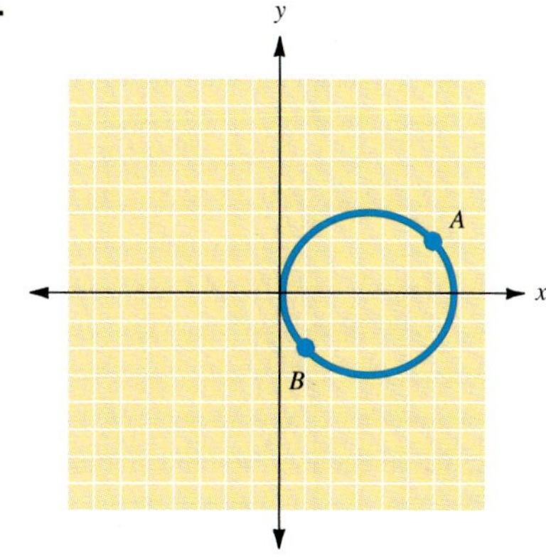

5.

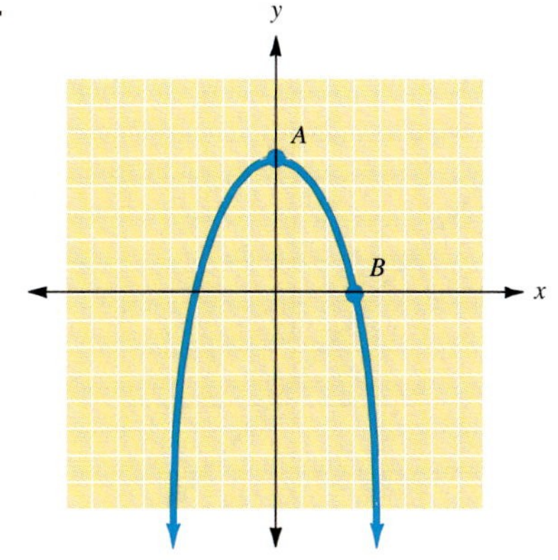

6.

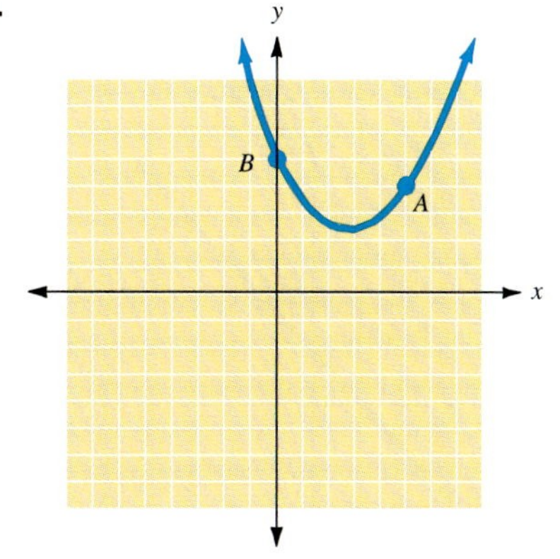

ANSWERS

7. ______

8. ______

9. ______

10. ______

11. ______

12. ______

13. ______

14. ______

7.

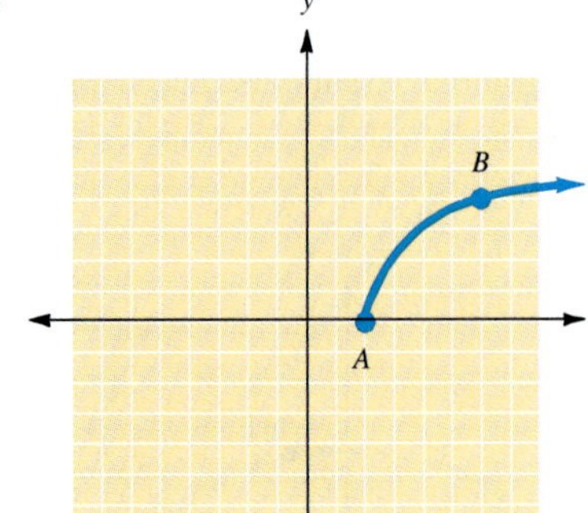

8.

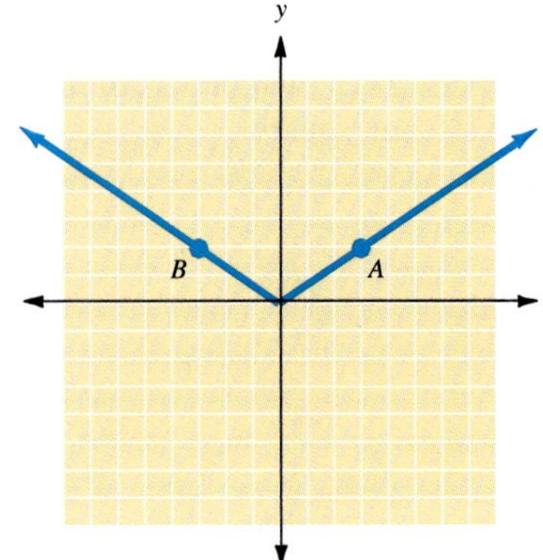

9.

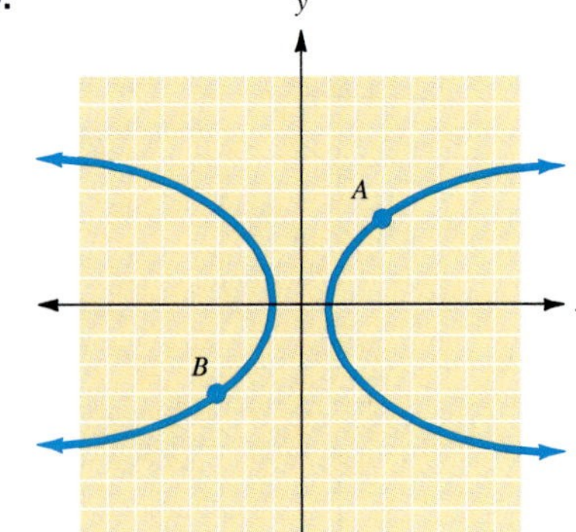

10.

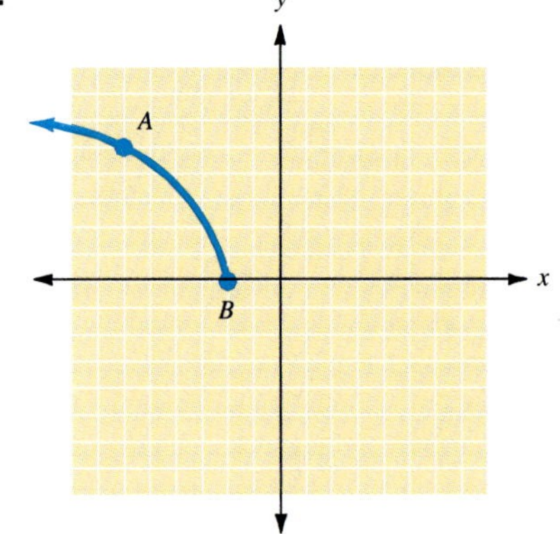

11.

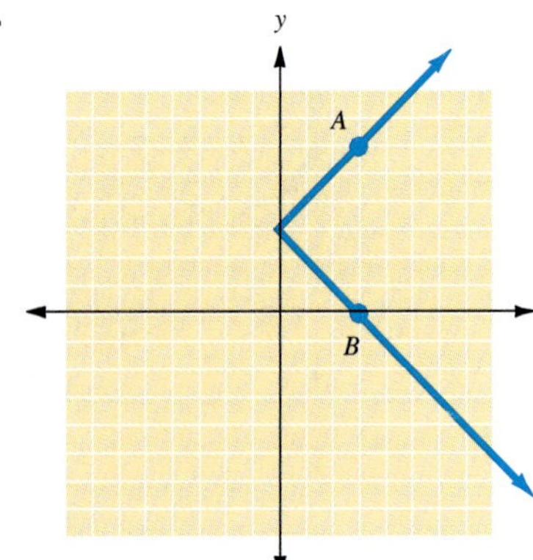

12.

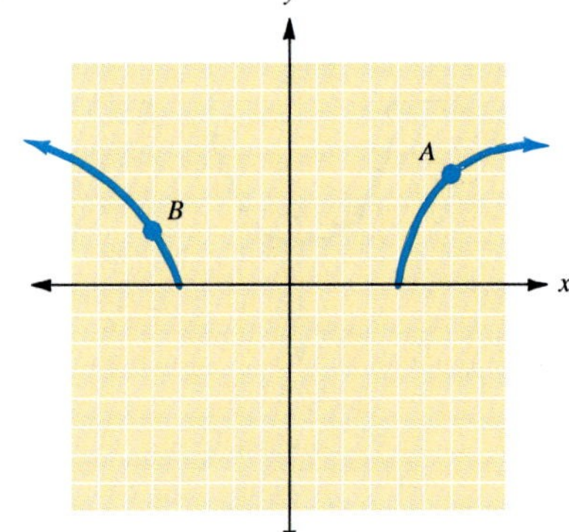

In exercises 13 to 20, use the graph of the function to estimate each of the following values: **(a)** $f(1)$, **(b)** $f(-1)$, **(c)** $f(0)$, **(d)** $f(3)$, and **(e)** $f(-2)$.

13.

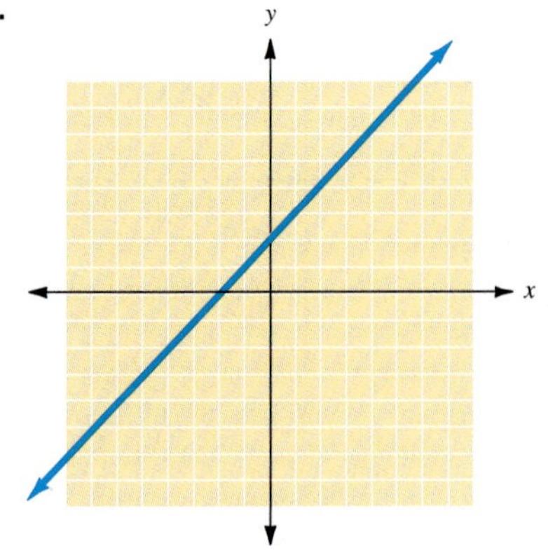

14.

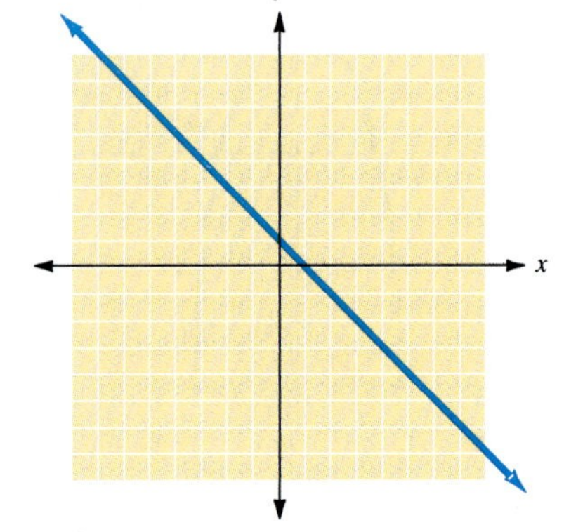

15.

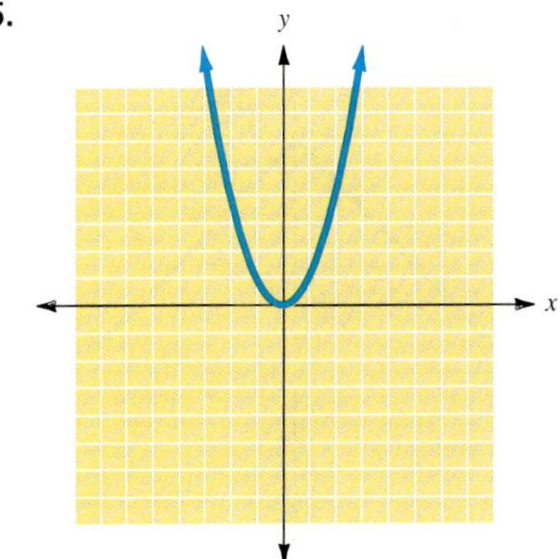

16.

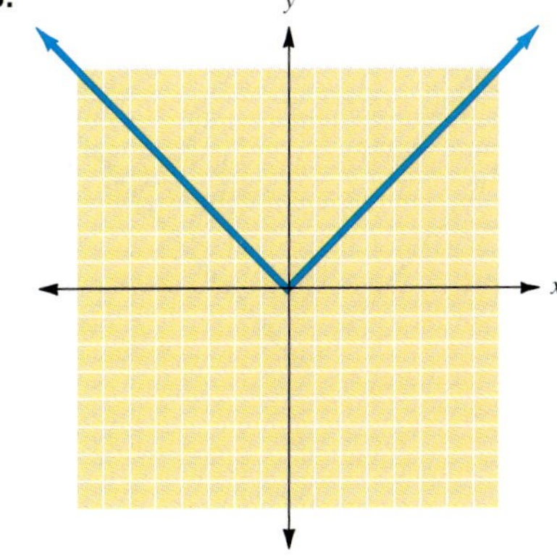

17.

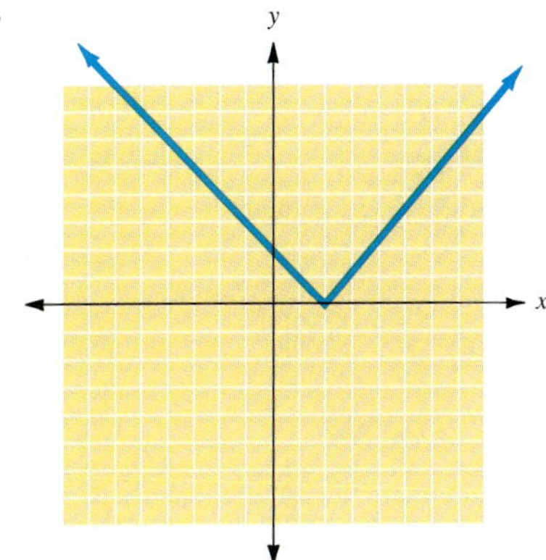

18.

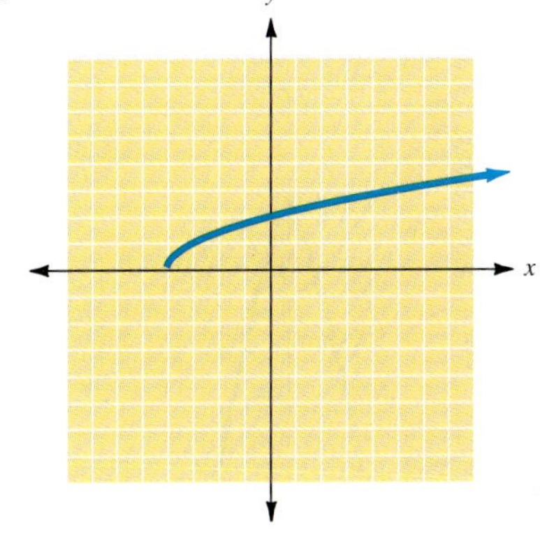

19.

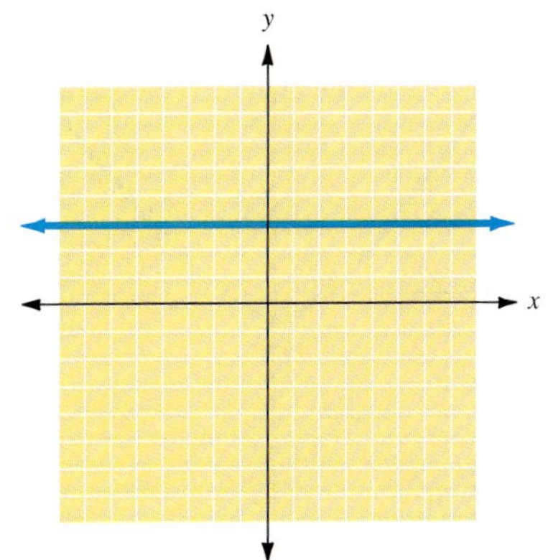

20.

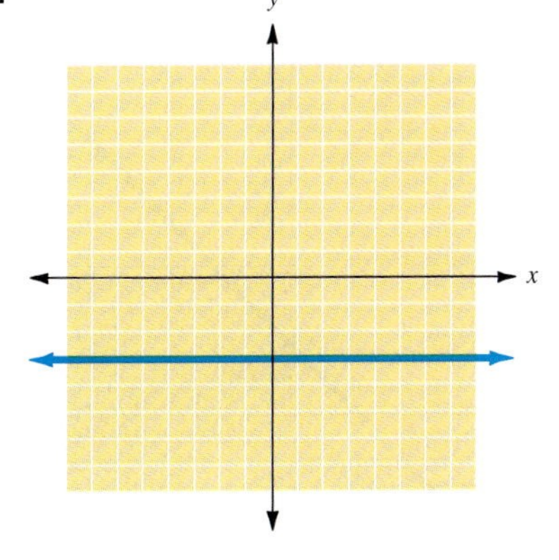

In exercises 21 to 28, use the graph of $f(x)$ to find all values of x such that **(a)** $f(x) = -1$, **(b)** $f(x) = 0$, and **(c)** $f(x) = 2$.

21.

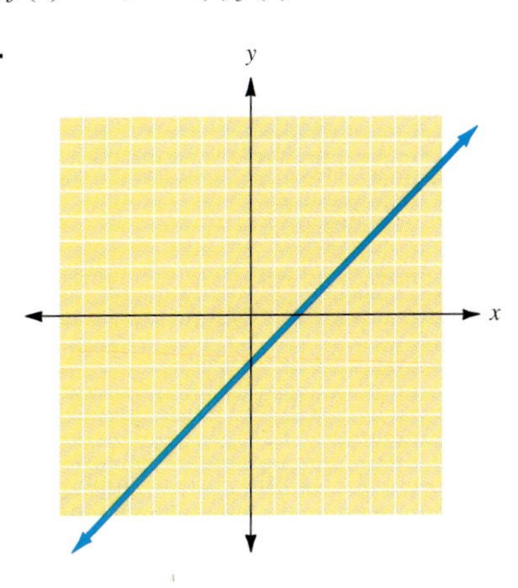

22.

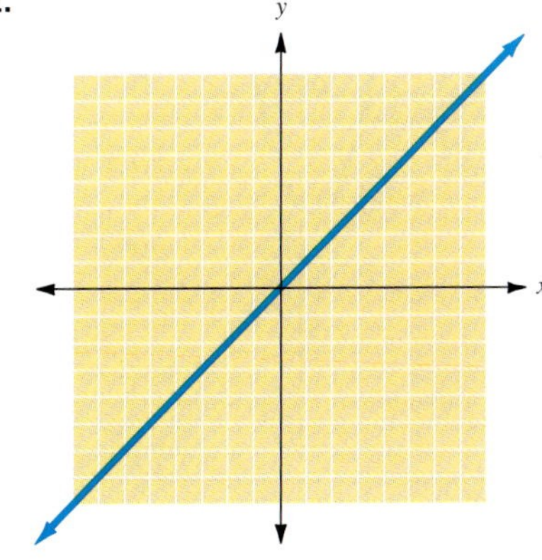

ANSWERS

15. ______________________

16. ______________________

17. ______________________

18. ______________________

19. ______________________

20. ______________________

21. ______________________

22. ______________________

23. ______

24. ______

25. ______

26. ______

27. ______

28. ______

29. ______

23.

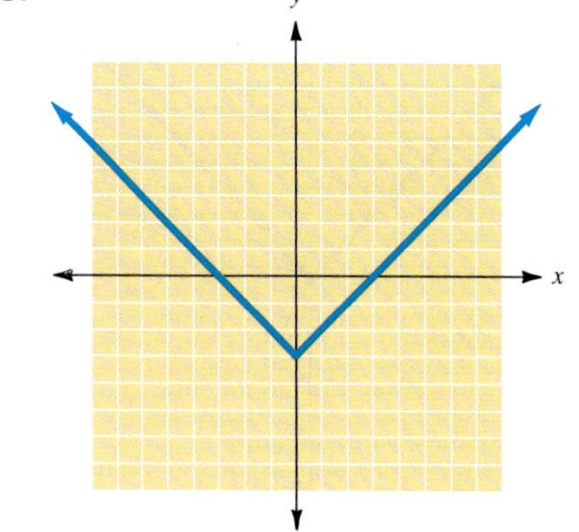

24.

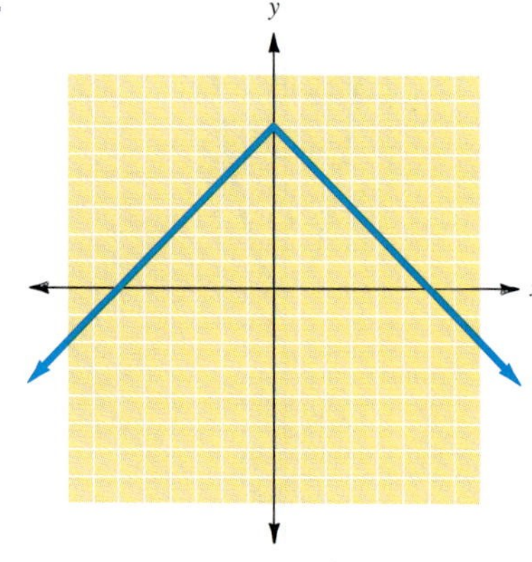

25.

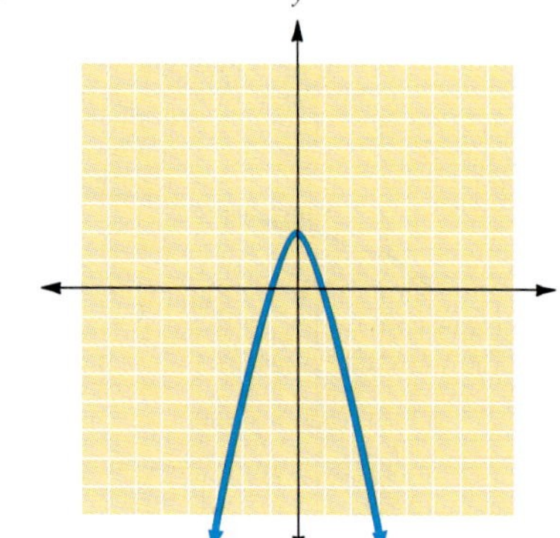

26.

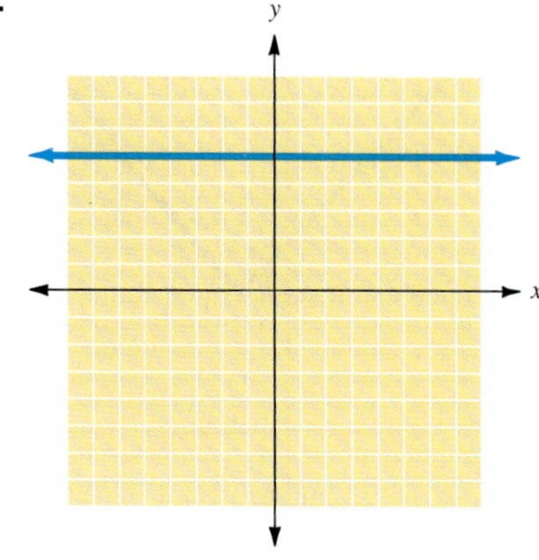

27.

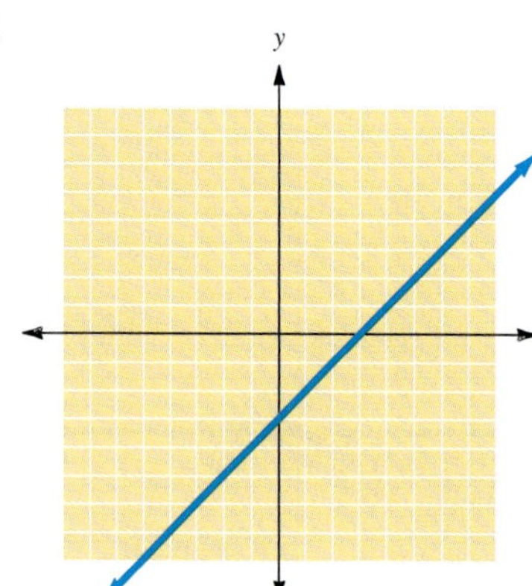

28.

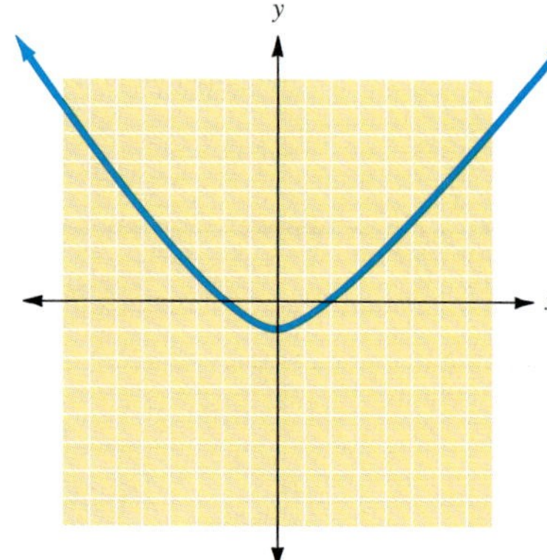

In exercises 29 to 32, find the x and y intercepts from the graph.

29.

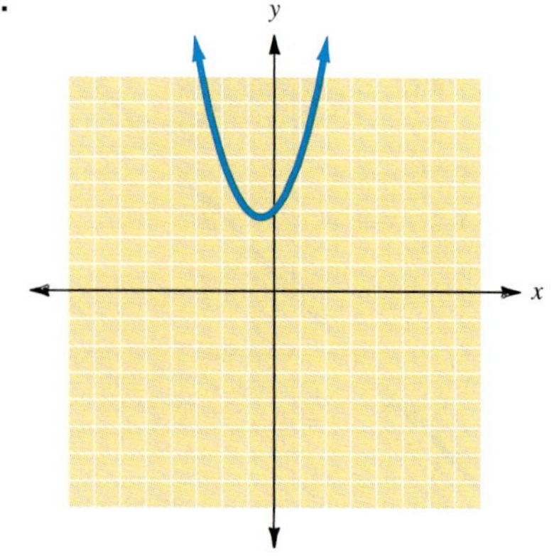

30.

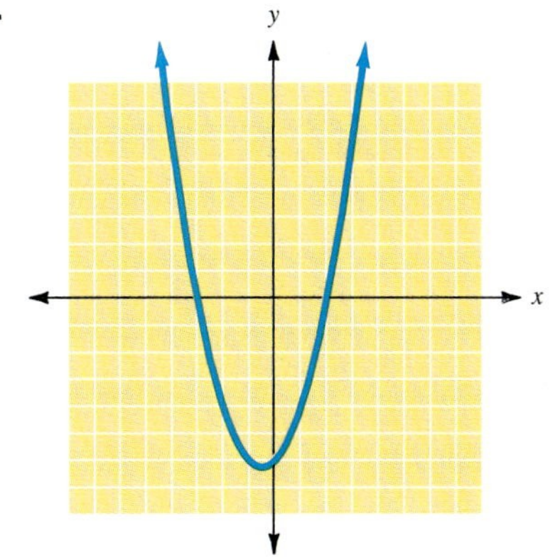

31.

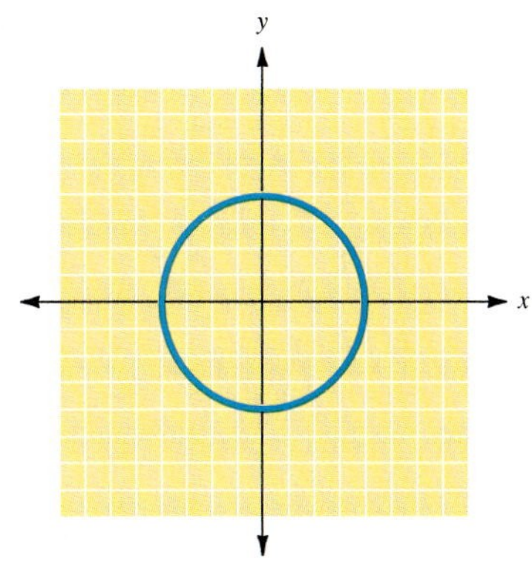

32.

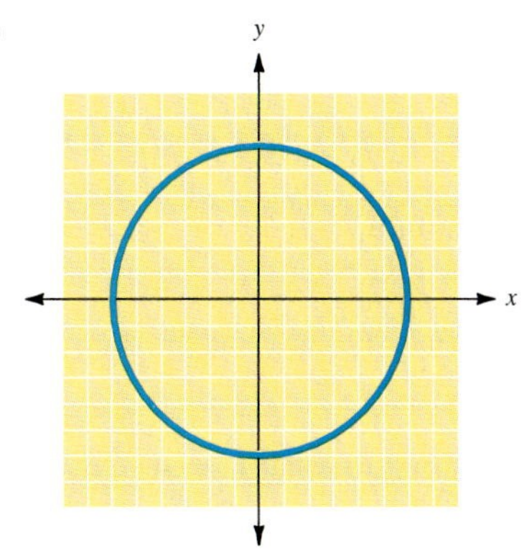

33. Your friend Sam Weatherby is a salesperson and has just been offered two new jobs, one for \$280 a month plus 9% of the amount of all his sales over \$20,000. The second job offer is \$280 a month plus 3% of the amount of his sales.

Sam and his spouse are about to have a child, and he feels that he has to make \$4000 a month just to make ends meet. He has called you to ask for your help in deciding which job to take. To help him picture his options, graph both offers on the same graph, and add a graph of the income of \$4000.

Next, write an explanation that answers these questions: How much does he have to sell in each position to earn \$4000? When is the first offer better? When is the second better? What sales would he have made to make less than \$3500 in each position? Which job should he take?

ANSWERS

30. ____________________

31. ____________________

32. ____________________

33. ____________________

34. ______________________

34. The cost of a taxi ride is shown in the graph below:

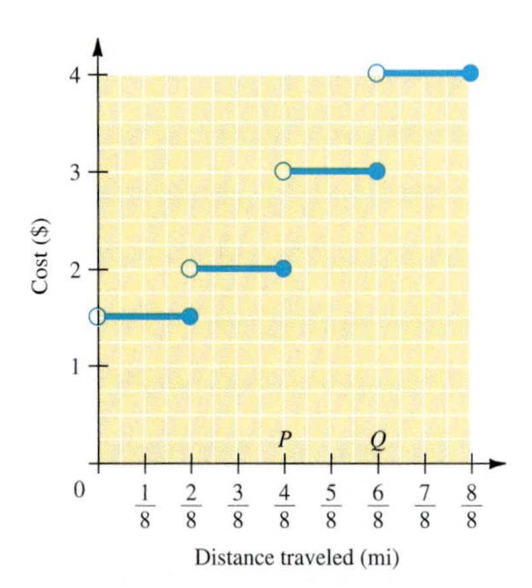

(a) How much does it cost to ride the taxi for $\frac{3}{4}$ of a mile?

(b) If the cost of the taxi ride is \$2, how far was the ride?

(c) What is the cost of the ride between points P and Q?

Answers

1. $A(-4, 1), B(1, -4)$ **3.** $A(2, 5), B(-2, -4)$ **5.** $A(0, 5), B(3, 0)$
7. $A(2, 0), B(6, 4)$ **9.** $A(3, 3), B(-3, -3)$ **11.** $A(3, 6), B(3, 0)$
13. **(a)** 3; **(b)** 1; **(c)** 2; **(d)** 5; **(e)** 0 **15.** **(a)** 1; **(b)** 1; **(c)** 0; **(d)** 9; **(e)** 4
17. **(a)** 1; **(b)** 3; **(c)** 2; **(d)** 1; **(e)** 4 **19.** **(a)** 3; **(b)** 3; **(c)** 3; **(d)** 3; **(e)** 3
21. **(a)** 1; **(b)** 2; **(c)** 4 **23.** **(a)** 2, -2; **(b)** 3, -3; **(c)** 5, -5
25. **(a)** -1.5, 1.5; **(b)** 1, -1; **(c)** 0 **27.** **(a)** 2; **(b)** 3; **(c)** 5
29. x int: none; y int: (0, 3) **31.** x int: $(-4, 0)$ and $(4, 0)$; y int: $(0, 4)$ and $(0, -4)$
33.

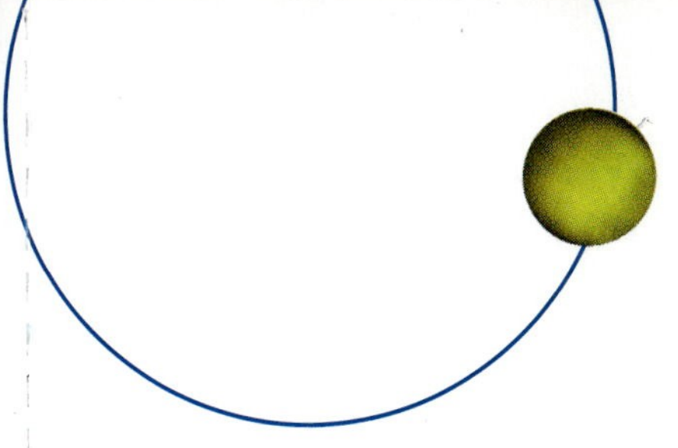

Cumulative Test for Chapters 1 to 3

Name ____________

Section ______ Date ______

This test is provided to help you in the process of reviewing the previous chapters. Answers are provided in the back of the book. If you missed any answers, be sure to go back and review the appropriate chapter sections.

1. Given the set of numbers $\left\{-8, \frac{\pi}{2}, 3.2, \sqrt{5}, 0\right\}$, identify the numbers that belong to **(a)** the integers; **(b)** the rational numbers.

2. Simplify the expressions: **(a)** $(x^3y^4)(x^5y^2)$ **(b)** $\frac{(xy)^5}{xy^2}$

3. Graph the set $\{x \mid x < -2 \text{ or } x > 4\}$.

4. Write 68,000,000 in scientific notation.

5. Evaluate the expression: $\frac{12 - 5 \cdot 2}{3 + 2^2}$

Solve each of the following equations.

6. $2(3x + 9) = 8(2 + x)$

7. $2x - 3(x + 2) = 4(5 - x) + 7$

8. $4(x - 5) = -6(x - 10)$

9. $|x - 4| = 5$

Solve each of the following inequalities.

10. $-8(2 - x) \geq 16$

11. $|2x - 6| < 4$

12. $3|2x - 5| > 3$

13. Solve for the indicated variable: $3x + 5y = 15$ for y

ANSWERS

1. ____________
2. ____________
3. ____________
4. ____________
5. ____________
6. ____________
7. ____________
8. ____________
9. ____________
10. ____________
11. ____________
12. ____________
13. ____________

14. ______
15. ______
16. ______
17. ______
18. ______
19. ______
20. ______

Solve the following application.

14. On her way to a business meeting, Connie drove on the freeway, and the trip took 3 h. Returning, she decided to take a side road, and her speed along that route averaged 8 mi/h slower than on the freeway. If her return trip took $3\frac{1}{2}$ h and the distance driven was the same each way, find her average speed in each direction.

15. You know that w is directly proportional to u and inversely proportional to v. If $w = 20$ when $u = 10$ and $v = 3$, find k, the constant of variation.

16. Identify the domain and range of the following set of ordered pairs.

$\{(2, 7), (3, 5), (-1, 1), (-2, 0), (4, 5)\}$

17. In which quadrant is the point $(-6, -7)$ located?

18. Find the coordinates of the labeled points.

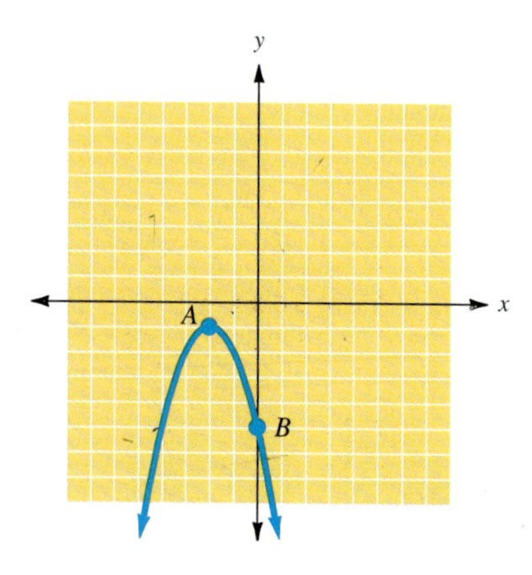

19. If $f(x) = 3x^2 + 5x - 9$, find $f(-1)$.

In each of the following, determine which relations are functions.

20. $\{(1, 2), (-1, 2), (3, 4), (5, 6)\}$

21.

x	y
−3	0
−2	1
−1	5
6	3

22. Use the vertical line test to determine whether the given graph represents a function.

(a)

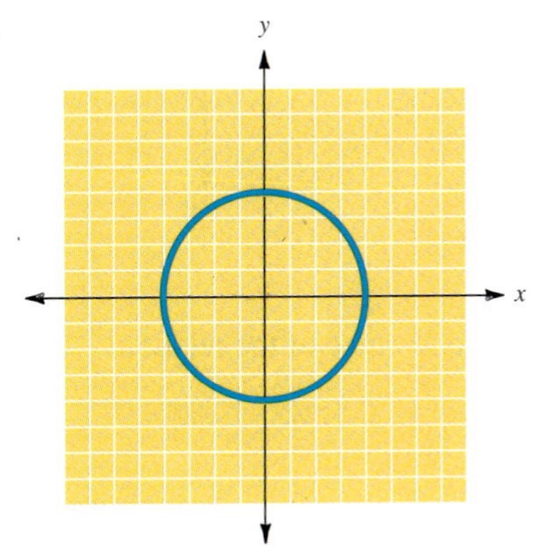

(b)

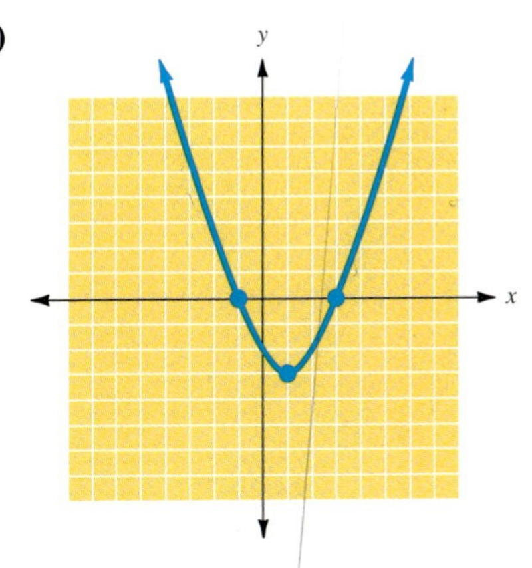

23. Use the given graph to estimate **(a)** $f(-3)$, **(b)** $f(0)$, and **(c)** $f(3)$.

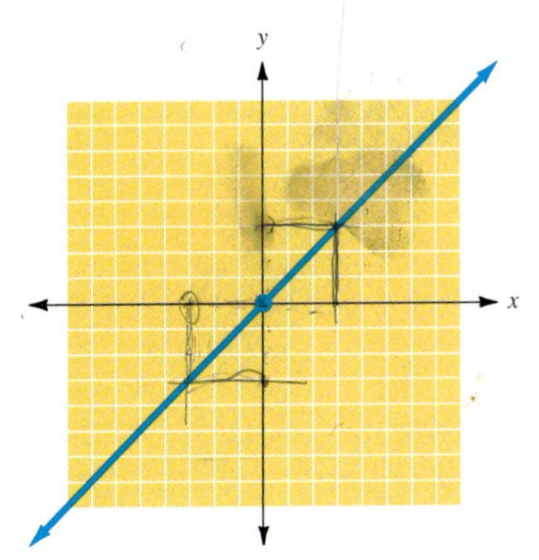

ANSWERS

21. ________________

22. ________________

23. ________________

24. ______

25. ______

24. Find the x and y intercepts of the given lines.

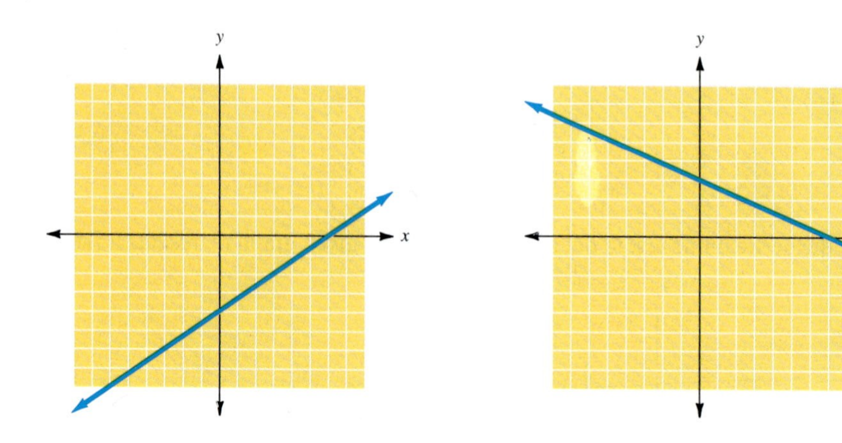

25. Use the given graph to determine the domain and range of the relation.

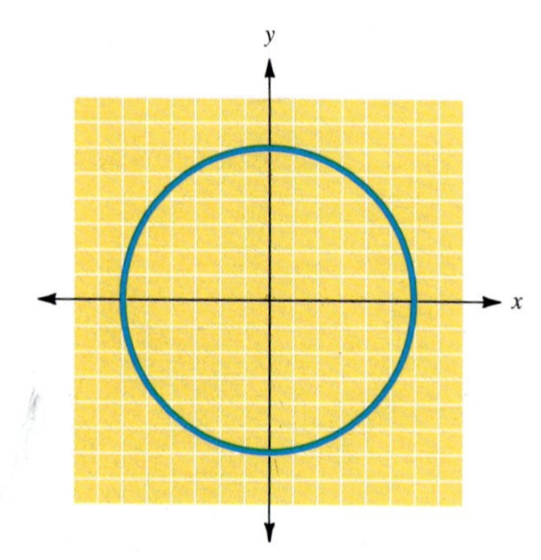

GRAPHS OF LINEAR EQUATIONS AND FUNCTIONS

4

INTRODUCTION

Graphs are used to discern patterns and trends that may be difficult to see when looking at a list of numbers or other kinds of data. The word "graph" comes from Latin and Greek roots and means "to draw a picture." This is just what a graph does in mathematics: It draws a picture of a relationship between two or more variables. But, as in art, these graphs can be difficult to interpret without a little practice and training. This chapter is the beginning of that training. And the training is important because graphs are used in every field in which numbers are used.

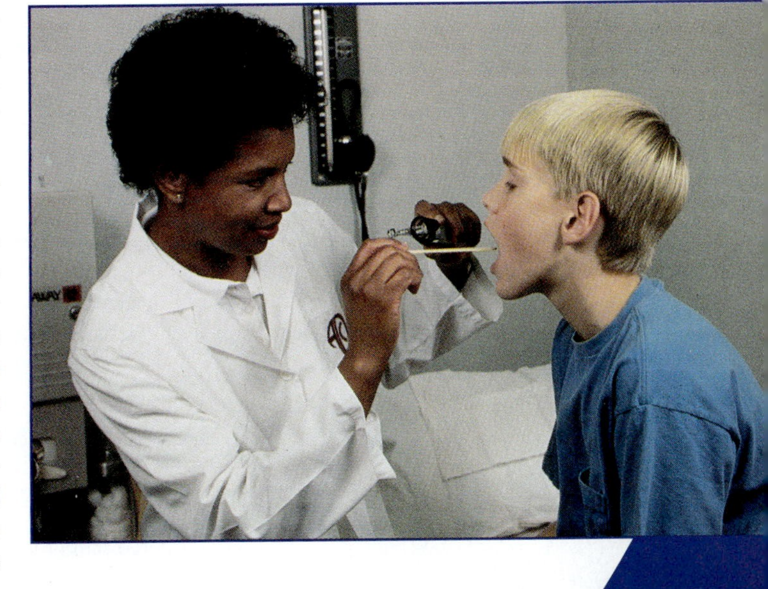

In the field of pediatric medicine, there has been controversy about the use of somatotropin (human growth hormone) to help children whose growth has been impeded by various health problems. The reason for the controversy is that many doctors are giving this expensive drug therapy to children who are simply shorter than average or shorter than their parents want them to be. The question of which children are not growing normally because of some serious health defect and need the therapy and which children are healthy and simply small of stature and thus should not be subjected to this treatment has been vigorously argued by professionals here and in Europe, where the therapy is being used.

Some of the measures used to distinguish between the two groups are blood tests and age and height measurements. The age and height measurements are graphed and monitored over several years of a child's life to monitor the rate of growth. If during a certain period the child's rate of growth slows to below 4.5 centimeters per year, this indicates that something may be seriously wrong. The graph can also indicate if the child's size fits within a range considered normal at each age of the child's life.

4.1 The Graph of a Linear Equation

4.1 OBJECTIVES

1. Find three ordered pairs for an equation in two variables
2. Graph a line from three points
3. Graph a line by the intercept method
4. Graph a line that passes through the origin
5. Determine domain and range
6. Graph horizontal and vertical lines

In previous algebra classes you have solved equations in one variable such as $3x - 2 = 5x + 4$.

Solving such an equation required finding the value of the variable, in this case x, that made the equation a true statement. In this case, that value is $x = -3$, because

$$3(-3) - 2 = 5(-3) + 4$$

This is a true statement because each side of the equation is equal to -11; no other value for x makes this statement true. The solution can be written in three different ways. We can write $x = -3$, $\{x|x = -3\}$ which is read "the set of all x such that x equals -3," or simply $\{-3\}$, which is the set containing the number -3.

What if we have an equation in two variables, such as $3x + y = 6$? The solution set is defined in a similar manner.

Definitions: Solution Set for an Equation in Two Variables

The **solution set** for an equation in two variables is the set containing all ordered pairs of real numbers (x, y) that will make the equation a true statement.

The solution set for an equation in two variables is a set of ordered pairs. Typically, there will be an infinite number of ordered pairs that make an equation a true statement. We can find some of these ordered pairs by substituting a value for x, then solving the remaining equation for y. We will use that technique in Example 1.

Example 1

Finding Ordered Pair Solutions

Find three ordered pairs that are solutions for each equation.

(a) $3x + y = 6$

We will pick three values for x, set up a table for ordered pairs, and then determine the related value for y.

x	y
-1	
0	
1	

Substituting -1 for x, we get

$$3(-1) + y = 6$$
$$-3 + y = 6$$
$$y = 9$$

The ordered pair $(-1, 9)$ is a solution to the equation $3x + y = 6$.
Substituting 0 for x, we get

$$3(0) + y = 6$$
$$0 + y = 6$$
$$y = 6$$

The ordered pair $(0, 6)$ is a solution to the equation $3x + y = 6$.
Substituting 1 for x, we get

$$3(1) + y = 6$$
$$3 + y = 6$$
$$y = 3$$

NOTE To indicate the set of all solutions to the equation, we write

$\{(x, y) \mid 3x + y = 6\}$

The ordered pair $(1, 3)$ is a solution to the equation $3x + y = 6$.
Completing the table gives us the following:

x	y
-1	9
0	6
1	3

(b) $2x - y = 1$

Let's try a different set of values for x. We will use the following table.

x	y
-5	
0	
5	

Substituting -5 for x, we get

$$2(-5) - y = 1$$
$$-10 - y = 1$$
$$-y = 11$$
$$y = -11$$

The ordered pair $(-5, -11)$ is a solution to the equation $2x - y = 1$.
Substituting 0 for x, we get

$$2(0) - y = 1$$
$$0 - y = 1$$
$$-y = 1$$
$$y = -1$$

NOTE Again, the set of all solutions is

$\{(x, y) \mid 2x - y = 1\}$

The ordered pair $(0, -1)$ is a solution to the equation $2x - y = 1$.

Substituting 5 for x, we get

$$2(5) - y = 1$$
$$10 - y = 1$$
$$-y = -9$$
$$y = 9$$

The ordered pair $(5, 9)$ is a solution to the equation $2x - y = 1$.

Completing the table gives us the following:

x	y
−5	−11
0	−1
5	9

CHECK YOURSELF 1

Find three ordered pairs that are solutions for each equation.

(a) $2x - y = 6$ **(b)** $3x + y = 2$

The graph of the solution set of an equation in two variables, usually called the **graph of the equation,** is the set of all points with coordinates (x, y) that satisfy the equation.

In this chapter, we are primarily interested in a particular kind of equation in x and y and the graph of that equation. The equations we refer to involve x and y to the first power, and they are called **linear equations.**

NOTE Why can A and B not both be zero? First, recall that, although x and y are variables, A, B, and C are constants. With that in mind, look at the equation if A and B are both zero.

$$(0)x + (0)y = C$$
$$0 + 0 = C$$
$$0 = C$$

Because zero must be a constant, we are left with the statement

$$0 = 0$$

This would be a true statement regardless of the values of x and y. Its graph would be every point in the plane.

Definitions: Linear Equations

An equation of the form

$Ax + By = C$

in which A and B cannot both be zero, is called the **standard form for a line.** Its graph is always a line.

Example 2

Graphing by Plotting Points

Graph the equation

$x + y = 5$

NOTE Because two points determine a line, technically two points are all that are needed to graph the equation. You may want to locate at least one other point as a check of your work.

This is a linear equation in two variables. To draw its graph, we can begin by assigning values to x and finding the corresponding values for y. For instance, if $x = 1$, we have

$$1 + y = 5$$
$$y = 4$$

Therefore, $(1, 4)$ satisfies the equation and is on the graph of $x + y = 5$.

Similarly, (2, 3), (3, 2), and (4, 1) are in the graph. Often these results are recorded in a table of values, as shown below. We then plot the points determined and draw a line through those points.

NOTE If you first rewrite an equation so that y is isolated on the left side, it can be easily entered and graphed with a graphing calculator. In this case, graph the equation

$y = -x + 5$

$x + y = 5$

x	y
1	4
2	3
3	2
4	1

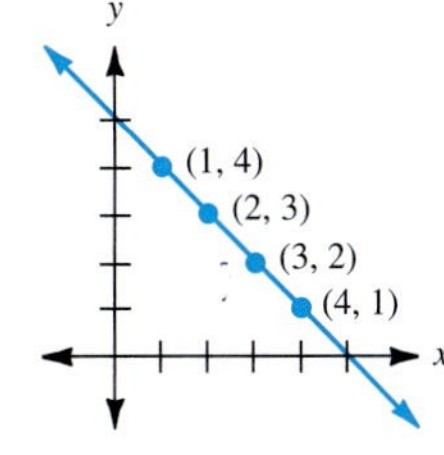

Every point on the graph of the equation $x + y = 5$ has coordinates that satisfy the equation, and every point with coordinates that satisfy the equation lies on the line.

CHECK YOURSELF 2

Graph the equation $2x - y = 6$.

NOTE An **algorithm** is a sequence of steps that solve a problem.

The following algorithm summarizes our first approach to **graphing a linear equation in two variables.**

Step by Step: To Graph a Linear Equation

Step 1 Find at least three solutions for the equation, and write your results in a table of values.

Step 2 Graph the points associated with the ordered pairs found in step 1.

Step 3 Draw a line through the points plotted above to form the graph of the equation.

Two particular points are often used in graphing an equation because they are very easy to find. The x *intercept* of a line is the point at which the line crosses the x axis. If the x intercept exists, it can be found by setting $y = 0$ in the equation and solving for x. The y *intercept* is the point at which the line crosses the y axis. If the y intercept exists, it is found by letting $x = 0$ and solving for y.

Example 3

Graphing by the Intercept Method

Use the intercepts to graph the equation

$x - 2y = 6$

NOTE Solving for y, we get

$y = \frac{1}{2}x - 3$

To graph this result on your calculator, you can enter

$Y_1 = (1 \div 2)x - 3$

using the $\boxed{x, T, \theta, n}$ key for x.

To find the x intercept, let $y = 0$.

$x - 2 \cdot 0 = 6$

$x = 6$

The x intercept is (6, 0).

To find the y intercept, let $x = 0$.

$$0 - 2y = 6$$
$$-2y = 6$$
$$y = -3$$

The y intercept is $(0, -3)$.

Graphing the intercepts and drawing the line through those intercepts, we have the desired graph.

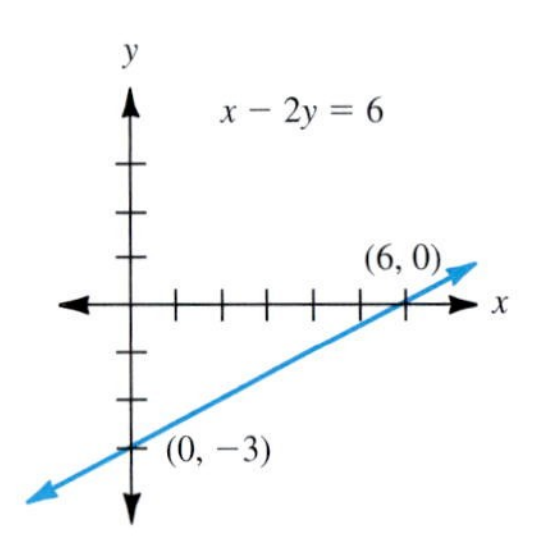

CHECK YOURSELF 3

Graph, using the intercept method.

$4x + 3y = 12$

The following algorithm summarizes the steps of graphing a line by the **intercept method.**

Step by Step: Graphing by the Intercept Method

Step 1 Find the x intercept. Let $y = 0$, and solve for x.
Step 2 Find the y intercept. Let $x = 0$, and solve for y.
Step 3 Plot the two intercepts determined in steps 1 and 2.
Step 4 Draw a line through the intercepts.

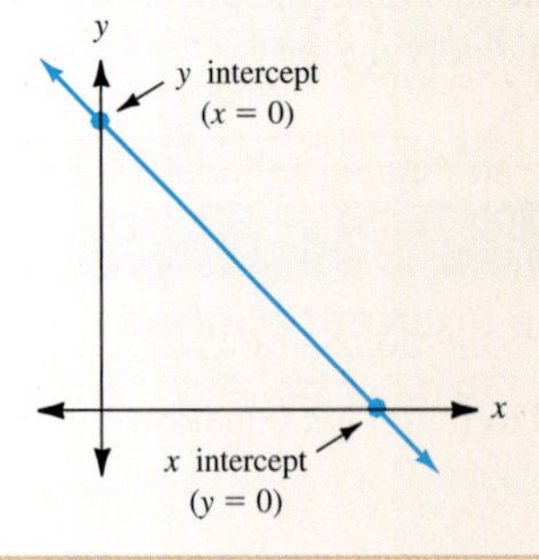

When can the intercept method not be used? Some lines have only one intercept. For instance, the graph of $x + 2y = 0$ passes through the origin. In this case, other points must be used to graph the equation.

Example 4

Graphing a Line That Passes Through the Origin

NOTE Graph the equation

$y = -\frac{1}{2}x$

Note that the line passes through the origin.

Graph $x + 2y = 0$.

Letting $y = 0$ gives

$$x + 2 \cdot 0 = 0$$
$$x = 0$$

Thus (0, 0) is a solution, and the line has only one intercept.

We continue by choosing any other convenient values for x. If $x = 2$:

$$2 + 2y = 0$$
$$2y = -2$$
$$y = -1$$

So $(2, -1)$ is a solution. You can easily verify that $(4, -2)$ is also a solution. Again, plotting the points and drawing the line through those points, we have the desired graph.

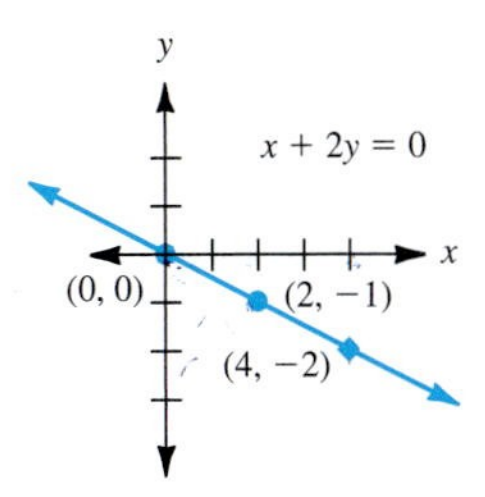

CHECK YOURSELF 4

Graph the equation $x - 3y = 0$.

In Section 3.1, we defined the terms *domain* and *range.* Recall that the domain of a relation is the set of all the first elements in the ordered pairs. The range is the set of all the second elements. Recall that a line is the graph of a set of ordered pairs. In Example 5, we will examine the domain and range for the graph of a line.

Example 5

Finding the Domain and Range

Find the domain and range for the relation described by the equation

$$x + y = 5$$

We can analyze the domain and range either graphically or algebraically. First, we will look at a graphical analysis. From Example 2, let's look at the graph of the equation.

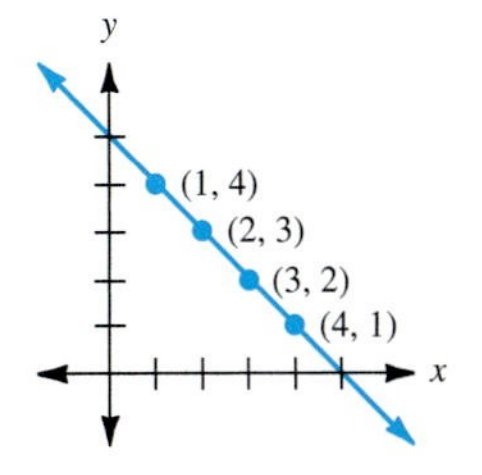

The graph continues forever at both ends. For every value of x, there is an associated point on the line. Therefore, the domain (D) is the set of all real numbers. In set notation, we write

$$D = \{x \mid x \in R\}$$

This is read, "The domain is the set of every x that is a real number."

To find the range (R), we look at the graph to see what values are associated with y. Note that every y is associated with some point. The range is written as

$$R = \{y \mid y \in R\}$$

This is read, "The range is the set of every y that is a real number."

Let's find the domain and range for the same relation by using an algebraic analysis. Look at the following equation.

$$x + y = 5$$

To determine the domain, we need to find every value of x that allows us to solve for y. That combination will result in an ordered pair (x, y). The set of all those x values is the domain of the relation.

We can find a value for y for *any* real value of x. For example, if $x = -5$,

$$-5 + y = 5$$
$$y = 10$$

The ordered pair $(-5, 10)$ is part of the relation. As in our graphical analysis, the domain is

$$D = \{x \mid x \in R\}$$

By a similar argument, we can substitute any value for y and solve the equation for x. The range is

$$R = \{y \mid y \in R\}$$

CHECK YOURSELF 5

Find the domain and range for the relation described by the following equation.

$$x - y = 4$$

Two types of linear equations are worthy of special attention. Their graphs are lines that are parallel to the x or y axis, and the equations are special cases of the general form

$$Ax + By = C$$

in which either $A = 0$ or $B = 0$.

Rules and Properties: Vertical or Horizontal Lines

1. A line with an equation of the form

   $y = k$

   is horizontal (parallel to the x axis).
2. A line with an equation of the form

   $x = h$

   is vertical (parallel to the y axis).

Example 6 illustrates both cases.

Example 6

Graphing Horizontal and Vertical Lines

NOTE Because part **(a)** is a function, it can be graphed on your calculator. Part **(b)** is not a function and cannot be graphed on your calculator.

(a) Graph the line with equation

$y = 3$

You can think of the equation in the equivalent form

$0 \cdot x + y = 3$

Note that any ordered pair of the form (__, 3) will satisfy the equation. Because x is multiplied by 0, y will always be equal to 3.

For instance, $(-2, 3)$ and $(3, 3)$ are on the graph. The graph, a horizontal line, is shown below.

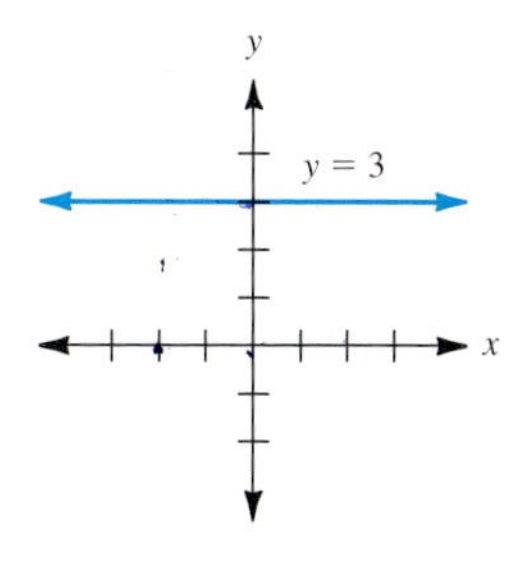

The domain for a horizontal line is every real number. The range is a single y value. We write

$D = \{x | x \in R\}$ and $R = \{3\}$

(b) Graph the line with equation

$x = -2$

In this case, you can think of the equation in the equivalent form

$x + 0 \cdot y = -2$

NOTE Notice that

$D = \{-2\}$

and

$R = \{y | y \in R\}$

Now any ordered pair of the form (−2, __) will satisfy the equation. Examples are $(-2, -1)$ and $(-2, 3)$. The graph, a vertical line, is shown below.

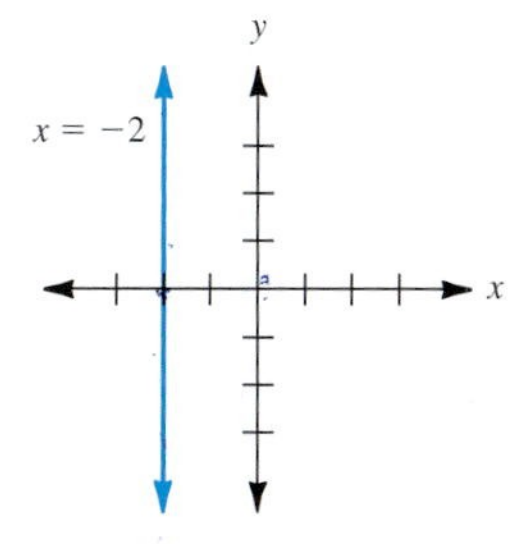

CHECK YOURSELF 6

Graph each equation and state the domain and range.

(a) $y = -3$ **(b)** $x = 5$

CHECK YOURSELF ANSWERS

1. **(a)** Answers will vary, but could include $(0, -6)$; **(b)** Answers will vary, but could include $(0, 2)$. **2.**

x	y
0	−6
1	−4
2	−2

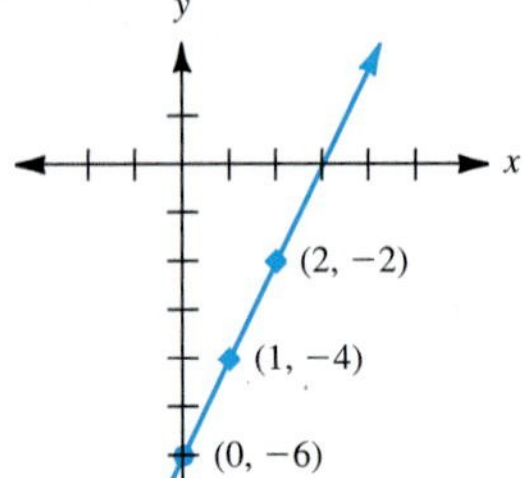

3.

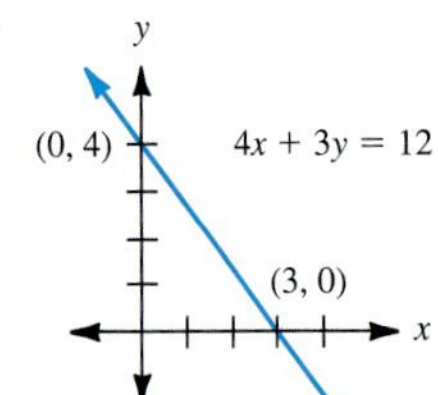

4.

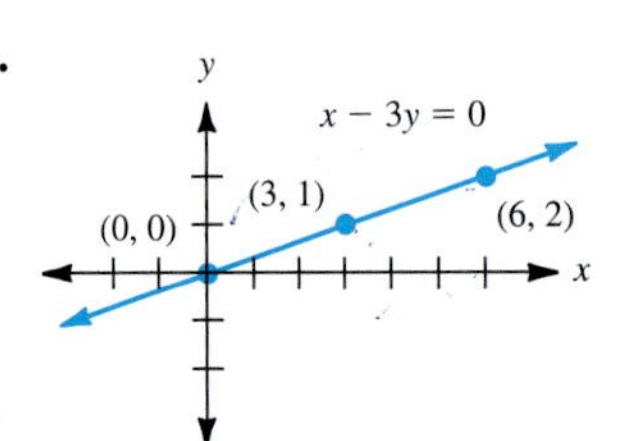

5. $D = \{x \mid x \in R\}$ and $R = \{y \mid y \in R\}$

6. **(a)**

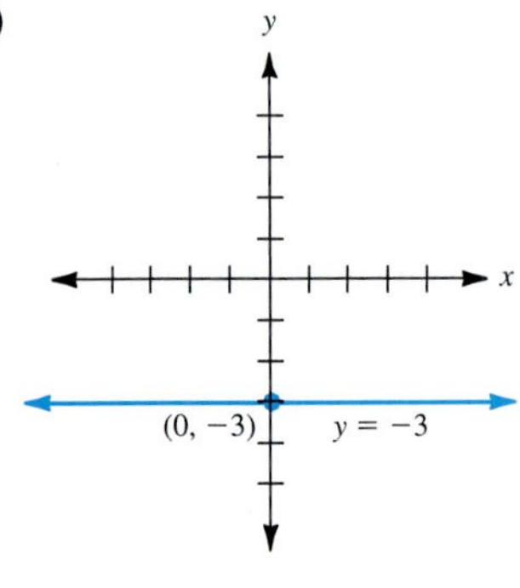

$D = \{x \mid x \in R\}$
$R = \{-3\}$

(b)

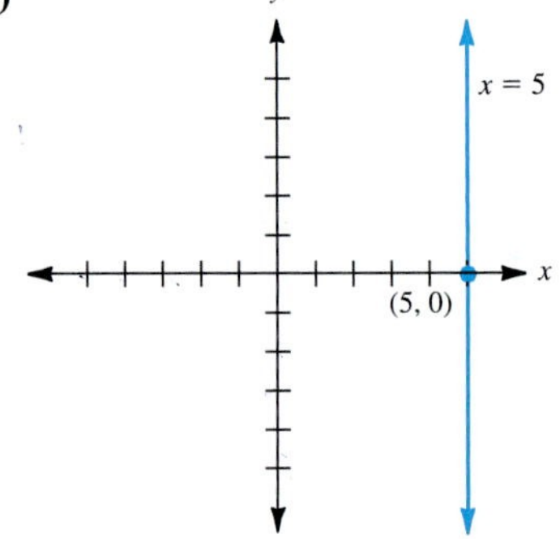

$D = \{5\}$
$R = \{y \mid y \in R\}$

Name ____________

Section ________ Date ________

4.1 Exercises

In exercises 1 to 8, find three ordered pairs that are solutions to the given equations.

1. $2x + y = 5$ **2.** $3x + y = 7$ **3.** $7x - y = 8$ **4.** $5x - y = 3$

5. $4x + 5y = 20$ **6.** $2x + 3y = 6$ **7.** $3x + y = 0$ **8.** $2x - y = 0$

In exercises 9 to 26, graph each of the equations.

9. $x + y = 6$ **10.** $x - y = 6$ **11.** $y = x - 2$

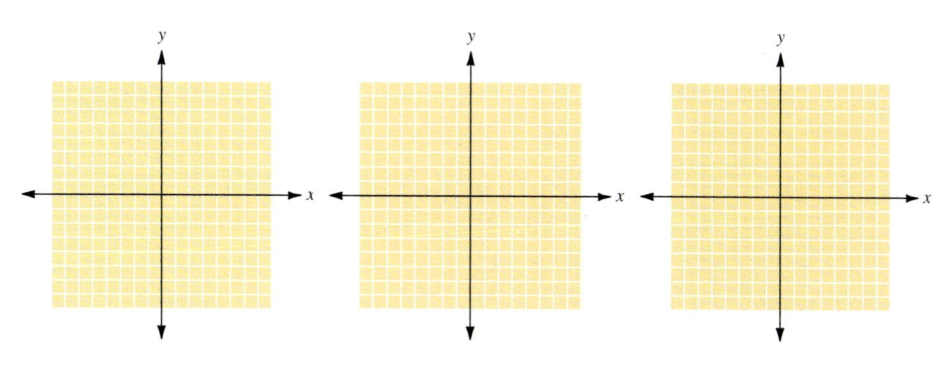

12. $y = x + 5$ **13.** $y = x + 1$ **14.** $y = 2x + 2$

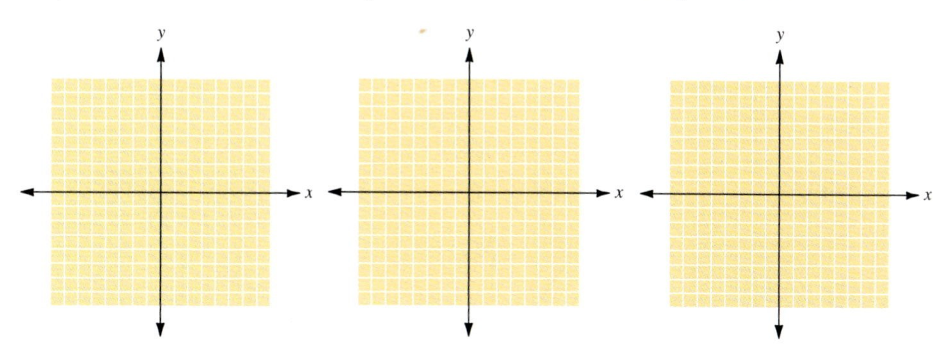

15. $y = -2x + 1$ **16.** $y = -3x + 1$ **17.** $y = \frac{1}{2}x - 3$

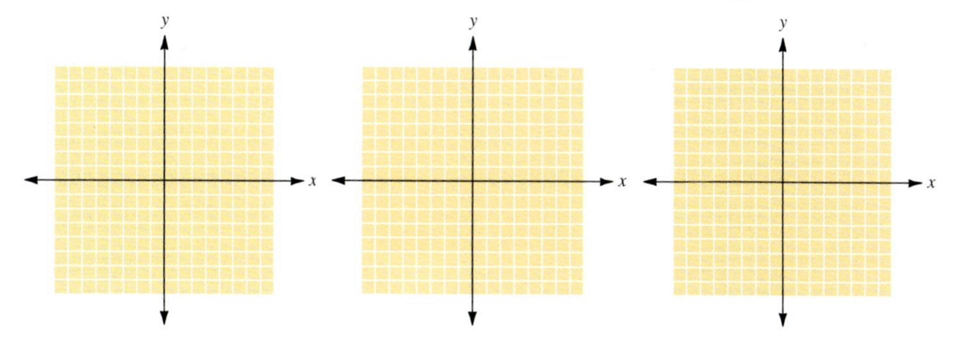

ANSWERS

1. ____________
2. ____________
3. ____________
4. ____________
5. ____________
6. ____________
7. ____________
8. ____________
9. ____________
10. ____________
11. ____________
12. ____________
13. ____________
14. ____________
15. ____________
16. ____________
17. ____________

ANSWERS

18. ______

19. ______

20. ______

21. ______

22. ______

23. ______

24. ______

25. ______

26. ______

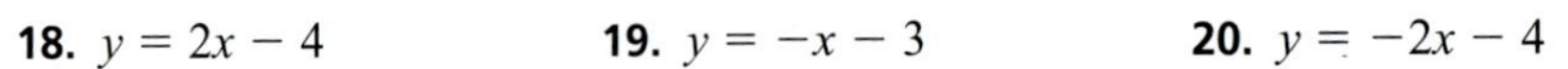

18. $y = 2x - 4$

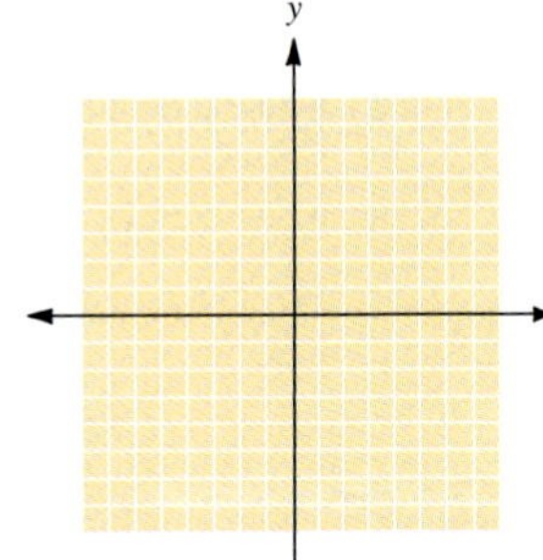

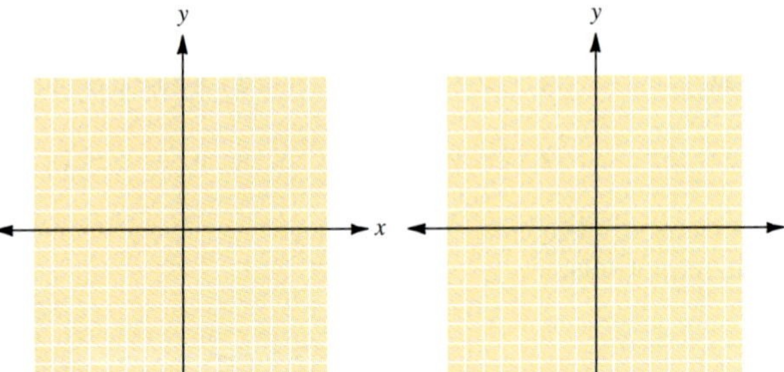

19. $y = -x - 3$

20. $y = -2x - 4$

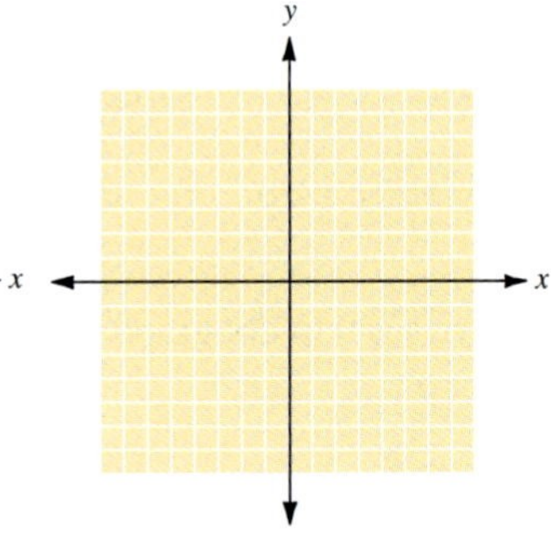

21. $x + 2y = 0$

22. $x - 2y = 0$

23. $x = 4$

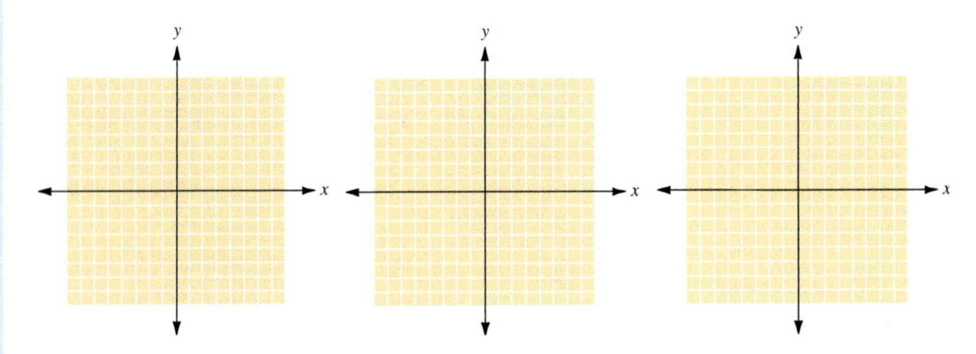

24. $x = -4$

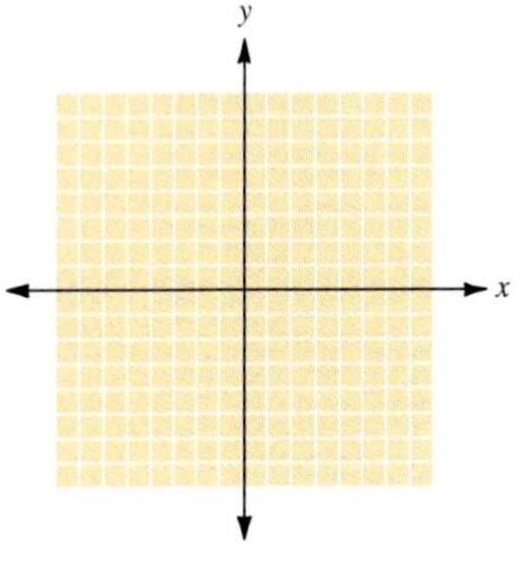

25. $y = 4$

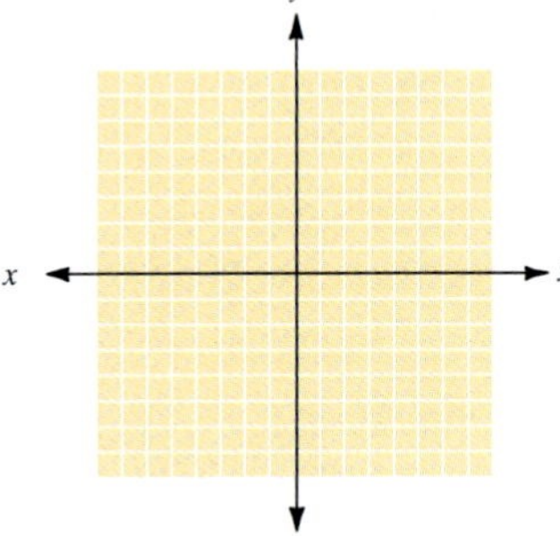

26. $y = -6$

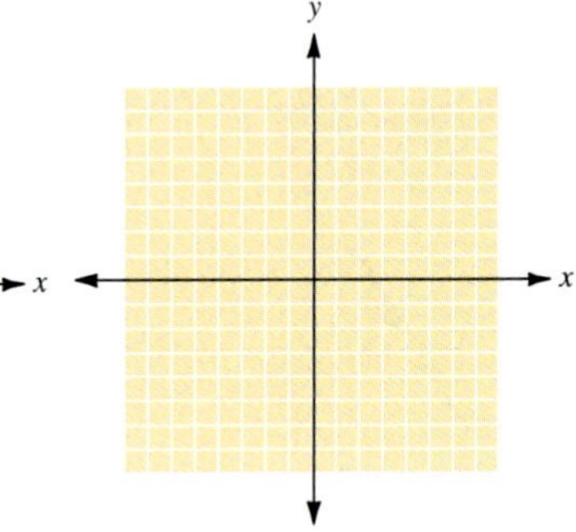

In exercises 27 to 38, find the x and y intercepts and then graph each equation.

27. $x - 2y = 4$ **28.** $x + 3y = 6$ **29.** $2x - y = 6$

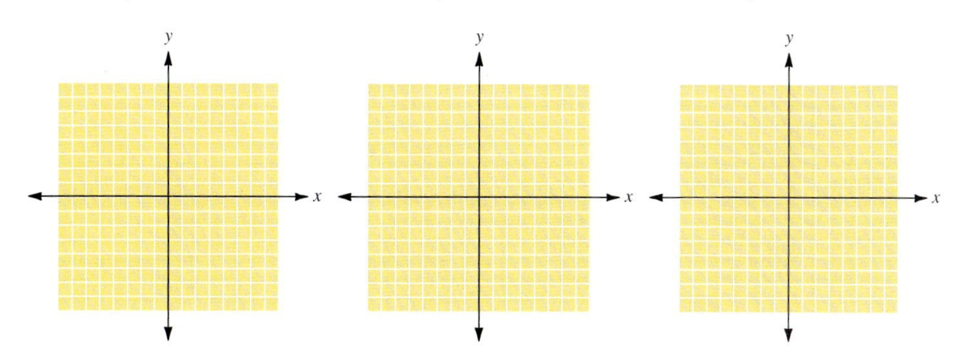

30. $3x + 2y = 12$ **31.** $2x + 5y = 10$ **32.** $2x - 3y = 6$

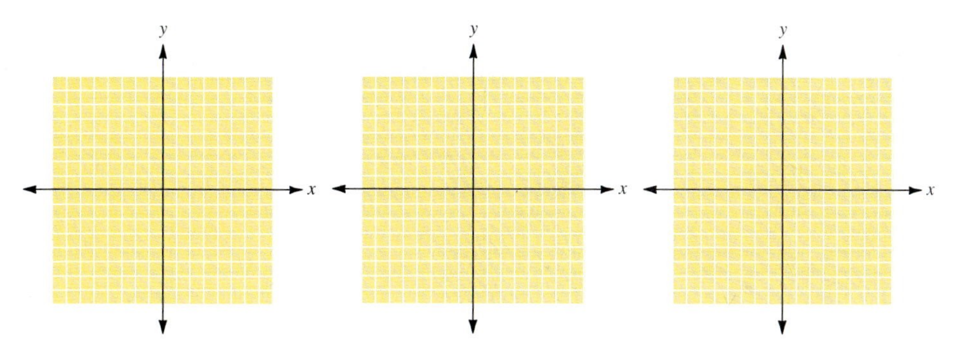

33. $5x - 6y = 0$ **34.** $2x + 7y = 0$ **35.** $x + 4y + 8 = 0$

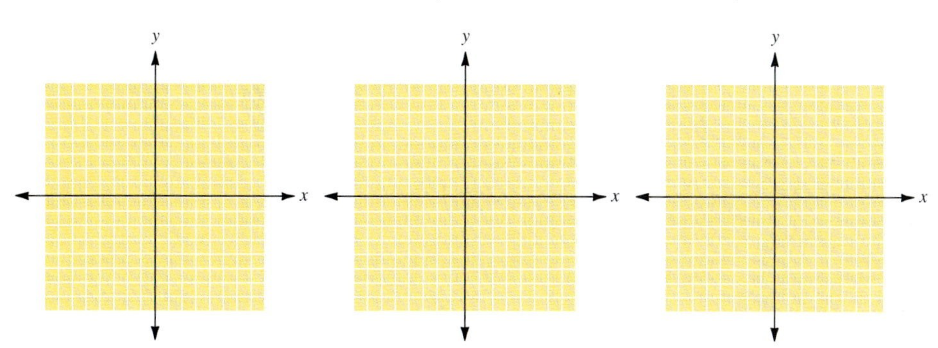

ANSWERS

27. ______

28. ______

29. ______

30. ______

31. ______

32. ______

33. ______

34. ______

35. ______

ANSWERS

36. ______

37. ______

38. ______

39. ______

40. ______

41. ______

42. ______

43. ______

44. ______

45. ______

46. ______

47. ______

48. ______

49. ______

50. ______

51. ______

52. ______

53. ______

54. ______

36. $2x - y + 6 = 0$ **37.** $8x = 4y$ **38.** $6x = -7y$

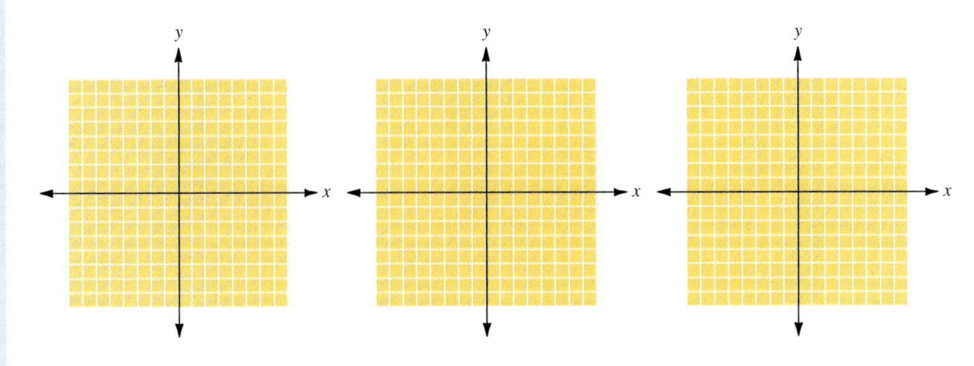

In exercises 39 to 46, find the domain and range of each of the relations.

39. $3x + 2y = 4$

40. $5x - 4y = 20$

41. $6x + 2y = 18$

42. $-x + 5y = 8$

43. $x = 4$

44. $2x - 10 = 0$

45. $y = 3$

46. $3y + 12 = 0$

For exercises 47 to 54, select a window that allows you to see both the x and y intercepts on your calculator. If that is not possible, explain why not.

47. $x + y = 40$

48. $x - y = 80$

49. $2x + 3y = 900$

50. $5x - 8y = 800$

51. $y = 5x + 90$

52. $y = 3x - 450$

53. $y = 30x$

54. $y = 200$

Two distinct lines in the plane either are parallel or they intersect. In exercises 55 to 58, graph each pair of equations on the same set of axes, and find the point of intersection, where possible.

55. $x + y = 6, x - y = 4$

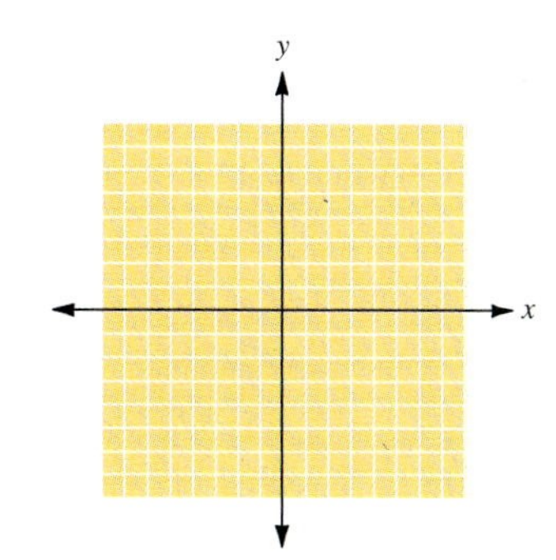

56. $y = x + 3, y = -x + 1$

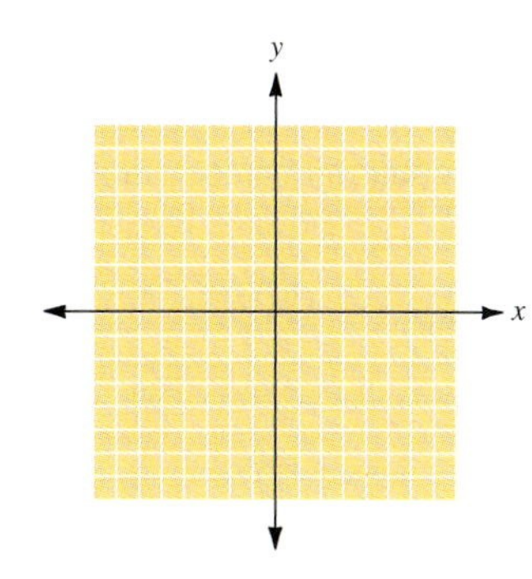

57. $y = 2x, y = x + 1$

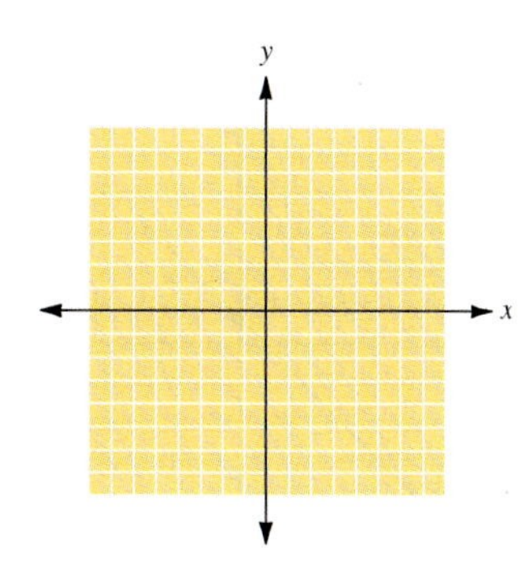

58. $2x + y = 3, 2x + y = 5$

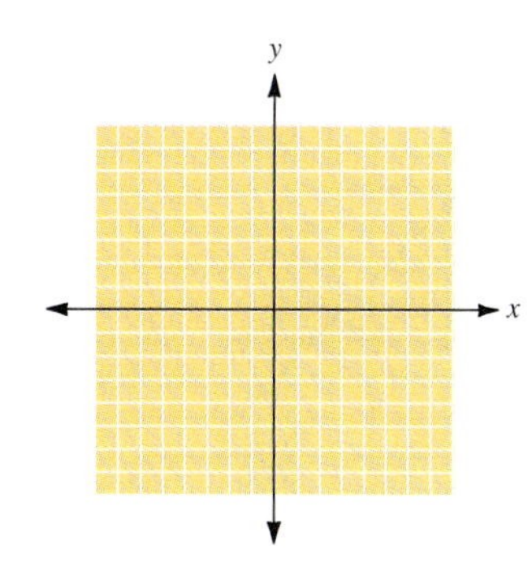

59. Graph $y = x$ and $y = 2x$ on the same set of axes. What do you observe?

60. Graph $y = 2x + 1$ and $y = -2x + 1$ on the same set of axes. What do you observe?

61. Graph $y = 2x$ and $y = 2x + 1$ on the same set of axes. What do you observe?

62. Graph $y = 3x + 1$ and $y = 3x - 1$ on the same set of axes. What do you observe?

63. Graph $y = 2x$ and $y = -\frac{1}{2}x$ on the same set of axes. What do you observe?

64. Graph $y = \frac{1}{3}x + \frac{7}{3}$ and $y = -3x + 2$ on the same set of axes. What do you observe?

ANSWERS

55. ______

56. ______

57. ______

58. ______

59. ______

60. ______

61. ______

62. ______

63. ______

64. ______

ANSWERS

65. ____________

66. ____________

67. ____________

68. ____________

69. ____________

70. ____________

Use your graphing utility to graph each of the following equations.

65. $y = -3$

66. $y = 2$

67. $y = 3x - 1$

68. $y = -2x + 2$

69. Write an equation whose graph will have no x intercept but will have a y intercept at (0, 6).

70. Write an equation whose graph will have no y intercept but will have an x intercept at $(-5, 0)$.

Answers

1. (0, 5), (1, 3), (−1, 7)

3. (0, −8), (1, −1), (−1, −15)

5. (0, 4), (5, 0), $\left(-1, \frac{24}{5}\right)$

7. (0, 0), (1, −3), (−1, 3)

9. $x + y = 6$

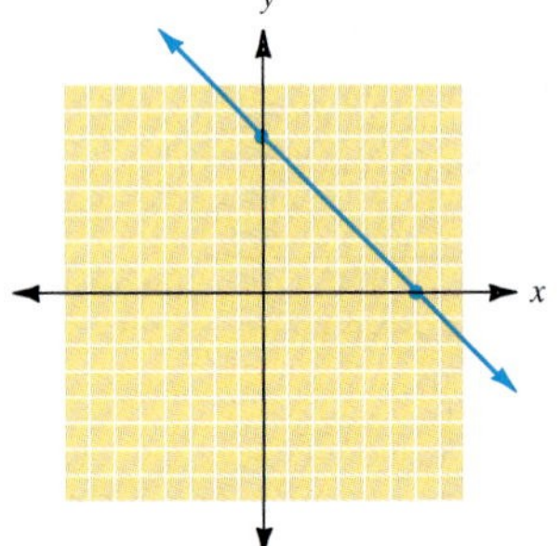

11. $y = x - 2$

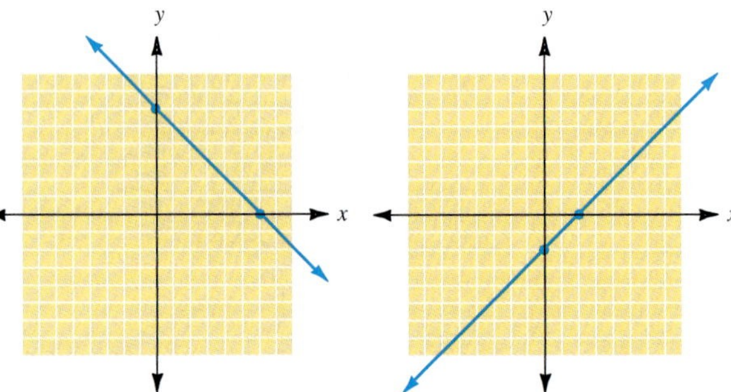

13. $y = x + 1$

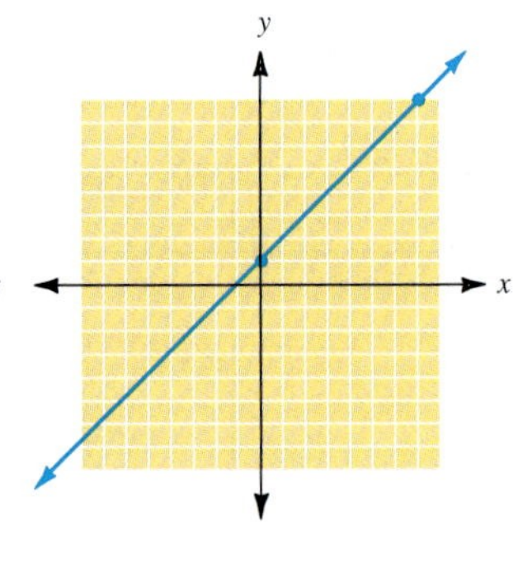

15. $y = -2x + 1$

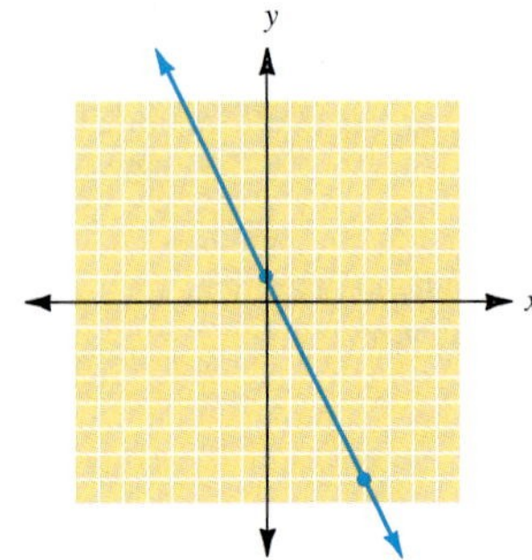

17. $y = \frac{1}{2}x - 3$

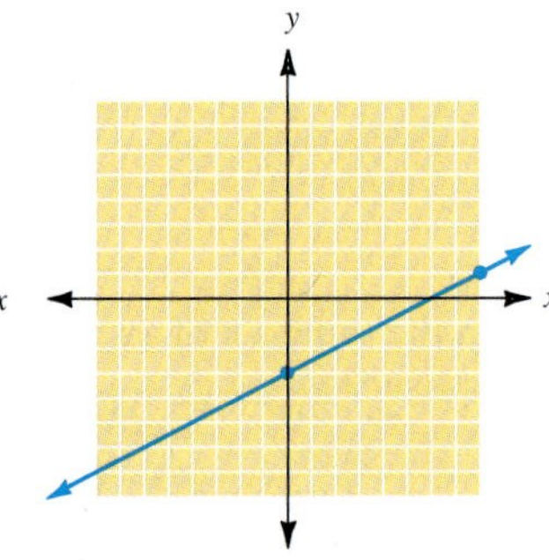

19. $y = -x - 3$

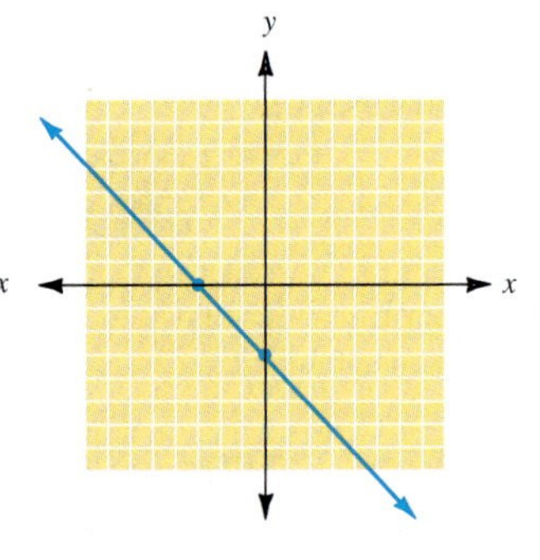

21. $x + 2y = 0$

23. $x = 4$

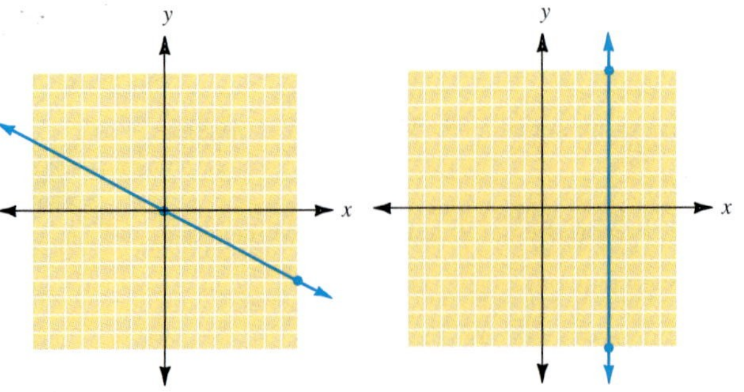

25. $y = 4$

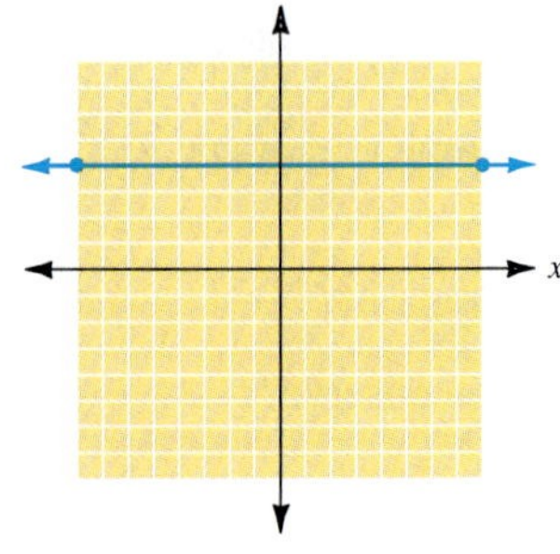

27. $x - 2y = 4$; y intercept $(0, -2)$; x intercept $(4, 0)$

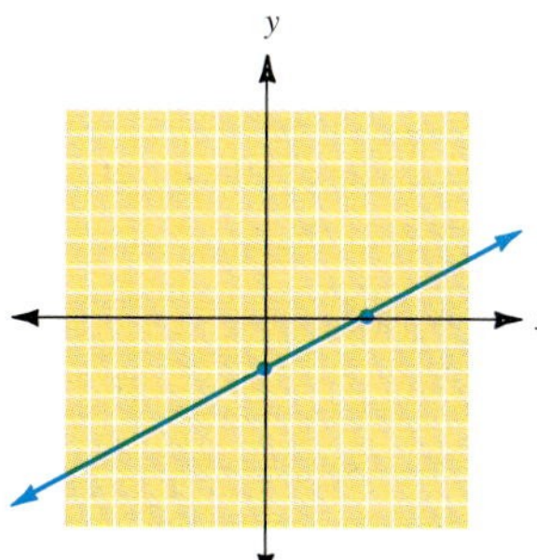

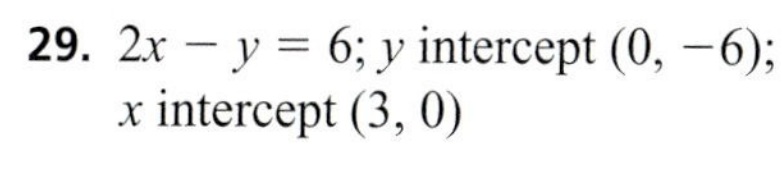
29. $2x - y = 6$; y intercept $(0, -6)$; x intercept $(3, 0)$

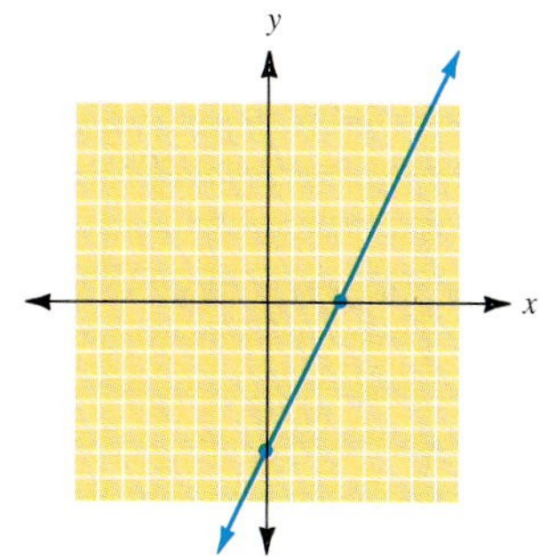

31. $2x + 5y = 10$; y intercept $(0, 2)$; x intercept $(5, 0)$

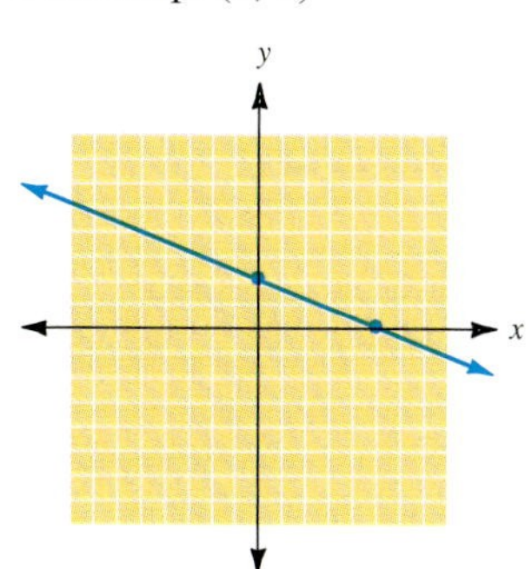

33. $5x - 6y = 0$; intercepts: $(0, 0)$

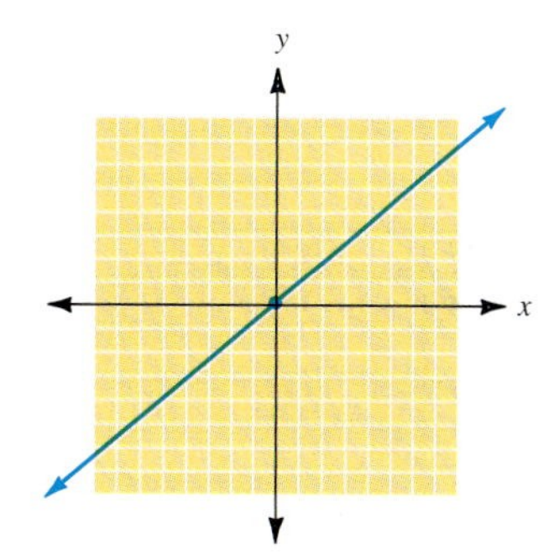

35. $x + 4y + 8 = 0$; y intercept $(0, -2)$; x intercept $(-8, 0)$

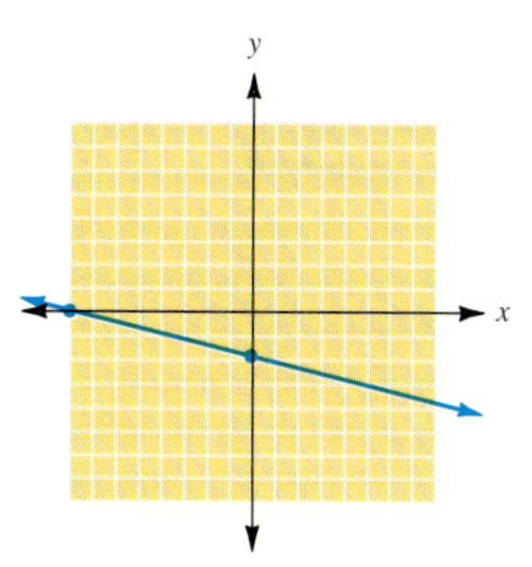

37. $8x = 4y$; intercepts: $(0, 0)$

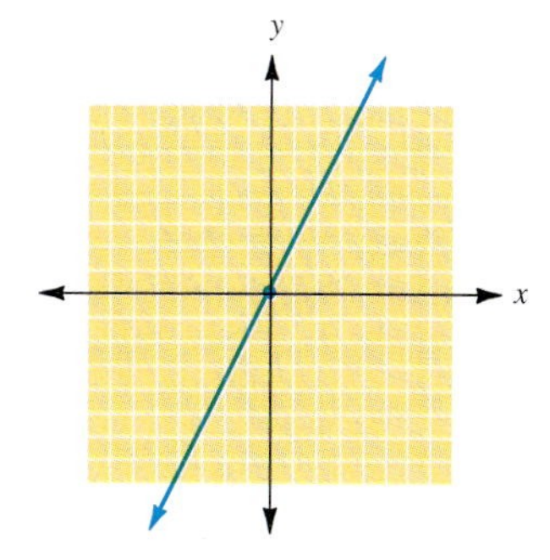

39. D: $\{x \mid x \in R\}$; R: $\{y \mid y \in R\}$
41. D: $\{x \mid x \in R\}$; R: $\{y \mid y \in R\}$
43. D: $\{4\}$; R: $\{y \mid y \in R\}$
45. D: $\{x \mid x \in R\}$; R: $\{3\}$
47. X max > 40, Y max > 40
49. X max > 450, Y max > 300
51. X min < -18, Y max > 90
53. Any viewing window that shows the origin

55. Intersection: (5, 1)

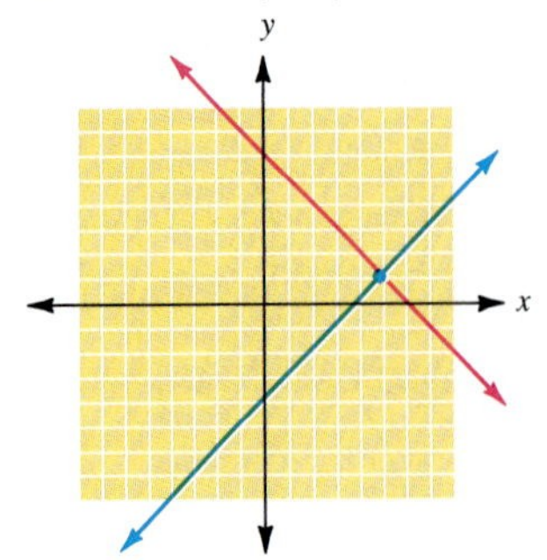

57. Intersection: (1, 2)

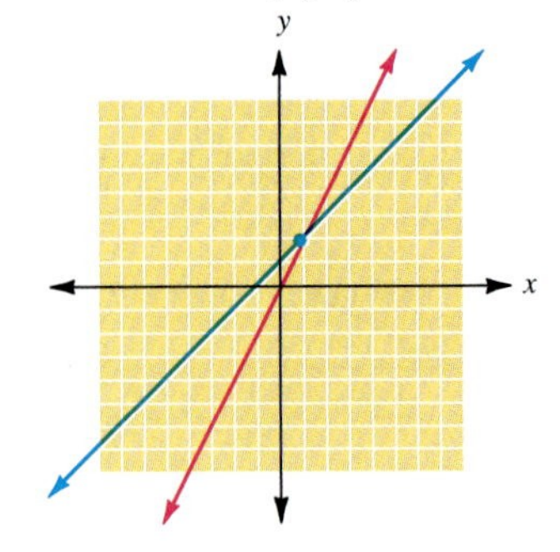

59. The line corresponding to $y = 2x$ is steeper than that corresponding to $y = x$.

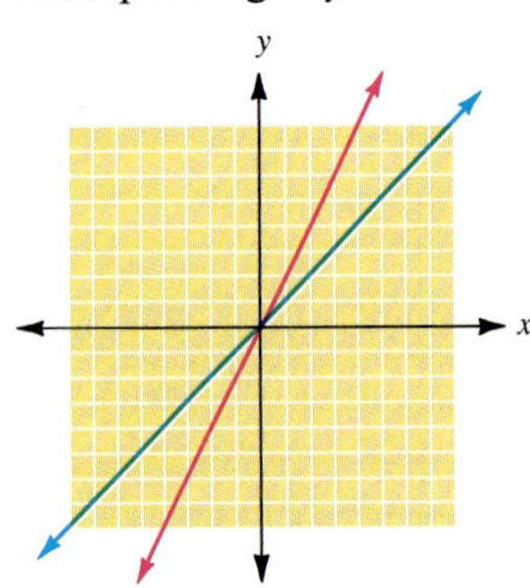

61. The two lines appear to be parallel.

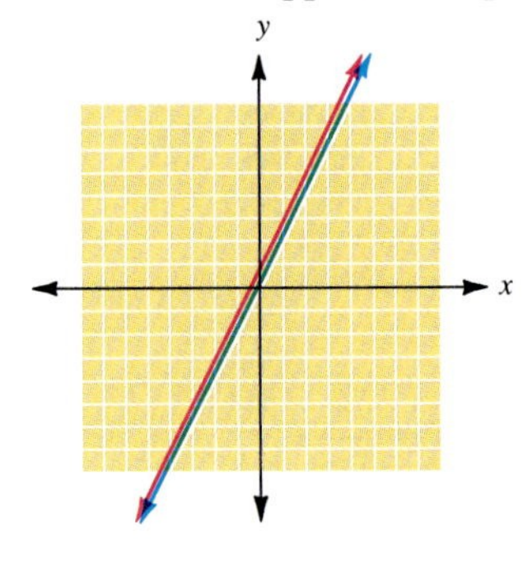

63. The lines appear to be perpendicular.

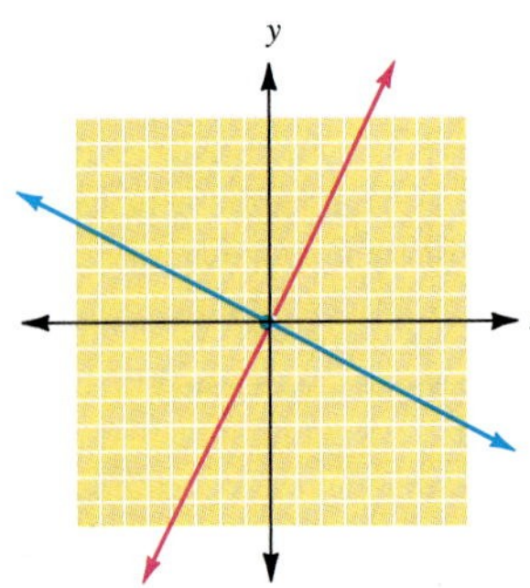

65.

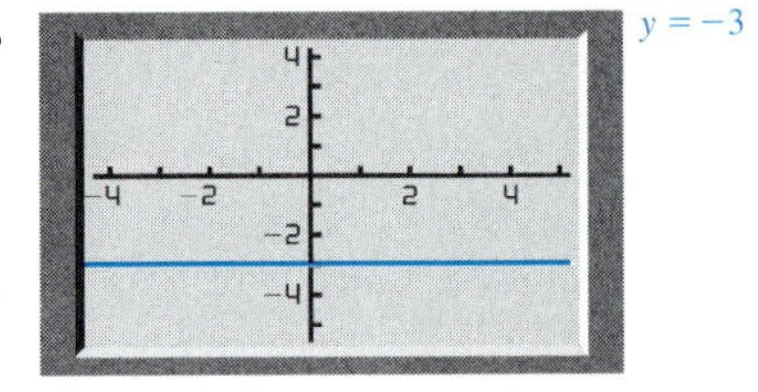

67.

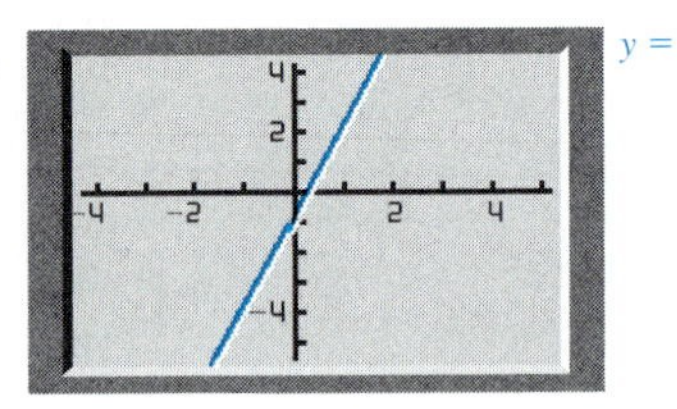

69. $y = 6$

The Slope of a Line

OBJECTIVES

1. Find the slope of a line
2. Find the slopes of parallel and perpendicular lines
3. Find the slope of a line given an equation
4. Find the slope given a graph
5. Graph a linear equation using the slope and a point

On the coordinate system below, plot a point, any point.

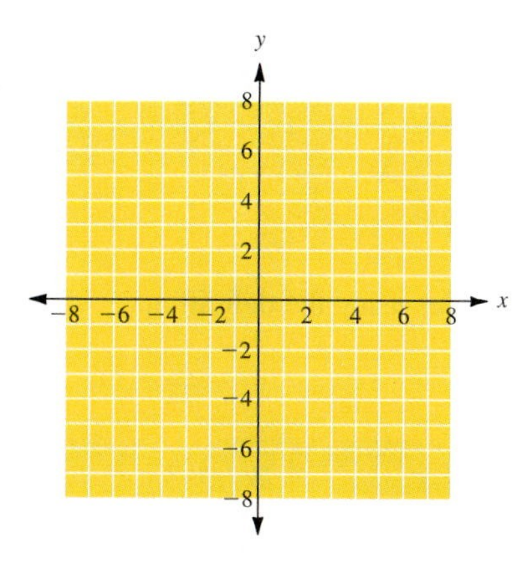

How many different lines can you draw through that point? Hundreds? Thousands? Millions? Actually, there is no limit to the number of different lines that pass through that point.

Now, on the coordinate system above, plot a second point.

How many different lines can you draw through those points? Only one! The two points were enough to define the line.

In Section 4.3, we will see how we can find the equation of a line if we are given two of its points. The first part of finding that equation is finding the **slope** of the line, which is a way of describing the steepness of a line.

The Definitions box contains a formula for slope. First, we choose any two distinct points on the line, say, P with coordinates (x_1, y_1) and Q with coordinates (x_2, y_2). As we move along the line from P to Q, the x value, or coordinate, changes from x_1 to x_2. That change in x, also called the **horizontal change,** is $x_2 - x_1$. Similarly, as we move from P to Q, the corresponding change in y, called the **vertical change,** is $y_2 - y_1$. The *slope* is then defined as the ratio of the vertical change to the horizontal change. The letter m is used to represent the slope, which we now define.

NOTE The difference, $x_2 - x_1$, is often called the **run.** The difference, $y_2 - y_1$, is the **rise.** So the slope can be thought of as "rise over run."

Note that $x_1 \neq x_2$ or $x_2 - x_1 \neq 0$ ensures that the denominator is nonzero, so that the slope is defined. It also means the line cannot be vertical.

Definitions: Slope of a Line

The *slope* of a line through two distinct points $P(x_1, y_1)$ and $Q(x_2, y_2)$ is given by

$$m = \frac{\text{change in } y}{\text{change in } x} = \frac{y_2 - y_1}{x_2 - x_1}$$

when $x_1 \neq x_2$.

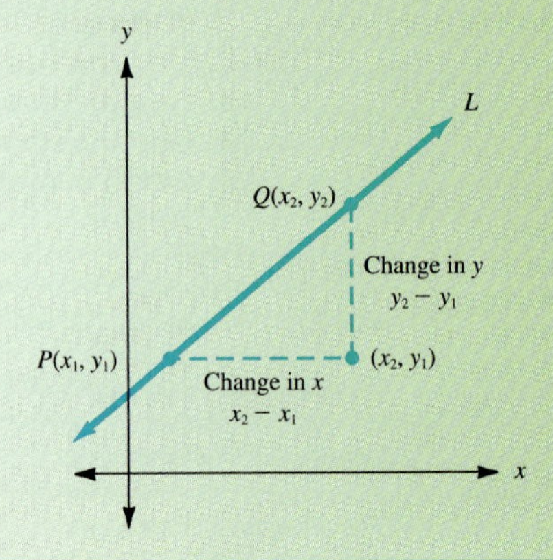

Let's look at some examples using the definition.

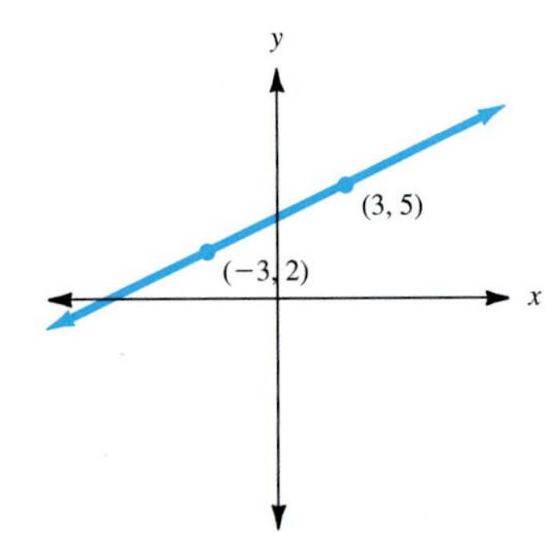

Example 1

Finding the Slope Through Two Points

Find the slope of the line through the points $(-3, 2)$ and $(3, 5)$.

Let $(x_1, y_1) = (-3, 2)$ and $(x_2, y_2) = (3, 5)$. From the definition we have

$$m = \frac{5 - 2}{3 - (-3)} = \frac{3}{6} = \frac{1}{2}$$

Note that if the pairs are reversed, so that

$(x_1, y_1) = (3, 5)$ and $(x_2, y_2) = (-3, 2)$

then we have

$$m = \frac{2 - 5}{-3 - 3} = \frac{-3}{-6} = \frac{1}{2}$$

The slope in either case is the same.

NOTE The work here suggests that no matter which point is chosen as (x_1, y_1) or (x_2, y_2), the slope formula will give the same result. Simply stay with your choice once it is made, and use the same order of subtraction in the numerator and the denominator.

CHECK YOURSELF 1

Find the slope of the line through the points $(-2, -1)$ *and* $(1, 1)$.

The slope indicates both the direction of a line and its steepness. First, we will compare the steepness of the lines in two examples.

Example 2

Finding the Slope

Find the slope of the line through $(-2, -3)$ and $(2, 5)$.

Again, by equation (1),

$$m = \frac{5 - (-3)}{2 - (-2)} = \frac{8}{4} = 2$$

Compare the lines in Examples 1 and 2. In Example 1 the line has slope $\frac{1}{2}$. The slope here is 2. Now look at the two lines. Do you see the idea of slope as measuring steepness? The greater the absolute value of the slope, the steeper the line.

CHECK YOURSELF 2

Find the slope of the line through the points $(-1, 2)$ and $(2, 7)$. Draw a sketch of this line and the line in the Check Yourself 1 exercise on the same coordinate axes. Compare the lines and the two slopes.

The sign of the slope indicates in which direction the line tilts, as Example 3 illustrates.

Example 3

Finding the Slope

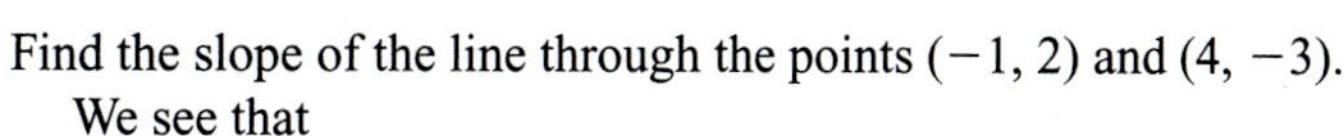

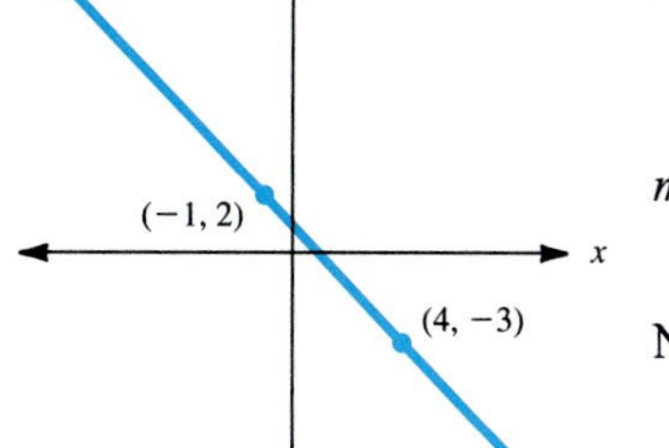

Find the slope of the line through the points $(-1, 2)$ and $(4, -3)$.

We see that

$$m = \frac{-3 - 2}{4 - (-1)} = \frac{-5}{5} = -1$$

Now the slope is negative.

Comparing this with our previous examples, we see that

1. In Examples 1 and 2, the lines were rising from left to right, and the slope was **positive.**

2. In this example, the line is falling from left to right, and the slope is **negative.**

CHECK YOURSELF 3

Find the slope of the line through the points (−2, 5) *and* (4, −1).

Let's continue by looking at the slopes of lines in two particular cases.

Example 4

Finding the Slope of a Horizontal Line

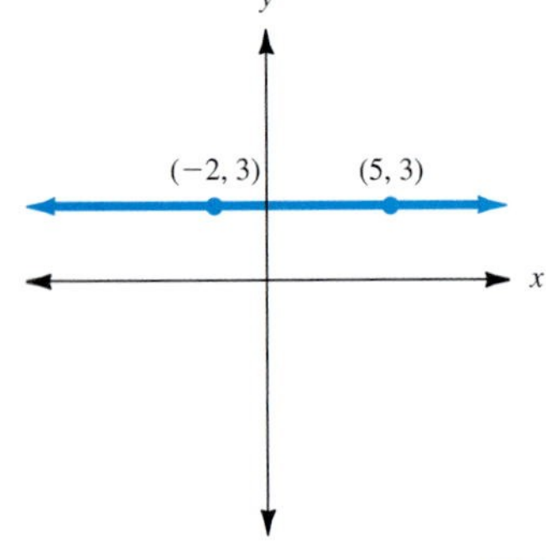

Find the slope of the line through (−2, 3) and (5, 3).

$$m = \frac{3 - 3}{5 - (-2)} = \frac{0}{7} = 0$$

The slope of the line is 0. Note that the line is parallel to the x axis and $y_2 - y_1 = 0$. *The slope of any horizontal line will be* 0.

CHECK YOURSELF 4

Find the slope of the line through the points (−2, −4) *and* (3, −4).

Example 5

Finding the Slope of a Vertical Line

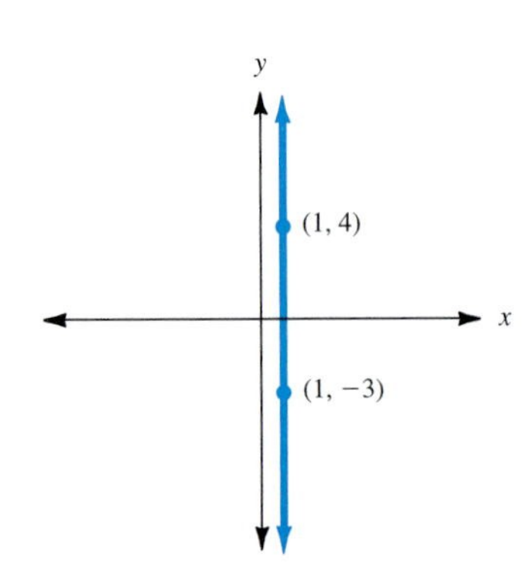

Find the slope of the line through the points (1, −3) and (1, 4).

$$m = \frac{4 - (-3)}{1 - 1} = \frac{7}{0}, \text{ which is undefined.}$$

Here the line is parallel to the y axis, and $x_2 - x_1$ (the denominator of the slope formula) is 0. Because division by 0 is undefined, we say that the slope is **undefined,** as will be the case for *any vertical line.*

Be very careful not to confuse a slope of 0 (in the case of a horizontal line) with an undefined slope (in the case of a vertical line).

CHECK YOURSELF 5

Find the slope of the line through the points (2, −3) *and* (2, 7).

Here is a summary of our work in the previous examples.

1. If the slope of a line is *positive,* the line is rising from left to right.

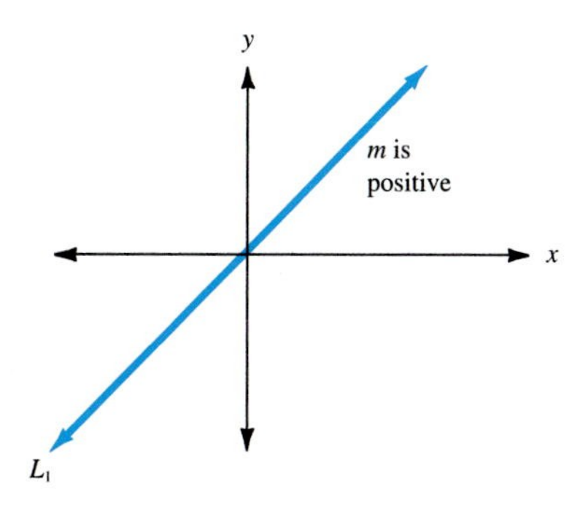

2. If the slope of a line is *negative,* the line is falling from left to right.

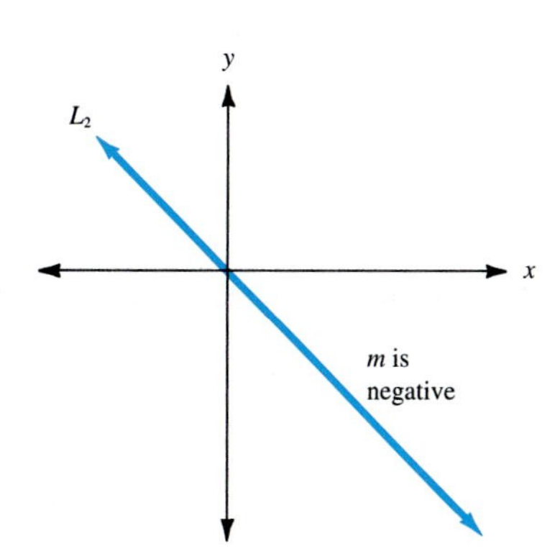

3. If the slope of a line is 0, the line is *horizontal.*

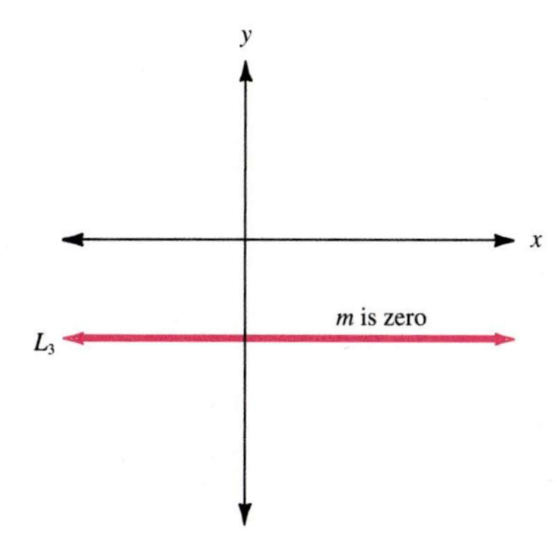

4. If the slope of a line is undefined, the line is *vertical.*

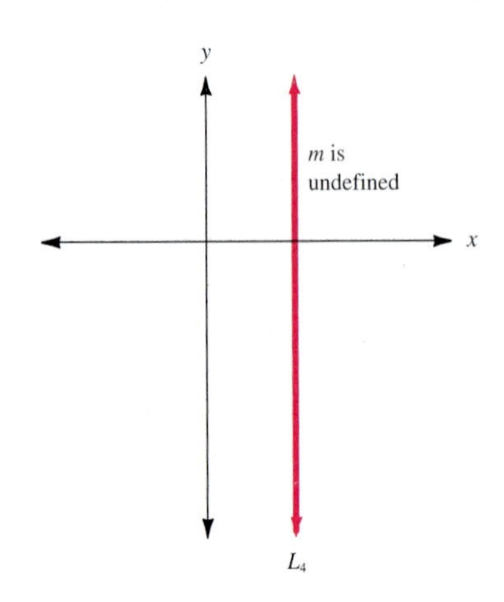

There are two more important results regarding the slope. Recall from geometry that two distinct lines in the plane either intersect at a point or never intersect. Two lines in the plane that do not intersect are called **parallel lines.** It can be shown that two distinct parallel lines will always have the same slope, and we can state the following.

NOTE This means that if the lines are parallel, then their slopes are equal. Conversely, if the slopes are equal, then the lines are parallel.

Mathematicians use the symbol ⇔ to represent "if and only if."

Definitions: Slopes of Parallel Lines

For nonvertical lines L_1 and L_2, if line L_1 has slope m_1 and line L_2 has slope m_2, then

L_1 is parallel to L_2 if and only if $m_1 = m_2$

Note: All vertical lines are parallel to each other.

Example 6

Parallel Lines

Are lines L_1 through (2, 3) and (4, 6) and L_2 through (−4, 2) and (0, 8) parallel, or do they intersect?

$$m_1 = \frac{6 - 3}{4 - 2} = \frac{3}{2}$$

$$m_2 = \frac{8 - 2}{0 - (-4)} = \frac{6}{4} = \frac{3}{2}$$

NOTE Unless, of course, L_1 and L_2 are actually the *same line.* In this case, a quick sketch will show that the lines are distinct.

Because the slopes of the lines are equal, the lines are parallel. They do *not* intersect.

CHECK YOURSELF 6

Are lines L_1 through (−2, −1) and (1, 4) and L_2 through (−3, 4) and (0, 8) parallel, or do they intersect?

Two lines are perpendicular if they intersect at right angles. Also, if two lines (which are not vertical or horizontal) are perpendicular, their slopes are the negative reciprocals of each other. We can then state the following result for perpendicular lines.

Definitions: Slopes of Perpendicular Lines

For nonvertical lines L_1 and L_2, if line L_1 has slope m_1 and line L_2 has slope m_2, then

L_1 is perpendicular to L_2 if and only if $m_1 = -\frac{1}{m_2}$

or, equivalently,

$m_1 \cdot m_2 = -1$

Note: Horizontal lines are perpendicular to vertical lines.

Example 7

Perpendicular Lines

Are lines L_1 through points $(-2, 3)$ and $(1, 7)$ and L_2 through points $(2, 4)$ and $(6, 1)$ perpendicular?

NOTE

$\left(\frac{4}{3}\right)\left(-\frac{3}{4}\right) = -1$

$$m_1 = \frac{7 - 3}{1 - (-2)} = \frac{4}{3}$$

$$m_2 = \frac{1 - 4}{6 - 2} = -\frac{3}{4}$$

Because the slopes are negative reciprocals of each other, the lines are perpendicular.

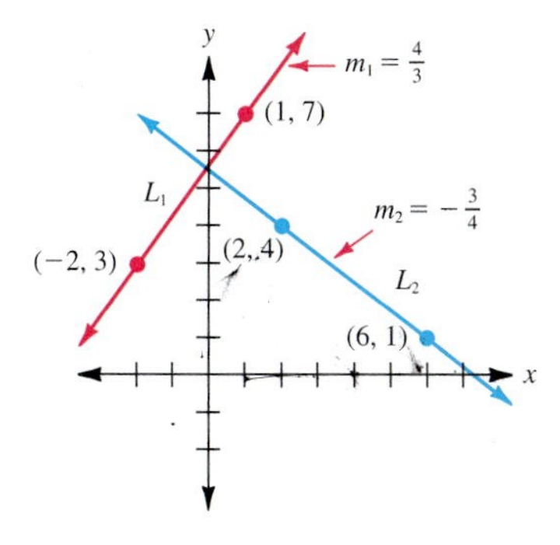

CHECK YOURSELF 7

Are lines L_1 through points (1, 3) and (4, 1) and L_2 through points (−2, 4) and (2, 10) perpendicular?

Given the equation of a line, we can also find its slope, as Example 8 illustrates.

NOTE Let's try solving the original equation for y:

$$3x + 2y = 6$$
$$2y = -3x + 6$$
$$y = -\frac{3}{2}x + 3$$

Consider the coefficient of x. What do you observe?

Example 8

Finding the Slope from an Equation

Find the slope of the line with equation $3x + 2y = 6$.

First, find any two points on the line. In this case, (2, 0) and (0, 3), the x and y intercepts, will work and are easy to find. From the slope formula,

$$m = \frac{0 - 3}{2 - 0} = \frac{-3}{2} = -\frac{3}{2}$$

The slope of the line with equation $3x + 2y = 6$ is $-\frac{3}{2}$.

CHECK YOURSELF 8

Find the slope of the line with equation $3x - 4y = 12$.

We can find the slope of a graphed line by identifying two points on the line. We will use that technique in Example 9.

Example 9

Finding the Slope from a Graph

Determine the slope of the line from its graph.

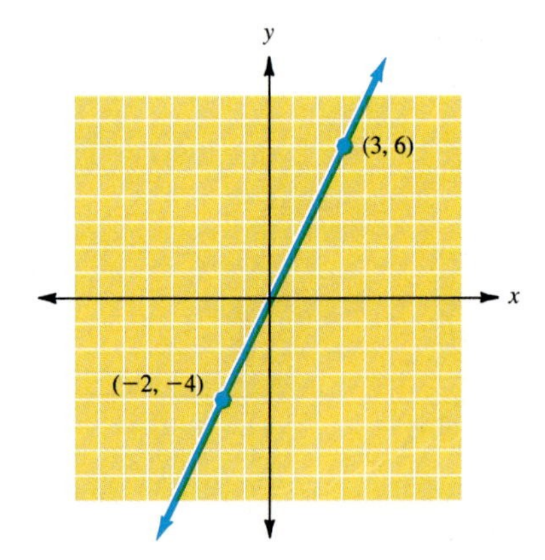

We can choose any two points to determine the slope of the line. Here we pick (3, 6) and (−2, −4). We find the slope by the usual method.

$$m = \frac{6 - (-4)}{3 - (-2)} = \frac{10}{5} = 2$$

The slope of the line is 2.

CHECK YOURSELF 9

Determine the slope of the line from its graph.

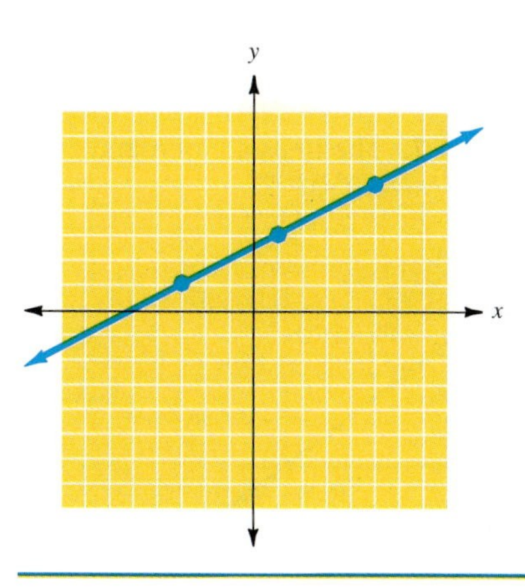

The slope of a line can also be useful in graphing a line. In Example 10, the slope of a line is used in sketching its graph.

Example 10

Graphing a Line with a Given Slope

Suppose a line has slope $\frac{3}{2}$ and passes through the point (5, 2). Graph the line.

First, locate the point (5, 2) in the coordinate system. Now, because the slope, $\frac{3}{2}$, is the ratio of the change in y to the change in x, move 2 units to the right in the x direction and then 3 units up in the y direction. This determines a second point, here (7, 5), and we can draw our graph.

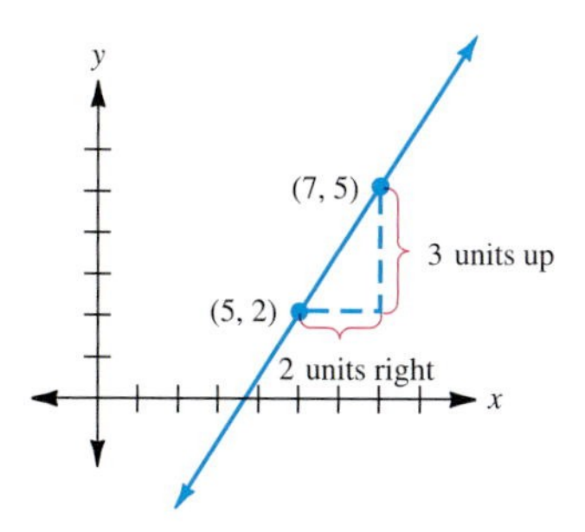

CHECK YOURSELF 10

Graph the line with slope $-\frac{3}{4}$ *that passes through the point* (2, 3). *Hint: Consider the x change as* 4 *units and the y change as* −3 *units (down).*

Because, given a point on a line and its slope, we can graph the line, we also should be able to write its equation. That is, in fact, the case, as we will see in Section 4.3.

CHECK YOURSELF ANSWERS

1. $m = \frac{2}{3}$ **2.** $m = \frac{5}{3}$. This line is steeper than the line of check yourself 1.

3. $m = -1$ **4.** $m = 0$ **5.** Undefined

6. The lines intersect. **7.** The lines are perpendicular. **8.** $m = \frac{3}{4}$

9. $m = \frac{1}{2}$ **10.**

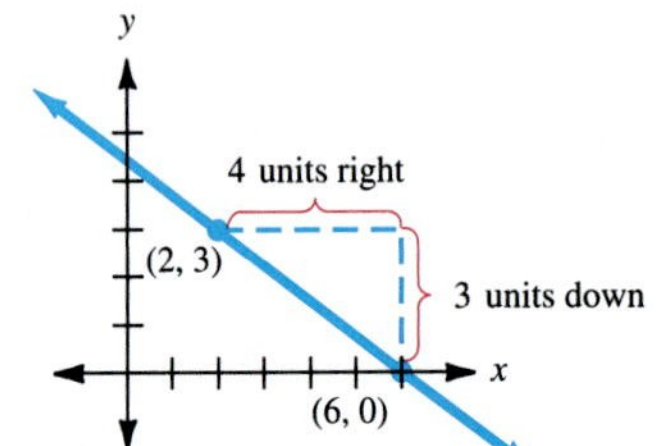

Name ______________________

Section ________ Date ________

4.2 Exercises

In exercises 1 to 12, find the slope (if it exists) of the line determined by the following pairs of points. Sketch each line so that you can compare the slopes.

1. $(2, 3)$ and $(4, 7)$ **2.** $(-1, 2)$ and $(5, 3)$ **3.** $(2, -3)$ and $(-2, -5)$

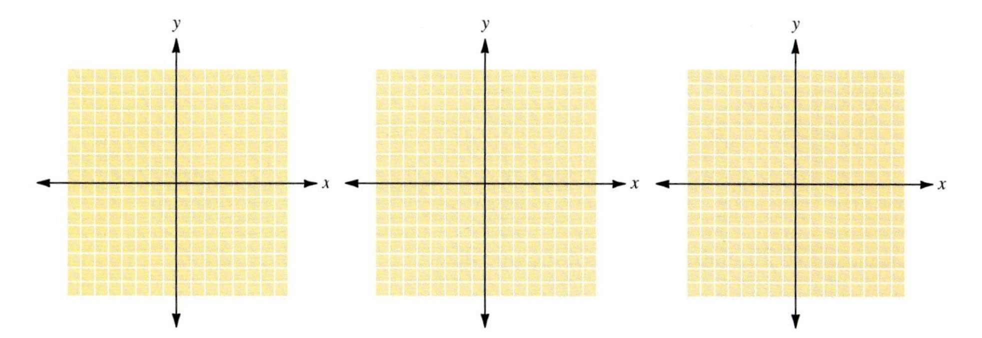

4. $(0, 0)$ and $(5, 7)$ **5.** $(2, 5)$ and $(-3, 5)$ **6.** $(-2, -4)$ and $(5, 3)$

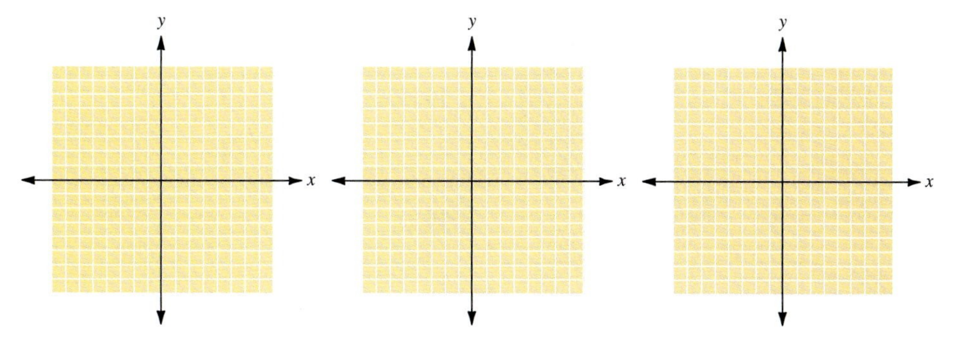

7. $(-1, 4)$ and $(-1, 7)$ **8.** $(4, 2)$ and $(-2, 5)$ **9.** $(8, -3)$ and $(-2, -5)$

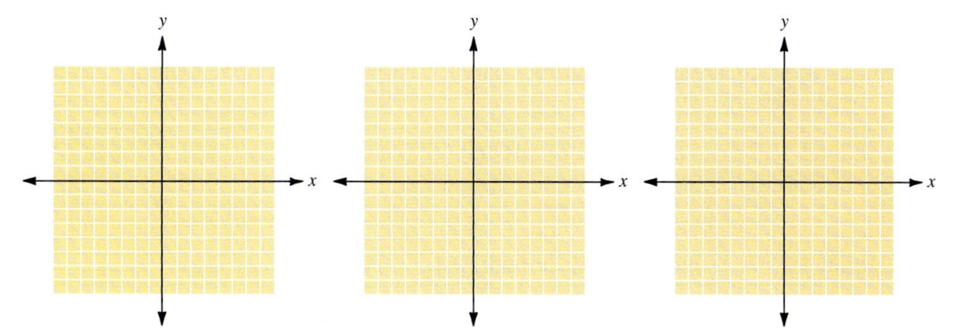

ANSWERS

1. ______________________
2. ______________________
3. ______________________
4. ______________________
5. ______________________
6. ______________________
7. ______________________
8. ______________________
9. ______________________

ANSWERS

10. ____________
11. ____________
12. ____________
13. ____________
14. ____________
15. ____________
16. ____________
17. ____________
18. ____________
19. ____________
20. ____________
21. ____________
22. ____________
23. ____________
24. ____________
25. ____________
26. ____________

10. (4, −3) and (−2, 7) **11.** (−4, −3) and (2, −7) **12.** (3, 6) and (3, −4)

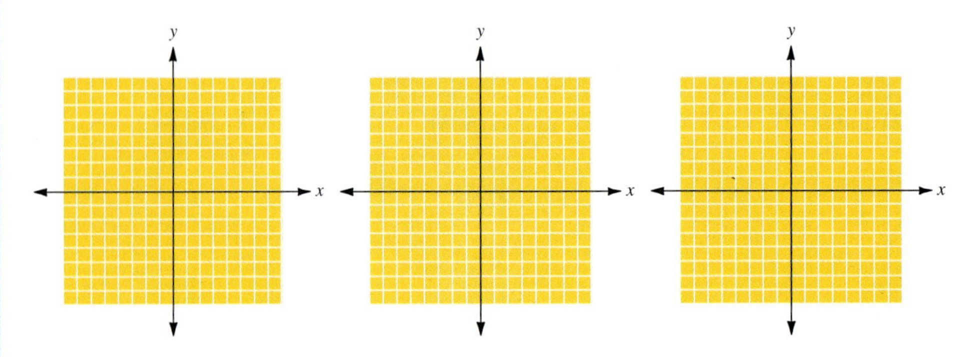

In exercises 13 to 20, find the slope of the line determined by each equation.

13. $y = -3x - \frac{1}{2}$

14. $y = \frac{1}{4}x + 3$

15. $y + \frac{1}{2}x = 2$

16. $2y - 3x + 5 = 0$

17. $2x - 3y = 6$

18. $x + 4y = 4$

19. $3x + 4y = 12$

20. $x - 3y = 9$

In exercises 21 to 26, are the pairs of lines parallel, perpendicular, or neither?

21. L_1 through (−2, −3) and (4, 3); L_2 through (3, 5) and (5, 7)

22. L_1 through (−2, 4) and (1, 8); L_2 through (−1, −1) and (−5, 2)

23. L_1 through (8, 5) and (3, −2); L_2 through (−2, 4) and (4, −1)

24. L_1 through (−2, −3) and (3, −1); L_2 through (−3, 1) and (7, 5)

25. L_1 with equation $x - 3y = 6$; L_2 with equation $3x + y = 3$

26. L_1 with equation $x + 2y = 4$; L_2 with equation $2x + 4y = 5$

ANSWERS

27. ______
28. ______
29. ______
30. ______
31. ______
32. ______
33. ______
34. ______
35. ______
36. ______
37. ______
38. ______
39. ______

27. Find the slope of any line parallel to the line through points $(-2, 3)$ and $(4, 5)$.

28. Find the slope of any line perpendicular to the line through points $(0, 5)$ and $(-3, -4)$.

29. A line passing through $(-1, 2)$ and $(4, y)$ is parallel to a line with slope 2. What is the value of y?

30. A line passing through $(2, 3)$ and $(5, y)$ is perpendicular to a line with slope $\frac{3}{4}$. What is the value of y?

If points P, Q, and R are collinear (lie on the same line), the slope of the line through P and Q must equal the slope of the line through Q and R. In exercises 31 to 36, use the slope concept to determine whether the sets of points are collinear.

31. $P(-2, -3)$, $Q(3, 2)$, and $R(4, 3)$

32. $P(-5, 1)$, $Q(-2, 4)$, and $R(4, 9)$

33. $P(0, 0)$, $Q(2, 4)$, and $R(-3, 6)$

34. $P(-2, 5)$, $Q(-5, 2)$, and $R(1, 12)$

35. $P(2, 4)$, $Q(-3, -6)$, and $R(-4, 8)$

36. $P(-1, 5)$, $Q(2, -4)$, and $R(-2, 8)$

In exercises 37 to 44, graph the lines through each of the specified points having the given slope.

37. $(0, 1)$, $m = 3$

38. $(0, -2)$, $m = -2$

39. $(3, -1)$, $m = 2$

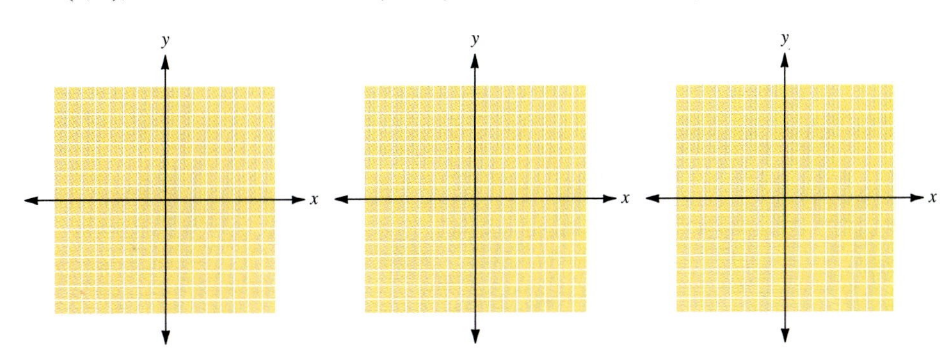

ANSWERS

40. ______

41. ______

42. ______

43. ______

44. ______

45. ______

46. ______

40. $(2, -3), m = -3$ **41.** $(2, 3), m = \frac{2}{3}$ **42.** $(-2, 1), m = -\frac{3}{4}$

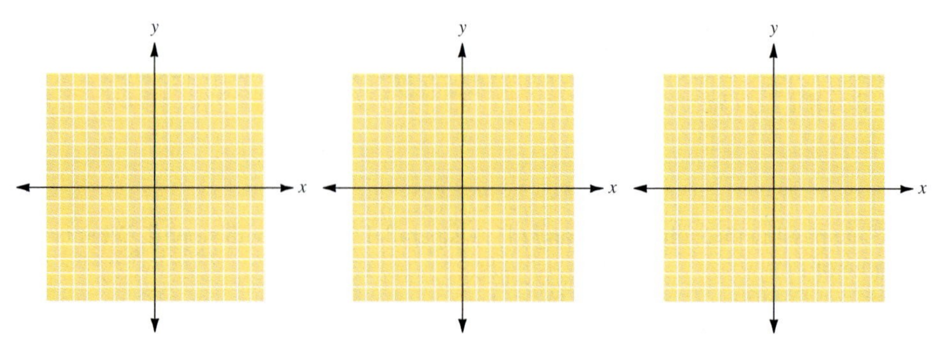

43. $(4, 2), m = 0$ **44.** $(3, 0)$, m is undefined

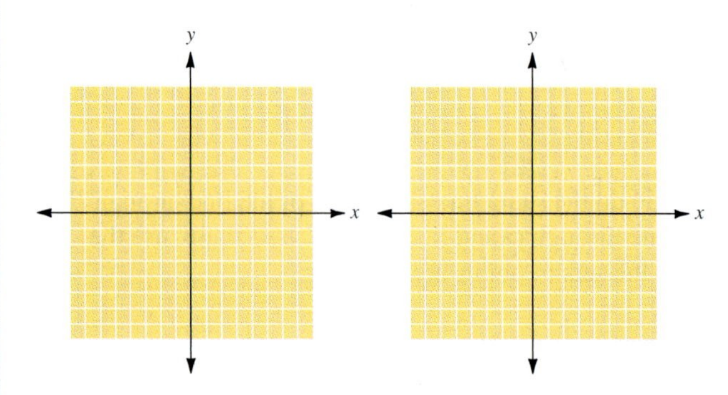

45. On the same graph, sketch lines with slope 2 through each of the following points: $(-1, 0)$, $(2, 0)$, and $(5, 0)$.

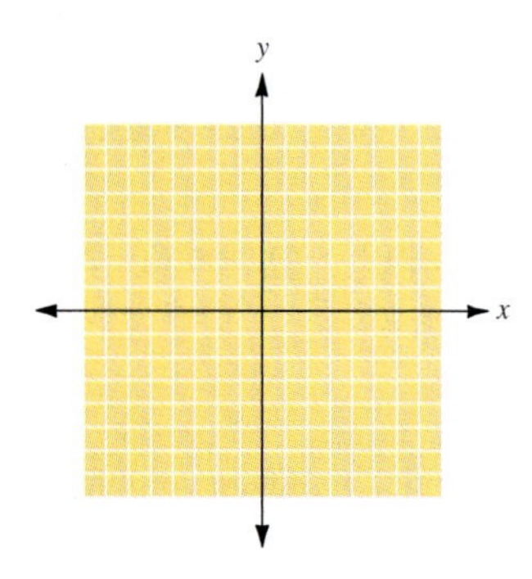

46. On the same graph, sketch one line with slope $\frac{1}{3}$ and one line with slope -3, having both pass through point $(2, 3)$.

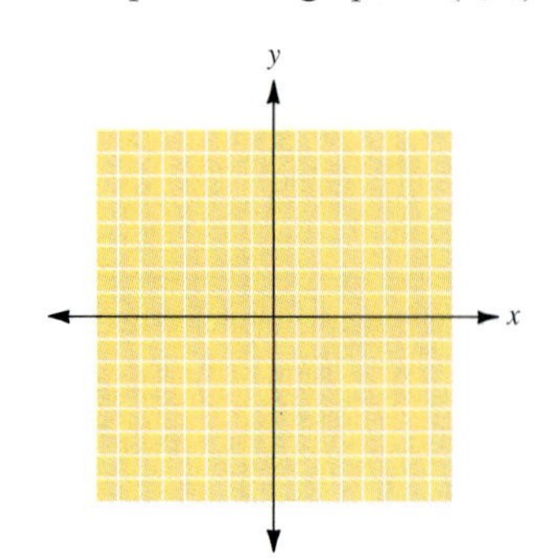

A four-sided figure (quadrilateral) is a parallelogram if the opposite sides have the same slope. If the adjacent sides are perpendicular, the figure is a rectangle. In exercises 47 to 50, for each quadrilateral *ABCD*, determine whether it is a parallelogram; then determine whether it is a rectangle.

47. $A(0, 0), B(2, 0), C(2, 3), D(0, 3)$

48. $A(-3, 2), B(1, -7), C(3, -4), D(-1, 5)$

49. $A(0, 0), B(4, 0), C(5, 2), D(1, 2)$

50. $A(-3, -5), B(2, 1), C(-4, 6), D(-9, 0)$

In exercises 51 to 54, solve each equation for y, then use your graphing utility to graph each equation.

51. $2x + 5y = 10$

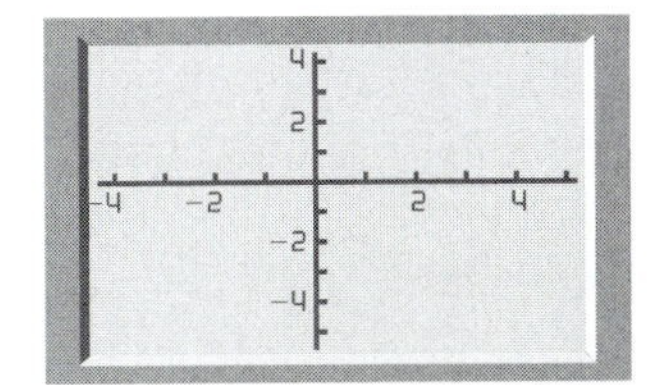

52. $5x - 3y = 12$

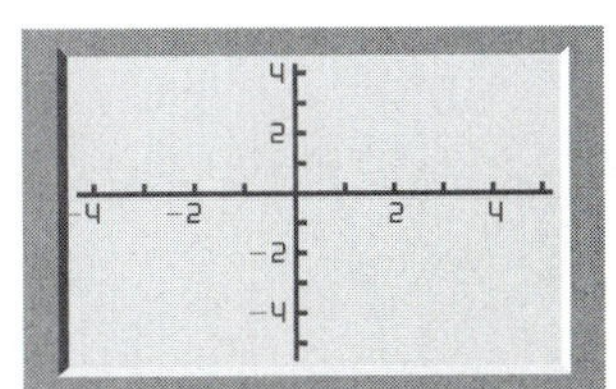

53. $x + 7y = 14$

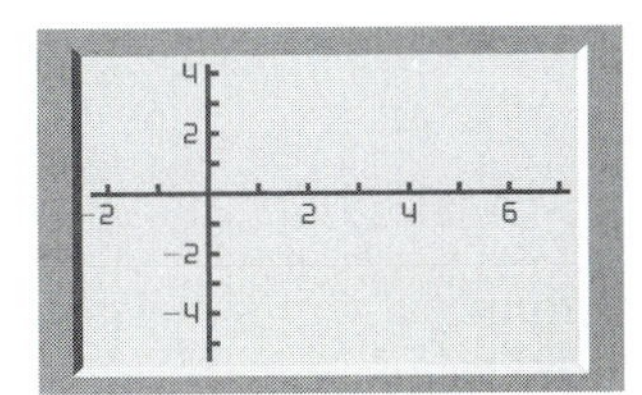

54. $-2x - 3y = 9$

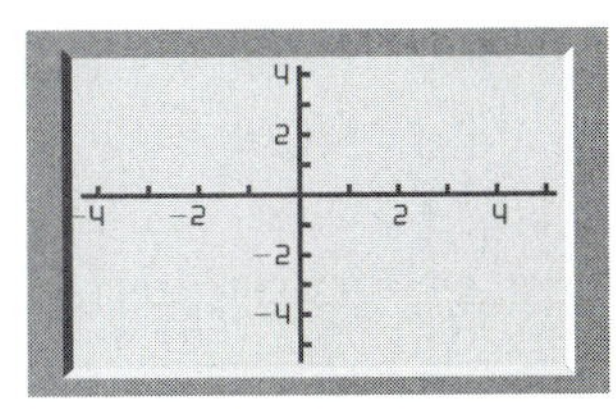

ANSWERS

47. ______

48. ______

49. ______

50. ______

51. ______

52. ______

53. ______

54. ______

ANSWERS

55. ______

56. ______

57. ______

58. ______

59. ______

60. ______

61. ______

62. ______

In exercises 55 to 62, use the graph to determine the slope of the line.

55.

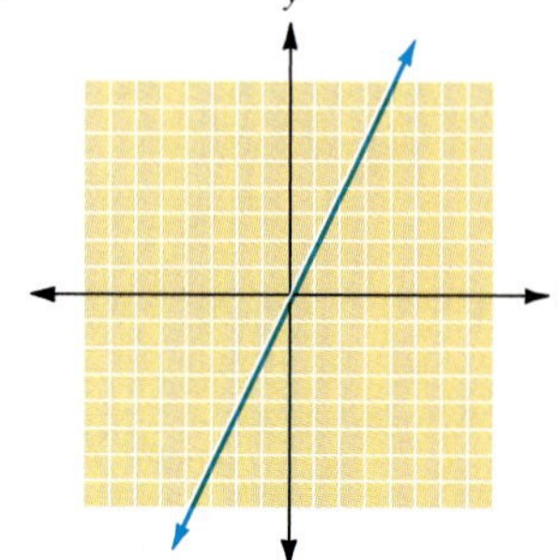

56.

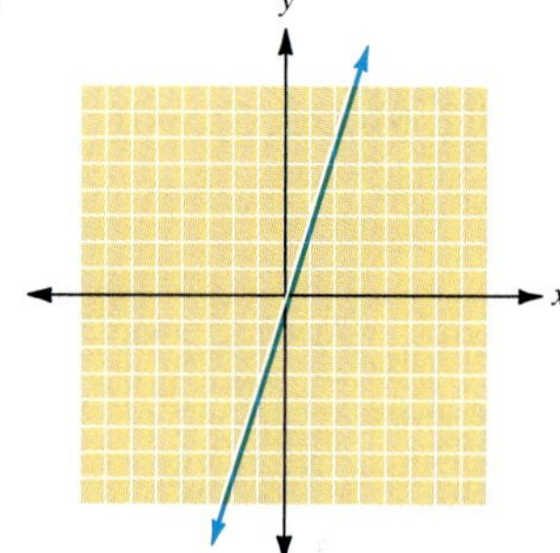

57.

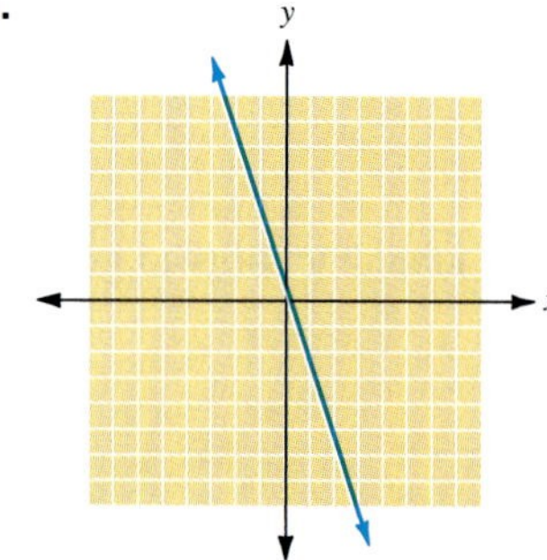

58.

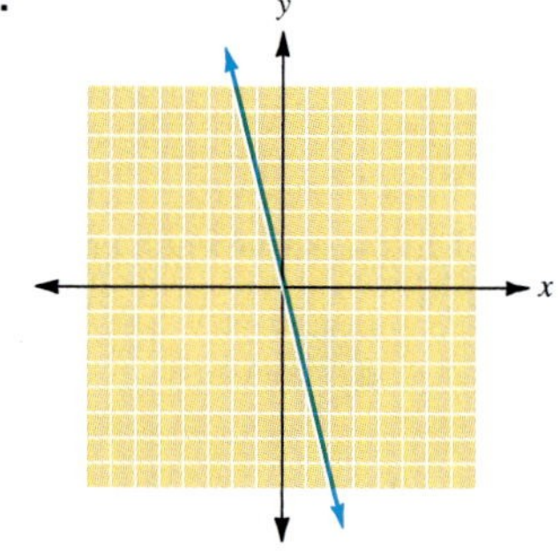

59.

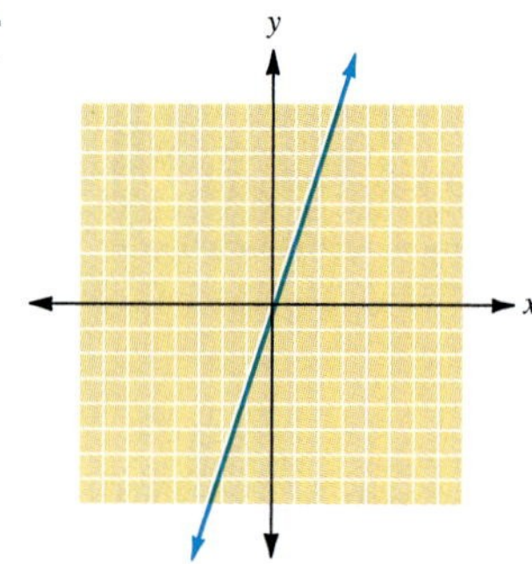

60.

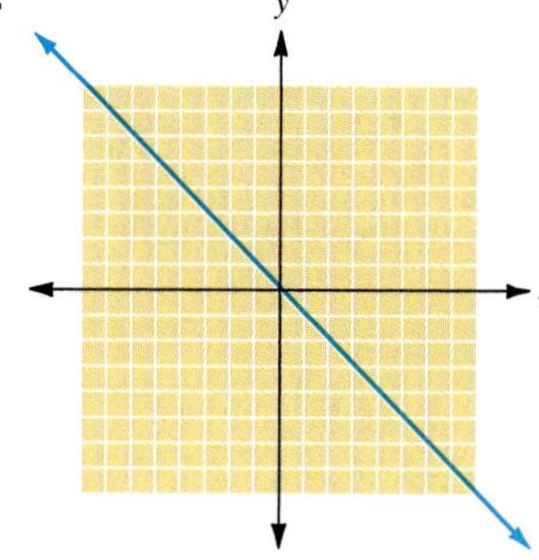

61.

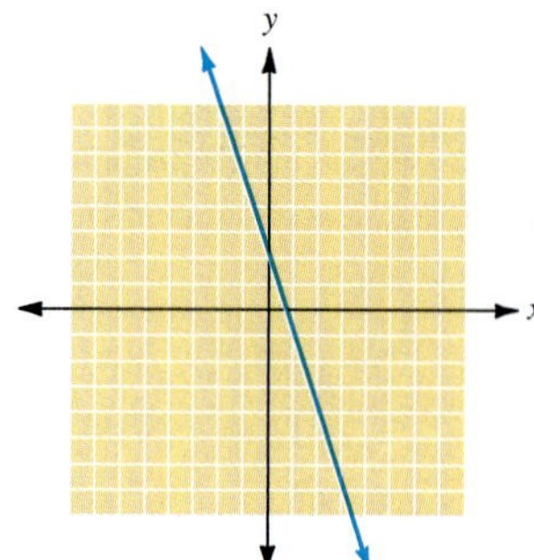

62.

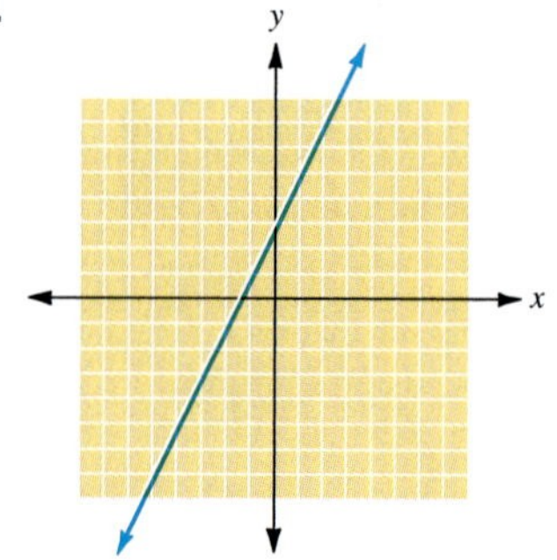

63. Consider the equation $y = 2x + 3$.

(a) Complete the following table of values, and plot the resulting points.

Point	x	y
A	5	
B	6	
C	7	
D	8	
E	9	

(b) As the x coordinates change by 1 (for example, as you move from point A to point B), by how much does the corresponding y coordinate change?

(c) Is your answer to part (b) the same if you move from B to C? from C to D? from D to E?

(d) Describe the "growth rate" of the line using these observations. Complete the following statement: When the x value grows by 1 unit, the y value _______ _________________.

64. Repeat exercise 63 using $y = 2x + 5$.

65. Repeat exercise 63 using $y = 3x - 2$.

66. Repeat exercise 63 using $y = 3x - 4$.

67. Repeat exercise 63 using $y = -4x + 50$.

68. Repeat exercise 63 using $y = -4x + 40$.

69. Summarize the results of exercises 63 to 68. In particular, how does the concept of "growth rate" connect to the concept of slope?

70. **Consumer affairs.** In 1995, the cost of a soft drink was 75¢. By 1999, the cost of the same soft drink had risen to \$1.25. During this period, what was the rate of change of the cost of the soft drink? (*Hint*: Assume a linear growth rate for the price as the years go by. Find the slope of the line.)

ANSWERS

63. (a)
(b)
(c)
(d)

64. (a)
(b)
(c)
(d)

65. (a)
(b)
(c)
(d)

66. (a)
(b)
(c)
(d)

67. (a)
(b)
(c)
(d)

68. (a)
(b)
(c)
(d)

69.

70.

ANSWERS

71. ______________

72. ______________

71. Science. On a certain February day in Philadelphia, the temperature at 6:00 A.M. was 10°F. By 2:00 P.M. the temperature was up to 26°F. What was the average hourly rate of temperature change? (*Hint:* Assume a linear growth rate for the temperature as the hours go by. Find the slope.)

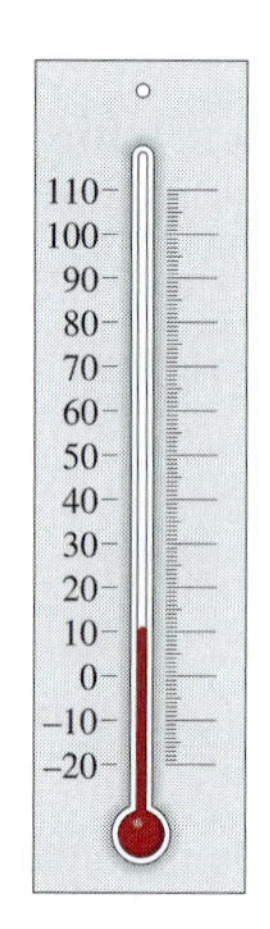

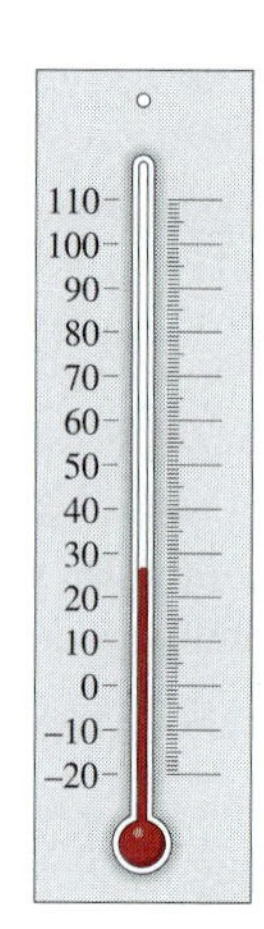

72. Construction. The rise-to-run ratio used to describe the steepness of the roof on a certain house is 4 to 12. Determine the maximum height of the attic if the house is 32 ft wide.

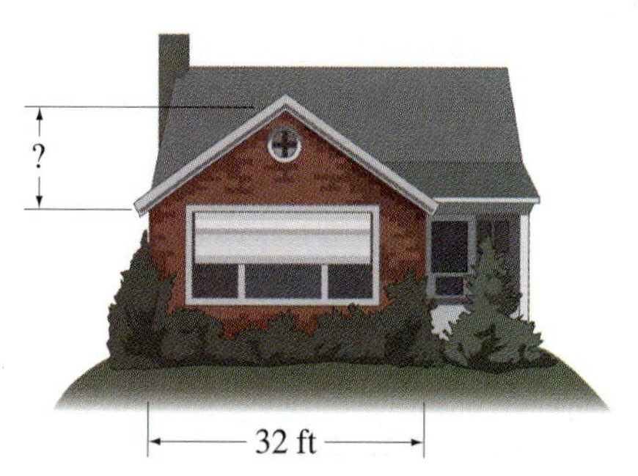

Answers

1. 2

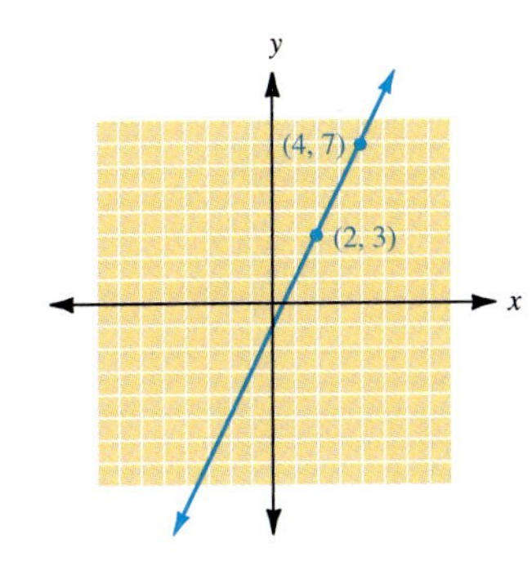

3. $\frac{1}{2}$

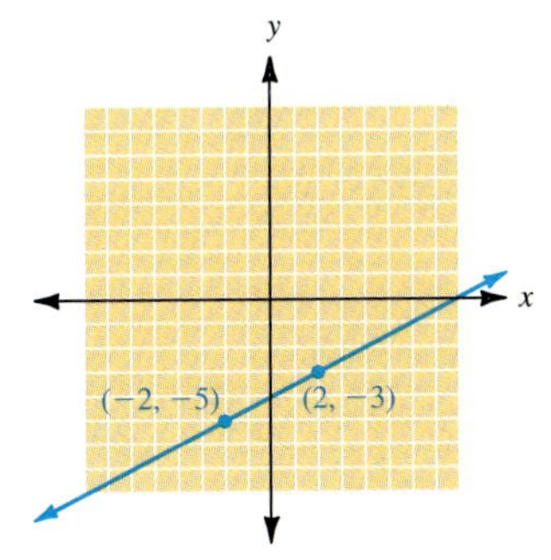

5. 0

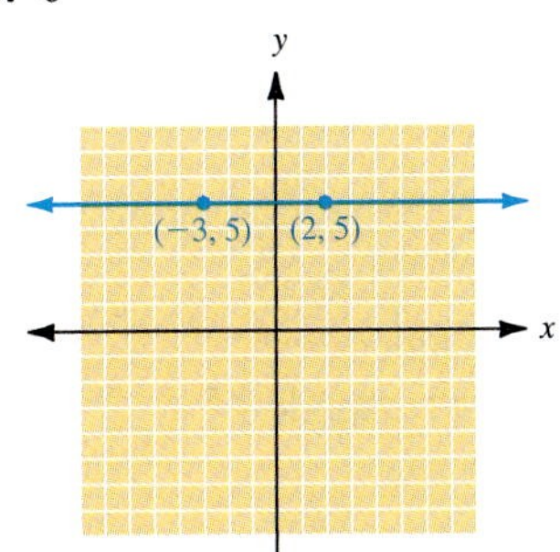

7. Undefined

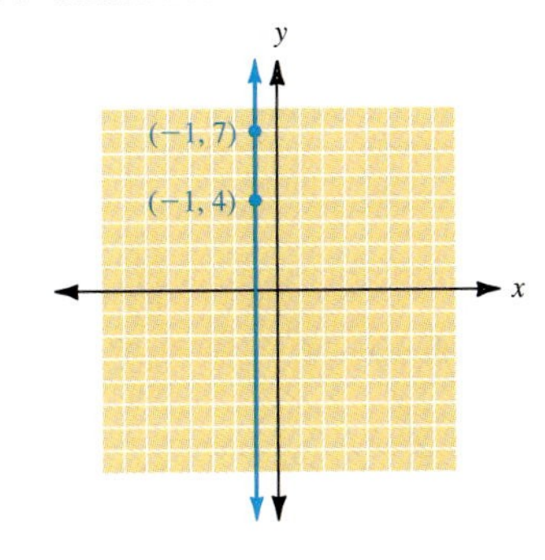

9. $\frac{1}{5}$

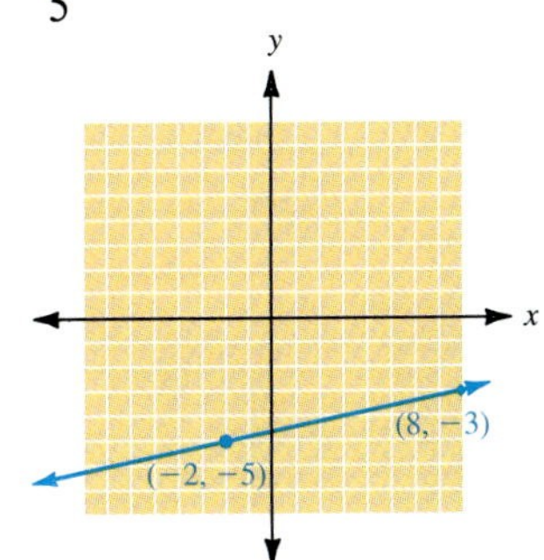

11. $-\frac{2}{3}$

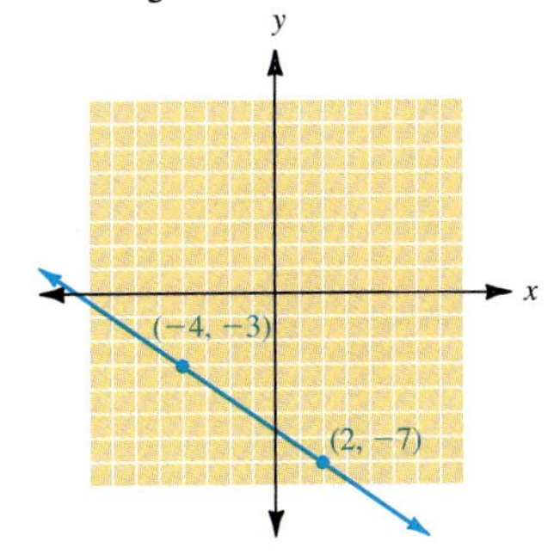

13. -3 **15.** $-\frac{1}{2}$ **17.** $\frac{2}{3}$ **19.** $-\frac{3}{4}$ **21.** Parallel

23. Neither **25.** Perpendicular **27.** $\frac{1}{3}$ **29.** 12

31. Collinear **33.** Not collinear **35.** Not collinear

37.

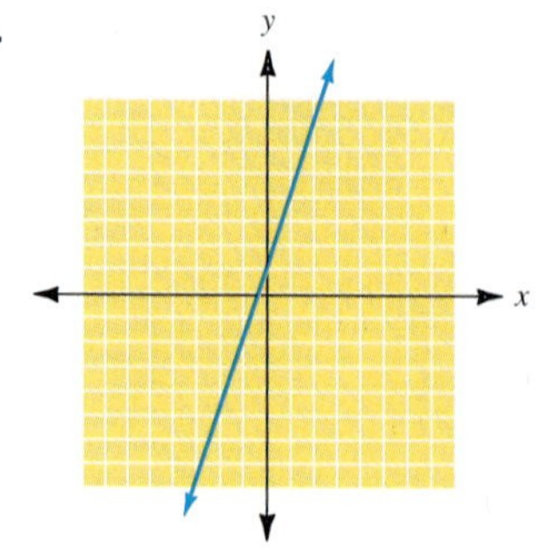

39.

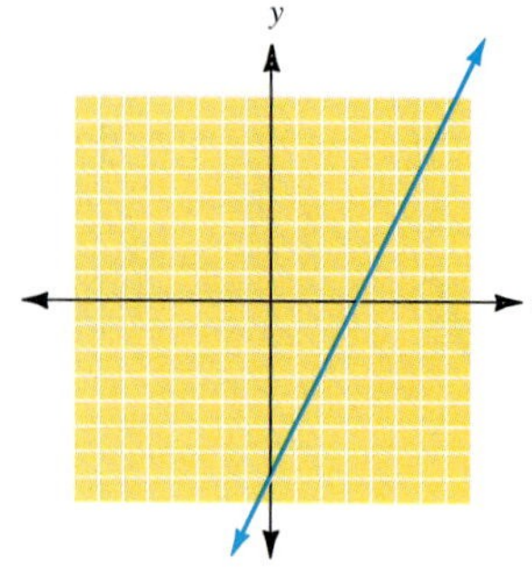

41.

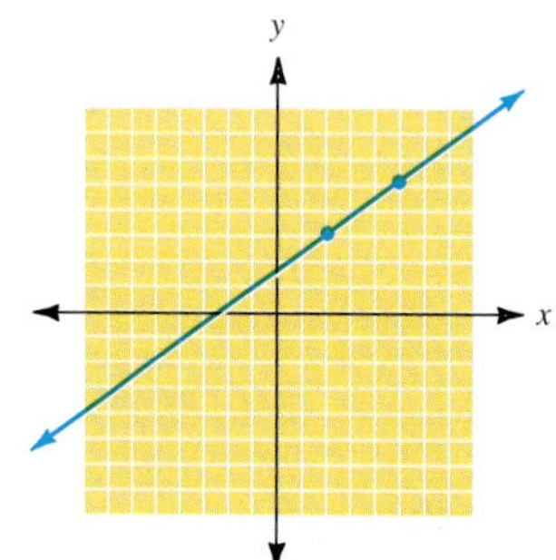

43.

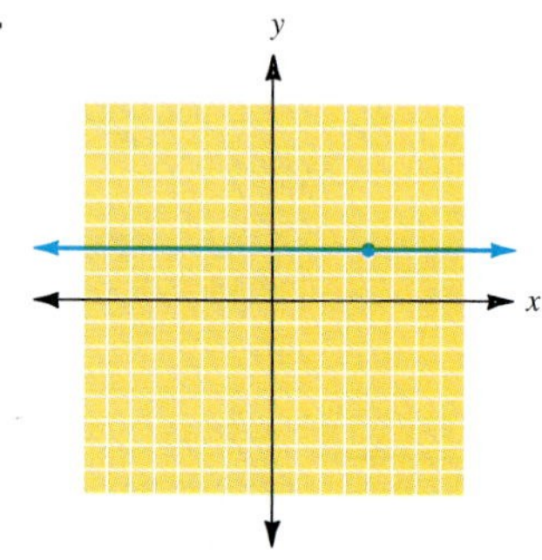

45.

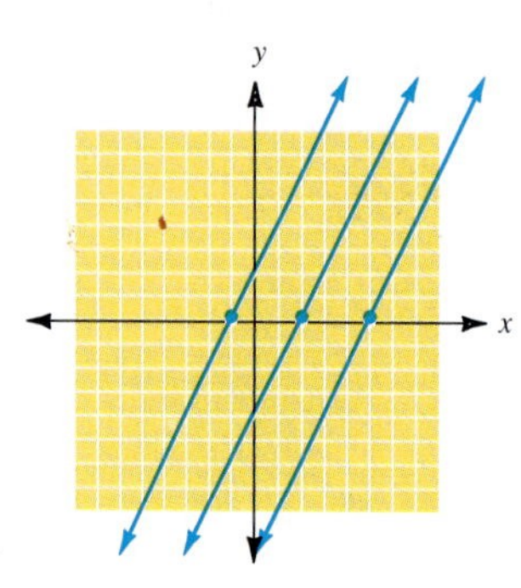

47. Parallelogram, rectangle
49. Parallelogram, not a rectangle

51. $y = -\frac{2}{5}x + 2$

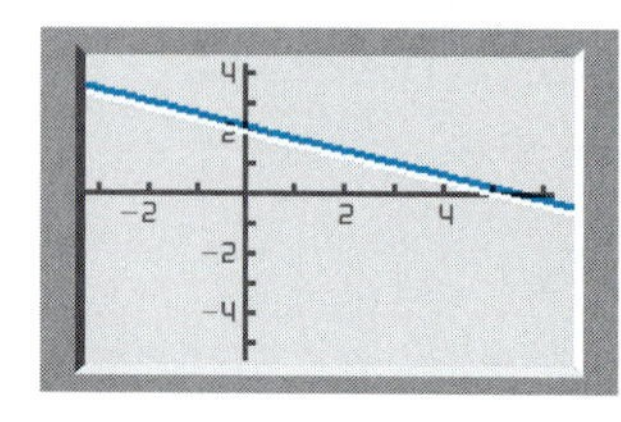

53. $y = -\frac{1}{7}x + 2$

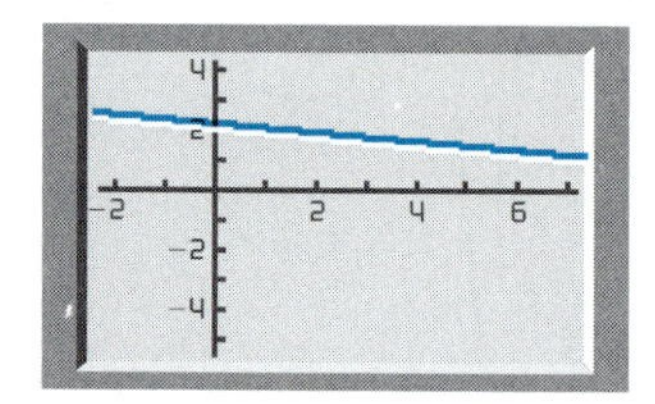

55. 2 **57.** -3 **59.** 3 **61.** -3
63. **(a)** y values: 13, 15, 17, 19, 21; **(b)** y increases by 2; **(c)** yes; **(d)** grows by 2 units
65. **(a)** y values: 13, 16, 19, 22, 25; **(b)** y increases by 3; **(c)** yes; **(d)** grows by 3 units
67. **(a)** y values: 30, 26, 22, 18, 14; **(b)** y decreases by 4; **(c)** yes; **(d)** drops by 4 units
69. **71.** 2°/hour

4.3 Forms of Linear Equations

OBJECTIVES

1. Write the equation of a line given its slope and y intercept
2. Find the slope and y intercept of a line from its equation
3. Graph the equation of a line
4. Write the equation of a line given a point and a slope
5. Write the equation of a line given two points
6. Express the equation of a line as a linear function
7. Graph a linear function

The special form

$Ax + By = C$

in which A and B cannot both be zero, is called the **standard form for a linear equation.** In Section 4.2, we determined the slope of a line from two ordered pairs. In this section, we will look at several forms for a linear equation. For the first of these forms, we use the concept of slope to write the equation of a line.

NOTE The coordinates are (0, b) because the x coordinate on the y axis is zero and the y coordinate of the intercept is b.

First, suppose we know the y intercept $(0, b)$ of a line L, and its slope m. Let $P(x, y)$ be any other point on that line. Using $(0, b)$ as (x_1, y_1) and (x, y) as (x_2, y_2) in the slope formula, we have

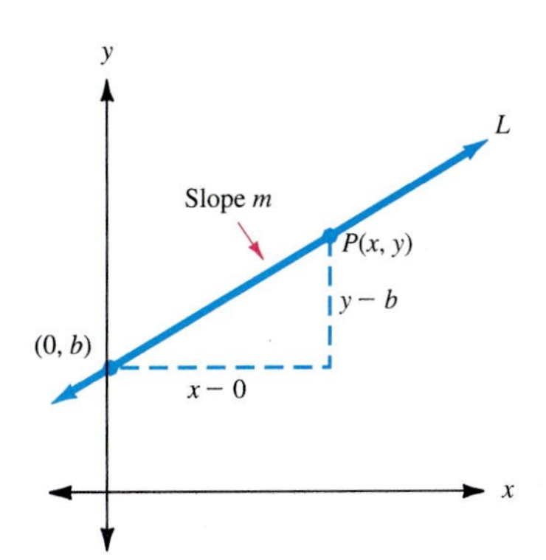

$$m = \frac{y - b}{x - 0} \qquad (1)$$

or

$$m = \frac{y - b}{x} \qquad (2)$$

Multiplying both sides of equation (2) by x gives

$mx = y - b$

or

$$y = mx + b \qquad (3)$$

Equation (3) will be satisfied by any point on line L, including $(0, b)$. It is called the **slope-intercept form** for a line, and we can state the following general result.

NOTE The slope-intercept form for the equation of a line is the most convenient form for entering an equation into the calculator.

Definitions: Slope-Intercept Form for the Equation of a Line

The equation of a line with y intercept $(0, b)$ and with slope m can be written as $y = mx + b$.

Example 1

Finding the Equation of a Line

Write the equation of the line with slope 2 and y intercept (0, 3).

Here $m = 2$ and $b = 3$. Applying the slope-intercept form, we have

NOTE The x coefficient is 2; the y intercept is (0, 3).

$$y = 2x + 3$$

as the equation of the specified line.

It is easy to see that whenever a linear equation is written in slope-intercept form, the slope of the line is simply the x coefficient and the y intercept is determined by the constant.

CHECK YOURSELF 1

Write the equation of the line with slope $-\frac{2}{3}$ *and y intercept* (0, −3).

Note that the slope-intercept form now gives us a second (and generally more efficient) means of finding the slope of a line whose equation is written in standard form. Recall that we determined two specific points on the line and then applied the slope formula. Now, rather than using specific points, we can simply solve the given equation for y to rewrite the equation in the slope-intercept form and identify the slope of the line as the x coefficient.

Example 2

Finding the Slope and y Intercept of a Line

Find the slope and y intercept of the line with equation

$$2x + 3y = 3$$

To write the equation in slope-intercept form, we solve for y.

$$2x + 3y = 3$$

$$3y = -2x + 3 \quad \text{Subtract } 2x \text{ from both sides.}$$

$$y = -\frac{2}{3}x + 1 \quad \text{Divide by 3.}$$

We now see that the slope of the line is $-\frac{2}{3}$ and the y intercept is (0, 1).

CHECK YOURSELF 2

Find the slope and y intercept of the line with equation.

$$3x - 4y = 8$$

−3x + 8

We can also use the slope-intercept form to determine whether the graphs of given equations will be parallel, intersecting, or perpendicular lines.

Example 3

Verifying That Two Lines Are Perpendicular

NOTE Two lines are perpendicular if their slopes are negative reciprocals, so

$m_1 = -\frac{1}{m_2}$

Show that the graphs of $3x + 4y = 4$ and $-4x + 3y = 12$ are perpendicular lines.

First, we solve each equation for y.

$$3x + 4y = 4$$
$$4y = -3x + 4$$
$$y = -\frac{3}{4}x + 1 \qquad (4)$$

$$-4x + 3y = 12$$
$$3y = 4x + 12$$
$$y = \frac{4}{3}x + 4 \qquad (5)$$

We now look at the product of the two slopes: $-\frac{3}{4} \cdot \frac{4}{3} = -1$. Any two lines whose slopes have a product of -1 are perpendicular lines. These two lines are perpendicular.

CHECK YOURSELF 3

Show that the graphs of the equations

$-3x + 2y = 4$ and $2x + 3y = 9$

are perpendicular lines.

The slope-intercept form can also be used in graphing a line, as Example 4 illustrates.

Example 4

Graphing the Equation of a Line

Graph the line $2x + 3y = 3$.

In Example 2, we found that the slope-intercept form for this equation is

$$y = -\frac{2}{3}x + 1$$

To graph the line, plot the y intercept at $(0, 1)$. Now, because the slope m is equal to $-\frac{2}{3}$, from $(0, 1)$ we move to the right 3 units and then *down* 2 units, to locate a second point on the graph of the line, here $(3, -1)$. We can now draw a line through the two points to complete the graph.

NOTE We treat $-\frac{2}{3}$ as $\frac{-2}{+3}$ to move to the right 3 units and down 2 units.

CHECK YOURSELF 4

Graph the line with equation

$3x - 4y = 8$

Hint: You worked with this equation in Check Yourself 2.

The following algorithm summarizes the use of graphing with the slope-intercept form.

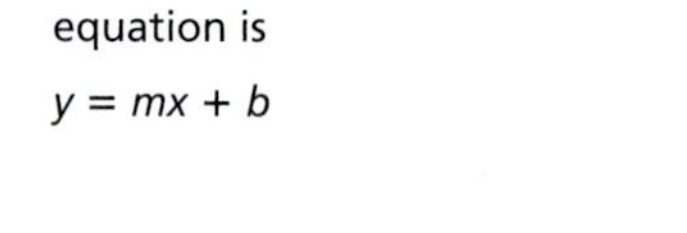

NOTE The desired form for the equation is

$y = mx + b$

Step by Step: Graphing by Using the Slope-Intercept Form

Step 1 Write the original equation of the line in slope-intercept form.

Step 2 Determine the slope m and the y intercept $(0, b)$.

Step 3 Plot the y intercept at $(0, b)$.

Step 4 Use m (the change in y over the change in x) to determine a second point on the desired line.

Step 5 Draw a line through the two points determined above to complete the graph.

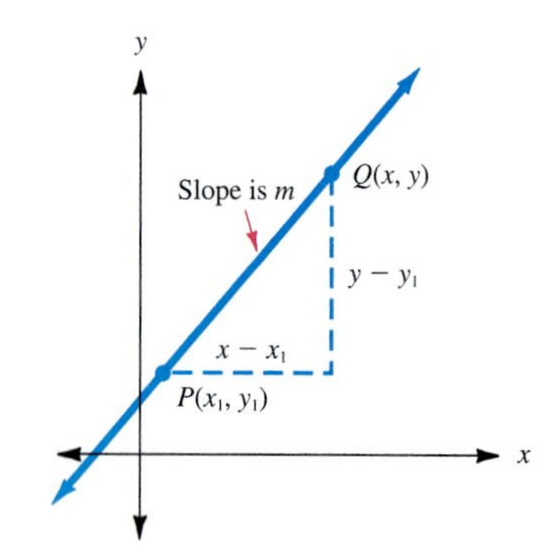

Often in mathematics it is useful to be able to write the equation of a line, given its slope and *any* point on the line. We will now derive a third special form for a line for this purpose.

Suppose a line has slope m and passes through the known point $P(x_1, y_1)$. Let $Q(x, y)$ be any other point on the line. Once again we can use the definition of slope and write

$$m = \frac{y - y_1}{x - x_1} \tag{6}$$

Multiplying both sides of equation (6) by $x - x_1$, we have

$$m(x - x_1) = y - y_1$$

or

$$y - y_1 = m(x - x_1) \tag{7}$$

Equation (7) is called the **point-slope form** for the equation of a line, and all points lying on the line [including (x_1, y_1)] will satisfy this equation. We can state the following general result.

NOTE Recall that a vertical line has undefined slope. The equation of a line with undefined slope passing through the point (x_1, y_1) is given by $x = x_1$. For example, the equation of a line with undefined slope passing through (3, 5) is $x = 3$.

Definitions: Point-Slope Form for the Equation of a Line

The equation of a line with slope m that passes through point (x_1, y_1) is given by

$$y - y_1 = m(x - x_1)$$

Example 5

Finding the Equation of a Line

Write the equation for the line that passes through point (3, −1) with a slope of 3.

Letting $(x_1, y_1) = (3, -1)$ and $m = 3$ in point-slope form, we have

$$y - (-1) = 3(x - 3)$$

or

$$y + 1 = 3x - 9$$

We can write the final result in slope-intercept form as

$$y = 3x - 10$$

CHECK YOURSELF 5

Write the equation of the line that passes through point (−2, 4) with a slope of $\frac{3}{2}$. Write your result in slope-intercept form.

Because we know that two points determine a line, it is natural that we should be able to write the equation of a line passing through two given points. Using the point-slope form together with the slope formula will allow us to write such an equation.

Example 6

Finding the Equation of a Line

Write the equation of the line passing through (2, 4) and (4, 7).

First, we find m, the slope of the line. Here

$$m = \frac{7 - 4}{4 - 2} = \frac{3}{2}$$

NOTE We could just as well have chosen to let

$(x_1, y_1) = (4, 7)$

The resulting equation will be the same in either case. Take time to verify this for yourself.

Now we apply the point-slope form with $m = \frac{3}{2}$ and $(x_1, y_1) = (2, 4)$:

$$y - 4 = \frac{3}{2}(x - 2) \qquad \text{Distribute the } \frac{3}{2}.$$

$$y - 4 = \frac{3}{2}x - 3 \qquad \text{Write the result in slope-intercept form.}$$

$$y = \frac{3}{2}x + 1$$

CHECK YOURSELF 6

Write the equation of the line passing through (−2, 5) and (1, 3). Write your result in slope-intercept form.

A line with slope zero is a horizontal line. A line with an undefined slope is vertical. Example 7 illustrates the equations of such lines.

Example 7

Finding the Equation of a Line

(a) Find the equation of a line passing through $(7, -2)$ with a slope of zero.

We could find the equation by letting $m = 0$. Substituting into the slope-intercept form, we can solve for b.

$$y = mx + b$$

$$-2 = 0(7) + b$$

$$-2 = b$$

So,

$$y = 0x - 2 \qquad y = -2$$

It is far easier to remember that any line with a zero slope is a horizontal line and has the form

$$y = b$$

The value for b will always be the y coordinate for the given point.

(b) Find the equation of a line with undefined slope passing through $(4, -5)$.

A line with undefined slope is vertical. It will always be of the form $x = a$, in which a is the x coordinate for the given point. The equation is

$$x = 4$$

CHECK YOURSELF 7

(a) Find the equation of a line with zero slope that passes through point $(-3, 5)$.

(b) Find the equation of a line passing through $(-3, -6)$ with undefined slope.

Alternate methods for finding the equation of a line through two points do exist and have particular significance in other fields of mathematics, such as statistics. Example 8 shows such an alternate approach.

Example 8

Finding the Equation of a Line

Write the equation of the line through points $(-2, 3)$ and $(4, 5)$.

First, we find m, as before.

$$m = \frac{5 - 3}{4 - (-2)} = \frac{2}{6} = \frac{1}{3}$$

We now make use of the slope-intercept equation, but in a slightly different form.

Because $y = mx + b$, we can write

$$b = y - mx$$

NOTE We substitute these values because the line must pass through (−2, 3).

Now, letting $x = -2$, $y = 3$, and $m = \frac{1}{3}$, we can calculate b.

$$b = 3 - \left(\frac{1}{3}\right)(-2)$$

$$= 3 + \frac{2}{3} = \frac{11}{3}$$

With $m = \frac{1}{3}$ and $b = \frac{11}{3}$, we can apply the slope-intercept form to write the equation of the desired line. We have

$$y = \frac{1}{3}x + \frac{11}{3}$$

CHECK YOURSELF 8

Repeat the Check Yourself 6 exercise, using the technique illustrated in Example 8.

We now know that we can write the equation of a line once we have been given appropriate geometric conditions, such as a point on the line and the slope of that line. In some applications, the slope may be given not directly but through specified parallel or perpendicular lines.

Example 9

Finding the Equation of a Line

Find the equation of the line passing through $(-4, -3)$ and parallel to the line determined by $3x + 4y = 12$.

First, we find the slope of the given parallel line, as before.

$$3x + 4y = 12$$

$$4y = -3x + 12$$

$$y = -\frac{3}{4}x + 3$$

NOTE The slope of the given line is $-\frac{3}{4}$.

Now, because the slope of the desired line must also be $-\frac{3}{4}$, we can use the point-slope form to write the required equation.

NOTE The line must pass through (−4, −3), so let $(x_1, y_1) = (-4, -3)$.

$$y - (-3) = -\frac{3}{4}[x - (-4)]$$

This simplifies to

$$y = -\frac{3}{4}x - 6$$

and we have our equation in slope-intercept form.

CHECK YOURSELF 9

Find the equation of the line passing through (5, 4) and perpendicular to the line with equation $2x - 5y = 10$. Hint: Recall that the slopes of perpendicular lines are negative reciprocals of each other.

NOTE Recall the *vertical line test,* which was used to determine whether a graph was the graph of a function.

Because a vertical line cannot pass through a line with defined slope more than once, any nonvertical line can be represented as a function.

Example 10

Writing Equations as Functions

Rewrite each linear equation as a function of x.

(a) $y = 3x - 4$

This can be rewritten as

$$f(x) = 3x - 4$$

(b) $2x - 3y = 6$

We must first solve the equation for y (recall that this will give us the slope-intercept form).

$$-3y = -2x + 6$$

$$y = \frac{2}{3}x - 2$$

This can be rewritten as

$$f(x) = \frac{2}{3}x - 2$$

CHECK YOURSELF 10

Rewrite each equation as a function of x.

(a) $y = -2x + 5$ **(b)** $3x + 5y = 15$

The process of finding the graph of a linear function is identical to the process of finding the graph of a linear equation.

Example 11

Graphing a Linear Function

Graph the function

$$f(x) = 3x - 5$$

We could use the slope and y intercept to graph the line, or we can find three points (the third is just a check point) and draw the line through them. We will do the latter.

$f(0) = -5 \qquad f(1) = -2 \qquad f(2) = 1$

We will use the three points $(0, -5)$, $(1, -2)$, and $(2, 1)$ to graph the line.

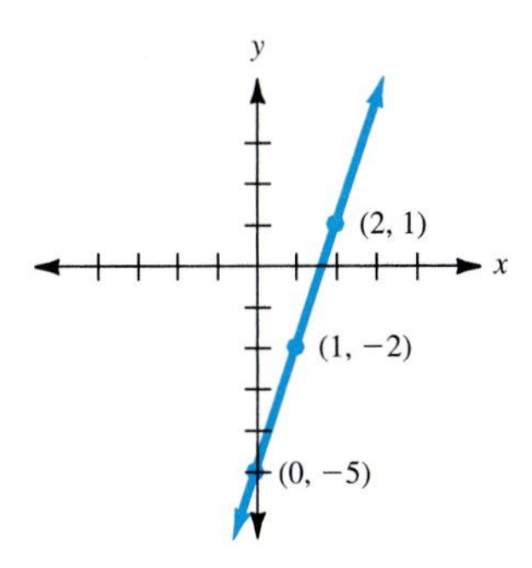

CHECK YOURSELF 11

Graph the function

$f(x) = 5x - 3$

One benefit of having a function written in $f(x)$ form is that it makes it fairly easy to substitute values for x. In Example 11, we substituted the values 0, 1, and 2. Sometimes it is useful to substitute nonnumeric values for x.

Example 12

Substituting Nonnumeric Values for *x*

Let $f(x) = 2x + 3$. Evaluate f as indicated.

(a) $f(a)$

Substituting a for x in our equation, we see that

$f(a) = 2a + 3$

(b) $f(2 + h)$

Substituting $2 + h$ for x in our equation, we get

$f(2 + h) = 2(2 + h) + 3$

Distributing the 2, then simplifying, we have

$$f(2 + h) = 4 + 2h + 3$$
$$= 2h + 7$$

CHECK YOURSELF 12

Let $f(x) = 4x - 2$. *Evaluate f as indicated.*

(a) $f(b)$ **(b)** $f(4 + h)$

CHECK YOURSELF ANSWERS

1. $y = -\frac{2}{3}x - 3$ **2.** $m = \frac{3}{4}$ and the y intercept is $(0, -2)$

3. $m_1 = \frac{3}{2}$ and $m_2 = -\frac{2}{3}$; $(m_1)(m_2) = -1$ **4.**

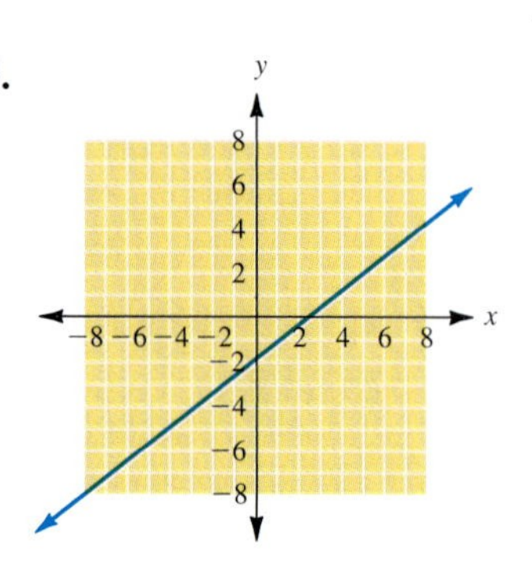

5. $y = \frac{3}{2}x + 7$ **6.** $y = -\frac{2}{3}x + \frac{11}{3}$ **7.** **(a)** $y = 5$; **(b)** $x = -3$

8. $y = -\frac{2}{3}x + \frac{11}{3}$ **9.** $y = -\frac{5}{2}x + \frac{33}{2}$

10. **(a)** $f(x) = -2x + 5$; **(b)** $f(x) = -\frac{3}{5}x + 3$ **11.**

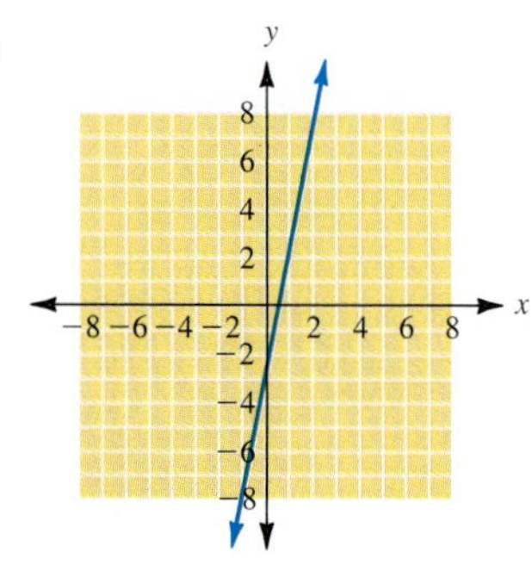

12. **(a)** $4b - 2$; **(b)** $4h + 14$

Name ______________

Section ________ Date ________

4.3 Exercises

In exercises 1 to 8, match the graph with one of these equations: **(a)** $y = 2x$; **(b)** $y = x + 1$; **(c)** $y = -x + 3$; **(d)** $y = 2x + 1$; **(e)** $y = -3x - 2$; **(f)** $y = \frac{2}{3}x + 1$; **(g)** $y = -\frac{4}{3}x + 1$; and **(h)** $y = -4x$.

1.

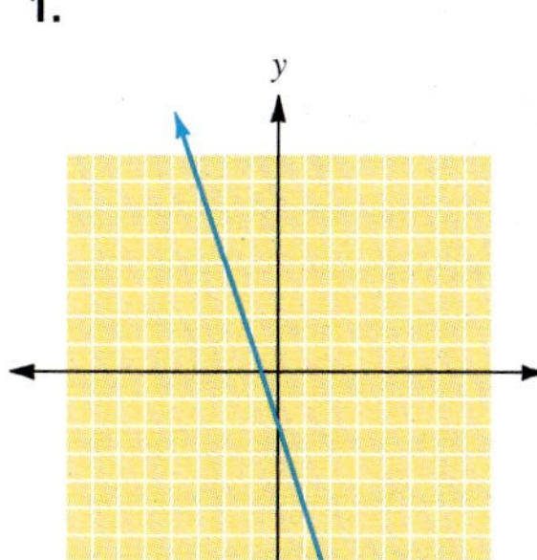

2.

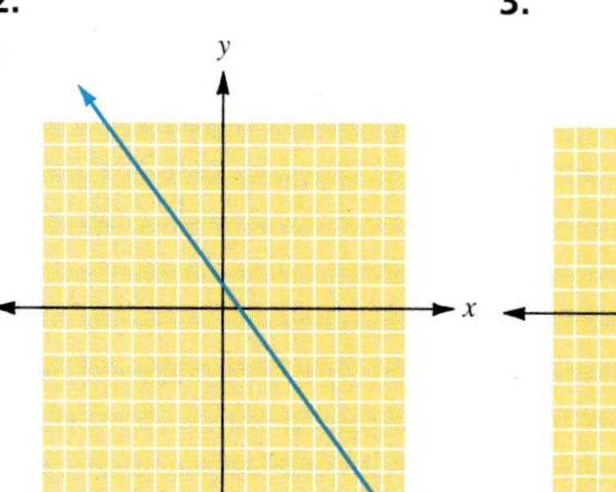

3.

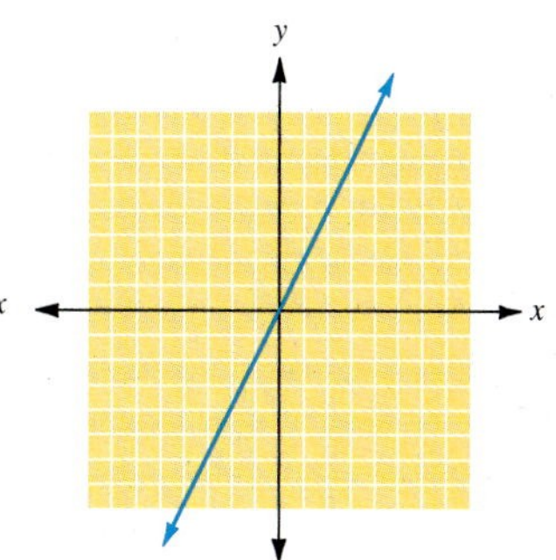

4.

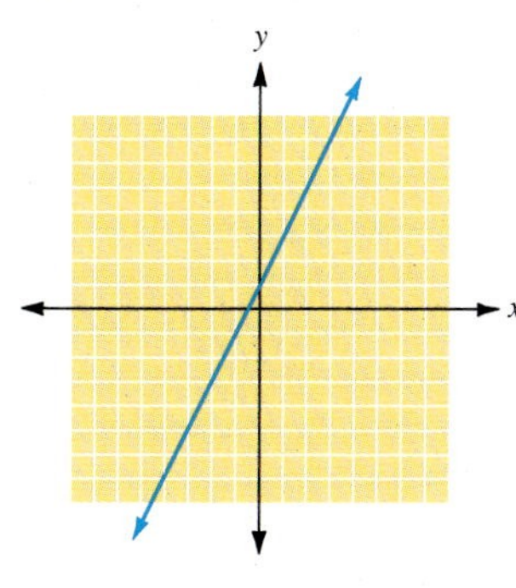

5. **6.**

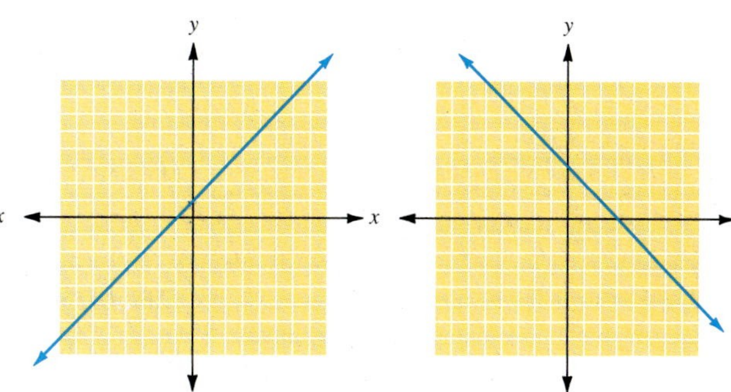

7.

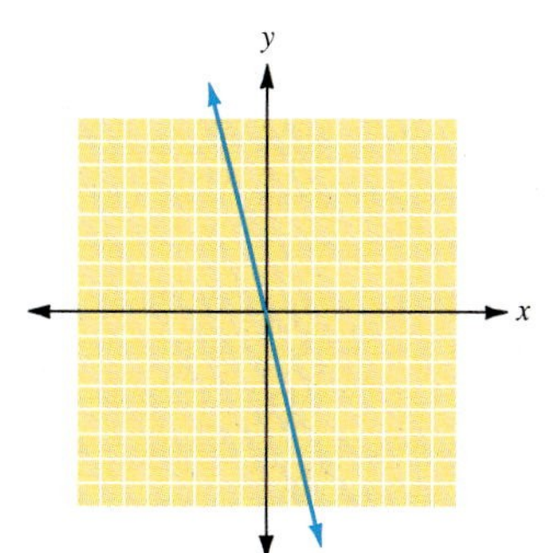

8.

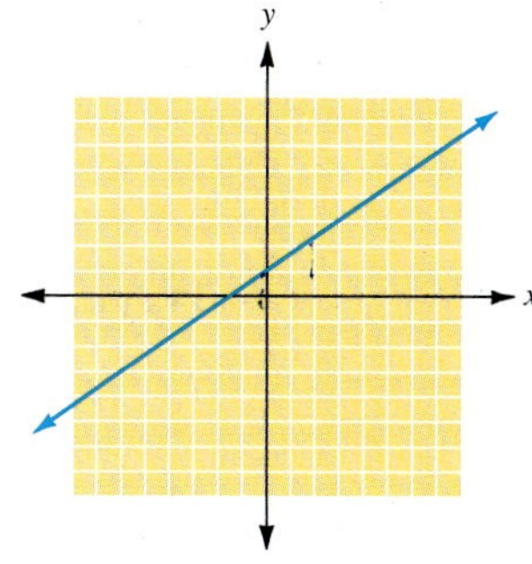

ANSWERS

1. ______________
2. ______________
3. ______________
4. ______________
5. ______________
6. ______________
7. ______________
8. ______________

ANSWERS

9. ______

10. ______

11. ______

12. ______

13. ______

14. ______

15. ______

16. ______

17. ______

18. ______

19. ______

20. ______

21. ______

22. ______

23. ______

24. ______

25. ______

26. ______

27. ______

28. ______

29. ______

30. ______

31. ______

32. ______

33. ______

In exercises 9 to 20, write each equation in slope-intercept form. Give its slope and y intercept.

9. $x + y = 5$

10. $2x + y = 3$

11. $2x - y = -2$

12. $x + 3y = 6$

13. $x + 3y = 9$

14. $4x - y = 8$

15. $2x - 3y = 6$

16. $3x - 4y = 12$

17. $2x - y = 0$

18. $3x + y = 0$

19. $y + 3 = 0$

20. $y - 2 = 0$

In exercises 21 to 38, write the equation of the line passing through each of the given points with the indicated slope. Give your results in slope-intercept form, when possible.

21. $(0, 2)$, $m = 3$

22. $(0, -4)$, $m = -2$

23. $(0, 2)$, $m = \frac{3}{2}$

24. $(0, -3)$, $m = -2$

25. $(0, 4)$, $m = 0$

26. $(0, 5)$, $m = -\frac{3}{5}$

27. $(0, -5)$, $m = \frac{5}{4}$

28. $(0, -4)$, $m = -\frac{3}{4}$

29. $(1, 2)$, $m = 3$

30. $(-1, 2)$, $m = 3$

31. $(-2, -3)$, $m = -3$

32. $(1, -4)$, $m = -4$

33. $(5, -3)$, $m = \frac{2}{5}$

34. $(4, 3)$, $m = 0$

35. $(2, -3)$, m is undefined

36. $(2, -5)$, $m = \frac{1}{4}$

37. $(5, 0)$, $m = -\frac{4}{5}$

38. $(-3, 0)$, m is undefined

In exercises 39 to 48, write the equation of the line passing through each of the given pairs of points. Write your result in slope-intercept form, when possible.

39. $(2, 3)$ and $(5, 6)$

40. $(3, -2)$ and $(6, 4)$

41. $(-2, -3)$ and $(2, 0)$

42. $(-1, 3)$ and $(4, -2)$

43. $(-3, 2)$ and $(4, 2)$

44. $(-5, 3)$ and $(4, 1)$

45. $(2, 0)$ and $(0, -3)$

46. $(2, -3)$ and $(2, 4)$

47. $(0, 4)$ and $(-2, -1)$

48. $(-4, 1)$ and $(3, 1)$

In exercises 49 to 58, write the equation of the line L satisfying the given geometric conditions.

49. L has slope 4 and y intercept $(0, -2)$.

50. L has slope $-\frac{2}{3}$ and y intercept $(0, 4)$.

51. L has x intercept $(4, 0)$ and y intercept $(0, 2)$.

52. L has x intercept $(-2, 0)$ and slope $\frac{3}{4}$.

53. L has y intercept $(0, 4)$ and a 0 slope.

54. L has x intercept $(-2, 0)$ and an undefined slope.

55. L passes through point $(3, 2)$ with a slope of 5.

56. L passes through point $(-2, -4)$ with a slope of $-\frac{3}{2}$.

57. L has y intercept $(0, 3)$ and is parallel to the line with equation $y = 3x - 5$.

58. L has y intercept $(0, -3)$ and is parallel to the line with equation $y = \frac{2}{3}x + 1$.

In exercises 59 to 70, write the equation of each line in function form.

59. L has y intercept $(0, 4)$ and is perpendicular to the line with equation $y = -2x + 1$.

60. L has y intercept $(0, 2)$ and is parallel to the line with equation $y = -1$.

61. L has y intercept $(0, 3)$ and is parallel to the line with equation $y = 2$.

ANSWERS

34. ______
35. ______
36. ______
37. ______
38. ______
39. ______
40. ______
41. ______
42. ______
43. ______
44. ______
45. ______
46. ______
47. ______
48. ______
49. ______
50. ______
51. ______
52. ______
53. ______
54. ______
55. ______
56. ______
57. ______
58. ______
59. ______
60. ______
61. ______

62. ________
63. ________
64. ________
65. ________
66. ________
67. ________
68. ________
69. ________
70. ________
71. ________
72. ________
73. ________
74. ________
75. ________
76. ________
77. ________
78. ________
79. ________
80. ________

62. L has y intercept (0, 2) and is perpendicular to the line with equation $2x - 3y = 6$.

63. L passes through point $(-3, 2)$ and is parallel to the line with equation $y = 2x - 3$.

64. L passes through point $(-4, 3)$ and is parallel to the line with equation $y = -2x + 1$.

65. L passes through point (3, 2) and is parallel to the line with equation $y = \frac{4}{3}x + 4$.

66. L passes through point $(-2, -1)$ and is perpendicular to the line with equation $y = 3x + 1$.

67. L passes through point $(5, -2)$ and is perpendicular to the line with equation $y = -3x - 2$.

68. L passes through point (3, 4) and is perpendicular to the line with equation $y = -\frac{3}{5}x + 2$.

69. L passes through $(-2, 1)$ and is parallel to the line with equation $x + 2y = 4$.

70. L passes through $(-3, 5)$ and is parallel to the x axis.

71. Geometry. Find the equation of the perpendicular bisector of the segment joining $(-3, -5)$ and $(5, 9)$. *Hint:* First determine the midpoint of the segment. The midpoint would be $\left(\frac{-3 + 5}{2}, \frac{-5 + 9}{2}\right) = (1, 2)$. The perpendicular bisector passes through that point and is perpendicular to the line segment connecting the points.

72. Geometry. Find the equation of the perpendicular bisector of the segment joining $(-2, 3)$ and $(8, 5)$.

In exercises 73 to 76, without graphing, compare and contrast the slopes and y intercepts of the equations.

73. **(a)** $y = 2x + 1$, **(b)** $y = 2x - 5$, **(c)** $y = 2x$

74. **(a)** $y = 3x + 2$, **(b)** $y = \frac{3}{4}x + 2$, **(c)** $y = -2x + 2$

75. **(a)** $y = 4x + 5$, **(b)** $y = -4x + 5$, **(c)** $y = 4x - 5$

76. **(a)** $y = 3x$, **(b)** $y = \frac{1}{3}x$, **(c)** $y = -3x$

In exercises 77 to 80, use your graphing utility to graph the following.

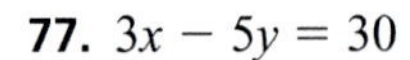

77. $3x - 5y = 30$

78. $2x + 7y = 14$

79. The line with slope $\frac{2}{3}$ and y intercept at (0, 7).

80. The line with slope $-\frac{1}{5}$ and y intercept at (0, 3).

In exercises 81 to 88, use the graph to determine the slope and y intercept of the line. Then write the equation of the line in slope-intercept form.

81. **82.** **83.**

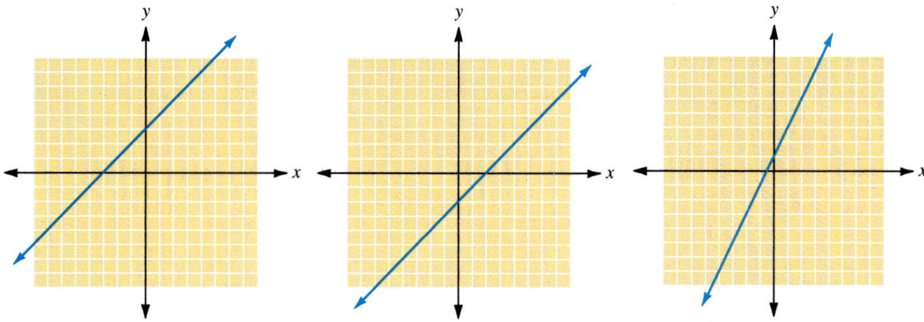

84. **85.** **86.**

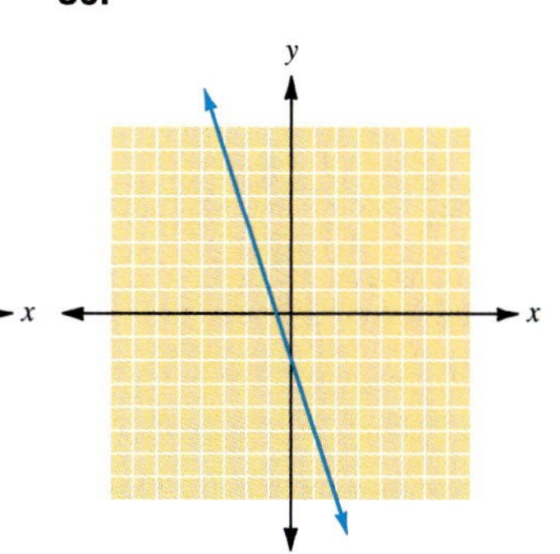

87. **88.**

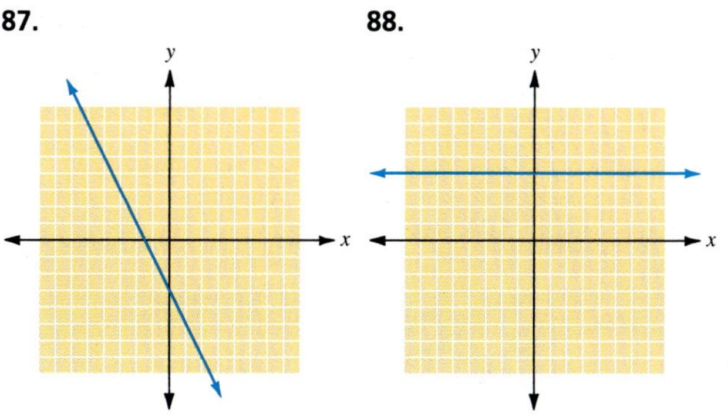

In exercises 89 to 94, graph the functions.

89. $f(x) = 3x + 7$ **90.** $f(x) = -2x - 5$ **91.** $f(x) = -2x + 7$

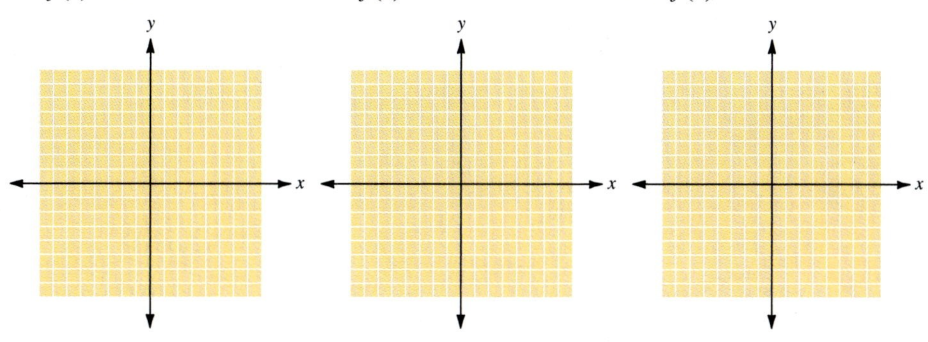

ANSWERS

81. ____________

82. ____________

83. ____________

84. ____________

85. ____________

86. ____________

87. ____________

88. ____________

89. ____________

90. ____________

91. ____________

ANSWERS

92. ______

93. ______

94. ______

95. ______

96. ______

97. ______

98. ______

99. ______

100. ______

101. ______

102. ______

103. ______

104. ______

105. ______

106. ______

92. $f(x) = -3x + 8$ **93.** $f(x) = -x - 1$ **94.** $f(x) = -2x - 5$

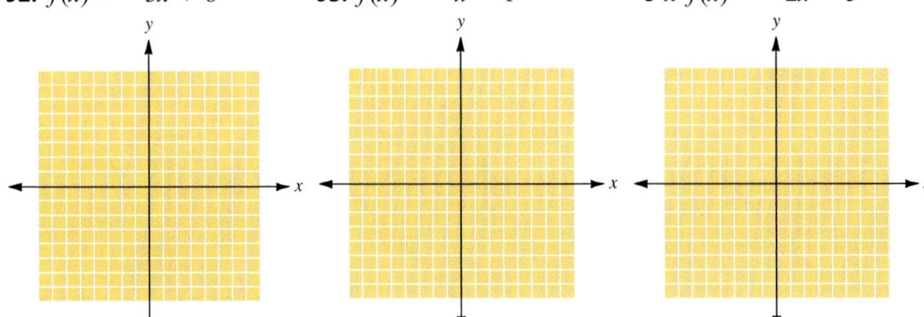

In exercises 95 to 100, if $f(x) = 4x - 3$, find the following.

95. $f(5)$

96. $f(0)$

97. $f(4)$

98. $f(-1)$

99. $f(-4)$

100. $f\left(\frac{1}{2}\right)$

In exercises 101 to 106, if $f(x) = 5x - 1$, find the following.

101. $f(a)$

102. $f(2r)$

103. $f(x + 1)$

104. $f(a - 2)$

105. $f(x + h)$

106. $\dfrac{f(x + h) - f(x)}{h}$

In exercises 107 to 110, if $g(x) = -3x + 2$, find the following.

107. $g(m)$

108. $g(5n)$

109. $g(x + 2)$

110. $g(s - 1)$

In exercises 111 to 114, let $f(x) = 2x + 3$.

111. Find $f(1)$

112. Find $f(3)$

113. Form the ordered pairs $(1, f(1))$ and $(3, f(3))$.

114. Write the equation of the line passing through the points determined by the ordered pairs in exercise 113.

ANSWERS

107. ______

108. ______

109. ______

110. ______

111. ______

112. ______

113. ______

114. ______

Answers

1. **(e)** **3.** **(a)** **5.** **(b)** **7.** **(h)** **9.** $y = -x + 5, m = -1, y$ int: $(0, 5)$

11. $y = 2x + 2, m = 2, y$ int: $(0, 2)$ **13.** $y = -\frac{1}{3}x + 3, m = -\frac{1}{3}, y$ int: $(0, 3)$

15. $y = \frac{2}{3}x - 2, m = \frac{2}{3}, y$ int: $(0, -2)$ **17.** $y = 2x, m = 2, y$ int: $(0, 0)$

19. $y = -3, m = 0, y$ int: $(0, -3)$ **21.** $y = 3x + 2$ **23.** $y = \frac{3}{2}x + 2$

25. $y = 4$ **27.** $y = \frac{5}{4}x - 5$ **29.** $y = 3x - 1$ **31.** $y = -3x - 9$

33. $y = \frac{2}{5}x - 5$ **35.** $x = 2$ **37.** $y = -\frac{4}{5}x + 4$ **39.** $y = x + 1$

41. $y = \frac{3}{4}x - \frac{3}{2}$ **43.** $y = 2$ **45.** $y = \frac{3}{2}x - 3$ **47.** $y = \frac{5}{2}x + 4$

49. $y = 4x - 2$ **51.** $y = -\frac{1}{2}x + 2$ **53.** $y = 4$ **55.** $y = 5x - 13$

57. $y = 3x + 3$ **59.** $f(x) = \frac{1}{2}x + 4$ **61.** $f(x) = 3$ **63.** $f(x) = 2x + 8$

65. $f(x) = \frac{4}{3}x - 2$ **67.** $f(x) = \frac{1}{3}x - \frac{11}{3}$ **69.** $f(x) = -\frac{1}{2}x$

71. $y = -\frac{4}{7}x + \frac{18}{7}$ **73.** Same slope but different y int.

75. **(a)** and **(b)** have the same y intercept but **(a)** rises and **(b)** falls, both at the same rate. **(a)** and **(c)** have the same slope but different y int.

77. 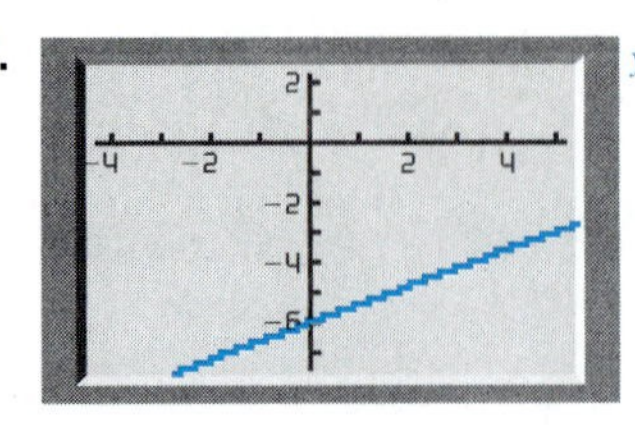

$y = \frac{3}{5}x - 6$

79. 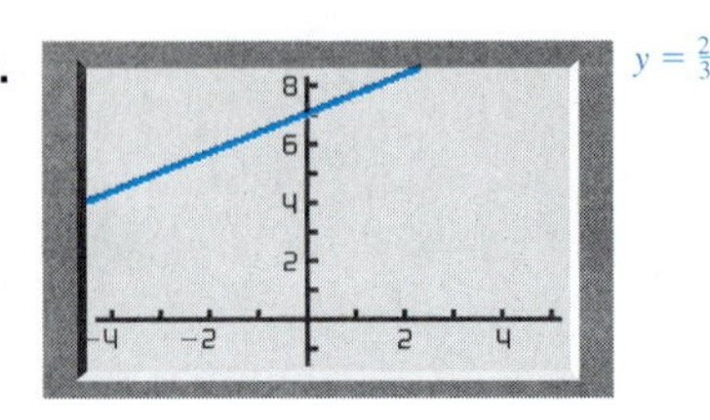

$y = \frac{2}{3}x + 7$

81. Slope = 1; y int (0, 3); $y = x + 3$

83. Slope = 2; y int (0, 1); $y = 2x + 1$

85. Slope = -3; y int (0, 1); $y = -3x + 1$

87. Slope = -2; y int (0, -3); $y = -2x - 3$

89. 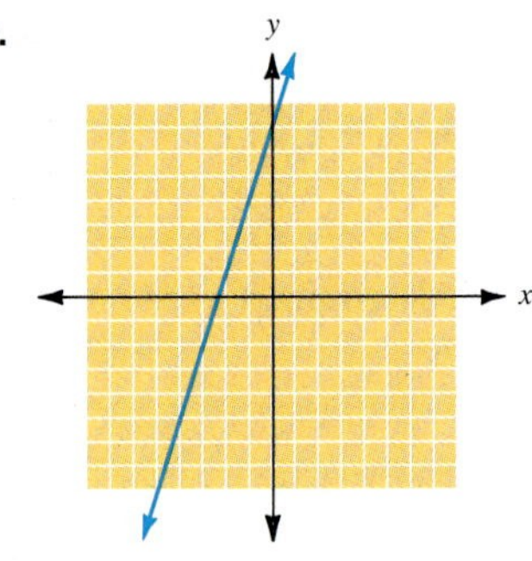

91. 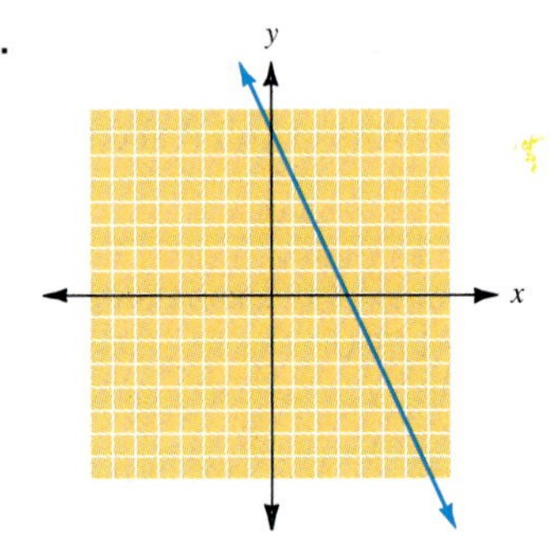

93. 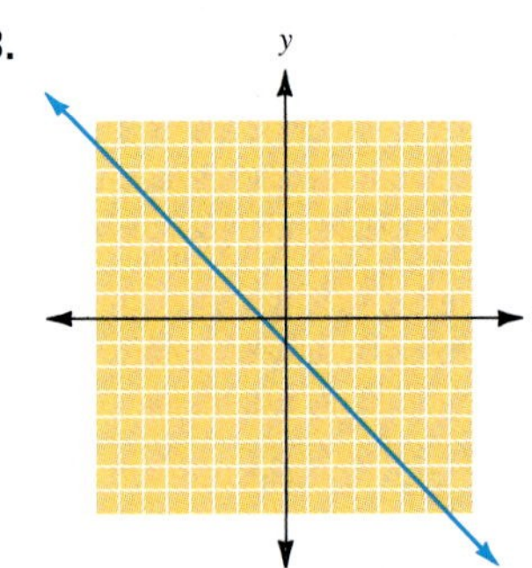

95. 17 **97.** 13 **99.** -19

101. $5a - 1$ **103.** $5x + 4$

105. $5x + 5h - 1$ **107.** $-3m + 2$

109. $-3x - 4$ **111.** 5

113. (1, 5), (3, 9)

Solving Equations and Inequalities Graphically

OBJECTIVES

1. Solve linear equations graphically
2. Solve linear inequalities graphically

In Chapter 2, we solved linear equations and inequalities. In this section, we will graphically demonstrate solutions for similar statements. In using this section, note that each graphical demonstration is accompanied by the algebraic solution, which appears in the margin. The techniques of this section are not designed as an alternative to the algebra. They are rather an introduction to the idea of "viewing a solution." This is a skill that will be very useful as you continue to study mathematics.

In our first example, we will solve a simple linear equation. The graphical method may seem cumbersome, but once you master it, you will find it quite helpful, particularly if you are a visual learner.

Example 1

A Graphical Approach to Solving a Linear Equation

Graphically solve the following equation.

NOTE Algebraically

$2x - 6 = 0$

$2x = 6$

$x = 3$

$$2x - 6 = 0$$

Step 1 Let each side of the equation represent a function of x.

$$f(x) = 2x - 6$$

$$g(x) = 0$$

Step 2 Graph the two functions on the same set of axes.

NOTE We ask the question, "when is the graph of f equal to the graph of g?" Specifically, for what values of x does this occur?

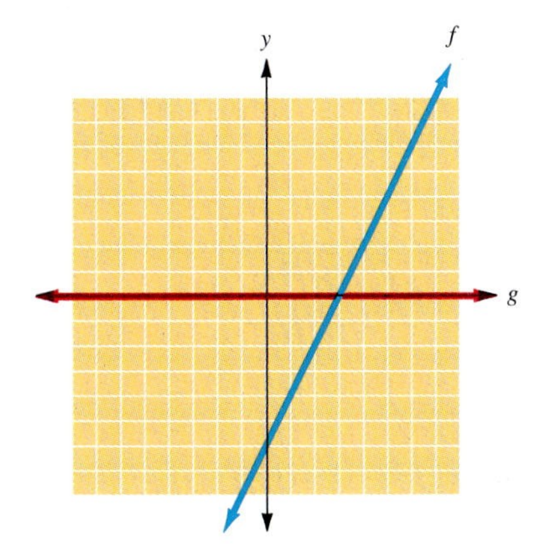

Step 3 Find the intersection of the two graphs. To do this, examine the two graphs closely to see where they intersect. Identify the coordinates of that point. This intersection determines the solution to the original equation.

The two lines intersect on the x axis at the point (3, 0). We are looking for the x value at the point of intersection, which is 3.

CHECK YOURSELF 1

Graphically solve the following equation.

$-3x + 6 = 0$

The graph of the equation is often used to check the algebraic solution. This concept is illustrated in Example 2.

Example 2

Solving Linear Equations Algebraically and Graphically

Solve the linear equation algebraically, then graphically display the solution.

$2(x + 3) = -3x - 4$

NOTE Algebraically

$2x + 6 = -3x - 4$

$5x + 6 = -4$

$5x = -10$

$x = -2$

The solution set is $\{-2\}$.

To graphically display the solution, let

$f(x) = 2(x + 3)$

$g(x) = -3x - 4$

Graphing both lines, we get

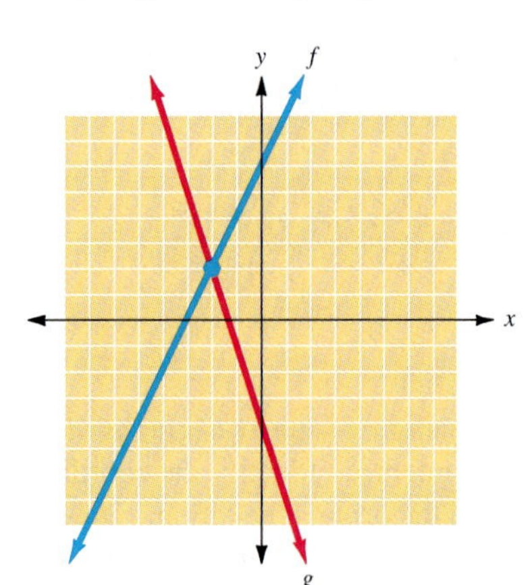

The point of intersection appears to be $(-2, 2)$, which confirms that -2 is a reasonable solution to the equation

$$2(x + 3) = -3x - 4$$

CHECK YOURSELF 2

First solve the linear equation algebraically, then graphically display the solution.

$$-3x + 4 = 2x - 1$$

The following algorithm summarizes our work in graphically solving an equation.

Step by Step: Graphically Solving an Equation

Step 1 Let each side of the equation represent a function of x.
Step 2 Graph the two functions on the same set of axes.
Step 3 Find the intersection of the two graphs. The x value at this intersection represents the solution to the original equation.

We will now use the graphs of linear functions to determine the solutions of a linear inequality.

Linear inequalities in one variable, x, are obtained from linear equations by replacing the symbol for equality ($=$) with one of the inequality symbols ($<, >, \le, \ge$).

The general form for a linear inequality in one variable is

$$x < a$$

in which the symbol $<$ can be replaced with $>$, $\le$, or $\ge$. Examples of linear inequalities in one variable include

$$x \ge -3 \qquad 2x + 5 > 7 \qquad 2x - 3 \le 5x + 6$$

Recall that the solution set for an equation is the set of all values for the variable (or ordered pair) that make the equation a true statement. Similarly, the solution set for an inequality is the set of all values that make the inequality a true statement. Example 3 looks at the graphical approach to solving an inequality.

NOTE Algebraic solution:

$$2x + 5 > 7$$
$$2x + 5 - 5 > 7 - 5$$
$$2x > 2$$
$$x > 1$$

Example 3

Solving an Inequality Graphically

Solve the inequality graphically.

$$2x + 5 > 7$$

First, rewrite the inequality as a comparison of two functions. Here, $f(x) > g(x)$, in which $f(x) = 2x + 5$ and $g(x) = 7$.

Now graph the two functions on a single set of axes.

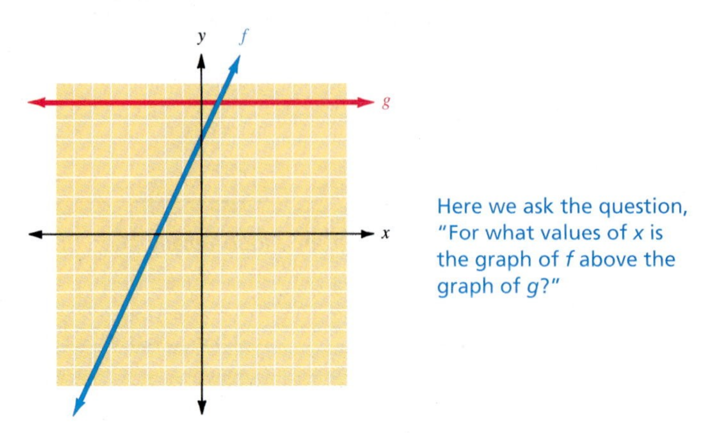

Here we ask the question, "For what values of x is the graph of f above the graph of g?"

Next, draw a vertical dotted line through the point of intersection of the two functions. In this case, there will be a vertical line through the point (1, 7).

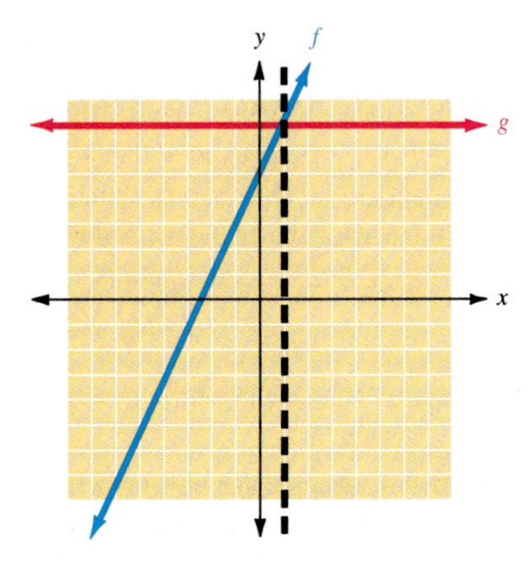

The solution set is every x value that results in $f(x)$ being greater than $g(x)$, which is every x value to the right of the dotted line.

NOTE The solution set will be all the x values that make the original statement, $2x + 5 > 7$, true.

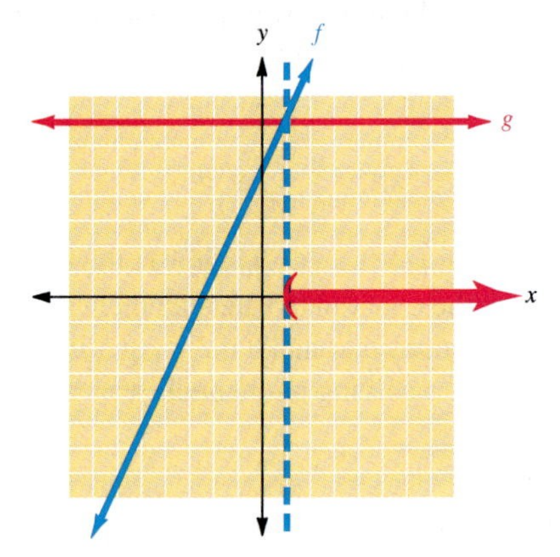

Finally, we express the solution set in set notation

$\{x \mid x > 1\}$

CHECK YOURSELF 3

Solve the inequality $3x - 2 < 4$ *graphically.*

In Example 3, the function $g(x) = 7$ resulted in a horizontal line. In Example 4, we see that the same method works when comparing any two functions.

Example 4

Solving an Inequality Graphically

NOTE Algebraic solution

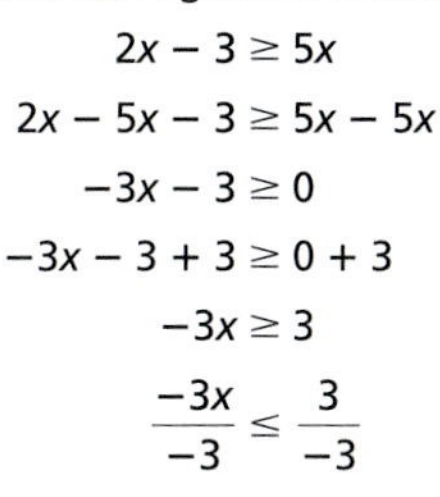

$$2x - 3 \geq 5x$$
$$2x - 5x - 3 \geq 5x - 5x$$
$$-3x - 3 \geq 0$$
$$-3x - 3 + 3 \geq 0 + 3$$
$$-3x \geq 3$$
$$\frac{-3x}{-3} \leq \frac{3}{-3}$$

(note what happens when we divide by a negative number)

$$x \leq -1$$

Solve the inequality graphically.

$2x - 3 \geq 5x$

First, rewrite the inequality as a comparison of two functions. Here, $f(x) \geq g(x)$, and $f(x) = 2x - 3$ and $g(x) = 5x$.

Now graph the two functions on a single set of axes.

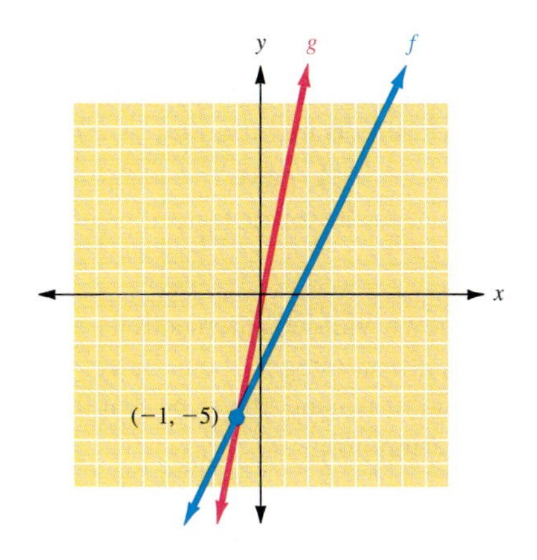

As in Example 3, draw a vertical line through the point of intersection of the two functions. The vertical line will go through the point $(-1, -5)$. In this case, the line is included (greater than or *equal to*), so the line is solid, not dotted.

Again, we need to mark every x value that makes the statement true. In this case, that is every x for which the line representing $f(x)$ is above or intersects the line representing $g(x)$. That is the region in which $f(x)$ is greater than or equal to $g(x)$. We mark the x values to the left of the line, but we also want to include the x value on the line, so we make it a bracket rather than a parenthesis.

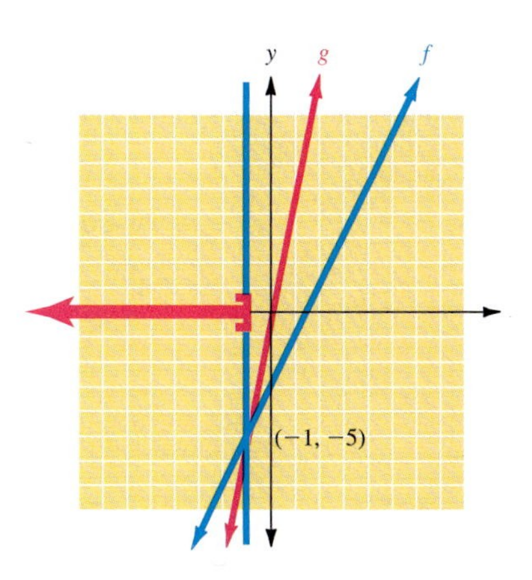

Finally, we express the solutions in set notation. We see that the solution set is every x value less than or equal to -1, so we write

$\{x \mid x \leq -1\}$

CHECK YOURSELF 4

Solve the inequality graphically.

$3x + 2 \geq -2x - 8$

The following algorithm summarizes our work in this section.

Step by Step: Solving an Inequality in One Variable Graphically

Step 1 Rewrite the inequality as a comparison of two functions.

$f(x) < g(x) \qquad f(x) > g(x) \qquad f(x) \leq g(x) \qquad f(x) \geq g(x)$

Step 2 Graph the two functions on a single set of axes.

Step 3 Draw a vertical line through the point of intersection of the two graphs. Use a dotted line if equality is not included ($<$ or $>$). Use a solid line if equality is included ($\leq$ or $\geq$).

Step 4 Mark the x values that make the inequality a true statement.

Step 5 Write the solutions in set notation.

The examples we have shown yielded intersections at x values that are integers. If the x value of the intersection is not an integer, it can be very difficult to read from a hand-drawn graph. If a graphing calculator is used, the trace feature can be used to get a very good approximation of the intersection point.

CHECK YOURSELF ANSWERS

1. $f(x) = -3x + 6$
$g(x) = 0$

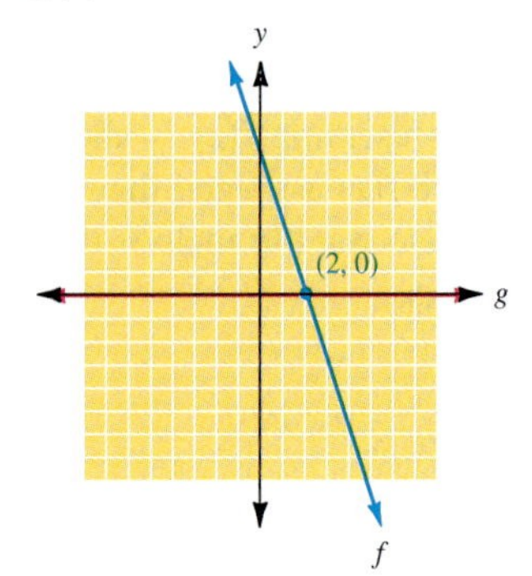

Solution set $\{2\}$

2. $f(x) = -3x + 4$
$g(x) = 2x - 1$

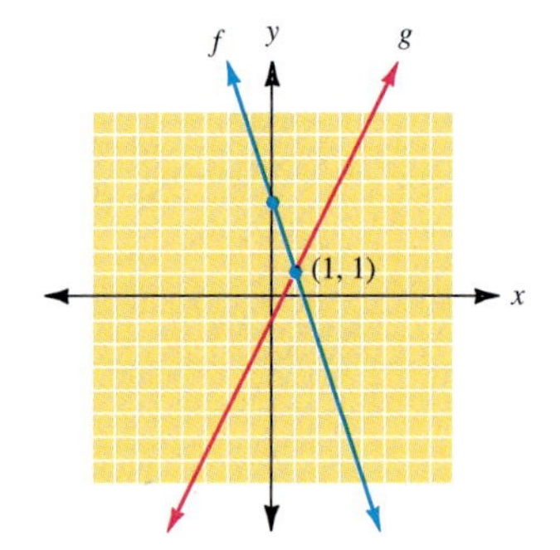

Solution set $\{1\}$

3.

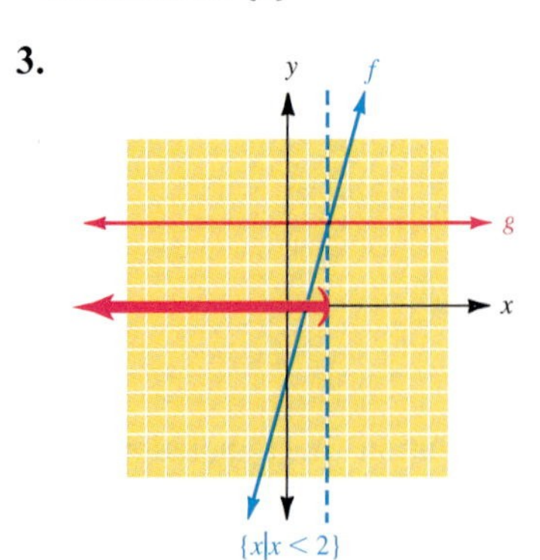

4.

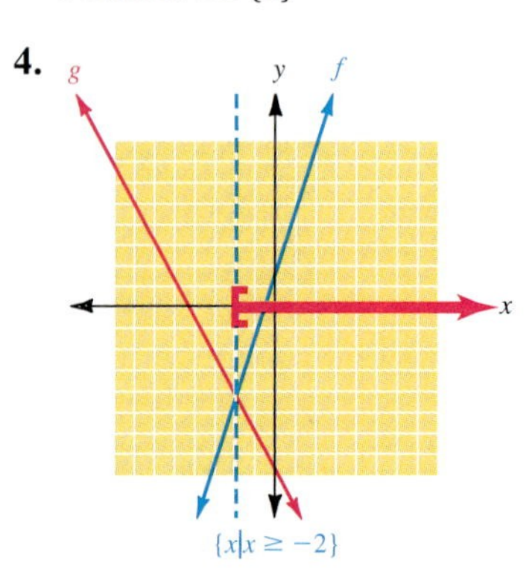

Name ____________________

Section ________ Date ________

4.4 Exercises

Graphically solve the following equations.

1. $2x - 8 = 0$

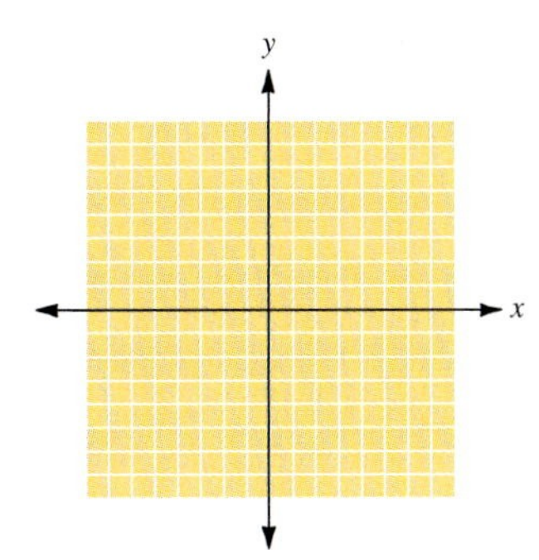

2. $4x + 12 = 0$

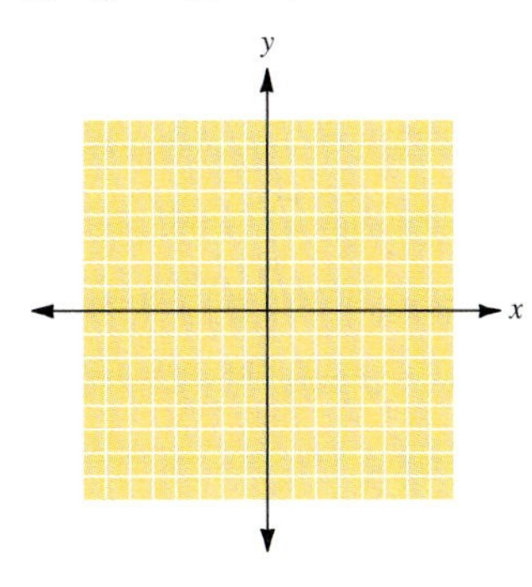

3. $7x - 7 = 0$

4. $2x - 6 = 0$

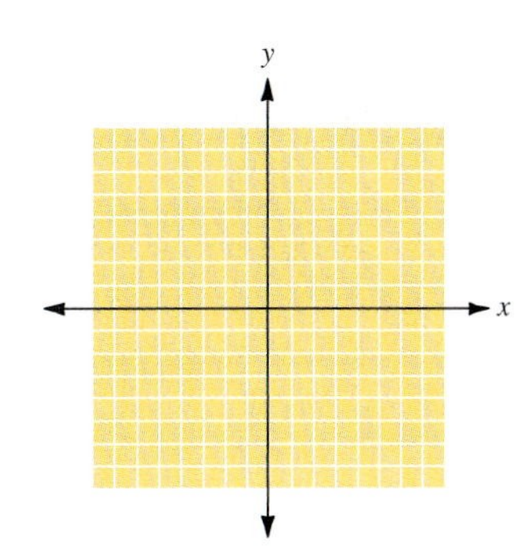

5. $5x - 8 = 2$

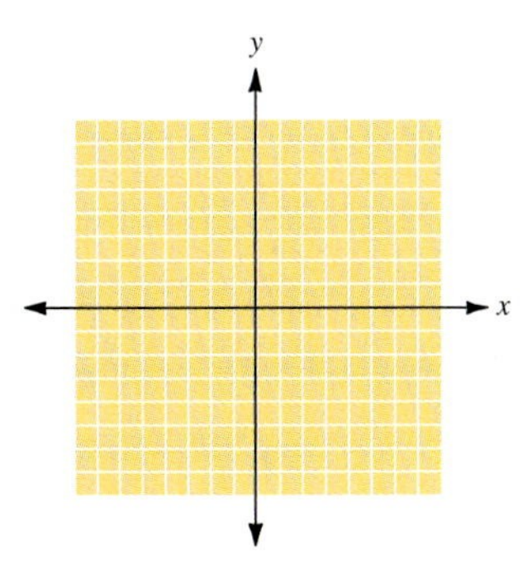

6. $4x + 5 = -3$

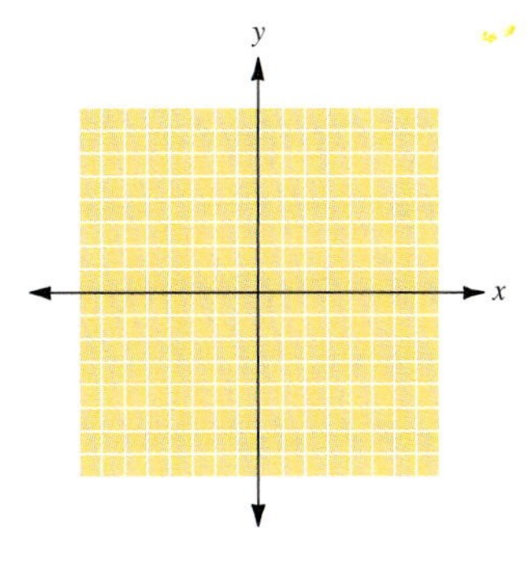

7. $2x - 3 = 7$

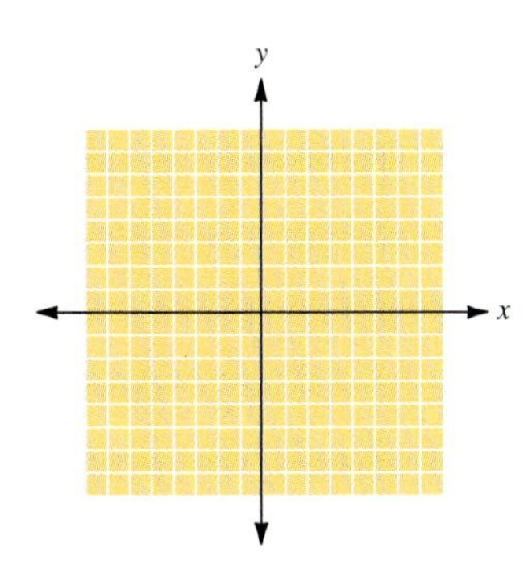

8. $5x + 9 = 4$

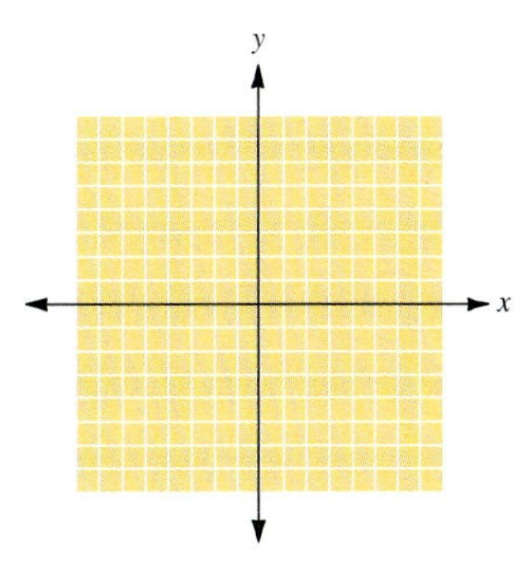

ANSWERS

1. ____________________
2. ____________________
3. ____________________
4. ____________________
5. ____________________
6. ____________________
7. ____________________
8. ____________________

ANSWERS

9. ______

10. ______

11. ______

12. ______

13. ______

14. ______

15. ______

16. ______

Solve the linear equations algebraically, then graphically display the solutions.

9. $3x - 2 = 2x + 1$

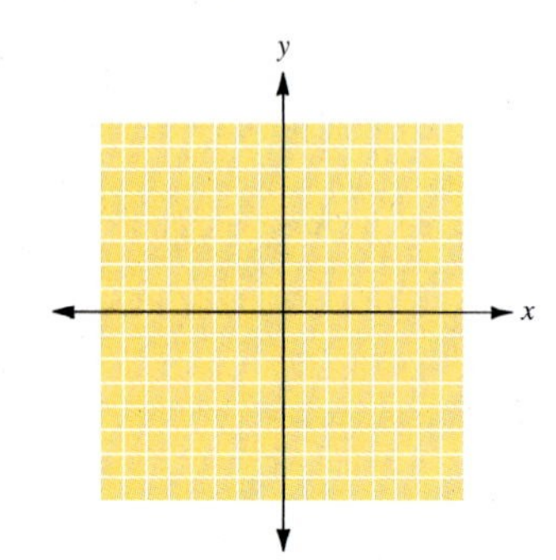

10. $4x + 3 = -x - 2$

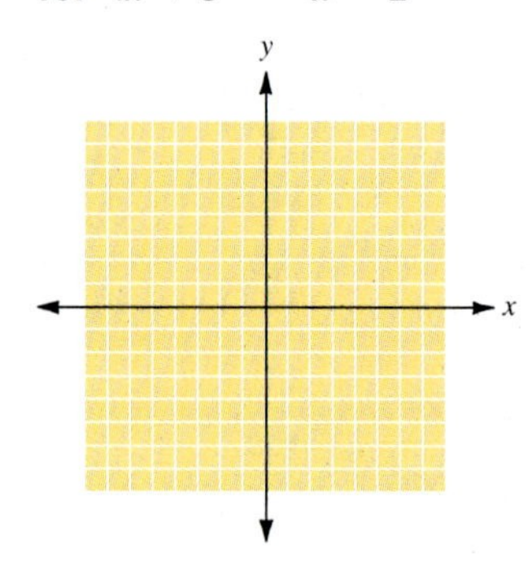

11. $\frac{7}{5}x - 3 = -\frac{2}{5}x + 6$

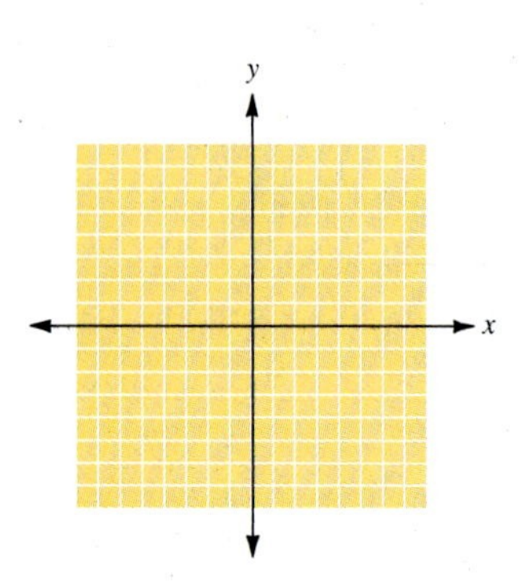

12. $2x - 3 = 3x - 2$

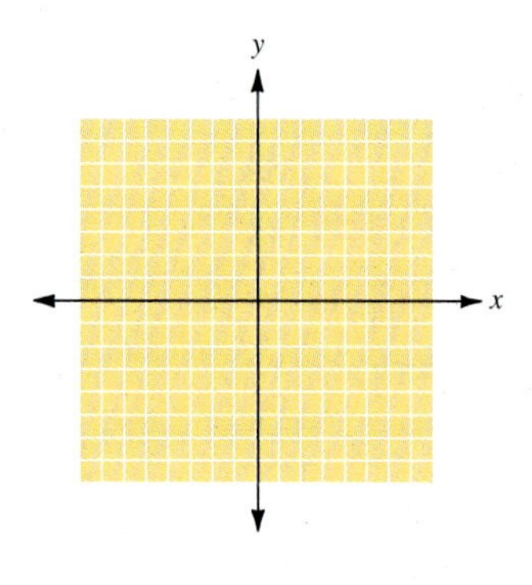

13. $3(x - 1) = 4x - 5$

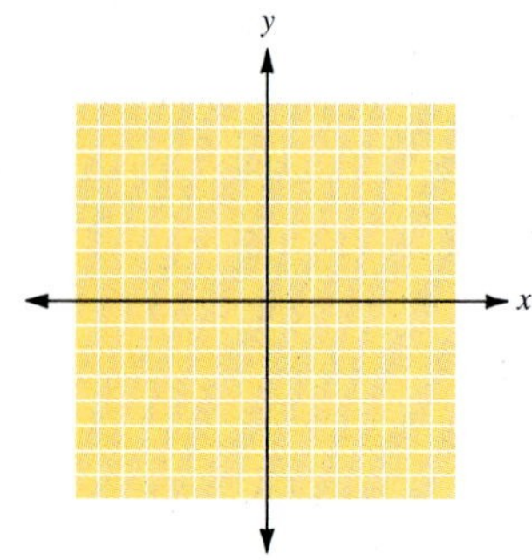

14. $2(x + 1) = 5x - 7$

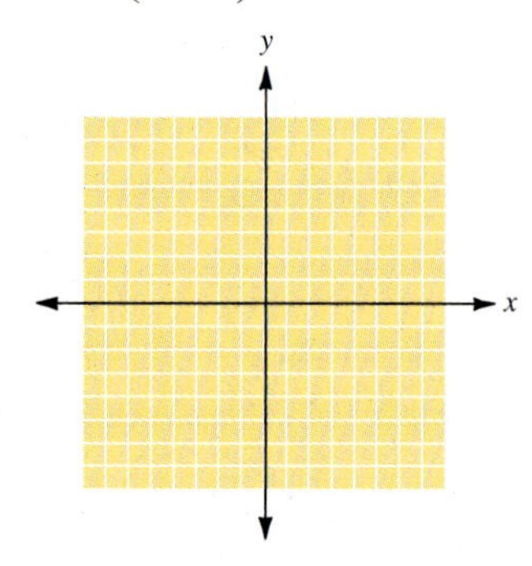

15. $7\left(\frac{1}{5}x - \frac{1}{7}\right) = x + 1$

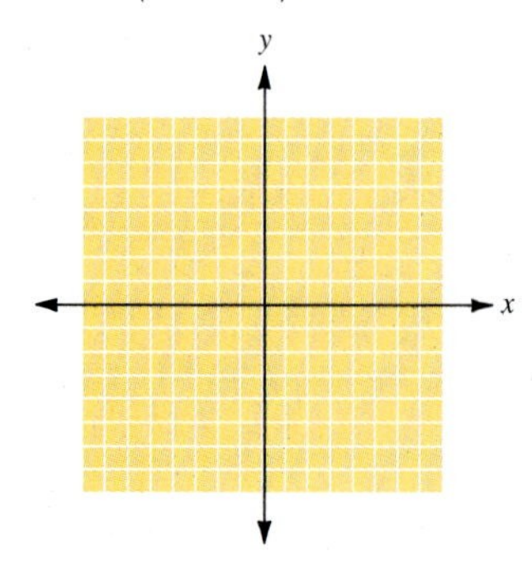

16. $2(2x - 1) = -2x + 10$

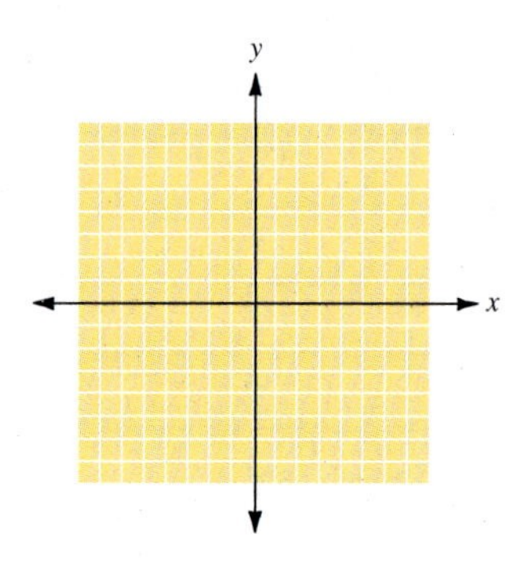

In exercises 17 to 32, solve each inequality graphically.

17. $2x < 8$

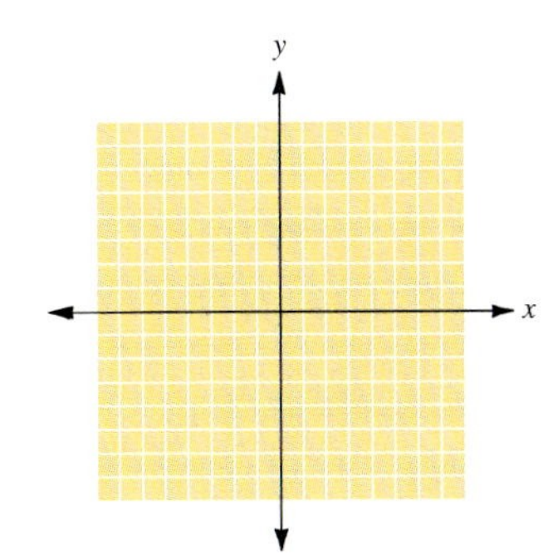

18. $-x < 4$

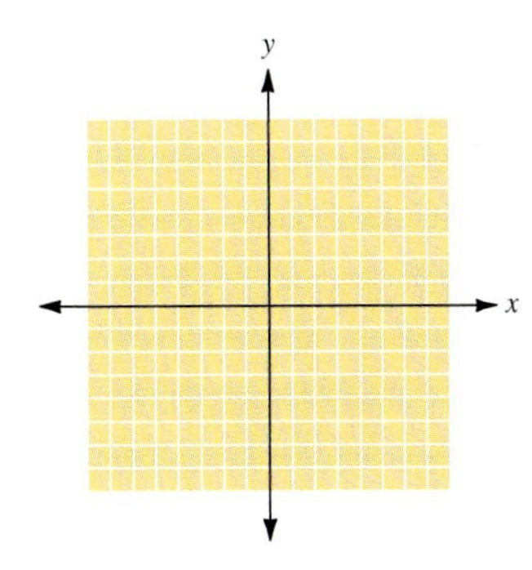

19. $\dfrac{x + 3}{2} < -1$

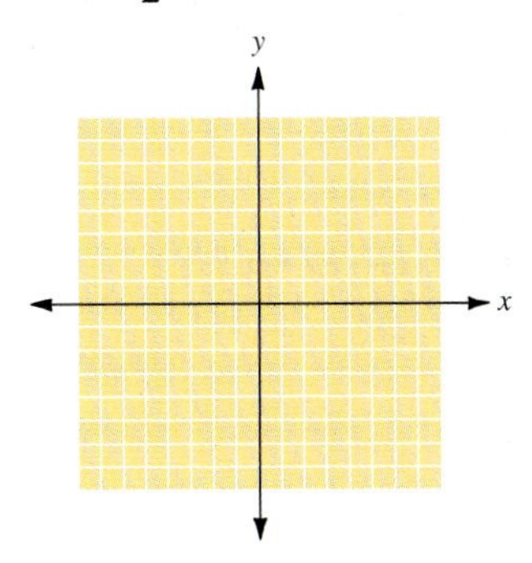

20. $\dfrac{-3x + 3}{4} > -3$

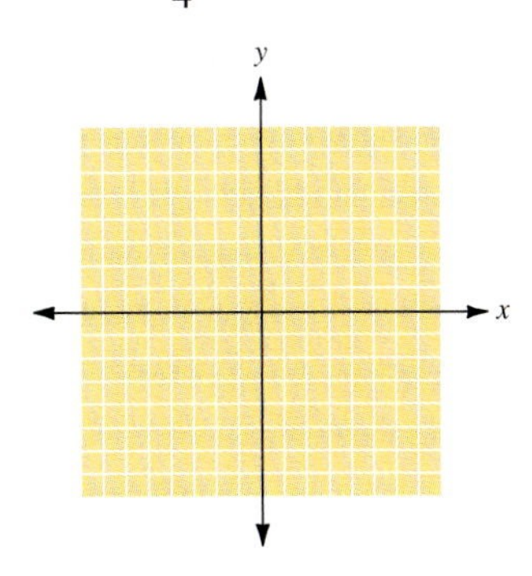

21. $6x \geq 6$

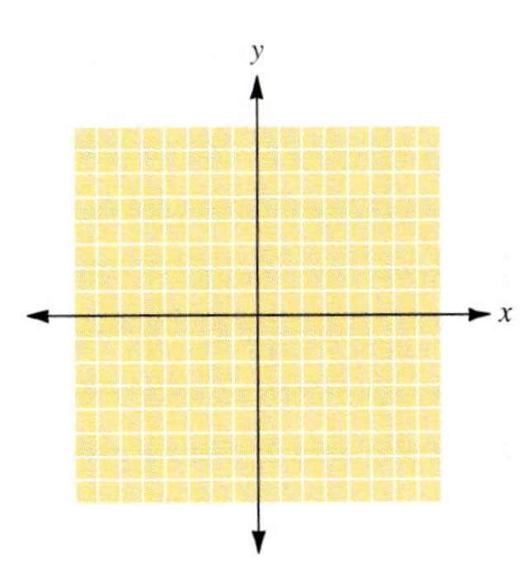

22. $-3x \leq 6$

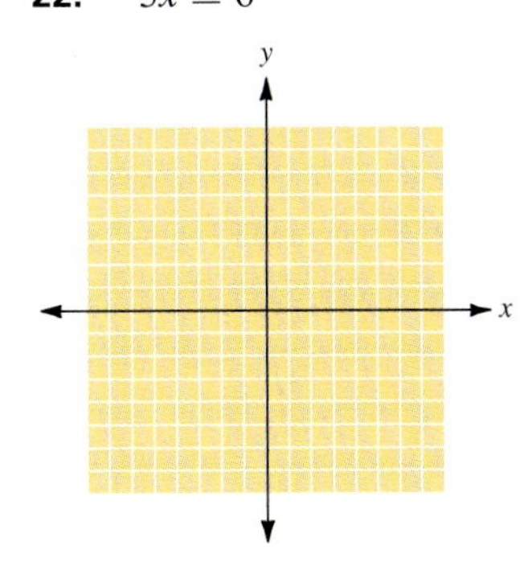

23. $7x - 7 < -2x + 2$

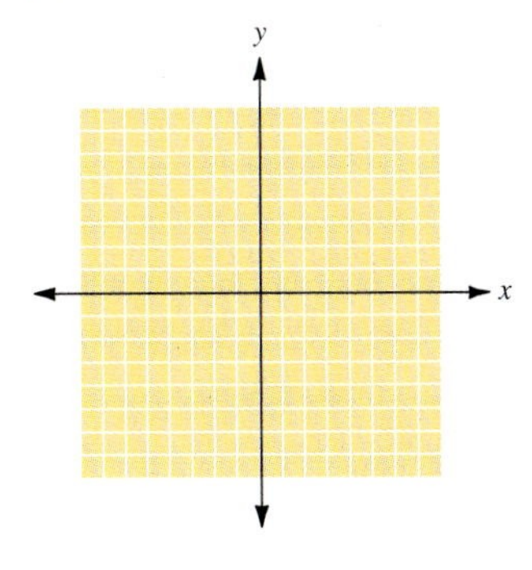

24. $7x + 2 > x - 4$

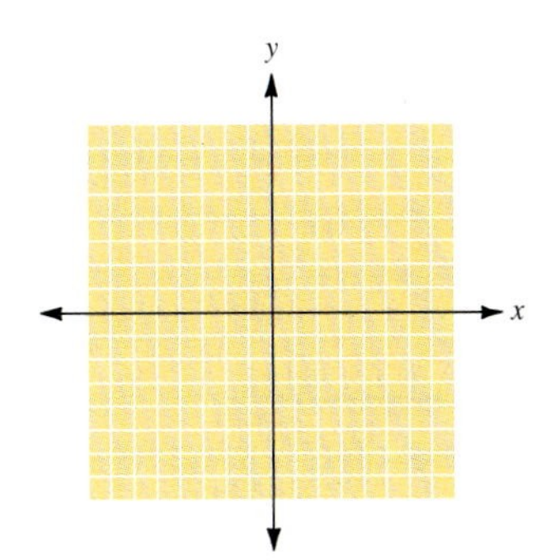

ANSWERS

17. ______

18. ______

19. ______

20. ______

21. ______

22. ______

23. ______

24. ______

ANSWERS

25. ______

26. ______

27. ______

28. ______

29. ______

30. ______

31. ______

32. ______

25. $2x - 7 \le -3(x - 1)$

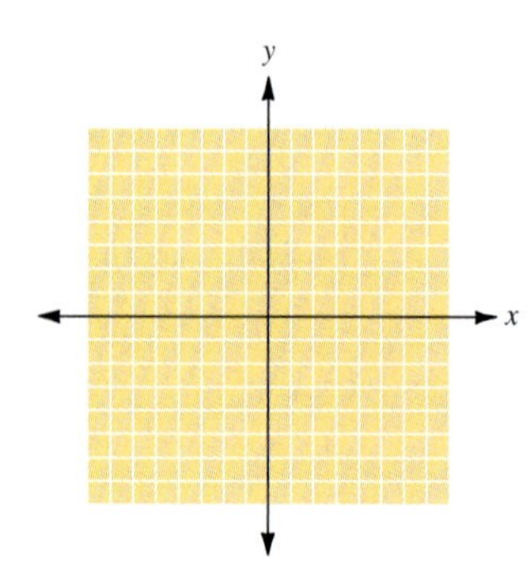

26. $2(3x + 1) < 4(x + 1)$

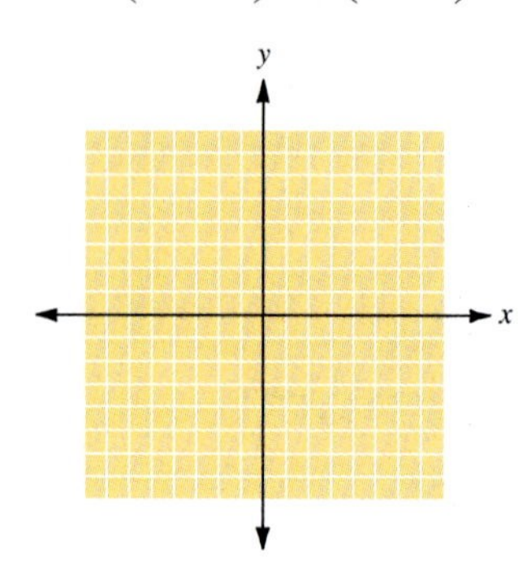

27. $6(1 + x) \ge 2(3x - 5)$

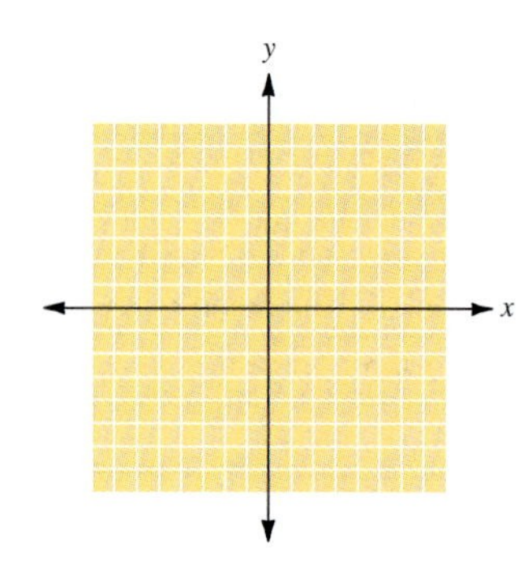

28. $2(x - 5) \ge 2x - 1$

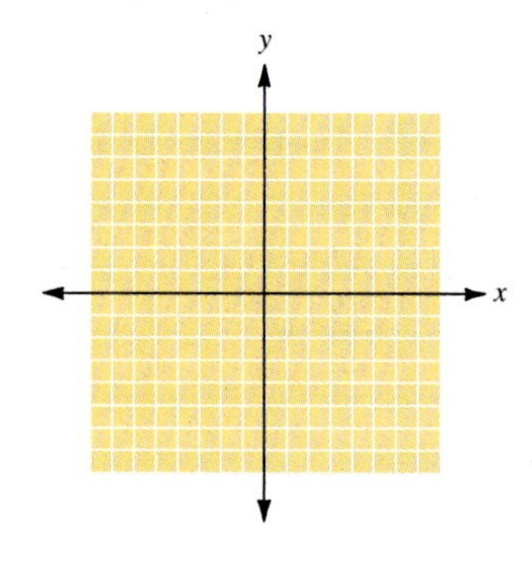

29. $3x < \frac{4x - 5}{3}$

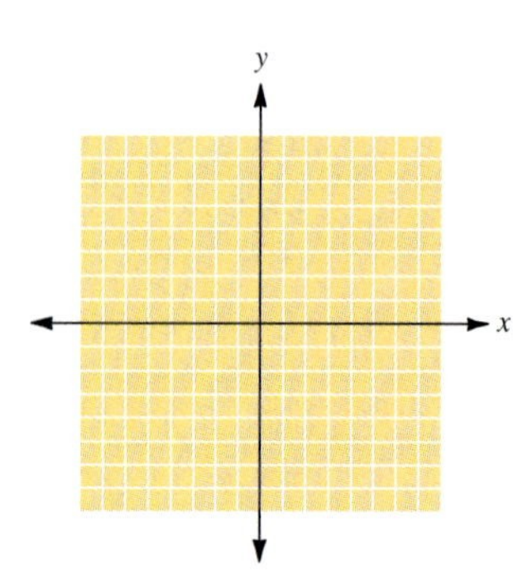

30. $-4x - 12 < x + 8$

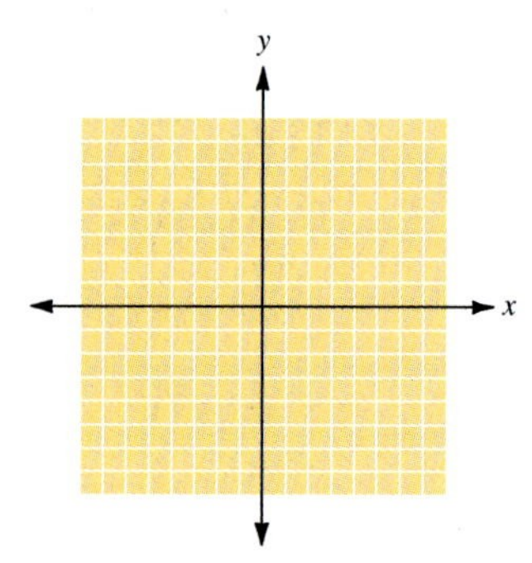

31. $4x - 6 \le 2x - 2(5x - 12)$

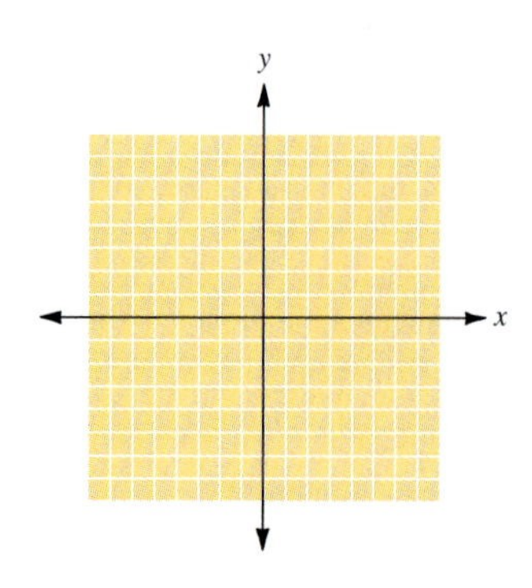

32. $5x + 3 > 2(4 - x) + 7x$

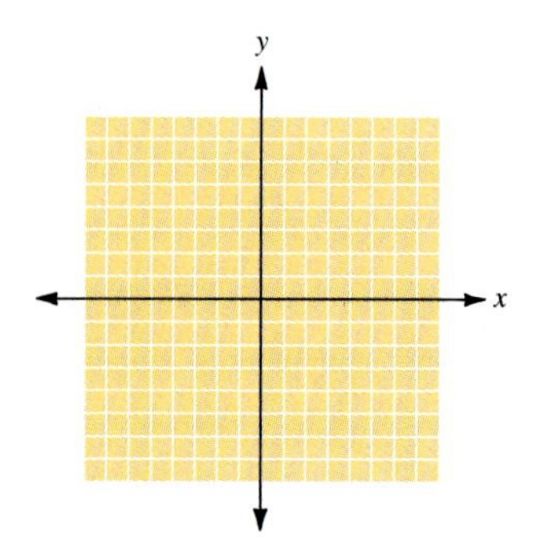

In exercises 33 to 38, solve the following applications.

33. Business. The cost to produce x units of wire is $C(x) = 50x + 5000$, and the revenue generated is $R(x) = 60x$. Find all values of x for which the product will at least break even.

34. Business. Find the values of x for which a product will at least break even if the cost is $C(x) = 85x + 900$ and the revenue is given by $R(x) = 105x$.

35. Car Rental. Tom and Jean went to Salem, Massachusetts, for 1 week. They needed to rent a car, so they checked out two rental firms. Wheels, Inc. wanted \$28 per day with no mileage fee. Downtown Edsel wanted \$98 per week and 14¢ per mile. Set up equations to express the rates of the two firms, and then decide when each deal should be taken.

36. Mileage. A fuel company has a fleet of trucks. The annual operating cost per truck is $C(x) = 0.58x + 7800$, in which x is the number of miles traveled by a truck per year. What number of miles will yield an operating cost that is less than \$25,000?

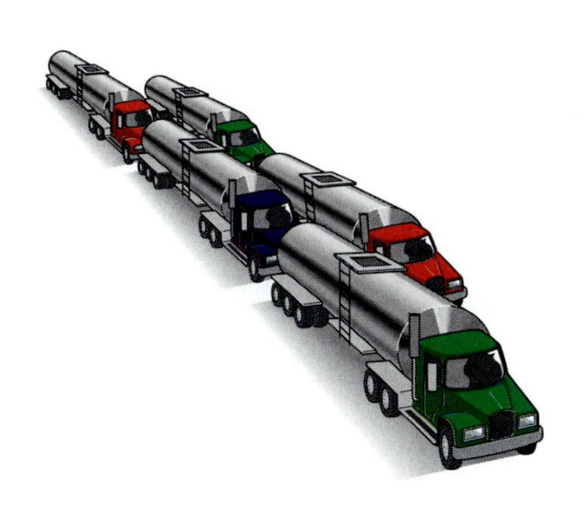

ANSWERS

33. ________

34. ________

35. ________

36. ________

ANSWERS

37. ______

38. ______

39. ______

40. ______

37. Wedding. Eileen and Tom are having their wedding reception at the Warrington Fire Hall. They can spend at the most \$3000 for the reception. If the hall charges a \$250 cleanup fee plus \$25 per person, find the largest number of people they can invite.

38. Tuition. A nearby college charges annual tuition of \$6440. Meg makes no more than \$1610 per year in her summer job. What is the smallest number of summers that she must work to make enough for 1 year's tuition?

39. Graphing. Explain to a relative how a graph is helpful in solving each inequality below. Be sure to include the significance of the point at which the lines meet (or what happens if the lines do not meet).

(a) $3x - 2 < 5$ **(b)** $3x - 2 \le 4 - x$ **(c)** $4(x - 1) \ge 2 + 4x$

40. College. Look at the data here about enrollment in college. Assume that the changes occurred at a constant rate over the years. Make one linear graph for men and one for women, but on the same set of axes. What conclusions could you draw from reading the graph?

Year	No., in Millions, of Men in the U.S. Enrolled in College	No., in Millions, of Women in the U.S. Enrolled in College
1960	2.3	1.2
1991	6.4	7.8

Answers

1.

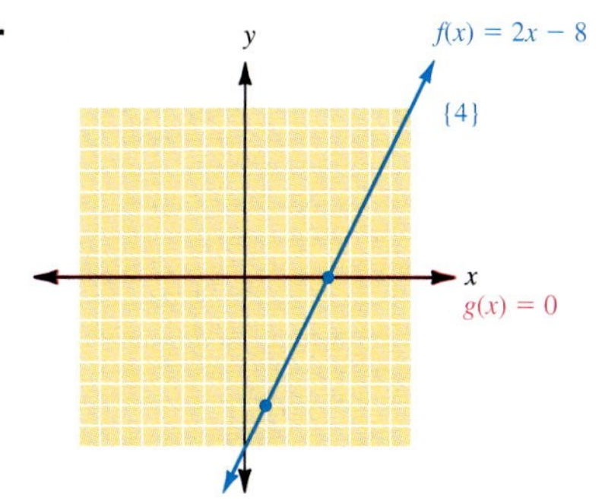

3.

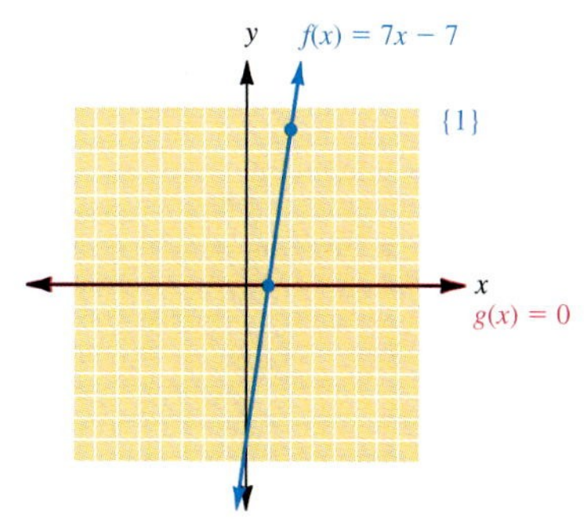

5.

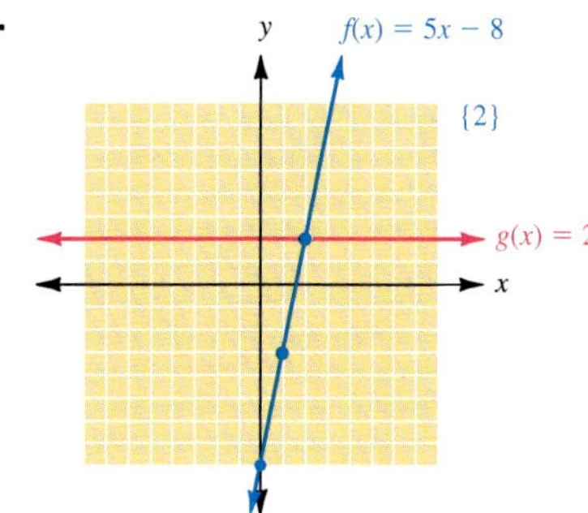

7.

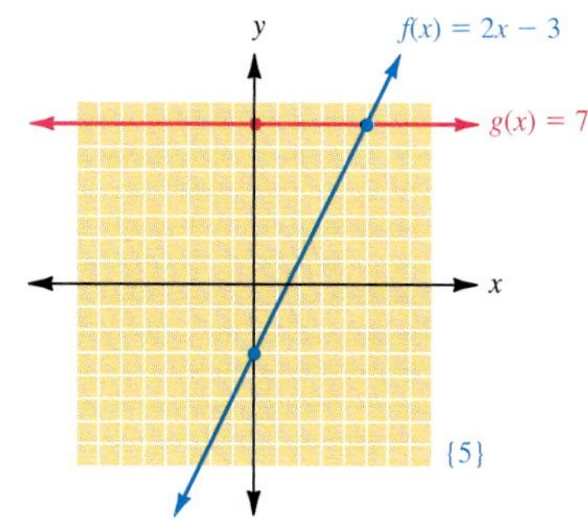

9.

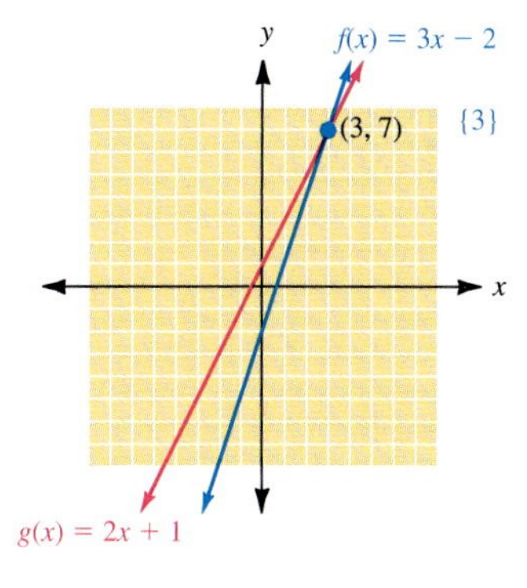

11.

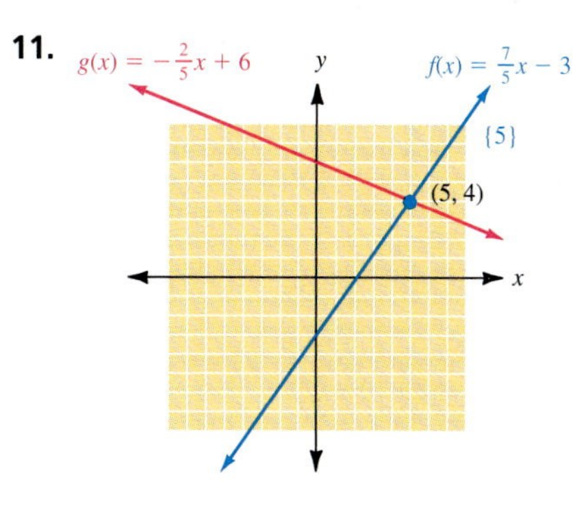

13.

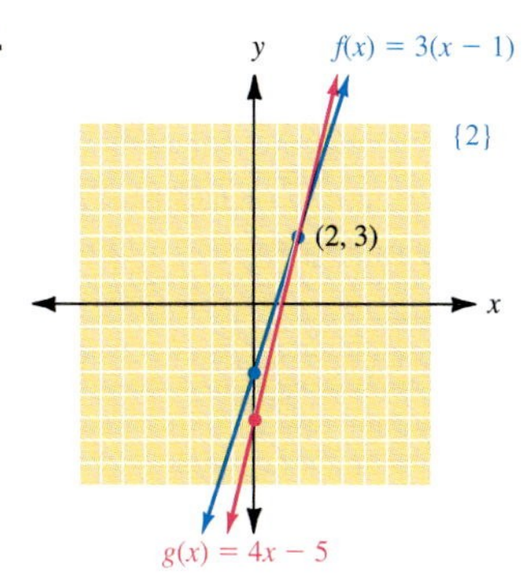

15.

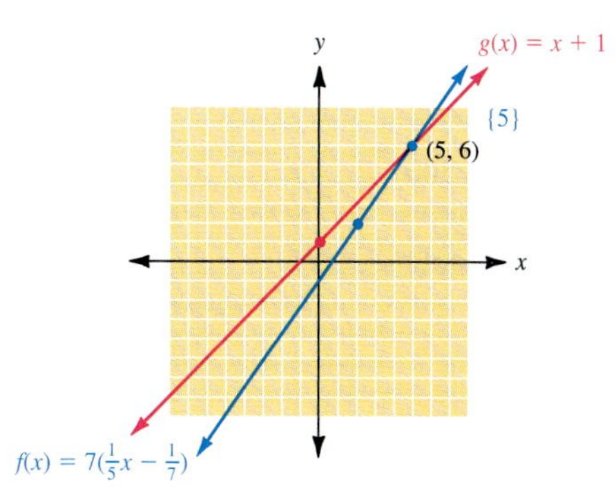

17.

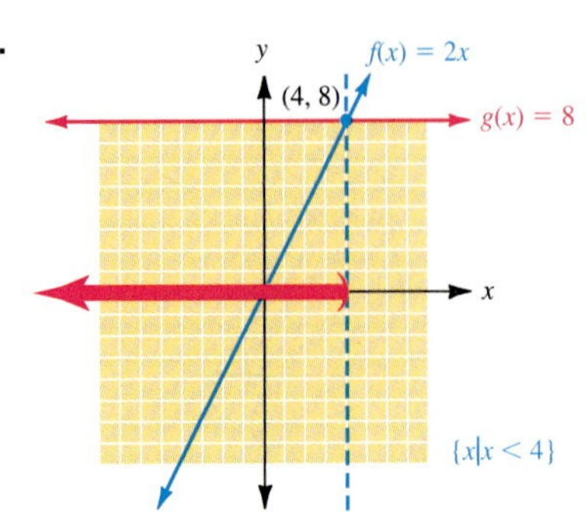

19.

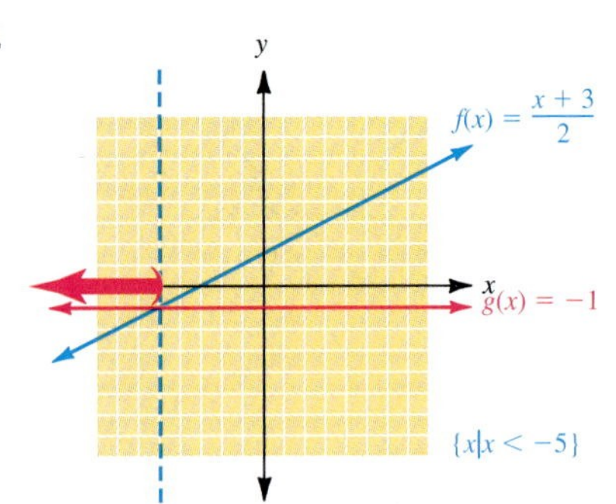

21.

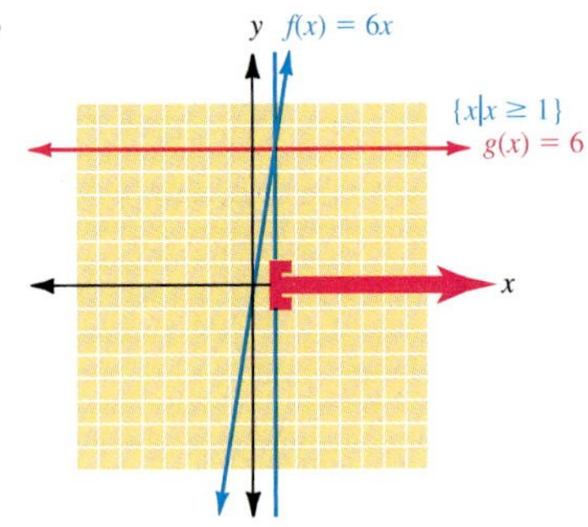

23.

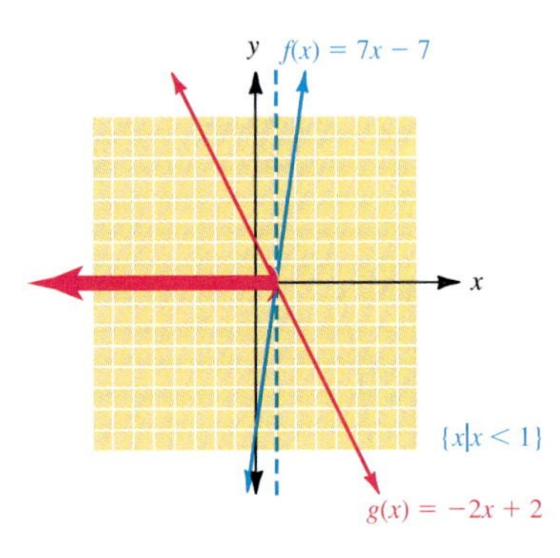

25.

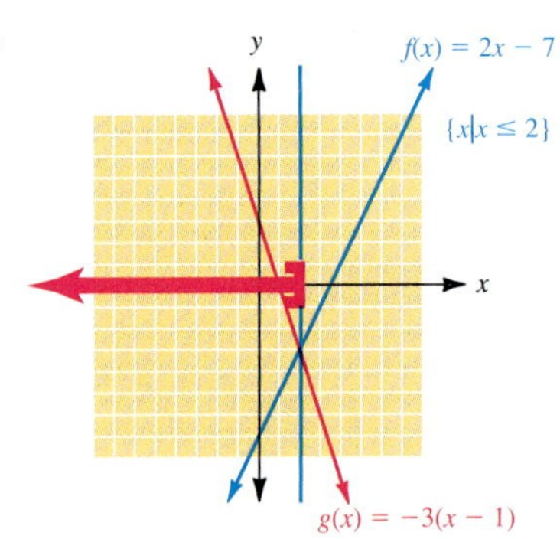

27.

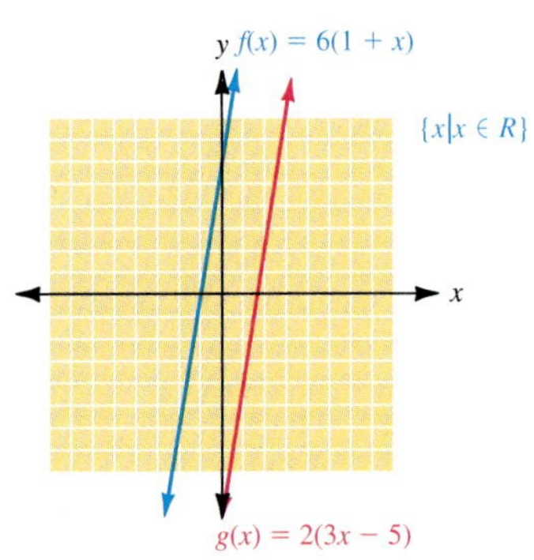

29.

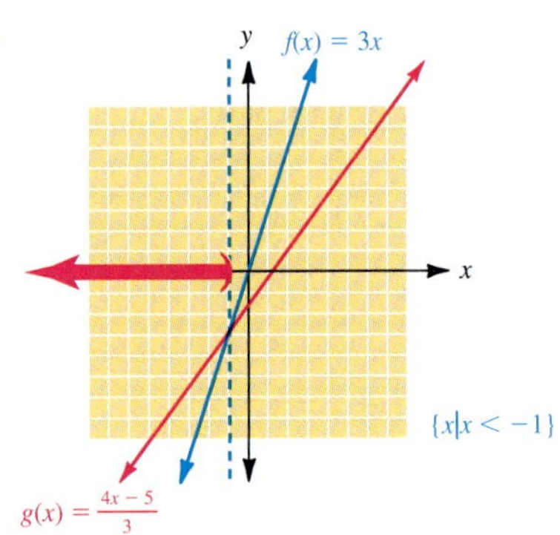

31.

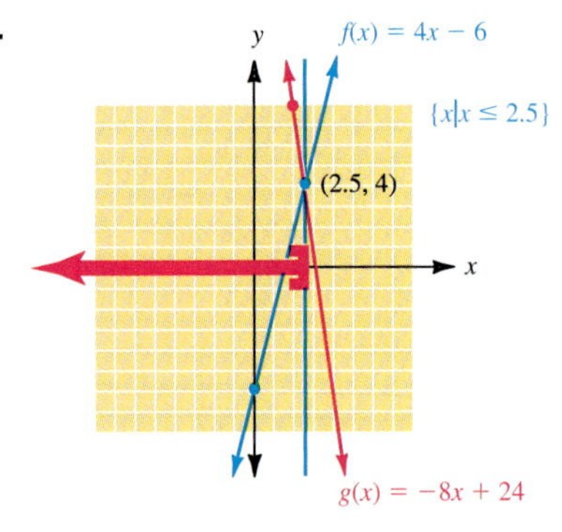

33. $\{x \mid x \geq 500\}$ **35.** If miles are under 700, Downtown Edsel; if over 700, Wheels, Inc.; $W = \$28 \times 7 = \$196$; $DE = 98 + 0.14x$ (x is number of miles)
37. 110 people **39.** 

Solving Absolute Value Equations and Inequalities Graphically

OBJECTIVES

1. Draw the graph of an absolute value function
2. Solve an absolute value equation graphically
3. Solve an absolute value inequality graphically

Equations may contain absolute value notation in their statements. In this section, we will look at graphically solving statements that include absolute values.

To look at a graphical approach to solving, we must first look at the graph of an absolute value function. We will start by looking at the graph of the function $f(x) = |x|$. All other graphs of absolute value functions are variations of this graph.

NOTE Graph the function $y = |x|$ as $Y_1 = \text{abs}(x)$

The graph can be found using a graphing calculator (most graphing calculators use **abs** to represent the absolute value). We will develop the graph from a table of values.

x	$f(x) = \|x\|$
−3	3
−2	2
−1	1
0	0
1	1
2	2

Plotting these ordered pairs, we see a pattern emerge. The graph is like a large V that has its vertex at the origin. The slope of the line to the right of 0 is 1, and the slope of the line to the left of 0 is -1.

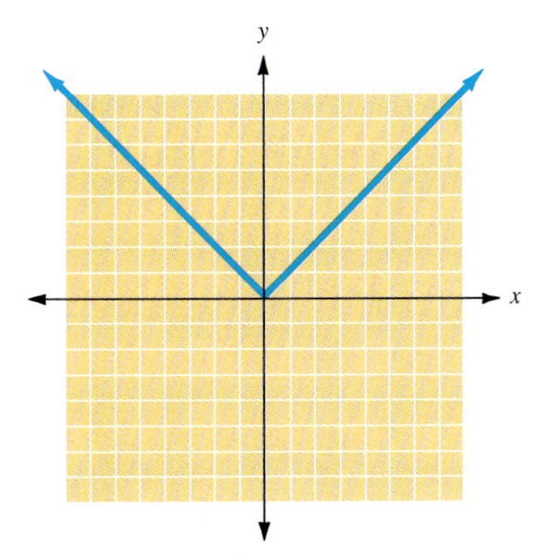

Let us now see what happens to the graph when we add or subtract some constant inside the absolute value bars.

Example 1

Graphing an Absolute Value Function

Graph each function.

NOTE $f(x) = |x - 3|$ would be entered as $Y_1 = \text{abs}(x - 3)$

(a) $f(x) = |x - 3|$

Again, we start with a table of values.

x	$f(x)$
−2	5
−1	4
0	3
1	2
2	1
3	0
4	1
5	2

Then, we plot the points associated with the set of ordered pairs.

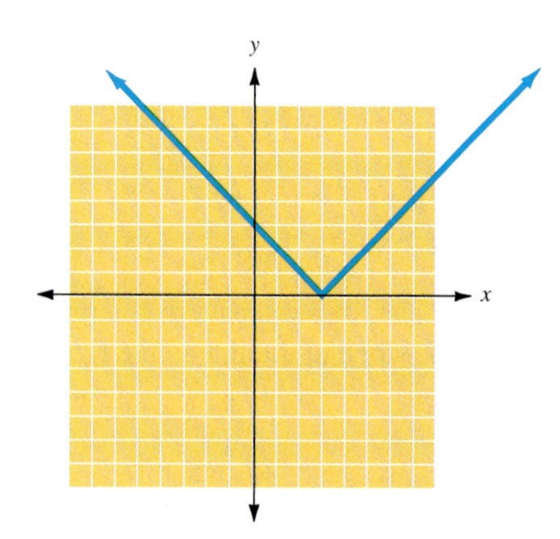

The graph of the function $f(x) = |x - 3|$ is the same shape as the graph of the function $f(x) = |x|$; it has just shifted to the right 3 units.

(b) $f(x) = |x + 1|$

We begin with a table of values.

x	$f(x)$
−2	1
−1	0
0	1
1	2
2	3
3	4

Then we graph.

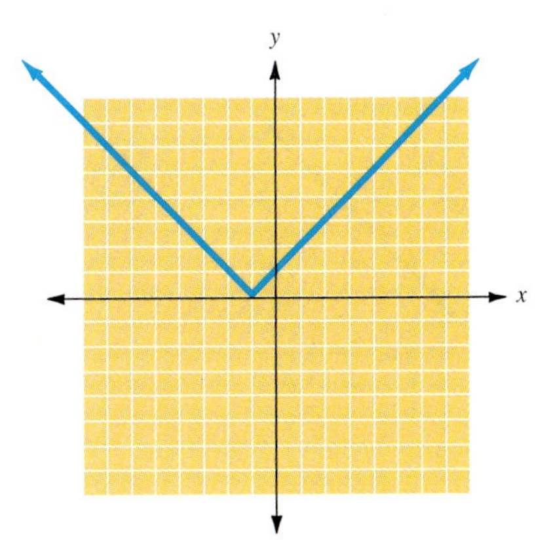

Note that the graph of $f(x) = |x + 1|$ is the same shape as the graph of the function $f(x) = |x|$, except that it is shifted 1 unit to the left.

CHECK YOURSELF 1

Graph each function.

(a) $f(x) = |x - 2|$ **(b)** $f(x) = |x + 3|$

We can summarize what we have discovered about the horizontal shift of the graph of an absolute value function.

Rules and Properties: Graphing Absolute Value Functions

The graph of the function $f(x) = |x - a|$ will be the same shape as the graph of $f(x) = |x|$ except that the graph will be shifted a units

to the right if a is positive
to the left if a is negative

NOTE If a is negative, $x - a$ will be x plus some positive number. For example, if $a = -2$,
$x - a = x - (-2) = x + 2$

We will now use these methods to solve equations that contain an absolute value expression.

Example 2

Solving an Absolute Value Equation Graphically

Graphically, find the solution set for the equation.

$|x - 3| = 4$

NOTE Algebraically
$|x - 3| = 4$
$x - 3 = 4$ or $x - 3 = -4$
$x = 7$ or $x = -1$

We graph the function associated with each side of the equation.

$f(x) = |x - 3|$ and $g(x) = 4$

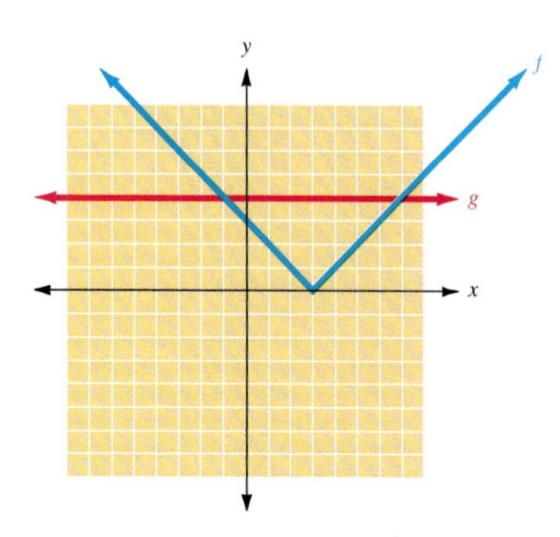

Then, we draw a vertical line through each of the intersection points.

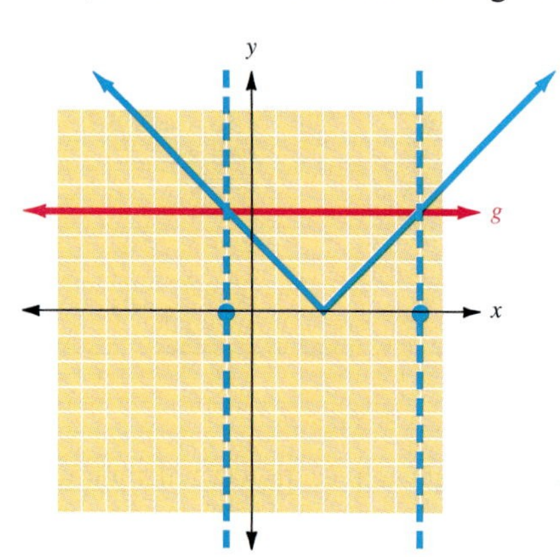

We ask the question, "For what values of x do f and g coincide?"

Looking at the x values of the two vertical lines, we find the solutions to the original equation. There are two x values that make the statement true: -1 and 7. The solution set is $\{-1, 7\}$.

CHECK YOURSELF 2

Graphically find the solution set for the equation.

$|x - 2| = 3$

Example 3 demonstrates a graphical approach to solving an absolute value inequality.

Example 3

Solving an Absolute Value Inequality Graphically

Graphically solve

$|x| < 6$

As we did in previous sections, we begin by letting each side of the inequality represent a function. Here

$f(x) = |x|$ and $g(x) = 6$

Now we graph both functions on the same set of axes.

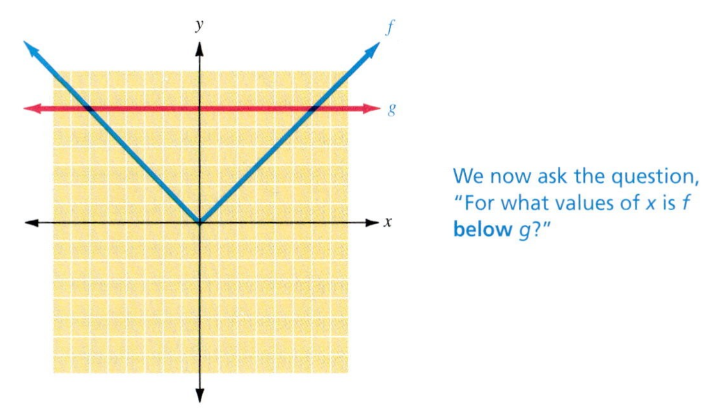

We now ask the question, "For what values of x is f **below** g?"

We next draw a dotted line (equality is not included) through the points of intersection of the two graphs.

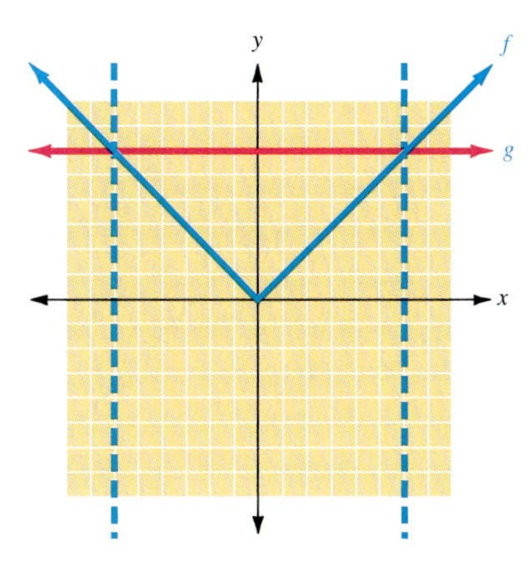

The solution set is any value of x for which the graph of $f(x)$ is below the graph of $g(x)$.

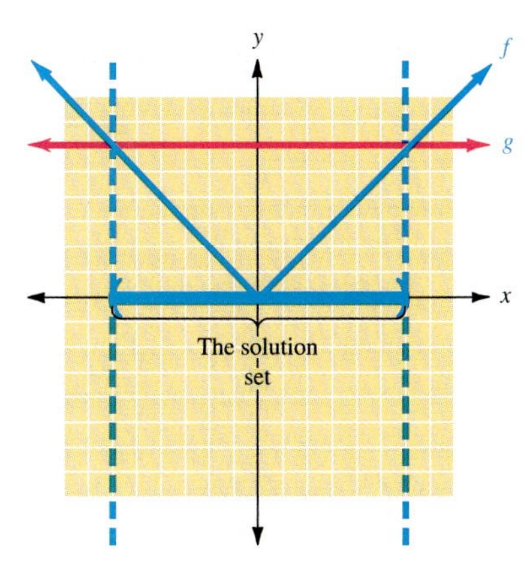

In set notation, we write $\{x \mid -6 < x < 6\}$.

CHECK YOURSELF 3

Graphically solve the inequality

$|x| > 3$

CHECK YOURSELF ANSWERS

1. (a)

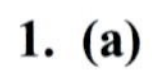

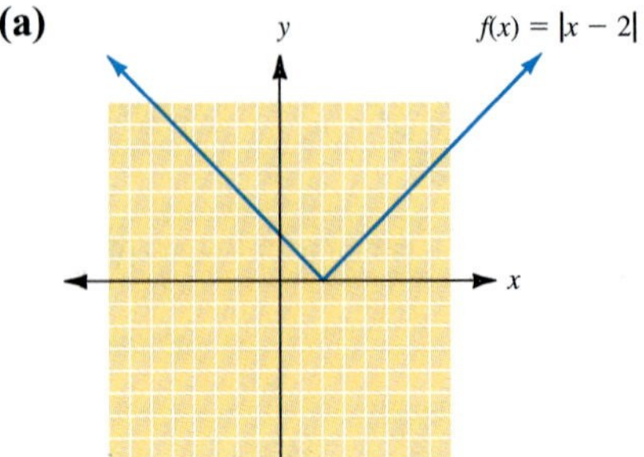

(b)

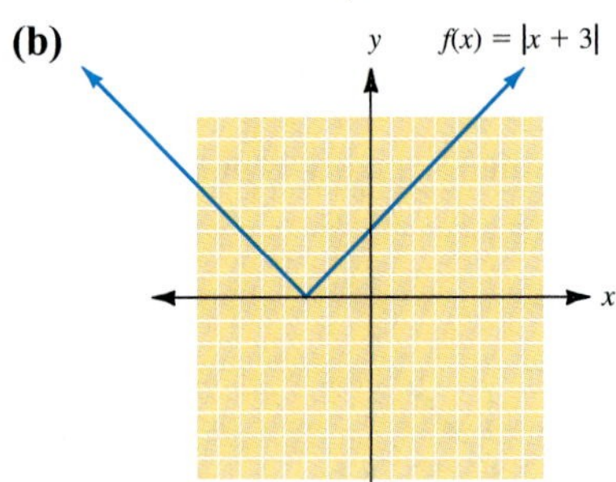

2.

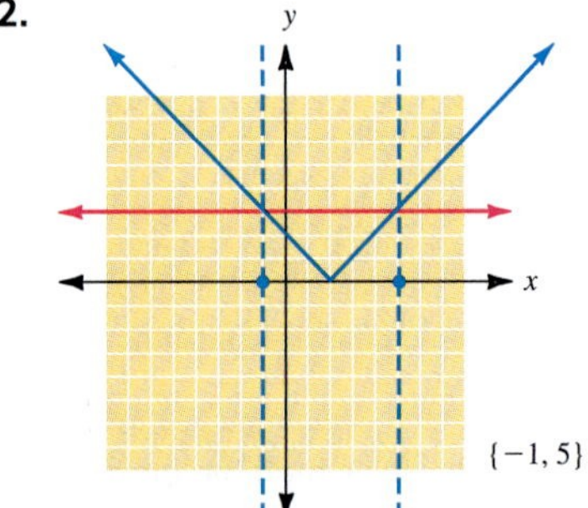

3.

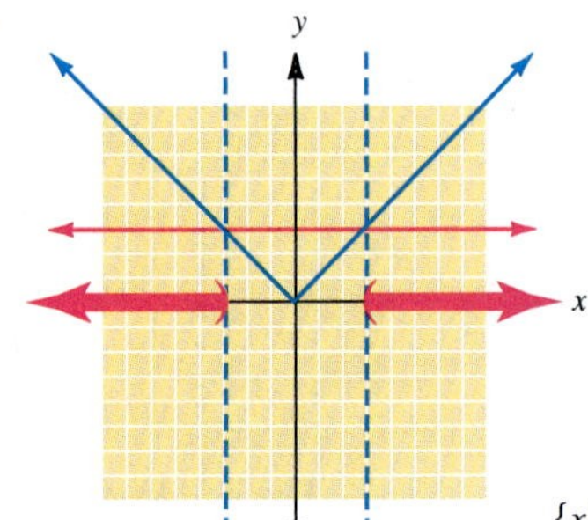

Name ____________________

Section __________ Date __________

4.5 Exercises

In exercises 1 to 6, graph each function.

1. $f(x) = |x - 3|$

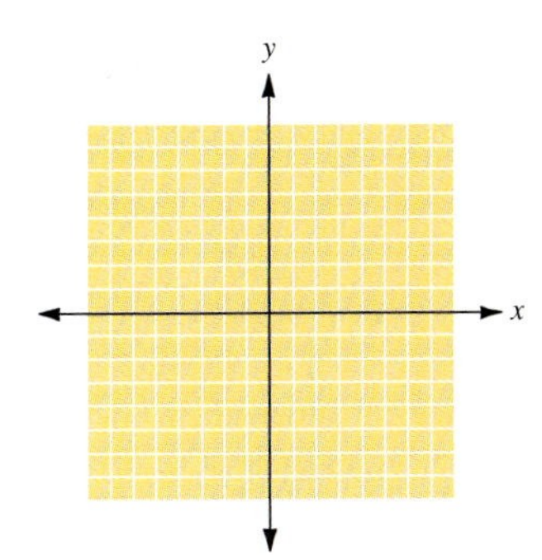

2. $f(x) = |x + 2|$

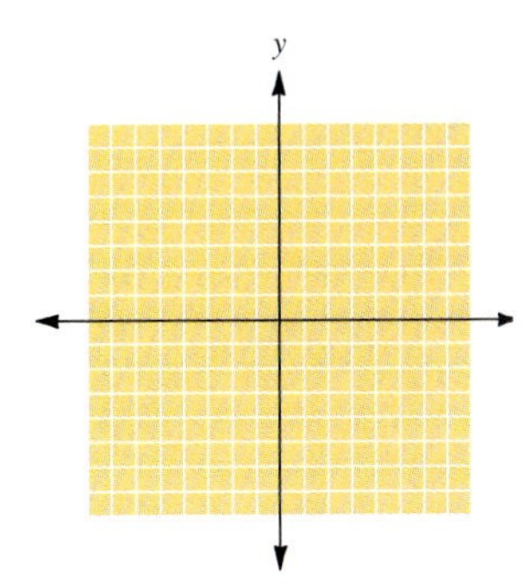

3. $f(x) = |x + 3|$

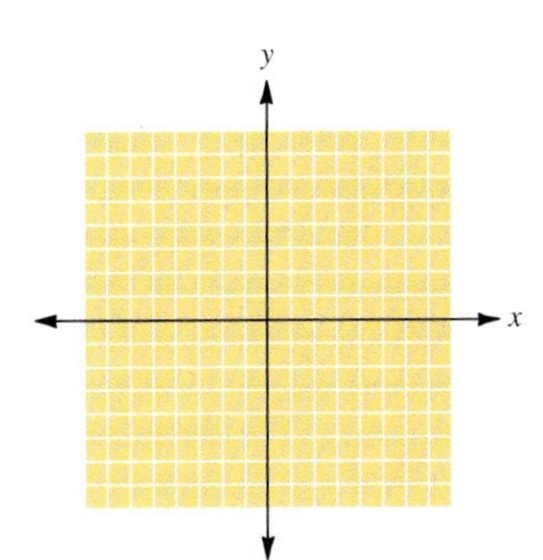

4. $f(x) = |x - 4|$

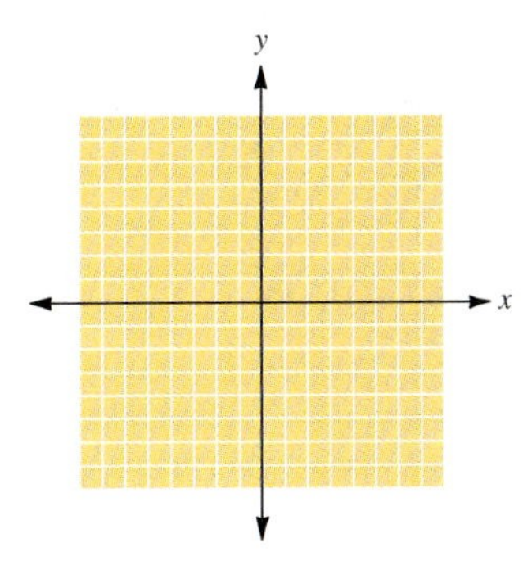

5. $f(x) = |x - (-3)|$

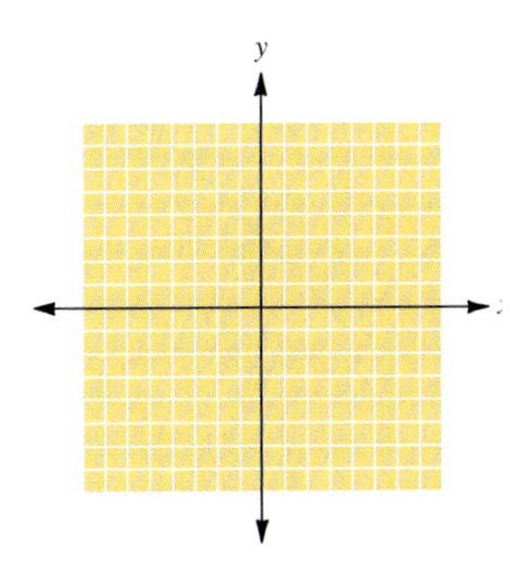

6. $f(x) = |x - (-5)|$

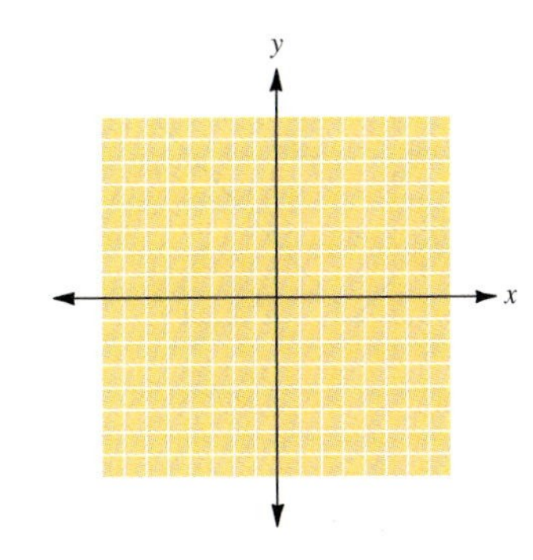

In exercises 7 to 12, solve the equations graphically.

7. $|x| = 3$

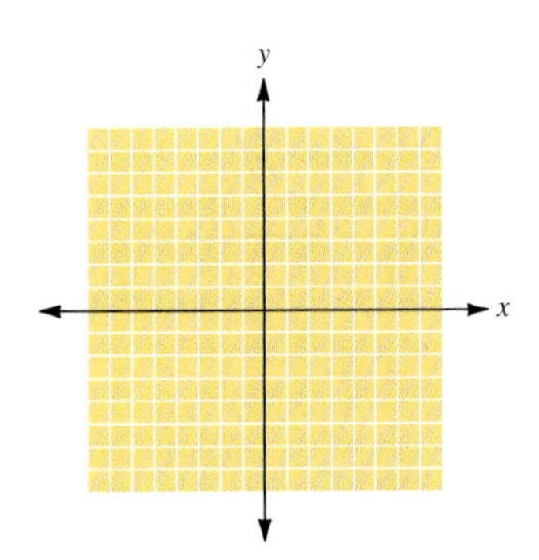

8. $|x| = 5$

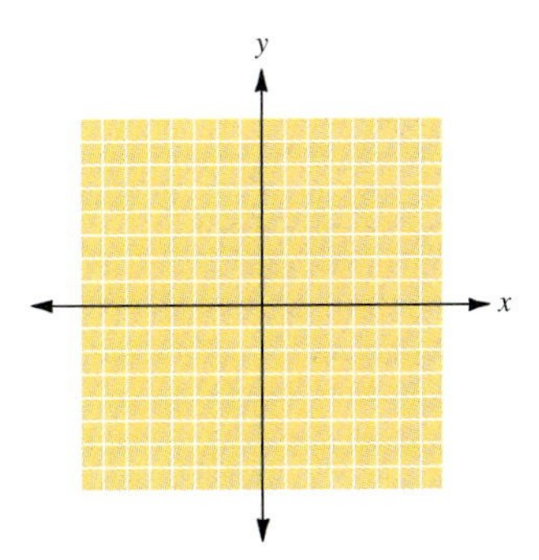

ANSWERS

1. ____________________
2. ____________________
3. ____________________
4. ____________________
5. ____________________
6. ____________________
7. ____________________
8. ____________________

ANSWERS

9. ______

10. ______

11. ______

12. ______

13. ______

14. ______

15. ______

16. ______

9. $|x - 2| = \frac{7}{2}$

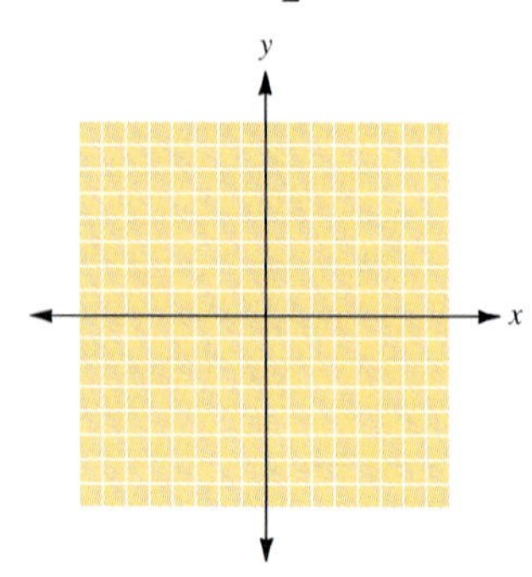

10. $|x - 5| = 3$

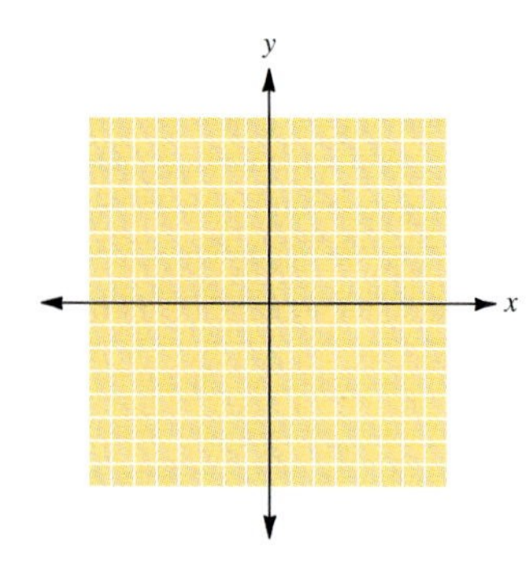

11. $|x + 2| = 4$

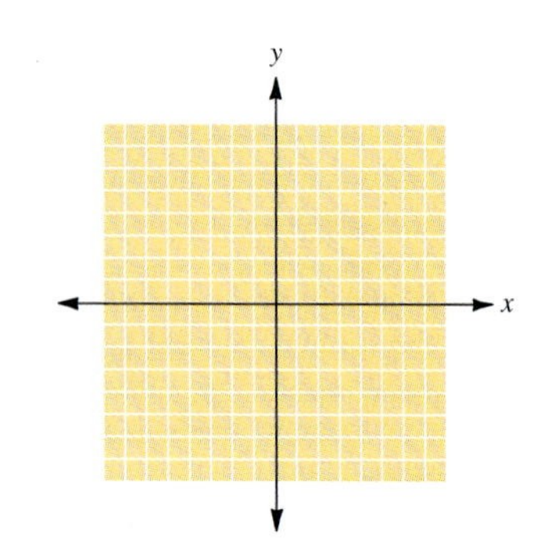

12. $|x + 4| = 2$

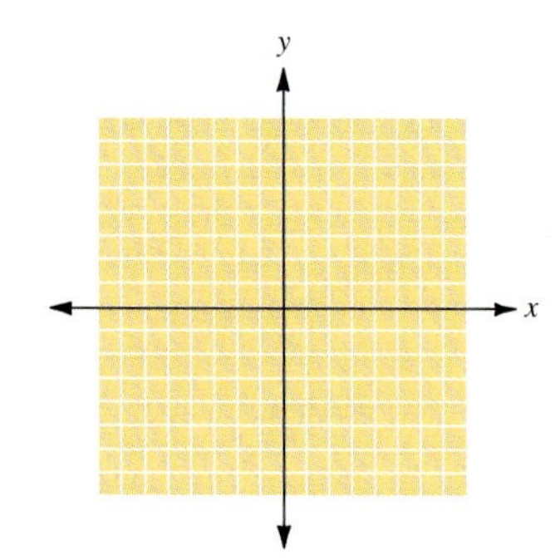

In exercises 13 to 16, determine the function represented by each graph.

13.

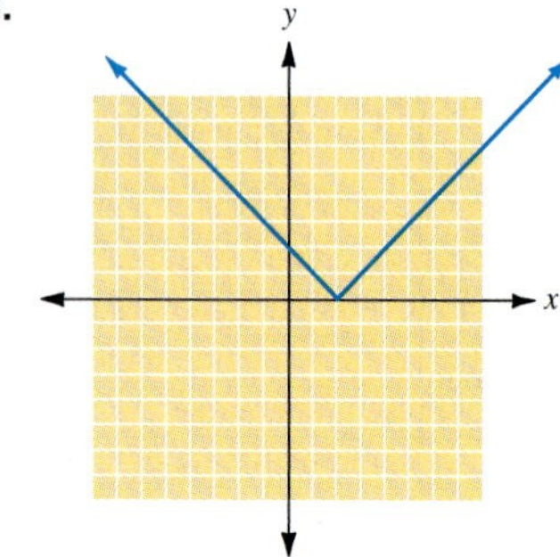

14.

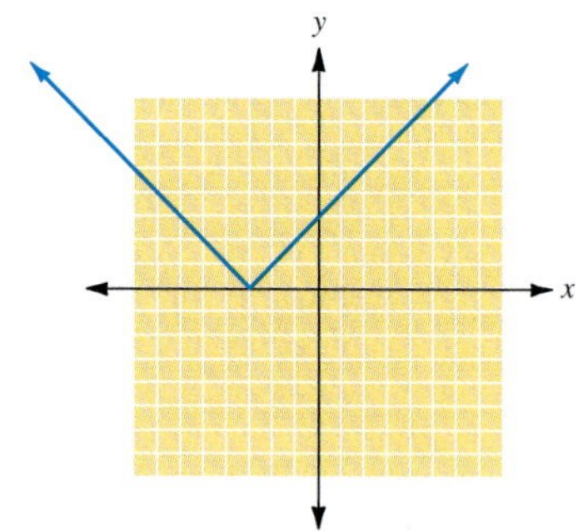

15.

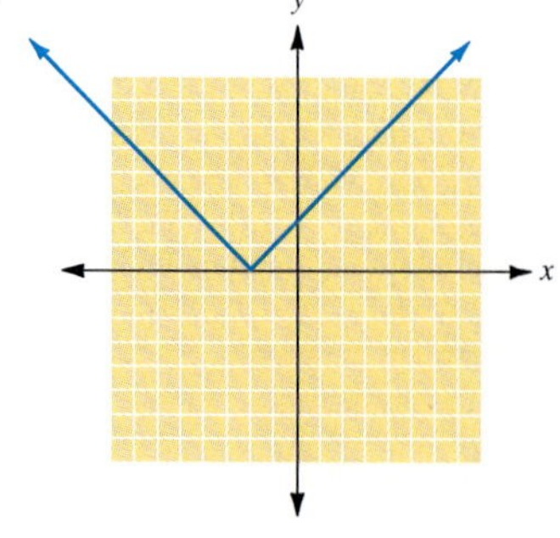

16.

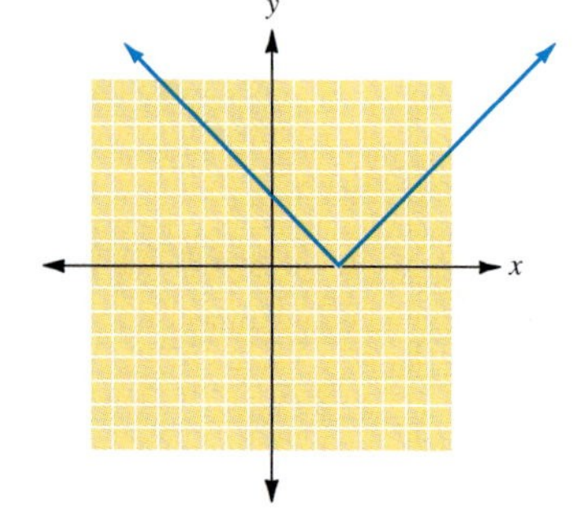

ANSWERS

17. ______

18. ______

19. ______

20. ______

21. ______

22. ______

23. ______

24. ______

In exercises 17 to 28, solve each inequality graphically.

17. $|x| < 4$

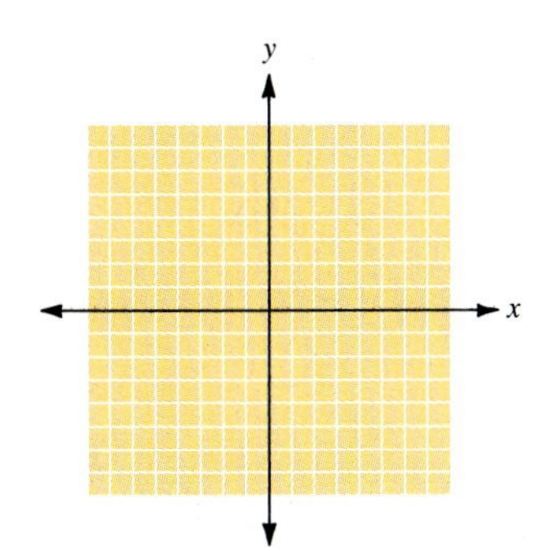

18. $|x| < 6$

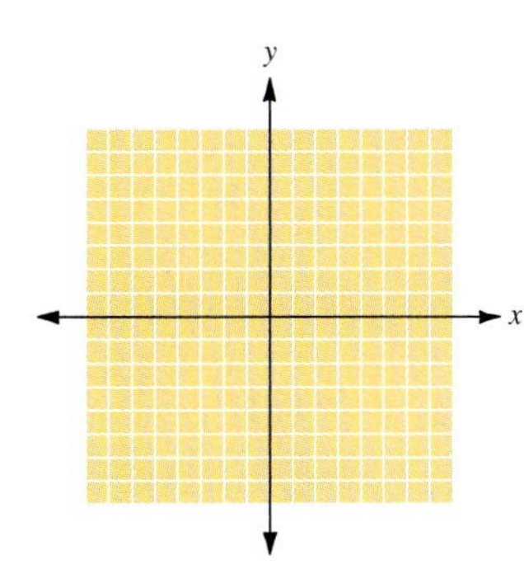

19. $|x| \geq 5$

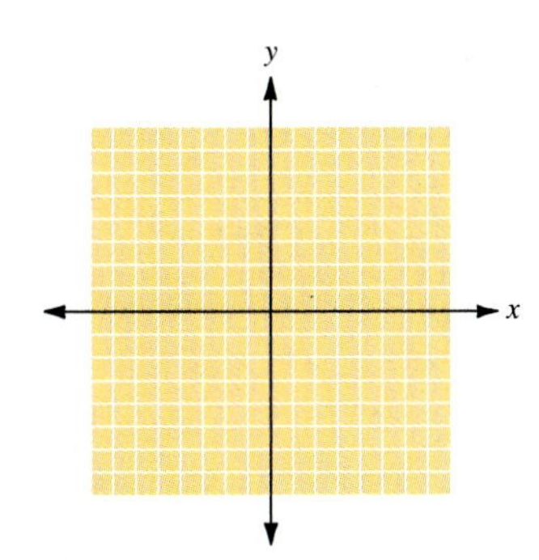

20. $|x| \geq 2$

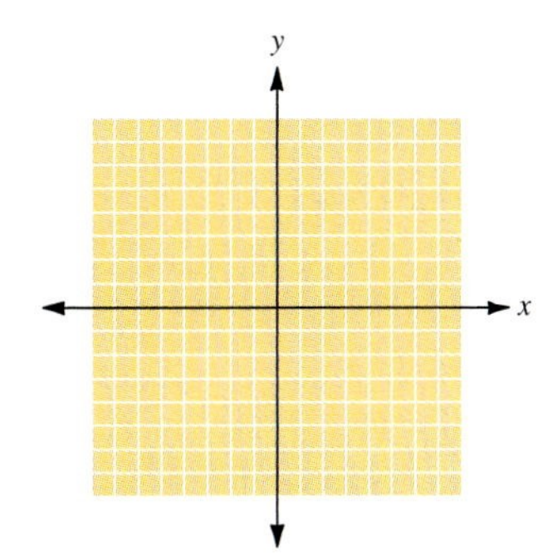

21. $|x - 3| < 4$

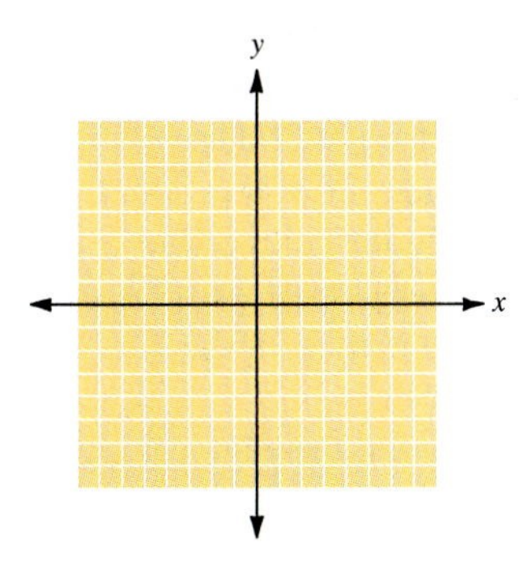

22. $|x - 1| < 5$

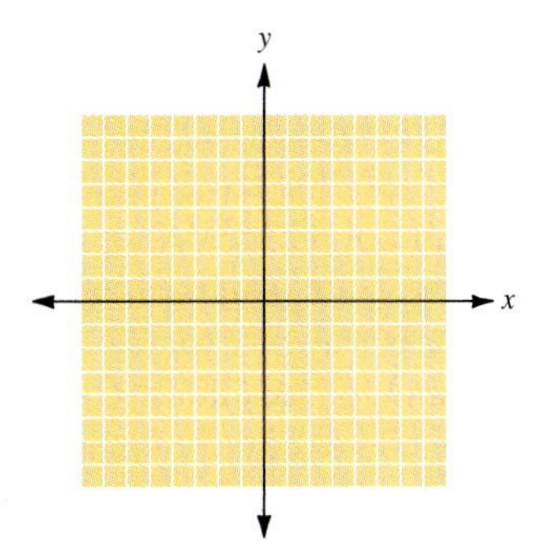

23. $|x - 2| \geq 5$

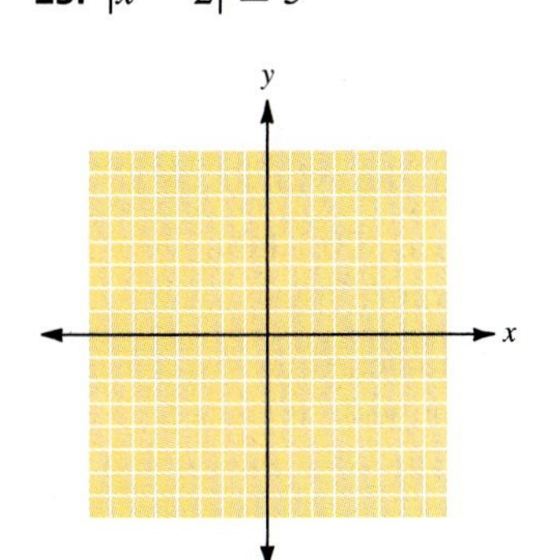

24. $|x + 2| > 4$

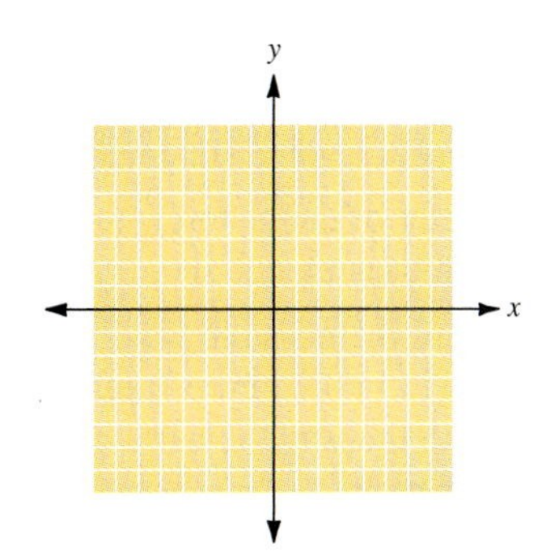

ANSWERS

25. ______

26. ______

27. ______

28. ______

29. ______

25. $|x + 1| \leq 5$

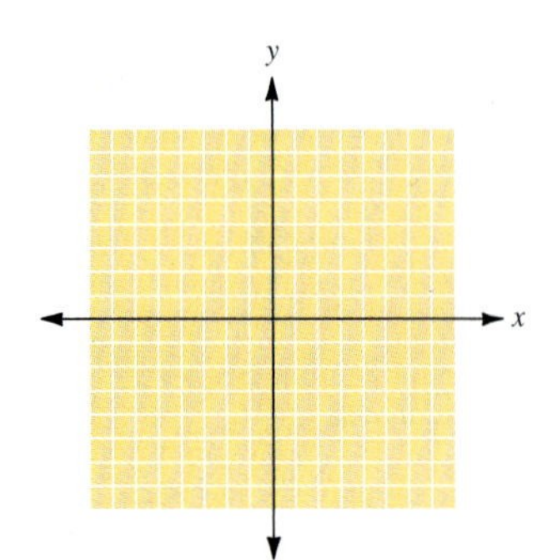

26. $|x + 4| > 1$

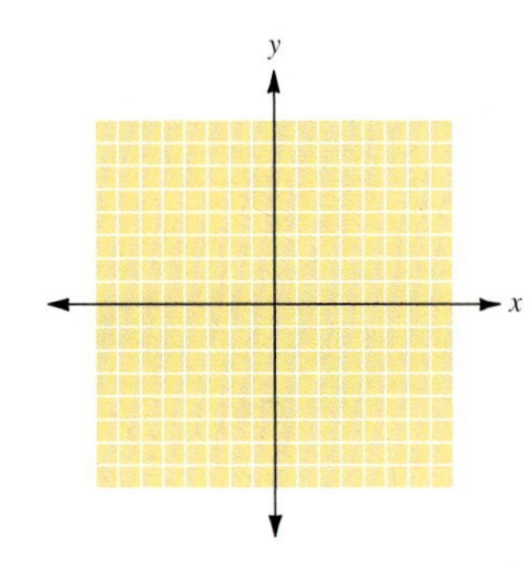

27. $|x + 2| \geq -2$

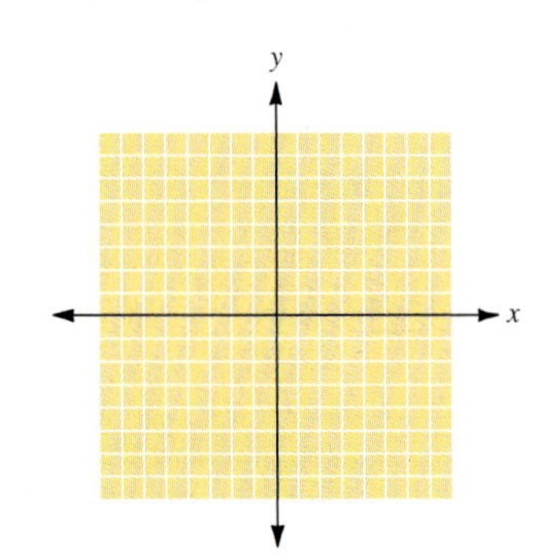

28. $|x - 4| > -1$

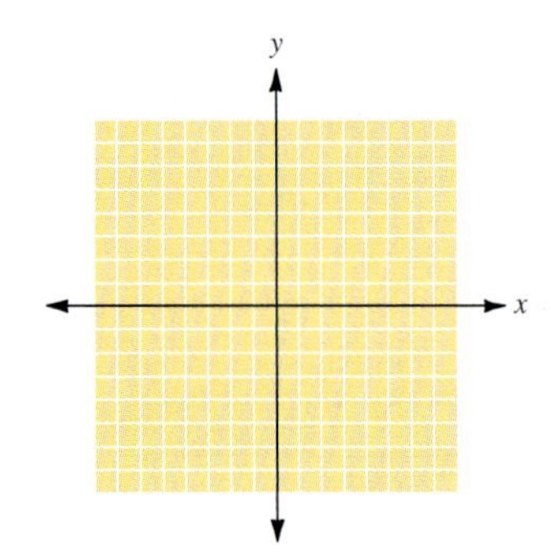

29. Assessing Piston Design. Combustion engines get their power from the force exerted by burning fuel on a piston inside a chamber. The piston is forced down out of the cylinder by the force of a small explosion caused by burning fuel mixed with air. The piston in turn moves a piston rod, which transfers the motion to the work of the engine. The rod is attached to a flywheel, which pushes the piston back into the cylinder to begin the process all over. Cars usually have four to eight of these cylinders and pistons. It is crucial that the piston and the cylinder fit well together, with just a thin film of oil separating the sides of the piston and the sides of the cylinder. When these are manufactured, the measurements for each part must be accurate. But, there is always some error. How much error is a matter for the engineers to set and for the quality control department to check.

Suppose the diameter of the cylinder is meant to be 7.6 cm, and the engineer specifies that this part must be manufactured to within 0.1 mm of that measurement. This figure is called the **tolerance.** As parts come off the assembly line, someone in quality control takes samples and measures the cylinders and the pistons. Given this information, complete the following.

1. Write an absolute value statement about the diameter, d_c, of the cylinder.
2. If the diameter of the piston is to be 7.59 cm with a tolerance of 0.1 mm, write an absolute value statement about the diameter, d_p, of the piston.
3. Investigate all the possible ways these two parts will fit together. If the two parts have to be within 0.1 mm of each other for the engine to run well, is there a problem with the way the parts may be paired together? Write your answer and use a graph to explain.

4. Accuracy in machining the parts is expensive, so the tolerance should be close enough to make sure the engine runs correctly, but not so close that the cost is prohibitive. If you think a tolerance of 0.1 mm is too large, find another that you think would work better. If it is too small, how much can it be enlarged and still have the engine run according to design? (That is, so $|d_c - d_p| \le 0.1$ mm.) Write the tolerance using absolute value signs. Explain your reasoning if you think a tolerance of 0.1 mm is not workable.

5. After you have decided on the appropriate tolerance for these parts, think about the quality control engineer's job. Hazard a few educated opinions to answer these questions: How many parts should be pulled off the line and measured? How often? How many parts can reasonably be expected to be outside the expected tolerance before the whole line is shut down and the tools corrected?

Answers

1.

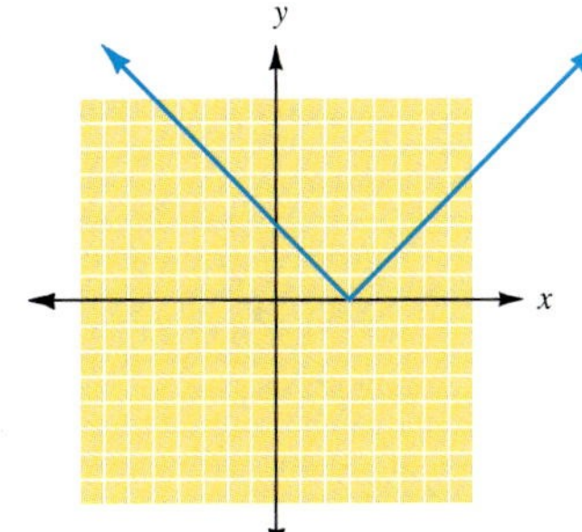

3.

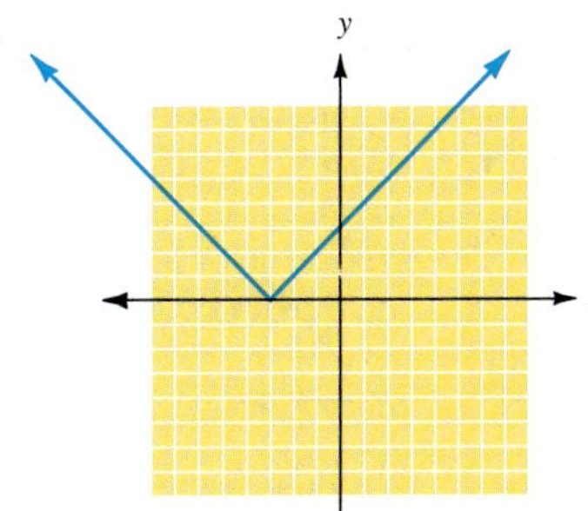

5.

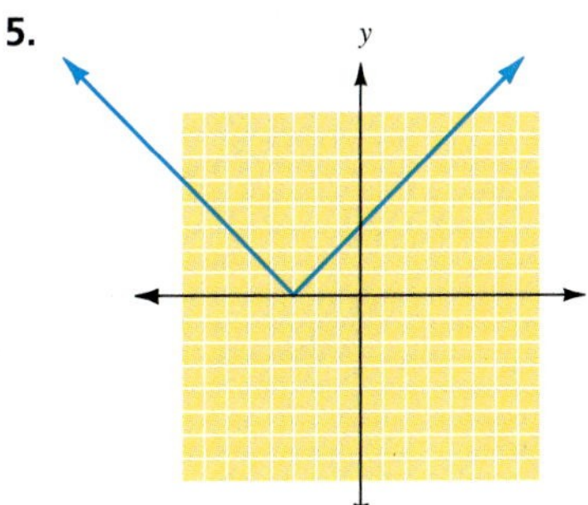

7.

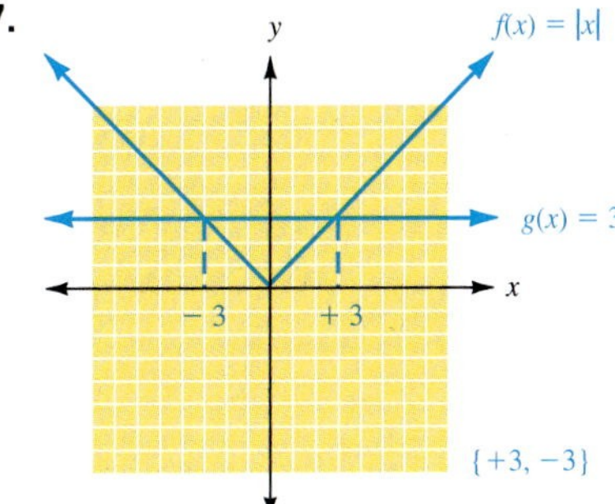

9.

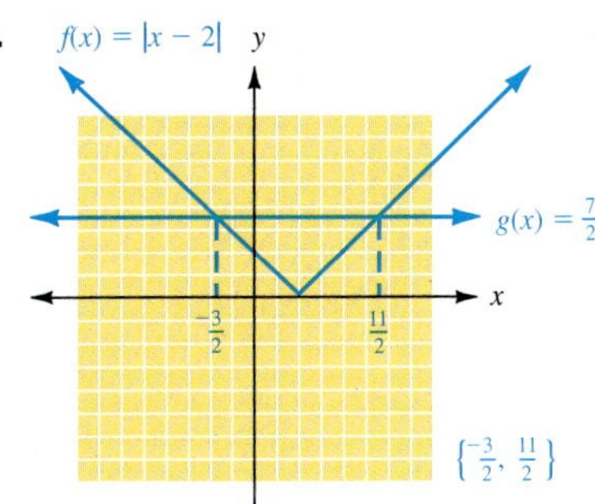

11.

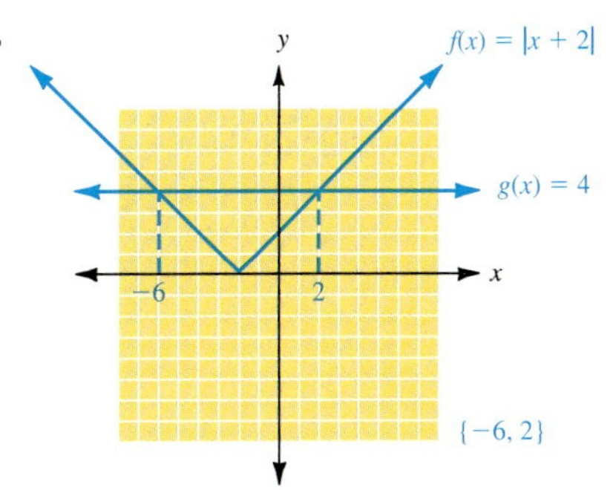

13. $f(x) = |x - 2|$ **15.** $f(x) = |x + 2|$

17.

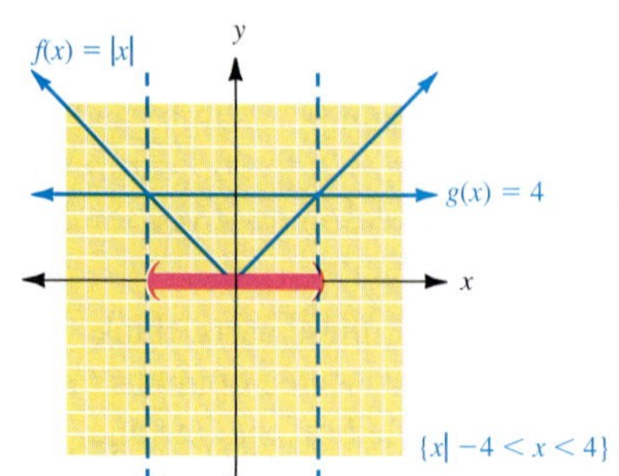

19.

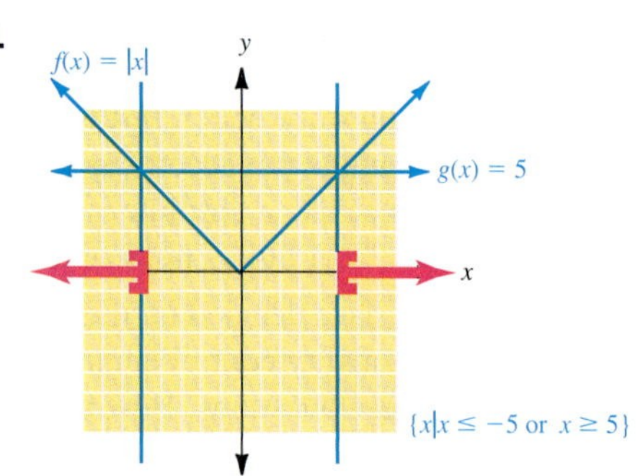

21.

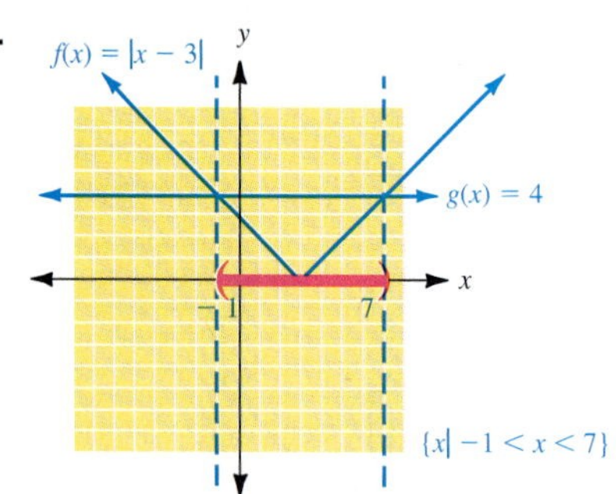

23.

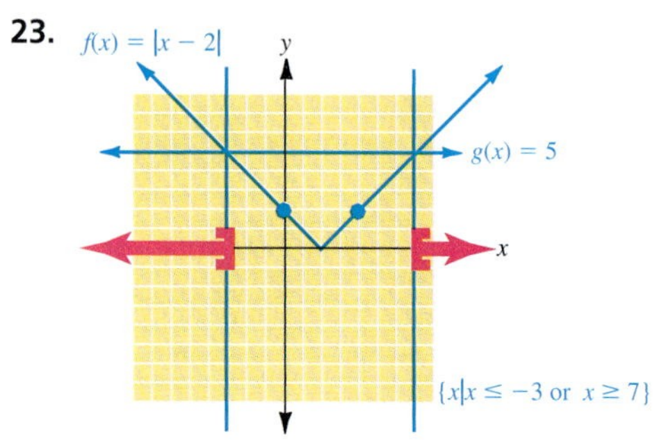

25.

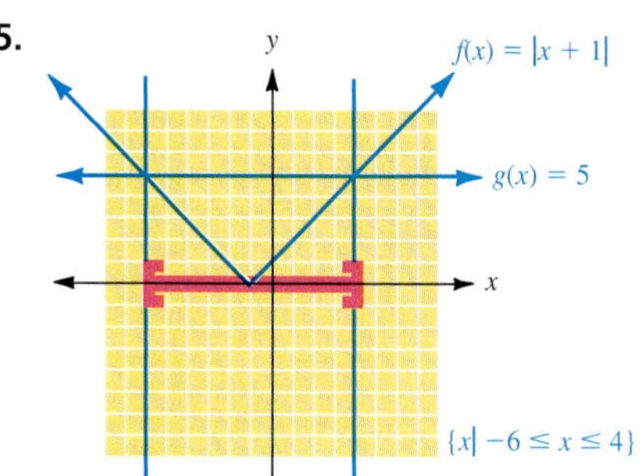

27.

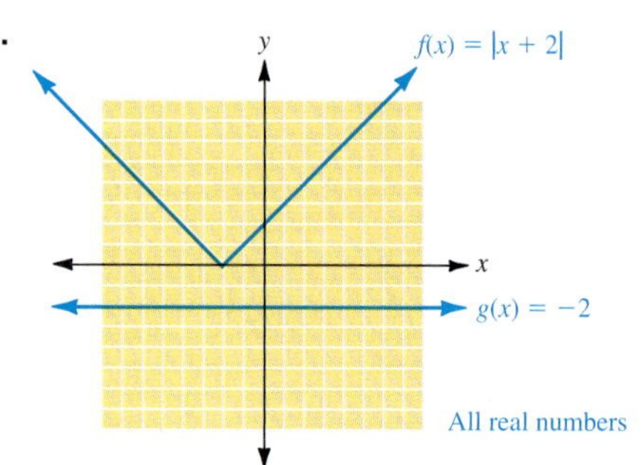

29.

Applications of Two-Variable Graphs

OBJECTIVES

1. Solve an application of a linear function
2. Identify the dependent and independent variables for an application
3. Create a scatter plot

It is unusual to find a newspaper that does not have several two-variable graphs. Each of the following graphs was found in a daily newspaper.

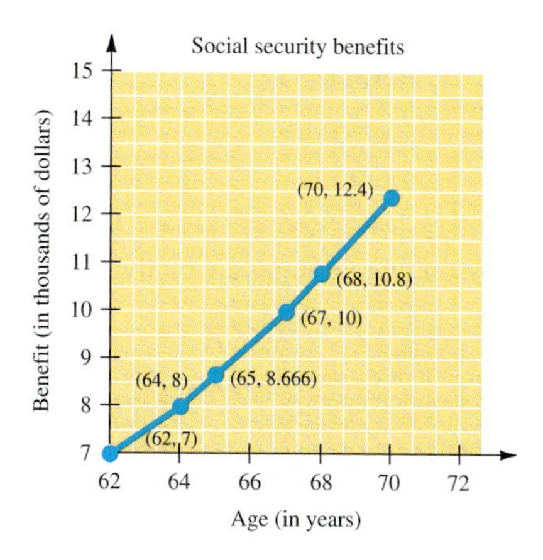

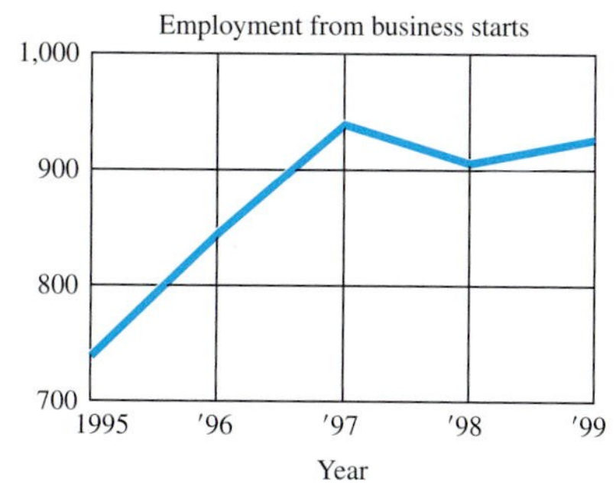

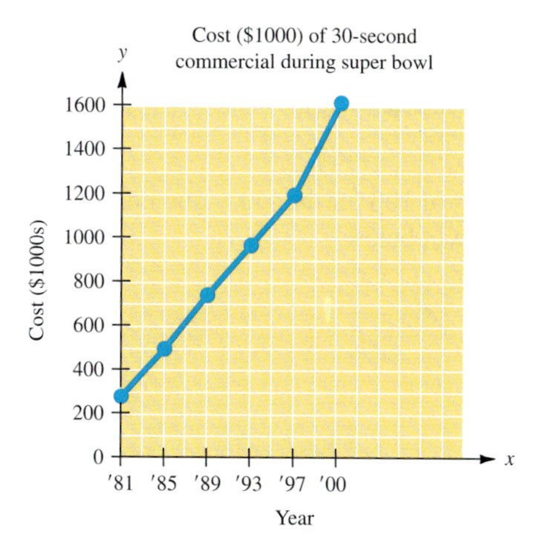

All three of the graphs above use the principles described in this chapter. There are, however, a few noticeable differences. Notice the units on the x and y axes for each graph. This leads to the following rule of graphing applications.

Rules and Properties: Graphing Applications

When graphing an application, each axis must be labeled with appropriate units.

Note also that none of the graphs above use equal units for the two axes. This is usually the case with a graphical application. That observation leads to our first example.

Example 1

Adjusting the Scale on the *x* and *y* Axes

A car rental agency advertises daily rates for a midsized automobile at $20 per day plus 10¢ per mile. The cost per day C and the distance driven in miles s are then related by the following linear equation:

$$C = 0.10s + 20 \qquad (1)$$

Graph the relationship between C and s.

First, we proceed by finding three points on the graph.

s	C
0	20
100	30
200	40

So as the distance s varies from 0 to 200 mi, the cost C changes from $20 to $40. To draw a "reasonable" graph, it makes sense to choose a different scale for the horizontal (or s) axis than for the vertical (or C) axis.

NOTE Before you graph this function on your calculator, adjust the scales on both axes. This is done from the WINDOW menu.

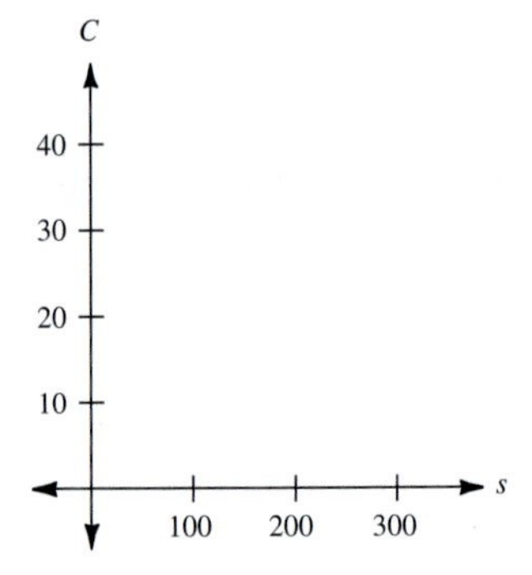

We have chosen units of 100 for the s axis and units of 10 for the C axis. The graph can then be completed, as shown below.

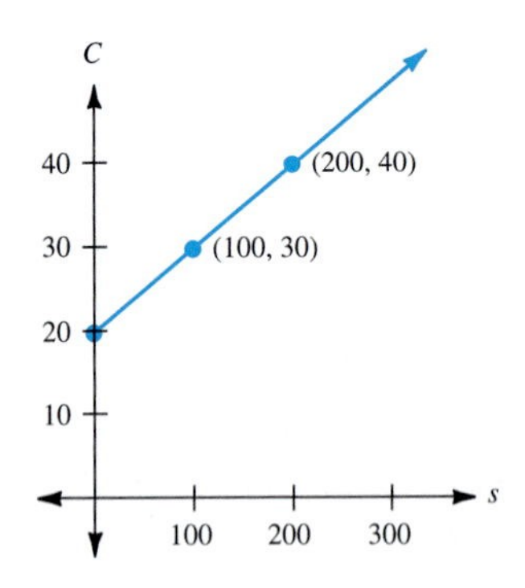

Note that the graph of equation (1) does not extend beyond the first quadrant because of the nature of our problem, in which solutions are only realistic when $s \geq 0$.

CHECK YOURSELF 1

A salesperson's monthly salary S is based on a fixed salary of $1200 plus 8% of all monthly sales x. The linear equation relating S and x is

$$S = 0.08x + 1200$$

Graph the relationship between S and x. Hint: Find the monthly salary for sales of $0, $10,000, and $20,000.

In our second example, we will find and graph a linear equation from just two points.

Example 2

An Application of a Linear Function

In producing a new product, a manufacturer predicts that the number of items produced x and the cost in dollars C of producing those items will be related by a linear equation.

Suppose that the cost of producing 100 items will be $5000 and the cost of producing 500 items will be $15,000. Find the linear equation relating x and C.

To solve this problem, we must find the equation of the line passing through points (100, 5000) and (500, 15,000).

Although the numbers are considerably larger than we have encountered thus far in this section, the process is exactly the same.

First, we find the slope:

$$m = \frac{15{,}000 - 5000}{500 - 100} = \frac{10{,}000}{400} = 25$$

We can now use the point-slope form as before to find the desired equation.

$$C - 5000 = 25(x - 100)$$

$$C - 5000 = 25x - 2500$$

$$C = 25x + 2500$$

To graph the equation we have just derived, we must choose the scaling on the x and C axes carefully to get a "reasonable" picture. Here we choose increments of 100 on the x axis and 2500 on the C axis because those seem appropriate for the given information.

NOTE Notice how the change in scaling "distorts" the slope of the line.

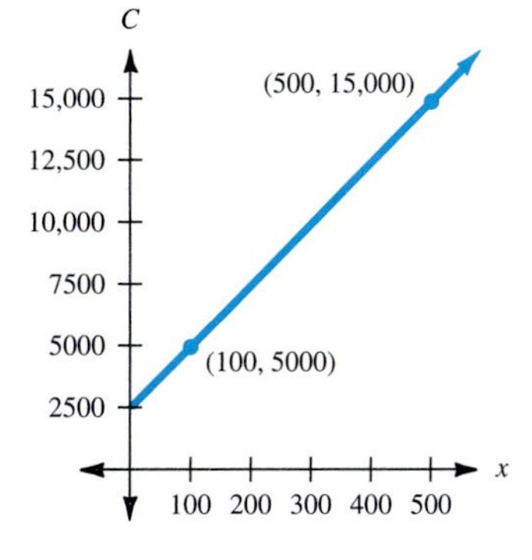

CHECK YOURSELF 2

A company predicts that the value in dollars V and the time that a piece of equipment has been in use t are related by a linear equation. If the equipment is valued at \$1500 after 2 years and at \$300 after 10 years, find the linear equation relating t and V.

When an equation is written such that the left side is y and the right side is an expression involving the variable x, such as

$$y = 3x - 2$$

we can rewrite the equation as a function. In this case, we have

$$f(x) = 3x - 2$$

This implies that

$$y = f(x)$$

We can say that y is a function of x, or y is dependent on x. That leads to the following definitions.

Definitions: Independent Variable and Dependent Variable

Given that $y = f(x)$,

x is called the **independent variable** and y is called the **dependent variable.**

Identifying which variable is independent and which is dependent is important in many applications.

Example 3

Identifying the Dependent Variable

From each pair, identify which variable is dependent on the other.

(a) The age of a car and its resale value.

The resale value depends on the age, so we would assign the age of the car the independent variable (x) and the resale value the dependent variable (y).

(b) The amount of interest earned in a bank account and the amount of time the money has been in the bank.

NOTE If you think about it, you will see that time will be the independent variable in most ordered pairs. Most everything depends on time rather than the reverse.

The interest depends on the time, so interest is the dependent variable (y) and time is the independent variable (x).

(c) The number of cigarettes you have smoked and the probability of dying from a smoking-related disease.

The number of cigarettes is the independent variable (x), and the probability of dying from a smoking-related disease is the dependent variable (y).

CHECK YOURSELF 3

From each pair, identify which variable is dependent on the other.

(a) The number of credits taken and the amount of tuition paid.

(b) The temperature of a cup of coffee and the length of time since it was poured.

In the next example, you will combine the skills you have learned to this point of the section.

Example 4

Modeling with a Function

Shaquille and Kobe are interested in renting a gym for summer basketball. They are told that they must pay a flat rate of \$200 plus \$75 per hour.

(a) Identify the dependent and independent variables.

(b) Find the equation of the relationship.

(c) Scale the axes and graph the relationship.

(d) Find $f(2)$ and $f(5)$.

(a) The independent variable is the number of hours of use. The dependent variable is the total cost.

(b) The equation is $y = 200 + 75x$, or we could write $f(x) = 200 + 75x$.

(c) Using only the first quadrant (why?), we get the graph

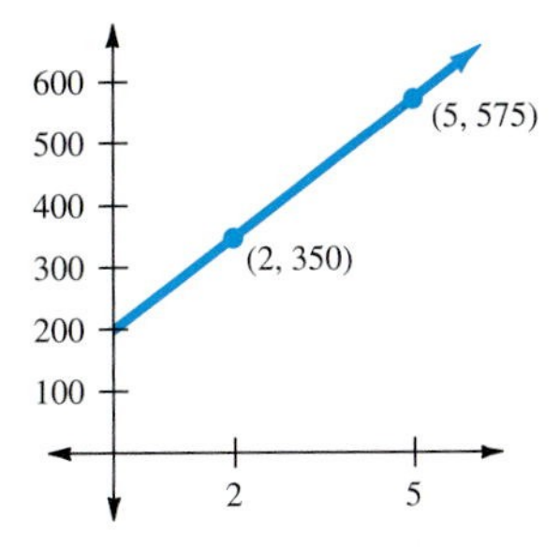

NOTE $f(2)$ is the cost of a 2-h rental.

(d) $f(2) = \$350$, $f(5) = \$575$

CHECK YOURSELF 4

Tiger and Sergio are selling instructional videos. Their contract gives them $10,000 *plus* $1.25 *per video (think of this as* $1250 *for each* 1000 *videos sold)*

(a) Identify the dependent and independent variables.
(b) Find the equation of the relationship.
(c) Scale the axes and graph the relationship.
(b) Find $f(3000)$ and $f(50{,}000)$.

In Example 2, we found a graph given two points. Although this is occasionally useful, it is far more common that we have many points. To graph several points we use a scatter plot.

A **scatter plot** is a graph of a set of ordered pairs. Scatter plots help us see the relationship between two sets of data. For example, the following graph represents the relationship between the number of wins and the number of losses that a professional football team might have in a full season. We can use this graph to determine the number of losses that a team with 10 wins would have.

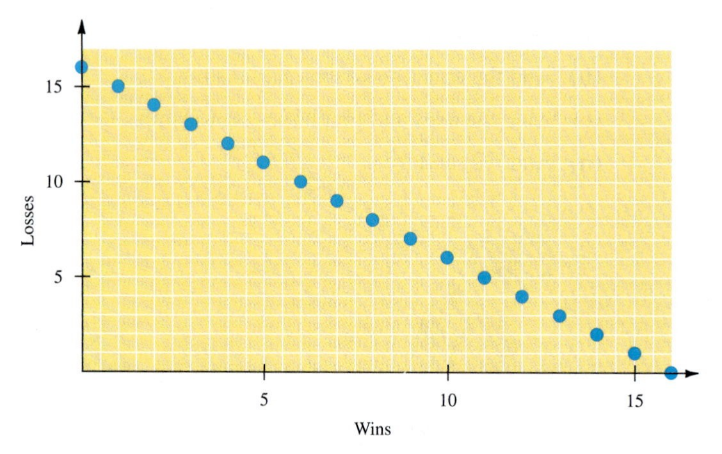

The ordered pair (10, 6) indicates that a team with 10 wins would have 6 losses. Notice that the ordered pairs form a perfect line with slope -1.

The set of ordered pairs graphed below shows the relationship between the number of miles driven and the amount of gas purchased the last 12 times that Allie filled her gas tank. Notice that the points almost form a straight line.

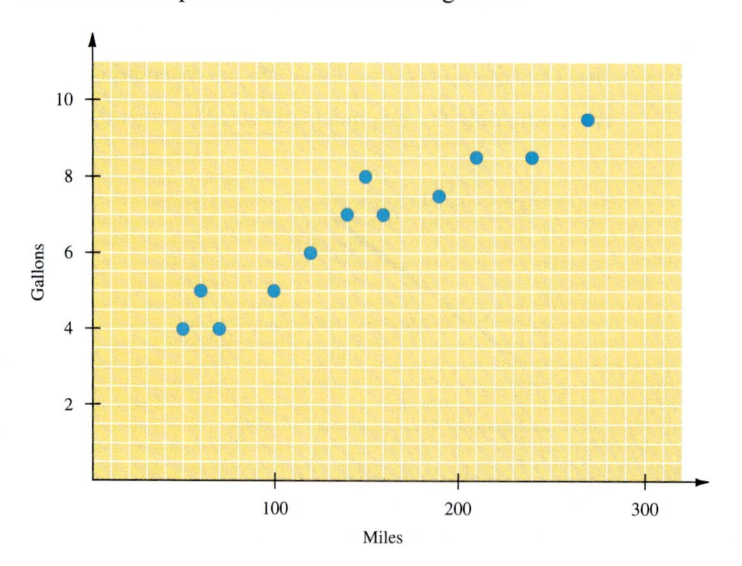

NOTE A *prediction line* is a line that gives us a "reasonable" estimation for y when we have a given x. We will reserve definition of the word "reasonable" for future mathematics classes.

Suppose that you were asked to estimate the amount of gas Allie will need to buy if she drives 250 miles. Even though there is no ordered pair associated with 250 miles, you can comfortably guess that Allie will need about 9 gallons of gas. You arrived at this answer by noting that the points fell in a fairly straight line, and you estimated where that line would be when the x was 250. Essentially, you created a **prediction line,** which is a line that is used to estimate the y value when you are given a value for x.

In a subsequent mathematics or statistics class, you will learn how to find a prediction line. The first step in finding a prediction line is to create and sketch a scatter plot, as Example 5 illustrates.

Example 5

Creating a Scatter Plot

Carlotta kept the following chart next to her treadmill. Create a scatter plot for the ordered pairs.

Minutes	Miles
53	6.4
48	5.7
55	6.8
30	4.5
40	5.2
62	7.0
35	4.9
50	6.0
65	7.2

Each combination of minutes and miles makes an ordered pair. The first ordered pair is (53, 6.4). The scatter plot is the graph of all nine ordered pairs.

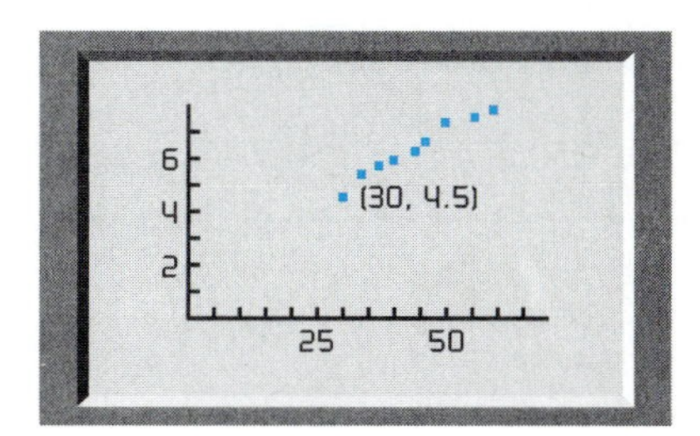

CHECK YOURSELF 5

Whitney keeps track of use of the copy machine in the library. She created the following chart:

Month	School Days	Duplication Count
September	9	1230
October	21	3268
December	8	1124
January	15	2253
February	19	2872
March	17	2597
April	21	3410
May	22	3502
June	10	1470

Create a scatter plot for the ordered pairs.

CHECK YOURSELF ANSWERS

1.

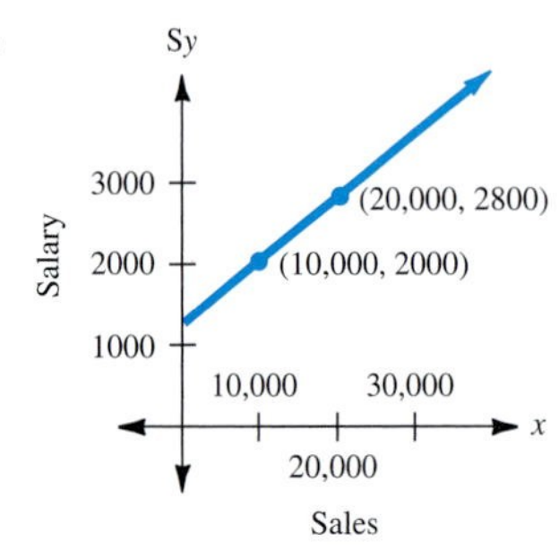

2. $V = -150t + 1800$

3. **(a)** Tuition is dependent, number of credits taken is independent; **(b)** The temperature is dependent, the time since the coffee was poured is independent.

4. **(a)** The number of videos is independent. The total amount is dependent; **(b)** $y = 10{,}000 + 1.25x$;

(c)

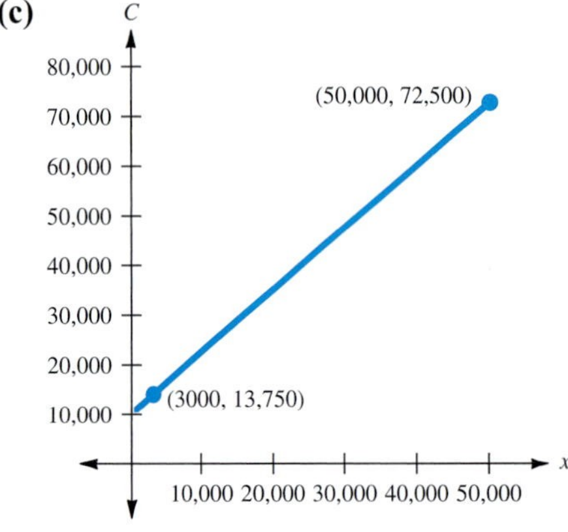

(d) $f(3000) = 13{,}750$, $f(50{,}000) = 72{,}500$

5.

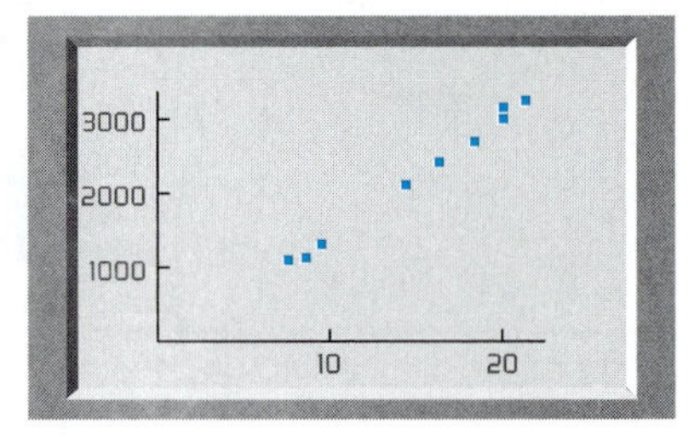

Name ______________________

Section ________ Date ________

4.6 Exercises

ANSWERS

1. ______________________

2. ______________________

1. **Consumer Affairs.** A car rental agency charges \$20 per day and 16¢ per mile for the use of a compact automobile. The cost of the rental C and the number of miles driven per day s are related by the equation

$$C = 0.16s + 20$$

Graph the relationship between C and s. Be sure to select appropriate scaling for the C and s axes.

2. **Checking Account Charges.** A bank has the following structure for charges on checking accounts. The monthly charge consists of a fixed amount of \$8 and an additional charge of 12¢ per check. The monthly cost of an account C and the number of checks written per month n are related by the equation

$$C = 0.12n + 8$$

Graph the relationship between C and n.

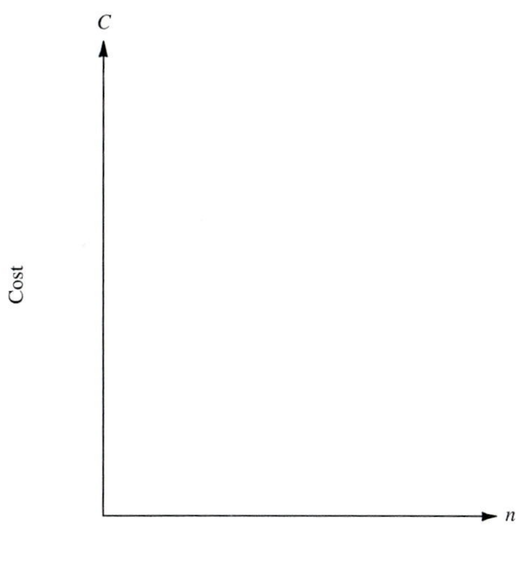

ANSWERS

3. ______

4. ______

5. ______

3. Tuition Charges. A college has tuition charges based on the following pattern. Tuition is \$35 per credit-hour plus a fixed student fee of \$75.

(a) Write a linear equation that shows the relationship between the total tuition charge T and the number of credit-hours taken h.

(b) Graph the relationship between T and h.

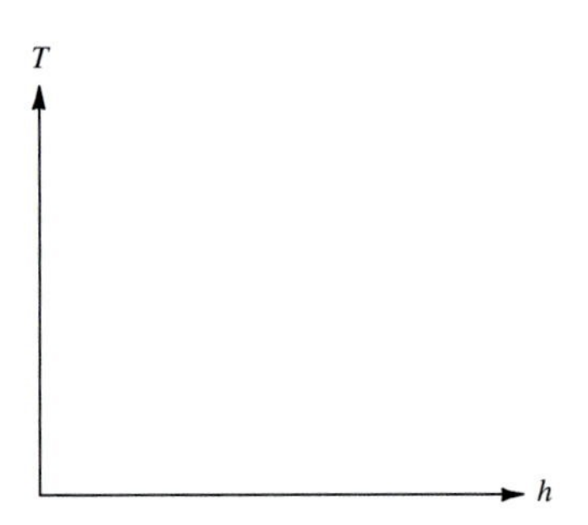

4. Weekly Salary. A salesperson's weekly salary is based on a fixed amount of \$200 plus 10% of the total amount of weekly sales.

(a) Write an equation that shows the relationship between the weekly salary S and the amount of weekly sales x (in dollars).

(b) Graph the relationship between S and x.

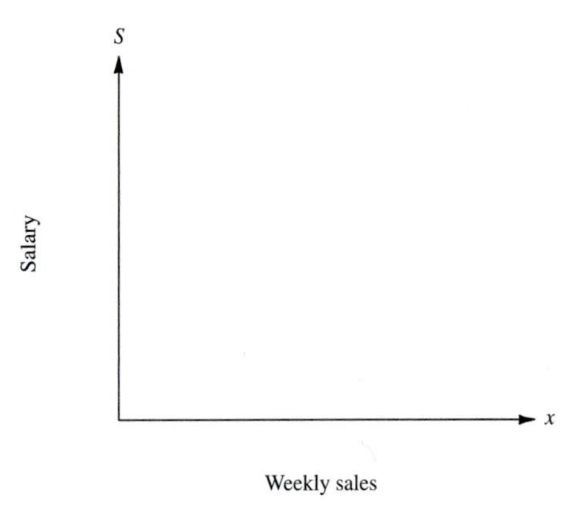

5. Science. A temperature of 10°C corresponds to a temperature of 50°F. Also 40°C corresponds to 104°F. Find the linear equation relating F and C.

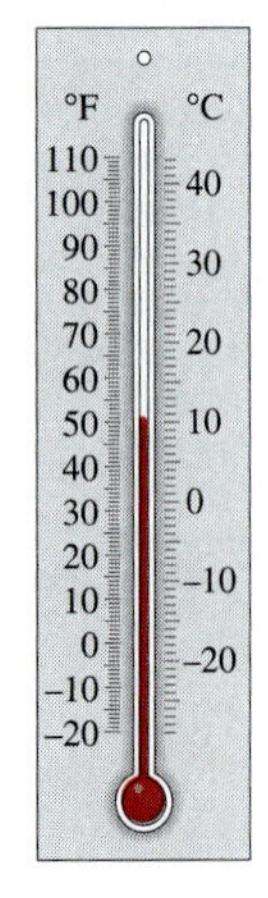

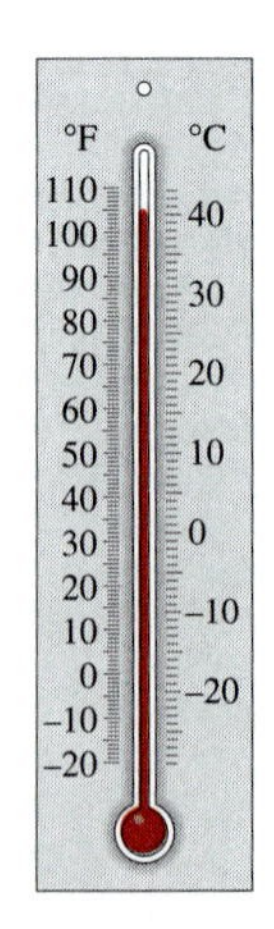

ANSWERS

6. ______

7. ______

8. ______

6. A realtor receives \$500 a month plus 4% commission on sales. The equation that describes the total monthly income, I (in dollars), of the realtor is $I = .04s + 500$, in which s is the amount of sales. **(a)** Graph this equation for $0 \le s \le 160$, where s is the sales in thousands. **(b)** Plot the point whose coordinates are (90, 4100) on the graph. Write a sentence to describe the meaning of this ordered pair.

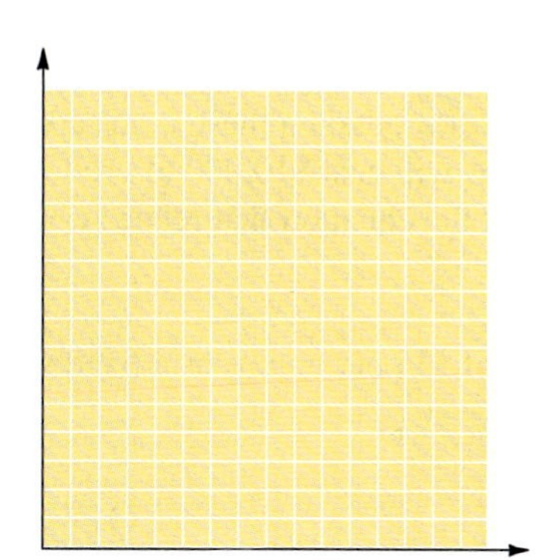

7. An electrician charges \$55 plus \$1 per minute to wire an addition to a house. The equation that describes the total cost, C, of the job is $C = t + 55$ in which t is the number of minutes the electrician works. **(a)** Graph the equation for $0 \le t \le 80$. **(b)** Plot the point whose coordinates are (30, 85) on the graph. Write a sentence to describe the meaning of this ordered pair.

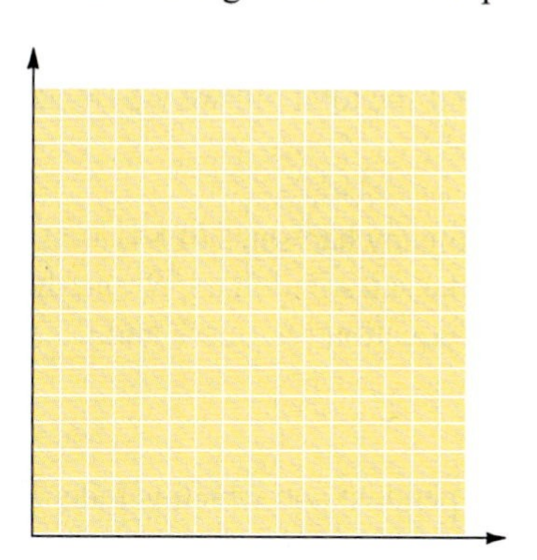

8. A business purchases a new duplicating machine for \$5,000. The depreciated value, v, after t years is given by $v = 5000 - 250t$. Sketch a graph of this equation.

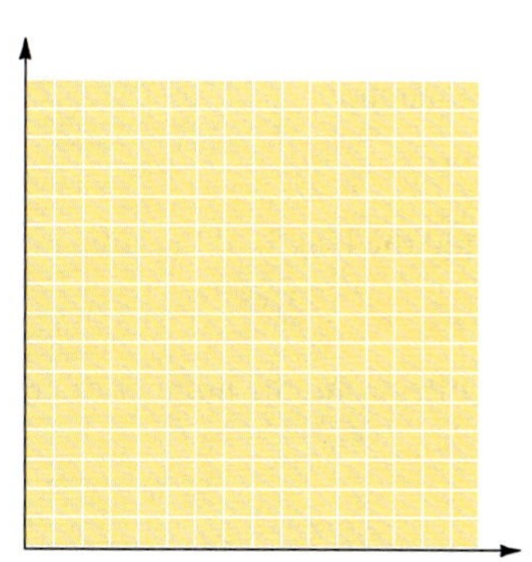

ANSWERS

9. ____
10. ____
11. ____
12. ____
13. ____
14. ____
15. ____
16. ____
17. ____
18. ____

9. Trac Hunyh's weekly cost of operating a taxi is \$100 plus 15 cents a mile. The equation that describes Trac's cost is $C = 100 + 0.15m$, in which m is the number of miles driven in a week. **(a)** Graph the equation for $0 \leq m \leq 200$. **(b)** How many miles would Trac have to drive for the weekly cost to be \$127?

10. Business. In planning for a new item, a manufacturer assumes that the number of items produced, x, and the cost in dollars, C, of producing these items are related by a linear equation. Projections are that 100 items will cost \$10,000 to produce and that 300 items will cost \$22,000 to produce. Find the equation that relates C and x.

11. Business. Mike bills a customer at the rate of \$35 per hour plus a fixed service call charge of \$50.

(a) Write an equation that will allow you to compute the total bill for any number of hours, x, that it takes to complete a job.

(b) What will the total cost of a job be if it takes 3.5 hours to complete?

(c) How many hours would a job have to take if the total bill were \$160.25?

12. Business. Two years after an expansion, a company had sales of \$42,000. Four years later the sales were \$102,000. Assuming that the sales in dollars, S, and the time, t, in years are related by a linear equation, find the equation relating S and t.

In exercises 13 to 18, identify which variable is dependent and which is independent.

13. The amount of a phone bill and the length of the call.

14. The cost of filling a car's gas tank and the size of the tank.

15. The height of a ball thrown in the air and the time in the air.

16. The amount of penalty on an unpaid tax bill and the length of the time unpaid.

17. The length of time needed to graduate from college and the number of credits taken per semester.

18. The amount of snowfall in Boston and the length of the winter.

In exercises 19 to 24, create a scatter plot from the given information.

19. In a local industrial plant, the number of work-hours in safety training and the number of work-hours lost as a result of accidents have been recorded for 10 divisions.

Division	No. of Work-Hours in Safety Training	No. of Work-Hours Lost from Accidents
1	10	80
2	15	75
3	20	72
4	25	70
5	30	60
6	40	53
7	45	50
8	50	48
9	60	42
10	65	35

20. In a statistics class, the mid-term and final exam scores were collected for 10 students. Each exam was worth a total of 100 points.

Mid-Term Exam Scores	Final Exam Scores
71	80
79	85
84	88
76	81
62	75
93	90
88	87
91	96
68	82
77	83

21. A rental car agency has collected data relating the number of miles traveled and the total cost in dollars.

Miles Traveled (in thousands)	Cost (in $)
2	60
6	100
10	200
14	275
8	175
5	90
12	290
16	400
3	75
18	450
21	475

ANSWERS

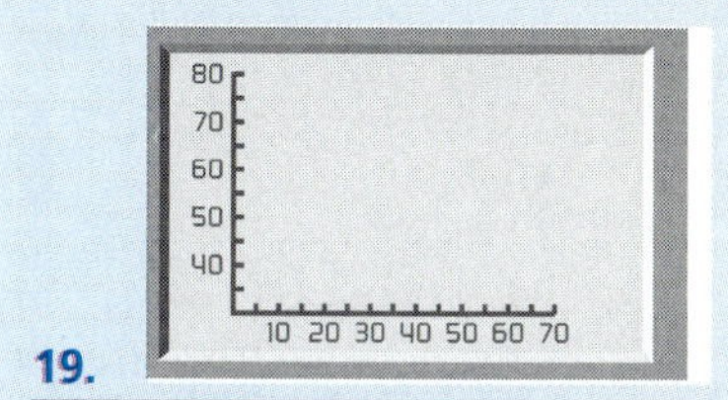

19.

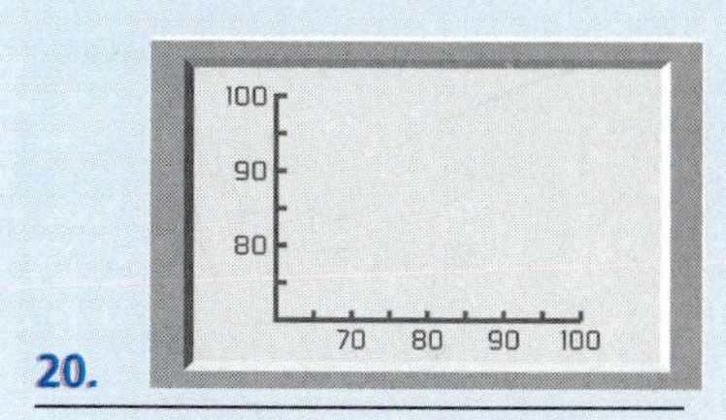

20.

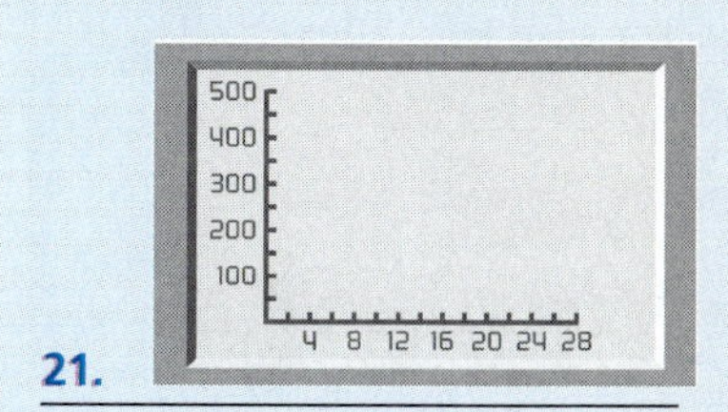

21.

ANSWERS

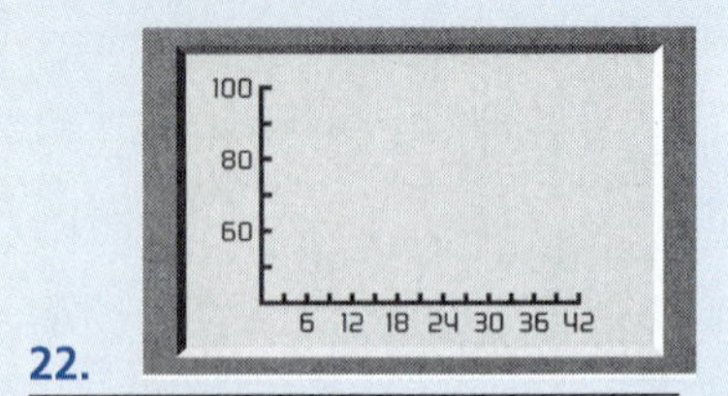

22.

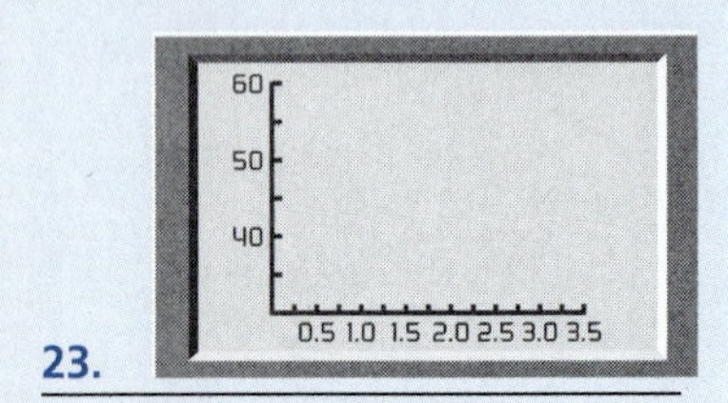

23.

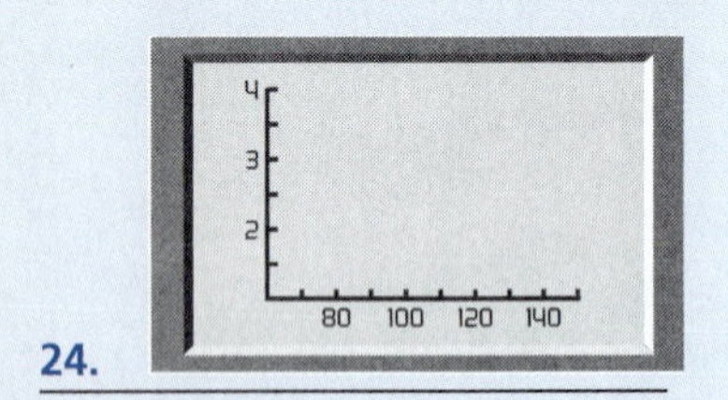

24.

22. A math placement test was given to all entering freshmen at Bucks County Community College. The placement test scores and the score on the first test were recorded for students in a college algebra class.

Placement Test Scores (max. of 40)	First Test Score
25	78
18	75
30	88
14	65
10	62
32	85
12	68
16	73
22	78
27	82
38	93

23. Students claim they can tell the cost of a textbook by the thickness of the book. They picked nine books of roughly the same height and weight. The following data were collected.

Thickness (in cm)	Cost (in $)
1.0	44
0.8	43
3.0	53
2.4	50
1.6	46
1.9	48
0.5	42
1.2	45
3.2	54

24. The following table shows the IQ of 12 students along with their cumulative grade point average (GPA) after 4 years of college.

IQ	GPA
117	3.2
93	2.6
102	2.9
110	3.1
88	2.4
75	1.9
107	3.1
111	3.2
120	3.5
95	2.7
115	3.4
99	2.9

25. Exercise and Age. Aerobic exercise requires that your heartbeat be at a certain rate for 12 minutes or more for full physical benefit. To determine the proper heart rate for a healthy person, start with the number 220 and subtract the person's age. Then multiply by 0.70. The result is the target aerobic heart rate, the rate to maintain during exercise.

1. Write a formula for the relation between a person's age (A) and the person's target aerobic heart rate (R).
2. Using at least 10 different ages, construct a table of target heart rates by age.
3. Draw a graph of this table of values.
4. What are reasonable limits for the person's age that you would use with your formula? Would it make sense to use $A = 2$? Or $A = 150$? In other words, what is a reasonable domain for A?
5. What are the benefits of aerobic exercise over other types of exercise?
6. List some different types of exercise that are nonaerobic. Describe the differences between the two different types of exercise.

ANSWERS

25. ______________________

Answers

1. $C = 0.16s + 20$

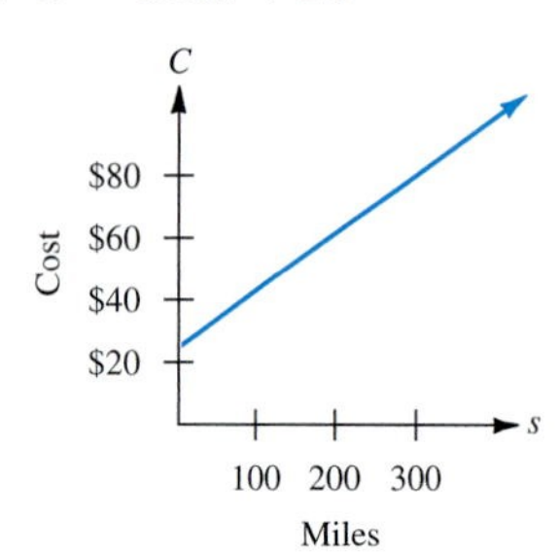

3. **(a)** $T = 35h + 75$ and **(b)** see graph

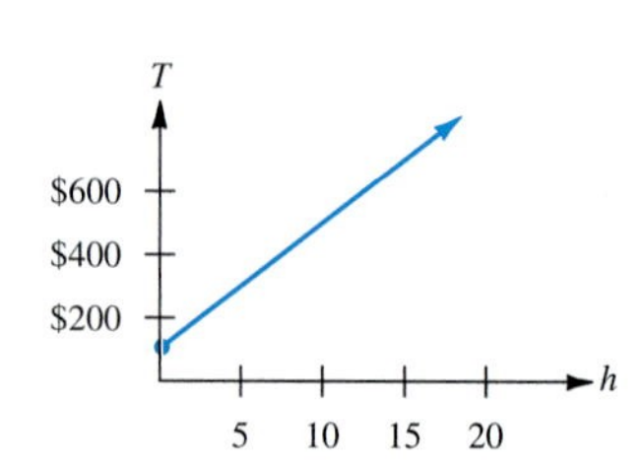

5. $F = \frac{9}{5}C + 32$

7. **(a)**

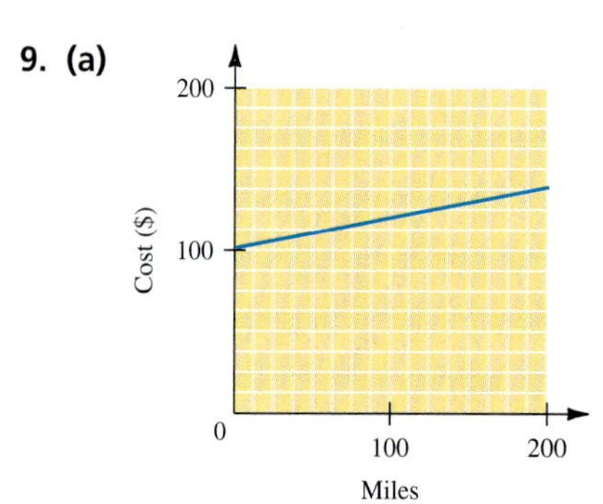

(b) The electrician charges $85 for 30 minutes of work.

9. **(a)**

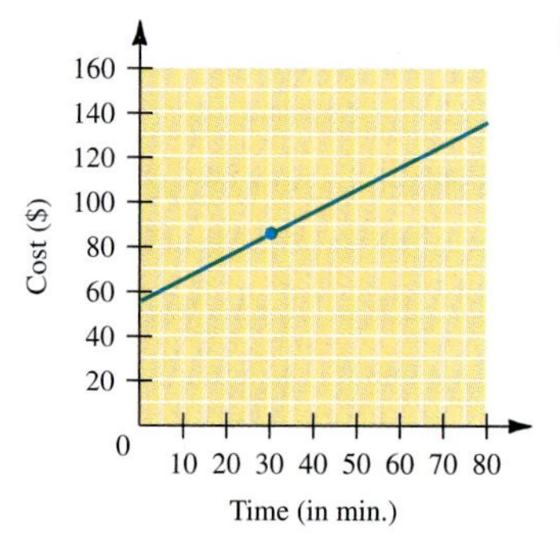

(b) 180 mi

11. **(a)** $C = 35x + 50$; **(b)** $172.50; **(c)** 3.15 h

13. Independent: length of call; dependent: amount of bill

15. Independent: time in air; dependent: height of ball

17. Independent: number of credits; dependent: time to graduate

19. 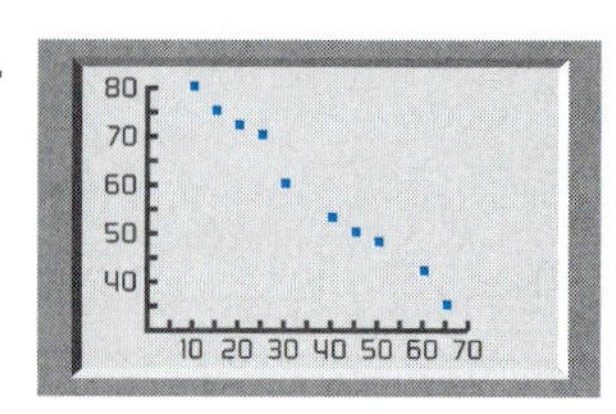

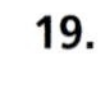

21. 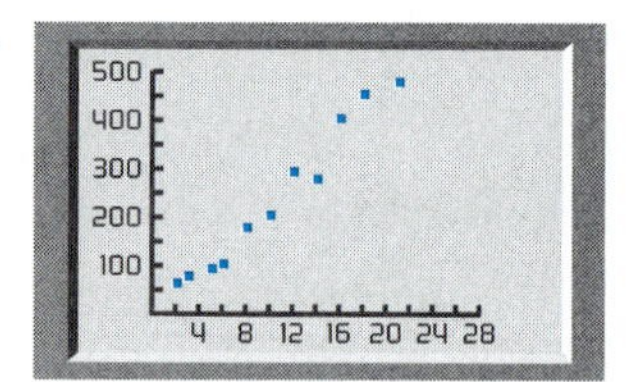

23. 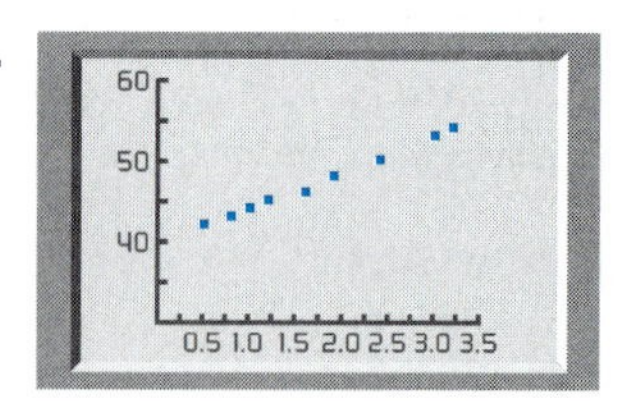

25. 

Summary

DEFINITION/PROCEDURE	EXAMPLE	REFERENCE
The Graph of a Linear Equation		Section 4.1
The **solution set** for an equation in two variables is the set containing all ordered pairs of real numbers (x, y) that will make the equation a true statement. An equation of the form $Ax + By = C \quad (1)$ in which A and B cannot both be zero, is a linear equation in two variables. The graph of such an equation is always a *line*. An equation in form (1) is called the **standard form of the equation of a line.** To graph a linear equation: **1.** Find at least three solutions for the equation, and write your results in a table of values. **2.** Graph the points associated with the ordered pairs found in step 1. **3.** Draw a line through the points plotted above to form the graph of the equation. A second approach to graphing linear equations uses the ***x*** and ***y*** **intercepts** of the line. The x intercept is the point at which the line intersects the x axis. The y intercept is the point at which the line intersects the y axis. To graph by the intercept method: **1.** Find the x intercept. Let $y = 0$ and solve for x. **2.** Find the y intercept. Let $x = 0$ and solve for y. **3.** Plot the two intercepts determined in steps 1 and 2. **4.** Draw a line through the intercepts.	(1, 3) is a solution for $3x + 2y = 9$ because $3 \cdot 1 + 2 \cdot 3 = 9$ is a true statement. To graph $y = 2x - 3$ $(0, -3)$, $(1, -1)$, and $(2, 1)$ are solutions.	pp. 190, 192, 193
Vertical or Horizontal Lines **1.** A line with an equation of the form $y = k$ is horizontal (parallel to the x axis). **2.** A line with an equation of the form $x = h$ is vertical (parallel to the y axis).		p. 196
The Slope of a Line		Section 4.2
The **slope** of a line gives a numerical measure of the direction and steepness, or inclination, of the line. The slope m of a line containing the distinct points in the plane (x_1, y_1) and (x_2, y_2) is given by $m = \frac{y_2 - y_1}{x_2 - x_1} \quad x_2 \neq x_1$ The slopes of two nonvertical parallel lines are equal. The slopes of two nonvertical perpendicular lines are the negative reciprocals of each other.	Given a line through $(2, -1)$ and $(-4, 5)$, the slope is $m = \frac{-1 - 5}{2 - (-4)} = \frac{-6}{6} = -1$	p. 208

Continued

DEFINITION/PROCEDURE	EXAMPLE	REFERENCE
Forms of Linear Equations		**Section 4.3**
There are two useful special forms for the equation of a line. The **slope-intercept form** of the equation of a line is $y = mx + b$, in which the line has slope m and y intercept $(0, b)$. The **point-slope form** of the equation of a line is $y - y_1 = m(x - x_1)$, in which the line has slope m and passes through the point (x_1, y_1). And $x = x_1$ is the equation of a line through (x, y) with undefined slope. The slope-intercept form can also be expressed as a function $f(x) = mx + b$	$y = \frac{2}{3}x + 4$ is in slope-intercept form. The slope m is $\frac{2}{3}$, and the y intercept is $(0, 4)$. If line l has slope $m = -2$ and passes through $(-2, 3)$, its equation is $y - 3 = -2[x - (-2)]$ $y - 3 = -2(x + 2)$ $y - 3 = -2x - 4$ $y = -2x - 1$	**pp. 228, 230**
Solving Equations and Inequalities Graphically		**Section 4.4**
Solving an Inequality in One Variable Graphically **1.** Rewrite the inequality as a comparison of two functions. $f(x) < g(x)$ $f(x) > g(x)$ $f(x) \le g(x)$ $f(x) \ge g(x)$ **2.** Graph the two functions on a single set of axes. **3.** Draw a vertical line through each point of intersection of the two functions. Use a dotted line if equality is not included ($<$ or $>$). Use a solid line if equality is included ($\le$ or $\ge$). **4.** Mark the x values that make the inequality a true statement. **5.** Write the solution in set notation.	To solve $2x - 6 > 8x$, let $f(x) = 2x - 6$ $g(x) = 8x$ then graph both lines. Draw a dotted line where they intersect. 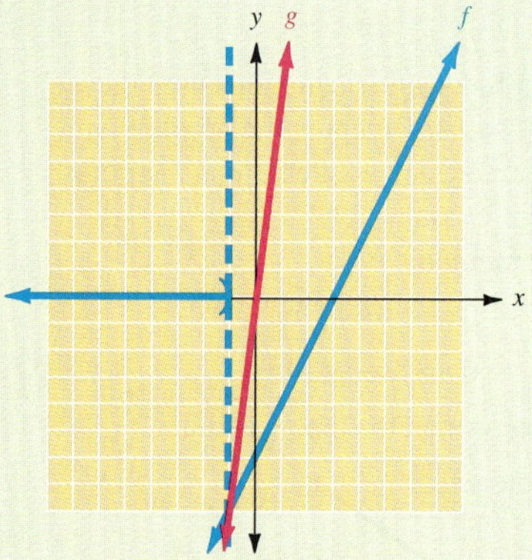 y g f x The solution is $x < -1$. The solution set is $\{x \mid x < -1\}$.	**p. 250**
Absolute Value Equations and Inequalities		**Section 4.4**
Finding a Graphical Solution for an Absolute Value Equation **1.** Let each side of the equation represent a function of x. **2.** Graph the two functions on the same set of axes. **3.** Find the intersection of the two graphs. The x value(s) at these intersection points represent the solutions to the original equation.		**p. 261**

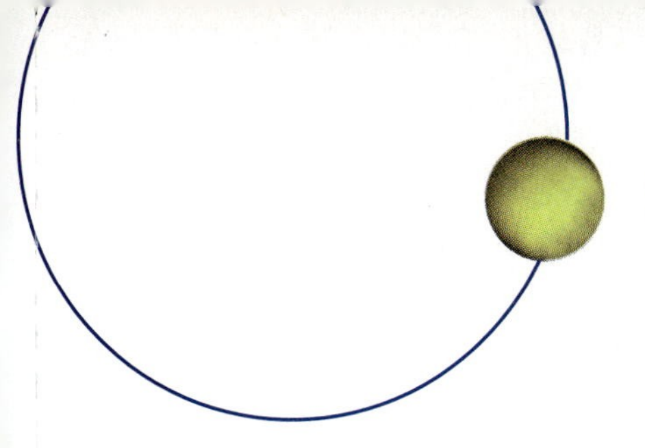

Summary Exercises

This summary exercise set is provided to give you practice with each of the objectives in the chapter. Each exercise is keyed to the appropriate chapter section. The answers are provided in the *Instructor's Manual*.

[4.1] In exercises 1 to 3, find three ordered pairs that are solutions for the given equations.

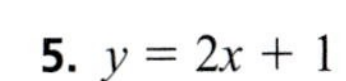

1. $2x - y = 8$

2. $3x + y = 7$

3. $x + 2y = 5$

In exercises 4 to 7, graph each equation.

4. $y = x + 1$

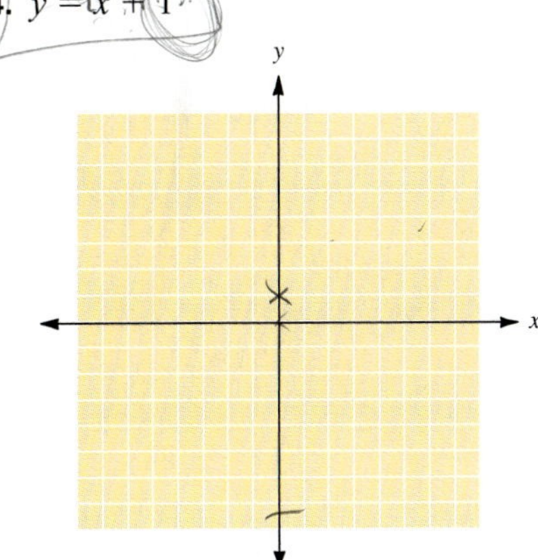

5. $y = 2x + 1$

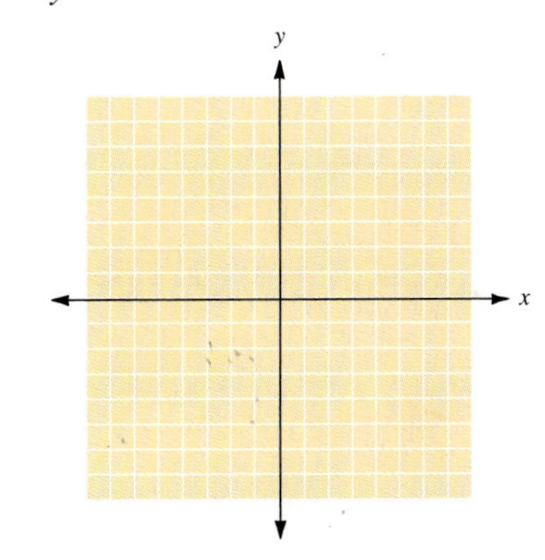

6. $y = -2x + 1$

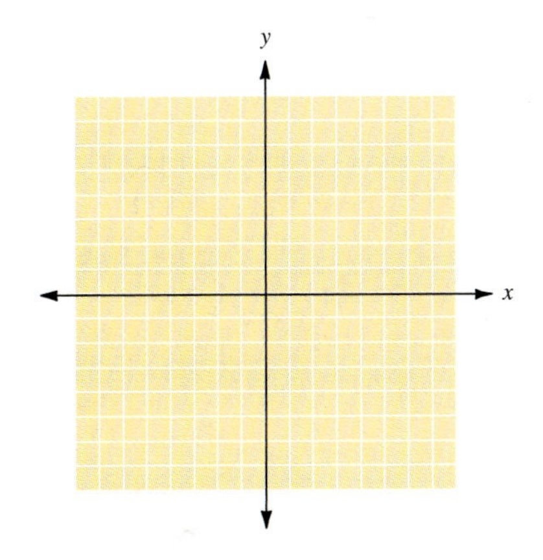

7. $y = -3x + 1$

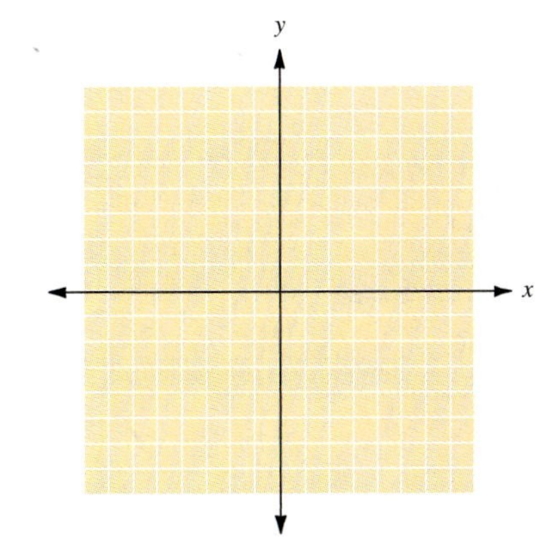

[4.1] Find the x and y intercepts, and then graph each of the following equations.

8. $x - 2y = 4$

9. $x + 3y = 6$

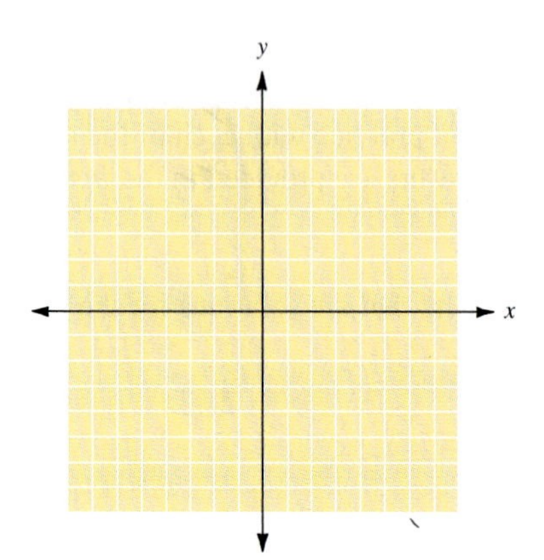

10. $2x - y = 6$

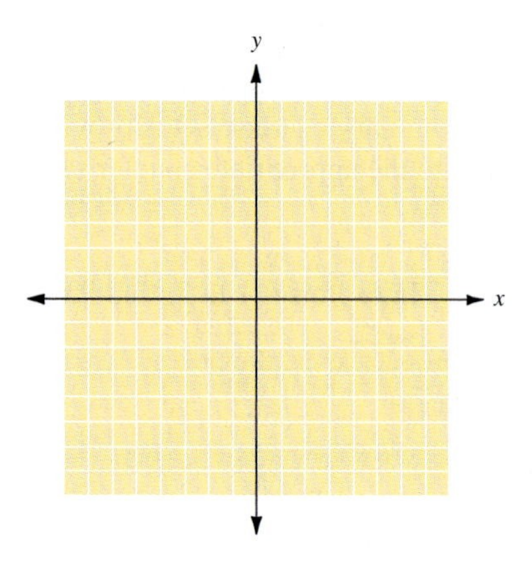

11. $3x + 2y = 12$

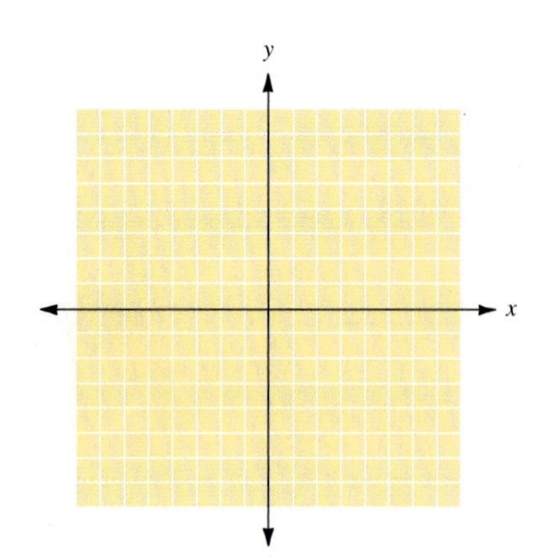

12. $2x + 5y = 10$

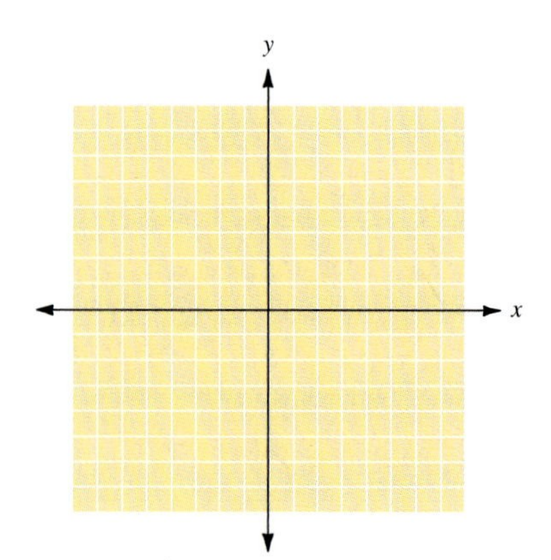

13. $2x - 3y = 6$

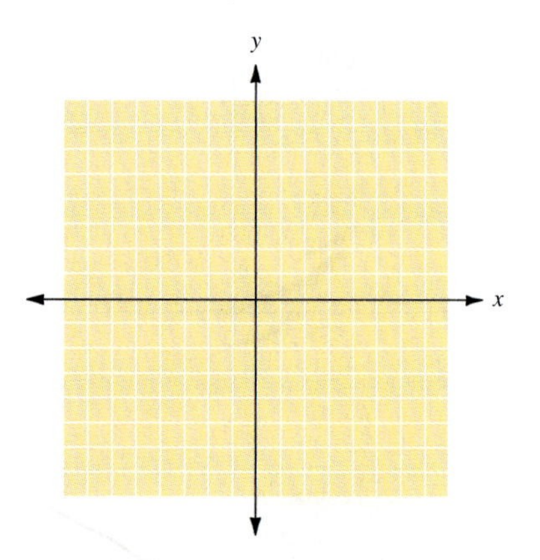

14. $y = 4$

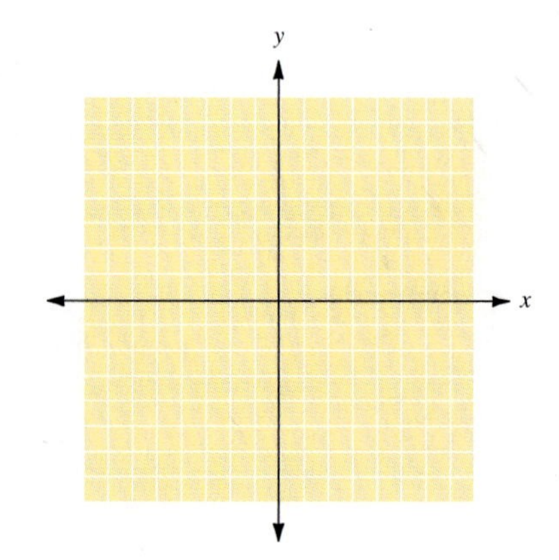

15. $x = -3$

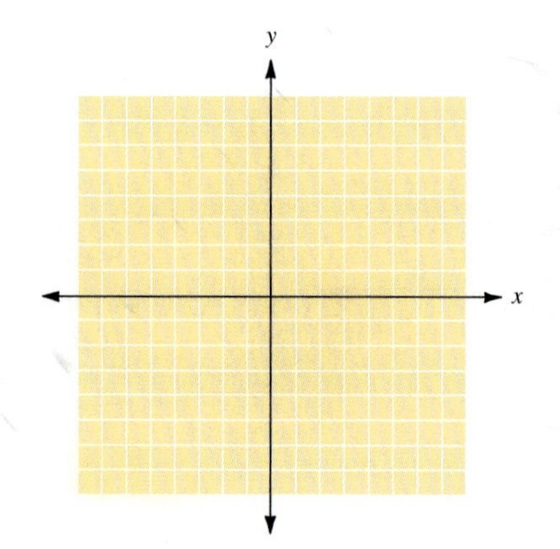

[4.1] In exercises 16 to 19, find the domain and range of each relation.

16. $4x - 5y = 40$

17. $-3x + 15y = 45$

18. $2x + 18 = 0$

19. $7y - 28 = 0$

[4.2] In exercises 20 to 27, find the slope (if it exists) of the line determined by the given pair of points.

20. (3, 2) and (5, 8)

21. (−2, 5) and (1, −1)

22. (2, −4) and (3, 5)

23. (−3, −4) and (3, 0)

24. (4, −3) and (4, 4)

25. (4, −2) and (−2, −2)

26. (2, −4) and (−1, −3)

27. (5, −2) and (5, 3)

In exercises 28 and 29, find the slope of the line determined by the given equation.

28. $3x + 2y = 6$

29. $x - 4y = 8$

In exercises 30 to 33, are the pairs of lines parallel, perpendicular, or neither?

30. L_1 through (−3, −2) and (1, 3)
L_2 through (0, 3) and (4, 8)

31. L_1 through (−4, 1) and (2, −3)
L_2 through (0, −3) and (2, 0)

32. L_1 with equation $x + 2y = 6$
L_2 with equation $x + 3y = 9$

33. L_1 with equation $4x + 6y = 18$
L_2 with equation $2x - 3y = 6$

[4.3] In exercises 34 to 43, write the equation of the line passing through the given point with the indicated slope. Give your results in slope-intercept form, when possible.

34. $(0, -5), m = \frac{2}{3}$

35. $(0, -3), m = 0$

36. $(2, 3), m = 3$

37. $(4, 3)$, m is undefined

38. $(3, -2), m = \frac{5}{3}$

39. $(-2, -3), m = 0$

40. $(-2, -4), m = -\frac{5}{2}$

41. $(-3, 2), m = -\frac{4}{3}$

42. $\left(\frac{2}{3}, -5\right), m = 0$

43. $\left(-\frac{5}{2}, -1\right)$, m is undefined

[4.3] In exercises 44 to 51, write the equation of the line L satisfying the given set of geometric conditions.

44. L passes through (−3, −1) and (3, 3).

45. L passes through (0, 4) and (5, 3).

46. L has slope $\frac{3}{4}$ and y intercept (0, 3).

47. L passes through $(4, -3)$ with a slope of $-\frac{5}{4}$.

48. L has y intercept $(0, -4)$ and is parallel to the line with equation $3x - y = 6$.

49. L passes through $(3, -2)$ and is perpendicular to the line with equation $3x - 5y = 15$.

50. L passes through $(2, -1)$ and is perpendicular to the line with equation $3x - 2y = 5$.

51. L passes through the point $(-5, -2)$ and is parallel to the line with equation $4x - 3y = 9$.

In exercises 52 to 55, use the graph to determine the slope and y intercept of the line.

52.

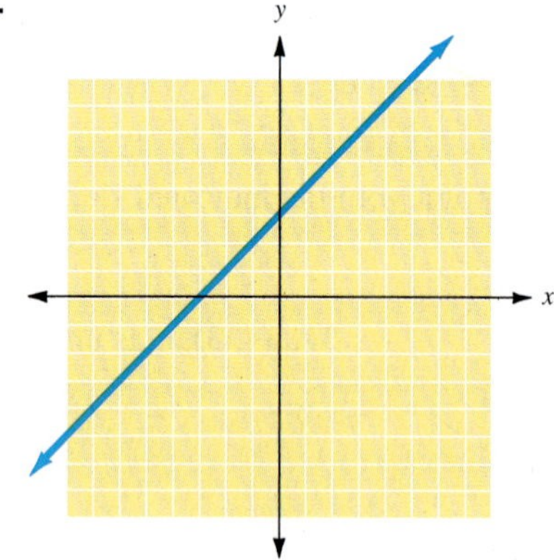

53.

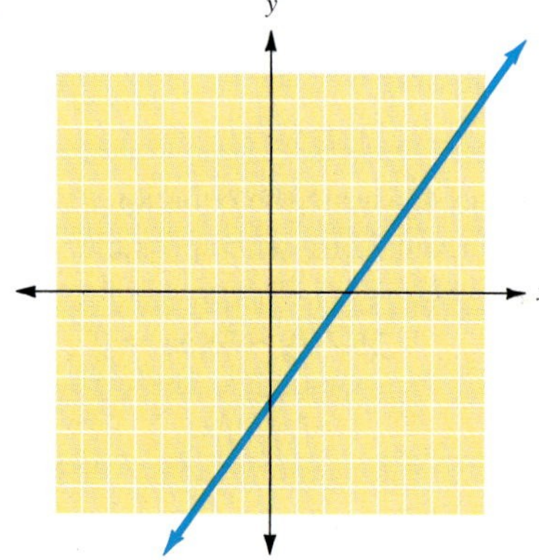

54.

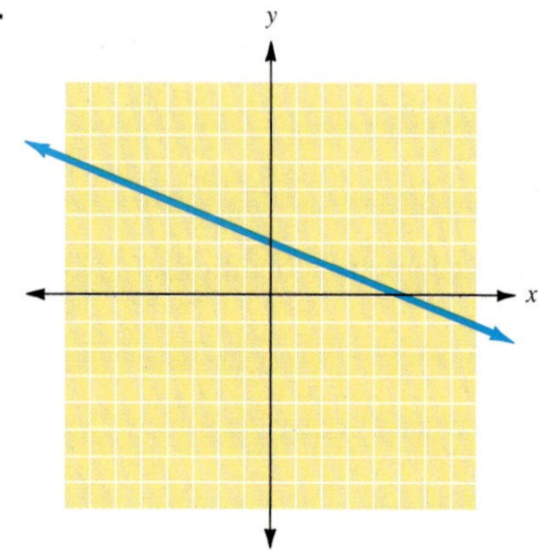

55.

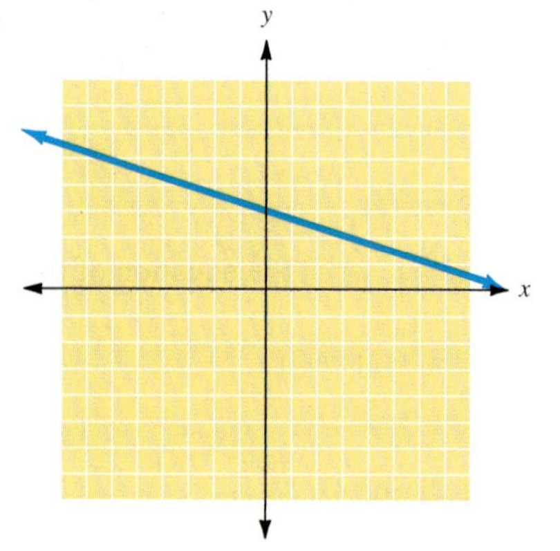

[4.4] Solve the following equations graphically. Check your solution by solving algebraically.

56. $3x - 6 = 0$

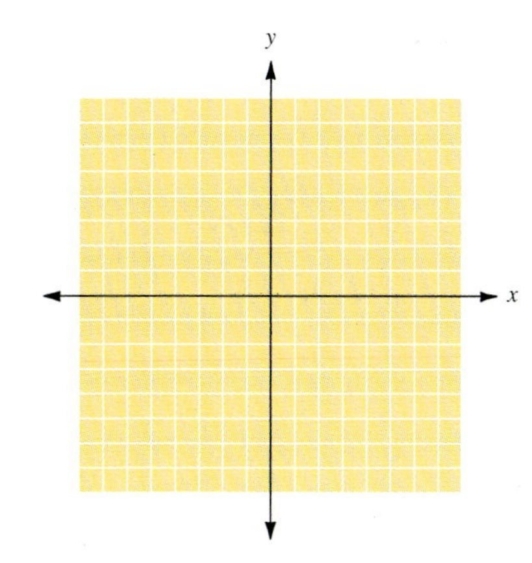

57. $4x + 3 = 7$

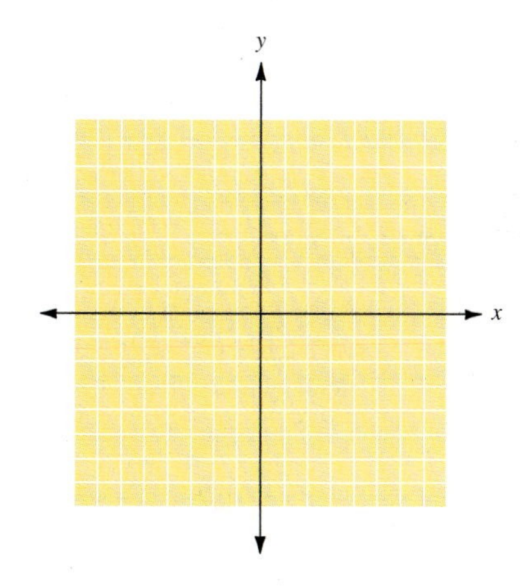

58. $3x + 5 = x + 7$

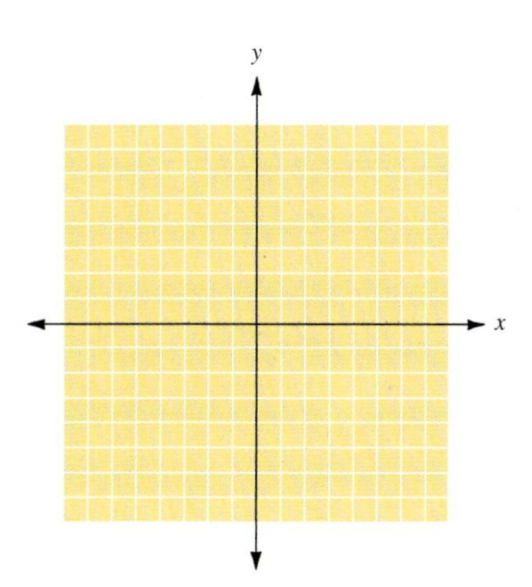

59. $4x - 3 = x - 6$

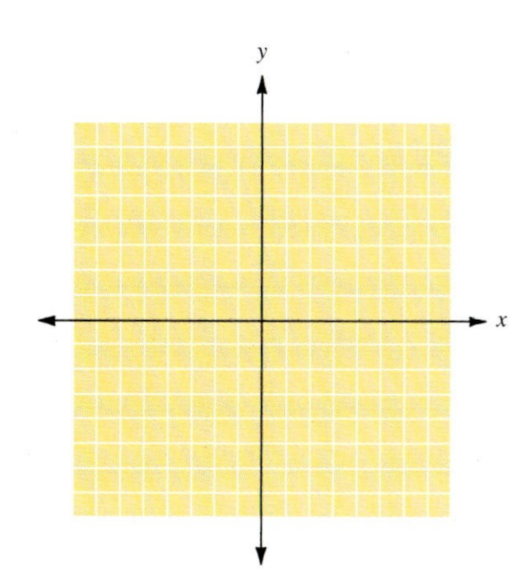

[4.4] Solve each of the following inequalities graphically.

60. $2x + 1 > 5$

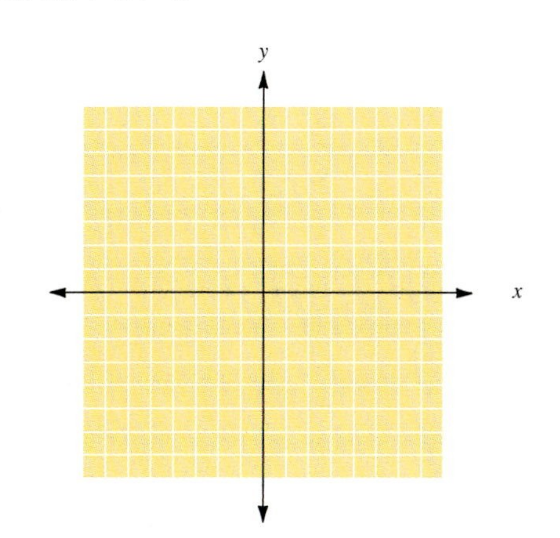

61. $-3 \geq -2x + 3$

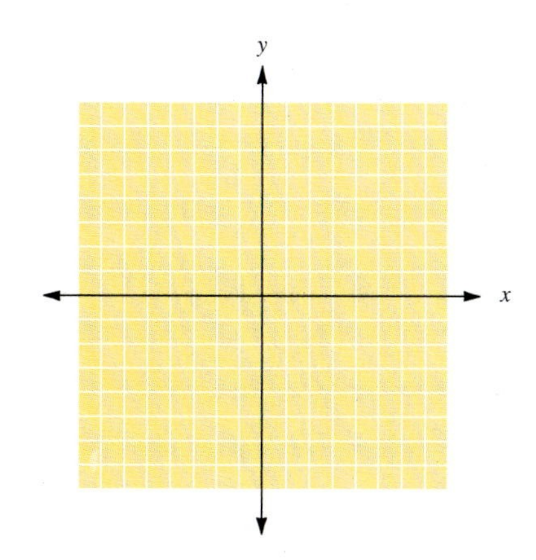

62. $3x + 2 \geq 6 - x$

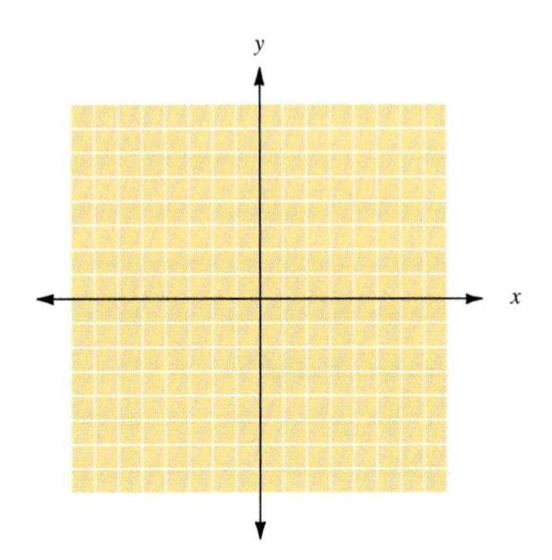

63. $3x - 5 < 15 - 2x$

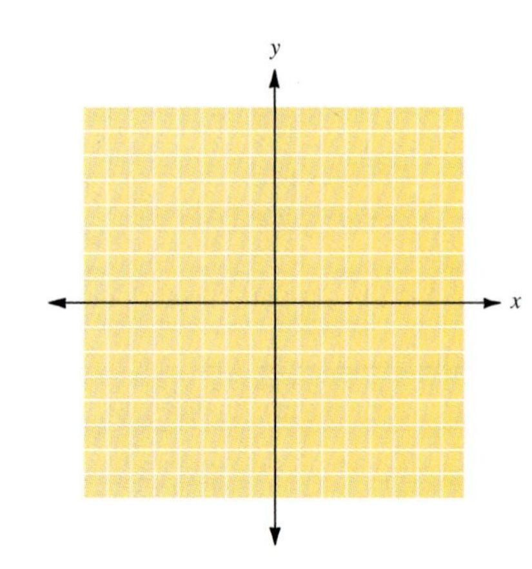

64. $x + 6 < -2x - 3$

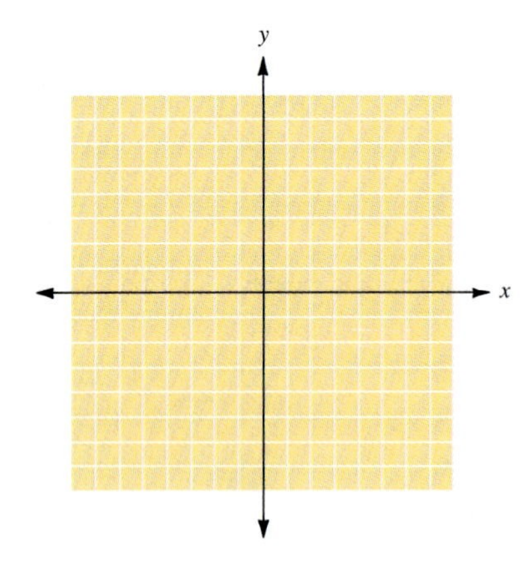

65. $4x - 3 \geq 7 - x$

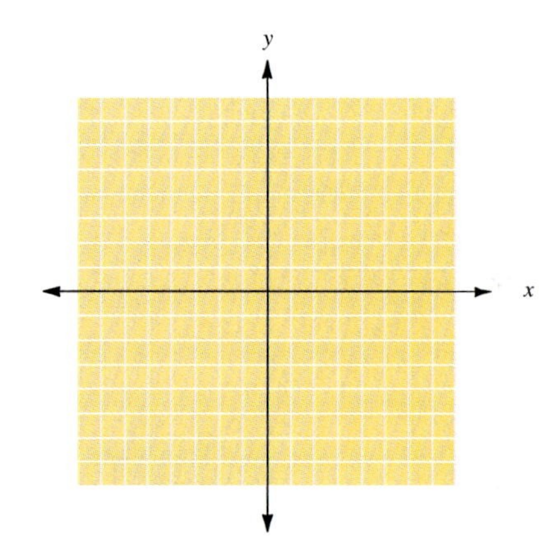

66. $x \geq -3 - 5x$

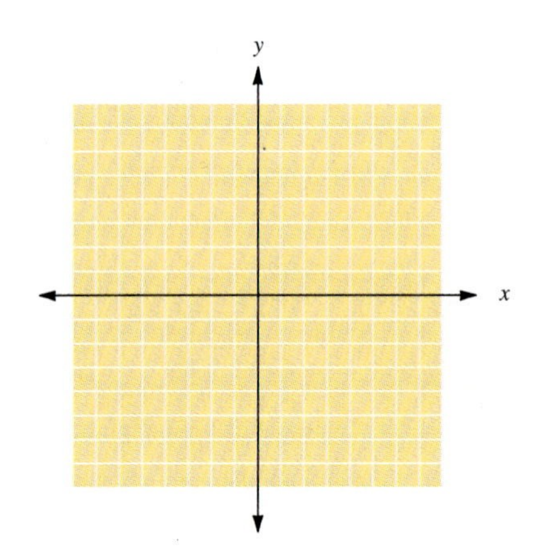

67. $x < 4 + 2x$

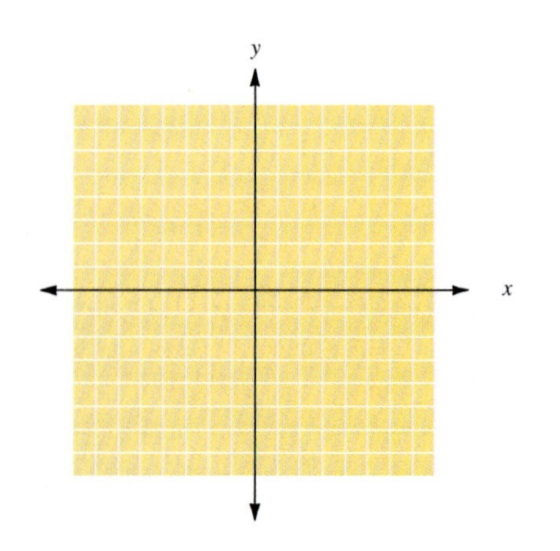

[4.5] Solve the following equations graphically.

68. $|x + 3| = 5$

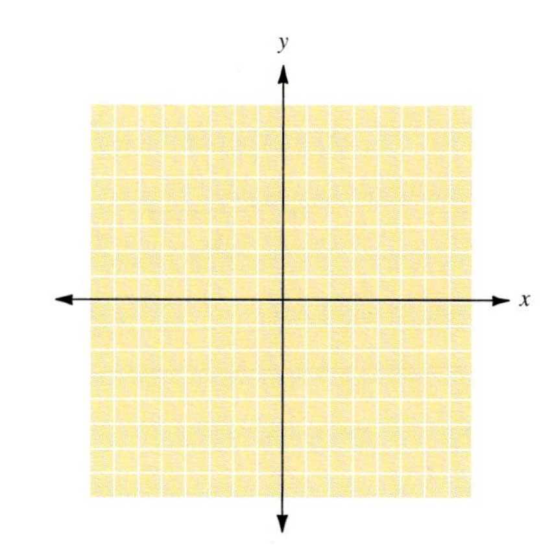

69. $|x - 2| = -2$

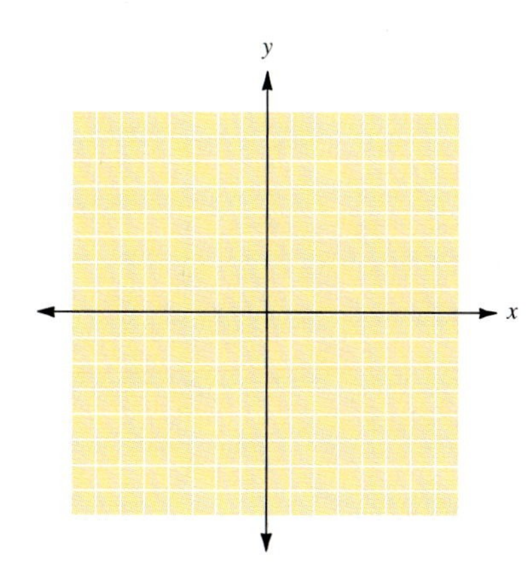

70. $|x - 2| = 3$

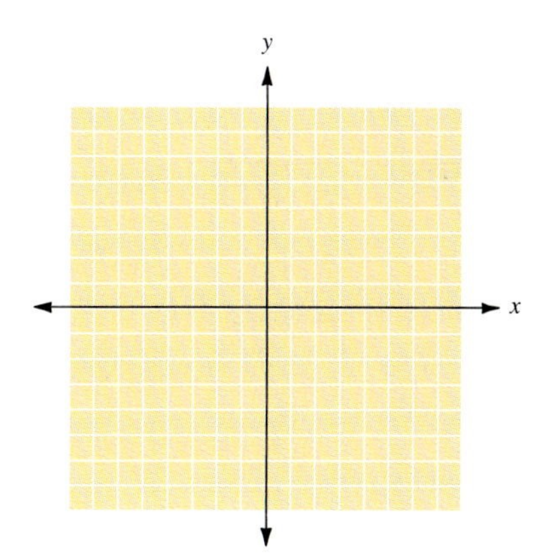

71. $|x - 4| = 2$

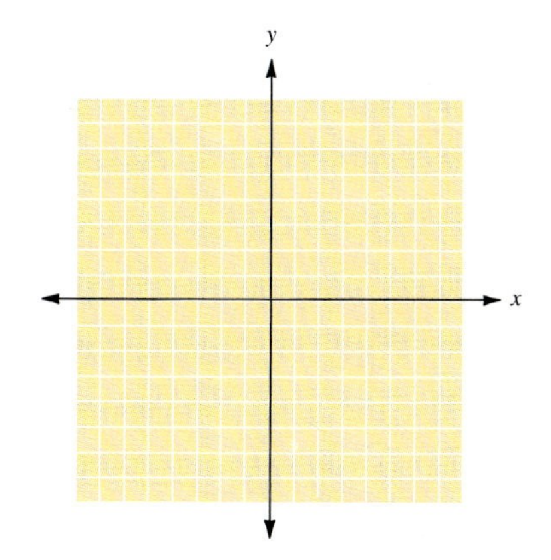

[4.5] Solve the following inequalities graphically.

72. $|x| < 7$

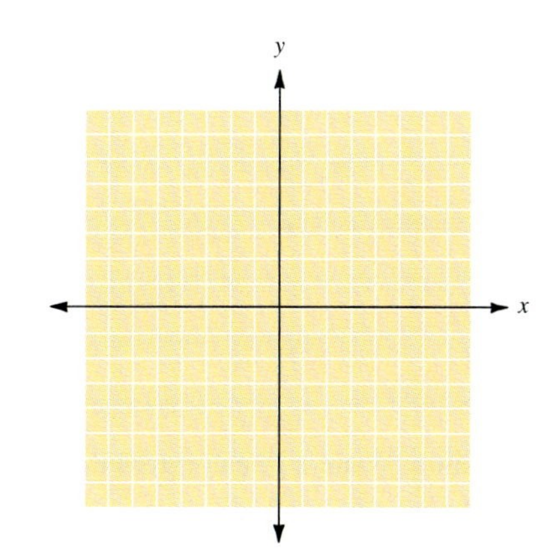

73. $|x| \leq 6$

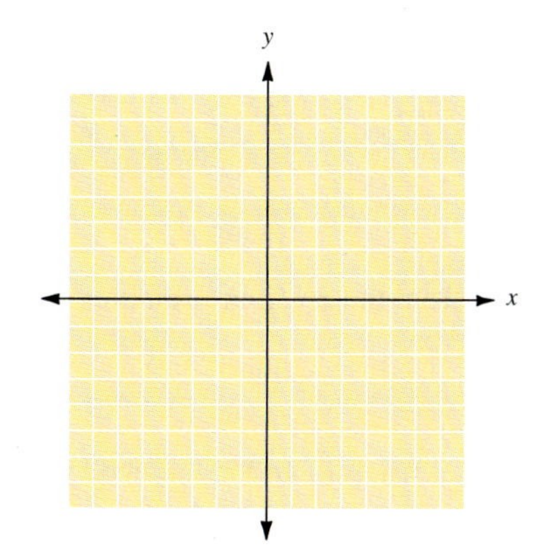

74. $|x - 1| > 6$

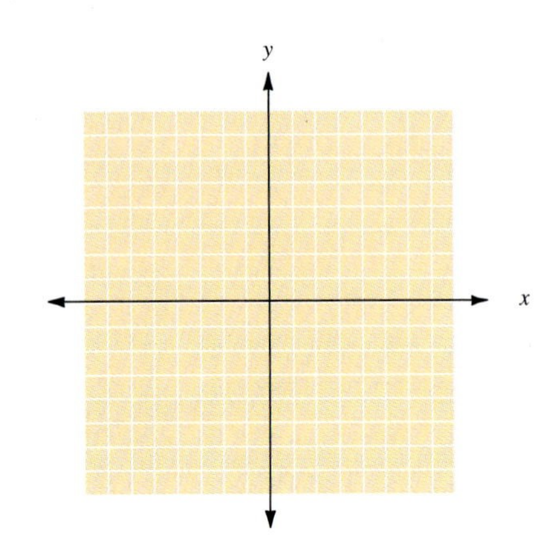

75. $|x + 3| \leq 4$

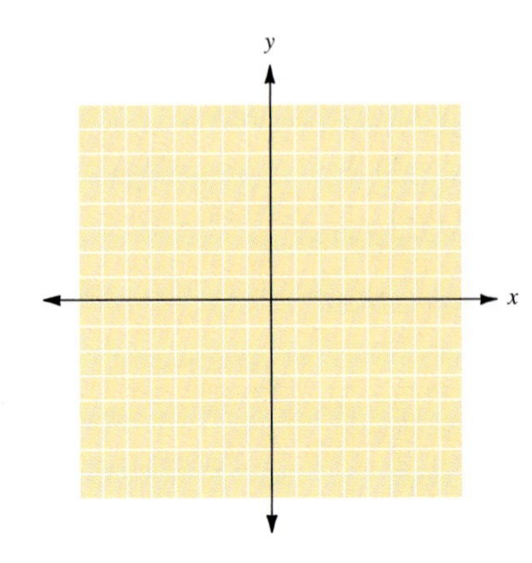

76. $|x + 5| \geq 2$

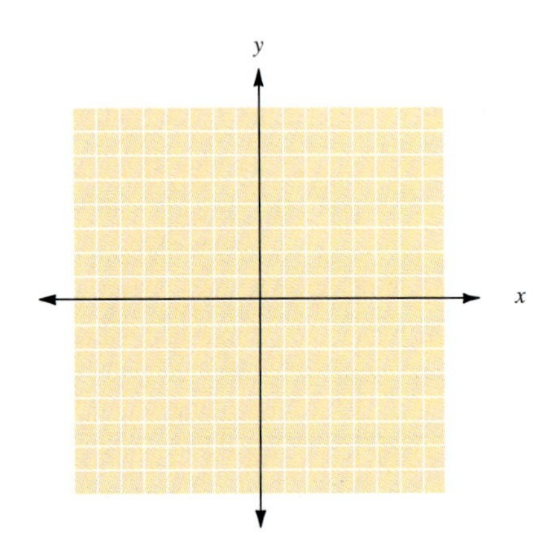

77. $|x - 1| < 8$

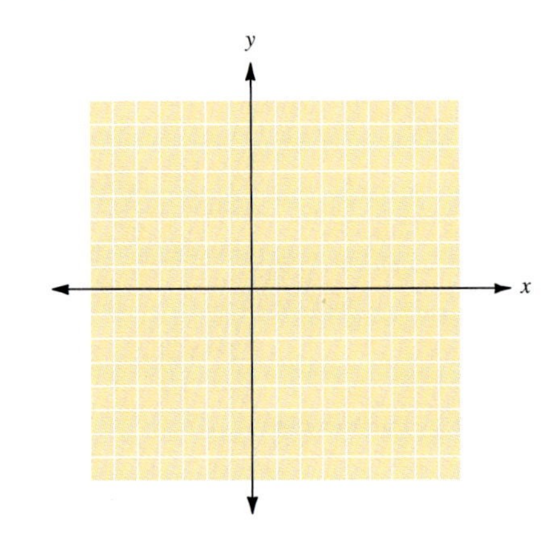

[4.6]

78. A public relations firm has developed a table that relates the number of years of experience and annual salary.

No. of Years	Salary (in thousands)
1	20
2	24
15	48
11	37
9	32
6	29

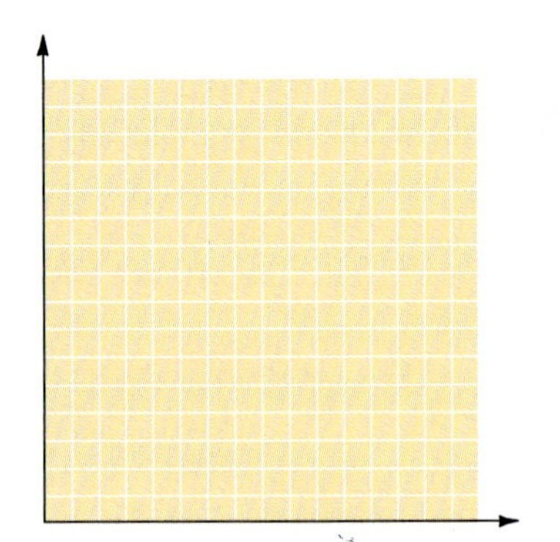

Create a scatter plot from the given information.

79. A local band charges a fixed fee of \$250 plus \$15 per hour for performing at a wedding. Write a linear equation that shows the relationship between the total cost, C, and the number of hours, h, played.

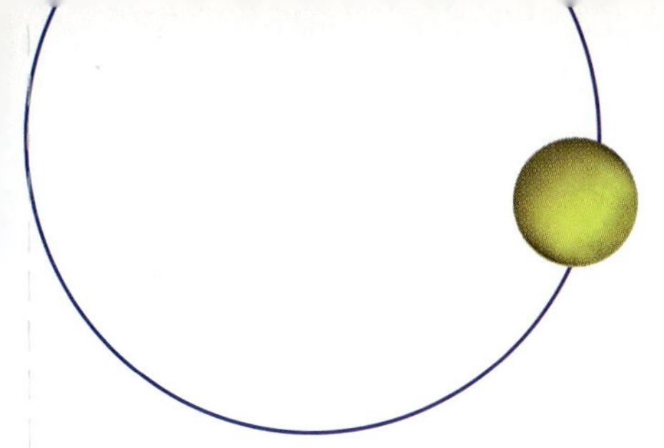

Self-Test for Chapter 4

Name ______________

Section ________ Date ________

ANSWERS

1. ______________

2. ______________

3. ______________

4. ______________

5. ______________

6. ______________

7. ______________

In exercises 1 and 2, identify the x and y intercepts and graph the equation.

1. $5x + 6y = 30$

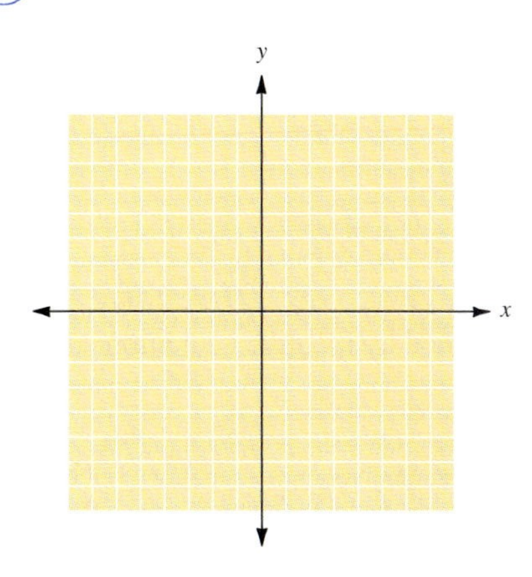

2. $48 - 16x = 0$

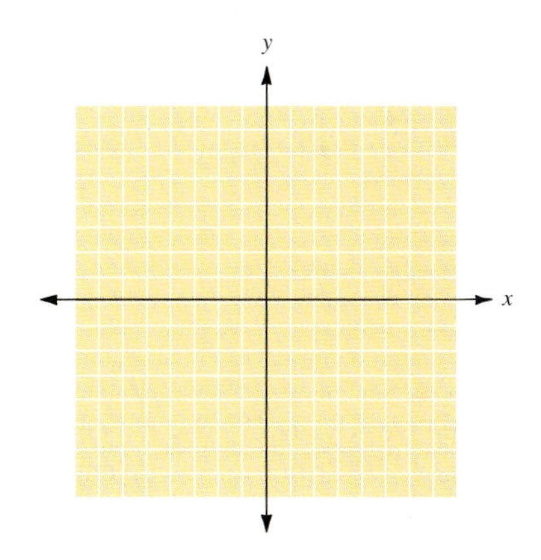

In exercises 3 to 5, find the domain and range of each relation.

3. $4x + 5y = 20$ 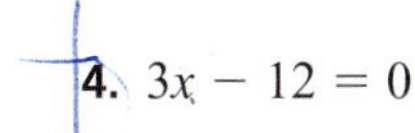 **4.** $3x - 12 = 0$

5. $4y + 28 = 0$

Write the equation of the line L that satisfies the given set of geometric conditions.

6. L passes through $(2, 3)$ and $(4, 7)$.

7. L has slope $\frac{2}{3}$ and y intercept of $(0, -4)$.

ANSWERS

8. ______
9. ______
10. ______
11. ______
12. ______
13. ______
14. ______
15. ______
16. ______
17. ______
18. ______
19. ______
20. ______

8. L passes through $(-1, 2)$ and has slope of -3.

9. L has y intercept $(0, -2)$ and is parallel to $4x - y = 8$.

10. L passes through the point $(3, 5)$ and is perpendicular to $3x + 2y = 4$.

In exercises 11 to 13, find the slope of the line connecting the given points.

11. $(5, 6)$ and $(-3, -10)$

12. $(9, 5)$ and $(9, -3)$

13. $(3, 6)$ and $(-2, 6)$

In exercises 14 to 16, find the slope of the line determined by the given equation.

14. $2x + 5y = 10$

15. $3x - 6 = 0$

16. $4y + 24 = 0$

In exercises 17 to 19, determine if the given lines are parallel, perpendicular or neither.

17. $3x + 5y = 10$
$5x + 3y = 10$

18. $2y - 3x = 12$
$3y + 2x = 15$

19. $5x + 6y = 2$
$15x + 18y = 2$

20. A study was made to determine the relation between weekly advertising expenditures and sales. The following data were recorded. Create a scatterplot for the given information.

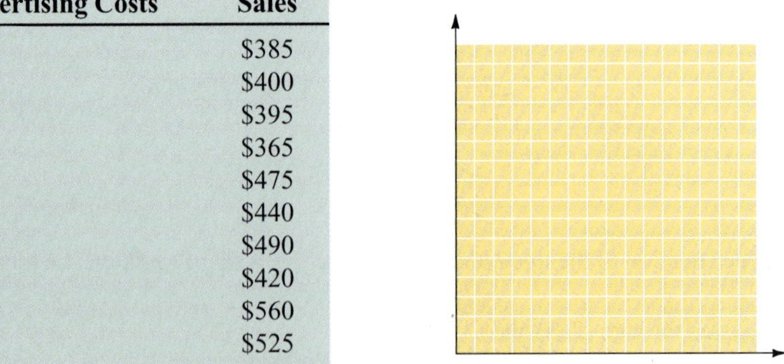

Advertising Costs	Sales
$40	$385
$20	$400
$25	$395
$20	$365
$30	$475
$50	$440
$40	$490
$20	$420
$50	$560
$40	$525
$25	$480
$50	$510

Solve the following inequalities graphically.

21. $2x + 1 > 5$

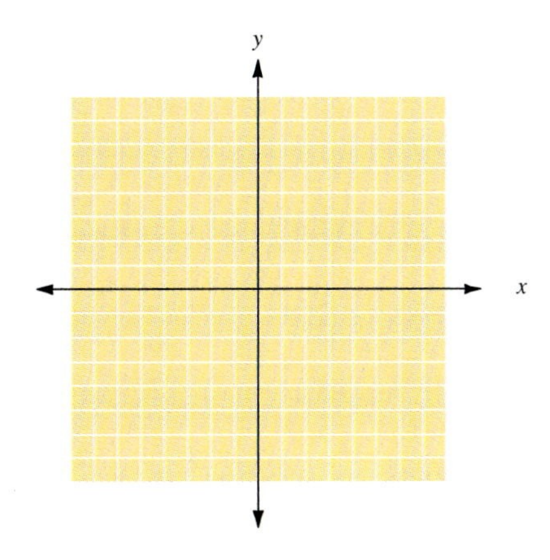

22. $-3 \geq -2x + 3$

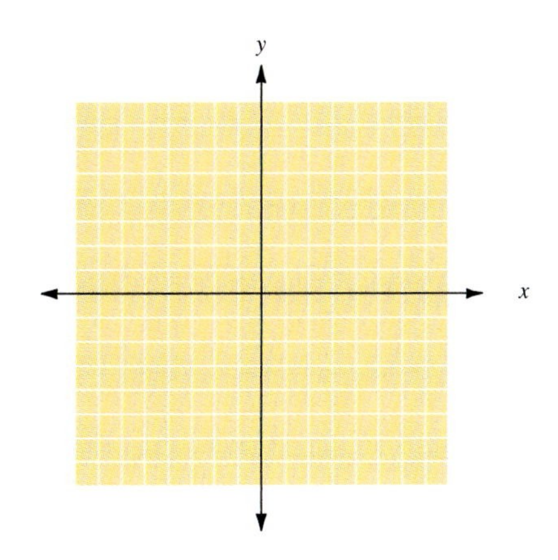

Solve the following equation graphically.

23. $|x - 2| = 5$

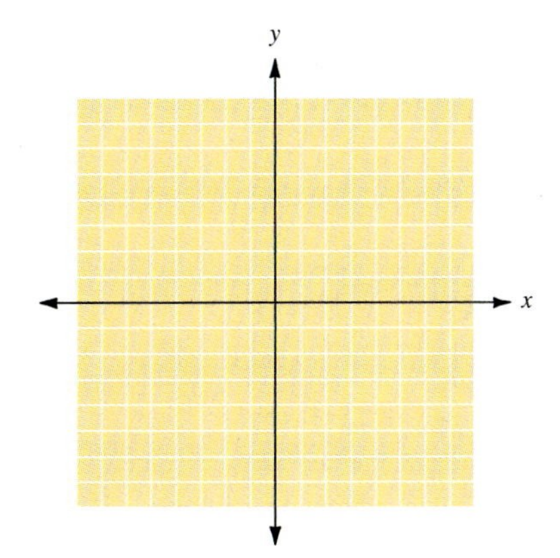

ANSWERS

21. ____________

22. ____________

23. ____________

ANSWERS

24. ____________________

25. ____________________

Solve the following inequalities graphically.

24. $|x| \leq 3$

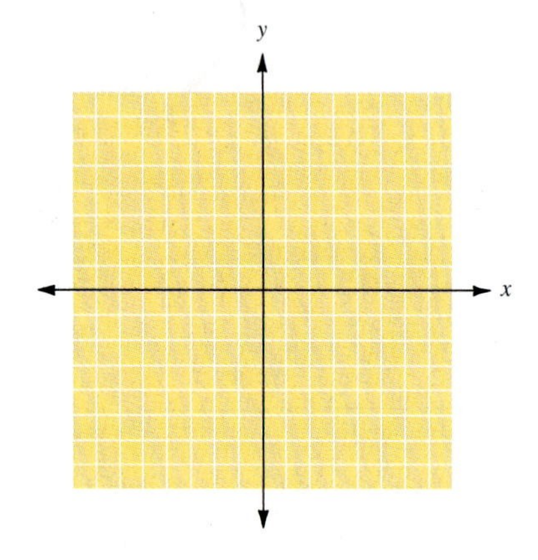

25. $|x + 1| \geq 4$

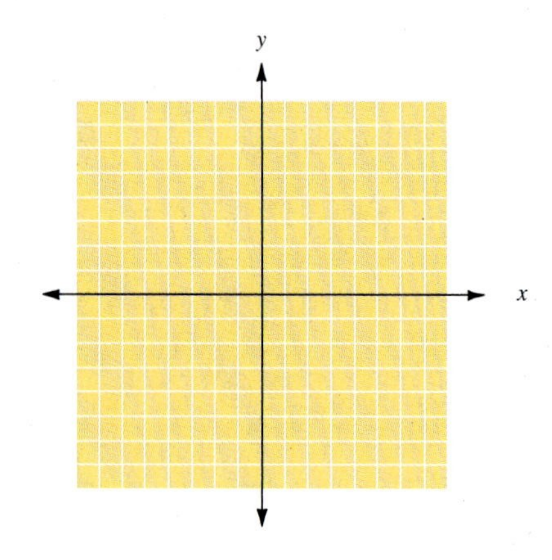

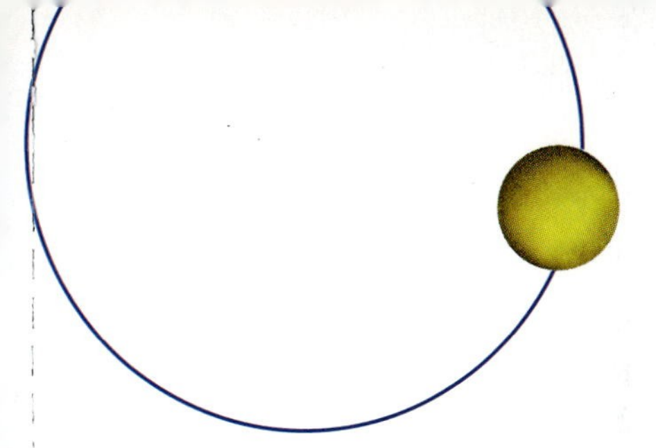

Cumulative Test for Chapters 1 to 4

Name ______________

Section ______ Date ______

ANSWERS

1. ______
2. ______
3. ______
4. ______ (0)
5. ______ (0)
6. ______
7. ______
8. ______
9. ______
10. ______
11. ______ (0)
12. ______ (0)
13. ______ (0)
14. ______
15. ______
16. ______

This test is provided to help you in the process of review of the previous chapters. Answers are provided in the back of the book. If you missed any problems, be sure to go back and review the appropriate section.

Identify the property of real numbers that justifies each statement.

1. $(5x)y = 5(xy)$

2. $a + 3b = 3b + a$

3. $2c(x + d) = 2cx + 2cd$

Graph each of the following sets.

4. $\{x | x < -2\}$

5. $\{x | -3 \le x \le 4\}$

Solve each of the following equations and check the results.

6. $5c - 7 = 3c - 2$

7. $3x + 11 = 5(x - 2)$

8. $\frac{1}{3}x - 5 = 8 - \frac{5}{6}x$

Solve each of the following equations for the indicated variable.

9. $A = \frac{1}{2}h(B + b)$ for B

10. $7p - 4y + 12 = 0$ for p

Solve and graph the solution to each inequality.

11. $3x - 5 > 5x + 9$

12. $-11 \le -3x + 4 \le -2$

13. $|5 - 2x| > 5$

Solve each of the following applications. Be sure to show the equation used for the solution.

14. The sum of two consecutive odd integers is -16. What are the two integers?

15. Joshua received grades of 85 and 91 on his first two mathematics tests. What score does he need on the next test to have an average of 90 for the three tests?

16. Carla and Jake leave town at the same time, traveling in opposite directions. If Carla travels at 55 mi/h and Jake travels at 35 mi/h, how long will it take for them to be 330 mi apart?

ANSWERS

17. ______
18. ______
19. ______
20. ______
21. ______
22. ______
23. ______
24. ______
25. ______
26. ______
27. ______
28. ______
29. ______
30. ______

17. At a school dance, admission was \$4 for tickets bought in advance and \$5 for those purchased at the door. If the 172 people attending paid a total of \$748, how many of each type of ticket was purchased?

Graph each of the following equations.

18. $x - y = 4$ **19.** $y = 3x - 1$ **20.** $3x + 2y = 12$

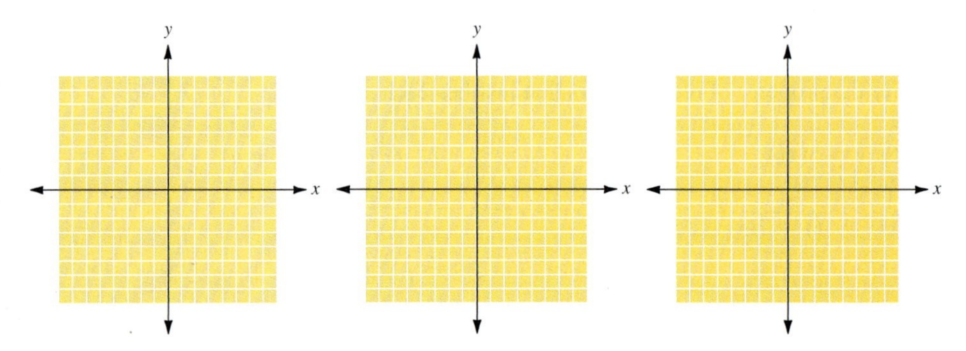

21. $x + 2y = 0$ **22.** $x = -4$

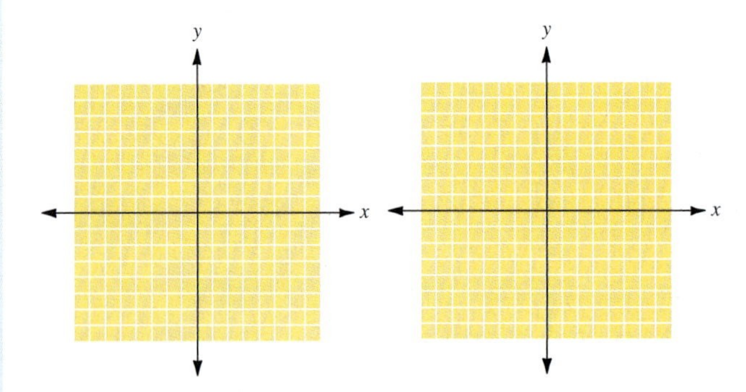

Find the slope (if it exists) of the line determined by each of the following pairs of points.

23. $(3, -4)$ and $(-3, 8)$

24. $(-2, 4)$ and $(-2, 7)$

25. $(-4, 3)$ and $(4, -3)$

26. $(5, -3)$ and $(-2, -3)$

Are the following pairs of lines parallel, perpendicular, or neither?

27. L_1 through $(-3, 7)$ and $(1, -5)$
L_2 through $(2, 3)$ and $(-4, 1)$

28. L_1 with equation $x - 2y = 7$
L_2 with equation $2x + 4y = 3$

Write the equation of the line L satisfying each of the following sets of conditions.

29. L passes through $(3, 2)$ and $(-1, 10)$.

30. L passes through $(3, 1)$ and is perpendicular to the line with equation $x - 2y = 5$.

5 Systems of Linear Relations

INTRODUCTION

Successful businesses juggle many factors, including workers' schedules, available machine time, and costs of raw material and storage. Getting the most efficient mix of these factors is crucial if the business is to make money. Systems of linear equations can be used to find the most efficient ways to combine these costs.

For example, the owner of a small bakery must decide how much of several kinds of bread to make based on the time it takes to produce each kind and the profit to be made on each one. The owner knows the bakery requires 0.8 hours of oven time and 1.25 hours of preparation time for every 8 loaves of bread, and that a coffee cake takes 1 hour of oven time and 1.25 hours of preparation time for every 6 coffee cakes. In 1 day the bakery has 12 hours of oven time and 16 hours of preparation time available. The owner knows that she clears a profit of \$0.50 for every loaf of bread sold and \$1.75 for every coffee cake sold. But, she has found that on a regular day she sells no more than 12 coffee cakes.

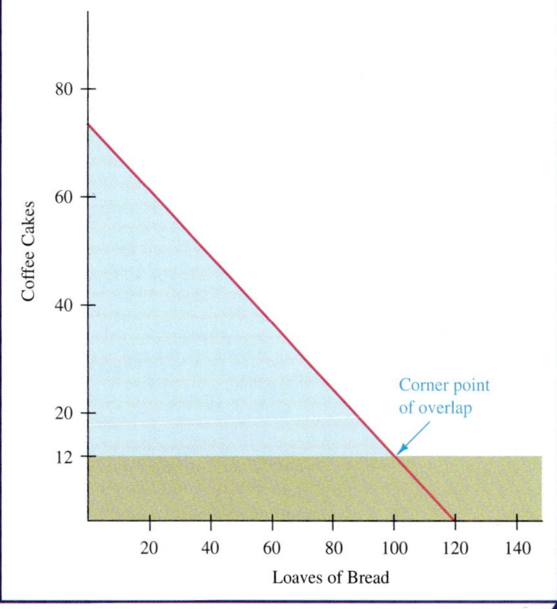

Given these constraints, how many of each type of product can be made in a day? Which combination of products would give the highest profit if all the products are sold?

These questions can be answered by graphing a system of inequalities that model the constraints, in which b is the number of loaves of bread, and c is the number of coffee cakes. We begin by writing inequalities for baking and preparation times, and coffee cake sales.

The graph of these three inequalities shows where all of these constraints overlap and indicate what is possible.

Baking time:

$$b\left(\frac{1}{10}\right) + c\left(\frac{1}{6}\right) \le 12$$

Preparation time:

$$b\left(\frac{1.25}{8}\right) + c\left(\frac{1.25}{6}\right) \leq 16$$

Coffee cake sales:

$$c < 12$$

There is one point of particular interest; the point (100, 12), indicates that if no more than 12 coffee cakes can be sold, then 100 loaves of bread and 12 coffee cakes is the combination that fits the constraints and makes the most profit. In this case, the profit will be 100($.50) + 12($1.75) or $71.00 for 1 day.

This graph not only tells the baker what combination gives the most profit but also indicates that she has room to market more of the coffee cakes. She might consider a sales promotion. To solve this problem, she would use a system of equations, which is the topic of this chapter.

5.1 Systems of Linear Equations in Two Variables

OBJECTIVES

1. Find ordered pairs associated with two equations
2. Solve a system by graphing
3. Solve a system by the addition method
4. Solve a system by the substitution method

Our work in this chapter focuses on systems of equations and the various solution techniques available for your work with such systems. First, let's consider what we mean by a system of equations.

In many applications, you will find it helpful to use two variables when labeling the quantities involved. Often this leads to a **linear equation in two variables.** A typical equation might be

$$x - 2y = 6$$

NOTE Of course, there are an infinite number of solutions for an equation of this type. You might want to verify that (2, −2) and (6, 0) are also solutions.

A solution for such an equation is any ordered pair of real numbers (x, y) that satisfies the equation. For example, the ordered pair $(4, -1)$ is a solution for the equation because substituting 4 for x and -1 for y results in a true statement.

$$4 - 2(-1) \stackrel{?}{=} 6$$

$$4 + 2 \stackrel{?}{=} 6$$

$$6 \stackrel{?}{=} 6 \qquad \text{True}$$

Whenever two or more equations are considered together, they form a **system of equations.** If the equations of the system are linear, the system is called a **linear system.** Our work here involves finding solutions for such systems. We present three methods for solving such systems: the graphing method, the addition method, and the substitution method.

We begin our discussion with a definition.

Definitions: Solution

A **solution** for a linear system of equations in two variables is an ordered pair of real numbers (x, y) that satisfies *both* equations in the system.

For instance, given the linear system

$$x - 2y = -1$$

$$2x + y = 8$$

the pair (3, 2) is a solution because after substituting 3 for x and 2 for y in the two equations of the system, we have the *two* true statements

NOTE Both equations are satisfied by (3, 2).

$$3 - 2(2) \stackrel{?}{=} -1 \quad \text{and} \quad 2(3) + 2 \stackrel{?}{=} 8$$

$$-1 = -1 \quad \text{and} \quad 8 = 8$$

The solution set for a linear equation in two variables may be graphed as a line. Because a solution to a system of equations represents a point on both lines, one approach to finding

NOTE It is helpful at this point to review Section 4.1 on graphing linear equations.

the solution for a system is to **graph** each equation on the same set of coordinate axes and then identify the point of intersection. This is shown in Example 1.

Example 1

Solving a System by Graphing

NOTE Solve each equation for y and then graph.

$y = -2x + 4$

and

$y = x - 5$

We can *approximate* the solution by tracing the curves near their intersection.

Solve the system by graphing.

$$2x + y = 4$$
$$x - y = 5$$

We graph the lines corresponding to the two equations of the system.

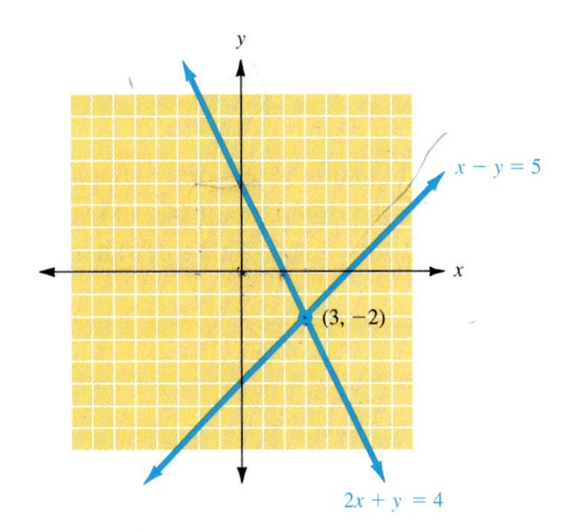

Each equation has an infinite number of solutions (ordered pairs) corresponding to points on a line. The point of intersection, here $(3, -2)$, is the *only* point lying on both lines, and so $(3, -2)$ is the only ordered pair satisfying both equations. Thus, $(3, -2)$ is the solution for the system.

CHECK YOURSELF 1

Solve the system by graphing.

$$3x - y = 2$$
$$x + y = 6$$

In Example 1, the two lines are nonparallel and intersect at only one point.

Definitions: Consistent System

A system of equations that has a unique solution is called a **consistent system.**

In Example 2, we examine a system representing two lines that have no point of intersection.

Example 2

Solving a System by Graphing

Solve the system by graphing.

$$2x - y = 4$$
$$6x - 3y = 18$$

The lines corresponding to the two equations are graphed below.

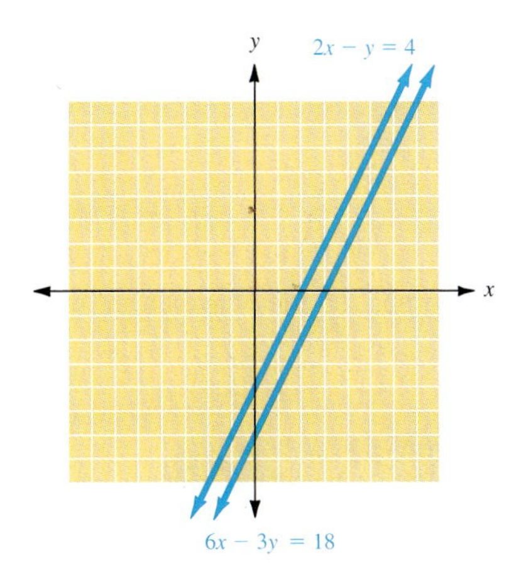

The lines are distinct and parallel. There is no point at which they intersect, so the system has no solution.

Definitions: Inconsistent System

A system of equations with no solution is called an **inconsistent system.**

CHECK YOURSELF 2

Solve the system, if possible.

$$3x - y = 1$$
$$6x - 2y = 3$$

Sometimes the equations in a system have the same graph.

Example 3

Solving a System by Graphing

Solve the system by graphing.

$$2x - y = 2$$
$$4x - 2y = 4$$

The equations are graphed as follows.

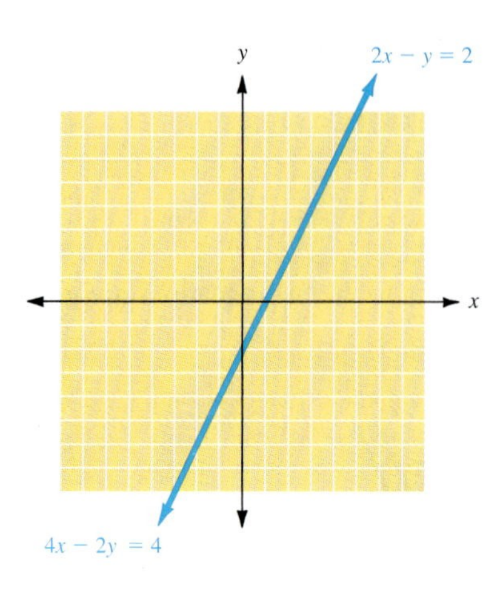

The lines have the same graph, so they have an infinite number of solutions in common.

Definitions: Dependent System

A system with an infinite number of solutions is called a **dependent system.**

CHECK YOURSELF 3

Solve the system by graphing.

$$6x - 3y = 12$$
$$y = 2x - 4$$

A second method for solving systems of linear equations in two variables is by the **addition method.**

Example 4 illustrates the addition method of solution.

Example 4

Solving a System by the Addition Method

Solve the system by the addition method.

$$5x - 2y = 12 \quad (1)$$
$$3x + 2y = 12 \quad (2)$$

NOTE The addition method is sometimes called **solution by elimination** for this reason.

In this case, adding the equations will eliminate variable y, and we have

$$8x = 24$$
$$x = 3 \quad (3)$$

Now equation (3) can be paired with either of the original equations. We let $x = 3$ in equation (1):

NOTE The solution should be checked by substituting these values into equation (2). Here

$$3(3) + 2\left(\frac{3}{2}\right) \stackrel{?}{=} 12$$
$$9 + 3 \stackrel{?}{=} 12$$
$$12 = 12$$

is a true statement.

$$5(3) - 2y = 12$$
$$15 - 2y = 12$$
$$-2y = -3$$
$$y = \frac{3}{2}$$

and $\left(3, \frac{3}{2}\right)$ is the solution for our system.

The solution set is $\left\{\left(3, \frac{3}{2}\right)\right\}$.

CHECK YOURSELF 4

Solve the system by the addition method.

$$4x - 3y = 19$$
$$-4x + 5y = -25$$

Example 4 and Check Yourself 4 were straightforward because adding the equations of the system immediately eliminated one of the variables. Example 5 illustrates a common situation in which we must multiply one or both of the equations by a nonzero constant before the addition method is applied. This multiplication results in a new system that is equivalent to the original system.

Procedure: Equivalent System

An **equivalent system** is formed whenever

1. One of the equations is multiplied by a nonzero number.
2. One of the equations is replaced by the sum of a constant multiple of another equation and that equation.

NOTE All these solutions can be approximated by graphing the lines and tracing near the intersection.This is particularly useful when the solutions are not integers (the technical term for such solutions is "ugly").

Example 5

Solving a System by the Addition Method

Solve the system by the addition method.

$$3x - 5y = 19 \quad (4)$$
$$5x + 2y = 11 \quad (5)$$

It is clear that adding the equations of the given system will *not* eliminate one of the variables. Therefore, we must use multiplication to form an equivalent system. The choice of multipliers depends on which variable we decide to eliminate. Here we have decided to eliminate y. We multiply equation (4) by 2 and equation (5) by 5. We then have

NOTE Note that the coefficients of y are now *opposites* of each other.

$$6x - 10y = 38$$
$$25x + 10y = 55$$

Adding now eliminates y and yields

$$31x = 93$$

$$x = 3 \qquad (6)$$

Pairing equation (6) with equation (4) gives an equivalent system, and we can substitute 3 for x in equation (4):

$$3 \cdot 3 - 5y = 19$$
$$9 - 5y = 19$$
$$-5y = 10$$
$$y = -2$$

NOTE Again, the solution should be checked by substitution in equation (5).

The solution set for the system is $\{(3, -2)\}$.

CHECK YOURSELF 5

Solve the system by the addition method.

$$2x + 3y = -18$$
$$6x - 10y = 22$$

The following algorithm summarizes the addition method of solving linear systems of two equations in two variables.

Step by Step: Solving by the Addition Method

Step 1 If necessary, multiply one or both of the equations by a constant so that one of the variables can be eliminated by addition.
Step 2 Add the equations of the equivalent system formed in step 1.
Step 3 Solve the equation found in step 2.
Step 4 Substitute the value found in step 3 into either of the equations of the original system to find the corresponding value of the remaining variable. The ordered pair formed is the solution to the system.
Step 5 Check the solution by substituting the pair of values found in step 4 into the other equation of the original system.

Example 6 illustrates two special situations you may encounter while applying the addition method.

Example 6

Solving a System by the Addition Method

Solve each system by the addition method.

(a) $4x + 5y = 20 \qquad (7)$
$8x + 10y = 19 \qquad (8)$

Multiply equation (7) by -2. Then

$$\begin{aligned} -8x - 10y &= -40 \\ 8x + 10y &= 19 \\ \hline 0 &= -21 \end{aligned}$$

We add the two left sides to get 0 and the two right sides to get -21.

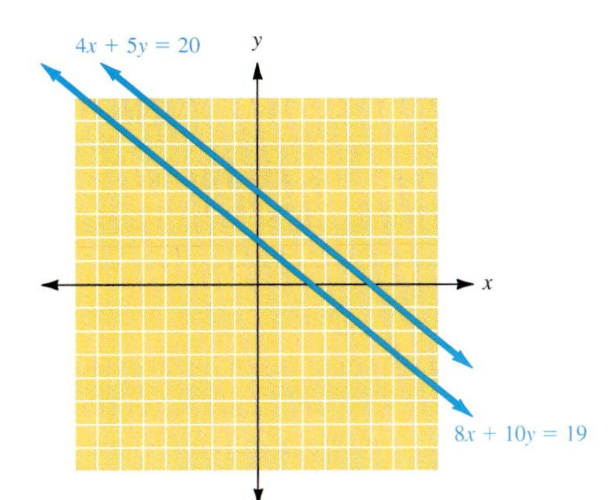

The result $0 = -21$ is a *false* statement, which means that there is no point of intersection. Therefore, the system is inconsistent, and there is no solution.

(b) $5x - 7y = 9$ (9)

$15x - 21y = 27$ (10)

Multiply equation (9) by -3. We then have

$$\begin{aligned} -15x + 21y &= -27 \\ 15x - 21y &= 27 \\ \hline 0 &= 0 \end{aligned}$$

We add the two equations.

NOTE The solution set could be written $\{(x, y) | 5x - 7y = 9\}$. This means the set of all ordered pairs (x, y) that make $5x - 7y = 9$ a true statement.

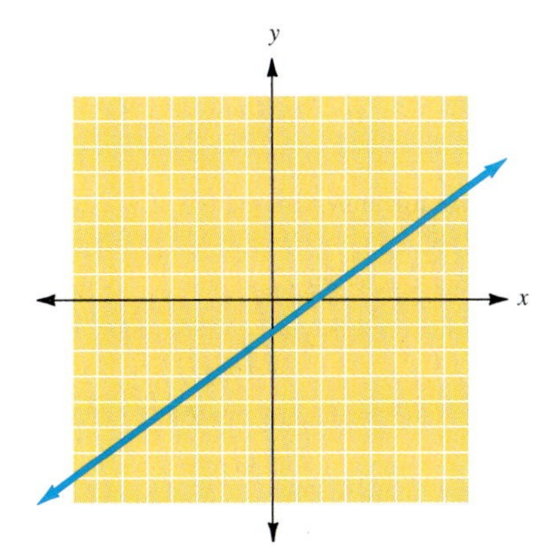

Both variables have been eliminated, and the result is a *true* statement. The two lines coincide, and there are an infinite number of solutions, one for each point on that line. We have a **dependent system.**

CHECK YOURSELF 6

Solve each system by the addition method.

(a) $3x + 2y = 8$
$9x + 6y = 11$

(b) $x - 2y = 8$
$3x - 6y = 24$

The results of Example 6 can be summarized as follows.

Rules and Properties: Solving a System of Two Linear Equations

When a system of two linear equations is solved:

1. If a false statement such as $3 = 4$ is obtained, then the system is inconsistent and has no solution.
2. If a true statement such as $8 = 8$ is obtained, then the system is dependent and has an infinite number of solutions.

A third method for finding the solutions of linear systems in two variables is called the **substitution method.** You may very well find the substitution method more difficult to apply in solving certain systems than the addition method, particularly when the equations involved in the substitution lead to fractions. However, the substitution method does have important extensions to systems involving higher degree equations, as you will see in later mathematics classes.

To outline the technique, we solve one of the equations from the original system for one of the variables. That expression is then substituted into the *other* equation of the system to provide an equation in a single variable. That equation is solved, and the corresponding value for the other variable is found as before, as Example 7 illustrates.

Example 7

Solving a System by the Substitution Method

(a) Solve the system by the substitution method.

$$2x - 3y = -3 \qquad (11)$$

$$y = 2x - 1 \qquad (12)$$

Because equation (12) is already solved for y, we substitute the expression $2x - 1$ for y in equation (11).

NOTE We now have an equation in the single variable x.

$$2x - 3(2x - 1) = -3$$

Solving for x gives

$$2x - 6x + 3 = -3$$

$$-4x = -6$$

$$x = \frac{3}{2}$$

NOTE To check this result, we substitute these values in equation (11) and have

$$2\left(\frac{3}{2}\right) - 3 \cdot 2 \stackrel{?}{=} -3$$

$$3 - 6 \stackrel{?}{=} -3$$

$$-3 = -3$$

A true statement!

We now substitute $\frac{3}{2}$ for x in equation (12).

$$y = 2\left(\frac{3}{2}\right) - 1$$

$$= 3 - 1 = 2$$

The solution set for our system is $\left\{\left(\frac{3}{2}, 2\right)\right\}$.

(b) Solve the system by the substitution method.

$$2x + 3y = 16 \qquad (13)$$

$$3x - y = 2 \qquad (14)$$

We start by solving equation (14) for y.

NOTE Why did we choose to solve for y in equation (14)? We could have solved for x, so that

$$x = \frac{y + 2}{3}$$

We simply chose the easier case to avoid fractions.

$$3x - y = 2$$

$$-y = -3x + 2 \qquad (15)$$

$$y = 3x - 2$$

Substituting in equation (13) yields

$$2x + 3(3x - 2) = 16$$

$$2x + 9x - 6 = 16$$

$$11x = 22$$

$$x = 2$$

We now substitute 2 for x in equation (15).

$$y = 3 \cdot 2 - 2$$

$$= 6 - 2 = 4$$

The solution set for the system is $\{(2, 4)\}$. We leave the check of this result to you.

NOTE The solution should be checked in *both* equations of the original system.

CHECK YOURSELF 7

Solve each system by the substitution method.

(a) $2x + 3y = 6$
$x = 3y + 6$

(b) $3x + 4y = -3$
$x + 4y = 1$

The following algorithm summarizes the substitution method for solving linear systems of two equations in two variables.

Step by Step: Solving by the Substitution Method

Step 1 If necessary, solve one of the equations of the original system for one of the variables.

Step 2 Substitute the expression obtained in step 1 into the *other* equation of the system to write an equation in a single variable.

Step 3 Solve the equation found in step 2.

Step 4 Substitute the value found in step 3 into the equation derived in step 1 to find the corresponding value of the remaining variable. The ordered pair formed is the solution for the system.

Step 5 Check the solution by substituting the pair of values found in step 4 into *both* equations of the original system.

A natural question at this point is, How do you decide which solution method to use? First, the graphical method can generally provide only approximate solutions. When exact solutions are necessary, one of the algebraic methods must be applied. Which method to use depends totally on the given system.

If you can easily solve for a variable in one of the equations, the substitution method should work well. However, if solving for a variable in either equation of the system leads to fractions, you may find the addition approach more efficient.

CHECK YOURSELF ANSWERS

1.

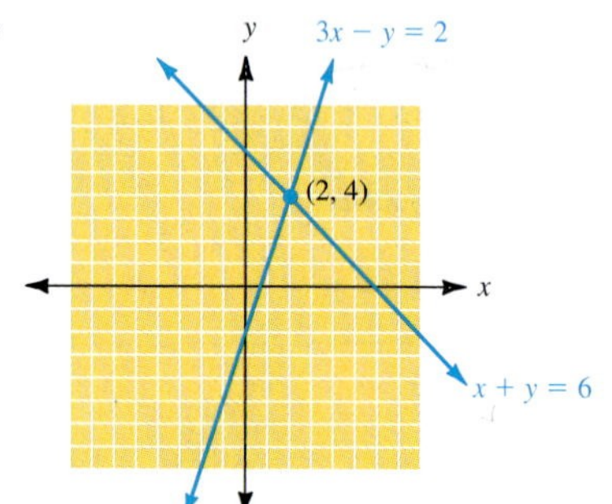

2.

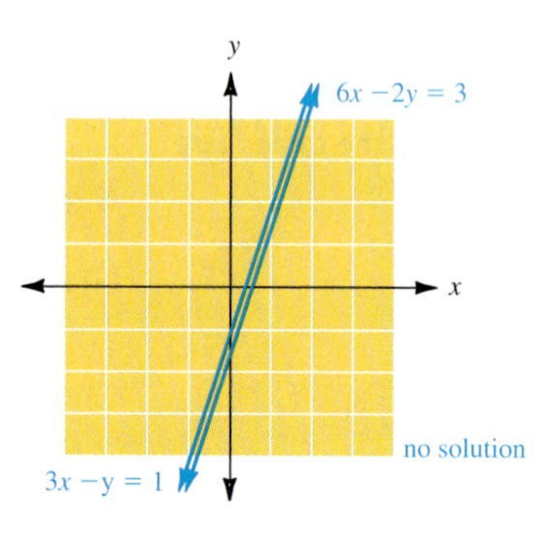

3.

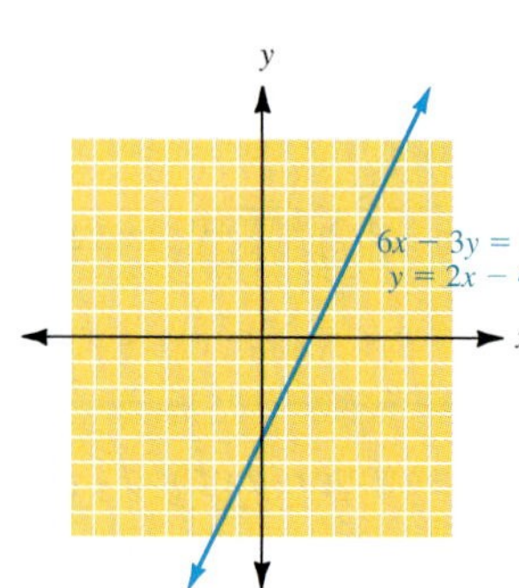

An infinite number of solutions.

4. $\left\{\left(\frac{5}{2}, -3\right)\right\}$ **5.** $\{(-3, -4)\}$

6. **(a)** Inconsistent system: no solution; **(b)** dependent system: an infinite number of solutions.

7. **(a)** $\left\{\left(4, -\frac{2}{3}\right)\right\}$; **(b)** $\left\{\left(-2, \frac{3}{4}\right)\right\}$

5.1 Exercises

Name ______________________

Section ________ Date ________

In exercises 1 to 8, solve each system by graphing. If a unique solution does not exist, state whether the system is dependent or inconsistent.

1. $x + y = 6$
$x - y = 4$

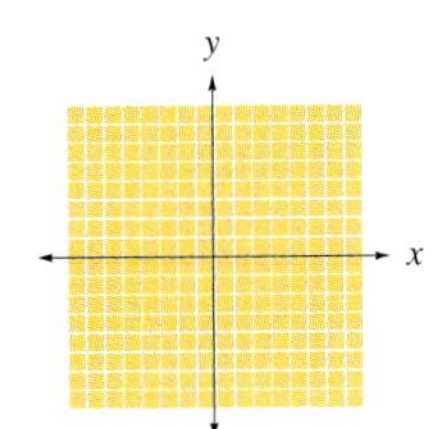

2. $x - y = 8$
$x + y = 2$

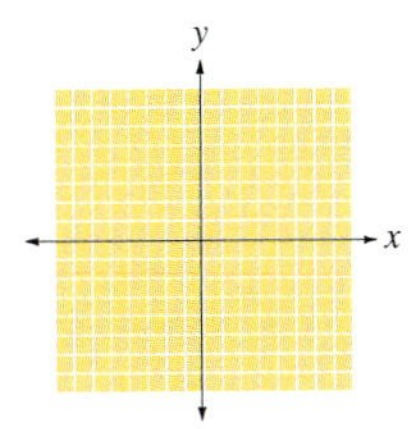

3. $x + 2y = 4$
$x - y = 1$

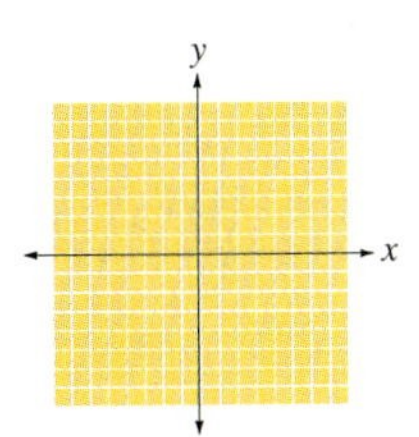

4. $x - 2y = 2$
$x + 2y = 6$

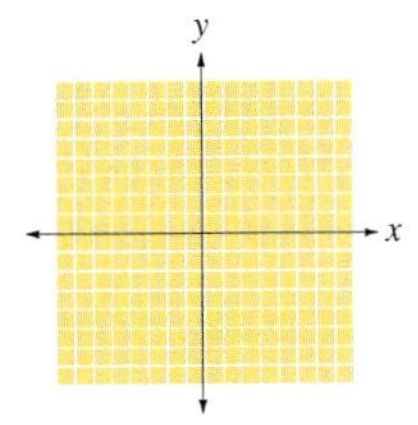

5. $3x - y = 3$
$3x - y = 6$

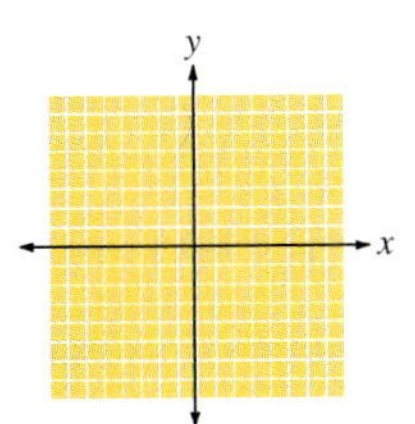

6. $3x + 2y = 12$
$y = 3$

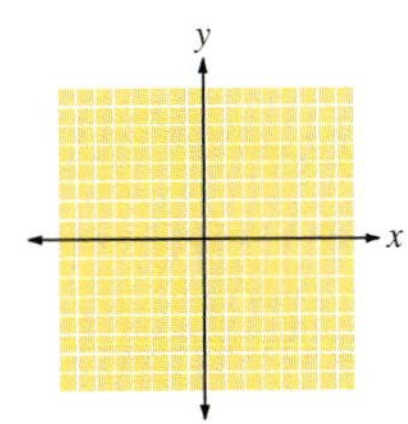

7. $x + 3y = 12$
$2x - 3y = 6$

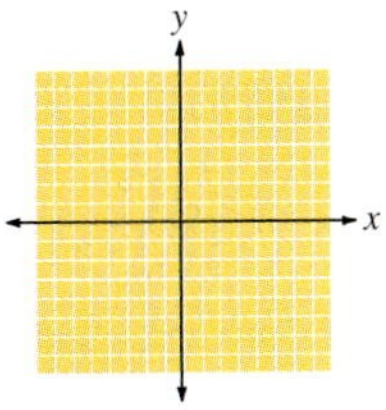

8. $3x - 6y = 9$
$x - 2y = 3$

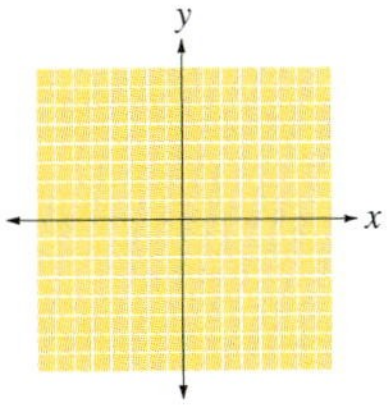

ANSWERS

1. ______________________
2. ______________________
3. ______________________
4. ______________________
5. ______________________
6. ______________________
7. ______________________
8. ______________________

ANSWERS

9. ______
10. ______
11. ______
12. ______
13. ______
14. ______
15. ______
16. ______
17. ______
18. ______
19. ______
20. ______
21. ______
22. ______
23. ______
24. ______
25. ______
26. ______
27. ______
28. ______

In exercises 9 to 22, solve each system by the addition method. If a unique solution does not exist, state whether the system is inconsistent or dependent.

9. $2x - y = 1$
$-2x + 3y = 5$

10. $x + 3y = 12$
$2x - 3y = 6$

11. $x + 2y = -2$
$3x + 2y = -12$

12. $2x + 3y = 1$
$5x + 3y = 16$

13. $x + y = 3$
$3x - 2y = 4$

14. $x - y = -2$
$2x + 3y = 21$

15. $2x + y = 8$
$-4x - 2y = -16$

16. $3x - 4y = 2$
$4x - y = 20$

17. $5x - 2y = 31$
$4x + 3y = 11$

18. $2x - y = 4$
$6x - 3y = 10$

19. $3x - 2y = 7$
$-6x + 4y = -15$

20. $3x + 4y = 0$
$5x - 3y = -29$

21. $-2x + 7y = 2$
$3x - 5y = -14$

22. $5x - 2y = 3$
$10x - 4y = 6$

In exercises 23 to 34, solve each system by the substitution method. If a unique solution does not exist, state whether the system is inconsistent or dependent.

23. $x - y = 7$
$y = 2x - 12$

24. $x - y = 4$
$x = 2y - 2$

25. $3x + 2y = -18$
$x = 3y + 5$

26. $3x - 18y = 4$
$x = 6y + 2$

27. $10x - 2y = 4$
$y = 5x - 2$

28. $4x + 5y = 6$
$y = 2x - 10$

ANSWERS

29. ______

30. ______

31. ______

32. ______

33. ______

34. ______

35. ______

36. ______

37. ______

38. ______

39. ______

40. ______

41. ______

42. ______

43. ______

44. ______

29. $3x + 4y = 9$
$y = 3x + 1$

30. $6x - 5y = 27$
$x = 5y + 2$

31. $x - 7y = 3$
$2x - 5y = 15$

32. $4x + 3y = -11$
$5x + y = -11$

33. $4x - 12y = 5$
$-x + 3y = -1$

34. $5x - 6y = 21$
$x - 2y = 5$

In exercises 35 to 40, solve each system by any method discussed in this section.

35. $2x - 3y = 4$
$x = 3y + 6$

36. $5x + y = 2$
$5x - 3y = 6$

37. $4x - 3y = 0$
$5x + 2y = 23$

38. $7x - 2y = -17$
$x + 4y = 4$

39. $3x - y = 17$
$5x + 3y = 5$

40. $7x + 3y = -51$
$y = 2x + 9$

In exercises 41 to 44, solve each system by any method discussed in this section. *Hint:* You should multiply to clear fractions as your first step.

41. $\frac{1}{2}x - \frac{1}{3}y = 8$
$\frac{1}{3}x + y = -2$

42. $\frac{1}{5}x - \frac{1}{2}y = 0$
$x - \frac{3}{2}y = 4$

43. $\frac{2}{3}x + \frac{3}{5}y = -3$
$\frac{1}{3}x + \frac{2}{5}y = -3$

44. $\frac{3}{8}x - \frac{1}{2}y = -5$
$\frac{1}{4}x + \frac{3}{2}y = 4$

Answers

1. Solution: (5, 1)

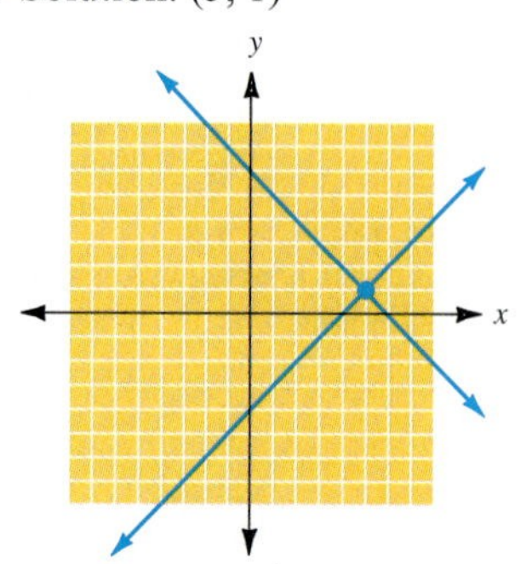

3. Solution: (2, 1)

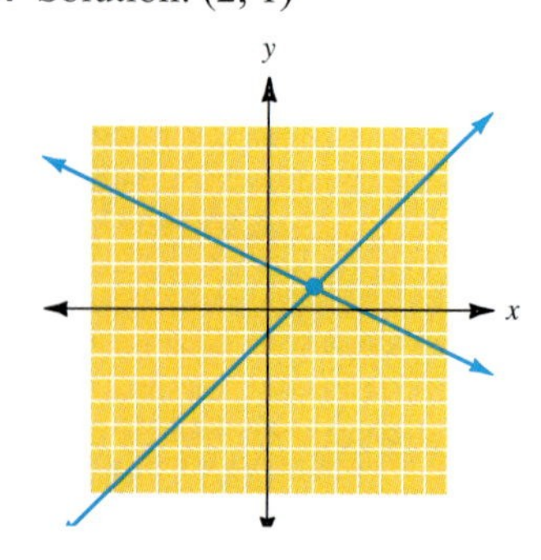

5. Inconsistent system

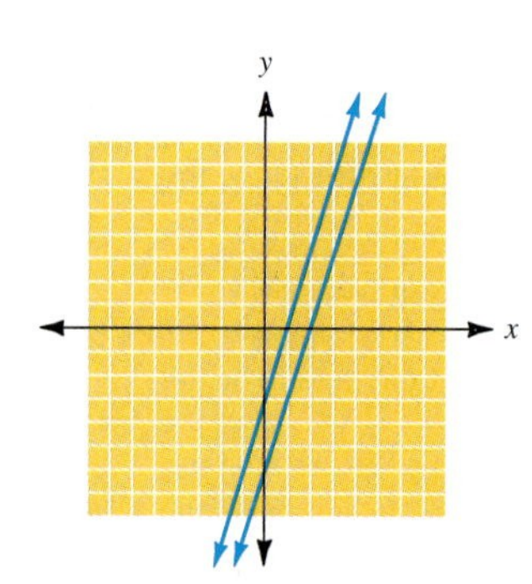

7. Solution: (6, 2)

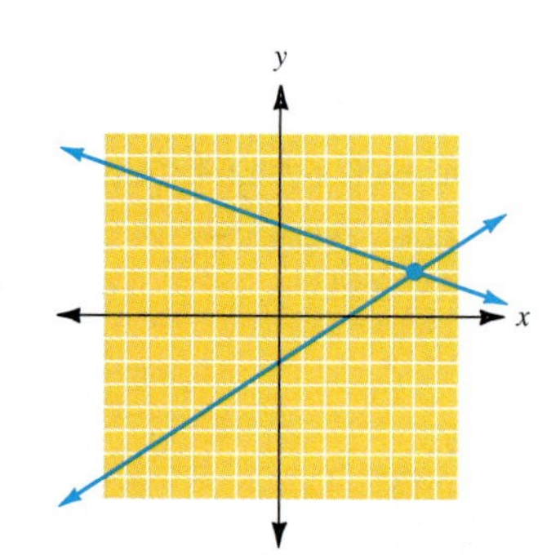

9. $\{(2, 3)\}$ **11.** $\left\{\left(-5, \frac{3}{2}\right)\right\}$ **13.** $\{(2, 1)\}$ **15.** Dependent system

17. $\{(5, -3)\}$ **19.** Inconsistent system **21.** $\{(-8, -2)\}$ **23.** $\{(5, -2)\}$

25. $\{(-4, -3)\}$ **27.** Dependent system **29.** $\left\{\left(\frac{1}{3}, 2\right)\right\}$ **31.** $\{(10, 1)\}$

33. Inconsistent system **35.** $\left\{\left(-2, -\frac{8}{3}\right)\right\}$ **37.** $\{(3, 4)\}$ **39.** $\{(4, -5)\}$

41. $\{(12, -6)\}$ **43.** $\{(9, -15)\}$

Applications of Systems of Linear Equations

OBJECTIVE

1. Use a system of equations to solve an application

We are now ready to apply our equation-solving skills to solving various applications or word problems. Being able to extend these skills to problem solving is an important goal, and the procedures developed here are used throughout the rest of the book.

Although we consider applications from a variety of areas in this section, all are approached with the same five-step strategy presented here to begin the discussion.

Step by Step: Solving Applications

Step 1 Read the problem carefully to determine the unknown quantities.

Step 2 Choose a variable to represent any unknown.

Step 3 Translate the problem to the language of algebra to form a system of equations.

Step 4 Solve the system of equations, and answer the question of the original problem.

Step 5 Verify your solution by returning to the original problem.

Example 1

Solving a Mixture Problem

A coffee merchant has two types of coffee beans, one selling for \$9 per pound and the other for \$15 per pound. The beans are to be mixed to provide 100 lb of a mixture selling for \$13.50 per pound. How much of each type of coffee bean should be used to form 100 lb of the mixture?

Step 1 The unknowns are the amounts of the two types of beans.

Step 2 We use two variables to represent the two unknowns. Let x be the amount of \$9 beans and y the amount of \$15 beans.

Step 3 We now want to establish a system of two equations. One equation will be based on the *total amount* of the mixture, the other on the mixture's *value.*

NOTE Because we use *two* variables, we must form *two* equations.

$$x + y = 100 \qquad \text{The mixture must weigh 100 lb.} \qquad (1)$$

$$9x + 15y = 1350 \qquad (2)$$

Step 4 An easy approach to the solution of the system is to multiply equation (1) by -9 and add to eliminate x.

$$
\begin{aligned}
-9x - 9y &= -900 \\
9x + 15y &= 1350 \\
6y &= 450 \\
y &= 75 \text{ lb}
\end{aligned}
$$

By substitution in equation (1), we have

$x = 25$ lb

Step 5 To check the result, show that the value of the \$9 beans, added to the value of the \$15 beans, equals the desired value of the mixture.

CHECK YOURSELF 1

Peanuts, which sell for \$4.80 per pound, and cashews, which sell for \$12 per pound, are to be mixed to form a 60-lb mixture selling for \$6 per pound. How much of each type of nut should be used?

A related problem is illustrated in Example 2.

Example 2

Solving a Mixture Problem

A chemist has a 25% and a 50% acid solution. How much of each solution should be used to form 200 mL of a 35% acid solution?

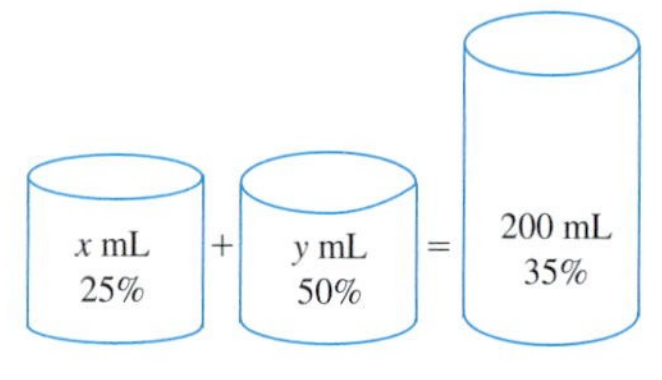

Drawing a sketch of a problem is often a valuable part of the problem-solving strategy.

Step 1 The unknowns in this case are the amounts of the 25% and 50% solutions to be used in forming the mixture.

Step 2 Again we use two variables to represent the two unknowns. Let x be the amount of the 25% solution and y the amount of the 50% solution. Let's draw a picture before proceeding to form a system of equations.

Step 3 Now, to form our two equations, we want to consider two relationships: the *total amounts* combined and the *amounts of acid* combined.

From our sketch of the problem, we have

NOTE Total amounts combined.

NOTE Amounts of acid combined.

$$x + y = 200 \qquad (3)$$

$$0.25x + 0.50y = 0.35(200) \qquad (4)$$

Step 4 Now, clear equation (4) of decimals by multiplying equation (4) by 100. The solution then proceeds as before, with the result

$x = 120$ mL (25% solution)

$y = 80$ mL (50% solution)

Step 5 To check, show that the amount of acid in the 25% solution, (0.25)(120), added to the amount in the 50% solution, (0.50)(80), equals the correct amount in the mixture, (0.35)(200). We leave that to you.

CHECK YOURSELF 2

A pharmacist wants to prepare 300 mL of a 20% alcohol solution. How much of a 30% solution and a 15% solution should be used to form the desired mixture?

Applications that involve a constant rate of travel, or speed, require the use of the distance formula

$d = rt$

in which d = distance traveled
r = rate, or speed
t = time

Example 3 illustrates this approach.

Example 3

Solving a Distance-Rate-Time Problem

A boat can travel 36 mi downstream in 2 h. Coming back upstream, the boat takes 3 h. What is the rate of the boat in still water? What is the rate of the current?

Step 1 We want to find the two rates.

Step 2 Let x be the rate of the boat in still water and y the rate of the current.

NOTE Downstream the rate is then

$x + y$

Upstream, the rate is

$x - y$

Step 3 To form a system, think about the following. Downstream, the rate of the boat is *increased* by the effect of the current. Upstream, the rate is *decreased.*

In many applications, it helps to lay out the information in tabular form. Let's try that strategy here.

	d	r	t
Downstream	36	$x + y$	2
Upstream	36	$x - y$	3

Because $d = rt$, from the table we can easily form two equations:

$$36 = (x + y)(2) \tag{5}$$

$$36 = (x - y)(3) \tag{6}$$

Step 4 We clear equations (5) and (6) of parentheses and simplify, to write the equivalent system

$$x + y = 18$$

$$x - y = 12$$

Solving, we have

$$x = 15 \text{ mi/h}$$

$$y = 3 \text{ mi/h}$$

Step 5 To check, verify the $d = rt$ equation in *both* the upstream and the downstream cases. We leave that to you.

CHECK YOURSELF 3

A plane flies 480 mi in an easterly direction, with the wind, in 4 h. Returning westerly along the same route, against the wind, the plane takes 6 h. What is the rate of the plane in still air? What is the rate of the wind?

The use of systems of equations in problem solving has many applications in a business setting. Example 4 illustrates one such application.

Example 4

Solving a Business-Based Application

A manufacturer produces a standard model and a deluxe model of a 25-inch (in.) television set. The standard model requires 12 h of labor to produce, and the deluxe model requires 18 h. The company has 360 h of labor available per week. The plant's capacity is a total of 25 sets per week. If all the available time and capacity are to be used, how many of each type of set should be produced?

Step 1 The unknowns in this case are the number of standard and deluxe models that can be produced.

NOTE The choices for x and y could have been reversed.

Step 2 Let x be the number of standard models and y the number of deluxe models.

Step 3 Our system will come from the two given conditions that fix the total number of sets that can be produced and the total labor hours available.

$$x + y = 25 \quad \leftarrow \text{Total number of sets}$$

$$12x + 18y = 360 \quad \leftarrow \text{Total labor hours available}$$

($12x$: Labor hours—standard sets; $18y$: Labor hours—deluxe sets)

Step 4 Solving the system in step 3, we have

$x = 15$ and $y = 10$

which tells us that to use all the available capacity, the plant should produce 15 standard sets and 10 deluxe sets per week.

Step 5 We leave the check of this result to the reader.

CHECK YOURSELF 4

A manufacturer produces standard cassette players and compact disc players. The cassette players require 2 h of electronic assembly and the CD players 3 h. The cassette players require 4 h of case assembly and the CD players 2 h. The company has 120 h of electronic assembly time available per week and 160 h of case assembly time. How many of each type of unit can be produced each week if all available assembly time is to be used?

Let's look at one final application that leads to a system of two equations.

Example 5

Solving a Business-Based Application

Two car rental agencies have the following rate structures for a subcompact car. Urent charges \$50 per day plus 15¢ per mile. Painz charges \$45 per day plus 20¢ per mile. If you rent a car for 1 day, for what number of miles will the two companies have the same total charge?

Letting c represent the total a company will charge and m the number of miles driven, we calculate the following.

For Urent:

NOTE You first saw this type of linear model in exercises in Section 4.3.

$$c = 50 + 0.15m \tag{7}$$

For Painz:

$$c = 45 + 0.20m \tag{8}$$

The system can be solved most easily by substitution. Substituting $45 + 0.20m$ for c in equation (7) gives

$$45 + 0.20m = 50 + 0.15m$$

$$0.05m = 5$$

$$m = 100 \text{ mi}$$

The graph of the system is shown below.

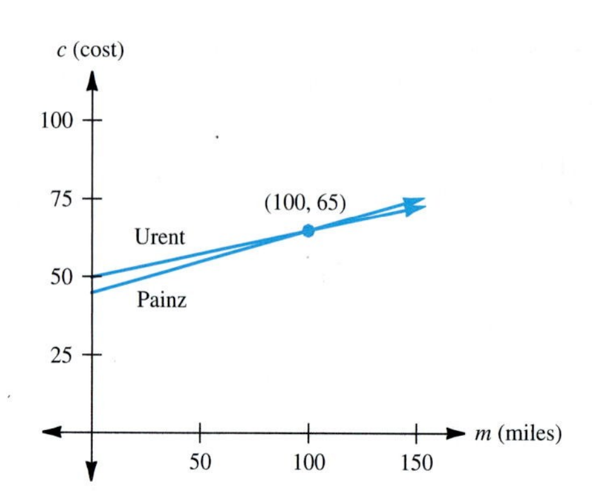

From the graph, how would you make a decision about which agency to use?

CHECK YOURSELF 5

For a compact car, the same two companies charge \$54 per day plus 20¢ per mile and \$51 per day plus 22¢ per mile. For a 2-day rental, when will the charges be the same?

CHECK YOURSELF ANSWERS

1. 50 lb of peanuts and 10 lb of cashews
2. 100 mL of the 30% and 200 mL of the 15%
3. 100 mi/h plane and 20 mi/h wind
4. 30 cassette players and 20 CD players
5. At 300 mi, \$168 charge

Name ______________

Section ________ Date ________

5.2 Exercises

Each application in exercises 1 to 8 can be solved by the use of a system of linear equations. Match the application with the appropriate system below.

(a) $12x + 5y = 116$
$8x + 12y = 112$

(b) $x + y = 8000$
$0.06x + 0.09y = 600$

(c) $x + y = 200$
$0.20x + 0.60y = 90$

(d) $x + y = 36$
$y = 3x - 4$

(e) $2(x + y) = 36$
$3(x - y) = 36$

(f) $x + y = 200$
$6.50x + 4.50y = 980$

(g) $L = 2W + 3$
$2L + 2W = 36$

(h) $x + y = 120$
$2.20x + 5.40y = 360$

1. **Number problem.** One number is 4 less than 3 times another. If the sum of the numbers is 36, what are the two numbers?

2. **Recreation.** Suppose a movie theater sold 200 adult and student tickets for a showing with a revenue of $980. If the adult tickets were $6.50 and the student tickets were $4.50, how many of each type of ticket were sold?

3. **Geometry.** The length of a rectangle is 3 cm more than twice its width. If the perimeter of the rectangle is 36 cm, find the dimensions of the rectangle.

4. **Business.** An order of 12 dozen roller-ball pens and 5 dozen ballpoint pens cost $116. A later order for 8 dozen roller-ball pens and 12 dozen ballpoint pens cost $112. What was the cost of 1 dozen of each type of pen?

5. **Mixture problem.** A candy merchant wants to mix peanuts selling at $2.20 per pound with cashews selling at $5.40 per pound to form 120 lb of a mixed-nut blend that will sell for $3 per pound. What amount of each type of nut should be used?

6. **Investment.** Donald has investments totaling $8000 in two accounts—one a savings account paying 6% interest, and the other a bond paying 9%. If the annual interest from the two investments was $600, how much did he have invested at each rate?

7. **Mixture problem.** A chemist wants to combine a 20% alcohol solution with a 60% solution to form 200 mL of a 45% solution. How much of each solution should be used to form the mixture?

8. **Motion problem.** Xian was able to make a downstream trip of 36 mi in 2 h. Returning upstream, he took 3 h to make the trip. How fast can his boat travel in still water? What was the rate of the river's current?

ANSWERS

1. ______
2. ______
3. ______
4. ______
5. ______
6. ______
7. ______
8. ______

9. ______

10. ______

11. ______

12. ______

13. ______

14. ______

15. ______

16. ______

In exercises 9 to 30, solve by choosing a variable to represent each unknown quantity and writing a system of equations.

9. **Mixture problem.** Suppose 750 tickets were sold for a concert with a total revenue of \$5300. If adult tickets were \$8 and student tickets were \$4.50, how many of each type of ticket were sold?

10. **Mixture problem.** Theater tickets sold for \$7.50 on the main floor and \$5 in the balcony. The total revenue was \$3250, and there were 100 more main-floor tickets sold than balcony tickets. Find the number of each type of ticket sold.

11. **Geometry.** The length of a rectangle is 3 in. less than twice its width. If the perimeter of the rectangle is 84 in., find the dimensions of the rectangle.

12. **Geometry.** The length of a rectangle is 5 cm more than 3 times its width. If the perimeter of the rectangle is 74 cm, find the dimensions of the rectangle.

13. **Mixture problem.** A garden store sold 8 bags of mulch and 3 bags of fertilizer for \$24. The next purchase was for 5 bags of mulch and 5 bags of fertilizer. The cost of that purchase was \$25. Find the cost of a single bag of mulch and a single bag of fertilizer.

14. **Mixture problem.** The cost of an order for 10 computer disks and 3 packages of paper was \$22.50. The next order was for 30 disks and 5 packages of paper, and its cost was \$53.50. Find the price of a single disk and a single package of paper.

15. **Mixture problem.** A coffee retailer has two grades of decaffeinated beans—one selling for \$4 per pound and the other for \$6.50 per pound. She wishes to blend the beans to form a 150-lb mixture that will sell for \$4.75 per pound. How many pounds of each grade of bean should be used in the mixture?

16. **Mixture problem.** A candy merchant sells jelly beans at \$3.50 per pound and gumdrops at \$4.70 per pound. To form a 200-lb mixture that will sell for \$4.40 per pound, how many pounds of each type of candy should be used?

17. **Investment.** Cheryl decided to divide \$12,000 into two investments—one a time deposit that pays 8% annual interest and the other a bond that pays 9%. If her annual interest was \$1010, how much did she invest at each rate?

18. **Investment.** Miguel has \$3000 more invested in a mutual fund paying 5% interest than in a savings account paying 3%. If he received \$310 in interest for 1 year, how much did he have invested in the two accounts?

19. **Science.** A chemist mixes a 10% acid solution with a 50% acid solution to form 400 mL of a 40% solution. How much of each solution should be used in the mixture?

20. **Science.** A laboratory technician wishes to mix a 70% saline solution and a 20% solution to prepare 500 mL of a 40% solution. What amount of each solution should be used?

21. **Motion.** A boat traveled 36 mi up a river in 3 h. Returning downstream, the boat took 2 h. What is the boat's rate in still water, and what is the rate of the river's current?

22. **Motion.** A jet flew east a distance of 1800 mi with the jetstream in 3 h. Returning west, against the jetstream, the jet took 4 h. Find the jet's speed in still air and the rate of the jetstream.

23. **Number problem.** The sum of the digits of a two-digit number is 8. If the digits are reversed, the new number is 36 more than the original number. Find the original number. *Hint:* If u represents the units digit of the number and t the tens digit, the original number can be represented by $10t + u$.

24. **Number problem.** The sum of the digits of a two-digit number is 10. If the digits are reversed, the new number is 54 less than the original number. What was the original number?

25. **Business.** A manufacturer produces a battery-powered calculator and a solar model. The battery-powered model requires 10 min of electronic assembly and the solar model 15 min. There are 450 min of assembly time available per day. Both models require 8 min for packaging, and 280 min of packaging time are available per day. If the manufacturer wants to use all the available time, how many of each unit should be produced per day?

ANSWERS

17. ______

18. ______

19. ______

20. ______

21. ______

22. ______

23. ______

24. ______

25. ______

ANSWERS

26. ______

27. ______

28. ______

29. ______

30. ______

26. **Business.** A small tool manufacturer produces a standard- and a cordless-model power drill. The standard model takes 2 h of labor to assemble and the cordless model 3 h. There are 72 h of labor available per week for the drills. Material costs for the standard drill are \$10, and for the cordless drill they are \$20. The company wishes to limit material costs to \$420 per week. How many of each model drill should be produced to use all the available resources?

27. **Economics.** In economics, a demand equation gives the quantity D that will be demanded by consumers at a given price p, in dollars. Suppose that $D = 210 - 4p$ for a particular product.

    A supply equation gives the supply S that will be available from producers at price p. Suppose also that for the same product $S = 10p$.

    The equilibrium point is that point at which the supply equals the demand (here, where $S = D$). Use the given equations to find the equilibrium point.

28. **Economics.** Suppose the demand equation for a product is $D = 150 - 3p$ and the supply equation is $S = 12p$. Find the equilibrium point for the product.

29. **Consumer affairs.** Two car rental agencies have the following rate structure for compact cars.

    Company A: \$30/day and 22¢/mi.

    Company B: \$28/day and 26¢/mi.

    For a 2-day rental, at what number of miles will the charges be the same?

30. **Construction.** Two construction companies submit the following bid.

    Company A: \$5000 plus \$15/square foot of building.

    Company B: \$7000 plus \$12.50/square foot of building.

    For what number of square feet of building will the bids of the two companies be the same?

Certain systems that are not linear can be solved with the methods of this section if we first substitute to change variables. For instance, the system

$$\frac{1}{x} + \frac{1}{y} = 4$$

$$\frac{1}{x} - \frac{3}{y} = -6$$

can be solved by the substitutions $u = \frac{1}{x}$ and $v = \frac{1}{y}$. That gives the system $u + v = 4$ and $u - 3v = -6$. The system is then solved for u and v, and the corresponding values for x and y are found. Use this method to solve the systems in exercises 31 to 34.

31. $\frac{1}{x} + \frac{1}{y} = 4$

$\frac{1}{x} - \frac{3}{y} = -6$

32. $\frac{1}{x} + \frac{3}{y} = 1$

$\frac{4}{x} + \frac{3}{y} = 3$

33. $\frac{2}{x} + \frac{3}{y} = 4$

$\frac{2}{x} - \frac{6}{y} = 10$

34. $\frac{4}{x} - \frac{3}{y} = -1$

$\frac{12}{x} - \frac{1}{y} = 1$

Writing the equation of a line through two points can be done by the following method. Given the coordinates of two points, substitute each pair of values into the equation $y = mx + b$. This gives a system of two equations in variables m and b, which can be solved as before.

In exercises 35 and 36, write the equation of the line through each of the following pairs of points, using the method outlined above.

35. (2, 1) and (4, 4)

36. (−3, 7) and (6, 1)

In exercises 37 and 38, use your calculator to approximate the solution to each system. Express your answer to the nearest tenth.

37. $y = 2x - 3$
$2x + 3y = 1$

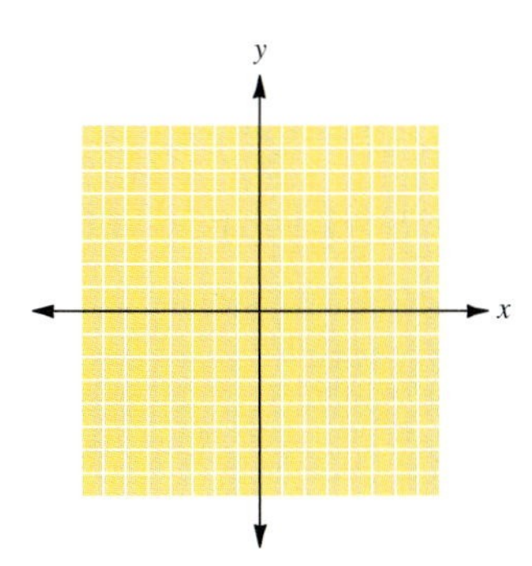

38. $3x - 4y = -7$
$2x + 3y = -1$

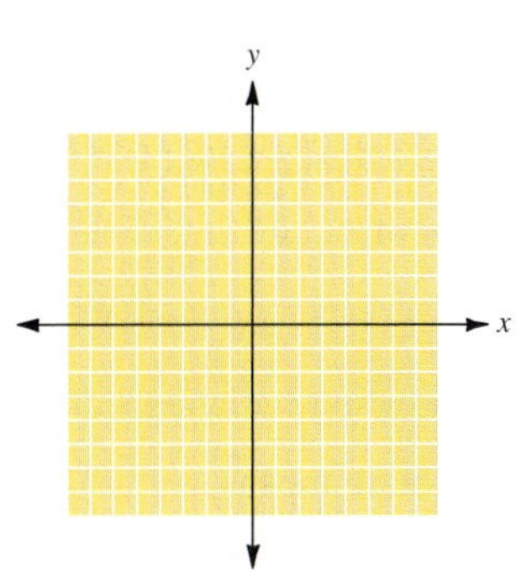

ANSWERS

31. ____________

32. ____________

33. ____________

34. ____________

35. ____________

36. ____________

37. ____________

38. ____________

ANSWERS

39. ______

40. ______

41. ______

42. ______

43. ______

44. ______

For exercises 39 and 40, adjust the viewing window on your calculator so that you can see the point of intersection for the two lines representing the equations in the system. Then approximate the solution.

39. $5x - 12y = 8$
$7x + 2y = 44$

40. $9x - 3y = 10$
$x + 5y = 58$

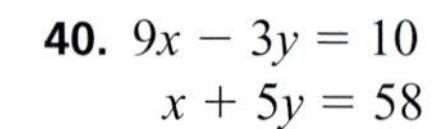

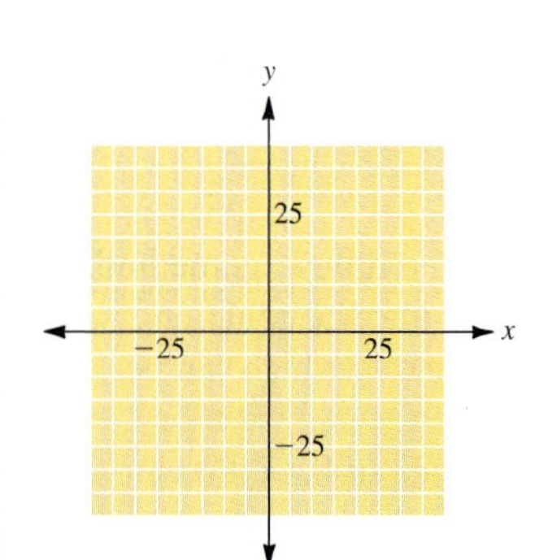

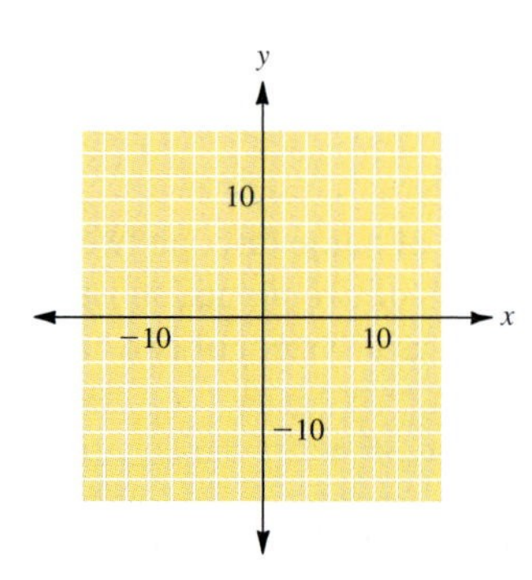

41. Find values for m and b in the following system so that the solution to the system is (1, 2).

$mx + 3y = 8$
$-3x + 4y = b$

42. Find values for m and b in the following system so that the solution to the system is (−3, 4).

$5x + 7y = b$
$mx + y = 22$

43. Complete the following statements in your own words:

"To solve an equation means to . . . ."

"To solve a system of equations means to . . . ."

44. A system of equations such as the one below is sometimes called a "2-by-2" system of linear equations.

$3x + 4y = 1$
$x - 2y = 6$

Explain this term.

45. Complete this statement in your own words: "All the points on the graph of the equation $2x + 3y = 6$ . . . ." Exchange statements with other students. Do you agree with other students' statements?

46. Does a system of linear equations always have a solution? How can you tell without graphing that a system of two equations will be graphed as two parallel lines? Give some examples to explain your reasoning.

47. Suppose we have the following linear system:

$$Ax + By = C \qquad (1)$$

$$Dx + Ey = F \qquad (2)$$

(a) Write the slope of the line determined by equation (1).

(b) Write the slope of the line determined by equation (2).

(c) What must be true about the given coefficients to guarantee that the system is consistent?

48. We have discussed three different methods of solving a system of two linear equations in two unknowns: the graphical method, the addition method, and the substitution method. Discuss the strengths and weaknesses of each method.

49. Determine a system of two linear equations for which the solution is (3, 4). Are there other systems that have the same solution? If so, determine at least one more and explain why this can be true.

50. Suppose we have the following linear system:

$$Ax + By = C \qquad (1)$$

$$Dx + Ey = F \qquad (2)$$

(a) Multiply equation (1) by $-D$, multiply equation (2) by A and add. This will allow you to eliminate x. Solve for y and indicate what must be true about the coefficients for a unique value for y to exist.

(b) Now return to the original system and eliminate y instead of x. (*Hint:* try multiplying equation (1) by E and equation (2) by $-B$.) Solve for x and again indicate what must be true about the coefficients for a unique value for x to exist.

ANSWERS

45. ______

46. ______

47. (a)

(b)

(c)

48. ______

49. ______

50. (a)

(b)

Answers

1. (d) **3.** (g) **5.** (h) **7.** (c) **9.** 550 adult, 200 student tickets
11. 27 in. × 15 in. **13.** Mulch: $1.80; fertilizer: $3.20
15. 105 lb of $4 beans, 45 lb of $6.50 beans **17.** $7000 time deposit, $5000 bond
19. 100 mL of 10%, 300 mL of 50% **21.** 15 mi/h boat, 3 mi/h current
23. 26 **25.** 15 battery powered, 20 solar models **27.** $p = 15$ **29.** 100 mi
31. $\left\{\left(\frac{2}{3}, \frac{2}{5}\right)\right\}$ **33.** $\left\{\left(\frac{1}{3}, -\frac{3}{2}\right)\right\}$ **35.** $y = \frac{3}{2}x - 2$
37. $\{(1.3, -0.5)\}$

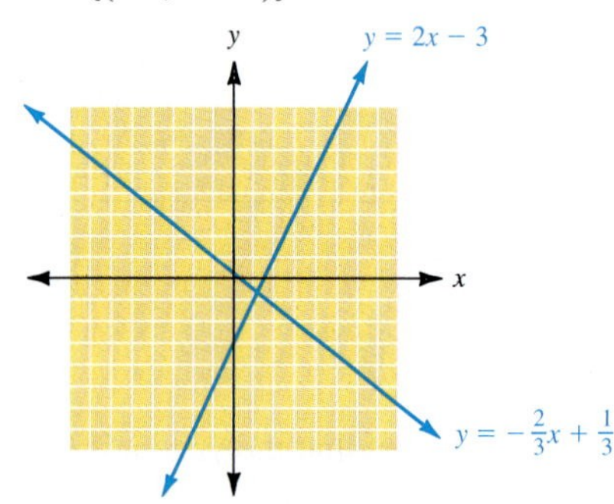

39. $\{(6, 2)\}$

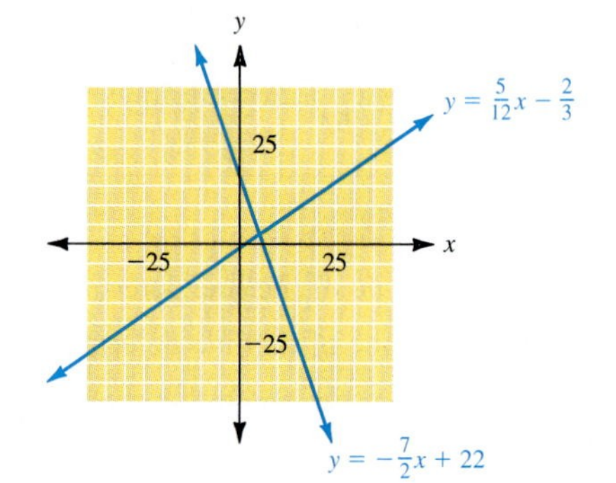

41. $m = 2, b = 5$ **43.** **45.**

47. (a) $\frac{-A}{B}$; **(b)** $\frac{-D}{E}$; **(c)** $AE - BD \neq 0$ **49.**

5.3 Systems of Linear Equations in Three Variables

5.3 OBJECTIVES

1. Find ordered triples associated with three equations
2. Solve a system by the addition method
3. Interpret a solution graphically
4. Use a system of three equations to solve an application

Suppose an application involves three quantities that we want to label x, y, and z. A typical equation used in solving the application might be

$$2x + 4y - z = 8$$

This is called a **linear equation in three variables.** The solution for such an equation is an **ordered triple** (x, y, z) of real numbers that satisfies the equation. For example, the ordered triple (2, 1, 0) is a solution for the equation above because substituting 2 for x, 1 for y, and 0 for z results in the following true statement.

$$2 \cdot 2 + 4 \cdot 1 - 0 \stackrel{?}{=} 8$$

$$4 + 4 \stackrel{?}{=} 8$$

$$8 = 8 \quad \text{True}$$

Of course, other solutions, in fact infinitely many, exist. You might want to verify that $(1, 1, -2)$ and $(3, 1, 2)$ are also solutions. To extend the concepts of the last section, we want to consider systems of three linear equations in three variables such as

$$x + y + z = 5$$

$$2x - y + z = 9$$

$$x - 2y + 3z = 16$$

NOTE For a unique solution to exist, when *three variables* are involved, we must have *three equations.*

The solution for such a system is the set of all ordered triples that satisfy each equation of the system. In this case, you can verify that $(2, -1, 4)$ is a solution for the system because that ordered triple makes each equation a true statement.

Let's turn now to the process of solving such a system. In this section, we will consider the addition method. We will then apply what we have learned to solving applications.

NOTE The choice of which variable to eliminate is yours. Generally, you should pick the variable that allows the easiest computation.

The central idea is to choose *two pairs* of equations from the system and, by the addition method, to eliminate the *same variable* from each of those pairs. The method is best illustrated by example. So let's proceed to see how the solution for the previous system was determined.

Example 1

Solving a Linear System in Three Variables

Solve the system.

$$x + y + z = 5 \qquad (1)$$

$$2x - y + z = 9 \qquad (2)$$

$$x - 2y + 3z = 16 \qquad (3)$$

First we choose two of the equations and the variable to eliminate. Variable y seems convenient in this case. Pairing equations (1) and (2) and then adding, we have

NOTE Any pair of equations could have been selected.

$$\begin{array}{r} x + y + z = 5 \\ 2x - y + z = 9 \\ \hline 3x \quad + 2z = 14 \end{array} \qquad (4)$$

We now want to choose a different pair of equations to eliminate y. Using equations (1) and (3) this time, we multiply equation (1) by 2 and then add the result to equation (3):

$$\begin{array}{r} 2x + 2y + 2z = 10 \\ x - 2y + 3z = 16 \\ \hline 3x \quad + 5z = 26 \end{array} \qquad (5)$$

We now have equations (4) and (5) in variables x and z.

$$3x + 2z = 14$$

$$3x + 5z = 26$$

Because we are now dealing with a system of two equations in two variables, any of the methods of the previous section apply. We have chosen to multiply equation (4) by -1 and then add that result to equation (5). This yields

$$3z = 12$$

$$z = 4$$

Substituting $z = 4$ in equation (4) gives

$$3x + 2 \cdot 4 = 14$$

$$3x + 8 = 14$$

$$3x = 6$$

$$x = 2$$

NOTE Any of the original equations could have been used.

Finally, letting $x = 2$ and $z = 4$ in equation (1) gives

$$2 + y + 4 = 5$$

$$y = -1$$

and $(2, -1, 4)$ is shown to be the solution for the system.

NOTE To check, substitute these values into the other equations of the original system.

CHECK YOURSELF 1

Solve the system.

$$x - 2y + z = 0$$

$$2x + 3y - z = 16$$

$$3x - y - 3z = 23$$

One or more of the equations of a system may already have a missing variable. The elimination process is simplified in that case, as Example 2 illustrates.

Example 2

Solving a Linear System in Three Variables

Solve the system.

$$2x + y - z = -3 \quad (6)$$

$$y + z = 2 \quad (7)$$

$$4x - y + z = 12 \quad (8)$$

Noting that equation (7) involves only y and z, we must simply find another equation in those same two variables. Multiply equation (6) by -2 and add the result to equation (8) to eliminate x.

NOTE We now have a *second* equation in y and z.

$$\begin{aligned} -4x - 2y + 2z &= 6 \\ 4x - y + z &= 12 \\ \hline -3y + 3z &= 18 \\ y - z &= -6 \quad (9) \end{aligned}$$

We now form a system consisting of equations (7) and (9) and solve as before.

$$\begin{aligned} y + z &= 2 \\ y - z &= -6 \\ \hline 2y &= -4 \\ y &= -2 \end{aligned}$$

Adding eliminates z.

From equation (7), if $y = -2$,

$$-2 + z = 2$$

$$z = 4$$

and from equation (6), if $y = -2$ and $z = 4$,

$$2x - 2 - 4 = -3$$

$$2x = 3$$

$$x = \frac{3}{2}$$

The solution for the system is

$$\left(\frac{3}{2}, -2, 4\right)$$

CHECK YOURSELF 2

Solve the system.

$$x + 2y - z = -3$$

$$x - y + z = 2$$

$$x - z = 3$$

The following algorithm summarizes the procedure for finding the solutions for a linear system of three equations in three variables.

Step by Step: Solving a System of Three Equations in Three Unknowns

Step 1 Choose a pair of equations from the system, and use the addition method to eliminate one of the variables.

Step 2 Choose a *different* pair of equations, and eliminate the *same* variable.

Step 3 Solve the system of two equations in two variables determined in steps 1 and 2.

Step 4 Substitute the values found above into one of the original equations, and solve for the remaining variable.

Step 5 The solution is the ordered triple of values found in steps 3 and 4. It can be checked by substituting into the other equations of the original system.

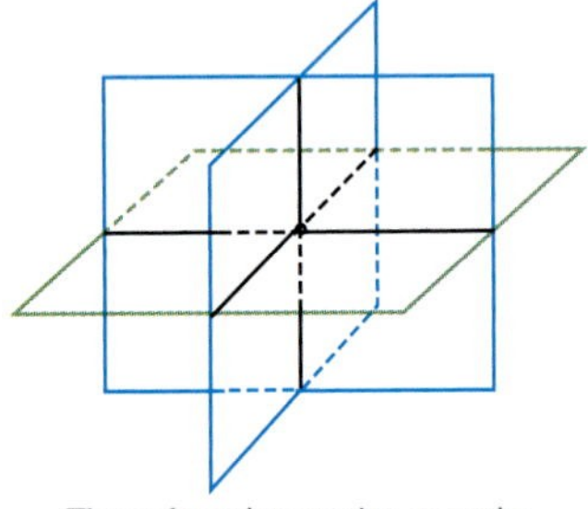
Three planes intersecting at a point

Systems of three equations in three variables may have (1) exactly one solution, (2) infinitely many solutions, or (3) no solution. Before we look at an algebraic approach in the second and third cases, let's discuss the geometry involved.

The graph of a linear equation in three variables is a plane (a flat surface) in three dimensions. Two distinct planes either will be parallel or will intersect in a line.

If three distinct planes intersect, that intersection will be either a single point (as in our first example) or a line (think of three pages in an open book—they intersect along the binding of the book).

Let's look at an example of how we might proceed in these cases.

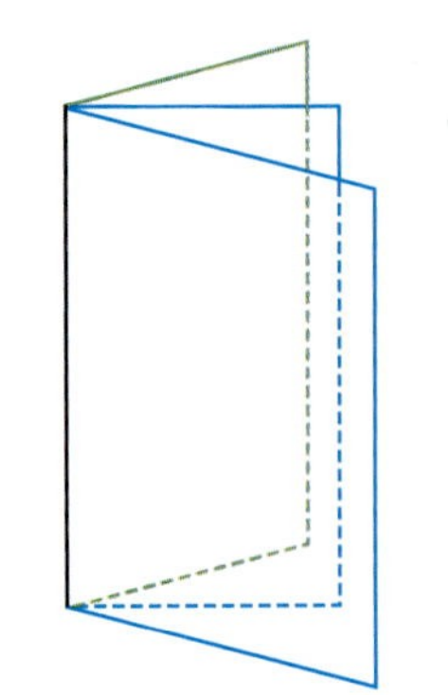
Three planes intersecting in a line

Example 3

Solving a Dependent Linear System in Three Variables

Solve the system.

$$x + 2y - z = 5 \qquad (10)$$

$$x - y + z = -2 \qquad (11)$$

$$-5x - 4y + z = -11 \qquad (12)$$

We begin as before by choosing two pairs of equations from the system and eliminating the same variable from each of the pairs. Adding equations (10) and (11) gives

$$2x + y = 3 \qquad (13)$$

Adding equations (10) and (12) gives

$$-4x - 2y = -6 \qquad (14)$$

Now consider the system formed by equations (13) and (14). We multiply equation (13) by 2 and add again:

$$\begin{aligned} 4x + 2y &= 6 \\ -4x - 2y &= -6 \\ \hline 0 &= 0 \end{aligned}$$

NOTE There are ways of representing the solutions, as you will see in later courses.

This true statement tells us that the system has an infinite number of solutions (lying along a straight line). Again, such a system is dependent.

CHECK YOURSELF 3

Solve the system.

$$2x - y + 3z = 3$$
$$-x + y - 2z = 1$$
$$y - z = 5$$

There is a third possibility for the solutions of systems in three variables, as Example 4 illustrates.

Example 4

Solving an Inconsistent Linear System in Three Variables

Solve the system.

$$3x + y - 3z = 1 \qquad (15)$$
$$-2x - y + 2z = 1 \qquad (16)$$
$$-x - y + z = 2 \qquad (17)$$

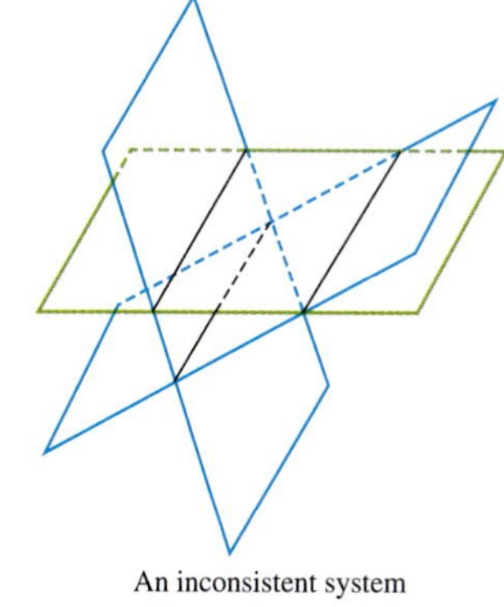

An inconsistent system

This time we eliminate variable y. Adding equations (15) and (16), we have

$$x - z = 2 \qquad (18)$$

Adding equations (15) and (17) gives

$$2x - 2z = 3 \qquad (19)$$

Now, multiply equation (18) by -2 and add the result to equation (19).

$$\begin{aligned} -2x + 2z &= -4 \\ 2x - 2z &= 3 \\ \hline 0 &= -1 \end{aligned}$$

All the variables have been eliminated, and we have arrived at a contradiction, $0 = -1$. This means that the system is *inconsistent* and has no solutions. There is *no* point common to all three planes.

CHECK YOURSELF 4

Solve the system.

$$x - y - z = 0$$
$$-3x + 2y + z = 1$$
$$3x - y + z = -1$$

As a closing note, we have by no means illustrated all possible types of inconsistent and dependent systems. Other possibilities involve either distinct parallel planes or planes that coincide. The solution techniques in these additional cases are, however, similar to those illustrated above.

In many instances, if an application involves three unknown quantities, you will find it useful to assign three variables to those quantities and then build a system of three equations from the given relationships in the problem. The extension of our problem-solving strategy is natural, as Example 5 illustrates.

Example 5

Solving a Number Problem

The sum of the digits of a three-digit number is 12. The tens digit is 2 less than the hundreds digit, and the units digit is 4 less than the sum of the other two digits. What is the number?

Step 1 The three unknowns are, of course, the three digits of the number.

NOTE Sometimes it helps to choose variable letters that relate to the words as is done here.

Step 2 We now want to assign variables to each of the three digits. Let u be the units digit, t be the tens digit, and h be the hundreds digit.

Step 3 There are three conditions given in the problem that allow us to write the necessary three equations. From those conditions

NOTE Take a moment now to go back to the original problem and pick out those conditions. That skill is a crucial part of the problem-solving strategy.

$$h + t + u = 12$$
$$t = h - 2$$
$$u = h + t - 4$$

Step 4 There are various ways to approach the solution. To use addition, write the system in the equivalent form

$$\begin{aligned} h + t + u &= 12 \\ -h + t \quad &= -2 \\ -h - t + u &= -4 \end{aligned}$$

and solve by our earlier methods. The solution, which you can verify, is $h = 5$, $t = 3$, and $u = 4$. The desired number is 534.

Step 5 To check, you should show that the digits of 534 meet each of the conditions of the original problem.

CHECK YOURSELF 5

The sum of the measures of the angles of a triangle is 180°. In a given triangle, the measure of the second angle is twice the measure of the first. The measure of the third angle is 30° less than the sum of the measures of the first two. Find the measure of each angle.

Let's continue with a financial application that will lead to a system of three equations.

Example 6

Solving an Investment Application

Monica decided to divide a total of \$42,000 into three investments: a savings account paying 5% interest, a time deposit paying 7%, and a bond paying 9%. Her total annual interest from the three investments was \$2600, and the interest from the savings account was \$200 less than the total interest from the other two investments. How much did she invest at each rate?

Step 1 The three amounts are the unknowns.

NOTE Again, we choose letters that suggest the unknown quantities—*s* for savings, *t* for time deposit, and *b* for bond.

Step 2 We let s be the amount invested at 5%, t the amount at 7%, and b the amount at 9%. Note that the interest from the savings account is then $0.05s$, and so on.

A table will help with the next step.

NOTE For 1 year, the interest formula is

$I = Pr$

(interest equals principal times rate).

	5%	7%	9%
Principal	s	t	b
Interest	$0.05s$	$0.07t$	$0.09b$

Step 3 Again there are three conditions in the given problem. By using the table above, they lead to the following equations.

NOTE Total invested.

NOTE Total interest.

NOTE The savings interest was \$200 *less than* that from the other two investments.

$$s + t + b = 42{,}000$$

$$0.05s + 0.07t + 0.09b = 2600$$

$$0.05s = 0.07t + 0.09b - 200$$

Step 4 We clear of decimals and solve as before, with the result

NOTE Find the interest earned from each investment, and verify that the conditions of the problem are satisfied.

$s = \$24{,}000 \qquad t = \$11{,}000 \qquad b = \$7000$

Step 5 We leave the check of these solutions to you.

CHECK YOURSELF 6

Glenn has a total of \$11,600 invested in three accounts: a savings account paying 6% interest, a stock paying 8%, and a mutual fund paying 10%. The annual interest from the stock and mutual fund is twice that from the savings account, and the mutual fund returned \$120 more than the stock. How much did Glenn invest in each account?

CHECK YOURSELF ANSWERS

1. $(5, 1, -3)$ **2.** $(1, -3, -2)$

3. The system is dependent (there are an infinite number of solutions).

4. The system is inconsistent (there are no solutions).

5. The three angles are 35°, 70°, and 75°.

6. \$5000 in savings, \$3000 in stocks, and \$3600 in mutual funds.

Exercises

Name ______________________

Section ________ Date ________

In exercises 1 to 20, solve each system of equations. If a unique solution does not exist, state whether the system is inconsistent or has an infinite number of solutions.

1. $$\begin{aligned} x - y + z &= 3 \\ 2x + y + z &= 8 \\ 3x + y - z &= 1 \end{aligned}$$

2. $$\begin{aligned} x - y - z &= 2 \\ 2x + y + z &= 8 \\ x + y + z &= 6 \end{aligned}$$

3. $$\begin{aligned} x + y + z &= 1 \\ 2x - y + 2z &= -1 \\ -x - 3y + z &= 1 \end{aligned}$$

4. $$\begin{aligned} x - y - z &= 6 \\ -x + 3y + 2z &= -11 \\ 3x + 2y + z &= 1 \end{aligned}$$

5. $$\begin{aligned} x + y + z &= 1 \\ -2x + 2y + 3z &= 20 \\ 2x - 2y - z &= -16 \end{aligned}$$

6. $$\begin{aligned} x + y + z &= -3 \\ 3x + y - z &= 13 \\ 3x + y - 2z &= 18 \end{aligned}$$

7. $$\begin{aligned} 2x + y - z &= 2 \\ -x - 3y + z &= -1 \\ -4x + 3y + z &= -4 \end{aligned}$$

8. $$\begin{aligned} x + 4y - 6z &= 8 \\ 2x - y + 3z &= -10 \\ 3x - 2y + 3z &= -18 \end{aligned}$$

9. $$\begin{aligned} 3x - y + z &= 5 \\ x + 3y + 3z &= -6 \\ x + 4y - 2z &= 12 \end{aligned}$$

10. $$\begin{aligned} 2x - y + 3z &= 2 \\ x - 2y + 3z &= 1 \\ 4x - y + 5z &= 5 \end{aligned}$$

11. $$\begin{aligned} x + 2y + z &= 2 \\ 2x + 3y + 3z &= -3 \\ 2x + 3y + 2z &= 2 \end{aligned}$$

12. $$\begin{aligned} x - 4y - z &= -3 \\ x + 2y + z &= 5 \\ 3x - 7y - 2z &= -6 \end{aligned}$$

13. $$\begin{aligned} x + 3y - 2z &= 8 \\ 3x + 2y - 3z &= 15 \\ 4x + 2y + 3z &= -1 \end{aligned}$$

14. $$\begin{aligned} x + y - z &= 2 \\ 3x + 5y - 2z &= -5 \\ 5x + 4y - 7z &= -7 \end{aligned}$$

15. $$\begin{aligned} x + y - z &= 2 \\ x - 2z &= 1 \\ 2x - 3y - z &= 8 \end{aligned}$$

16. $$\begin{aligned} x + y + z &= 6 \\ x - 2y &= -7 \\ 4x + 3y + z &= 7 \end{aligned}$$

17. $$\begin{aligned} x - 3y + 2z &= 1 \\ 16y - 9z &= 5 \\ 4x + 4y - z &= 8 \end{aligned}$$

18. $$\begin{aligned} x - 4y + 4z &= -1 \\ y - 3z &= 5 \\ 3x - 4y + 6z &= 1 \end{aligned}$$

19. $$\begin{aligned} x + 2y - 4z &= 13 \\ 3x + 4y - 2z &= 19 \\ 3x + 2z &= 3 \end{aligned}$$

20. $$\begin{aligned} x + 2y - z &= 6 \\ -3x - 2y + 5z &= -12 \\ x - 2z &= 3 \end{aligned}$$

ANSWERS

1. ____________
2. ____________
3. ____________
4. ____________
5. ____________
6. ____________
7. ____________
8. ____________
9. ____________
10. ____________
11. ____________
12. ____________
13. ____________
14. ____________
15. ____________
16. ____________
17. ____________
18. ____________
19. ____________
20. ____________

ANSWERS

21. ______

22. ______

23. ______

24. ______

25. ______

26. ______

27. ______

28. ______

Solve exercises 21 to 36 by choosing a variable to represent each unknown quantity and writing a system of equations.

21. Number problem. The sum of three numbers is 16. The largest number is equal to the sum of the other two, and 3 times the smallest number is 1 more than the largest. Find the three numbers.

22. Number problem. The sum of three numbers is 24. Twice the smallest number is 2 less than the largest number, and the largest number is equal to the sum of the other two. What are the three numbers?

23. Coin problem. A cashier has 25 coins consisting of nickels, dimes, and quarters with a value of \$4.90. If the number of dimes is 1 less than twice the number of nickels, how many of each type of coin does she have?

24. Recreation. A theater has tickets at \$6 for adults, \$3.50 for students, and \$2.50 for children under 12 years old. A total of 278 tickets were sold for one showing with a total revenue of \$1300. If the number of adult tickets sold was 10 less than twice the number of student tickets, how many of each type of ticket were sold for the showing?

25. Geometry. The perimeter of a triangle is 19 cm. If the length of the longest side is twice that of the shortest side and 3 cm less than the sum of the lengths of the other two sides, find the lengths of the three sides.

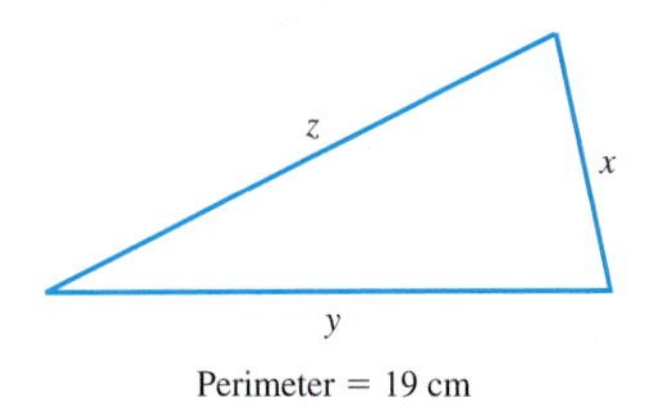

Perimeter = 19 cm

26. Geometry. The measure of the largest angle of a triangle is 10° more than the sum of the measures of the other two angles and 10° less than 3 times the measure of the smallest angle. Find the measures of the three angles of the triangle.

27. Investments. Jovita divides \$17,000 into three investments: a savings account paying 6% annual interest, a bond paying 9%, and a money market fund paying 11%. The annual interest from the three accounts is \$1540, and she has three times as much invested in the bond as in the savings account. What amount does she have invested in each account?

28. Investments. Adrienne has \$6000 invested among a savings account paying 3%, a time deposit paying 4%, and a bond paying 8%. She has \$1000 less invested in the bond than in her savings account, and she earned a total of \$260 in annual interest. What has she invested in each account?

29. Number problem. The sum of the digits of a three-digit number is 9, and the tens digit of the number is twice the hundreds digit. If the digits are reversed in order, the new number is 99 more than the original number. What is the original number?

30. Number problem. The sum of the digits of a three-digit number is 9. The tens digit is 3 times the hundreds digit. If the digits are reversed in order, the new number is 99 less than the original number. Find the original three-digit number.

31. Motion. Roy, Sally, and Jeff drive a total of 50 mi to work each day. Sally drives twice as far as Roy, and Jeff drives 10 mi farther than Sally. Use a system of three equations in three unknowns to find how far each person drives each day.

32. Consumer affairs. A parking lot has spaces reserved for motorcycles, cars, and vans. There are five more spaces reserved for vans than for motorcycles. There are three times as many car spaces as van and motorcycle spaces combined. If the parking lot has 180 total reserved spaces, how many of each type are there?

The solution process illustrated in this section can be extended to solving systems of more than three variables in a natural fashion. For instance, if four variables are involved, eliminate one variable in the system and then solve the resulting system in three variables as before. Substituting those three values into one of the original equations will provide the value for the remaining variable and the solution for the system.

In exercises 33 and 34, use this procedure to solve the system.

33.
$$\begin{aligned} x + 2y + 3z + w &= 0 \\ -x - y - 3z + w &= -2 \\ x - 3y + 2z + 2w &= -11 \\ -x + y - 2z + w &= 1 \end{aligned}$$

34.
$$\begin{aligned} x + y - 2z - w &= 4 \\ x - y + z + 2w &= 3 \\ 2x + y - z - w &= 7 \\ x - y + 2z + w &= 2 \end{aligned}$$

In some systems of equations there are more equations than variables. We can illustrate this situation with a system of three equations in two variables. To solve this type of system, pick any two of the equations and solve this system. Then substitute the solution obtained into the third equation. If a true statement results, the solution used is the solution to the entire system. If a false statement occurs, the system has no solution.

In exercises 35 and 36, use this procedure to solve each system.

35.
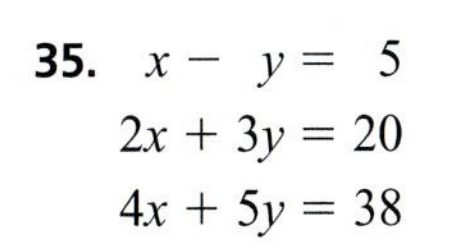
$$\begin{aligned} x - y &= 5 \\ 2x + 3y &= 20 \\ 4x + 5y &= 38 \end{aligned}$$

36.
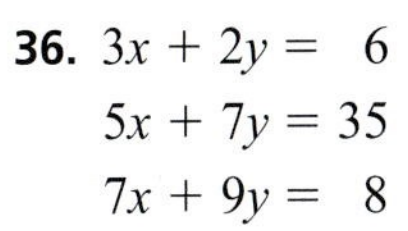
$$\begin{aligned} 3x + 2y &= 6 \\ 5x + 7y &= 35 \\ 7x + 9y &= 8 \end{aligned}$$

ANSWERS

29. ________
30. ________
31. ________
32. ________
33. ________
34. ________
35. ________
36. ________

37. ____

38. ____

39. (a)
(b) ____

37. Experiments have shown that cars (C), trucks (T), and buses (B) emit different amounts of air pollutants. In one such experiment, a truck emitted 1.5 pounds (lb) of carbon dioxide (CO_2) per passenger-mile and 2 grams (g) of nitrogen oxide (NO) per passenger-mile. A car emitted 1.1 lb of CO_2 per passenger-mile and 1.5 g of NO per passenger-mile. A bus emitted 0.4 lb of CO_2 per passenger-mile and 1.8 g of NO per passenger-mile. A total of 85 mi was driven by the three vehicles, and 73.5 lb of CO_2 and 149.5 g of NO were collected. Use the following system of equations to determine the miles driven by each vehicle.

$$T + C + B = 85.0$$
$$1.5T + 1.1C + 0.4B = 73.5$$
$$2T + 1.5C + 1.8B = 149.5$$

38. Experiments have shown that cars (C), trucks (T), and trains (R) emit different amounts of air pollutants. In one such experiment, a truck emitted 0.8 lb of carbon dioxide per passenger-mile and 1 g of nitrogen oxide per passenger-mile. A car emitted 0.7 lb of CO_2 per passenger-mile and 0.9 g of NO per passenger-mile. A train emitted 0.5 lb of CO_2 per passenger-mile and 4 g of NO per passenger-mile. A total of 141 mi was driven by the three vehicles, and 82.7 lb of CO_2 and 424.4 g of NO were collected. Use the following system of equations to determine the miles driven by each vehicle.

$$T + C + R = 141.0$$
$$0.8T + 0.7C + 0.5R = 82.7$$
$$T + 0.9C + 4R = 424.4$$

39. In Chapter 9 you will learn about quadratic functions and their graphs. A quadratic function has the form $y = ax^2 + bx + c$, in which a, b, and c are specific numbers and $a \neq 0$. Three distinct points on the graph are enough to determine the equation.

(a) Suppose that (1, 5), (2, 10), and (3, 19) are on the graph of $y = ax^2 + bx + c$. Substituting the pair (1, 5) into this equation (that is, let $x = 1$ and $y = 5$) yields $5 = a + b + c$. Substituting each of the other ordered pairs yields: $10 = 4a + 2b + c$ and $19 = 9a + 3b + c$. Solve the resulting system of equations to determine the values of a, b, and c. Then write the equation of the function.

(b) Repeat the work of part (a) using the following three points: (1, 2), (2, 9), and (3, 22).

Answers

1. $\{(1, 2, 4)\}$ **3.** $\{(-2, 1, 2)\}$ **5.** $\{(-4, 3, 2)\}$
7. Infinite number of solutions **9.** $\left\{\left(3, \frac{1}{2}, -\frac{7}{2}\right)\right\}$ **11.** $\{(3, 2, -5)\}$
13. $\{(2, 0, -3)\}$ **15.** $\left\{\left(4, -\frac{1}{2}, \frac{3}{2}\right)\right\}$ **17.** Inconsistent system
19. $\left\{\left(2, \frac{5}{2}, -\frac{3}{2}\right)\right\}$ **21.** 3, 5, 8 **23.** 3 nickels, 5 dimes, 17 quarters
25. 4 cm, 7 cm, 8 cm **27.** \$3000 savings, \$9000 bond, \$5000 money market
29. 243 **31.** Roy 8 mi, Sally 16 mi, Jeff 26 mi **33.** $(1, 2, -1, -2)$
35. (7, 2) **37.** $T = 20$, $C = 25$, $B = 40$
39. **(a)** $y = 2x^2 - x + 4$; **(b)** $y = 3x^2 - 2x + 1$

5.4 Graphing Linear Inequalities in Two Variables

5.4 OBJECTIVES

1. Graph linear inequalities in two variables
2. Graph a region defined by linear inequalities

CAUTION

Note that we have **two** variables. This is a different process than we discussed in Section 4.4, when we had only **one** variable.

What does the solution set look like when we are faced with an inequality in two variables? We will see that it is a set of ordered pairs best represented by a shaded region. Recall that the general form for a linear inequality in two variables is

$$ax + by < c$$

in which a and b cannot both be 0. The symbol $<$ can be replaced with $>$, $\leq$, or $\geq$. Some examples are

$$y < -2x + 6 \qquad x - 2y > 4 \qquad \text{or} \qquad 2x - 3y \geq x + 5y$$

As was the case with an equation, the solution set of a linear inequality in two variables is, a set of ordered pairs of real numbers. However, in the case of the linear inequalities, we will find that the solution sets will be all the points in an entire region of the plane, called a **half plane.**

To determine such a solution set, let's start with the first inequality listed above. To graph the solution set of

$$y < -2x + 6$$

we begin by writing the corresponding linear equation

$$y = -2x + 6$$

First, note that the graph of $y = -2x + 6$ is simply a straight line.

Now, to graph the solution set of $y < -2x + 6$, we must include all ordered pairs that satisfy that inequality. For instance, if $x = 1$, we have

$$y < -2 \cdot 1 + 6$$

$$y < 4$$

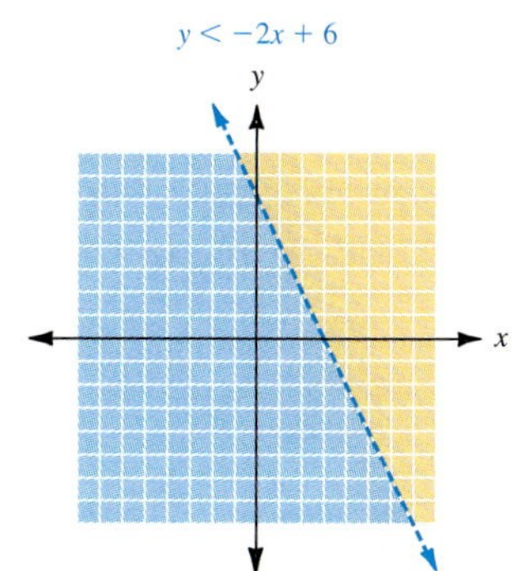

NOTE The line is dashed to indicate that the points do not represent solutions.

So we want to include all points of the form $(1, y)$, in which $y < 4$. Of course, because $(1, 4)$ is *on* the corresponding line, this means that we want all points *below* the line along the vertical line $x = 1$. The result will be similar for any choice of x, and our solution set will then contain all points below the line $y = -2x + 6$. We can then graph the solution set as the shaded region shown. We have the following definition.

NOTE We call the graph of the equation

$ax + by = c$

the **boundary line** of the half plane.

Definitions: Solution Set of an Inequality

In general, the solution set of an inequality of the form

$ax + by < c \qquad \text{or} \qquad ax + by > c$

will be a half plane either above or below the corresponding line determined by

$ax + by = c$

How do we decide which half plane represents the desired solution set? The use of a **test point** provides an easy answer. Choose any point *not* on the line. Then substitute the coordinates of that point into the given inequality. If the coordinates satisfy the inequality (result in a true statement), then shade the region or half plane that includes the test point; if not, shade the opposite half plane. Example 1 illustrates the process.

Example 1

Graphing a Linear Inequality

Graph the linear inequality

$x - 2y < 4$

First, we graph the corresponding equation

$x - 2y = 4$

to find the boundary line. Now to decide on the appropriate half plane, we need a test point *not* on the line. As long as the line *does not pass through the origin,* we can always use (0, 0) as a test point. It provides the easiest computation.

Here letting $x = 0$ and $y = 0$, we have

$0 - 2 \cdot 0 \overset{?}{<} 4$

$0 < 4$

Because this is a true statement, we proceed to shade the half plane including the origin (the test point), as shown.

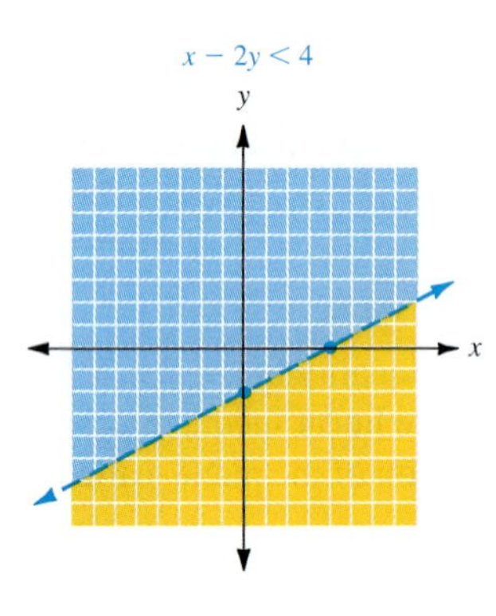

NOTE The boundary line is dashed to indicate that any point **on** the line itself is **not** a solution.

CHECK YOURSELF 1

Graph the solution set of $3x + 4y > 12$.

The graphs of some linear inequalities will include the boundary line. That will be the case whenever equality is included with the inequality statement, as illustrated in Example 2.

Example 2

Graphing a Linear Inequality

Graph the solution set of

$y \le 2x$

We proceed as before by graphing the boundary line (it is solid because equality is included). The only difference between this and previous examples is that we *cannot use the origin* as a test point. Do you see why?

Choosing (1, 1) as our test point gives the statement

$1 \overset{?}{\le} 2 \cdot 1$

$1 \le 2$

Because the statement is *true,* we shade the half plane *including* the test point (1, 1).

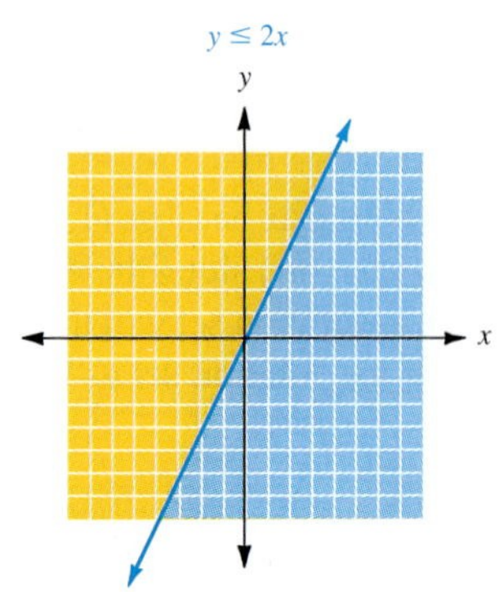

NOTE The choice of (1, 1) is arbitrary. We simply want *any* point *not* on the line.

CHECK YOURSELF 2

Graph the solution set of $3x + y > 0$.

Let's consider a special case of graphing linear inequalities in the rectangular coordinate system.

Example 3

Graphing a Linear Inequality

NOTE Here we specify the rectangular coordinate system to indicate we want a two-dimensional graph.

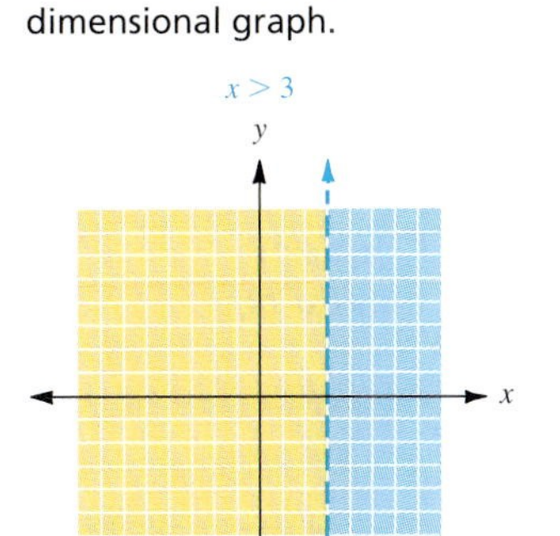

Graph the solution set of $x > 3$.

First, we draw the boundary line (a dashed line because equality is not included) corresponding to

$$x = 3$$

We can choose the origin as a test point in this case, and that results in the false statement

$$0 > 3$$

We then shade the half plane *not* including the origin. In this case, the solution set is represented by the half plane to the right of the vertical boundary line.

As you may have observed, in this special case choosing a test point is not really necessary. Because we want values of x that are *greater than* 3, we want those ordered pairs that are to the *right* of the boundary line.

CHECK YOURSELF 3

Graph the solution set of

$$y \le 2$$

in the rectangular coordinate system.

Applications of linear inequalities will often involve more than one inequality condition. Consider Example 4.

Example 4

Graphing a Region Defined by Linear Inequalities

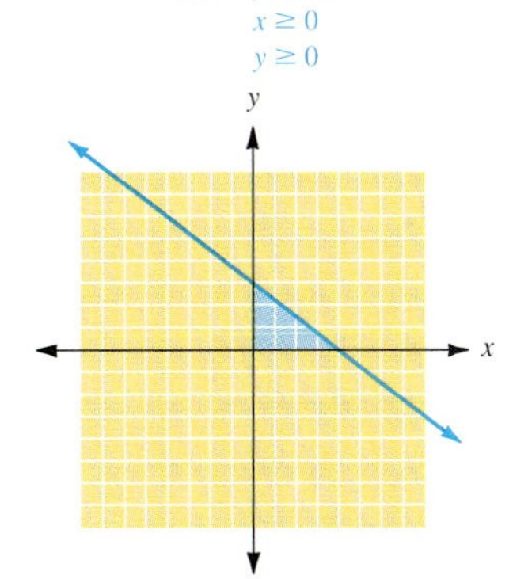

Graph the region satisfying the following conditions.

$$3x + 4y \le 12$$
$$x \ge 0$$
$$y \ge 0$$

The solution set in this case must satisfy *all three conditions.* As before, the solution set of the first inequality is graphed as the half plane *below* the boundary line. The second and third inequalities mean that x and y must also be nonnegative. Therefore, our solution set is restricted to the first quadrant (and the appropriate segments of the x and y axes), as shown.

CHECK YOURSELF 4

Graph the region satisfying the following conditions.

$$3x + 4y < 12$$
$$x \ge 0$$
$$y \ge 1$$

The following algorithm summarizes our work in graphing linear inequalities in two variables.

Step by Step: Graphing a Linear Inequality

Step 1 Replace the inequality symbol with an equality symbol to form the equation of the boundary line of the solution set.

Step 2 Graph the boundary line. Use a dashed line if equality is not included ($<$ or $>$). Use a solid line if equality is included ($\le$ or $\ge$).

Step 3 Choose any convenient test point *not* on the boundary line.

Step 4 If the inequality is *true* for the test point, shade the half plane *including* the test point. If the inequality is *false* for the test point, shade the half plane *not including* the test point.

CHECK YOURSELF ANSWERS

1.

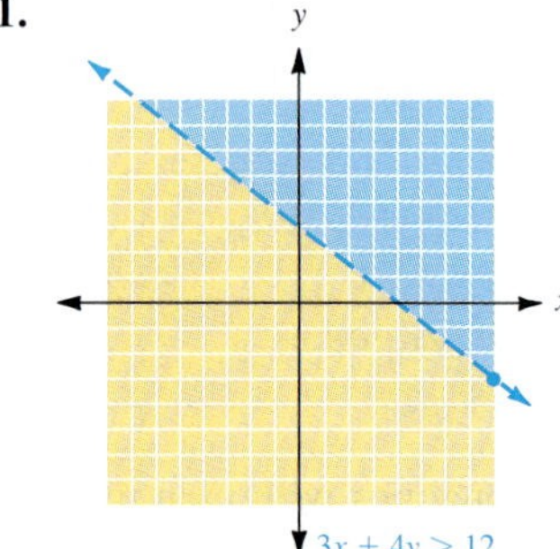

2.

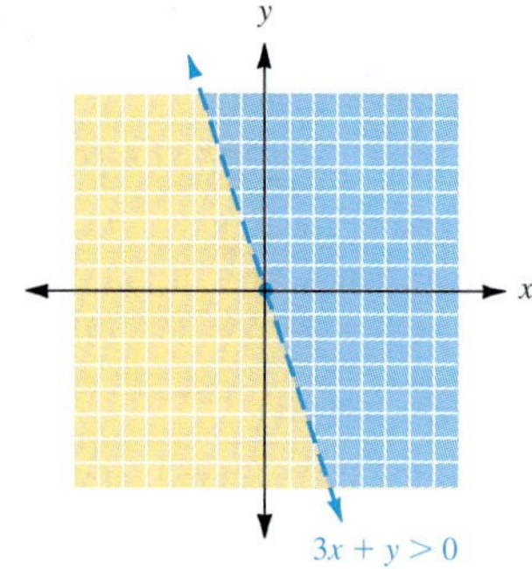

3.

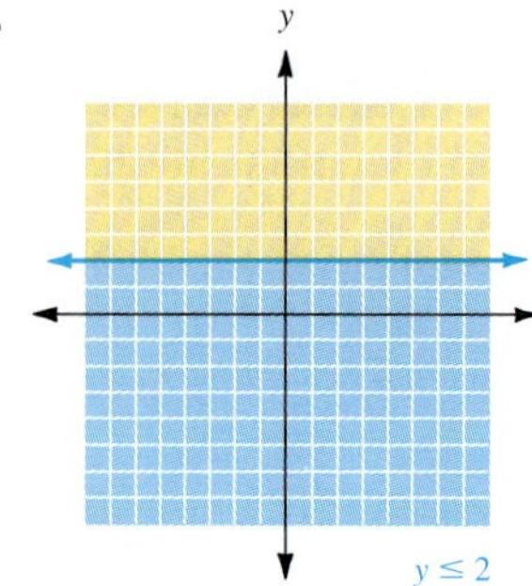

4.

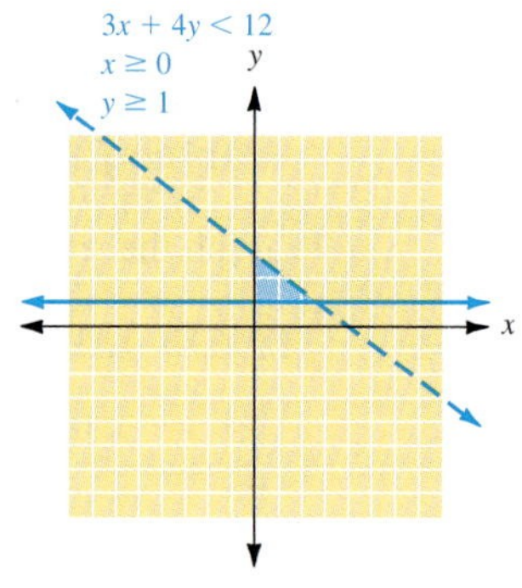

Name

Section Date

5.4 Exercises

In exercises 1 to 24, graph the solution sets of the linear inequalities.

1. $x + y < 4$
2. $x + y \geq 6$
3. $x - y \geq 3$

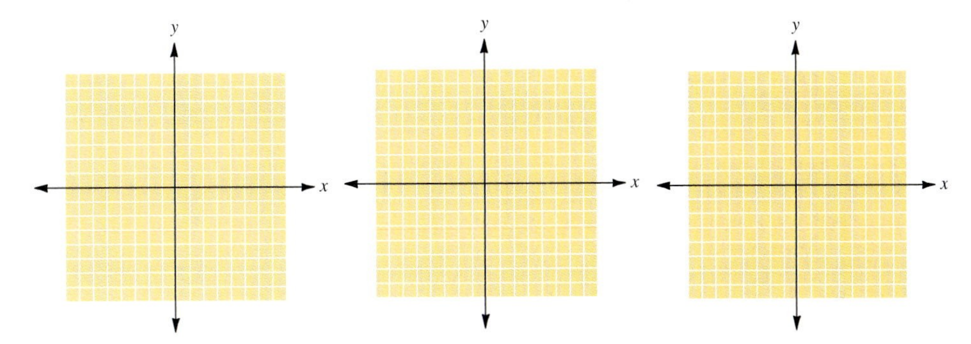

4. $x - y < 5$
5. $y \geq 2x + 1$
6. $y < 3x - 4$

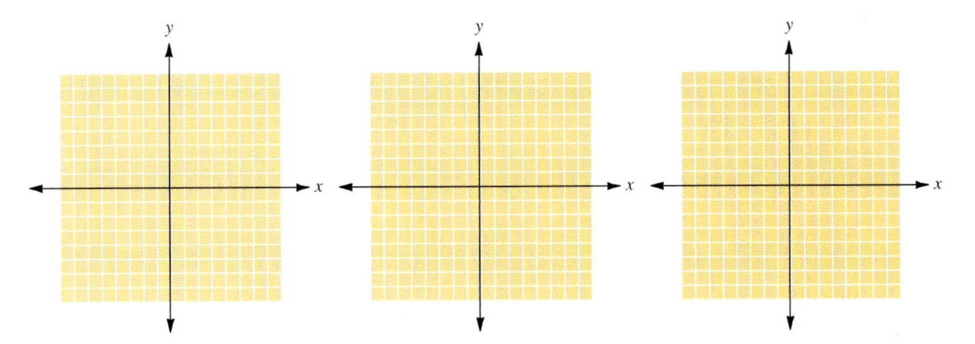

7. $2x + 3y < 6$
8. $3x - 4y \geq 12$
9. $x - 4y > 8$

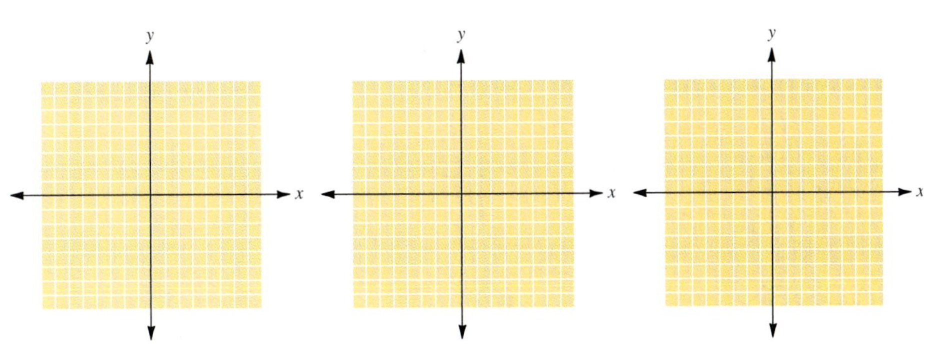

10. $2x + 5y \leq 10$
11. $y \geq 3x$
12. $y \leq -2x$

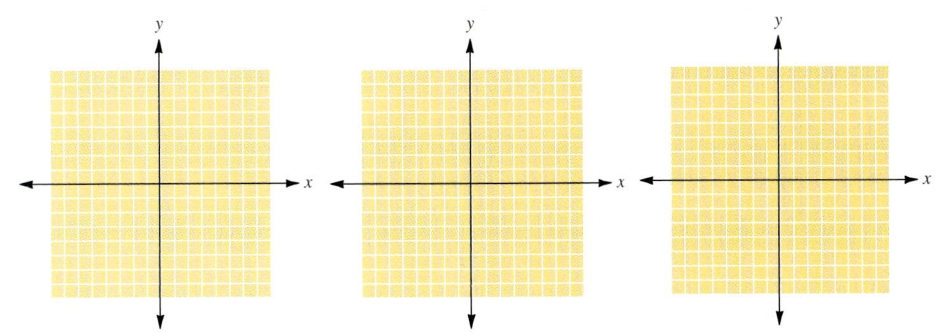

ANSWERS

1.
2.
3.
4.
5.
6.
7.
8.
9.
10.
11.
12.

ANSWERS

13. ______

14. ______

15. ______

16. ______

17. ______

18. ______

19. ______

20. ______

21. ______

22. ______

23. ______

24. ______

13. $x - 2y > 0$ **14.** $x + 4y \le 0$ **15.** $x < 3$

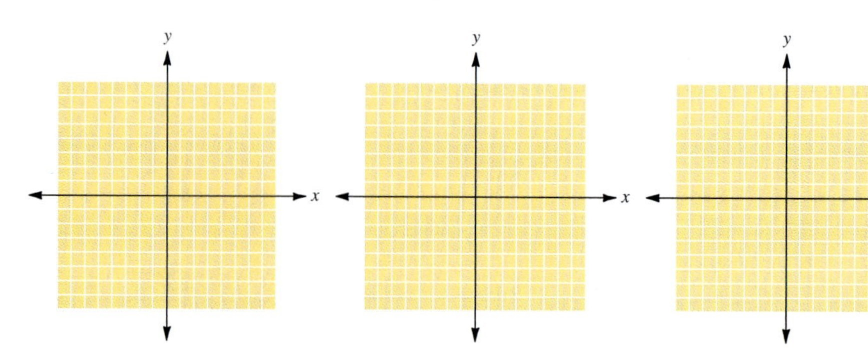

16. $y < -2$ **17.** $y > 3$ **18.** $x \le -4$

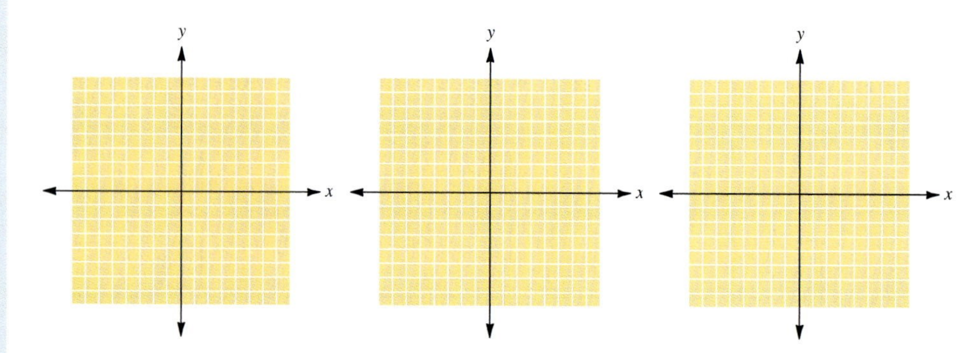

19. $3x - 6 \le 0$ **20.** $-2y > 6$ **21.** $0 < x < 1$

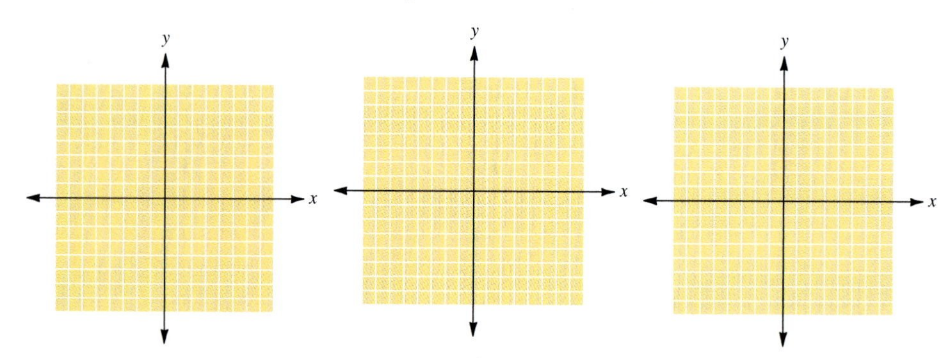

22. $-2 \le y \le 1$ **23.** $1 \le x \le 3$ **24.** $1 < y < 5$

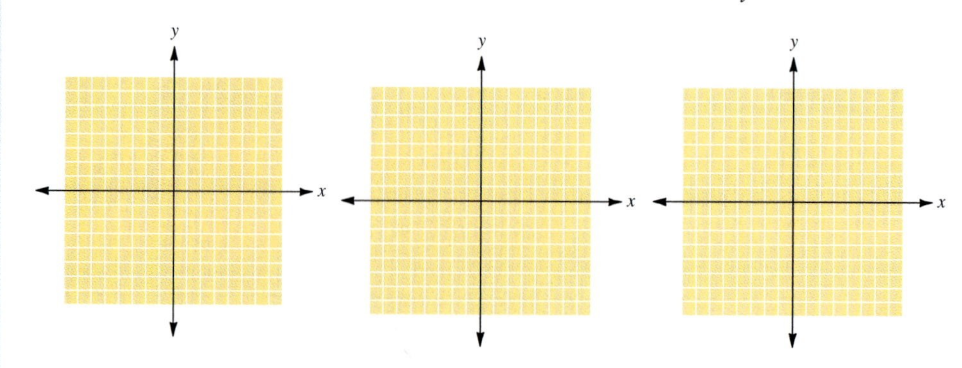

In exercises 25 to 28, graph the region satisfying each set of conditions.

25. $0 \le x \le 3$
$2 \le y \le 4$

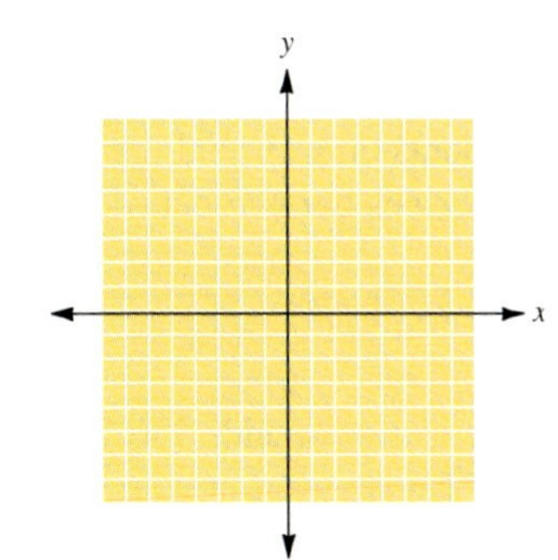

26. $1 \le x \le 5$
$0 \le y \le 3$

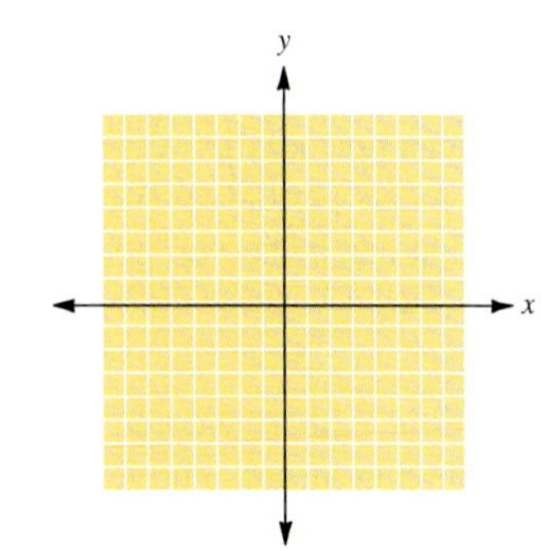

27. $x + 2y \le 4$
$x \ge 0$
$y \ge 0$

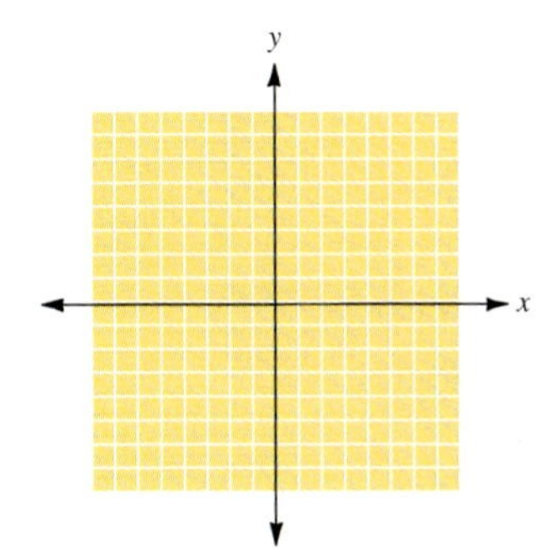

28. $2x + 3y \le 6$
$x \ge 0$
$y \ge 0$

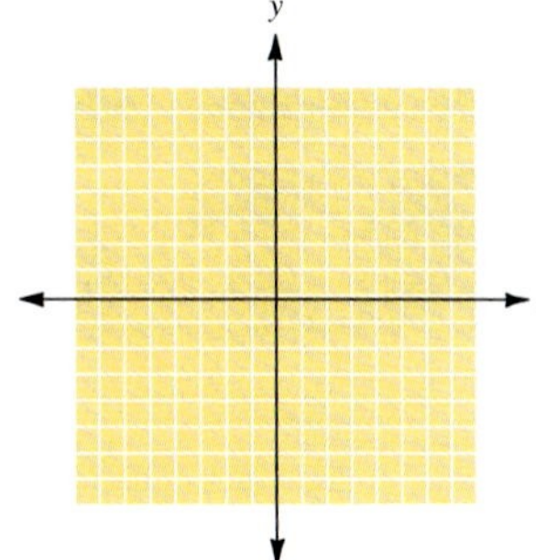

29. Assume that you are working only with the variable x. Describe the set of solutions for the statement $x > -1$.

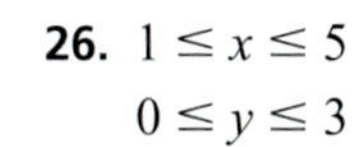

30. Now, assume that you are working in two variables, x and y. Describe the set of solutions for the statement $x > -1$.

31. Manufacturing. A manufacturer produces a standard model and a deluxe model of a 13-in. television set. The standard model requires 12 h to produce, and the deluxe model requires 18 h. The labor available is limited to 360 h per week.

If x represents the number of standard-model sets produced per week and y represents the number of deluxe models, draw a graph of the region representing the feasible values for x and y. (This region will be the solution set for the system of inequalities.) Keep in mind that the values for x and y must be nonnegative because they represent a quantity of items.

ANSWERS

25. ______

26. ______

27. ______

28. ______

29. ______

30. ______

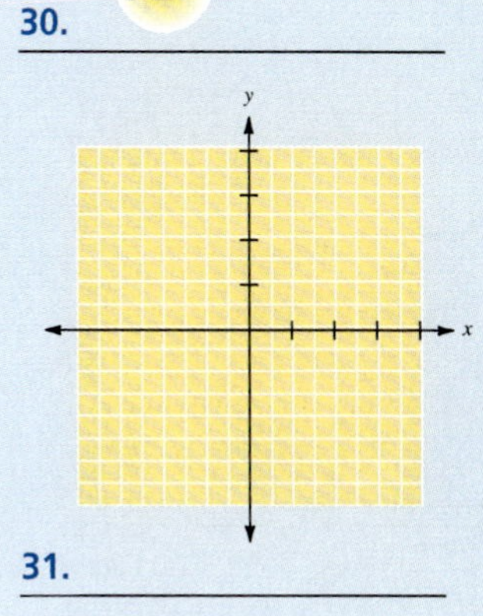

31. ______

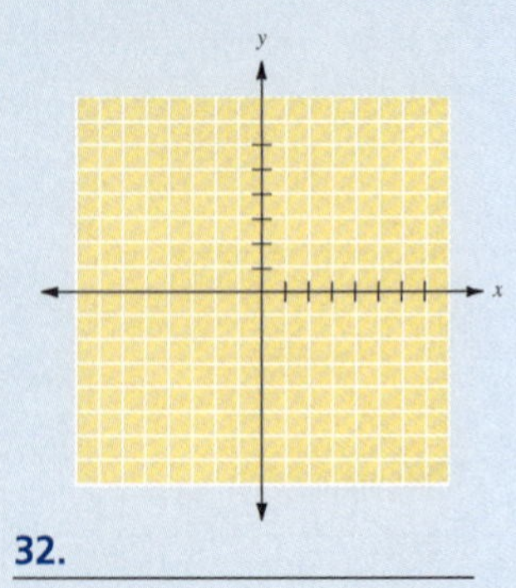

32. ______

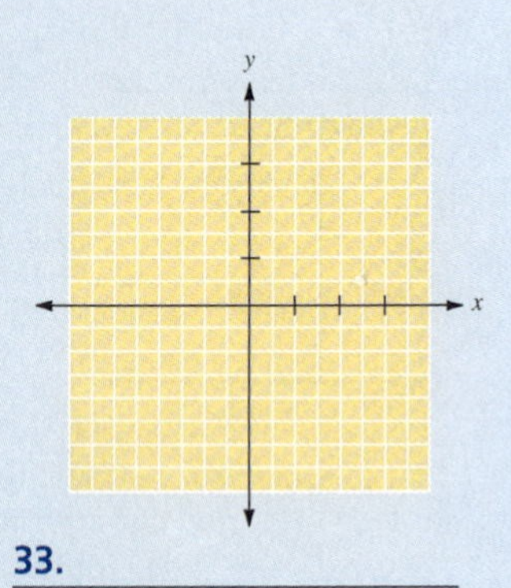

33. ______

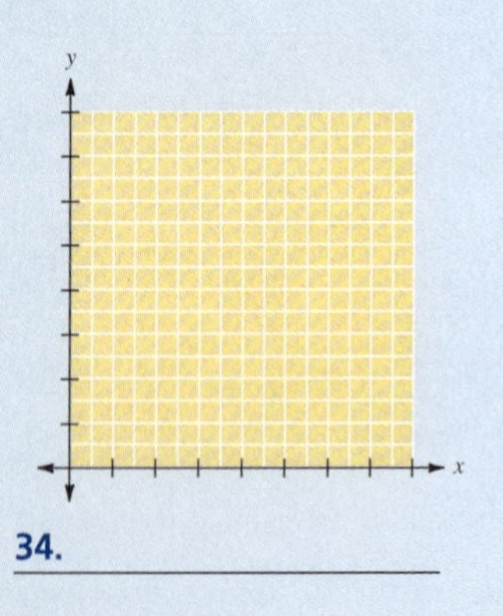

34. ______

35. ______

36. ______

32. Manufacturing. A manufacturer produces portable radios and CD players. The radios require 10 h of labor to produce and CD players require 20 h. Let x represent the number of radios produced and y the number of CD players.

If the labor hours available are limited to 300 h per week, graph the region representing the feasible values for x and y.

33. Serving capacity. A hospital food service can serve at most 1000 meals per day. Patients on a normal diet receive 3 meals per day and patients on a special diet receive 4 meals per day. Write a linear inequality that describes the number of patients that can be served per day and draw its graph.

34. Time on job. The movie and TV critic for the local radio station spends 3 to 7 hours daily reviewing movies and fewer than 4 hours reviewing TV shows. Let x represent the hours watching movies and y represent the time spent watching TV. Write two inequalities that model the situation, and graph their intersection.

In exercises 35 to 38, write an inequality for the shaded region shown in the figure.

35.

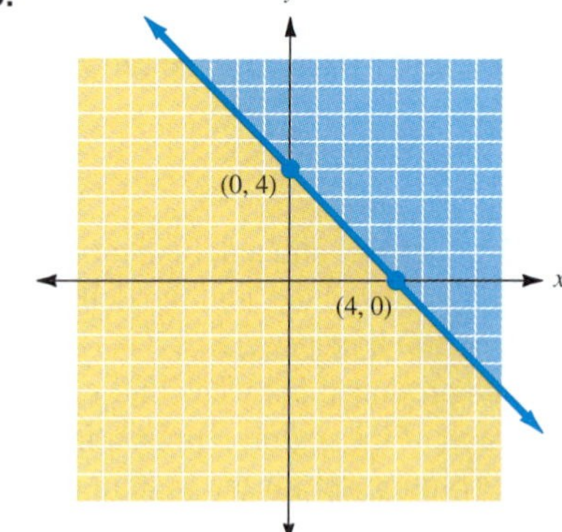

36.

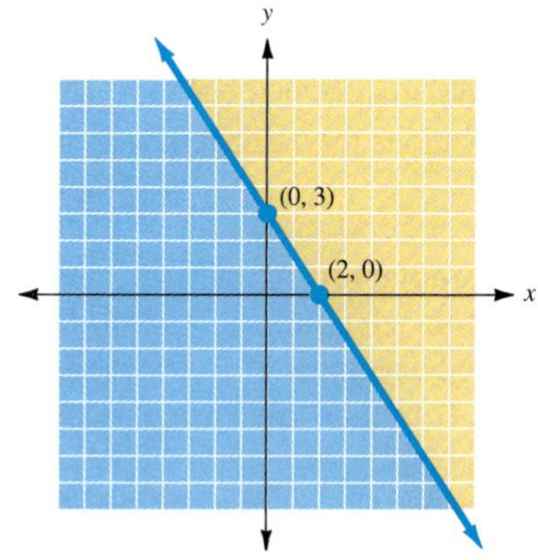

37.

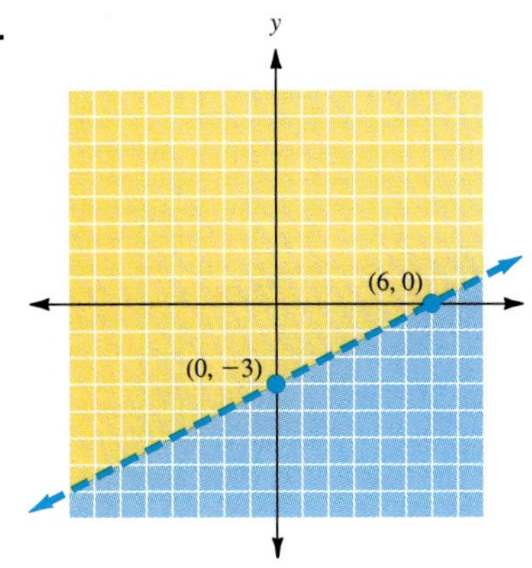

38.

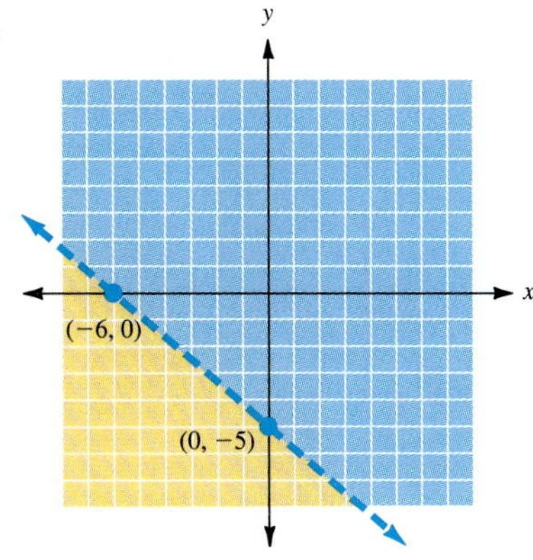

ANSWERS

37. ____________________

38. ____________________

Answers

1. $x + y < 4$

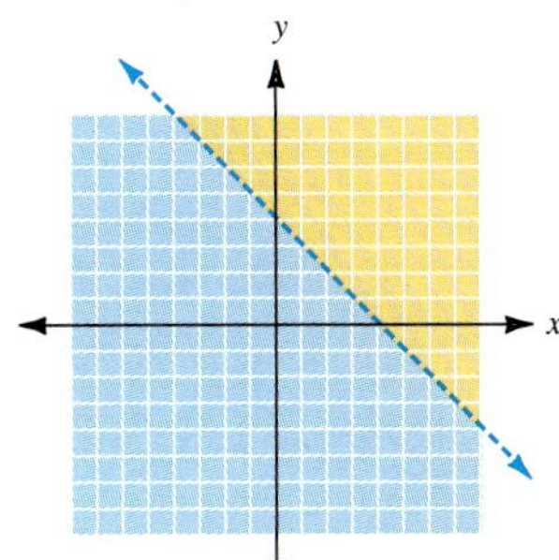

3. $x - y \geq 3$

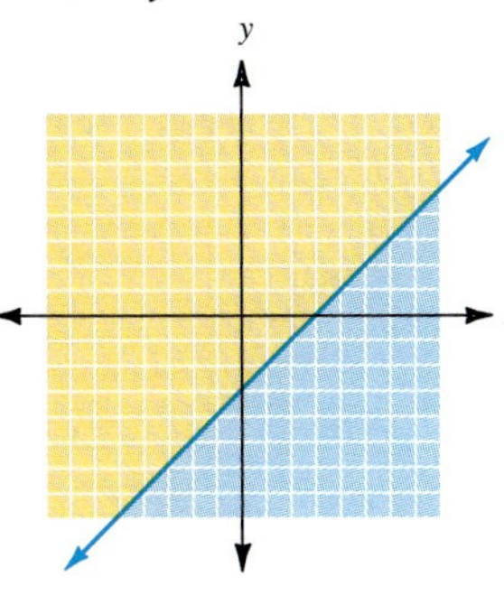

5. $y \geq 2x + 1$

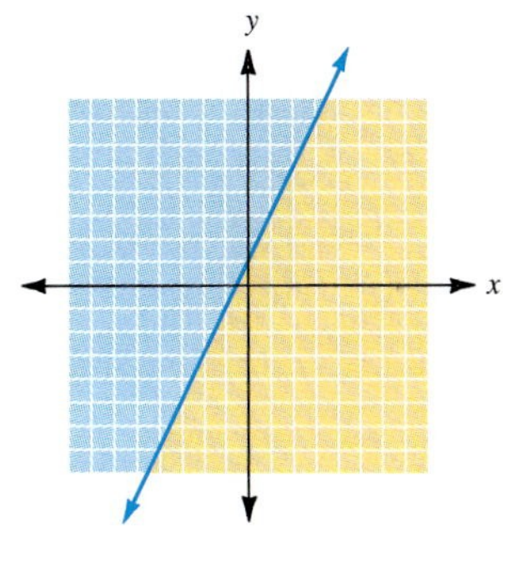

7. $2x + 3y < 6$

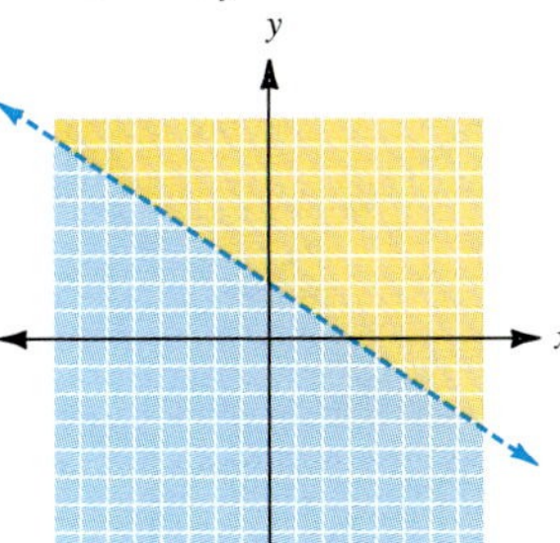

9. $x - 4y > 8$

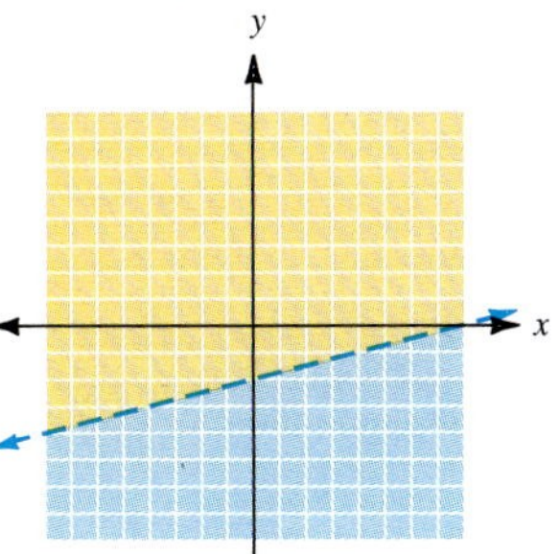

11. $y \geq 3x$

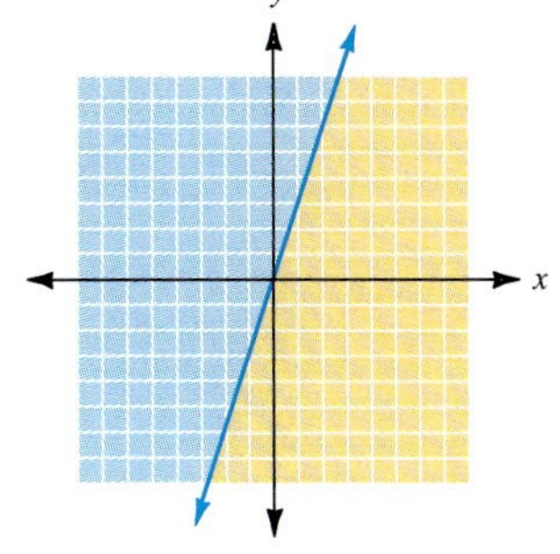

13. $x - 2y > 0$

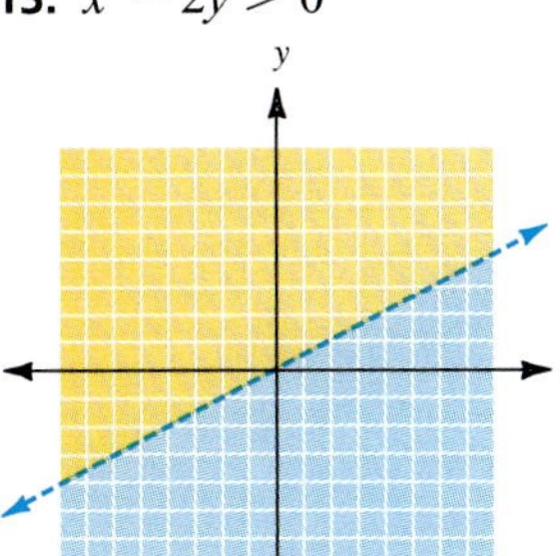

15. $x < 3$

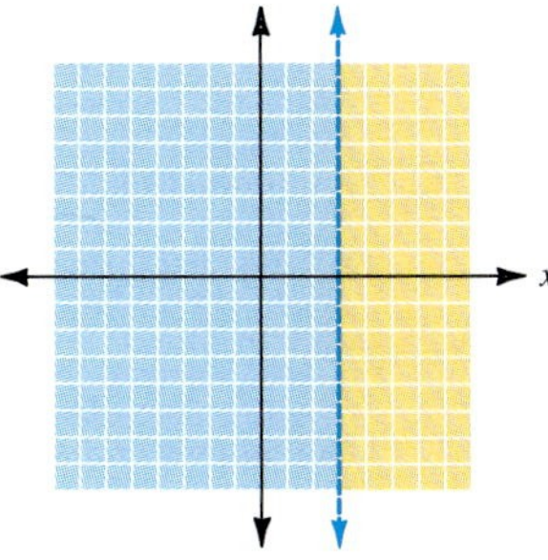

17. $y > 3$

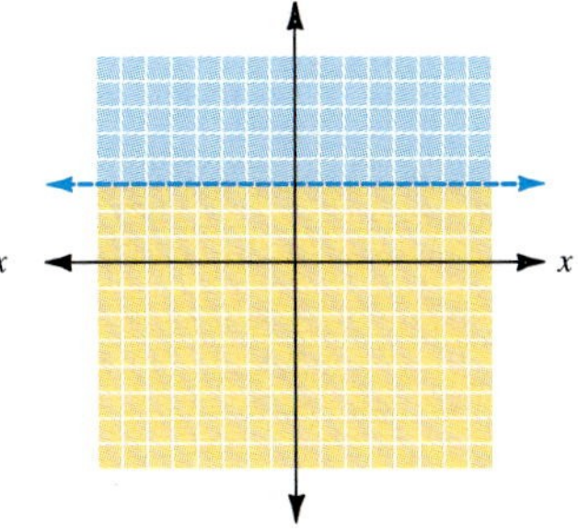

19. $3x - 6 \le 0$

21. $0 < x < 1$

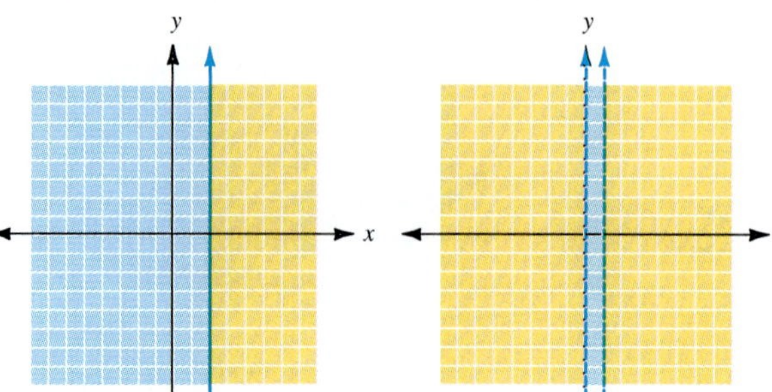

23. $1 \le x \le 3$

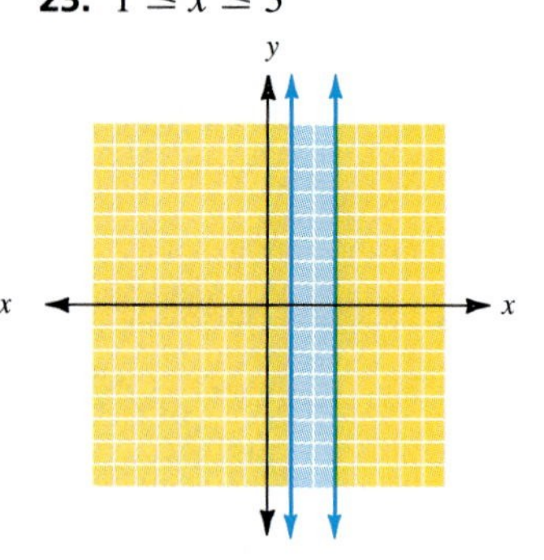

25. $0 \le x \le 3$
$2 \le y \le 4$

27. $x + 2y \le 4$
$x \ge 0$
$y \ge 0$

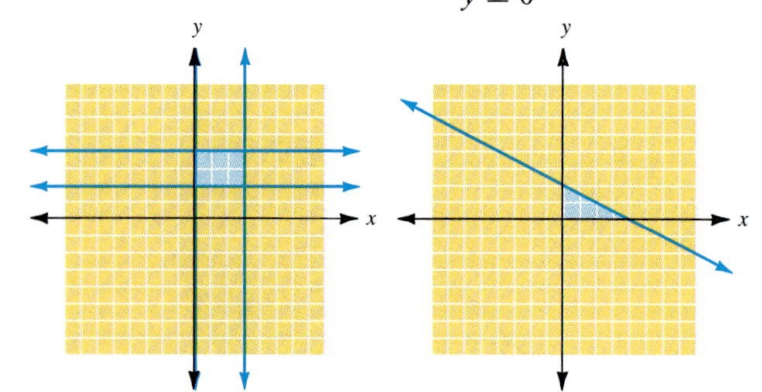

29.  

31.

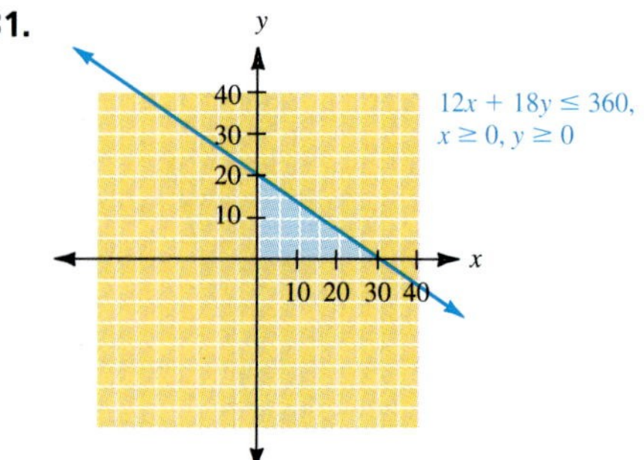

33.

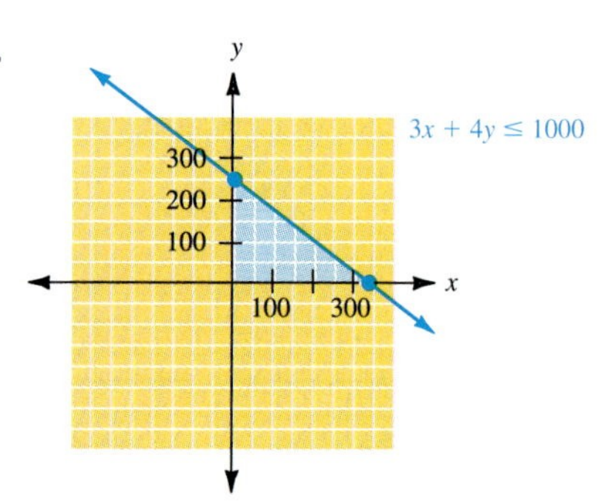

35. $y \ge -x + 4$

37. $y < \frac{1}{2}x - 3$

Graphing Systems of Linear Inequalities in Two Variables

OBJECTIVES

1. Graph a system of linear inequalities in two variables
2. Solve an application of a system of linear inequalities

In Section 5.1, we dealt with finding the solution set of a system of linear equations. That solution set represented the points of intersection of the graphs of the equations in the system. In this section, we extend that idea to include systems of linear inequalities.

In this case, the solution set is all ordered pairs that satisfy each inequality. **The graph of the solution set of a system of linear inequalities** is then the intersection of the graphs of the individual inequalities. Let's look at an example.

Example 1

Solving a System by Graphing

Solve the following system of linear inequalities by graphing

$x + y > 4$

$x - y < 2$

We start by graphing each inequality separately. The boundary line is drawn, and using (0, 0) as a test point, we see that we should shade the half plane above the line in both graphs.

NOTE Notice that the boundary line is dashed to indicate that points on the line do *not* represent solutions.

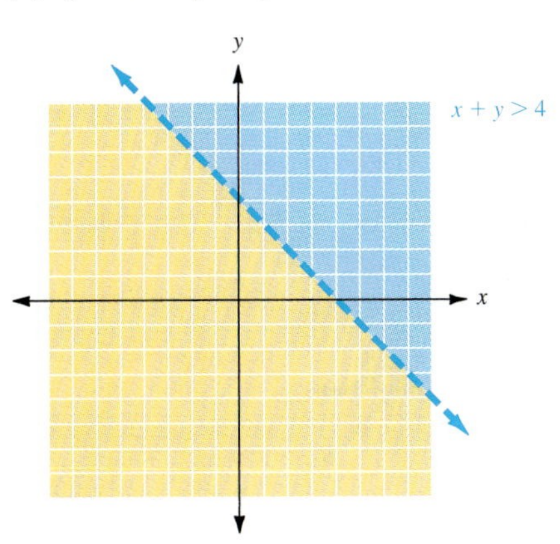

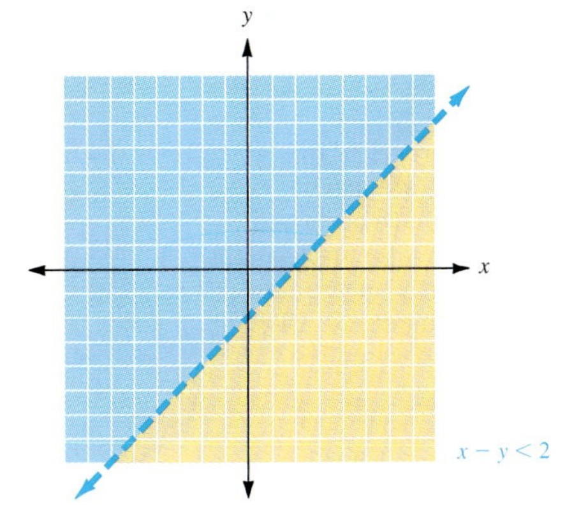

In practice, the graphs of the two inequalities are combined on the same set of axes, as is shown below. The graph of the solution set of the original system is the intersection of the graphs drawn above.

NOTE Points on the lines are not included in the solution set.

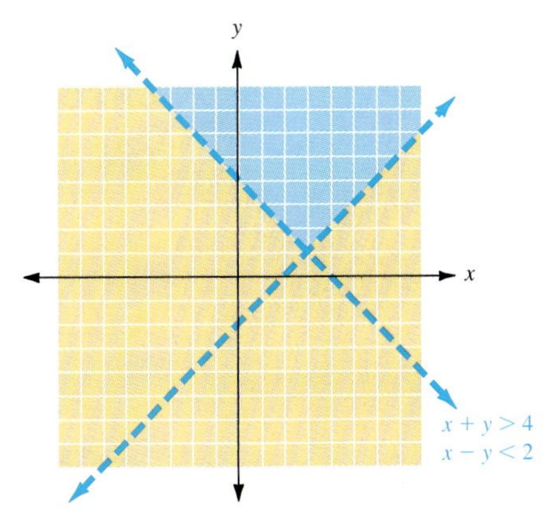

CHECK YOURSELF 1

Solve the following system of linear inequalities by graphing.

$$2x - y < 4$$
$$x + y < 3$$

Most applications of systems of linear inequalities lead to **bounded regions.** This requires a system of three or more inequalities, as shown in Example 2.

Example 2

Solving a System by Graphing

Solve the following system of linear inequalities by graphing.

$$x + 2y \leq 6$$
$$x + y \leq 5$$
$$x \geq 2$$
$$y \geq 0$$

On the same set of axes, we graph the boundary line of each of the inequalities. We then choose the appropriate half planes (indicated by the arrow that is perpendicular to the line) in each case, and we locate the intersection of those regions for our graph.

NOTE The vertices of the shaded region are given because they have particular significance in later applications of this concept. Can you see how the coordinates of the vertices were determined?

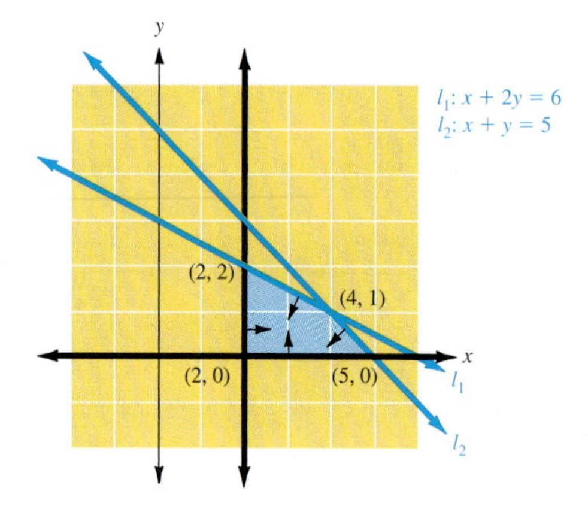

CHECK YOURSELF 2

Solve the following system of linear inequalities by graphing.

$$2x - y \leq 8 \qquad x \geq 0$$
$$x + y \leq 7 \qquad y \geq 0$$

Let's expand on Example 4, Section 5.2, to see an application of our work with systems of linear inequalities. Consider Example 3.

Example 3

Solving a Business-Based Application

A manufacturer produces a standard model and a deluxe model of a 25-in. television set. The standard model requires 12 h of labor to produce, and the deluxe model requires 18 h. The labor available is limited to 360 h per week. Also, the plant capacity is limited to producing a total of 25 sets per week. Draw a graph of the region representing the number of sets that can be produced, given these conditions.

As suggested earlier, we let x represent the number of standard-model sets produced and y the number of deluxe-model sets. Because the labor is limited to 360 h, we have

NOTE The total labor is limited to (or less than or equal to) 360 h.

$$12x + 18y \leq 360 \qquad (1)$$

12 h per standard set (under $12x$); 18 h per deluxe set (under $18y$)

The total production, here $x + y$ sets, is limited to 25, so we can write

$$x + y \leq 25 \qquad (2)$$

NOTE We have $x \geq 0$ and $y \geq 0$ because the number of sets produced cannot be negative.

For convenience in graphing, we divide both members of inequality (1) by 6, to write the equivalent system

$$\begin{aligned} 2x + 3y &\leq 60 \\ x + y &\leq 25 \\ x &\geq 0 \\ y &\geq 0 \end{aligned}$$

We now graph the system of inequalities as before. The shaded area represents all possibilities in terms of the number of sets that can be produced.

NOTE The shaded area is called the **feasible region.** All points in the region meet the given conditions of the problem and represent possible production options.

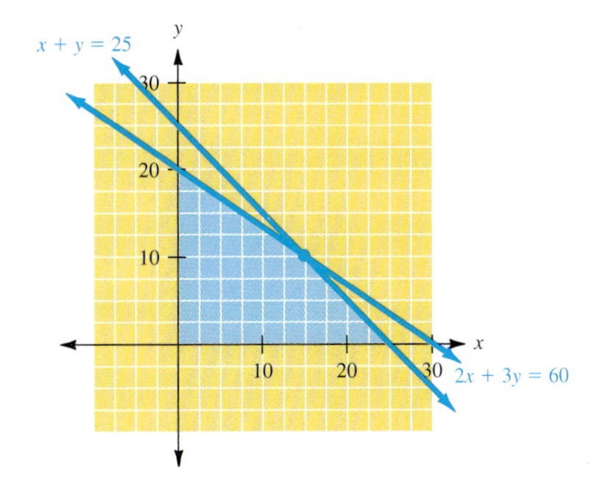

CHECK YOURSELF 3

A manufacturer produces TVs and CD players. The TVs require 10 h of labor to produce and the CD players require 20 h. The labor hours available are limited to 300 h per week. Existing orders require that at least 10 TVs and at least 5 CD players be produced per week. Draw a graph of the region representing the possible production options.

CHECK YOURSELF ANSWERS

1. $2x - y < 4$
$x + y < 3$

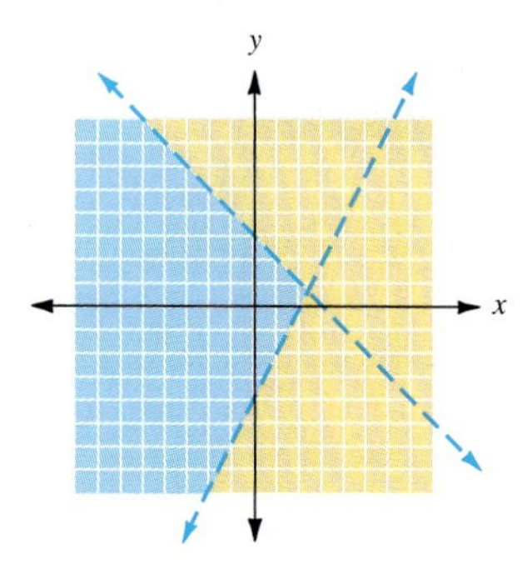

2. $2x - y \leq 8$
$x + y \leq 7$
$x \geq 0$
$y \geq 0$

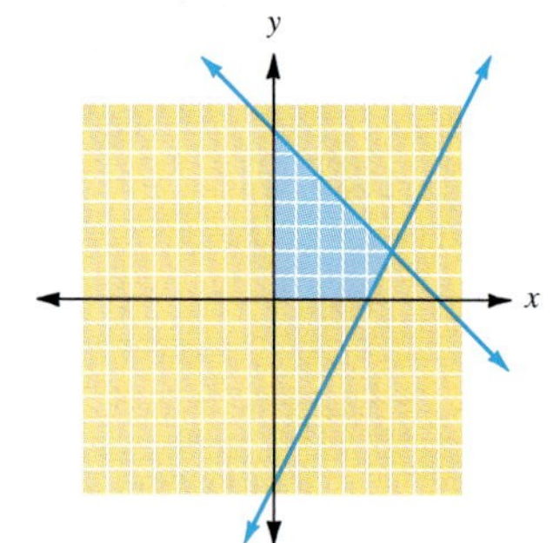

3. Let x be the number of TVs and y be the number of CD players. The system is

$$10x + 20y \leq 300$$
$$x \geq 10$$
$$y \geq 5$$

5.5 Exercises

Name ______________________

Section ________ Date ________

In exercises 1 to 18, solve each system of linear inequalities by graphing.

1. $x + 2y \le 4$
$x - y \ge 1$

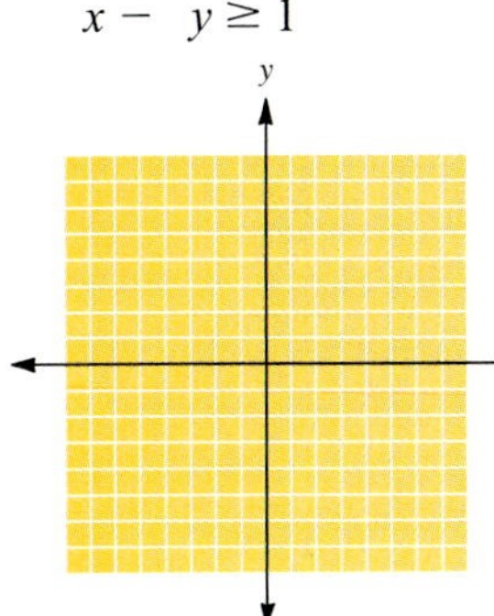

2. $3x - y > 6$
$x + y < 6$

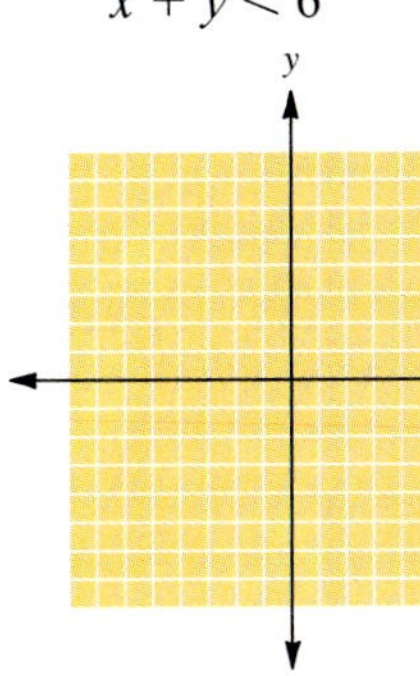

3. $3x + y < 6$
$x + y > 4$

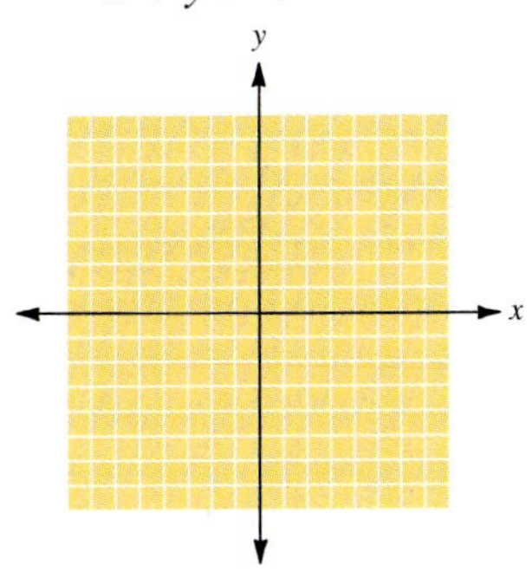

4. $2x + y \ge 8$
$x + y \ge 4$

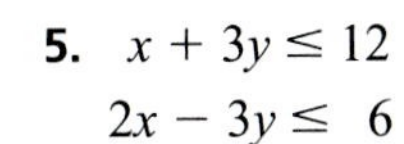

5. $x + 3y \le 12$
$2x - 3y \le 6$

6. $x - 2y > 8$
$3x - 2y > 12$

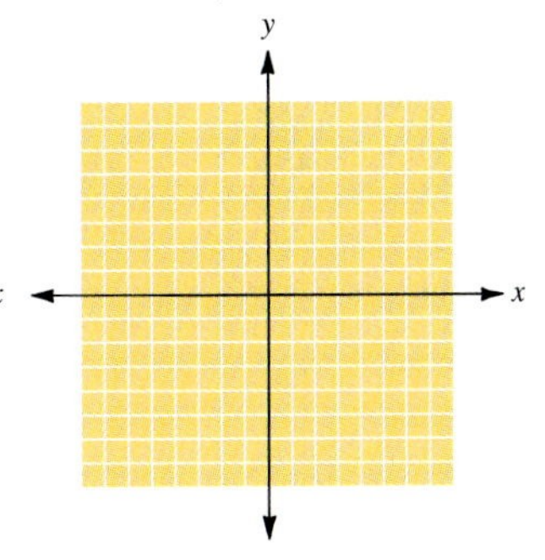

7. $3x + 2y \le 12$
$x \ge 2$

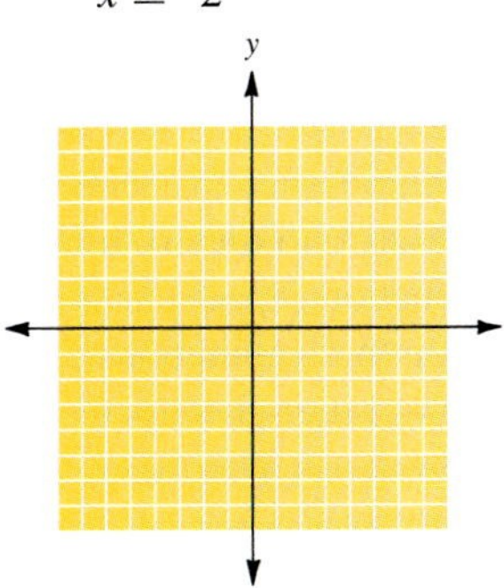

8. $2x + y \le 6$
$y \ge 1$

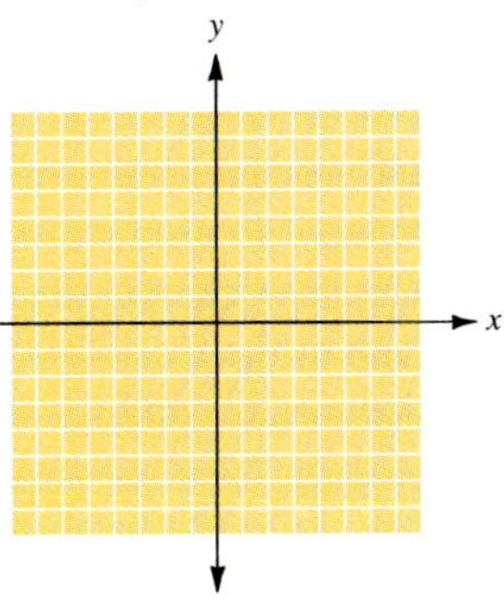

9. $2x + y < 8$
$x > 1$
$y > 2$

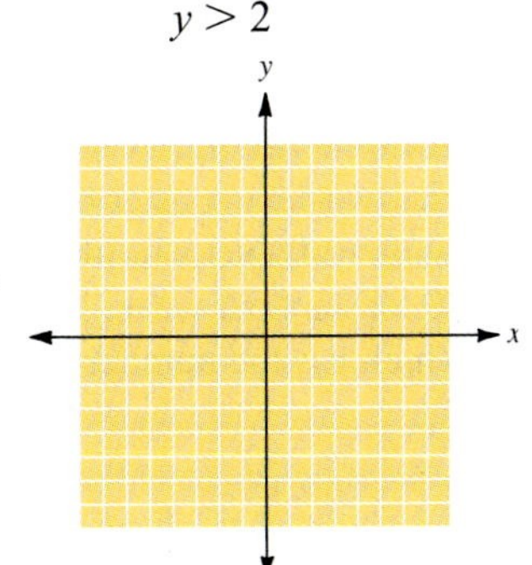

10. $3x - y \le 6$
$x \ge 1$
$y \le 3$

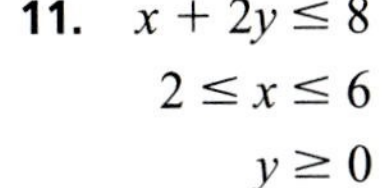

11. $x + 2y \le 8$
$2 \le x \le 6$
$y \ge 0$

12. $x + y < 6$
$0 \le y \le 3$
$x \ge 1$

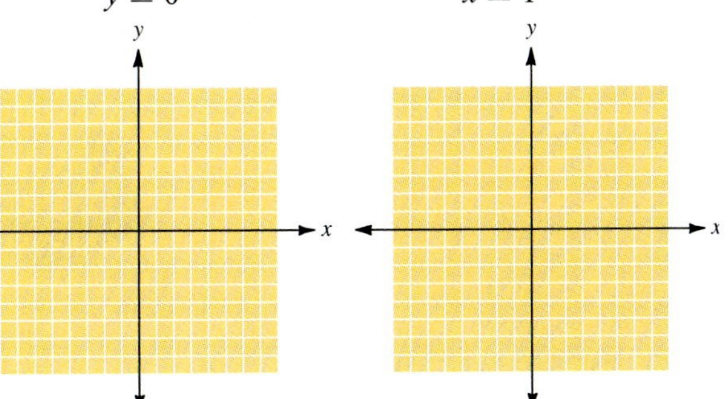

ANSWERS

1. ______________________
2. ______________________
3. ______________________
4. ______________________
5. ______________________
6. ______________________
7. ______________________
8. ______________________
9. ______________________
10. ______________________
11. ______________________
12. ______________________

ANSWERS

13. ________________

14. ________________

15. ________________

16. ________________

17. ________________

18. ________________

19. ________________

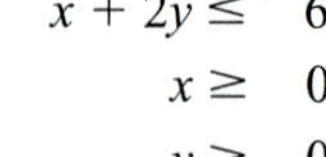
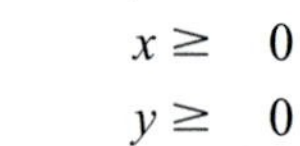

13. $3x + y \le 6$
$x + y \le 4$
$x \ge 0$
$y \ge 0$

14. $x - 2y \ge -2$
$x + 2y \le 6$
$x \ge 0$
$y \ge 0$

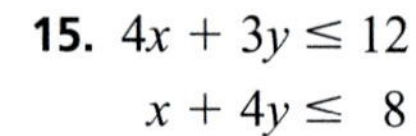

15. $4x + 3y \le 12$
$x + 4y \le 8$
$x \ge 0$
$y \ge 0$

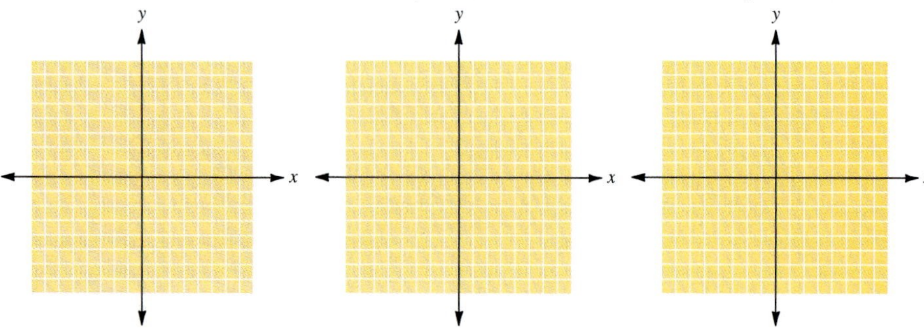

16. $2x + y \le 8$
$x + y \ge 3$
$x \ge 0$
$y \ge 0$

17. $x - 4y \le -4$
$x + 2y \le 8$
$x \ge 2$

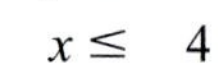

18. $x - 3y \ge -6$
$x + 2y \ge 4$
$x \le 4$

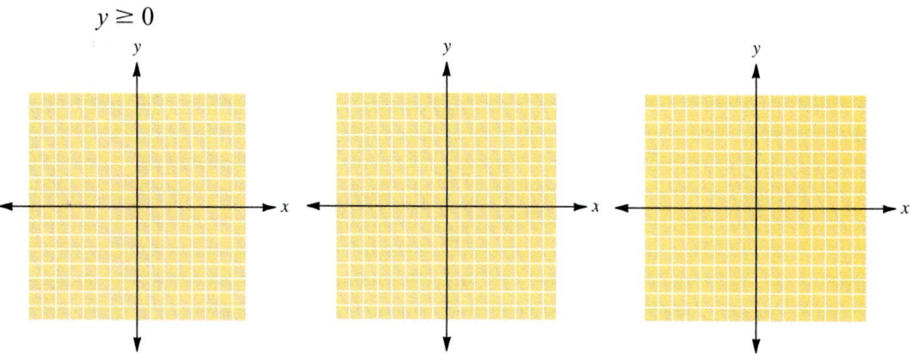

In exercises 19 and 20, draw the appropriate graph.

19. Manufacturing. A manufacturer produces both two-slice and four-slice toasters. The two slice toaster takes 6 h of labor to produce and the four-slice toaster 10 h. The labor available is limited to 300 h per week, and the total production capacity is 40 toasters per week. Draw a graph of the feasible region, given these conditions, in which x is the number of two-slice toasters and y is the number of four-slice toasters.

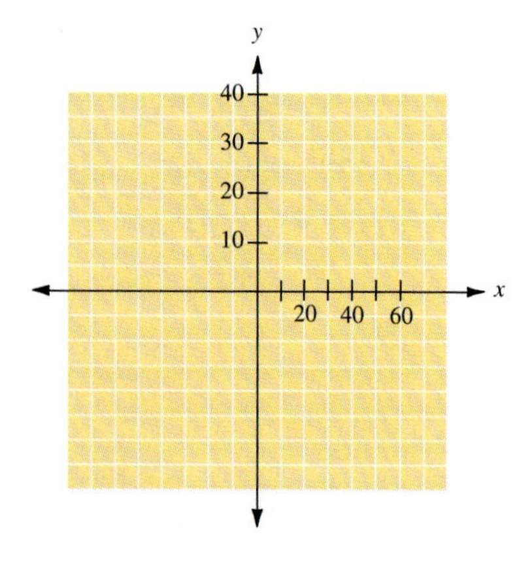

20. Production. A small firm produces both AM and AM/FM car radios. The AM radios take 15 h to produce, and the AM/FM radios take 20 h. The number of production hours is limited to 300 h per week. The plant's capacity is limited to a total of 18 radios per week, and existing orders require that at least 4 AM radios and at least 3 AM/FM radios be produced per week. Draw a graph of the feasible region, given these conditions, in which x is the number of AM radios and y the number of AM/FM radios.

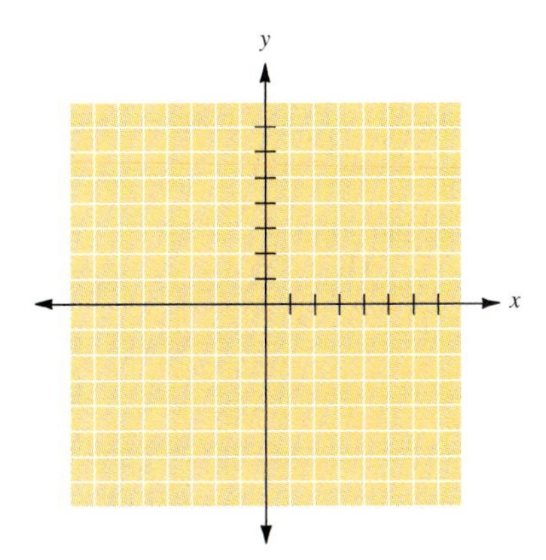

21. When you solve a system of linear inequalities, it is often easier to shade the region that is not part of the solution, rather than the region that is. Try this method, then describe its benefits.

22. Describe a system of linear inequalities for which there is no solution.

23. Write the system of inequalities whose graph is the shaded region.

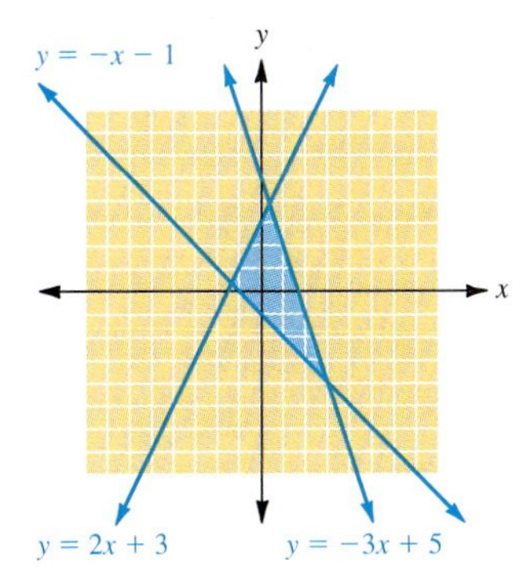

24. Write the system of inequalities whose graph is the shaded region.

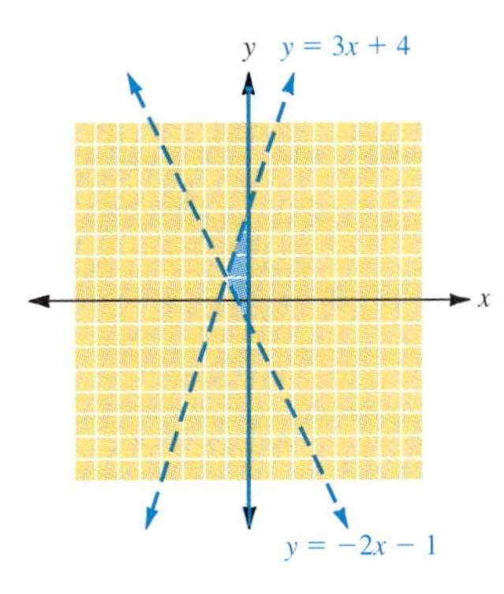

ANSWERS

20. ______

21. ______

22. ______

23. ______

24. ______

Answers

1. 

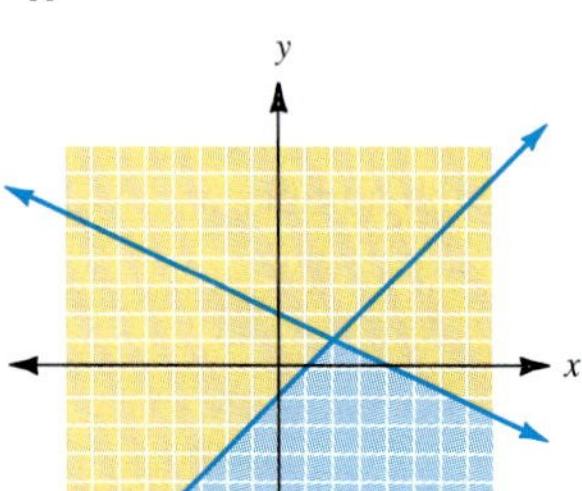

3.

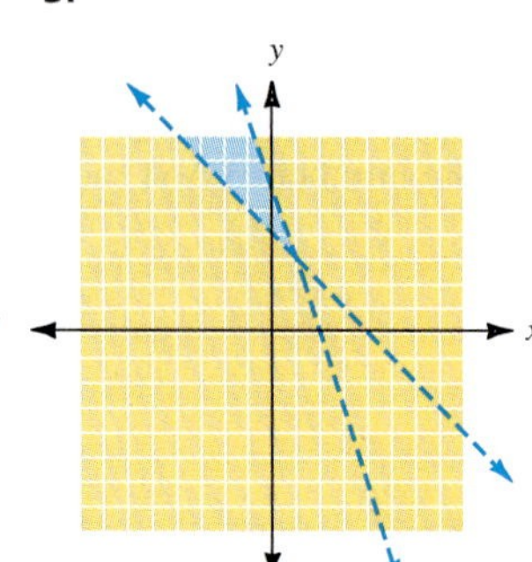

5.

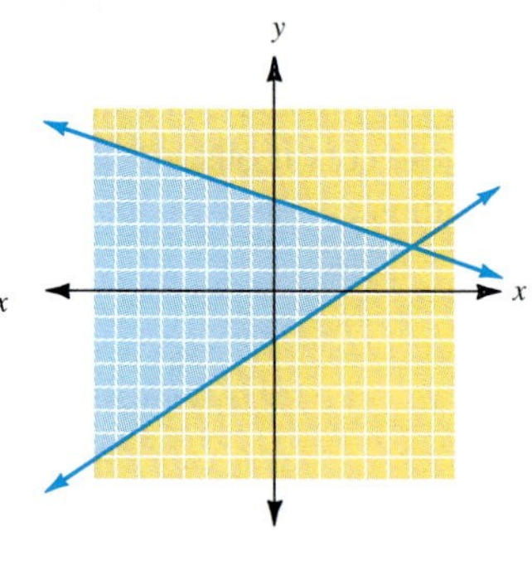

7.

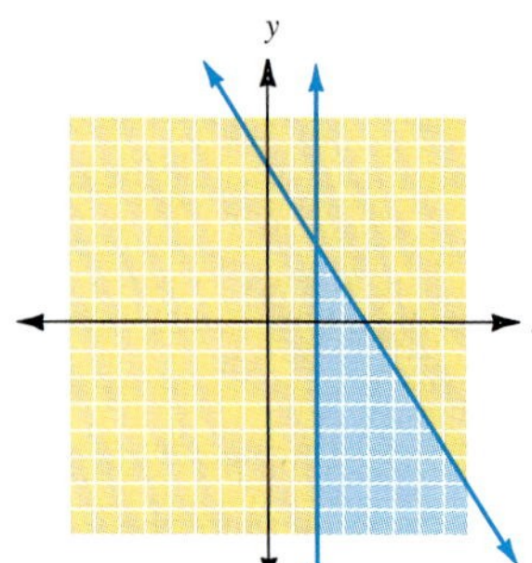

9.

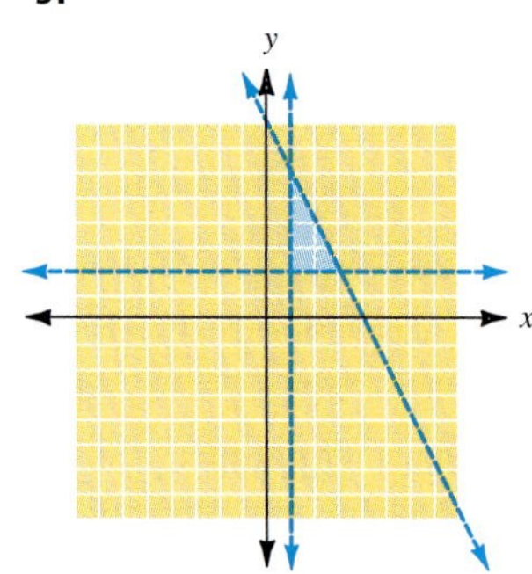

11.

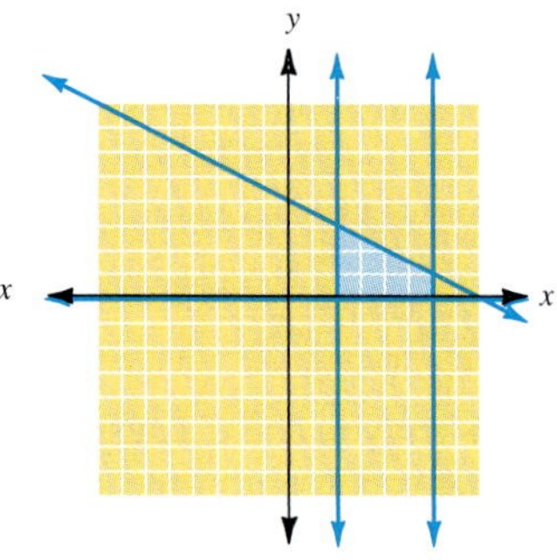

13.

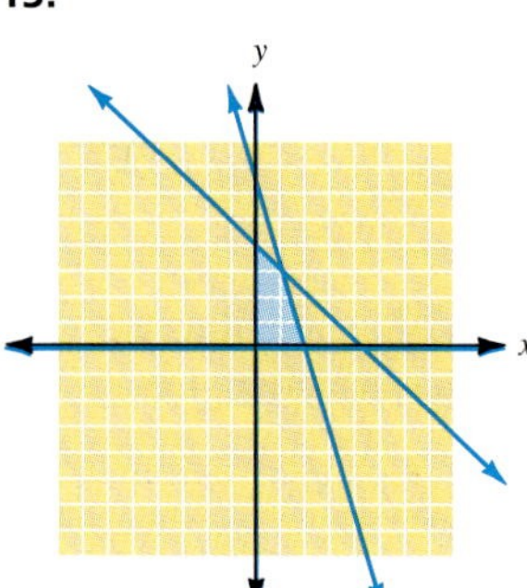

15.

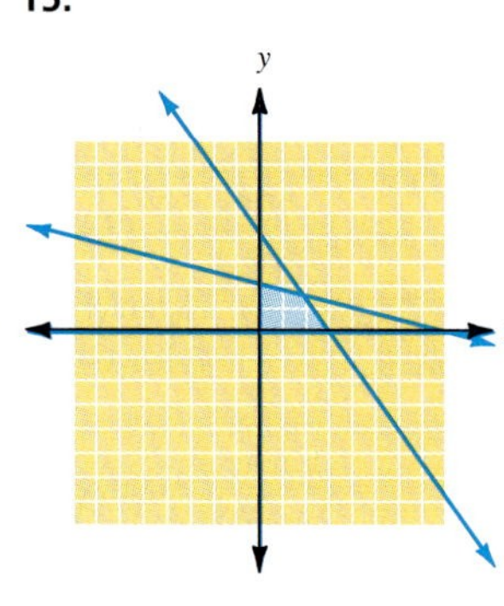

17.

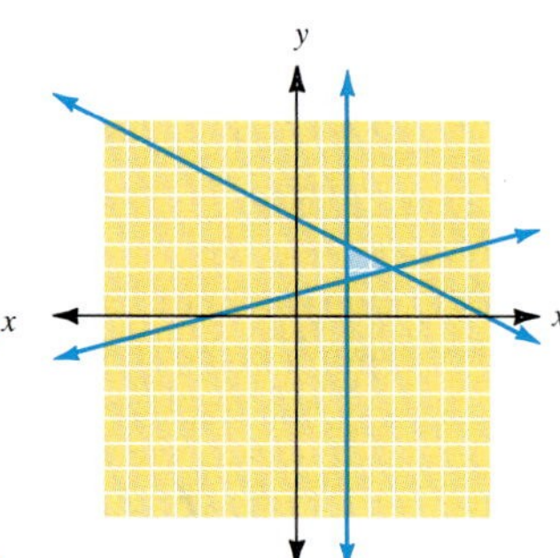

19.

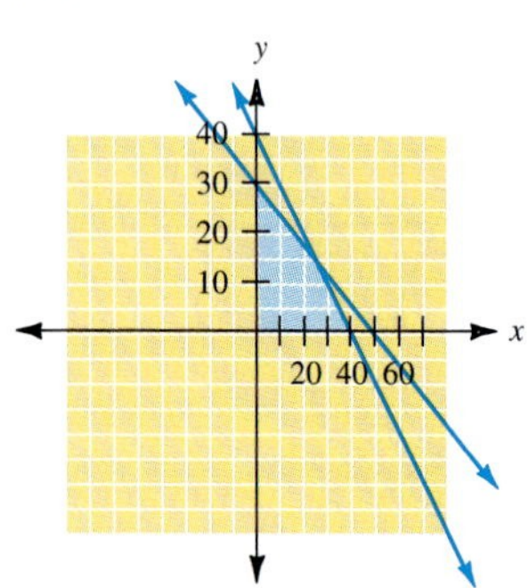

21. 

23. $y \leq 2x + 3$
$y \leq -3x + 5$
$y \geq -x - 1$

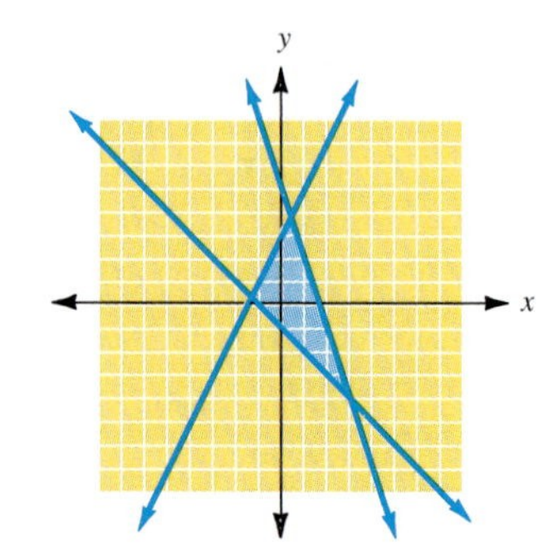

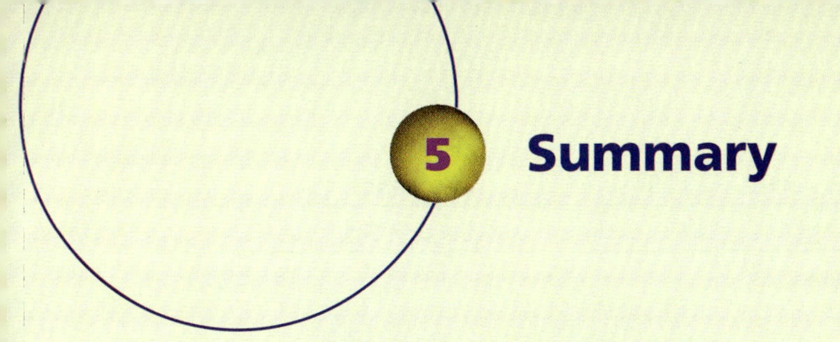

5 Summary

Definition/Procedure	Example	Reference
Systems of Linear Equations in Two Variables		**Section 5.1**
A **system of linear equations** is two or more linear equations considered together. A solution for a linear system in two variables is an ordered pair of real numbers (x, y) that satisfies both equations in the system. There are three solution techniques: the graphing method, the addition method, and the substitution method.	The solution for the system $2x - y = 7$ $x + y = 2$ is $(3, -1)$. It is the only ordered pair that will satisfy each equation.	p. 305
Solving by the Graphing Method Graph each equation of the system on the same set of coordinate axes. If a solution exists, it will correspond to the point of intersection of the two lines. Such a system is called a **consistent system.** If a solution does not exist, there is no point at which the two lines intersect. Such lines are parallel, and the system is called an **inconsistent system.** If there are an infinite number of solutions, the lines coincide. Such a system is called a **dependent system.** You may or may not be able to determine exact solutions for the system of equations with this method. Consistent system No solutions—an inconsistent system An infinite number of solutions—a dependent system	To solve $2x - y = 7$ $x + y = 2$ by graphing (3, −1) One solution—a consistent system No solutions—an inconsistent system An infinite number of solutions—a dependent system.	pp. 306–308

Continued

DEFINITION/PROCEDURE	EXAMPLE	REFERENCE
Systems of Linear Equations in Two Variables		**Section 5.1**
Solving by the Addition Method **1.** If necessary, multiply one or both of the equations by a constant so that one of the variables can be eliminated by addition. **2.** Add the equations of the equivalent system formed in step 1. **3.** Solve the equation found in step 2. **4.** Substitute the value found in step 3 into either of the equations of the original system to find the corresponding value of the remaining variable. The ordered pair formed is the solution to the system. **5.** Check the solution by substituting the pair of values found in step 4 into the other equation of the original system.	To solve $5x - 2y = 11 \quad (1)$ $2x + 3y = 12 \quad (2)$ Multiply equation (1) by 3 and equation (2) by 2. Then add to eliminate y. $19x = 57$ $x = 3$ Substituting 3 for x in equation (1), we have $15 - 2y = 11$ $y = 2$ (3, 2) is the solution.	**p. 310**
Solving by the Substitution Method **1.** If necessary, solve one of the equations of the original system for one of the variables. **2.** Substitute the expression obtained in step 1 into the other equation of the system to write an equation in a single variable. **3.** Solve the equation found in step 2. **4.** Substitute the value found in step 3 into the equation derived in step 1 to find the corresponding value of the remaining variable. The ordered pair formed is the solution for the system. **5.** Check the solution by substituting the pair of values found in step 4 in *both* equations of the original system.	To solve $3x - 2y = 6 \quad (3)$ $6x + y = 2 \quad (4)$ by substitution, solve (4) for y. $y = -6x + 2 \quad (5)$ Substituting in (3) gives $3x - 2(-6x + 2) = 6$ and $x = \frac{2}{3}$ Substituting $\frac{2}{3}$ for x in (5) gives $y = (-6)\left(\frac{2}{3}\right) + 2$ $= -4 + 2 = -2$ The solution is $\left(\frac{2}{3}, -2\right)$	**p. 313**
Applications of Systems of Linear Equations		**Section 5.2**
1. Read the problem carefully to determine the unknown quantities. **2.** Choose a variable to represent any unknown. **3.** Translate the problem to the language of algebra to form a system of equations. **4.** Solve the system of equations by any of the methods discussed, and answer the question in the original problem. **5.** Verify your solution by returning to the original problem.	Also determine the condition that relates the unknown quantities. Use a different letter for each variable. A table or a sketch often helps in writing the equations of the system.	**p. 319**

DEFINITION/PROCEDURE	EXAMPLE	REFERENCE

Systems of Linear Equations in Three Variables

Section 5.3

A solution for a linear system of three equations in three variables is an ordered triple of numbers (x, y, z) that satisfies each equation in the system.

Solving a System of Three Equations in Three Unknowns

1. Choose a pair of equations from the system, and use the addition method to eliminate one of the variables.
2. Choose a different pair of equations, and eliminate the same variable.
3. Solve the system of two equations in two variables determined in steps 1 and 2.
4. Substitute the values found above into one of the original equations, and solve for the remaining variable.
5. The solution is the ordered triple of values found in steps 3 and 4. It can be checked by substituting into the other equations of the original system.

To solve

$$x + y - z = 6 \quad (6)$$
$$2x - 3y + z = -9 \quad (7)$$
$$3x + y + 2z = 2 \quad (8)$$

Adding (6) and (7) gives

$$3x - 2y = -3 \quad (9)$$

Multiplying (6) by 2 and adding the result to (8) gives

$$5x + 3y = 14 \quad (10)$$

The system consisting of (9) and (10) is solved as before and

$$x = 1 \qquad y = 3$$

Substituting these values into (6) gives

$$z = -2$$

The solution is $(1, 3, -2)$.

p. 336

Graphing Linear Inequalities in Two Variables

Section 5.4

In general, the solution set of an inequality of the form

$$ax + by < c \qquad \text{or} \qquad ax + by > c$$

will be a **half plane** either above or below the **boundary line** determined by

$$ax + by = c$$

The boundary line is included in the graph if equality is included in the statement of the original inequality. Such a line is solid. The boundary line is dashed if it is not included in the graph.

To graph a linear inequality:

1. Replace the inequality symbol with an equality symbol to form the equation of the boundary line of the solution set.
2. Graph the boundary line. Use a dashed line if equality is not included ($<$ or $>$). Use a solid line if equality is included ($\le$ or $\ge$).
3. Choose any convenient test point *not* on the boundary line.
4. If the inequality is *true* for the test point, shade the half plane *including* the test point. If the inequality is *false* for the test point, shade the half plane *not including* the test point.

To graph

$$x - 2y < 4$$

p. 345

Graphing Systems of Linear Inequalities in Two Variables

Section 5.5

A **system of linear inequalities** is two or more linear inequalities considered together. The **graph of the solution set** of a system of linear inequalities is the intersection of the graphs of the individual inequalities.

To solve

$$x + 2y \le 8$$
$$x + y \le 6$$
$$x \ge 0$$
$$y \ge 0$$

Continued

DEFINITION/PROCEDURE	EXAMPLE	REFERENCE
Graphing Systems of Linear Inequalities in Two Variables		**Section 5.5**
Solving Systems of Linear Inequalities by Graphing **1.** Graph each inequality, shading the appropriate half plane, on the same set of coordinate axes. **2.** The graph of the system is the intersection of the regions shaded in step 1.	by graphing 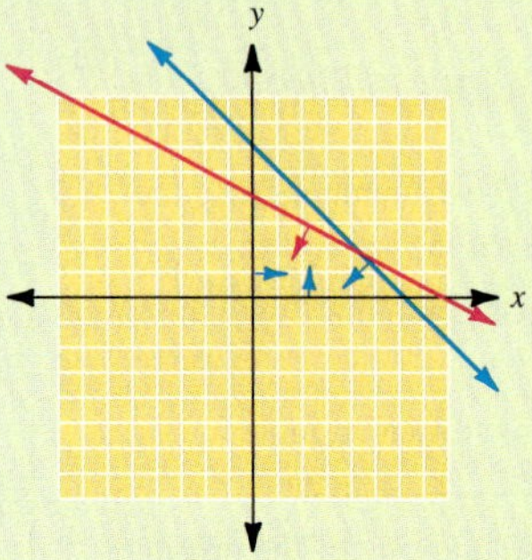	**pp. 355–356**

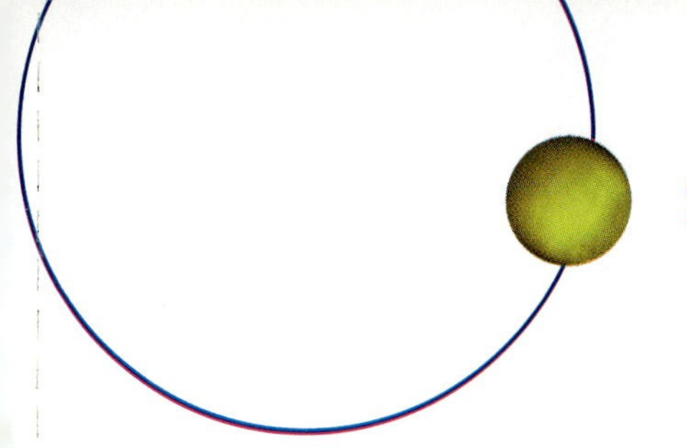

Summary Exercises

This summary exercise set will give you practice with each of the objectives in the chapter. The answers are provided in the *Instructor's Manual*.

[5.1] Solve each of the following systems by graphing.

1. $x + y = 8$
$x - y = 4$

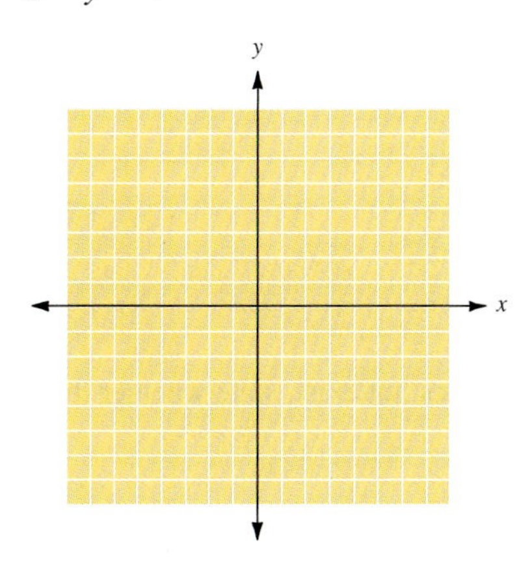

2. $x + 2y = 8$
$x - y = 5$

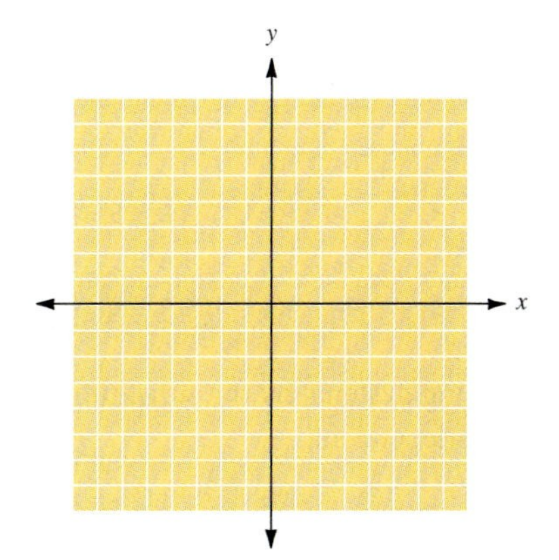

3. $2x + 3y = 12$
$2x + y = 8$

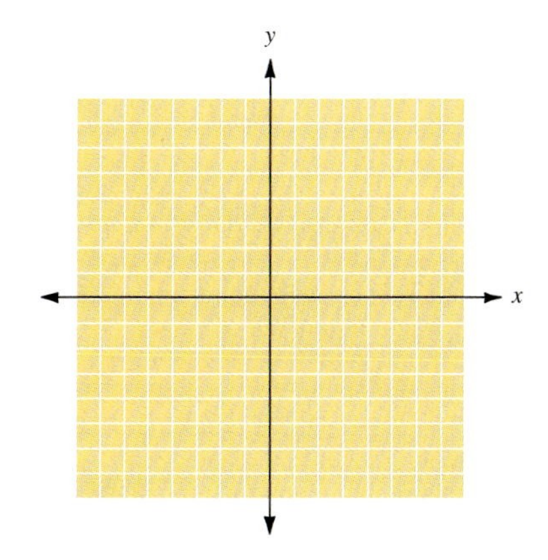

4. $x + 4y = 8$
$y = 1$

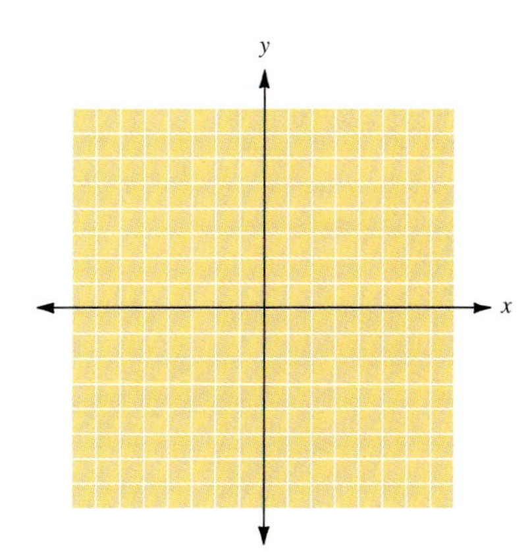

Solve each of the following systems by the addition method. If a unique solution does not exist, state whether the given system is inconsistent or dependent.

5. $x + 2y = 7$
$x - y = 1$

6. $x + 3y = 14$
$4x + 3y = 29$

7. $3x - 5y = 5$
$-x + y = -1$

8. $x - 4y = 12$
$2x - 8y = 24$

9. $6x + 5y = -9$
$-5x + 4y = 32$

10. $3x + y = -17$
$5x - 3y = -19$

11. $3x + y = 8$
$-6x - 2y = -10$

12. $5x - y = -17$
$4x + 3y = -6$

13. $7x - 4y = 27$
$5x + 6y = 6$

14. $4x - 3y = 1$
$6x + 5y = 30$

15. $x - \frac{1}{2}y = 8$
$\frac{2}{3}x + \frac{3}{2}y = -2$

16. $\frac{1}{5}x - 2y = 4$
$\frac{3}{5}x + \frac{2}{3}y = -8$

Solve each of the following systems by the substitution method. If a unique solution does not exist, state whether the given system is inconsistent or dependent.

17. $2x + y = 23$
$x = y + 4$

18. $x - 5y = 26$
$y = x - 10$

19. $3x + y = 7$
$y = -3x + 5$

20. $2x - 3y = 13$
$x = 3y + 9$

21. $5x - 3y = 13$
$x - y = 3$

22. $4x - 3y = 6$
$x + y = 12$

23. $3x - 2y = -12$
$6x + y = 1$

24. $x - 4y = 8$
$-2x + 8y = -16$

[5.2] Solve each of the following problems by choosing a variable to represent each unknown quantity. Then, write a system of equations that will allow you to solve for each variable.

25. **Number problem.** One number is 2 more than 3 times another. If the sum of the two numbers is 30, find the two numbers.

26. **Money value.** Suppose that a cashier has 78 \$5 and \$10 dollar bills with a value of \$640. How many of each type of bill does she have?

27. Ticket sales. Tickets for a basketball game sold at $7 for an adult ticket and $4.50 for a student ticket. If the revenue from 1200 tickets was $7400, how many of each type of ticket were sold?

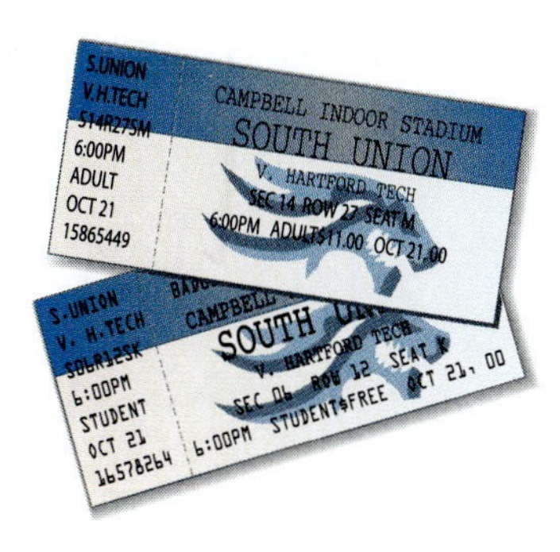

28. Purchase price. A purchase of eight blank cassette tapes and four blank videotapes costs $36. A second purchase of four cassette tapes and five videotapes costs $30. What is the price of a single cassette tape and of a single videotape?

29. Rectangles. The length of a rectangle is 4 cm less than twice its width. If the perimeter of the rectangle is 64 cm, find the dimensions of the rectangle.

30. Mixture. A grocer in charge of bulk foods wishes to combine peanuts selling for $2.25 per pound and cashews selling for $6 per pound. What amount of each nut should be used to form a 120-lb mixture selling for $3 per pound?

31. Investments. Reggie has two investments totaling $17,000—one a savings account paying 6%, the other a time deposit paying 8%. If his annual interest is $1200, what does he have invested in each account?

32. Mixture. A pharmacist mixes a 20% alcohol solution and a 50% alcohol solution to form 600 mL of a 40% solution. How much of each solution should she use in forming the mixture?

33. Motion. A jet flying east, with the wind, makes a trip of 2200 mi in 4 h. Returning, against the wind, the jet can travel only 1800 mi in 4 h. What is the plane's rate in still air? What is the rate of the wind?

34. Number problem. The sum of the digits of a two-digit number is 9. If the digits are reversed, the new number is 45 more than the original number. What is the original number?

35. Work. A manufacturer produces CD-Roms and $3\frac{1}{2}$-in. drives. The CD-Roms require 20 min of component assembly time; the $3\frac{1}{2}$-in. drives, 25 min. The manufacturer has 500 min of component assembly time available per day. Each drive requires 30 min for packaging and testing, and 690 min of that time is available per day. How many of each of the drives should be produced daily to use all the available time?

36. Equilibrium price. If the demand equation for a product is $D = 270 - 5p$ and the supply equation is $S = 13p$, find the equilibrium point.

37. Rental charges. Two car rental agencies have the following rates for the rental of a compact automobile:

Company A: $18 per day plus 22¢ per mile.

Company B: $22 per day plus 18¢ per mile.

For a 3-day rental, at what number of miles will the charges from the two companies be the same?

[5.3] Solve each of the following systems by the addition method. If a unique solution does not exist, state whether the given system is inconsistent or dependent.

38.
$$\begin{aligned} x - y + z &= 0 \\ x + 4y - z &= 14 \\ x + y - z &= 6 \end{aligned}$$

39.
$$\begin{aligned} x - y + z &= 3 \\ 3x + y + 2z &= 15 \\ 2x - y + 2z &= 7 \end{aligned}$$

40.
$$\begin{aligned} x - y - z &= 2 \\ -2x + 2y + z &= -5 \\ -3x + 3y + z &= -10 \end{aligned}$$

41.
$$\begin{aligned} x - y \quad &= 3 \\ 2y + z &= 5 \\ x \quad + 2z &= 7 \end{aligned}$$

42.
$$\begin{aligned} x + y + z &= 2 \\ x + 3y - 2z &= 13 \\ y - 2z &= 7 \end{aligned}$$

43.
$$\begin{aligned} x + y - z &= -1 \\ x - y + 2z &= 2 \\ -5x - y - z &= -1 \end{aligned}$$

44.
$$\begin{aligned} 2x + 3y + z &= 7 \\ -2x - 9y + 2z &= 1 \\ 4x - 6y + 3z &= 10 \end{aligned}$$

Solve each of the following problems by choosing a variable to represent each unknown quantity. Then, write a system of equations that will allow you to solve for each variable.

45. Number problem. The sum of three numbers is 15. The largest number is 4 times the smallest number, and it is also 1 more than the sum of the other two numbers. Find the three numbers.

46. Number problem. The sum of the digits of a three-digit number is 16. The tens digit is 3 times the hundreds digit, and the units digit is 1 more than the hundreds digit. What is the number?

47. Tickets sold. A theater has orchestra tickets at $10, box-seat tickets at $7, and balcony tickets at $5. For one performance, a total of 360 tickets were sold, and the total revenue was $3040. If the number of orchestra tickets sold was 40 more than that of the other two types combined, how many of each type of ticket were sold for the performance?

48. Triangles. The measure of the largest angle of a triangle is 15° less than 4 times the measure of the smallest angle and 30° more than the sum of the measures of the other two angles. Find the measures of the three angles of the triangle.

49. Investments. Rachel divided $12,000 into three investments: a savings account paying 5%, a stock paying 7%, and a mutual fund paying 9%. Her annual interest from the investments was $800, and the amount that she had invested at 5% was equal to the sum of the amounts invested in the other accounts. How much did she have invested in each type of account?

50. Number problem. The difference of two positive numbers is 3, and the sum of those numbers is 41. Find the two numbers.

51. Number problem. The sum of two integers is 144, and the difference 42. What are the two integers?

52. Rectangle. A rectangular building lot is $1\frac{1}{2}$ times as wide as it is long. The perimeter of the lot is 400 ft. Find the length and width of the lot.

53. Break-even analysis. A manufacturer's cost for producing x units of a product is given by

$$C = 10x + 3600$$

The revenue from selling x units of that product is given by

$$R = 100x$$

Find the break-even point for this product.

[5.4] Graph the solution set for each of the following linear inequalities.

54. $y < 2x + 1$

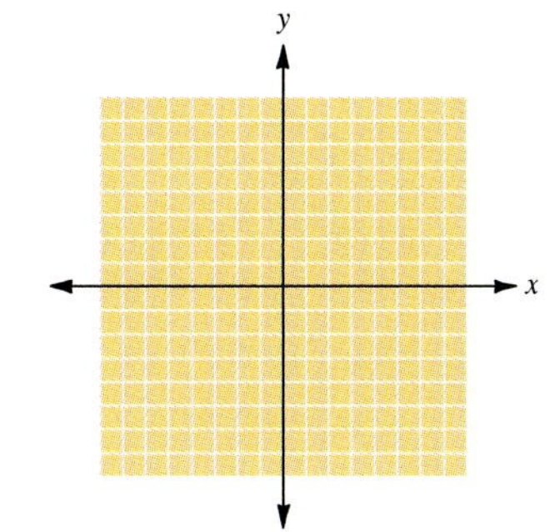

55. $y \geq -2x + 3$

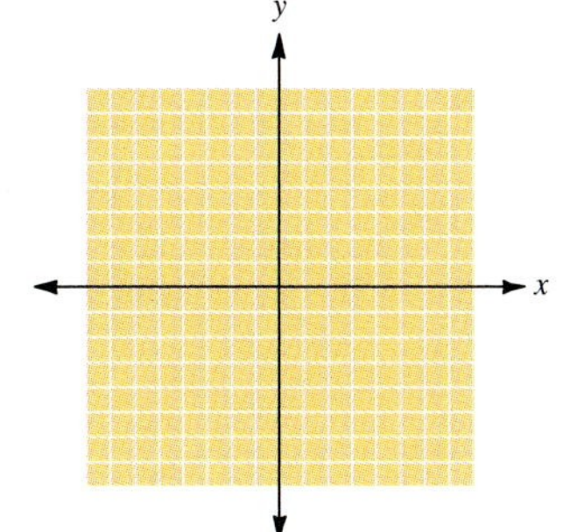

56. $3x + 2y \geq 6$

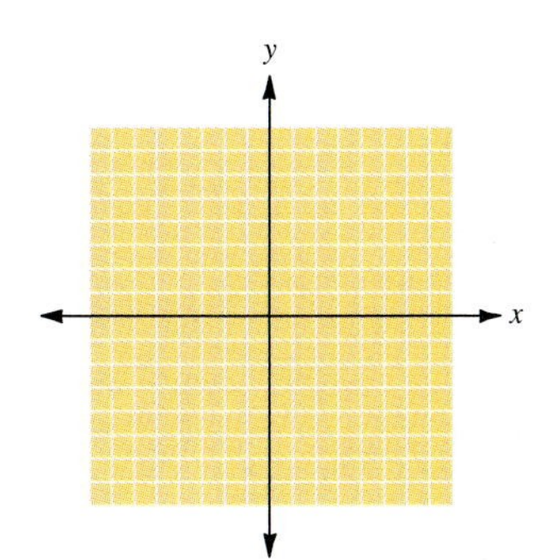

57. $3x - 5y < 15$

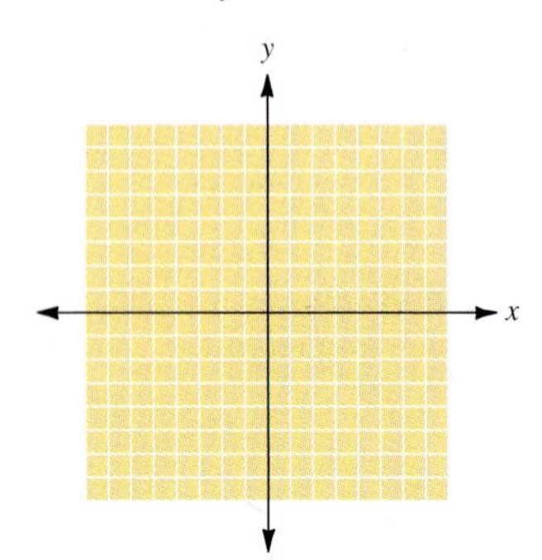

58. $y < -2x$

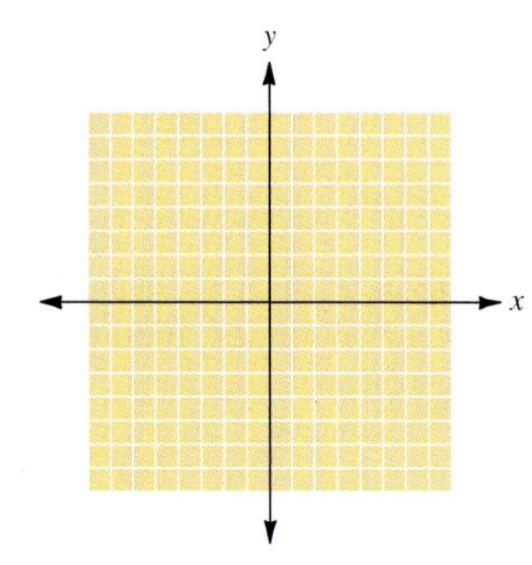

59. $4x - y \geq 0$

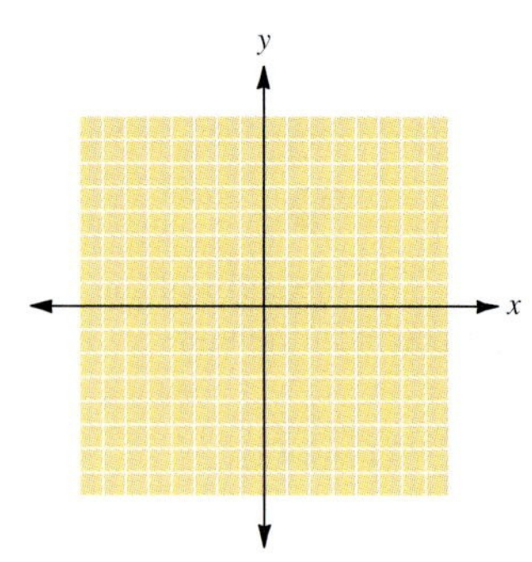

60. $y \geq -3$

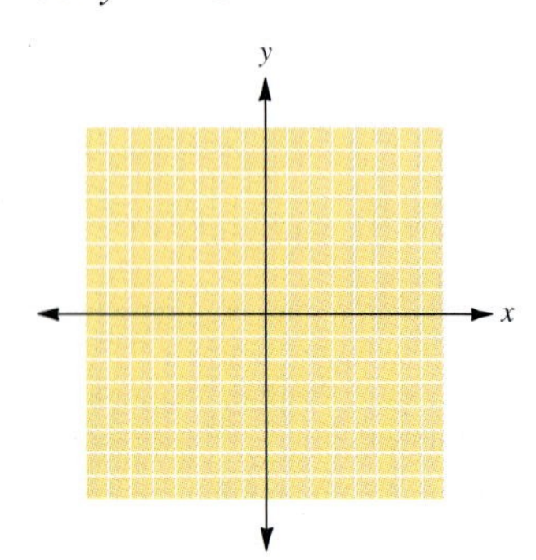

61. $x < 4$

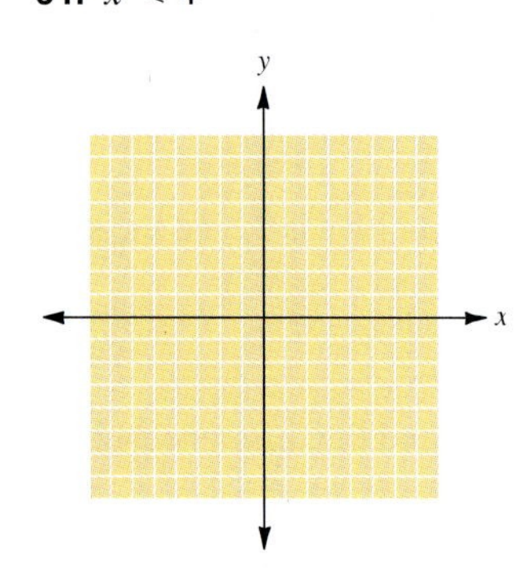

[5.5] Solve each of the following systems of linear inequalities graphically.

62. $x - y < 7$
$x + y > 3$

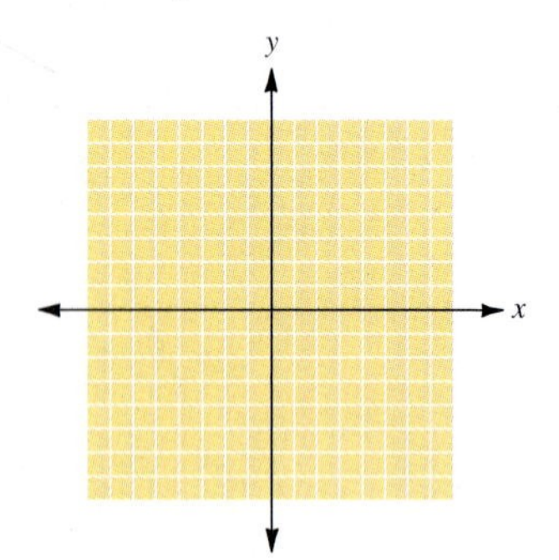

63. $x - 2y \leq -2$
$x + 2y \leq 6$

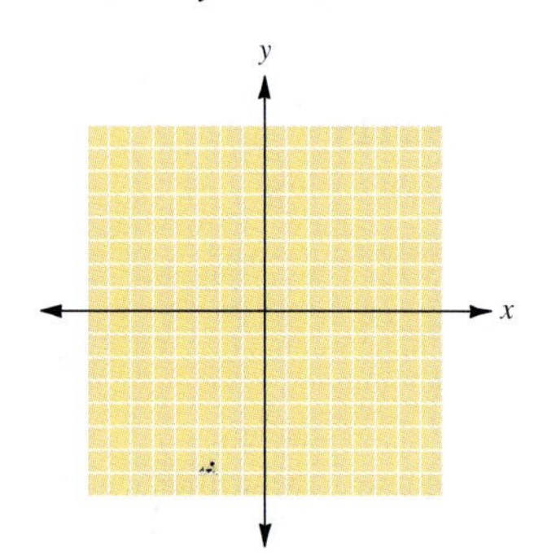

64. $x - 6y < 6$
$-x + y < 4$

65. $2x + y \leq 8$
$x \geq 1$
$y \geq 0$

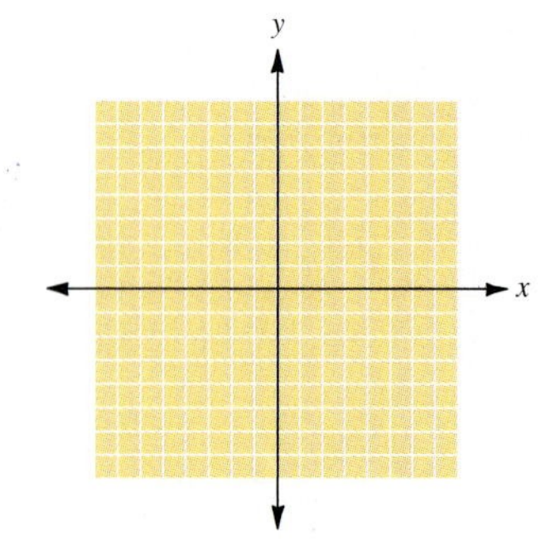

66. $2x + y \leq 6$
$x \geq 1$
$y \geq 0$

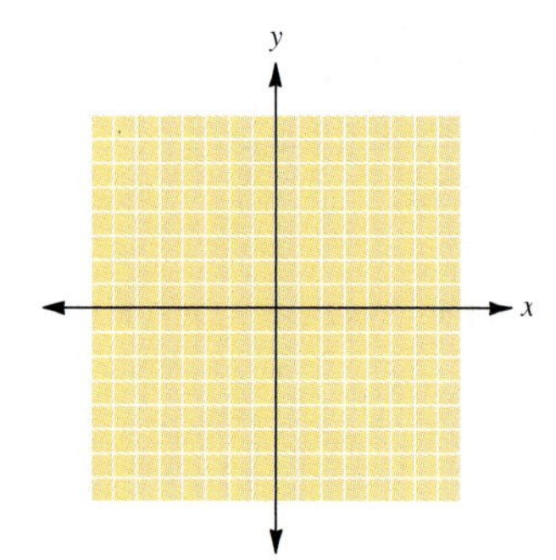

67. $4x + y \leq 8$
$x \geq 0$
$y \geq 2$

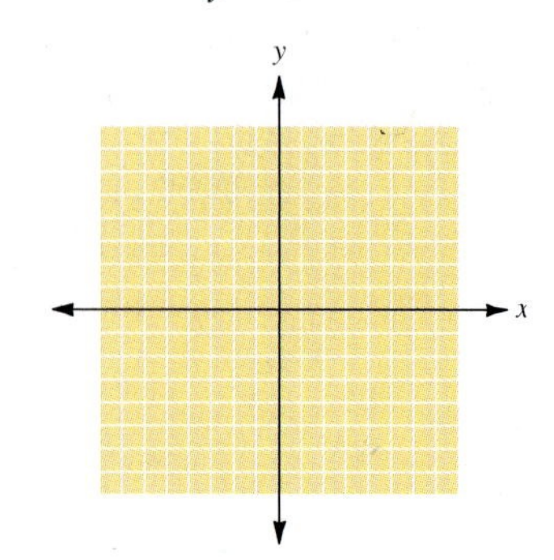

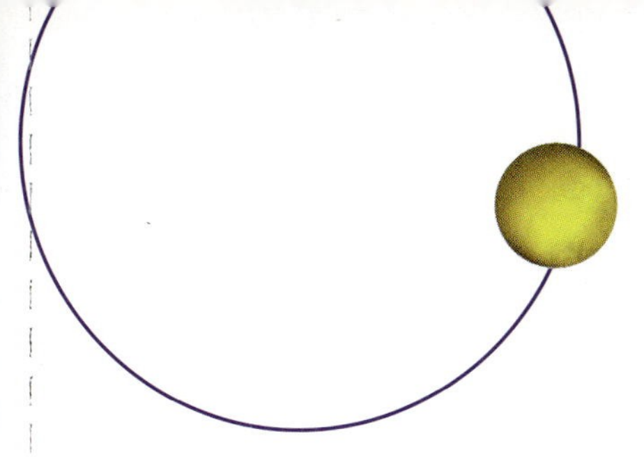

Self-Test for Chapter 5

Name ______________________

Section ________ Date ________

ANSWERS

1. ______________
2. ______________
3. ______________
4. ______________
5. ______________
6. ______________
7. ______________
8. ______________
9. ______________
10. ______________
11. ______________
12. ______________

The purpose of this self-test is to help you check your progress and to review for a chapter test in class. Allow yourself about 1 hour to take the test. When you are done, check your answers in the back of the book. If you missed any answers, be sure to go back and review the appropriate sections in the chapter and the exercises that are provided.

Solve each of the following systems. If a unique solution does not exist, state whether the given system is inconsistent or dependent.

1. $\begin{aligned} 3x + y &= -5 \\ 5x - 2y &= -23 \end{aligned}$

2. $\begin{aligned} 4x - 2y &= -10 \\ y &= 2x + 5 \end{aligned}$

3. $\begin{aligned} 9x - 3y &= 4 \\ -3x + y &= -1 \end{aligned}$

4. $\begin{aligned} 5x - 3y &= 5 \\ 3x + 2y &= -16 \end{aligned}$

5. $\begin{aligned} x - 2y &= 5 \\ 2x + 5y &= 10 \end{aligned}$

6. $\begin{aligned} 5x - 3y &= 20 \\ 4x + 9y &= -3 \end{aligned}$

Solve each of the following systems.

7. $\begin{aligned} x - y + z &= 1 \\ -2x + y + z &= 8 \\ x + 5z &= 19 \end{aligned}$

8. $\begin{aligned} x + 3y - 2z &= -6 \\ 3x - y + 2z &= 8 \\ -2x + 3y - 4z &= -11 \end{aligned}$

Solve each of the following problems by choosing a variable to represent each unknown quantity. Then write a system of equations that will allow you to solve for each variable.

9. An order for 30 computer disks and 12 printer ribbons totaled \$147. A second order for 12 more disks and 6 additional ribbons cost \$66. What was the cost per individual disk and ribbon?

10. A candy dealer wants to combine jawbreakers selling for \$2.40 per pound and licorice selling for \$3.90 per pound to form a 100-lb mixture that will sell for \$3 per pound. What amount of each type of candy should be used?

11. A small electronics firm assembles 5-in. portable television sets and 12-in. models. The 5-in. set requires 9 h of assembly time; the 12-in. set, 6 h. Each unit requires 5 h for packaging and testing. If 72 h of assembly time and 50 h of packaging and testing time are available per week, how many of each type of set should be finished if the firm wishes to use all its available capacity?

12. Hans decided to divide \$14,000 into three investments: a savings account paying 6% annual interest, a bond paying 9%, and a mutual fund paying 13%. His annual interest from the three investments was \$1100, and he had twice as much invested in the bond as in the mutual fund. What amount did he invest in each type?

ANSWERS

13. ______

14. ______

15. ______

16. ______

17. ______

18. ______

19. ______

20. ______

21. ______

22. ______

13. To fence around a rectangular yard requires 260 ft of fencing. The length of the yard is 20 ft less than twice the width. Find the dimensions of the yard.

Graph the solution set in each of the following.

14. $5x + 6y \leq 30$

15. $x + 3y > 6$

16. $4x - 8 \leq 0$

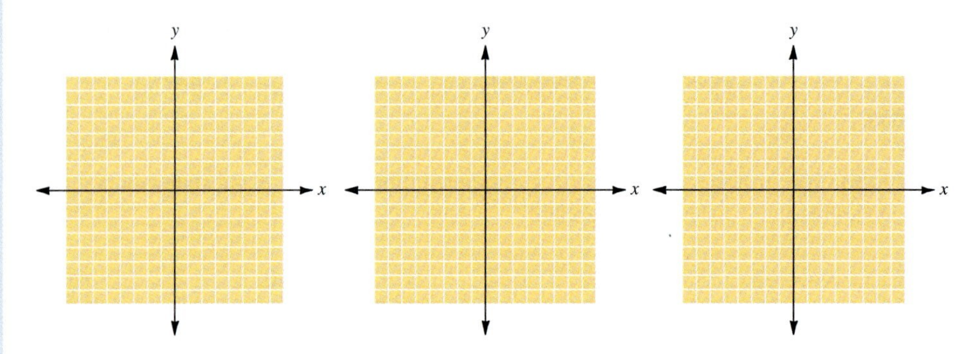

17. $2y + 4 > 0$

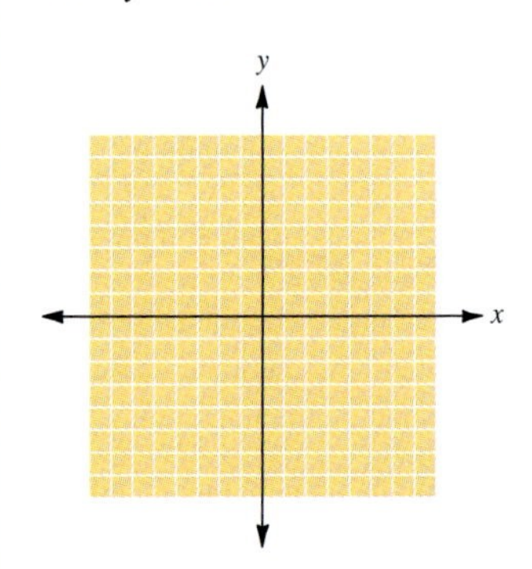

Solve each of the following systems of linear inequalities by graphing.

18. $x - 2y < 6$
$x + y < 3$

19. $3x + 4y \geq 12$
$x \geq 1$

20. $x + 2y \leq 8$
$x + y \leq 6$
$x \geq 0$
$y \geq 0$

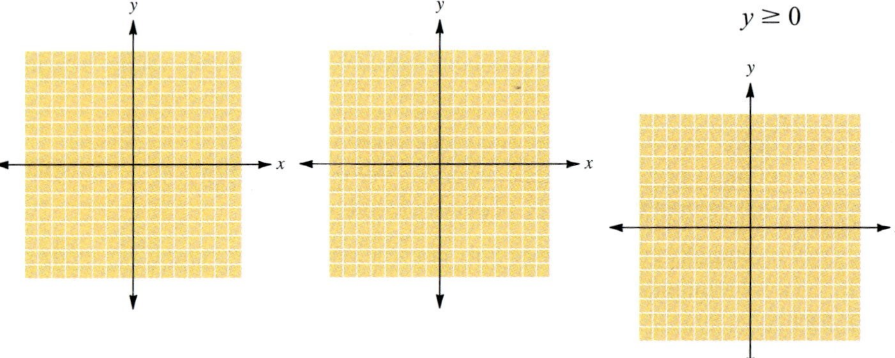

Solve each of the systems.

21. $6x - 3y = -1$
$-3x + 9y = 13$

22. $2x - z = 9$
$x - y = 7$
$-3x + 2z = -13$

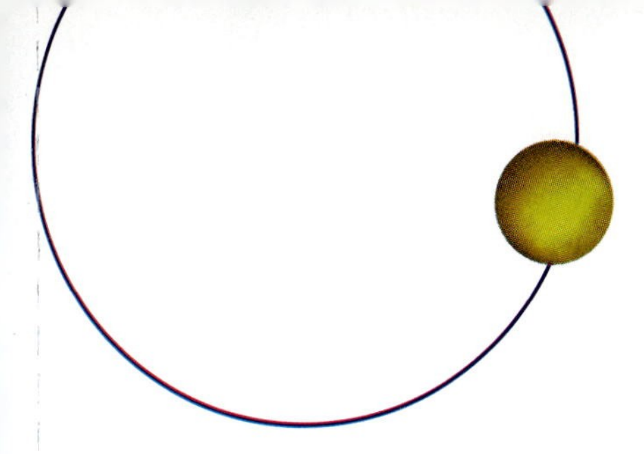

Cumulative Test for Chapters 1 to 5

Name ______________________

Section __________ Date __________

ANSWERS

1. ______________________
2. ______________________
3. ______________________
4. ______________________
5. ______________________
6. ______________________
7. ______________________
8. ______________________
9. ______________________
10. ______________________
11. ______________________
12. ______________________
13. ______________________
14. ______________________

This test is provided to help you in the process of reviewing the previous chapters. Answers are provided in the back of the book. If you missed any answers, be sure to go back and review the appropriate chapter sections.

1. Simplify the expression $(x^9y^3)^2$.

2. Perform the indicated operations: $56 \cdot 2 \div 4 - (-4^3 + 5 \cdot 6)$

3. If $f(x) = -4x^3 + 3x^2 - 5$, evaluate $f(-2)$.

Solve each equation.

4. $3x - 2(x - 5) + 8 = -3(4 - x)$

5. $|3x - 5| = 10$

6. $\frac{1}{R} = \frac{1}{R_1} + \frac{1}{R_2}$ for R

Solve the following inequalities.

7. $3x + 5 \le 8$

8. $2x - 9 > 4x - 5$

9. $|2x - 3| < 9$

10. $|x - 5| > 8$

Write the equation of the line L that satisfies the given conditions.

11. L has y intercept $(0, -3)$ and slope of 2

12. L passes through $(1, 3)$ and $(-2, 1)$

13. L has y intercept of $(0, -2)$ and is perpendicular to the line with equation $4x - 5y = 20$

In each of the following, determine which relations are functions.

14. $\{(1, 2), (-1, 2), (3, 4), (5, 6)\}$

ANSWERS

15. ______

16. ______

17. ______

18. ______

19. ______

20. ______

21. ______

22. ______

23. ______

24. ______

25. ______

15.

x	y
-3	0
-2	1
-1	5
-1	3

Use the vertical line test to determine whether each of the following graphs represents a function.

16.

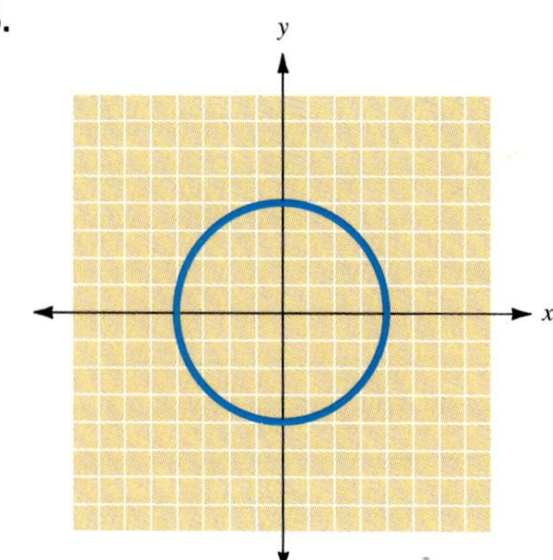

17.

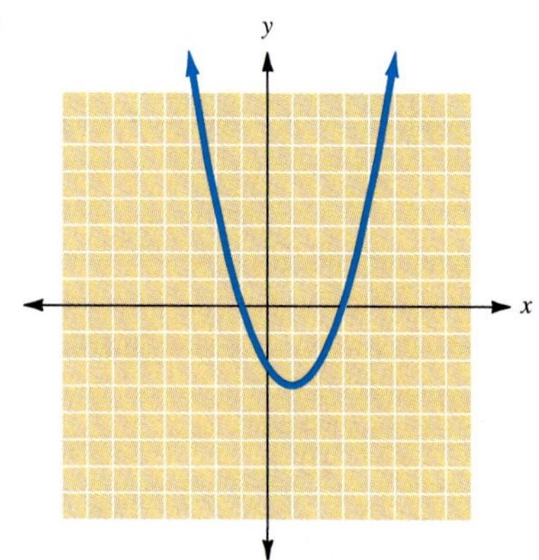

Solve each of the following systems. If a unique solution does not exist, state whether the given system is inconsistent or dependent.

18. $\begin{aligned} x + 3y &= 12 \\ 2x - 3y &= 6 \end{aligned}$

19. $\begin{aligned} 7x + 3y &= -51 \\ y &= 2x + 9 \end{aligned}$

20. $\begin{aligned} 4x - 12y &= 7 \\ -x + 3y &= 9 \end{aligned}$

21. $\begin{aligned} 8x - 4y &= 32 \\ -2x + y &= -8 \end{aligned}$

22. $\begin{aligned} 7x - 2y &= -17 \\ x + 4y &= 4 \end{aligned}$

23. $\begin{aligned} x + y + 2z &= 9 \\ -x - z &= -1 \\ 2x - 3y &= -17 \end{aligned}$

Solve each of the following problems by choosing a variable to represent each unknown quantity. Then write a system of equations that will allow you to solve for each variable.

24. An order for 3 three-ring binders and 10 packages of paper totaled \$34. A second order for 6 binders and 4 packages of paper cost \$28. What were the individual costs of a binder and a package of paper?

25. Josepha divided her \$10,000 inheritance into two investments: a bond paying 10% and a time deposit paying 7%. If her annual interest from the two investments was \$910, what amount did she have invested in each type?

6

Polynomials and Polynomial Functions

INTRODUCTION

When developing aircraft, autos, and boats, engineers use computer design programs. To ensure that the smooth, curved shapes created by the design fit together when manufactured, engineers use "polynomial splines." These splines are also useful for civil engineers who design tunnels and highways. These splines help in the design of a roadway, ensuring that changes in direction and altitude occur smoothly and gradually. In some cases, roads make transitions from, say, a valley floor to a mountain pass while covering a distance of only a few kilometers. For example: a road passes through a valley at 50 meters altitude and then climbs through some hills, reaching an altitude of 350 meters before descending again. The following graph shows the change in altitude for 19 km of the roadway.

The road seems very steep in this graph, but remember that the y axis is the altitude measured in meters, and the x axis is horizontal distance measured in kilometers. To get a true feeling for the vertical change, the horizontal axis would have to be stretched by a factor of 1070.

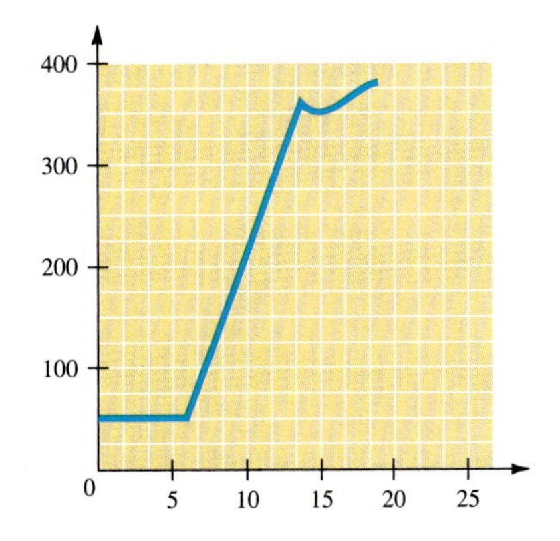

Based on measurements of the distance, altitude, and change in the slope taken at intervals along the planned path of the road, formulas are developed that model the roadway for short sections. The formulas are then pieced together to form a model of the road over several kilometers. Such formulas are found by using algebraic methods to solve systems of linear equations. The resulting splines can be linear, quadratic, or even cubic equations. Here are some of the splines that could be fit together to create the roadway needed in our graph; y is the altitude measured in meters and x is the horizontal distance, measured in kilometers.

First 6 km: $y = 50$

Next 7.5 km: $y = 40x - 190$

Final 5.5 km: $y = -0.8x^3 + 41x^2 - 690x + 4176$

By entering these equations in your calculator, you can see part of the design for the road. Be certain that you adjust your graphing window appropriately.

Polynomials are useful in many areas. In this chapter, you will learn how to solve problems involving polynomials.

6.1 Introduction to Polynomials and Polynomial Functions

6.1 OBJECTIVES

1. Identify like terms
2. Find the degree of a polynomial
3. Find an ordered pair associated with a given polynomial function

In Chapter 4, we looked at a class of functions called *linear functions*. In this section, we examine polynomial functions. We begin by defining some important words.

Definitions: Term

A **term** is a number or the product of a number and one or more variables, raised to a power.

Example 1

Identifying Terms

Which of the following are terms?

$5x^3$ $\quad 2x^2 + 3x$ $\quad 4xy$

$5x^3$ and $4xy$ are terms. $2x^2 + 3x$ is not a term; it is the sum of two terms.

CHECK YOURSELF 1

Which of the following are terms?

(a) $5xy$ **(b)** $4x^3 - 2y$ **(c)** $2x^3y^2$ **(d)** x^7

If terms contain exactly the same variables raised to the *same powers,* they are called **like terms.** Examples include $6s$ and $7s$, $4x^2$ and $9x^2$, and $7xy^2z^3$ and $10xy^2z^3$. The following are *not* like terms

$6s$ and $7t$ — Different variables

$4x^2$ and $9x^3$ — Different exponents

$7x^2y^3z^3$ and $10xy^2z^3$ — Different exponents

Example 2

Identifying Like Terms

For each of the following pairs of terms, decide whether they are like terms.

(a) $5x^3$ and $5x^2$

Not like terms—different exponents

(b) $3xy$ and $2xy$

Like terms

(c) $4xy^2z$ and $9xy^2z$

Like terms

(d) $2xy^2$ and $7x^2y$

Not like terms—different exponents

(e) $3x^2y$ and $2x^2z$

Not like terms—different variables

CHECK YOURSELF 2

For each of the following pairs of terms, decide whether they are like terms.

(a) $2ab^2c$ and $3ab^2d$ **(b)** $4xy^3$ and $\frac{1}{2}xy^3$

(c) $5x^2y^3z^2$ and $7x^3y^2z^2$ **(d)** $3x$ and $4xy$

A **polynomial** consists of one or more terms in which the only allowable exponents are the whole numbers, 0, 1, 2, 3, . . . and so on. The terms are connected by addition or subtraction signs.

Certain polynomials are given special names according to the number of terms that they have.

NOTE The prefix "mono" means one.

A polynomial with one term is called a **monomial.** For example,

$7x^3 \qquad 2x^2y^3 \qquad 4xy \qquad -12$

are all polynomials. But,

$\frac{9}{x} \qquad 5\sqrt{x}$

are not polynomials because the variable in the first term is in the denominator and the variable in the second term is under the radical (we will see in Chapter 8 that the exponent in this case is $\frac{1}{2}$).

NOTE The prefix "bi" means two and "tri" means three.

A polynomial with two terms is a **binomial.** A polynomial with three terms is called a **trinomial.**

The numerical factor in a term is called the **numerical coefficient,** or more simply the **coefficient,** of that term. For example, in the terms

$8x^4 \qquad 9x^3y^4 \qquad 6x^2y^7 \qquad -10xy$

the numerical coefficients are 8, 9, 6, and -10.

Example 3

Classifying Polynomials

Which of the following are polynomials? Classify the polynomials as monomial, binomial, or trinomial.

(a) $5x^2y$

Monomial

(b) $3m + 5n$

Binomial

(c) $4a^3 + 3a - 2$

Trinomial

NOTE Remember that $\frac{1}{x} = x^{-1}$

(d) $5y^2 - \frac{2}{x}$

$5y^2 - \frac{2}{x}$ is not a polynomial because the variable x is in the denominator.

CHECK YOURSELF 3

Which of the following are polynomials? Classify the polynomials as monomial, binomial, or trinomial.

(a) $5x^2 - 6x$ **(b)** $8x^5$

(c) $5x^3 - 3xy + 7y^2$ **(d)** $9x - \frac{3}{x}$

It is also useful to classify polynomials by their **degree.**

Definitions: Degree of Monomials

The **degree** of a monomial is the sum of the exponents of the variable factors.

Example 4

Determining the Degree of a Monomial

(a) $5x^2$ has degree 2.

(b) $7n^5$ has degree 5.

(c) $3a^2b^4$ has degree 6. (The sum of the powers, 2 and 4, is 6.)

(d) 9 has degree 0 (because $9 = 9 \cdot 1 = 9x^0$).

CHECK YOURSELF 4

Give the degree of each monomial.

(a) $4x^2$ **(b)** $7x^3y^2$

(c) $8p^2s$ **(d)** 5

Definitions: Degree of Polynomials

The **degree** of a polynomial is that of the term with the highest degree.

Example 5

Determining the Degree of a Polynomial

(a) $7x^3 - 5x^2 + 5$ has degree 3.

(b) $5y^7 - 3y^2 + 5y - 7$ has degree 7.

(c) $4a^2b^3 - 5abc^2$ has degree 5 because the sum of the variable powers in the term with highest degree ($4a^2b^3$) is 5.

Polynomials such as those in Examples 5(a) and 5(b) are called **polynomials in one variable**, and they are usually written in *descending form* so that the power of the variable decreases from left to right. In that case, the coefficient of the first term is called the **leading coefficient**.

CHECK YOURSELF 5

Give the degree of each polynomial. For those polynomials in one variable, write in descending form and give the leading coefficient.

(a) $7x^4 - 5xy + 2$ **(b)** 5 **(c)** $4x^2 - 7x^3 - 8x + 5$

A **polynomial function** is a function in which the expression on the right-hand side is a polynomial expression. For example,

$f(x) = 3x^3 + 2x^2 - x + 5$

$g(x) = 2x^3 - 5x - 1$

$P(x) = 2x^7 + 2x^6 - x^3 + 2x^4 + 5x^2 - 6$

are all polynomial functions. Because they are functions, every x value determines a unique ordered pair.

Example 6

Finding Ordered Pairs

Given $f(x) = 3x^3 + 2x^2 - x + 5$ and $g(x) = 2x^3 - 5x - 1$, find the following ordered pairs.

(a) $(0, f(0))$

To find $f(0)$, we substitute 0 for x in the function $f(x) = 3x^3 + 2x^2 - x + 5$.

$$f(0) = 3(0)^3 + 2(0)^2 - (0) + 5$$
$$= 0 + 0 - 0 + 5$$
$$= 5$$

Therefore, $(0, f(0)) = (0, 5)$.

Name ______________

Section ________ Date ________

6.1 Exercises

Which of the following expressions are polynomials?

1. $7x^3$

2. $5x^3 - \frac{3}{x}$

3. $4x^4y^2 - 3x^3y$

4. 7

5. -7

6. $4x^3 + x$

7. $\frac{3 + x}{x^2}$

8. $5a^2 - 2a + 7$

For each of the following polynomials, list the terms and the coefficients.

9. $2x^2 - 3x$

10. $-5x^3 + x$

11. $4x^3 - 3x + 2$

12. $7x^2$

For each of the following pairs of terms, decide whether they are like terms.

13. $3xy^2c$ and $2xy^2c$

14. $5xy^3$ and $4xy^3$

15. $-6xy$ and $2x^2y^2$

16. $-3x^3y^4$ and $2x^4y^3$

Classify each of the following as a monomial, binomial, or trinomial where possible.

17. $7x^3 - 3x^2$

18. $4x^7$

19. $7y^2 + 4y + 5$

20. $2x^2 + 3xy + y^2$

21. $-2x^4 - 3x^2 + 5x - 2$

22. $x^4 + \frac{5}{x} + 7$

ANSWERS

1. ______________
2. ______________
3. ______________
4. ______________
5. ______________
6. ______________
7. ______________
8. ______________
9. ______________
10. ______________
11. ______________
12. ______________
13. ______________
14. ______________
15. ______________
16. ______________
17. ______________
18. ______________
19. ______________
20. ______________
21. ______________
22. ______________

ANSWERS

23. ______
24. ______
25. ______
26. ______
27. ______
28. ______
29. ______
30. ______
31. ______
32. ______
33. ______
34. ______
35. ______
36. ______
37. ______
38. ______
39. ______
40. ______
41. ______
42. ______
43. ______
44. ______
45. ______
46. ______

23. $6y^8$

24. $4x^4 - 2x^2 + 5x - 7$

25. $x^5 - \frac{3}{x^2}$

26. $4x^2 - 9$

Arrange in descending-exponent form if necessary, and give the degree of each polynomial.

27. $4x^5 - 3x^2$

28. $5x^2 - 3x^3 + 4$

29. $7x^7 - 5x^9 + 4x^3$

30. $2 + x$

31. $4x$

32. $x^{17} - 3x^4$

33. $5x^2 - 3x^5 + x^6 - 7$

34. 5

Find the values of each of the following polynomials for the given values of the variable.

35. $6x + 1$; $x = 1$ and $x = -1$

36. $5x - 5$; $x = 2$ and $x = -2$

37. $x^3 - 2x$; $x = 2$ and $x = -2$

38. $3x^2 + 7$; $x = 3$ and $x = -3$

39. $3x^2 + 4x - 2$; $x = 4$ and $x = -4$

40. $2x^2 - 5x + 1$; $x = 2$ and $x = -2$

41. $-x^2 - 2x + 3$; $x = 1$ and $x = -3$

42. $-x^2 - 5x - 6$; $x = -3$ and $x = -2$

Indicate whether each of the following statements is always true, sometimes true, or never true.

43. A monomial is a polynomial.

44. A binomial is a trinomial.

45. The degree of a trinomial is 3.

46. A trinomial has three terms.

47. A polynomial has four or more terms.

48. A binomial must have two coefficients.

49. If x equals 0, the value of a polynomial in x equals 0.

50. The coefficient of the leading term in a polynomial is the largest coefficient of the polynomial.

In exercises 51 to 58, the polynomial functions $f(x)$ and $g(x)$ are given. Find the ordered pairs **(a)** $(0, f(0))$, **(b)** $(2, f(2))$, **(c)** $(-2, f(-2))$, **(d)** $(0, g(0))$, **(e)** $(2, g(2))$, and **(f)** $(-2, g(-2))$

51. $f(x) = 2x^2 + 3x - 5$ and $g(x) = 4x^2 - 5x - 7$

52. $f(x) = 3x^2 + 7x - 9$ and $g(x) = 7x^2 + 8x - 9$

53. $f(x) = x^3 + 8x^2 - 4x + 10$ and $g(x) = 3x^3 + 4x^2 - 5x - 3$

54. $f(x) = -x^3 + 3x^2 - 7x + 8$ and $g(x) = -2x^3 + 4x^2 - 9x - 2$

55. $f(x) = -4x^3 - 5x^2 + 8x - 12$ and $g(x) = -x^3 + 5x^2 + 7x - 14$

56. $f(x) = 7x^3 + 5x - 9$ and $g(x) = 4x^3 + 3x^2 + 8$

57. $f(x) = 10x^3 + 5x^2 - 3x$ and $g(x) = 8x^3 + 4x^2 + 9$

58. $f(x) = 7x^2 - 5$ and $g(x) = 3x^3 + 5x^2 + 6x + 9$

ANSWERS

47. ______

48. ______

49. ______

50. ______

51. (a) (b) (c) (d) (e) (f)

52. (a) (b) (c) (d) (e) (f)

53. (a) (b) (c) (d) (e) (f)

54. (a) (b) (c) (d) (e) (f)

55. (a) (b) (c) (d) (e) (f)

56. (a) (b) (c) (d) (e) (f)

57. (a) (b) (c) (d) (e) (f)

58. (a) (b) (c) (d) (e) (f)

ANSWERS

59. ____

60. ____

61. ____

62. ____

63. ____

64. ____

59. Cost of typing. The cost, in dollars, of typing a term paper is given as three times the number of pages plus 20. Use x as the number of pages to be typed and write a polynomial to describe this cost. Find the cost of typing a 50-page paper.

60. Manufacturing. The cost, in dollars, of making suits is described as 20 times the number of suits plus 150. Use x as the number of suits and write a polynomial to describe this cost. Find the cost of making seven suits.

61. Revenue. The revenue, in dollars, when x pairs of shoes are sold is given by $R(x) = 3x^2 - 95$. Find the revenue when 12 pairs of shoes are sold.

62. Manufacturing. The cost in dollars of manufacturing x wing nuts is given by $C(x) = 0.07x + 13.3$. Find the cost when 375 wing nuts are made.

Let $P(x) = 2x^3 - 3x^2 + 5x - 5$ and $Q(x) = -x^2 + 2x - 2$. In exercises 63 and 64, find each of the following.

63. $P[Q(1)]$

64. $Q[P(1)]$

Answers

1. Polynomial **3.** Polynomial **5.** Polynomial **7.** Not a polynomial
9. $2x^2, -3x; 2, -3$ **11.** $4x^3, -3x, 2; 4, -3, 2$ **13.** Like terms
15. Not like terms **17.** Binomial **19.** Trinomial **21.** Not classified
23. Monomial **25.** Not a polynomial **27.** $4x^5 - 3x^2; 5$
29. $-5x^9 + 7x^7 + 4x^3; 9$ **31.** $4x; 1$ **33.** $x^6 - 3x^5 + 5x^2 - 7; 6$
35. $7, -5$ **37.** $4, -4$ **39.** 62, 30 **41.** 0, 0 **43.** Always
45. Sometimes **47.** Sometimes **49.** Sometimes
51. **(a)** $(0, -5)$; **(b)** $(2, 9)$; **(c)** $(-2, -3)$; **(d)** $(0, -7)$; **(e)** $(2, -1)$; **(f)** $(-2, 19)$
53. **(a)** $(0, 10)$; **(b)** $(2, 42)$; **(c)** $(-2, 42)$; **(d)** $(0, -3)$; **(e)** $(2, 27)$; **(f)** $(-2, -1)$
55. **(a)** $(0, -12)$; **(b)** $(2, -48)$; **(c)** $(-2, -16)$; **(d)** $(0, -14)$; **(e)** $(2, 12)$; **(f)** $(-2, 0)$
57. **(a)** $(0, 0)$; **(b)** $(2, 94)$; **(c)** $(-2, -54)$; **(d)** $(0, 9)$; **(e)** $(2, 89)$; **(f)** $(-2, -39)$
59. $C(x) = 3x + 20$; \$170 **61.** \$337 **63.** -15

Combining Functions

OBJECTIVES

1. Find the sum or difference of two functions
2. Find the product of two functions
3. Find the quotient of two functions
4. Find the domain of the sum, difference, product, or quotient of two functions

The profit that a company makes on an item is determined by subtracting the cost of making the item from the total revenue the company receives from selling the item. This is an example of **combining functions.** It can be written as

$$P(x) = R(x) - C(x)$$

Many applications of functions involve the combining of two or more component functions. In this section, we will look at several properties that allow for the addition, subtraction, multiplication, and division of functions.

Definitions: Sum of Two Functions

The **sum of two functions** f and g is written as $f + g$ and can be defined as

$$(f + g)(x) = f(x) + g(x)$$

for every value of x that is in the domain of both functions f and g.

Definitions: Difference of Two Functions

The **difference of two functions** f and g is written as $f - g$ and can be defined as

$$(f - g)(x) = f(x) - g(x)$$

for every value of x that is in the domain of both functions f and g.

Example 1

Finding the Sum or Difference of Two Functions

Given the functions $f(x) = 2x - 1$ and $g(x) = -3x + 4$

(a) Find $(f + g)(x)$.

$$(f + g)(x) = f(x) + g(x)$$
$$= (2x - 1) + (-3x + 4) = -x + 3$$

(b) Find $(f - g)(x)$.

$$(f - g)(x) = f(x) - g(x)$$
$$= (2x - 1) - (-3x + 4) = 5x - 5$$

(c) Find $(f + g)(2)$.

If we use the definition of the sum of two functions, we find that

$$(f + g)(2) = f(2) + g(2)$$
$$= 3 + (-2) = 1$$

As an alternative, we could use part (a) and say

$$(f + g)(x) = -x + 3$$

Therefore,

$$(f + g)(2) = -2 + 3$$
$$= 1$$

CHECK YOURSELF 1

Given the functions $f(x) = -2x - 3$ and $g(x) = 5x - 1$,

(a) Find $(f + g)(x)$. **(b)** Find $(f - g)(x)$. **(c)** Find $(f + g)(2)$.

In defining the sum of two functions, we indicated that the domain was determined by the domain of both functions. We will find the domain in Example 2.

Example 2

Finding the Domain of the Sum or Difference of Functions

Given $f(x) = 2x - 4$ and $g(x) = \frac{1}{x}$,

(a) Find $(f + g)(x)$.

$$(f + g)(x) = (2x - 4) + \frac{1}{x} = 2x - 4 + \frac{1}{x}$$

(b) Find the domain of $f + g$.

The domain of $f + g$ is the set of all numbers in the domain of f and also in the domain of g. The domain of f consists of all real numbers. The domain of g consists of all real numbers except 0 because we cannot divide by 0. The domain of $f + g$ is the set of all real numbers except 0. We write $D = \{x \mid x \neq 0\}$.

CHECK YOURSELF 2

Given $f(x) = -3x + 1$ and $g(x) = \frac{1}{(x - 2)}$,

(a) Find $(f + g)(x)$. **(b)** Find the domain of $f + g$.

Definitions: Product of Two Functions

The **product of two functions** f and g is written as $f \cdot g$ and can be defined as

$$(f \cdot g)(x) = f(x) \cdot g(x)$$

for every value of x that is in the domain of both functions f and g.

Example 3

Finding the Product of Two Functions

NOTE From earlier work in algebra you should recall that the product of two binomials, $(x + a)(x + b)$ is

$x^2 + bx + ax + ab$

Given $f(x) = x - 1$ and $g(x) = x + 5$, find $(f \cdot g)(x)$.

$$(f \cdot g)(x) = f(x) \cdot g(x) = (x - 1)(x + 5) = x^2 + 5x - x - 5 = x^2 + 4x - 5$$

CHECK YOURSELF 3

Given $f(x) = x - 3$ and $g(x) = x + 2$, find $(f \cdot g)(x)$.

The final operation on functions that we will look at involves the division of two functions.

Definitions: Quotient of Two Functions

The **quotient of two functions** f and g is written as $f \div g$ and can be defined as

$$(f \div g)(x) = f(x) \div g(x)$$

for every value of x that is in the domain of both functions f and g, such that $g(x) \neq 0$.

Example 4

Finding the Quotient of Two Functions

Given $f(x) = x - 1$ and $g(x) = x + 5$,

(a) Find $(f \div g)(x)$.

$$(f \div g)(x) = f(x) \div g(x) = (x - 1) \div (x + 5) = \frac{x - 1}{x + 5}$$

(b) Find the domain of $f \div g$.

The domain is the set of all real numbers except -5 because $g(-5) = 0$, and division by 0 is undefined. We write $D = \{x \mid x \neq -5\}$.

CHECK YOURSELF 4

Given $f(x) = x - 3$ and $g(x) = x + 2$,

(a) Find $(f \div g)(x)$.

(b) Find the domain of $f \div g$.

CHECK YOURSELF ANSWERS

1. **(a)** $3x - 4$; **(b)** $-7x - 2$; **(c)** 2

2. **(a)** $-3x + 1 + \dfrac{1}{(x - 2)}$; **(b)** $D = \{x | x \neq 2\}$ **3.** $(x - 3)(x + 2) = x^2 - x - 6$

4. **(a)** $\dfrac{(x - 3)}{(x + 2)}$; **(b)** $D = \{x | x \neq -2\}$

Exercises

Name ______________________

Section ________ Date ________

In exercises 1 to 8, find **(a)** $(f + g)(x)$, **(b)** $(f - g)(x)$, **(c)** $(f + g)(3)$, and **(d)** $(f - g)(2)$.

1. $f(x) = -4x + 5 \quad g(x) = 7x - 4$

2. $f(x) = 9x - 3 \quad g(x) = -3x + 5$

3. $f(x) = 8x - 2 \quad g(x) = -5x + 6$

4. $f(x) = -7x + 9 \quad g(x) = 2x - 1$

5. $f(x) = x^2 + x - 1 \quad g(x) = -3x^2 - 2x + 5$

6. $f(x) = -3x^2 - 2x + 5 \quad g(x) = 5x^2 + 3x - 6$

7. $f(x) = -x^3 - 5x + 8 \quad g(x) = 2x^2 + 3x - 4$

8. $f(x) = 2x^3 + 3x^2 - 5 \quad g(x) = -4x^2 + 5x - 7$

In exercises 9 to 14, find **(a)** $(f + g)(x)$ and **(b)** the domain of $f + g$.

9. $f(x) = -9x + 11 \quad g(x) = 15x - 7$

10. $f(x) = -11x + 3 \quad g(x) = 8x - 5$

11. $f(x) = 3x + 2 \quad g(x) = \dfrac{1}{x - 2}$

12. $f(x) = -2x + 5 \quad g(x) = \dfrac{3}{x + 1}$

13. $f(x) = x^2 + x - 5 \quad g(x) = \dfrac{2}{3x + 1}$

14. $f(x) = 3x^2 - 5x + 1 \quad g(x) = \dfrac{-2}{2x - 3}$

ANSWERS

1. (a)
(b)
(c) (d)

2. (a)
(b)
(c) (d)

3. (a)
(b)
(c) (d)

4. (a)
(b)
(c) (d)

5. (a)
(b)
(c) (d)

6. (a)
(b)
(c) (d)

7. (a)
(b)
(c) (d)

8. (a)
(b)
(c) (d)

9. (a)
(b)

10. (a)
(b)

11. (a)
(b)

12. (a)
(b)

13. (a)
(b)

14. (a)
(b)

ANSWERS

15. (a)

(b)

(c)

16. (a)

(b)

(c)

17. (a)

(b)

(c)

18. (a)

(b)

(c)

19. (a)

(b)

(c)

20. (a)

(b)

(c)

21.

22.

23.

24.

25.

26.

In exercises 15 to 20, find **(a)** $(f \cdot g)(x)$, **(b)** $(f \div g)(x)$, and **(c)** the domain of $f \div g$.

15. $f(x) = 2x - 1 \quad g(x) = x - 3$

16. $f(x) = -x + 3 \quad g(x) = x + 4$

17. $f(x) = 3x + 2 \quad g(x) = 2x - 1$

18. $f(x) = -3x + 5 \quad g(x) = -x + 2$

19. $f(x) = 2 - x \quad g(x) = 5 + 2x$

20. $f(x) = x + 5 \quad g(x) = 1 - 3x$

In business, the profit, $P(x)$, obtained from selling x units of a product is equal to the revenue, $R(x)$, minus the cost, $C(x)$. In exercises 21 and 22, find the profit, $P(x)$, for selling x units.

21. $R(x) = 25x \quad C(x) = x^2 + 4x + 50$

22. $R(x) = 20x \quad C(x) = x^2 + 2x + 30$

Let $V(t)$ be the velocity of an object that has been thrown in the air. It can be shown that $V(t)$ is the combination of three functions: the initial velocity, V_0 (this is a constant); the acceleration due to gravity, g (this is also a constant); and the time that has elapsed, t. We have:

$$V(t) = V_0 + g \cdot t$$

In exercises 23 and 24, find the velocity as a function of time t.

23. $V_0 = 10$ m/s, $\quad g = -9.8$ m/s^2

24. $V_0 = 64$ ft/s, $\quad g = -32$ ft/s^2

The revenue produced from the sale of an item can be found by multiplying the price, $p(x)$ by the quantity sold, x. In exercises 25 and 26, find the revenue produced from selling x items.

25. $p(x) = 119 - 6x$

26. $p(x) = 1190 - 36x$

Answers

1. **(a)** $3x + 1$; **(b)** $-11x + 9$; **(c)** 10; **(d)** -13 **3.** **(a)** $3x + 4$; **(b)** $13x - 8$; **(c)** 13; **(d)** 18 **5.** **(a)** $-2x^2 - x + 4$; **(b)** $4x^2 + 3x - 6$; **(c)** -17; **(d)** 16
7. **(a)** $-x^3 + 2x^2 - 2x + 4$; **(b)** $-x^3 - 2x^2 - 8x + 12$; **(c)** -11; **(d)** -20
9. **(a)** $6x^2 + 4$; **(b)** all real numbers **11.** **(a)** $3x + 2 + \frac{1}{x - 2}$; **(b)** $\{x \mid x \neq 2\}$
13. **(a)** $x^2 + x - 5 + \frac{2}{3x + 1}$; **(b)** $\left\{x \mid x \neq -\frac{1}{3}\right\}$
15. **(a)** $(2x - 1)(x - 3) = 2x^2 - 7x + 3$; **(b)** $\frac{2x - 1}{x - 3}$; **(c)** $\{x \mid x \neq 3\}$
17. **(a)** $6x^2 + x - 2$; **(b)** $\frac{3x + 2}{2x - 1}$; **(c)** $\left\{x \mid x \neq \frac{1}{2}\right\}$
19. **(a)** $-2x^2 - x + 10$; **(b)** $\frac{2 - x}{5 + 2x}$; **(c)** $\left\{x \mid x \neq -\frac{5}{2}\right\}$ **21.** $P(x) = -x^2 + 21x - 50$
23. $V(t) = 10 - 9.8t$ **25.** $R(x) = 119x - 6x^2$

6.3 Addition and Subtraction of Polynomials

6.3 OBJECTIVES

1. Find the sum of two polynomial functions
2. Find the difference of two polynomial functions

To find the sum of two polynomials, we add the like terms. Given

$(3x^3 + 2x - 5) + (x^3 - x^2 + 5x - 1)$

we could use the associative and commutative properties of addition to rewrite the sum as

$(3x^3 + x^3) + (-x^2) + (2x + 5x) + (-5 - 1)$

This simplifies to

$4x^3 - x^2 + 7x - 6$

The process for adding polynomial functions is nearly identical, as the next example illustrates.

Example 1

Adding Two Polynomial Functions

Given $f(x) = 3x^3 + 2x^2 - x + 5$ and $g(x) = 2x^3 - 5x - 1$, and letting $h(x) = f(x) + g(x)$, find $h(x)$.

$h(x) = f(x) + g(x)$

$= (3x^3 + 2x^2 - x + 5) + (2x^3 - 5x - 1)$ Substitute the polynomial functions.

$= (3x^3 + 2x^3) + 2x^2 + (-5x - x) + (5 - 1)$ Collect like terms.

$= 5x^3 + 2x^2 - 6x + 4$

CHECK YOURSELF 1

Given $f(x) = x^3 - 7x + 1$ and $g(x) = x^3 + 2x^2 - 3x - 4$, and letting $h(x) = f(x) + g(x)$, find $h(x)$.

Example 2 demonstrates a method by which we can check our results when we add two polynomials.

Example 2

Adding Two Polynomial Functions

Given $f(x) = 3x^3 + 2x^2 - x + 5$ and $g(x) = 2x^3 - 5x - 1$, and letting $h(x) = f(x) + g(x)$, find the following.

(a) $f(2) + g(2)$

$f(2) = 3(2)^3 + 2(2)^2 - (2) + 5$

$= 24 + 8 - 2 + 5$

$= 35$

$$g(2) = 2(2)^3 - 5(2) - 1$$
$$= 16 - 10 - 1$$
$$= 5$$
$$f(2) + g(2) = 35 + 5$$
$$= 40$$

(b) Use the result from Example 1 to find $h(2)$

From Example 1, we have

$$h(x) = 5x^3 + 2x^2 - 6x + 4$$

Therefore,

$$h(2) = 5(2)^3 + 2(2)^2 - 6(2) + 4$$
$$= 40 + 8 - 12 + 4$$
$$= 40$$

Note that $h(2) = f(2) + g(2)$. This helps us confirm that we have correctly added the polynomials.

CHECK YOURSELF 2

Given $f(x) = x^3 - 7x + 1$ *and* $g(x) = x^3 + 2x^2 - 3x - 4$, *and letting* $h(x) = f(x) + g(x)$, *complete the following.*

(a) Find $f(2) + g(2)$.

(b) Use the result of Check Yourself 1 to find $h(2)$.

(c) Compare the results of **(a)** and **(b).**

Subtracting polynomials proceeds in a similar fashion. We view the subtraction of a quantity as adding the opposite of that quantity, so that

$$a - b = a + (-b)$$

When we are subtracting a polynomial, we apply the distributive property before combining like terms.

NOTE Both statements are just applications of the distributive property. This shows us that the *opposite* of $a + b$ is $-a - b$ and that the *opposite* of $a - b$ is $-a + b$.

Rules and Properties: Distribution of the Negative

$$-(a + b) = -a - b$$

and

$$-(a - b) = -a + b$$

We can now go on to subtracting polynomials.

Example 3

Subtracting Polynomial Functions

Given $f(x) = 7x^2 - 2x$ and $g(x) = 4x^2 + 5x$, and letting $h(x) = f(x) - g(x)$, find the following.

(a) $h(x)$

$$h(x) = f(x) - g(x)$$
$$= (7x^2 - 2x) - (4x^2 + 5x)$$
$$= 7x^2 - 2x - 4x^2 - 5x$$
$$= 7x^2 - 4x^2 - 2x - 5x$$
$$= 3x^2 - 7x$$

(b) $f(2) - g(2)$

$$f(2) - g(2) = (7(2)^2 - 2(2)) - (4(2)^2 + 5(2))$$
$$= (28 - 4) - (16 + 10)$$
$$= (24) - (26)$$
$$= -2$$

(c) Use the results of part (a) to find $h(2)$.

$$h(2) = 3(2)^2 - 7(2)$$
$$= 12 - 14$$
$$= -2$$

As was the case with addition, we have found an easy way to check our work when we are subtracting polynomial functions. We have found that $(2, h(2)) = (2, -2) = (2, f(2) - g(2))$.

CHECK YOURSELF 3

Given $f(x) = 8x^2 - x$ *and* $g(x) = x^2 - 2x$, *and letting* $h(x) = f(x) - g(x)$, *find the following.*

(a) $h(x)$

(b) $f(2) - g(2)$

(c) Use the result of part (a) to find $h(2)$.

Example 4

Subtracting Polynomial Functions

Given $f(x) = 5x^2 + 2x$ and $g(x) = 3x^2 - 5x$, and letting $h(x) = f(x) - g(x)$, find the following.

(a) $h(x)$

$$\begin{aligned} h(x) &= f(x) - g(x) \\ &= (5x^2 + 2x) - (3x^2 - 5x) \\ &= 5x^2 + 2x - 3x^2 + 5x \\ &= 5x^2 - 3x^2 + 2x + 5x \\ &= 2x^2 + 7x \end{aligned}$$

(b) $f(1) - g(1)$

$$\begin{aligned} f(1) - g(1) &= (5(1)^2 + 2(1)) - (3(1)^2 - 5(1)) \\ &= (5 + 2) - (3 - 5) \\ &= (7) - (-2) \\ &= 9 \end{aligned}$$

(c) Use the result of part (a) to find $h(1)$.

$$\begin{aligned} h(1) &= 2(1)^2 + 7(1) \\ &= 2 + 7 \\ &= 9 \end{aligned}$$

We see that $(1, h(1)) = (1, 9) = (1, f(1) - g(1))$.

CHECK YOURSELF 4

Given $f(x) = 3x^2 - 4x$ and $g(x) = 6x^2 - x$, and letting $h(x) = f(x) - g(x)$, find the following.

(a) $h(x)$
(b) $f(1) - g(1)$
(c) Use the results of part (a) to find $h(1)$.

CHECK YOURSELF ANSWERS

1. $2x^3 + 2x^2 - 10x - 3$ **2.** **(a)** 1; **(b)** 1; **(c)** $h(2) = f(2) + g(2)$
3. **(a)** $h(x) = 7x^2 + x$; **(b)** 30; **(c)** 30
4. **(a)** $h(x) = -3x^2 - 3x$; **(b)** -6; **(c)** -6

Exercises

Name ________________

Section ________ Date ________

In exercises 1 to 8, $f(x)$ and $g(x)$ are given. Let $h(x) = f(x) + g(x)$. Find **(a)** $h(x)$, **(b)** $f(1) + g(1)$, and **(c)** use the results of part **(a)** to find $h(1)$.

1. $f(x) = 5x - 3$ and $g(x) = 4x + 7$

2. $f(x) = 7x^2 + 3x$ and $g(x) = 5x^2 - 7x$

3. $f(x) = 5x^2 + 3x$ and $g(x) = 4x + 2x^2$

4. $f(x) = 8x^2 - 3x + 10$ and $g(x) = 7x^2 + 2x - 12$

5. $f(x) = -3x^2 - 5x - 7$ and $g(x) = 2x^2 + 3x + 5$

6. $f(x) = 2x^3 + 5x^2 + 8$ and $g(x) = -5x^3 - 2x^2 + 7x$

7. $f(x) = -5x^2 - 3x - 15$ and $g(x) = 5x^3 - 8x - 10$

8. $f(x) = 5x^2 + 12x - 5$ and $g(x) = 3x^3 + 7x - 9$

In exercises 9 to 16, $f(x)$ and $g(x)$ are given. Let $h(x) = f(x) - g(x)$. Find **(a)** $h(x)$, **(b)** $f(1) - g(1)$, and **(c)** use the results of part **(a)** to find $h(1)$.

9. $f(x) = 7x + 10$ and $g(x) = 5x - 3$

10. $f(x) = 5x - 12$ and $g(x) = 8x - 7$

11. $f(x) = 7x^2 - 3x$ and $g(x) = -5x^2 - 2x$

12. $f(x) = -10x^2 + 3x$ and $g(x) = -3x^2 - 3x$

13. $f(x) = 8x^2 - 5x - 7$ and $g(x) = 5x^2 - 3x$

14. $f(x) = 5x^3 - 2x^2 - 8x$ and $g(x) = 7x^3 - 3x^2 - 8$

15. $f(x) = 5x^2 - 5$ and $g(x) = 8x^2 - 7x$

16. $f(x) = 5x^2 - 7x$ and $g(x) = 9x^2 - 3$

In exercises 17 to 28, simplify each polynomial function.

17. $f(x) = (2x - 3) + (4x + 7) - (2x - 3)$

18. $f(x) = (5x - 2) - (2x + 3) + (7x - 7)$

ANSWERS

1. ________________
2. ________________
3. ________________
4. ________________
5. ________________
6. ________________
7. ________________
8. ________________
9. ________________
10. ________________
11. ________________
12. ________________
13. ________________
14. ________________
15. ________________
16. ________________
17. ________________
18. ________________

ANSWERS

19. ______
20. ______
21. ______
22. ______
23. ______
24. ______
25. ______
26. ______
27. ______
28. ______
29. ______
30. ______
31. ______
32. ______
33. ______
34. ______
35. ______
36. ______
37. ______

19. $f(x) = (5x^2 - 2x + 7) - (2x^2 + 3x + 1) - (4x^2 + 3x + 3)$

20. $f(x) = (8x^2 - 8x + 5) + (3x^2 + 2x - 7) - (x^2 - 6x + 4)$

21. $f(x) = (8x^2 - 3) + (5x^2 + 7x) - (7x - 8)$

22. $f(x) = (9x^2 + 7x) - (5x^2 - 8x) - (4x^2 + 3)$

23. $f(x) = x - [5x - (x - 3)]$

24. $f(x) = x^2 - [7x^2 - (x^2 + 7)]$

25. $f(x) = (2x - 3) - [x - (2x + 7)]$

26. $f(x) = (3x^2 + 5x) - [x^2 - (2x^2 - 3x)]$

27. $f(x) = 2x - (3x + 2[x - 2(x - 3)])$

28. $f(x) = 3x - (5x - 2[x - 3(x + 4)])$

Suppose that revenue is given by the polynomial $R(x)$ and cost is given by the polynomial $C(x)$. Profit $P(x)$ can then be found with the formula

$P(x) = R(x) - C(x)$

In exercises 29 to 32, find the polynomial representing profit in each expression.

29. $R(x) = 100x$
$C(x) = 2000 + 50x$

30. $R(x) = 250x$
$C(x) = 5000 + 175x$

31. $R(x) = 100x + 2x^2$
$C(x) = 2000 + 50x + 5x^2$

32. $R(x) = 250x + 5x^2$
$C(x) = 5000 + 175x + 10x^2$

33. If $P(x) = -2x + 1$, find $P(a + h) - P(a)$.

34. If $P(x) = 5x$, find $\dfrac{P(a + h) - P(a)}{h}$.

35. Find the difference when $4x^2 + 2x + 1$ is subtracted from the sum of $x^2 - 2x - 3$ and $3x^2 + 5x - 7$.

36. Subtract $7x^2 + 5x - 3$ from the sum of $2x^2 - 5x + 7$ and $-9x^2 - 2x + 5$.

37. Find the difference when $7x^2 - 3x + 2$ is subtracted from the sum of $2x^2 + 5x - 4$ and $6x^2 - 7x + 9$.

ANSWERS

38. ____
39. ____
40. ____
41. ____
42. ____
43. (a) (b) (c) (d) ____
44. (a) (b) (c) (d) ____

38. Subtract $4x^2 + 3x - 8$ from the sum of $2x^2 + 4x + 3$ and $5x^2 - 3x - 7$.

39. Subtract $8x^2 - 2x$ from the sum of $x^2 - 5x$ and $7x^2 + 5$.

40. Find the difference when $9a^2 - 7$ is subtracted from the sum of $5a^2 - 5$ and $-2a^2 - 2$.

41. The length of a rectangle is 1 cm more than twice its width. Represent the width of the rectangle by w, and write a polynomial to express the perimeter of the rectangle in terms of w. Be sure to simplify your result.

42. One integer is 2 more than twice the first. Another is 3 less than 3 times the first. Represent the first integer by x, and then write a polynomial to express the sum of the three integers in terms of x. Be sure to simplify your result.

43. The number 3078 can be written as the polynomial

$3(10)^3 + 0(10)^2 + 7(10)^1 + 8(10)^0$

because 10 is the *base* of the number system we commonly use. All numbers can be written as a polynomial. Interpret the following polynomials by writing them the way they would normally appear.

(a) $7(10)^4 + 5(10)^3 + 0(10)^2 + 2(10) + 0(10)^0 =$ ____________

(b) $4(10)^2 + 2(10) + 3(10)^{-1} + 2(10)^{-2} + 5(10)^{-3} =$ ____________

Write these numbers as polynomials:

(c) $6525 =$ ____________

(d) $99.95 =$ ____________

44. In exercise 43, the first number in the list could be written with a variable in place of the 10: $7(n)^4 + 5(n)^3 + 0(n)^2 + 2(n) + 0(n)^0$. The number could still be written as 75020, but the value of the number would be very different from 75 thousand 20 if the *base* were different. Try $n = 8$ and calculate the value of 75020_8. This is read "75020 base 8." Did you get 31248?

The number 11011 in base 10 is eleven thousand eleven. Written as a polynomial:

$1(n)^4 + 1(n)^3 + 0(n)^2 + 1(n) + 1(n)^0$

If $n = 2$, we have another value for 11011 but this time in *base 2,* another very widely used number system because it is used by computers. Evaluate 11011_2 in base 10.

Write the following as polynomials and then evaluate the numbers in base 10:

(a) $546302_7 =$ ____________ = ____________ in base 10

(b) $111100111_2 =$ ____________ = ____________ in base 10

(c) $21112_3 =$ ____________ = ____________ in base 10

(d) $21112_5 =$ ____________ = ____________ in base 10

You may want to find out more about base 2 numbers or the binary number system because versions of it are widely used in computers, bar code scanners, and other electronic devices.

ANSWERS

45. ______

46. ______

47. ______

48. ______

49. ______

50. ______

Find values for a, b, c, and d so that the following equations are true.

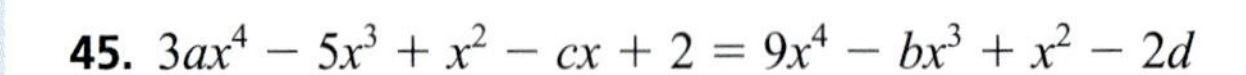

45. $3ax^4 - 5x^3 + x^2 - cx + 2 = 9x^4 - bx^3 + x^2 - 2d$

46. $(4ax^3 - 3bx^2 - 10) - 3(x^3 + 4x^2 - cx - d) = x^3 - 6x + 8$

47. Geometry. A rectangle has sides of $8x + 9$ and $6x - 7$. Find the polynomial that represents its perimeter.

48. Geometry. A triangle has sides $3x + 7$, $4x - 9$, and $5x + 6$. Find the polynomial that represents its perimeter.

49. For the given figure, write the polynomial that represents the perimeter.

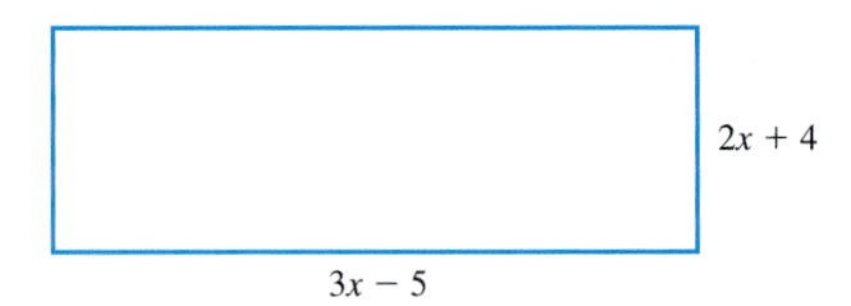

50. For the given figure, the perimeter is given by $14x + 5$. Write the polynomial that represents the missing side.

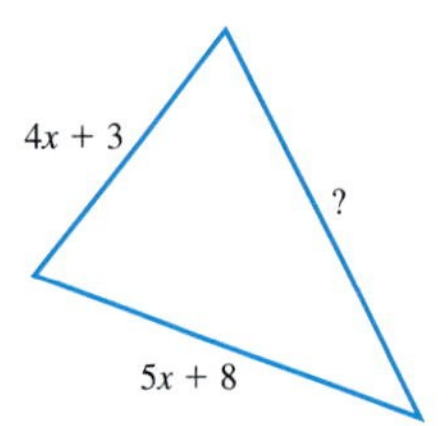

Answers

1. (a) $9x + 4$; **(b)** 13; **(c)** 13 **3. (a)** $7x^2 + 7x$; **(b)** 14; **(c)** 14 **5. (a)** $-x^2 - 2x - 2$; **(b)** -5; **(c)** -5 **7. (a)** $5x^3 - 5x^2 - 11x - 25$; **(b)** -36; **(c)** -36; **9. (a)** $2x + 13$; **(b)** 15; **(c)** 15 **11. (a)** $12x^2 - x$; **(b)** 11; **(c)** 11 **13. (a)** $3x^2 - 2x - 7$; **(b)** -6; **(c)** -6 **15. (a)** $-3x^2 + 7x - 5$; **(b)** -1; **(c)** -1 **17.** $4x + 7$ **19.** $-x^2 - 8x + 3$ **21.** $13x^2 + 5$ **23.** $-3x - 3$ **25.** $3x + 4$ **27.** $x - 12$ **29.** $50x - 2000$ **31.** $-3x^2 + 50x - 2000$ **33.** $-2h$ **35.** $x - 11$ **37.** $x^2 + x + 3$ **39.** $-3x + 5$ **41.** $6w + 2$ **43. (a)** 75,020; **(b)** 420.325; **(c)** $6(10)^3 + 5(10)^2 + 2(10) + 5(10)^0$; **(d)** $9(10)^1 + 9(10)^0 + 9(10)^{-1} + 5(10)^{-2}$ **45.** $a = 3$; $b = 5$; $c = 0$; $d = -1$ **47.** $28x + 4$ **49.** $10x - 2$

6.4 Multiplication of Polynomials and Special Products

6.4 OBJECTIVES

1. Evaluate $f(x) \cdot g(x)$ for a given x
2. Multiply two polynomial functions
3. Square a binomial
4. Find the product of two binomials as a difference of squares

In Section 1.4, you saw the first exponent property and used that property to multiply monomials. Let's review.

NOTE

$a^m a^n = a^{m+n}$

Notice the use of the associative and commutative properties to "regroup" and "reorder" the factors.

Example 1

Multiplying Monomials

Multiply.

$$(8x^2y)(4x^3y^4) = (8 \cdot 4)(x^{2+3})(y^{1+4})$$

Add exponents.

Multiply.

$$= 32x^5y^5$$

CHECK YOURSELF 1

Multiply.

(a) $(4a^3b)(9a^3b^2)$ **(b)** $(-5m^3n)(7mn^5)$

We now want to extend the process to multiplying polynomial functions.

Example 2

Multiplying a Monomial and a Binomial Function

Given $f(x) = 5x^2$ and $g(x) = 3x^2 - 5x$, and letting $h(x) = f(x) \cdot g(x)$, find $h(x)$.

$$h(x) = f(x) \cdot g(x)$$
$$= 5x^2 \cdot (3x^2 - 5x)$$ Apply the distributive property.
$$= 5x^2 \cdot 3x^2 - 5x^2 \cdot 5x$$
$$= 15x^4 - 25x^3$$

CHECK YOURSELF 2

Given $f(x) = 3x^2$ and $g(x) = 4x^2 + x$, and letting $h(x) = f(x) \cdot g(x)$, find $h(x)$.

We can check this result by comparing the values of $h(x)$ and of $f(x) \cdot g(x)$ for a specific value of x. This is illustrated in Example 3.

Example 3

Multiplying a Monomial and a Binomial Function

Given $f(x) = 5x^2$ and $g(x) = 3x^2 - 5x$, and letting $h(x) = f(x) \cdot g(x)$, compare $f(1) \cdot g(1)$ with $h(1)$.

$$\begin{aligned} f(1) \cdot g(1) &= 5(1)^2 \cdot (3(1)^2 - 5(1)) \\ &= 5(3 - 5) \\ &= 5(-2) \\ &= -10 \end{aligned}$$

From Example 2, we know that

$$h(x) = 15x^4 - 25x^3$$

So

$$\begin{aligned} h(1) &= 15(1)^4 - 25(1)^3 \\ &= 15 - 25 \\ &= -10 \end{aligned}$$

Therefore, $h(1) = f(1) \cdot g(1)$.

CHECK YOURSELF 3

Given $f(x) = 3x^2$ and $g(x) = 4x^2 + x$, and letting $h(x) = f(x) \cdot g(x)$, compare $f(1) \cdot g(1)$ with $h(1)$.

The distributive property is also used to multiply two polynomial functions. To consider the pattern, let's start with the product of two binomial functions.

Example 4

Multiplying Binomial Functions

Given $f(x) = x + 3$ and $g(x) = 2x + 5$, and letting $h(x) = f(x) \cdot g(x)$, find the following.

(a) $h(x)$

$$h(x) = f(x) \cdot g(x)$$

$= (x + 3)(2x + 5)$ Apply the distributive property.

$= (x + 3)(2x) + (x + 3)(5)$ Apply the distributive property again.

$= (x)(2x) + (3)(2x) + (x)(5) + (3)(5)$

$= 2x^2 + 6x + 5x + 15$

$= 2x^2 + 11x + 15$

Notice that this ensures that each term in the first polynomial is multiplied by each term in the second polynomial.

(b) $f(1) \cdot g(1)$

$$f(1) \cdot g(1) = (1 + 3)(2(1) + 5)$$
$$= 4(7)$$
$$= 28$$

(c) $h(1)$

From part (a), we have $h(x) = 2x^2 + 11x + 15$, so

$$h(1) = 2(1)^2 + 11(1) + 15$$
$$= 2 + 11 + 15$$
$$= 28$$

Again, we see that $h(1) = f(1) \cdot g(1)$.

CHECK YOURSELF 4

Given $f(x) = 3x - 2$ and $g(x) = x + 3$, and letting $h(x) = f(x) \cdot g(x)$, find the following.

(a) $h(x)$ **(b)** $f(1) \cdot g(1)$ **(c)** $h(1)$

Certain products occur frequently enough in algebra that it is worth learning special formulas for dealing with them. Consider these products of two equal binomial factors.

NOTE
$a^2 + 2ab + b^2$
and
$a^2 - 2ab + b^2$
are called **perfect-square trinomials.**

$$(a + b)^2 = (a + b)(a + b)$$
$$= a^2 + 2ab + b^2 \quad (1)$$
$$(a - b)^2 = (a - b)(a - b)$$
$$= a^2 - 2ab + b^2 \quad (2)$$

We can summarize these statements as follows.

Rules and Properties: Squaring a Binomial

The square of a binomial has three terms, (1) the square of the first term, (2) twice the product of the two terms, and (3) the square of the last term.

$(a + b)^2 = a^2 + 2ab + b^2$

and

$(a - b)^2 = a^2 - 2ab + b^2$

Example 5

Squaring a Binomial

Find each of the following binomial squares.

NOTE Be sure to write out the expansion in detail.

(a) $(x + 5)^2 = x^2 + 2(x)(5) + 5^2$

Square of first term — Twice the product of the two terms — Square of last term

$$= x^2 + 10x + 25$$

CAUTION

Be Careful! A very common mistake in squaring binomials is to forget *the middle* term!

$(y + 7)^2$

is not equal to

$y^2 + (7)^2$

The correct square is

$y^2 + 14y + 49$

The square of a binomial is *always* a trinomial.

(b) $(2a - 7)^2 = (2a)^2 - 2(2a)(7) + (-7)^2$

$$= 4a^2 - 28a + 49$$

CHECK YOURSELF 5

Find each of the following binomial squares.

(a) $(x + 8)^2$ **(b)** $(3x - 5)^2$

Another special product involves binomials that differ only in sign. It will be extremely important in your work later in this chapter on factoring. Consider the following:

$$(a + b)(a - b) = a^2 - ab + ab + b^2$$
$$= a^2 - b^2$$

Rules and Properties: Product of Binomials Differing in Sign

$(a + b)(a - b) = a^2 - b^2$

In words, the product of two binomials that differ only in the signs of their second terms is the difference of the squares of the two terms of the binomials.

Example 6

Finding a Special Product

Multiply.

(a) $(x - 3)(x + 3) = x^2 - (3)^2$

$$= x^2 - 9$$

CAUTION

The entire term $2x$ is squared, not just the x.

(b) $(2x - 3y)(2x + 3y) = (2x)^2 - (3y)^2$

$$= 4x^2 - 9y^2$$

(c) $(5a + 4b^2)(5a - 4b^2) = (5a)^2 - (4b^2)^2$

$$= 25a^2 - 16b^4$$

CHECK YOURSELF 6

Find each of the following products.

(a) $(y + 5)(y - 5)$ **(b)** $(2x - 3)(2x + 3)$ **(c)** $(4r + 5s^2)(4r - 5s^2)$

NOTE This format ensures that each term of one polynomial multiplies each term of the other.

When multiplying two polynomials that don't fit one of the special product patterns, there are two different ways to set up the multiplication. Example 7 will illustrate the vertical approach.

Example 7

Multiplying Polynomials

Multiply $3x^3 - 2x^2 + 5$ and $3x + 2$.

Step 1

$$\begin{array}{rrrr} 3x^3 - & 2x^2 & & + 5 \\ & & 3x + & 2 \\ \hline 6x^3 - & 4x^2 & & + 10 \end{array}$$

Multiply by 2.

Step 2

$$\begin{array}{rrrrr} & 3x^3 & - 2x^2 & & + 5 \\ & & & 3x & + 2 \\ \hline & 6x^3 & - 4x^2 & & + 10 \\ 9x^4 & - 6x^3 & & + 15x & \\ \hline \end{array}$$

Multiply by $3x$. Note that we align the terms in the partial product.

Step 3

$$\begin{array}{rrrrr} & 3x^3 & - 2x^2 & & + 5 \\ & & & 3x & + 2 \\ \hline & 6x^3 & - 4x^2 & & + 10 \\ 9x^4 & - 6x^3 & & + 15x & \\ \hline 9x^4 & & - 4x^2 & + 15x & + 10 \end{array}$$

Add the partial products.

CHECK YOURSELF 7

Find the following product, using the vertical method.

$(4x^3 - 6x - 7)(3x - 2)$

A horizontal approach to the multiplication in Example 7 is also possible by the distributive property. As we see in Example 8, we first distribute $3x$ over the trinomial and then we distribute 2 over the trinomial.

Example 8

Multiplying Polynomials

Multiply $(3x + 2)(3x^3 - 2x^2 + 5)$, using a horizontal format.

Step 1

$(3x + 2)(3x^3 - 2x^2 + 5)$

Step 2

NOTE Again, this ensures that each term of one polynomial multiplies each term of the other.

$$= \underbrace{9x^4 - 6x^3 + 15x}_{\text{Step 1}} + \underbrace{6x^3 - 4x^2 + 10}_{\text{Step 2}}$$

Combine like terms.

$$= 9x^4 - 4x^2 + 15x + 10$$

Write the product in descending form.

CHECK YOURSELF 8

Find the product of Check Yourself 7, using a horizontal format.

Multiplication sometimes involves the product of more than two polynomials. In such cases, the associative property of multiplication allows us to regroup the factors to make the multiplication easier. Generally, we choose to start with the product of binomials. Example 9 illustrates this approach.

Example 9

Multiplying Polynomials

Find the products.

(a) $x(x + 3)(x - 3) = x(x^2 - 9)$ — Find the product $(x + 3)(x - 3)$.

$= x^3 - 9x$ — Then distribute x as the last step.

(b) $2x(x + 3)(2x - 1) = 2x(2x^2 + 5x - 3)$ — Find the product of the binomials.

$= 4x^3 + 10x^2 - 6x$ — Then distribute $2x$.

CHECK YOURSELF 9

Find each of the following products.

(a) $m(2m + 3)(2m - 3)$ **(b)** $3a(2a + 5)(a - 3)$

CHECK YOURSELF ANSWERS

1. **(a)** $36a^6b^3$; **(b)** $-35m^4n^6$ **2.** $h(x) = 12x^4 + 3x^3$ **3.** $f(1) \cdot g(1) = 15 = h(1)$
4. **(a)** $h(x) = 3x^2 + 7x - 6$; **(b)** $f(1) \cdot g(1) = 4$; **(c)** $h(1) = 4$
5. **(a)** $x^2 + 16x + 64$; **(b)** $9x^2 - 30x + 25$
6. **(a)** $y^2 - 25$; **(b)** $4x^2 - 9$; **(c)** $16r^2 - 25s^4$ **7.** $12x^4 - 8x^3 - 18x^2 - 9x + 14$
8. $12x^4 - 8x^3 - 18x^2 - 9x + 14$ **9.** **(a)** $4m^3 - 9m$; **(b)** $6a^3 - 3a^2 - 45a$

Name ____________

Section ______ Date ______

6.4 Exercises

In exercises 1 to 6, find each product.

1. $(4x)(5y)$

2. $(-3m)(5n)$

3. $(6x^2)(-3x^3)$

4. $(5y^4)(3y^2)$

5. $(5r^2s)(6r^3s^4)$

6. $(-8a^2b^5)(-3a^3b^2)$

In exercises 7 to 14, $f(x)$ and $g(x)$ are given. Let $h(x) = f(x) \cdot g(x)$. Find **(a)** $h(x)$, **(b)** $f(1) \cdot g(1)$, and **(c)** use the result of **(a)** to find $h(1)$.

7. $f(x) = 3x$ and $g(x) = 2x^2 - 3x$

8. $f(x) = 4x$ and $g(x) = 2x^2 - 7x$

9. $f(x) = -5x$ and $g(x) = -3x^2 - 5x + 8$

10. $f(x) = 2x^2$ and $g(x) = -7x^2 + 2x$

11. $f(x) = 4x^3$ and $g(x) = 9x^2 + 3x - 5$

12. $f(x) = 2x^3$ and $g(x) = 2x^3 - 4x$

13. $f(x) = 3x^3$ and $g(x) = 5x^2 - 4x$

14. $f(x) = -x^2$ and $g(x) = -7x^3 - 5x^2$

In exercises 15 to 24, find each product.

15. $(x + y)(x + 3y)$

16. $(x - 3y)(x + 5y)$

17. $(x - 2y)(x + 7y)$

18. $(x + 7y)(x - 3y)$

19. $(5x - 7y)(5x - 9y)$

20. $(3x - 5y)(7x + 2y)$

21. $(7x - 5y)(7x - 4y)$

22. $(9x + 7y)(3x - 2y)$

23. $(5x^2 - 2y)(3x + 2y^2)$

24. $(6x^2 - 5y^2)(3x^2 - 2y)$

ANSWERS

1. ______ 2. ______
3. ______ 4. ______
5. ______
6. ______
7. ______
8. ______
9. ______
10. ______
11. ______
12. ______
13. ______
14. ______
15. ______
16. ______
17. ______
18. ______
19. ______
20. ______
21. ______
22. ______
23. ______
24. ______

ANSWERS

25. ______
26. ______
27. ______
28. ______
29. ______
30. ______
31. ______
32. ______
33. ______
34. ______
35. ______
36. ______
37. ______
38. ______
39. ______
40. ______
41. ______
42. ______
43. ______
44. ______
45. ______
46. ______
47. ______
48. ______
49. ______
50. ______

In exercises 25 to 38, multiply polynomial expressions using the special product formulas.

25. $(x + 5)^2$

26. $(x - 7)^2$

27. $(2x - 3)^2$

28. $(5x + 3)^2$

29. $(4x - 3y)^2$

30. $(7x - 5y)^2$

31. $(4x + 3y^2)^2$

32. $(3x^3 - 7y)^2$

33. $(x - 3y)(x + 3y)$

34. $(x + 5y)(x - 5y)$

35. $(2x - 3y)(2x + 3y)$

36. $(5x + 3y)(5x - 3y)$

37. $(4x^2 + 3y)(4x^2 - 3y)$

38. $(7x - 6y^2)(7x + 6y^2)$

In exercises 39 to 42, multiply using the vertical format.

39. $(3x - y)(x^2 + 3xy - y^2)$

40. $(5x + y)(x^2 - 3xy + y^2)$

41. $(x - 2y)(x^2 + 2xy + 4y^2)$

42. $(x + 3y)(x^2 - 3xy + 9y^2)$

In exercises 43 to 46, simplify each function.

43. $f(x) = x(x - 3)(x + 1)$

44. $f(x) = x(x + 4)(x - 2)$

45. $f(x) = 2x(x - 5)(x + 4)$

46. $f(x) = x^2(x - 4)(x^2 + 5)$

Multiply the following.

47. $\left(\frac{x}{2} + \frac{2}{3}\right)\left(\frac{2x}{3} - \frac{2}{5}\right)$

48. $\left(\frac{x}{3} + \frac{3}{4}\right)\left(\frac{3x}{4} - \frac{3}{5}\right)$

49. $[x + (y - 2)][x - (y - 2)]$

50. $[x + (3 - y)][x - (3 - y)]$

If the polynomial $p(x)$ represents the selling price of an object, then the polynomial $R(x)$, in which $R(x) = x \cdot p(x)$, is the revenue produced by selling x objects. Use this information to solve exercises 51 and 52.

51. If $p(x) = 100 - 0.2x$, find $R(x)$. Find $R(50)$.

52. If $p(x) = 250 - 0.5x$, find $R(x)$. Find $R(20)$.

In exercises 53 to 56, label the statements as true or false.

53. $(x + y)^2 = x^2 + y^2$

54. $(x - y)^2 = x^2 - y^2$

55. $(x + y)^2 = x^2 + 2xy + y^2$

56. $(x - y)^2 = x^2 - 2xy + y^2$

57. Area. The length of a rectangle is given by $3x + 5$ centimeters (cm) and the width is given by $2x - 7$ cm. Express the area of the rectangle in terms of x.

58. Area. The base of a triangle measures $3y + 7$ inches (in.) and the height is $2y - 3$ in. Express the area of the triangle in terms of y.

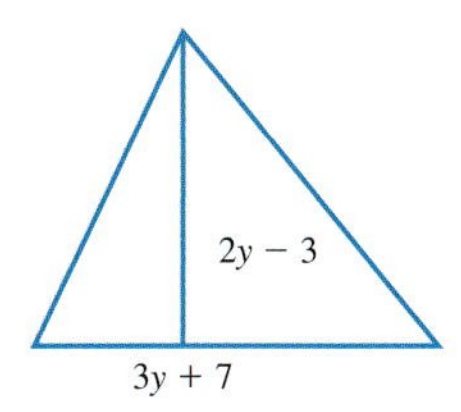

59. Revenue. The price of an item is given by $p = 10 - 3x$. If the revenue generated is found by multiplying the number of items x sold by the price of an item, find the polynomial that represents the revenue.

60. Revenue. The price of an item is given by $p = 100 - 2x^2$. Find the polynomial that represents the revenue generated from the sale of x items.

61. Tree planting. Suppose an orchard is planted with trees in straight rows. If there are $5x - 4$ rows with $5x - 4$ trees in each row, how many trees are there in the orchard?

62. Area of a square. A square has sides of length $3x - 2$ centimeters (cm). Express the area of the square as a polynomial.

ANSWERS

51. ______
52. ______
53. ______
54. ______
55. ______
56. ______
57. ______
58. ______
59. ______
60. ______
61. ______
62. ______

ANSWERS

63. ______

64. ______

65. ______

66. ______

67. ______

63. Area of a rectangle. The length and width of a rectangle are given by two consecutive odd integers. Write an expression for the area of the rectangle.

64. Area of a rectangle. The length of a rectangle is 6 less than three times the width. Write an expression for the area of the rectangle.

65. Work with another student to complete this table and write the polynomial. A paper box is to be made from a piece of cardboard 20 inches (in.) wide and 30 in. long. The box will be formed by cutting squares out of each of the four corners and folding up the sides to make a box.

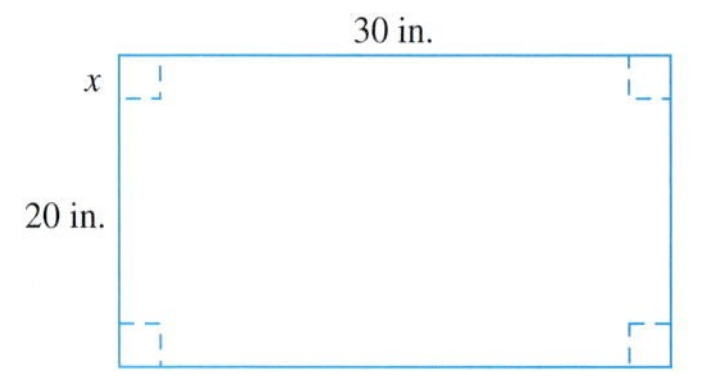

If x is the dimension of the side of the square cut out of the corner, when the sides are folded up, the box will be x inches tall. You should use a piece of paper to try this to see how the box will be made. Complete the following chart.

Length of Side of Corner Square	Length of Box	Width of Box	Depth of Box	Volume of Box
1 in.				
2 in.				
3 in.				
n in.				

Write general formulas for the width, length, and height of the box and a general formula for the *volume* of the box, and simplify it by multiplying. The variable will be the height, the side of the square cut out of the corners. What is the highest power of the variable in the polynomial you have written for the volume? Extend the table to decide what the dimensions are for a box with maximum volume. Draw a sketch of this box and write in the dimensions.

66. Complete the following statement: $(a + b)^2$ is not equal to $a^2 + b^2$ because. . . . But, wait! Isn't $(a + b)^2$ *sometimes* equal to $a^2 + b^2$? What do you think?

67. Is $(a + b)^3$ ever equal to $a^3 + b^3$? Explain.

68. ____________

69. ____________

70. ____________

71. ____________

72. ____________

73. ____________

68. In the following figures, identify the length and the width of the square, and then find the area.

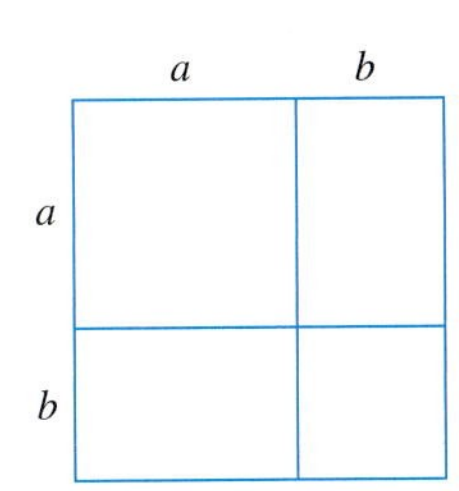

Length = _______

Width = _______

Area = _______

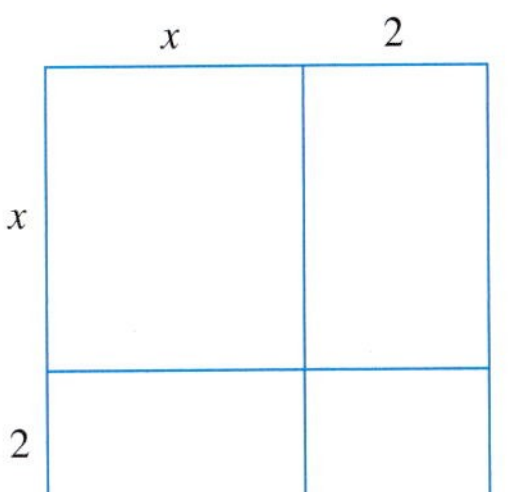

Length = _______

Width = _______

Area = _______

69. The square shown is x units on a side. The area is _______.

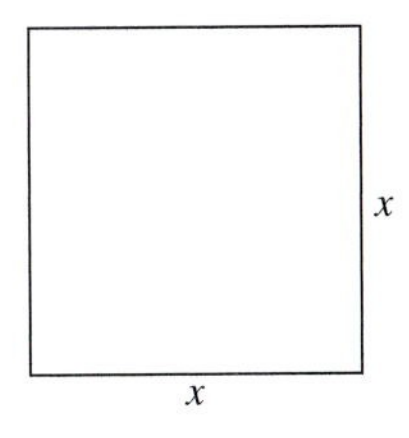

Draw a picture of what happens when the sides are doubled. The area is _______.

Continue the picture to show what happens when the sides are tripled. The area is _______.

If the sides are quadrupled, the area is _______.

In general, if the sides are multiplied by n, the area is _______.

If each side is increased by 3, the area is increased by _______.

If each side is decreased by 2, the area is decreased by _______.

In general, if each side is increased by n, the area is increased by _______, and if each side is decreased by n, the area is decreased by _______.

For each of the following problems, let x represent the number, then write an expression for the product.

70. The product of 6 more than a number and 6 less than that number.

71. The square of 5 more than a number.

72. The square of 4 less than a number.

73. The product of 5 less than a number and 5 more than that number.

ANSWERS

74. ____

75. ____

76. ____

77. ____

78. ____

79. ____

Note that $(28)(32) = (30 - 2)(30 + 2) = 900 - 4 = 896$. Use this pattern to find each of the following products.

74. $(49)(51)$

75. $(27)(33)$

76. $(34)(26)$

77. $(98)(102)$

78. $(55)(65)$

79. $(64)(56)$

Answers

1. $20xy$ **3.** $-18x^5$ **5.** $30r^5s^5$ **7.** **(a)** $6x^3 - 9x^2$; **(b)** -3; **(c)** -3
9. **(a)** $15x^3 + 25x^2 - 40x$; **(b)** 0; **(c)** 0 **11.** **(a)** $36x^5 + 12x^4 - 20x^3$; **(b)** 28; **(c)** 28
13. **(a)** $15x^5 - 12x^4$; **(b)** 3; **(c)** 3 **15.** $x^2 + 4xy + 3y^2$
17. $x^2 + 5xy - 14y^2$ **19.** $25x^2 - 80xy + 63y^2$ **21.** $49x^2 - 63xy + 20y^2$
23. $15x^3 + 10x^2y^2 - 6xy - 4y^3$ **25.** $x^2 + 10x + 25$ **27.** $4x^2 - 12x + 9$
29. $16x^2 - 24xy + 9y^2$ **31.** $16x^2 + 24xy^2 + 9y^4$ **33.** $x^2 - 9y^2$
35. $4x^2 - 9y^2$ **37.** $16x^4 - 9y^2$ **39.** $3x^3 + 8x^2y - 6xy^2 + y^3$
41. $x^3 - 8y^3$ **43.** $x^3 - 2x^2 - 3x$ **45.** $2x^3 - 2x^2 - 40x$
47. $\frac{x^2}{3} + \frac{11x}{45} - \frac{4}{15}$ **49.** $x^2 - y^2 + 4y - 4$ **51.** $100x - 0.2x^2$, 4500
53. False **55.** True **57.** $6x^2 - 11x - 35\ \text{cm}^2$ **59.** $10x - 3x^2$
61. $25x^2 - 40x + 16$ **63.** $x(x + 2)$ or $x^2 + 2x$ **65.**

67.  **69.** x^2; $4x^2$; $9x^2$; $16x^2$; n^2x^2; $6x + 9$; $4x - 4$; $2xn + n^2$; $2xn - n^2$

71. $x^2 + 10x + 25$ **73.** $x^2 - 25$ **75.** 891 **77.** 9996 **79.** 3584

6.5 Common Factors and Factoring by Grouping

OBJECTIVES

1. Remove the greatest common factor (GCF)
2. Factor by grouping

When the integers 3 and 5 are multiplied, the product is 15. We call 3 and 5 **factors** of 15.

Writing $3 \cdot 5 = 15$ indicates multiplication, but when we write $15 = 3 \cdot 5$, we say we have **factored** 15. In general, factoring is the reverse of multiplication. We can extend this idea to algebra.

From the last section on multiplying polynomials and special products, we know that

$$(2x + 3)(x - 2) = 2x^2 - x - 6$$

But what if we begin with $2x^2 - x - 6$? How do we write the polynomial as a product of other polynomials?

NOTE Recall that a prime number is any integer greater than 1 that has only itself and 1 as factors. Writing

$15 = 3 \cdot 5$

as a product of prime factors is called the **completely factored form** for 15.

There are a number of methods we can use to factor polynomials: factoring out the greatest common factor (GCF), factoring by grouping, factoring the difference of two squares, and factoring the sum or difference of two cubes.

The first step in factoring is always to factor out the **greatest common factor (GCF),** if any.

NOTE In fact, we will see that factoring out the GCF is the *first* method to try in any of the factoring problems we will discuss.

Definitions: Greatest Common Factor

The **greatest common factor (GCF)** of a polynomial is the monomial with the highest degree and the largest numerical coefficient that is a factor of each term of the polynomial.

Once the GCF is found, we apply the distributive property to write the original polynomial as a product of the GCF and the polynomial formed by dividing each term by that GCF. Example 1 illustrates this approach.

Example 1

Factoring Out a Monomial

(a) Factor $4x^3 - 12x$.

NOTE Here 4 is the GCF of the numerical coefficients, and the highest common power of x is 1, so the GCF of $4x^3 - 12x$ is $4x$.

Note that the numerical coefficient of the GCF is 4 and the variable factor is x (the highest power common to each term). So

$$4x^3 - 12x = 4x \cdot x^2 - 4x \cdot 3$$
$$= 4x(x^2 - 3)$$

(b) Factor $6a^3b^2 - 12a^2b^3 + 24a^4b^4$.

NOTE Here 6 is the GCF of the numerical coefficients, the highest common power of a is 2, and the highest common power of b is 2.

Here the GCF is $6a^2b^2$, and we can write

$$6a^3b^2 - 12a^2b^3 + 24a^4b^4$$
$$= 6a^2b^2 \cdot a - 6a^2b^2 \cdot 2b + 6a^2b^2 \cdot 4a^2b^2$$
$$= 6a^2b^2 (a - 2b + 4a^2b^2)$$

(c) Factor $8m^4n^2 - 16m^2n^2 + 24mn^3 - 32mn^4$.

Here the GCF is $8mn^2$, and we have

$8m^4n^2 - 16m^2n^2 + 24mn^3 - 32mn^4$

$= 8mn^2 \cdot m^3 - 8mn^2 \cdot 2m + 8mn^2 \cdot 3n - 8mn^2 \cdot 4n^2$

$= 8mn^2(m^3 - 2m + 3n - 4n^2)$

Notice that in Example 1(b) it is also true that

$6a^3b^2 - 12a^2b^3 + 24a^4b^4 = 3ab(2a^2b - 4ab^2 + 8a^3b^3)$

However, this is not in *completely factored form* because we agree that this means factoring out the GCF (that monomial with the largest possible coefficient and degree). In this case, we must remove $6a^2b^2$.

CHECK YOURSELF 1

Write each of the following in completely factored form.

(a) $7x^3y - 21x^2y^2 + 28xy^3$

(b) $15m^4n^4 - 5mn^3 + 20mn^2 - 25m^2n^2$

A related factoring method is called **factoring by grouping.** We introduce this method in Example 2.

Example 2

Finding a Common Factor

(a) Factor $3x(x + y) + 2(x + y)$.

We see that *the binomial* $x + y$ is a common factor and can be removed.

NOTE Because of the commutative property, the factors can be written in either order.

$3x(x + y) + 2(x + y)$

$= (x + y) \cdot 3x + (x + y) \cdot 2$

$= (x + y)(3x + 2)$

(b) Factor $3x^2(x - y) + 6x(x - y) + 9(x - y)$.

We note that here the GCF is $3(x - y)$. Factoring as before, we have

$3(x - y)(x^2 + 2x + 3)$

CHECK YOURSELF 2

Completely factor each of the polynomials.

(a) $7a(a - 2b) + 3(a - 2b)$

(b) $4x^2(x + y) - 8x(x + y) - 16(x + y)$

If the terms of a polynomial have no common factor (other than 1), factoring by grouping is the preferred method, as illustrated in Example 3.

Example 3

Factoring by Grouping Terms

Suppose we want to factor the polynomial

$ax - ay + bx - by$

NOTE Notice that our example has *four* terms. That is the clue for trying the factoring by grouping method.

As you can see, the polynomial has no common factors. However, look at what happens if we separate the polynomial into *two groups* of *two terms*.

$$ax - ay + bx - by$$

$$= \underbrace{ax - ay}_{(1)} + \underbrace{bx - by}_{(2)}$$

Now *each* group has a common factor, and we can write the polynomial as

$a(x - y) + b(x - y)$

In this form, we can see that $x - y$ is the GCF. Factoring out $x - y$, we get

$a(x - y) + b(x - y) = (x - y)(a + b)$

CHECK YOURSELF 3

Use the factoring by grouping method.

$x^2 - 2xy + 3x - 6y$

Be particularly careful of your treatment of algebraic signs when applying the factoring by grouping method. Consider Example 4.

Example 4

Factoring by Grouping Terms

Factor $2x^3 - 3x^2 - 6x + 9$.

We group the polynomial as follows.

$$\underbrace{2x^3 - 3x^2}_{(1)} - \underbrace{6x + 9}_{(2)}$$ Remove the common factor of -3 from the second two terms.

$$= x^2(2x - 3) - 3(2x - 3)$$

$$= (2x - 3)(x^2 - 3)$$

NOTE Notice that $9 = (-3)(-3)$.

CHECK YOURSELF 4

Factor by grouping.

$3y^3 + 2y^2 - 6y - 4$

It may also be necessary to change the order of the terms as they are grouped. Look at Example 5.

Example 5

Factoring by Grouping Terms

Factor $x^2 - 6yz + 2xy - 3xz$.

Grouping the terms as before, we have

$$\underbrace{x^2 - 6yz}_{(1)} + \underbrace{2xy - 3xz}_{(2)}$$

Do you see that we have accomplished nothing because there are no common factors in the first group?

We can, however, rearrange the terms to write the original polynomial as

$$\underbrace{x^2 + 2xy}_{(1)} - \underbrace{3xz - 6yz}_{(2)}$$

$= x(x + 2y) - 3z(x + 2y)$

$= (x + 2y)(x - 3z)$

We can now remove the common factor of $x + 2y$ in group (1) and group (2).

Note: It is often true that the grouping can be done in more than one way. The factored form will be the same.

CHECK YOURSELF 5

We can write the polynomial of Example 5 *as*

$x^2 - 3xz + 2xy - 6yz$

Factor, and verify that the factored form is the same in either case.

CHECK YOURSELF ANSWERS

1. **(a)** $7xy(x^2 - 3xy + 4y^2)$; **(b)** $5mn^2(3m^3n^2 - n + 4 - 5m)$
2. **(a)** $(a - 2b)(7a + 3)$; **(b)** $4(x + y)(x^2 - 2x - 4)$ **3.** $(x - 2y)(x + 3)$
4. $(3y + 2)(y^2 - 2)$ **5.** $(x - 3z)(x + 2y)$

Name ______________________

Section ________ Date ________

6.5 Exercises

Find the greatest common factor for each of the following sets of terms.

1. 10, 12

2. 15, 35

3. 16, 32, 88

4. 55, 33, 132

5. x^2, x^5

6. y^7, y^9

7. a^3, a^6, a^9

8. b^4, b^6, b^8

9. $5x^4, 10x^5$

10. $8y^9, 24y^3$

11. $8a^4, 6a^6, 10a^{10}$

12. $9b^3, 6b^5, 12b^4$

13. $9x^2y, 12xy^2, 15x^2y^2$

14. $12a^3b^2, 18a^2b^3, 6a^4b^4$

15. $15ab^3, 10a^2bc, 25b^2c^3$

16. $9x^2, 3xy^3, 6y^3$

17. $15a^2bc^2, 9ab^2c^2, 6a^2b^2c^2$

18. $18x^3y^2z^3, 27x^4y^2z^3, 81xy^2z$

19. $(x + y)^2, (x + y)^3$

20. $12(a + b)^4, 4(a + b)^3$

In exercises 21 to 38, completely factor each polynomial.

21. $6x + 9y$

22. $7a - 21b$

23. $4x^2 - 12x$

24. $5a^2 + 25a$

ANSWERS

1. ______
2. ______
3. ______
4. ______
5. ______
6. ______
7. ______
8. ______
9. ______
10. ______
11. ______
12. ______
13. ______
14. ______
15. ______
16. ______
17. ______
18. ______
19. ______
20. ______
21. ______
22. ______
23. ______
24. ______

ANSWERS

25. ______
26. ______
27. ______
28. ______
29. ______
30. ______
31. ______
32. ______
33. ______
34. ______
35. ______
36. ______
37. ______
38. ______
39. ______
40. ______
41. ______
42. ______
43. ______
44. ______
45. ______
46. ______
47. ______
48. ______
49. ______
50. ______

25. $18m^2n + 27mn^2$

26. $24c^2d^3 - 30c^3d^2$

27. $5x^3 - 15x^2 + 25x$

28. $28r^3 - 21r^2 + 7r$

29. $12m^3n - 6mn + 18mn^2$

30. $18w^2z + 27wz - 36wz^2$

31. $4a^3b^2 - 8a^2b + 12ab^2 - 4ab$

32. $9r^2s^3 + 27r^3s^2 - 6r^2s^2 + 3rs$

33. $x(y - z) + 3(y - z)$

34. $2a(c - d) - b(c - d)$

35. $3(m - n) + 5(m - n)^2$

36. $4(r + 2s) - 3(r + 2s)^2$

37. $5x^2(x - y) - 10x(x - y) + 15(x - y)$

38. $7a^2(a + 2b) + 21a(a + 2b) - 14(a + 2b)$

Determine if the factoring in each of the following is correct.

39. $x^2 - x - 6 = (x - 3)(x + 2)$

40. $x^2 + x - 12 = (x + 4)(x - 3)$

41. $x^2 + x - 12 = (x + 6)(x - 2)$

42. $x^2 + 2x - 8 = (x + 8)(x - 1)$

43. $2x^2 - 5x - 3 = (2x + 1)(x - 3)$

44. $6x^2 - 13x + 6 = (3x - 2)(2x - 3)$

In exercises 45 to 50, factor each polynomial by grouping.

45. $ab - ac + b^2 - bc$

46. $ax + 2a + bx + 2b$

47. $6r^2 + 12rs - r - 2s$

48. $2mn - 4m^2 + 3n - 6m$

49. $ab^2 - 2b^2 + 3a - 6$

50. $r^2s^2 - 3s^2 - 2r^2 + 6$

In exercises 51 to 54, factor each polynomial by grouping. *Hint:* Consider a rearrangement of terms.

51. $x^2 - 10y - 5xy + 2x$

52. $a^2 - 12b + 3ab - 4a$

53. $m^2 - 6n^3 + 2mn^2 - 3mn$

54. $r^2 - 3rs^2 - 12s^3 + 4rs$

In exercises 55 to 58, factor. *Hint:* Consider *three* groups of *two* terms.

55. $x^3 - x^2 + 3x + x^2y - xy + 3y$

56. $m^3 - m^2 - 4m + 2m^2n - 2mn - 8n$

57. $a^3 - a^2b - 3a^2 + 3ab + 3a - 3b$

58. $r^3 + 2r^2s + r^2 + 2rs - 3r - 6s$

59. The GCF of $2x - 6$ is 2. The GCF of $5x + 10$ is 5. Find the greatest common factor of the product $(2x - 6)(5x + 10)$.

60. The GCF of $3z + 12$ is 3. The GCF of $4z + 8$ is 4. Find the GCF of the product $(3z + 12)(4z + 8)$.

61. The GCF of $2x^3 - 4x$ is $2x$. The GCF of $3x + 6$ is 3. Find the GCF of the product $(2x^3 - 4x)(3x + 6)$.

62. State, in a sentence, the rule that the previous three exercises illustrated.

63. For the monomials x^4y^2, x^8y^6, and x^9y^4, explain how you can determine the GCF by inspecting exponents.

64. It is not possible to use the grouping method to factor $2x^3 + 6x^2 + 8x + 4$. Is it correct to conclude that the polynomial is prime? Justify your answer.

65. **Area of a rectangle.** The area of a rectangle with width t is given by $33t - t^2$. Factor the expression and determine the length of the rectangle in terms of t.

66. **Area of a rectangle.** The area of a rectangle of length x is given by $3x^2 + 5x$. Find the width of the rectangle.

67. For centuries, mathematicians have found factoring numbers into prime factors a fascinating subject. A prime number is a number that cannot be written as a product of any whole numbers but 1 and itself. The list of positive primes begins with 2 because 1 is not considered a prime number and then goes on: 3, 5, 7, 11, ... What are the first 10 primes? What are the primes less than 100? If you list the numbers from 1 to 100 and then cross out all numbers that are factors of 2, 3, 5, and 7, what is left? Are all the numbers not crossed out prime? Write a paragraph to explain why this might be so. You might want to investigate the sieve of Eratosthenes, a system from 230 B.C. for finding prime numbers.

ANSWERS

51. ______

52. ______

53. ______

54. ______

55. ______

56. ______

57. ______

58. ______

59. ______

60. ______

61. ______

62. ______

63. ______

64. ______

65. ______

66. ______

67. ______

ANSWERS

68.  

69. 

68. If we could make a list of all the prime numbers, what number would be at the end of the list? Because there are an infinite number of prime numbers, there is no "largest prime number." But is there some formula that will give us all the primes? Here are some formulas proposed over the centuries:

$n^2 + n + 17 \qquad 2n^2 + 29 \qquad n^2 - n + 11$

In all these expressions, $n = 1, 2, 3, 4, \ldots$, that is, a positive integer beginning with 1. Investigate these expressions with a partner. Do the expressions give prime numbers when they are evaluated for these values of n? Do the expressions give *every* prime in the range of resulting numbers? Can you put in *any* positive number for n?

69. How are primes used in coding messages and for security? Work together to decode the messages. The messages are coded using this code: After the numbers are factored into prime factors, the power of 2 gives the number of the letter in the alphabet. This code would be easy for a code breaker to figure out, but you might make up a code that would be more difficult to break.

(a) 1310720, 229376, 1572864, 1760, 460, 2097152, 336

(b) 786432, 143, 4608, 278528, 1344, 98304, 1835008, 352, 4718592, 5242880

(c) Code a message using this rule. Exchange your message with a partner to decode it.

Answers

1. 2 **3.** 8 **5.** x^2 **7.** a^3 **9.** $5x^4$ **11.** $2a^4$ **13.** $3xy$ **15.** $5b$ **17.** $3abc^2$ **19.** $(x + y)^2$ **21.** $3(2x + 3y)$ **23.** $4x(x - 3)$ **25.** $9mn(2m + 3n)$ **27.** $5x(x^2 - 3x + 5)$ **29.** $6mn(2m^2 - 1 + 3n)$ **31.** $4ab(a^2b - 2a + 3b - 1)$ **33.** $(y - z)(x + 3)$ **35.** $(m - n)(3 + 5m - 5n)$ **37.** $5(x - y)(x^2 - 2x + 3)$ **39.** Correct **41.** Incorrect **43.** Correct **45.** $(b - c)(a + b)$ **47.** $(r + 2s)(6r - 1)$ **49.** $(a - 2)(b^2 + 3)$ **51.** $(x + 2)(x - 5y)$ **53.** $(m - 3n)(m + 2n^2)$ **55.** $(x + y)(x^2 - x + 3)$ **57.** $(a - b)(a^2 - 3a + 3)$ **59.** 10 **61.** $6x$ **63.** **65.** $33 - t$ **67.**  **69.** 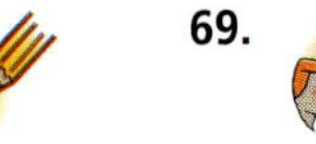 

6.6 Factoring Special Polynomials

OBJECTIVES

1. Factor the difference of two squares
2. Factor the sum or difference of two cubes

In this section, we will look at several special polynomials. These polynomials are special because they fit a recognizable pattern. Pattern recognition is an important element of mathematics. Many mathematical discoveries were made because somebody recognized a pattern.

The first pattern, which we saw in Section 6.4, is called the difference of two squares.

CAUTION

What about the sum of two squares, such as

$x^2 + 25$

In general, it is *not possible* to factor (using real numbers) a sum of two squares. So

$(x^2 + 25) \neq (x + 5)(x + 5)$

Rules and Properties: The Difference of Two Squares

$$a^2 - b^2 = (a + b)(a - b) \quad (1)$$

In words: The product of the sum and difference of two terms gives the *difference of two squares.*

Equation (1) is easy to apply in factoring. It is just a matter of recognizing a binomial as the difference of two squares.

To confirm this identity, use the FOIL method to multiply

$(a + b)(a - b)$

Example 1

Factoring the Difference of Two Squares

NOTE We are looking for perfect squares—the exponents must be multiples of 2 and the coefficients perfect squares—1, 4, 9, 16, and so on.

(a) Factor $x^2 - 25$.

Note that our example has two terms—a clue to try factoring as the difference of two squares.

$$x^2 - 25 = (x)^2 - (5)^2$$
$$= (x + 5)(x - 5)$$

(b) Factor $9a^2 - 16$.

$$9a^2 - 16 = (3a)^2 - (4)^2$$
$$= (3a + 4)(3a - 4)$$

(c) Factor $25m^4 - 49n^2$.

$$25m^4 - 49n^2 = (5m^2)^2 - (7n)^2$$
$$= (5m^2 + 7n)(5m^2 - 7n)$$

CHECK YOURSELF 1

Factor each of the following binomials.

(a) $y^2 - 36$ **(b)** $25m^2 - n^2$ **(c)** $16a^4 - 9b^2$

We mentioned earlier that factoring out a common factor should always be considered your first step. Then other steps become obvious. Consider Example 2.

Example 2

Factoring the Difference of Two Squares

Factor $a^3 - 16ab^2$.

First note the common factor of a. Removing that factor, we have

$$a^3 - 16ab^2 = a(a^2 - 16b^2)$$

We now see that the binomial factor is a difference of squares, and we can continue to factor as before. So

$$a^3 - 16ab^2 = a(a + 4b)(a - 4b)$$

CHECK YOURSELF 2

Factor $2x^3 - 18xy^2$.

You may also have to apply the difference of two squares method *more than once* to completely factor a polynomial.

Example 3

Factoring the Difference of Two Squares

Factor $m^4 - 81n^4$.

$$m^4 - 81n^4 = (m^2 + 9n^2)(m^2 - 9n^2)$$

Do you see that we are not done in this case? Because $m^2 - 9n^2$ is still factorable, we can continue to factor as follows.

$$m^4 - 81n^4 = (m^2 + 9n^2)(m + 3n)(m - 3n)$$

NOTE The other binomial factor, $m^2 + 9n^2$, is a *sum of two squares,* which cannot be factored further.

CHECK YOURSELF 3

Factor $x^4 - 16y^4$.

Two additional patterns for factoring certain binomials include the sum or difference of two cubes.

Rules and Properties: The Sum or Difference of Two Cubes

$$a^3 + b^3 = (a + b)(a^2 - ab + b^2) \quad (2)$$

$$a^3 - b^3 = (a - b)(a^2 + ab + b^2) \quad (3)$$

NOTE Be sure you take the time to expand the product on the right-hand side to confirm the identity.

Example 4

Factoring the Sum or Difference of Two Cubes

NOTE We are now looking for perfect cubes—the exponents must be multiples of 3 and the coefficients perfect cubes—1, 8, 27, 64, and so on.

(a) Factor $x^3 + 27$.

The first term is the cube of x, and the second is the cube of 3, so we can apply equation (2). Letting $a = x$ and $b = 3$, we have

$$x^3 + 27 = (x + 3)(x^2 - 3x + 9)$$

(b) Factor $8w^3 - 27z^3$.

This is a difference of cubes, so use equation (3).

$$\begin{aligned} 8w^3 - 27z^3 &= (2w - 3z)[(2w)^2 + (2w)(3z) + (3z)^2] \\ &= (2w - 3z)(4w^2 + 6wz + 9z^2) \end{aligned}$$

NOTE Again, looking for a *common factor* should be your first step.

(c) Factor $5a^3b - 40b^4$.

First note the common factor of $5b$. The binomial is the difference of cubes, so use equation (3).

NOTE Remember to write the GCF as a part of the final factored form.

$$\begin{aligned} 5a^3b - 40b^4 &= 5b(a^3 - 8b^3) \\ &= 5b(a - 2b)(a^2 + 2ab + 4b^2) \end{aligned}$$

CHECK YOURSELF 4

Factor completely.

(a) $27x^3 + 8y^3$ **(b)** $3a^4 - 24ab^3$

In each example in this section, we factored a polynomial expression. If we are given a polynomial function to factor, there is no change in the ordered pairs represented by the function after it is factored.

Example 5

Factoring a Polynomial Function

Given the function $f(x) = 9x^2 + 15x$, complete the following.

(a) Find $f(1)$.

$$\begin{aligned} f(1) &= 9(1)^2 + 15(1) \\ &= 9 + 15 \\ &= 24 \end{aligned}$$

(b) Factor $f(x)$.

$$\begin{aligned} f(x) &= 9x^2 + 15x \\ &= 3x(3x + 5) \end{aligned}$$

(c) Find $f(1)$ from the factored form of $f(x)$.

$$\begin{aligned} f(1) &= 3(1)(3(1) + 5) \\ &= 3(8) \\ &= 24 \end{aligned}$$

CHECK YOURSELF 5

Given the function $f(x) = 16x^5 + 10x^2$, complete the following.

(a) Find $f(1)$.

(b) Factor $f(x)$.

(c) Find $f(1)$ from the factored form of $f(x)$.

CHECK YOURSELF ANSWERS

1. **(a)** $(y + 6)(y - 6)$; **(b)** $(5m + n)(5m - n)$; **(c)** $(4a^2 + 3b)(4a^2 - 3b)$
2. $2x(x + 3y)(x - 3y)$ **3.** $(x^2 + 4y^2)(x + 2y)(x - 2y)$
4. **(a)** $(3x + 2y)(9x^2 - 6xy + 4y^2)$; **(b)** $3a(a - 2b)(a^2 + 2ab + 4b^2)$
5. **(a)** 26; **(b)** $2x^2(8x^3 + 5)$; **(c)** 26

Name ______________

Section ________ Date ________

6.6 Exercises

For each of the following binomials, state whether the binomial is a difference of squares.

1. $3x^2 + 2y^2$

2. $5x^2 - 7y^2$

3. $16a^2 - 25b^2$

4. $9n^2 - 16m^2$

5. $16r^2 + 4$

6. $p^2 - 45$

7. $16a^2 - 12b^3$

8. $9a^2b^2 - 16c^2d^2$

9. $a^2b^2 - 25$

10. $4a^3 - b^3$

Factor the following binomials.

11. $x^2 - 49$

12. $m^2 - 64$

13. $a^2 - 81$

14. $b^2 - 36$

15. $9p^2 - 1$

16. $4x^2 - 9$

17. $25a^2 - 16$

18. $16m^2 - 49$

19. $x^2y^2 - 25$

20. $m^2n^2 - 9$

21. $4c^2 - 25d^2$

22. $9a^2 - 49b^2$

23. $49p^2 - 64q^2$

24. $25x^2 - 36y^2$

25. $x^4 - 16y^2$

26. $a^2 - 25b^4$

ANSWERS

1. ______ 2. ______
3. ______
4. ______
5. ______
6. ______
7. ______
8. ______
9. ______
10. ______
11. ______
12. ______
13. ______
14. ______
15. ______
16. ______
17. ______
18. ______
19. ______
20. ______
21. ______
22. ______
23. ______
24. ______
25. ______
26. ______

ANSWERS

27. ______
28. ______
29. ______
30. ______
31. ______
32. ______
33. ______
34. ______
35. ______
36. ______
37. ______
38. ______
39. ______
40. ______
41. ______
42. ______
43. ______
44. ______
45. ______
46. ______
47. ______
48. ______
49. ______
50. ______

27. $a^3 - 4ab^2$

28. $9p^2q - q^3$

29. $a^4 - 16b^4$

30. $81x^4 - y^4$

31. $x^3 + 64$

32. $y^3 - 8$

33. $m^3 - 125$

34. $b^3 + 27$

35. $a^3b^3 - 27$

36. $p^3q^3 - 64$

37. $8w^3 + z^3$

38. $c^3 - 27d^3$

39. $r^3 - 64s^3$

40. $125x^3 + y^3$

41. $8x^3 - 27y^3$

42. $64m^3 - 27n^3$

43. $8x^3 + y^6$

44. $m^6 - 27n^3$

45. $4x^3 - 32y^3$

46. $3a^3 + 81b^3$

47. $18x^3 - 2xy^2$

48. $50a^2b - 2b^3$

49. $12m^3n - 75mn^3$

50. $63p^4 - 7p^2q^2$

ANSWERS

51. ______
52. ______
53. ______
54. ______
55. ______
56. ______
57. ______
58. ______
59. ______
60. ______
61. ______
62. ______
63. ______
64. ______
65. ______
66. ______
67. ______
68. ______

For each of the functions in exercises 51 to 56, **(a)** find $f(1)$, **(b)** factor $f(x)$, and **(c)** find $f(1)$ from the factored form of $f(x)$.

51. $f(x) = 12x^5 + 21x^2$

52. $f(x) = -6x^3 - 10x$

53. $f(x) = -8x^5 + 20x$

54. $f(x) = 5x^5 - 35x^3$

55. $f(x) = x^5 + 3x^2$

56. $f(x) = 6x^6 - 16x^5$

Factor each expression.

57. $x^2(x + y) - y^2(x + y)$

58. $a^2(b - c) - 16b^2(b - c)$

59. $2m^2(m - 2n) - 18n^2(m - 2n)$

60. $3a^3(2a + b) - 27ab^2(2a + b)$

61. Find a value for k so that $kx^2 - 25$ will have the factors $2x + 5$ and $2x - 5$.

62. Find a value for k so that $9m^2 - kn^2$ will have the factors $3m + 7n$ and $3m - 7n$.

63. Find a value for k so that $2x^3 - kxy^2$ will have the factors $2x$, $x - 3y$, and $x + 3y$.

64. Find a value for k so that $20a^3b - kab^3$ will have the factors $5ab$, $2a - 3b$, and $2a + 3b$.

65. Complete the following statement in complete sentences: "To factor a number you . . . ."

66. Complete this statement: "To factor an algebraic expression into prime factors means . . . ."

67. Verify the formula for factoring the sum of two cubes by finding the product $(a + b)(a^2 - ab + b^2)$.

68. Verify the formula for factoring the difference of two cubes by finding the product $(a - b)(a^2 + ab + b^2)$.

ANSWERS

69. ____________________

70. ____________________

69. What are the characteristics of a monomial that is a perfect cube?

70. Suppose you factored the polynomial $4x^2 - 16$ as follows:

$4x^2 - 16 = (2x + 4)(2x - 4)$

Would this be in completely factored form? If not, what would be the final form?

Answers

1. No **3.** Yes **5.** No **7.** No **9.** Yes **11.** $(x + 7)(x - 7)$
13. $(a + 9)(a - 9)$ **15.** $(3p + 1)(3p - 1)$ **17.** $(5a + 4)(5a - 4)$
19. $(xy + 5)(xy - 5)$ **21.** $(2c + 5d)(2c - 5d)$ **23.** $(7p + 8q)(7p - 8q)$
25. $(x^2 + 4y)(x^2 - 4y)$ **27.** $a(a + 2b)(a - 2b)$
29. $(a^2 + 4b^2)(a + 2b)(a - 2b)$ **31.** $(x + 4)(x^2 - 4x + 16)$
33. $(m - 5)(m^2 + 5m + 25)$ **35.** $(ab - 3)(a^2b^2 + 3ab + 9)$
37. $(2w + z)(4w^2 - 2wz + z^2)$ **39.** $(r - 4s)(r^2 + 4rs + 16s^2)$
41. $(2x - 3y)(4x^2 + 6xy + 9y^2)$ **43.** $(2x + y^2)(4x^2 - 2xy^2 + y^4)$
45. $4(x - 2y)(x^2 + 2xy + 4y^2)$ **47.** $2x(3x + y)(3x - y)$
49. $3mn(2m + 5n)(2m - 5n)$ **51.** **(a)** 33; **(b)** $3x^2(4x^3 + 7)$; **(c)** 33
53. **(a)** 12; **(b)** $4x(-2x^4 + 5)$; **(c)** 12 **55.** **(a)** 4; **(b)** $x^2(x^3 + 3)$; **(c)** 4
57. $(x + y)^2(x - y)$ **59.** $2(m - 2n)(m + 3n)(m - 3n)$ **61.** 4 **63.** 18
65.  **67.**  **69.** 

6.7 Factoring Trinomials: The *ac* Method

6.7 OBJECTIVES

1. Use the *ac* test to determine whether a trinomial is factorable over the integers
2. Use the results of the *ac* test to factor a trinomial
3. For a given value of x, evaluate $f(x)$ before and after factoring

The product of two binomials of the form

$$(\_x + \_)(\_x + \_)$$

will be a trinomial. In your earlier mathematics classes, you used the FOIL method to find the product of two binomials. In this section, we will use the factoring by grouping method to find the binomial factors for a trinomial.

First, let's look at some factored trinomials.

Example 1

Matching Trinomials and Their Factors

Determine which of the following are true statements.

(a) $x^2 - 2x - 8 = (x - 4)(x + 2)$

This is a true statement. Using the FOIL method, we see that

$$(x - 4)(x + 2) = x^2 + 2x - 4x - 8$$
$$= x^2 - 2x - 8$$

(b) $x^2 - 6x + 5 = (x - 2)(x - 3)$

Not a true statement, because

$$(x - 2)(x - 3) = x^2 - 3x - 2x + 6$$
$$= x^2 - 5x + 6$$

(c) $x^2 + 5x - 14 = (x - 2)(x + 7)$

True, because

$$(x - 2)(x + 7) = x^2 + 7x - 2x - 14$$
$$= x^2 + 5x - 14$$

(d) $x^2 - 8x - 15 = (x - 5)(x - 3)$

False, because

$$(x - 5)(x - 3) = x^2 - 3x - 5x + 15$$
$$= x^2 - 8x + 15$$

CHECK YOURSELF 1

Determine which of the following are true statements.

(a) $2x^2 - 2x - 3 = (2x - 3)(x + 1)$

(b) $3x^2 + 11x - 4 = (3x - 1)(x + 4)$

(c) $2x^2 - 7x + 3 = (x - 3)(2x - 1)$

The first step in learning to factor a trinomial is to identify its coefficients. To be consistent, we first write the trinomial in standard $ax^2 + bx + c$ form, then label the three coefficients as a, b, and c.

Example 2

Identifying the Coefficients of $ax^2 + bx + c$

When necessary, rewrite the trinomial in $ax^2 + bx + c$ form. Then label a, b, and c.

(a) $x^2 - 3x - 18$

$a = 1 \qquad b = -3 \qquad c = -18$

(b) $x^2 - 24x + 23$

$a = 1 \qquad b = -24 \qquad c = 23$

(c) $x^2 + 8 - 11x$

First rewrite the trinomial in descending order.

$x^2 - 11x + 8$

Then,

$a = 1 \qquad b = -11 \qquad c = 8$

CHECK YOURSELF 2

When necessary, rewrite the trinomial in $ax^2 + bx + c$ form. Then label a, b, and c.

(a) $x^2 + 5x - 14$ **(b)** $x^2 - 18x + 17$ **(c)** $x - 6 + 2x^2$

Not all trinomials can be factored. To discover if a trinomial is factorable, we try the *ac* test.

Rules and Properties: The *ac* Test

A trinomial of the form $ax^2 + bx + c$ is factorable if (and only if) there are two integers, m and n, such that

$ac = mn$ and $b = m + n$

In Example 3, we will determine whether each trinomial is factorable by finding the values of m and n.

Example 3

Using the *ac* Test

Use the *ac* test to determine which of the following trinomials can be factored. Find the values of m and n for each trinomial that can be factored.

(a) $x^2 - 3x - 18$

First, we note that $a = 1$, $b = -3$, and $c = -18$, so $ac = 1(-18) = -18$.

Then, we look for two numbers, m and n, such that $mn = ac$ and $m + n = b$. In this case, that means

$mn = -18 \qquad m + n = -3$

We will look at every pair of integers with a product of -18. We then look at the sum of each pair.

mn	$m + n$
$1(-18) = -18$	$1 + (-18) = -17$
$2(-9) = -18$	$2 + (-9) = -7$
$3(-6) = -18$	$3 + (-6) = -3$

We need look no further because we have found two integers whose mn product is -18 and $m + n$ sum is -3.

$m = 3 \qquad n = -6$

(b) $x^2 - 24x + 23$

We see that $a = 1$, $b = -24$, and $c = 23$. So, $ac = 23$ and $b = -24$. Therefore, we want

$mn = 23 \qquad m + n = -24$

We now work with integer pairs, looking for two integers with a product of 23 and a sum of -24.

mn	$m + n$
$1(23) = 23$	$1 + 23 = 24$
$-1(-23) = 23$	$-1 + (-23) = -24$

We find that $m = -1$ and $n = -23$.

(c) $x^2 - 11x + 8$

We see that $a = 1$, $b = -11$, and $c = 8$. So, $ac = 8$ and $b = -11$. Therefore, we want

$mn = 8 \qquad m + n = -11$

mn	$m + n$
$1(8) = 8$	$1 + 8 = 9$
$2(4) = 8$	$2 + 4 = 6$
$-1(-8) = 8$	$-1 + (-8) = -9$
$-2(-4) = 8$	$-2 + (-4) = -6$

There is no other pair of integers with a product of 8, and none has a sum of -11. The trinomial $x^2 - 11x + 8$ is not factorable.

(d) $2x^2 + 7x - 15$

We see that $a = 2$, $b = 7$, and $c = -15$. So, $ac = -30$ and $b = 7$. Therefore, we want

$mn = -30 \qquad m + n = 7$

mn	$m + n$
$1(-30) = -30$	$1 + (-30) = -29$
$2(-15) = -30$	$2 + (-15) = -13$
$3(-10) = -30$	$3 + (-10) = -7$
$5(-6) = -30$	$5 + (-6) = -1$
$6(-5) = -30$	$6 + (-5) = 1$
$10(-3) = -30$	$10 + (-3) = 7$

There is no need to go further. We have found two integers with a product of -30 and a sum of 7. So $m = 10$ and $n = -3$.

In this example, you may have noticed patterns and shortcuts that make it easier to find m and n. By all means, use those patterns. This is essential in mathematical thinking. You are taught a step-by-step process that will always work for solving a problem; this process is called an *algorithm*. It is very easy to teach a computer an algorithm. It is very difficult (some would say impossible) for a computer to have insight. Shortcuts that you discover are insights. They may be the most important part of your mathematical education.

CHECK YOURSELF 3

Use the ac test to determine which of the following trinomials can be factored. Find the values of m and n for each trinomial that can be factored.

(a) $x^2 - 7x + 12$
(b) $x^2 + 5x - 14$
(c) $2x^2 + x - 6$
(d) $3x^2 - 6x + 7$

So far we have used the results of the *ac* test only to determine whether a trinomial is factorable. The results can also be used to help factor the trinomial.

Example 4

Using the Results of the *ac* Test to Factor a Trinomial

Rewrite the middle term as the sum of two terms, then factor by grouping.

(a) $x^2 - 3x - 18$

We see that $a = 1$, $b = -3$, and $c = -18$, so

$ac = -18 \qquad b = -3$

We are looking for two numbers, m and n, so that

$mn = -18 \qquad m + n = -3$

In Example 3, we found that the two integers were 3 and -6 because $3(-6) = -18$ and $3 + (-6) = -3$. That result is used to rewrite the middle term (here $-3x$) as the sum of two terms. We now rewrite the middle term as the sum of $3x$ and $-6x$.

$x^2 + 3x - 6x - 18$

Then, we factor by grouping:

$$x^2 + 3x - 6x - 18 = x(x + 3) - 6(x + 3)$$
$$= (x + 3)(x - 6)$$

(b) $x^2 - 24x + 23$

We use the results of Example 3(b), in which we found $m = -1$ and $n = -23$, to rewrite the middle term of the expression.

$x^2 - 24x + 23 = x^2 - x - 23x + 23$

Then, we factor by grouping:

$$x^2 - x - 23x + 23 = x(x - 1) - 23(x - 1)$$
$$= (x - 1)(x - 23)$$

(c) $2x^2 + 7x - 15$

From example 3(d), we know that this trinomial is factorable and that $m = 10$ and $n = -3$. We use that result to rewrite the middle term of the trinomial.

$$\begin{aligned} 2x^2 + 7x - 15 &= 2x^2 + 10x - 3x - 15 \\ &= 2x(x + 5) - 3(x + 5) \\ &= (x + 5)(2x - 3) \end{aligned}$$

CHECK YOURSELF 4

Rewrite the middle term as the sum of two terms, then factor by grouping.

(a) $x^2 - 7x + 12$
(b) $x^2 + 5x - 14$
(c) $2x^2 - x - 6$
(d) $3x^2 - 7x - 6$

Not all product pairs need to be tried to find m and n. A look at the sign pattern will eliminate many of the possibilities. Assuming the lead coefficient to be positive, there are four possible sign patterns.

Pattern	Example	Conclusion
1. b and c are both positive.	$2x^2 + 13x + 15$	m and n must be positive.
2. b is negative and c is positive.	$x^2 - 3x + 2$	m and n must both be negative.
3. b is positive and c is negative.	$x^2 + 5x - 14$	m and n are of opposite signs. (The value with the larger absolute value is positive.)
4. b and c are both negative.	$x^2 - 4x - 4$	m and n are of opposite signs. (The value with the larger absolute value is negative.)

Sometimes the factors of a trinomial seem obvious. At other times you might be certain that there are only a couple of possible sets of factors for a trinomial. It is perfectly acceptable to check these proposed factors to see if they work. If you find the factors in this manner, we say that you have used the **trial and error method.** This method is discussed in Section 6.7*, which follows this section.

To this point we have been factoring polynomial expressions. When a function is defined by a polynomial expression, we can factor that expression without affecting any of the ordered pairs associated with the function. Factoring the expression makes it easier to find some of the ordered pairs.

In particular, we will be looking for values of x that cause $f(x)$ to be 0. We do this by using the **zero product rule.**

Rules and Properties: Zero Product Rule

If $0 = ab$, then either $a = 0$, $b = 0$, or both are zero.

Another way to say this is, if the product of two numbers is zero, then at least one of those numbers must be zero.

Example 5

Factoring Polynomial Functions

Given the function $f(x) = 2x^2 + 7x - 15$, complete the following.

(a) Rewrite the function in factored form.

From Example 4(c) we have

$f(x) = (x + 5)(2x - 3)$

(b) Find the ordered pair associated with $f(0)$.

$f(0) = (0 + 5)(0 - 3) = -15$

The ordered pair is $(0, -15)$.

(c) Find all ordered pairs $(x, 0)$.

We are looking for the x value for which $f(x) = 0$, so

$0 = (x + 5)(2x - 3)$

By the zero product rule, we know that either

$(x + 5) = 0$ or $(2x - 3) = 0$

which means that

$x = -5$ or $x = \dfrac{3}{2}$

The ordered pairs are $(-5, 0)$ and $\left(\dfrac{3}{2}, 0\right)$. Check the original function to see that these ordered pairs are associated with that function.

CHECK YOURSELF 5

Given the function $f(x) = 2x^2 - x - 6$, complete the following.

(a) Rewrite the function in factored form.
(b) Find the ordered pair associated with $f(0)$.
(c) Find all ordered pairs $(x, 0)$.

CHECK YOURSELF ANSWERS

1. **(a)** False; **(b)** true; **(c)** true
2. **(a)** $a = 1, b = 5, c = -14$; **(b)** $a = 1, b = -18, c = 17$; **(c)** $a = 2, b = 1, c = -6$
3. **(a)** Factorable, $m = -4, n = -3$; **(b)** factorable, $m = 7, n = -2$; **(c)** factorable, $m = 4, n = -3$; **(d)** not factorable
4. **(a)** $x^2 - 3x - 4x + 12 = (x - 3)(x - 4)$; **(b)** $x^2 + 7x - 2x - 14 = (x + 7)(x - 2)$; **(c)** $2x^2 - x - 6 = 2x^2 - 4x + 3x - 6 = (2x + 3)(x - 2)$; **(d)** $3x^2 - 7x - 6 = 3x^2 - 9x + 2x - 6 = (3x + 2)(x - 3)$
5. **(a)** $f(x) = (2x + 3)(x - 2)$; **(b)** $(0, -6)$; **(c)** $\left(-\dfrac{3}{2}, 0\right)$ and $(2, 0)$

6.7 Exercises

Name ______________

Section ______ Date ______

In exercises 1 to 8, determine which are true statements.

1. $x^2 - 2x - 3 = (x + 1)(x - 3)$

2. $x^2 - 2x - 8 = (x - 2)(x + 4)$

3. $2x^2 - 5x + 4 = (2x - 1)(x - 4)$

4. $3x^2 - 13x - 10 = (3x + 2)(x - 5)$

5. $x^2 - x - 6 = (x - 5)(x + 1)$

6. $6x^2 + 7x - 3 = (3x - 1)(2x + 3)$

7. $-2x^2 + 11x - 5 = (-x + 5)(2x + 1)$

8. $-6x^2 + 13x - 6 = (2x - 3)(-3x + 2)$

In exercises 9 to 16, when necessary, rewrite the trinomial in $ax^2 + bx + c$ form, then label a, b, and c.

9. $x^2 + 3x - 5$

10. $x^2 - 2x - 1$

11. $2x^2 + 5x + 3$

12. $-3x^2 - x + 2$

13. $x - 1 + 2x^2$

14. $4 - 5x - 3x^2$

15. $2x + 3x^2 - 5$

16. $x - x^2 + 4$

In exercises 17 to 24, use the ac test to determine which trinomials can be factored. Find the values of m and n for each trinomial that can be factored.

17. $x^2 + 3x - 10$

18. $x^2 - x - 12$

19. $x^2 - 2x + 3$

20. $6x^2 - 7x + 2$

ANSWERS

1. ______________
2. ______________
3. ______________
4. ______________
5. ______________
6. ______________
7. ______________
8. ______________
9. ______________
10. ______________
11. ______________
12. ______________
13. ______________
14. ______________
15. ______________
16. ______________
17. ______________
18. ______________
19. ______________
20. ______________

ANSWERS

21. ______
22. ______
23. ______
24. ______
25. ______
26. ______
27. ______
28. ______
29. ______
30. ______
31. ______
32. ______
33. ______
34. ______
35. ______
36. ______
37. ______
38. ______
39. ______
40. ______
41. ______
42. ______
43. ______
44. ______
45. ______
46. ______

21. $2x^2 - 3x + 2$

22. $3x^2 - 10x - 8$

23. $2x^2 + 5x + 2$

24. $3x^2 + x - 2$

In exercises 25 to 70, completely factor each polynomial expression.

25. $x^2 + 7x + 12$

26. $x^2 + 9x + 20$

27. $x^2 - 9x + 8$

28. $x^2 - 11x + 10$

29. $x^2 - 15x + 50$

30. $x^2 - 13x + 40$

31. $x^2 + 7x - 30$

32. $x^2 - 7x - 18$

33. $x^2 - 10x + 24$

34. $x^2 + 13x - 30$

35. $x^2 - 7x - 44$

36. $x^2 - 15x - 54$

37. $x^2 + 8xy + 15y^2$

38. $x^2 - 9xy + 20y^2$

39. $x^2 - 16xy + 55y^2$

40. $x^2 - 9xy - 22y^2$

41. $3x^2 + 11x - 20$

42. $2x^2 + 9x - 18$

43. $5x^2 + 18x - 8$

44. $3x^2 - 20x - 7$

45. $12x^2 + 23x + 5$

46. $8x^2 + 30x + 7$

47. $4x^2 + 20x + 25$

48. $9x^2 - 24x + 16$

49. $5x^2 + 19x - 30$

50. $3x^2 + 17x - 28$

51. $5x^2 + 24x - 36$

52. $3x^2 - 14x - 24$

53. $10x^2 - 7x - 12$

54. $6x^2 + 5x - 21$

55. $16x^2 + 40x + 25$

56. $18x^2 + 45x + 7$

57. $7x^2 - 17xy + 6y^2$

58. $5x^2 + 17xy - 12y^2$

59. $8x^2 - 30xy + 7y^2$

60. $8x^2 - 14xy - 15y^2$

61. $3x^2 - 24x + 45$

62. $2x^2 + 10x - 28$

63. $2x^2 - 26x + 72$

64. $3x^2 + 39x + 120$

65. $6x^3 - 31x^2 + 5x$

66. $8x^3 + 25x^2 + 3x$

67. $5x^3 + 14x^2 - 24x$

68. $3x^4 + 17x^3 - 28x^2$

69. $3x^3 - 15x^2y - 18xy^2$

70. $2x^3 - 10x^2y - 72xy^2$

ANSWERS

47. ______
48. ______
49. ______
50. ______
51. ______
52. ______
53. ______
54. ______
55. ______
56. ______
57. ______
58. ______
59. ______
60. ______
61. ______
62. ______
63. ______
64. ______
65. ______
66. ______
67. ______
68. ______
69. ______
70. ______

ANSWERS

71. (a)
(b)
(c)

72. (a)
(b)
(c)

73. (a)
(b)
(c)

74. (a)
(b)
(c)

75. (a)
(b)
(c)

76. (a)
(b)
(c)

77.

78.

79.

80.

81.

82.

83.

84.

85.

86.

87.

88.

89.

90.

In exercises 71 to 76, for each function, **(a)** rewrite the function in factored form, **(b)** find the ordered pair associated with $f(0)$, and **(c)** find all ordered pairs $(x, 0)$.

71. $f(x) = x^2 - 2x - 3$

72. $f(x) = x^2 - 3x - 10$

73. $f(x) = 2x^2 + 3x - 2$

74. $f(x) = 3x^2 - 11x + 6$

75. $f(x) = 3x^2 + 5x - 28$

76. $f(x) = 10x^2 + 13x - 3$

Certain trinomials in quadratic form can be factored with similar techniques. For instance, we can factor $x^4 - 5x^2 - 6$ as $(x^2 - 6)(x^2 + 1)$. In exercises 77 to 88, apply a similar method to completely factor each polynomial.

77. $x^4 + 3x^2 + 2$

78. $x^4 - 7x^2 + 10$

79. $x^4 - 8x^2 - 33$

80. $x^4 + 5x^2 - 14$

81. $y^6 - 2y^3 - 15$

82. $x^6 + 10x^3 + 21$

83. $x^5 - 6x^3 - 16x$

84. $x^6 - 8x^4 + 15x^2$

85. $x^4 - 5x^2 - 36$

86. $x^4 - 5x^2 + 4$

87. $x^6 - 6x^3 - 16$

88. $x^6 - 2x^3 - 3$

In exercises 89 to 96, determine a value of the number k so that the polynomial can be factored.

89. $x^2 + 5x + k$

90. $x^2 + 3x + k$

91. $6x^2 + x + k$

92. $4x^2 - x + k$

93. $x^2 + kx - 6$

94. $x^2 + kx - 15$

95. $6x^2 + kx - 3$

96. $2x^2 + kx - 15$

97. The product of three numbers is $x^3 + 6x^2 + 8x$. Show that the numbers are consecutive even integers. (*Hint:* Factor the expression.)

98. The product of three numbers is $x^3 + 3x^2 + 2x$. Show that the numbers are consecutive integers.

In each of the following, **(a)** factor the given function, **(b)** identify the values of x for which $f(x) = 0$, **(c)** graph $f(x)$ using the graphing calculator and determine where the graph crosses the x axis, and **(d)** compare the results of (b) and (c).

99. $f(x) = x^2 - 2x - 8$

100. $f(x) = x^2 - 3x - 10$

101. $f(x) = 2x^2 - x - 3$

102. $f(x) = 3x^2 - x - 2$

In exercises 103 and 104, determine the binomials that represent the dimensions of the given figure.

103.

Area = $2x^2 + 7x - 15$?

?

104.

Area = $3x^2 + 11x + 10$?

?

ANSWERS

91.
92.
93.
94.
95.
96.
97.
98.
99. (a) (b)
100. (a) (b)
101. (a) (b)
102. (a) (b)
103.
104.

Answers

1. True **3.** False **5.** False **7.** False **9.** $a = 1, b = 3, c = -5$
11. $a = 2, b = 5, c = 3$ **13.** $a = 2, b = 1, c = -1$ **15.** $a = 3, b = 2, c = -5$
17. Factorable, $m = 5, n = -2$ **19.** Not factorable **21.** Not factorable
23. Factorable, $m = 4, n = 1$ **25.** $(x + 3)(x + 4)$ **27.** $(x - 8)(x - 1)$
29. $(x - 10)(x - 5)$ **31.** $(x + 10)(x - 3)$ **33.** $(x - 6)(x - 4)$
35. $(x - 11)(x + 4)$ **37.** $(x + 3y)(x + 5y)$ **39.** $(x - 11y)(x - 5y)$
41. $(3x - 4)(x + 5)$ **43.** $(5x - 2)(x + 4)$ **45.** $(3x + 5)(4x + 1)$
47. $(2x + 5)^2$ **49.** $(5x - 6)(x + 5)$ **51.** $(5x - 6)(x + 6)$
53. $(2x - 3)(5x + 4)$ **55.** $(4x + 5)^2$ **57.** $(7x - 3y)(x - 2y)$
59. $(4x - y)(2x - 7y)$ **61.** $3(x - 5)(x - 3)$ **63.** $2(x - 4)(x - 9)$
65. $x(6x - 1)(x - 5)$ **67.** $x(x + 4)(5x - 6)$ **69.** $3x(x - 6y)(x + y)$
71. **(a)** $(x - 3)(x + 1)$; **(b)** $(0, -3)$; **(c)** $(3, 0)$ and $(-1, 0)$
73. **(a)** $(2x - 1)(x + 2)$; **(b)** $(0, -2)$; **(c)** $\left(\frac{1}{2}, 0\right)$ and $(-2, 0)$
75. **(a)** $(x + 4)(3x - 7)$; **(b)** $(0, -28)$; **(c)** $(-4, 0)$ and $\left(\frac{7}{3}, 0\right)$
77. $(x^2 + 1)(x^2 + 2)$ **79.** $(x^2 - 11)(x^2 + 3)$ **81.** $(y^3 - 5)(y^3 + 3)$
83. $x(x^2 + 2)(x^2 - 8)$ **85.** $(x + 3)(x - 3)(x^2 + 4)$
87. $(x - 2)(x^2 + 2x + 4)(x^3 + 2)$ **89.** 4 **91.** $-2, -1$, or -5
93. $-5, 5, -1$, or 1 **95.** $7, -7, 17, -17, -3$, or 3 **97.** $x(x + 2)(x + 4)$
99. **(a)** $(x - 4)(x + 2)$; **(b)** $4, -2$ **101.** **(a)** $(2x - 3)(x + 1)$; **(b)** $\frac{3}{2}, -1$
103. $(2x - 3)$ and $(x + 5)$

6.7* Factoring Trinomials: Trial and Error

6.7* OBJECTIVES

1. Factor a trinomial of the form $ax^2 + bx + c$
2. Completely factor a trinomial

Recall that the product of two binomials may be a trinomial of the form

$$ax^2 + bx + c$$

This suggests that some trinomials may be factored as the product of two binomials. And, in fact, factoring trinomials in this way is probably the most common type of factoring that you will encounter in algebra. One process for factoring a trinomial into a product of two binomials is called *trial and error*.

As before, let's introduce the factoring technique with an example from multiplication. Consider

$$(x + 3)(x + 4) = x^2 + 4x + 3x + 12$$
$$= x^2 + 7x + 12$$

Product of first terms, x and x

Sum of inner and outer products, $3x$ and $4x$

Product of last terms, 3 and 4

To reverse the multiplication process to one of factoring, we see that the product of the *first* terms of the binomial factors is the *first* term of the given trinomial, the product of the *last* terms of the binomial factors is the *last* term of the trinomial, and the *middle* term of the trinomial must equal the sum of the *outer* and *inner* products. That leads us to the following sign patterns in factoring a trinomial.

Example	Pattern	Factoring Sign Pattern
$x^2 + bx + c$	Both signs are positive.	$(x + \ )(x + \ )$
$x^2 - bx + c$	The constant is positive, and the x coefficient is negative.	$(x - \ )(x - \ )$
$x^2 + bx - c$ or $x^2 - bx - c$	The constant is negative.	$(x + \ )(x - \ )$

Given the above information let's work through an example.

Example 1

Factoring Trinomials, $a = 1$

To factor

$$x^2 + 7x + 10$$

the desired sign pattern is

$$(x + \_\_)(x + \_\_)$$

From the constant, 10, and the x coefficient of our original trinomial, 7, we want two numbers whose product is 10 and whose sum is 7.

Consider the following:

NOTE With practice, you will do much of this work mentally. We show the factors and their sums here, and in later examples, to emphasize the process.

Factors of 10	Sum
1, 10	11
2, 5	7

We can see that the correct factorization is

$$x^2 + 7x + 10 = (x + 2)(x + 5)$$

NOTE To check, multiply the factors using the method of Section 6.4.

CHECK YOURSELF 1

Factor $x^2 + 8x + 15$.

Example 2

Factoring Trinomials, $a = 1$

Factor $x^2 - 9x + 14$. Do you see that the sign pattern must be as follows?

$(x - \_\_)(x - \_\_)$

We then want two factors of 14 whose sum is -9.

Factors of 14	Sum
$-1, -14$	-15
$-2, -7$	-9

NOTE Here we use two negative factors of 14 because the coefficient of the x term is negative and the constant is positive.

Because the desired middle term is $-9x$, the correct factors are

$$x^2 - 9x + 14 = (x - 2)(x - 7)$$

CHECK YOURSELF 2

Factor $x^2 - 12x + 32$.

Let's turn now to applying our factoring technique to a trinomial whose constant term is negative. Consider the following example.

Example 3

Factoring Trinomials, $a = 1$

Factor the function.

$$f(x) = x^2 + 4x - 12$$

In this case, the sign pattern is

$(x - \_\_)(x + \_\_)$

NOTE Because the constant is now negative, the signs in the binomial factors must be *opposite*.

Here we want two numbers whose product is -12 and whose sum is 4. Again let's look at the possible factors:

Factors of -12	Sum
$1, -12$	-11
$-1, 12$	11
$3, -4$	-1
$-3, 4$	1
$2, -6$	-4
$-2, 6$	4

From the information above, we see that the correct factors are

$$f(x) = x^2 + 4x - 12 = (x - 2)(x + 6)$$

CHECK YOURSELF 3

Factor the function $f(x) = x^2 - 7x - 18$.

Thus far we have considered only trinomials of the form $x^2 + bx + c$. Suppose that the leading coefficient is *not* 1. In general, to factor the trinomial $ax^2 + bx + c$ (with $a \neq 1$), we must consider binomial factors of the form

$$(\_\_x + \_\_)(\_\_x + \_\_)$$

in which one or both of the coefficients of x in the binomial factors is greater than 1. Again let's look at a multiplication example for some clues to the technique. Consider

$$(2x + 3)(3x + 5) = 6x^2 + 19x + 15$$

Product of $2x$ and $3x$

Sum of outer and inner products, $10x$ and $9x$

Product of 3 and 5

Now, to reverse the process to factoring, we can proceed as in the following example.

Example 4

Factoring Trinomials, $a \neq 1$

To factor the function $f(x) = 5x^2 + 9x + 4$, we must have the pattern

$$(\_\_x + \_\_)(\_\_x + \_\_)$$

This product must be 4.

This product must be 5.

NOTE Now that the lead coefficient is no longer 1, we must be prepared to try both 1, 4 and 4, 1.

Factors of 5	Factors of 4
1, 5	1, 4
	4, 1
	2, 2

Therefore the possible binomial factors are

$(x + 1)(5x + 4)$

$(x + 4)(5x + 1)$

$(x + 2)(5x + 2)$

Checking the middle terms of each product, we see that the proper factorization is

$f(x) = 5x^2 + 9x + 4 = (x + 1)(5x + 4)$

CHECK YOURSELF 4

Factor the function $f(x) = 6x^2 - 17x + 7$.

The sign patterns discussed before remain the same when the leading coefficient is not 1. Look at the following example involving a trinomial with a negative constant.

Example 5

Factoring Trinomials, $a \neq 1$

(a) Factor the function $f(x) = 6x^2 + 7x - 3$. The sign patterns are

$(\_x + \_)(\_x - \_)$ or $(\_x - \_)(\_x + \_)$

Factors of 6	Factors of −3
1, 6	1, −3
2, 3	−1, 3

NOTE Again, as the number of factors for the first coefficient and the constant increase, the number of possible factors becomes larger. Can we reduce the search? One clue: If the trinomial has no common factors (other than 1), then a binomial factor can have no common factor. This means that $6x - 3$, $6x + 3$, $3x - 3$, and $3x + 3$ need not be considered. They are shown here to completely illustrate the possibilities.

There are eight possible binomial factors:

$(x + 1)(6x - 3)$

$(x - 1)(6x + 3)$

$(x + 3)(6x - 1)$

$(x - 3)(6x + 1)$

$(2x + 1)(3x - 3)$

$(2x - 1)(3x + 3)$

$(3x + 1)(2x - 3)$

$(3x - 1)(2x + 3)$

Again, checking the middle terms, we have the correct factors:

$f(x) = 6x^2 + 7x - 3 = (3x - 1)(2x + 3)$

Factoring certain trinomials in more than one variable involves similar techniques, as is illustrated below.

(b) Factor.

$4x^2 - 16xy + 7y^2$

From the first term of the trinomial we see that possible first terms for our binomial factors are $4x$ and x or $2x$ and $2x$. The last term of the trinomial tells us that the only choices for the last terms of our binomial factors are y and $7y$. So given the sign of the middle and last terms, the only possible factors are

NOTE Find the middle term of each product.

$(4x - 7y)(x - y)$

$(4x - y)(x - 7y)$

$(2x - 7y)(2x - y)$

From the middle term of our original trinomial we see that $2x - 7y$ and $2x - y$ are the proper factors.

$(2x - 7y)(2x - y)$

CHECK YOURSELF 5

Factor $6a^2 + 11ab - 10b^2$.

Recall our earlier comment that the *first step* in any factoring problem is to remove any existing common factors. As before, it may be necessary to combine common-term factoring with other methods (such as factoring a trinomial into a product of binomials) to completely factor a polynomial. Look at the following example.

Example 6

Factoring Trinomials, $a \neq 1$

(a) Factor.

$f(x) = 2x^2 - 16x + 30$

First note the common factor of 2. So we can write

NOTE "Remove" the common factor of 2.

$f(x) = 2x^2 - 16x + 30 = 2(x^2 - 8x + 15)$

Now, as the second step, examine the trinomial factor. By our earlier methods we know that

$x^2 - 8x + 15 = (x - 3)(x - 5)$

and we have

$f(x) = 2x^2 - 16x + 30 = 2(x - 3)(x - 5)$

in completely factored form.

(b) Factor.

$6x^3 + 15x^2y - 9xy^2$

As before, note the common factor of $3x$ in each term of the trinomial. Removing that common factor, we have

$$6x^3 + 15x^2y - 9xy^2 = 3x(2x^2 + 5xy - 3y^2)$$

Again, considering the trinomial factor, we see that $2x^2 + 5xy - 3y^2$ has factors of $2x - y$ and $x + 3y$. And our original trinomial becomes

$$3x(2x - y)(x + 3y)$$

in completely factored form.

CHECK YOURSELF 6

Factor.

(a) $f(x) = 9x^2 - 39x + 36$ **(b)** $24a^3 + 4a^2b - 8ab^2$

Note: When factoring, we require that all coefficients be integers. Given this restriction, not all polynomials are factorable over the integers. The following illustrates.

To factor $x^2 - 9x + 12$, we know that the only possible binomial factors (using integers as coefficients) are

$$(x - 1)(x - 12)$$

$$(x - 2)(x - 6)$$

$$(x - 3)(x - 4)$$

You can easily verify that *none* of these pairs gives the correct middle term of $-9x$. We then say that the original trinomial is not factorable using integers as coefficients.

CHECK YOURSELF ANSWERS

1. $(x + 3)(x + 5)$ **2.** $(x - 4)(x - 8)$ **3.** $f(x) = (x - 9)(x + 2)$
4. $f(x) = (2x - 1)(3x - 7)$ **5.** $(3a - 2b)(2a + 5b)$
6. **(a)** $f(x) = 3(x - 3)(3x - 4)$; **(b)** $4a(3a + 2b)(2a - b)$

Name ______________________

Section ________ Date ________

6.7* Exercises

In exercises 1 to 8, determine which are true statements.

1. $x^2 - 2x - 3 = (x + 1)(x - 3)$

2. $x^2 - 2x - 8 = (x - 2)(x + 4)$

3. $2x^2 - 5x + 4 = (2x - 1)(x - 4)$

4. $3x^2 - 13x - 10 = (3x + 2)(x - 5)$

5. $x^2 - x - 6 = (x - 5)(x + 1)$

6. $6x^2 + 7x - 3 = (3x - 1)(2x + 3)$

7. $-2x^2 + 11x - 5 = (-x + 5)(2x + 1)$

8. $-6x^2 + 13x - 6 = (2x - 3)(-3x + 2)$

In exercises 9 to 20, when necessary, rewrite the trinomial in $ax^2 + bx + c$ form, then label a, b, and c.

9. $x^2 + 3x - 5$

10. $x^2 - 2x - 1$

11. $2x^2 + 5x + 3$

12. $-3x^2 - x + 2$

13. $x - 1 + 2x^2$

14. $4 - 5x - 3x^2$

15. $2x + 3x^2 - 5$

16. $x - x^2 + 4$

17. $-7x^2 + 9x - 18$

18. $5x^2 + 7x - 9$

19. $-3x^2 + 5x - 10$

20. $x^2 + 4$

In exercises 21 to 70, completely factor each polynomial expression.

21. $x^2 + 6x + 8$

22. $x^2 - 11x + 24$

23. $x^2 - 17x + 60$

24. $x^2 - 14x + 49$

ANSWERS

1. ______
2. ______
3. ______
4. ______
5. ______
6. ______
7. ______
8. ______
9. ______
10. ______
11. ______
12. ______
13. ______
14. ______
15. ______
16. ______
17. ______
18. ______
19. ______
20. ______
21. ______
22. ______
23. ______
24. ______

ANSWERS

25. ______
26. ______
27. ______
28. ______
29. ______
30. ______
31. ______
32. ______
33. ______
34. ______
35. ______
36. ______
37. ______
38. ______
39. ______
40. ______
41. ______
42. ______
43. ______
44. ______
45. ______
46. ______
47. ______
48. ______
49. ______
50. ______
51. ______
52. ______

25. $x^2 + 7x + 12$

26. $x^2 + 9x + 20$

27. $x^2 - 9x + 8$

28. $x^2 - 11x + 10$

29. $x^2 - 15x + 50$

30. $x^2 - 13x + 40$

31. $x^2 + 7x - 30$

32. $x^2 - 7x - 18$

33. $x^2 - 10x + 24$

34. $x^2 + 13x - 30$

35. $x^2 - 7x - 44$

36. $x^2 - 15x - 54$

37. $x^2 + 8xy + 15y^2$

38. $x^2 - 9xy + 20y^2$

39. $x^2 - 16xy + 55y^2$

40. $x^2 - 9xy - 22y^2$

41. $3x^2 + 11x - 20$

42. $2x^2 + 9x - 18$

43. $5x^2 + 18x - 8$

44. $3x^2 - 20x - 7$

45. $12x^2 + 23x + 5$

46. $8x^2 + 30x + 7$

47. $4x^2 + 20x + 25$

48. $9x^2 - 24x + 16$

49. $5x^2 + 19x - 30$

50. $3x^2 + 17x - 28$

51. $5x^2 + 24x - 36$

52. $3x^2 - 14x - 24$

53. $10x^2 - 7x - 12$

54. $6x^2 + 5x - 21$

55. $16x^2 + 40x + 25$

56. $18x^2 + 45x + 7$

57. $7x^2 - 17xy + 6y^2$

58. $5x^2 + 17xy - 12y^2$

59. $8x^2 - 30xy + 7y^2$

60. $8x^2 - 14xy - 15y^2$

61. $3x^2 - 24x + 45$

62. $2x^2 + 10x - 28$

63. $2x^2 - 26x + 72$

64. $3x^2 + 39x + 120$

65. $6x^3 - 31x^2 + 5x$

66. $8x^3 + 25x^2 + 3x$

67. $5x^3 + 14x^2 - 24x$

68. $3x^4 + 17x^3 - 28x^2$

69. $3x^3 - 15x^2y - 18xy^2$

70. $2x^3 - 10x^2y - 72xy^2$

In exercises 71 to 78, determine a value of the number k so that the polynomial can be factored.

71. $x^2 + 5x + k$

72. $x^2 + 3x + k$

73. $6x^2 + x + k$

74. $4x^2 - 2x + k$

75. $x^2 + kx - 6$

76. $x^2 + kx - 15$

77. $6x^2 + kx - 3$

78. $2x^2 + kx - 15$

79. The product of three numbers is $x^3 + 6x^2 + 8x$. Show that the numbers are consecutive even integers. (*Hint:* Factor the expression.)

80. The product of three numbers is $x^3 + 3x^2 + 2x$. Show that the numbers are consecutive integers.

ANSWERS

53. ______
54. ______
55. ______
56. ______
57. ______
58. ______
59. ______
60. ______
61. ______
62. ______
63. ______
64. ______
65. ______
66. ______
67. ______
68. ______
69. ______
70. ______
71. ______
72. ______
73. ______
74. ______
75. ______
76. ______
77. ______
78. ______
79. ______
80. ______

Answers

1. True **3.** False **5.** False **7.** False **9.** $a = 1, b = 3, c = -5$
11. $a = 2, b = 5, c = 3$ **13.** $a = 2, b = 1, c = -1$ **15.** $a = 3, b = 2, c = -5$
17. $a = -7, b = 9, c = -18$ **19.** $a = -3, b = 5, c = -10$
21. $(x + 2)(x + 4)$ **23.** $(x - 12)(x - 5)$ **25.** $(x + 3)(x + 4)$
27. $(x - 8)(x - 1)$ **29.** $(x - 10)(x - 5)$ **31.** $(x + 10)(x - 3)$
33. $(x - 6)(x - 4)$ **35.** $(x - 11)(x + 4)$ **37.** $(x + 3y)(x + 5y)$
39. $(x - 11y)(x - 5y)$ **41.** $(3x - 4)(x + 5)$ **43.** $(5x - 2)(x + 4)$
45. $(3x + 5)(4x + 1)$ **47.** $(2x + 5)^2$ **49.** $(5x - 6)(x + 5)$
51. $(5x - 6)(x + 6)$ **53.** $(2x - 3)(5x + 4)$ **55.** $(4x + 5)^2$
57. $(7x - 3y)(x - 2y)$ **59.** $(4x - y)(2x - 7y)$ **61.** $3(x - 5)(x - 3)$
63. $2(x - 4)(x - 9)$ **65.** $x(6x - 1)(x - 5)$ **67.** $x(x + 4)(5x - 6)$
69. $3x(x - 6y)(x + y)$ **71.** 4 **73.** $-2, -1$, or -5 **75.** $-5, 5, -1$, or 1
77. $7, -7, 17, -17, -3$, or 3 **79.** $x(x + 2)(x + 4)$

6.8 Division of Polynomials

6.8 OBJECTIVES

1. Divide a polynomial by a monomial
2. Use long division to divide a polynomial by a binomial

In Section 1.4, we introduced the quotient rule for exponents, which was used to divide one monomial by another monomial. Let's review that process.

Step by Step: To Divide a Monomial by a Monomial

Step 1 Divide the coefficients.
Step 2 Use the quotient rule for exponents to combine the variables.

Example 1

Dividing Monomials

NOTE The quotient rule says: If x is not zero and $m > n$,

$$\frac{x^m}{x^n} = x^{m-n}$$

(a) $\dfrac{8x^4}{2x^2} = 4x^{4-2}$ Divide: $\dfrac{8}{2} = 4$ Subtract the exponents.

$= 4x^2$

(b) $\dfrac{45a^5b^3}{9a^2b} = 5a^3b^2$

CHECK YOURSELF 1

Divide.

(a) $\dfrac{16a^5}{8a^3}$ (b) $\dfrac{28m^4n^3}{7m^3n}$

Now let's look at how this can be extended to divide any polynomial by a monomial. For example, to divide $12a^3 + 8a^2$ by $4a$, proceed as follows:

NOTE Technically, this step depends on the distributive property and the definition of division.

$$\frac{12a^3 + 8a^2}{4a} = \frac{12a^3}{4a} + \frac{8a^2}{4a}$$

Divide each term in the numerator by the denominator, $4a$.

Now do each division.

$= 3a^2 + 2a$

The preceding work leads us to the following rule.

Step by Step: To Divide a Polynomial by a Monomial

Step 1 Divide each term of the polynomial by the monomial.

Step 2 Simplify the results.

Example 2

Dividing by Monomials

Divide each term by 2.

(a) $$\frac{4a^2 + 8}{2} = \frac{4a^2}{2} + \frac{8}{2}$$
$$= 2a^2 + 4$$

Divide each term by 6y.

(b) $$\frac{24y^3 - 18y^2}{6y} = \frac{24y^3}{6y} - \frac{18y^2}{6y}$$
$$= 4y^2 - 3y$$

Remember the rules for signs in division.

(c) $$\frac{15x^2 + 10x}{-5x} = \frac{15x^2}{-5x} + \frac{10x}{-5x}$$
$$= -3x - 2$$

NOTE With practice you can write just the quotient.

(d) $$\frac{14x^4 + 28x^3 - 21x^2}{7x^2} = \frac{14x^4}{7x^2} + \frac{28x^3}{7x^2} - \frac{21x^2}{7x^2}$$
$$= 2x^2 + 4x - 3$$

(e) $$\frac{9a^3b^4 - 6a^2b^3 + 12ab^4}{3ab} = \frac{9a^3b^4}{3ab} - \frac{6a^2b^3}{3ab} + \frac{12ab^4}{3ab}$$
$$= 3a^2b^3 - 2ab^2 + 4b^3$$

CHECK YOURSELF 2

Divide.

(a) $\frac{20y^3 - 15y^2}{5y}$

(b) $\frac{8a^3 - 12a^2 + 4a}{-4a}$

(c) $\frac{16m^4n^3 - 12m^3n^2 + 8mn}{4mn}$

We are now ready to look at dividing one polynomial by another polynomial (with more than one term). The process is very much like long division in arithmetic, as Example 3 illustrates.

Example 3

Dividing by Binomials

Divide $x^2 + 7x + 10$ by $x + 2$.

NOTE The first term in the dividend, x^2, is divided by the first term in the divisor, x.

Step 1 $x + 2\overline{)x^2 + 7x + 10}$ with quotient x

Divide x^2 by x to get x.

Step 2

$$\begin{array}{r} x \\ x + 2\overline{)x^2 + 7x + 10} \\ \underline{x^2 + 2x} \end{array}$$

Multiply the divisor, $x + 2$, by x.

NOTE Remember: To subtract $x^2 + 2x$, mentally change each sign to $-x^2 - 2x$, and add. Take your time and be careful here. It's where most errors are made.

Step 3

$$\begin{array}{r} x \\ x + 2\overline{)x^2 + 7x + 10} \\ \underline{x^2 + 2x} \\ 5x + 10 \end{array}$$

Subtract and bring down 10.

Step 4

$$\begin{array}{r} x + 5 \\ x + 2\overline{)x^2 + 7x + 10} \\ \underline{x^2 + 2x} \\ 5x + 10 \end{array}$$

Divide $5x$ by x to get 5.

NOTE Note that we repeat the process until the degree of the remainder is less than that of the divisor or until there is no remainder.

Step 5

$$\begin{array}{r} x + 5 \\ x + 2\overline{)x^2 + 7x + 10} \\ \underline{x^2 + 2x} \\ 5x + 10 \\ \underline{5x + 10} \\ 0 \end{array}$$

Multiply $x + 2$ by 5 and then subtract.

The quotient is $x + 5$.

CHECK YOURSELF 3

Divide $x^2 + 9x + 20$ by $x + 4$.

In Example 3, we showed all the steps separately to help you see the process. In practice, the work can be shortened.

Example 4

Dividing by Binomials

Divide $x^2 + x - 12$ by $x - 3$.

NOTE You might want to write out a problem like $408 \div 17$, to compare the steps.

$$\begin{array}{r} x + 4 \\ x - 3\overline{)x^2 + x - 12} \\ \underline{x^2 - 3x} \\ 4x - 12 \\ \underline{4x - 12} \\ 0 \end{array}$$

1. Divide x^2 by x to get x, the first term of the quotient.
2. Multiply $x - 3$ by x.
3. Subtract. Remember to mentally change the signs to $-x^2 + 3x$ and add. Bring down -12.
4. Divide $4x$ by x to get 4, the second term of the quotient.
5. Multiply $x - 3$ by 4 and subtract.

The quotient is $x + 4$.

CHECK YOURSELF 4

Divide.

$(x^2 + 2x - 24) \div (x - 4)$

You may have a remainder in algebraic long division just as in arithmetic. Consider Example 5.

Example 5

Dividing by Binomials

Divide $4x^2 - 8x + 11$ by $2x - 3$.

$$\begin{array}{r|l} & 2x - 1 \\ \hline 2x - 3 & 4x^2 - 8x + 11 \\ & \underline{4x^2 - 6x} \\ & -2x + 11 \\ & \underline{-2x + 3} \\ & 8 \end{array}$$

Quotient: $2x - 1$; Divisor: $2x - 3$; Remainder: 8

This result can be written as

$$\frac{4x^2 - 8x + 11}{2x - 3}$$

$$= 2x - 1 + \frac{8}{2x - 3}$$

Quotient: $2x - 1$; Remainder: 8; Divisor: $2x - 3$

CHECK YOURSELF 5

Divide.

$(6x^2 - 7x + 15) \div (3x - 5)$

The division process shown in our previous examples can be extended to dividends of a higher degree. The steps involved in the division process are exactly the same, as Example 6 illustrates.

Example 6

Dividing by Binomials

Divide $6x^3 + x^2 - 4x - 5$ by $3x - 1$.

$$
\begin{array}{r}
2x^2 + x - 1 \\
3x - 1 \overline{)\, 6x^3 + x^2 - 4x - 5} \\
\underline{6x^3 - 2x^2} \\
3x^2 - 4x \\
\underline{3x^2 - x} \\
-3x - 5 \\
\underline{-3x + 1} \\
-6
\end{array}
$$

This result can be written as

$$\frac{6x^3 + x^2 - 4x - 5}{3x - 1} = 2x^2 + x - 1 + \frac{-6}{3x - 1}$$

CHECK YOURSELF 6

Divide $4x^3 - 2x^2 + 2x + 15$ *by* $2x + 3$.

Suppose that the dividend is "missing" a term in some power of the variable. You can use 0 as the coefficient for the missing term. Consider Example 7.

Example 7

Dividing by Binomials

Divide $x^3 - 2x^2 + 5$ by $x + 3$.

$$
\begin{array}{r}
x^2 - 5x + 15 \\
x + 3 \overline{)\, x^3 - 2x^2 + 0x + 5} \\
\underline{x^3 + 3x^2} \\
-5x^2 + 0x \\
\underline{-5x^2 - 15x} \\
15x + 5 \\
\underline{15x + 45} \\
-40
\end{array}
$$

Write $0x$ for the "missing" term in x.

This result can be written as

$$\frac{x^3 - 2x^2 + 5}{x + 3} = x^2 - 5x + 15 + \frac{-40}{x + 3}$$

CHECK YOURSELF 7

Divide.

$(4x^3 + x + 10) \div (2x - 1)$

You should always arrange the terms of the divisor and the dividend in descending-exponent form before starting the long division process, as illustrated in Example 8.

Example 8

Dividing by Binomials

Divide $5x^2 - x + x^3 - 5$ by $-1 + x^2$.

Write the divisor as $x^2 - 1$ and the dividend as $x^3 + 5x^2 - x - 5$.

$$
\begin{array}{r|l}
 & x + 5 \\
x^2 - 1 & x^3 + 5x^2 - x - 5 \\
 & \underline{x^3 \qquad\quad - x} \\
 & 5x^2 \qquad - 5 \\
 & \underline{5x^2 \qquad - 5} \\
 & 0
\end{array}
$$

Write $x^3 - x$, the product of x and $x^2 - 1$, so that like terms fall in the same columns.

CHECK YOURSELF 8

Divide.

$(5x^2 + 10 + 2x^3 + 4x) \div (2 + x^2)$

CHECK YOURSELF ANSWERS

1. **(a)** $2a^2$; **(b)** $4mn^2$ **2.** **(a)** $4y^2 - 3y$; **(b)** $-2a^2 + 3a - 1$; **(c)** $4m^3n^2 - 3m^2n + 2$

3. $x + 5$ **4.** $x + 6$ **5.** $2x + 1 + \dfrac{20}{3x - 5}$ **6.** $2x^2 - 4x + 7 + \dfrac{-6}{2x + 3}$

7. $2x^2 + x + 1 + \dfrac{11}{2x - 1}$ **8.** $2x + 5$

Name ______________________

Section ________ Date ________

6.8 Exercises

Divide.

1. $\frac{18x^6}{9x^2}$

2. $\frac{20a^7}{5a^5}$

3. $\frac{35m^3n^2}{7mn^2}$

4. $\frac{42x^5y^2}{6x^3y}$

5. $\frac{3a + 6}{3}$

6. $\frac{4x - 8}{4}$

7. $\frac{9b^2 - 12}{3}$

8. $\frac{10m^2 + 5m}{5}$

9. $\frac{16a^3 - 24a^2}{4a}$

10. $\frac{9x^3 + 12x^2}{3x}$

11. $\frac{12m^2 + 6m}{-3m}$

12. $\frac{20b^3 - 25b^2}{-5b}$

13. $\frac{18a^4 + 12a^3 - 6a^2}{6a}$

14. $\frac{21x^5 - 28x^4 + 14x^3}{7x}$

15. $\frac{20x^4y^2 - 15x^2y^3 + 10x^3y}{5x^2y}$

16. $\frac{16m^3n^3 + 24m^2n^2 - 40mn^3}{8mn^2}$

Perform the indicated divisions.

17. $\frac{x^2 + 5x + 6}{x + 2}$

18. $\frac{x^2 + 8x + 15}{x + 3}$

19. $\frac{x^2 - x - 20}{x + 4}$

20. $\frac{x^2 - 2x - 35}{x + 5}$

21. $\frac{2x^2 + 5x - 3}{2x - 1}$

22. $\frac{3x^2 + 20x - 32}{3x - 4}$

23. $\frac{2x^2 - 3x - 5}{x - 3}$

24. $\frac{3x^2 + 17x - 12}{x + 6}$

ANSWERS

1. ______
2. ______
3. ______
4. ______
5. ______
6. ______
7. ______
8. ______
9. ______
10. ______
11. ______
12. ______
13. ______
14. ______
15. ______
16. ______
17. ______
18. ______
19. ______
20. ______
21. ______
22. ______
23. ______
24. ______

ANSWERS

25. ______
26. ______
27. ______
28. ______
29. ______
30. ______
31. ______
32. ______
33. ______
34. ______
35. ______
36. ______
37. ______
38. ______
39. ______
40. ______
41. ______
42. ______
43. ______
44. ______
45. ______
46. ______
47. ______
48. ______

25. $\dfrac{4x^2 - 18x - 15}{x - 5}$

26. $\dfrac{3x^2 - 18x - 32}{x - 8}$

27. $\dfrac{6x^2 - x - 10}{3x - 5}$

28. $\dfrac{4x^2 + 6x - 25}{2x + 7}$

29. $\dfrac{x^3 + x^2 - 4x - 4}{x + 2}$

30. $\dfrac{x^3 - 2x^2 + 4x - 21}{x - 3}$

31. $\dfrac{4x^3 + 7x^2 + 10x + 5}{4x - 1}$

32. $\dfrac{2x^3 - 3x^2 + 4x + 4}{2x + 1}$

33. $\dfrac{x^3 - x^2 + 5}{x - 2}$

34. $\dfrac{x^3 + 4x - 3}{x + 3}$

35. $\dfrac{25x^3 + x}{5x - 2}$

36. $\dfrac{8x^3 - 6x^2 + 2x}{4x + 1}$

37. $\dfrac{2x^2 - 8 - 3x + x^3}{x - 2}$

38. $\dfrac{x^2 - 18x + 2x^3 + 32}{x + 4}$

39. $\dfrac{x^4 - 1}{x - 1}$

40. $\dfrac{x^4 + x^2 - 16}{x + 2}$

41. $\dfrac{x^3 - 3x^2 - x + 3}{x^2 - 1}$

42. $\dfrac{x^3 + 2x^2 + 3x + 6}{x^2 + 3}$

43. $\dfrac{x^4 + 2x^2 - 2}{x^2 + 3}$

44. $\dfrac{x^4 + x^2 - 5}{x^2 - 2}$

45. $\dfrac{y^3 + 1}{y + 1}$

46. $\dfrac{y^3 - 8}{y - 2}$

47. $\dfrac{x^4 - 1}{x^2 - 1}$

48. $\dfrac{x^6 - 1}{x^3 - 1}$

49. Find the value of c so that $\dfrac{y^2 - y + c}{y + 1} = y - 2$

50. Find the value of c so that $\dfrac{x^3 + x^2 + x + c}{x^2 + 1} = x + 1$

51. Write a summary of your work with polynomials. Explain how a polynomial is recognized, and explain the rules for the arithmetic of polynomials—how to add, subtract, multiply, and divide. What parts of this chapter do you feel you understand very well, and what part(s) do you still have questions about, or feel unsure of? Exchange papers with another student and compare your answers.

52. An interesting (and useful) thing about division of polynomials: To find out about this interesting thing, do this division. Compare your answer with another student.

$(x - 2)\overline{)\,2x^2 + 3x - 5}$ Is there a remainder?

Now, evaluate the polynomial $2x^2 + 3x - 5$ when $x = 2$. Is this value the same as the remainder?

Try $(x + 3)\overline{)\,5x^2 - 2x + 1}$. Is there a remainder?

Evaluate the polynomial $5x^2 - 2x + 1$ when $x = -3$. Is this value the same as the remainder?

What happens when there is no remainder?

Try $(x - 6)\overline{)\,3x^3 - 14x^2 - 23x - 6}$. Is the remainder zero?

Evaluate the polynomial $3x^3 - 14x^2 - 23x + 6$ when $x = 6$. Is this value zero? Write a description of the patterns you see. When does the pattern hold? Make up several more examples, and test your conjecture.

In exercises 53 to 56, **(a)** divide the polynomial $f(x)$ by the given linear factor and **(b)** factor the quotient obtained in part (a). Then, **(c)** using your graphing calculator, graph the polynomial $f(x)$ and determine where the graph passes through the x axis and **(d)** compare the results of parts (b) and (c).

53. $f(x) = x^3 + 5x^2 + 2x - 8;\ x + 2$

54. $f(x) = x^3 - 2x^2 - 11x + 12;\ x + 3$

55. $f(x) = x^3 + x^2 - 4x - 4;\ x + 1$

56. $f(x) = x^3 - 3x^2 - 16x + 48;\ x - 3$

ANSWERS

49. ____

50. ____

51. ____

52. ____

53. (a) (b) (c) ____

54. (a) (b) (c) ____

55. (a) (b) (c) ____

56. (a) (b) (c) ____

Answers

1. $2x^4$ **3.** $5m^2$ **5.** $a + 2$ **7.** $3b^2 - 4$ **9.** $4a^2 - 6a$ **11.** $-4m - 2$
13. $3a^3 + 2a^2 - a$ **15.** $4x^2y - 3y^2 + 2x$ **17.** $x + 3$ **19.** $x - 5$
21. $x + 3$ **23.** $2x + 3 + \frac{4}{x - 3}$ **25.** $4x + 2 + \frac{-5}{x - 5}$
27. $2x + 3 + \frac{5}{3x - 5}$ **29.** $x^2 - x - 2$ **31.** $x^2 + 2x + 3 + \frac{8}{4x - 1}$
33. $x^2 + x + 2 + \frac{9}{x - 2}$ **35.** $5x^2 + 2x + 1 + \frac{2}{5x - 2}$
37. $x^2 + 4x + 5 + \frac{2}{x - 2}$ **39.** $x^3 + x^2 + x + 1$ **41.** $x - 3$
43. $x^2 - 1 + \frac{1}{x^2 + 3}$ **45.** $y^2 - y + 1$ **47.** $x^2 + 1$ **49.** $c = -2$
51. **53.** **(a)** $(x^2 + 3x - 4)$; **(b)** $(x + 4)(x - 1)$; **(c)** $-2, -4, 1$

55. **(a)** $x^2 - 4$; **(b)** $(x + 2)(x - 2)$; **(c)** $-1, 2, -2$

Solving Quadratic Equations by Factoring

OBJECTIVES

1. Solve a quadratic equation by factoring
2. Find the zeros of a quadratic function

The factoring techniques you have learned provide us with tools for solving equations that can be written in the form

$$ax^2 + bx + c = 0 \qquad a \neq 0$$

This is a quadratic equation in one variable, here x. You can recognize such a quadratic equation by the fact that the highest power of the variable x is the second power.

in which a, b, and c are constants.

An equation written in the form $ax^2 + bx + c = 0$ is called a **quadratic equation in standard form.** Using factoring to solve quadratic equations requires the **zero-product rule,** which says that if the product of two factors is 0, then one or both of the factors must be equal to 0. In symbols:

Rules and Properties: Zero-Product Rule

If $a \cdot b = 0$, then $a = 0$ or $b = 0$ or $a = b = 0$.

Let's see how the rule is applied to solving quadratic equations.

Example 1

Solving Equations by Factoring

Solve.

$$x^2 - 3x - 18 = 0$$

Factoring on the left, we have

$$(x - 6)(x + 3) = 0$$

NOTE To use the zero-product rule, 0 must be on one side of the equation.

By the zero-product rule, we know that one or both of the factors must be zero. We can then write

$$x - 6 = 0 \qquad \text{or} \qquad x + 3 = 0$$

Solving each equation gives

$$x = 6 \qquad \text{or} \qquad x = -3$$

NOTE Graph the function

$y = x^2 - 3x - 18$

on your graphing calculator. The solutions to the equation $0 = x^2 - 3x - 18$ will be the x coordinates of the points on the curve at which $y = 0$. Those are the points at which the graph intercepts the x axis.

The two solutions are 6 and -3 and the solution set is written as $\{x|x = -3, 6\}$ or simply $\{-3, 6\}$.

The solutions are sometimes called the **zeros,** or **roots,** of the equation. They represent the x coordinates of the points where the graph of the equation $y = x^2 - 3x - 18$ crosses the x axis. In this case, the graph crosses the x axis at $(-3, 0)$ and $(6, 0)$.

Quadratic equations can be checked in the same way as linear equations were checked: by substitution. For instance, if $x = 6$, we have

$$6^2 - 3 \cdot 6 - 18 \stackrel{?}{=} 0$$

$$36 - 18 - 18 \stackrel{?}{=} 0$$

$$0 = 0$$

which is a true statement. We leave it to you to check the solution of -3.

CHECK YOURSELF 1

Solve $x^2 - 9x + 20 = 0$.

Other factoring techniques are also used in solving quadratic equations. Example 2 illustrates this concept.

Example 2

Solving Equations by Factoring

(a) Solve $x^2 - 5x = 0$.

Again, factor the left side of the equation and apply the zero-product principle.

CAUTION

A *common mistake* is to forget the statement $x = 0$ when you are solving equations of this type. Be sure to include both solutions in the solution set.

$$x(x - 5) = 0$$

Now

$$x = 0 \quad \text{or} \quad x - 5 = 0$$

$$x = 5$$

The two solutions are 0 and 5. The solution set is $\{0, 5\}$.

(b) Solve $x^2 - 9 = 0$.

Factoring yields

$$(x + 3)(x - 3) = 0$$

$$x + 3 = 0 \quad \text{or} \quad x - 3 = 0$$

$$x = -3 \qquad\qquad x = 3$$

The solution set is $\{-3, 3\}$, which may be written as $\{\pm 3\}$.

NOTE The symbol ± is read "plus or minus."

CHECK YOURSELF 2

Solve by factoring.

(a) $x^2 + 8x = 0$ **(b)** $x^2 - 16 = 0$

Example 3 illustrates a crucial point. Our solution technique depends on the zero-product rule, which means that the product of factors *must be equal to* 0. The importance of this is shown now.

Example 3

Solving Equations by Factoring

Solve $2x^2 - x = 3$.

The first step in the solution is to write the equation in standard form (that is, when one side of the equation is 0). So start by adding -3 to both sides of the equation. Then,

CAUTION

Consider the equation

$x(2x - 1) = 3$

Students are sometimes tempted to write

$x = 3$ or $2x - 1 = 3$

This is *not correct.* Instead, subtract 3 from both sides of the equation as *the first step* to write

$2x^2 - x - 3 = 0$

in standard form. Only *now* can you factor and proceed as before.

$2x^2 - x - 3 = 0$ Make sure all terms are on one side of the equation. The other side will be 0.

You can now factor and solve by using the zero-product rule.

$$(2x - 3)(x + 1) = 0$$

$$2x - 3 = 0 \quad \text{or} \quad x + 1 = 0$$

$$2x = 3 \qquad\qquad x = -1$$

$$x = \frac{3}{2}$$

The solution set is $\left\{\frac{3}{2}, -1\right\}$.

CHECK YOURSELF 3

Solve $3x^2 = 5x + 2$.

Always examine the quadratic member of an equation for common factors. It will make your work much easier, as Example 4 illustrates.

Example 4

Solving Equations by Factoring

Solve $3x^2 - 3x - 60 = 0$.

First, note the common factor 3 in the quadratic member of the equation. Factoring out the 3, we have

$$3(x^2 - x - 20) = 0$$

Now divide both sides of the equation by 3.

NOTE Notice the advantage of dividing both members by 3. The coefficients in the quadratic member become smaller, and that member is much easier to factor.

$$\frac{3(x^2 - x - 20)}{3} = \frac{0}{3}$$

or

$$x^2 - x - 20 = 0$$

We can now factor and solve as before.

$$(x - 5)(x + 4) = 0$$

$$x - 5 = 0 \quad \text{or} \quad x + 4 = 0$$

$$x = 5 \qquad\qquad x = -4 \quad \text{or} \quad \{5, -4\}$$

CHECK YOURSELF 4

Solve $2x^2 - 10x - 48 = 0$.

In Chapter 3, we introduced the concept of a function and later expressed the equation of a line in function form. Another type of function is called a quadratic function.

Definitions: Quadratic Function

A **quadratic function** is a function that can be written in the form

$f(x) = ax^2 + bx + c$

in which a, b, and c are real numbers and $a \neq 0$.

For example, $f(x) = 3x^2 - 2x - 1$ and $g(x) = x^2 - 2$ are quadratic functions. In working with functions, we often want to find the values of x for which $f(x) = 0$. As in quadratic equations, these values are called the **zeros of the function.** They represent the x coordinates of the points where the graph of the function crosses the x axis. To find the zeros of a quadratic function, a quadratic equation must be solved.

Example 5

Finding the Zeros of a Function

NOTE The graph of $f(x) = x^2 - x - 2$ intercepts the x axis at the points (2, 0) and (−1, 0), so 2 and −1 are the *zeros* of the function.

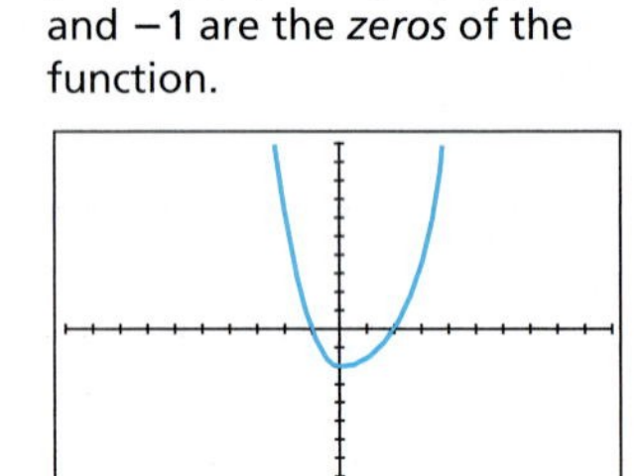

Find the zeros of $f(x) = x^2 - x - 2$.

To find the zeros of $f(x) = x^2 - x - 2$, we must solve the quadratic equation $f(x) = 0$.

$$f(x) = 0$$

$$x^2 - x - 2 = 0$$

$$(x - 2)(x + 1) = 0$$

$$x - 2 = 0 \quad \text{or} \quad x + 1 = 0$$

$$x = 2 \qquad x = -1 \quad \text{or} \quad \{-1, 2\}$$

The zeros of the function are −1 and 2.

CHECK YOURSELF 5

Find the zeros of $f(x) = 2x^2 - x - 3$.

CHECK YOURSELF ANSWERS

1. $\{4, 5\}$ **2. (a)** $\{0, -8\}$; **(b)** $\{4, -4\}$ **3.** $\left\{-\frac{1}{3}, 2\right\}$ **4.** $\{-3, 8\}$ **5.** $\left\{-1, \frac{3}{2}\right\}$

Name ________________

Section ________ Date ________

6.9 Exercises

In exercises 1 to 46, solve the quadratic equations by factoring.

1. $x^2 + 4x + 3 = 0$

2. $x^2 - 5x + 4 = 0$

3. $x^2 - 2x - 15 = 0$

4. $x^2 + 4x - 32 = 0$

5. $x^2 - 11x + 30 = 0$

6. $x^2 + 14x + 48 = 0$

7. $x^2 - 4x - 21 = 0$

8. $x^2 + 5x - 36 = 0$

9. $x^2 - 5x = 50$

10. $x^2 + 14x = -33$

11. $x^2 = 2x + 35$

12. $x^2 = 6x + 27$

13. $x^2 - 8x = 0$

14. $x^2 + 7x = 0$

15. $x^2 + 10x = 0$

16. $x^2 - 9x = 0$

17. $x^2 = 5x$

18. $4x = x^2$

19. $x^2 - 25 = 0$

20. $x^2 - 49 = 0$

21. $x^2 = 64$

22. $x^2 = 36$

23. $4x^2 + 12x + 9 = 0$

24. $9x^2 - 30x + 25 = 0$

ANSWERS

1. ________________
2. ________________
3. ________________
4. ________________
5. ________________
6. ________________
7. ________________
8. ________________
9. ________________
10. ________________
11. ________________
12. ________________
13. ________________
14. ________________
15. ________________
16. ________________
17. ________________
18. ________________
19. ________________
20. ________________
21. ________________
22. ________________
23. ________________
24. ________________

ANSWERS

25. ______
26. ______
27. ______
28. ______
29. ______
30. ______
31. ______
32. ______
33. ______
34. ______
35. ______
36. ______
37. ______
38. ______
39. ______
40. ______
41. ______
42. ______
43. ______
44. ______
45. ______
46. ______

25. $2x^2 - 17x + 36 = 0$

26. $5x^2 + 17x - 12 = 0$

27. $3x^2 - x = 4$

28. $6x^2 = 13x - 6$

29. $6x^2 = 7x - 2$

30. $4x^2 - 3 = x$

31. $2m^2 = 12m + 54$

32. $5x^2 - 55x = 60$

33. $4x^2 - 24x = 0$

34. $6x^2 - 9x = 0$

35. $5x^2 = 15x$

36. $7x^2 = -49x$

37. $x(x - 2) = 15$

38. $x(x + 3) = 28$

39. $x(2x - 3) = 9$

40. $x(3x + 1) = 52$

41. $2x(3x + 1) = 28$

42. $3x(2x - 1) = 30$

43. $(x - 3)(x - 1) = 15$

44. $(x + 4)(x + 1) = 18$

45. $(2x + 1)(x - 4) = 11$

46. $(3x - 5)(x + 2) = 14$

In exercises 47 to 50, find the zeros of the functions.

47. $f(x) = 3x^2 - 24x + 36$

48. $f(x) = 2x^2 - 6x - 56$

49. $f(x) = 4x^2 + 16x - 20$

50. $f(x) = 3x^2 - 33x + 54$

51. Explain the differences between solving the equations $3(x - 2)(x + 5) = 0$ and $3x(x - 2)(x + 5) = 0$.

52. How can a graphing calculator be used to determine the zeros of a quadratic function?

In exercises 53 to 56, write an equation that has the following solutions. *Hint:* Write the binomial factors and then the quadratic member of the equation.

53. $\{2, -3\}$

54. $\{0, 5\}$

55. $\{6, 2\}$

56. $\{-4, 4\}$

The zero-product rule can be extended to three or more factors. If $a \cdot b \cdot c = 0$, then at least one of these factors is 0. In exercises 57 to 60, use this information to solve the equations.

57. $x^3 - 3x^2 - 10x = 0$

58. $x^3 + 8x^2 + 15x = 0$

59. $x^3 - 9x = 0$

60. $x^3 = 16x$

In exercises 61 to 64, extend the ideas in the previous exercises to find solutions for the following equations. *Hint:* Apply factoring by grouping.

61. $x^3 + x^2 - 4x - 4 = 0$

62. $x^3 - 5x^2 - x + 5 = 0$

63. $x^4 - 10x^2 + 9 = 0$

64. $x^4 - 5x^2 + 4 = 0$

The net productivity of a forested wetland as related to the amount of water moving through the wetland can be expressed by a quadratic equation. In exercises 65 to 68, if y represents the amount of wood produced, in grams per square meter, and x represents the amount of water present, in centimeters, determine where the productivity is zero in each wetland represented by the equations.

65. $y = -3x^2 + 300x$

66. $y = -4x^2 + 500x$

67. $y = -6x^2 + 792x$

68. $y = -7x^2 + 1022x$

ANSWERS

47. ______
48. ______
49. ______
50. ______
51. ______
52. ______
53. ______
54. ______
55. ______
56. ______
57. ______
58. ______
59. ______
60. ______
61. ______
62. ______
63. ______
64. ______
65. ______
66. ______
67. ______
68. ______

69. ______

70. ______

71. ______

72. ______

69. Break-even analysis. The manager of a bicycle shop knows that the cost of selling x bicycles is $C = 20x + 60$ and the revenue from selling x bicycles is $R = x^2 - 8x$. Find the break-even value of x.

70. Break-even analysis. A company that produces computer games has found that its operating cost in dollars is $C = 40x + 150$ and its revenue in dollars is $R = 65x - x^2$. For what value(s) of x will the company break even?

Use your calculator to graph $f(x)$.

71. $f(x) = 3x^2 - 24x + 36$. Note the x values at which the graph crosses the x axis. Compare your answer to the solution for exercise 47.

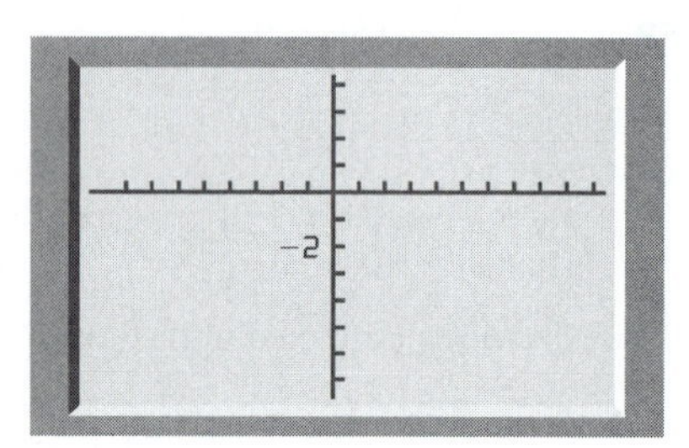

72. $f(x) = 2x^2 - 6x - 56$. Note the x values at which the graph crosses the x axis. Compare your answer to the solution for exercise 48.

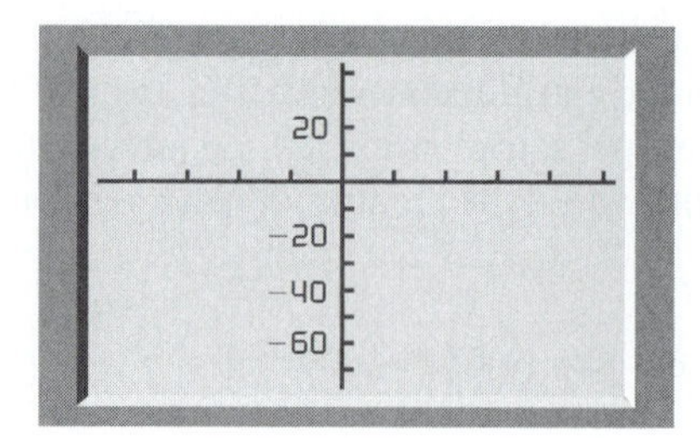

ANSWERS

73. ____________________

73. Work with another student and use your calculator to solve this exercise. Find an equation for each graph given. There could be more than one equation for some of the graphs. Remember the connection between the x intercepts and the zeros.

(a)

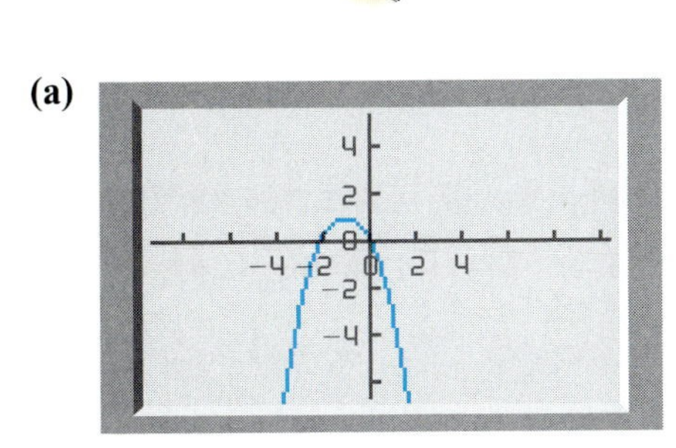

(b)

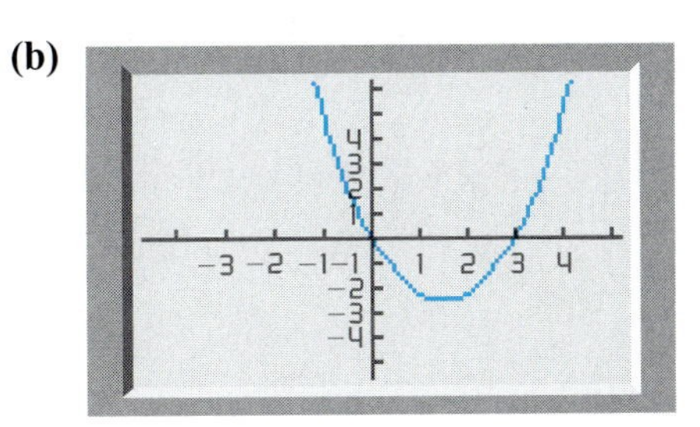

(c)

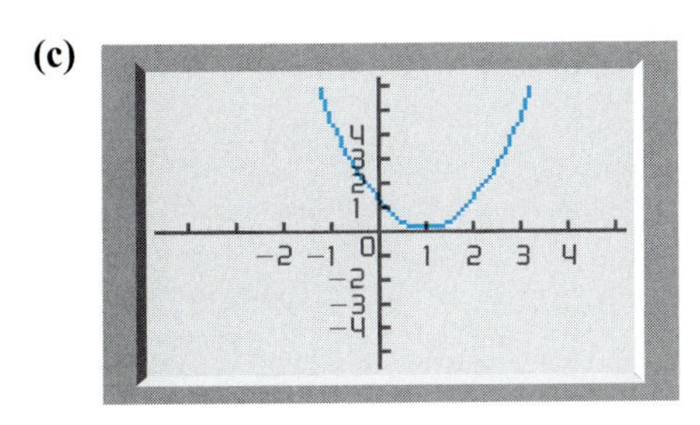

(d)

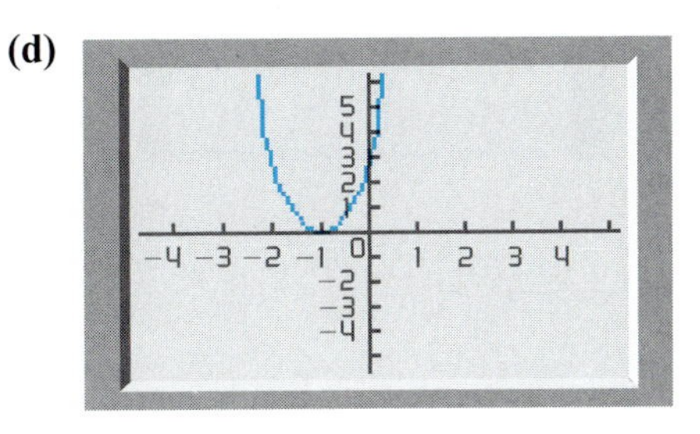

(e)

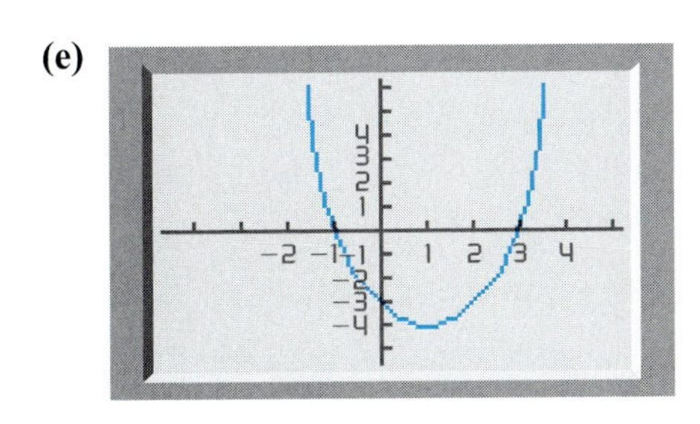

(f)

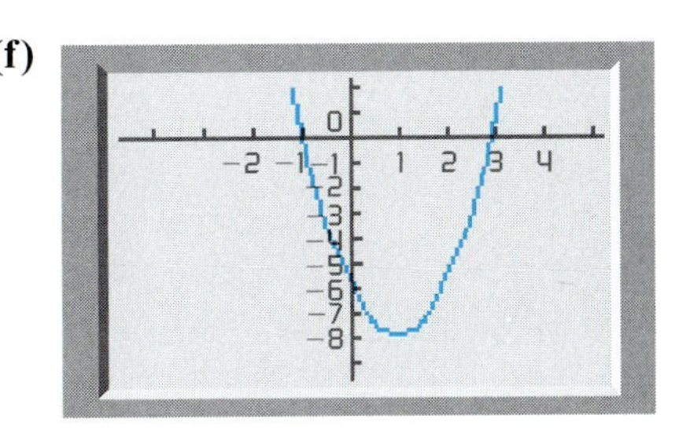

(g)

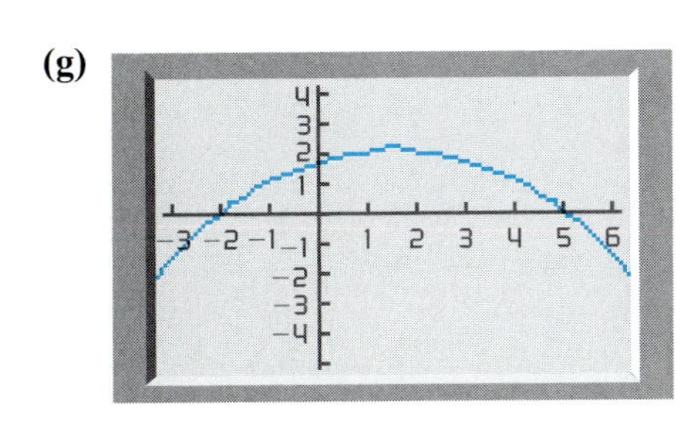

(h)

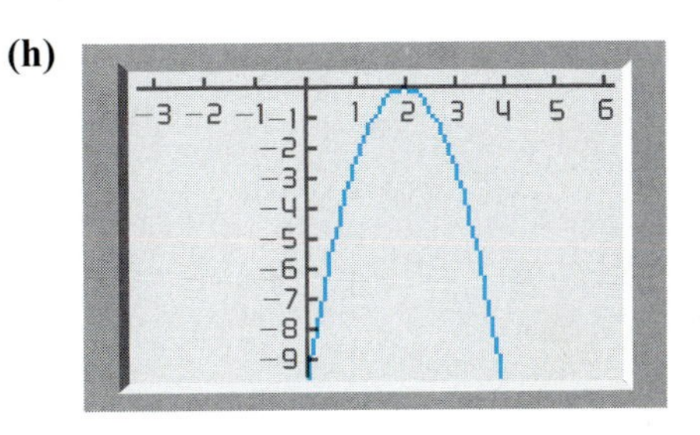

Answers

1. $\{-3, -1\}$ **3.** $\{-3, 5\}$ **5.** $\{5, 6\}$ **7.** $\{-3, 7\}$ **9.** $\{-5, 10\}$
11. $\{-5, 7\}$ **13.** $\{0, 8\}$ **15.** $\{0, -10\}$ **17.** $\{0, 5\}$ **19.** $\{-5, 5\}$
21. $\{-8, 8\}$ **23.** $\left\{-\frac{3}{2}\right\}$ **25.** $\left\{4, \frac{9}{2}\right\}$ **27.** $\left\{-1, \frac{4}{3}\right\}$ **29.** $\left\{\frac{1}{2}, \frac{2}{3}\right\}$
31. $\{-3, 9\}$ **33.** $\{0, 6\}$ **35.** $\{0, 3\}$ **37.** $\{-3, 5\}$ **39.** $\left\{-\frac{3}{2}, 3\right\}$
41. $\left\{-\frac{7}{3}, 2\right\}$ **43.** $\{-2, 6\}$ **45.** $\left\{-\frac{3}{2}, 5\right\}$ **47.** $\{2, 6\}$ **49.** $\{-5, 1\}$
51. **53.** $x^2 + x - 6 = 0$ **55.** $x^2 - 8x + 12 = 0$ **57.** $\{-2, 0, 5\}$
59. $\{-3, 0, 3\}$ **61.** $\{-2, -1, 2\}$ **63.** $\{-3, -1, 1, 3\}$ **65.** $\{0 \text{ cm}, 100 \text{ cm}\}$
67. $\{0 \text{ cm}, 132 \text{ cm}\}$ **69.** $\{30\}$
71. (2, 0) (6, 0) −2 — 2, 6 are the zeros of exercise 47 **73.**

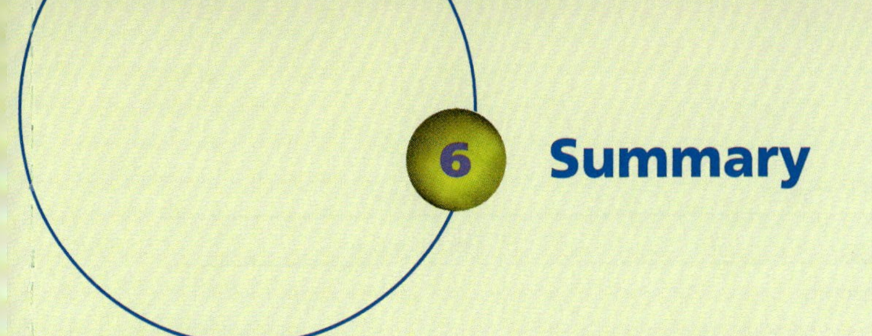

Summary

Definition/Procedure	Example	Reference
Introduction to Polynomials and Polynomial Functions		**Section 6.1**
A **term** is a number or the product of a number and one or more variables, raised to a power. If terms contain exactly the same variables raised to the same powers, they are called **like terms.** A **monomial** is a term in which only whole numbers can appear as exponents. A **polynomial** may be a monomial or any finite sum (or difference) of monomials. We call a polynomial with two terms a **binomial,** one with three terms a **trinomial.**	$3x$ and $7x^3$ are terms. $3x^2 + 2x + 1$ is a polynomial. $2x^2 + 5$ is a binomial. $f(x) = 3x^3 + 2x^2 - 5x + 1$ is a polynomial function.	pp. 379–380
Combining Functions		**Section 6.2**
The **sum of two functions** f and g is written $f + g$. It is defined as $(f + g)(x) = f(x) + g(x)$ The **difference of two functions** f and g is written $f - g$. It is defined as $(f - g)(x) = f(x) - g(x)$ The **product of two functions** f and g is written $f \cdot g$. It is defined as $(f \cdot g)(x) = f(x) \cdot g(x)$ The **quotient of two functions** f and g is written $f \div g$. It is defined as $(f \div g)(x) = f(x) \div g(x)$	If $f(x) = x + 3$ and $g(x) = 2x^2$ $(f \cdot g)(x) = (x + 3)(2x^2)$ $= 2x^3 + 6x^2$	pp. 389, 391
Addition and Subtraction of Polynomials		**Section 6.3**
Adding Polynomials To add polynomials, simply combine any like terms by using the distributive property. *Subtracting Polynomials* To subtract polynomials, enclose the polynomial being subtracted in parentheses, preceded by a subtraction symbol. Then remove the parentheses and combine like terms.	$(4x^2 - 6x + 2) - (3x^2 - 2x - 5)$ $= 4x^2 - 6x + 2 - 3x^2 + 2x + 5$ $= x^2 - 4x + 7$	pp. 395–396
Multiplication of Polynomials and Special Products		**Section 6.4**
Multiplying Polynomials To multiply two polynomials, multiply each term of the first polynomial by each term of the second polynomial. *Special Products* Certain special products can be found by applying the following formulas. $(a + b)(a - b) = a^2 - b^2$ Product of binomials differing in sign. $(a + b)^2 = a^2 + 2ab + b^2$ Squaring a binomial. $(a - b)^2 = a^2 - 2ab + b^2$ Squaring a binomial.	To multiply: $\begin{array}{r} 3x^2 - 2x + 5 \\ 5x - 2 \\ \hline -6x^2 + 4x - 10 \\ 15x^3 - 10x^2 + 25x \quad \\ \hline 15x^3 - 16x^2 + 29x - 10 \end{array}$ $(2x - 3)(2x + 3) = 4x^2 - 9$ $(2x + 5)^2 = 4x^2 + 20x + 25$ $(3a - b)^2 = 9a^2 - 6ab + b^2$	pp. 403, 405, 406
Common Factors and Factoring by Grouping		**Section 6.5**
The **greatest common factor (GCF)** of a polynomial is the monomial with highest degree and largest numerical coefficient that is a factor of each term of the polynomial.	$5x^2y^2 - 10xy^2 + 35x^2y$ has GCF $5xy$.	p. 415

Continued

DEFINITION/PROCEDURE	EXAMPLE	REFERENCE
Factoring Special Polynomials		**Section 6.6**
To Factor a Polynomial In general, you can apply the following steps. **1.** If a polynomial has a GCF other than 1, factor out that *greatest common factor.* **2.** If a polynomial is a binomial, try factoring it as a **difference of two squares** or as a **sum or difference of two cubes.**	Factor: $5x^2y^2 - 10xy^2 + 35x^2y$ $= 5xy(xy - 2y + 7x)$ Factor: $16x^2 - 9y^2$ $= (4x + 3y)(4x - 3y)$ Factor: $x^3 + 8y^3$ $= (x + 2y)(x^2 - 2xy + 4y^2)$	**pp. 423, 424**
Factoring Trinomials: The *ac* Method		**Section 6.7**
The ac Test A trinomial of the form $ax^2 + bx + c$ is factorable if (and only if) there are two integers, m and n, such that $ac = mn \qquad b = m + n$ *To Factor a Trinomial* In general, you can apply the following steps. **1.** Write the trinomial in standard $ax^2 + bx + c$ form. **2.** Then, label the three coefficients a, b, and c. **3.** Find two integers, m and n, such that $ac = mn$ and $b = m + n$ **4.** Rewrite the trinomial as $ax^2 + mx + nx + c$ **5.** Factor by grouping. Not all trinomials are factorable. To discover if a trinomial is factorable, try the ***ac* test.**	Given $2x^2 - 5x - 3$ $a = 2, b = -5, c = -3$ $ac = -6, b = -5$ $m = -6 \quad n = 1$ $mn = -6 \quad m + n = -5$ $2x^2 - 6x + x - 3$ $2x(x - 3) + 1(x - 3)$ $(x - 3)(2x + 1)$	**p. 432**
Division of Polynomials		**Section 6.8**
To Divide a Polynomial by a Monomial Divide each term of the polynomial by the monomial. Then simplify the results.	$\frac{9x^4 + 6x^3 - 15x^2}{3x}$ $= 3x^3 + 2x^2 - 5x$	**p. 454**
Solving Quadratic Equations by Factoring		**Section 6.9**
Zero Product Rule If $0 = ab$, then either $a = 0$, $b = 0$, or both are zero.		**p. 463**
Solving Quadratic Equations by Factoring **1.** Add or subtract the necessary terms on both sides of the equation so that the equation is in standard form (set equal to 0). **2.** Factor the quadratic expression. **3.** Set each factor equal to 0. **4.** Solve the resulting equations to find the solutions. **5.** Check each solution by substituting in the original equation.	To solve: $x^2 + 7x = 30$ $x^2 + 7x - 30 = 0$ $(x + 10)(x - 3) = 0$ $x + 10 = 0$ or $x - 3 = 0$ $x = -10$ and $x = 3$ are solutions.	**p. 464**

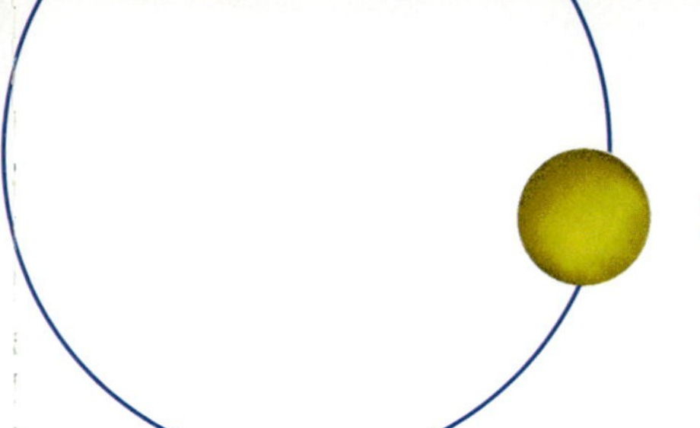

Summary Exercises

This summary exercise set is provided to give you practice with each of the objectives in the chapter. Each exercise is keyed to the appropriate chapter section. The answers are provided in the *Instructor's Manual.*

[6.1] Classify each of the following as a monomial, binomial, or trinomial, if possible.

1. $5x^3 - 2x^2$

2. $7x^5$

3. $4x^5 - 8x^3 + 5$

4. $x^3 + 2x^2 - 5x + 3$

5. $9a^2 - 18a$

6. $4x^2 + \dfrac{2}{x}$

Arrange in descending-exponent form, and give the degree of each polynomial.

7. $5x^5 + 3x^2$

8. $9x$

9. $6x^2 + 4x^4 + 6$

10. $5 + x$

11. -8

12. $9x^4 - 3x + 7x^6$

13. Given $f(x) = 4x^2 + 5x - 6$, find $f(-2)$.

14. Given $f(x) = 5x^2 - 4x + 8$, find $f(3)$.

15. Given $f(x) = -2x^3 + 4x^2 - 5x + 8$, find $f(3)$.

16. Given $f(x) = 3x^3 + 5x^2 - 2x + 6$, find $f(-1)$.

17. Given $f(x) = -2x^4 - 5x^2 + 3x - 6$, find $f(2)$.

18. Given $f(x) = -x^5 - 3x^4 + 5x^3 - x - 4$, find $f(-2)$.

In each of the following, the polynomial functions $f(x)$ and $g(x)$ are given. Find the following ordered pairs: **(a)** $(0, f(0))$, **(b)** $(2, f(2))$, **(c)** $(0, g(0))$, and **(d)** $(2, g(2))$.

19. $f(x) = 5x^2 - 3x - 4$ and $g(x) = 3x^2 + 6x - 5$

20. $f(x) = 2x^2 + 4x + 6$ and $g(x) = 3x^2 - 4x + 3$

21. $f(x) = x^3 - 3x^2 + 7x$ and $g(x) = -2x^3 - x^2 + 3x + 1$

22. $f(x) = -x^4 - 3x^2 + 2x$ and $g(x) = -2x^4 + 3x^3 - 2x^2 + 2$

[6.2]–[6.3] In each of the following, $f(x)$ and $g(x)$ are given. Let $h(x) = f(x) + g(x)$. Find **(a)** $h(x)$, **(b)** $f(1) + g(1)$, and **(c)** $h(1)$.

23. $f(x) = 4x^2 + 5x - 3$ and $g(x) = -2x^2 + x - 5$

24. $f(x) = -3x^3 + 2x^2 - 5$ and $g(x) = 4x^3 - 4x^2 + 5x + 6$

25. $f(x) = 2x^4 + 4x^2 + 5$ and $g(x) = x^3 - 5x^2 + 6x$

26. $f(x) = 3x^3 + 5x - 5$ and $g(x) = -2x^3 + 2x^2 + 5x$

In each of the following, $f(x)$ and $g(x)$ are given. Let $h(x) = f(x) - g(x)$. Find **(a)** $h(x)$, **(b)** $f(1) - g(1)$, and **(c)** $h(1)$.

27. $f(x) = 7x^2 - 2x + 3$ and $g(x) = 2x^2 - 5x - 7$.

28. $f(x) = 9x^2 - 4x$ and $g(x) = 5x^2 + 3$.

29. $f(x) = 8x^2 + 5x$ and $g(x) = 4x^2 - 3x$.

30. $f(x) = -2x^2 - 3x$ and $g(x) = -3x^2 + 4x - 5$.

Simplify each of the following polynomial functions.

31. $f(x) = [(3x^2 + 4x) - 3] - (2x^2 - 6x - 2)$

32. $f(x) = (2x^3 + 4x) + (-3x^3 - 2x^2 + 5) - (4x^2 - 6x - 7)$

33. $f(x) = (3x - 7) - [5x^2 - (8x + 6)] + (-6x^2 - 9)$

34. $f(x) = (-3x^2 + 6x + 9) - (7x^2 + 5x - 8) + [2 - (-3x^2 + 7x + 9)]$

35. $f(x) = -4x^2(-6x + 5) - [(3x^3 - 5x^2 + 6) - (x^3 + 6x^2 + 4x - 2)]$

36. $f(x) = -5x^2(3x^2 + 6x - 2) + [-2x(-4x + 5) + (-3x^2 - x - 1)]$

[6.4] In each of the following, $f(x)$ and $g(x)$ are given. Let $h(x) = f(x) \cdot g(x)$. Find **(a)** $h(x)$, **(b)** $f(1) \cdot g(1)$, and **(c)** $h(1)$.

37. $f(x) = 4x$ and $g(x) = 3x^2 - 5x$

38. $f(x) = -6x$ and $g(x) = -4x^2 - 5x + 7$

39. $f(x) = 3x^3$ and $g(x) = -2x^2 - 8x + 1$

40. $f(x) = -7x^3$ and $g(x) = 5x^2 - 4x + 10$

Multiply each polynomial expression.

41. $5x(3x^2 - 4x)$

42. $5y^2(2y^3 - 3y^2 + 5y)$

43. $(x - 2y)(x + 3y)$

44. $(a - 5b)(a - 6b)$

45. $(3c - 5d)(5c + 2d)$

46. $(4x^2 - y)(2x + 3y^2)$

47. $x(x - 3)(x + 2)$

48. $2y(2y + 3)(3y + 2)$

Multiply the following polynomial expressions using the special product formulas.

49. $(x + 8)^2$

50. $(y - 5)^2$

51. $(2a - 3b)^2$

52. $(5x + 2y)^2$

53. $(x - 4y)(x + 4y)$

54. $(2c - 3d)(2c + 3d)$

Multiply each polynomial expression using the vertical method.

55. $(2x - 3)(x^2 - 5x + 2)$

56. $(5a - b)(2a^2 - 3ab - 2b^2)$

[6.5] Factor each of the following polynomials completely.

57. $18x^2y + 24xy^2$

58. $35a^3 - 28a^2 + 7a$

59. $18m^2n^2 - 27m^2n + 45m^2n^3$

60. $x(2x - y) + y(2x - y)$

61. $5(w - 3z) - 10(w - 3z)^2$

62. $x^2 - 4x + 5x - 20$

63. $x^2 + 7x - 2x - 14$

64. $6x^2 + 4x - 15x - 10$

65. $12x^2 - 9x - 28x + 21$

[6.6] Factor each of the following binomials completely.

66. $x^2 - 64$

67. $25a^2 - 16$

68. $16m^2 - 49n^2$

69. $3w^3 - 12wz^2$

70. $a^4 - 16b^4$

71. $m^3 - 64$

72. $8x^3 + 1$

73. $8c^3 - 27d^3$

74. $125m^3 + 64n^3$

75. $2x^4 + 54x$

For each of the following functions, **(a)** find $f(1)$, **(b)** factor $f(x)$, and **(c)** find $f(1)$ from the factored form.

76. $f(x) = 3x^3 + 15x^2$

77. $f(x) = -6x^3 - 2x^2$

78. $f(x) = 12x^6 - 8x^4$

79. $f(x) = 2x^5 - 2x$

[6.6]–[6.7*] Completely factor each of the following polynomial expressions.

80. $x^2 + 12x + 20$

81. $a^2 - a - 12$

82. $w^2 - 13w + 40$

83. $r^2 - 9r - 36$

84. $x^2 - 8xy - 48y^2$

85. $a^2 + 17ab + 30b^2$

86. $5x^2 + 13x - 6$

87. $2a^2 + 3a - 35$

88. $4r^2 + 20r + 21$

89. $6c^2 - 19c + 10$

90. $6m^2 - 19mn + 10n^2$

91. $8x^2 + 14xy - 15y^2$

92. $9x^2 - 15x - 6$

93. $5w^2 - 25wz + 30z^2$

94. $3c^3 + 18c^2 + 15c$

95. $2a^3 + 4a^2b - 6ab^2$

96. $x^4 + 6x^2 + 5$

97. $a^4 - 3a^2b^2 - 4b^4$

[6.8] Divide.

98. $\frac{9a^5}{3a^2}$

99. $\frac{24m^4n^2}{6m^2n}$

100. $\frac{15a - 10}{5}$

101. $\frac{32a^3 + 24a}{8a}$

102. $\frac{9r^2s^3 - 18r^3s^2}{-3rs^2}$

103. $\frac{35x^3y^2 - 21x^2y^3 + 14x^3y}{7x^2y}$

Perform the indicated long division.

104. $\frac{x^2 - 2x - 15}{x + 3}$

105. $\frac{2x^2 + 9x - 35}{2x - 5}$

106. $\frac{x^2 - 8x + 17}{x - 5}$

107. $\frac{6x^2 - x - 10}{3x + 4}$

108. $\frac{6x^3 + 14x^2 - 2x - 6}{6x + 2}$

109. $\frac{4x^3 + x + 3}{2x - 1}$

110. $\frac{3x^2 + x^3 + 5 + 4x}{x + 2}$

111. $\frac{2x^4 - 2x^2 - 10}{x^2 - 3}$

[6.9] Solve each of the following equations by factoring.

112. $x^2 + 5x - 6 = 0$

113. $x^2 - 2x - 8 = 0$

114. $x^2 + 7x = 30$

115. $x^2 - 6x = 40$

116. $x^2 = 11x - 24$

117. $x^2 = 28 - 3x$

118. $x^2 - 10x = 0$

119. $x^2 = 12x$

120. $x^2 - 25 = 0$

121. $x^2 = 144$

122. $2x^2 - x - 3 = 0$

123. $3x^2 - 4x = 15$

124. $3x^2 + 9x - 30 = 0$

125. $4x^2 + 24x = -32$

126. $x(x - 3) = 18$

127. $(x - 2)(2x + 1) = 33$

128. $x^3 - 2x^2 - 15x = 0$

129. $x^3 + x^2 - 4x - 4 = 0$

130. Suppose that the cost, in dollars, of producing x stereo systems is given by

$$C(x) = 3000 - 60x + 3x^2$$

How many systems can be produced for $7500?

131. The demand equation for a certain type of computer paper is predicted to be

$$D = -4p + 72$$

The supply equation is predicted to be

$$S = -p^2 + 24p - 3$$

Find the equilibrium price.

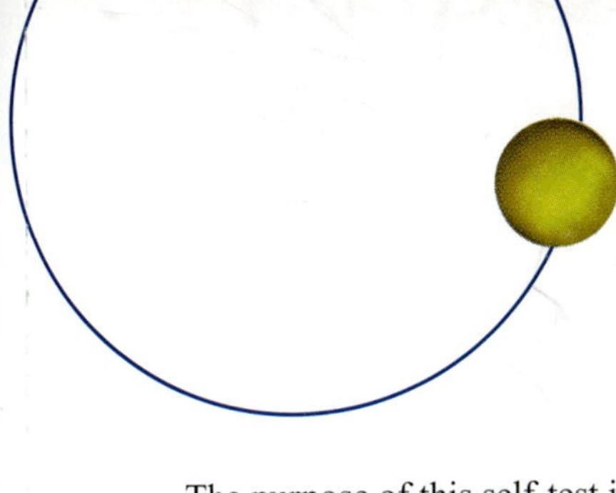

Self-Test for Chapter 6

Name ______________

Section ________ Date ________

The purpose of this self-test is to help you check your progress and to review for a chapter test in class. Allow yourself about 1 hour to take the test. When you are done, check your answers in the back of the book. If you missed any answers, be sure to go back and review the appropriate sections in the chapter and the exercises that are provided.

Classify each of the following as a monomial, binomial, or trinomial, if possible.

1. $6x^2 + 7x$

2. $5x^2 + 8x - 8$

3. $-3x + \frac{1}{x}$

Arrange in descending-exponent form, and give the coefficients and degree.

4. $-3x^2 + 8x^4 - 7$

5. Given $f(x) = 3x^2 - 4x + 5$, find $f(-2)$.

In the following, $f(x)$ and $g(x)$ are given. Find **(a)** $h(x) = f(x) + g(x)$, **(b)** $p(x) = f(x) - g(x)$, **(c)** $f(1) + g(1)$, **(d)** $f(1) - g(1)$, **(e)** $h(1)$, **(f)** $p(1)$.

6. $f(x) = 4x^2 - 3x + 7$ and $g(x) = 2x^2$

7. $f(x) = -3x^3 + 5x^2 - 2x - 7$ and $g(x) = -2x^2 + 7x - 2$

Simplify the following polynomial function.

8. $f(x) = 5x - \{(4x + 2[x - 3(x + 2)]\}$

Multiply each of the following polynomials.

9. $(2a - 5b)(3a + 7b)$

10. $(5m - 3n)(5m + 3n)$

11. $(2a + 3b)^2$

12. $(2x - 5)(x^2 - 4x + 3)$

Factor each of the following polynomials completely.

13. $14a^2b^2 - 21a^2b + 35ab^2$

14. $x^2 - 3xy + 5x - 15y$

15. $25c^2 - 64d^2$

16. $27x^3 - 1$

17. $16a^4 + 2ab^3$

18. $x^2 - 2x - 48$

19. $10x^2 - 39x + 14$

20. $6x^3 + 3x^2 - 45x$

ANSWERS

1. ______
2. ______
3. ______
4. ______
5. ______
6. (a) (b) (c) (d) (e) (f) ______
7. (a) (b) (c) (d) (e) (f) ______
8. ______
9. ______
10. ______
11. ______
12. ______
13. ______
14. ______
15. ______
16. ______
17. ______
18. ______
19. ______
20. ______

ANSWERS

21. ______

22. ______

23. ______

24. ______

25. ______

Perform the indicated long division.

21. $\dfrac{3x^2 - 2x - 4}{3x + 1}$

22. $\dfrac{4x^3 - 5x^2 + 7x - 9}{x - 2}$

23. $\dfrac{3x^4 - 2x^2 - 5}{x^2 + 1}$

Solve each of the following equations by factoring.

24. $2x^2 + 7x + 3 = 0$

25. $6x^2 = 10 - 11x$

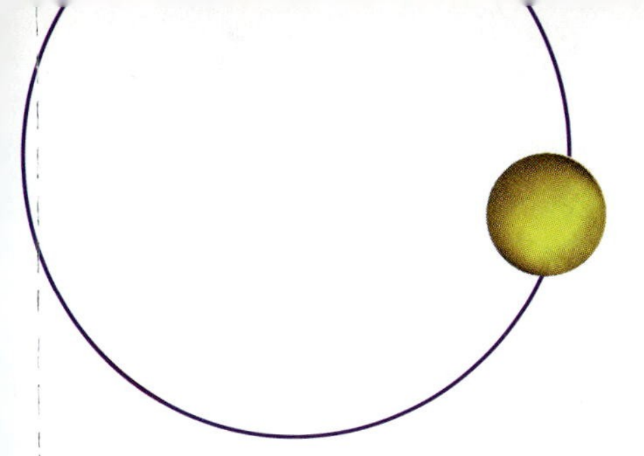

Cumulative Test for Chapters 1 to 6

Name ______________________

Section __________ Date __________

ANSWERS

1. ______________
2. ______________
3. ______________
4. ______________
5. ______________
6. ______________
7. ______________
8. (a) (b) (c) ______________

This test is provided to help you in the process of reviewing previous chapters. Answers are provided in the back of the book. If you missed any answers, be sure to go back and review the appropriate chapter sections.

1. Solve the equation $4x - 2(x + 1) = 3(5 - x) - 7$.

2. If $f(x) = 4x^3 - 5x^2 + 7x - 11$, find $f(-2)$.

3. Find the x and y intercepts and graph the equation $2x + 3y = 12$.

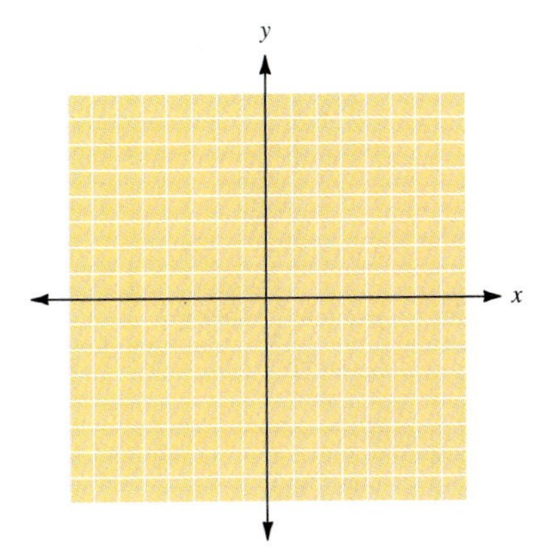

4. Find the equation of the line that passes through the point $(-1, -2)$ and is perpendicular to the line $4x + 5y = 15$.

5. Simplify the expression $(4x^3y^2)^3(-2x^2y^3)^2$.

6. Evaluate the expression $18 \div 3 \cdot 2 - (3 - 4^2)$.

7. Find the domain and range of the relation $4x - 3y = 15$.

8. Use the given graph to estimate **(a)** $f(-2)$, **(b)** $f(0)$, and **(c)** $f(3)$.

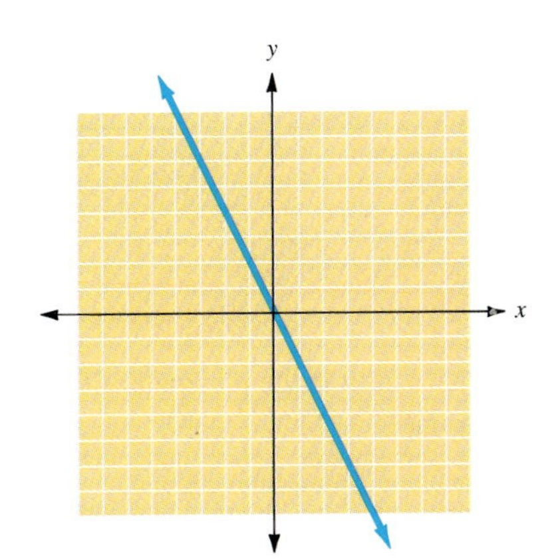

ANSWERS

9. ______
10. ______
11. ______
12. ______
13. ______
14. (a)
(b)
(c)
(d)
15. ______
16. ______
17. ______
18. ______
19. ______
20. ______
21. ______
22. ______
23. ______
24. ______
25. ______

Simplify each of the following polynomial functions.

9. $f(x) = (2x^2 - 3x + 4) - (3x^2 + 2x - 5)$

10. $f(x) = (2x + 3)(5x - 4)$.

Factor each of the following completely.

11. $3x^3 - x^2 - 2x$

12. $16x^2 - 25y^2$

13. $3x^2 - 3xy + x - y$

14. If $f(x) = -5x + 1$ and $g(x) = 8x + 6$, find **(a)** $(f + g)(x)$, **(b)** $(f - g)(x)$, **(c)** $(f \cdot g)(x)$, and **(d)** $\left(\frac{f}{g}\right)(x)$.

Solve the following equations.

15. $-6x - 6(2x - 9) = -3(x + 7)$

16. $|2x + 46| = 10$

17. $4(3x - 6) = -(8x - 2)$

Solve the following inequalities.

18. $4(3 + x) \geq 8$

19. $|x + 9| \leq 4$

20. $-3|x + 5| > -3$

21. Solve the following system of equations

$$2x + 3y = 6$$
$$5x + 3y = -24$$

22. Solve the equation $P = P_0 + IRT$ for R.

23. Solve the equation $2x^2 - 5x - 3 = 0$.

Solve the following applications.

24. The length of a rectangle is 3 cm more than twice its width. If the perimeter of the rectangle is 54 cm, find the dimensions of the rectangle.

25. The sum of two integers is 30 and the difference is 2. What are the two integers?

7

RATIONAL EXPRESSIONS AND FUNCTIONS

INTRODUCTION

The House of Representatives is made up of officials elected from congressional districts within each state. The number of representatives a state sends to the U.S. House of Representatives depends on the state's population. The total number of representatives to the House has grown from 106 in 1790 to 435, the maximum number established in 1930. These 435 representatives are apportioned to the 50 states on the basis of population. This apportionment is revised after every decennial (10-year) census.

If a particular state has a population, A, and its number of representatives is equal to a, then $\frac{A}{a}$ represents the ratio of people in the state to their total number of representatives in the U.S. House. It follows that the total population of the country, P, and its total number of representatives, r, is represented by the ratio $\frac{P}{r}$. If another state, with population, E, has e number of representatives, then $\frac{E}{e}$ should also be equal to $\frac{P}{r}$ and to $\frac{A}{a}$, if the apportionment is fair.

A comparison of these ratios for states in 1990 finds Pennsylvania with 546,880 people per representative and Arizona with 610,800—Arizona was above the national average of 571,750 people per representative for 1990, and Pennsylvania below. This is not so much a result of political backroom bargaining as it is a result of ratios that do not divide out evenly—there are remainders when the proportions are solved. Should the numbers be rounded up or down? If they are all rounded down, the total is too small, if rounded up, the total number of representatives would be more than the 435 seats in the House. So, because all the states cannot be treated equally, the question of what is fair and how to decide who gets the additional representatives has been debated in Congress since its inception.

Simplification of Rational Expressions and Functions

7.1 OBJECTIVES

1. Simplify a rational expression
2. Identify a rational function
3. Simplify a rational function
4. Graph a rational function

Our work in this chapter will expand your experience with algebraic expressions to include algebraic fractions or **rational expressions.** We consider the four basic operations of addition, subtraction, multiplication, and division in the next sections. Fortunately, you will observe many parallels to your previous work with arithmetic fractions.

First, let's define what we mean by a rational expression. Recall that a rational number is the ratio of two integers. Similarly, a rational expression can be written as the ratio of two polynomials, in which the denominator cannot have the value 0.

NOTE The word "rational" comes from "ratio."

Definitions: Rational Expression

A **rational expression** is the ratio of two polynomials. It can be written as

$$\frac{P}{Q}$$

in which P and Q are polynomials and Q cannot have the value 0.

The expressions

$$\frac{x-3}{x+1} \qquad \frac{x^2+5}{x-3} \qquad \text{and} \qquad \frac{x^2-2x}{x^2+3x+1}$$

are all rational expressions. The restriction that the denominator of the expressions not be 0 means that certain values for the variable may have to be excluded because division by 0 is undefined.

Example 1

Precluding Division by Zero

(a) For what values of x is the following expression undefined?

$$\frac{x}{x-5}$$

NOTE A fraction is undefined when its denominator is equal to 0.

NOTE Note that when $x = 5$, $\frac{x}{x-5}$ becomes $\frac{5}{5-5}$, or $\frac{5}{0}$.

To answer this question, we must find where the denominator is 0. Set

$$x - 5 = 0$$

or

$$x = 5$$

The expression $\frac{x}{x-5}$ is undefined for $x = 5$.

(b) For what values of x is the following expression undefined?

$$\frac{3}{x + 5}$$

Again, set the denominator equal to 0:

$$x + 5 = 0$$

or

$$x = -5$$

The expression $\frac{3}{x + 5}$ is undefined for $x = -5$.

CHECK YOURSELF 1

For what values of the variable are the following expressions undefined?

(a) $\frac{1}{r + 7}$ **(b)** $\frac{5}{2x - 9}$

Generally, we want to write rational expressions in the simplest possible form. To begin our discussion of simplifying rational expressions, let's review for a moment. As we pointed out previously, there are many parallels to your work with arithmetic fractions. Recall that

$$\frac{3}{5} = \frac{3 \cdot 2}{5 \cdot 2} = \frac{6}{10}$$

so

$$\frac{3}{5} \quad \text{and} \quad \frac{6}{10}$$

name equivalent fractions. In a similar fashion,

$$\frac{10}{15} = \frac{5 \cdot 2}{5 \cdot 3} = \frac{2}{3}$$

so

$$\frac{10}{15} \quad \text{and} \quad \frac{2}{3}$$

name equivalent fractions.

We can always multiply or divide the numerator and denominator of a fraction by the same nonzero number. The same pattern is true in algebra.

Rules and Properties: Fundamental Principle of Rational Expressions

For polynomials P, Q, and R,

$$\frac{P}{Q} = \frac{PR}{QR} \quad \text{when } Q \neq 0 \text{ and } R \neq 0$$

This principle can be used in two ways. We can multiply or divide the numerator and denominator of a rational expression by the same nonzero polynomial. The result will always be an expression that is equivalent to the original one.

In simplifying arithmetic fractions, we used this principle to divide the numerator and denominator by all common factors. With arithmetic fractions, those common factors are generally easy to recognize. Given rational expressions in which the numerator and denominator are polynomials, we must determine those factors as our first step. The most important tools for simplifying expressions are the factoring techniques in Chapter 6.

NOTE In fact, you will see that most of the methods in this chapter depend on factoring polynomials.

Example 2

Simplifying Rational Expressions

Simplify each rational expression. Assume denominators are not 0.

NOTE We find the common factors 4, x, and y in the numerator and denominator. We divide the numerator and denominator by the common factor $4xy$. Note that

$$\frac{4xy}{4xy} = 1$$

(a) $\dfrac{4x^2y}{12xy^2} = \dfrac{4xy \cdot x}{4xy \cdot 3y}$

$= \dfrac{x}{3y}$

NOTE We have *divided* the numerator and denominator by the common factor $x - 2$. Again note that

$$\frac{x-2}{x-2} = 1$$

(b) $\dfrac{3x - 6}{x^2 - 4} = \dfrac{3(x - 2)}{(x + 2)(x - 2)}$ Factor the numerator and the denominator.

We can now divide the numerator and denominator by the common factor $x - 2$:

$$\frac{3(x-2)}{(x+2)(x-2)} = \frac{3}{x+2}$$

and the rational expression is in simplest form.

CAUTION

Pick any value other than 0 for the variable x, and substitute. You will quickly see that

$$\frac{x+2}{x+3} \neq \frac{2}{3}$$

Be Careful! Given the expression

$$\frac{x+2}{x+3}$$

students are often tempted to divide by variable x, as in

$$\frac{x+2}{x+3} \stackrel{?}{=} \frac{2}{3}$$

This is not a valid operation. We can only divide by common *factors,* and in the expression above, the variable x is a *term* in both the numerator and the denominator. The numerator and denominator of a rational expression must be factored *before* common factors are divided out. Therefore,

$$\frac{x+2}{x+3}$$

is in its simplest possible form.

CHECK YOURSELF 2

Simplify each expression.

(a) $\dfrac{36a^3b}{9ab^2}$ **(b)** $\dfrac{x^2 - 25}{4x + 20}$

The same techniques are used when trinomials need to be factored. Example 3 further illustrates the simplification of rational expressions.

Example 3

Simplifying Rational Expressions

Simplify each rational expression.

NOTE Divide by the common factor $x + 1$, using the fact that

$$\frac{x+1}{x+1} = 1$$

when $x \neq -1$

(a) $$\frac{5x^2 - 5}{x^2 - 4x - 5} = \frac{5(x^2 - 1)}{x^2 - 4x - 5} = \frac{5(x-1)(x+1)}{(x-5)(x+1)}$$

$$= \frac{5(x-1)}{x-5}$$

NOTE In part (c) we factor by grouping in the numerator and use the sum of cubes in the denominator. Note that

$x^3 + 2x^2 - 3x - 6$

$= x^2(x + 2) - 3(x + 2)$

$= (x + 2)(x^2 - 3)$

(b) $$\frac{2x^2 + x - 6}{2x^2 - x - 3} = \frac{(x+2)(2x-3)}{(x+1)(2x-3)}$$

$$= \frac{x+2}{x+1}$$

(c) $$\frac{x^3 + 2x^2 - 3x - 6}{x^3 + 8} = \frac{(x+2)(x^2-3)}{(x+2)(x^2-2x+4)} = \frac{x^2 - 3}{x^2 - 2x + 4}$$

CHECK YOURSELF 3

Simplify each rational expression.

(a) $\dfrac{x^2 - 5x + 6}{3x^2 - 6x}$ **(b)** $\dfrac{3x^2 + 14x - 5}{3x^2 + 2x - 1}$

Simplifying certain algebraic expressions involves recognizing a particular pattern. Verify for yourself that

$$3 - 9 = -(9 - 3)$$

In general, it is true that

$$a - b = a + (-b) = -b + a = -1(b - a)$$

or, by dividing the left and right sides of the equation by $b - a$,

NOTE Notice that

$$\frac{a-b}{a-b} = 1$$

but

$$\frac{a-b}{b-a} = -1$$

when $a \neq b$.

$$\frac{a-b}{b-a} = \frac{-(b-a)}{b-a} = -1$$

Example 4 makes use of this result.

Example 4

Simplifying Rational Expressions

Simplify each rational expression.

NOTE From the margin note above, we get

$$\frac{x-2}{2-x} = -1$$

(a) $$\frac{2x-4}{4-x^2} = \frac{2\overset{-1}{(x-2)}}{(2+x)(2-x)}$$

$$= \frac{2(-1)}{2+x} = \frac{-2}{2+x}$$

(b) $\dfrac{9 - x^2}{x^2 + 2x - 15} = \dfrac{(3 + x)\overset{-1}{\cancel{(3 - x)}}}{(x + 5)\cancel{(x - 3)}}$

$= \dfrac{(3 + x)(-1)}{x + 5}$

$= \dfrac{-x - 3}{x + 5}$

CHECK YOURSELF 4

Simplify each rational expression.

(a) $\dfrac{5x - 20}{16 - x^2}$ **(b)** $\dfrac{x^2 - 6x - 27}{81 - x^2}$

The following algorithm summarizes our work with simplifying rational expressions.

Step by Step: Simplifying Rational Expressions

Step 1 Completely factor both the numerator and denominator of the expression.

Step 2 Divide the numerator and denominator by *all* common factors.

Step 3 The resulting expression will be in simplest form (or in lowest terms).

To identify rational functions, we begin with a definition.

Definitions: Rational Function

A **rational function** is a function that is defined by a rational expression. It can be written as

$$f(x) = \frac{P}{Q}$$

in which P and Q are polynomials and $Q(x) \neq 0$ for all x.

Example 5

Identifying Rational Functions

Which of the following are rational functions?

(a) $f(x) = 3x^3 - 2x + 5$ This is a rational function; it could be written over the denominator 1, and 1 is a polynomial.

(b) $f(x) = \dfrac{3x^2 - 5x + 2}{2x - 1}$ This is a rational function; it is the ratio of two polynomials.

(c) $f(x) = 3x^3 + 3\sqrt{x}$ This is not a rational function; it is not the ratio of two polynomials.

NOTE Recall from Chapter 6 that there are no square roots of variables in a polynomial.

CHECK YOURSELF 5

Which of the following are rational functions?

(a) $f(x) = x^5 - 2x^4 + 3x - 1$ **(b)** $f(x) = \dfrac{x^2 - x + 7}{\sqrt{x} - 1}$

(c) $f(x) = \dfrac{3x^3 + 3x}{2x + 1}$

When we simplify a rational function, it is important that we note the x values that need to be excluded, particularly when we are trying to draw the graph of a function. The set of ordered pairs of the simplified function will be exactly the same as the set of ordered pairs of the original function. If an excluded value for x yields ordered pair $(x, f(x))$, that ordered pair represents a "hole" in the graph. These holes are breaks in the curve. We use an open circle to designate them on a graph.

Example 6

Simplifying a Rational Function

Given the function

$$f(x) = \frac{x^2 + 2x + 1}{x + 1}$$

complete the following.

(a) Simplify the rational expression on the right.

$$\frac{x^2 + 2x + 1}{x + 1} = \frac{(x + 1)(x + 1)}{(x + 1)}$$

$$= (x + 1) \qquad x \neq -1$$

NOTE Notice that $f(-1)$ is undefined.

(b) Rewrite the function in simplified form.

$$f(x) = x + 1 \qquad x \neq -1$$

(c) Find the ordered pair associated with the hole in the graph of the original function.

Plugging -1 into the simplified function yields the ordered pair $(-1, 0)$. This represents the hole in the graph of the function

$$f(x) = \frac{x^2 + 2x + 1}{x + 1}$$

CHECK YOURSELF 6

Given the function

$$f(x) = \frac{5x^2 - 10x}{5x}$$

complete the following.

(a) Rewrite the function in simplified form.

(b) Find the ordered pair associated with the hole in the graph of the original function.

Certain rational functions can be graphed as a line with a "hole" in it. One such function is examined in Example 7.

Example 7

Graphing a Rational Function

Graph the following function.

$$f(x) = \frac{x^2 + 2x + 1}{x + 1}$$

From Example 6, we know that

$$\frac{x^2 + 2x + 1}{x + 1} = x + 1 \qquad x \neq -1$$

Therefore,

$$f(x) = x + 1 \qquad x \neq -1$$

Because an x value of -1 results in division by 0, there can be no point on the graph with an x value of -1. The graph will be the graph of the line $f(x) = x + 1$, with an open circle at the point $(-1, 0)$.

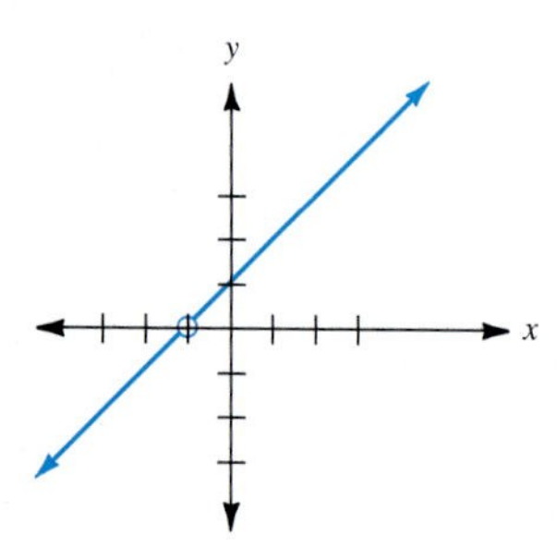

CHECK YOURSELF 7

Graph the function $f(x) = \dfrac{5x^2 - 10x}{5x}$.

CHECK YOURSELF ANSWERS

1. (a) $r = -7$; **(b)** $x = \dfrac{9}{2}$ **2(a)** $\dfrac{4a^2}{b}$; **(b)** $\dfrac{x - 5}{4}$ **3. (a)** $\dfrac{x - 3}{3x}$; **(b)** $\dfrac{x + 5}{x + 1}$

4. (a) $\dfrac{-5}{x + 4}$; **(b)** $\dfrac{-x - 3}{x + 9}$

5. (a) A rational function; **(b)** not a rational function; and **(c)** a rational function

6. (a) $f(x) = x - 2, x \neq 0$; **(b)** $(0, -2)$ **7.**

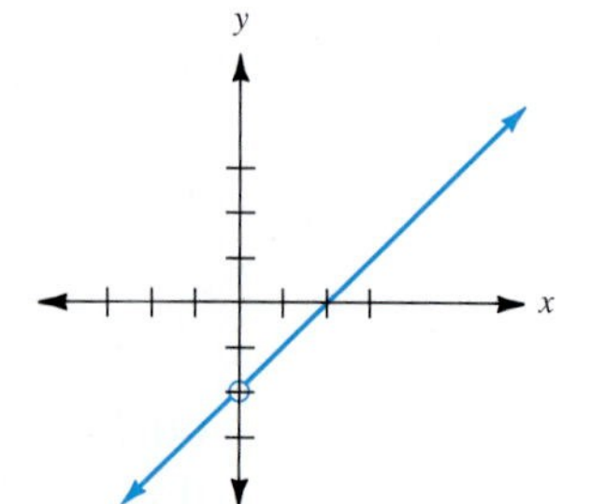

Name ____________

Section ________ Date ________

7.1 Exercises

In exercises 1 to 12, for what values of the variable is each rational expression undefined?

1. $\dfrac{x}{x - 5}$ **2.** $\dfrac{y}{y + 7}$ **3.** $\dfrac{x + 5}{3}$

4. $\dfrac{x - 6}{4}$ **5.** $\dfrac{2x - 3}{2x - 1}$ **6.** $\dfrac{3x - 2}{3x + 1}$

7. $\dfrac{2x + 5}{x}$ **8.** $\dfrac{3x - 7}{x}$ **9.** $\dfrac{x(x + 1)}{x + 2}$

10. $\dfrac{x + 2}{3x - 7}$ **11.** $\dfrac{4 - x}{x}$ **12.** $\dfrac{2x + 7}{3x + \frac{1}{3}}$

In exercises 13 to 16, evaluate each expression, using a calculator.

13. $\dfrac{3x}{2x - 1}$ for $x = 2$ **14.** $\dfrac{5x}{4x - 3}$ for $x = 2$

15. $\dfrac{2x + 3}{x + 3}$ for $x = -6$ **16.** $\dfrac{4x - 7}{2x - 1}$ for $x = -2$

In exercises 17 to 48, simplify each expression. Assume the denominators are not 0.

17. $\dfrac{14}{21}$ **18.** $\dfrac{45}{75}$ **19.** $\dfrac{4x^5}{6x^2}$

20. $\dfrac{25x^6}{20x^2}$ **21.** $\dfrac{10x^2y^5}{25xy^2}$ **22.** $\dfrac{18a^2b^3}{24a^4b^3}$

23. $\dfrac{-42x^3y}{14xy^3}$ **24.** $\dfrac{-15x^3y^3}{-20xy^2}$ **25.** $\dfrac{28a^5b^3c^2}{84a^2bc^4}$

26. $\dfrac{-52p^5q^3r^2}{39p^3q^5r^2}$ **27.** $\dfrac{6x - 24}{x^2 - 16}$ **28.** $\dfrac{x^2 - 25}{3x - 15}$

29. $\dfrac{x^2 + 2x + 1}{6x + 6}$ **30.** $\dfrac{5y^2 - 10y}{y^2 + y - 6}$ **31.** $\dfrac{x^2 - 5x - 14}{x^2 - 49}$

ANSWERS

1. ______ 2. ______
3. ______
4. ______
5. ______ 6. ______
7. ______ 8. ______
9. ______ 10. ______
11. ______ 12. ______
13. ______ 14. ______
15. ______ 16. ______
17. ______ 18. ______
19. ______ 20. ______
21. ______ 22. ______
23. ______ 24. ______
25. ______ 26. ______
27. ______ 28. ______
29. ______ 30. ______
31. ______

ANSWERS

32. ______ 33. ______
34. ______ 35. ______
36. ______ 37. ______
38. ______
39. ______
40. ______
41. ______ 42. ______
43. ______ 44. ______
45. ______ 46. ______
47. ______
48. ______
49. ______
50. ______
51. ______
52. ______
53. ______
54. ______
55. ______
56. ______
57. ______
58. ______
59. ______
60. ______

32. $\dfrac{2m^2 + 11m - 21}{4m^2 - 9}$ **33.** $\dfrac{3b^2 - 14b - 5}{b - 5}$ **34.** $\dfrac{a^2 - 9b^2}{a^2 + 8ab + 15b^2}$

35. $\dfrac{2y^2 + 3yz - 5z^2}{2y^2 + 11yz + 15z^2}$ **36.** $\dfrac{6x^2 - x - 2}{3x^2 - 5x + 2}$ **37.** $\dfrac{x^3 - 64}{x^2 - 16}$

38. $\dfrac{r^2 - rs - 6s^2}{r^3 + 8s^3}$ **39.** $\dfrac{a^4 - 81}{a^2 + 5a + 6}$ **40.** $\dfrac{c^4 - 16}{c^2 - 3c - 10}$

41. $\dfrac{xy - 2x + 3y - 6}{x^2 + 8x + 15}$ **42.** $\dfrac{cd - 3c + 5d - 15}{d^2 - 7d + 12}$ **43.** $\dfrac{x^2 + 3x - 18}{x^3 - 3x^2 - 2x + 6}$

44. $\dfrac{y^2 + 2y - 35}{y^2 - 5y - 3y + 15}$ **45.** $\dfrac{2m - 10}{25 - m^2}$ **46.** $\dfrac{5x - 20}{16 - x^2}$

47. $\dfrac{49 - x^2}{2x^2 - 13x - 7}$ **48.** $\dfrac{2x^2 - 7x + 3}{9 - x^2}$

In exercises 49 to 54, identify which functions are rational functions.

49. $f(x) = 4x^2 - 5x + 6$ **50.** $f(x) = \dfrac{x^3 - 2x^2 + 7}{\sqrt{x} + 2}$

51. $f(x) = \dfrac{x^2 - x - 1}{x + 2}$ **52.** $f(x) = \dfrac{\sqrt{x} - x + 3}{x - 2}$

53. $f(x) = 5x^2 - \sqrt[3]{x}$ **54.** $f(x) = \dfrac{x^2 - x + 5}{x}$

For the given functions in exercises 55 to 60, **(a)** rewrite the function in simplified form, and **(b)** find the ordered pair associated with the hole in the graph of the original function.

55. $f(x) = \dfrac{x^2 - x - 2}{x + 1}$ **56.** $f(x) = \dfrac{x^2 + x - 12}{x + 4}$

57. $f(x) = \dfrac{3x^2 + 5x - 2}{x + 2}$ **58.** $f(x) = \dfrac{2x^2 - 7x + 5}{2x - 5}$

59. $f(x) = \dfrac{x^2 + 4x + 4}{5(x + 2)}$ **60.** $f(x) = \dfrac{x^2 - 6x + 9}{7(x - 3)}$

In exercises 61 to 66, graph the rational functions. Indicate the coordinates of the hole in the graph.

61. $f(x) = \dfrac{x^2 - 2x - 8}{x + 2}$

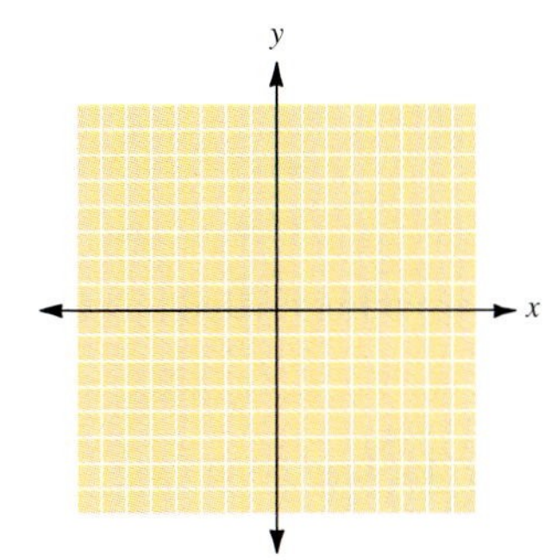

62. $f(x) = \dfrac{x^2 + 4x - 5}{x + 5}$

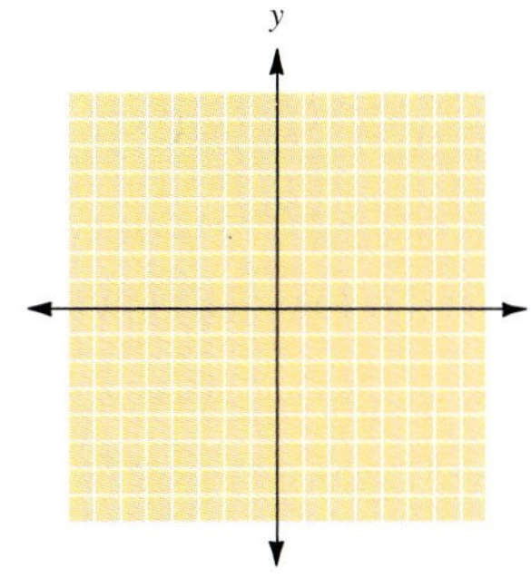

63. $f(x) = \dfrac{x^2 + 4x + 3}{x + 1}$

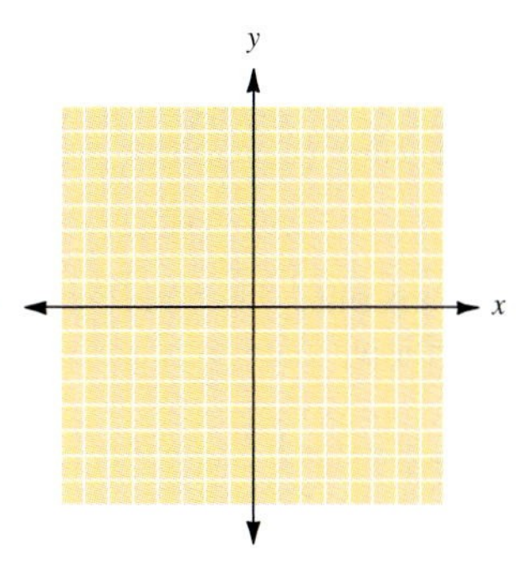

64. $f(x) = \dfrac{x^2 + 7x + 10}{x + 2}$

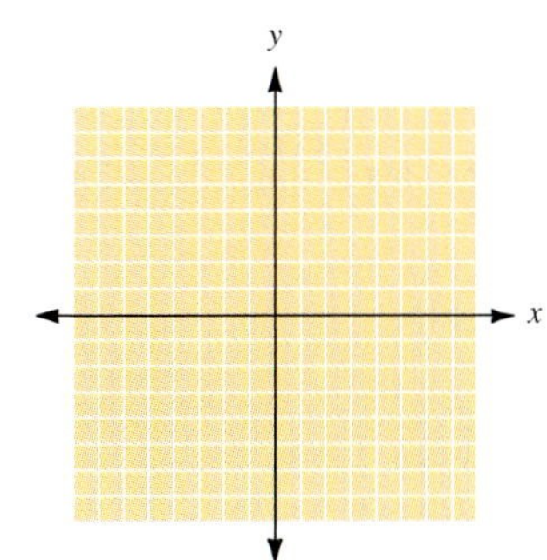

65. $f(x) = \dfrac{x^2 - 4x + 3}{x - 1}$

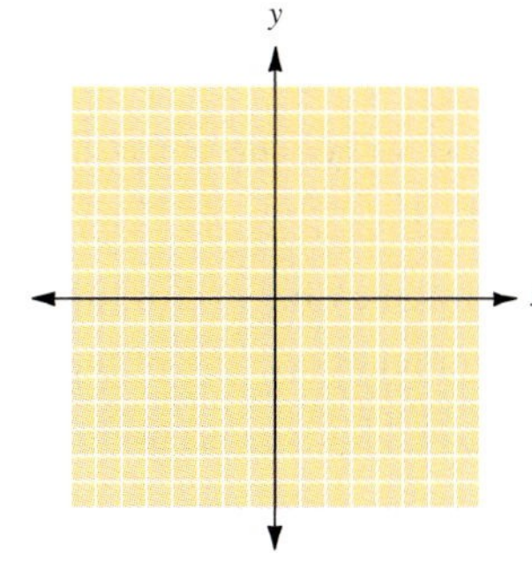

66. $f(x) = \dfrac{x^2 - 6x + 8}{x - 4}$

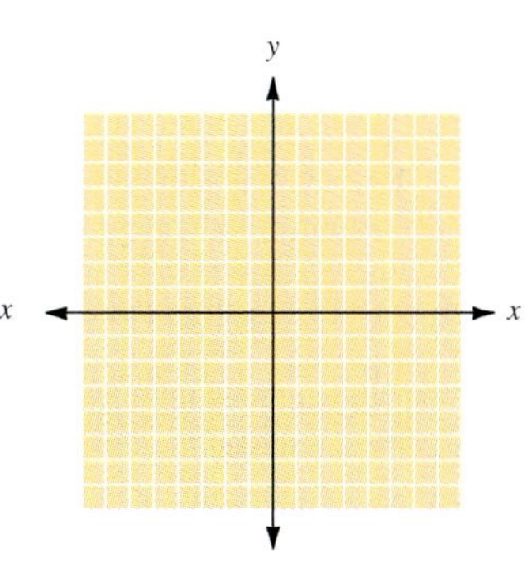

67. Explain why the following statement is false.

$$\frac{6m^2 + 2m}{2m} = 6m^2 + 1$$

68. State and explain the Fundamental Principle of Rational Expressions.

69. The rational expression $\dfrac{x^2 - 4}{x + 2}$ can be simplified to $x - 2$. Is this reduction true for all values of x? Explain.

70. What is meant by a rational expression in lowest terms?

In exercises 71 to 76, simplify.

71. $\dfrac{2(x + h) - 2x}{(x + h) - x}$

72. $\dfrac{-3(x + h) - (-3x)}{(x + h) - x}$

73. $\dfrac{3(x + h) - 3 - (3x - 3)}{(x + h) - x}$

74. $\dfrac{2(x + h) + 5 - (2x + 5)}{(x + h) - x}$

75. $\dfrac{(x + h)^2 - x^2}{(x + h) - x}$

76. $\dfrac{(x + h)^3 - x^3}{(x + h) - x}$

ANSWERS

61. ____
62. ____
63. ____
64. ____
65. ____
66. ____

67. ____

68. ____
69. ____

70. ____
71. ____
72. ____
73. ____
74. ____
75. ____
76. ____

ANSWERS

77. ______

78. ______

79. ______

80. ______

81. (a)
(b)
(c)

(d)
(e)

(f)

Given $f(x) = \frac{P(x)}{Q(x)}$, if the graphs of $P(x)$ and $Q(x)$ intersect at $(a, 0)$, then $x - a$ is a factor of both $P(x)$ and $Q(x)$. Use a graphing calculator to find the common factor for the expressions in exercises 77 and 78.

77. $f(x) = \frac{x^2 + 4x - 5}{x^2 + 3x - 10}$

78. $f(x) = \frac{2x^2 + 11x - 21}{2x^2 + 15x + 7}$

79. Revenue. The total revenue from the sale of a popular video is approximated by the rational function

$$R(x) = \frac{300x^2}{x^2 + 9}$$

in which x is the number of months since the video has been released and $R(x)$ is the total revenue in hundreds of dollars.

(a) Find the total revenue generated by the end of the first month.

(b) Find the total revenue generated by the end of the second month.

(c) Find the total revenue generated by the end of the third month.

(d) Find the revenue in the second month only.

80. Cost. A company has a set-up cost of \$3500 for the production of a new product. The cost to produce a single unit is \$8.75.

(a) Define a rational function that gives the average cost per unit when x units are produced.

(b) Find the average cost when 50 units are produced.

81. Besides holes, we sometimes encounter a different sort of "break" in the graph of a rational function. Consider the rational function

$$f(x) = \frac{1}{x - 3}$$

(a) For what value(s) of x is the function undefined?

(b) Complete the following table.

x	$f(x)$
4	
3.1	
3.01	
3.001	
3.0001	

(c) What do you observe concerning $f(x)$ if x is chosen close to 3 (but slightly larger than 3)?

(d) Complete the table.

x	$f(x)$
2	
2.9	
2.99	
2.999	
2.9999	

(e) What do you observe concerning $f(x)$ if x is chosen close to 3 (but slightly smaller than 3)?

(f) Graph the function on your graphing calculator. Describe the behavior of the graph of $f(x)$ near $x = 3$.

82. Consider the rational function

$$f(x) = \frac{1}{x + 2}$$

(a) For what value(s) of x is the function undefined?

(b) Complete the following table.

x	$f(x)$
-3	
-2.1	
-2.01	
-2.001	
-2.0001	

(c) What do you observe concerning $f(x)$ if x is chosen close to -2? (but slightly smaller than -2)?

(d) Complete the following table.

x	$f(x)$
-1	
-1.9	
-1.99	
-1.999	
-1.9999	

(e) What do you observe concerning $f(x)$ if x is chosen close to -2 (but slightly larger than -2)?

(f) Graph the function on your graphing calculator. Describe the behavior of the graph near $x = -2$.

83. Age Ratios. Your friend has a 4-year-old cousin, Amy, who has a 9-year-old brother. The younger child is upset because not only does her brother refuse to let her play with him and his friends, he teases her because she can never catch up in age! You explain to the child that as she gets older, this age difference will not seem like such a big deal.

Write an expression for the ratio of the younger child's age to her brother's age.

ANSWERS

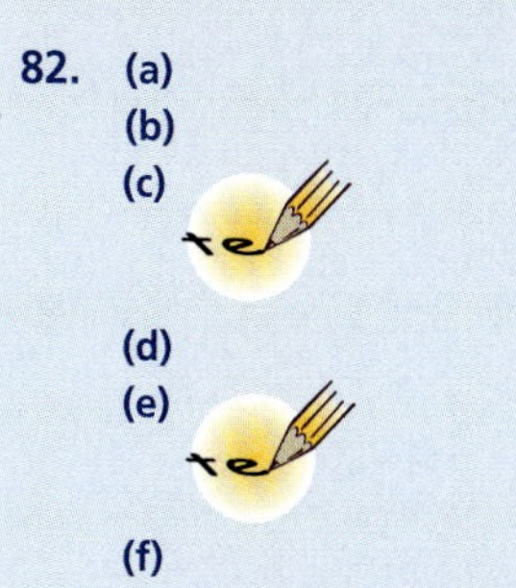

82. (a)
(b)
(c)
(d)
(e)
(f)

83.

1. What happens to the ratio as the children grow older?
2. Draw a graph of the ratio as a function of Amy's age to show how this ratio changes.
3. Assume that Amy and her brother live to be 100 and 105 years old, respectively. What will the ratio between their ages be?
4. Draw a graph of the ratio of Amy's brother's age to Amy's age. Does this graph have anything in common with the first graph? Explain.
5. Write a short explanation for Amy (who is only 4, remember!) of your conclusions about how their age ratios change over time.

Answers

1. 5 **3.** Never undefined **5.** $\frac{1}{2}$ **7.** 0 **9.** -2 **11.** 0 **13.** 2
15. 3 **17.** $\frac{2}{3}$ **19.** $\frac{2x^3}{3}$ **21.** $\frac{2xy^3}{5}$ **23.** $\frac{-3x^2}{y^2}$ **25.** $\frac{a^3b^2}{3c^2}$
27. $\frac{6}{x+4}$ **29.** $\frac{x+1}{6}$ **31.** $\frac{x+2}{x+7}$ **33.** $3b + 1$ **35.** $\frac{y-z}{y+3z}$
37. $\frac{x^2+4x+16}{x+4}$ **39.** $\frac{(a^2+9)(a-3)}{a+2}$ **41.** $\frac{y-2}{x+5}$ **43.** $\frac{x+6}{x^2-2}$
45. $\frac{-2}{m+5}$ **47.** $-\frac{x+7}{2x+1}$ **49.** Rational **51.** Rational
53. Not rational **55.** **(a)** $f(x) = x - 2$; **(b)** $(-1, -3)$
57. **(a)** $f(x) = 3x - 1$; **(b)** $(-2, -7)$ **59.** **(a)** $f(x) = \frac{x+2}{5}$; **(b)** $(-2, 0)$

61.

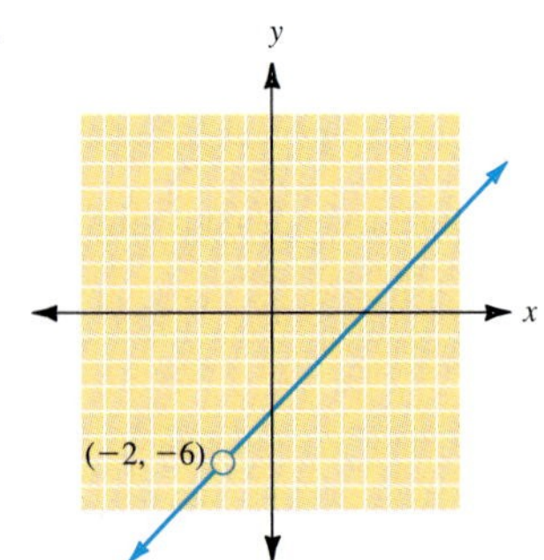

63.

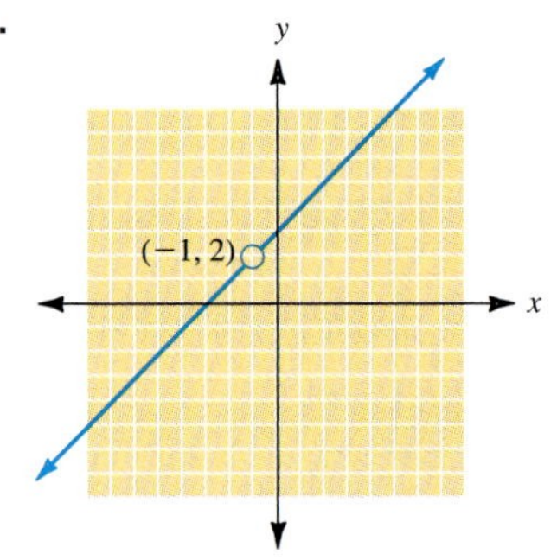

65.

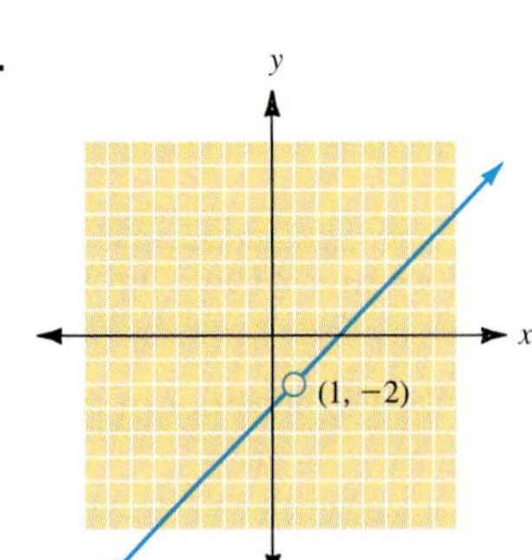

67.  **69.** 

71. 2 **73.** 3 **75.** $2x + h$ **77.** $x + 5$
79. **(a)** \$3000; **(b)** \$9231; **(c)** \$15,000; **(d)** \$6231
81. **(a)** 3; **(b)** 1, 10, 100, 1000, 10,000; **(c)** ; **(d)** $-1, -10, -100, -1000, -10,000$
83. 

7.2 Multiplication and Division of Rational Expressions and Functions

7.2 OBJECTIVES

1. Multiply two rational expressions
2. Divide two rational expressions
3. Multiply two rational functions
4. Divide two rational functions

Once again, let's turn to an example from arithmetic to begin our discussion of multiplying rational expressions. Recall that to multiply two fractions, we multiply the numerators and multiply the denominators. For instance,

$$\frac{2}{5} \cdot \frac{3}{7} = \frac{2 \cdot 3}{5 \cdot 7} = \frac{6}{35}$$

In algebra, the pattern is exactly the same.

Rules and Properties: Multiplying Rational Expressions

For polynomials P, Q, R, and S,

$$\frac{P}{Q} \cdot \frac{R}{S} = \frac{PR}{QS} \quad \text{when } Q \neq 0 \quad \text{and} \quad S \neq 0$$

NOTE For all problems with rational expressions, assume denominators are not 0.

Example 1

Multiplying Rational Expressions

Multiply.

$$\frac{2x^3}{5y^2} \cdot \frac{10y}{3x^2} = \frac{20x^3y}{15x^2y^2}$$

$$= \frac{5x^2y \cdot 4x}{5x^2y \cdot 3y}$$

Divide by the common factor $5x^2y$ to simplify.

$$= \frac{4x}{3y}$$

CHECK YOURSELF 1

Multiply.

$$\frac{9a^2b^3}{5ab^4} \cdot \frac{20ab^2}{27ab^3}$$

NOTE The factoring methods in Chapter 6 are used to simplify rational expressions.

Generally, you will find it best to divide by any common factors before you multiply, as Example 2 illustrates.

Example 2

Multiplying Rational Expressions

Multiply as indicated.

(a) $\dfrac{x}{x^2 - 3x} \cdot \dfrac{6x - 18}{9x}$

Factor.

$= \dfrac{\overset{1}{\cancel{x}}}{\underset{1}{\cancel{x}}\underset{1}{\cancel{(x-3)}}} \cdot \dfrac{\overset{2}{\cancel{6}}\cancel{(x-3)}}{\underset{3}{\cancel{9}}x}$

Divide by the common factors of 3, x, and $x - 3$.

$= \dfrac{2}{3x}$

(b) $\dfrac{x^2 - y^2}{5x^2 - 5xy} \cdot \dfrac{10xy}{x^2 + 2xy + y^2}$

Factor and divide by the common factors of 5, x, $x - y$, and $x + y$.

$= \dfrac{\overset{1}{\cancel{(x+y)}}\overset{1}{\cancel{(x-y)}}}{\underset{1}{\cancel{5}}x\underset{1}{\cancel{(x-y)}}} \cdot \dfrac{\overset{2}{\cancel{10}}xy}{\underset{1}{\cancel{(x+y)}}(x + y)}$

$= \dfrac{2y}{x + y}$

NOTE From Section 7.1, recall that

$\dfrac{2 - x}{x - 2} = -1$

(c) $\dfrac{4}{x^2 - 2x} \cdot \dfrac{10x - 5x^2}{8x + 24}$

$= \dfrac{\overset{1}{\cancel{4}}}{x\cancel{(x-2)}} \cdot \dfrac{5x\overset{-1}{\cancel{(2-x)}}}{\underset{2}{\cancel{8}}(x + 3)}$

$= \dfrac{-5}{2(x + 3)}$

CHECK YOURSELF 2

Multiply as indicated.

(a) $\dfrac{x^2 - 5x - 14}{4x^2} \cdot \dfrac{8x + 56}{x^2 - 49}$ **(b)** $\dfrac{x}{2x - 6} \cdot \dfrac{3x - x^2}{2}$

The following algorithm summarizes our work in multiplying rational expressions.

Step by Step: Multiplying Rational Expressions

Step 1 Write each numerator and denominator in completely factored form.

Step 2 Divide by any common factors appearing in both the numerator and denominator.

Step 3 Multiply as needed to form the product.

In dividing rational expressions, you can again use your experience from arithmetic. Recall that

NOTE We invert the *divisor* (the second fraction) and multiply.

$$\frac{3}{5} \div \frac{2}{3} = \frac{3}{5} \cdot \frac{3}{2} = \frac{9}{10}$$

Once more, the pattern in algebra is identical.

Rules and Properties: Dividing Rational Expressions

For polynomials P, Q, R, and S,

$$\frac{P}{Q} \div \frac{R}{S} = \frac{P}{Q} \cdot \frac{S}{R} = \frac{PS}{QR}$$

when $Q \neq 0 \quad R \neq 0 \quad$ and $\quad S \neq 0$

To divide rational expressions, invert the divisor and multiply as before, as Example 3 illustrates.

Example 3

Dividing Rational Expressions

Divide as indicated.

NOTE Invert the divisor and multiply.

(a) $\dfrac{3x^2}{8x^3y} \div \dfrac{9x^2y^2}{4y^4} = \dfrac{3x^2}{8x^3y} \cdot \dfrac{4y^4}{9x^2y^2} = \dfrac{y}{6x^3}$

Be Careful! Invert the divisor, then factor.

(b) $\dfrac{2x^2 + 4xy}{9x - 18y} \div \dfrac{4x + 8y}{3x - 6y} = \dfrac{2x^2 + 4xy}{9x - 18y} \cdot \dfrac{3x - 6y}{4x + 8y}$

$$= \frac{2x(x + 2y)}{9(x - 2y)} \cdot \frac{3(x - 2y)}{4(x + 2y)} = \frac{x}{6}$$

(c) $\dfrac{2x^2 - x - 6}{4x^2 + 6x} \div \dfrac{x^2 - 4}{4x} = \dfrac{2x^2 - x - 6}{4x^2 + 6x} \cdot \dfrac{4x}{x^2 - 4}$

$$= \frac{(2x + 3)(x - 2)}{2x(2x + 3)} \cdot \frac{4x}{(x + 2)(x - 2)} = \frac{2}{x + 2}$$

CHECK YOURSELF 3

Divide and simplify.

(a) $\dfrac{5xy}{7x^3} \div \dfrac{10y^2}{14x^3}$

(b) $\dfrac{3x - 9y}{2x + 10y} \div \dfrac{x^2 - 3xy}{4x^2 + 20xy}$

(c) $\dfrac{x^2 - 9}{x^3 - 27} \div \dfrac{x^2 - 2x - 15}{2x^2 - 10x}$

We summarize our work in dividing fractions with the following algorithm.

Step by Step: Dividing Rational Expressions

Step 1 Invert the divisor (the *second* rational expression) to write the problem as one of multiplication.

Step 2 Proceed as in the algorithm for the multiplication of rational expressions.

The product of two rational functions is always a rational function. Given two rational functions, $f(x)$ and $g(x)$, we can rename the product, so

$$h(x) = f(x) \cdot g(x)$$

This will always be true for values of x for which both f and g are defined. So, for example, $h(1) = f(1) \cdot g(1)$ as long as both $f(1)$ and $g(1)$ exist.

Example 4 illustrates this concept.

Example 4

Multiplying Rational Functions

Given the rational functions

$$f(x) = \frac{x^2 - 3x - 10}{x + 1} \quad \text{and} \quad g(x) = \frac{x^2 - 4x - 5}{x - 5}$$

find the following.

(a) $f(0) \cdot g(0)$

Because $f(0) = -10$ and $g(0) = 1$, then $f(0) \cdot g(0) = (-10)(1) = -10$.

(b) $f(5) \cdot g(5)$

Although we can find $f(5)$, $g(5)$ is undefined. 5 is an excluded value for the domain of the function g. Therefore, $f(5) \cdot g(5)$ is undefined.

(c) $h(x) = f(x) \cdot g(x)$

$$h(x) = f(x) \cdot g(x)$$

$$= \frac{x^2 - 3x - 10}{x + 1} \cdot \frac{x^2 - 4x - 5}{x - 5}$$

$$= \frac{(x - 5)(x + 2)}{\cancel{(x + 1)}_{1}} \cdot \frac{\cancel{(x + 1)}^{1}\cancel{(x - 5)}^{1}}{\cancel{(x - 5)}_{1}}$$

$$= (x - 5)(x + 2) \qquad x \neq -1, x \neq 5$$

(d) $h(0)$

$$h(0) = (0 - 5)(0 + 2) = -10$$

(e) $h(5)$

Although the temptation is to substitute 5 for x in part (c), notice that the function is undefined when x is -1 or 5. As was true in part (b), the function is undefined at that point.

CHECK YOURSELF 4

Given the rational functions

$$f(x) = \frac{x^2 - 2x - 8}{x + 2} \quad \text{and} \quad g(x) = \frac{x^2 - 3x - 10}{x - 4}$$

find the following.

(a) $f(0) \cdot g(0)$ **(b)** $f(4) \cdot g(4)$ **(c)** $h(x) = f(x) \cdot g(x)$ **(d)** $h(0)$ **(e)** $h(4)$

When we divide two rational functions to create a third rational function, we must be certain to exclude values for which the denominator is equal to zero, as Example 5 illustrates.

Example 5

Dividing Polynomial Functions

Given the rational functions

$$f(x) = \frac{x^3 - 2x^2}{x + 2} \quad \text{and} \quad g(x) = \frac{x^2 - 3x + 2}{x - 4}$$

complete the following.

(a) Find $\frac{f(0)}{g(0)}$.

Because $f(0) = 0$ and $g(0) = -\frac{1}{2}$, then

$$\frac{f(0)}{g(0)} = \frac{0}{-\frac{1}{2}} = 0$$

(b) Find $\frac{f(1)}{g(1)}$.

Although we can find both $f(1)$ and $g(1)$, $g(1) = 0$, so division is undefined when $x = 1$. 1 is an excluded value for the domain of the quotient.

(c) Find $h(x) = \frac{f(x)}{g(x)}$.

$$h(x) = \frac{f(x)}{g(x)}$$

$$= \frac{\frac{x^3 - 2x^2}{x + 2}}{\frac{x^2 - 3x + 2}{x - 4}} \qquad \text{Invert and multiply.}$$

$$= \frac{x^3 - 2x^2}{x + 2} \cdot \frac{x - 4}{x^2 - 3x + 2}$$

$$= \frac{x^2\overset{1}{\cancel{(x - 2)}}}{x + 2} \cdot \frac{x - 4}{(x - 1)\underset{1}{\cancel{(x - 2)}}}$$

$$= \frac{x^2(x - 4)}{(x + 2)(x - 1)} \qquad x \neq -2, 1, 2, 4$$

(d) For which values of x is $h(x)$ undefined?

$h(x)$ will be undefined for any value of x for which $f(x)$ is undefined, $g(x)$ is undefined, or $g(x) = 0$.

$h(x)$ is undefined for the values -2, 1, 2, and 4.

CHECK YOURSELF 5

Given the rational functions

$$f(x) = \frac{x^2 - 2x + 1}{x + 3} \quad \text{and} \quad g(x) = \frac{x^2 - 5x + 4}{x - 2}$$

complete the following.

(a) Find $\frac{f(0)}{g(0)}$. **(b)** Find $\frac{f(1)}{g(1)}$. **(c)** Find $h(x) = \frac{f(x)}{g(x)}$.

(d) For which values of x is $h(x)$ undefined?

CHECK YOURSELF ANSWERS

1. $\frac{4a}{3b^2}$ **2. (a)** $\frac{2(x + 2)}{x^2}$; **(b)** $\frac{-x^2}{4}$ **3. (a)** $\frac{x}{y}$; **(b)** 6; **(c)** $\frac{2x}{x^2 + 3x + 9}$

4. (a) -10; **(b)** undefined; **(c)** $h(x) = (x - 5)(x + 2) \quad x \neq 4, x \neq -2$; **(d)** -10; and **(e)** undefined **5. (a)** $-\frac{1}{6}$; **(b)** undefined; **(c)** $h(x) = \frac{(x - 1)(x - 2)}{(x + 3)(x - 4)}$; and **(d)** $x \neq -3, 1, 2, 4$

Name ______________

Section ________ Date ________

7.2 Exercises

In exercises 1 to 36, multiply or divide as indicated. Express your result in simplest form.

1. $\frac{x^2}{3} \cdot \frac{6x}{x^4}$

2. $\frac{-y^3}{10} \cdot \frac{15y}{y^6}$

3. $\frac{a}{7a^3} \div \frac{a^2}{21}$

4. $\frac{p^5}{8} \div \frac{-p^2}{12p}$

5. $\frac{4xy^2}{15x^3} \cdot \frac{25xy}{16y^3}$

6. $\frac{3x^3y}{10xy^3} \cdot \frac{5xy^2}{-9xy^3}$

7. $\frac{8b^3}{15ab} \div \frac{2ab^2}{20ab^3}$

8. $\frac{4x^2y^2}{9x^3} \div \frac{-8y^2}{27xy}$

9. $\frac{m^3n}{2mn} \cdot \frac{6mn^2}{m^3n} \div \frac{3mn}{5m^2n}$

10. $\frac{4cd^2}{5cd} \cdot \frac{3c^3d}{2c^2d} \div \frac{9cd}{20cd^3}$

11. $\frac{5x + 15}{3x} \cdot \frac{9x^2}{2x + 6}$

12. $\frac{a^2 - 3a}{5a} \cdot \frac{20a^2}{3a - 9}$

13. $\frac{3b - 15}{6b} \div \frac{4b - 20}{9b^2}$

14. $\frac{7m^2 + 28m}{4m} \div \frac{5m + 20}{12m^2}$

15. $\frac{x^2 - 3x - 10}{5x} \cdot \frac{15x^2}{3x - 15}$

16. $\frac{y^2 - 8y}{4y} \cdot \frac{12y^2}{y^2 - 64}$

17. $\frac{c^2 + 2c - 8}{6c} \div \frac{5c + 20}{18c}$

18. $\frac{m^2 - 49}{5m} \div \frac{3m + 21}{20m^2}$

ANSWERS

1. ______
2. ______
3. ______
4. ______
5. ______
6. ______
7. ______
8. ______
9. ______
10. ______
11. ______
12. ______
13. ______
14. ______
15. ______
16. ______
17. ______
18. ______

ANSWERS

19. ______
20. ______
21. ______
22. ______
23. ______
24. ______
25. ______
26. ______
27. ______
28. ______
29. ______
30. ______
31. ______
32. ______
33. ______
34. ______
35. ______
36. ______
37. (a) (b) (c) (d) (e) ______

19. $\dfrac{x^2 - 2x - 8}{4x - 16} \cdot \dfrac{10x}{x^2 - 4}$

20. $\dfrac{y^2 + 7y + 10}{y^2 + 5y} \cdot \dfrac{2y}{y^2 - 4}$

21. $\dfrac{d^2 - 3d - 18}{16d - 96} \div \dfrac{d^2 - 9}{20d}$

22. $\dfrac{b^2 + 6b + 8}{b^2 + 4b} \div \dfrac{b^2 - 4}{2b}$

23. $\dfrac{2x^2 - x - 3}{3x^2 + 7x + 4} \cdot \dfrac{3x^2 - 11x - 20}{4x^2 - 9}$

24. $\dfrac{4p^2 - 1}{2p^2 - 9p - 5} \cdot \dfrac{3p^2 - 13p - 10}{9p^2 - 4}$

25. $\dfrac{a^2 - 9}{2a^2 - 6a} \div \dfrac{2a^2 + 5a - 3}{4a^2 - 1}$

26. $\dfrac{2x^2 - 5x - 7}{4x^2 - 9} \div \dfrac{5x^2 + 5x}{2x^2 + 3x}$

27. $\dfrac{2w - 6}{w^2 + 2w} \cdot \dfrac{3w}{3 - w}$

28. $\dfrac{3y - 15}{y^2 + 3y} \cdot \dfrac{4y}{5 - y}$

29. $\dfrac{a - 7}{2a + 6} \div \dfrac{21 - 3a}{a^2 + 3a}$

30. $\dfrac{x - 4}{x^2 + 2x} \div \dfrac{16 - 4x}{3x + 6}$

31. $\dfrac{x^2 - 9y^2}{2x^2 - xy - 15y^2} \cdot \dfrac{4x + 10y}{x^2 + 3xy}$

32. $\dfrac{2a^2 - 7ab - 15b^2}{2ab - 10b^2} \cdot \dfrac{2a^2 - 3ab}{4a^2 - 9b^2}$

33. $\dfrac{3m^2 - 5mn + 2n^2}{9m^2 - 4n^2} \div \dfrac{m^3 - m^2n}{9m^2 + 6mn}$

34. $\dfrac{2x^2y - 5xy^2}{4x^2 - 25y^2} \div \dfrac{4x^2 + 20xy}{2x^2 + 15xy + 25y^2}$

35. $\dfrac{x^3 + 8}{x^2 - 4} \cdot \dfrac{5x - 10}{x^3 - 2x^2 + 4x}$

36. $\dfrac{a^3 - 27}{a^2 - 9} \div \dfrac{a^3 + 3a^2 + 9a}{3a^3 + 9a^2}$

37. Let $f(x) = \dfrac{x^2 - 3x - 4}{x + 2}$ and $g(x) = \dfrac{x^2 - 2x - 8}{x - 4}$. Find **(a)** $f(0) \cdot g(0)$, **(b)** $f(4) \cdot g(4)$, **(c)** $h(x) = f(x) \cdot g(x)$, **(d)** $h(0)$, and **(e)** $h(4)$.

38. Let $f(x) = \dfrac{x^2 - 4x + 3}{x + 5}$ and $g(x) = \dfrac{x^2 + 7x + 10}{x - 3}$. Find **(a)** $f(1) \cdot g(1)$, **(b)** $f(3) \cdot g(3)$, **(c)** $h(x) = f(x) \cdot g(x)$, **(d)** $h(1)$, and **(e)** $h(3)$.

39. Let $f(x) = \dfrac{2x^2 - 3x - 5}{x + 2}$ and $g(x) = \dfrac{3x^2 + 5x - 2}{x + 1}$. Find **(a)** $f(1) \cdot g(1)$, **(b)** $f(-2) \cdot g(-2)$, **(c)** $h(x) = f(x) \cdot g(x)$, **(d)** $h(1)$, and **(e)** $h(-2)$.

40. Let $f(x) = \dfrac{x^2 - 1}{x - 3}$ and $g(x) = \dfrac{x^2 - 9}{x - 1}$. Find **(a)** $f(2) \cdot g(2)$, **(b)** $f(3) \cdot g(3)$, **(c)** $h(x) = f(x) \cdot g(x)$, **(d)** $h(2)$, and **(e)** $h(3)$.

41. Let $f(x) = \dfrac{3x^2 + x - 2}{x - 2}$ and $g(x) = \dfrac{x^2 - 4x - 5}{x + 4}$. Find **(a)** $\dfrac{f(0)}{g(0)}$, **(b)** $\dfrac{f(1)}{g(1)}$, **(c)** $h(x) = \dfrac{f(x)}{g(x)}$, and **(d)** the values of x for which $h(x)$ is undefined.

42. Let $f(x) = \dfrac{x^2 + x}{x - 5}$ and $g(x) = \dfrac{x^2 - x - 6}{x - 5}$. Find **(a)** $\dfrac{f(0)}{g(0)}$, **(b)** $\dfrac{f(2)}{g(2)}$, **(c)** $h(x) = \dfrac{f(x)}{g(x)}$, and **(d)** the values of x for which $h(x)$ is undefined.

The results from multiplying and dividing rational expressions can be checked by using a graphing calculator. To do this, define one expression in Y_1 and the other in Y_2. Then define the operation in Y_3 as $Y_1 \cdot Y_2$ or $Y_1 \div Y_2$. Put your simplified result in Y_4 (sorry, you still must simplify algebraically). Deselect the graphs for Y_1 and Y_2. If you have correctly simplified the expression, the graphs of Y_3 and Y_4 will be identical. Use this technique in exercises 43 to 46.

43. $\dfrac{x^3 - 3x^2 + 2x - 6}{x^2 - 9} \cdot \dfrac{5x^2 + 15x}{20x}$

44. $\dfrac{3a^3 + a^2 - 9a - 3}{15a^2 + 5a} \cdot \dfrac{3a^2 + 9}{a^4 - 9}$

45. $\dfrac{x^4 - 16}{x^2 + x - 6} \div (x^3 + 4x)$

46. $\dfrac{w^3 + 27}{w^2 + 2w - 3} \div (w^3 - 3w^2 + 9w)$

ANSWERS

38. (a) (b) (c) (d) (e)

39. (a) (b) (c) (d) (e)

40. (a) (b) (c) (d) (e)

41. (a) (b) (c) (d)

42. (a) (b) (c) (d)

43.

44.

45.

46.

Answers

1. $\frac{2}{x}$ **3.** $\frac{3}{a^4}$ **5.** $\frac{5}{12x}$ **7.** $\frac{16b^3}{3a}$ **9.** $5mn$ **11.** $\frac{15x}{2}$ **13.** $\frac{9b}{8}$

15. $x^2 + 2x$ **17.** $\frac{3(c - 2)}{5}$ **19.** $\frac{5x}{2(x - 2)}$ **21.** $\frac{5d}{4(d - 3)}$ **23.** $\frac{x - 5}{2x + 3}$

25. $\frac{2a + 1}{2a}$ **27.** $\frac{-6}{w + 2}$ **29.** $\frac{-a}{6}$ **31.** $\frac{2}{x}$ **33.** $\frac{3}{m}$ **35.** $\frac{5}{x}$

37. **(a)** -4; **(b)** undefined; **(c)** $(x + 1)(x - 4)$ $x \neq -2, 4$; **(d)** -4; **(e)** undefined

39. **(a)** -6; **(b)** undefined; **(c)** $(2x - 5)(3x - 1)$ $x \neq -2, -1$; **(d)** -6; **(e)** undefined

41. **(a)** $-\frac{4}{5}$; **(b)** $\frac{5}{4}$; **(c)** $\frac{(3x - 2)(x + 4)}{(x - 2)(x - 5)}$ $x \neq -4, -1, 2, 5$; **(d)** $2, -4, -1, 5$

43. $\frac{x^2 + 2}{4}$ **45.** $\frac{x + 2}{x(x + 3)}$

7.3 Addition and Subtraction of Rational Expressions and Functions

7.3 OBJECTIVES

1. Add two rational expressions
2. Subtract two rational expressions
3. Add two rational functions
4. Subtract two rational functions

Recall that adding or subtracting two arithmetic fractions with the same denominator is straightforward. The same is true in algebra. To add or subtract two rational expressions with the same denominator, we add or subtract their numerators and then write that sum or difference over the common denominator.

Rules and Properties: Adding or Subtracting Rational Expressions

$$\frac{P}{R} + \frac{Q}{R} = \frac{P + Q}{R}$$

and

$$\frac{P}{R} - \frac{Q}{R} = \frac{P - Q}{R}$$

when $R \neq 0$.

Example 1

Adding and Subtracting Rational Expressions

Perform the indicated operations.

NOTE Because we have common denominators, we simply perform the indicated operations on the numerators.

$$\frac{3}{2a^2} - \frac{1}{2a^2} + \frac{5}{2a^2} = \frac{3 - 1 + 5}{2a^2}$$

$$= \frac{7}{2a^2}$$

CHECK YOURSELF 1

Perform the indicated operations.

$$\frac{5}{3y^2} + \frac{4}{3y^2} - \frac{7}{3y^2}$$

The sum or difference of rational expressions should always be expressed in simplest form. Consider Example 2.

Example 2

Adding and Subtracting Rational Expressions

Add or subtract as indicated.

(a) $\dfrac{5x}{x^2 - 9} + \dfrac{15}{x^2 - 9}$ — Add the numerators.

$$= \frac{5x + 15}{x^2 - 9}$$

$$= \frac{5(x + 3)}{(x - 3)(x + 3)} = \frac{5}{x - 3}$$

Factor and divide by the common factor.

(b) $\dfrac{3x + y}{2x} - \dfrac{x - 3y}{2x} = \dfrac{(3x + y) - (x - 3y)}{2x}$

Be sure to *enclose the second numerator* in parentheses.

$$= \frac{3x + y - x + 3y}{2x}$$

Remove the parentheses by *changing each sign.*

$$= \frac{2x + 4y}{2x} = \frac{2(x + 2y)}{2x}$$

Factor and divide by the common factor of 2.

$$= \frac{x + 2y}{x}$$

CHECK YOURSELF 2

Perform the indicated operations.

(a) $\dfrac{6a}{a^2 - 2a - 8} + \dfrac{12}{a^2 - 2a - 8}$ **(b)** $\dfrac{5x - y}{3y} - \dfrac{2x - 4y}{3y}$

Now, what if our rational expressions *do not* have common denominators? In that case, we must use the least common denominator (LCD). The **least common denominator** is the simplest polynomial that is divisible by each of the individual denominators. Each expression in the desired sum or difference is then "built up" to an equivalent expression having that LCD as a denominator. We can then add or subtract as before.

Although in many cases we can find the LCD by inspection, we can state an algorithm for finding the LCD that is similar to the one used in arithmetic.

NOTE By **inspection,** we mean you look at the denominators and find that the LCD is obvious (as in Example 2).

NOTE Again, we see the key role that factoring plays in the process of working with rational expressions.

Step by Step: Finding the Least Common Denominator

Step 1 Write each of the denominators in completely factored form.

Step 2 Write the LCD as the product of each prime factor to the highest power to which it appears in the factored form of any individual denominator.

Example 3 illustrates the procedure.

Example 3

Finding the LCD for Two Rational Expressions

Find the LCD for each of the following pairs of rational expressions.

(a) $\frac{3}{4x^2}$ and $\frac{5}{6xy}$

Factor the denominators.

NOTE You may very well be able to find this LCD by inspecting the numerical coefficients and the variable factors.

$4x^2 = 2^2 \cdot x^2$

$6xy = 2 \cdot 3 \cdot x \cdot y$

The LCD must have the factors

$2^2 \cdot 3 \cdot x^2 \cdot y$

and so $12x^2y$ is the desired LCD.

(b) $\frac{7}{x-3}$ and $\frac{2}{x+5}$

Here, neither denominator can be factored. The LCD must have the factors $x - 3$ and $x + 5$. So the LCD is

NOTE It is generally best to leave the LCD in this factored form.

$(x - 3)(x + 5)$

CHECK YOURSELF 3

Find the LCD for the following pairs of rational expressions.

(a) $\frac{3}{8a^3}$ and $\frac{5}{6a^2}$ **(b)** $\frac{4}{x+7}$ and $\frac{3}{x-5}$

Let's see how factoring techniques are applied in Example 4.

Example 4

Finding the LCD for Two Rational Expressions

Find the LCD for the following pairs of rational expressions.

(a) $\frac{2}{x^2 - x - 6}$ and $\frac{1}{x^2 - 9}$

Factoring, we have

$x^2 - x - 6 = (x + 2)(x - 3)$

and

$x^2 - 9 = (x + 3)(x - 3)$

NOTE The LCD must contain *each* of the factors appearing in the original denominators.

The LCD of the given expressions is then

$(x + 2)(x - 3)(x + 3)$

NOTE Because $(x - 3)$ appears only once in each denominator, it appears only once in the LCD.

(b) $\frac{5}{x^2 - 4x + 4}$ and $\frac{3}{x^2 + 2x - 8}$

Again, we factor:

$x^2 - 4x + 4 = (x - 2)^2$

$x^2 + 2x - 8 = (x - 2)(x + 4)$

The LCD is then

$(x - 2)^2 (x + 4)$

NOTE The LCD must contain $(x - 2)^2$ as a factor because $x - 2$ appears *twice* as a factor in the first denominator.

CHECK YOURSELF 4

Find the LCD for the following pairs of rational expressions.

(a) $\dfrac{3}{x^2 - 2x - 15}$ and $\dfrac{5}{x^2 - 25}$

(b) $\dfrac{5}{y^2 + 6y + 9}$ and $\dfrac{3}{y^2 - y - 12}$

Let's look at Example 5, in which the concept of the LCD is applied in adding and subtracting rational expressions.

Example 5

Adding and Subtracting Rational Expressions

Add or subtract as indicated.

(a) $\dfrac{5}{4xy} + \dfrac{3}{2x^2}$

The LCD for $2x^2$ and $4xy$ is $4x^2y$. We rewrite each of the rational expressions with the LCD as a denominator.

$$\frac{5}{4xy} + \frac{3}{2x^2} = \frac{5 \cdot x}{4xy \cdot x} + \frac{3 \cdot 2y}{2x^2 \cdot 2y}$$

$$= \frac{5x}{4x^2y} + \frac{6y}{4x^2y} = \frac{5x + 6y}{4x^2y}$$

Multiply the first rational expression by $\frac{x}{x}$ and the second by $\frac{2y}{2y}$ to form the LCD of $4x^2y$.

NOTE Notice that in each case we are multiplying by 1: $\frac{x}{x}$ in the first fraction and $\frac{2y}{2y}$ in the second fraction, which is why the resulting fractions are equivalent to the original ones.

(b) $\dfrac{3}{a - 3} - \dfrac{2}{a}$

The LCD for a and $a - 3$ is $a(a - 3)$. We rewrite each of the rational expressions with that LCD as a denominator.

$$\frac{3}{a - 3} - \frac{2}{a}$$

$$= \frac{3a}{a(a - 3)} - \frac{2(a - 3)}{a(a - 3)}$$

Subtract the numerators.

$$= \frac{3a - 2(a - 3)}{a(a - 3)}$$

Remove the parentheses, and combine like terms.

$$= \frac{3a - 2a + 6}{a(a - 3)} = \frac{a + 6}{a(a - 3)}$$

CHECK YOURSELF 5

Perform the indicated operations.

(a) $\frac{3}{2ab} + \frac{4}{5b^2}$ **(b)** $\frac{5}{y+2} - \frac{3}{y}$

Let's proceed to Example 6, in which factoring will be required in forming the LCD.

Example 6

Adding and Subtracting Rational Expressions

Add or subtract as indicated.

(a) $\frac{-5}{x^2 - 3x - 4} + \frac{8}{x^2 - 16}$

We first factor the two denominators.

$$x^2 - 3x - 4 = (x + 1)(x - 4)$$
$$x^2 - 16 = (x + 4)(x - 4)$$

We see that the LCD must be

$$(x + 1)(x + 4)(x - 4)$$

Again, rewriting the original expressions with factored denominators gives

NOTE We use the facts that $\frac{x+4}{x+4} = 1$ and $\frac{x+1}{x+1} = 1$

$$\frac{-5}{(x+1)(x-4)} + \frac{8}{(x-4)(x+4)}$$

$$= \frac{-5(x+4)}{(x+1)(x-4)(x+4)} + \frac{8(x+1)}{(x-4)(x+4)(x+1)}$$

$$= \frac{-5(x+4) + 8(x+1)}{(x+1)(x-4)(x+4)}$$ Now add the numerators.

$$= \frac{-5x - 20 + 8x + 8}{(x+1)(x-4)(x+4)}$$ Combine like terms in the numerator.

$$= \frac{3x - 12}{(x+1)(x-4)(x+4)}$$ Factor.

$$= \frac{3(x-4)}{(x+1)(x-4)(x+4)}$$ Divide by the common factor $x - 4$.

$$= \frac{3}{(x+1)(x+4)}$$

(b) $\frac{5}{x^2 - 5x + 6} - \frac{3}{4x - 12}$

Again, factor the denominators.

$$x^2 - 5x + 6 = (x - 2)(x - 3)$$
$$4x - 12 = 4(x - 3)$$

The LCD is $4(x - 2)(x - 3)$, and proceeding as before, we have

$$\frac{5}{(x-2)(x-3)} - \frac{3}{4(x-3)}$$

$$= \frac{5 \cdot 4}{4(x-2)(x-3)} - \frac{3(x-2)}{4(x-2)(x-3)}$$

$$= \frac{20 - 3(x-2)}{4(x-2)(x-3)}$$

$$= \frac{20 - 3x + 6}{4(x-2)(x-3)} = \frac{-3x + 26}{4(x-2)(x-3)}$$

Simplify the numerator and combine like terms.

CHECK YOURSELF 6

Add or subtract as indicated.

(a) $\frac{-4}{x^2 - 4} + \frac{7}{x^2 - 3x - 10}$ **(b)** $\frac{5}{3x - 9} - \frac{2}{x^2 - 9}$

Example 7 looks slightly different from those you have seen thus far, but the reasoning involved in performing the subtraction is exactly the same.

Example 7

Subtracting Rational Expressions

Subtract.

$$3 - \frac{5}{2x - 1}$$

To perform the subtraction, remember that 3 is equivalent to the fraction $\frac{3}{1}$, so

$$3 - \frac{5}{2x-1} = \frac{3}{1} - \frac{5}{2x-1}$$

The LCD for 1 and $2x - 1$ is just $2x - 1$. We now rewrite the first expression with that denominator.

$$3 - \frac{5}{2x-1} = \frac{3(2x-1)}{(2x-1)} - \frac{5}{2x-1}$$

$$= \frac{3(2x-1) - 5}{2x-1} = \frac{6x - 8}{2x - 1}$$

CHECK YOURSELF 7

Subtract.

$$\frac{4}{3x + 1} - 3$$

Example 8 uses an observation from Section 7.1. Recall that

$$a - b = -(b - a)$$
$$= -1(b - a)$$

Let's see how this is used in adding rational expressions.

Example 8

Adding and Subtracting Rational Expressions

Add.

$$\frac{x^2}{x - 5} + \frac{3x + 10}{5 - x}$$

Your first thought might be to use a denominator of $(x - 5)(5 - x)$. However, we can simplify our work considerably if we multiply the numerator and denominator of the second fraction by -1 to find a common denominator.

NOTE Use

$$\frac{-1}{-1} = 1$$

NOTE Notice that

$(-1)(5 - x) = x - 5$

The fractions now have a common denominator, and we can add as before.

$$\frac{x^2}{x - 5} + \frac{3x + 10}{5 - x}$$
$$= \frac{x^2}{x - 5} + \frac{(-1)(3x + 10)}{(-1)(5 - x)}$$
$$= \frac{x^2}{x - 5} + \frac{-3x - 10}{x - 5}$$
$$= \frac{x^2 - 3x - 10}{x - 5}$$
$$= \frac{(x + 2)(x - 5)}{x - 5}$$
$$= x + 2$$

CHECK YOURSELF 8

Add.

$$\frac{x^2}{x - 7} + \frac{10x - 21}{7 - x}$$

The sum of two rational functions is always a rational function. Given two rational functions, $f(x)$ and $g(x)$, we can rename the sum, so $h(x) = f(x) + g(x)$. This will always be true for values of x for which both f and g are defined. So, for example, $h(-2) = f(-2) + g(-2)$, so long as both $f(-2)$ and $g(-2)$ exist.

Example 9 illustrates this approach.

Example 9

Adding Two Rational Functions

Given

$$f(x) = \frac{3x}{x + 5} \quad \text{and} \quad g(x) = \frac{x}{x - 4}$$

complete the following.

(a) Find $f(1) + g(1)$.

Because $f(1) = \frac{1}{2}$ and $g(1) = -\frac{1}{3}$, then

$$f(1) + g(1) = \frac{1}{2} + \left(-\frac{1}{3}\right)$$
$$= \frac{3}{6} + \left(-\frac{2}{6}\right)$$
$$= \frac{1}{6}$$

(b) Find $h(x) = f(x) + g(x)$.

$h(x) = f(x) + g(x)$

$$= \frac{3x}{x+5} + \frac{x}{x-4}$$
$$= \frac{3x(x-4) + x(x+5)}{(x+5)(x-4)}$$
$$= \frac{3x^2 - 12x + x^2 + 5x}{(x+5)(x-4)}$$
$$= \frac{4x^2 - 7x}{(x+5)(x-4)} \qquad x \neq -5, 4$$

(c) Find the ordered pair $(1, h(1))$.

$$h(1) = \frac{-3}{-18} = \frac{1}{6}$$

The ordered pair is $\left(1, \frac{1}{6}\right)$.

CHECK YOURSELF 9

Given

$$f(x) = \frac{x}{2x-5} \qquad \text{and} \qquad g(x) = \frac{2x}{3x-1}$$

complete the following.

(a) Find $f(1) + g(1)$.
(b) Find $h(x) = f(x) + g(x)$.
(c) Find the ordered pair $(1, h(1))$.

When subtracting rational functions, you must take particular care with the signs in the numerator of the expression being subtracted.

Example 10

Subtracting Rational Functions

Given

$$f(x) = \frac{3x}{x + 5} \quad \text{and} \quad g(x) = \frac{x - 2}{x - 4}$$

complete the following.

(a) Find $f(1) - g(1)$.

Because $f(1) = \frac{1}{2}$ and $g(1) = \frac{1}{3}$, then

$$f(1) - g(1) = \frac{1}{2} - \frac{1}{3}$$

$$= \frac{3}{6} - \frac{2}{6}$$

$$= \frac{1}{6}$$

(b) Find $h(x) = f(x) - g(x)$.

$$h(x) = \frac{3x}{x + 5} - \frac{x - 2}{x - 4}$$

$$= \frac{3x(x - 4) - (x - 2)(x + 5)}{(x + 5)(x - 4)}$$

$$= \frac{(3x^2 - 12x) - (x^2 + 3x - 10)}{(x + 5)(x - 4)}$$

$$= \frac{2x^2 - 15x + 10}{(x + 5)(x - 4)} \qquad x \neq -5, 4$$

(c) Find the ordered pair $(1, h(1))$.

$$h(1) = \frac{-3}{-18} = \frac{1}{6}$$

The ordered pair is $\left(1, \frac{1}{6}\right)$.

CHECK YOURSELF 10

Given

$$f(x) = \frac{x}{2x - 5} \quad \text{and} \quad g(x) = \frac{2x - 1}{3x - 1}$$

complete the following.

(a) Find $f(1) - g(1)$.
(b) Find $h(x) = f(x) - g(x)$.
(c) Find the ordered pair $(1, h(1))$.

CHECK YOURSELF ANSWERS

1. $\dfrac{2}{3y^2}$ **2. (a)** $\dfrac{6}{a-4}$; **(b)** $\dfrac{x+y}{y}$ **3. (a)** $24a^3$; **(b)** $(x+7)(x-5)$
4. (a) $(x-5)(x+5)(x+3)$; **(b)** $(y+3)^2(y-4)$
5. (a) $\dfrac{8a+15b}{10ab^2}$; **(b)** $\dfrac{2y-6}{y(y+2)}$ **6. (a)** $\dfrac{3}{(x-2)(x-5)}$; **(b)** $\dfrac{5x+9}{3(x+3)(x-3)}$
7. $\dfrac{-9x+1}{3x+1}$ **8.** $x-3$ **9. (a)** $\dfrac{2}{3}$; **(b)** $h(x) = \dfrac{7x^2-11x}{(2x-5)(3x-1)}$ $x \neq \dfrac{1}{3}, \dfrac{5}{2}$;
(c) $\left(1, \dfrac{2}{3}\right)$ **10. (a)** $-\dfrac{5}{6}$; **(b)** $h(x) = \dfrac{-x^2+11x-5}{(2x-5)(3x-1)}$ $x \neq \dfrac{1}{3}, \dfrac{5}{2}$; **(c)** $\left(1, -\dfrac{5}{6}\right)$

Probability and Pari-Mutual Betting. In most gambling games, payoffs are determined by the **odds.** At horse and dog tracks, the odds (D) are a ratio that is calculated by taking into account the total amount wagered (A), the amount wagered on a particular animal (a), and the government share, called the take-out (f). The ratio is then rounded down to a comparison of integers like 99 to 1, 3 to 1, or 5 to 2. Below is the formula that tracks use to find odds.

$$D = \frac{A(1-f)}{a} - 1$$

Work with a partner to complete the following.

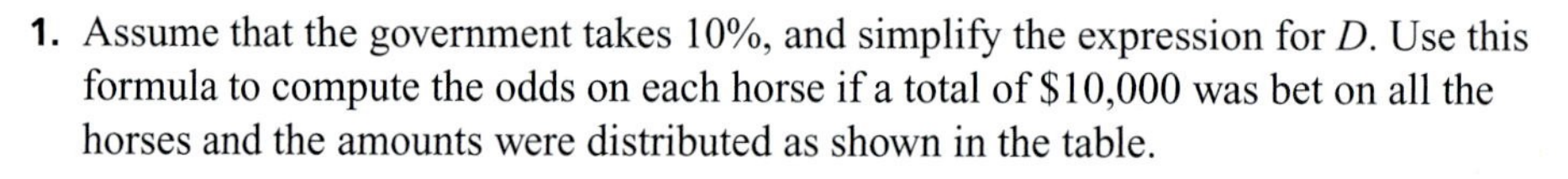

1. Assume that the government takes 10%, and simplify the expression for D. Use this formula to compute the odds on each horse if a total of \$10,000 was bet on all the horses and the amounts were distributed as shown in the table.

Horse	Total Amount Wagered on This Horse to Win	Odds: Amount Paid on Each Dollar Bet If Horse Wins
1	\$5000	
2	\$1000	
3	\$2000	
4	\$1500	
5	\$ 500	

2. Odds can be used as a guide in determining the chance that a given horse will win. The probability of a horse winning is related to many variables, such as track condition, how the horse is feeling, and weather. However, the odds do reflect the consensus opinion of racing fans and can be used to give some idea of the probability.
   The relationship between odds and probability is given by the equations

   $$P(\text{win}) = \frac{1}{D+1}$$

   and $P(\text{loss}) = 1 - P(\text{win})$

   or $P(\text{loss}) = 1 - \dfrac{1}{D+1}$

   Solve this equation for D, the odds against the horse winning. Do the probabilities for each horse winning all add up to 1? Should they add to 1?

Name ______________

Section ________ Date ________

7.3 Exercises

In exercises 1 to 36, perform the indicated operations. Express your results in simplest form.

1. $\frac{7}{2x^2} + \frac{5}{2x^2}$

2. $\frac{11}{3b^3} - \frac{2}{3b^3}$

3. $\frac{5}{3a + 7} + \frac{2}{3a + 7}$

4. $\frac{6}{5x + 3} - \frac{3}{5x + 3}$

5. $\frac{2x}{x - 3} - \frac{6}{x - 3}$

6. $\frac{7w}{w + 3} + \frac{21}{w + 3}$

7. $\frac{y^2}{2y + 8} + \frac{3y - 4}{2y + 8}$

8. $\frac{x^2}{4x - 12} - \frac{9}{4x - 12}$

9. $\frac{4m - 7}{m - 5} - \frac{2m + 3}{m - 5}$

10. $\frac{3b - 8}{b - 6} + \frac{b - 16}{b - 6}$

11. $\frac{x - 7}{x^2 - x - 6} + \frac{2x - 2}{x^2 - x - 6}$

12. $\frac{5x - 12}{x^2 - 8x + 15} - \frac{3x - 2}{x^2 - 8x + 15}$

13. $\frac{5}{3x} + \frac{3}{2x}$

14. $\frac{4}{5w} - \frac{3}{4w}$

15. $\frac{6}{a} + \frac{3}{a^2}$

16. $\frac{3}{p} - \frac{7}{p^2}$

17. $\frac{2}{m} - \frac{2}{n}$

18. $\frac{3}{x} + \frac{3}{y}$

19. $\frac{3}{4b^2} - \frac{5}{3b^3}$

20. $\frac{4}{5x^3} - \frac{3}{2x^2}$

ANSWERS

1. ______________
2. ______________
3. ______________
4. ______________
5. ______________
6. ______________
7. ______________
8. ______________
9. ______________
10. ______________
11. ______________
12. ______________
13. ______________
14. ______________
15. ______________
16. ______________
17. ______________
18. ______________
19. ______________
20. ______________

ANSWERS

21. ______
22. ______
23. ______
24. ______
25. ______
26. ______
27. ______
28. ______
29. ______
30. ______
31. ______
32. ______
33. ______
34. ______
35. ______
36. ______
37. ______
38. ______
39. ______
40. ______
41. ______
42. ______

21. $\frac{2}{a} - \frac{1}{a - 2}$

22. $\frac{4}{c} + \frac{3}{c + 1}$

23. $\frac{2}{x + 1} + \frac{3}{x + 2}$

24. $\frac{4}{y - 1} + \frac{2}{y + 3}$

25. $4 - \frac{3}{3x + 2}$

26. $-6 - \frac{1}{2x + 3}$

27. $\frac{2w}{w - 7} + \frac{w}{w - 2}$

28. $\frac{3n}{n + 5} + \frac{n}{n - 4}$

29. $\frac{3x}{3x - 2} - \frac{2x}{2x + 1}$

30. $\frac{4}{x - 1} - \frac{3}{1 - x}$

31. $\frac{6}{m - 7} + \frac{2}{7 - m}$

32. $\frac{5}{a - 5} - \frac{3}{5 - a}$

33. $\frac{2x}{2x - 3} + \frac{3}{3 - 2x}$

34. $\frac{5}{y^2 + 5y + 6} + \frac{2}{y + 2}$

35. $\frac{4m}{m^2 - 3m + 2} - \frac{1}{m - 2}$

36. $\frac{x}{x^2 - 1} - \frac{2}{x - 1}$

As we saw in Section 7.2 exercises, the graphing calculator can be used to check our work. In exercises 37 to 42, enter the first rational expression in Y_1 and the second in Y_2. In Y_3, you will enter either $Y_1 + Y_2$ or $Y_1 - Y_2$. Enter your algebraically simplified rational expression in Y_4. The graphs of Y_3 and Y_4 will be identical if you have correctly simplified the expression.

37. $\frac{6y}{y^2 - 8y + 15} + \frac{9}{y - 3}$

38. $\frac{8a}{a^2 - 8a + 12} + \frac{4}{a - 2}$

39. $\frac{6x}{x^2 - 10x + 24} - \frac{18}{x - 6}$

40. $\frac{21p}{p^2 - 3p - 10} - \frac{15}{p - 5}$

41. $\frac{2}{z^2 - 4} + \frac{3}{z^2 + 2z - 8}$

42. $\frac{5}{x^2 - 3x - 10} + \frac{2}{x^2 - 25}$

In exercises 43 to 46, find **(a)** $f(1) + g(1)$, **(b)** $h(x) = f(x) + g(x)$, and **(c)** the ordered pair $(1, h(1))$.

43. $f(x) = \dfrac{3x}{x + 1}$ and $g(x) = \dfrac{2x}{x - 3}$

44. $f(x) = \dfrac{4x}{x - 4}$ and $g(x) = \dfrac{x + 4}{x + 1}$

45. $f(x) = \dfrac{x}{x + 1}$ and $g(x) = \dfrac{1}{x^2 + 2x + 1}$

46. $f(x) = \dfrac{x + 2}{x - 4}$ and $g(x) = \dfrac{x + 3}{x + 4}$

In exercises 47 to 50, find **(a)** $f(1) - g(1)$, **(b)** $h(x) = f(x) - g(x)$, and **(c)** the ordered pair $(1, h(1))$.

47. $f(x) = \dfrac{x + 5}{x - 5}$ and $g(x) = \dfrac{x - 5}{x + 5}$

48. $f(x) = \dfrac{2x}{x - 4}$ and $g(x) = \dfrac{3x}{x + 7}$

49. $f(x) = \dfrac{x + 9}{4x - 36}$ and $g(x) = \dfrac{x - 9}{x^2 - 18x + 81}$

50. $f(x) = \dfrac{4x + 1}{x + 5}$ and $g(x) = -\dfrac{2}{x}$

In exercises 51 to 60, evaluate each expression at the given variable value(s).

51. $\dfrac{5x + 5}{x^2 + 3x + 2} - \dfrac{x - 3}{x^2 + 5x + 6}, x = -4$

52. $\dfrac{y - 3}{y^2 - 6y + 8} + \dfrac{2y - 6}{y^2 - 4}, y = 3$

53. $\dfrac{2m + 2n}{m^2 - n^2} + \dfrac{m - 2n}{m^2 + 2mn - n^2}, m = 3, n = 2$

54. $\dfrac{w - 3z}{w^2 - 2wz + z^2} - \dfrac{w + 2z}{w^2 - z^2}, w = 2, z = 1$

ANSWERS

43. (a)
(b)
(c)

44. (a)
(b)
(c)

45. (a)
(b)
(c)

46. (a)
(b)
(c)

47. (a)
(b)
(c)

48. (a)
(b)
(c)

49. (a)
(b)
(c)

50. (a)
(b)
(c)

51. 52.

53. 54.

ANSWERS

55. ______

56. ______

57. ______

58. ______

59. ______

60. ______

55. $\dfrac{1}{a-3} - \dfrac{1}{a+3} + \dfrac{2a}{a^2-9}$, $a = 4$

56. $\dfrac{1}{m+1} + \dfrac{1}{m-3} - \dfrac{4}{m^2-2m-3}$, $m = -2$

57. $\dfrac{3w^2+16w-8}{w^2+2w-8} + \dfrac{w}{w+4} - \dfrac{w-1}{w-2}$, $w = 3$

58. $\dfrac{4x^2-7x-45}{x^2-6x+5} - \dfrac{x+2}{x-1} - \dfrac{x}{x-5}$, $x = -3$

59. $\dfrac{a^2-9}{2a^2-5a-3} \cdot \left(\dfrac{1}{a-2} + \dfrac{1}{a+3}\right)$, $a = -3$

60. $\dfrac{m^2-2mn+n^2}{m^2+2mn-3n^2} \cdot \left(\dfrac{2}{m-n} - \dfrac{1}{m+n}\right)$, $m = 4, n = -3$

Answers

1. $\dfrac{6}{x^2}$ **3.** $\dfrac{7}{3a+7}$ **5.** 2 **7.** $\dfrac{y-1}{2}$ **9.** 2 **11.** $\dfrac{3}{x+2}$
13. $\dfrac{19}{6x}$ **15.** $\dfrac{3(2a+1)}{a^2}$ **17.** $\dfrac{2(n-m)}{mn}$
19. $\dfrac{9b-20}{12b^3}$ **21.** $\dfrac{a-4}{a(a-2)}$ **23.** $\dfrac{5x+7}{(x+1)(x+2)}$
25. $\dfrac{12x+5}{3x+2}$ **27.** $\dfrac{w(3w-11)}{(w-7)(w-2)}$ **29.** $\dfrac{7x}{(3x-2)(2x+1)}$ **31.** $\dfrac{4}{m-7}$
33. 1 **35.** $\dfrac{3m+1}{(m-1)(m-2)}$ **37.** $\dfrac{15}{y-5}$
39. $\dfrac{-12}{x-4}$ **41.** $\dfrac{5z+14}{(z+2)(z-2)(z+4)}$
43. **(a)** $\dfrac{1}{2}$; **(b)** $\dfrac{5x^2-7x}{(x+1)(x-3)}$; **(c)** $\left(1, \dfrac{1}{2}\right)$ **45.** **(a)** $\dfrac{3}{4}$; **(b)** $\dfrac{x^2+x+1}{(x+1)^2}$; **(c)** $\left(1, \dfrac{3}{4}\right)$
47. **(a)** $-\dfrac{5}{6}$; **(b)** $\dfrac{20x}{(x-5)(x+5)}$; **(c)** $\left(1, -\dfrac{5}{6}\right)$
49. **(a)** $-\dfrac{3}{16}$; **(b)** $\dfrac{x+5}{4(x-9)}$; **(c)** $\left(1, -\dfrac{3}{16}\right)$ **51.** 1 **53.** $\dfrac{49}{25}$ **55.** 2
57. 8 **59.** Undefined

7.4 Complex Fractions

7.4 OBJECTIVES

1. Use the fundamental principle to simplify a complex fraction
2. Use division to simplify a complex fraction

Our work in this section deals with two methods for simplifying complex fractions. We begin with a definition. A **complex fraction** is a fraction that has a fraction in its numerator or denominator (or both). Some examples are

$$\frac{\frac{5}{6}}{\frac{3}{4}} \qquad \frac{\frac{4}{x}}{\frac{3}{x+1}} \qquad \text{and} \qquad \frac{1+\frac{1}{x}}{1-\frac{1}{x}}$$

Two methods can be used to simplify complex fractions. Method 1 involves the fundamental principle, and Method 2 involves inverting and multiplying.

NOTE Fundamental principle:

$$\frac{P}{Q} = \frac{PR}{QR}$$

when $Q \neq 0$ and $R \neq 0$

Recall that by the *fundamental principle* we can always multiply the numerator and denominator of a fraction by the same nonzero quantity. In simplifying a complex fraction, we multiply the numerator and denominator by the LCD of all fractions that appear within the complex fraction.

Here the denominators are 5 and 10, so we can write

NOTE Again, we are multiplying by $\frac{10}{10}$ or 1.

$$\frac{\frac{3}{5}}{\frac{7}{10}} = \frac{\frac{3}{5}\cdot 10}{\frac{7}{10}\cdot 10} = \frac{6}{7}$$

Our second approach interprets the complex fraction as indicating division and applies our earlier work in dividing fractions in which we *invert and multiply.*

$$\frac{\frac{3}{5}}{\frac{7}{10}} = \frac{3}{5} \div \frac{7}{10} = \frac{3}{5} \cdot \frac{10}{7} = \frac{6}{7} \qquad \text{Invert and multiply.}$$

Which method is better? The answer depends on the expression you are trying to simplify. Both approaches are effective, and you should be familiar with both. With practice you will be able to tell which method may be easier to use in a particular situation.

Let's look at the same two methods applied to the simplification of an algebraic complex fraction.

Example 1

Simplifying Complex Fractions

Simplify.

$$\frac{1+\frac{2x}{y}}{2-\frac{x}{y}}$$

Method 1 The LCD of 1, $\frac{2x}{y}$, 2, and $\frac{x}{y}$ is y. So we multiply the numerator and denominator by y.

$$\frac{1 + \frac{2x}{y}}{2 - \frac{x}{y}} = \frac{\left(1 + \frac{2x}{y}\right) \cdot y}{\left(2 - \frac{x}{y}\right) \cdot y}$$

Distribute y over the numerator and denominator.

$$= \frac{1 \cdot y + \frac{2x}{y} \cdot y}{2 \cdot y - \frac{x}{y} \cdot y}$$

Simplify.

$$= \frac{y + 2x}{2y - x}$$

Method 2 In this approach, we must *first work separately* in the numerator and denominator to form single fractions.

NOTE Make sure you understand the steps in forming a single fraction in the numerator and denominator.

$$\frac{1 + \frac{2x}{y}}{2 - \frac{x}{y}} = \frac{\frac{y}{y} + \frac{2x}{y}}{\frac{2y}{y} - \frac{x}{y}} = \frac{\frac{y + 2x}{y}}{\frac{2y - x}{y}}$$

$$= \frac{y + 2x}{y} \cdot \frac{y}{2y - x}$$

Invert the divisor and multiply.

$$= \frac{y + 2x}{2y - x}$$

CHECK YOURSELF 1

Simplify.

$$\frac{\frac{x}{y} - 1}{\frac{2x}{y} + 2}$$

Again, simplifying a complex fraction means writing an equivalent simple fraction in lowest terms, as Example 2 illustrates.

Example 2

Simplifying Complex Fractions

Simplify.

$$\frac{1 - \frac{2y}{x} + \frac{y^2}{x^2}}{1 - \frac{y^2}{x^2}}$$

We choose the first method of simplification in this case. The LCD of all the fractions that appear is x^2. So we multiply the numerator and denominator by x^2.

$$\frac{1 - \frac{2y}{x} + \frac{y^2}{x^2}}{1 - \frac{y^2}{x^2}} = \frac{\left(1 - \frac{2y}{x} + \frac{y^2}{x^2}\right) \cdot x^2}{\left(1 - \frac{y^2}{x^2}\right) \cdot x^2}$$

Distribute x^2 over the numerator and denominator, and simplify.

$$= \frac{x^2 - 2xy + y^2}{x^2 - y^2}$$

Factor the numerator and denominator.

$$= \frac{(x - y)(x - y)}{(x + y)(x - y)} = \frac{x - y}{x + y}$$

Divide by the common factor $x - y$.

CHECK YOURSELF 2

Simplify.

$$\frac{1 + \frac{5}{x} + \frac{6}{x^2}}{1 - \frac{9}{x^2}}$$

In Example 3, we will illustrate the second method of simplification for purposes of comparison.

Example 3

Simplifying Complex Fractions

Simplify.

$$\frac{1 - \frac{1}{x + 2}}{x - \frac{2}{x - 1}}$$

NOTE Again, take time to make sure you understand how the numerator and denominator are rewritten as single fractions.

NOTE Method 2 is probably the more efficient in this case. The LCD of the denominators would be $(x + 2)(x - 1)$, leading to a somewhat more complicated process if method 1 were used.

$$\frac{1 - \frac{1}{x + 2}}{x - \frac{2}{x - 1}} = \frac{\frac{x + 2}{x + 2} - \frac{1}{x + 2}}{\frac{x(x - 1)}{x - 1} - \frac{2}{x - 1}} = \frac{\frac{x + 1}{x + 2}}{\frac{x^2 - x - 2}{x - 1}}$$

$$= \frac{x + 1}{x + 2} \cdot \frac{x - 1}{x^2 - x - 2}$$

$$= \frac{x + 1}{x + 2} \cdot \frac{x - 1}{(x - 2)(x + 1)}$$

$$= \frac{x - 1}{(x + 2)(x - 2)}$$

CHECK YOURSELF 3

Simplify.

$$\frac{2 + \dfrac{5}{x - 3}}{x - \dfrac{1}{2x + 1}}$$

The following algorithm summarizes our work with complex fractions.

Step by Step: Simplifying Complex Fractions

Method 1

Step 1 Multiply the numerator and denominator of the complex fraction by the LCD of all the fractions that appear within the numerator and denominator.

Step 2 Simplify the resulting rational expression, writing the expression in lowest terms.

Method 2

Step 1 Write the numerator and denominator of the complex fraction as single fractions, if necessary.

Step 2 Invert the denominator and multiply as before, writing the result in lowest terms.

CHECK YOURSELF ANSWERS

1. $\dfrac{x - y}{2x + 2y}$ **2.** $\dfrac{x + 2}{x - 3}$ **3.** $\dfrac{2x + 1}{(x - 3)(x + 1)}$

Exercises

Name ______ Section ______ Date ______

In exercises 1 to 39, simplify each complex fraction.

1. $\dfrac{\frac{2}{3}}{\frac{6}{8}}$

2. $\dfrac{\frac{5}{6}}{\frac{10}{15}}$

3. $\dfrac{\frac{2}{3} + \frac{1}{2}}{\frac{3}{4} - \frac{1}{3}}$

4. $\dfrac{\frac{3}{4} + \frac{1}{2}}{\frac{7}{8} - \frac{1}{4}}$

5. $\dfrac{2 + \frac{1}{3}}{3 - \frac{1}{5}}$

6. $\dfrac{1 + \frac{3}{4}}{2 - \frac{1}{8}}$

7. $\dfrac{\frac{x}{8}}{\frac{x^2}{4}}$

8. $\dfrac{\frac{a^2}{10}}{\frac{a^3}{15}}$

9. $\dfrac{\frac{3}{m}}{\frac{6}{m^2}}$

10. $\dfrac{\frac{15}{x^2}}{\frac{20}{x^3}}$

11. $\dfrac{\frac{y + 1}{y}}{\frac{y - 1}{2y}}$

12. $\dfrac{\frac{x + 3}{4x}}{\frac{x - 3}{2x}}$

13. $\dfrac{\frac{a + 2b}{3a}}{\frac{a^2 + 2ab}{9b}}$

14. $\dfrac{\frac{m - 3n}{4m}}{\frac{m^2 - 3mn}{8n}}$

ANSWERS

1. ______
2. ______
3. ______
4. ______
5. ______
6. ______
7. ______
8. ______
9. ______
10. ______
11. ______
12. ______
13. ______
14. ______

ANSWERS

15. ______

16. ______

17. ______

18. ______

19. ______

20. ______

21. ______

22. ______

23. ______

24. ______

25. ______

26. ______

27. ______

28. ______

29. ______

30. ______

15. $\dfrac{\dfrac{x-2}{x^2-9}}{\dfrac{x^2-4}{x^2+3x}}$

16. $\dfrac{\dfrac{x+5}{x^2-6x}}{\dfrac{x^2-25}{x^2-36}}$

17. $\dfrac{2-\dfrac{1}{x}}{2+\dfrac{1}{x}}$

18. $\dfrac{3+\dfrac{1}{b}}{3-\dfrac{1}{b}}$

19. $\dfrac{\dfrac{1}{x}-\dfrac{1}{y}}{\dfrac{1}{xy}}$

20. $\dfrac{\dfrac{1}{ab}}{\dfrac{1}{a}+\dfrac{1}{b}}$

21. $\dfrac{\dfrac{x^2}{y^2}-1}{\dfrac{x}{y}+1}$

22. $\dfrac{\dfrac{m}{n}+2}{\dfrac{m^2}{n^2}-4}$

23. $\dfrac{1+\dfrac{3}{a}-\dfrac{4}{a^2}}{1+\dfrac{2}{a}-\dfrac{3}{a^2}}$

24. $\dfrac{1-\dfrac{2}{x}-\dfrac{8}{x^2}}{1-\dfrac{1}{x}-\dfrac{6}{x^2}}$

25. $\dfrac{\dfrac{x^2}{y}+2x+y}{\dfrac{1}{y^2}-\dfrac{1}{x^2}}$

26. $\dfrac{\dfrac{a}{b}+1-\dfrac{2b}{a}}{\dfrac{1}{b^2}-\dfrac{4}{a^2}}$

27. $\dfrac{1+\dfrac{1}{x-1}}{1-\dfrac{1}{x-1}}$

28. $\dfrac{2-\dfrac{1}{m-2}}{2+\dfrac{1}{m-2}}$

29. $\dfrac{1-\dfrac{1}{y-1}}{y-\dfrac{8}{y+2}}$

30. $\dfrac{1+\dfrac{1}{x+2}}{x-\dfrac{18}{x-3}}$

31. $\dfrac{\dfrac{1}{x-3}+\dfrac{1}{x+3}}{\dfrac{1}{x-3}-\dfrac{1}{x+3}}$

32. $\dfrac{\dfrac{2}{m-2}+\dfrac{1}{m-3}}{\dfrac{2}{m-2}-\dfrac{1}{m-3}}$

33. $\dfrac{\dfrac{x}{x+1}+\dfrac{1}{x-1}}{\dfrac{x}{x-1}-\dfrac{1}{x+1}}$

34. $\dfrac{\dfrac{y}{y-4}+\dfrac{1}{y+2}}{\dfrac{4}{y-4}-\dfrac{1}{y+2}}$

35. $\dfrac{\dfrac{a+1}{a-1}-\dfrac{a-1}{a+1}}{\dfrac{a+1}{a-1}+\dfrac{a-1}{a+1}}$

36. $\dfrac{\dfrac{x+2}{x-2}-\dfrac{x-2}{x+2}}{\dfrac{x+2}{x-2}+\dfrac{x-2}{x+2}}$

37. $1+\dfrac{1}{1+\dfrac{1}{x}}$

38. $1+\dfrac{1}{1-\dfrac{1}{y}}$

39. $1+\dfrac{1}{1+\dfrac{1}{1+\dfrac{1}{x}}}$

40. **(a)** Extend the "continued fraction" patterns in exercises 37 and 39 to write the next complex fraction. **(b)** Simplify the complex fraction obtained in (a).

41. Compare your results in exercises 37, 39, and 40. Could you have predicted the result?

42. Outline the two different methods used to simplify a complex fraction. What are the advantages of each method?

ANSWERS

31. ____________

32. ____________

33. ____________

34. ____________

35. ____________

36. ____________

37. ____________

38. ____________

39. ____________

40. (a)

(b) ____________

41. ____________

42. ____________

43. ______

44. ______

45. ______

46. ______

47. ______

43. Can the expression $\frac{x^{-1} + y^{-1}}{x^{-2} + y^{-2}}$ be written as $\frac{x^2 + y^2}{x + y}$? If not, what is the correct simplified form?

44. Write and simplify a complex fraction that is the reciprocal of $x + \frac{6}{x - 1}$.

45. Let $f(x) = \frac{3}{x}$. Write and simplify a complex fraction whose numerator is $f(3 + h) - f(3)$ and whose denominator is h.

46. Write and simplify a complex fraction that is the arithmetic mean of $\frac{1}{x}$ and $\frac{1}{x - 1}$.

47. Write and simplify a complex fraction that is the reciprocal of $\frac{1}{x} + \frac{1}{y}$.

Suppose you drive at 40 mi/h from city A to city B. You then return along the same route from city B to city A at 50 mi/h. What is your average rate for the round trip? Your obvious guess would be 45 mi/h, but you are in for a surprise.

Suppose that the cities are 200 mi apart. Your time from city A to city B is the distance divided by the rate, or

$$\frac{200 \text{ mi}}{40 \text{ mi/h}} = 5 \text{ h}$$

Similarly, your time from city B to city A is

$$\frac{200 \text{ mi}}{50 \text{ mi/h}} = 4 \text{ h}$$

The total time is then 9 h, and now using *rate equals distance divided by time,* we have

$$\frac{400 \text{ mi}}{9 \text{ h}} = \frac{400}{9} \text{ mi/h} = 44\frac{4}{9} \text{ mi/h}$$

Note that the rate for the round trip is independent of the distance involved. For instance, try the same computations above if cities A and B are 400 mi apart.

The answer to the problem above is the complex fraction

$$R = \frac{2}{\frac{1}{R_1} + \frac{1}{R_2}}$$

in which R_1 = rate going

R_2 = rate returning

R = rate for round trip

48. ______

49. ______

50. ______

51. ______

52. ______

Use this information to solve exercises 48 to 51.

48. Verify that if $R_1 = 40$ mi/h and $R_2 = 50$ mi/h, then $R = 44\frac{4}{9}$ mi/h, by simplifying the complex fraction *after* substituting those values.

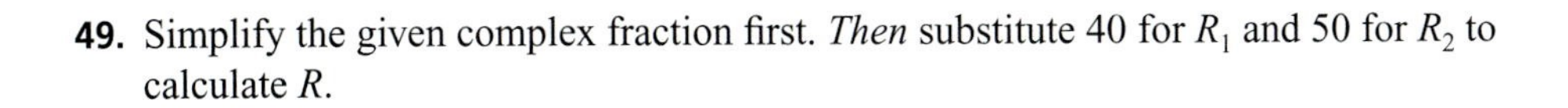

49. Simplify the given complex fraction first. *Then* substitute 40 for R_1 and 50 for R_2 to calculate R.

50. Repeat exercise 48, with $R_1 = 50$ mi/h and $R_2 = 60$ mi/h.

51. Use the procedure in exercise 49 with the above values for R_1 and R_2.

52. The following inequality is used when the U.S. House of Representatives seats are apportioned (see the chapter opener for more information).

$$\frac{\dfrac{E}{e} - \dfrac{A}{a+1}}{\dfrac{A}{a+1}} < \frac{\dfrac{A}{a} - \dfrac{E}{e+1}}{\dfrac{E}{e+1}}$$

Show that this inequality can be simplified to

$$\frac{A}{\sqrt{a(a+1)}} > \frac{E}{\sqrt{e(e+1)}}.$$

Here, A and E represent the populations of two states of the United States, and a and e are the number of representatives each of these two states have in the U.S. House of Representatives.

53. Mathematicians have shown that there are situations in which the method for apportionment described in the chapter's introduction does not work, and a state may not even get its basic quota of representatives. They give the table below of a hypothetical seven states and their populations as an example.

State	Population	Exact Quota	Number of Reps.
A	325	1.625	2
B	788	3.940	4
C	548	2.740	3
D	562	2.810	3
E	4263	21.315	21
F	3219	16.095	15
G	295	1.475	2
Total	10,000	50	50

53. ______

54. ______

In this case, the total population of all states is 10,000, and there are 50 representatives in all, so there should be no more than 10,000/50 or 200 people per representative. The quotas are found by dividing the population by 200. Whether a state, A, should get an additional representative before another state, E, should get one is decided in this method by using the simplified inequality below. If the ratio

$$\frac{A}{\sqrt{a(a+1)}} > \frac{E}{\sqrt{e(e+1)}}$$

is true, then A gets an extra representative before E does.

(a) If you go through the process of comparing the inequality above for each pair of states, state F loses a representative to state G. Do you see how this happens? Will state F complain?

(b) Alexander Hamilton, one of the signers of the Constitution, proposed that the extra representative positions be given one at a time to states with the largest remainder until all the "extra" positions were filled. How would this affect the table? Do you agree or disagree?

54. In Italy in the 1500s, Pietro Antonio Cataldi expressed square roots as infinite, continued fractions. It is not a difficult process to follow. For instance, if you want the square root of 5, then let

$$x + 1 = \sqrt{5}$$

Squaring both sides gives

$$(x+1)^2 = 5 \qquad \text{or} \qquad x^2 + 2x + 1 = 5$$

which can be written

$$x(x+2) = 4$$

$$x = \frac{4}{x+2}$$

One can continue replacing x with $\frac{4}{x+2}$:

$$x = \cfrac{4}{2 + \cfrac{4}{2 + \cfrac{4}{2 + \cfrac{4}{2 + \ldots}}}}$$

to obtain

$$\sqrt{5} - 1$$

(a) Evaluate this complex fraction and then add 1 and see how close it is to the square root of 5. What should you put where the ellipses (. . .) are? Try a number you feel is close to $\sqrt{5}$. How far would you have to go to get the square root correct to the nearest hundredth?

(b) Develop an infinite complex fraction for $\sqrt{10}$.

ANSWERS

55. ______

56. ______

57. ______

58. ______

In exercises 55 and 56, use the table utility on your graphing calculator to complete the table. Comment on the equivalence of the two expressions.

55.

x	-3	-2	-1	0	1	2	3
$\dfrac{1 - \dfrac{2}{x}}{1 - \dfrac{4}{x^2}}$							
$\dfrac{x}{x + 2}$							

56.

x	-3	-2	-1	0	1	2	3
$\dfrac{-8 + \dfrac{20}{x}}{4 - \dfrac{25}{x^2}}$							
$\dfrac{-4x}{2x + 5}$							

57. Here is yet another method for simplifying a complex fraction. Suppose we want to simplify

$$\frac{\frac{3}{5}}{\frac{7}{10}}$$

Multiply the numerator and denominator of the complex fraction by $\frac{10}{7}$.

(a) What principle allows you to do this?

(b) Why was $\frac{10}{7}$ chosen?

(c) When learning to divide fractions, you may have heard the saying "Yours is not to reason why . . . just invert and multiply." How does this method serve to explain the "reason why" we invert and multiply?

58. Suppose someone wrote a fraction as follows $\frac{\frac{1}{2}}{3}$. Give two ways that this fraction can be interpreted, and simplify each. Do you see why it is important to clearly indicate the "main fraction line"? On your graphing calculator, type 1/2/3 and press ENTER. Which way is your calculator interpreting the fraction?

Answers

1. $\frac{8}{9}$ **3.** $\frac{14}{5}$ **5.** $\frac{5}{6}$ **7.** $\frac{1}{2x}$ **9.** $\frac{m}{2}$ **11.** $\frac{2(y+1)}{y-1}$ **13.** $\frac{3b}{a^2}$
15. $\frac{x}{(x+2)(x-3)}$ **17.** $\frac{2x-1}{2x+1}$ **19.** $y-x$ **21.** $\frac{x-y}{y}$ **23.** $\frac{a+4}{a+3}$
25. $\frac{x^2y(x+y)}{x-y}$ **27.** $\frac{x}{x-2}$ **29.** $\frac{y+2}{(y-1)(y+4)}$ **31.** $\frac{x}{3}$ **33.** 1
35. $\frac{2a}{a^2+1}$ **37.** $\frac{2x+1}{x+1}$ **39.** $\frac{3x+2}{2x+1}$ **41.**

43. No; $\frac{xy(x+y)}{y^2+x^2}$ **45.** $\frac{-1}{3+h}$ **47.** $\frac{xy}{x+y}$ **49.** $44\frac{4}{9}$ mi/h
51. $54\frac{6}{11}$ mi/h **53.**

55.

x	−3	−2	−1	0	1	2	3
$\dfrac{1-\frac{2}{x}}{1-\frac{4}{x^2}}$	3	error	−1	error	$\frac{1}{3}$	error	$\frac{3}{5}$
$\dfrac{x}{x+2}$	3	error	−1	0	$\frac{1}{3}$	$\frac{1}{2}$	$\frac{3}{5}$

57.

Solving Rational Equations

OBJECTIVES

1. Rewrite a rational equation by clearing the fractions
2. Solve an equation that contains a rational expression
3. Find the zeros of a rational function

Applications of your work in algebra will often result in equations involving rational expressions. Our objective in this section is to develop methods to find the solutions of such equations.

The usual technique is to multiply both sides of the equation by the lowest common denominator (LCD) of all the rational expressions appearing in the equation. The resulting equation will be cleared of fractions, and we can then proceed to solve the equation as before. Example 1 illustrates the process.

Example 1

Clearing Equations of Fractions

Solve.

$$\frac{2x}{3} + \frac{x}{5} = 13$$

NOTE The LCM for 3 and 5 is 15.

The LCD for $\frac{2x}{3}$ and $\frac{x}{5}$ is 15.

The LCM for 3 and 5 is 15. Multiplying both sides of the equation by 15, we have

$$15\left(\frac{2x}{3} + \frac{x}{5}\right) = 15 \cdot 13 \qquad \text{Distribute 15 on the left.}$$

$$15 \cdot \frac{2x}{3} + 15 \cdot \frac{x}{5} = 15 \cdot 13$$

$$10x + 3x = 195 \qquad \text{Simplify. The equation is now cleared of fractions.}$$

$$13x = 195$$

$$x = 15$$

To check, substitute 15 in the original equation.

$$\frac{2 \cdot 15}{3} + \frac{15}{5} \stackrel{?}{=} 13$$

$$10 + 3 \stackrel{?}{=} 13$$

$$13 = 13 \qquad \text{A true statement.}$$

So 15 is the solution for the equation.

Be Careful! A common mistake is to confuse an *equation* such as

$$\frac{2x}{3} + \frac{x}{5} = 13$$

and an *expression* such as

$$\frac{2x}{3} + \frac{x}{5}$$

Let's compare.

Equation: $\frac{2x}{3} + \frac{x}{5} = 13$

Here we want to *solve the equation for x,* as in Example 1. We multiply both sides by the LCD to clear fractions and proceed as before.

Expression: $\frac{2x}{3} + \frac{x}{5}$

Here we want to find *a third fraction* that is equivalent to the given expression. We write each fraction as an equivalent fraction with the LCD as a common denominator.

$$\frac{2x}{3} + \frac{x}{5} = \frac{2x \cdot 5}{3 \cdot 5} + \frac{x \cdot 3}{5 \cdot 3}$$

$$= \frac{10x}{15} + \frac{3x}{15} = \frac{10x + 3x}{15}$$

$$= \frac{13x}{15}$$

CHECK YOURSELF 1

Solve.

$$\frac{3x}{2} - \frac{x}{3} = 7$$

The process is similar when variables are in the denominators. Consider Example 2.

Example 2

Solving an Equation Involving Rational Expressions

Solve.

NOTE We assume that x cannot have the value 0. Do you see why?

$$\frac{7}{4x} - \frac{3}{x^2} = \frac{1}{2x^2}$$

The LCM of $4x$, x^2, and $2x^2$ is $4x^2$. So, the LCD for the equation is $4x^2$. Multiplying both sides by $4x^2$, we have

$$4x^2\left(\frac{7}{4x} - \frac{3}{x^2}\right) = 4x^2 \cdot \frac{1}{2x^2}$$ Distribute $4x^2$ on the left side.

$$4x^2 \cdot \frac{7}{4x} - 4x^2 \cdot \frac{3}{x^2} = 4x^2 \cdot \frac{1}{2x^2}$$ Simplify.

$$7x - 12 = 2$$

$$7x = 14$$

$$x = 2$$

We leave the check of the solution, $x = 2$, to you. Be sure to return to the original equation and substitute 2 for x.

CHECK YOURSELF 2

Solve.

$$\frac{5}{2x} - \frac{4}{x^2} = \frac{7}{2x^2}$$

Example 3 illustrates the same solution process when there are binomials in the denominators.

Example 3

Solving an Equation Involving Rational Expressions

Solve.

NOTE Here we assume that x *cannot* have the value -2 or 3.

$$\frac{4}{x+2} + 3 = \frac{3x}{x-3}$$

The LCD is $(x + 2)(x - 3)$. Multiplying by that LCD, we have

NOTE Notice that multiplying *each term* by the LCD is the same as multiplying both sides of the equation by the LCD.

$$(x+2)(x-3)\left(\frac{4}{x+2}\right) + (x+2)(x-3)(3) = (x+2)(x-3)\left(\frac{3x}{x-3}\right)$$

Or, simplifying each term, we have

$$4(x-3) + 3(x+2)(x-3) = 3x(x+2)$$

We now clear the parentheses and proceed as before.

$$4x - 12 + 3x^2 - 3x - 18 = 3x^2 + 6x$$

$$3x^2 + x - 30 = 3x^2 + 6x$$

$$x - 30 = 6x$$

$$-5x = 30$$

$$x = -6$$

Again, we leave the check of this solution to you.

CHECK YOURSELF 3

Solve.

$$\frac{5}{x-4} + 2 = \frac{2x}{x-3}$$

Factoring plays an important role in solving equations containing rational expressions.

Example 4

Solving an Equation Involving Rational Expressions

Solve.

$$\frac{3}{x-3} - \frac{7}{x+3} = \frac{2}{x^2-9}$$

In factored form, the denominator on the right side is $(x - 3)(x + 3)$, which forms the LCD, and we multiply each term by that LCD.

$$(x-3)(x+3)\left(\frac{3}{x-3}\right) - (x-3)(x+3)\left(\frac{7}{x+3}\right) = (x-3)(x+3)\left[\frac{2}{(x-3)(x+3)}\right]$$

Again, simplifying each term on the right and left sides, we have

$$3(x+3) - 7(x-3) = 2$$
$$3x + 9 - 7x + 21 = 2$$
$$-4x = -28$$
$$x = 7$$

Be sure to check this result by substitution in the original equation.

CHECK YOURSELF 4

Solve $\frac{4}{x-4} - \frac{3}{x+1} = \frac{5}{x^2-3x-4}$.

Whenever we multiply both sides of an equation by an expression containing a variable, there is the possibility that a proposed solution may make that multiplier 0. As we pointed out earlier, multiplying by 0 does not give an equivalent equation, and therefore verifying solutions by substitution serves not only as a check of our work but also as a check for extraneous solutions. Consider Example 5.

Example 5

Solving an Equation Involving Rational Expressions

Solve.

NOTE Notice that we must assume that $x \neq 2$.

$$\frac{x}{x-2} - 7 = \frac{2}{x-2}$$

The LCD is $x - 2$, and multiplying, we have

NOTE Notice that each of the three terms gets multiplied by $(x - 2)$.

$$\left(\frac{x}{x-2}\right)(x-2) - 7(x-2) = \left(\frac{2}{x-2}\right)(x-2)$$

Simplifying yields

$$x - 7(x-2) = 2$$
$$x - 7x + 14 = 2$$
$$-6x = -12$$
$$x = 2$$

CAUTION

Because division by 0 is undefined, we conclude that 2 is *not a solution* for the original equation. It is an extraneous solution. The original equation has no solution.

To check this result, by substituting 2 for x, we have

$$\frac{2}{2-2} - 7 \stackrel{?}{=} \frac{2}{2-2}$$

$$\frac{2}{0} - 7 \stackrel{?}{=} \frac{2}{0}$$

CHECK YOURSELF 5

Solve $\frac{x-3}{x-4} = 4 + \frac{1}{x-4}$.

Equations involving rational expressions may also lead to quadratic equations, as illustrated in Example 6.

Example 6

Solving an Equation Involving Rational Expressions

Solve.

NOTE Assume $x \neq 3$ and $x \neq 4$.

$$\frac{x}{x-4} = \frac{15}{x-3} - \frac{2x}{x^2 - 7x + 12}$$

After factoring the trinomial denominator on the right, the LCD of $x - 3$, $x - 4$, and $x^2 - 7x + 12$ is $(x - 3)(x - 4)$. Multiplying by that LCD, we have

$$(x-3)(x-4)\left(\frac{x}{x-4}\right) = (x-3)(x-4)\left(\frac{15}{x-3}\right) - (x-3)(x-4)\left[\frac{2x}{(x-3)(x-4)}\right]$$

Simplifying yields

$$x(x-3) = 15(x-4) - 2x \qquad \text{Remove the parentheses.}$$

$$x^2 - 3x = 15x - 60 - 2x \qquad \text{Write in standard form and factor.}$$

$$x^2 - 16x + 60 = 0$$

$$(x-6)(x-10) = 0$$

So

$$x = 6 \qquad \text{or} \qquad x = 10$$

Verify that 6 and 10 are both solutions for the original equation.

CHECK YOURSELF 6

Solve $\frac{3x}{x+2} - \frac{2}{x+3} = \frac{36}{x^2 + 5x + 6}$.

The following algorithm summarizes our work in solving equations containing rational expressions.

Step by Step: Solving Equations Containing Rational Expressions

Step 1 Clear the equation of fractions by multiplying both sides of the equation by the LCD of all the fractions that appear.
Step 2 Solve the equation resulting from step 1.
Step 3 Check all solutions by substitution in the original equation.

The techniques we have just discussed can also be used to find the zeros of rational functions. Remember that a zero of a function is a value of x for which $f(x) = 0$.

Example 7

Finding the Zeros of a Function

CAUTION

This is different from finding values for which the rational function does not exist. In that case, we found x values that caused division by zero. Here, we set the function equal to zero and find the x values that make the statement true.

Find the zeros of

$$f(x) = \frac{1}{x} - \frac{3}{7x} - \frac{4}{21}$$

Set the function equal to 0, and solve the resulting equation for x.

$$f(x) = \frac{1}{x} - \frac{3}{7x} - \frac{4}{21} = 0$$

The LCM for x, $7x$, and 21 is $21x$. Multiplying both sides by $21x$, we have

$$21x\left(\frac{1}{x} - \frac{3}{7x} - \frac{4}{21}\right) = 21x \cdot 0 \quad \text{Distribute } 21x \text{ on the left side.}$$

$$21 - 9 - 4x = 0 \quad \text{Simplify.}$$

$$12 - 4x = 0$$

$$12 = 4x$$

$$3 = x$$

So 3 is the value of x for which $f(x) = 0$, that is, 3 is a zero of $f(x)$.

CHECK YOURSELF 7

Find the zeros of the function.

$$f(x) = \frac{5x + 2}{x - 6} - \frac{11}{4}$$

CHECK YOURSELF ANSWERS

1. $\{6\}$ **2.** $\{3\}$ **3.** $\{9\}$ **4.** $\{-11\}$ **5.** No solution **6.** $\left\{-5, \frac{8}{3}\right\}$
7. $\left\{-\frac{74}{9}\right\}$

Exercises

Name ______________________

Section ________ Date ________

In exercises 1 to 8, decide whether each of the following is an expression or an equation. If it is an equation, solve it. If it is an expression, write it as a single fraction.

1. $\frac{x}{2} - \frac{x}{3} = 6$

2. $\frac{x}{4} - \frac{x}{7} = 3$

3. $\frac{x}{2} - \frac{x}{5}$

4. $\frac{x}{6} - \frac{x}{8}$

5. $\frac{3x + 1}{4} = x - 1$

6. $\frac{3x - 1}{2} - \frac{x}{5} - \frac{x + 3}{4}$

7. $\frac{x}{4} = \frac{x}{12} + \frac{1}{2}$

8. $\frac{2x - 1}{3} + \frac{x}{2}$

In exercises 9 to 50, solve each equation.

9. $\frac{x}{3} + \frac{3}{2} = \frac{x}{6} + \frac{7}{3}$

10. $\frac{x}{10} - \frac{1}{5} = \frac{x}{5} + \frac{1}{2}$

11. $\frac{4}{x} + \frac{3}{4} = \frac{10}{x}$

12. $\frac{3}{x} = \frac{5}{3} - \frac{7}{x}$

13. $\frac{5}{4x} - \frac{1}{2} = \frac{1}{2x}$

14. $\frac{7}{6x} - \frac{1}{3} = \frac{1}{2x}$

15. $\frac{3}{x + 4} = \frac{2}{x + 3}$

16. $\frac{5}{x - 2} = \frac{4}{x - 1}$

17. $\frac{9}{x} + 2 = \frac{2x}{x + 3}$

18. $\frac{6}{x} + 3 = \frac{3x}{x + 1}$

ANSWERS

1. ______________
2. ______________
3. ______________
4. ______________
5. ______________
6. ______________
7. ______________
8. ______________
9. ______________
10. ______________
11. ______________
12. ______________
13. ______________
14. ______________
15. ______________
16. ______________
17. ______________
18. ______________

ANSWERS

19. ______
20. ______
21. ______
22. ______
23. ______
24. ______
25. ______
26. ______
27. ______
28. ______
29. ______
30. ______
31. ______
32. ______
33. ______
34. ______
35. ______
36. ______
37. ______
38. ______
39. ______

19. $\frac{3}{x+2} - \frac{5}{x} = \frac{13}{x+2}$

20. $\frac{7}{x} - \frac{2}{x-3} = \frac{6}{x}$

21. $\frac{3}{2} + \frac{2}{2x-4} = \frac{1}{x-2}$

22. $\frac{2}{x-1} + \frac{5}{2x-2} = \frac{3}{4}$

23. $\frac{x}{3x+12} + \frac{x-1}{x+4} = \frac{5}{3}$

24. $\frac{x}{4x-12} - \frac{x-4}{x-3} = \frac{1}{8}$

25. $\frac{x-1}{x+3} - \frac{x-3}{x} = \frac{3}{x^2+3x}$

26. $\frac{x+1}{x-2} - \frac{x+3}{x} = \frac{6}{x^2-2x}$

27. $\frac{1}{x-2} - \frac{2}{x+2} = \frac{2}{x^2-4}$

28. $\frac{1}{x+4} + \frac{1}{x-4} = \frac{12}{x^2-16}$

29. $\frac{7}{x+5} - \frac{1}{x-5} = \frac{x}{x^2-25}$

30. $\frac{2}{x-2} = \frac{3}{x+2} + \frac{x}{x^2-4}$

31. $\frac{11}{x+2} - \frac{5}{x^2-x-6} = \frac{1}{x-3}$

32. $\frac{5}{x-4} = \frac{1}{x+2} - \frac{2}{x^2-2x-8}$

33. $\frac{5}{x-2} - \frac{3}{x+3} = \frac{24}{x^2+x-6}$

34. $\frac{3}{x+1} - \frac{5}{x+6} = \frac{2}{x^2+7x+6}$

35. $\frac{x}{x-3} - 2 = \frac{3}{x-3}$

36. $\frac{x}{x-5} + 2 = \frac{5}{x-5}$

37. $\frac{2}{x^2-3x} - \frac{1}{x^2+2x} = \frac{2}{x^2-x-6}$

38. $\frac{2}{x^2-x} - \frac{4}{x^2+5x-6} = \frac{3}{x^2+6x}$

39. $\frac{2}{x^2-4x+3} - \frac{3}{x^2-9} = \frac{2}{x^2+2x-3}$

40. $\dfrac{2}{x^2 - 4} - \dfrac{1}{x^2 + x - 2} = \dfrac{3}{x^2 - 3x + 2}$

41. $2 - \dfrac{6}{x^2} = \dfrac{1}{x}$

42. $3 - \dfrac{7}{x} - \dfrac{6}{x^2} = 0$

43. $1 - \dfrac{7}{x - 2} + \dfrac{12}{(x - 2)^2} = 0$

44. $1 + \dfrac{3}{x + 1} = \dfrac{10}{(x + 1)^2}$

45. $1 + \dfrac{3}{x^2 - 9} = \dfrac{10}{x + 3}$

46. $3 - \dfrac{7}{x^2 - x - 6} = \dfrac{5}{x - 3}$

47. $\dfrac{2x}{x - 3} + \dfrac{2}{x - 5} = \dfrac{3x}{x^2 - 8x + 15}$

48. $\dfrac{x}{x - 4} = \dfrac{5x}{x^2 - x - 12} - \dfrac{3}{x + 3}$

49. $\dfrac{2x}{x + 2} = \dfrac{5}{x^2 - x - 6} - \dfrac{1}{x - 3}$

50. $\dfrac{3x}{x - 1} = \dfrac{2}{x - 2} - \dfrac{2}{x^2 - 3x + 2}$

In exercises 51 to 58, find the zeros of each function.

51. $f(x) = \dfrac{x}{10} - \dfrac{12}{5}$

52. $f(x) = \dfrac{4x}{3} - \dfrac{x}{6}$

53. $f(x) = \dfrac{12}{x + 5} - \dfrac{5}{x}$

54. $f(x) = \dfrac{1}{x - 2} - \dfrac{3}{x}$

55. $f(x) = \dfrac{1}{x - 3} + \dfrac{2}{x} - \dfrac{5}{3x}$

56. $f(x) = \dfrac{2}{x} - \dfrac{1}{x + 1} - \dfrac{3}{x^2 + x}$

57. $f(x) = 1 + \dfrac{39}{x^2} - \dfrac{16}{x}$

58. $f(x) = x - \dfrac{72}{x} + 1$

ANSWERS

40. ______
41. ______
42. ______
43. ______
44. ______
45. ______
46. ______
47. ______
48. ______
49. ______
50. ______
51. ______
52. ______
53. ______
54. ______
55. ______
56. ______
57. ______
58. ______

Answers

1. Equation, {36} **3.** Expression, $\frac{3x}{10}$ **5.** Equation, {5}

7. Equation, {3} **9.** {5} **11.** {8} **13.** $\left\{\frac{3}{2}\right\}$ **15.** {−1}

17. $\left\{-\frac{9}{5}\right\}$ **19.** $\left\{-\frac{2}{3}\right\}$ **21.** No solution **23.** {−23} **25.** {6}

27. {4} **29.** {8} **31.** {4} **33.** $\left\{\frac{3}{2}\right\}$ **35.** No solution **37.** {7}

39. {5} **41.** $\left\{2, -\frac{3}{2}\right\}$ **43.** {5, 6} **45.** {4, 6} **47.** $\left\{-\frac{1}{2}, 6\right\}$

49. $\left\{-\frac{1}{2}\right\}$ **51.** 24 **53.** $\frac{25}{7}$ **55.** $\frac{3}{4}$ **57.** 3, 13

7.6 Applications of Rational Equations

7.6 OBJECTIVES

1. Solve a literal equation that involves a rational expression
2. Solve a distance problem that involves a rational equation
3. Solve a work problem
4. Convert units

In the previous section, we solved rational equations. In this section we will see several applications of those techniques. Given a literal equation such as

$$\frac{1}{R} = \frac{1}{R_1} + \frac{1}{R_2}$$

we can use the techniques of Section 7.5 to solve for one of the variables.

Example 1

Solving a Literal Equation

NOTE This is a parallel electric circuit. The symbol for a resistor is —⩘⩘—.

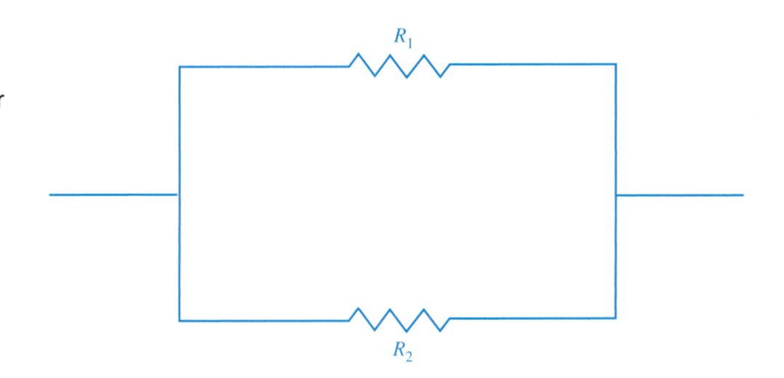

NOTE Recall that the numbers 1 and 2 are *subscripts*. We read R_1 as "R sub 1" and R_2 as "R sub 2."

If two resistors with resistances R_1 and R_2 are connected in parallel, the combined resistance R can be found from

$$\frac{1}{R} = \frac{1}{R_1} + \frac{1}{R_2}$$

Solve the formula for R.

First, the LCD is RR_1R_2, and we multiply:

$$RR_1R_2 \cdot \frac{1}{R} = RR_1R_2 \cdot \frac{1}{R_1} + RR_1R_2 \cdot \frac{1}{R_2}$$

Simplifying yields

$$R_1R_2 = RR_2 + RR_1 \qquad \text{Factor out } R \text{ on the right.}$$

$$R_1R_2 = R(R_2 + R_1) \qquad \text{Divide by } R_2 + R_1 \text{ to isolate } R.$$

NOTE Reversing the left and right sides of an equation uses the symmetric property of equality.

$$\frac{R_1R_2}{R_2 + R_1} = R \quad \text{or} \quad R = \frac{R_1R_2}{R_1 + R_2}$$

CHECK YOURSELF 1

Solve for D_1.

$$\frac{1}{F} = \frac{1}{D_1} + \frac{1}{D_2}$$

NOTE This formula involves the focal length of a convex lens.

Many distance problems also lead to rational equations. You will recall that distance (d), rate (r), and time (t) are related by the following formula

Rules and Properties: The Distance Relationship I

$$d = r \cdot t$$

Treating this as a literal equation, we can produce two variations that frequently lead to rational equations.

Rules and Properties: The Distance Relationship II

$$r = \frac{d}{t}$$

and

Rules and Properties: The Distance Relationship III

$$t = \frac{d}{r}$$

Example 2

Finding a Rate

A boat, which moves at 36 mi/h in still water, travels 28 mi downstream in the same amount of time that it takes to travel 20 mi upstream. Find the speed of the current.

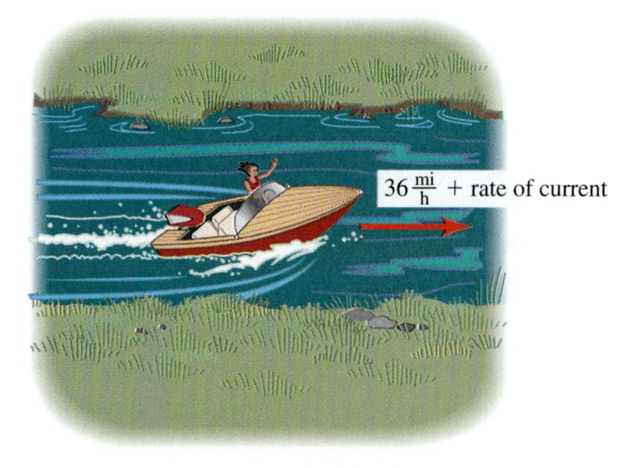

When solving an application that involves the distance relationship, it is usually best to begin by completing the following table.

	Distance	Rate	Time
Upstream			
Downstream			

Letting r represent the rate of the current, we have the following information:

	Distance	Rate	Time
Upstream	20	$36 - r$	
Downstream	28	$36 + r$	

Using relationship III above, we can complete the last column of the table.

	Distance	Rate	Time
Upstream	20	$36 - r$	$\frac{20}{(36 - r)}$
Downstream	28	$36 + r$	$\frac{28}{(36 + r)}$

Having completed the table, we go back to the original problem. The key phrase is "in the same amount of time." That means that the time going upstream is the same as the time going downstream. This leads to the equation

$$\frac{28}{(36 + r)} = \frac{20}{(36 - r)}$$

The LCD is $(36 + r)(36 - r)$. Clearing the fractions, we get

$$28(36 - r) = 20(36 + r)$$

$$1008 - 28r = 720 + 20r$$

$$288 = 48r$$

$$r = 6$$

The rate of the current is 6 mi/h.

CHECK YOURSELF 2

A boat, which moves at 30 mi/h in still water, travels 3 mi downstream in the same amount of time that it takes to travel 2 mi upstream. Find the speed of the current.

Another type of application that frequently leads to a rational equation is something called a **work problem.** Solving a work problem is simplified by using the following work principles.

Rules and Properties: Work Principle I

If a job takes t hours to complete, then, for each hour,

$$\frac{1}{t}$$

represents the portion of the job that has been completed.

This principle confirms that, if a job takes 2 h to do, then $\frac{1}{2}$ of the job is done each hour. If a job takes 10 h to do, then $\frac{1}{10}$ of the job is done each hour.

Extending the idea behind the first principle, we find the second work principle.

Rules and Properties: Work Principle II

If two entities are working on the same job, and the first would take a hours to complete the job alone and the second b hours to complete the job alone, then the expression

$$\frac{1}{a} + \frac{1}{b}$$

represents the portion of the job completed each hour if both entities are at work.

Combining these two principles, we get an equation that will allow us to find the total time it takes to complete a job.

Rules and Properties: Work Principle III

If two entities are working on the same job, and the first would take a hours to complete the job alone and the second b hours to complete the job alone, then the equation

$$\frac{1}{a} + \frac{1}{b} = \frac{1}{t}$$

can be used to find t, the time it will take to complete the job.

We will use this equation in the next example.

Example 3

Solving a Work Problem

Jason and Hilger are required to paint over the graffiti on a wall. If Jason worked alone, it would take him 20 h to repaint. Working alone, Hilger could do the job in 15 h. How long will it take them to do the painting if they work together?

Using the work principle, we get the equation

$$\frac{1}{20} + \frac{1}{15} = \frac{1}{t}$$

Clearing the fractions (LCD = $60t$) yields

$$3t + 4t = 60$$

$$7t = 60$$

$$t = \frac{60}{7}$$

Working together, they will finish the job in $8\frac{4}{7}$ h. Note that, when working together, it will always take less time to complete a job than it will take either individual!

CHECK YOURSELF 3

Filling a hot tub with a hose will take 4 h. Draining the same tub takes 7 h. If the drain is open, how long will it take to fill the tub? (Hint: Treat the draining time as a negative number.)

The techniques used with rational equations are also used when performing **unit conversions.** The following table contains some of the common conversion factors.

Common Conversion Factors
1 mile = 5280 feet
1 hour = 60 minutes
1 minute = 60 seconds
1 kilogram = 1000 grams
1 second = 1,000,000 microseconds
1 gigabyte = 1,000,000,000 bytes

Each of the conversion factors can be rewritten as one of two fractions that are equal to one. For example, we can rewrite the statement 1 kilogram = 1000 grams as either

$$\frac{1 \text{ kilogram}}{1000 \text{ grams}} = 1 \quad \text{or} \quad \frac{1000 \text{ grams}}{1 \text{ kilogram}} = 1$$

Example 4

Converting Units

An SST can fly over 2000 mi/h. To the nearest tenth, how many miles does it cover each second?

First, note that 2000 mi/h can be written in several different forms. When doing unit conversion, it is best to write a rate (like 2000 mi/h) as a ratio, so we write it as

$$\frac{2000 \text{ mi}}{1 \text{ h}}$$

This is now multiplied by the appropriate conversion ratios. To figure out what those ratios should be, it is best to first write the units of the desired result.

$$\frac{2000 \text{ mi}}{1 \text{ h}} = \qquad\qquad = \frac{\text{mi}}{1 \text{ s}}$$

To convert hours to seconds, use the conversion ratio $\frac{1 \text{ h}}{60 \text{ min}} \cdot \frac{1 \text{ min}}{60 \text{ s}} = \frac{1 \text{ h}}{3600 \text{ s}}$

$$\frac{2000 \text{ mi}}{1 \text{ h}} \cdot \frac{1 \text{ h}}{3600 \text{ s}} = \frac{2000 \text{ mi}}{3600 \text{ s}} = \frac{5 \text{ mi}}{9 \text{ s}} \approx \frac{0.55 \text{ mi}}{1 \text{ s}}$$

When traveling at 2000 mi/h, the SST covers 0.55 mi every second.

CHECK YOURSELF 4

A roller coaster reaches a maximum speed of 75 ft/s. To the nearest tenth, convert the speed to miles per hour.

CHECK YOURSELF ANSWERS

1. $\frac{FD_2}{D_2 - F}$ **2.** 6 mi/h **3.** $\frac{28}{3}$ h **4.** 51.1 mi/h

Name ____________

Section ______ Date ______

7.6 Exercises

In exercises 1 to 8, solve each equation for the indicated variable.

1. $\frac{1}{x} = \frac{1}{a} - \frac{1}{b}$ for x

2. $\frac{1}{x} = \frac{1}{a} + \frac{1}{b}$ for a

3. $\frac{1}{R} = \frac{1}{R_1} + \frac{1}{R_2}$ for R_1

4. $\frac{1}{F} = \frac{1}{D_1} + \frac{1}{D_2}$ for D_2

5. $y = \frac{x + 1}{x - 1}$ for x

6. $y = \frac{x - 3}{x - 2}$ for x

7. $t = \frac{A - P}{Pr}$ for P

8. $I = \frac{nE}{R + nr}$ for n

Solve the following problems.

9. **Motion.** A motorboat can travel 20 mi/h in still water. If the boat can travel 3 mi downstream on a river in the same time it takes to travel 2 mi upstream, what is the rate of the river's current?

10. **Motion.** A small jet has an airspeed (the rate in still air) of 300 mi/h. During one day's flights, the pilot noted that the plane could fly 85 mi with a tailwind in the same time it took to fly 65 mi against that same wind. What was the rate of the wind?

11. **Motion.** A plane flew 720 mi with a steady 30-mi/h tailwind. The pilot then returned to the starting point, flying against that same wind. If the round-trip flight took 10 h, what was the plane's airspeed?

12. **Motion.** Janet and Michael took a canoeing trip, traveling 6 mi upstream along a river, against a 2 mi/h current. They then returned downstream to the starting point of their trip. If their entire trip took 4 h, what was their rate in still water?

13. **Motion.** Po Ling can bicycle 75 mi in the same time it takes her to drive 165 mi. If her driving rate is 30 mi/h faster than her rate on the bicycle, find each rate.

ANSWERS

1. ____________
2. ____________
3. ____________
4. ____________
5. ____________
6. ____________
7. ____________
8. ____________
9. ____________
10. ____________
11. ____________
12. ____________
13. ____________

ANSWERS

14. ______

15. ______

16. ______

17. ______

18. ______

19. ______

20. ______

21. ______

22. ______

14. Motion. A passenger train can travel 275 mi in the same time a freight train takes to travel 225 mi. If the speed of the passenger train is 10 mi/h more than that of the freight train, find the speed of each train.

15. Motion. A light plane took 1 h longer to fly 540 mi on the first portion of a trip than to fly 360 mi on the second. If the rate was the same for each portion, what was the flying time for each leg of the trip?

16. Motion. Gilbert took 2 h longer to drive 240 mi on the first day of a business trip than to drive 144 mi on the second day. If his rate was the same both days, what was his driving time for each day?

17. Motion. An express train and a passenger bus leave the same city, at the same time, for a destination 350 mi away. The rate of the train is 20 mi/h faster than the rate of the bus. If the train arrives at its destination 2 h ahead of the bus, find each rate.

18. Motion. A private plane and a commercial plane take off from an airport at the same time for a city 720 mi away. The rate of the private plane is 180 mi/h less than that of the commercial plane. If the commercial plane arrives 2 h ahead of the private plane, find each plane's rate.

19. Work. One road crew can pave a section of highway in 15 h. A second crew, working with newer equipment, can do the same job in 10 h. How long whould it take to pave that same section of highway if both crews worked together?

20. Work. One computer printer can print a company's weekly payroll checks in 60 min. A second printer would take 90 min to complete the job. How long would it take the two printers, operating together, to print the checks?

21. Work. An inlet pipe can fill a tank in 10 h. An outlet pipe can drain that same tank in 30 h. The inlet valve is opened, but the outlet valve is accidentally left open. How long will it take to fill the tank with both valves open?

22. Work. A bathtub can be filled in 8 min. It takes 12 min for the bathtub to drain. If the faucet is turned on but the drain is also left open, how long will it take to fill the tub?

ANSWERS

23. ____

24. ____

25. ____

26. ____

27. ____

28. ____

29. ____

30. ____

31. ____

32. ____

23. Work. An electrician can wire a house in 20 h. If she works with an apprentice, the same job can be completed in 12 h. How long would it take the apprentice, working alone, to wire the house?

24. Work. A landscaper can prepare and seed a new lawn in 12 h. If he works with an assistant, the job takes 8 h. How long would it take the assistant, working alone, to complete the job?

25. Work. An experienced roofer can work twice as fast as her helper. Working together, they can shingle a new section of roof in 4 h. How long would it take the experienced roofer, working alone, to complete the same job?

26. Work. One model copier operates at 3 times the speed of another. Working together, the copiers can copy a report in 8 min. The faster model breaks down, and the other model must be used. How long will the job take with the machine that is available?

27. Work. A college uses two optical scanners to grade multiple-choice tests. One model takes 12 min longer to complete the scoring of a test than the other model. If by both models working together the test can be scored in 8 min, how long would each model take to score the same test, used by itself?

28. Work. Virginia can complete her company's monthly report in 5 h less time than Carl. If they work together, the report will take them 6 h to finish. How long would it take Virginia, working alone?

29. Number analysis. The sum of the reciprocals of two consecutive integers is equal to 11 times the reciprocal of the product of those integers. What are the two integers?

30. Number analysis. The sum of the reciprocals of two consecutive even integers is equal to 10 times the reciprocal of the product of those integers. Find the two integers.

31. Number analysis. If the same number is added to the numerator and denominator of $\frac{2}{5}$, the result is $\frac{4}{5}$. What is that number?

32. Number analysis. If the same number is subtracted from the numerator and denominator of $\frac{11}{15}$, the result is $\frac{1}{3}$. Find that number.

ANSWERS

33. ______

34. ______

35. ______

36. ______

37. ______

38. ______

39. ______

40. ______

33. **Number analysis.** One positive number is 2 more than another. If the sum of the reciprocals of the two numbers is $\frac{7}{24}$, what are those numbers?

34. **Number analysis.** One integer is 3 less than another. If the sum of the reciprocals of the two numbers is $\frac{7}{10}$, find the two integers.

35. A car is traveling at the rate of 60 ft/s. To the nearest tenth, convert the speed to miles per hour.

36. How many seconds are there in the month of June?

37. Arlene's hard drive has 2 gigabytes of memory. How many bytes is this?

38. Mr. Fernandez can walk 5 mi in 2 h. What is his speed in feet per second, to the nearest tenth?

39. Trucks in Sam's home state are taxed at the rate of 0.15¢ per gram. His truck weighs 1620 kg. What is his annual tax?

40. Suppose you travel to a certain destination at an average speed of 30 mi/h. What should your speed be on the return trip to bring your overall average speed to 60 mi/h?

Answers

1. $\frac{ab}{b - a}$ **3.** $\frac{RR_2}{R_2 - R}$ **5.** $\frac{y + 1}{y - 1}$ **7.** $\frac{A}{1 + rt}$ **9.** 4 mi/h
11. 150 mi/h **13.** Bicycling 25 mi/h, driving 55 mi/h **15.** 3 h, 2 h
17. Bus 50 mi/h, train 70 mi/h **19.** 6 h **21.** 15 h **23.** 30 h **25.** 6 h
27. 12 min, 24 min **29.** 5, 6 **31.** 10 **33.** 6, 8 **35.** 40.9 mi/h
37. 2,000,000,000 **39.** $2430

7.7 Negative Integer Exponents

7.7 OBJECTIVES

1. Define the zero exponent
2. Use the definition of a negative exponent to simplify an expression
3. Use the properties of exponents to simplify expressions that contain negative exponents

In Section 1.4, all the exponents we looked at were positive integers. In this section, we look at the meaning of zero and negative integer exponents. First, let's look at an application of the quotient rule that will yield a zero exponent.

Recall that, in the quotient rule, to divide expressions with the same base, keep the base and subtract the exponents.

$$\frac{a^m}{a^n} = a^{m-n}$$

Now, suppose that we allow *m to equal n.* We then have

$$\frac{a^m}{a^m} = a^{m-m} = a^0 \qquad (1)$$

But we know that it is also true that

$$\frac{a^m}{a^m} = 1 \qquad (2)$$

Comparing equations (1) and (2), we see that the following definition is reasonable.

NOTE We must have $a \neq 0$. The form 0^0 is called **indeterminate** and is considered in later mathematics classes.

Rules and Properties: The Zero Exponent

For any real number a when $a \neq 0$,

$a^0 = 1$

Example 1

The Zero Exponent

Use the above definition to simplify each expression.

(a) $17^0 = 1$

(b) $(a^3b^2)^0 = 1$

(c) $6x^0 = 6 \cdot 1 = 6$

(d) $-3y^0 = -3$

NOTE Notice that in $6x^0$ the exponent 0 applies *only* to x.

CHECK YOURSELF 1

Simplify each expression.

(a) 25^0 **(b)** $(m^4n^2)^0$ **(c)** $8s^0$ **(d)** $-7t^0$

Recall that, in the product rule, to multiply expressions with the same base, keep the base and add the exponents.

$$a^m \cdot a^n = a^{m+n}$$

Now, what if we allow one of the exponents to be negative and apply the product rule? Suppose, for instance, that $m = 3$ and $n = -3$. Then

$$a^m \cdot a^n = a^3 \cdot a^{-3} = a^{3+(-3)}$$
$$= a^0 = 1$$

so

$$a^3 \cdot a^{-3} = 1$$

Dividing both sides by a^3, we get

$$a^{-3} = \frac{1}{a^3}$$

NOTE John Wallis (1616–1702), an English mathematician, was the first to fully discuss the meaning of 0, negative, and rational exponents.

Rules and Properties: Negative Integer Exponents

For any nonzero real number a and whole number n,

$$a^{-n} = \frac{1}{a^n}$$

and a^{-n} is the **multiplicative inverse** of a^n.

Example 2 illustrates this definition.

Example 2

Using Properties of Exponents

NOTE From this point on, to *simplify* will mean to write the expression with *positive exponents only.*

NOTE Also, we will restrict all variables so that they represent nonzero real numbers.

Simplify the following expressions.

(a) $y^{-5} = \frac{1}{y^5}$

(b) $4^{-2} = \frac{1}{4^2} = \frac{1}{16}$

(c) $(-3)^3 = \frac{1}{(-3)^3} = \frac{1}{-27} = -\frac{1}{27}$

(d) $\left(\frac{2}{3}\right)^{-3} = \frac{1}{\left(\frac{2}{3}\right)^3} = \frac{1}{\frac{8}{27}} = \frac{27}{8}$

CHECK YOURSELF 2

Simplify each of the following expressions.

(a) a^{-10} **(b)** 2^{-4} **(c)** $(-4)^{-2}$ **(d)** $\left(\frac{5}{2}\right)^{-2}$

Example 3 illustrates the case in which coefficients are involved in an expression with negative exponents. As will be clear, some caution must be used.

Example 3

Using Properties of Exponents

Simplify each of the following expressions.

CAUTION

The expressions

$4w^{-2}$ and $(4w)^{-2}$

are *not* the same. Do you see why?

(a) $2x^{-3} = 2 \cdot \frac{1}{x^3} = \frac{2}{x^3}$

The exponent −3 applies only to the variable *x*, and *not* to the coefficient 2.

(b) $4w^{-2} = 4 \cdot \frac{1}{w^2} = \frac{4}{w^2}$

(c) $(4w)^{-2} = \frac{1}{(4w)^2} = \frac{1}{16w^2}$

CHECK YOURSELF 3

Simplify each of the following expressions.

(a) $3w^{-4}$ **(b)** $10x^{-5}$ **(c)** $(2y)^{-4}$ **(d)** $-5t^{-2}$

Suppose that a variable with a negative exponent appears in the denominator of an expression. Our previous definition can be used to write a complex fraction that can then be simplified. For instance,

$$\frac{1}{a^{-2}} = \frac{1}{\frac{1}{a^2}} = 1 \cdot \frac{a^2}{1} = a^2$$

Positive exponent in numerator.

Negative exponent in denominator.

To divide, we invert and multiply.

To avoid the intermediate steps, we can write that, in general,

Rules and Properties: Negative Exponents in a Denominator

For any nonzero real number *a* and integer *n*,

$$\frac{1}{a^{-n}} = a^n$$

Example 4

Using Properties of Exponents

Simplify each of the following expressions.

(a) $\frac{1}{y^{-3}} = y^3$

(b) $\frac{1}{2^{-5}} = 2^5 = 32$

(c) $\frac{3}{4x^{-2}} = \frac{3x^2}{4}$ The exponent −2 applies only to x, not to 4.

(d) $\frac{a^{-3}}{b^{-4}} = \frac{b^4}{a^3}$

CHECK YOURSELF 4

Simplify each of the following expressions.

(a) $\frac{1}{x^{-4}}$ **(b)** $\frac{1}{3^{-3}}$ **(c)** $\frac{2}{3a^{-2}}$ **(d)** $\frac{c^{-5}}{d^{-7}}$

NOTE To review these properties, return to Section 1.4.

The product and quotient rules for exponents apply to expressions that involve any integer exponent—positive, negative, or 0. Example 5 illustrates this concept.

Example 5

Using Properties of Exponents

Simplify each of the following expressions, and write the result, using positive exponents only.

(a) $x^3 \cdot x^{-7} = x^{3+(-7)}$

$= x^{-4} = \frac{1}{x^4}$

Add the exponents by the product rule.

(b) $\frac{m^{-5}}{m^{-3}} = m^{-5-(-3)} = m^{-5+3}$

$= m^{-2} = \frac{1}{m^2}$

Subtract the exponents by the quotient rule.

(c) $\frac{x^5x^{-3}}{x^{-7}} = \frac{x^{5+(-3)}}{x^{-7}} = \frac{x^2}{x^{-7}} = x^{2-(-7)} = x^9$

We apply first the product rule and then the quotient rule.

NOTE Notice that m^{-5} in the numerator becomes m^5 in the denominator, and m^{-3} in the denominator becomes m^3 in the numerator. We then simplify as before.

In simplifying expressions involving negative exponents, there are often alternate approaches. For instance, in Example 5(b), we could have made use of our earlier work to write

$$\frac{m^{-5}}{m^{-3}} = \frac{m^3}{m^5} = m^{3-5} = m^{-2} = \frac{1}{m^2}$$

CHECK YOURSELF 5

Simplify each of the following expressions.

(a) $x^9 \cdot x^{-5}$ **(b)** $\dfrac{y^{-7}}{y^{-3}}$ **(c)** $\dfrac{a^{-3}a^2}{a^{-5}}$

The properties of exponents can be extended to include negative exponents. One of these properties, the quotient-power rule, is particularly useful when rational expressions are raised to a negative power. Let's look at the rule and apply it to negative exponents.

Rules and Properties: Quotient-Power Rule

$$\left(\frac{a}{b}\right)^n = \frac{a^n}{b^n}, \quad b \neq 0$$

Rules and Properties: Raising Quotients to a Negative Power

$$\left(\frac{a}{b}\right)^{-n} = \frac{a^{-n}}{b^{-n}} = \frac{b^n}{a^n} = \left(\frac{b}{a}\right)^n \qquad a \neq 0,\ b \neq 0$$

Example 6

Extending the Properties of Exponents

Simplify each expression.

(a) $\left(\dfrac{s^3}{t^2}\right)^{-2} = \left(\dfrac{t^2}{s^3}\right)^2 = \dfrac{t^4}{s^6}$

(b) $\left(\dfrac{m^2}{n^{-2}}\right)^{-3} = \left(\dfrac{n^{-2}}{m^2}\right)^3 = \dfrac{n^{-6}}{m^6} = \dfrac{1}{n^6m^6}$

CHECK YOURSELF 6

Simplify each expression.

(a) $\left(\dfrac{s^3}{3t^2}\right)^{-3}$ **(b)** $\left(\dfrac{x^5}{y^{-2}}\right)^{-3}$

As you might expect, more complicated expressions require the use of more than one of the properties for simplification. Example 7 illustrates such cases.

Example 7

Using Properties of Exponents

Simplify each of the following expressions.

(a) $\dfrac{(a^2)^{-3}(a^3)^4}{(a^{-3})^3} = \dfrac{a^{-6} \cdot a^{12}}{a^{-9}}$ Apply the power rule to each factor.

$= \dfrac{a^{-6+12}}{a^{-9}} = \dfrac{a^6}{a^{-9}}$ Apply the product rule.

$= a^{6-(-9)} = a^{6+9} = a^{15}$ Apply the quotient rule.

NOTE It may help to separate the problem into three fractions, one for the coefficients and one for each of the variables.

(b) $$\frac{8x^{-2}y^{-5}}{12x^{-4}y^{3}} = \frac{8}{12} \cdot \frac{x^{-2}}{x^{-4}} \cdot \frac{y^{-5}}{y^{3}}$$

$$= \frac{2}{3} \cdot x^{-2-(-4)} \cdot y^{-5-3}$$

$$= \frac{2}{3} \cdot x^{2} \cdot y^{-8} = \frac{2x^{2}}{3y^{8}}$$

(c) $$\left(\frac{pr^{3}s^{-5}}{p^{3}r^{-3}s^{-2}}\right)^{-2} = \left(\frac{p^{3}r^{-3}s^{-2}}{pr^{3}s^{-5}}\right)^{2}$$

$$= \frac{p^{6}r^{-6}s^{-4}}{p^{2}r^{6}s^{-10}}$$

$$= p^{4}r^{-12}s^{6} = \frac{p^{4}s^{6}}{r^{12}}$$

Be Careful! Another possible first step (and generally an efficient one) is to rewrite an expression by using our earlier definitions.

$$a^{-n} = \frac{1}{a^{n}} \quad \text{and} \quad \frac{1}{a^{-n}} = a^{n}$$

For instance, in Example 8(b), we would *correctly* write

$$\frac{8x^{-2}y^{-5}}{12x^{-4}y^{3}} = \frac{8x^{4}}{12x^{2}y^{3}y^{5}}$$

A *common error* is to write

$$\frac{8x^{-2}y^{-5}}{12x^{-4}y^{3}} = \frac{12x^{4}}{8x^{2}y^{3}y^{5}}$$ This is *not* correct.

The coefficients should not be moved along with the factors in x. Keep in mind that the negative exponents apply *only* to the variables. The coefficients remain *where they were* in the original expression when the expression is rewritten by using this approach.

CHECK YOURSELF 7

Simplify each of the following expressions.

(a) $\dfrac{(x^{5})^{-2}(x^{2})^{3}}{(x^{-4})^{3}}$ (b) $\dfrac{12a^{-3}b^{-2}}{16a^{-2}b^{3}}$ (c) $\left(\dfrac{xy^{-3}z^{-5}}{x^{-4}y^{-2}z^{3}}\right)^{-3}$

CHECK YOURSELF ANSWERS

1. **(a)** 1; **(b)** 1; **(c)** 8; **(d)** -7 **2.** **(a)** $\frac{1}{a^{10}}$; **(b)** $\frac{1}{16}$; **(c)** $\frac{1}{16}$; **(d)** $\frac{4}{25}$

3. **(a)** $\frac{3}{w^{4}}$; **(b)** $\frac{10}{x^{5}}$; **(c)** $\frac{1}{16y^{4}}$; **(d)** $-\frac{5}{t^{2}}$ **4.** **(a)** x^{4}; **(b)** 27; **(c)** $\frac{2a^{2}}{3}$; **(d)** $\frac{d^{7}}{c^{5}}$

5. **(a)** x^{4}; **(b)** $\frac{1}{y^{4}}$; **(c)** a^{4} **6.** **(a)** $\frac{27t^{6}}{s^{9}}$; **(b)** $\frac{1}{x^{15}y^{6}}$ **7.** **(a)** x^{8}; **(b)** $\frac{3}{4ab^{5}}$; **(c)** $\frac{y^{3}z^{24}}{x^{15}}$

Name ____________

Section ______ Date ______

Exercises

In exercises 1 to 22, simplify each expression.

1. x^{-5}

2. 3^{-3}

3. 5^{-2}

4. x^{-8}

5. $(-5)^{-2}$

6. $(-3)^{-3}$

7. $(-2)^{-3}$

8. $(-2)^{-4}$

9. $\left(\frac{2}{3}\right)^{-3}$

10. $\left(\frac{3}{4}\right)^{-2}$

11. $3x^{-2}$

12. $4x^{-3}$

13. $-5x^{-4}$

14. $(-2x)^{-4}$

15. $(-3x)^{-2}$

16. $-5x^{-2}$

17. $\frac{1}{x^{-3}}$

18. $\frac{1}{x^{-5}}$

19. $\frac{2}{5x^{-3}}$

20. $\frac{3}{4x^{-4}}$

21. $\frac{x^{-3}}{y^{-4}}$

22. $\frac{x^{-5}}{y^{-3}}$

In exercises 23 to 32, use the properties of exponents to simplify the expressions.

23. $x^5 \cdot x^{-3}$

24. $y^{-4} \cdot y^5$

25. $a^{-9} \cdot a^6$

26. $w^{-5} \cdot w^3$

27. $z^{-2} \cdot z^{-8}$

28. $b^{-7} \cdot b^{-1}$

29. $a^{-5} \cdot a^5$

30. $x^{-4} \cdot x^4$

31. $\frac{x^{-5}}{x^{-2}}$

32. $\frac{x^{-3}}{x^{-6}}$

ANSWERS

1. ______ 2. ______
3. ______ 4. ______
5. ______ 6. ______
7. ______ 8. ______
9. ______ 10. ______
11. ______ 12. ______
13. ______ 14. ______
15. ______ 16. ______
17. ______ 18. ______
19. ______ 20. ______
21. ______ 22. ______
23. ______ 24. ______
25. ______ 26. ______
27. ______ 28. ______
29. ______ 30. ______
31. ______ 32. ______

ANSWERS

33. ______ 34. ______
35. ______ 36. ______
37. ______ 38. ______
39. ______ 40. ______
41. ______ 42. ______
43. ______ 44. ______
45. ______ 46. ______
47. ______ 48. ______
49. ______ 50. ______
51. ______ 52. ______
53. ______ 54. ______
55. ______ 56. ______
57. ______ 58. ______
59. ______ 60. ______
61. ______ 62. ______
63. ______ 64. ______
65. ______ 66. ______
67. ______

In exercises 33 to 58, use the properties of exponents to simplify the following.

33. $(x^5)^3$ **34.** $(w^4)^6$ **35.** $(2x^{-3})(x^2)^4$

36. $(p^4)(3p^3)^2$ **37.** $(3a^{-4})(a^3)(a^2)$ **38.** $(5y^{-2})(2y)(y^5)$

39. $(x^4y)(x^2)^3(y^3)^0$ **40.** $(r^4)^2(r^2s)(s^3)^2$ **41.** $(ab^2c)(a^4)^4(b^2)^3(c^3)^4$

42. $(p^2qr^2)(p^2)(q^3)^2(r^2)^0$ **43.** $(x^5)^{-3}$ **44.** $(x^{-2})^{-3}$

45. $(b^{-4})^{-2}$ **46.** $(a^0b^{-4})^3$ **47.** $(x^5y^{-3})^2$

48. $(p^{-3}q^2)^{-2}$ **49.** $(x^{-4}y^{-2})^{-3}$ **50.** $(3x^{-2}y^{-2})^3$

51. $(2x^{-3}y^0)^{-5}$ **52.** $\dfrac{a^{-6}}{b^{-4}}$ **53.** $\dfrac{x^{-2}}{y^{-4}}$

54. $\left(\dfrac{x^{-3}}{y^2}\right)^{-3}$ **55.** $\dfrac{x^{-4}}{y^{-2}}$ **56.** $\dfrac{(3x^{-4})^2(2x^2)}{x^6}$

57. $(4x^{-2})^2(3x^{-4})$ **58.** $(5x^{-4})^{-4}(2x^3)^{-5}$

In exercises 59 to 90, simplify each expression.

59. $(2x^5)^4(x^3)^2$ **60.** $(3x^2)^3(x^2)^4(x^2)$ **61.** $(2x^{-3})^3(3x^3)^2$

62. $(x^2y^3)^4(xy^3)^0$ **63.** $(xy^5z)^4(xyz^2)^8(x^6yz)^5$ **64.** $(x^2y^2z^2)^0(xy^2z)^2(x^3yz^2)$

65. $(3x^{-2})(5x^2)^2$ **66.** $(2a^3)^2(a^0)^5$ **67.** $(2w^3)^4(3w^{-5})^2$

68. $(3x^3)^2(2x^4)^5$

69. $\dfrac{3x^6}{2y^9} \cdot \dfrac{y^5}{x^3}$

70. $\dfrac{x^8}{y^6} \cdot \dfrac{2y^9}{x^3}$

71. $(-7x^2y)(-3x^5y^6)^4$

72. $\left(\dfrac{2w^5z^3}{3x^3y^9}\right)\left(\dfrac{x^5y^4}{w^4z^0}\right)^2$

73. $(2x^2y^{-3})(3x^{-4}y^{-2})$

74. $(-5a^{-2}b^{-4})(2a^5b^0)$

75. $\dfrac{(x^{-3})(y^2)}{y^{-3}}$

76. $\dfrac{6x^3y^{-4}}{24x^{-2}y^{-2}}$

77. $\dfrac{15x^{-3}y^2z^{-4}}{20x^{-4}y^{-3}z^2}$

78. $\dfrac{24x^{-5}y^{-3}z^2}{36x^{-2}y^3z^{-2}}$

79. $\dfrac{x^{-5}y^{-7}}{x^0y^{-4}}$

80. $\left(\dfrac{xy^3z^{-4}}{x^{-3}y^{-2}z^2}\right)^{-2}$

81. $\dfrac{x^{-2}y^2}{x^3y^{-2}} \cdot \dfrac{x^{-4}y^2}{x^{-2}y^{-2}}$

82. $\left(\dfrac{x^{-3}y^3}{x^{-4}y^2}\right)^3 \cdot \left(\dfrac{x^{-2}y^{-2}}{xy^4}\right)^{-1}$

83. $x^{2n} \cdot x^{3n}$

84. $x^{n+1} \cdot x^{3n}$

85. $\dfrac{x^{n+3}}{x^{n+1}}$

86. $\dfrac{x^{n-4}}{x^{n-1}}$

87. $(y^n)^{3n}$

88. $(x^{n+1})^n$

89. $\dfrac{x^{2n} \cdot x^{n+2}}{x^{3n}}$

90. $\dfrac{x^n \cdot x^{3n+5}}{x^{4n}}$

91. Can $(a + b)^{-1}$ be written as $\dfrac{1}{a} + \dfrac{1}{b}$ by using the properties of exponents? If not, why not? Explain.

92. Write a short description of the difference between $(-4)^{-3}$, -4^{-3}, $(-4)^3$, and -4^3. Are any of these equal?

93. If $n > 0$, which of the following expressions are negative?

$(-n)^{-3}$, $-n^{-3}$, n^{-3}, $(-n)^{-3}$, $(-n)^3$, $-n^3$

If $n < 0$, which of these expressions are negative? Explain what effect a negative in the exponent has on the sign of the result when an exponential expression is simplified.

ANSWERS

68. ______
69. ______
70. ______
71. ______
72. ______ 73. ______
74. ______ 75. ______
76. ______ 77. ______
78. ______ 79. ______
80. ______ 81. ______
82. ______
83. ______
84. ______
85. ______
86. ______
87. ______
88. ______
89. ______
90. ______
91. ______
92. ______
93. ______

Answers

1. $\frac{1}{x^5}$ **3.** $\frac{1}{25}$ **5.** $\frac{1}{25}$ **7.** $-\frac{1}{8}$ **9.** $\frac{27}{8}$ **11.** $\frac{3}{x^2}$ **13.** $\frac{-5}{x^4}$

15. $\frac{1}{9x^2}$ **17.** x^3 **19.** $\frac{2x^3}{5}$ **21.** $\frac{y^4}{x^3}$ **23.** x^2 **25.** $\frac{1}{a^3}$ **27.** $\frac{1}{z^{10}}$ **29.** 1 **31.** $\frac{1}{x^3}$

33. x^{15} **35.** $2x^5$ **37.** $3a$ **39.** $x^{10}y$

41. $a^{17}b^8c^{13}$ **43.** $\frac{1}{x^{15}}$ **45.** b^8 **47.** $\frac{x^{10}}{y^6}$ **49.** $x^{12}y^6$ **51.** $\frac{x^{15}}{32}$

53. $\frac{y^4}{x^2}$ **55.** $\frac{y^2}{x^4}$ **57.** $\frac{48}{x^8}$ **59.** $16x^{26}$ **61.** $\frac{72}{x^3}$ **63.** $x^{42}y^{33}z^{25}$

65. $75x^2$ **67.** $144w^2$ **69.** $\frac{3x^3}{2y^4}$ **71.** $-567x^{22}y^{25}$ **73.** $\frac{6}{x^2y^5}$

75. $\frac{y^5}{x^3}$ **77.** $\frac{3xy^5}{4z^6}$ **79.** $\frac{1}{x^5y^3}$ **81.** $\frac{y^8}{x^7}$ **83.** x^{5n} **85.** x^2

87. y^{3n^2} **89.** x^2 **91.** **93.**

7 Summary

Definition/Procedure	Example	Reference
Simplification of Rational Expressions and Functions		**Section 7.1**
Rational expressions have the form $\dfrac{P}{Q}$ in which P and Q are polynomials and $Q(x) \neq 0$ for all x	$\dfrac{x^2 - 5x}{x - 3}$ is a rational expression. The variable x cannot have the value 3.	p. 486
Fundamental Principle of Rational Expressions For polynomials P, Q, and R, $\dfrac{P}{Q} = \dfrac{PR}{QR}$ when $Q \neq 0$ and $R \neq 0$. This principle can be used in two ways. We can multiply or divide the numerator and denominator of a rational expression by the same nonzero polynomial.	This uses the fact that $\dfrac{R}{R} = 1$ when $R \neq 0$.	p. 487
Simplifying Rational Expressions To simplify a rational expression, use the following algorithm. **1.** Completely factor both the numerator and denominator of the expression. **2.** Divide the numerator and denominator by *all* common factors. **3.** The resulting expression will be in simplest form (or in lowest terms).	$\dfrac{x^2 - 4}{x^2 - 2x - 8} = \dfrac{(x - 2)(x + 2)}{(x - 4)(x + 2)} = \dfrac{x - 2}{x - 4}$	p. 488
Identifying Rational Functions A rational function is a function that is defined by a rational expression. It can be written as $f(x) = \dfrac{P}{Q}$ in which P and Q are polynomials, $Q \neq 0$		p. 490
Simplifying Rational Functions When we simplify a rational function, it is important that we note the x values that need to be excluded, particularly when we are trying to draw the graph of a function. The set of ordered pairs of the simplified function will be exactly the same as the set of ordered pairs of the original function. If we plug the excluded value(s) for x into the simplified expression, we get a set of ordered pairs that represent "holes" in the graph. These holes are breaks in the curve. We use an open circle to designate them on a graph.		p. 490

Continued

DEFINITION/PROCEDURE	EXAMPLE	REFERENCE
Multiplication and Division of Rational Expressions and Functions		**Section 7.2**
Multiplying Rational Expressions For polynomials P, Q, R, and S, $\frac{P}{Q} \cdot \frac{R}{S} = \frac{PR}{QS}$ when $Q \neq 0, S \neq 0$ In practice, we apply the following algorithm to multiply two rational expressions. **1.** Write each numerator and denominator in completely factored form. **2.** Divide by any common factors appearing in both the numerator and denominator. **3.** Multiply as needed to form the desired product.	$\frac{2x - 6}{x^2 - 9} \cdot \frac{x^2 + 3x}{6x + 24}$ $= \frac{2(x - 3)}{(x - 3)(x + 3)} \cdot \frac{x(x + 3)}{6(x + 4)}$ $= \frac{x}{3(x + 4)}$	**pp. 499–500**
Dividing Rational Expressions For polynomials P, Q, R, and S, $\frac{P}{Q} \div \frac{R}{S} = \frac{P}{Q} \cdot \frac{S}{R} = \frac{PS}{QR}$ when $Q \neq 0, R \neq 0, S \neq 0$ To divide two rational expressions, you can apply the following algorithm. **1.** Invert the divisor (the *second* rational expression) to write the problem as one of multiplication. **2.** Proceed as in the algorithm for the multiplication of rational expressions.	$\frac{5y}{2y - 8} \div \frac{10y^2}{y^2 - y - 12}$ $= \frac{5y}{2y - 8} \cdot \frac{y^2 - y - 12}{10y^2}$ $= \frac{5y}{2(y - 4)} \cdot \frac{(y - 4)(y + 3)}{10y^2}$ $= \frac{y + 3}{4y}$	**pp. 501–502**
Addition and Subtraction of Rational Expressions and Functions		**Section 7.3**
Adding and Subtracting Rational Expressions To add or subtract rational expressions with the same denominator, add or subtract their numerators and then write that sum over the common denominator. The result should be written in lowest terms. In symbols, $\frac{P}{R} + \frac{Q}{R} = \frac{P + Q}{R}$ and $\frac{P}{R} - \frac{Q}{R} = \frac{P - Q}{R}$ when $R \neq 0$.	$\frac{5w}{w^2 - 16} - \frac{20}{w^2 - 16}$ $= \frac{5w - 20}{w^2 - 16}$ $= \frac{5(w - 4)}{(w + 4)(w - 4)}$ $= \frac{5}{w + 4}$	**p. 509**
Least Common Denominator The **least common denominator (LCD)** of a group of rational expressions is the simplest polynomial that is divisible by each of the individual denominators of the rational expressions. To find the LCD, use the following algorithm. **1.** Write each of the denominators in completely factored form. **2.** Write the LCD as the product of each prime factor, to the highest power to which it appears in the factored form of any individual denominators.	To find the LCD for $\frac{2}{x^2 + 2x + 1}$ and $\frac{3}{x^2 + x}$ write $x^2 + 2x + 1 = (x + 1)(x + 1)$ $x^2 + x = x(x + 1)$ The LCD is $x(x + 1)(x + 1)$	**p. 510**

DEFINITION/PROCEDURE	EXAMPLE	REFERENCE
Addition and Subtraction of Rational Expressions and Functions		Section 7.3
Now to add or subtract rational expressions with different denominators, we first find the LCD by the procedure outlined above. We then rewrite each of the rational expressions with that LCD as a common denominator. Then we can add or subtract as before.	$\frac{2}{(x+1)^2} - \frac{3}{x(x+1)}$ $= \frac{2 \cdot x}{(x+1)^2 x} - \frac{3(x+1)}{x(x+1)(x+1)}$ $= \frac{2x - 3(x+1)}{x(x+1)(x+1)}$ $= \frac{-x-3}{x(x+1)(x+1)}$	p. 515
Complex Fractions		Section 7.4
Complex fractions are fractions that have a fraction in their numerator or denominator (or both). There are two commonly used methods for simplifying complex fractions: methods 1 and 2. *Method 1* 1. Multiply the numerator and denominator of the complex fraction by the LCD of all the fractions that appear within the numerator and denominator. 2. Simplify the resulting rational expression, writing the result in lowest terms. *Method 2* 1. Write the numerator and denominator of the complex fraction as single fractions, if necessary. 2. Invert the denominator and multiply as before, writing the result in lowest terms.	Simplify $\frac{1 - \frac{2}{x}}{1 - \frac{4}{x^2}}$ Method 1: $\frac{\left(1 - \frac{2}{x}\right)x^2}{\left(1 - \frac{4}{x^2}\right)x^2}$ $= \frac{x^2 - 2x}{x^2 - 4} = \frac{x(x-2)}{(x+2)(x-2)}$ $= \frac{x}{x+2}$ Method 2: $\frac{\frac{x-2}{x}}{\frac{x^2-4}{x^2}}$ $= \frac{x-2}{x} \cdot \frac{x^2}{x^2-4}$ $= \frac{x-2}{x} \cdot \frac{x^2}{(x+2)(x-2)}$ $= \frac{x}{x+2}$	p. 526
Solving Rational Equations		Section 7.5
To solve an equation involving rational expressions, you should apply the following algorithm. 1. Clear the equation of fractions by multiplying both sides of the equation by the LCD of all the fractions that appear. 2. Solve the equation resulting from step 1.	Solve $\frac{3}{x-3} - \frac{2}{x+2} = \frac{19}{x^2 - x - 6}$ Multiply by the LCD $(x-3)(x+2)$: $3(x+2) - 2(x-3) = 19$ $3x + 6 - 2x + 6 = 19$ $x = 7$	

Continued

DEFINITION/PROCEDURE	EXAMPLE	REFERENCE
Solving Rational Equations		**Section 7.5**
3. Check all solutions by substitution in the original equation.	Check: $\frac{3}{4} - \frac{2}{9} \stackrel{?}{=} \frac{19}{36}$ $\frac{19}{36} = \frac{19}{36}$ True	**p. 540**
Negative Integer Exponents		**Section 7.7**
Zero Exponent For any real number a when $a \neq 0$, $a^0 = 1$ *Negative Integer Exponents* For any nonzero real number a and whole number n, $a^{-n} = \frac{1}{a^n}$ and a^{-n} is the **multiplicative inverse** of a^n. *Raising Quotients to a Negative Power* $\left(\frac{a}{b}\right)^{-n} = \frac{a^{-n}}{b^{-n}} = \frac{b^n}{a^n} = \left(\frac{b}{a}\right)^n \quad a \neq 0 \quad b \neq 0$	$5^0 = 1$ $x^{-3} = \frac{1}{x^3}$ $2^{-4} = \frac{1}{2^4} = \frac{1}{16}$ $2y^{-5} = \frac{2}{y^5}$ $(2y)^{-5} = \frac{1}{(2y)^5} = \frac{1}{32y^5}$ $\frac{1}{x^{-4}} = x^4$ $\frac{1}{3^{-3}} = 3^3 = 27$	**pp. 555–557**

Summary Exercises

This summary exercise set is provided to give you practice with each of the objectives in the chapter. Each exercise is keyed to the appropriate chapter section. The answers are provided in the *Instructor's Manual*.

[7.1] For what value of the variable will each of the following rational expressions be undefined?

1. $\frac{x}{2}$

2. $\frac{3}{y}$

3. $\frac{2}{x - 5}$

4. $\frac{3x}{2x - 5}$

Simplify each of the following rational expressions.

5. $\frac{18x^5}{24x^3}$

6. $\frac{15m^3n}{-5mn^2}$

7. $\frac{8y - 64}{y - 8}$

8. $\frac{5x - 20}{x^2 - 16}$

9. $\frac{9 - x^2}{x^2 + 2x - 15}$

10. $\frac{3w^2 + 8w - 35}{2w^2 + 13w + 15}$

11. $\frac{6a^2 - ab - b^2}{9a^2 - b^2}$

12. $\frac{6w - 3z}{8w^3 - z^3}$

Graph the following rational functions. Indicate the coordinates of the hole in the graph.

13. $f(x) = \frac{x^2 - 3x - 4}{x + 1}$

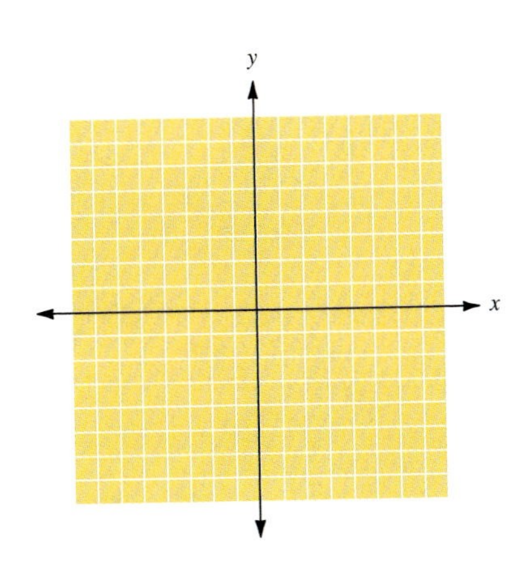

14. $f(x) = \frac{x^2 + x - 6}{x - 2}$

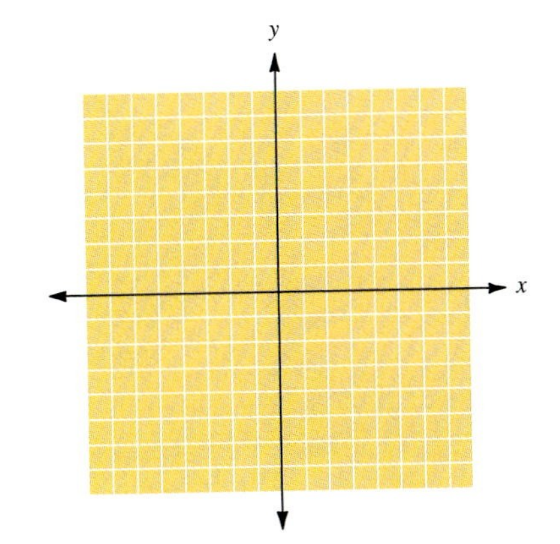

[7.2] Multiply or divide as indicated. Express your results in simplest form.

15. $\dfrac{x^5}{24} \cdot \dfrac{20}{x^3}$

16. $\dfrac{a^3b}{4ab^2} \div \dfrac{ab}{12ab^2}$

17. $\dfrac{6y - 18}{9y} \cdot \dfrac{10}{5y - 15}$

18. $\dfrac{m^2 - 3m}{m^2 - 5m + 6} \cdot \dfrac{m^2 - 4}{m^2 + 7m + 10}$

19. $\dfrac{a^2 - 2a}{a^2 - 4} \div \dfrac{2a^2}{3a + 6}$

20. $\dfrac{r^2 + 2rs}{r^3 - r^2s} \div \dfrac{5r + 10s}{r^2 - 2rs + s^2}$

21. $\dfrac{x^2 - 2xy - 3y^2}{x^2 - xy - 2y^2} \cdot \dfrac{x^2 - 4y^2}{x^2 - 8xy + 15y^2}$

22. $\dfrac{w^3 + 3w^2 + 2w + 6}{w^4 - 4} \div (w^3 + 27)$

23. Let $f(x) = \dfrac{x^2 - 16}{x - 5}$ and $g(x) = \dfrac{x^2 - 25}{x + 4}$. Find **(a)** $f(3) \cdot g(3)$, **(b)** $h(x) = f(x) \cdot g(x)$, **(c)** $h(3)$.

24. Let $f(x) = \dfrac{2x^2 - 5x - 3}{x - 4}$ and $g(x) = \dfrac{x^2 - 3x - 4}{2x^2 + 5x + 2}$. Find **(a)** $f(3) \cdot g(3)$, **(b)** $h(x) = f(x) \cdot g(x)$, **(c)** $h(3)$.

[7.3] Perform the indicated operations. Express your results in simplified form.

25. $\dfrac{5x + 7}{x + 4} - \dfrac{2x - 5}{x + 4}$

26. $\dfrac{3}{4x^2} + \dfrac{5}{6x}$

27. $\dfrac{2}{x - 5} - \dfrac{1}{x}$

28. $\dfrac{2}{y + 5} + \dfrac{3}{y + 4}$

29. $\dfrac{2}{3m - 3} - \dfrac{5}{2m - 2}$

30. $\dfrac{7}{x - 3} - \dfrac{5}{3 - x}$

31. $\dfrac{5}{4x + 4} + \dfrac{5}{2x - 2}$

32. $\dfrac{2a}{a^2 - 9a + 20} + \dfrac{8}{a - 4}$

33. $\dfrac{2}{s-1} - \dfrac{6s}{s^2+s-2}$

34. $\dfrac{4}{x^2-9} - \dfrac{3}{x^2-4x+3}$

35. $\dfrac{x^2-14x-8}{x^2-2x-8} + \dfrac{2x}{x-4} - \dfrac{3}{x+2}$

36. $\dfrac{w^2+2wz+z^2}{w^2-wz-2z^2} \cdot \left(\dfrac{3}{w+z} - \dfrac{1}{w-z}\right)$

37. Let $f(x) = \dfrac{2x}{x-2}$ and $g(x) = \dfrac{x}{x-3}$. Find **(a)** $f(4) + g(4)$, **(b)** $h(x) = f(x) + g(x)$, **(c)** the ordered pair $(4, h(4))$.

38. Let $f(x) = \dfrac{x+2}{x-2}$ and $g(x) = \dfrac{x+1}{x-7}$. Find **(a)** $f(3) - g(3)$, **(b)** $h(x) = f(x) - g(x)$, **(c)** the ordered pair $(3, h(3))$.

[7.4] Simplify each of the following complex fractions.

39. $\dfrac{\dfrac{x^2}{12}}{\dfrac{x^3}{8}}$

40. $\dfrac{\dfrac{y-1}{y^2-4}}{\dfrac{y^2-1}{y^2-y-2}}$

41. $\dfrac{1+\dfrac{a}{b}}{1-\dfrac{a}{b}}$

42. $\dfrac{2-\dfrac{x}{y}}{4-\dfrac{x^2}{y^2}}$

43. $\dfrac{\dfrac{1}{r}-\dfrac{1}{s}}{\dfrac{1}{r^2}-\dfrac{1}{s^2}}$

44. $\dfrac{1-\dfrac{1}{x+2}}{1+\dfrac{1}{x+2}}$

45. $\dfrac{1-\dfrac{2}{x-1}}{x+\dfrac{3}{x-4}}$

46. $\dfrac{\dfrac{w}{w+1}-\dfrac{1}{w-1}}{\dfrac{w}{w-1}+\dfrac{1}{w+1}}$

47. $\dfrac{1}{1 - \dfrac{1}{1 - \dfrac{1}{y - 1}}}$

48. $1 - \dfrac{1}{1 + \dfrac{1}{1 - \dfrac{1}{x}}}$

49. $\dfrac{1 - \dfrac{1}{x - 1}}{x - \dfrac{8}{x + 2}}$

50. $\dfrac{1}{1 - \dfrac{1}{1 + \dfrac{1}{y + 1}}}$

[7.5] Solve each of the following equations.

51. $\dfrac{1}{2x} + \dfrac{1}{3x} = \dfrac{1}{6}$

52. $\dfrac{5}{2x^2} - \dfrac{1}{4x} = \dfrac{1}{x}$

53. $\dfrac{x}{x - 2} + 1 = \dfrac{x + 4}{x - 2}$

54. $\dfrac{2x - 1}{x - 3} - \dfrac{5}{x - 3} = 1$

55. $\dfrac{2}{3x + 1} = \dfrac{1}{x + 2}$

56. $\dfrac{5}{x + 1} + \dfrac{1}{x - 2} = \dfrac{7}{x + 1}$

57. $\dfrac{4}{x - 1} - \dfrac{5}{3x - 7} = \dfrac{3}{x - 1}$

58. $\dfrac{7}{x} - \dfrac{1}{x - 3} = \dfrac{9}{x^2 - 3x}$

59. $\dfrac{2}{x - 3} - \dfrac{11}{x^2 - 9} = \dfrac{3}{x + 3}$

60. $\dfrac{5}{x + 3} + \dfrac{1}{x - 5} = 1$

61. $\dfrac{2}{x - 4} = \dfrac{x}{x - 2} - \dfrac{x + 4}{x^2 - 6x + 8}$

62. $\dfrac{x}{x - 5} = \dfrac{3x}{x^2 - 7x + 10} + \dfrac{8}{x - 2}$

[7.6] Solve each equation for the indicated variable.

63. $\dfrac{1}{T} + \dfrac{1}{T_1} = \dfrac{1}{T_2}$ for T_1

64. $R = \dfrac{R_1R_2}{R_1 + R_2}$ for R_1

65. $\dfrac{1}{F} = \dfrac{1}{D_1} + \dfrac{1}{D_2}$ for D_2

66. $\dfrac{1}{x} = \dfrac{1}{a} + \dfrac{1}{b}$ for x

[7.6]

67. **Motion.** On the first day of a business trip, Min Yeh drove 225 mi. On her second day, it took her 2 h longer to drive 315 mi. If her rate was the same each day, what was her driving time each day?

68. **Motion.** Brett made a trip of 240 mi using the freeway. Returning by a different route, he found that the distance was only 200 mi but that traffic slowed his speed by 8 mi/h. If the trip took the same time in both directions, what was Brett's rate each way?

69. **Work.** A painter could paint an office complex in 10 h although it would take his helper 15 h. How long would it take to complete the job if the two worked together?

70. **Work.** A water tank can be filled through an inlet pipe in 10 h. The tank will take 15 h to drain through an outlet pipe. The inlet pipe is opened to begin filling the tank, but the outlet valve is also inadvertently left open. How long will it take to fill the tank?

71. **Work.** Salvatore and Susan can construct a fence in 6 h. If Susan could complete the same job by herself in 9 h, how long would it take Salvatore, working alone?

72. **Work.** One model printer can print a company's monthly billings three times as fast as another model. If the two printers, working together, can complete the job in 9 h, how long will it take the faster model, working alone?

73. **Number analysis.** The sum of the reciprocal of an integer and the square of the reciprocal of that integer is $\frac{4}{9}$. Find the integer.

74. **Number analysis.** The difference between the reciprocal of an integer and the square of its reciprocal is $\frac{3}{16}$. What is the integer?

75. **Number analysis.** The same number is added to the numerator and denominator of $\frac{7}{10}$. If the resulting fraction is equal to $\frac{5}{6}$, what number was added to the numerator and denominator?

76. **Number analysis.** One integer is 2 more than another. If the sum of the reciprocals of the integers is $\frac{5}{12}$, find the two integers.

[7.7] Simplify each expression, using the properties of exponents.

77. $4x^{-5}$

78. $(2w)^{-3}$

79. $\frac{3}{m^{-4}}$

80. $\frac{a^{-5}}{b^{-4}}$

81. $y^{-5} \cdot y^2$

82. $\frac{w^{-7}}{w^{-3}}$

83. $(m^{-6})^{-2}$

84. $(m^3n^{-5})^{-2}$

85. $\left(\frac{a^{-4}}{b^{-2}}\right)^3$

86. $\left(\frac{r^{-5}}{s^4}\right)^{-2}$

87. $(5w^{-2})^2(2w^{-2})$

88. $(5a^2b^{-3})(2a^{-2}b^{-6})$

89. $\frac{7a^{-4}b^4}{28a^{-3}b^{-3}}$

90. $\left(\frac{m^{-3}n^{-3}}{m^{-4}n^4}\right)^3$

91. $\left(\frac{x^{-4}y^{-3}z^2}{x^{-3}y^2z^{-4}}\right)^{-2}$

Self-Test for Chapter 7

Name ______________________

Section __________ Date __________

ANSWERS

1. ______________________
2. ______________________
3. ______________________
4. ______________________
5. ______________________
6. ______________________
7. ______________________
8. ______________________
9. ______________________

The purpose of this self-test is to help you check your progress and to review for a chapter test in class. Allow yourself about 1 hour to take the test. When you are done, check your answers in the back of the book. If you missed any answers, be sure to go back and review the appropriate sections in the chapter and the exercises that are provided.

Simplify each of the following rational expressions.

1. $\dfrac{-21x^5y^3}{28xy^5}$

2. $\dfrac{3w^2 + w - 2}{3w^2 - 8w + 4}$

3. $\dfrac{x^3 + 2x^2 - 3x}{x^3 - 3x^2 + 2x}$

4. Graph the following. Indicate the coordinates of the hole in the graph.

$$f(x) = \frac{x^2 - 5x + 4}{x - 4}$$

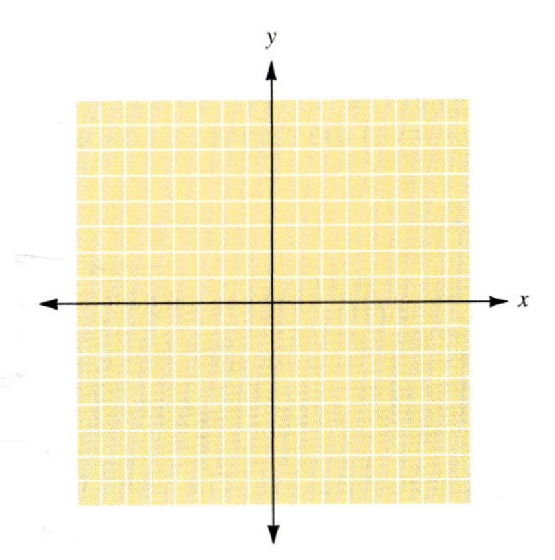

Multiply or divide as indicated.

5. $\dfrac{3ab^2}{5ab^3} \cdot \dfrac{20a^2b}{21b}$

6. $\dfrac{m^2 - 3m}{m^2 - 9} \div \dfrac{4m}{m^2 - m - 12}$

7. $\dfrac{x^2 - 3x}{5x^2} \cdot \dfrac{10x}{x^2 - 4x + 3}$

8. $\dfrac{x^2 + 3xy}{2x^3 - x^2y} \div \dfrac{x^2 + 6xy + 9y^2}{4x^2 - y^2}$

9. $\dfrac{9x^2 - 9x - 4}{6x^2 - 11x + 3} \cdot \dfrac{15 - 10x}{3x - 4}$

10. ______
11. ______
12. ______
13. ______
14. ______
15. ______
16. ______
17. ______
18. ______
19. ______
20. ______
21. ______
22. ______
23. ______
24. ______
25. ______

Add or subtract as indicated.

10. $\dfrac{5}{x-2} - \dfrac{1}{x}$

11. $\dfrac{2}{x+3} + \dfrac{12}{x^2-9}$

12. $\dfrac{6x}{x^2-x-2} - \dfrac{2}{x+1}$

13. $\dfrac{3}{x^2-3x-4} + \dfrac{5}{x^2-16}$

Simplify each of the following complex fractions.

14. $\dfrac{3 - \dfrac{x}{y}}{9 - \dfrac{x^2}{y^2}}$

15. $\dfrac{1 - \dfrac{10}{z+3}}{2 - \dfrac{12}{z-1}}$

16. $\dfrac{\dfrac{1}{x} + \dfrac{1}{y}}{x^2 - y^2}$

Solve the following equations.

17. $\dfrac{x}{x+3} + 1 = \dfrac{3x-6}{x+3}$

18. $\dfrac{2x}{x+1} = \dfrac{3}{x-2} + \dfrac{1}{x^2-x-2}$

Solve each of the following applications.

19. One positive number is 3 more than another. If the sum of the reciprocals of the two numbers is $\frac{1}{2}$, find the two numbers.

20. Stephen drove 250 mi to visit Sandra. Returning by a shorter route, he found that the trip was only 225 mi, but traffic slowed his speed by 5 mi/h. If the two trips took exactly the same time, what was his rate each way?

21. Juan can paint the neighbor's house four times as fast as Ariel can. The year they worked together it took them 8 days. How long would it take each to paint the house alone?

22. Given the fraction $\frac{4}{7}$, the numerator is multiplied by a certain number, and the same number is added to the denominator. The result is $\frac{6}{5}$. What is the number?

Simplify each expression.

23. $(x^4y^{-5})^2$

24. $\dfrac{9c^{-5}d^3}{18c^{-7}d^{-4}}$

25. $\dfrac{(x^{-2}y)^{-3}}{(x^{-1}y^{-4})^{-2}}$

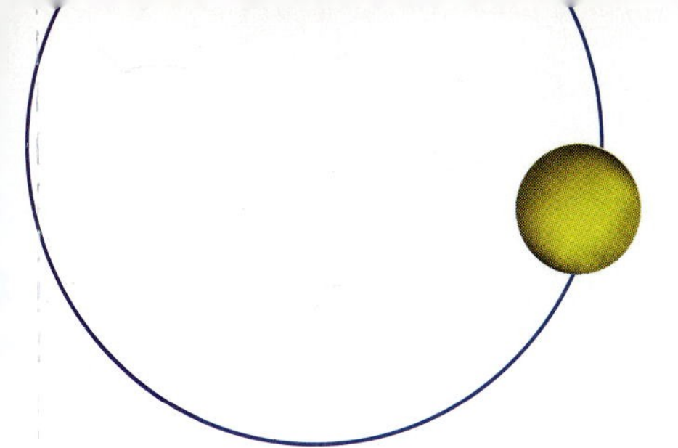

Cumulative Test for Chapters 1 to 7

Name ____________________

Section ________ Date ________

ANSWERS

1. ____________________
2. ____________________
3. ____________________
4. ____________________
5. ____________________
6. ____________________
7. ____________________
8. ____________________
9. ____________________
10. ____________________
11. ____________________
12. ____________________
13. ____________________

This test is provided to help you in the process of reviewing the previous chapters. Answers are provided in the back of the book. If you missed any answers, be sure to go back and review the appropriate chapter section.

1. Solve the equation $5x - 3(2x + 6) = 4 - (3x - 2)$.

2. If $f(x) = 5x^4 - 3x^2 + 7x - 9$, find $f(-1)$.

3. Find the equation of the line that is parallel to the line $6x + 7y = 42$ and has a y intercept of $(0, -3)$.

4. Find the x and y intercepts of the equation: $7x - 6y = -42$.

Simplify each of the following polynomial functions.

5. $f(x) = 3x - 2(x - (3x - 1)) + 6x(x - 2)$

6. $f(x) = x(2x - 1)(x + 3)$

7. Find the domain and range of the relation $7x - 14 = 0$.

8. Evaluate the expression: $6^2 - (16 \div 8 \cdot 2) - 4^2$

Factor each of the following completely.

9. $6x^3 + 7x^2 - 3x$

10. $16x^{16} - 9y^8$

Simplify each of the following rational expressions.

11. $\dfrac{5}{x - 1} - \dfrac{2x + 6}{x^2 + 2x - 3}$

12. $\dfrac{x + 1}{x^2 - 5x - 6} \div \dfrac{x^2 - 1}{x - 6}$

13. $\dfrac{1 - \dfrac{3}{x + 3}}{\dfrac{1}{x^2 - 9}}$

ANSWERS

14. (a)

(b)

(c)

15.

16.

17.

18.

19.

20.

21.

22.

23.

24.

25.

14. If $f(x) = \frac{1}{x - 5}$ and $g(x) = 8x + 6$, find **(a)** $f(x) + g(x)$, **(b)** $\frac{f(x)}{g(x)}$, **(c)** domain of $\frac{f(x)}{g(x)}$.

Solve the following equations.

15. $7x + (x - 10) = -12(x - 5)$

16. $|-9x - 6| = 2$

17. $-4(7x + 6) = 8(5x + 12)$

Solve the following inequalities.

18. $-4(-2x - 7) > -6x$

19. $|5x - 4| < 3$

20. $-6|2x + 6| \leq -12$

21. Solve the equation $\frac{5}{x} = \frac{2}{x + 3}$ for x.

22. Solve the system of equations

$$\begin{aligned} 4x - 3y &= 15 \\ x + y &= 2 \end{aligned}$$

23. Simplify the expression

$$\left(\frac{a^{-2}b}{a^3b^{-2}}\right)^2$$

Solve the following applications.

24. When each works alone, Barry can mow a lawn in 3 h less time than Don. When they work together, it takes 2 h. How long does it take each to do the job by himself?

25. The length of a rectangle is 2 cm less than twice the width. The area of the rectangle is 180 cm^2. Find the length and width of the rectangle.

8 RADICAL EXPRESSIONS

INTRODUCTION

As we have all experienced firsthand, many consumer goods increase in price from year to year. This increase is usually measured by the Consumer Price Index (CPI), which measures a change in the prices of such everyday goods and services as energy, food, shelter, apparel, transportation, medical care, and utilities. The percent change in the CPI is a reflection of the purchasing power of the dollar and indicates the rate of inflation.

Because many labor contracts and government programs such as Social Security increase or decrease along with the CPI, the method used to calculate this index is hotly debated by economists and statisticians. Beginning in April 1997, the Bureau of Labor Statistics began releasing an experimental CPI that uses a **geometric mean formula.** This new method may more accurately reflect the true cost-of-living increase or decrease because it takes into consideration that consumers' buying habits change as prices fluctuate. For instance, consumers may switch from romaine lettuce to iceberg lettuce or spinach if the price of romaine lettuce is too high.

Item	Cost Last Month	Cost This Month	Relative Price Change: $\frac{\textit{Cost this month}}{\textit{Cost last month}}$
1 pound of iceberg lettuce	\$0.50	\$0.62	$\frac{0.62}{0.50}$
1 pound of romaine lettuce	\$0.75	\$0.80	$\frac{0.80}{0.75}$
1 pound of spinach	\$0.70	\$0.78	$\frac{0.78}{0.70}$

To compute the CPI, prices of individual items are averaged together to produce relative price changes or indices for 9108 item-area categories. The number 9108 is found by multiplying 207 items by 44 geographic areas from around the United States. For example, the cost to consumers of lettuce and spinach in one city is reflected in the table.

The new formula computes the relative price change, R, for salad greens the following way:

$$R = \left(\frac{0.62}{0.50} \cdot \frac{0.80}{0.75} \cdot \frac{0.78}{0.70}\right)^{1/3} = \sqrt[3]{\frac{0.62}{0.50}} \cdot \sqrt[3]{\frac{0.80}{0.75}} \cdot \sqrt[3]{\frac{0.78}{0.70}}$$

$$\approx 1.14 \text{ or a } 14\% \text{ increase}$$

From February 1990 through February 2000, the Bureau of Labor Statistics computed an inflation rate of 28.1%, which is equivalent to an annual growth rate of 2.51%. The old method, which is also still being used for computing the index, computed an annual growth rate for this same period of 2.80%. In this chapter, we will work with the exponents and radicals used to compute these rates.

8.1 Roots and Radicals

8.1 OBJECTIVES

1. Use radical notation
2. Evaluate expressions that contain radicals
3. Use a calculator to estimate or evaluate radical expressions
4. Simplify expressions that contain radicals

In Chapters 1 and 7 we discussed the properties of integer exponents. Over the next six sections, we will be working toward an extension of those properties. To achieve that objective, we must develop a notation that "reverses" the power process.

A statement such as

$$x^2 = 9$$

is read as "x squared equals 9."

In this section we are concerned with the relationship between the base x and the number 9. Equivalently, we can say that "x is the square root of 9."

NOTE We will see later that a negative number has *no* real square roots.

We know from experience that x must be 3 (because $3^2 = 9$) or -3 [because $(-3)^2 = 9$]. We see that 9 has the two square roots, 3 and -3. In fact, every positive number has *two* square roots, one positive and one negative. In general,

If $x^2 = a$, we say x is a *square root* of a.

We also know that

$$3^3 = 27$$

and similarly we call 3 a *cube root* of 27. Here 3 is the *only* real number with that property. Every real number (positive or negative) has *one* real cube root.

Definitions: Roots

In general, we can state that if

$$x^n = a$$

then x is an *nth root* of a.

NOTE The symbol $\sqrt{}$ first appeared in print in 1525. In Latin, "radix" means root, and this was contracted to a small *r*. The present symbol may have been used because it resembled the manuscript form of that small *r*.

We are now ready for new notation. The symbol $\sqrt{}$ is called a *radical sign*. We saw above that 3 was the positive square root of 9.

We call 3 the *principal square root* of 9, and we write

$$\sqrt{9} = 3$$

In some applications we will want to indicate the negative square root; to do so we must write

$$-\sqrt{9} = -3$$

to indicate the negative root.

NOTE You will see this used later in our work with quadratic equations in Chapter 9.

If both square roots need to be indicated, we can write

$$\pm\sqrt{9} = \pm 3$$

Every radical expression contains three parts, as shown below. The principal nth root of a is written as

NOTE The index of 2 for square roots is generally not written. We understand that

$\sqrt{a}$

is the principal square root of a.

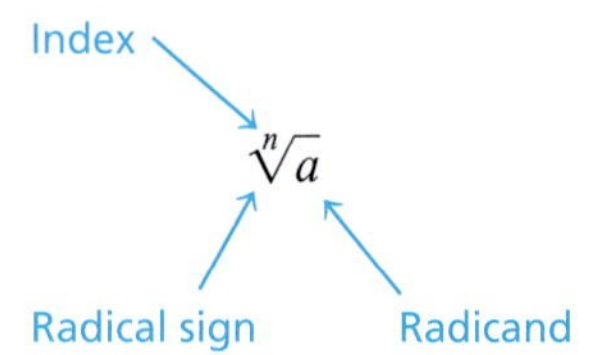

Example 1

Evaluating Radical Expressions

Evaluate, if possible.

(a) $\sqrt{49} = 7$

(b) $-\sqrt{49} = -7$

(c) $\pm\sqrt{49} = \pm 7$

(d) $\sqrt{-49}$ is not a real number.

NOTE Notice that neither 7^2 nor $(-7)^2$ equals -49.

Let's examine part (d) more carefully. Suppose that for some real number x,

$x = \sqrt{-49}$

By our earlier definition, this means that

$x^2 = -49$

which is impossible. There is no real square root for -49. We call $\sqrt{-49}$ an *imaginary number*.

NOTE We consider imaginary numbers in detail in Section 8.7.

CHECK YOURSELF 1

Evaluate, if possible.

(a) $\sqrt{64}$ **(b)** $-\sqrt{64}$ **(c)** $\pm\sqrt{64}$ **(d)** $\sqrt{-64}$

Our next example considers cube roots.

Example 2

Evaluating Radical Expressions

Evaluate, if possible.

(a) $\sqrt[3]{64} = 4$ Because $4^3 = 64$

(b) $-\sqrt[3]{64} = -4$

(c) $\sqrt[3]{-64} = -4$ Because $(-4)^3 = -64$

NOTE Notice that the cube root of a *negative* number is *negative.*

CHECK YOURSELF 2

Evaluate.

(a) $\sqrt[3]{125}$ **(b)** $-\sqrt[3]{125}$ **(c)** $\sqrt[3]{-125}$

NOTE The word "indices" is the plural of "index."

Let's consider radicals with other indices in the next example.

Example 3

Evaluating Radical Expressions

Evaluate, if possible.

(a) $\sqrt[4]{81} = 3$ Because $3^4 = 81$

NOTE In general, an *even* root of a *negative* number is *not real;* it is *imaginary.*

(b) $\sqrt[4]{-81}$ is not a real number.

(c) $\sqrt[5]{32} = 2$ Because $2^5 = 32$

(d) $\sqrt[5]{-32} = -2$ Because $(-2)^5 = -32$

CHECK YOURSELF 3

Evaluate, if possible.

(a) $\sqrt[4]{16}$ **(b)** $\sqrt[5]{243}$ **(c)** $\sqrt[4]{-16}$ **(d)** $\sqrt[5]{-243}$

Note: All the numbers of our previous examples and exercises were chosen so that the results would be *rational numbers.* That is, our radicands were

Perfect squares: 1, 4, 9, 16, 25, . . .

Perfect cubes: 1, 8, 27, 64, 125, . . .

and so on.

The square root of a number that is *not* a perfect square (or the cube root of a number that is *not* a perfect cube) is not a rational number.

Expressions such as $\sqrt{2}$, $\sqrt{3}$, and $\sqrt{5}$ are *irrational numbers.* A calculator with a square root key [√] will give decimal approximations for such numbers.

Example 4

Estimating Radical Expressions

NOTE On some calculators, the square root is shown as the "2nd function" or "inverse" of x^2. If that is the case, press the [2nd function] key and then the [x^2] key. On graphing calculators, you press the [√] key, then enclose the radicand in parentheses.

Using a calculator, find the decimal approximation for each of the following. Round all answers to three decimal places.

(a) $\sqrt{17}$

Enter 17 in your calculator and press the [√] key. The display will read 4.123105626 (if your calculator displays 10 digits). If this is rounded to three decimal places, the result is 4.123.

(b) $\sqrt{28}$

The display should read 5.291502622. Rounded to three decimal places, the result is 5.292.

(c) $\sqrt{-11}$

Enter -11 by first entering 11 and then pressing the [+/−] key. Take the square root by pressing the [√] key. The display will read ERROR. This indicates that -11 does not have a real square root.

CHECK YOURSELF 4

Use a calculator to find the decimal approximation for each of the following. Round each answer to three decimal places.

(a) $\sqrt{13}$ **(b)** $\sqrt{38}$ **(c)** $\sqrt{-21}$

NOTE Not all scientific calculators have this key.

To evaluate roots other than square roots by using scientific calculators, the key marked [y^x] can be used together with the [INV] key. (On some calculators, the [INV] key is [2nd F].)

Example 5

Estimating Radical Expressions

Using a calculator, find a decimal approximation for each of the following. Round each answer to three decimal places.

(a) $\sqrt[4]{12}$

NOTE Again, depending upon your calculator, you may have only an eight-digit display.

Enter 12 and press [INV] [y^x]. Then enter 4 and press [=]. The display will read 1.861209718. Rounded to three decimal places, the result is 1.861.

(b) $\sqrt[5]{27}$

Enter 27 and press [INV] [y^x]. Then enter 5 and press [=]. The display will read 1.933182045. Rounded to three decimal places, the result is 1.933.

CHECK YOURSELF 5

Use a calculator to find the decimal approximation for each of the following. Round each answer to three decimal places.

(a) $\sqrt[4]{35}$ **(b)** $\sqrt[5]{29}$

A certain amount of caution should be exercised in dealing with principal even roots. For example, consider the statement

$$\sqrt{x^2} = x \qquad (1)$$

First, let $x = 2$ in Equation (1).

NOTE Because $x = 2$, $\sqrt{x^2} = x$

$$\sqrt{2^2} = \sqrt{4} = 2 \qquad (2)$$

NOTE Because here $x = -2$, $\sqrt{x^2} \neq x$

Now let $x = -2$.

$$\sqrt{(-2)^2} = \sqrt{4} = 2 \qquad (3)$$

We see that statement (1) is not true when x is negative, but we can write

$$\sqrt{x^2} = \begin{cases} x & \text{when } x \geq 0 \\ -x & \text{when } x < 0 \end{cases}$$

From your earlier work with absolute values you will remember that

$$|x| = \begin{cases} x & \text{when } x \geq 0 \\ -x & \text{when } x < 0 \end{cases}$$

and we can summarize the discussion by writing

NOTE Statement (4) can be extended to $\sqrt[n]{x^n} = |x|$ when n is *even*.

$$\sqrt{x^2} = |x| \qquad (4)$$

Example 6

Evaluating Radical Expressions

Evaluate.

(a) $\sqrt{5^2} = 5$

(b) $\sqrt{(-4)^2} = |-4| = 4$

NOTE Alternately we could write $\sqrt{(-4)^2} = \sqrt{16} = 4$

(c) $\sqrt[4]{2^4} = 2$

(d) $\sqrt[4]{(-3)^4} = |-3| = 3$

CHECK YOURSELF 6

Evaluate.

(a) $\sqrt{6^2}$ **(b)** $\sqrt{(-6)^2}$ **(c)** $\sqrt[4]{3^4}$ **(d)** $\sqrt[4]{(-3)^4}$

Note: The case for roots with indices that are odd does *not* require the use of absolute value, as illustrated in Example 6. For instance,

$$\sqrt[3]{3^3} = \sqrt[3]{27} = 3$$

$$\sqrt[3]{(-3)^3} = \sqrt[3]{-27} = -3$$

and we see that

$$\sqrt[n]{x^n} = x \qquad \text{when } n \text{ is odd.}$$

To summarize, we can write

$$\sqrt[n]{x^n} = \begin{cases} |x| & \text{when } n \text{ is even} \\ x & \text{when } n \text{ is odd} \end{cases}$$

Let's turn now to a final example in which variables are involved in the radicand.

Example 7

Simplifying Radical Expressions

Simplify the following.

(a) $\sqrt[3]{a^3} = a$

(b) $\sqrt{16m^2} = 4|m|$

(c) $\sqrt[5]{32x^5} = 2x$

(d) $\sqrt[4]{x^8} = x^2$ Because $(x^2)^4 = x^8$.

(e) $\sqrt[3]{27y^6} = 3y^2$ Do you see why?

NOTE Notice that we can determine the power of the variable in our root by dividing the power in the radicand by the index. In Example 7*d*, $8 \div 4 = 2$.

CHECK YOURSELF 7

Simplify.

(a) $\sqrt[4]{x^4}$ **(b)** $\sqrt{49w^2}$ **(c)** $\sqrt[5]{a^{10}}$ **(d)** $\sqrt[3]{8y^9}$

CHECK YOURSELF ANSWERS

1. **(a)** 8; **(b)** -8; **(c)** ± 8; **(d)** not a real number **2.** **(a)** 5; **(b)** -5; **(c)** -5
3. **(a)** 2; **(b)** 3; **(c)** not a real number; **(d)** -3
4. **(a)** 3.606; **(b)** 6.164; **(c)** not a real number
5. **(a)** 2.432; **(b)** 1.961 **6.** **(a)** 6; **(b)** 6; **(c)** 3; **(d)** 3
7. **(a)** $|x|$; **(b)** $7|w|$; **(c)** a^2; **(d)** $2y^3$

Name ______________

Section ________ Date ________

8.1 Exercises

Evaluate each of the following roots where possible.

1. $\sqrt{49}$
2. $\sqrt{36}$
3. $-\sqrt{36}$
4. $-\sqrt{81}$
5. $\pm\sqrt{81}$
6. $\pm\sqrt{49}$
7. $\sqrt{-49}$
8. $\sqrt{-25}$
9. $\sqrt[3]{27}$
10. $\sqrt[3]{64}$
11. $\sqrt[3]{-64}$
12. $-\sqrt[3]{125}$
13. $-\sqrt[3]{216}$
14. $\sqrt[3]{-27}$
15. $\sqrt[4]{81}$
16. $\sqrt[5]{32}$
17. $\sqrt[5]{-32}$
18. $\sqrt[4]{-81}$
19. $-\sqrt[4]{16}$
20. $\sqrt[5]{-243}$
21. $\sqrt[4]{-16}$
22. $-\sqrt[5]{32}$
23. $-\sqrt[5]{243}$
24. $-\sqrt[4]{625}$
25. $\sqrt{\frac{4}{9}}$
26. $\sqrt{\frac{9}{25}}$
27. $\sqrt[3]{\frac{8}{27}}$
28. $\sqrt[3]{-\frac{27}{64}}$

ANSWERS

1. ______ 2. ______
3. ______ 4. ______
5. ______ 6. ______
7. ______
8. ______
9. ______
10. ______
11. ______
12. ______
13. ______
14. ______
15. ______
16. ______
17. ______
18. ______
19. ______
20. ______
21. ______
22. ______
23. ______
24. ______
25. ______
26. ______
27. ______
28. ______

ANSWERS

29. ______ 30. ______
31. ______ 32. ______
33. ______ 34. ______
35. ______ 36. ______
37. ______
38. ______
39. ______
40. ______
41. ______
42. ______
43. ______
44. ______
45. ______
46. ______
47. ______
48. ______
49. ______
50. ______
51. ______
52. ______
53. ______
54. ______
55. ______
56. ______
57. ______
58. ______
59. ______
60. ______

29. $\sqrt{6^2}$

30. $\sqrt{9^2}$

31. $\sqrt{(-3)^2}$

32. $\sqrt{(-5)^2}$

33. $\sqrt[3]{4^3}$

34. $\sqrt[3]{(-5)^3}$

35. $\sqrt[4]{3^4}$

36. $\sqrt[4]{(-2)^4}$

Simplify each of the following roots.

37. $\sqrt{x^2}$

38. $\sqrt[3]{w^3}$

39. $\sqrt[5]{y^5}$

40. $\sqrt[7]{z^7}$

41. $\sqrt{9x^2}$

42. $\sqrt{81y^2}$

43. $\sqrt{a^4b^6}$

44. $\sqrt{w^6z^{10}}$

45. $\sqrt{16x^4}$

46. $\sqrt{49y^6}$

47. $\sqrt[4]{y^{20}}$

48. $\sqrt[3]{a^{18}}$

49. $\sqrt[4]{m^8n^{12}}$

50. $\sqrt[3]{a^6b^9}$

51. $\sqrt[3]{125a^3}$

52. $\sqrt[3]{-27x^3}$

53. $\sqrt[5]{32x^5y^{15}}$

54. $\sqrt[5]{-32m^{10}n^5}$

Using a calculator, evaluate the following. Round each answer to three decimal places.

55. $\sqrt{15}$

56. $\sqrt{29}$

57. $\sqrt{156}$

58. $\sqrt{213}$

59. $\sqrt{-15}$

60. $\sqrt{-79}$

61. $\sqrt[3]{83}$

62. $\sqrt[3]{97}$

63. $\sqrt[5]{123}$

64. $\sqrt[5]{283}$

65. $\sqrt[3]{-15}$

66. $\sqrt[5]{-29}$

Label each of the following statements as true or false.

67. $\sqrt{16x^{16}} = 4x^4$

68. $\sqrt{36c^2} = 6c$

69. $\sqrt[3]{(4x^6y^9)^3} = 4x^6y^9$

70. $\sqrt[4]{(x - 4)^4} = x - 4$

71. $\sqrt{x^4 + 16} = x^2 + 4$

72. $\sqrt[3]{x^8 + 27} = x^2 + 3$

73. $\sqrt{16x^{-4}y^{-4}}$ is not a real number

74. $\sqrt[3]{-8x^6y^6}$ is not a real number.

75. Is there any prime number whose square root is an integer? Explain your answer.

76. Determine two consecutive integers whose square roots are also consecutive integers.

77. Try the following using your calculator.

(a) Choose a number greater than 1 and find its square root. Then find the square root of the result and continue in this manner, observing the successive square roots. Do these numbers seem to be approaching a certain value? If so, what?

(b) Choose a number greater than 0 but less than 1 and find its square root. Then find the square root of the result, and continue in this manner, observing successive square roots. Do these numbers seem to be approaching a certain value? If so, what?

78. (a) Can a number be equal to its own square root?

(b) Other than the number(s) found in part a, is a number always greater than its square root? Investigate.

ANSWERS

61. ______
62. ______
63. ______
64. ______
65. ______
66. ______
67. ______
68. ______
69. ______
70. ______
71. ______
72. ______
73. ______
74. ______
75. ______
76. ______

77. ______

78. ______

ANSWERS

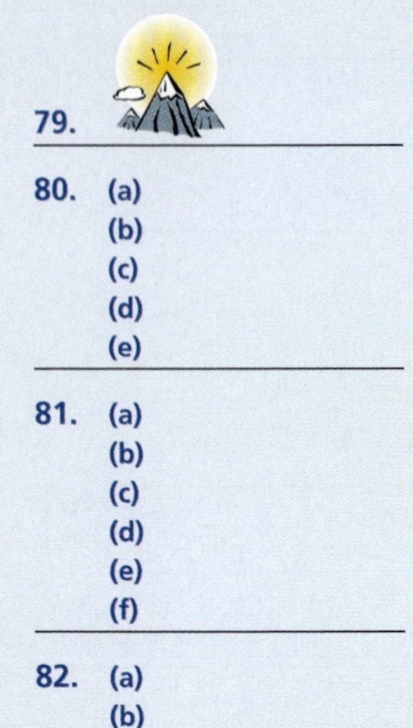

79.

80. (a) (b) (c) (d) (e)

81. (a) (b) (c) (d) (e) (f)

82. (a) (b) (c) (d)

79. Let a and b be positive numbers. If a is greater than b, is it always true that the square root of a is greater than the square root of b? Investigate.

80. Suppose that a weight is attached to a string of length L, and the other end of the string is held fixed. If we pull the weight and then release it, allowing the weight to swing back and forth, we can observe the behavior of a simple pendulum. The period, T, is the time required for the weight to complete a full cycle, swinging forward and then back. The following formula may be used to describe the relationship between T and L.

$$T = 2\pi\sqrt{\frac{L}{g}}$$

If L is expressed in centimeters, then $g = 980 \text{ cm/s}^2$. For each of the following string lengths, calculate the corresponding period. Round to the nearest tenth of a second.

(a) 30 cm **(b)** 50 cm **(c)** 70 cm **(d)** 90 cm **(e)** 110 cm

81. In parts (a) through (f), evaluate when possible.

(a) $\sqrt{4 \cdot 9}$ **(b)** $\sqrt{4} \cdot \sqrt{9}$ **(c)** $\sqrt{9 \cdot 16}$

(d) $\sqrt{9} \cdot \sqrt{16}$ **(e)** $\sqrt{(-4)(-25)}$ **(f)** $\sqrt{-4} \cdot \sqrt{-25}$

(g) Based on parts (a) through (f), make a general conjecture concerning $\sqrt{ab}$. Be careful to specify any restrictions on possible values for a and b.

82. In parts (a) through (d), evaluate when possible.

(a) $\sqrt{9 + 16}$ **(b)** $\sqrt{9} + \sqrt{16}$ **(c)** $\sqrt{36 + 64}$ **(d)** $\sqrt{36} + \sqrt{64}$

(e) Based on parts (a) through (d), what can you say about $\sqrt{a + b}$ and $\sqrt{a} + \sqrt{b}$?

Answers

1. 7 **3.** -6 **5.** ± 9 **7.** Not a real number **9.** 3 **11.** -4 **13.** -6 **15.** 3 **17.** -2 **19.** -2 **21.** Not a real number **23.** -3 **25.** $\frac{2}{3}$ **27.** $\frac{2}{3}$ **29.** 6 **31.** 3 **33.** 4 **35.** 3 **37.** $|x|$ **39.** y **41.** $3|x|$ **43.** $|a^2b^3|$ **45.** $4x^2$ **47.** $|y^5|$ **49.** $|m^2n^3|$ **51.** $5a$ **53.** $2xy^3$ **55.** 3.873 **57.** 12.490 **59.** Not a real number **61.** 4.362 **63.** 2.618 **65.** -2.466 **67.** False **69.** True **71.** False **73.** False **75.** No **77.** **79.**

81. **(a)** 6; **(b)** 6; **(c)** 12; **(d)** 12; **(e)** 10; **(f)** Not possible

8.2 Simplification of Radical Expressions

8.2 OBJECTIVES

1. Simplify a radical expression by using the product property
2. Simplify a radical expression by using the quotient property

NOTE A precise set of conditions for a radical to be in simplified form will follow in this section.

In the last section, we introduced the radical notation. For some applications, we will want to make sure that all radical expressions are written in *simplified form.* To accomplish this objective, we will need two basic properties. In stating these properties, and in our subsequent examples, we will assume that all variables represent positive real numbers whenever the index of a radical is even. To develop our first property, consider an expression such as

$$\sqrt{25 \cdot 4}$$

One approach to simplify the expression would be

$$\sqrt{25 \cdot 4} = \sqrt{100} = 10$$

Now what happens if we separate the original radical as follows?

$$\sqrt{25 \cdot 4} = \sqrt{25} \cdot \sqrt{4}$$

$$= 5 \cdot 2 = 10$$

The result in either case is the same, and this suggests our first property for radicals.

NOTE As we stated in the first paragraph, a and b are assumed to be positive real numbers when n is an even integer.

Rules and Properties: Product Property for Radicals

$$\sqrt[n]{ab} = \sqrt[n]{a} \cdot \sqrt[n]{b}$$

In words, the radical of a product is equal to the product of the radicals.

The second property we will need is similar.

NOTE To convince yourself that this must be the case, at least for square roots, let $a = 100$ and $b = 4$ and evaluate both sides of the equation.

Rules and Properties: Quotient Property for Radicals

$$\sqrt[n]{\frac{a}{b}} = \frac{\sqrt[n]{a}}{\sqrt[n]{b}}$$

In words, the radical of a quotient is the quotient of the radicals.

CAUTION

Be Careful! Students sometimes assume that because

$$\sqrt{ab} = \sqrt{a} \cdot \sqrt{b}$$

it should also be true that

$$\sqrt{a + b} = \sqrt{a} + \sqrt{b}$$ This is *not* true.

NOTE You can easily see that this is *not* true. Let $a = 9$ and $b = 16$ in the statement.

With these two properties, we are now ready to define the simplified form for a radical expression. A radical is in simplified form if the following three conditions are satisfied.

Definitions: Simplified Form for a Radical Expression

1. The radicand has no factor raised to a power greater than or equal to the index.
2. No fraction appears in the radical.
3. No radical appears in a denominator.

Our initial example deals with satisfying the first of the above conditions. Essentially, we want to find the largest perfect-square factor (in the case of a square root) in the radicand and then apply the product property to simplify the expression.

Example 1

Simplifying Radical Expressions

Write each expression in simplified form.

NOTE The largest perfect-square factor of 18 is 9.

(a) $\sqrt{18} = \sqrt{9 \cdot 2}$

$= \sqrt{9} \cdot \sqrt{2}$ Apply the product property.

$= 3\sqrt{2}$

NOTE The largest perfect-square factor of 75 is 25.

(b) $\sqrt{75} = \sqrt{25 \cdot 3}$

$= \sqrt{25} \cdot \sqrt{3}$

$= 5\sqrt{3}$

NOTE The largest perfect-square factor of $27x^3$ is $9x^2$. Note that the exponent must be *even* in a perfect square.

(c) $\sqrt{27x^3} = \sqrt{9x^2 \cdot 3x}$

$= \sqrt{9x^2} \cdot \sqrt{3x} = 3x\sqrt{3x}$

(d) $\sqrt{72a^3b^4} = \sqrt{36a^2b^4 \cdot 2a}$

$= \sqrt{36a^2b^4} \cdot \sqrt{2a}$

$= 6ab^2\sqrt{2a}$

CHECK YOURSELF 1

Write each expression in simplified form.

(a) $\sqrt{45}$ **(b)** $\sqrt{200}$ **(c)** $\sqrt{75p^5}$ **(d)** $\sqrt{98m^3n^4}$

Writing a cube root in simplest form involves finding factors of the radicand that are perfect cubes, as illustrated in Example 2. The process illustrated in this example is extended in an identical fashion to simplify radical expressions with any index.

Example 2

Simplifying Radical Expressions

Write each expression in simplified form.

(a) $\sqrt[3]{48} = \sqrt[3]{8 \cdot 6}$

$= \sqrt[3]{8} \cdot \sqrt[3]{6} = 2\sqrt[3]{6}$

NOTE In a perfect cube, the exponent must be a *multiple of* 3.

(b) $\sqrt[3]{24x^4} = \sqrt[3]{8x^3 \cdot 3x}$

$= \sqrt[3]{8x^3} \cdot \sqrt[3]{3x} = 2x\sqrt[3]{3x}$

(c) $\sqrt[3]{54a^7b^4} = \sqrt[3]{27a^6b^3 \cdot 2ab}$

$= \sqrt[3]{27a^6b^3} \cdot \sqrt[3]{2ab} = 3a^2b\sqrt[3]{2ab}$

CHECK YOURSELF 2

Write each expression in simplified form.

(a) $\sqrt[3]{128w^4}$ (b) $\sqrt[3]{40x^5y^7}$ (c) $\sqrt[4]{48a^8b^5}$

Satisfying our second condition for a radical to be in simplified form (no fractions should appear inside the radical) requires the second property for radicals. Consider the following example.

Example 3

Simplifying Radical Expressions

Write each expression in simplified form.

NOTE Apply the quotient property.

(a) $\sqrt{\dfrac{5}{9}} = \dfrac{\sqrt{5}}{\sqrt{9}}$

$= \dfrac{\sqrt{5}}{3}$

(b) $\sqrt{\dfrac{a^4}{25}} = \dfrac{\sqrt{a^4}}{\sqrt{25}} = \dfrac{a^2}{5}$

(c) $\sqrt[3]{\dfrac{5x^2}{8}} = \dfrac{\sqrt[3]{5x^2}}{\sqrt[3]{8}} = \dfrac{\sqrt[3]{5x^2}}{2}$

CHECK YOURSELF 3

Write each expression in simplified form.

(a) $\sqrt{\dfrac{7}{16}}$ (b) $\sqrt{\dfrac{3}{25a^2}}$ (c) $\sqrt[3]{\dfrac{5x}{27}}$

Our next example also begins with the application of the quotient property for radicals. However, an additional step is required because, as we will see, the third condition (no radicals can appear in a denominator) must also be satisfied during the process.

Example 4

Rationalizing the Denominator

Write $\sqrt{\frac{3}{5}}$ in simplified form.

$$\sqrt{\frac{3}{5}} = \frac{\sqrt{3}}{\sqrt{5}}$$

The application of the quotient property satisfies the second condition—there are now no fractions *inside* a radical. However, we now have a radical in the denominator, violating the third condition. The expression will not be simplified until that radical is removed.

To remove the radical in the denominator, we multiply the numerator and denominator by the *same* expression, here $\sqrt{5}$. This is called *rationalizing the denominator.*

NOTE The value of the expression is *not* changed as we multiply by $\frac{\sqrt{5}}{\sqrt{5}}$, or 1.

$$\frac{\sqrt{3}}{\sqrt{5}} = \frac{\sqrt{3} \cdot \sqrt{5}}{\sqrt{5} \cdot \sqrt{5}}$$

$$= \frac{\sqrt{15}}{\sqrt{25}} = \frac{\sqrt{15}}{5}$$

NOTE The point here is to arrive at a perfect square inside the radical in the denominator. This is done by multiplying the numerator and denominator by $\sqrt{5}$ because

$\sqrt{5} \cdot \sqrt{5} = \sqrt{5^2} = \sqrt{25}$

CHECK YOURSELF 4

Simplify $\sqrt{\frac{3}{7}}$

Let's look at some further examples that involve rationalizing the denominator of an expression.

Example 5

Rationalizing the Denominator

Write each expression in simplified form.

NOTE We multiply numerator and denominator by $\sqrt{2}$. Why did we choose $\sqrt{2}$? Note that

$\sqrt{8} = \sqrt{2^3}$

so

$$\sqrt{8}\sqrt{2} = \sqrt{2^3}\sqrt{2} = \sqrt{2^4} = \sqrt{4^2} = 4$$

(a) $\frac{3}{\sqrt{8}} = \frac{3 \cdot \sqrt{2}}{\sqrt{8} \cdot \sqrt{2}}$

$$= \frac{3\sqrt{2}}{\sqrt{16}} = \frac{3\sqrt{2}}{4}$$

(b) $\sqrt[3]{\dfrac{5}{4}} = \dfrac{\sqrt[3]{5}}{\sqrt[3]{4}}$

Now note that

$$\sqrt[3]{4} \cdot \sqrt[3]{2} = \sqrt[3]{8} = 2$$

so multiplying the numerator and denominator by $\sqrt[3]{2}$ will produce a perfect cube inside the radical in the denominator. Continuing, we have

$$\frac{\sqrt[3]{5}}{\sqrt[3]{4}} = \frac{\sqrt[3]{5} \cdot \sqrt[3]{2}}{\sqrt[3]{4} \cdot \sqrt[3]{2}}$$

$$= \frac{\sqrt[3]{10}}{\sqrt[3]{8}} = \frac{\sqrt[3]{10}}{2}$$

NOTE Why did we use $\sqrt[3]{2}$? Note that

$$\sqrt[3]{4} \cdot \sqrt[3]{2} = \sqrt[3]{2^2} \cdot \sqrt[3]{2}$$
$$= \sqrt[3]{2^3}$$

and the exponent is a multiple of 3.

CHECK YOURSELF 5

Simplify each expression.

(a) $\dfrac{5}{\sqrt{12}}$ **(b)** $\sqrt[3]{\dfrac{2}{9}}$

As our final example, we illustrate the process of rationalizing a denominator when variables are involved in a rational expression.

Example 6

Rationalizing Variable Denominators

Simplify each expression.

(a) $\sqrt{\dfrac{8x^3}{3y}}$

By the quotient property we have

$$\sqrt{\frac{8x^3}{3y}} = \frac{\sqrt{8x^3}}{\sqrt{3y}}$$

Because the numerator can be simplified in this case, let's start with that procedure.

$$\frac{\sqrt{8x^3}}{\sqrt{3y}} = \frac{\sqrt{4x^2} \cdot \sqrt{2x}}{\sqrt{3y}} = \frac{2x\sqrt{2x}}{\sqrt{3y}}$$

Multiplying the numerator and denominator by $\sqrt{3y}$ will rationalize the denominator.

$$\frac{2x\sqrt{2x} \cdot \sqrt{3y}}{\sqrt{3y} \cdot \sqrt{3y}} = \frac{2x\sqrt{6xy}}{\sqrt{9y^2}} = \frac{2x\sqrt{6xy}}{3y}$$

(b) $\dfrac{2}{\sqrt[3]{3x}}$

To satisfy the third condition, we must remove the radical from the denominator. For this we need a perfect cube inside the radical in the denominator. Multiplying the numerator

NOTE
$\sqrt[3]{9x^2} = \sqrt[3]{3^2x^2}$
so
$\sqrt[3]{3x} \cdot \sqrt[3]{9x^2} = \sqrt[3]{3^3x^3}$
and each exponent is a multiple of 3.

and denominator by $\sqrt[3]{9x^2}$ will provide the perfect cube. So

$$\frac{2\sqrt[3]{9x^2}}{\sqrt[3]{3x} \cdot \sqrt[3]{9x^2}} = \frac{2\sqrt[3]{9x^2}}{\sqrt[3]{27x^3}}$$

$$= \frac{2\sqrt[3]{9x^2}}{3x}$$

CHECK YOURSELF 6

Simplify each expression.

(a) $\sqrt{\frac{12a^3}{5b}}$ **(b)** $\frac{3}{\sqrt[3]{2w^2}}$

The following algorithm summarizes our work in simplifying radical expressions.

NOTE In the case of a cube root, steps 1 and 2 would refer to perfect cubes, etc.

Step by Step: Simplifying Radical Expressions

Step 1 To satisfy the first condition: Determine the largest perfect-square factor of the radicand. Apply the product property to "remove" that factor from inside the radical.

Step 2 To satisfy the second condition: Use the quotient property to write the expression in the form

$$\frac{\sqrt{a}}{\sqrt{b}}$$

If b is a perfect square, remove the radical in the denominator. If that is not the case, proceed to step 3.

Step 3 Multiply the numerator and denominator of the radical expression by an appropriate radical to remove the radical in the denominator. Simplify the resulting expression when necessary.

CHECK YOURSELF ANSWERS

1. **(a)** $3\sqrt{5}$; **(b)** $10\sqrt{2}$; **(c)** $5p^2\sqrt{3p}$; **(d)** $7mn^2\sqrt{2m}$
2. **(a)** $4w\sqrt[3]{2w}$; **(b)** $2xy^2\sqrt[3]{5x^2y}$; **(c)** $2a^2b\sqrt[4]{3b}$ **3.** **(a)** $\frac{\sqrt{7}}{4}$; **(b)** $\frac{\sqrt{3}}{5a}$; **(c)** $\frac{\sqrt[3]{5x}}{3}$
4. $\frac{\sqrt{21}}{7}$ **5.** **(a)** $\frac{5\sqrt{3}}{6}$; **(b)** $\frac{\sqrt[3]{6}}{3}$ **6.** **(a)** $\frac{2a\sqrt{15ab}}{5b}$; **(b)** $\frac{3\sqrt[3]{4w}}{2w}$

8.2 Exercises

Name ______ Section ______ Date ______

Use the product property to write each expression in simplified form.

1. $\sqrt{12}$
2. $\sqrt{24}$
3. $\sqrt{50}$
4. $\sqrt{28}$
5. $-\sqrt{108}$
6. $\sqrt{32}$
7. $\sqrt{52}$
8. $-\sqrt{96}$
9. $\sqrt{60}$
10. $\sqrt{150}$
11. $-\sqrt{125}$
12. $\sqrt{128}$
13. $\sqrt{288}$
14. $-\sqrt{300}$
15. $\sqrt{450}$
16. $\sqrt{432}$
17. $\sqrt[3]{16}$
18. $\sqrt[3]{-54}$
19. $\sqrt[3]{-48}$
20. $\sqrt[3]{250}$
21. $\sqrt[3]{135}$
22. $\sqrt[3]{-160}$
23. $\sqrt[4]{32}$
24. $\sqrt[4]{96}$
25. $\sqrt[4]{243}$
26. $\sqrt[4]{1250}$

Use the product property to write each expression in simplified form. Assume that all variables represent positive real numbers.

27. $\sqrt{18z^2}$
28. $\sqrt{45a^2}$
29. $\sqrt{63x^4}$
30. $\sqrt{54w^4}$
31. $\sqrt{98m^3}$
32. $\sqrt{75a^5}$
33. $\sqrt{80x^2y^3}$
34. $\sqrt{108p^5q^2}$
35. $\sqrt[3]{40b^3}$
36. $\sqrt[3]{16x^3}$
37. $\sqrt[3]{48p^9}$
38. $\sqrt[3]{-80a^6}$

ANSWERS

1. ______ 2. ______
3. ______ 4. ______
5. ______ 6. ______
7. ______ 8. ______
9. ______ 10. ______
11. ______ 12. ______
13. ______ 14. ______
15. ______ 16. ______
17. ______ 18. ______
19. ______ 20. ______
21. ______ 22. ______
23. ______ 24. ______
25. ______ 26. ______
27. ______
28. ______
29. ______
30. ______
31. ______
32. ______
33. ______
34. ______
35. ______
36. ______
37. ______
38. ______

ANSWERS

39. ______
40. ______
41. ______
42. ______
43. ______
44. ______
45. ______
46. ______
47. ______
48. ______
49. ______
50. ______
51. ______
52. ______
53. ______
54. ______
55. ______
56. ______
57. ______ 58. ______
59. ______ 60. ______
61. ______ 62. ______
63. ______ 64. ______
65. ______ 66. ______
67. ______ 68. ______
69. ______ 70. ______

39. $\sqrt[3]{54m^7}$ **40.** $\sqrt[3]{250x^{13}}$ **41.** $\sqrt[3]{-24a^5b^4}$

42. $\sqrt[3]{128r^6s^2}$ **43.** $\sqrt[3]{56x^6y^5z^4}$ **44.** $-\sqrt[3]{250a^4b^{15}c^9}$

45. $\sqrt[4]{32x^8}$ **46.** $\sqrt[4]{162y^{12}}$ **47.** $\sqrt[4]{243a^{15}}$

48. $\sqrt[4]{80p^{11}}$ **49.** $\sqrt[4]{96w^5z^{13}}$ **50.** $\sqrt[4]{128a^{12}b^{17}}$

51. $\sqrt[5]{64w^{10}}$ **52.** $\sqrt[5]{96a^5b^{12}}$

Use the quotient property to write each expression in simplified form. Assume that all variables represent positive real numbers.

53. $\sqrt{\frac{5}{16}}$ **54.** $\sqrt{\frac{11}{36}}$ **55.** $\sqrt{\frac{x^4}{25}}$

56. $\sqrt{\frac{a^6}{49}}$ **57.** $\sqrt{\frac{5}{9y^4}}$ **58.** $\sqrt{\frac{7}{25x^2}}$

59. $\sqrt[3]{\frac{5}{8}}$ **60.** $\sqrt[3]{\frac{3}{64}}$ **61.** $\sqrt[3]{\frac{4x^2}{27}}$

62. $\sqrt[4]{\frac{5x^3}{16}}$ **63.** $\sqrt[4]{\frac{3}{81a^8}}$ **64.** $\sqrt[3]{\frac{7}{8y^6}}$

Write each expression in simplified form. Assume that all variables represent positive real numbers.

65. $\sqrt{\frac{4}{5}}$ **66.** $\sqrt{\frac{7}{3}}$ **67.** $\frac{3}{\sqrt{10}}$

68. $\frac{5}{\sqrt{7}}$ **69.** $\sqrt{\frac{5}{8}}$ **70.** $\frac{7}{\sqrt{12}}$

71. $\dfrac{\sqrt{6}}{\sqrt{7}}$

72. $\dfrac{\sqrt{5}}{\sqrt{11}}$

73. $\dfrac{2\sqrt{3}}{\sqrt{10}}$

74. $\dfrac{3\sqrt{5}}{\sqrt{3}}$

75. $\sqrt[3]{\dfrac{7}{4}}$

76. $\sqrt[3]{\dfrac{5}{9}}$

77. $\dfrac{5}{\sqrt[3]{16}}$

78. $\dfrac{\sqrt[3]{3}}{\sqrt[3]{4}}$

79. $\sqrt{\dfrac{3}{x}}$

80. $\sqrt{\dfrac{7}{y}}$

81. $\sqrt{\dfrac{12}{w}}$

82. $\dfrac{\sqrt{18}}{\sqrt{a}}$

83. $\dfrac{\sqrt{8m^3}}{\sqrt{5n}}$

84. $\sqrt{\dfrac{24x^5}{7y}}$

85. $\sqrt[3]{\dfrac{5}{y}}$

86. $\sqrt[3]{\dfrac{7}{x^2}}$

87. $\dfrac{3}{\sqrt[3]{2x}}$

88. $\dfrac{5}{\sqrt[3]{3a}}$

89. $\sqrt[3]{\dfrac{2}{5x^2}}$

90. $\sqrt[3]{\dfrac{5}{7w^2}}$

91. $\dfrac{\sqrt[3]{5}}{\sqrt[3]{4a^2}}$

92. $\dfrac{\sqrt[3]{2}}{\sqrt[3]{9m^2}}$

93. $\sqrt[3]{\dfrac{a^5}{b^7}}$

94. $\sqrt[3]{\dfrac{w^7}{z^{10}}}$

Label each of the following statements as true or false.

95. $\sqrt{16x^{16}} = 4x^8$

96. $\sqrt{x^2 + y^2} = x + y$

97. $\dfrac{\sqrt{x^2 - 25}}{\sqrt{x - 5}} = \sqrt{x + 5}$

98. $\sqrt[3]{x^6} \cdot \sqrt[3]{x^3 - 1} = x^2\sqrt[3]{x - 1}$

99. $\sqrt[3]{(8b^6)^2} = \left(\sqrt[3]{8b^6}\right)^2$

100. $\dfrac{\sqrt[3]{8x^3}}{\sqrt[3]{2x}} = \sqrt[3]{4x^2}$

ANSWERS

71. ______ 72. ______

73. ______ 74. ______

75. ______ 76. ______

77. ______ 78. ______

79. ______ 80. ______

81. ______ 82. ______

83. ______

84. ______

85. ______

86. ______

87. ______

88. ______

89. ______

90. ______

91. ______

92. ______

93. ______

94. ______

95. ______

96. ______

97. ______

98. ______

99. ______ 100. ______

ANSWERS

101.

102.

103.

104.

105. (a)
(b)
(c)
(d)

Simplify.

101. $\dfrac{7\sqrt{x^2y^4} \cdot \sqrt{36xy}}{6\sqrt{x^{-6}y^{-2}} \cdot \sqrt{49x^{-1}y^{-3}}}$

102. $\dfrac{3\sqrt[3]{32c^{12}d^2} \cdot \sqrt[3]{2c^5d^4}}{4\sqrt[3]{9c^8d^{-2}} \cdot \sqrt[3]{3c^{-3}d^{-4}}}$

103. Explain the difference between a pair of binomials in which the middle sign is changed, and the opposite of a binomial. To illustrate, use $4 - \sqrt{7}$.

104. Determine the missing binomial in the following:

$(\sqrt{3} - 2)(\qquad) = -1.$

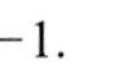

105. Use a calculator to evaluate the following expressions in parts (a) through (d). Round your answer to the nearest hundredth.

(a) $3\sqrt{5} + 4\sqrt{5}$ **(b)** $7\sqrt{5}$ **(c)** $2\sqrt{6} + 3\sqrt{6}$ **(d)** $5\sqrt{6}$

(e) Based on parts (a) through (d) make a conjecture concerning $a\sqrt{m} + b\sqrt{m}$. Check your conjecture on an example of your own similar to parts (a) through (d).

Answers

1. $2\sqrt{3}$ **3.** $5\sqrt{2}$ **5.** $-6\sqrt{3}$ **7.** $2\sqrt{13}$ **9.** $2\sqrt{15}$ **11.** $-5\sqrt{5}$
13. $12\sqrt{2}$ **15.** $15\sqrt{2}$ **17.** $2\sqrt[3]{2}$ **19.** $-2\sqrt[3]{6}$ **21.** $3\sqrt[3]{5}$
23. $2\sqrt[4]{2}$ **25.** $3\sqrt[4]{3}$ **27.** $3z\sqrt{2}$ **29.** $3x^2\sqrt{7}$ **31.** $7m\sqrt{2m}$
33. $4xy\sqrt{5y}$ **35.** $2b\sqrt[3]{5}$ **37.** $2p^3\sqrt[3]{6}$ **39.** $3m^2\sqrt[3]{2m}$
41. $-2ab\sqrt[3]{3a^2b}$ **43.** $2x^2yz\sqrt[3]{7y^2z}$ **45.** $2x^2\sqrt[4]{2}$ **47.** $3a^3\sqrt[4]{3a^3}$
49. $2wz^3\sqrt[4]{6wz}$ **51.** $2w^2\sqrt[5]{2}$ **53.** $\dfrac{\sqrt{5}}{4}$ **55.** $\dfrac{x^2}{5}$ **57.** $\dfrac{\sqrt{5}}{3y^2}$ **59.** $\dfrac{\sqrt[3]{5}}{2}$
61. $\dfrac{\sqrt[3]{4x^2}}{3}$ **63.** $\dfrac{\sqrt[4]{3}}{3a^2}$ **65.** $\dfrac{2\sqrt{5}}{5}$ **67.** $\dfrac{3\sqrt{10}}{10}$ **69.** $\dfrac{\sqrt{10}}{4}$ **71.** $\dfrac{\sqrt{42}}{7}$
73. $\dfrac{\sqrt{30}}{5}$ **75.** $\dfrac{\sqrt[3]{14}}{2}$ **77.** $\dfrac{5\sqrt[3]{4}}{4}$ **79.** $\dfrac{\sqrt{3x}}{x}$ **81.** $\dfrac{2\sqrt{3w}}{w}$
83. $\dfrac{2m\sqrt{10mn}}{5n}$ **85.** $\dfrac{\sqrt[3]{5y^2}}{y}$ **87.** $\dfrac{3\sqrt[3]{4x^2}}{2x}$ **89.** $\dfrac{\sqrt[3]{50x}}{5x}$ **91.** $\dfrac{\sqrt[3]{10a}}{2a}$
93. $\dfrac{a\sqrt[3]{a^2b^2}}{b^3}$ **95.** True **97.** True **99.** True **101.** x^5y^5
103. **105.** **(a)** 15.65; **(b)** 15.65; **(c)** 12.25; **(d)** 12.25

8.3 Operations on Radical Expressions

OBJECTIVES

1. Add two radical expressions
2. Subtract two radical expressions
3. Multiply two radical expressions
4. Divide two radical expressions

The addition and subtraction of radical expressions exactly parallel our earlier work with polynomials containing like terms. Let's review for a moment.

To add $3x^2 + 4x^2$, we have

NOTE This uses the distributive property.

$$3x^2 + 4x^2 = (3 + 4)x^2$$
$$= 7x^2$$

Keep in mind that we were able to simplify or combine the above expressions because of like terms in x^2. (Recall that like terms have the same variable factor raised to the same power.)

We *cannot* combine terms such as

$$4a^3 + 3a^2 \qquad \text{or} \qquad 3x - 5y$$

By extending these ideas, radical expressions can be combined *only* if they are *similar*, that is, if the expressions contain the same radicand with the same index. This is illustrated in the first example.

Example 1

Adding or Subtracting Radical Expressions

Add or subtract as indicated.

NOTE Apply the distributive property again.

(a) $3\sqrt{7} + 2\sqrt{7} = (3 + 2)\sqrt{7}$
$= 5\sqrt{7}$

(b) $7\sqrt{3} - 4\sqrt{3} = (7 - 4)\sqrt{3} = 3\sqrt{3}$

(c) $5\sqrt{10} - 3\sqrt{10} + 2\sqrt{10} = (5 - 3 + 2)\sqrt{10}$
$= 4\sqrt{10}$

NOTE The expressions have different radicands, $\sqrt{5}$ and $\sqrt{3}$.

(d) $2\sqrt{5} + 3\sqrt{3}$ cannot be combined or simplified.

NOTE The expressions have different indices, 2 and 3.

(e) $\sqrt{7} + \sqrt[3]{7}$ cannot be simplified.

(f) $5\sqrt{x} + 2\sqrt{x} = (5 + 2)\sqrt{x}$
$= 7\sqrt{x}$

(g) $5\sqrt{3ab} - 2\sqrt{3ab} + 3\sqrt{3ab} = (5 - 2 + 3)\sqrt{3ab} = 6\sqrt{3ab}$

NOTE The radicands are *not* the same.

(h) $\sqrt[3]{3x^2} + \sqrt[3]{3x}$ cannot be simplified.

CHECK YOURSELF 1

Add or subtract as indicated.

(a) $5\sqrt{3} + 2\sqrt{3}$

(b) $7\sqrt{5} - 2\sqrt{5} + 3\sqrt{5}$

(c) $2\sqrt{3} + 3\sqrt{2}$

(d) $\sqrt{2y} + 5\sqrt{2y} - 3\sqrt{2y}$

(e) $2\sqrt[3]{3m} - 5\sqrt[3]{3m}$

(f) $\sqrt{5x} - \sqrt[3]{5x}$

Often it is necessary to simplify radical expressions by the methods of Section 8.2 before they can be combined. The following example illustrates how the product property is applied.

Example 2

Adding or Subtracting Radical Expressions

Add or subtract as indicated.

(a) $\sqrt{48} + 2\sqrt{3}$

In this form, the radicals cannot be combined. However, note that the first radical can be simplified by our earlier methods because 48 has the perfect-square factor 16.

$$\sqrt{48} = \sqrt{16 \cdot 3} = 4\sqrt{3}$$

With this result we can proceed as before.

$$\sqrt{48} + 2\sqrt{3} = 4\sqrt{3} + 2\sqrt{3}$$
$$= (4 + 2)\sqrt{3} = 6\sqrt{3}$$

NOTE Notice that each of the radicands has a perfect-square factor. The reader should provide the details for the simplification of each radical.

(b) $\sqrt{50} - \sqrt{32} + \sqrt{98} = 5\sqrt{2} - 4\sqrt{2} + 7\sqrt{2}$

$$= (5 - 4 + 7)\sqrt{2} = 8\sqrt{2}$$

(c) $x\sqrt{2x} + 3\sqrt{8x^3}$

Note that

$$3\sqrt{8x^3} = 3\sqrt{4x^2 \cdot 2x}$$
$$= 3\sqrt{4x^2} \cdot \sqrt{2x}$$
$$= 3 \cdot 2x\sqrt{2x} = 6x\sqrt{2x}$$

So

NOTE We can now combine the similar radicals.

$$x\sqrt{2x} + 3\sqrt{8x^3} = x\sqrt{2x} + 6x\sqrt{2x}$$
$$= (x + 6x)\sqrt{2x} = 7x\sqrt{2x}$$

(d) $\sqrt[3]{2a} - \sqrt[3]{16a} + \sqrt[3]{54a} = \sqrt[3]{2a} - 2\sqrt[3]{2a} + 3\sqrt[3]{2a}$

$$= (1 - 2 + 3)\sqrt[3]{2a}$$
$$= 2\sqrt[3]{2a}$$

CHECK YOURSELF 2

Add or subtract as indicated.

(a) $\sqrt{125} + 3\sqrt{5}$

(b) $\sqrt{75} - \sqrt{27} + \sqrt{48}$

(c) $5\sqrt{24y^3} - y\sqrt{6y}$

(d) $\sqrt[3]{81x} - \sqrt[3]{3x} + \sqrt[3]{24x}$

It may also be necessary to apply the quotient property before combining rational expressions. Consider the following example.

Example 3

Adding or Subtracting Radical Expressions

Add or subtract as indicated.

(a) $2\sqrt{6} + \sqrt{\dfrac{2}{3}}$

We apply the quotient property to the *second term* and rationalize the denominator.

NOTE Multiply by $\dfrac{\sqrt{3}}{\sqrt{3}}$, or 1.

$$\sqrt{\frac{2}{3}} = \frac{\sqrt{2}}{\sqrt{3}} = \frac{\sqrt{2}\cdot\sqrt{3}}{\sqrt{3}\cdot\sqrt{3}} = \frac{\sqrt{6}}{3}$$

So

$$2\sqrt{6} + \sqrt{\frac{2}{3}} = 2\sqrt{6} + \frac{\sqrt{6}}{3}$$

$$= \left(2 + \frac{1}{3}\right)\sqrt{6} = \frac{7}{3}\sqrt{6}$$

NOTE Note that $\dfrac{\sqrt{6}}{3}$ and $\dfrac{1}{3}\sqrt{6}$ are equivalent.

(b) $\sqrt{20x} - \sqrt{\dfrac{x}{5}}$

Again we first simplify the two expressions. So

$$\sqrt{20x} - \sqrt{\frac{x}{5}} = 2\sqrt{5x} - \frac{\sqrt{x}\cdot\sqrt{5}}{\sqrt{5}\cdot\sqrt{5}}$$

$$= 2\sqrt{5x} - \frac{\sqrt{5x}}{5}$$

$$= \left(2 - \frac{1}{5}\right)\sqrt{5x} = \frac{9}{5}\sqrt{5x}$$

CHECK YOURSELF 3

Add or subtract as indicated.

(a) $3\sqrt{7} + \sqrt{\dfrac{1}{7}}$

(b) $\sqrt{40x} - \sqrt{\dfrac{2x}{5}}$

Our next example illustrates how our earlier methods for adding fractions may have to be applied in working with radical expressions.

Example 4

Adding Radical Expressions

Add $\frac{\sqrt{5}}{3} + \frac{2}{\sqrt{5}}$.

Our first step will be to rationalize the denominator of the second fraction, to write the sum as

$$\frac{\sqrt{5}}{3} + \frac{2\sqrt{5}}{\sqrt{5} \cdot \sqrt{5}}$$

or

$$\frac{\sqrt{5}}{3} + \frac{2\sqrt{5}}{5}$$

The LCD of the fractions is 15, and rewriting each fraction with that denominator, we have

$$\frac{\sqrt{5} \cdot 5}{3 \cdot 5} + \frac{2\sqrt{5} \cdot 3}{5 \cdot 3} = \frac{5\sqrt{5} + 6\sqrt{5}}{15}$$

$$= \frac{11\sqrt{5}}{15}$$

CHECK YOURSELF 4

Subtract $\frac{3}{\sqrt{10}} - \frac{\sqrt{10}}{5}$.

In Section 8.2 we introduced the product and quotient properties for radical expressions. At that time they were used for simplifying radicals.

If we turn those properties around, we have our rules for the multiplication and division of radical expressions. For multiplication:

Rules and Properties: Multiplying Radical Expressions

$$\sqrt[n]{a} \cdot \sqrt[n]{b} = \sqrt[n]{a \cdot b}$$

In words, the product of two roots is the root of the product of the radicands.

The use of this multiplication rule is illustrated in our next example. Again we assume that all variables represent positive real numbers.

Example 5

Multiplying Radical Expressions

Multiply.

NOTE Just multiply the radicands.

(a) $\sqrt{7} \cdot \sqrt{5} = \sqrt{7 \cdot 5} = \sqrt{35}$

(b) $\sqrt{3x} \cdot \sqrt{10y} = \sqrt{3x \cdot 10y}$
$= \sqrt{30xy}$

(c) $\sqrt[3]{4x} \cdot \sqrt[3]{7x} = \sqrt[3]{4x \cdot 7x}$
$= \sqrt[3]{28x^2}$

CHECK YOURSELF 5

Multiply.

(a) $\sqrt{6} \cdot \sqrt{7}$ **(b)** $\sqrt{5a} \cdot \sqrt{11b}$
(c) $\sqrt[3]{3y} \cdot \sqrt[3]{5y}$

Keep in mind that all radical expressions should be written in simplified form. Often we have to apply the methods of Section 8.2 to simplify a product once it has been formed.

Example 6

Multiplying Radical Expressions

Multiply and simplify.

NOTE $\sqrt{18}$ is *not* in simplified form. 9 is a perfect-square factor of 18.

(a) $\sqrt{3} \cdot \sqrt{6} = \sqrt{18}$
$= \sqrt{9 \cdot 2} = \sqrt{9}\sqrt{2}$
$= 3\sqrt{2}$

(b) $\sqrt{5x} \cdot \sqrt{15x} = \sqrt{75x^2}$
$= \sqrt{25x^2 \cdot 3} = \sqrt{25x^2} \cdot \sqrt{3}$
$= 5x\sqrt{3}$

NOTE Now we want a factor that is a *perfect cube.*

(c) $\sqrt[3]{4a^2b} \cdot \sqrt[3]{10a^2b^2} = \sqrt[3]{40a^4b^3} = \sqrt[3]{8a^3b^3 \cdot 5a}$
$= \sqrt[3]{8a^3b^3} \cdot \sqrt[3]{5a} = 2ab\sqrt[3]{5a}$

CHECK YOURSELF 6

Multiply and simplify.

(a) $\sqrt{10} \cdot \sqrt{20}$ **(b)** $\sqrt{6x} \cdot \sqrt{15x}$
(c) $\sqrt[3]{9p^2q^2} \cdot \sqrt[3]{6pq^2}$

We are now ready to combine multiplication with the techniques for the addition and subtraction of radicals. This will allow us to multiply radical expressions with more than one term. Consider the following examples.

Example 7

Using the Distributive Property

Multiply and simplify.

NOTE We distribute $\sqrt{2}$ over the sum $\sqrt{5} + \sqrt{7}$ to multiply.

(a) $\sqrt{2}(\sqrt{5} + \sqrt{7})$

Distributing $\sqrt{2}$, we have

$$\sqrt{2} \cdot \sqrt{5} + \sqrt{2} \cdot \sqrt{7} = \sqrt{10} + \sqrt{14}$$

The expression cannot be simplified further.

NOTE Distribute $\sqrt{3}$.

(b)
$$\begin{aligned} \sqrt{3}(\sqrt{6} + 2\sqrt{15}) &= \sqrt{3} \cdot \sqrt{6} + \sqrt{3} \cdot 2\sqrt{15} \\ &= \sqrt{18} + 2\sqrt{45} \\ &= 3\sqrt{2} + 6\sqrt{5} \end{aligned}$$

NOTE Alternately we could choose to simplify $\sqrt{8x}$ in the original expression as our first step. We leave it to the reader to verify that the result would be the same.

(c)
$$\begin{aligned} \sqrt{x}(\sqrt{2x} + \sqrt{8x}) &= \sqrt{x} \cdot \sqrt{2x} + \sqrt{x} \cdot \sqrt{8x} \\ &= \sqrt{2x^2} + \sqrt{8x^2} \\ &= x\sqrt{2} + 2x\sqrt{2} = 3x\sqrt{2} \end{aligned}$$

CHECK YOURSELF 7

Multiply and simplify.

(a) $\sqrt{3}(\sqrt{10} + \sqrt{2})$ **(b)** $\sqrt{2}(3 + 2\sqrt{6})$ **(c)** $\sqrt{a}(\sqrt{3a} + \sqrt{12a})$

If both of the radical expressions involved in a multiplication statement have two terms, we must apply the patterns for multiplying polynomials developed in Chapter 6. The following example illustrates.

Example 8

Multiplying Radical Binomials

Multiply and simplify.

(a) $(\sqrt{3} + 1)(\sqrt{3} + 5)$

To write the desired product, we use the FOIL pattern for multiplying binomials.

$$(\sqrt{3} + 1)(\sqrt{3} + 5)$$

First, Outer, Inner, Last

$$= \underbrace{\sqrt{3} \cdot \sqrt{3}}_{\text{First}} + \underbrace{5 \cdot \sqrt{3}}_{\text{Outer}} + \underbrace{1 \cdot \sqrt{3}}_{\text{Inner}} + \underbrace{1 \cdot 5}_{\text{Last}}$$

NOTE Combine the outer and inner products.

$$= 3 + 6\sqrt{3} + 5$$

NOTE Combine the first and last products.

$$= 8 + 6\sqrt{3}$$

(b) $(\sqrt{6} + \sqrt{2})(\sqrt{6} - \sqrt{2})$

Multiplying as before, we have

NOTE Notice that sum of the outer and inner products $-\sqrt{12} + \sqrt{12}$ is 0.

$$\sqrt{6} \cdot \sqrt{6} - \sqrt{6} \cdot \sqrt{2} + \sqrt{6} \cdot \sqrt{2} - \sqrt{2} \cdot \sqrt{2} = 6 - 2 = 4$$

Note: Two binomial radical expressions that differ *only* in the sign of the second term are called *conjugates* of each other. So

NOTE Notice the form of the product

$(a + b)(a - b)$

gives

$a^2 - b^2$

when a and b are square roots. The product will be rational.

$$\sqrt{6} + \sqrt{2} \qquad \text{and} \qquad \sqrt{6} - \sqrt{2}$$

are conjugates, and their product does *not* contain a radical—the product is a rational number. That will always be the case with two conjugates. This will have particular significance later in this section.

NOTE We write the square as a product and apply the multiplication pattern for binomials.

(c) $(\sqrt{2} + \sqrt{5})^2 = (\sqrt{2} + \sqrt{5})(\sqrt{2} + \sqrt{5})$

Multiplying as before, we have

$$\begin{aligned}(\sqrt{2} + \sqrt{5})^2 &= \sqrt{2} \cdot \sqrt{2} + \sqrt{2} \cdot \sqrt{5} + \sqrt{2} \cdot \sqrt{5} + \sqrt{5} \cdot \sqrt{5} \\ &= 2 + \sqrt{10} + \sqrt{10} + 5 \\ &= 7 + 2\sqrt{10}\end{aligned}$$

Note: This square can also be evaluated by using our earlier formula for the square of a binomial

$$(a + b)^2 = a^2 + 2ab + b^2$$

in which $a = \sqrt{2}$ and $b = \sqrt{5}$.

CHECK YOURSELF 8

Multiply and simplify.

(a) $(\sqrt{2} + 3)(\sqrt{2} + 5)$ **(b)** $(\sqrt{5} - \sqrt{3})(\sqrt{5} + \sqrt{3})$

(c) $(\sqrt{7} - \sqrt{3})^2$

We are now ready to state our basic rule for the division of radical expressions. Again, it is simply a restatement of our earlier quotient property.

Rules and Properties: Dividing Radical Expressions

$$\frac{\sqrt[n]{a}}{\sqrt[n]{b}} = \sqrt[n]{\frac{a}{b}}$$

In words, the quotient of two roots is the root of the quotient of the radicands.

Although we illustrate the use of this property in one of the examples that follow, the division of rational expressions is most often carried out by rationalizing the denominator. This process can be divided into two types of problems, those with a monomial divisor and those with binomial divisors. The following series of examples will illustrate.

Example 9

Dividing Radical Expressions

Simplify each expression. Again assume that all variables represent positive real numbers.

NOTE Use $\frac{\sqrt{5}}{\sqrt{5}} = 1$

(a) $\frac{3}{\sqrt{5}} = \frac{3 \cdot \sqrt{5}}{\sqrt{5} \cdot \sqrt{5}} = \frac{3\sqrt{5}}{5}$ We multiply numerator and denominator by $\sqrt{5}$ to rationalize the denominator.

(b) $\frac{\sqrt{7x}}{\sqrt{10y}} = \frac{\sqrt{7x} \cdot \sqrt{10y}}{\sqrt{10y} \cdot \sqrt{10y}}$

$= \frac{\sqrt{70xy}}{10y}$

NOTE Notice that

$\sqrt[3]{2} \cdot \sqrt[3]{4} = \sqrt[3]{2} \cdot \sqrt[3]{2^2}$

$= \sqrt[3]{2^3}$

$= 2$

(c) $\frac{3}{\sqrt[3]{2}} = \frac{3\sqrt[3]{4}}{\sqrt[3]{2} \cdot \sqrt[3]{4}}$ In this case we want a perfect cube in the denominator, and so we multiply numerator and denominator by $\sqrt[3]{4}$.

$= \frac{3\sqrt[3]{4}}{2}$

Note that these division problems are identical to those we saw earlier in Section 8.2 when we were simplifying radical expressions. They are shown here to illustrate this case of division with radicals.

CHECK YOURSELF 9

Simplify each expression.

(a) $\frac{5}{\sqrt{7}}$ **(b)** $\frac{\sqrt{3a}}{\sqrt{5b}}$ **(c)** $\frac{5}{\sqrt[3]{9}}$

Our division rule can be particularly useful if the radicands in the numerator and denominator have common factors. Consider the following example.

Example 10

Dividing Radical Expressions

Simplify

NOTE Notice the common factor of 5 in the radicand of the numerator and denominator.

$$\frac{\sqrt{10}}{\sqrt{15a}}$$

We apply the division rule so that the radicand can be reduced as a fraction:

$$\frac{\sqrt{10}}{\sqrt{15a}} = \sqrt{\frac{10}{15a}} = \sqrt{\frac{2}{3a}}$$

Now we use the quotient property and rationalize the denominator:

$$\sqrt{\frac{2}{3a}} = \frac{\sqrt{2}}{\sqrt{3a}} = \frac{\sqrt{2} \cdot \sqrt{3a}}{\sqrt{3a} \cdot \sqrt{3a}}$$

$$= \frac{\sqrt{6a}}{3a}$$

CHECK YOURSELF 10

Simplify $\dfrac{\sqrt{15}}{\sqrt{18x}}$.

We now turn our attention to a second type of division problem involving radical expressions. Here the divisors (the denominators) are binomials. This will use the idea of conjugates that we saw in Example 8, earlier in this section. The following example illustrates.

Example 11

Rationalizing Radical Denominators

Rationalize each denominator.

(a) $\dfrac{6}{\sqrt{6} + \sqrt{2}}$

NOTE If a radical expression has a sum or difference in the denominator, multiply the numerator and denominator by the *conjugate* of the denominator to rationalize.

NOTE See Example 8(b) for the details of the multiplication in the denominator.

Recall that $\sqrt{6} - \sqrt{2}$ is the conjugate of $\sqrt{6} + \sqrt{2}$, and the product of conjugates is *always a rational number*. Therefore, to rationalize the denominator, we multiply by $\sqrt{6} - \sqrt{2}$.

$$\frac{6}{\sqrt{6} + \sqrt{2}} = \frac{6(\sqrt{6} - \sqrt{2})}{(\sqrt{6} + \sqrt{2})(\sqrt{6} - \sqrt{2})}$$

$$= \frac{6(\sqrt{6} - \sqrt{2})}{4}$$

$$= \frac{3(\sqrt{6} - \sqrt{2})}{2}$$

(b) $\dfrac{\sqrt{5} + \sqrt{3}}{\sqrt{5} - \sqrt{3}}$

Multiply numerator and denominator by $\sqrt{5} + \sqrt{3}$

NOTE Combine like terms, factor, and divide the numerator and denominator by 2 to simplify.

$$\frac{(\sqrt{5} + \sqrt{3})(\sqrt{5} + \sqrt{3})}{(\sqrt{5} - \sqrt{3})(\sqrt{5} + \sqrt{3})} = \frac{5 + \sqrt{15} + \sqrt{15} + 3}{5 - 3}$$

$$= \frac{8 + 2\sqrt{15}}{2} = \frac{2(4 + \sqrt{15})}{2}$$

$$= 4 + \sqrt{15}$$

CHECK YOURSELF 11

Rationalize the denominator.

(a) $\dfrac{4}{\sqrt{3} - \sqrt{2}}$ **(b)** $\dfrac{\sqrt{6} + \sqrt{3}}{\sqrt{6} - \sqrt{3}}$

CHECK YOURSELF ANSWERS

1. (a) $7\sqrt{3}$; **(b)** $8\sqrt{5}$; **(c)** cannot be simplified; **(d)** $3\sqrt{2y}$; **(e)** $-3\sqrt[3]{3m}$; **(f)** cannot be simplified

2. (a) $8\sqrt{5}$; **(b)** $6\sqrt{3}$; **(c)** $9y\sqrt{6y}$; **(d)** $4\sqrt[3]{3x}$

3. (a) $\frac{22}{7}\sqrt{7}$; **(b)** $\frac{9}{5}\sqrt{10x}$ **4.** $\frac{\sqrt{10}}{10}$ **5. (a)** $\sqrt{42}$; **(b)** $\sqrt{55ab}$; **(c)** $\sqrt[3]{15y^2}$

6. (a) $10\sqrt{2}$; **(b)** $3x\sqrt{10}$; **(c)** $3pq\sqrt[3]{2q}$

7. (a) $\sqrt{30} + \sqrt{6}$; **(b)** $3\sqrt{2} + 4\sqrt{3}$; **(c)** $3a\sqrt{3}$

8. (a) $17 + 8\sqrt{2}$; **(b)** 2; **(c)** $10 - 2\sqrt{21}$ **9. (a)** $\frac{5\sqrt{7}}{7}$; **(b)** $\frac{\sqrt{15ab}}{5b}$; **(c)** $\frac{5\sqrt[3]{3}}{3}$

10. $\frac{\sqrt{30x}}{6x}$ **11. (a)** $4(\sqrt{3} + \sqrt{2})$; **(b)** $3 + 2\sqrt{2}$

8.3 Exercises

Add or subtract as indicated. Assume that all variables represent positive real numbers.

1. $3\sqrt{5} + 4\sqrt{5}$
2. $5\sqrt{6} + 3\sqrt{6}$
3. $5\sqrt{x} + 3\sqrt{x}$
4. $9\sqrt{y} - 5\sqrt{y}$
5. $11\sqrt{3a} - 8\sqrt{3a}$
6. $2\sqrt{5w} + 3\sqrt{5w}$
7. $7\sqrt{m} + 6\sqrt{n}$
8. $8\sqrt{a} - 6\sqrt{b}$
9. $2\sqrt[3]{2} + 7\sqrt[3]{2}$
10. $5\sqrt[4]{3} - 2\sqrt[4]{3}$
11. $8\sqrt{6} - 2\sqrt{6} + 3\sqrt{6}$
12. $8\sqrt{3} + 2\sqrt{3} - 7\sqrt{3}$

Simplify the radical expressions when necessary. Then add or subtract as indicated. Again assume that all variables represent positive real numbers.

13. $\sqrt{20} + \sqrt{5}$
14. $\sqrt{27} + \sqrt{3}$
15. $\sqrt{18} + \sqrt{50}$
16. $\sqrt{28} + \sqrt{63}$
17. $4\sqrt{28} - \sqrt{63}$
18. $2\sqrt{40} + \sqrt{90}$
19. $\sqrt{98} - \sqrt{18} + \sqrt{8}$
20. $\sqrt{108} - \sqrt{27} + \sqrt{75}$
21. $\sqrt[3]{81} + \sqrt[3]{3}$
22. $\sqrt[3]{16} - \sqrt[3]{2}$
23. $2\sqrt[3]{128} - 3\sqrt[3]{2}$
24. $3\sqrt[3]{81} - 2\sqrt[3]{3}$
25. $\sqrt{54w} - \sqrt{24w}$
26. $\sqrt{27p} + \sqrt{75p}$
27. $\sqrt{18x^3} + \sqrt{8x^3}$
28. $\sqrt{125y^3} - \sqrt{20y^3}$
29. $\sqrt[3]{54x^4} - \sqrt[3]{16x^4} + \sqrt[3]{128x^4}$
30. $\sqrt[3]{81a^5} + \sqrt[3]{24a^5} - \sqrt[3]{192a^5}$
31. $\sqrt[3]{16w^5} + 2w\sqrt[3]{2w^2} - \sqrt[3]{2w^5}$
32. $\sqrt[4]{2z^7} - z\sqrt[4]{32z^3} + \sqrt[4]{162z^7}$

Name ____________

Section ______ Date ______

ANSWERS

1. ____
2. ____
3. ____
4. ____
5. ____
6. ____
7. ____
8. ____
9. ____
10. ____
11. ____
12. ____
13. ____
14. ____
15. ____
16. ____
17. ____
18. ____
19. ____
20. ____
21. ____ 22. ____
23. ____ 24. ____
25. ____ 26. ____
27. ____ 28. ____
29. ____ 30. ____
31. ____ 32. ____

ANSWERS

33. ______ 34. ______
35. ______ 36. ______
37. ______ 38. ______
39. ______ 40. ______
41. ______
42. ______
43. ______
44. ______
45. ______
46. ______
47. ______
48. ______
49. ______
50. ______
51. ______
52. ______
53. ______
54. ______
55. ______
56. ______
57. ______
58. ______
59. ______
60. ______
61. ______
62. ______

33. $\sqrt{3} + \sqrt{\frac{1}{3}}$

34. $\sqrt{6} - \sqrt{\frac{1}{6}}$

35. $\sqrt{6} - \sqrt{\frac{2}{3}}$

36. $\sqrt{10} + \sqrt{\frac{2}{5}}$

37. $\sqrt[3]{48} - \sqrt[3]{\frac{3}{4}}$

38. $\sqrt[3]{96} + \sqrt[3]{\frac{4}{9}}$

39. $\frac{\sqrt{6}}{2} + \frac{1}{\sqrt{6}}$

40. $\frac{\sqrt{10}}{2} - \frac{1}{\sqrt{10}}$

41. $\frac{\sqrt{12}}{3} - \frac{1}{\sqrt{3}}$

42. $\frac{\sqrt{20}}{5} + \frac{2}{\sqrt{5}}$

Multiply each of the following expressions.

43. $\sqrt{7} \cdot \sqrt{6}$

44. $\sqrt{3} \cdot \sqrt{10}$

45. $\sqrt{a} \cdot \sqrt{11}$

46. $\sqrt{10} \cdot \sqrt{w}$

47. $\sqrt{3} \cdot \sqrt{7} \cdot \sqrt{2}$

48. $\sqrt{5} \cdot \sqrt{7} \cdot \sqrt{3}$

49. $\sqrt[3]{4} \cdot \sqrt[3]{9}$

50. $\sqrt[3]{5} \cdot \sqrt[3]{7}$

Multiply and simplify each of the following expressions.

51. $\sqrt{3} \cdot \sqrt{12}$

52. $\sqrt{5} \cdot \sqrt{20}$

53. $\sqrt{7} \cdot \sqrt{7}$

54. $\sqrt{10} \cdot \sqrt{10}$

55. $\sqrt[3]{9p^2} \cdot \sqrt[3]{6p}$

56. $\sqrt[3]{25x^2} \cdot \sqrt[3]{10x^2}$

57. $\sqrt[3]{4x^2y} \cdot \sqrt[3]{10xy^3}$

58. $\sqrt[3]{18r^2s^2} \cdot \sqrt[3]{9r^2s}$

59. $\sqrt{2}(\sqrt{3} + 5)$

60. $\sqrt{3}(\sqrt{5} - 7)$

61. $\sqrt{5}(\sqrt{3} - \sqrt{2})$

62. $\sqrt{7}(\sqrt{5} + \sqrt{2})$

63. $\sqrt{3}(5\sqrt{2} - \sqrt{18})$

64. $\sqrt{2}(2\sqrt{10} + \sqrt{40})$

65. $\sqrt{x}(\sqrt{3x} + \sqrt{27x})$

66. $\sqrt{y}(\sqrt{8y} - \sqrt{2y})$

67. $\sqrt[3]{4}(\sqrt[3]{4} + \sqrt[3]{32})$

68. $\sqrt[3]{6}(\sqrt[3]{32} - \sqrt[3]{4})$

69. $(\sqrt{2} + 3)(\sqrt{2} - 4)$

70. $(\sqrt{3} - 1)(\sqrt{3} + 5)$

71. $(\sqrt{3} - 2)(\sqrt{3} - 5)$

72. $(\sqrt{5} + 1)(\sqrt{5} + 4)$

73. $(\sqrt{2} + 3\sqrt{5})(\sqrt{2} - 2\sqrt{5})$

74. $(\sqrt{6} - 2\sqrt{3})(\sqrt{6} - 3\sqrt{3})$

75. $(\sqrt{5} + 3)(\sqrt{5} - 3)$

76. $(\sqrt{10} + 2)(\sqrt{10} - 2)$

77. $(\sqrt{a} + \sqrt{3})(\sqrt{a} - \sqrt{3})$

78. $(\sqrt{m} + \sqrt{7})(\sqrt{m} - \sqrt{7})$

79. $(\sqrt{3} - 5)^2$

80. $(\sqrt{5} + \sqrt{2})^2$

81. $(\sqrt{a} + 3)^2$

82. $(\sqrt{x} - 4)^2$

83. $(\sqrt{x} + \sqrt{y})^2$

84. $(\sqrt{r} - \sqrt{s})^2$

Rationalize the denominator in each of the following expressions. Simplify when necessary.

85. $\dfrac{\sqrt{3}}{\sqrt{7}}$

86. $\dfrac{\sqrt{5}}{\sqrt{3}}$

87. $\dfrac{\sqrt{2a}}{\sqrt{3b}}$

88. $\dfrac{\sqrt{5x}}{\sqrt{6y}}$

89. $\dfrac{3}{\sqrt[3]{4}}$

90. $\dfrac{2}{\sqrt[3]{9}}$

91. $\dfrac{1}{2 + \sqrt{3}}$

92. $\dfrac{2}{3 - \sqrt{2}}$

ANSWERS

63. ______ 64. ______
65. ______ 66. ______
67. ______ 68. ______
69. ______
70. ______
71. ______
72. ______
73. ______
74. ______
75. ______ 76. ______
77. ______
78. ______
79. ______
80. ______
81. ______
82. ______
83. ______
84. ______
85. ______
86. ______
87. ______
88. ______
89. ______
90. ______
91. ______
92. ______

ANSWERS

93. ______
94. ______
95. ______
96. ______
97. ______
98. ______
99. ______
100. ______
101. ______
102. ______
103. ______
104. ______
105. ______
106. ______
107. ______
108. ______

93. $\dfrac{8}{3 - \sqrt{5}}$

94. $\dfrac{20}{4 + \sqrt{6}}$

95. $\dfrac{\sqrt{7} - 2}{\sqrt{7} + 2}$

96. $\dfrac{\sqrt{5} + 3}{\sqrt{5} - 3}$

97. $\dfrac{\sqrt{6} + \sqrt{3}}{\sqrt{6} - \sqrt{3}}$

98. $\dfrac{\sqrt{7} - \sqrt{5}}{\sqrt{7} + \sqrt{5}}$

99. $\dfrac{\sqrt{w} + 3}{\sqrt{w} - 3}$

100. $\dfrac{\sqrt{x} - 5}{\sqrt{x} + 5}$

101. $\dfrac{\sqrt{x} - \sqrt{y}}{\sqrt{x} + \sqrt{y}}$

102. $\dfrac{\sqrt{m} + \sqrt{n}}{\sqrt{m} - \sqrt{n}}$

Simplify each of the following radical expressions.

103. $x\sqrt[3]{8x^4} + 4\sqrt[3]{27x^7}$

104. $\sqrt[3]{8x^2} - \sqrt[3]{27x^2}$

105. $\sqrt[4]{16a^5} + \sqrt[4]{81a^5}$

106. $\sqrt[4]{256x^7} - 2\sqrt[4]{81x^7}$

107. $\dfrac{\sqrt{2x^2 + 3x}}{\sqrt{x}}$

108. $\dfrac{\sqrt{x^2 - 9}}{\sqrt{x + 3}}$

Answers

1. $7\sqrt{5}$ **3.** $8\sqrt{x}$ **5.** $3\sqrt{3a}$ **7.** Cannot be simplified **9.** $9\sqrt[3]{2}$ **11.** $9\sqrt{6}$ **13.** $3\sqrt{5}$ **15.** $8\sqrt{2}$ **17.** $5\sqrt{7}$ **19.** $6\sqrt{2}$ **21.** $4\sqrt[3]{3}$ **23.** $5\sqrt[3]{2}$ **25.** $\sqrt{6w}$ **27.** $5x\sqrt{2x}$ **29.** $5x\sqrt[3]{2x}$ **31.** $3w\sqrt[3]{2w^2}$ **33.** $\frac{4}{3}\sqrt{3}$ **35.** $\frac{2}{3}\sqrt{6}$ **37.** $\frac{3}{2}\sqrt[3]{6}$ **39.** $\frac{2}{3}\sqrt{6}$ **41.** $\frac{1}{3}\sqrt{3}$ **43.** $\sqrt{42}$ **45.** $\sqrt{11a}$ **47.** $\sqrt{42}$ **49.** $\sqrt[3]{36}$ **51.** 6 **53.** 7 **55.** $3p\sqrt[3]{2}$ **57.** $2xy\sqrt[3]{5y}$ **59.** $\sqrt{6} + 5\sqrt{2}$ **61.** $\sqrt{15} - \sqrt{10}$ **63.** $2\sqrt{6}$ **65.** $4x\sqrt{3}$ **67.** $6\sqrt[3]{2}$ **69.** $-10 - \sqrt{2}$ **71.** $13 - 7\sqrt{3}$ **73.** $-28 + \sqrt{10}$ **75.** -4 **77.** $a - 3$ **79.** $28 - 10\sqrt{3}$ **81.** $a + 6\sqrt{a} + 9$ **83.** $x + 2\sqrt{xy} + y$ **85.** $\frac{\sqrt{21}}{7}$ **87.** $\frac{\sqrt{6ab}}{3b}$ **89.** $\frac{3\sqrt[3]{2}}{2}$ **91.** $2 - \sqrt{3}$ **93.** $6 + 2\sqrt{5}$ **95.** $\frac{11 - 4\sqrt{7}}{3}$ **97.** $3 + 2\sqrt{2}$ **99.** $\frac{w + 6\sqrt{w} + 9}{w - 9}$ **101.** $\frac{x - 2\sqrt{xy} + y}{x - y}$ **103.** $14x^2\sqrt[3]{x}$ **105.** $5a\sqrt[4]{a}$ **107.** $\sqrt{2x + 3}$

8.4 Solving Radical Equations

8.4 OBJECTIVES

1. Solve an equation containing a radical expression
2. Solve an equation containing two radical expressions
3. Solve an application that involves a radical equation

In this section, we wish to establish procedures for solving equations involving radicals. The basic technique we will use involves raising both sides of an equation to some power. However, doing so requires some caution.

NOTE

$$x^2 = 1$$
$$x^2 - 1 = 0$$
$$(x + 1)(x - 1) = 0$$

so the solutions are 1 and -1.

For example, let's begin with the equation $x = 1$. Squaring both sides gives us $x^2 = 1$, which has two solutions, 1 and -1. Clearly -1 is not a solution to the original equation. We refer to -1 as an *extraneous solution*.

We must be aware of the possibility of extraneous solutions any time we raise both sides of an equation to any *even power*. Having said that, we are now prepared to introduce the power property of equality.

Rules and Properties: The Power Property of Equality

Given any two expressions a and b and any positive integer n,

If $a = b$, then $a^n = b^n$.

Note that although in applying the power property you will never lose a solution, you will often find an extraneous one as a result of raising both sides of an equation to some power. Because of this, it is very important that you *check all solutions*.

Example 1

Solving a Radical Equation

NOTE Notice that

$(\sqrt{x + 2})^2 = x + 2$

That is why squaring both sides of the equation removes the radical.

Solve $\sqrt{x + 2} = 3$.

Squaring each side, we have

$$(\sqrt{x + 2})^2 = 3^2$$
$$x + 2 = 9$$
$$x = 7$$

Substituting 7 into the original equation, we find

$$\sqrt{7 + 2} \stackrel{?}{=} 3$$
$$\sqrt{9} \stackrel{?}{=} 3$$
$$3 = 3$$

Because this is a true statement, we have found the solution for the equation, $x = 7$.

CHECK YOURSELF 1

Solve the equation $\sqrt{x - 5} = 4$.

Example 2

Solving a Radical Equation

NOTE Applying the power property will only remove the radical if that radical is isolated on one side of the equation.

Solve $\sqrt{4x + 5} + 1 = 0$.

We must *first isolate the radical* on the left side:

$$\sqrt{4x + 5} = -1$$

NOTE Notice that on the right $(-1)^2 = 1$.

Then, squaring both sides, we have

$$(\sqrt{4x + 5})^2 = (-1)^2$$

$$4x + 5 = 1$$

and solving for x, we find that

$$x = -1$$

Now we will check the solution by substituting -1 for x in the original equation:

NOTE $\sqrt{1} = 1$, the principal root.

$$\sqrt{4(-1) + 5} + 1 \stackrel{?}{=} 0$$

$$\sqrt{1} + 1 \stackrel{?}{=} 0$$

NOTE This is clearly a false statement, so -1 is *not* a solution for the original equation.

and $\qquad 2 \neq 0$

Because -1 is an extraneous solution, there are *no solutions* to the original equation.

CHECK YOURSELF 2

Solve $\sqrt{3x - 2} + 2 = 0$.

Let's consider an example in which the procedure we have described will involve squaring a binomial.

Example 3

Solving a Radical Equation

NOTE These problems can also be solved graphically. With a graphing utility, plot the two graphs $Y_1 = \sqrt{x + 3}$ and $Y_2 = x + 1$. Note that the graphs have one point of intersection, where $x = 1$.

Solve $\sqrt{x + 3} = x + 1$.

We can square each side, as before.

$$(\sqrt{x + 3})^2 = (x + 1)^2$$

$$x + 3 = x^2 + 2x + 1$$

Simplifying this gives us the quadratic equation

NOTE We solved similar equations in Section 6.9.

$$x^2 + x - 2 = 0$$

Factoring, we have

$$(x - 1)(x + 2) = 0$$

which gives us the possible solutions

NOTE Verify this for yourself by substituting 1 and then −2 for x in the original equation.

$$x = 1 \qquad \text{or} \qquad x = -2$$

Now we check for extraneous solutions and find that $x = 1$ is a valid solution, but that $x = -2$ does not yield a true statement.

CAUTION

Be Careful! Sometimes (as in this example), one side of the equation contains a binomial. In that case, we must remember the middle term when we square the binomial. The square of a binomial *is always a trinomial.*

CHECK YOURSELF 3

Solve $\sqrt{x - 5} = x - 7$.

It is not always the case that one of the solutions is extraneous. We may have zero, one, or two valid solutions when we generate a quadratic from a radical equation.

In the following example we see a case in which both of the solutions derived will satisfy the equation.

Example 4

Solving a Radical Equation

Solve $\sqrt{7x + 1} - 1 = 2x$.

First, *we must isolate the term involving the radical.*

NOTE Again, with a graphing utility plot $Y_1 = \sqrt{7x + 1}$ and $Y_2 = 2x + 1$. Where do they intersect?

$$\sqrt{7x + 1} = 2x + 1$$

We can now square both sides of the equation.

$$7x + 1 = 4x^2 + 4x + 1$$

Now we write the quadratic equation in standard form.

$$4x^2 - 3x = 0$$

Factoring, we have

$$x(4x - 3) = 0$$

which yields two possible solutions

$$x = 0 \qquad \text{or} \qquad x = \frac{3}{4}$$

Checking the solutions by substitution, we find that both values for x give true statements, as follows.

Letting x be 0, we have

$$\sqrt{7(0) + 1} - 1 \stackrel{?}{=} 2(0)$$

$$\sqrt{1} - 1 \stackrel{?}{=} 0$$

$$\text{or} \qquad 0 = 0 \qquad \text{A true statement.}$$

Letting x be $\frac{3}{4}$, we have

$$\sqrt{7\left(\frac{3}{4}\right) + 1} - 1 \stackrel{?}{=} 2\left(\frac{3}{4}\right)$$

$$\sqrt{\frac{25}{4}} - 1 \stackrel{?}{=} \frac{3}{2}$$

$$\frac{5}{2} - 1 \stackrel{?}{=} \frac{3}{2}$$

$$\frac{3}{2} = \frac{3}{2}$$ Again a true statement.

CHECK YOURSELF 4

Solve $\sqrt{5x + 1} - 1 = 3x$.

Sometimes when an equation involves more than one radical, we must apply the power property more than once. In such a case, it is generally best to avoid having to work with two radicals on the same side of the equation. The following example illustrates one approach to the solution of such equations.

Example 5

Solving an Equation Containing Two Radicals

Solve $\sqrt{x - 2} - \sqrt{2x - 6} = 1$.

First we isolate $\sqrt{x - 2}$ by adding $\sqrt{2x - 6}$ to both sides of the equation. This gives

NOTE $1 + \sqrt{2x - 6}$ is a binomial of the form $a + b$, in which a is 1 and b is $\sqrt{2x - 6}$. The square on the right then has the form $a^2 + 2ab + b^2$.

$$\sqrt{x - 2} = 1 + \sqrt{2x - 6}$$

Then squaring each side, we have

$$x - 2 = 1 + 2\sqrt{2x - 6} + 2x - 6$$

We now isolate the radical that remains on the right side.

$$-x + 3 = 2\sqrt{2x - 6}$$

We must square again to remove that radical.

$$x^2 - 6x + 9 = 4(2x - 6)$$

Now solve the quadratic equation that results.

$$x^2 - 14x + 33 = 0$$

$$(x - 3)(x - 11) = 0$$

So

$x = 3$ or $x = 11$ are the possible solutions.

Checking the possible solutions, you will find that $x = 3$ yields the only valid solution. You should verify that for yourself.

CHECK YOURSELF 5

Solve $\sqrt{x+3} - \sqrt{2x+4} + 1 = 0$.

Earlier in this section, we noted that extraneous roots were possible whenever we raised both sides of the equation to an *even power*. In the following example, we will raise both sides of the equation to an odd power. We will still check the solutions, but in this case it will simply be a check of our work and not a search for extraneous solutions.

Example 6

Solving a Radical Equation

NOTE Because a *cube root* is involved, we *cube* both sides to remove the radical.

Solve $\sqrt[3]{x^2 + 23} = 3$.

Cubing each side, we have

$$x^2 + 23 = 27$$

which results in the quadratic equation

$$x^2 - 4 = 0$$

This has two solutions

$$x = 2 \qquad \text{or} \qquad x = -2$$

Checking the solutions, we find that both result in true statements. Again you should verify this result.

CHECK YOURSELF 6

Solve $\sqrt[3]{x^2 - 8} - 2 = 0$.

We summarize our work in this section in the following algorithm for solving equations involving radicals.

Step by Step: Solving Equations Involving Radicals

Step 1 Isolate a radical on one side of the equation.

Step 2 Raise each side of the equation to the smallest power that will eliminate the isolated radical.

Step 3 If any radicals remain in the equation derived in step 2, return to step 1 and continue the solution process.

Step 4 Solve the resulting equation to determine any possible solutions.

Step 5 Check all solutions to determine whether extraneous solutions may have resulted from step 2.

Did you ever stand on a beach and wonder how far out into the ocean you could see? Or have you wondered how close a ship has to be to spot land? In either case, the function

$$d(h) = \sqrt{2h}$$

can be used to estimate the distance to the horizon (in miles) from a given height (in feet).

Example 7

Estimating a Distance

Cordelia stood on a cliff gazing out at the ocean. Her eyes were 100 ft above the ocean. She saw a ship on the horizon. Approximately how far was she from that ship?

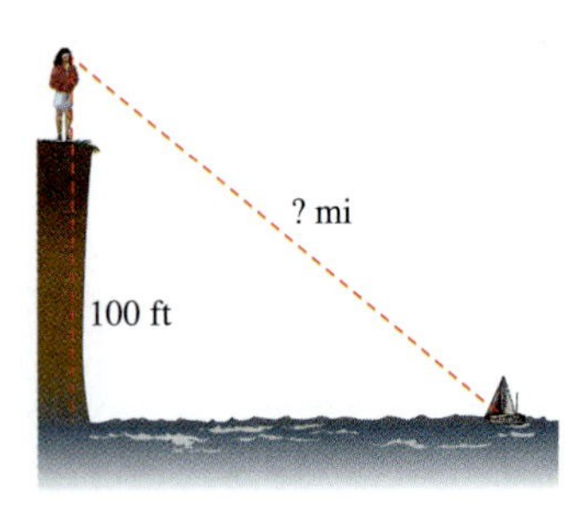

Substituting 100 for h in the equation, we get

$$d(h) = \sqrt{2(100)}$$

$$d(h) = \sqrt{200}$$

$$d(h) \approx 14 \text{ mi}$$

CHECK YOURSELF 7

From a plane flying at 35,000 ft, how far away is the horizon?

CHECK YOURSELF ANSWERS

1. $\{21\}$ **2.** No solution **3.** $\{9\}$ **4.** $\left\{0, -\frac{1}{9}\right\}$ **5.** $\{6\}$ **6.** $\{4, -4\}$

7. $d(h) \approx 265$ mi

8.4 Exercises

Name ____________

Section ______ Date ______

Solve each of the following equations. Be sure to check your solutions.

1. $\sqrt{x} = 2$

2. $\sqrt{x} - 3 = 0$

3. $2\sqrt{y} - 1 = 0$

4. $3\sqrt{2z} = 9$

5. $\sqrt{m + 5} = 3$

6. $\sqrt{y + 7} = 5$

7. $\sqrt{2x + 4} - 4 = 0$

8. $\sqrt{3x + 3} - 6 = 0$

9. $\sqrt{3x - 2} + 2 = 0$

10. $\sqrt{4x + 1} + 3 = 0$

11. $\sqrt{x - 1} = \sqrt{1 - x}$

12. $\sqrt{x + 1} = \sqrt{1 + x}$

13. $\sqrt{w + 3} = \sqrt{3 + w}$

14. $\sqrt{w - 3} = \sqrt{3 - w}$

15. $\sqrt{2x - 3} + 1 = 3$

16. $\sqrt{3x + 1} - 2 = -1$

17. $2\sqrt{3z + 2} - 1 = 5$

18. $3\sqrt{4q - 1} - 2 = 7$

19. $\sqrt{15 - 2x} = x$

20. $\sqrt{48 - 2y} = y$

21. $\sqrt{x + 5} = x - 1$

22. $\sqrt{2x - 1} = x - 8$

ANSWERS

1. ____________
2. ____________
3. ____________
4. ____________
5. ____________
6. ____________
7. ____________
8. ____________
9. ____________
10. ____________
11. ____________
12. ____________
13. ____________
14. ____________
15. ____________
16. ____________
17. ____________
18. ____________
19. ____________
20. ____________
21. ____________
22. ____________

ANSWERS

23. ______

24. ______

25. ______

26. ______

27. ______

28. ______

29. ______

30. ______

31. ______

32. ______

33. ______

34. ______

35. ______

36. ______

37. ______

38. ______

39. ______

40. ______

41. ______

42. ______

43. ______

44. ______

23. $\sqrt{3m - 2} + m = 10$

24. $\sqrt{2x + 1} + x = 7$

25. $\sqrt{t + 9} + 3 = t$

26. $\sqrt{2y + 7} + 4 = y$

27. $\sqrt{6x + 1} - 1 = 2x$

28. $\sqrt{7x + 1} - 1 = 3x$

29. $\sqrt[3]{x - 5} = 3$

30. $\sqrt[3]{x + 6} = 2$

31. $\sqrt[3]{x^2 - 1} = 2$

32. $\sqrt[3]{x^2 + 11} = 3$

Solve each of the following equations. Be sure to check your solutions.

33. $\sqrt{2x} = \sqrt{x + 1}$

34. $\sqrt{3x} = \sqrt{5x - 1}$

35. $2\sqrt{3r} = \sqrt{r + 11}$

36. $5\sqrt{2q - 7} = \sqrt{15q}$

37. $\sqrt{x + 2} + 1 = \sqrt{x + 4}$

38. $\sqrt{x + 5} - 1 = \sqrt{x + 3}$

39. $\sqrt{4m - 3} - 2 = \sqrt{2m - 5}$

40. $\sqrt{2c - 1} = \sqrt{3c + 1} - 1$

41. $\sqrt{x + 1} + \sqrt{x} = 1$

42. $\sqrt{z - 1} - \sqrt{6 - z} = 1$

43. $\sqrt{5x + 6} - \sqrt{x + 3} = 3$

44. $\sqrt{5y + 6} - \sqrt{3y + 4} = 2$

45. $\sqrt{y^2 + 12y} - 3\sqrt{5} = 0$

46. $\sqrt{x^2 + 2x} - 2\sqrt{6} = 0$

47. $\sqrt{\dfrac{x - 3}{x + 2}} = \dfrac{2}{3}$

48. $\dfrac{\sqrt{x - 2}}{x - 2} = \dfrac{x - 5}{\sqrt{x - 2}}$

49. $\sqrt{\sqrt{t} + 5} = 3$

50. $\sqrt{\sqrt{s} - 1} = \sqrt{s - 7}$

51. For what values of x is $\sqrt{(x - 1)^2} = x - 1$ a true statement?

52. For what values of x is $\sqrt[3]{(x - 1)^3} = x - 1$ a true statement?

Solve for the indicated variable.

53. $h = \sqrt{pq}$ for q

54. $c = \sqrt{a^2 + b^2}$ for a

55. $v = \sqrt{2gR}$ for R

56. $v = \sqrt{2gR}$ for g

57. $r = \sqrt{\dfrac{S}{2\pi}}$ for S

58. $r = \sqrt{\dfrac{3V}{4\pi}}$ for V

59. $r = \sqrt{\dfrac{2V}{\pi h}}$ for V

60. $r = \sqrt{\dfrac{2V}{\pi h}}$ for h

61. $d = \sqrt{(x - 1)^2 + (y - 2)^2}$ for x

62. $d = \sqrt{(x - 1)^2 + (y - 2)^2}$ for y

ANSWERS

45. __________

46. __________

47. __________

48. __________

49. __________

50. __________

51. __________

52. __________

53. __________

54. __________

55. __________

56. __________

57. __________

58. __________

59. __________

60. __________

61. __________

62. __________

ANSWERS

63. ______

64. ______

65. ______

66. ______

67. ______

68. ______

69. ______

70. ______

A weight suspended on the end of a string is a *pendulum.* The most common example of a pendulum (this side of Edgar Allen Poe) is the kind found in many clocks.

The regular back-and-forth motion of the pendulum is *periodic,* and one such cycle of motion is called a *period.* The time, in seconds, that it takes for one period is given by the radical equation

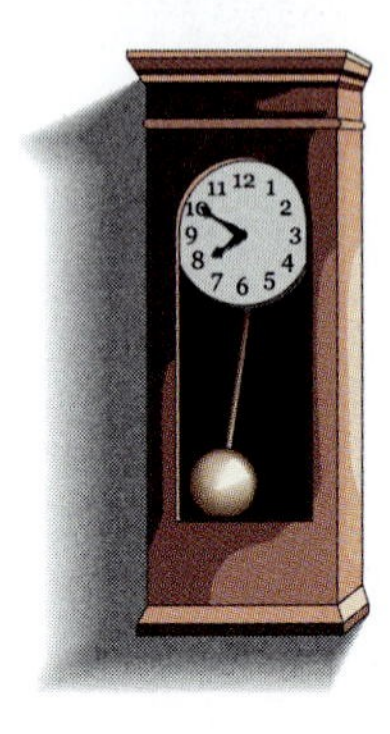

$$t = 2\pi\sqrt{\frac{l}{g}}$$

in which g is the force of gravity (10 m/s^2) and l is the length of the pendulum.

63. Find the period (to the nearest hundredth of a second) if the pendulum is 0.9 m long.

64. Find the period if the pendulum is 0.049 m long.

65. Solve the equation for length l.

66. How long would the pendulum be if the period were exactly 1 s?

Solve each of the following applications.

67. The sum of an integer and its square root is 12. Find the integer.

68. The difference between an integer and its square root is 12. What is the integer?

69. The sum of an integer and twice its square root is 24. What is the integer?

70. The sum of an integer and 3 times its square root is 40. Find the integer.

ANSWERS

71. ______
72. ______
73. ______
74. ______
75. ______
76. ______
77. ______
78. ______
79. ______
80. ______
81. ______
82. ______
83. ______

71. If a plane flies at 30,000 ft, how far away is the horizon?

72. Janine was looking out across the ocean from her hotel room on the beach. Her eyes were 250 ft above the ground. She saw a ship on the horizon. Approximately how far was the ship from her?

73. Given a distance, d, to the horizon, what altitude would allow you to see that far?

When a car comes to a sudden stop, you can determine the skidding distance (in feet) for a given speed (in miles per hour) using the formula $s(x) = 2\sqrt{5x}$, in which s is skidding distance and x is speed. Calculate the skidding distance for the following speeds.

74. 55 mi/h

75. 65 mi/h

76. 75 mi/h

77. 40 mi/h

78. Given the skidding distance s, what formula would allow you to calculate the speed in miles per hour?

79. Use the formula obtained in exercise 78 to determine the speed of a car in miles per hour if the skid marks were 35 ft long.

For each given equation, use a graphing calculator to solve. Express solutions to the nearest hundredth. (*Hint:* Define Y_1 by the expression on the left side of the equation and define Y_2 by the expression on the right side. Graph these functions and locate any intersection points. For each such point, the x value represents a solution.)

80. $\sqrt{x + 4} = x - 3$

81. $\sqrt{2 - x} = x + 4$

82. $3 - 2\sqrt{x + 4} = 2x - 5$

83. $5 - 3\sqrt{2 - x} = 3 - 4x$

Answers

1. 4 **3.** $\frac{1}{4}$ **5.** 4 **7.** 6 **9.** No solution **11.** 1

13. All real numbers **15.** $\frac{7}{2}$ **17.** $\frac{7}{3}$ **19.** 3 **21.** 4 **23.** 6

25. 7 **27.** $0, \frac{1}{2}$ **29.** 32 **31.** ± 3 **33.** 1 **35.** 1 **37.** $-\frac{7}{4}$

39. 3, 7 **41.** 0 **43.** 6 **45.** $-15, 3$ **47.** 7 **49.** 16 **51.** $x \geq 1$

53. $q = \frac{h^2}{p}$ **55.** $R = \frac{v^2}{2g}$ **57.** $S = 2\pi r^2$ **59.** $V = \frac{\pi h r^2}{2}$

61. $x = 1 \pm \sqrt{d^2 - (y - 2)^2}$ **63.** 1.88 s **65.** $l = \frac{t^2 g}{4\pi^2}$ **67.** 9

69. 16 **71.** $\approx$245 mi **73.** $\frac{d^2}{2}$ **75.** $\approx$36 ft **77.** $\approx$28 ft

79. $\approx$60 mi/h **81.** $\{-2\}$ **83.** $\{0.44\}$

8.5 Geometric and Other Applications of Radical Expressions

8.5 OBJECTIVES

1. Solve applications of the Pythagorean theorem
2. Solve applications of the golden ratio

Perhaps the most famous of all mathematical theorems is the **Pythagorean theorem** (sometimes called Pythagoras's theorem). Although the Greek mathematician, Pythagoras, and his followers gave name to the theorem, it was known to the Babylonians around 2000 B.C.E., about 1500 years before the time of the Pythagoreans.

Rules and Properties: Pythagorean Theorem

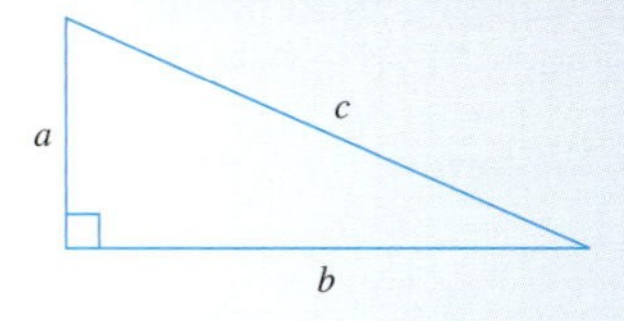

$a^2 + b^2 = c^2$

In a right triangle, the sum of the squares of the two sides is equal to the square of the hypotenuse.

NOTE In the movie *The Wizard of Oz*, the scarecrow misstates this theorem.

Example 1

Applying the Pythagorean Theorem

Fredrica and Thanh are building a set for the school play. They need to put together several wooden 8 ft × 3 ft rectangles, each with a diagonal piece. How long is each diagonal piece?

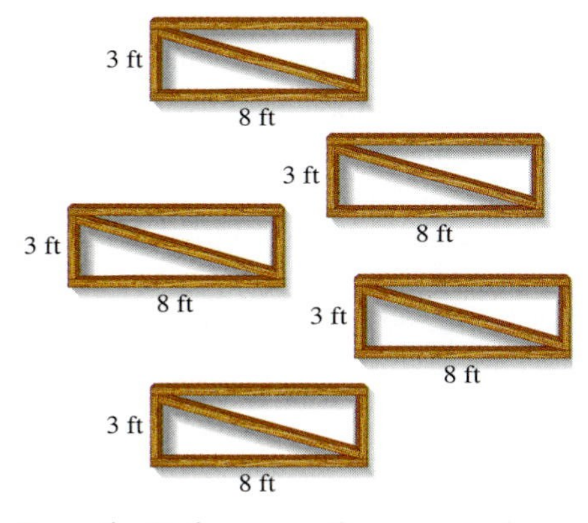

From the Pythagorean theorem, we have

$$a^2 + b^2 = c^2$$
$$8^2 + 3^2 = c^2$$
$$64 + 9 = c^2$$
$$73 = c^2$$
$$c = \sqrt{73}$$
$$c \approx 8.54 \text{ ft}$$

Each diagonal should be about $8\frac{1}{2}$ ft long.

CHECK YOURSELF 1

Fernando needs to build a handicap-access ramp into the gym. The Americans with Disabilities Act requires such a ramp to have a rise of no more than 1 in. for each 12 in. of run. With this in mind, Fernando has established a run of 19 ft for a rise of 18 in. Find the length of the ramp.

We can also use the Pythagorean theorem to find the length of a missing side, as in the next example.

Example 2

Applying the Pythagorean Theorem

A flag pole has a 20-ft rope hanging from the top. When extended to the ground, the rope touches a spot 5 ft from the pole. How high is the flag pole?

Let the pole have height h. We have

$$h^2 = 20^2 - 5^2$$

$$h^2 = 400 - 25$$

$$h^2 = 375$$

$$h = \sqrt{375}$$

$$h \approx 19.4 \text{ ft}$$

CHECK YOURSELF 2

A 10-ft tent pole is to be stabilized by a 12-ft guy wire. How far will the wire be staked from the pole?

The Pythagoreans were interested in mathematics, music, and art. They were particularly interested in areas in which mathematics could be connected to music or art. One of their discoveries was the **golden rectangle.** This became an important element in painting and sculpture. It is still used today in those fields.

What follows is the method that Euclid used around 300 B.C.E. to develop the golden rectangle.

Start with a square with sides of length 2.

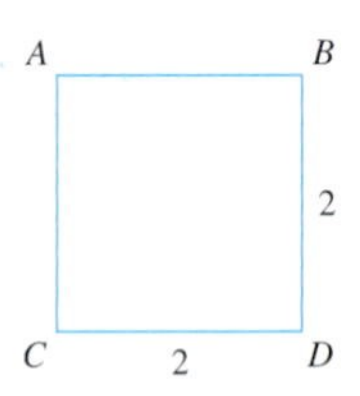

Connect corner B to the midpoint of side CD, which we label as O.

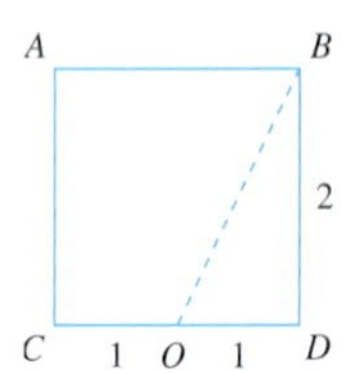

Pivot the length OB on point O so that it passes through D. Label the end of that segment as E.

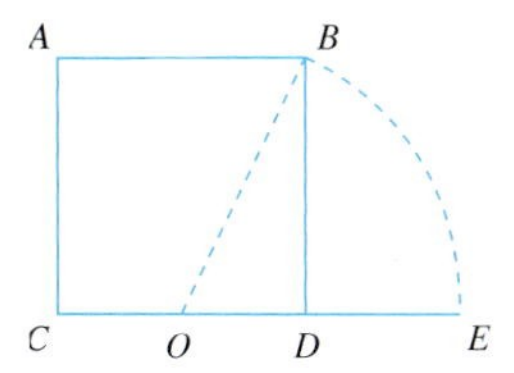

Create rectangle $AFEC$. This is a golden rectangle.

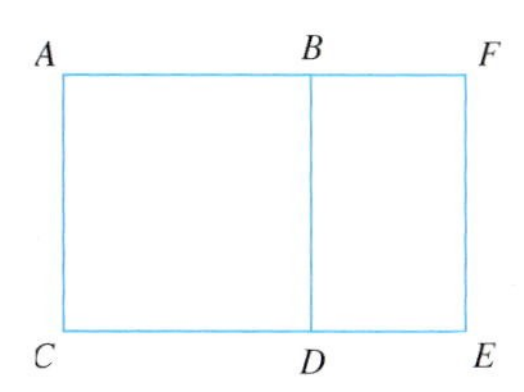

Example 3

Finding the Golden Ratio

The ratio of the length of a golden rectangle to its width is called the **golden ratio.** Use the rectangle given to find the golden ratio.

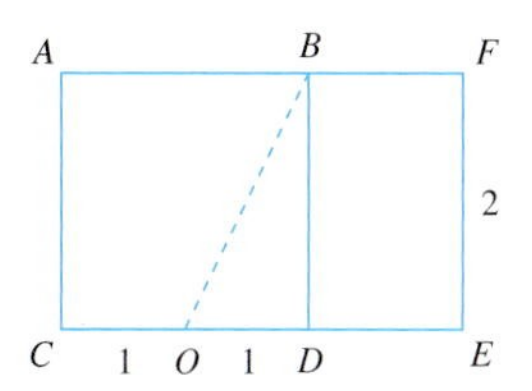

We want $\frac{CE}{FE}$.

$CE = CO + OE$

$= 1 + OE$

OE is the same as OB, which is the hypotenuse of triangle BOD.

$OB^2 = 2^2 + 1^2$

$= 4 + 1$

$= 5$

$OB = \sqrt{5} = OE$

$CE = 1 + \sqrt{5}$

$\frac{CE}{FE} = \frac{1 + \sqrt{5}}{2}$

This is the golden ratio.

CHECK YOURSELF 3

The reciprocal of the golden ratio is $\dfrac{2}{1 + \sqrt{5}}$. *Rewrite this ratio, rationalizing the denominator.*

The golden ratio is also called the **divine section,** a term first used by Franciscan monk and mathematician Luca Pacioli in the sixteenth century. It was believed to be the most important ratio in all of art and architecture. The next example illustrates one such case.

Example 4

Applying the Golden Ratio

The public façade for almost all Greek buildings used the golden ratio. Pictured below is such a façade, this one with a base of approximately 20 m. What is the height of the façade?

The ratio of the shorter side to the longer must be $\dfrac{2}{1 + \sqrt{5}}$. In this case,

$$\frac{2}{1 + \sqrt{5}} = \frac{x}{20}$$

Clearing the fractions, we have

$$(1 + \sqrt{5})x = 40$$

$$x = \frac{40}{1 + \sqrt{5}}$$

$$x \approx 12.4 \text{ m}$$

The height of the façade is just over 12 m.

CHECK YOURSELF 4

In Greek sculpture, the human torso was always sculpted using the golden ratio. If the height of a torso was to be 3 ft, what would the width have been?

CHECK YOURSELF ANSWERS

1. 228.7 in., just over 19 ft **2.** Approximately 6.6 ft

3. $\dfrac{\sqrt{5} - 1}{2}$ **4.** $h \approx 1.85$ ft

Name ______________________

Section ________ Date ________

8.5 Exercises

Using the Pythagorean theorem, find the length x in each triangle. Write answers to the nearest tenth.

1.

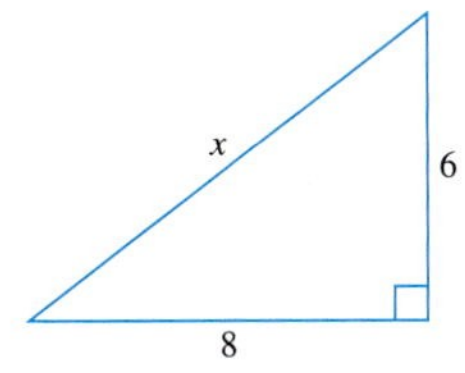

2.

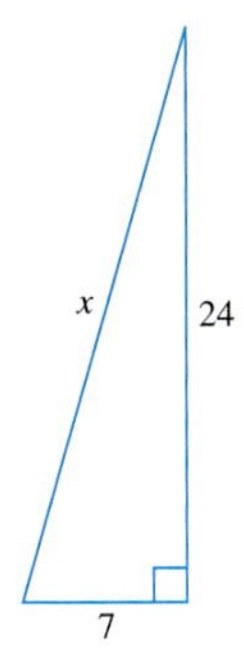

3.

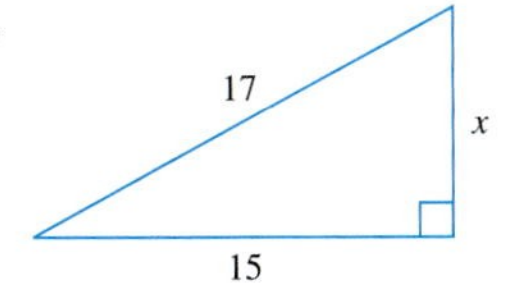

4.

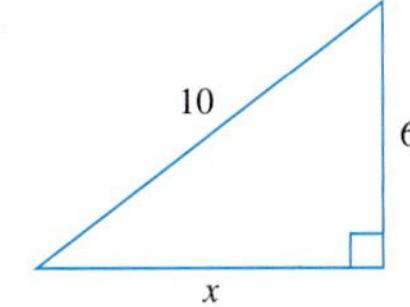

5.

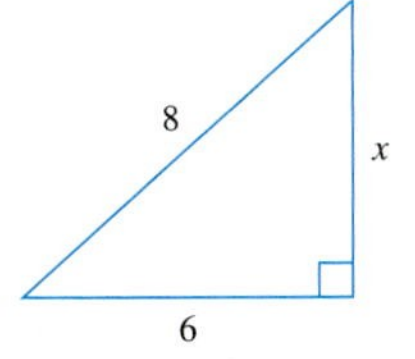

6.

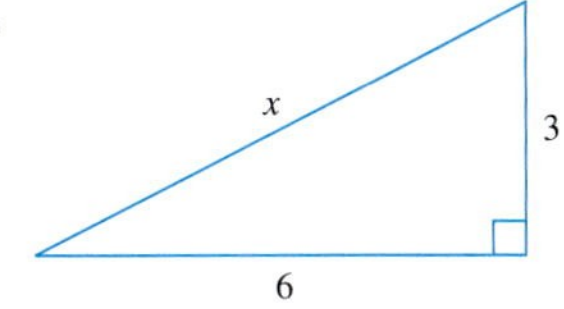

7. Find the length of the diagonal of a rectangle whose length is 11 cm and whose width is 7 cm.

8. Find the length of the diagonal of a rectangle whose width is 4 in. and whose length is 7 in.

9. Find the width of a rectangle whose diagonal is 13 ft and whose length is 10 ft.

10. Find the length of a rectangle whose diagonal is 9 in. and whose width is 5 in.

11. How long must a wire be to run from the top of a 20-ft pole to a point on the ground 7 ft from the base of the pole?

ANSWERS

1. ______________
2. ______________
3. ______________
4. ______________
5. ______________
6. ______________
7. ______________
8. ______________
9. ______________
10. ______________
11. ______________

ANSWERS

12. ______

13. ______

14. ______

15. ______

16. ______

12. The base of a 14-ft ladder is 5 ft away from a wall. How high from the floor is the top of the ladder?

13. The length of one leg of a right triangle is 2 in. more than the other. If the length of the hypotenuse is 10 in., what are the lengths of the two legs?

14. The length of one side of a rectangle is 3 cm more than the other side. If the diagonal is 15 cm, find the lengths of the two sides.

15. To "square" the wall of a house being built by a carpenter, measurements are taken, as shown in the accompanying figure. Is the wall square?

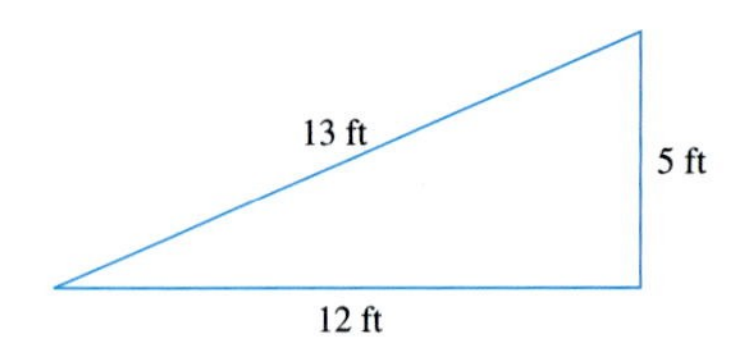

16. A television camera at ground level is filming the lift-off of a space shuttle that is rising vertically. The camera is 1500 ft from the launch pad. How far apart are the camera and the shuttle when the shuttle is 1 mi directly above the launch pad? (*Hint:* 1 mi = 5280 ft)

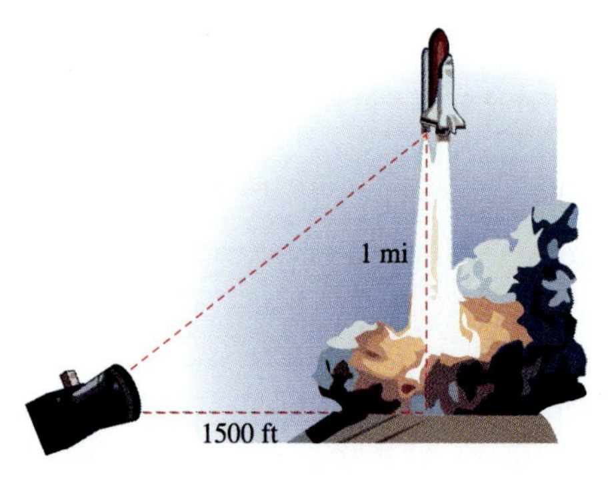

17. An air traffic controller is tracking two planes at the same altitude converging on a point in space as they fly at right angles to each other. One plane is 75 mi from the point when the other plane is 100 mi from the point. How far apart are the two planes?

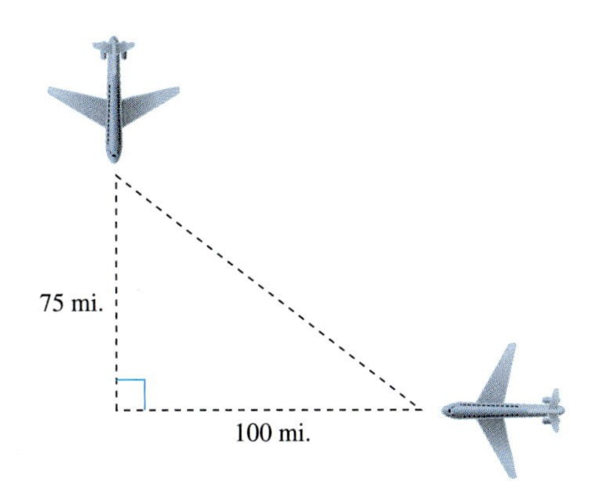

18. A baseball diamond is in the shape of a square. The vertices of the square are the three bases and home plate. If the distance from first base to second base is 90 ft, what is the distance between home plate and the second base?

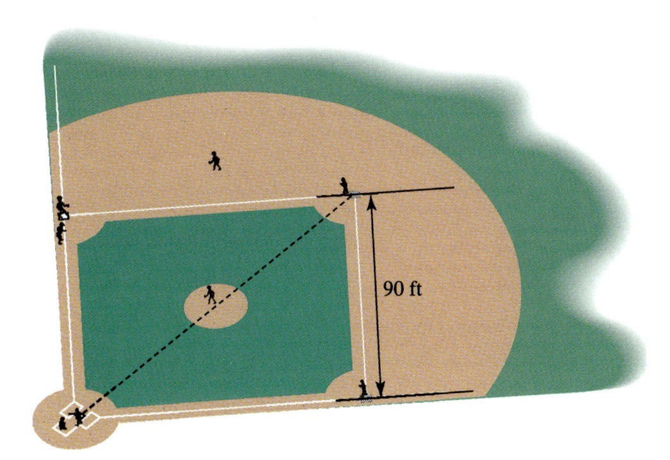

19. The Parthenon in Athens is an example of a building with a width to height ratio that is almost equal to the golden ratio. If the Parthenon is 101 ft wide, what is its height (to the nearest foot) if we assume the dimensions are in the golden ratio?

20. The Great Pyramid of Giza has dimensions as follows: height, $h = 481$ ft; base, $b = 756$ ft; and slant height, $s = 612$ ft. Find the ratio of s to $\frac{1}{2}b$ and b to h. Are either of these ratios equal to the golden ratio?

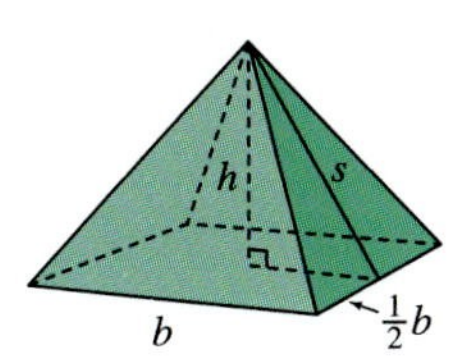

ANSWERS

17. ______

18. ______

19. ______

20. ______

ANSWERS

21. (a)
    (b)

22.

23.

24.

21. **(a)** What are the length and width of a standard index card?

    (b) Find the ratio of length to width and compare it with the golden ratio.

22. If a rectangle is to have sides related by the golden ratio, what is the length if the width (the shorter side) is 2 units?

23. If a window is 5 ft wide, how high should it be, to the nearest tenth of a foot, to be a golden rectangle?

24. A photograph is to be printed on a rectangle in the golden ratio. If it is 9 cm on its shorter side, how wide is it, to the nearest centimeter?

Answers

1. 10 **3.** 8 **5.** 5.3 **7.** 13.0 cm **9.** 8.3 ft **11.** 21.2 ft
13. 6 in.; 8 in. **15.** Yes **17.** 125 mi **19.** 62 ft
21. **(a)** 5 in. by 3in.; **(b)** 1.67 **23.** 3.1 ft or 8.1 ft

8.6 Rational Exponents

OBJECTIVES

1. Define rational exponents
2. Simplify expressions containing rational exponents
3. Use a calculator to estimate the value of an expression containing rational exponents
4. Write an expression in radical or exponential form

In Section 8.1, we discussed the radical notation, along with the concept of roots. In this section, we use that concept to develop a new notation, using exponents that provide an alternate way of writing these roots.

That new notation involves **rational numbers as exponents.** To start the development, we extend all the previous properties of exponents to include rational exponents.

Given that extension, suppose that

$$a = 4^{1/2} \tag{1}$$

Squaring both sides of the equation yields

$$a^2 = (4^{1/2})^2$$

or

NOTE We will see later in this chapter that the property $(x^m)^n = x^{mn}$ holds for rational numbers m and n.

$$a^2 = 4^{(1/2)(2)}$$

$$a^2 = 4^1$$

$$a^2 = 4 \tag{2}$$

From equation (2) we see that a is the number whose square is 4; that is, a is the principal square root of 4. Using our earlier notation, we can write

$$a = \sqrt{4}$$

But from (1)

$$a = 4^{1/2}$$

and to be consistent, we must have

NOTE $4^{1/2}$ indicates the *principal square root* of 4.

$$4^{1/2} = \sqrt{4}$$

This argument can be repeated for any exponent of the form $\frac{1}{n}$, so it seems reasonable to make the following definition.

Definitions: Rational Exponents

If a is any real number and n is a positive integer ($n > 1$), then

$$a^{1/n} = \sqrt[n]{a}$$

We restrict a so that a is nonnegative when n is even. In words, $a^{1/n}$ indicates the principal nth root of a.

Example 1 illustrates the use of rational exponents to represent roots.

Example 1

Writing Expressions in Radical Form

Write each expression in radical form and then simplify.

(a) $25^{1/2} = \sqrt{25} = 5$

NOTE $27^{1/3}$ is the *cube root* of 27.

(b) $27^{1/3} = \sqrt[3]{27} = 3$

(c) $-36^{1/2} = -\sqrt{36} = -6$

(d) $(-36)^{1/2} = \sqrt{-36}$ is not a real number.

NOTE $32^{1/5}$ is the *fifth root* of 32.

(e) $32^{1/5} = \sqrt[5]{32} = 2$

CHECK YOURSELF 1

Write each expression in radical form and simplify.

(a) $8^{1/3}$ **(b)** $-64^{1/2}$ **(c)** $81^{1/4}$

We are now ready to extend our exponent notation to allow *any* rational exponent, again assuming that our previous exponent properties must still be valid. Note that

NOTE This is because $\frac{m}{n} = (m)\left(\frac{1}{n}\right) = \left(\frac{1}{n}\right)(m)$

$$a^{m/n} = (a^{1/n})^m = (a^m)^{1/n}$$

From our earlier work, we know that $a^{1/n} = \sqrt[n]{a}$, and combining this with the above observation, we offer the following definition for $a^{m/n}$.

Definitions:

For any real number a and positive integers m and n with $n > 1$,

$$a^{m/n} = (\sqrt[n]{a})^m = \sqrt[n]{a^m}$$

NOTE The two radical forms for $a^{m/n}$ are equivalent, and the choice of which form to use generally depends on whether we are evaluating numerical expressions or rewriting expressions containing variables in radical form.

This new extension of our rational exponent notation is applied in Example 2.

Example 2

Simplifying Expressions with Rational Exponents

Simplify each expression.

(a) $9^{3/2} = (9^{1/2})^3 = (\sqrt{9})^3$

$= 3^3 = 27$

(b) $\left(\frac{16}{81}\right)^{3/4} = \left(\left(\frac{16}{81}\right)^{1/4}\right)^3 = \left(\sqrt[4]{\frac{16}{81}}\right)^3$

$= \left(\frac{2}{3}\right)^3 = \frac{8}{27}$

(c) $(-8)^{2/3} = ((-8)^{1/3})^2 = (\sqrt[3]{-8})^2$

$= (-2)^2 = 4$

NOTE This illustrates why we use $(\sqrt[n]{a})^m$ for $a^{m/n}$ when evaluating numerical expressions. The numbers involved will be smaller and easier to work with.

In (a) we could also have evaluated the expression as

$$9^{3/2} = \sqrt{9^3} = \sqrt{729}$$
$$= 27$$

CHECK YOURSELF 2

Simplify each expression.

(a) $16^{3/4}$ **(b)** $\left(\frac{8}{27}\right)^{2/3}$ **(c)** $(-32)^{3/5}$

Now we want to extend our rational exponent notation. Using the definition of negative exponents, we can write

$$a^{-m/n} = \frac{1}{a^{m/n}}$$

Example 3 illustrates the use of negative rational exponents.

Example 3

Simplifying Expressions with Rational Exponents

Simplify each expression.

(a) $16^{-1/2} = \frac{1}{16^{1/2}} = \frac{1}{4}$

(b) $27^{-2/3} = \frac{1}{27^{2/3}} = \frac{1}{(\sqrt[3]{27})^2} = \frac{1}{3^2} = \frac{1}{9}$

CHECK YOURSELF 3

Simplify each expression.

(a) $16^{-1/4}$ **(b)** $81^{-3/4}$

Graphing calculators can be used to evaluate expressions that contain rational exponents by using the [^] key and the parentheses keys.

Example 4

Estimating Powers Using a Calculator

Using a graphing calculator, evaluate each of the following. Round all answers to three decimal places.

(a) $45^{2/5}$

NOTE If you are using a scientific calculator, try using the [y^x] key in place of the [^] key.

Enter 45 and press the [^] key. Then use the following keystrokes:

[(] 2 [÷] 5) [)]

Press [ENTER], and the display will read 4.584426407. Rounded to three decimal places, the result is 4.584.

(b) $38^{-2/3}$

Enter 38 and press the [∧] key. Then use the following keystrokes:

NOTE The [(−)] key changes the sign of the exponent to minus.

[(] [(−)] 2 [÷] 3 [)]

Press [ENTER], and the display will read 0.088473037. Rounded to three decimal places, the result is 0.088.

CHECK YOURSELF 4

Evaluate each of the following by using a calculator. Round each answer to three decimal places.

(a) $23^{3/5}$ **(b)** $18^{-4/7}$

As we mentioned earlier in this section, we assume that all our previous exponent properties will continue to hold for rational exponents. Those properties are restated here.

Rules and Properties: Properties of Exponents

For any nonzero real numbers a and b and rational numbers m and n,

1. Product rule $a^m \cdot a^n = a^{m+n}$
2. Quotient rule $\frac{a^m}{a^n} = a^{m-n}$
3. Power rule $(a^m)^n = a^{mn}$
4. Product-power rule $(ab)^m = a^m b^m$
5. Quotient-power rule $\left(\frac{a}{b}\right)^m = \frac{a^m}{b^m}$

We restrict a and b to being nonnegative real numbers when m or n indicates an even root.

Example 5 illustrates the use of our extended properties to simplify expressions involving rational exponents. Here, we assume that all variables represent positive real numbers.

Example 5

Simplifying Expressions

Simplify each expression.

NOTE Product rule—add the exponents.

(a) $x^{2/3} \cdot x^{1/2} = x^{2/3+1/2}$

$= x^{4/6+3/6} = x^{7/6}$

NOTE Quotient rule—subtract the exponents.

(b) $\frac{w^{3/4}}{w^{1/2}} = w^{3/4-1/2}$

$= w^{3/4-2/4} = w^{1/4}$

NOTE Power rule—multiply the exponents.

(c) $(a^{2/3})^{3/4} = a^{(2/3)(3/4)}$

$= a^{1/2}$

CHECK YOURSELF 5

Simplify each expression.

(a) $z^{3/4} \cdot z^{1/2}$ **(b)** $\dfrac{x^{5/6}}{x^{1/3}}$ **(c)** $(b^{5/6})^{2/5}$

As you would expect from your previous experience with exponents, simplifying expressions often involves using several exponent properties.

Example 6

Simplifying Expressions

Simplify each expression.

(a) $(x^{2/3} \cdot y^{5/6})^{3/2}$

$= (x^{2/3})^{3/2} \cdot (y^{5/6})^{3/2}$ Product power rule.

$= x^{(2/3)(3/2)} \cdot y^{(5/6)(3/2)} = xy^{5/4}$ Power rule.

(b) $\left(\dfrac{r^{-1/2}}{s^{1/3}}\right)^6 = \dfrac{(r^{-1/2})^6}{(s^{1/3})^6}$ Quotient-power rule.

$= \dfrac{r^{-3}}{s^2} = \dfrac{1}{r^3 s^2}$ Power rule.

(c) $\left(\dfrac{4a^{-2/3} \cdot b^2}{a^{1/3} \cdot b^{-4}}\right)^{1/2} = \left(\dfrac{4b^2 \cdot b^4}{a^{1/3} \cdot a^{2/3}}\right)^{1/2} = \left(\dfrac{4b^6}{a}\right)^{1/2}$ We simplify inside the parentheses as the first step.

$= \dfrac{(4b^6)^{1/2}}{a^{1/2}} = \dfrac{4^{1/2}(b^6)^{1/2}}{a^{1/2}}$

$= \dfrac{2b^3}{a^{1/2}}$

CHECK YOURSELF 6

Simplify each expression.

(a) $(a^{3/4} \cdot b^{1/2})^{2/3}$ **(b)** $\left(\dfrac{w^{1/2}}{z^{-1/4}}\right)^4$ **(c)** $\left(\dfrac{8x^{-3/4}y}{x^{1/4} \cdot y^{-5}}\right)^{1/3}$

We can also use the relationships between rational exponents and radicals to write expressions involving rational exponents as radicals and vice versa.

Example 7

Writing Expressions in Radical Form

Write each expression in radical form.

NOTE Here we use $a^{m/n} = \sqrt[n]{a^m}$, which is generally the preferred form in this situation.

(a) $a^{3/5} = \sqrt[5]{a^3}$

(b) $(mn)^{3/4} = \sqrt[4]{(mn)^3}$

$= \sqrt[4]{m^3n^3}$

NOTE Notice that the exponent applies *only* to the variable *y*.

(c) $2y^{5/6} = 2\sqrt[6]{y^5}$

NOTE Now the exponent applies to 2*y* because of the parentheses.

(d) $(2y)^{5/6} = \sqrt[6]{(2y)^5}$

$= \sqrt[6]{32y^5}$

CHECK YOURSELF 7

Write each expression in radical form.

(a) $(ab)^{2/3}$ **(b)** $3x^{3/4}$ **(c)** $(3x)^{3/4}$

Example 8

Writing Expressions in Exponential Form

Using rational exponents, write each expression and simplify.

(a) $\sqrt[3]{5x} = (5x)^{1/3}$

(b) $\sqrt{9a^2b^4} = (9a^2b^4)^{1/2}$

$= 9^{1/2}(a^2)^{1/2}(b^4)^{1/2} = 3ab^2$

(c) $\sqrt[4]{16w^{12}z^8} = (16w^{12}z^8)^{1/4}$

$= 16^{1/4}(w^{12})^{1/4}(z^8)^{1/4} = 2w^3z^2$

CHECK YOURSELF 8

Using rational exponents, write each expression and simplify.

(a) $\sqrt{7a}$ **(b)** $\sqrt[3]{27p^6q^9}$ **(c)** $\sqrt[4]{81x^8y^{16}}$

CHECK YOURSELF ANSWERS

1. (a) 2; **(b)** −8; **(c)** 3 **2. (a)** 8; **(b)** $\frac{4}{9}$; **(c)** −8 **3. (a)** $\frac{1}{2}$; **(b)** $\frac{1}{27}$

4. (a) 6.562; **(b)** 0.192 **5. (a)** $z^{5/4}$; **(b)** $x^{1/2}$; **(c)** $b^{1/3}$

6. (a) $a^{1/2}b^{1/3}$; **(b)** w^2z; **(c)** $\frac{2y^2}{x^{1/3}}$ **7. (a)** $\sqrt[3]{a^2b^2}$; **(b)** $3\sqrt[4]{x^3}$; **(c)** $\sqrt[4]{27x^3}$

8. (a) $(7a)^{1/2}$; **(b)** $3p^2q^3$; **(c)** $3x^2y^4$

Name ____________________

Section ________ Date ________

8.6 Exercises

In exercises 1 to 12, use the definition of $a^{1/n}$ to evaluate each expression.

1. $36^{1/2}$

2. $100^{1/2}$

3. $-25^{1/2}$

4. $(-64)^{1/2}$

5. $(-49)^{1/2}$

6. $-49^{1/2}$

7. $27^{1/3}$

8. $(-64)^{1/3}$

9. $81^{1/4}$

10. $-32^{1/5}$

11. $\left(\frac{4}{9}\right)^{1/2}$

12. $\left(\frac{27}{8}\right)^{1/3}$

In exercises 13 to 22, use the definition of $a^{m/n}$ to evaluate each expression.

13. $27^{2/3}$

14. $16^{3/2}$

15. $(-8)^{4/3}$

16. $125^{2/3}$

17. $32^{2/5}$

18. $-81^{3/4}$

19. $81^{3/2}$

20. $(-243)^{3/5}$

21. $\left(\frac{8}{27}\right)^{2/3}$

22. $\left(\frac{9}{4}\right)^{3/2}$

In exercises 23 to 32, use the definition of $a^{-m/n}$ to evaluate the following expressions. Use your calculator to check each answer.

23. $25^{-1/2}$

24. $27^{-1/3}$

25. $81^{-1/4}$

26. $121^{-1/2}$

27. $9^{-3/2}$

28. $16^{-3/4}$

ANSWERS

1. ____
2. ____
3. ____
4. ____
5. ____
6. ____
7. ____
8. ____
9. ____
10. ____
11. ____
12. ____
13. ____
14. ____
15. ____
16. ____
17. ____
18. ____
19. ____ 20. ____
21. ____ 22. ____
23. ____ 24. ____
25. ____ 26. ____
27. ____ 28. ____

ANSWERS

29. ______
30. ______
31. ______
32. ______
33. ______
34. ______
35. ______
36. ______
37. ______
38. ______
39. ______
40. ______
41. ______
42. ______
43. ______
44. ______
45. ______
46. ______
47. ______
48. ______
49. ______ 50. ______
51. ______ 52. ______
53. ______ 54. ______
55. ______ 56. ______
57. ______ 58. ______
59. ______ 60. ______

29. $64^{-5/6}$

30. $16^{-3/2}$

31. $\left(\frac{4}{25}\right)^{-1/2}$

32. $\left(\frac{27}{8}\right)^{-2/3}$

In exercises 33 to 76, use the properties of exponents to simplify each expression. Assume all variables represent positive real numbers.

33. $x^{1/2} \cdot x^{1/2}$

34. $a^{2/3} \cdot a^{1/3}$

35. $y^{3/5} \cdot y^{1/5}$

36. $m^{1/4} \cdot m^{5/4}$

37. $b^{2/3} \cdot b^{3/2}$

38. $p^{5/6} \cdot p^{2/3}$

39. $\frac{x^{2/3}}{x^{1/3}}$

40. $\frac{a^{5/6}}{a^{1/6}}$

41. $\frac{s^{7/5}}{s^{2/5}}$

42. $\frac{z^{9/2}}{z^{3/2}}$

43. $\frac{w^{5/4}}{w^{1/2}}$

44. $\frac{b^{7/6}}{b^{2/3}}$

45. $(x^{3/4})^{4/3}$

46. $(y^{4/3})^{3/4}$

47. $(a^{2/5})^{3/2}$

48. $(p^{3/4})^{2/3}$

49. $(y^{-3/4})^{8}$

50. $(w^{-2/3})^{6}$

51. $(a^{2/3} \cdot b^{3/2})^{6}$

52. $(p^{3/4} \cdot q^{5/2})^{4}$

53. $(2x^{1/5} \cdot y^{3/5})^{5}$

54. $(3m^{3/4} \cdot n^{5/4})^{4}$

55. $(s^{3/4} \cdot t^{1/4})^{4/3}$

56. $(x^{5/2} \cdot y^{5/7})^{2/5}$

57. $(8p^{3/2} \cdot q^{5/2})^{2/3}$

58. $(16a^{1/3} \cdot b^{2/3})^{3/4}$

59. $(x^{3/5} \cdot y^{3/4} \cdot z^{3/2})^{2/3}$

60. $(p^{5/6} \cdot q^{2/3} \cdot r^{5/3})^{3/5}$

61. $\dfrac{a^{5/6} \cdot b^{3/4}}{a^{1/3} \cdot b^{1/2}}$

62. $\dfrac{x^{2/3} \cdot y^{3/4}}{x^{1/2} \cdot y^{1/2}}$

63. $\dfrac{(r^{-1} \cdot s^{1/2})^3}{r \cdot s^{-1/2}}$

64. $\dfrac{(w^{-2} \cdot z^{-1/4})^6}{w^{-8}z^{1/2}}$

65. $\left(\dfrac{x^{12}}{y^8}\right)^{1/4}$

66. $\left(\dfrac{p^9}{q^6}\right)^{1/3}$

67. $\left(\dfrac{m^{-1/4}}{n^{1/2}}\right)^4$

68. $\left(\dfrac{r^{1/5}}{s^{-1/2}}\right)^{10}$

69. $\left(\dfrac{r^{-1/2} \cdot s^{3/4}}{t^{1/4}}\right)^4$

70. $\left(\dfrac{a^{1/3} \cdot b^{-1/6}}{c^{-1/6}}\right)^6$

71. $\left(\dfrac{8x^3 \cdot y^{-6}}{z^{-9}}\right)^{1/3}$

72. $\left(\dfrac{16p^{-4} \cdot q^6}{r^2}\right)^{-1/2}$

73. $\left(\dfrac{16m^{-3/5} \cdot n^2}{m^{1/5} \cdot n^{-2}}\right)^{1/4}$

74. $\left(\dfrac{27x^{5/6} \cdot y^{-4/3}}{x^{-7/6} \cdot y^{5/3}}\right)^{1/3}$

75. $\left(\dfrac{x^{3/2} \cdot y^{1/2}}{z^2}\right)^{1/2}\left(\dfrac{x^{3/4} \cdot y^{3/2}}{z^{-3}}\right)^{1/3}$

76. $\left(\dfrac{p^{1/2} \cdot q^{4/3}}{r^{-4}}\right)^{3/4}\left(\dfrac{p^{15/8} \cdot q^{-3}}{r^6}\right)^{1/3}$

In exercises 77 to 84, write each expression in radical form. Do not simplify.

77. $a^{3/4}$

78. $m^{5/6}$

79. $2x^{2/3}$

80. $3m^{-2/5}$

81. $3x^{2/5}$

82. $2y^{-3/4}$

83. $(3x)^{2/5}$

84. $(2y)^{-3/4}$

In exercises 85 to 88, write each expression using rational exponents, and simplify when necessary.

85. $\sqrt{7a}$

86. $\sqrt{25w^4}$

87. $\sqrt[3]{8m^6n^9}$

88. $\sqrt[5]{32r^{10}s^{15}}$

ANSWERS

61. ______
62. ______
63. ______
64. ______
65. ______
66. ______
67. ______
68. ______
69. ______
70. ______
71. ______
72. ______
73. ______
74. ______
75. ______
76. ______
77. ______
78. ______
79. ______
80. ______
81. ______
82. ______
83. ______ 84. ______
85. ______ 86. ______
87. ______ 88. ______

ANSWERS

89. ______
90. ______
91. ______
92. ______
93. ______
94. ______
95. ______
96. ______
97. ______
98. ______
99. ______
100. ______
101. ______
102. ______
103. ______
104. ______
105. ______
106. ______
107. ______
108. ______
109. ______
110. ______

In exercises 89 to 92, evaluate each expression, using a calculator. Round each answer to three decimal places.

89. $46^{3/5}$

90. $23^{2/7}$

91. $12^{-2/5}$

92. $36^{-3/4}$

93. Describe the difference between x^{-2} and $x^{1/2}$.

94. Some rational exponents, like $\frac{1}{2}$, can easily be rewritten as terminating decimals (0.5). Others, like $\frac{1}{3}$, cannot. What is it that determines which rational numbers can be rewritten as terminating decimals?

In exercises 95 to 104, apply the appropriate multiplication patterns. Then simplify your result.

95. $a^{1/2}(a^{3/2} + a^{3/4})$

96. $2x^{1/4}(3x^{3/4} - 5x^{-1/4})$

97. $(a^{1/2} + 2)(a^{1/2} - 2)$

98. $(w^{1/3} - 3)(w^{1/3} + 3)$

99. $(m^{1/2} + n^{1/2})(m^{1/2} - n^{1/2})$

100. $(x^{1/3} + y^{1/3})(x^{1/3} - y^{1/3})$

101. $(x^{1/2} + 2)^2$

102. $(a^{1/3} - 3)^2$

103. $(r^{1/2} + s^{1/2})^2$

104. $(p^{1/2} - q^{1/2})^2$

As is suggested by several of the preceding exercises, certain expressions containing rational exponents are factorable. For instance, to factor $x^{2/3} - x^{1/3} - 6$, let $u = x^{1/3}$. Note that $x^{2/3} = (x^{1/3})^2 = u^2$.

Substituting, we have $u^2 - u - 6$, and factoring yields $(u - 3)(u + 2)$ or $(x^{1/3} - 3)(x^{1/3} + 2)$.

In exercises 105 to 110, use this technique to factor each expression.

105. $x^{2/3} + 4x^{1/3} + 3$

106. $y^{2/5} - 2y^{1/5} - 8$

107. $a^{4/5} - 7a^{2/5} + 12$

108. $w^{4/3} + 3w^{2/3} - 10$

109. $x^{4/3} - 4$

110. $x^{2/5} - 16$

In exercises 111 to 120, perform the indicated operations. Assume that n represents a positive integer and that the denominators are not zero.

111. $x^{3n} \cdot x^{2n}$

112. $p^{1-n} \cdot p^{n+3}$

113. $(y^2)^{2n}$

114. $(a^{3n})^3$

115. $\dfrac{r^{n+2}}{r^n}$

116. $\dfrac{w^n}{w^{n-3}}$

117. $(a^3 \cdot b^2)^{2n}$

118. $(c^4 \cdot d^2)^{3m}$

119. $\left(\dfrac{x^{n+2}}{x^n}\right)^{1/2}$

120. $\left(\dfrac{b^n}{b^{n-3}}\right)^{1/3}$

In exercises 121 to 124, write each expression in exponent form, simplify, and give the result as a single radical.

121. $\sqrt{\sqrt{x}}$

122. $\sqrt[3]{\sqrt{a}}$

123. $\sqrt[4]{\sqrt{y}}$

124. $\sqrt{\sqrt[3]{w}}$

In exercises 125 to 130, simplify each expression. Write your answer in scientific notation.

125. $(4 \times 10^8)^{1/2}$

126. $(8 \times 10^6)^{1/3}$

127. $(16 \times 10^{-12})^{1/4}$

128. $(9 \times 10^{-4})^{1/2}$

129. $(16 \times 10^{-8})^{1/2}$

130. $(16 \times 10^{-8})^{3/4}$

131. While investigating rainfall runoff in a region of semiarid farmland, a researcher encounters the following formula:

$$t = C\left(\frac{L}{xy^2}\right)^{1/3}$$

Evaluate t when $C = 20$, $L = 600$, $x = 3$, and $y = 5$.

132. The average velocity of water in an open irrigation ditch is given by the formula

$$V = \frac{1.5x^{2/3}y^{1/2}}{z}$$

Evaluate V when $x = 27$, $y = 16$, and $z = 12$.

133. Use the properties of exponents to decide what x should be to make each statement true. Explain your choices regarding which properties of exponents you decide to use.

(a) $(a^{2/3})^x = a$

(b) $(a^{5/6})^x = \dfrac{1}{a}$

(c) $a^{2x} \cdot a^{3/2} = 1$

(d) $(\sqrt{a^{2/3}})^x = a$

ANSWERS

111. ______
112. ______
113. ______
114. ______
115. ______
116. ______
117. ______
118. ______
119. ______
120. ______
121. ______
122. ______
123. ______
124. ______
125. ______
126. ______
127. ______
128. ______
129. ______
130. ______
131. ______
132. ______

133. ______

ANSWERS

134. ______

135. ______

134. The geometric mean is used to measure average inflation rates or interest rates. If prices increased by 15% over 5 years, then the average *annual* rate of inflation is obtained by taking the 5th root of 1.15:

$(1.15)^{1/5} = 1.0283$ or ~2.8%

The 1 is added to 0.15 because we are taking the original price and adding 15% of that price. We could write that as

$P + 0.15P$

Factoring, we get

$$\begin{aligned} P + 0.15P &= P(1 + 0.15) \\ &= P(1.15) \end{aligned}$$

In the introduction to this chapter, the following statement was made: "From February 1990 through February 2000, the Bureau of Labor Statistics computed an inflation rate of 68.1%, which is equivalent to an annual growth rate of 2.51%." From February 1990 through February 2000 is 12 months. To what exponent was 1.281 raised to obtain this average annual growth rate?

135. On your calculator, try evaluating $(-9)^{4/2}$ in the following two ways:

(a) $((-9)^4)^{1/2}$ **(b)** $((-9)^{1/2})^4$

Discuss the results.

Answers

1. 6 **3.** -5 **5.** Not a real number **7.** 3 **9.** 3 **11.** $\frac{2}{3}$ **13.** 9 **15.** 16 **17.** 4 **19.** 729 **21.** $\frac{4}{9}$ **23.** $\frac{1}{5}$ **25.** $\frac{1}{3}$ **27.** $\frac{1}{27}$ **29.** $\frac{1}{32}$ **31.** $\frac{5}{2}$ **33.** x **35.** $y^{4/5}$ **37.** $b^{13/6}$ **39.** $x^{1/3}$ **41.** s **43.** $w^{3/4}$ **45.** x **47.** $a^{3/5}$ **49.** $\frac{1}{y^6}$ **51.** a^4b^9 **53.** $32xy^3$ **55.** $st^{1/3}$ **57.** $4pq^{5/3}$ **59.** $x^{2/5}y^{1/2}z$ **61.** $a^{1/2}b^{1/4}$ **63.** $\frac{s^2}{r^4}$ **65.** $\frac{x^3}{y^2}$ **67.** $\frac{1}{mn^2}$ **69.** $\frac{s^3}{r^2t}$ **71.** $\frac{2xz^3}{y^2}$ **73.** $\frac{2n}{m^{1/5}}$ **75.** $xy^{3/4}$ **77.** $\sqrt[4]{a^3}$ **79.** $2\sqrt[3]{x^2}$ **81.** $3\sqrt[5]{x^2}$ **83.** $\sqrt[5]{9x^2}$ **85.** $(7a)^{1/2}$ **87.** $2m^2n^3$ **89.** 9.946 **91.** 0.370 **93.** **95.** $a^2 + a^{5/4}$ **97.** $a - 4$ **99.** $m - n$ **101.** $x + 4x^{1/2} + 4$ **103.** $r + 2r^{1/2}s^{1/2} + s$ **105.** $(x^{1/3} + 1)(x^{1/3} + 3)$ **107.** $(a^{2/5} - 3)(a^{2/5} - 4)$ **109.** $(x^{2/3} - 2)(x^{2/3} + 2)$ **111.** x^{5n} **113.** y^{4n} **115.** r^2 **117.** $a^{6n}b^{4n}$ **119.** x **121.** $\sqrt[4]{x}$ **123.** $\sqrt[8]{y}$ **125.** 2×10^4 **127.** 2×10^{-3} **129.** 4×10^{-4} **131.** 40 **133.** **135.** **(a)** 81; **(b)** not defined

8.7 Complex Numbers

8.7 OBJECTIVES

1. Define a complex number
2. Add and subtract complex numbers
3. Multiply and divide complex numbers

Radicals such as

$$\sqrt{-4} \quad \text{and} \quad \sqrt{-49}$$

are *not* real numbers because no real number squared produces a negative number. Our work in this section will extend our number system to include these **imaginary numbers,** which will allow us to consider radicals such as $\sqrt{-4}$.

First we offer a definition.

Definitions: The Imaginary Number *i*

The number *i* is defined as

$$i = \sqrt{-1}$$

Note that this means that

$$i^2 = -1$$

This definition of the number i gives us an alternate means of indicating the square root of a negative number.

Definitions: Rules and Procedures: Writing an Imaginary Number

When *a* is a positive real number,

$$\sqrt{-a} = \sqrt{a}i \quad \text{or} \quad i\sqrt{a}$$

Example 1

Using the Number *i*

Write each expression as a multiple of i.

(a) $\sqrt{-4} = \sqrt{4}i = 2i$

(b) $-\sqrt{-9} = -\sqrt{9}i = -3i$

(c) $\sqrt{-8} = \sqrt{8}i = 2\sqrt{2}i$ or $2i\sqrt{2}$

(d) $\sqrt{-7} = \sqrt{7}i$ or $i\sqrt{7}$

NOTE We simplify $\sqrt{8}$ as $2\sqrt{2}$. Note that we write the *i in front of* the radical to make it clear that *i* is *not part of* the radicand.

CHECK YOURSELF 1

Write each radical as a multiple of i.

(a) $\sqrt{-25}$ **(b)** $\sqrt{-24}$

We are now ready to define complex numbers in terms of the number i.

Definitions: Complex Number

A **complex number** is any number that can be written in the form

$a + bi$

in which a and b are real numbers and

$i = \sqrt{-1}$

NOTE The term "imaginary number" was introduced by René Descartes in 1637. Euler used i to indicate $\sqrt{-1}$ in 1748, but it was not until 1832 that Gauss used the term "complex number."

NOTE The first application of these numbers was made by Charles Steinmetz (1865–1923) in explaining the behavior of electric circuits.

NOTE Also, $5i$ is called a **pure imaginary** number.

NOTE The real numbers can be considered a subset of the set of complex numbers.

The form $a + bi$ is called the **standard form** of a complex number. We call a the **real part** of the complex number and b the **imaginary part.** Some examples follow.

$3 + 7i$ is an example of a complex number with real part 3 and imaginary part 7.

$5i$ is also a complex number because it can be written as $0 + 5i$.

-3 is a complex number because it can be written as $-3 + 0i$.

The basic operations of addition and subtraction on complex numbers are defined here.

Rules and Properties: Adding and Subtracting Complex Numbers

For the complex numbers $a + bi$ and $c + di$,

$(a + bi) + (c + di) = (a + c) + (b + d)i$

$(a + bi) - (c + di) = (a - c) + (b - d)i$

In words, we add or subtract the real parts and the imaginary parts of the complex numbers.

Example 2 illustrates the use of these definitions.

Example 2

Adding and Subtracting Complex Numbers

Perform the indicated operations.

NOTE The regrouping is essentially a matter of combining like terms.

(a) $(5 + 3i) + (6 - 7i) = (5 + 6) + (3 - 7)i$
$= 11 - 4i$

(b) $5 + (7 - 5i) = (5 + 7) + (-5i)$
$= 12 - 5i$

(c) $(8 - 2i) - (3 - 4i) = (8 - 3) + [-2 - (-4)]i$
$= 5 + 2i$

CHECK YOURSELF 2

Perform the indicated operations.

(a) $(4 - 7i) + (3 - 2i)$ **(b)** $-7 + (-2 + 3i)$ **(c)** $(-4 + 3i) - (-2 - i)$

Because complex numbers are binomial in form, the product of two complex numbers is found by applying our earlier multiplication pattern for binomials, as Example 3 illustrates.

Example 3

Multiplying Complex Numbers

Multiply.

(a) $(2 + 3i)(3 - 4i)$

$$= 2 \cdot 3 + 2(-4i) + (3i)3 + (3i)(-4i)$$
$$= 6 + (-8i) + 9i + (-12i^2)$$
$$= 6 - 8i + 9i + (-12)(-1)$$
$$= 6 + i + 12$$
$$= 18 + i$$

NOTE We can replace i^2 with -1 because of the definition of i, and we usually do so because of the resulting simplification.

(b) $(1 - 2i)(3 - 4i)$

$$= 1 \cdot 3 + 1(-4i) + (-2i)3 + (-2i)(-4i)$$
$$= 3 + (-4i) + (-6i) + 8i^2$$
$$= 3 - 10i + 8(-1)$$
$$= 3 + 10i - 8$$
$$= -5 - 10i$$

CHECK YOURSELF 3

Multiply $(2 - 5i)(3 - 2i)$.

Example 3 suggests the following pattern for multiplication on complex numbers.

Rules and Properties: Multiplying Complex Numbers

For the complex numbers $a + bi$ and $c + di$,

$$(a + bi)(c + di) = ac + adi + bci + bdi^2$$
$$= ac + adi + bci - bd$$
$$= (ac - bd) + (ad + bc)i$$

This formula for the general product of two complex numbers can be memorized. However, you will find it much easier to get used to the multiplication pattern as it is applied to complex numbers than to memorize this formula.

There is one particular product form that will seem very familiar. We call $a + bi$ and $a - bi$ **complex conjugates.** For instance,

$3 + 2i$ and $3 - 2i$

are complex conjugates.

Consider the product

$$\begin{aligned}(3 + 2i)(3 - 2i) &= 3^2 - (2i)^2 \\ &= 9 - 4i^2 = 9 - 4(-1) \\ &= 9 + 4 = 13\end{aligned}$$

The product of $3 + 2i$ and $3 - 2i$ is a real number. In general, we can write the product of two complex conjugates as

$$(a + bi)(a - bi) = a^2 + b^2$$

The fact that this product is always a real number will be very useful when we consider the division of complex numbers later in this section.

Example 4

Multiplying Complex Numbers

Multiply.

NOTE We could get the same result by applying the formula above with $a = 7$ and $b = 4$.

$$\begin{aligned}(7 - 4i)(7 + 4i) &= 7^2 - (4i)^2 \\ &= 7^2 - 4^2(-1) \\ &= 7^2 + 4^2 \\ &= 49 + 16 = 65\end{aligned}$$

CHECK YOURSELF 4

Multiply $(5 + 3i)(5 - 3i)$.

We are now ready to discuss the division of complex numbers. Generally, we find the quotient by multiplying the numerator and denominator by the conjugate of the denominator, as Example 5 illustrates.

Example 5

Dividing Complex Numbers

Divide.

NOTE Think of $3i$ as $0 + 3i$ and of its conjugate as $0 - 3i$, or $-3i$.

(a) $\dfrac{6 + 9i}{3i}$

$$\frac{6 + 9i}{3i} = \frac{(6 + 9i)(-3i)}{(3i)(-3i)}$$

The conjugate of $3i$ is $-3i$, and so we multiply the numerator and denominator by $-3i$.

NOTE Multiplying the numerator and denominator in the original expression by i would yield the same result. Try it yourself.

$$= \frac{-18i - 27i^2}{-9i^2}$$

$$= \frac{-18i - 27(-1)}{(-9)(-1)}$$

$$= \frac{27 - 18i}{9} = 3 - 2i$$

NOTE We multiply by $\dfrac{3 - 2i}{3 - 2i}$, which equals 1.

(b) $\dfrac{3 - i}{3 + 2i} = \dfrac{(3 - i)(3 - 2i)}{(3 + 2i)(3 - 2i)}$

$$= \frac{9 - 6i - 3i + 2i^2}{9 - 4i^2}$$

$$= \frac{9 - 9i - 2}{9 + 4}$$

NOTE To write a complex number in standard form, we separate the real component from the imaginary.

$$= \frac{7 - 9i}{13} = \frac{7}{13} - \frac{9}{13}i$$

(c) $\dfrac{2 + i}{4 - 5i} = \dfrac{(2 + i)(4 + 5i)}{(4 - 5i)(4 + 5i)}$

$$= \frac{8 + 4i + 10i + 5i^2}{16 - 25i^2}$$

$$= \frac{8 + 14i - 5}{16 + 25}$$

$$= \frac{3 + 14i}{41} = \frac{3}{41} + \frac{14}{41}i$$

CHECK YOURSELF 5

Divide.

(a) $\dfrac{5 + i}{5 - 3i}$ (b) $\dfrac{4 + 10i}{2i}$

We conclude this section with the following diagram, which summarizes the structure of the system of complex numbers.

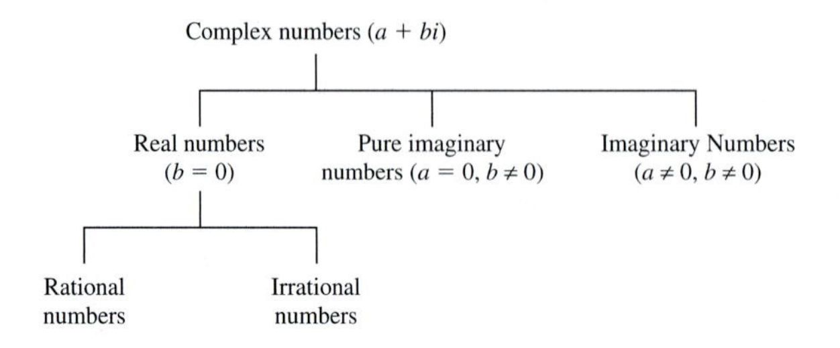

CHECK YOURSELF ANSWERS

1. **(a)** $5i$; **(b)** $2i\sqrt{6}$ **2.** **(a)** $7 - 9i$; **(b)** $-9 + 3i$; **(c)** $-2 + 4i$ **3.** $-4 - 19i$

4. 34 **5.** **(a)** $\frac{11}{17} + \frac{10}{17}i$; **(b)** $5 - 2i$

Name ______________________

Section ________ Date ________

8.7 Exercises

In exercises 1 to 10, write each root as a multiple of i. Simplify your results when necessary.

1. $\sqrt{-16}$

2. $\sqrt{-36}$

3. $-\sqrt{-64}$

4. $-\sqrt{-25}$

5. $\sqrt{-21}$

6. $\sqrt{-19}$

7. $\sqrt{-12}$

8. $\sqrt{-24}$

9. $-\sqrt{-108}$

10. $-\sqrt{-192}$

In exercises 11 to 26, perform the indicated operations.

11. $(3 + i) + (5 + 2i)$

12. $(2 + 3i) + (4 + 5i)$

13. $(3 - 2i) + (-2 + 7i)$

14. $(-5 + 3i) + (-2 + 7i)$

15. $(5 + 4i) - (3 + 2i)$

16. $(7 + 6i) - (3 + 5i)$

17. $(8 - 5i) - (3 + 2i)$

18. $(7 - 3i) - (-2 - 5i)$

19. $(5 + i) + (2 + 3i) + 7i$

20. $(3 - 2i) + (2 + 3i) + 7i$

21. $(2 + 3i) - (3 - 5i) + (4 + 3i)$

22. $(5 - 7i) + (7 + 3i) - (2 - 7i)$

23. $(7 + 3i) - [(3 + i) - (2 - 5i)]$

24. $(8 - 2i) - [(4 + 3i) - (-2 + i)]$

25. $(5 + 3i) + (-5 - 3i)$

26. $(8 - 7i) + (-8 + 7i)$

In exercises 27 to 42, find each product. Write your answer in standard form.

27. $3i(3 + 5i)$

28. $2i(7 + 3i)$

29. $4i(3 - 7i)$

30. $2i(6 + 3i)$

ANSWERS

1. ______ 2. ______

3. ______ 4. ______

5. ______ 6. ______

7. ______ 8. ______

9. ______ 10. ______

11. ______

12. ______

13. ______

14. ______

15. ______

16. ______

17. ______

18. ______

19. ______

20. ______

21. ______

22. ______

23. ______

24. ______

25. ______

26. ______

27. ______

28. ______

29. ______

30. ______

ANSWERS

31. ________
32. ________
33. ________
34. ________
35. ________
36. ________
37. ________
38. ________
39. ________
40. ________
41. ________
42. ________
43. ________
44. ________
45. ________
46. ________
47. ________
48. ________
49. ________
50. ________
51. ________
52. ________
53. ________
54. ________
55. ________
56. ________

31. $-2i(4 - 3i)$

32. $-5i(2 - 7i)$

33. $6i\left(\frac{2}{3} + \frac{5}{6}i\right)$

34. $4i\left(\frac{1}{2} + \frac{3}{4}i\right)$

35. $(3 + 2i)(2 + 3i)$

36. $(5 - 2i)(3 - i)$

37. $(4 - 3i)(2 + 5i)$

38. $(7 + 2i)(3 - 2i)$

39. $(-2 - 3i)(-3 + 4i)$

40. $(-5 - i)(-3 - 4i)$

41. $(5 - 2i)^2$

42. $(3 + 7i)^2$

In exercises 43 to 50, write the conjugate of each complex number. Then find the product of the given number and the conjugate.

43. $3 - 2i$

44. $5 + 2i$

45. $2 + 3i$

46. $7 - i$

47. $-3 - 2i$

48. $-5 - 7i$

49. $5i$

50. $-3i$

In exercises 51 to 62, find each quotient, and write your answer in standard form.

51. $\dfrac{3 + 2i}{i}$

52. $\dfrac{5 - 3i}{-i}$

53. $\dfrac{6 - 4i}{2i}$

54. $\dfrac{8 + 12i}{-4i}$

55. $\dfrac{3}{2 + 5i}$

56. $\dfrac{5}{2 - 3i}$

57. $\dfrac{13}{2 + 3i}$

58. $\dfrac{-17}{3 + 5i}$

59. $\dfrac{2 + 3i}{4 + 3i}$

60. $\dfrac{4 - 2i}{5 - 3i}$

61. $\dfrac{3 - 4i}{3 + 4i}$

62. $\dfrac{7 + 2i}{7 - 2i}$

63. The first application of complex numbers was suggested by the Norwegian surveyor Caspar Wessel in 1797. He found that complex numbers could be used to represent distance and direction on a two-dimensional grid. Why would a surveyor care about such a thing?

64. To what sets of numbers does 1 belong?

In this section, we defined $\sqrt{-4} = \sqrt{4}i = 2i$ in the process of expressing the square root of a negative number as a multiple of i.

Particular care must be taken with products in which two negative radicands are involved. For instance,

$$\sqrt{-3} \cdot \sqrt{-12} = (i\sqrt{3})(i\sqrt{12})$$
$$= i^2\sqrt{36} = (-1)\sqrt{36} = -6$$

is correct. However, if we try to apply the product property for radicals, we have

$$\sqrt{-3} \cdot \sqrt{-12} \stackrel{?}{=} \sqrt{(-3)(-12)} = \sqrt{36} = 6$$

which is *not* correct. The property $\sqrt{a} \cdot \sqrt{b} = \sqrt{ab}$ is not applicable in the case in which a and b are both negative. Radicals such as $\sqrt{-a}$ must be written in the standard form $i\sqrt{a}$ *before* multiplying, to use the rules for real-valued radicals.

In exercises 65 to 72, find each product.

65. $\sqrt{-5} \cdot \sqrt{-7}$

66. $\sqrt{-3} \cdot \sqrt{-10}$

67. $\sqrt{-2} \cdot \sqrt{-18}$

68. $\sqrt{-4} \cdot \sqrt{-25}$

69. $\sqrt{-6} \cdot \sqrt{-15}$

70. $\sqrt{-5} \cdot \sqrt{-30}$

71. $\sqrt{-10} \cdot \sqrt{-10}$

72. $\sqrt{-11} \cdot \sqrt{-11}$

ANSWERS

57. ______
58. ______
59. ______
60. ______
61. ______
62. ______
63. ______
64. ______
65. ______
66. ______
67. ______
68. ______
69. ______
70. ______
71. ______
72. ______

ANSWERS

73. ______

74. ______

75. ______

76. ______

77. ______

78. ______

79. ______

80. ______

81. ______

Because $i^2 = -1$, the positive integral powers of i form an interesting pattern. Consider the following.

$i = i$

$i^2 = -1$

$i^3 = i^2 \cdot i = (-1)i = -i$

$i^4 = i^2 \cdot i^2 = (-1)(-1) = 1$

$i^5 = i^4 \cdot i = 1 \cdot i = i$

$i^6 = i^4 \cdot i^2 = 1(-1) = -1$

$i^7 = i^4 \cdot i^3 = 1(-i) = -i$

$i^8 = i^4 \cdot i^4 = 1 \cdot 1 = 1$

Given the pattern above, do you see that any power of i will simplify to i, -1, $-i$, or 1? The easiest approach to simplifying higher powers of i is to write that power in terms of i^4 (because $1^4 = 1$). As an example,

$$i^{18} = i^{16} \cdot i^2 = (i^4)^4 \cdot i^2 = 1^4(-1) = -1$$

In exercises 73 to 80, use these comments to simplify each power of i.

73. i^{10}

74. i^9

75. i^{20}

76. i^{15}

77. i^{38}

78. i^{40}

79. i^{51}

80. i^{61}

81. Show that a square root of i is $\frac{\sqrt{2}}{2} + \frac{\sqrt{2}}{2}i$. That is, $\left(\frac{\sqrt{2}}{2} + \frac{\sqrt{2}}{2}i\right)^2 = i$.

Answers

1. $4i$ **3.** $-8i$ **5.** $i\sqrt{21}$ **7.** $2i\sqrt{3}$ **9.** $-6i\sqrt{3}$ **11.** $8 + 3i$ **13.** $1 + 5i$ **15.** $2 + 2i$ **17.** $5 - 7i$ **19.** $7 + 11i$ **21.** $3 + 11i$ **23.** $6 - 3i$ **25.** $0 + 0i$ **27.** $-15 + 9i$ **29.** $28 + 12i$ **31.** $-6 - 8i$ **33.** $-5 + 4i$ **35.** $13i$ **37.** $23 + 14i$ **39.** $18 + i$ **41.** $21 - 20i$ **43.** $3 + 2i, 13$ **45.** $2 - 3i, 13$ **47.** $-3 + 2i, 13$ **49.** $-5i, 25$ **51.** $2 - 3i$ **53.** $-2 - 3i$ **55.** $\frac{6}{29} - \frac{15}{29}i$ **57.** $2 - 3i$ **59.** $\frac{17}{25} + \frac{6}{25}i$ **61.** $-\frac{7}{25} - \frac{24}{25}i$ **63.** **65.** $-\sqrt{35}$ **67.** -6 **69.** $-3\sqrt{10}$ **71.** -10 **73.** -1 **75.** 1 **77.** -1 **79.** $-i$ **81.**

8 Summary

DEFINITION/PROCEDURE	EXAMPLE	REFERENCE
Roots and Radicals		**Section 8.1**
Square Roots Every positive number has two square roots. The positive or principal square root of a number a is denoted $\sqrt{a}$. The negative square root is written as $-\sqrt{a}$	$\sqrt{25} = 5$ 5 is the principal square root of 25 because $5^2 = 25$. $-\sqrt{49} = -7$ because $(-7)^2 = 49$.	
Higher Roots Cube roots, fourth roots, and so on are denoted by using an index and a radical. The principal nth root of a is written as $\sqrt[n]{a}$ (Index; Radical sign; Radicand) **Radicals Containing Variables** In general, $\sqrt[n]{x^n} = \begin{cases} \|x\| & \text{if } n \text{ is even} \\ x & \text{if } n \text{ is odd} \end{cases}$	$\sqrt[3]{27} = 3$ $\sqrt[3]{-64} = -4$ $\sqrt[4]{81} = 3$ $\sqrt{4^2} = 4$ $\sqrt{(-5)^2} = 5$ $\sqrt[3]{(-3)^3} = -3$ $\sqrt{m^2} = \|m\|$ $\sqrt[3]{27x^3} = 3x$	p. 581
Simplification of Radical Expressions		**Section 8.2**
Simplifying radical expressions entails applying two properties for radicals.		
Product Property $\sqrt[n]{ab} = \sqrt[n]{a} \cdot \sqrt[n]{b}$	$\sqrt{35} = \sqrt{5 \cdot 7}$ $= \sqrt{5} \cdot \sqrt{7}$	p. 591
Quotient Property $\sqrt[n]{\frac{a}{b}} = \frac{\sqrt[n]{a}}{\sqrt[n]{b}}, b \neq 0$		p. 591
Simplified Form for Radicals A radical is in *simplified form* if the following three conditions are satisfied. 1. The radicand has no factor raised to a power greater than or equal to the index. 2. No fraction appears in the radical. 3. No radical appears in a denominator. **Note:** Satisfying the third condition may require *rationalizing the denominator.*	$\sqrt{\frac{2}{5}} = \frac{\sqrt{2}}{\sqrt{5}}$ $\sqrt{18x^3} = \sqrt{9x^2 \cdot 2x}$ $= \sqrt{9x^2} \cdot \sqrt{2x}$ $= 3x\sqrt{2x}$ $\sqrt{\frac{5}{9}} = \frac{\sqrt{5}}{\sqrt{9}} = \frac{\sqrt{5}}{3}$ $\sqrt{\frac{3}{7x}} = \frac{\sqrt{3}}{\sqrt{7x}} = \frac{\sqrt{3} \cdot \sqrt{7x}}{\sqrt{7x} \cdot \sqrt{7x}}$ $= \frac{\sqrt{21x}}{\sqrt{49x^2}} = \frac{\sqrt{21x}}{7x}$	p. 592

Continued

DEFINITION/PROCEDURE	EXAMPLE	REFERENCE
Operations on Radical Expressions		**Section 8.3**
Radical expressions may be combined only if they are *similar,* that is, if they have the same radicand with the same index. Similar radicals are combined by application of the distributive property.	$8\sqrt{5} + 3\sqrt{5} = (8 + 3)\sqrt{5}$ $= 11\sqrt{5}$ $2\sqrt{18} - 4\sqrt{2}$ $= 2\sqrt{9 \cdot 2} - 4\sqrt{2}$ $= 2\sqrt{9} \cdot \sqrt{2} - 4\sqrt{2}$ $= 2 \cdot 3\sqrt{2} - 4\sqrt{2}$ $= 6\sqrt{2} - 4\sqrt{2} = (6 - 4)\sqrt{2}$ $= 2\sqrt{2}$	p. 601
Multiplication To multiply two radical expressions, we use $\sqrt[n]{a} \cdot \sqrt[n]{b} = \sqrt[n]{ab}$ and simplify the product. If binomial expressions are involved, we use the distributive property or the FOIL method.	$\sqrt{3x} \cdot \sqrt{6x^2} = \sqrt{18x^3}$ $= \sqrt{9x^2 \cdot 2x}$ $= \sqrt{9x^2} \cdot \sqrt{2x}$ $= 3x\sqrt{2x}$ $\sqrt{2}(5 + \sqrt{8}) = \sqrt{2} \cdot 5 + \sqrt{2} \cdot \sqrt{8}$ $= 5\sqrt{2} + 4$ $(3 + \sqrt{2})(5 - \sqrt{2})$ $= 15 - 3\sqrt{2} + 5\sqrt{2} - 2$ $= 13 + 2\sqrt{2}$	p. 604
Division To divide two radical expressions, rationalize the denominator by multiplying the numerator and denominator by the appropriate radical. If the divisor (the denominator) is a binomial, multiply the numerator and denominator by the conjugate of the denominator.	$\frac{5}{\sqrt{8}} = \frac{5 \cdot \sqrt{2}}{\sqrt{8} \cdot \sqrt{2}} = \frac{5\sqrt{2}}{\sqrt{16}}$ $= \frac{5\sqrt{2}}{4}$ **Note:** $3 + \sqrt{5}$ is the conjugate of $3 - \sqrt{5}$. $\frac{2}{3 - \sqrt{5}} = \frac{2(3 + \sqrt{5})}{(3 - \sqrt{5})(3 + \sqrt{5})}$ $= \frac{2(3 + \sqrt{5})}{4}$ $= \frac{3 + \sqrt{5}}{2}$	p. 607
Solving Radical Equations		**Section 8.4**
Power Property of Equality If $a = b$ then $a^n = b^n$	If $\sqrt{x + 1} = 5$ then $(\sqrt{x + 1})^2 = 5^2$ $x + 1 = 25$ $x = 24$	p. 615
If an equation involves two radicals, rewrite the equation so that there is one radical on each side and then use the power property to solve it.	Given $\sqrt{x} + \sqrt{x + 7} = 7$ $\sqrt{x} = 7 - \sqrt{x + 7}$ $x = 49 - 14\sqrt{x + 7} + (x + 7)$ $x = 56 + x - 14\sqrt{x + 7}$ $-56 = -14\sqrt{x + 7}$ $4 = \sqrt{x + 7}$ $16 = x + 7$ $x = 9$	p. 618

DEFINITION/PROCEDURE	EXAMPLE	REFERENCE
Geometric and Other Applications of Radical Expressions		**Section 8.5**
Pythagorean Theorem 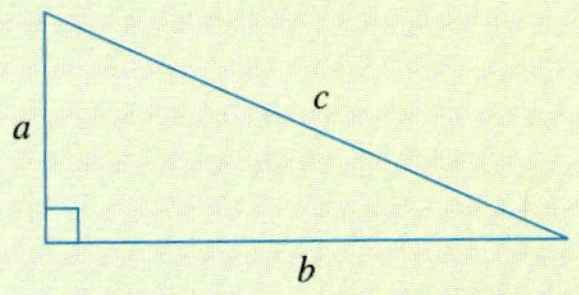 $a^2 + b^2 = c^2$ In a right triangle, the sum of the squares of the two sides is equal to the square of the hypotenuse.	Given triangle 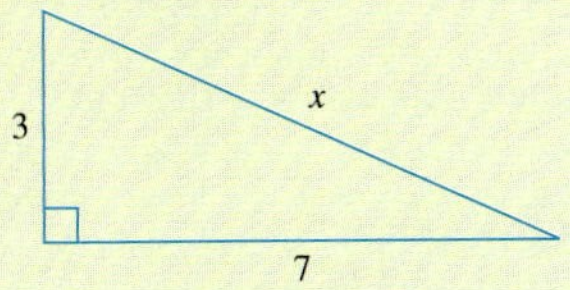 $x^2 = 7^2 + 3^2$ $x^2 = 49 + 9$ $x = \sqrt{58}$ ≈ 7.6	**p. 627**
Rational Exponents		**Section 8.6**
Rational exponents are an alternate way of indicating roots. We use the following definition. If a is any real number and n is a positive integer ($n > 1$), $a^{1/n} = \sqrt[n]{a}$ We restrict a so that a is nonnegative when n is even. We also define the following. For any real number a and positive integers m and n, with $n > 1$, then $a^{m/n} = (\sqrt[n]{a})^m = \sqrt[n]{a^m}$	$36^{1/2} = \sqrt{36} = 6$ $-27^{1/3} = -\sqrt[3]{27} = -3$ $243^{1/5} = \sqrt[5]{243} = 3$ $25^{-1/2} = \dfrac{1}{\sqrt{25}} = \dfrac{1}{5}$ $27^{2/3} = (\sqrt[3]{27})^2$ $= 3^2 = 9$ $(a^4b^8)^{3/4} = \sqrt[4]{(a^4b^8)^3}$ $= \sqrt[4]{a^{12}b^{24}} = a^3b^6$	**p. 635**
Properties of Exponents The following five properties for exponents continue to hold for rational exponents. *Product Rule* $a^m \cdot a^n = a^{m+n}$	$x^{1/2} \cdot x^{1/3} = x^{1/2+1/3} = x^{5/6}$	**p. 638**
Quotient Rule $\dfrac{a^m}{a^n} = a^{m-n}$	$\dfrac{x^{3/2}}{x^{1/2}} = x^{3/2-1/2} = x^{2/2} = x$	**p. 638**
Power Rule $(a^m)^n = a^{m \cdot n}$	$(x^{1/3})^5 = x^{1/3 \cdot 5} = x^{5/3}$	**p. 638**
Product-Power Rule $(ab)^m = a^mb^m$	$(2xy)^{1/2} = 2^{1/2}x^{1/2}y^{1/2}$	**p. 638**
Quotient-Power Rule $\left(\dfrac{a}{b}\right)^m = \dfrac{a^m}{b^m}$	$\left(\dfrac{x^{1/3}}{3}\right)^2 = \dfrac{(x^{1/3})^2}{3^2}$ $= \dfrac{x^{2/3}}{9}$	**p. 638**
Complex Numbers		**Section 8.7**
The number i is defined as $i = \sqrt{-1}$	$\sqrt{-16} = 4i$ $\sqrt{-8} = 2i\sqrt{2}$	

Continued

DEFINITION/PROCEDURE	EXAMPLE	REFERENCE
Complex Numbers		**Section 8.7**
Note that this means that $i^2 = -1$ A **complex number** is any number that can be written in the form $a + bi$ in which a and b are real numbers and $i = \sqrt{-1}$		p. 647
Addition and Subtraction For the complex numbers $a + bi$ and $c + di$, $(a + bi) + (c + di) = (a + c) + (b + d)i$ and $(a + bi) - (c + di) = (a - c) + (b - d)i$	$(2 + 3i) + (-3 - 5i)$ $= (2 - 3) + (3 - 5)i$ $= -1 - 2i$ $(5 - 2i) - (3 - 4i)$ $= (5 - 3) + (-2 - (-4))i$ $= 2 + 2i$	p. 648
Multiplication For the complex numbers $a + bi$ and $c + di$, $(a + bi)(c + di) = (ac - bd) + (ad + bc)i$ **Note:** It is generally easier to use the FOIL multiplication pattern and the definition of i, rather than to apply the above formula.	$(2 + 5i)(3 - 4i)$ $= 6 - 8i + 15i - 20i^2$ $= 6 + 7i - 20(-1)$ $= 26 + 7i$	p. 649
Division To divide two complex numbers, we multiply the numerator and denominator by the complex conjugate of the denominator and write the result in standard form.	$\frac{3 + 2i}{3 - 2i} = \frac{(3 + 2i)(3 + 2i)}{(3 - 2i)(3 + 2i)}$ $= \frac{9 + 6i + 6i + 4i^2}{9 - 4i^2}$ $= \frac{9 + 12i + 4(-1)}{9 - 4(-1)}$ $= \frac{5 + 12i}{13} = \frac{5}{13} + \frac{12}{13}i$	p. 651

Summary Exercises

This summary exercise set is provided to give you practice with each of the objectives of the chapter. Each exercise is keyed to the appropriate chapter section. The answers are provided in the instructor's manual that accompanies this text. Your instructor will provide guidelines on how to best use these exercises in your instructional program.

[8.1] Evaluate each of the following roots over the set of real numbers.

1. $\sqrt{121}$

2. $-\sqrt{64}$

3. $\sqrt{-81}$

4. $\sqrt[3]{64}$

5. $\sqrt[3]{-64}$

6. $\sqrt[4]{81}$

7. $\sqrt{\dfrac{9}{16}}$

8. $\sqrt[3]{-\dfrac{8}{27}}$

9. $\sqrt{8^2}$

Simplify each of the following expressions. Assume that all variables represent positive real numbers for all subsequent exercises in this exercise set.

10. $\sqrt{4x^2}$

11. $\sqrt{a^4}$

12. $\sqrt{36y^2}$

13. $\sqrt{49w^4z^6}$

14. $\sqrt[3]{x^9}$

15. $\sqrt[3]{-27b^6}$

16. $\sqrt[3]{8r^3s^9}$

17. $\sqrt[4]{16x^4y^8}$

18. $\sqrt[5]{32p^5q^{15}}$

[8.2] Use the product property to write each of the following expressions in simplified form.

19. $\sqrt{45}$

20. $-\sqrt{75}$

21. $\sqrt{60x^2}$

22. $\sqrt{108a^3}$

23. $\sqrt[3]{32}$

24. $\sqrt[3]{-80w^4z^3}$

Use the quotient property to write each of the following expressions in simplified form.

25. $\sqrt{\dfrac{9}{16}}$

26. $\sqrt{\dfrac{7}{36}}$

27. $\sqrt{\dfrac{y^4}{49}}$

28. $\sqrt{\dfrac{2x}{9}}$

29. $\sqrt{\dfrac{5}{16x^2}}$

30. $\sqrt[3]{\dfrac{5a^2}{27}}$

[8.3] Simplify each of the following expressions if necessary. Then add or subtract as indicated.

31. $7\sqrt{10} + 4\sqrt{10}$

32. $5\sqrt{3x} - 2\sqrt{3x}$

33. $7\sqrt[3]{2x} + 3\sqrt[3]{2x}$

34. $8\sqrt{10} - 3\sqrt{10} + 2\sqrt{10}$

35. $\sqrt{72} + \sqrt{50}$

36. $\sqrt{54} - \sqrt{24}$

37. $9\sqrt{7} - 2\sqrt{63}$

38. $\sqrt{20} - \sqrt{45} + 2\sqrt{125}$

39. $2\sqrt[3]{16} + 3\sqrt[3]{54}$

40. $\sqrt{27w^3} - w\sqrt{12w}$

41. $\sqrt[3]{128a^5} + 6a\sqrt[3]{2a^2}$

42. $\sqrt{20} + \frac{3}{\sqrt{5}}$

43. $\sqrt{72x} - \sqrt{\frac{x}{2}}$

44. $\sqrt[3]{81a^4} - a\sqrt[3]{\frac{a}{9}}$

45. $\frac{\sqrt{15}}{3} - \frac{1}{\sqrt{15}}$

[8.3] Multiply and simplify each of the following expressions.

46. $\sqrt{3x} \cdot \sqrt{7y}$

47. $\sqrt{6x^2} \cdot \sqrt{18}$

48. $\sqrt[3]{4a^2b} \cdot \sqrt[3]{ab^2}$

49. $\sqrt{5}(\sqrt{3} + 2)$

50. $\sqrt{6}(\sqrt{8} - \sqrt{2})$

51. $\sqrt{a}(\sqrt{5a} + \sqrt{125a})$

52. $(\sqrt{3} + 5)(\sqrt{3} - 7)$

53. $(\sqrt{7} - \sqrt{2})(\sqrt{7} + \sqrt{3})$

54. $(\sqrt{5} - 2)(\sqrt{5} + 2)$

55. $(\sqrt{7} - \sqrt{3})(\sqrt{7} + \sqrt{3})$

56. $(2 + \sqrt{3})^2$

57. $(\sqrt{5} - \sqrt{2})^2$

Rationalize the denominator, and write each of the following expressions in simplified form.

58. $\sqrt{\frac{3}{7}}$

59. $\frac{\sqrt{12}}{\sqrt{x}}$

60. $\frac{\sqrt{10a}}{\sqrt{5b}}$

61. $\sqrt[3]{\frac{3}{a^2}}$

62. $\frac{2}{\sqrt[3]{3x}}$

63. $\frac{\sqrt[3]{x^2}}{\sqrt[3]{y^5}}$

Divide and simplify each of the following expressions.

64. $\frac{1}{3 + \sqrt{2}}$

65. $\frac{11}{5 - \sqrt{3}}$

66. $\frac{\sqrt{5} - 2}{\sqrt{5} + 2}$

67. $\frac{\sqrt{x} - 3}{\sqrt{x} + 3}$

[8.4] Solve each of the following equations. Be sure to check your solutions.

68. $\sqrt{x - 5} = 4$

69. $\sqrt{3x - 2} + 2 = 5$

70. $\sqrt{y + 7} = y - 5$

71. $\sqrt{2x - 1} + x = 8$

72. $\sqrt[3]{5x + 2} = 3$

73. $\sqrt[3]{x^2 + 2} - 3 = 0$

74. $\sqrt{z + 7} = 1 + \sqrt{z}$

75. $\sqrt{4x + 5} - \sqrt{x - 1} = 3$

Solve each of the following equations for the indicated variable.

76. $r = \sqrt{x^2 + y^2}$ for x

77. $t = 2\pi\sqrt{\frac{l}{10}}$ for l

[8.5] Use the Pythagorean theorem to find the length x in each triangle. Write answers to the nearest tenth.

78.

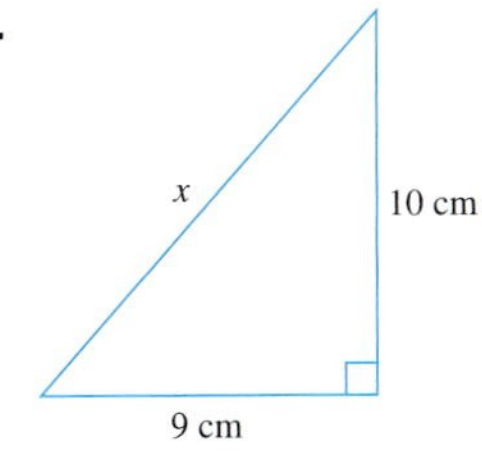

79.

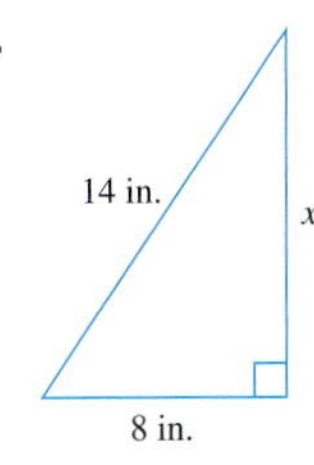

80. Find the width of a rectangle whose diagonal is 12 ft and whose length is 6 ft.

81. If a window is 3 ft on its shorter side, how high should it be, to the nearest tenth of a foot, to be a golden rectangle?

82. The base of a ladder that is 16 ft long is placed 4 ft from a wall of a building. How high up the wall is the top of the ladder?

83. How long must a ladder be to reach 8 m up the side of a building when the base of the ladder is 2 m from the building?

[8.6] Evaluate each of the following expressions.

84. $49^{1/2}$

85. $-100^{1/2}$

86. $(-27)^{1/3}$

87. $16^{1/4}$

88. $64^{2/3}$

89. $25^{3/2}$

90. $\left(\frac{4}{9}\right)^{3/2}$

91. $49^{-1/2}$

92. $81^{-3/4}$

Use the properties of exponents to simplify each of the following expressions.

93. $x^{3/2} \cdot x^{5/2}$

94. $b^{2/3} \cdot b^{3/2}$

95. $\frac{r^{8/5}}{r^{3/5}}$

96. $\frac{a^{5/4}}{a^{1/2}}$

97. $(x^{3/5})^{2/3}$

98. $(y^{-4/3})^{6}$

99. $(x^{4/5}y^{3/2})^{10}$

100. $(16x^{1/3} \cdot y^{2/3})^{3/4}$

101. $\left(\frac{x^{-2}y^{-1/6}}{x^{-4}y}\right)^{3}$

102. $\left(\frac{27y^{3}z^{-6}}{x^{-3}}\right)^{1/3}$

Write each of the following expressions in radical form.

103. $x^{3/4}$

104. $(w^2z)^{2/5}$

105. $3a^{2/3}$

106. $(3a)^{2/3}$

Write each of the following expressions, using rational exponents, and simplify when necessary.

107. $\sqrt[5]{7x}$

108. $\sqrt{16w^4}$

109. $\sqrt[3]{27p^3q^9}$

110. $\sqrt[4]{16a^8b^{16}}$

[8.7] Write each of the following roots as a multiple of i. Simplify your result.

111. $\sqrt{-49}$

112. $\sqrt{-13}$

113. $-\sqrt{-60}$

Perform the indicated operations.

114. $(2 + 3i) + (3 - 5i)$

115. $(7 - 3i) + (-3 - 2i)$

116. $(5 - 3i) - (2 + 5i)$

117. $(-4 + 2i) - (-1 - 3i)$

Find each of the following products.

118. $4i(7 - 2i)$

119. $(5 - 2i)(3 + 4i)$

120. $(3 - 4i)^2$

121. $(2 - 3i)(2 + 3i)$

Find each of the following quotients, and write your answer in standard form.

122. $\dfrac{5 - 15i}{5i}$

123. $\dfrac{10}{3 - 4i}$

124. $\dfrac{3 - 2i}{3 + 2i}$

125. $\dfrac{5 + 10i}{2 + i}$

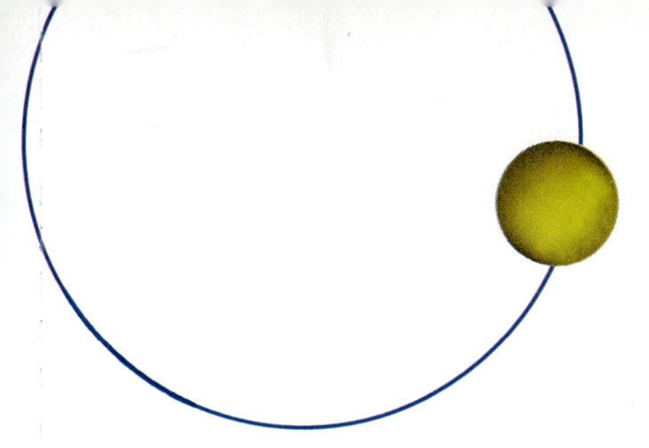

Self-Test for Chapter 8

Name ____________________

Section ________ Date ________

ANSWERS

1. ________
2. ________
3. ________
4. ________
5. ________
6. ________
7. ________
8. ________
9. ________
10. ________
11. ________
12. ________
13. ________
14. ________
15. ________

The purpose of this self-test is to help you check your progress and to review for a chapter test in class. Allow yourself about 1 hour to take the test. When you are done, check your answers in the back of the book. If you missed any problems, be sure to go back and review the appropriate sections in the chapter and the exercises that are provided.

Simplify each expression. Assume that all variables represent positive real numbers in all subsequent problems.

1. $\sqrt{49a^4}$

2. $\sqrt[3]{-27w^6z^9}$

Use the product or quotient properties to write each expression in simplified form.

3. $\sqrt[3]{9p^7q^5}$

4. $\dfrac{7x}{\sqrt{64y^2}}$

Rationalize the denominator, and write each expression in simplified form.

5. $\sqrt{\dfrac{5x}{8y}}$

6. $\dfrac{3}{\sqrt[3]{9x}}$

Simplify each expression if necessary. Then add or subtract as indicated.

7. $\sqrt{3x^3} + x\sqrt{75x} - \sqrt{27x^3}$

8. $\sqrt[3]{54m^4} + m\sqrt[3]{16m}$

Multiply or divide as indicated. Then simplify your result.

9. $\sqrt{6x}(\sqrt{18x} - \sqrt{2x})$

10. $\dfrac{\sqrt{6} - \sqrt{3}}{\sqrt{6} + \sqrt{3}}$

Solve the following equations. Be sure to check your solutions.

11. $\sqrt{x - 7} - 2 = 0$

12. $\sqrt{3w + 4} + w = 8$

Use the properties of exponents to simplify each expression.

13. $(16x^4)^{3/2}$

14. $(27m^{3/2}n^{-6})^{2/3}$

15. $\left(\dfrac{16r^{-1/3}s^{5/3}}{rs^{-7/3}}\right)^{3/4}$

ANSWERS

16. ______

17. ______

18. ______

19. ______

20. ______

21. ______

22. ______

23. ______

24. ______

25. ______

16. Use the Pythagorean theorem to find the length x in the given triangle. Write your answer to the nearest tenth.

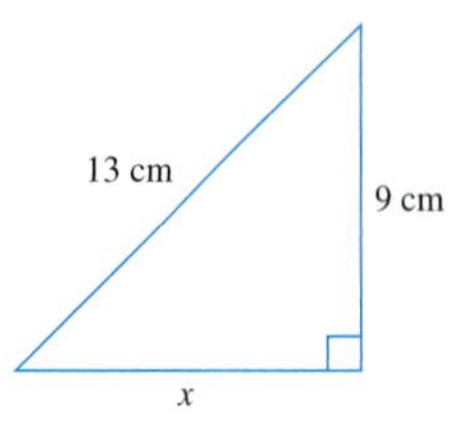

17. Find the width of a rectangle whose diagonal is 25 ft and whose length is 12 ft.

18. If the longer side of a rectangle is 6 cm, find the width (to the nearest tenth of a cm) to make the rectangle a golden rectangle.

19. What is the length of a diagonal in a rectangle with sides 4.5 cm and 7.8 cm? Give your answer to the nearest tenth of a centimeter.

20. In a right triangle, one leg has a length of 40 cm and the hypotenuse is 80 cm. Find the length of the other leg.

Write the expression in radical form and simplify.

21. $(a^7b^3)^{2/5}$

Write the expression, using rational exponents. Then simplify.

22. $\sqrt[3]{125p^9q^6}$

Perform the indicated operations.

23. $(-2 + 3i) - (-5 - 7i)$

24. $(5 - 3i)(-4 + 2i)$

25. $\dfrac{10 - 20i}{3 - i}$

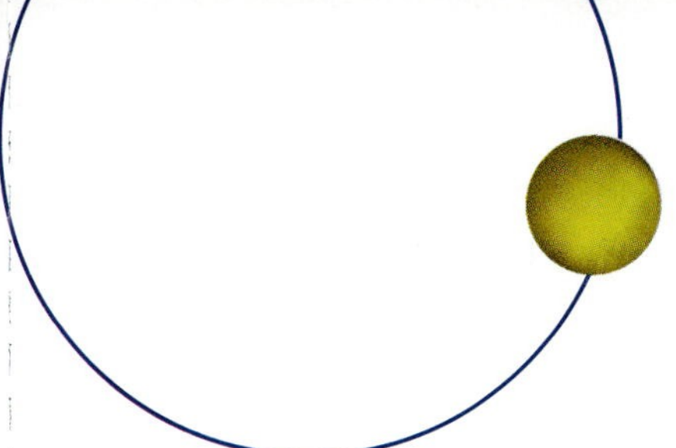

Cumulative Test for Chapters 1 to 8

Name ______________

Section ________ Date ________

ANSWERS

1. ______
2. ______
3. ______
4. ______
5. ______
6. ______
7. ______
8. ______
9. ______
10. ______
11. ______
12. ______
13. ______
14. ______
15. ______
16. ______

This test is provided to help you in the process of reviewing the previous chapters. Answers are provided in the back of the book. If you missed any answers, be sure to go back and review the appropriate chapter section.

1. Solve the equation $7x - 6(x - 1) = 2(5 + x) + 11$.

2. If $f(x) = 3x^6 - 4x^3 + 9x^2 - 11$, find $f(-1)$.

3. Find the equation of the line that has a y intercept of $(0, -6)$ and is parallel to the line $6x - 4y = 18$.

4. Solve the equation $|3x - 5| = 4$.

Simplify each of the following polynomial functions.

5. $f(x) = 5x^2 - 8x + 11 - (-3x^2 - 2x + 8) - (-2x^2 - 4x + 3)$

6. $f(x) = (5x + 3)(2x - 9)$

Factor each of the following completely.

7. $2x^3 + x^2 - 3x$

8. $9x^4 - 36y^4$

9. $4x^2 + 8xy - 5x - 10y$

Simplify each of the following rational expressions.

10. $\dfrac{2x^2 + 13x + 15}{6x^2 + 7x - 3}$

11. $\dfrac{3}{x - 5} - \dfrac{2}{x - 1}$

12. $\dfrac{a^2 - 4a}{a^2 - 6a + 8} \cdot \dfrac{a^2 - 4}{2a^2}$

13. $\dfrac{a^2 - 9}{a^2 - a - 12} - \dfrac{a^2 - a - 6}{a^2 - 2a - 8}$

Simplify each of the following radical expressions.

14. $\sqrt{3x^3y}\,\sqrt{4x^5y^6}$

15. $(\sqrt{3} - 5)(\sqrt{2} + 3)$

16. $\sqrt{50x^3} - x\sqrt{32x}$

17. ______

18. ______

19. ______

20. ______

21. ______

22. ______

23. ______

24. ______

25. ______

17. Use the properties of exponents to simplify the expression $\left(\dfrac{a^{-2}b^{-1/4}c^{3}}{a^{-1/2}b^{3/4}c^{0}}\right)^{4}$.

18. Graph the equation $2x - 3y = 12$.

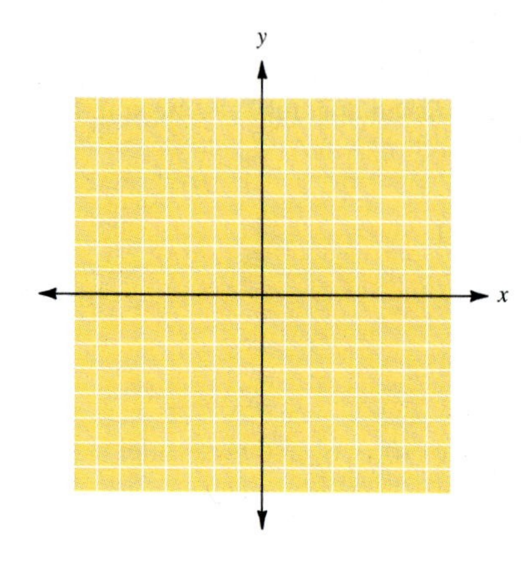

Find each of the following products.

19. $3i(6 - 5i)$

20. $(5 + 2i)(3 - 4i)$

21. Solve the equation $\sqrt{x + 14} - x + 6 = 0$

22. Solve the system

$$2x - 5y = 13$$
$$x + 3y = 12$$

Solve the following inequalities.

23. $3x - 5 \le 4x - 2(x - 6)$

24. $|x - 4| \le 5$

25. The difference between an integer and its square root is 12. Find the integer.

Quadratic Equations, Functions, and Inequalities

9

INTRODUCTION

Running power lines from a power plant, wind farm, or hydroelectric plant to a city is a very costly enterprise. Land must be cleared, towers built, and conducting wires strung from tower to tower across miles of countryside. Typical construction designs run from about 300- to 1200-ft spans, with towers about 75 to 200 ft high. Of course, if a lot of towers are needed, and many of them must be tall, the construction costs skyrocket.

Power line construction carries a unique set of problems. Towers must be built tall enough and close together enough to keep the conducting lines well above the ground. The sag of these wires (how much they droop from the towers) is a function of the weight of the conductor, the span length, and the tension in the wires. The amount of this sag, measured in feet, is approximated by the following formula:

$$\text{Sag} = \frac{wS^2}{8T}$$

in which w = weight of wires in pounds/foot

S = span length in feet

T = tension in wires measured in pounds

If the weight of the conducting wires is 2074 lb per 1000 ft, and the tension on the wires is 5000 lb, the relationship between the sag and the span is given by the following equation:

$$\text{Sag} = \frac{2.074S^2}{8(5000)}$$

The following graph shows how the sag increases as the distance between the towers increases.

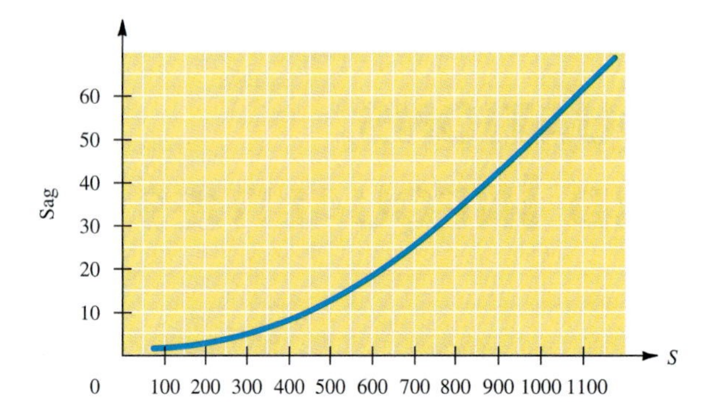

There would be a different graph for every combination of wire weight and tension. This graph represents a wire weight of 2.074 lb/ft and a 5000-lb tension on the conductor.

In this graph, when the towers are 600 ft apart, the conducting wires will sag 18.7 ft; if the towers are 1000 ft apart, the wires will sag 51.8 ft. Which will be the best for a particular area? The building planners must know what kind of clearance is needed over the terrain to decide which costs less: very tall towers spaced far apart or shorter towers placed closer together.

The actual curve of the power lines is called a **catenary curve.** The curve that we use to approximate the sag is a **parabola,** one of the conic sections we will study in this chapter.

9.1 Graphing a Factorable Quadratic Function

9.1 OBJECTIVES

1. Find the zeros for a factorable quadratic function
2. Find the vertex for the graph of a factorable quadratic function
3. Sketch the graph of a factorable quadratic function

If you know how fast a ball is thrown (its **initial velocity**) straight up into the air, you can find its maximum height and predict the number of seconds it will be in the air. If you know how long it was in the air, you can find both its maximum height and its initial velocity.

To analyze such a problem, we use a quadratic function.

Definitions: Quadratic Function

A function that can be written in the form

$f(x) = ax^2 + bx + c, \quad a \neq 0$

is called a **quadratic function.**

The height of a ball (in feet) thrown up from the ground is determined by the quadratic function

$$h(x) = -16x^2 + v_o x$$

in which v_o represents the initial velocity and x represents the number of seconds that have passed since the ball was thrown.

The function

$$h(x) = -16x^2 + 64x$$

gives us the height, after x seconds, of a ball thrown with an initial velocity of 64 ft/s.

Example 1

Finding the Zeros of a Quadratic Function

Find the zeros for the function

$$h(x) = -16x^2 + 64x$$

The zeros of the function are the values of x for which $h(x) = 0$. We wish to solve the equation

$$0 = -16x^2 + 64x \quad \text{or}$$

$$16x^2 - 64x = 0$$

In Section 6.9, we solved quadratic equations by the method of factoring. Using that technique, we find

$$16x(x - 4) = 0$$

The zero product rule tells us that there are two solutions, $x = 0$ or $x = 4$. Applying this answer to our ball-throwing example tells us that the ball is at ground level twice, at 0 s (just before the ball is thrown) and after 4 s (when the ball lands back on the ground).

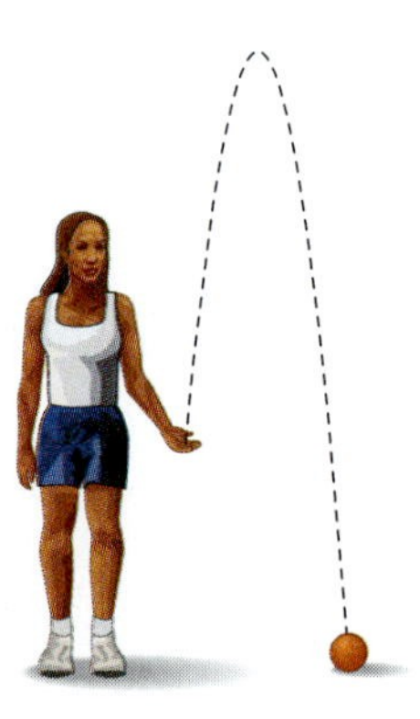

CHECK YOURSELF 1

Find the zeros for the function

$$h(x) = -16x^2 + 48x$$

In our next example, we will sketch the graph of the function $h(x)$.

Example 2

Graphing a Quadratic Function

NOTE This should not be confused with a "sketch" of the flight of the ball!

Sketch the graph of the function

$$h(x) = -16x^2 + 64x$$

From the table below, we can plot five points

x	$h(x)$
0	0
1	48
2	64
3	48
4	0

We will sketch our graph using only the first quadrant. We will plot the five points, then connect them with a smooth curve.

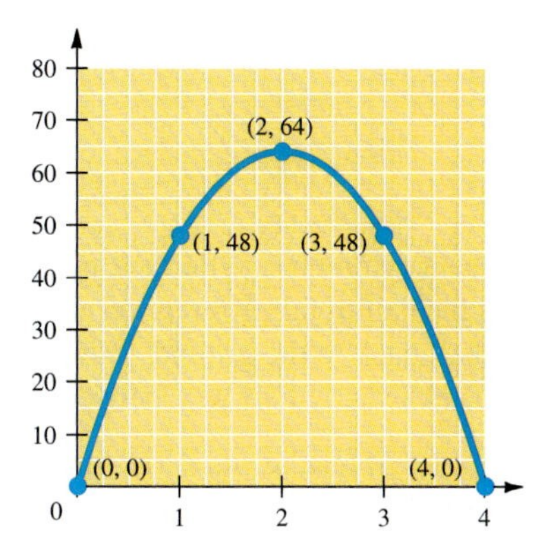

The shape of the graph of a quadratic function is called a **parabola.** Note that the curve at the vertex is not angled or pointed.

CHECK YOURSELF 2

Sketch the graph of the function

$h(x) = -16x^2 + 48x$

An important point for many mathematical applications is called the **vertex.**

Definitions: Vertex

The highest or lowest point on the graph for a quadratic function is called its **vertex.**

In the case of the ball thrown into the air, the vertex is the highest point the ball reaches.

Example 3

Finding the Vertex

Find the vertex for the function

$h(x) = -16x^2 + 64x$

The vertex of the function is the ordered pair $(x, h(x))$ for which $h(x)$ has the greatest value. We will look at a table of values that will help us identify the vertex.

x	$h(x)$
0	0
1	48
2	64
3	48
4	0

Note that there seems to be a symmetric pattern. It takes the ball exactly as much time to reach its vertex as it takes the ball to fall from its vertex to the ground. (This is because the thing that is bringing the ball back to the ground, gravity, is constant).

The ball reaches its vertex exactly half way between its time of release and the time it falls to the ground. In this case, it reaches its vertex after $\frac{(0 + 4)}{2} = 2$ s. To find its height at the vertex, substitute 2 for the x.

$$h(2) = -16(2)^2 + 64(2)$$
$$= -64 + 128$$
$$= 64$$

The vertex is represented by the ordered pair (2, 64).

CHECK YOURSELF 3

Find the vertex for the function

$h(x) = -16x^2 + 48x$

The preceding example demonstrated a method for sketching the graph of any factorable quadratic function. This method is summarized in the following algorithm.

Step by Step: Sketching the Graph of a Quadratic Function

Step 1 Factor the quadratic.

Step 2 Using the zero product rule, plot the points associated with the zeros of the function on the x axis.

Step 3 Find the vertex (the mean of the two x values in step 2 and the function value for that x) and plot the associated point.

Step 4 Draw a smooth curve connecting the three plotted points.

Example 4

Sketching the Graph of a Quadratic Function

Sketch the graph of the function

$f(x) = x^2 + 4x - 5$

Factoring, we have

$f(x) = (x - 1)(x + 5)$

Using the zero product rule, we find the two points (1, 0) and (−5, 0). We plot those points.

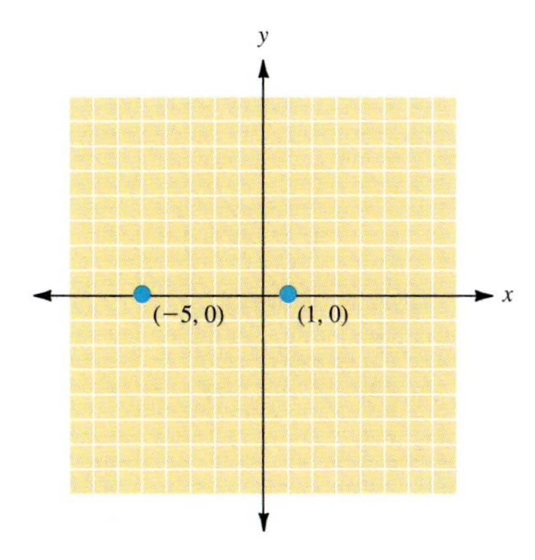

To find the vertex, we find the mean of 1 and −5, which is −2, then we find $f(-2)$.

$$\begin{aligned} f(-2) &= (-2)^2 + 4(-2) - 5 \\ &= 4 - 8 - 5 \\ &= -9 \end{aligned}$$

Plotting the vertex, (−2, −9), and connecting the points with a smooth curve, we get

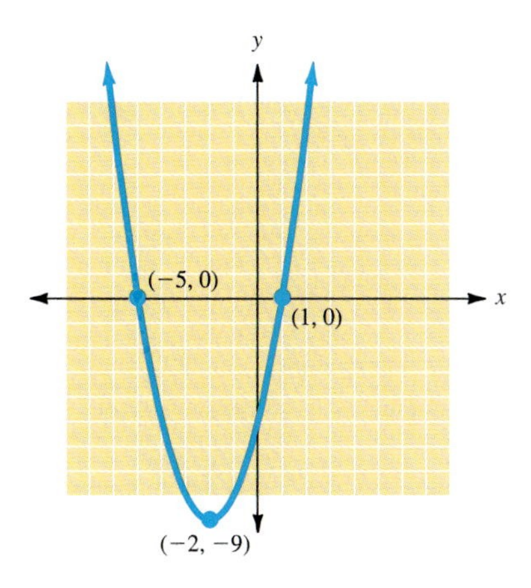

CHECK YOURSELF 4

Sketch the graph of the function

$f(x) = x^2 - 2x - 8$

CHECK YOURSELF ANSWERS

1. $(x = 0)$ and $(x = 3)$

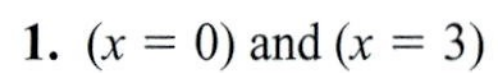

2.

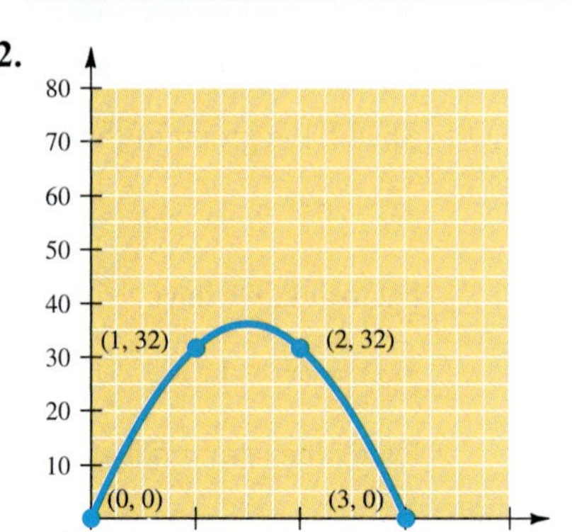

3. $\left(\frac{3}{2}, 36\right)$ **4.**

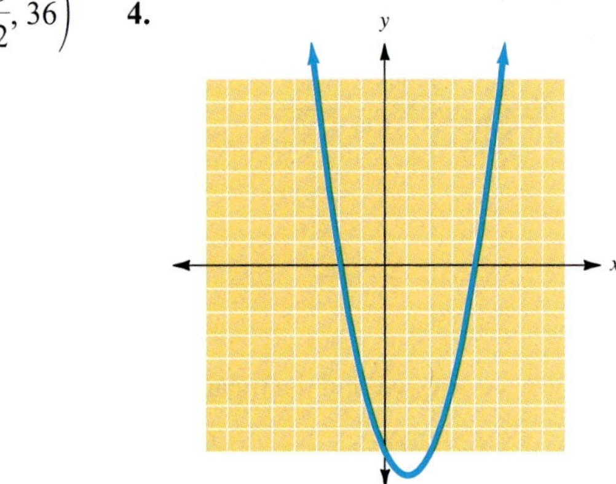

9.1 Exercises

Name ______________

Section ________ Date ________

In exercises 1 to 12, find the zeros and the vertex of the given quadratic function.

1. $f(x) = x^2 + 4x + 3$

2. $f(x) = x^2 - 2x - 24$

3. $f(x) = x^2 - 4x - 12$

4. $f(x) = 3x^2 - 24x + 36$

5. $f(x) = 4x^2 + 16x - 20$

6. $f(x) = 2x^2 - 10x + 12$

7. $f(x) = 8x^2 - 16x$

8. $f(x) = 9x^2 + 36x$

9. $f(x) = -5x^2 - 20x$

10. $f(x) = -6x^2 + 24x$

11. $f(x) = -x^2 + 2x + 3$

12. $f(x) = -x^2 + 8x - 12$

In exercises 13 to 20, sketch the graph of the given quadratic function.

13. $f(x) = x^2 + 4x + 3$

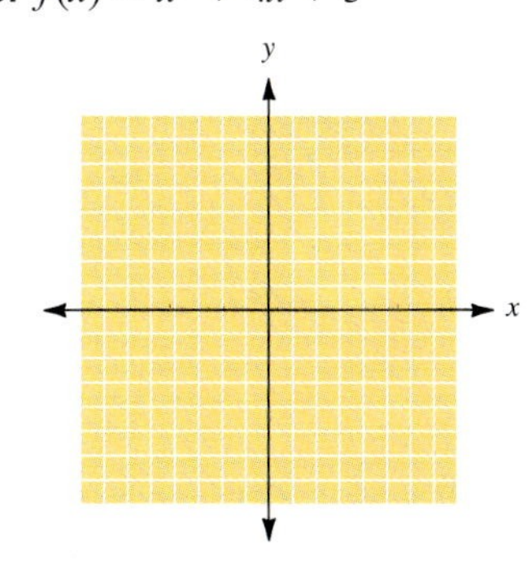

14. $f(x) = x^2 + x - 2$

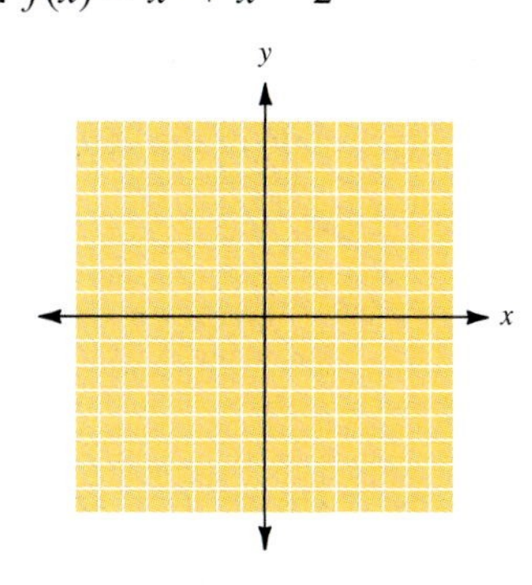

15. $f(x) = x^2 - 2x - 8$

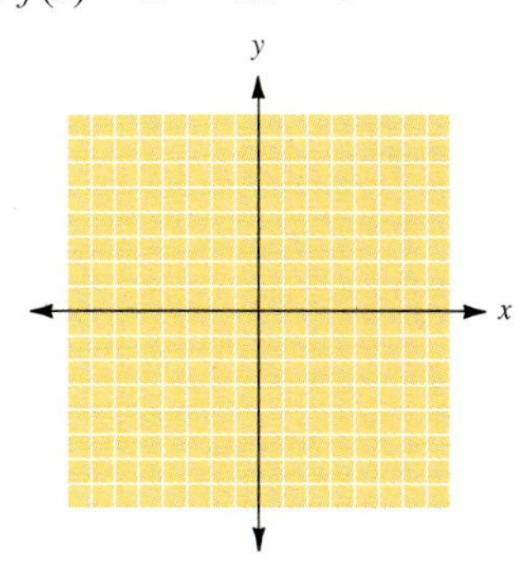

16. $f(x) = -3x^2 + 9x$

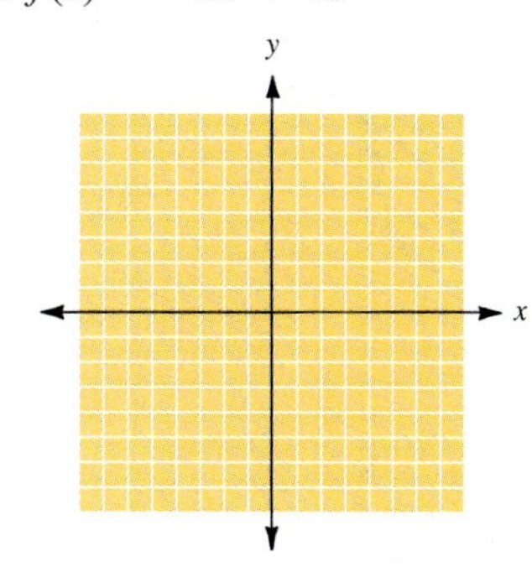

17. $f(x) = x^2 - x - 6$

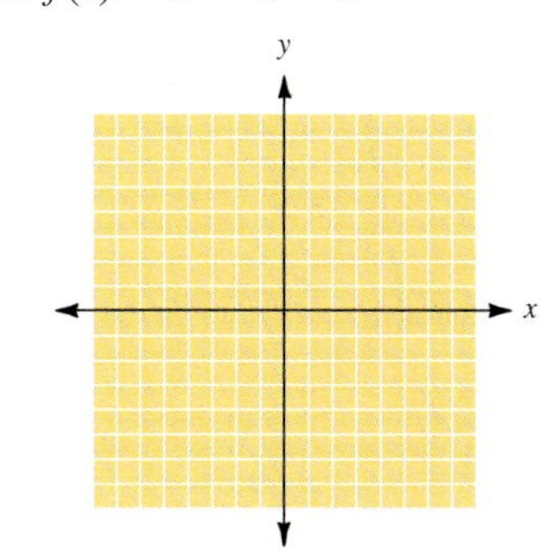

18. $f(x) = -x^2 + 2x + 3$

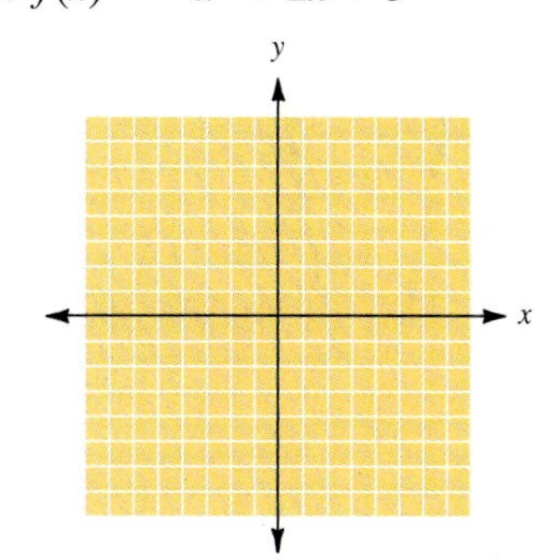

ANSWERS

1. ______
2. ______
3. ______
4. ______
5. ______
6. ______
7. ______
8. ______
9. ______
10. ______
11. ______
12. ______
13. ______
14. ______
15. ______
16. ______
17. ______
18. ______

ANSWERS

19.

20.

21. (a)

(b)

22. (a)

(b)

(c)

19. $f(x) = x(x - 6)$

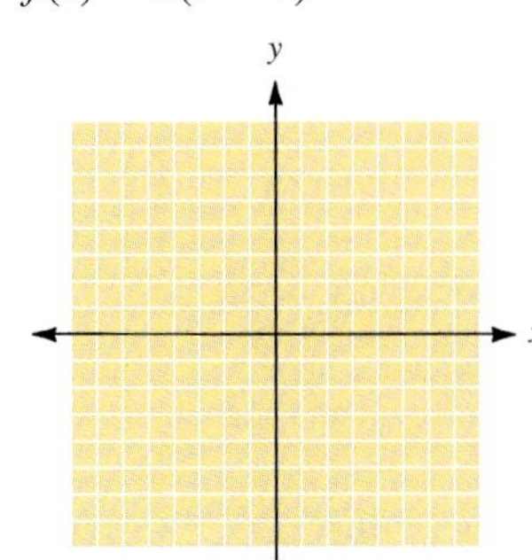

20. $f(x) = -2x^2 + 6x$

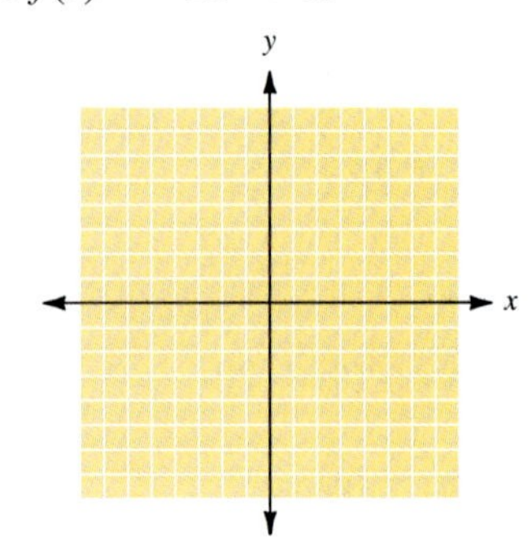

21. **(a)** Define a quadratic function that has zeros of 2 and 3. Express your function first in factored form, and then in standard form ($f(x) = ax^2 + bx + c$).

(b) Define another quadratic function with the same zeros.

22. Let $f(x) = ax^2 + bx$

(a) Write $f(x)$ in factored form.

(b) Find the zeros of $f(x)$.

(c) Find the x value of the vertex.

Answers

1. $-3, -1; (-2, -1)$ **3.** $-2, 6; (2, -16)$ **5.** $-5, 1; (-2, -36)$
7. $0, 2; (1, -8)$ **9.** $0, -4; (-2, 20)$ **11.** $3, -1; (1, 4)$

13.

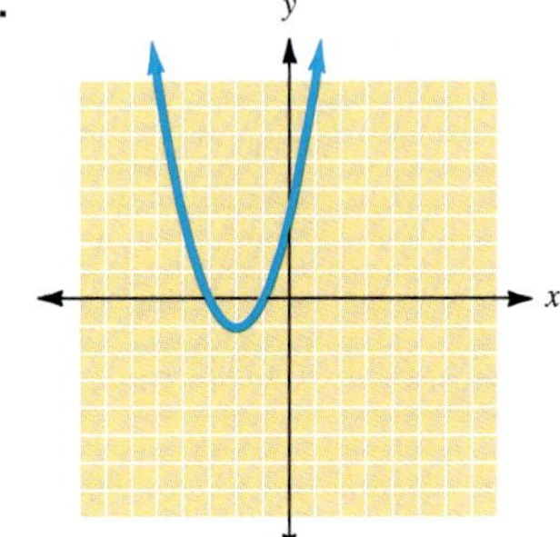

15.

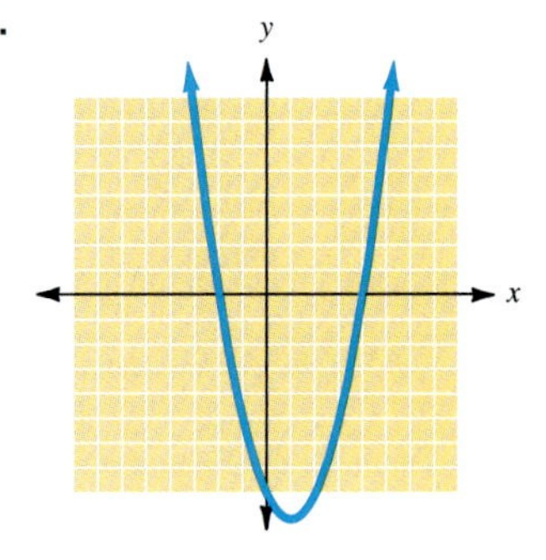

17.

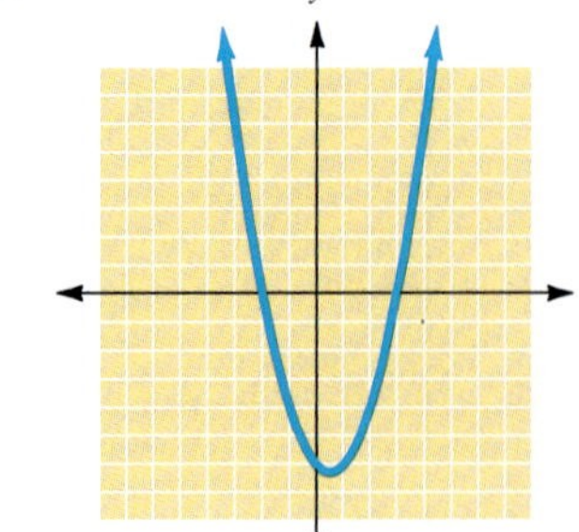

19.

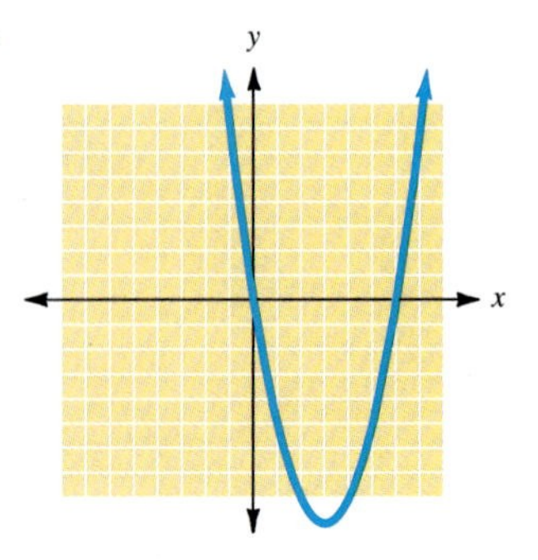

21. **(a)** $f(x) = (x - 2)(x - 3)$, $f(x) = x^2 - 5x + 6$; **(b)** $f(x) = 2x^2 - 10x + 12$

9.2 Solving Quadratic Equations by Completing the Square

OBJECTIVES

1. Solve a quadratic equation by the square root method
2. Solve a quadratic equation by completing the square
3. Solve a geometric application involving a quadratic equation

In Section 6.9, we solved quadratic equations by factoring and using the zero product rule. However, not all equations are factorable over the integers. In this section, we will look at another method that can be used to solve a quadratic equation, called the **square root method.** First, we will solve a special type of equation using the factoring method of Chapter 6.

Example 1

Solving Equations by Factoring

Solve the quadratic equation $x^2 = 16$ by factoring.

We write the equation in standard form:

$$x^2 - 16 = 0$$

Factoring, we have

NOTE Here, we factor the quadratic member of the equation as a difference of squares.

$$(x + 4)(x - 4) = 0$$

Finally, the solutions are

$$x = -4 \quad \text{or} \quad x = 4 \quad \text{or} \quad \{\pm 4\}$$

CHECK YOURSELF 1

Solve each of the following quadratic equations.

(a) $5x^2 = 180$ **(b)** $x^2 = 25$

The equation in Example 1 could have been solved in an alternative fashion. We could have used what is called the **square root method.** Again, given the equation

$$x^2 = 16$$

we can write the equivalent statement

$$x = \sqrt{16} \quad \text{or} \quad x = -\sqrt{16}$$

This yields the solutions

NOTE Be sure to include *both* the positive and the negative square roots when you use the square root method.

$$x = 4 \quad \text{or} \quad x = -4 \quad \text{or} \quad \{\pm 4\}$$

This discussion leads us to the following general result.

Rules and Properties: Square Root Property

If $x^2 = k$, when k is a complex number, then

$x = \sqrt{k}$ or $x = -\sqrt{k}$

Example 2 further illustrates the use of this property.

Example 2

Using the Square Root Method

Solve each equation by using the square root method.

(a) $x^2 = 9$

By the square root property,

$$x = \sqrt{9} \quad \text{or} \quad x = -\sqrt{9}$$
$$= 3 \qquad\qquad = -3 \quad \text{or} \quad \{\pm 3\}$$

NOTE If a calculator were used, $\sqrt{17} = 4.123$ (rounded to three decimal places).

(b) $x^2 - 17 = 0$

Add 17 to both sides of the equation.

$$x^2 = 17$$

so $x = \pm\sqrt{17}$ or $\{\pm\sqrt{17}\}$ or $\{-\sqrt{17}, \sqrt{17}\}$

(c) $2x^2 - 3 = 0$

$$2x^2 = 3$$
$$x^2 = \frac{3}{2}$$
$$x = \pm\sqrt{\frac{3}{2}}$$
$$x = \pm\frac{\sqrt{6}}{2} \quad \text{or} \quad \left\{\pm\frac{\sqrt{6}}{2}\right\}$$

NOTE In Example 2(d) we see that complex-number solutions may result.

(d) $x^2 + 1 = 0$

$$x^2 = -1$$
$$x = \pm\sqrt{-1} \qquad x = \pm i \quad \text{or} \quad \{\pm i\}$$

CHECK YOURSELF 2

Solve each equation.

(a) $x^2 = 5$ **(b)** $x^2 - 2 = 0$ **(c)** $3x^2 - 8 = 0$ **(d)** $x^2 + 9 = 0$

We can also use the approach in Example 2 to solve an equation of the form

$$(x + 3)^2 = 16$$

As before, by the square root property we have

$x + 3 = \pm 4$ Subtract 3 from both sides of the equation.

Solving for x yields

$x = -3 \pm 4$

which means that there are two solutions:

$x = -3 + 4$ or $x = -3 - 4$

$= 1$ $= -7$ or $\{1, -7\}$

Example 3

Using the Square Root Method

Use the square root method to solve each equation.

(a) $(x - 5)^2 - 5 = 0$

$$(x - 5)^2 = 5$$

$$x - 5 = \pm\sqrt{5}$$

$$x = 5 \pm \sqrt{5} \quad \text{or} \quad \{5 \pm \sqrt{5}\}$$

NOTE The two solutions $5 + \sqrt{5}$ and $5 - \sqrt{5}$ are abbreviated as $5 \pm \sqrt{5}$. Using a calculator, we find the approximate solutions $\{2.76, 7.24\}$.

(b) $3(y + 1)^2 - 2 = 0$

$$3(y + 1)^2 = 2$$

$$(y + 1)^2 = \frac{2}{3}$$

$$y + 1 = \pm\sqrt{\frac{2}{3}}$$

$$y = -1 \pm \frac{\sqrt{6}}{3}$$

$$= \frac{-3 \pm \sqrt{6}}{3} \quad \text{or} \quad \left\{\frac{-3 \pm \sqrt{6}}{3}\right\}$$

NOTE We have solved for y and rationalized the denominator.

$$\sqrt{\frac{2}{3}} = \frac{\sqrt{2}}{\sqrt{3}} = \frac{\sqrt{2} \cdot \sqrt{3}}{\sqrt{3} \cdot \sqrt{3}} = \frac{\sqrt{6}}{3}$$

Then we combine the terms on the right, using the common denominator of 3.

The approximate solutions are $\{-1.82, -0.18\}$.

CHECK YOURSELF 3

Using the square root method, solve each equation.

(a) $(x - 2)^2 - 3 = 0$ **(b)** $2(x - 1)^2 = 1$

Not all quadratic equations can be solved directly by factoring or using the square root method. We must extend our techniques.

The square root method is useful in this process because any quadratic equation can be written in the form

$$(x + h)^2 = k$$

which yields the solution

$$x = -h \pm \sqrt{k}$$

NOTE If $(x + h)^2 = k$, then

$x + h = \pm\sqrt{k}$

and

$x = -h \pm \sqrt{k}$

The process of changing an equation in standard form

$$ax^2 + bx + c = 0$$

to the form

$$(x + h)^2 = k$$

is called the method of **completing the square,** and it is based on the relationship between the middle term and the last term of any perfect-square trinomial.

Let's look at three perfect-square trinomials to see whether we can detect a pattern:

$$x^2 + 4x + 4 = (x + 2)^2 \quad (1)$$

$$x^2 - 6x + 9 = (x - 3)^2 \quad (2)$$

$$x^2 + 8x + 16 = (x + 4)^2 \quad (3)$$

NOTE Notice that this relationship is true *only* if the leading, or x^2, coefficient is 1. That will be important later.

Note that in each case the last (or constant) term is the square of one-half of the coefficient of x in the middle (or linear) term. For example, in equation (2),

$$x^2 - 6x + 9 = (x - 3)^2$$

$\frac{1}{2}$ of this coefficient is -3, and $(-3)^2 = 9$, the constant.

Verify this relationship for yourself in equation (3). To summarize, in perfect-square trinomials, the constant is always the square of one-half the coefficient of x.

We are now ready to use the above observation in the solution of quadratic equations by completing the square. Consider Example 4.

Example 4

Completing the Square to Solve an Equation

Solve $x^2 + 8x - 7 = 0$ by completing the square.

First, we rewrite the equation with the constant on the *right-hand side:*

$$x^2 + 8x = 7$$

NOTE Remember that if $(x + h)^2 = k$, then $x = -h \pm \sqrt{k}$.

Our objective is to have a perfect-square trinomial on the left-hand side. We know that we must add the square of one-half of the x coefficient to complete the square. In this case, that value is 16, so now we add 16 to each side of the equation.

NOTE $\frac{1}{2} \cdot 8 = 4$ and $4^2 = 16$

$$x^2 + 8x + 16 = 7 + 16$$

Factor the perfect-square trinomial on the left, and combine like terms on the right to yield

$$(x + 4)^2 = 23$$

Now the square root property yields

$$x + 4 = \pm\sqrt{23}$$

Subtracting 4 from both sides of the equation gives

$$x = -4 \pm \sqrt{23} \quad \text{or} \quad \{-4 \pm \sqrt{23}\}$$

As decimals, these solutions are approximated by $\{-8.8, 0.8\}$.

NOTE When you graph the related function, $y = x^2 + 8x - 7$, you will note that the x values for the x intercepts are just below 1 and just above -9. Be certain that you see how these points relate to the exact solutions, $-4 + \sqrt{23}$ and $-4 - \sqrt{23}$.

CHECK YOURSELF 4

Solve $x^2 - 6x - 2 = 0$ by completing the square.

Example 5

Completing the Square to Solve an Equation

Solve $x^2 + 5x - 3 = 0$ by completing the square.

$$x^2 + 5x - 3 = 0 \quad \text{Add 3 to both sides.}$$

$$x^2 + 5x = 3 \quad \text{Make the left-hand side a perfect square.}$$

NOTE Add the square of one-half of the x coefficient to both sides of the equation. Note that $\frac{1}{2} \cdot 5 = \frac{5}{2}$

$$x^2 + 5x + \left(\frac{5}{2}\right)^2 = 3 + \left(\frac{5}{2}\right)^2$$

$$\left(x + \frac{5}{2}\right)^2 = \frac{37}{4} \quad \text{Take the square root of both sides.}$$

$$x + \frac{5}{2} = \pm\frac{\sqrt{37}}{2} \quad \text{Solve for } x.$$

$$x = \frac{-5 \pm \sqrt{37}}{2} \quad \text{or} \quad \left\{\frac{-5 \pm \sqrt{37}}{2}\right\}$$

The approximate solutions are $\{-5.54, 0.54\}$.

CHECK YOURSELF 5

Solve $x^2 + 3x - 7 = 0$ by completing the square.

Some equations have nonreal complex solutions, as Example 6 illustrates.

Example 6

Completing the Square to Solve an Equation

Solve $x^2 + 4x + 13 = 0$ by completing the square.

NOTE Notice that the graph of $y = x^2 + 4x + 13$ does not intercept the x axis.

$$x^2 + 4x + 13 = 0 \quad \text{Subtract 13 from both sides.}$$

$$x^2 + 4x = -13 \quad \text{Add } \left[\frac{1}{2}(4)\right]^2 \text{ to both sides.}$$

$$x^2 + 4x + 4 = -13 + 4 \quad \text{Factor the left-hand side.}$$

$$(x + 2)^2 = -9 \quad \text{Take the square root of both sides.}$$

$$x + 2 = \pm\sqrt{-9} \quad \text{Simplify the radical.}$$

$$x + 2 = \pm\sqrt{9}i$$

$$x + 2 = \pm 3i$$

$$x = -2 \pm 3i \quad \text{or} \quad \{-2 \pm 3i\}$$

CHECK YOURSELF 6

Solve $x^2 + 10x + 41 = 0$.

Example 7 illustrates a situation in which the leading coefficient of the quadratic member is not equal to 1. As you will see, an extra step is required.

Example 7

Completing the Square to Solve an Equation

CAUTION

Before you can complete the square on the left, the coefficient of x^2 must be equal to 1. Otherwise, we must *divide* both sides of the equation by that coefficient.

Solve $3x^2 + 6x - 7 = 0$ by completing the square.

$3x^2 + 6x - 7 = 0$ Add 7 to both sides.

$3x^2 + 6x = 7$ Divide both sides by 3.

$x^2 + 2x = \frac{7}{3}$ Now, complete the square on the left.

$x^2 + 2x + 1 = \frac{7}{3} + 1$ The left side is now a perfect square.

$$(x + 1)^2 = \frac{10}{3}$$

$$x + 1 = \pm\sqrt{\frac{10}{3}}$$

$$x = -1 \pm \sqrt{\frac{10}{3}}$$

$$= \frac{-3 \pm \sqrt{30}}{3}$$

NOTE We have rationalized the denominator and combined the terms on the right side.

CHECK YOURSELF 7

Solve $2x^2 - 8x + 3 = 0$ by completing the square.

The following algorithm summarizes our work in this section with solving quadratic equations by completing the square.

Step by Step: Completing the Square

Step 1 Isolate the constant on the right side of the equation.

Step 2 Divide both sides of the equation by the coefficient of the x^2 term if that coefficient is not equal to 1.

Step 3 Add the square of one-half of the coefficient of the linear term to both sides of the equation. This will give a perfect-square trinomial on the left side of the equation.

Step 4 Write the left side of the equation as the square of a binomial, and simplify on the right side.

Step 5 Use the square root property, and then solve the resulting linear equations.

Let's proceed now to applications involving geometry.

Example 8

Applying the Completing of a Square

The length of a rectangle is 4 cm greater than its width. If the area of the rectangle is 108 cm^2, what are the approximate dimensions of the rectangle?

Step 1 You are asked to find the dimensions (the length and the width) of the rectangle.

Step 2 Whenever geometric figures are involved in an application, start by drawing, and *then labeling,* a sketch of the problem. Letting x represent the width and $x + 4$ the length, we have

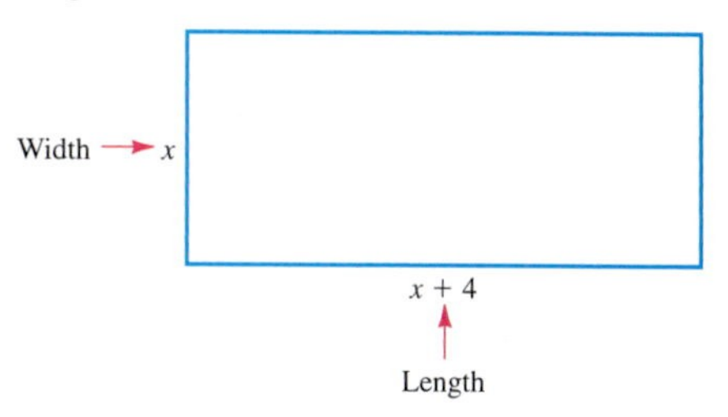

Step 3 The area of a rectangle is the product of its length and width, so

$x(x + 4) = 108$

NOTE Multiply and complete the square.

Step 4

$$\begin{aligned} x(x + 4) &= 108 \\ x^2 + 4x &= 108 \\ x^2 + 4x + 4 &= 108 + 4 \\ (x + 2)^2 &= 112 \\ x + 2 &= \pm\sqrt{112} \\ x &= -2 \pm \sqrt{112} \end{aligned}$$

Step 5 We reject $-2 - \sqrt{112}$ (cm) as a solution. A length cannot be negative, and so we must consider only $-2 + \sqrt{112}$ (cm) in finding the required dimensions.

The width x is approximately 8.6 cm, and the length $x + 4$ is 12.6 cm. Because (8.6 cm) (12.6 cm) gives a rectangle of area 108.36 cm^2, the solution is verified.

CHECK YOURSELF 8

In a triangle, the base is 4 in. less than its height. If its area is 35 in.^{2}, find the length of the base and the height of the triangle.

Example 9

Applying the Completing of a Square

An open box is formed from a rectangular piece of cardboard, whose length is 2 in. more than its width, by cutting 2-in. squares from each corner and folding up the sides. If the volume of the box is to be 100 in.^{3}, what must be the size of the original piece of cardboard?

Step 1 We are asked for the dimensions of the sheet of cardboard.

Step 2 Again sketch the problem.

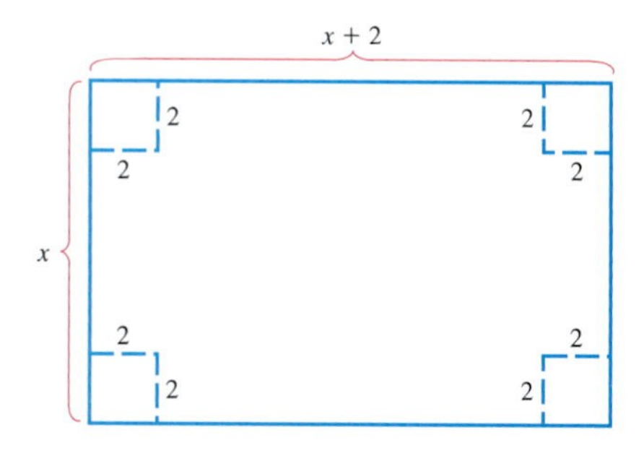

Step 3 To form an equation for volume, we sketch the completed box.

NOTE The original width of the cardboard was x. Removing two 2-in. squares leaves $x - 4$ for the width of the box. Similarly, the length of the box is $x - 2$. Do you see why?

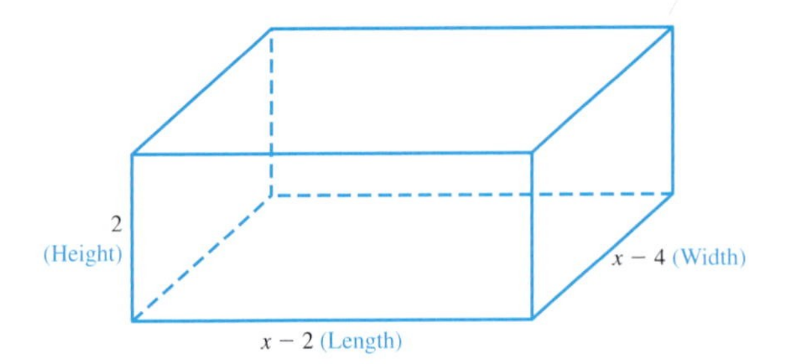

Because volume is the product of height, length, and width,

$$2(x - 2)(x - 4) = 100$$

Step 4

$$2(x - 2)(x - 4) = 100$$

Divide both sides by 2, and multiply on the left. Then solve as before.

$$(x - 2)(x - 4) = 50$$
$$x^2 - 6x + 8 = 50$$
$$x^2 - 6x = 42$$
$$x^2 - 6x + 9 = 42 + 9$$
$$(x - 3)^2 = 51$$
$$x - 3 = \pm\sqrt{51}$$
$$x = 3 \pm \sqrt{51}$$

Step 5 Again, we need consider only the positive solution. The width x of the original piece of cardboard is approximately 10.14 in., and its length $x + 2$ is 12.14 in. The dimensions of the completed box will be 6.14 by 8.14 by 2 in., which gives volume of an approximate 100 in.^{3}

CHECK YOURSELF 9

A similar box is to be made by cutting 3-cm squares from a piece of cardboard that is 4 cm longer than it is wide. If the required volume is 300 cm^3, find the dimensions of the original sheet of cardboard.

CHECK YOURSELF ANSWERS

1. **(a)** $\{-6, 6\}$; **(b)** $\{-5, 5\}$
2. **(a)** $\{\sqrt{5}, -\sqrt{5}\}$; **(b)** $\{\sqrt{2}, -\sqrt{2}\}$; **(c)** $\left\{\frac{2\sqrt{6}}{3}, -\frac{2\sqrt{6}}{3}\right\}$; **(d)** $\{3i, -3i\}$
3. **(a)** $\{2 \pm \sqrt{3}\}$; **(b)** $\left\{\frac{2 \pm \sqrt{2}}{2}\right\}$ **4.** $\{3 \pm \sqrt{11}\}$ **5.** $\left\{\frac{-3 \pm \sqrt{37}}{2}\right\}$
6. $\{-5 \pm 4i\}$ **7.** $\left\{\frac{4 \pm \sqrt{10}}{2}\right\}$ **8.** Base $\approx$ 6.6 in.; height $\approx$ 10.6 in.
9. 14.2 cm × 18.2 cm

9.2 Exercises

Name ______________

Section ________ Date ________

In exercises 1 to 8, solve by factoring or completing the square.

1. $x^2 + 6x + 5 = 0$

2. $x^2 + 5x + 6 = 0$

3. $z^2 - 2z - 35 = 0$

4. $q^2 - 5q - 24 = 0$

5. $2x^2 - 5x - 3 = 0$

6. $3x^2 + 10x - 8 = 0$

7. $6y^2 - y - 2 = 0$

8. $10z^2 + 3z - 1 = 0$

In exercises 9 to 20, use the square root method to find solutions for the equations.

9. $x^2 = 36$

10. $x^2 = 144$

11. $y^2 = 7$

12. $p^2 = 18$

13. $2x^2 - 12 = 0$

14. $3x^2 - 66 = 0$

15. $2t^2 + 12 = 4$

16. $3u^2 - 5 = -32$

17. $(x + 1)^2 = 12$

18. $(2x - 3)^2 = 5$

19. $(2z + 1)^2 - 3 = 0$

20. $(3p - 4)^2 + 9 = 0$

ANSWERS

1. ______
2. ______
3. ______
4. ______
5. ______
6. ______
7. ______
8. ______
9. ______
10. ______
11. ______
12. ______
13. ______
14. ______
15. ______
16. ______
17. ______
18. ______
19. ______
20. ______

ANSWERS

21. ______ 22. ______
23. ______ 24. ______
25. ______ 26. ______
27. ______
28. ______
29. ______
30. ______
31. ______
32. ______
33. ______
34. ______
35. ______
36. ______
37. ______
38. ______
39. ______
40. ______
41. ______
42. ______
43. ______
44. ______
45. ______
46. ______
47. ______
48. ______

In exercises 21 to 32, find the constant that must be added to each binomial expression to form a perfect-square trinomial.

21. $x^2 + 12x$

22. $r^2 - 14r$

23. $y^2 - 8y$

24. $w^2 + 16w$

25. $x^2 - 3x$

26. $z^2 + 5z$

27. $n^2 + n$

28. $x^2 - x$

29. $x^2 + \frac{1}{2}x$

30. $x^2 - \frac{1}{3}x$

31. $x^2 + 2ax$

32. $y^2 - 4ay$

In exercises 33 to 54, solve each equation by completing the square.

33. $x^2 + 12x - 2 = 0$

34. $x^2 - 14x - 7 = 0$

35. $y^2 - 2y = 8$

36. $z^2 + 4z - 72 = 0$

37. $x^2 - 2x - 5 = 0$

38. $x^2 - 2x = 3$

39. $x^2 + 10x + 13 = 0$

40. $x^2 + 3x - 17 = 0$

41. $z^2 - 5z - 7 = 0$

42. $q^2 - 8q + 20 = 0$

43. $m^2 - m - 3 = 0$

44. $y^2 + y - 5 = 0$

45. $x^2 + \frac{1}{2}x = 1$

46. $x^2 - \frac{1}{3}x = 2$

47. $2x^2 + 2x - 1 = 0$

48. $3x^2 - 3x = 1$

49. $3x^2 - 6x = 2$

50. $4x^2 + 8x - 1 = 0$

51. $3x^2 - 2x + 12 = 0$

52. $7y^2 - 2y + 3 = 0$

53. $x^2 + 8x + 20 = 0$

54. $x^2 - 2x + 10 = 0$

In exercises 55 to 60, find the constant that must be added to each binomial to form a perfect-square trinomial. Let x be the variable; other letters represent constants.

55. $x^2 + 2ax$

56. $x^2 + 2abx$

57. $x^2 + 3ax$

58. $x^2 + abx$

59. $a^2x^2 + 2ax$

60. $a^2x^2 + 4abx$

In exercises 61 and 62, solve each equation by completing the square.

61. $x^2 + 2ax = 4$

62. $x^2 + 2ax - 8 = 0$

Solve the following applications.

63. The width of a rectangle is 3 ft less than its length. If the area of the rectangle is 70 ft^2, what are the dimensions of the rectangle?

64. The length of a rectangle is 5 cm more than its width. If the area of the rectangle is 84 cm^2, find the dimensions of the rectangle.

65. The length of a rectangle is 2 cm more than 3 times its width. If the area of the rectangle is 85 cm^2, find the dimensions of the rectangle.

66. If the length of a rectangle is 3 ft less than twice its width and the area of the rectangle is 54 ft^2, what are the dimensions of the rectangle?

67. The length of a rectangle is 1 cm more than its width. If the length of the rectangle is doubled, the area of the rectangle is increased by 30 cm^2. What were the dimensions of the original rectangle?

ANSWERS

49. ______
50. ______
51. ______
52. ______
53. ______
54. ______
55. ______
56. ______
57. ______
58. ______
59. ______
60. ______
61. ______
62. ______
63. ______
64. ______
65. ______
66. ______
67. ______

68. ____

69. ____

70. ____

71. ____

72. ____

73. ____

74. ____

68. A box is to be made from a rectangular piece of tin that is twice as long as it is wide. To accomplish this, a 10-cm square is cut from each corner, and the sides are folded up. The volume of the finished box is to be 5000 cm^3. Find the dimensions of the original piece of tin.

Hint 1: To solve this equation, you will want to use the following sketch of the piece of tin. Note that the original dimensions are represented by x and $2x$. Do you see why? Also recall that the volume of the resulting box will be the product of the length, width, and height.

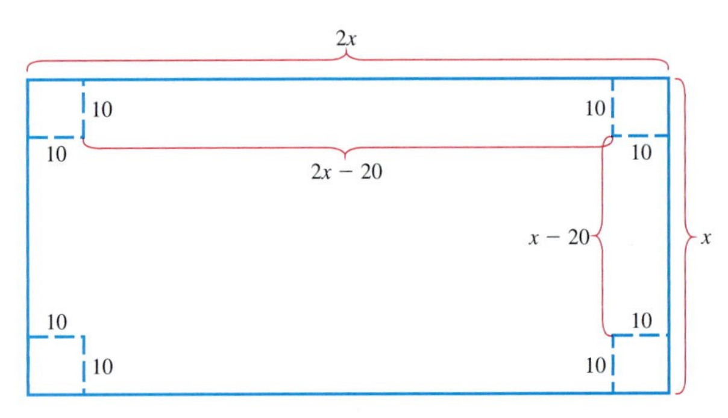

Hint 2: From this sketch, you can see that the equation that results from $V = LWH$ will be

$(2x - 20)(x - 20)10 = 5000$

69. An open box is formed from a square piece of material by cutting 2-in. squares from each corner of the material and folding up the sides. If the volume of the box that is formed is 72 in.^{3}, what was the size of the original piece of material?

70. An open carton is formed from a rectangular piece of cardboard that is 4 ft longer than it is wide, by removing 1-ft squares from each corner and folding up the sides. If the volume of the carton is then 12 ft^3, what were the dimensions of the original piece of cardboard?

71. A box that has a volume of 2000 in.^{3} was made from a square piece of tin. The square piece cut from each corner had sides of length 4 in. What were the original dimensions of the square?

72. A square piece of cardboard is to be formed into a box. After 5-cm squares are cut from each corner and the sides are folded up, the resulting box will have a volume of 400 cm^3. Find the length of a side of the original piece of cardboard.

73. Why must the leading coefficient of the quadratic member be set equal to 1 before using the technique of completing the square?

74. What relationship exists between the solutions of a quadratic equation and the graph of a quadratic function?

In exercises 75 to 78, use your graphing utility to find the graph. For each graph, approximate the x intercepts to the nearest tenth. (You may have to adjust the viewing window to see both intercepts.)

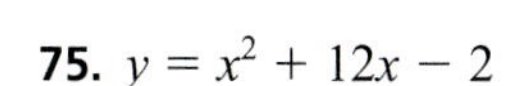

75. $y = x^2 + 12x - 2$

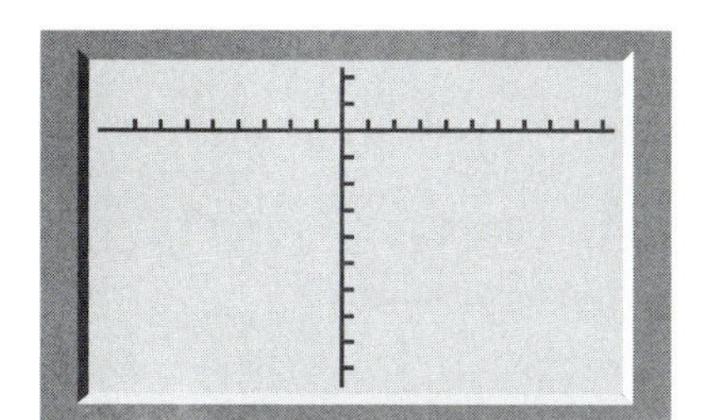

76. $y = x^2 - 14x - 7$

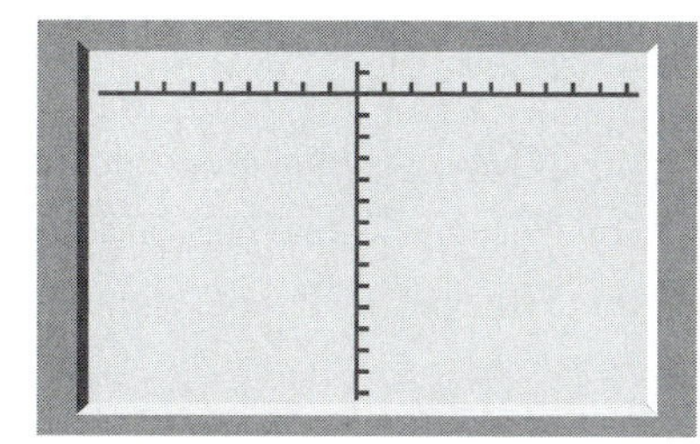

77. $y = x^2 - 2x - 8$

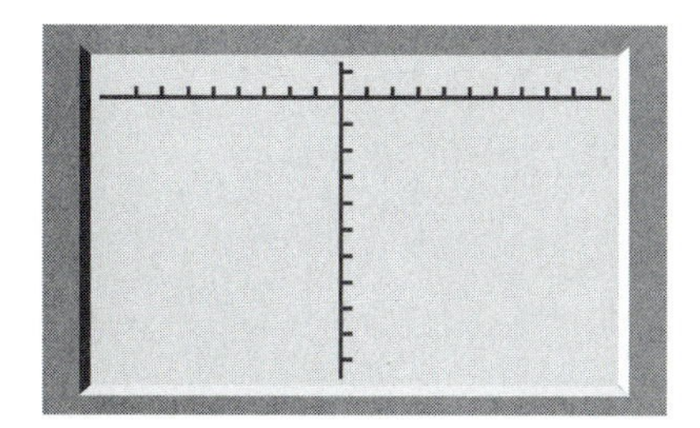

78. $y = x^2 + 4x - 72$

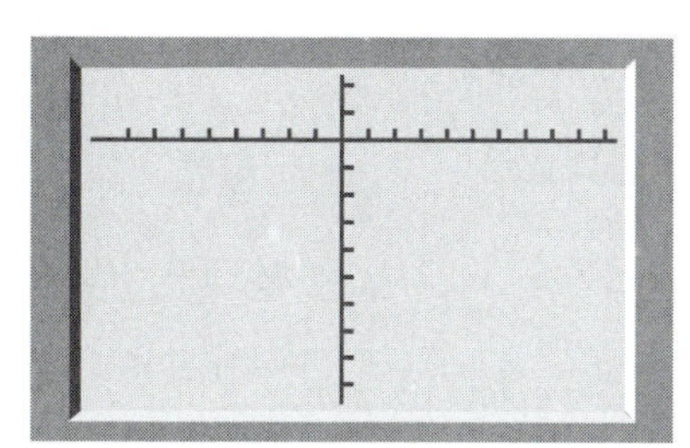

79. On your graphing calculator, view the graph of $f(x) = x^2 + 1$.

(a) What can you say about the x intercepts of the graph?

(b) Determine the zeros of the function, using the square root method.

(c) How does your answer to part (a) relate to your answer to part (b)?

80. Consider the following representation of "completing the square": Suppose we wish to complete the square for $x^2 + 10x$. A square with dimensions x by x has area equal to x^2.

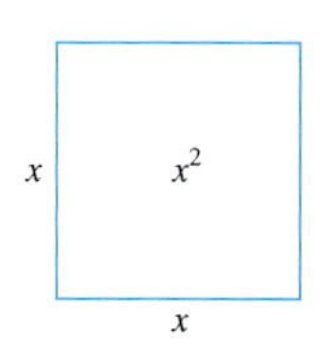

We divide the quantity $10x$ by 2 and get $5x$. If we extend the base x by 5 units, and draw the rectangle attached to the square, the rectangle's dimensions are 5 by x with an area of $5x$.

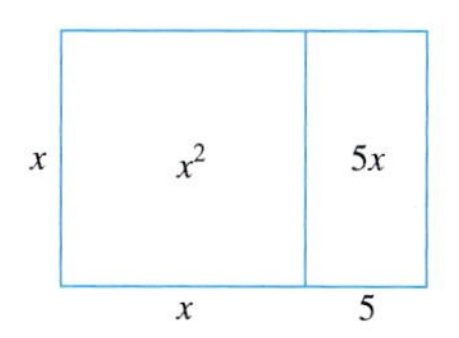

ANSWERS

75. ______

76. ______

77. ______

78. ______

79. (a)
(b)
(c)

ANSWERS

80. (a)
(b)
(c)

81. (a)
(b)
(c)

Now we extend the height by 5 units, and draw another rectangle whose area is $5x$.

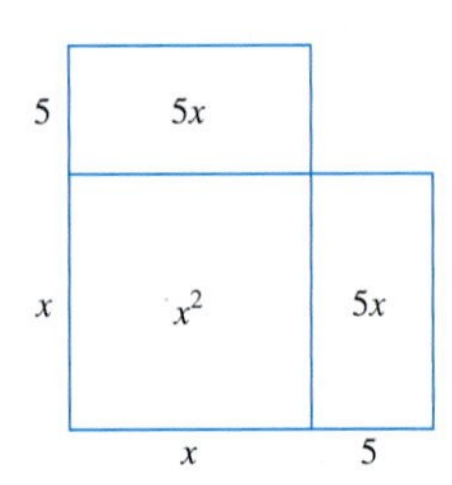

(a) What is the total area represented in the figure so far?

(b) How much area must be added to the figure to "complete the square"?

(c) Write the area of the completed square as a binomial squared.

81. Repeat the process described in exercise 80 with $x^2 + 16x$.

Answers

1. $\{-5, -1\}$ **3.** $\{-5, 7\}$ **5.** $\left\{-\frac{1}{2}, 3\right\}$ **7.** $\left\{-\frac{1}{2}, \frac{2}{3}\right\}$ **9.** $\{\pm 6\}$ **11.** $\{-\sqrt{7}, \sqrt{7}\}$ **13.** $\{-\sqrt{6}, \sqrt{6}\}$ **15.** $\{\pm 2i\}$ **17.** $\{-1 \pm 2\sqrt{3}\}$ **19.** $\left\{\frac{-1 \pm \sqrt{3}}{2}\right\}$ **21.** 36 **23.** 16 **25.** $\frac{9}{4}$ **27.** $\frac{1}{4}$ **29.** $\frac{1}{16}$ **31.** a^2 **33.** $\{-6 \pm \sqrt{38}\}$ **35.** $\{-2, 4\}$ **37.** $\{1 \pm \sqrt{6}\}$ **39.** $\{-5 \pm 2\sqrt{3}\}$ **41.** $\left\{\frac{5 \pm \sqrt{53}}{2}\right\}$ **43.** $\left\{\frac{1 \pm \sqrt{13}}{2}\right\}$ **45.** $\left\{\frac{-1 \pm \sqrt{17}}{4}\right\}$ **47.** $\left\{\frac{-1 \pm \sqrt{3}}{2}\right\}$ **49.** $\left\{\frac{3 \pm \sqrt{15}}{3}\right\}$ **51.** $\left\{\frac{1 \pm i\sqrt{35}}{3}\right\}$ **53.** $\{-4 \pm 2i\}$ **55.** a^2 **57.** $\frac{9}{4}a^2$ **59.** 1 **61.** $\{-a \pm \sqrt{4 + a^2}\}$ **63.** Width 7 ft, length 10 ft **65.** Width 5 cm, length 17 cm **67.** Width 5 cm, length 6 cm **69.** 10 in. by 10 in. **71.** $8 + 10\sqrt{5}$ in. or 30.4 in. **73.**

75.

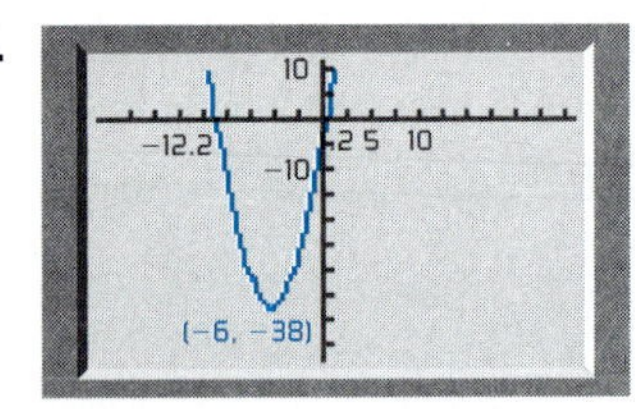

77.

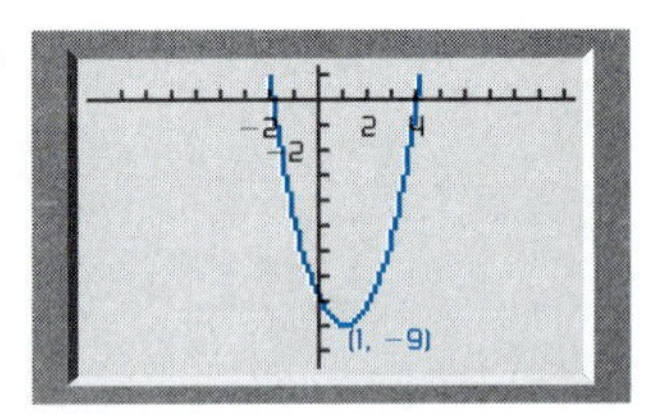

79. **(a)** There are none; **(b)** $\pm i$; **(c)** If the graph of $f(x)$ has no x intercepts, the zeros of the function are not real.

81. **(a)** $x^2 + 16x$; **(b)** 64; **(c)** $x^2 + 16x + 64 = (x + 8)^2$

Solving Quadratic Equations by Using the Quadratic Formula

OBJECTIVES

1. Solve a quadratic equation by using the quadratic formula
2. Determine the nature of the solutions of a quadratic equation by using the discriminant
3. Use the Pythagorean theorem to solve a geometric application
4. Solve applications involving quadratic equations

Every quadratic equation can be solved by using the quadratic formula. In this section, we will first describe how the quadratic formula is derived, then we will examine its use. Recall that a quadratic equation is any equation that can be written in the form

$$ax^2 + bx + c = 0 \qquad \text{when } a \neq 0$$

Step by Step: Deriving the Quadratic Formula

Step 1 Isolate the constant on the right side of the equation.
$$ax^2 + bx = -c$$

Step 2 Divide both sides by the coefficient of the x^2 term.
$$x^2 + \frac{b}{a}x = -\frac{c}{a}$$

Step 3 Add the square of one-half the x coefficient to both sides.
$$x^2 + \frac{b}{a}x + \frac{b^2}{4a^2} = -\frac{c}{a} + \frac{b^2}{4a^2}$$

Step 4 Factor the left side as a perfect-square binomial. Then apply the square root property.
$$\left(x + \frac{b}{2a}\right)^2 = \frac{-4ac + b^2}{4a^2}$$
$$x + \frac{b}{2a} = \pm\sqrt{\frac{b^2 - 4ac}{4a^2}}$$

Step 5 Solve the resulting linear equations.
$$x = -\frac{b}{2a} \pm \frac{\sqrt{b^2 - 4ac}}{2a}$$

Step 6 Simplify.
$$= \frac{-b \pm \sqrt{b^2 - 4ac}}{2a}$$

We now use the result derived above to state the **quadratic formula,** a formula that allows us to find the solutions for any quadratic equation.

Rules and Properties: The Quadratic Formula

Given any quadratic equation in the form

$$ax^2 + bx + c = 0 \qquad \text{when } a \neq 0$$

the two solutions to the equation are found using the formula

$$x = \frac{-b \pm \sqrt{b^2 - 4ac}}{2a}$$

Our first example uses an equation in standard form.

Example 1

Using the Quadratic Formula

NOTE Notice that the equation is in standard form.

Solve, using the quadratic formula.

$$6x^2 - 7x - 3 = 0$$

First, we determine the values for a, b, and c. Here,

$$a = 6 \qquad b = -7 \qquad c = -3$$

Substituting those values into the quadratic formula, we have

NOTE Because $b^2 - 4ac = 121$ is a perfect square, the two solutions in this case are rational numbers.

$$x = \frac{-(-7) \pm \sqrt{(-7)^2 - 4(6)(-3)}}{2(6)}$$

Simplifying inside the radical gives us

$$x = \frac{7 \pm \sqrt{121}}{12}$$

$$= \frac{7 \pm 11}{12}$$

NOTE Compare these solutions to the graph of $y = 6x^2 - 7x - 3$

This gives us the solutions

$$x = \frac{3}{2} \quad \text{or} \quad x = -\frac{1}{3} \quad \text{or} \quad \left\{\frac{3}{2}, -\frac{1}{3}\right\}$$

Note that because the solutions for the equation of this example are rational, the original equation could have been solved by our earlier method of factoring.

CHECK YOURSELF 1

Solve, using the quadratic formula.

$$3x^2 + 2x - 8 = 0$$

To use the quadratic formula, we often must write the equation in standard form. Example 2 illustrates this approach.

Example 2

Using the Quadratic Formula

NOTE The equation *must be in standard form* to determine *a*, *b*, and *c*.

Solve by using the quadratic formula.

$$9x^2 = 12x - 4$$

First, we must write the equation in standard form.

$$9x^2 - 12x + 4 = 0$$

Second, we find the values of a, b, and c. Here,

$$a = 9 \qquad b = -12 \qquad c = 4$$

Substituting these values into the quadratic formula, we find

$$x = \frac{-(-12) \pm \sqrt{(-12)^2 - 4(9)(4)}}{2(9)}$$

$$= \frac{12 \pm \sqrt{0}}{18}$$

NOTE The graph of $y = 9x^2 - 12x + 4$ intercepts the x axis only at the point $\left(\frac{2}{3}, 0\right)$.

and simplifying yields

$$x = \frac{2}{3} \quad \text{or} \quad \left\{\frac{2}{3}\right\}$$

CHECK YOURSELF 2

Use the quadratic formula to solve the equation.

$$4x^2 - 4x = -1$$

Thus far our examples and exercises have led to rational solutions. That is not always the case, as Example 3 illustrates.

Example 3

Using the Quadratic Formula

Using the quadratic formula, solve

$$x^2 - 3x = 5$$

Once again, to use the quadratic formula, we write the equation in standard form.

$$x^2 - 3x - 5 = 0$$

NOTE $a = 1$, $b = -3$, and $c = -5$

We now determine values for a, b, and c and substitute.

$$x = \frac{-(-3) \pm \sqrt{(-3)^2 - 4(1)(-5)}}{2(1)}$$

Simplifying as before, we have

$$x = \frac{3 \pm \sqrt{29}}{2} \quad \text{or} \quad \left\{\frac{3 \pm \sqrt{29}}{2}\right\}$$

CHECK YOURSELF 3

Using the quadratic equation, solve $2x^2 = x + 7$.

Example 4 requires some special care in simplifying the solution.

Example 4

Using the Quadratic Formula

Using the quadratic formula, solve

$$3x^2 - 6x + 2 = 0$$

CAUTION

Students are sometimes tempted to reduce this result to

$$\frac{6 \pm 2\sqrt{3}}{6} \stackrel{?}{=} 1 \pm 2\sqrt{3}$$

This is *not a valid step.* We must divide *each of the terms* in the numerator by 2 when simplifying the expression.

Here, we have $a = 3$, $b = -6$, and $c = 2$. Substituting gives

$$x = \frac{-(-6) \pm \sqrt{(-6)^2 - 4(3)(2)}}{2(3)}$$

$$= \frac{6 \pm \sqrt{12}}{6}$$

We now look for the largest perfect-square factor of 12, the radicand.

Simplifying, we note that $\sqrt{12}$ is equal to $\sqrt{4 \cdot 3}$, or $2\sqrt{3}$. We can then write the solutions as

$$x = \frac{6 \pm 2\sqrt{3}}{6} = \frac{2(3 \pm \sqrt{3})}{6} = \frac{3 \pm \sqrt{3}}{3}$$

CHECK YOURSELF 4

Solve by using the quadratic formula.

$$x^2 - 4x = 6$$

Let's examine a case in which the solutions are nonreal complex numbers.

Example 5

Using the Quadratic Formula

Solve by using the quadratic formula.

$$x^2 - 2x = -2$$

Rewriting in standard form, we have

$$x^2 - 2x + 2 = 0$$

Labeling the coefficients, we find that

$$a = 1 \qquad b = -2 \qquad c = 2$$

NOTE The solutions will have an imaginary part any time $\sqrt{b^2 - 4ac}$ is negative.

NOTE The graph of
$y = x^2 - 2x + 2$
does not intercept the x axis, so there are no real solutions.

Applying the quadratic formula, we have

$$x = \frac{2 \pm \sqrt{-4}}{2}$$

and noting that $\sqrt{-4}$ is $2i$, we can simplify to

$$x = 1 \pm i \quad \text{or} \quad \{1 \pm i\}$$

CHECK YOURSELF 5

Solve by using the quadratic formula.

$$x^2 - 4x + 6 = 0$$

In attempting to solve a quadratic equation, you should first try the factoring method. If this method does not work, you can apply the quadratic formula or the square root method to find the solution. The following algorithm outlines the steps for solving equations using the quadratic formula.

Step by Step: Solving a Quadratic Equation by Using the Quadratic Formula

Step 1 Write the equation in standard form (one side is equal to 0).

$$ax^2 + bx + c = 0$$

Step 2 Determine the values for a, b, and c.

Step 3 Substitute those values into the quadratic formula.

$$x = \frac{-b \pm \sqrt{b^2 - 4ac}}{2a}$$

Step 4 Simplify.

NOTE Although not necessarily distinct or real, every second-degree equation has two solutions.

Given a quadratic equation, the radicand $b^2 - 4ac$ determines the number of real solutions. Because of this, we call it the discriminant. Because the value of the discriminant is a real number, there are three possibilities.

NOTE Graphically, we can see the number of real solutions as the number of times the related quadratic function intercepts the x axis.

Rules and Properties: The Discriminant

If $b^2 - 4ac$ $\begin{cases} < 0 & \text{there are } \textit{no real solutions,} \text{ but two imaginary solutions} \\ = 0 & \text{there is } \textit{one real solution} \text{ (a double solution)} \\ > 0 & \text{there are } \textit{two distinct real solutions} \end{cases}$

Example 6

Analyzing the Discriminant

How many real solutions are there for each of the following quadratic equations?

(a) $x^2 + 7x - 15 = 0$

The value of the discriminant is $[49 - 4(1)(-15)]$ or 109. This indicates that there are two real solutions.

NOTE We could find two imaginary solutions by using the quadratic formula.

(b) $3x^2 - 5x + 7 = 0$

The value of the discriminant is negative (-59). There are no real solutions.

(c) $9x^2 - 12x + 4 = 0$

The value of the discriminant is 0. There is exactly one real solution (a double solution).

CHECK YOURSELF 6

How many real solutions are there for each of the following quadratic equations?

(a) $2x^2 - 3x + 2 = 0$ **(b)** $3x^2 + x - 11 = 0$

(c) $4x^2 - 4x + 1 = 0$ **(d)** $x^2 = -5x - 7$

Consider the following two applications involving thrown balls that can be solved by using the quadratic formula.

Example 7

Solving a Thrown-Ball Application

If a ball is thrown upward from the ground, the equation to find the height h of such a ball thrown with an initial velocity of 80 ft/s is

NOTE Here h measures the height above the ground, in feet, t seconds (s) after the ball is thrown upward.

$$h(t) = 80t - 16t^2$$

Find the time it takes the ball to reach a height of 48 ft.

First we substitute 48 for h, and then we rewrite the equation in standard form.

$$16t^2 - 80t + 48 = 0$$

To simplify the computation, we divide both sides of the equation by the common factor, 16. This yields

NOTE Notice that the result of dividing by 16

$$\frac{0}{16} = 0$$

is 0 on the right.

$$t^2 - 5t + 3 = 0$$

We solve for t as before, using the quadratic equation, with the result

$$t = \frac{5 \pm \sqrt{13}}{2}$$

NOTE There are two solutions because the ball reaches the height *twice*, once on the way up and once on the way down.

This gives us two solutions, $\frac{5 + \sqrt{13}}{2}$ and $\frac{5 - \sqrt{13}}{2}$. But, because we have specified units of time, we generally estimate the answer to the nearest tenth or hundredth of a second.

In this case, estimating to the nearest tenth of a second gives solutions of 0.7 and 4.3 s.

CHECK YOURSELF 7

The equation to find the height h of a ball thrown upward with an initial velocity of 64 ft/s is

$$h(t) = 64t - 16t^2$$

Find the time it takes the ball to reach a height of 32 ft.

Example 8

Solving a Thrown-Ball Application

NOTE The graph of $h(t) = 240 - 64t - 16t^2$ shows the height, h, at any time t.

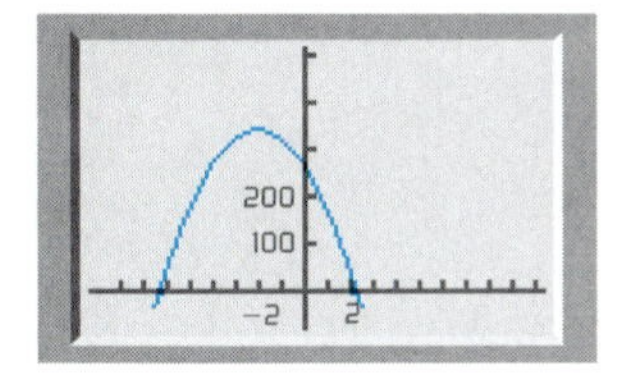

The height, h, of a ball thrown downward from the top of a 240-ft building with an initial velocity of 64 ft/s is given by

$$h(t) = 240 - 64t - 16t^2$$

At what time will the ball reach a height of 176 ft?

Let $h(t) = 176$, and write the equation in standard form.

$$176 = 240 - 64t - 16t^2$$

$$0 = 64 - 64t - 16t^2$$

$$16t^2 + 64t - 64 = 0$$

or

$$t^2 + 4t - 4 = 0$$

NOTE Again, we divide both sides of the equation by 16 to simplify the computation.

Applying the quadratic formula with $a = 1$, $b = 4$, and $c = -4$ yields

$$t = -2 \pm 2\sqrt{2}$$

NOTE The ball has a height of 176 ft at approximately 0.8 s.

Estimating these solutions, we have $t = -4.8$ and $t = 0.8$ s, but of these two values only the *positive value* makes any sense. (To accept the negative solution would be to say that the ball reached the specified height before it was thrown.)

CHECK YOURSELF 8

The height h of a ball thrown upward from the top of a 96-ft building with an initial velocity of 16 ft/s is given by

$$h(t) = 96 + 16t - 16t^2$$

When will the ball have a height of 32 ft above the ground? (Estimate your answer to the nearest tenth of a second.)

Another geometric result that generates quadratic equations in applications is the **Pythagorean theorem,** which we discussed in Section 8.5.

In Example 9, the solution of the quadratic equation contains radicals. Substituting a pair of solutions such as $\frac{3 \pm \sqrt{5}}{2}$ is a very difficult process. As in our thrown-ball applications, always check the "reasonableness" of the answer.

Example 9

A Triangular Application

One leg of a right triangle is 4 cm longer than the other leg. The length of the hypotenuse of the triangle is 12 cm. Find the length of the two legs.

As in any geometric problem, a sketch of the information will help us visualize.

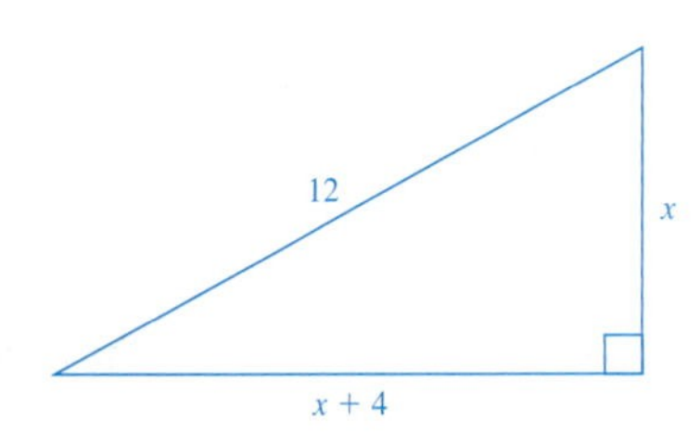

We assign variable x to the shorter leg and $x + 4$ to the other leg.

NOTE The sum of the squares of the legs of the triangle is equal to the square of the hypotenuse.

Now we apply the Pythagorean theorem to write an equation.

$$x^2 + (x + 4)^2 = (12)^2$$

$$x^2 + x^2 + 8x + 16 = 144$$

or

$$2x^2 + 8x - 128 = 0$$

Dividing both sides by 2, we have the equivalent equation

NOTE Dividing both sides of a quadratic equation by a common factor is always a prudent step. It simplifies your work with the quadratic formula.

$$x^2 + 4x - 64 = 0$$

Using the quadratic formula, we get

$$x = -2 + 2\sqrt{17} \quad \text{or} \quad x = -2 - 2\sqrt{17}$$

NOTE $\sqrt{17}$ is just slightly *more* than $\sqrt{16}$ or 4.

Now, we check our answers for reasonableness. We can reject $-2 - 2\sqrt{17}$ (do you see why?), but we should still check the reasonableness of the value $-2 + 2\sqrt{17}$. We could substitute $-2 + 2\sqrt{17}$ into the original equation, but it seems more prudent to simply check that it "makes sense" as a solution. Remembering that $\sqrt{16} = 4$, we estimate $-2 + 2\sqrt{17}$ as

$$-2 + 2(4) = 6$$

Our equation,

$$x^2 + (x + 4)^2 \overset{?}{\approx} (12)^2$$

when x equals 6, becomes

$$36 + 100 \approx 144$$

This indicates that our answer is at least reasonable.

CHECK YOURSELF 9

One leg of a right triangle is 2 cm longer than the other. The hypotenuse is 1 cm less than twice the length of the shorter leg. Find the length of each side of the triangle.

An important economic application involves supply and demand. Our last example illustrates that application.

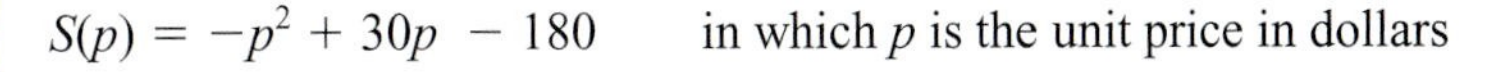

Example 10

An Economic Application

The number of intermediate algebra workbooks that a publisher is willing to produce is determined by the supply curve

$S(p) = -p^2 + 30p - 180$ in which p is the unit price in dollars

The demand for these workbooks is determined by the equation

$D(p) = -10p + 130$

Find the equilibrium price (the price at which supply and demand are equal).

Because supply equals demand ($S = D$ at equilibrium), we can write

$-p^2 + 30p - 180 = -10p + 130$

Rewriting this statement as a quadratic equation in standard form yields

$p^2 - 40p + 310 = 0$

When we apply the quadratic formula, we find the solutions

$p = 20 \pm 3\sqrt{10}$

$p \approx 10.51$ or $p \approx 29.49$

Although you might assume that the publisher will choose the higher price, it will, in fact, choose \$10.51. If you want to discover why, try substituting the two solutions into the original demand equation.

CHECK YOURSELF 10

The demand equation for CDs that accompany a text is predicted to be

$D = -6p + 30$ in which p is the unit price in dollars

The supply equation is predicted to be

$S = -p^2 + 12p - 20$

Find the equilibrium price.

CHECK YOURSELF ANSWERS

1. $\left\{-2, \frac{4}{3}\right\}$ **2.** $\left\{\frac{1}{2}\right\}$ **3.** $\left\{\frac{1 \pm \sqrt{57}}{4}\right\}$ **4.** $\{2 \pm \sqrt{10}\}$
5. $\{2 \pm i\sqrt{2}\}$ **6.** **(a)** None; **(b)** two; **(c)** one; **(d)** none **7.** 0.6 and 3.4 s
8. 2.6 s **9.** Approximately 4.3, 6.3, and 7.7 cm **10.** ≈$3.43

Name ______________________

Section __________ Date __________

9.3 Exercises

In exercises 1 to 8, solve each quadratic equation by first factoring and then using the quadratic formula.

1. $x^2 - 5x - 14 = 0$

2. $x^2 + 7x - 18 = 0$

3. $t^2 + 8t - 65 = 0$

4. $q^2 + 3q - 130 = 0$

5. $5x^2 + 4x - 1 = 0$

6. $3x^2 + 2x - 1 = 0$

7. $16t^2 - 24t + 9 = 0$

8. $6m^2 - 23m + 10 = 0$

In exercises 9 to 20, solve each quadratic equation by **(a)** completing the square, and **(b)** using the quadratic formula.

9. $x^2 - 2x - 5 = 0$

10. $x^2 + 6x - 1 = 0$

11. $x^2 + 3x - 27 = 0$

12. $t^2 + 4t - 7 = 0$

13. $2x^2 - 6x - 3 = 0$

14. $2x^2 - 6x + 1 = 0$

15. $2q^2 - 4q + 1 = 0$

16. $4r^2 - 2r + 1 = 0$

17. $3x^2 - x - 2 = 0$

18. $2x^2 - 8x + 3 = 0$

19. $2y^2 - y - 5 = 0$

20. $3m^2 + 2m - 1 = 0$

ANSWERS

1. ______________________
2. ______________________
3. ______________________
4. ______________________
5. ______________________
6. ______________________
7. ______________________
8. ______________________
9. ______________________
10. ______________________
11. ______________________
12. ______________________
13. ______________________
14. ______________________
15. ______________________
16. ______________________
17. ______________________
18. ______________________
19. ______________________
20. ______________________

ANSWERS

21. ____

22. ____

23. ____

24. ____

25. ____

26. ____

27. ____

28. ____

29. ____

30. ____

31. ____

32. ____

33. ____

34. ____

35. ____

36. ____

37. ____

38. ____

39. ____

40. ____

41. ____

42. ____

In exercises 21 to 42, solve each equation by using the quadratic formula.

21. $x^2 - 4x + 3 = 0$

22. $x^2 - 7x + 3 = 0$

23. $p^2 - 8p + 16 = 0$

24. $u^2 + 7u - 30 = 0$

25. $2x^2 - 2x - 3 = 0$

26. $2x^2 - 3x - 7 = 0$

27. $-3s^2 + 2s - 1 = 0$

28. $5t^2 - 2t - 2 = 0$

Hint: Clear each of the following equations of fractions or parentheses first.

29. $2x^2 - \frac{1}{2}x - 5 = 0$

30. $3x^2 + \frac{1}{3}x - 3 = 0$

31. $5t^2 - 2t - \frac{2}{3} = 0$

32. $3y^2 + 2y + \frac{3}{4} = 0$

33. $(x - 2)(x + 3) = 4$

34. $(x + 1)(x - 8) = 3$

35. $(t + 1)(2t - 4) - 7 = 0$

36. $(2w + 1)(3w - 2) = 1$

37. $3x - 5 = \frac{1}{x}$

38. $x + 3 = \frac{1}{x}$

39. $2t - \frac{3}{t} = 3$

40. $4p - \frac{1}{p} = 6$

41. $\frac{5}{y^2} + \frac{2}{y} - 1 = 0$

42. $\frac{6}{x^2} - \frac{2}{x} = 1$

In exercises 43 to 50, for each quadratic equation, find the value of the discriminant and give the number of real solutions.

43. $2x^2 - 5x = 0$

44. $3x^2 + 8x = 0$

45. $m^2 - 8m + 16 = 0$

46. $4p^2 + 12p + 9 = 0$

47. $3x^2 - 7x + 1 = 0$

48. $2x^2 - x + 5 = 0$

49. $2w^2 - 5w + 11 = 0$

50. $7q^2 - 3q + 1 = 0$

In exercises 51 to 62, find all the solutions of each quadratic equation. Use any applicable method.

51. $x^2 - 8x + 16 = 0$

52. $4x^2 + 12x + 9 = 0$

53. $3t^2 - 7t + 1 = 0$

54. $2z^2 - z + 5 = 0$

55. $5y^2 - 2y = 0$

56. $7z^2 - 6z - 2 = 0$

57. $(x - 1)(2x + 7) = -6$

58. $4x^2 - 3 = 0$

59. $x^2 + 9 = 0$

60. $(4x - 5)(x + 2) = 1$

61. $x - 3 - \dfrac{10}{x} = 0$

62. $1 + \dfrac{2}{x} + \dfrac{2}{x^2} = 0$

ANSWERS

43. ______

44. ______

45. ______

46. ______

47. ______

48. ______

49. ______

50. ______

51. ______

52. ______

53. ______

54. ______

55. ______

56. ______

57. ______

58. ______

59. ______

60. ______

61. ______

62. ______

ANSWERS

63. ______

64. ______

65. ______

66. ______

67. ______

68. ______

69. ______

70. ______

71. ______

The equation

$h(t) = 112t - 16t^2$

is the equation for the height of an arrow, shot upward from the ground with an initial velocity of 112 ft/s, in which t is the time, in seconds, after the arrow leaves the ground. Use this information to solve exercises 63 and 64. Your answers should be expressed to the nearest tenth of a second.

63. Find the time it takes for the arrow to reach a height of 112 ft.

64. Find the time it takes for the arrow to reach a height of 144 ft.

The equation

$h(t) = 320 - 32t - 16t^2$

is the equation for the height of a ball, thrown downward from the top of a 320-ft building with an initial velocity of 32 ft/s, in which t is the time after the ball is thrown down from the top of the building. Use this information to solve exercises 65 and 66. Express your results to the nearest tenth of a second.

65. Find the time it takes for the ball to reach a height of 240 ft.

66. Find the time it takes for the ball to reach a height of 96 ft.

67. Number problem. The product of two consecutive integers is 72. What are the two integers?

68. Number problem. The sum of the squares of two consecutive whole numbers is 61. Find the two whole numbers.

69. Rectangles. The width of a rectangle is 3 ft less than its length. If the area of the rectangle is 70 ft^2, what are the dimensions of the rectangle?

70. Rectangles. The length of a rectangle is 5 cm more than its width. If the area of the rectangle is 84 cm^2, find the dimensions.

71. Rectangles. The length of a rectangle is 2 cm more than 3 times its width. If the area of the rectangle is 85 cm^2, find the dimensions of the rectangle.

ANSWERS

72. ______

73. ______

74. ______

75. ______

76. ______

77. ______

78. ______

79. ______

80. ______

72. Rectangles. If the length of a rectangle is 3 ft less than twice its width, and the area of the rectangle is 54 ft^2, what are the dimensions of the rectangle?

73. Triangles. One leg of a right triangle is twice the length of the other. The hypotenuse is 6 m long. Find the length of each leg.

74. Triangles. One leg of a right triangle is 2 ft longer than the shorter side. If the length of the hypotenuse is 14 ft, how long is each leg?

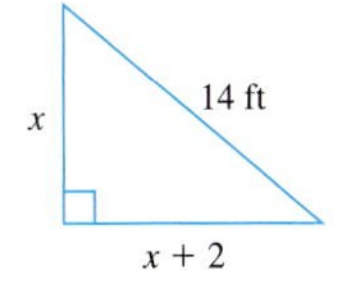

75. Triangles. One leg of a right triangle is 1 in. shorter than the other leg. The hypotenuse is 3 in. longer than the shorter side. Find the length of each side.

76. Triangles. The hypotenuse of a given right triangle is 5 cm longer than the shorter leg. The length of the shorter leg is 2 cm less than that of the longer leg. Find the length of the three sides.

77. Triangles. The sum of the lengths of the two legs of a right triangle is 25 m. The hypotenuse is 22 m long. Find the length of the two legs.

78. Triangles. The sum of the lengths of one side of a right triangle and the hypotenuse is 15 cm. The other leg is 5 cm shorter than the hypotenuse. Find the length of each side.

79. Thrown ball. If a ball is thrown vertically upward from the ground, its height, h, after t seconds is given by

$h(t) = 64t - 16t^2$

(a) How long does it take the ball to return to the ground? [*Hint:* Let $h(t) = 0$.]

(b) How long does it take the ball to reach a height of 48 ft on the way up?

80. Thrown ball. If a ball is thrown vertically upward from the ground, its height, h, after t seconds is given by

$h(t) = 96t - 16t^2$

(a) How long does it take the ball to return to the ground?

(b) How long does it take the ball to pass through a height of 128 ft on the way back down to the ground?

ANSWERS

81. ______

82. ______

83. ______

84. ______

85. ______

86. ______

81. Cost. Suppose that the cost $C(x)$, in dollars, of producing x chairs is given by

$$C(x) = 2400 - 40x + 2x^2$$

How many chairs can be produced for \$5400?

82. Profit. Suppose that the profit $T(x)$, in dollars, of producing and selling x appliances is given by

$$T(x) = -3x^2 + 240x - 1800$$

How many appliances must be produced and sold to achieve a profit of \$3000?

If a ball is thrown upward from the roof of a building 70 m tall with an initial velocity of 15 m/s, its approximate height, h, after t seconds is given by

$$h(t) = 70 + 15t - 5t^2$$

Note: The difference between this equation and the one we used in Example 8 has to do with the units used. When we used feet, the t^2 coefficient was -16 (from the fact that the acceleration due to gravity is approximately 32 ft/s^2). When we use meters as the height, the t^2 coefficient is -5 (that same acceleration becomes approximately 10 m/s^2). Use this information to solve exercises 83 and 84.

83. Thrown ball. How long does it take the ball to fall back to the ground?

84. Thrown ball. When will the ball reach a height of 80 m?

Changing the initial velocity to 25 m/s will only change the t coefficient. Our new equation becomes

$$h(t) = 70 + 25t - 5t^2$$

85. Thrown ball. How long will it take the ball to return to the thrower?

86. Thrown ball. When will the ball reach a height of 85 m?

The only part of the height equation that we have not discussed is the constant. You have probably noticed that the constant is always equal to the initial height of the ball (70 m in our previous problems). Now, let's have *you* develop an equation.

A ball is thrown upward from the roof of a 100-m building with an initial velocity of 20 m/s. Use this information to solve exercises 87 to 90.

87. Thrown ball. Find the equation for the height, h, of the ball after t seconds.

88. Thrown ball. How long will it take the ball to fall back to the ground?

89. Thrown ball. When will the ball reach a height of 75 m?

90. Thrown ball. Will the ball ever reach a height of 125 m? (*Hint:* Check the discriminant.)

A ball is thrown upward from the roof of a 100-ft building with an initial velocity of 20 ft/s. Use this information to solve exercises 91 to 94.

91. Thrown ball. Find the equation for the height, h, of the ball after t seconds.

92. Thrown ball. How long will it take the ball to fall back to the ground?

93. Thrown ball. When will the ball reach a height of 80 ft?

94. Thrown ball. Will the ball ever reach a height of 120 ft? Explain.

95. Profit. A small manufacturer's weekly profit in dollars is given by

$P(x) = -3x^2 + 270x$

Find the number of items x that must be produced to realize a profit of \$5100.

96. Profit. Suppose the profit in dollars is given by

$P(x) = -2x^2 + 240x$

Now how many items must be sold to realize a profit of \$5100?

97. Equilibrium price. The demand equation for a certain computer chip is given by

$D = -2p + 14$

The supply equation is predicted to be

$S = -p^2 + 16p - 2$

Find the equilibrium price.

ANSWERS

87. ____

88. ____

89. ____

90. ____

91. ____

92. ____

93. ____

94. ____

95. ____

96. ____

97. ____

ANSWERS

98. ______

99. ______

100. ______

101. ______

102. ______

103. ______

104. ______

105. ______

106. ______

107. ______

108. ______

109. ______

110. ______

111. ______

98. Equilibrium price. The demand equation for a certain type of print is predicted to be

$$D = -200p + 36{,}000$$

The supply equation is predicted to be

$$S = -p^2 + 400p - 24{,}000$$

Find the equilibrium price.

99. Can the solution of a quadratic equation with integer coefficients include one real and one imaginary number? Justify your answer.

100. Explain how the discriminant is used to predict the nature of the solutions of a quadratic equation.

In exercises 101 to 108, solve each equation for x.

101. $x^2 + y^2 = z^2$

102. $2x^2y^2z^2 = 1$

103. $x^2 - 36a^2 = 0$

104. $ax^2 - 9b^2 = 0$

105. $2x^2 + 5ax - 3a^2 = 0$

106. $3x^2 - 16bx + 5b^2 = 0$

107. $2x^2 + ax - 2a^2 = 0$

108. $3x^2 - 2bx - 2b^2 = 0$

109. Given that the polynomial $x^3 - 3x^2 - 15x + 25 = 0$ has as one of its solutions $x = 5$, find the other two solutions. (*Hint:* If you divide the given polynomial by $x - 5$ the quotient will be a quadratic equation. The remaining solutions will be the solutions for *that* equation.)

110. Given that $2x^3 + 2x^2 - 5x - 2 = 0$ has as one of its solutions $x = -2$, find the other two solutions. (*Hint:* In this case, divide the original polynomial by $x + 2$.)

111. Find all the zeros of the function $f(x) = x^3 + 1$.

112. Find the zeros of the function $f(x) = x^2 + x + 1$.

113. Find all six solutions to the equation $x^6 - 1 = 0$. (*Hint:* Factor the left-hand side of the equation first as the difference of squares, then as the sum and difference of cubes.)

114. Find all six solutions to $x^6 = 64$.

115. **(a)** Use the quadratic formula to solve $x^2 - 3x - 5 = 0$. For each solution give a decimal approximation to the nearest tenth.

(b) Graph the function $f(x) = x^2 - 3x - 5$ on your graphing calculator. Use a zoom utility and estimate the x intercepts to the nearest tenth.

(c) Describe the connection between parts (a) and (b).

116. **(a)** Solve the following equation using any appropriate method:

$$x^2 - 2x = 3$$

(b) Graph the following functions on your graphing calculator:

$f(x) = x^2 - 2x$ and $g(x) = 3$

Estimate the points of intersection of the graphs of f and g. In particular note the x coordinates of these points.

(c) Describe the connection between parts (a) and (b).

ANSWERS

112. ______

113. ______

114. ______

115. (a)
(b)
(c)
______

116. (a)
(b)
(c)
______

Answers

1. $\{-2, 7\}$ **3.** $\{-13, 5\}$ **5.** $\left\{-1, \frac{1}{5}\right\}$ **7.** $\left\{\frac{3}{4}\right\}$ **9.** $\{1 \pm \sqrt{6}\}$

11. $\left\{\frac{-3 \pm 3\sqrt{13}}{2}\right\}$ **13.** $\left\{\frac{3 \pm \sqrt{15}}{2}\right\}$ **15.** $\left\{\frac{2 \pm \sqrt{2}}{2}\right\}$ **17.** $\left\{-\frac{2}{3}, 1\right\}$

19. $\left\{\frac{1 \pm \sqrt{41}}{4}\right\}$ **21.** $\{1, 3\}$ **23.** $\{4\}$ **25.** $\left\{\frac{1 \pm \sqrt{7}}{2}\right\}$

27. $\left\{\frac{1 \pm i\sqrt{2}}{3}\right\}$ **29.** $\left\{\frac{1 \pm \sqrt{161}}{8}\right\}$ **31.** $\left\{\frac{3 \pm \sqrt{39}}{15}\right\}$

33. $\left\{\frac{-1 \pm \sqrt{41}}{2}\right\}$ **35.** $\left\{\frac{1 \pm \sqrt{23}}{2}\right\}$ **37.** $\left\{\frac{5 \pm \sqrt{37}}{6}\right\}$

39. $\left\{\frac{3 \pm \sqrt{33}}{4}\right\}$ **41.** $\{1 \pm \sqrt{6}\}$ **43.** 25, two **45.** 0, one

47. 37, two **49.** -63, none **51.** $\{4\}$ **53.** $\left\{\frac{7 \pm \sqrt{37}}{6}\right\}$ **55.** $\left\{0, \frac{2}{5}\right\}$

57. $\left\{\frac{-5 \pm \sqrt{33}}{4}\right\}$ **59.** $\{-3i, 3i\}$ **61.** $\{-2, 5\}$ **63.** {1.2 or 5.8 s}

65. 1.4 s **67.** $-9, -8$, or 8, 9 **69.** 7 by 10 ft **71.** 5 by 17 cm

73. 2.7 m, 5.4 m **75.** 5.5, 6.5, 8.5 in. **77.** 3.2 m, 21.8 m

79. **(a)** 4 s; **(b)** 1 s **81.** 50 chairs **83.** 5.5 s **85.** 5 s

87. $h(t) = 100 + 20t - 5t^2$ **89.** 5 s **91.** $h(t) = 100 + 20t - 16t^2$

93. 1.9 s **95.** 63 or 27 **97.** \$0.94 **99.** **101.** $\{\pm\sqrt{z^2 - y^2}\}$

103. $\{-6a, 6a\}$ **105.** $\left\{-3a, \frac{a}{2}\right\}$ **107.** $\left\{\frac{-a \pm a\sqrt{17}}{4}\right\}$ **109.** $\{-1 \pm \sqrt{6}\}$

111. $\left\{-1, \frac{1 \pm i\sqrt{3}}{2}\right\}$ **113.** $\left\{-1, 1, \frac{1 \pm i\sqrt{3}}{2}, \frac{-1 \pm i\sqrt{3}}{2}\right\}$

115. **(a)** $\{-1.2, 4.2\}$; **(b)** $-1.2, 4.2$; **(c)** Solutions to quadratic equation are the x intercepts of the graph

9.4 Equations that Are Quadratic in Form

OBJECTIVES

1. Solve a radical equation that is quadratic in form
2. Solve a fourth degree equation that is quadratic in form

Consider the following equations:

$$2x - 5\sqrt{x} + 3 = 0 \tag{1}$$

$$x^4 - 4x^2 + 3 = 0 \tag{2}$$

$$(x^2 - x)^2 - 8(x^2 - x) + 12 = 0 \tag{3}$$

None of these equations are quadratic, yet each can be readily solved by using quadratic methods.

Compare the following quadratic equations to the original three equations.

NOTE Let $u = \sqrt{x}$ in (1).

NOTE Let $u = x^2$ in (2).

NOTE Let $u = x^2 - x$ in (3).

$$2u^2 - 5u + 3 = 0 \tag{4}$$

$$u^2 - 4u + 3 = 0 \tag{5}$$

$$u^2 - 8u + 12 = 0 \tag{6}$$

In each case, a simple substitution has been made that resulted in a quadratic equation. Equations that can be rewritten in this manner are said to be *equations in quadratic form.*

Example 1

Solving a Radical Equation

Solve.

$$2x - 5\sqrt{x} + 3 = 0$$

By substituting u for $\sqrt{x}$, we have

NOTE Notice that $u^2 = x$ because $u = \sqrt{x}$.

$$2u^2 - 5u + 3 = 0$$

Factoring yields

$$(2u - 3)(u - 1) = 0$$

NOTE By intermediate solutions we mean values for u rather than for the original variable x.

which gives the intermediate solutions

$$u = \frac{3}{2} \quad \text{or} \quad u = 1$$

We must now solve for x and then check our solutions. Because $\sqrt{x} = u$, we can write

$$\sqrt{x} = \frac{3}{2} \quad \text{or} \quad \sqrt{x} = 1$$

NOTE We square both sides in each equation.

$$x = \frac{9}{4} \qquad\qquad x = 1$$

To check these solutions, we again simply substitute these values into the original equation. You should verify that each is a valid solution.

CHECK YOURSELF 1

Solve $3x - 8\sqrt{x} + 4 = 0$.

For certain equations in quadratic form, we can either solve by substitution (as we have done above) or solve directly by treating the equation as quadratic in some other power of the variable (in the case of the equation of the following example, x^2). In our next example, we show both methods of solution.

Example 2

Solving a Fourth Degree Equation

(a) Solve $x^4 - 4x^2 + 3 = 0$ by substitution:

Let $u = x^2$. Then

NOTE Notice that $u^2 = (x^2)^2 = x^4$

$u^2 - 4u + 3 = 0$

Factoring, we have

$(u - 1)(u - 3) = 0$

so

$u = 1 \quad \text{or} \quad u = 3$

Given these intermediate solutions, because $u = x^2$, we can write

$x^2 = 1 \quad \text{or} \quad x^2 = 3$

which, by using the square root method, yields the four solutions

NOTE There are *four* solutions to the original *fourth-degree* equation.

$x = \pm 1 \quad \text{or} \quad x = \pm\sqrt{3}$

We can check each of these solutions by substituting into the original equation. When we do so, we find that all four are valid solutions to the original equation.

(b) Solve $x^4 - 4x^2 + 3 = 0$ directly.

By treating the equation as quadratic in x^2, we can factor the left member, to write

$(x^2 - 1)(x^2 - 3) = 0$

This gives us the two equations

$x^2 - 1 = 0 \quad \text{or} \quad x^2 - 3 = 0$

Now

$x^2 = 1 \quad \text{or} \quad x^2 = 3$

Again, we have the four possible solutions

NOTE We apply the square root property.

$x = \pm 1 \quad \text{or} \quad x = \pm\sqrt{3}$

All check when they are substituted into the original equation.

CHECK YOURSELF 2

Solve $x^4 - 9x^2 + 20 = 0$ *by substitution and by factoring directly.*

In the following example, a binomial is replaced with u to make it easier to proceed with the solution.

Example 3

Solving a Fourth Degree Equation

Solve.

$(x^2 - x)^2 - 8(x^2 - x) + 12 = 0$

Because of the repeated factor $x^2 - x$, we substitute u for $x^2 - x$. Factoring the resulting equation

$u^2 - 8u + 12 = 0$

gives

$(u - 6)(u - 2) = 0$

So

$u = 6 \quad \text{or} \quad u = 2$

We now have two intermediate solutions to work with. Because $u = x^2 - x$, we have two cases:

If $u = 6$, then	If $u = 2$, then
$x^2 - x = 6$	$x^2 - x = 2$
$x^2 - x - 6 = 0$	$x^2 - x - 2 = 0$
$(x - 3)(x + 2) = 0$	$(x - 2)(x + 1) = 0$
$x = 3$ or $x = -2$	$x = 2$ or $x = -1$

NOTE Write in standard form.

NOTE Factor the quadratic member.

The quadratic equations now yield four solutions that we must check. Substituting into the original equation, you will find that all four are valid solutions.

CHECK YOURSELF 3

Solve for x:

$(x^2 - 2x)^2 - 11(x^2 - 2x) + 24 = 0$

To summarize our work with equations in quadratic form, two approaches are commonly used. The first involves substitution of a new intermediate variable to make the original equation quadratic. The second solves the original equation directly by treating the equation as quadratic in some other power of the original variable. The following algorithms outline the two approaches.

Step by Step: Solving by Substitution

Step 1 Make an appropriate substitution so that the equation becomes quadratic.

Step 2 Solve the resulting equation for the intermediate variable.

Step 3 Use the intermediate values found in step 2 to find possible solutions for the original variable.

Step 4 Check the solutions of step 3 by substitution into the original equation.

Step by Step: Solving by Factoring

Step 1 Treat the original equation as quadratic in some power of the variable, and factor.

Step 2 Solve the resulting equations.

Step 3 Check the solutions of step 2 by substitution into the original equation.

CHECK YOURSELF ANSWERS

1. $\left\{\frac{4}{9}, 4\right\}$ **2.** $\{\pm 2, \pm\sqrt{5}\}$ **3.** $\{-1, -2, 3, 4\}$

Name ____________

Section ______ Date ______

9.4 Exercises

Solve each of the following equations by factoring directly and then applying the zero product rule.

1. $x^4 - 9x^2 + 20 = 0$

2. $t^4 - 7t^2 + 12 = 0$

3. $x^4 + x^2 - 12 = 0$

4. $x^4 - 7x^2 - 18 = 0$

5. $2w^4 - 9w^2 + 4 = 0$

6. $3x^4 - 5x^2 + 2 = 0$

7. $x^4 - 4x^2 + 4 = 0$

8. $y^4 - 6y^2 + 9 = 0$

9. $3x^4 + 16x^2 - 12 = 0$

10. $2x^4 + 9x^2 - 5 = 0$

11. $2z^4 + 4z^2 - 70 = 0$

12. $3y^4 + 10y^2 - 8 = 0$

13. $4t^4 - 20t^2 = 0$

14. $r^4 - 81 = 0$

Solve each of the following equations.

15. $x^4 - x^2 - 12 = 0$

16. $w^4 + w^2 - 12 = 0$

17. $2y^4 + y^2 - 15 = 0$

18. $4x^4 - 5x^2 - 9 = 0$

19. $b - 20\sqrt{b} + 64 = 0$

20. $z - 6\sqrt{z} + 8 = 0$

ANSWERS

1. ____________
2. ____________
3. ____________
4. ____________
5. ____________
6. ____________
7. ____________
8. ____________
9. ____________
10. ____________
11. ____________
12. ____________
13. ____________
14. ____________
15. ____________
16. ____________
17. ____________
18. ____________
19. ____________
20. ____________

ANSWERS

21. ______
22. ______
23. ______
24. ______
25. ______
26. ______
27. ______
28. ______
29. ______
30. ______
31. ______
32. ______
33. ______
34. ______
35. ______
36. ______
37. ______
38. ______

21. $t - 8\sqrt{t} - 9 = 0$

22. $y - 24\sqrt{y} - 25 = 0$

23. $(x - 2)^2 - 3(x - 2) - 10 = 0$

24. $(w + 1)^2 - 5(w + 1) + 6 = 0$

25. $(x^2 + 2x)^2 + 3(x^2 + 2x) + 2 = 0$

26. $(x^2 - 4x)^2 - (x^2 - 4x) - 12 = 0$

Solve each of the following equations by any method.

27. $7m - 41\sqrt{m} - 6 = 0$

28. $(x + 1) - 6\sqrt{x + 1} + 8 = 0$

29. $(w - 3)^2 - 2(w - 3) = 15$

30. $(x^2 - 4x)^2 + 7(x^2 - 4x) + 12 = 0$

31. $2y^4 - 5y^2 = 12$

32. $4t^4 - 29t^2 + 25 = 0$

An equation involving rational exponents may sometimes be solved by substitution. For instance, to solve an equation of the form $ax^{1/2} + bx^{1/4} + c = 0$, make the substitution $u = x^{1/4}$. Note that $u^2 = (x^{1/4})^2 = x^{2/4} = x^{1/2}$.

Use the suggestion above to solve each of the following equations. Be sure to check your solutions.

33. $x^{1/2} - 4x^{1/4} + 3 = 0$

34. $x^{1/2} - 5x^{1/4} + 6 = 0$

35. $x^{1/2} - x^{1/4} = 2$

36. $2x^{1/2} + x^{1/4} - 1 = 0$

37. $x^{2/3} + 2x^{1/3} - 3 = 0$

38. $x^{2/5} - x^{1/5} = 6$

Certain equations involving rational expressions can also be solved by the method of substitution. For instance, to solve an equation of the form

$$\frac{a}{x^2} + \frac{b}{x} + c = 0$$

make the substitution $u = \frac{1}{x}$. Note that $u^2 = \left(\frac{1}{x}\right)^2 = \frac{1}{x^2}$.

Use the suggestion above to solve the following equations.

39. $\frac{1}{x^2} - \frac{6}{x} + 8 = 0$

40. $\frac{2}{x^2} - \frac{1}{x} = 3$

41. $\frac{3}{x^2} - \frac{5}{x} = 2$

42. $\frac{1}{(x + 1)^2} - \frac{5}{x + 1} + 4 = 0$

43. $\frac{1}{(x - 2)^2} + \frac{1}{x - 2} = 6$

44. $\frac{8}{(x - 3)^2} - \frac{2}{x - 3} = 1$

Solve each of the following applications.

45. Number problem. The sum of an integer and twice its square root is 24. What is the integer?

46. Number problem. The sum of an integer and 3 times its square root is 40. Find the integer.

47. Number problem. The sum of the reciprocal of an integer and the square of its reciprocal is $\frac{3}{4}$. What is the integer?

48. Number problem. The difference between the reciprocal of an integer and the square of its reciprocal is $\frac{2}{9}$. Find the integer.

ANSWERS

39. ______
40. ______
41. ______
42. ______
43. ______
44. ______
45. ______
46. ______
47. ______
48. ______

Answers

1. $\{\pm\sqrt{5}, \pm 2\}$ **3.** $\{\pm\sqrt{3}, \pm 2i\}$ **5.** $\left\{\frac{\pm\sqrt{2}}{2}, \pm 2\right\}$ **7.** $\{\pm\sqrt{2}\}$

9. $\left\{\frac{\pm\sqrt{6}}{3}, \pm i\sqrt{6}\right\}$ **11.** $\{\pm\sqrt{5}, \pm i\sqrt{7}\}$ **13.** $\{0, \pm\sqrt{5}\}$

15. $\{\pm 2, \pm i\sqrt{3}\}$ **17.** $\left\{\frac{\pm\sqrt{10}}{2}, \pm i\sqrt{3}\right\}$ **19.** $\{16, 256\}$ **21.** $\{81\}$

23. $\{7, 0\}$ **25.** $\{-1, -1 \pm i\}$ **27.** $\{36\}$ **29.** $\{0, 8\}$ **31.** $\left\{\pm 2, \frac{\pm i\sqrt{6}}{2}\right\}$

33. $\{1, 81\}$ **35.** $\{16\}$ **37.** $\{-27, 1\}$ **39.** $\left\{\frac{1}{4}, \frac{1}{2}\right\}$ **41.** $\left\{-3, \frac{1}{2}\right\}$

43. $\left\{\frac{5}{3}, \frac{5}{2}\right\}$ **45.** 16 **47.** 2

Quadratic Inequalities

9.5 OBJECTIVES

1. Solve a quadratic inequality graphically
2. Solve a quadratic inequality algebraically
3. Solve a quadratic inequality in two variables

A **quadratic inequality** is an inequality that can be written in the form

$ax^2 + bx + c < 0 \qquad \text{when } a \neq 0$

Note that the inequality symbol, < can be replaced by the symbol >, ≤, or ≥ in the above definition.

In Chapter 4, solutions to linear inequalities such as $4x + 2 < 0$ were analyzed graphically. Recall that, given the graph of the function $f(x) = 4x + 2$

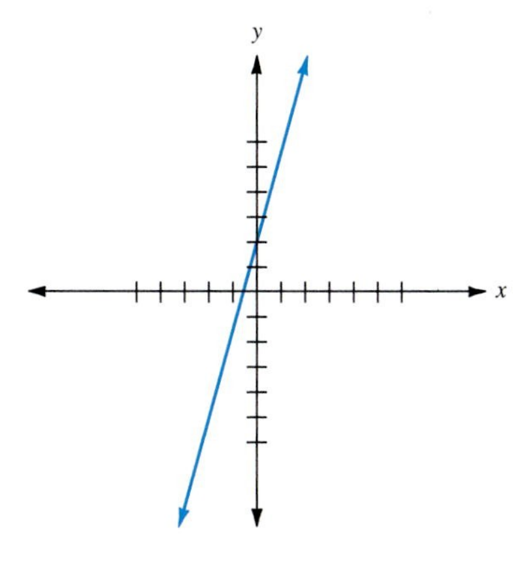

the solution set for the inequality is the set of all x values associated with points on the line that were *below* the x axis.

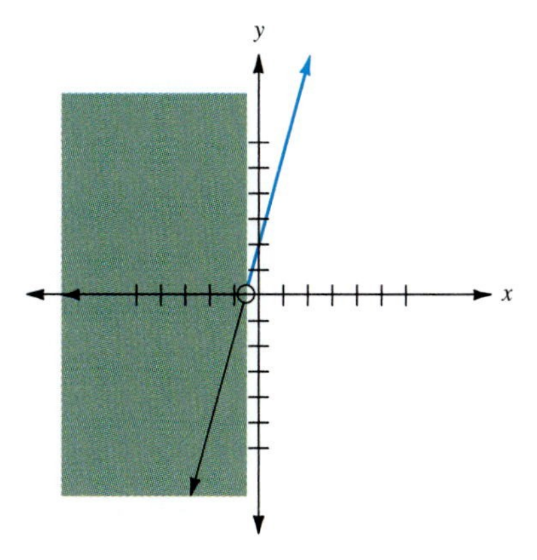

In this case, each x to the left of $-\frac{1}{2}$ is associated with a point on the line that is below the x axis. The solution set for the inequality $4x + 2 < 0$ is the set $\left\{x \middle| x < -\frac{1}{2}\right\}$.

The same principle can be applied to solving quadratic inequalities. Example 1 illustrates this concept.

Example 1

Solving a Quadratic Inequality Graphically

Solve the inequality

$x^2 - x - 12 \leq 0$

First, use the techniques in Section 9.1 to graph the function $f(x) = x^2 - x - 12$.

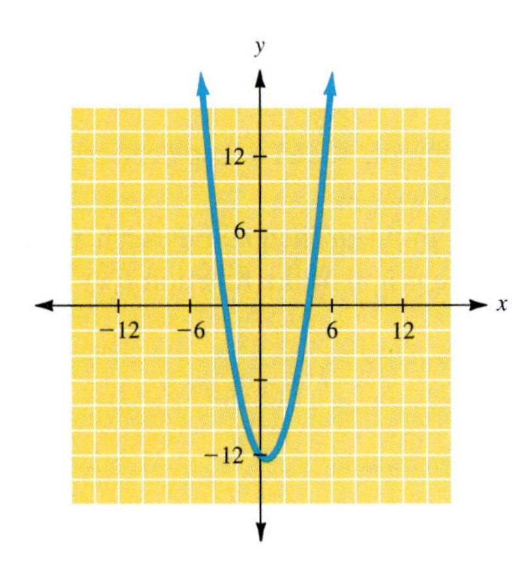

Next, looking at the graph, determine the values of x that make $x^2 - x - 12 \leq 0$ a true statement.

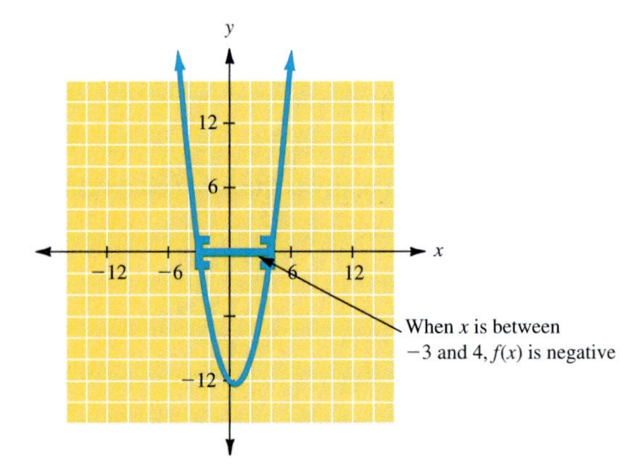

Notice that the graph is below the x axis for values of x between -3 and 4. The graph intercepts the x axis when x is -3 or 4. The solution set for the inequality is $\{x \mid -3 \leq x \leq 4\}$.

CHECK YOURSELF 1

Use a graph to solve the inequality

$x^2 - 3x - 10 \geq 0$

Algebraic methods can also be used to find the solutions to a quadratic inequality. Subsequent examples in this section will discuss algebraic methods. When solving an equation or inequality algebraically, it is always a good idea to compare the graph to the algebraic solutions.

NOTE If we expand the binomial product, we get

$x^2 - 2x - 3 < 0$

Looking at the graph of

$y = x^2 - 2x - 3$,

where is y less than 0 on the graph?

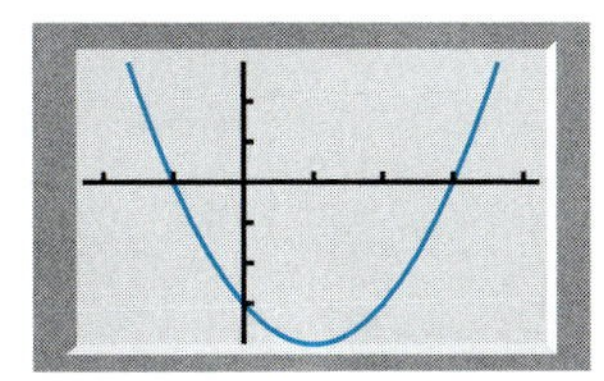

Example 2

Solving a Quadratic Inequality Algebraically

Solve $(x - 3)(x + 1) < 0$.

We start by finding the solutions of the corresponding quadratic equation. So

$$(x - 3)(x + 1) = 0$$

has solutions 3 and -1, called the *critical values* of x.

Our process depends on determining where each factor is positive or negative. To help visualize that process, we start with a number line and label it as shown below. We begin with our first critical value of -1.

NOTE $x + 1$ is negative if x is less than -1. $x + 1$ is positive if x is greater than -1.

Sign of $x + 1$ ----|++++++++++++

-1

We now continue in the same manner with the second critical value 3.

NOTE $x - 3$ is negative if x is less than 3. $x - 3$ is positive if x is greater than 3.

Sign of $x - 3$ -----------|++++

3

In practice, we combine the two steps above for the following result.

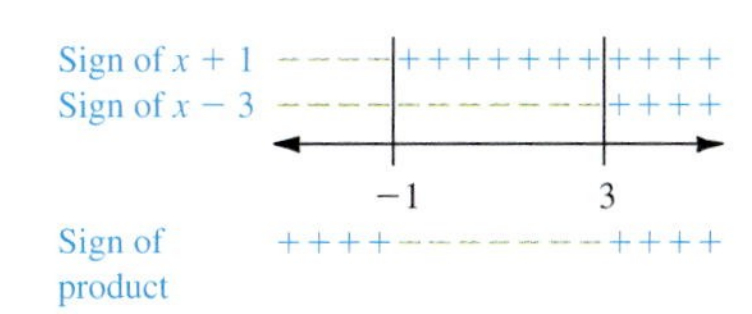

Examining the signs of the factors, we see that in this case:

NOTE Both factors are negative.

For any x less than -1, the product is positive.

NOTE The factors have opposite signs.

For any x between -1 and 3, the product is negative.

NOTE Both factors are positive.

For any x greater than 3, the product is again positive.

We return to the original inequality:

NOTE The product of the two binomials must be negative.

$$(x - 3)(x + 1) < 0$$

We can see that this is true only between -1 and 3. In set notation, the solution set can be written as

$$\{x \mid -1 < x < 3\}$$

On a number line, the graph of the solution set is

CHECK YOURSELF 2

Solve and graph the solution set.

$(x - 2)(x + 4) < 0$

We now consider an example in which the quadratic member of the inequality must be factored.

Example 3

Solving a Quadratic Inequality Algebraically

NOTE Examine the graph of $y = x^2 - 5x + 4$. For what values of x is y (the graph) greater than zero?

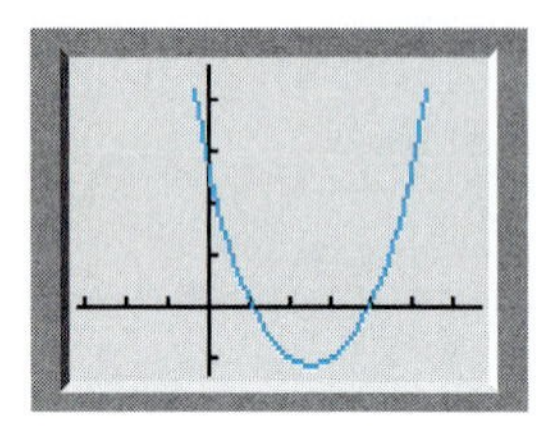

Solve $x^2 - 5x + 4 > 0$.

Factoring the quadratic member, we have

$(x - 1)(x - 4) > 0$

The critical values are 1 and 4, and we form the sign graph as before.

Sign of $x - 4$ ---- | -------- | ++++
Sign of $x - 1$ ---- | +++++++ | ++++
1 4
Sign of product ++++ -------- ++++

In this case, we want those values of x for which the product is *positive,* and we can see from the sign graph above that the solution set is

$\{x | x < 1 \text{ or } x > 4\}$

The graph of the solution set is shown below.

CHECK YOURSELF 3

Solve and graph the solution set.

$2x^2 - x - 3 > 0$

The method used in the previous examples works *only* when one side of the inequality is factorable and the other is 0. It is sometimes necessary to rewrite the inequality in an equivalent form, as Example 4 illustrates.

Example 4

Solving a Quadratic Inequality Algebraically

NOTE Use a calculator to graph both $f(x) = x^2 - 3x - 4$ and $g(x) = 6$. Where is $f(x)$ above $g(x)$? Compare this to the algebraic solution.

Solve $(x + 1)(x - 4) \geq 6$.

First, we multiply to clear the parentheses.

$x^2 - 3x - 4 \geq 6$

Now we subtract 6 from both sides so that the inequality is *related to 0:*

$x^2 - 3x - 10 \geq 0$

Factoring the quadratic member, we have

$(x - 5)(x + 2) \geq 0$

We can now proceed with the sign graph method as before.

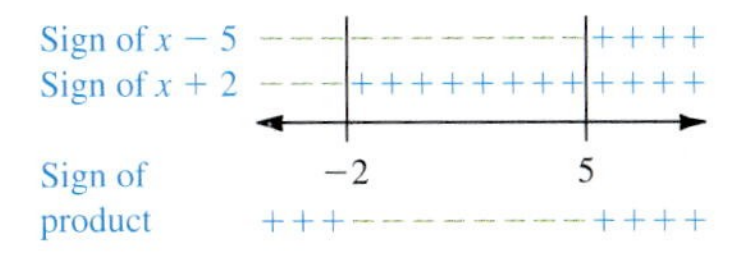

NOTE Both factors are negative if x is less than -2. Both factors are positive if x is greater than 5.

From the sign graph we see that the solution set is

$\{x | x \leq -2 \text{ or } x \geq 5\}$

The graph of the solution set is shown below.

CHECK YOURSELF 4

Solve and graph the solution set.

$(x - 5)(x + 7) \leq -11$

The following algorithm summarizes our work to this point in solving a quadratic inequality.

Step by Step: Solving a Quadratic Inequality Algebraically

Step 1 Clear the inequality of parentheses and fractions.
Step 2 Rewrite the inequality such that the expression is related to zero.
Step 3 Factor the quadratic member.
Step 4 Use a sign graph to find the values for x that make the inequality a true statement.
Step 5 Use a number line to graph the solution set.

Each of the quadratic inequalities we have investigated in this section has included only one variable. What does the solution set for a quadratic inequality in two variables look like? The next example illustrates.

Example 5

Solving a Quadratic Inequality in Two Variables

Use a graphing calculator to find the solution set for the inequality

$y \leq x^2 - 3x - 8$

Recall from Section 5.4 that the solution set for a linear inequality in two variables was a set of points in the plane. Further recall that, to find that set of points, we first graphed the line of the related equation. We will use a graphing calculator to graph

$f(x) = x^2 - 3x - 8$

Because the graph of $f(x)$ is a parabola, the solution set has a parabolic boundary.

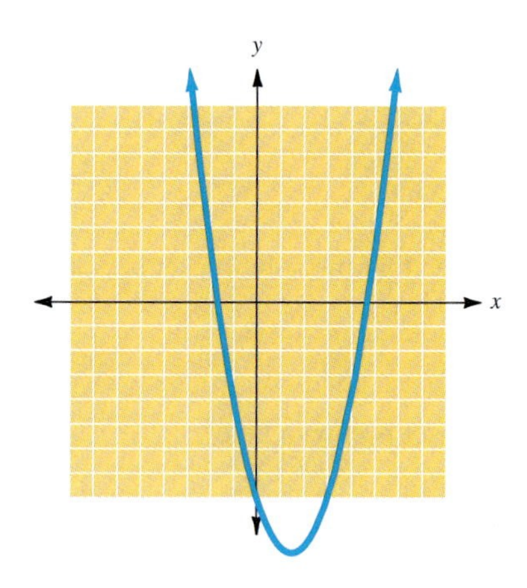

Now we decide which half-plane to shade. As we did with linear inequalities, we select a test point and find whether the test point is part of the solution set. The origin is usually the easiest test point to use.

$0 \leq (0)^2 - 3(0) - 8$

$0 \leq -8$ is not a true statement. We will shade the half-plane that does not include the origin.

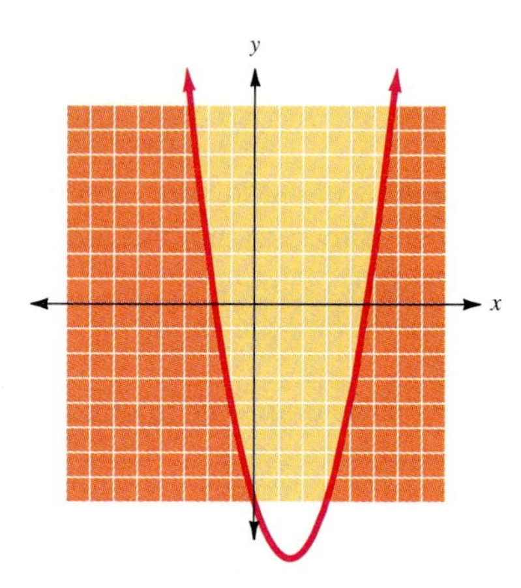

CHECK YOURSELF 5

Find the solution set for the inequality.

$y \geq x^2 - 7x + 12$

CHECK YOURSELF ANSWERS

1. $\{x \mid x \le -2 \text{ or } x \ge 5\}$

2. 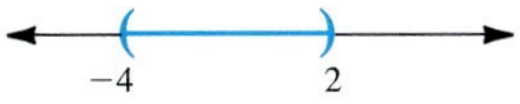

$\{x \mid -4 < x < 2\}$

3. 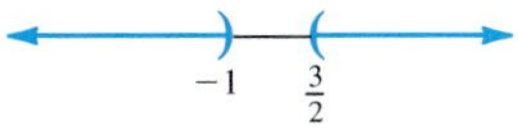

$\left\{x \mid x < -1 \text{ or } x > \frac{3}{2}\right\}$

4. 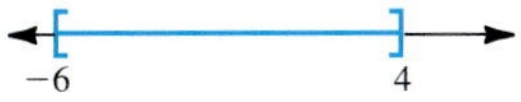

$\{x \mid -6 \le x \le 4\}$

5. 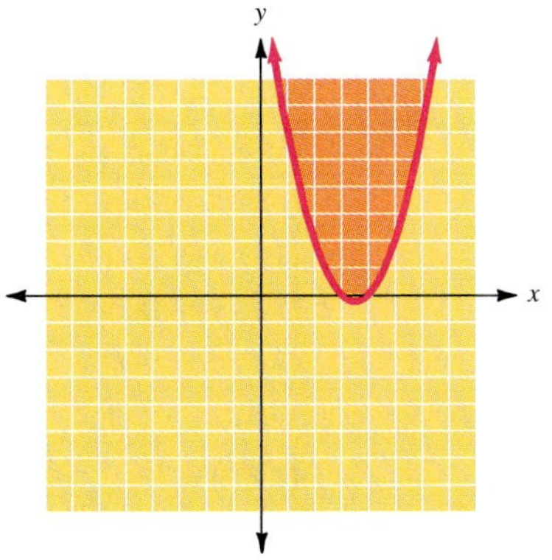

Trajectory and Height. So far in this chapter you have done many exercises involving balls that have been thrown upward with varying velocities. How does trajectory—the angle at which the ball is thrown—affect its height and time in the air?

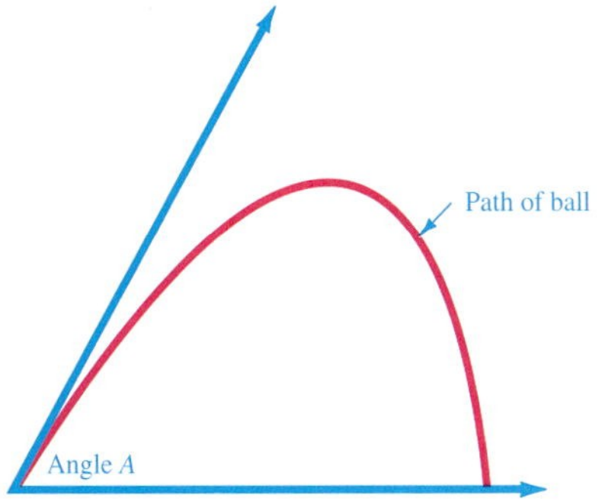

If you throw a ball from ground level with an initial upward velocity of 70 ft/s, the equation

$$h = -16t^2 + \partial(70)t$$

gives you the height, h, in feet t seconds after the ball has been thrown at a certain angle, A. The value of ∂ in the equation depends on the angle A and is given in the accompanying table.

Measure of Angle A in Degrees	Value of ∂
0	0.000
5	0.087
10	0.174
15	0.259
20	0.342
25	0.423
30	0.500
35	0.574
40	0.643
45	0.707
50	0.766
55	0.819
60	0.866
65	0.906
70	0.940
75	0.966
80	0.985
85	0.996
90 (straight up)	1.000

Investigate the following questions and write your conclusions to each one in complete sentences, showing all charts and graphs. Indicate what initial velocity you are using in each case.

1. Suppose an object is thrown from ground level with an initial upward velocity of 70 ft/s and at an angle of 45 degrees. What will be the height of the ball in 1 s (nearest tenth of a foot)? How long is the ball in the air?
2. Does the ball stay in the air longer if the angle of the throw is greater?
3. If you double the angle of the throw, will the ball stay in the air double the length of time?
4. If you double the angle of the throw, will the ball go twice as high?
5. Is the height of the ball directly related to the angle at which you throw it? That is, does the ball go higher if the angle of the throw is larger?
6. Repeat this exercise using another initial upward velocity.

Name ______________

Section ______ Date ______

9.5 Exercises

In exercises 1 to 8, solve each inequality, and graph the solution set.

1. $(x - 3)(x + 4) < 0$

0

2. $(x - 2)(x + 5) > 0$

0

3. $(x - 3)(x + 4) > 0$

0

4. $(x - 2)(x + 5) < 0$

0

5. $(x - 3)(x + 4) \le 0$

0

6. $(x - 2)(x + 5) \ge 0$

0

7. $(x - 3)(x + 4) \ge 0$

0

8. $(x - 2)(x + 5) \le 0$

0

In exercises 9 to 38, solve each inequality, and graph the solution set.

9. $x^2 - 3x - 4 > 0$

0

10. $x^2 - 2x - 8 < 0$

0

11. $x^2 + x - 12 \le 0$

0

12. $x^2 - 2x - 15 \ge 0$

0

13. $x^2 - 5x + 6 \ge 0$

0

14. $x^2 + 7x + 10 \le 0$

0

15. $x^2 + 2x \le 24$

0

16. $x^2 - 3x > 18$

0

ANSWERS

1. ______
2. ______
3. ______
4. ______
5. ______
6. ______
7. ______
8. ______
9. ______
10. ______
11. ______
12. ______
13. ______
14. ______
15. ______
16. ______

ANSWERS

17. ______
18. ______
19. ______
20. ______
21. ______
22. ______
23. ______
24. ______
25. ______
26. ______
27. ______
28. ______
29. ______
30. ______
31. ______
32. ______

17. $x^2 > 27 - 6x$

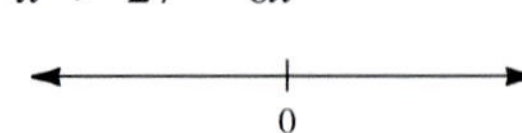

18. $x^2 \leq 7x - 12$

19. $2x^2 + x - 6 \leq 0$

20. $3x^2 - 10x - 8 < 0$

21. $4x^2 + x < 3$

22. $5x^2 - 13x \geq 6$

23. $x^2 - 16 \leq 0$

24. $x^2 - 9 > 0$

25. $x^2 \geq 25$

26. $x^2 < 49$

27. $4 - x^2 < 0$

28. $36 - x^2 \geq 0$

29. $x^2 - 4x \leq 0$

30. $x^2 + 5x > 0$

31. $x^2 \geq 6x$

32. $x^2 < 3x$

33. $4x > x^2$

34. $6x \le x^2$

35. $x^2 - 4x + 4 \le 0$

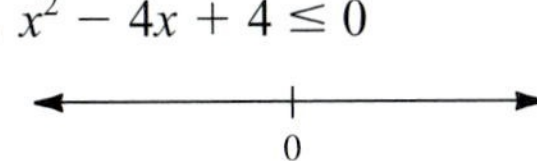

36. $x^2 + 6x + 9 \ge 0$

37. $(x + 3)(x - 6) \le 10$

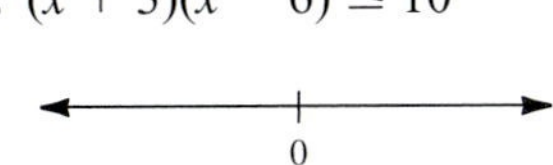

38. $(x + 4)(x - 5) > 22$

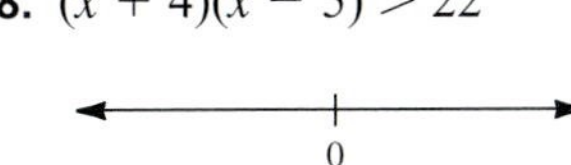

39. Can a quadratic inequality be solved if the quadratic member of the inequality is not factorable? If so, explain how the solution can be found. If not, explain why not.

40. Is it necessary to relate a quadratic inequality to 0 to solve it? Why or why not?

An inequality of the form

$(x - a)(x - b)(x - c) < 0$

can be solved by using a sign graph to consider the signs of *all three factors.* In exercises 41 to 46, use this suggestion to solve each inequality. Then graph the solution set.

41. $x(x - 2)(x + 1) < 0$

42. $x(x + 3)(x - 2) \ge 0$

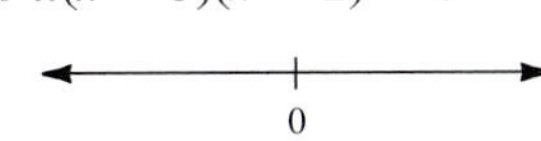

43. $(x - 3)(x + 2)(x - 1) \ge 0$

44. $(x - 5)(x + 1)(x - 4) < 0$

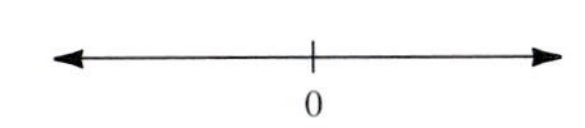

45. $x^3 - 2x^2 - 15x \le 0$

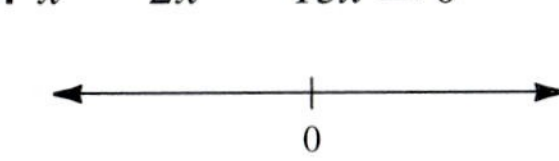

46. $x^3 + 2x^2 - 24x > 0$

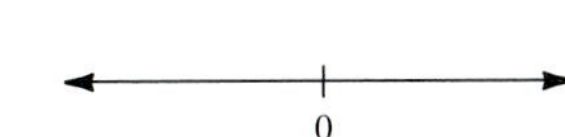

ANSWERS

33. ____
34. ____
35. ____
36. ____
37. ____
38. ____

39. ____

40. ____
41. ____
42. ____
43. ____
44. ____
45. ____
46. ____

47. ______

48. ______

49. ______

50. ______

51. ______

52. ______

Solve exercises 47 to 50.

47. A small manufacturer's weekly profit is given by

$$P(x) = -2x^2 + 220x$$

in which x is the number of items manufactured and sold. Find the number of items that must be manufactured and sold if the profit is to be greater than or equal to \$6000.

48. Suppose that a company's profit is given by

$$P(x) = -2x^2 + 360x$$

How many items must be produced and sold so that the profit will be at least \$16,000?

49. If a ball is thrown vertically upward from the ground with an initial velocity of 80 ft/s, its approximate height is given by

$$h(t) = -16t^2 + 80t$$

in which t is the time (in seconds) after the ball is released. When will the ball have a height of at least 96 ft?

50. Suppose a ball's height (in meters) is given by

$$h(t) = -5t^2 + 20t$$

When will the ball have a height of at least 15 m?

In exercises 51 to 54, use a graphing calculator to find the solution set for each inequality.

51. $y < x^2 + 6x$

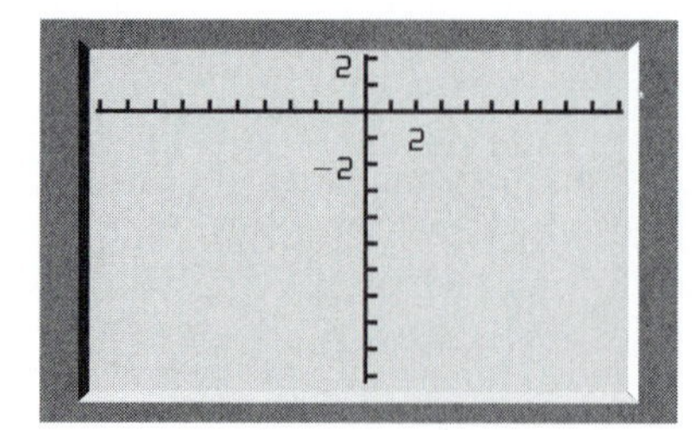

52. $y \geq x^2 - 49$

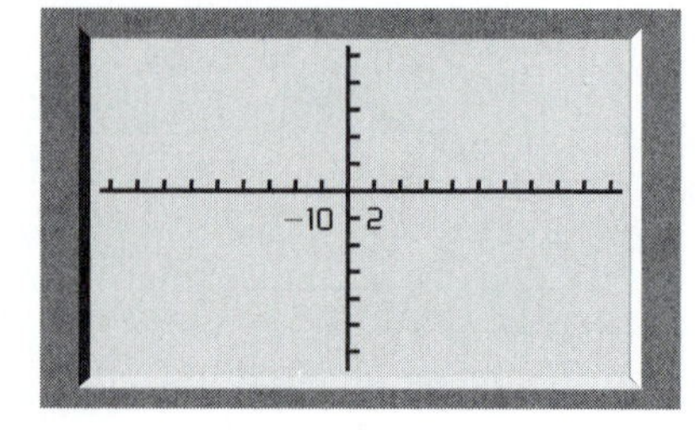

53. $y > x^2 - 5x - 6$

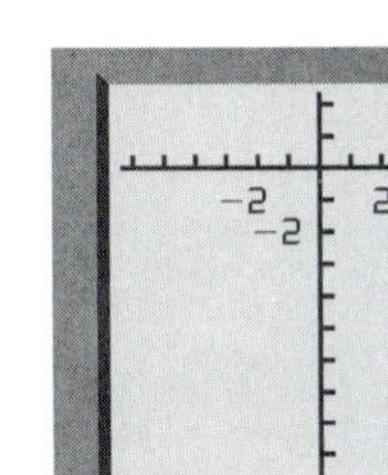

54. $y < 2x^2 + 7x - 15$

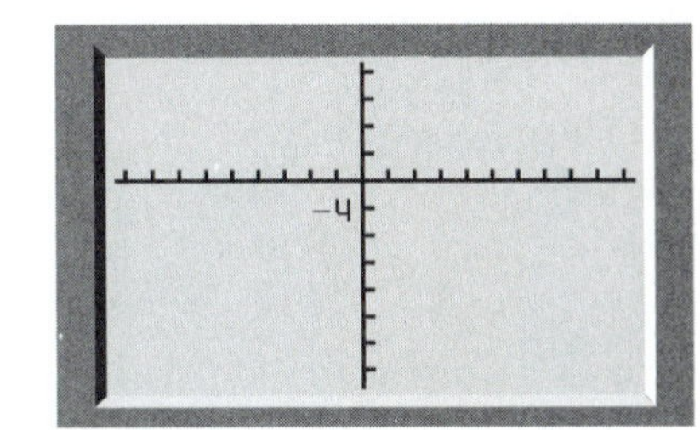

ANSWERS

53. ______

54. ______

Answers

1. 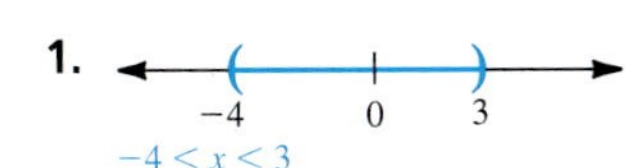
$-4 < x < 3$

3. $x < -4$ or $x > 3$

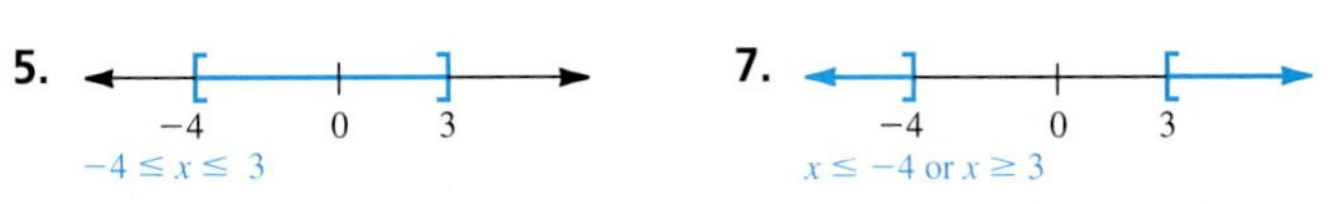

5. $-4 \le x \le 3$

7. $x \le -4$ or $x \ge 3$

9. $x < -1$ or $x > 4$

11. $-4 \le x \le 3$

13. $x \le 2$ or $x \ge 3$

15. $-6 \le x \le 4$

17. $x < -9$ or $x > 3$

19. $-2 \le x \le \frac{3}{2}$

21. $-1 < x < \frac{3}{4}$

23. $-4 \le x \le 4$

25. $x \le -5$ or $x \ge 5$

27. $x < -2$ or $x > 2$

29. $0 \le x \le 4$

31. $x \le 0$ or $x \ge 6$

33. $0 < x < 4$

35. $x = 2$

37. $-4 \le x \le 7$

39.

41. $x < -1$ or $0 < x < 2$

43. $-2 \le x \le 1$ or $x \ge 3$

45. $x \le -3$ or $0 \le x \le 5$

47. $50 \le x \le 60$

49. $\{t \mid 2 \le t \le 3\}$

51. $\{x \mid 6 < x < 0\}$

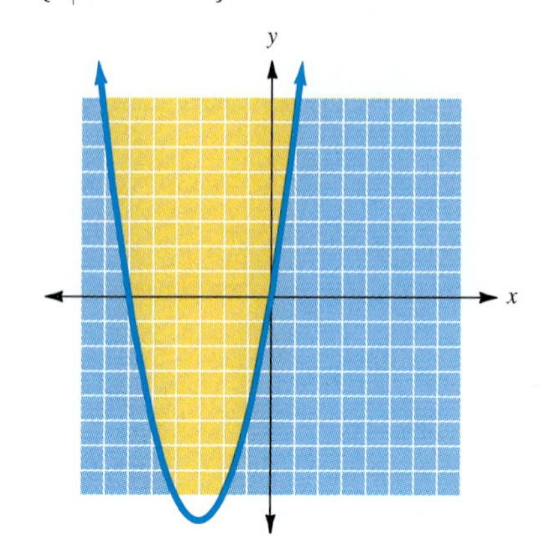

53. $\{x \mid -1 \le x \le 6\}$

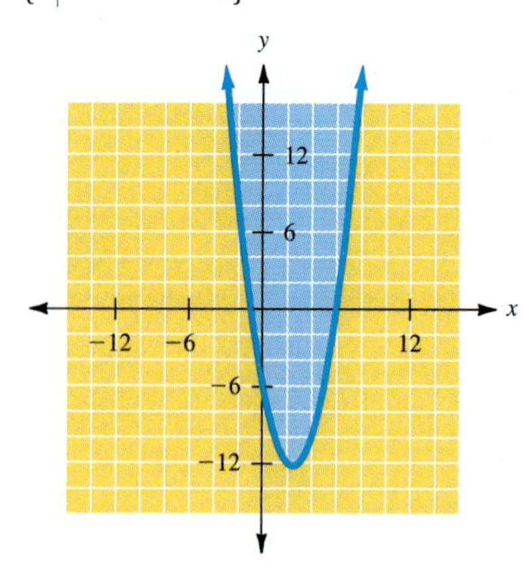

9 Summary

DEFINITION/PROCEDURE	EXAMPLE	REFERENCE
Graphing a Factorable Quadratic Function		**Section 9.1**
A function that can be written in the form $f(x) = ax^2 + bx + c$ is called a **quadratic function.**	$f(x) = 2x^2 - 3x + 2$ is a quadratic function.	**p. 671**
The height of a ball (in feet) thrown up from the ground is determined by the quadratic function $h(x) = -16x^2 + v_0x$ in which v_0 represents the initial velocity and x represents the number of seconds that have passed since the ball was thrown.	If a ball has an initial velocity of 64 ft/sec, its height is determined by $h(x) = -16x^2 + 64x$	**p. 671**
The zeros of the function are the values of x for which $h(x) = 0$.	If $0 = -16x^2 + 64x$ $0 = -16x(x - 4)$ $x = 0$ or $x = 4$	**p. 672**
Solving Quadratic Equations by Completing the Square		**Section 9.2**
Square Root Property If $x^2 = k$, when k is a complex number, then $x = \sqrt{k}$ or $x = -\sqrt{k}$	To solve: $(x - 3)^2 = 5$ $x - 3 = \pm\sqrt{5}$ $x = 3 \pm \sqrt{5}$	**p. 680**
Completing the Square **1.** Isolate the constant on the right side of the equation. **2.** Divide both sides of the equation by the coefficient of the x^2 term if that coefficient is not equal to 1. **3.** Add the square of one-half of the coefficient of the linear term to both sides of the equation. This will give a perfect-square trinomial on the left side of the equation. **4.** Write the left side of the equation as the square of a binomial, and simplify on the right side. **5.** Use the square root property, and then solve the resulting linear equations.	To solve: $x^2 + x = \frac{1}{2}$ $x^2 + x + \left(\frac{1}{2}\right)^2 = \frac{1}{2} + \left(\frac{1}{2}\right)^2$ $\left(x + \frac{1}{2}\right)^2 = \frac{3}{4}$ $x + \frac{1}{2} = \pm\sqrt{\frac{3}{4}}$ $x = \frac{-1 \pm \sqrt{3}}{2}$	**p. 684**
Solving Quadratic Equations by Using the Quadratic Formula		**Section 9.3**
Any quadratic equation can be solved by using the following algorithm. **1.** Write the equation in standard form (set it equal to 0). $ax^2 + bx + c = 0$ **2.** Determine the values for a, b, and c. **3.** Substitute those values into the quadratic formula $x = \frac{-b \pm \sqrt{b^2 - 4ac}}{2a}$ **4.** Write the solutions in simplest form.	To solve $x^2 - 2x = 4$ Write the equation as $x^2 - 2x - 4 = 0$ $a = 1 \quad b = -2 \quad c = -4$ $x = \frac{-(-2) \pm \sqrt{(-2)^2 - 4(1)(-4)}}{2 \cdot 1}$ $= \frac{2 \pm \sqrt{20}}{2}$ $= \frac{2 \pm 2\sqrt{5}}{2}$ $= 1 \pm \sqrt{5}$	**p. 697**

Continued

DEFINITION/PROCEDURE	EXAMPLE	REFERENCE
Solving Quadratic Equations by Using the Quadratic Formula		**Section 9.3**
The Discriminant The expression $b^2 - 4ac$ is called the **discriminant** for a quadratic equation. There are three possibilities: **1.** If $b^2 - 4ac < 0$, there are no real solutions (but two imaginary solutions). **2.** If $b^2 - 4ac = 0$, there is one real solution (a double solution). **3.** If $b^2 - 4ac > 0$, there are two distinct real solutions.	Given $2x^2 - 5x + 3 = 0$ $a = 2, b = -5, c =$ $b^2 - 4ac = 25 - 4(2)(3)$ $= 25 - 24$ $= 1$ There are two distinct solutions.	p. 697
Equations that are Quadratic in Form		**Section 9.4**
There are a variety of equations that are quadratic in form. $2x - 11\sqrt{x} + 12 = 0$ is quadratic in $\sqrt{x}$. $x^4 - 3x^2 - 10 = 0$ is quadratic in x^2. $(x^2 - 1)^2 + 2(x^2 - 1) - 8 = 0$ is quadratic in $x^2 - 1$. These equations can be solved by one of two techniques. *Solving by Substitution* **1.** Make an appropriate substitution so that the equation becomes quadratic. **2.** Solve the resulting equation for the intermediate variable. **3.** Use the intermediate values found in step 2 to find possible solutions for the original variable. **4.** Check the solutions of step 3 by substitution into the original equation.	To solve $2x - 11\sqrt{x} + 12 = 0$ let $u = \sqrt{x}$. Then $2u^2 - 11u + 12 = 0$ $(u - 4)(2u - 3) = 0$ $u = 4$ or $u = \frac{3}{2}$ Because $u = \sqrt{x}$, $\sqrt{x} = 4$ or $\sqrt{x} = \frac{3}{2}$ $x = 16$ $x = \frac{9}{4}$ Both solutions are valid.	p. 715
Solving by Factoring **1.** Treat the original equation as quadratic in some power of the variable, and factor. **2.** Solve the resulting equations. **3.** Check the solutions of step 2 by substitution into the original equation.	To solve $x^4 - 3x^2 - 10 = 0$ $(x^2 - 5)(x^2 + 2) = 0$ $x^2 = 5$ or $x^2 = -2$ $x = \pm\sqrt{5}$ $x = \pm i\sqrt{2}$	p. 716
Quadratic Inequalities		**Section 9.5**
Solving a Quadratic Inequality Algebraically **1.** Clear the inequality of parentheses and fractions. **2.** Rewrite the inequality such that the expression is related to zero. **3.** Factor the quadratic member and find the critical values of x. **4.** Use a sign graph to find the values for x that make the inequality a true statement. **5.** Use a number line to graph the solution set.	To solve $x^2 - 3x < 18$ $x^2 - 3x - 18 < 0$ $(x - 6)(x + 3) < 0$ Critical values of x are -3 and 6 Sign of $x + 3$: – – – + + + + + + + + Sign of $x - 6$: – – – – – – – – + + + -3 6 Sign of product: + + + – – – – – + + + The solution set is $\{x \mid -3 < x < 6\}$ -3 0 6	p. 725

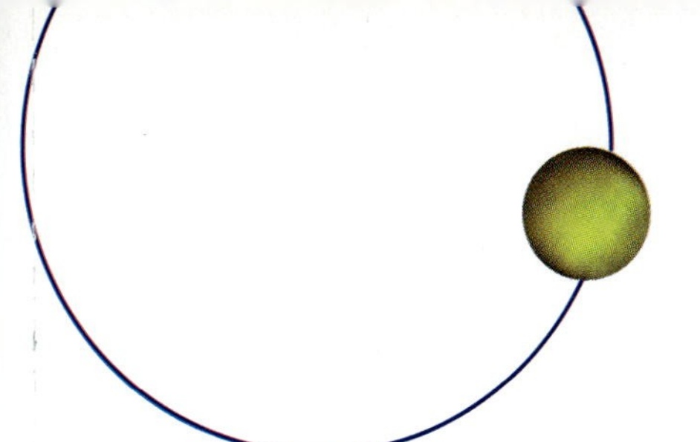

Summary Exercises

This summary exercise set is provided to give you practice with each of the objectives in the chapter. Each exercise is keyed to the appropriate chapter section. The answers are provided in the *Instructor's Manual*.

[9.1] Find the zeros of the following functions.

1. $f(x) = x^2 - x - 2$

2. $f(x) = 6x^2 + 7x + 2$

3. $f(x) = -2x^2 - 7x - 6$

4. $f(x) = -x^2 - 1$

[9.1] Find the coordinates for the vertex of each of the following.

5. $f(x) = x^2 + 2x - 8$

6. $f(x) = x^2 - 2x$

7. $f(x) = 8x^2 + 16x$

8. $f(x) = x^2 - 6x + 9$

9. $f(x) = x^2 + 4x + 4$

10. $f(x) = x^2 - 2x - 8$

11. $f(x) = x^2 - 4x - 5$

12. $f(x) = -x^2 - 2x - 15$

13. $f(x) = x^2 - 4x + 3$

14. $f(x) = x^2 + 6x + 8$

15. $f(x) = x^2 + 2x - 3$

16. $f(x) = -x^2 - 8x$

17. $f(x) = -x^2 - 2x + 8$

18. $f(x) = x^2 + 4x - 5$

[9.1] Graph each of the following.

19. $f(x) = x^2 + 5x$

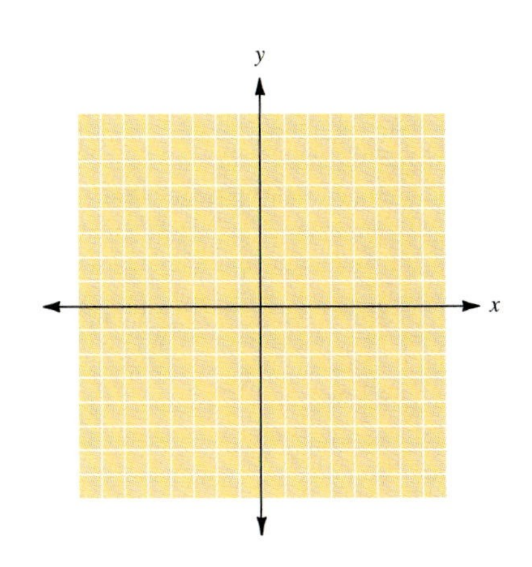

20. $f(x) = x^2 + 3x$

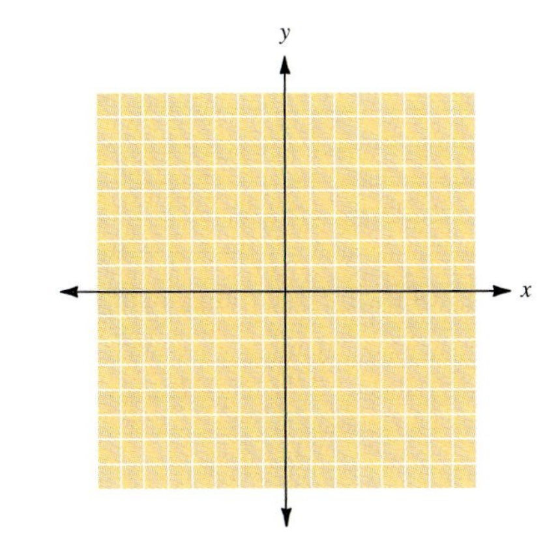

21. $f(x) = x^2 - 4x$

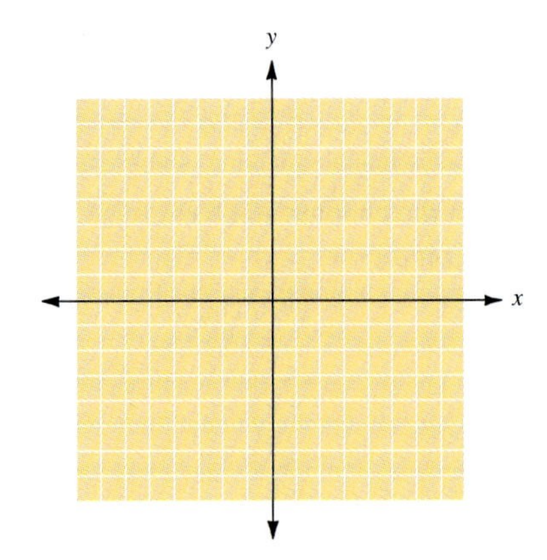

22. $f(x) = -x^2 + 2x$

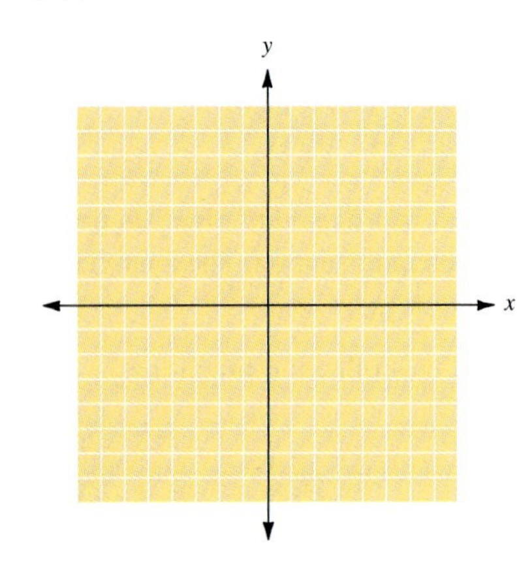

23. $f(x) = x^2 + 2x - 3$

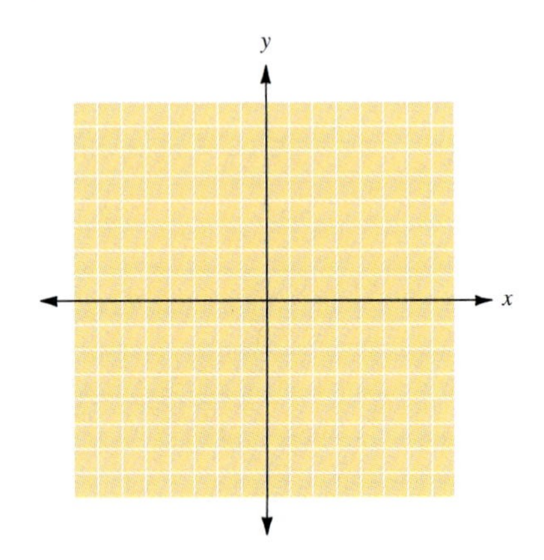

24. $f(x) = x^2 - 4x + 3$

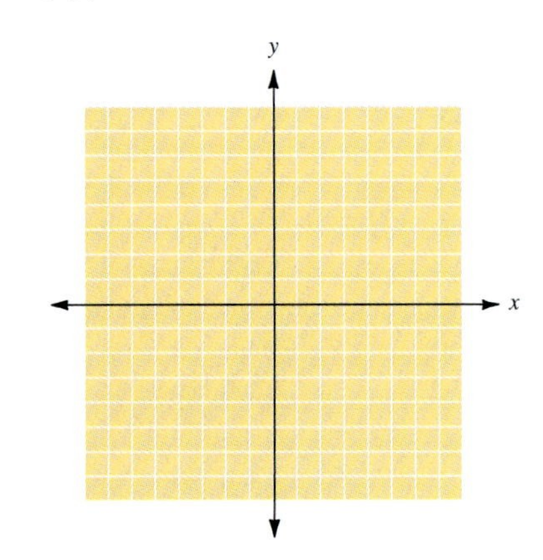

25. $f(x) = -x^2 - x + 6$

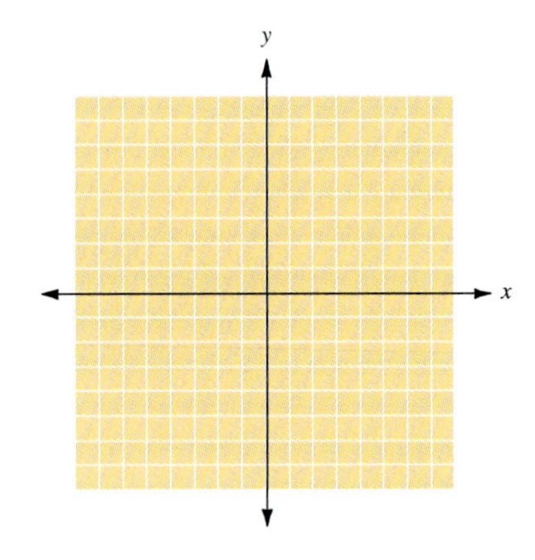

26. $f(x) = -x^2 + 3x + 4$

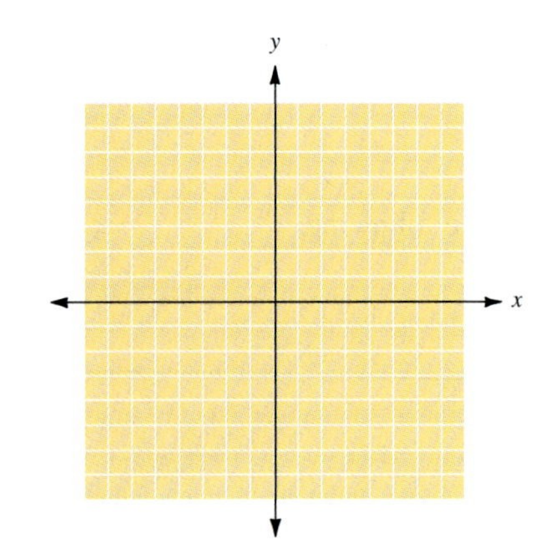

[9.2] Solve each of the following equations, using the square root method.

27. $x^2 - 8 = 0$

28. $3y^2 - 15 = 0$

29. $(x - 2)^2 = 20$

30. $(2x + 1)^2 - 10 = 0$

Find the constant that must be added to each of the following binomials to form a perfect-square trinomial.

31. $x^2 - 12x$

32. $y^2 + 3y$

Solve the following equations by completing the square.

33. $x^2 - 4x - 5 = 0$

34. $x^2 + 8x + 12 = 0$

35. $w^2 - 10w - 3 = 0$

36. $y^2 + 3y - 1 = 0$

37. $2x^2 - 8x - 5 = 0$

38. $3x^2 + 3x - 1 = 0$

[9.3] Solve each of the following equations using the quadratic formula.

39. $x^2 - 5x - 24 = 0$

40. $w^2 + 10w + 25 = 0$

41. $x^2 = 3x + 3$

42. $2y^2 - 5y + 2 = 0$

43. $3y^2 + 4y = 1$

44. $2y^2 + 5y + 4 = 0$

45. $(x - 5)(x + 3) = 13$

46. $\frac{1}{x^2} - \frac{4}{x} + 1 = 0$

47. $3x^2 + 2x + 5 = 0$

48. $(x - 1)(2x + 3) = -5$

[9.4] Solve each of the following equations.

49. $x^4 - 11x^2 + 18 = 0$

50. $x^4 + x^2 = 20$

51. $w^4 = 9w^2$

52. $p^4 - 16 = 0$

53. $m - \sqrt{m} - 12 = 0$

54. $(x - 3)^2 + 5(x - 3) = 14$

55. $(t^2 - 2t)^2 - 9(t^2 - 2t) + 8 = 0$

56. $x^{1/2} - 2x^{1/4} - 3 = 0$

57. $x^{2/3} + x^{1/3} = 2$

58. $\frac{10}{p^2} + \frac{3}{p} - 1 = 0$

[9.5] Solve the following inequalities and graph the solution set.

59. $(x - 2)(x + 5) > 0$

0

60. $(x - 1)(x - 6) < 0$

0

61. $(x + 1)(x + 3) \leq 0$

0

62. $(x + 4)(x - 5) \geq 0$

0

63. $x^2 - 5x - 24 \leq 0$

0

64. $x^2 + 4x \geq 21$

0

65. $x^2 \geq 64$

0

66. $x^2 + 5x \geq 0$

0

67. $(x + 2)(x - 6) < 9$

0

68. $(x - 1)(x + 2) \geq 4$

0

[9.2–9.5]

69. Thrown ball. If a ball is thrown vertically upward from the ground with an initial velocity of 64 ft/s its approximate height is given by $h(t) = -16t^2 + 64t$. When will the ball reach a height of at least 48 ft?

70. Cost. Suppose that the cost, in dollars, of producing x stereo systems is given by the equation $C(x) = 3000 - 60x + 3x^2$. How many systems can be produced if the cost cannot exceed $7500?

71. Rectangles. The length of a rectangle is 2 ft more than its width. If the area of the rectangle is 80 ft^2, what are the dimensions of the rectangle?

72. Rectangles. The length of a rectangle is 3 cm less than twice its width. The area of the rectangle is 35 cm^2. Find the length and width of the rectangle.

73. Rectangles. An open box is formed by cutting 3-in. squares from each corner of a rectangular piece of cardboard that is 3 in. longer than it is wide. If the box is to have a volume of 120 $in.^3$, what must be the size of the original piece of cardboard?

74. Profit. Suppose that a manufacturer's weekly profit P is given by

$$P(x) = -3x^2 + 240x$$

in which x is the number of items manufactured and sold. Find the number of items that must be manufactured and sold if the profit is to be at least $4500.

75. Thrown ball. If a ball is thrown vertically upward from the ground with an initial velocity of 96 ft/s, its approximate height is given by

$$h(t) = -16t^2 + 96t$$

When will the ball reach a height of at least 80 ft?

76. Rectangle. The length of a rectangle is 1 cm more than twice its width. If the length is doubled, the area of the new rectangle is 36 cm^2 more than that of the old. Find the dimensions of the original rectangle.

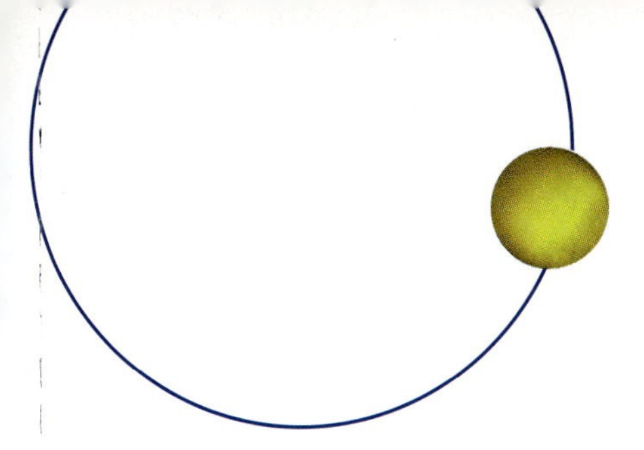

Self-Test for Chapter 9

Name ________

Section ________ Date ________

ANSWERS

1. ________
2. ________
3. ________
4. ________
5. ________
6. ________
7. ________
8. ________
9. ________
10. ________

The purpose of this self-test is to help you check your progress and to review for a chapter test in class. Allow yourself about 1 hour to take the test. When you are done, check your answers in the back of the book. If you missed any answers, be sure to go back and review the appropriate sections in the chapter and the exercises that are provided.

1. Find the zeros of the function $f(x) = 3x^2 - 10x - 8$.

Find the coordinates of the vertex of each of the following.

2. $f(x) = x^2 + x - 2$

3. $f(x) = x^2 - 4x - 5$

Graph the following functions.

4. $f(x) = x^2 + x - 6$

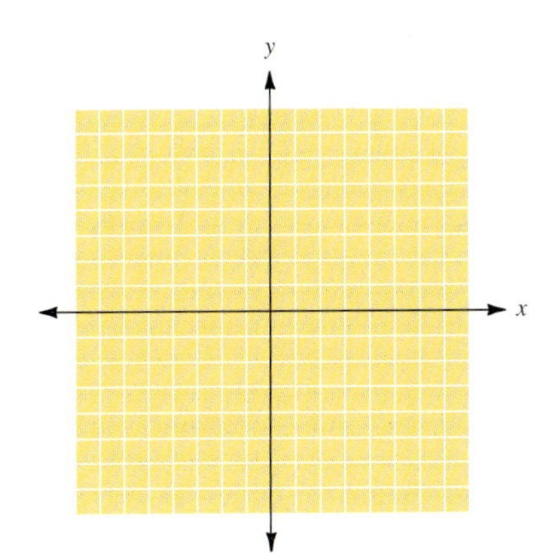

5. $f(x) = 2x^2 + 4x - 3$

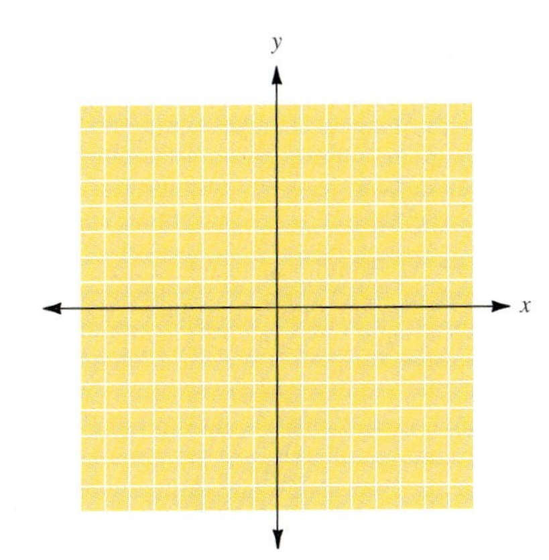

6. The length of a rectangle is 2 cm more than 3 times its width. If the area of the rectangle is 85 cm^2, what are the dimensions of the rectangle?

7. Suppose that the height (in feet) of a ball thrown upward from a raised platform is approximated by

$h(t) = -16t^2 + 32t + 32$

t seconds after the ball has been released. How long will it take the ball to hit the ground?

Solve each of the following equations, using the square root method.

8. $4w^2 - 20 = 0$

9. $(x - 1)^2 = 10$

10. $4(x - 1)^2 = 23$

ANSWERS

11. ______

12. ______

13. ______

14. ______

15. ______

16. ______

17. ______

18. ______

19. ______

20. ______

Solve each of the following equations by completing the square.

11. $m^2 + 3m - 1 = 0$

12. $2x^2 - 10x + 3 = 0$

Solve each of the following equations, using the quadratic formula.

13. $x^2 - 5x - 3 = 0$

14. $x^2 + 4x = 7$

Solve the following equations either by factoring directly or by substitution.

15. $x^4 - 12x^2 + 27 = 0$

16. $y - 11\sqrt{y} + 18 = 0$

17. $(m - 2)^2 + 2(m - 2) = 15$

18. $\frac{6}{x^2} + \frac{1}{x} - 1 = 0$

Solve each of the following inequalities, and graph the solution set.

19. $x^2 + 5x - 14 < 0$

20. $x^2 - 3x \geq 18$

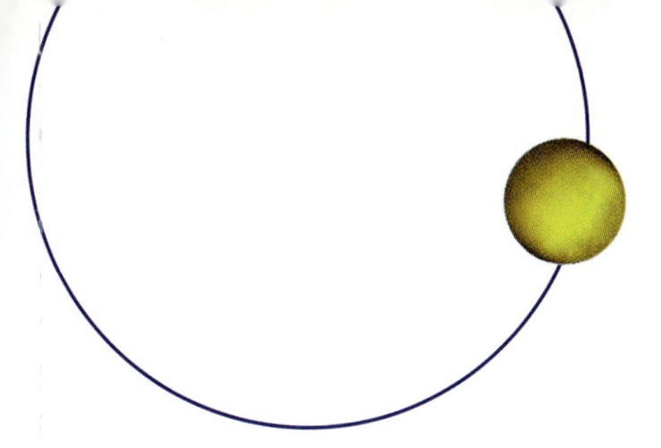

Cumulative Test for Chapters 1 to 9

Name ______________________

Section ________ Date ________

ANSWERS

1. ______________________

2. ______________________

3. ______________________

4. ______________________

5. ______________________

6. ______________________

7. ______________________

8. ______________________

9. ______________________

10. ______________________

11. ______________________

12. ______________________

13. ______________________

14. ______________________

15. ______________________

This test is provided to help you in the process of reviewing the previous chapters. Answers are provided in the back of the book. If you missed any answers, be sure to go back and review the appropriate chapter section.

1. Graph the following equation.

$2x - 3y = 6$

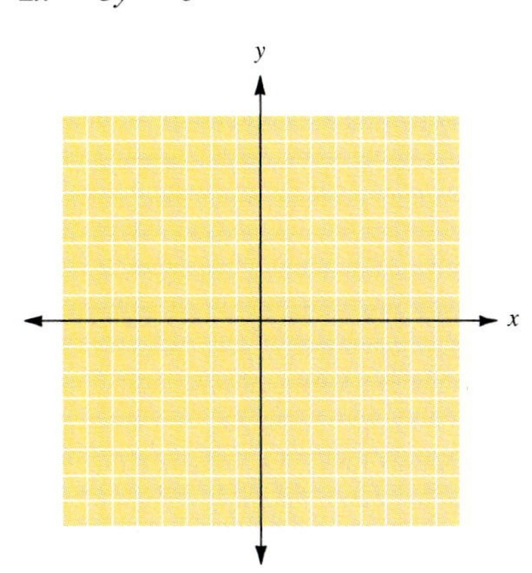

2. Find the slope of the line determined by the set of points.

$(-4, 7)$ and $(-3, 4)$

3. Let $f(x) = 6x^2 - 5x + 1$. Evaluate $f(-2)$.

4. Simplify the function $f(x) = (x^2 - 1)(x + 3)$.

5. Completely factor the expression $x^3 + x^2 - 6x$.

6. Simplify the expression $\frac{2}{x + 2} - \frac{3x - 2}{x^2 - x - 6}$.

7. Simplify the expression $(\sqrt{7} - \sqrt{2})(\sqrt{3} + \sqrt{6})$.

Solve each equation.

8. $2x - 7 = 0$

9. $3x - 5 = 5x + 3$

10. $0 = (x - 3)(x + 5)$

11. $x^2 - 3x + 2 = 0$

12. $x^2 + 7x - 30 = 0$

13. $x^2 - 3x - 3 = 0$

14. $(x - 3)^2 = 5$

15. $|2x - 3| = 7$

ANSWERS

16. ______
17. ______
18. ______
19. ______
20. ______
21. ______
22. ______
23. ______
24. ______
25. ______

16. $\dfrac{6}{x+3} + \dfrac{2}{x-3} = \dfrac{20}{x^2-9}$

17. $\sqrt{4x-3} - 3 = 2$

18. Find the equation of the line, L, that has y intercept of (0, 3) and is perpendicular to the line $2y - 4x = 5$.

Solve the following inequalities.

19. $2x - 3(x - 5) \geq 6$

20. $|x - 4| \geq 9$

21. Simplify the expression $\left(\dfrac{x^{-3}y^{11/8}}{x^{-5}y^{-21/8}}\right)^{1/2}$.

22. Simplify the expression $(3 - i)^2 - (5 - 2i)$.

Solve the following word problems.

23. Five times a number decreased by 7 is -72. Find the number.

24. One leg of a right triangle is 4 ft longer than the shorter leg. If the hypotenuse is 28 ft, how long is each leg?

25. Suppose that a manufacturer's weekly profit P is given by

$P(x) = -4x^2 + 320x$

in which x is the number of units manufactured and sold. Find the number of items that must be manufactured and sold to guarantee a profit of at least \$4956.

GRAPHS OF CONIC SECTIONS

10

INTRODUCTION

Large cities often commission fireworks artists to choreograph elaborate displays on holidays. Such displays look like beautiful paintings in the sky, in which the fireworks seem to dance to well-known popular and classical music. The displays are feats of engineering and very accurate timing. Suppose the designer wants a set of rockets of a certain color and shape to be released after a set of a different color and shape reaches a specific height and explodes. He must know the strength of the initial liftoff and use a quadratic equation to determine the proper time for setting off the second round.

The equation $h = -16t^2 + 100t$ gives the height in feet t seconds after the rockets are shot into the air if the initial velocity is 100 ft/s. Using this equation, the designer knows how high the rocket will ascend and when it will begin to fall. He can time the next round to achieve the effect he wishes. Displays that involve large banks of fireworks in shows that last up to an hour are programmed using computers, but quadratic equations are at the heart of the mechanism that creates the beautiful effects.

10.1 Graphing Parabolas

10.1 OBJECTIVES

1. Find the axis of symmetry
2. Find the vertex
3. Graph a parabola by the method of translation
4. Complete the square of a quadratic function and graph the parabola
5. Graph a parabola that opens to the left or right

In Chapter 4 we discussed the graphs of linear equations in two variables of the form

$ax + by = c$ in which a and b cannot both be 0

The graphs of all the linear equations were straight lines. Suppose that we now allow the terms in x and/or y to be quadratic; that is, we will allow squares in one or both of those terms. The graphs of such equations will form a family of curves called the **conic sections.** Conic sections are curves formed when a plane cuts through, or forms a section of, a cone. The conic sections include four curves—the **parabola, circle, ellipse,** and **hyperbola.** Examples of how these curves are formed are shown below.

NOTE The inclination of the plane determines which of the sections is formed.

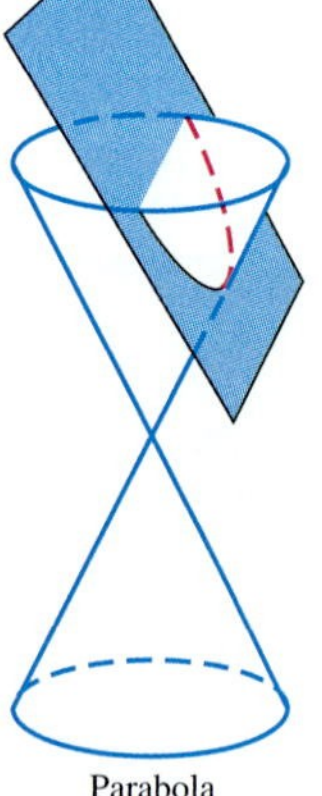

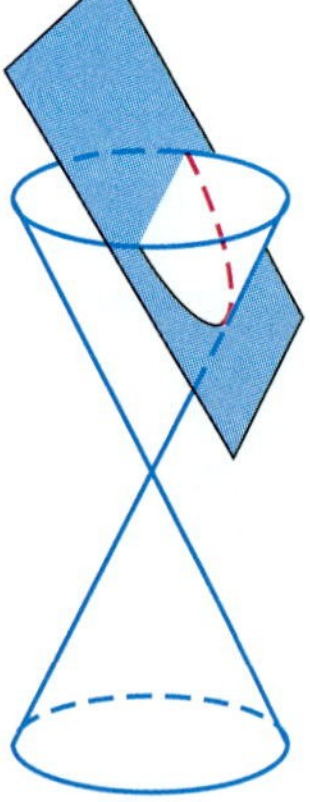

Parabola

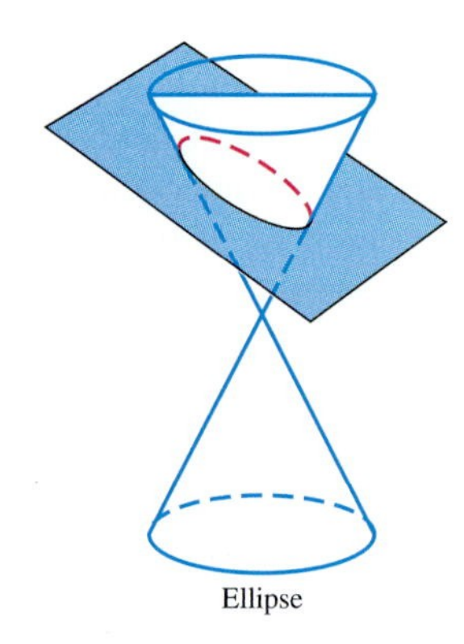

Ellipse

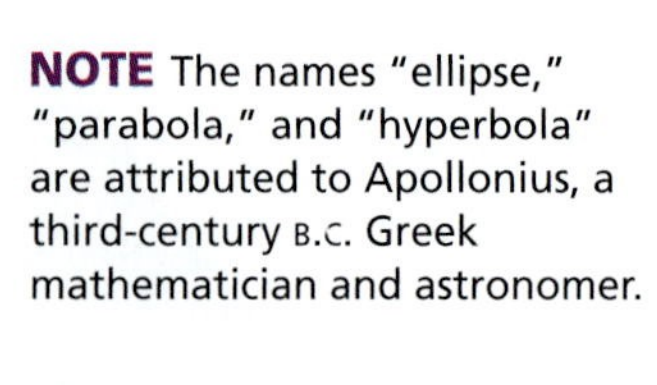

NOTE The names "ellipse," "parabola," and "hyperbola" are attributed to Apollonius, a third-century B.C. Greek mathematician and astronomer.

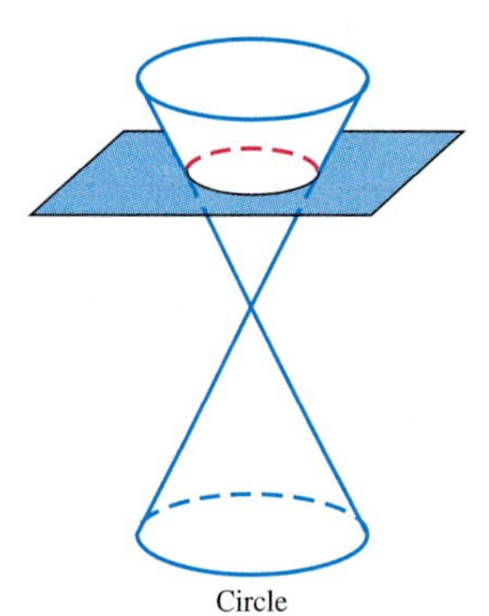

Circle

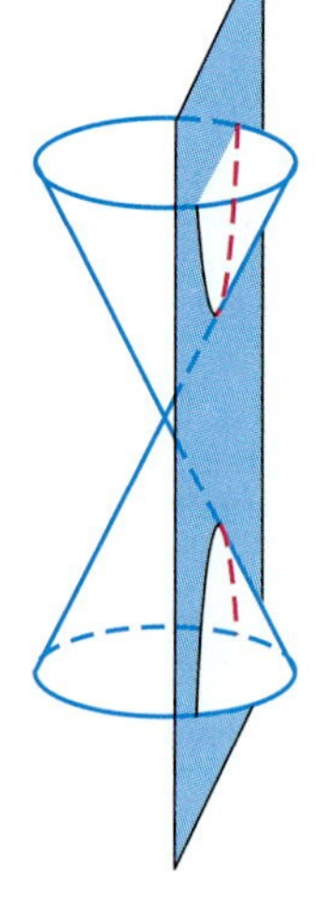

Hyperbola

Let's examine the first of these sections, the parabola. Consider the equation $y = x^2$. This equation is quadratic in x and linear in y. Its graph is a parabola. Let's look at that graph.

NOTE Notice that the parabola is rounded, and not pointed at the bottom.

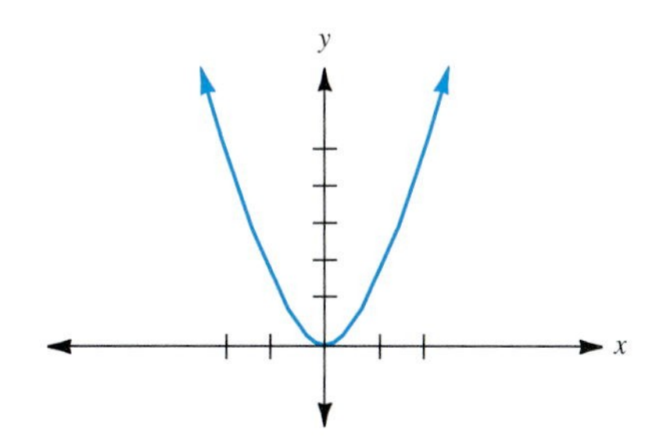

We could plot points, then connect them.

x	y
−2	4
−1	1
0	0
1	1
2	4

There are three elements of the graph that should be noted.

1. The graph opens upward.
2. The y axis cuts the graph into two equal parts. A line that does this is called an **axis of symmetry.**
3. The graph has a minimum point, called the **vertex.**

Let's compare that graph to the graph of the equation $y = -x^2$.

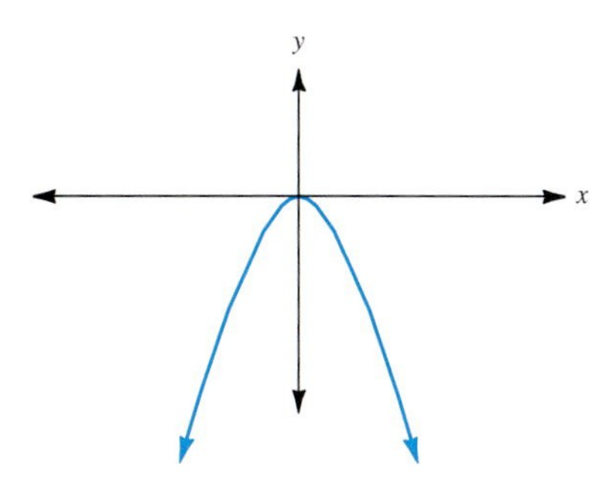

Looking at the three elements we examined earlier, we make three observations:

1. The graph opens downward.
2. The y axis is the *axis of symmetry.*
3. The graph has a maximum point, called the *vertex.*

It will always be the case that the sign of the coefficient of the x^2 term will determine which way the parabola opens. It will also be the case that a parabola opening upward has a minimum, and one opening downward has a maximum.

For every equation that is quadratic in x and linear in y, we will look for three things:

1. Does the graph open upward or downward?
2. Where is the axis of symmetry?
3. What are the coordinates of the vertex?

Example 1

Graphing with a Vertical Translation

Graph the equation $y = x^2 - 3$.

The difference between this graph and that of equation $y = x^2$ is that each y value has been decreased by 3. This results in a **translation** of 3 units in the negative direction on the y axis.

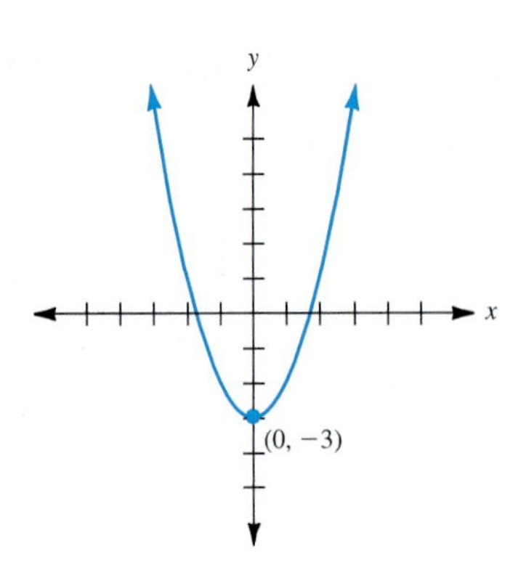

Note that the curve opens upward, the axis of symmetry is $x = 0$ (the y axis), and the vertex is $(0, -3)$.

CHECK YOURSELF 1

Graph the equation $y = -x^2 + 2$.

An equation of the form $y = (x - h)^2$ will be translated along the x axis with the axis of symmetry at $x = h$.

CAUTION

When you enter the equation in your graphing calculator, be certain that the negative is *in front of* the parentheses.

Example 2

Graphing with a Horizontal Translation

Graph the equation $y = -(x - 3)^2$.

Because the coefficient of the x^2 term is negative, the parabola opens downward and has a maximum point. Notice that when $x = 3$, y is 0. If x is more than 3, y will be negative. Try substituting values for x in the equation, then evaluate y. When x is less than 3, y is also negative. Thus, the vertex is at $(3, 0)$, and the axis of symmetry is at $x = 3$.

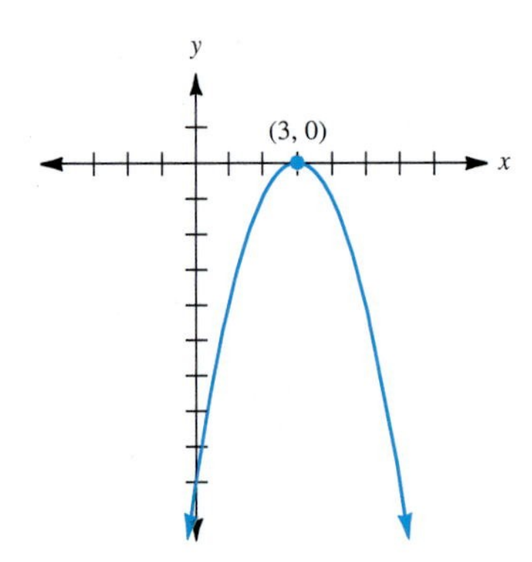

CHECK YOURSELF 2

Graph the equation $y = (x + 2)^2$.

Combining the lessons of the last two examples, we see that the graph of an equation of the form $y = (x - h)^2 + k$ is simply the parabola $y = x^2$ translated horizontally h units and vertically k units.

Example 3

Graphing a Parabola

NOTE Graph
$Y_1 =$ (x + 3) ^ 2 + 1

Graph the equation $y = (x + 3)^2 + 1$.

The parabola will be translated to the left 3 units and up 1 unit. The parabola opens upward, with an axis of symmetry at $x = -3$ and a vertex at $(-3, 1)$.

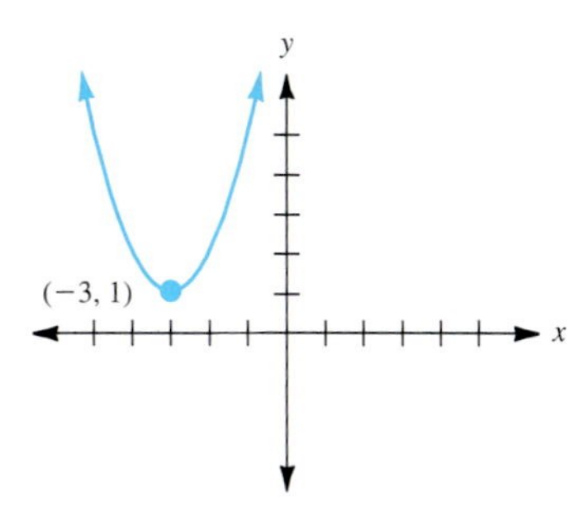

CHECK YOURSELF 3

Graph the equation $y = -(x - 2)^2 - 3$.

The rate at which the sides of a parabola rise (or fall) is determined by the coefficient of the x^2 term.

Example 4

Graphing Parabolas

Graph the parabolas $y = 2x^2$, $y = x^2$, and $y = \frac{1}{2}x^2$ on the same axes.

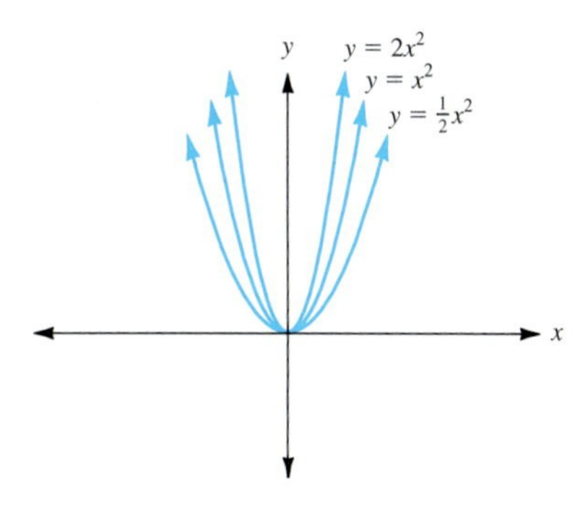

Notice that the larger the coefficient of x^2, the faster the parabola rises, thus the thinner the parabola appears.

CHECK YOURSELF 4

Graph the parabolas $y = -2x^2$, $y = -x^2$, *and* $y = -\frac{1}{2}x^2$ *on the same axes.*

We can now graph any equation of the form $y = a(x - h)^2 + k$.

Example 5

Graphing a Parabola

Graph the equation $y = -2(x - 3)^2 - 4$.

This parabola will open downward, the axis of symmetry is at $x = 3$, the vertex is at $(3, -4)$, and it has the shape of $y = 2x^2$.

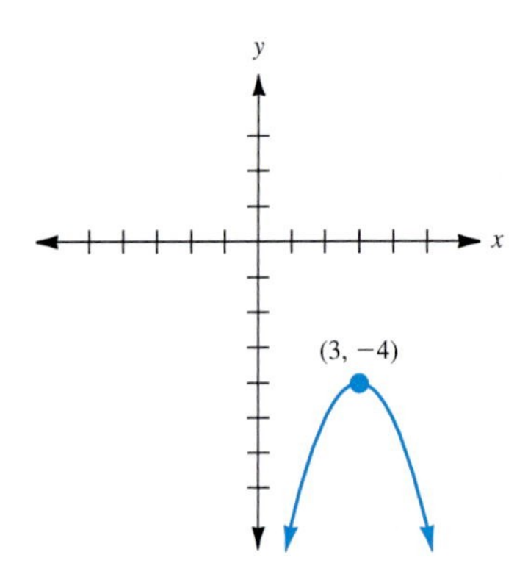

CHECK YOURSELF 5

Graph the equation $y = \frac{1}{4}(x + 3)^2 - 1$.

The standard form for a quadratic equation in two variables is $y = ax^2 + bx + c$. Any quadratic equation can be written in either this form or as $y = a(x - h)^2 + k$. The following example illustrates.

Example 6

Graphing an Equation of the Form $y = x^2 + bx + c$

Graph the equation

$$y = x^2 - 2x - 3$$

Using the techniques of Section 9.2, we will first complete the square.

NOTE One-half the middle term squared will complete the square.

$$y = x^2 - 2x - 3$$

$$= x^2 - 2x + 1 - 3 - 1$$ We can do this because $+1 - 1 = 0$.

$$= (x - 1)^2 - 4$$

From our work in the previous section, we know the parabola opens upward, the axis of symmetry is $x = 1$, and the vertex is at $(1, -4)$.

To improve the sketch, we can find the x intercepts. These are the x values for which $y = 0$, so

$$0 = (x - 1)^2 - 4$$

$$4 = (x - 1)^2$$

$$\pm 2 = x - 1$$

$$x = 1 + 2 \quad \text{or} \quad x = 1 - 2$$

$$= 3 \qquad\qquad = -1$$

The x intercepts are $(3, 0)$ and $(-1, 0)$.

Now draw a smooth curve connecting the vertex and the x intercepts.

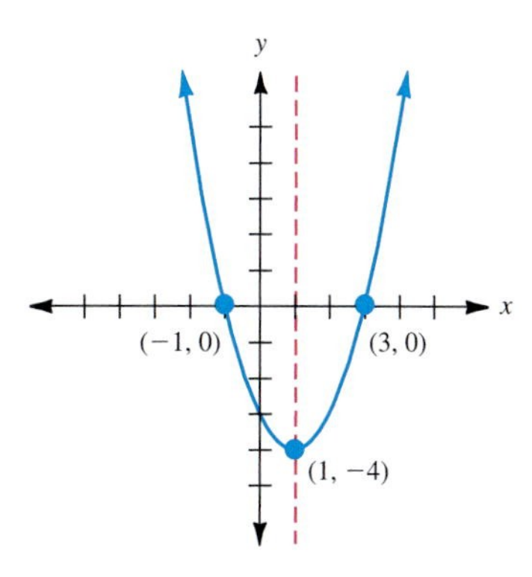

CHECK YOURSELF 6

Graph the equation

$$y = -x^2 + 6x - 5$$

Hint: Rewrite this as $y = -(x^2 - 6x + 9) - 5 + 9$.

A similar process will work if the quadratic member of the given equation is *not* factorable. In that case, one of two things happens:

1. The x intercepts are irrational. In this case, a calculator can be used to estimate the intercepts.
2. The x intercepts do not exist.

Consider the following example.

Example 7

Graphing an Equation of the Form $y = x^2 + bx + c$

Graph the equation

$$y = x^2 - 4x + 2$$

$$= x^2 - 4x + 2$$

$$= x^2 - 4x + 4 + 2 - 4$$

$$= (x - 2)^2 - 2$$

NOTE To keep the equation balanced, we both add and subtract 4.

The parabola opens upward, the axis of symmetry is $x = 2$, and the vertex is $(2, -2)$.

Again, we can improve the sketch if we find two symmetric points. Here the quadratic member is not factorable, and the x intercepts are irrational, so we would prefer to find another pair of symmetric points.

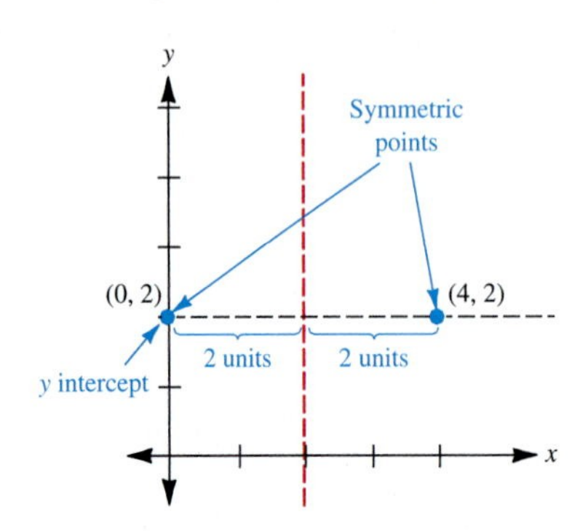

Note that (0, 2) is the y intercept of the parabola. We found the axis of symmetry at $x = 2$ earlier. Note that the symmetric point to (0, 2) lies along the horizontal line through the y intercept at the same distance (2 units) from the axis of symmetry, or at $x = 4$. Hence, (4, 2) is our symmetric point.

Draw a smooth curve connecting the points found above to form the parabola.

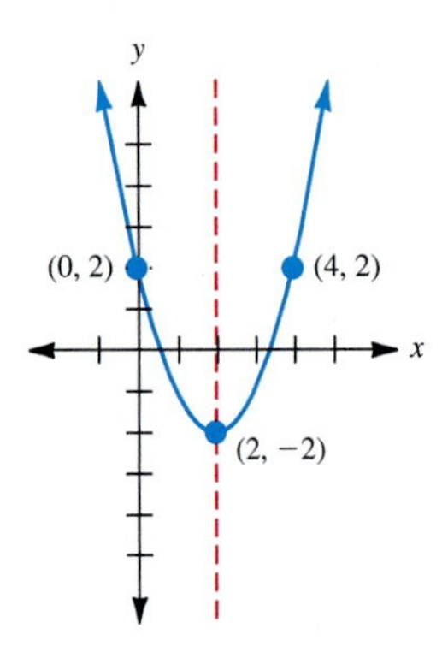

CHECK YOURSELF 7

Graph the equation $y = x^2 + 2x + 3$.

The coefficient of x^2 was 1 or -1 in our previous example and exercises. The following example shows the effect of different coefficients of the x^2 term.

Example 8

Graphing an Equation of the Form $y = ax^2 + bx + c$

Graph the equation

$$y = 2x^2 - 4x + 3$$

Step 1 Complete the square.

$$\begin{aligned} y &= 2(x^2 - 2x) + 3 \\ &= 2(x^2 - 2x + 1) + 3 - 2 \\ &= 2(x - 1)^2 + 1 \end{aligned}$$

NOTE We have added 2 times 1, so we must also subtract 2.

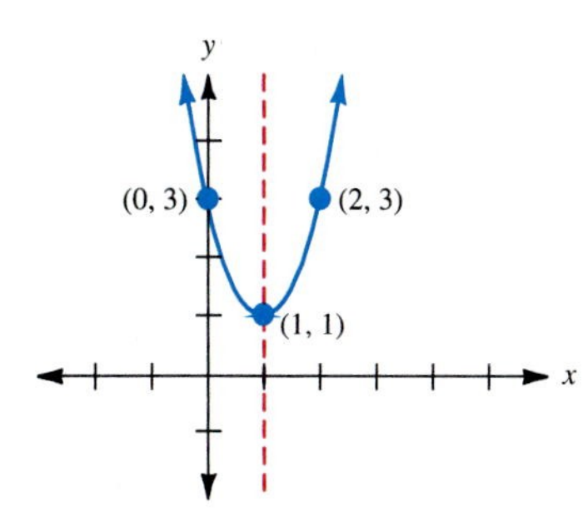

Step 2 The axis of symmetry is $x = 1$, the vertex is at (1, 1).

Step 3 Find symmetric points. Again the quadratic member is not factorable, and we use the y intercept (0, 3) and its symmetric point (2, 3).

Step 4 Connect the points with a smooth curve to form the parabola.

Compare this curve to those in previous examples. Note that the parabola is "tighter" about the axis of symmetry. That is the effect of the larger x^2 coefficient.

CHECK YOURSELF 8

Graph the equation $y = \frac{1}{2}x^2 - 2x - 2$.

The following algorithm summarizes our work thus far in this section.

Step by Step: Graphing a Quadratic Equation

Step 1 Complete the square for the quadratic variable.
Step 2 Find the axis of symmetry and the vertex.
Step 3 Determine two symmetric points.

Note: You can use the x intercepts if the quadratic member of the given equation is factorable. Otherwise use the y intercept and its symmetric point.

Step 4 Draw a smooth curve connecting the points found above, to form the parabola. You may choose to find additional pairs of symmetric points at this time.

If we use the algorithm to find the vertex of the equation $y = ax^2 + bx + c$, we get a useful result. The axis of symmetry will always occur at $x = \frac{-b}{2a}$.

So far we have dealt with equations of the form

$$y = ax^2 + bx + c$$

Suppose we reverse the role of x and y. We then have

$$x = ay^2 + by + c$$

which is quadratic in y but not in x. The graph of such an equation is once again a parabola, but this time the parabola will be horizontally oriented—opening to the right or left as a is positive or negative, respectively.

For $x = ay^2 + by + c$,

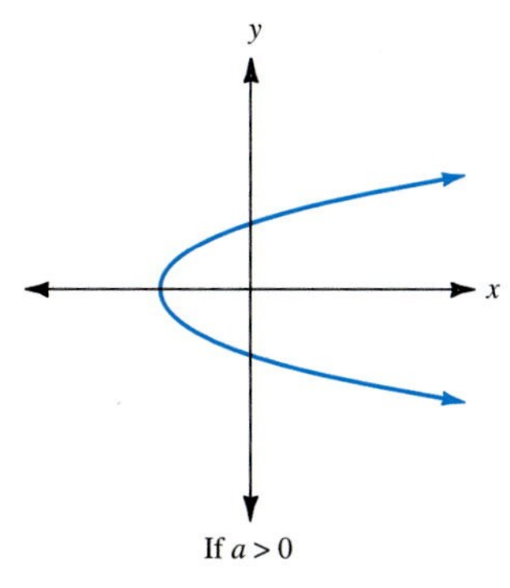

If $a > 0$

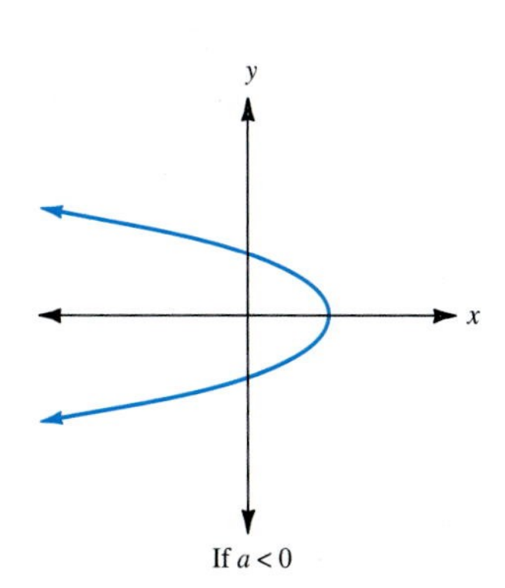

If $a < 0$

Much of what we did earlier is easily extended to this new case. The following example will illustrate the changes in the process.

Example 9

Graphing an Equation of the Form $x = ay^2 + by + c$

Graph the equation

$$x = y^2 + 2y - 3$$

Step 1 Complete the square

$$x = y^2 + 2y + 1 - 3 - 1$$

$$x = (y + 1)^2 - 4$$

Step 2 Find the axis of symmetry and the vertex. Because the y term is squared, the axis of symmetry is a horizontal line. Here, it is $y = -1$. Substituting -1 for y in the original equation, $(-4, -1)$ is the vertex.

Step 3 Sketch the parabola.

NOTE Note that, to find the y intercepts we let $x = 0$

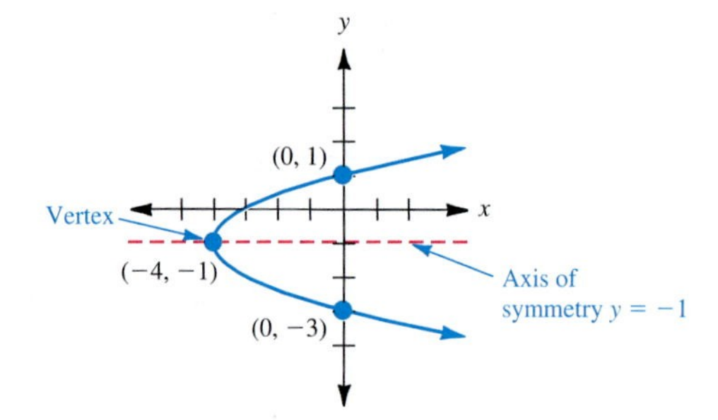

CHECK YOURSELF 9

Graph the equation $x = y^2 - 5y + 4$.

The vertex of a parabola can be found directly from the function

$$f(x) = ax^2 + bx + c$$

as the coordinates

$$\left(\frac{-b}{2a}, f\left(\frac{-b}{2a}\right)\right).$$

That technique will be used in the next example.

Given equations of the form $y = ax^2 + bx + c$, we know that if $a > 0$, then the vertex is the lowest point on the graph (the minimum value). Also, if $a < 0$, then the vertex is the highest point on the graph (the maximum value). We can use this result to solve a variety of problems in which we want to find the maximum or minimum value of a variable.

Example 10

An Application of a Quadratic Function

A software company sells a word processing program for personal computers. They have found that their monthly profit in dollars P from selling x copies of the program is approximated by

$$P(x) = -0.2x^2 + 80x - 1200$$

Find the number of copies of the program that should be sold to maximize the profit.

Because the function is quadratic, the graph must be a parabola. Also because the coefficient of x^2 is negative, the parabola must open downward and thus the vertex will give the maximum value for the profit P. To find the vertex,

$$x = \frac{-b}{2a} = \frac{-80}{2(-0.2)} = \frac{-80}{-0.4} = 200$$

The maximum profit must then occur when $x = 200$, and we substitute that value into the original equation:

$$P(x) = -0.2(200)^2 + (80)(200) - 1200$$
$$= \$6800$$

The maximum profit will occur when 200 copies are sold per month, and that profit will be \$6800.

CHECK YOURSELF 10

A company that sells portable radios finds that its weekly profit in dollars P and the number of radios sold x are related by

$$P(x) = -0.1x^2 + 20x - 200$$

Find the number of radios that should be sold to have the largest weekly profit and the amount of that profit.

Example 11

An Application of a Quadratic Function

A farmer has 1000 ft of fence and wishes to enclose the largest possible rectangular area with that fencing. Find the length and width of the largest possible area that can be enclosed.

NOTE As usual, when dealing with geometric figures, we start by drawing a sketch of the problem.

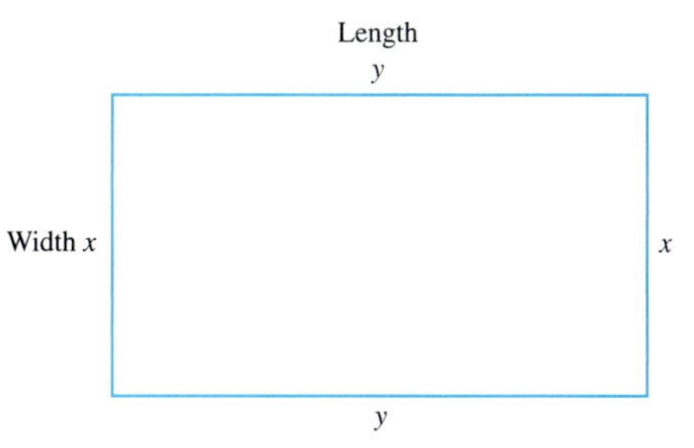

First, we can write the area A as

NOTE Area = length × width

$$A = xy \qquad (1)$$

Also because 1000 ft of fence is to be used, we know that

NOTE The perimeter of the region is
$2x + 2y$

$$2x + 2y = 1000$$
$$2y = 1000 - 2x$$
$$y = 500 - x \qquad (2)$$

Substituting for y in Equation (1), we have

$$A = x(500 - x) = 500x - x^2$$
$$= -x^2 + 500x \qquad (3)$$

Again, the graph for A is a parabola opening downward, and the largest possible area will occur at the vertex. As before,

NOTE The width x is 250 ft. From (2)
$y = 500 - 250$
$= 250$ ft

$$x = \frac{-500}{2(-1)} = \frac{-500}{-2} = 250$$

and the largest possible area is

NOTE The length is also 250 ft. The desired region is a square.

$$A = -(250)^2 + 500(250) = 62{,}500 \text{ ft}^2$$

CHECK YOURSELF 11

We want to enclose the largest possible rectangular area by using 400 ft of fence. The fence will be connected to the house, so only three sides of fence will be needed. What should be the dimensions of the rectangle and what is the largest possible area?

There are many physical models involving a parabolic path or design. The fact that a projectile follows a parabolic path was known in the time of Galileo. The reflective properties of the parabola also are important. If light rays enter a parabola along lines parallel to the axis of symmetry, the rays will reflect off the parabola and will pass through what is called the "focus" of the parabola. That fact accounts for the design of parabolic radar antennas and solar collectors.

The reflective surface of an automobile headlight is also a parabolic surface. The light source in the headlight is placed approximately at the focus, so that the light rays will reflect off the surface as nearly parallel rays.

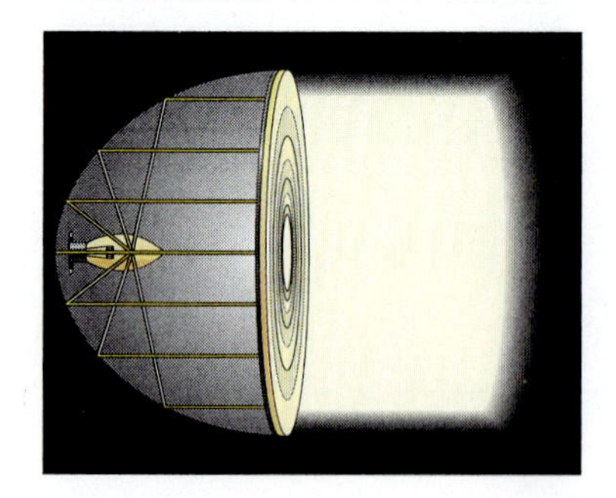

CHECK YOURSELF ANSWERS

1. $y = -x^2 + 2$

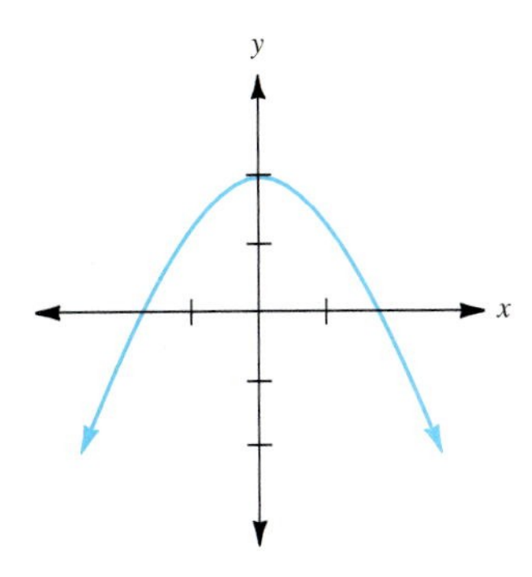

2. $y = (x + 2)^2$

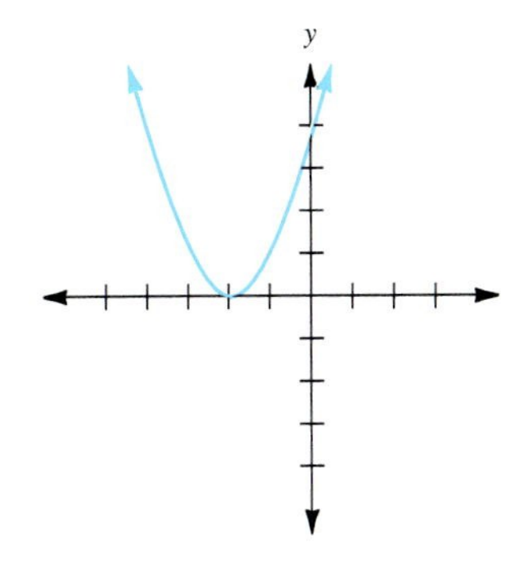

3. $y = -(x - 2)^2 - 3$

(2, −3)

4. 

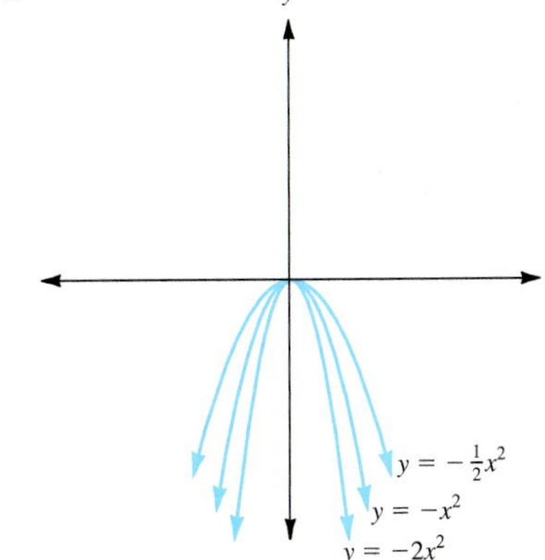

5. $y = \frac{1}{4}(x + 3)^2 - 1$

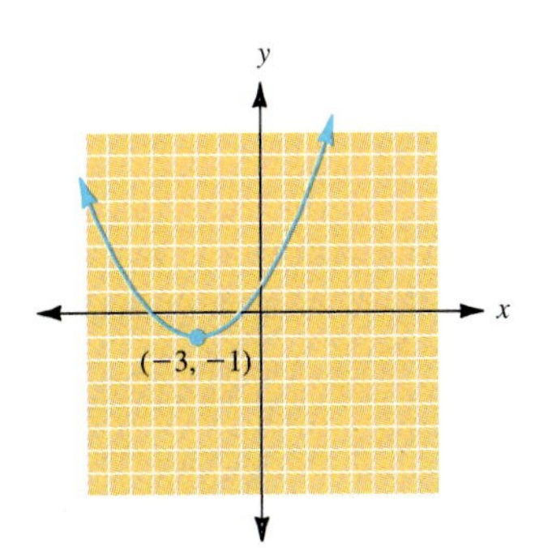

6. $y = -x^2 + 6x - 5$

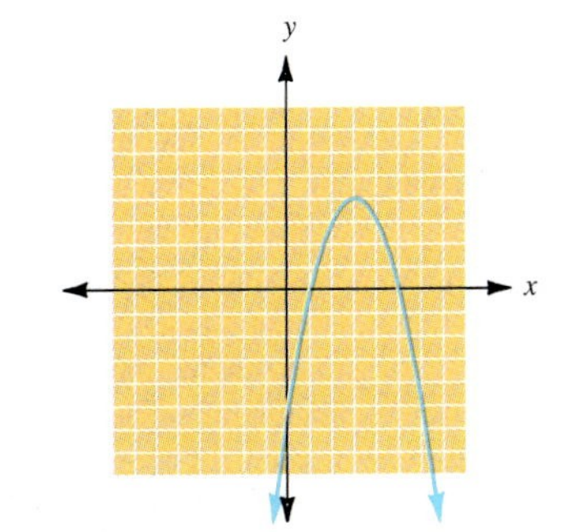

7. $y = x^2 + 2x + 3$

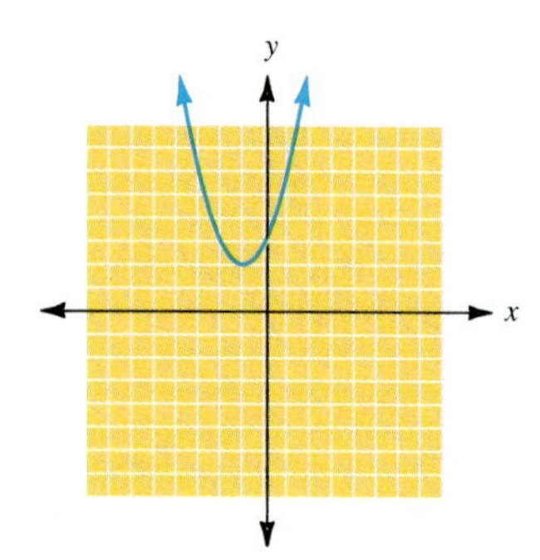

8. $y = \frac{1}{2}x^2 - 2x - 2$

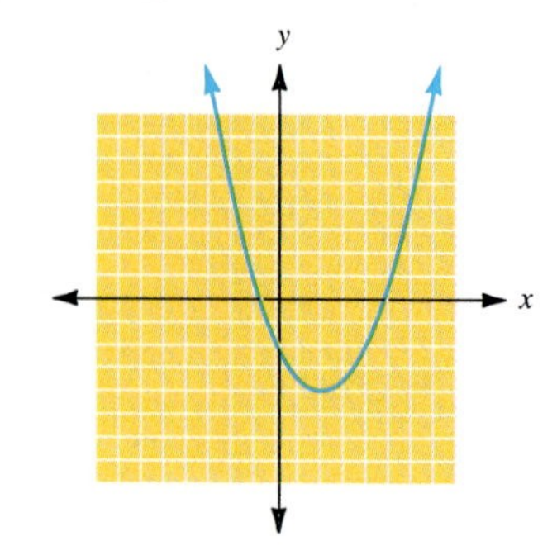

9. $x = y^2 - 5y + 4$

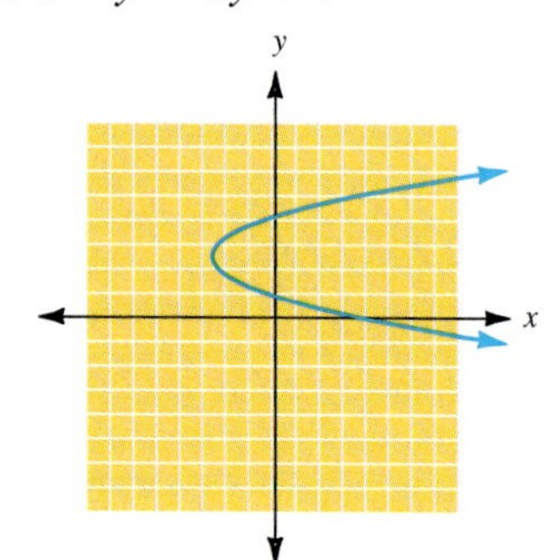

10. 100 radios, \$800

11. Width 100 ft, length 200 ft, area 20,000 ft^2

Name ____________________

Section ________ Date ________

10.1 Exercises

Match each graph with one of the equations below.

1.

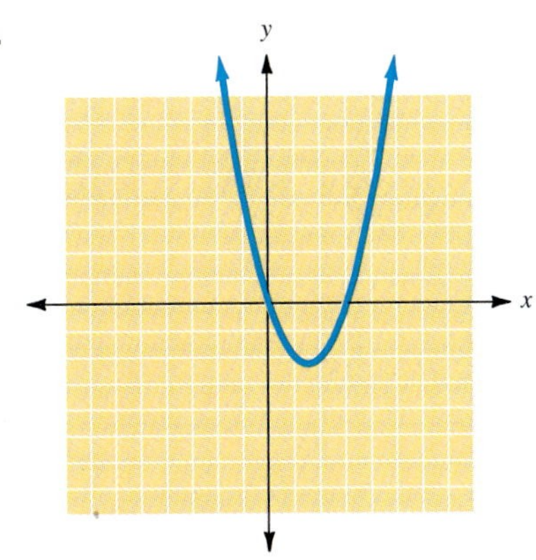

Equations

(a) $y = x^2 + 2$ **(b)** $y = 2x^2 - 1$

(c) $y = 2x + 1$ **(d)** $y = \left(x - \frac{3}{2}\right)^2 - \frac{9}{4}$

(e) $y = -(x + 2)^2 + 4$

(f) $y = -2x + 1$

(g) $y = (x + 1)^2 - 4$

(h) $y = -(x - 3)^2 + 1$

(i) $y = \left(x - \frac{1}{2}\right)^2 - \frac{9}{4}$

(j) $x = -(y - 1)^2 + 4$

ANSWERS

1. ________ 2. ________

3. ________ 4. ________

5. ________ 6. ________

7. ________ 8. ________

9. ________ 10. ________

2.

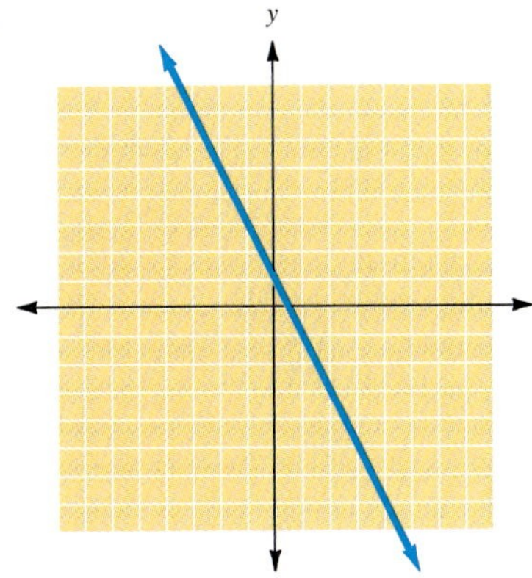

3.

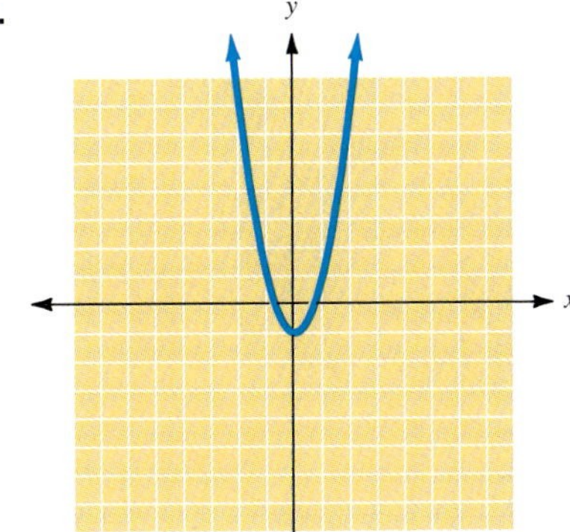

4.

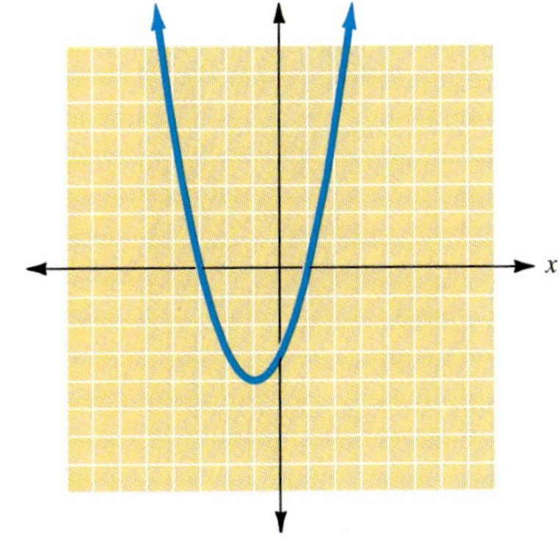

5.

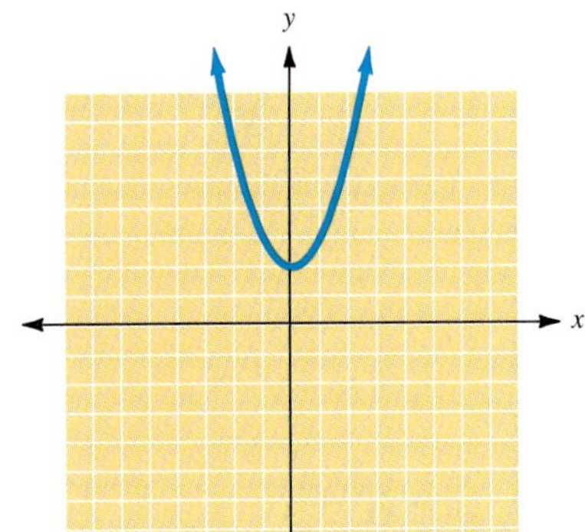

6.

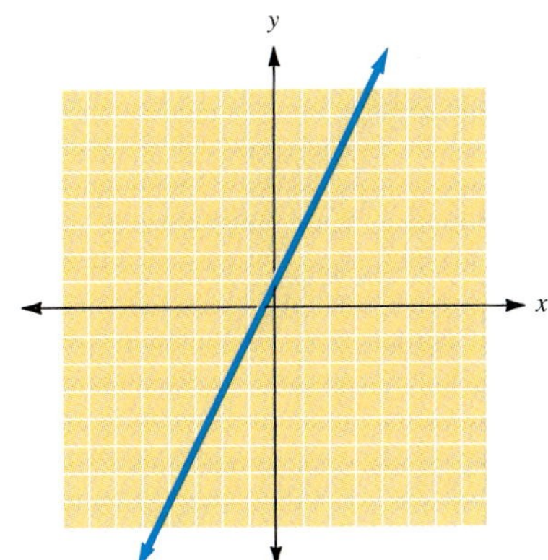

7.

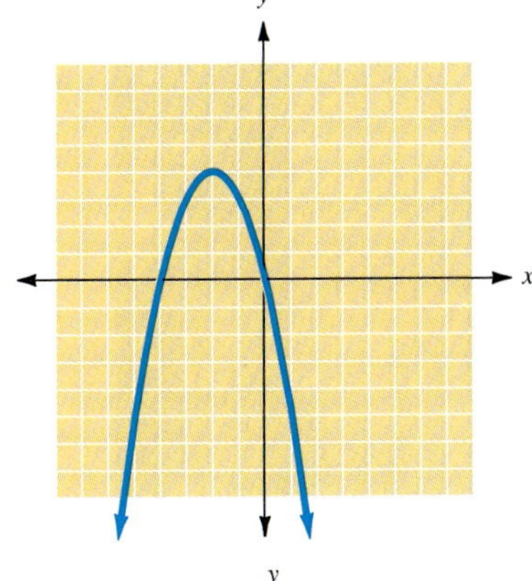

8.

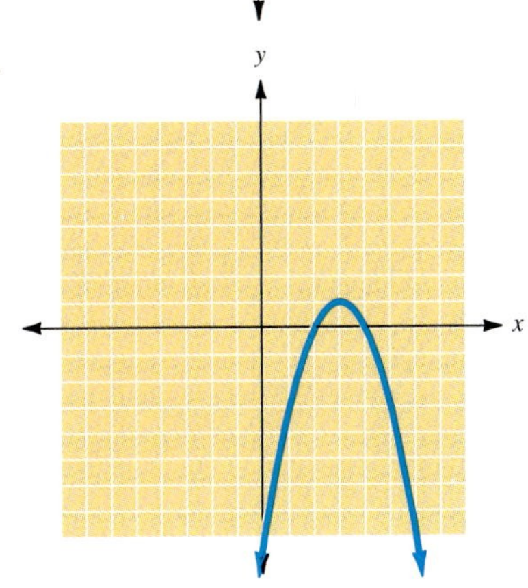

9.

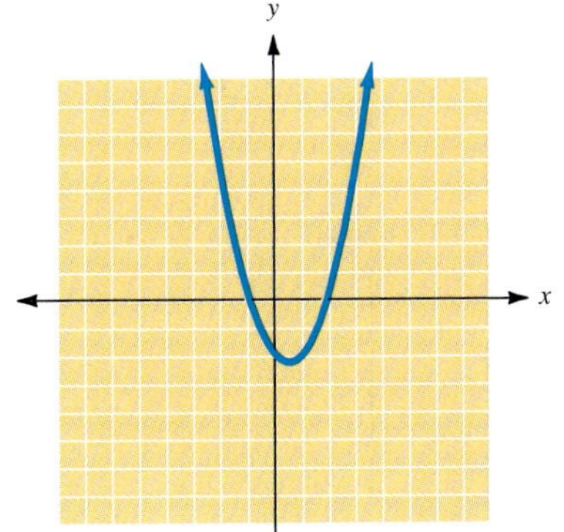

10.

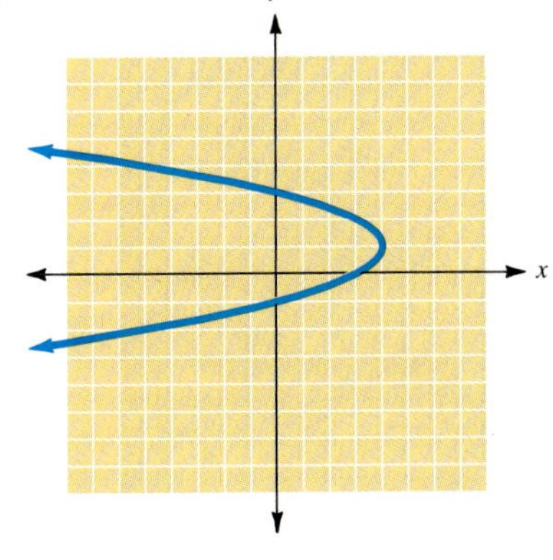

ANSWERS

11. ______

12. ______

13. ______

14. ______

15. ______

16. ______

17. ______

18. ______

19. ______

20. ______

21. ______

22. ______

23. ______

24. ______

25. ______

26. ______

Determine whether each parabola is translated from the origin to the left, to the right, upward, or downward.

11. $y = (x + 5)^2$

12. $y = x^2 + 7$

13. $y = (x - 2)^2$

14. $y = (x - 5)^2$

15. $x = y^2 - 4$

16. $x = (y + 2)^2$

Find the equation of the axis of symmetry and the coordinates for the vertex of each of the following.

17. $f(x) = x^2$

18. $f(x) = x^2 + 2$

19. $f(x) = x^2 - 5$

20. $f(x) = (x - 3)^2$

21. $f(x) = (x + 3)^2 + 1$

22. $f(x) = -(x + 2)^2 - 3$

23. $f(x) = -x^2 + 2x$

24. $f(x) = x^2 + 4x + 5$

Graph the following.

25. $y = -\frac{1}{2}(x + 3)^2$

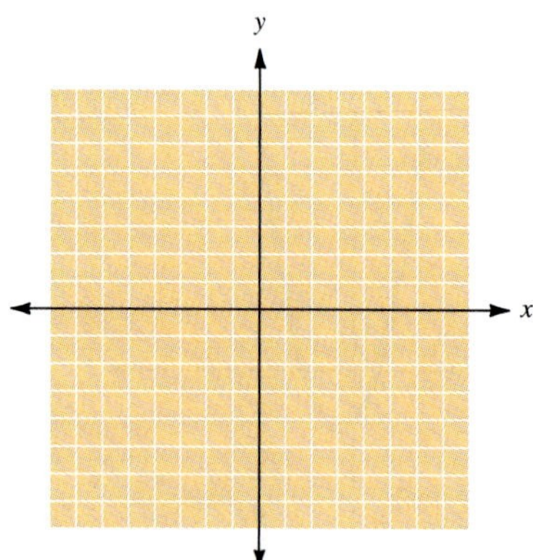

26. $y = (x - 3)^2 + 2$

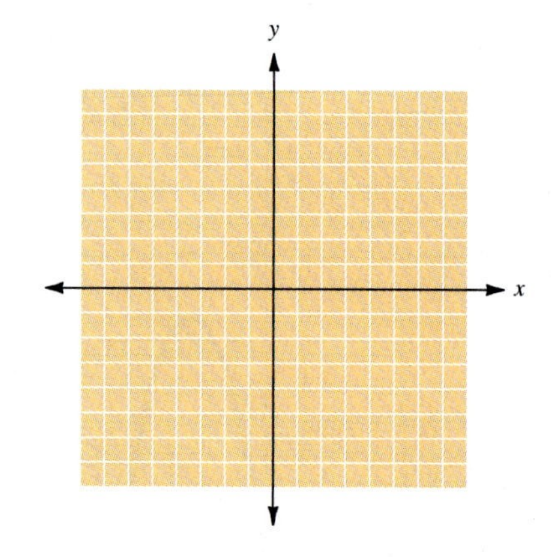

27. $y = -(x + 2)^2 - 1$

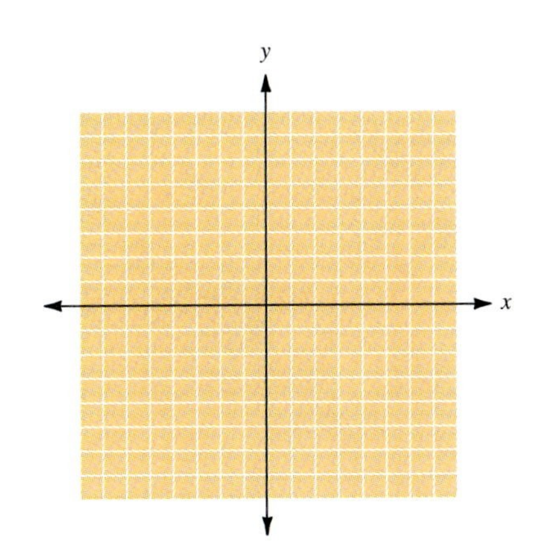

28. $y = 2(x + 3)^2 - 3$

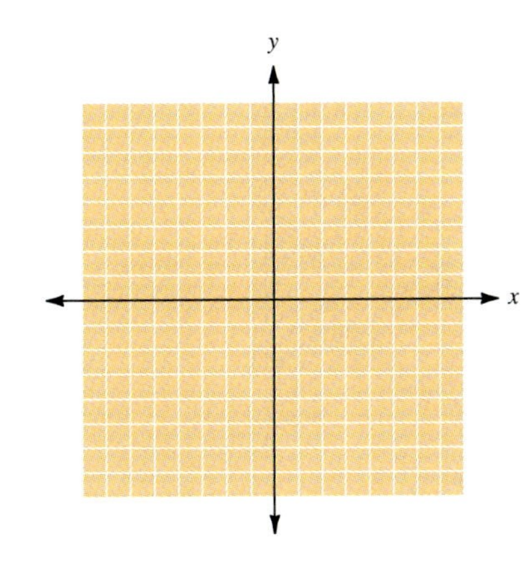

29. $y = x^2 - 2x$

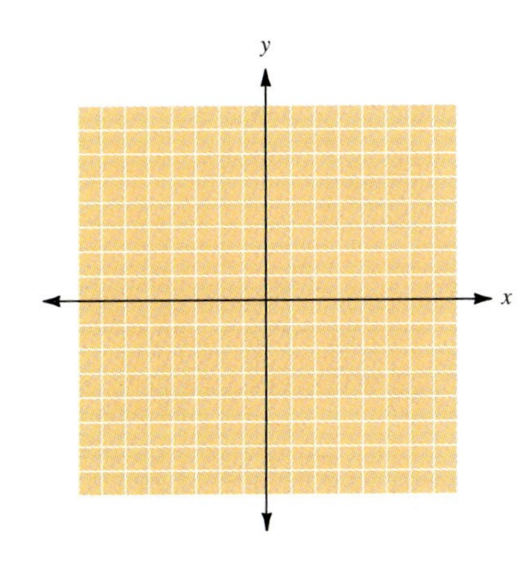

30. $y = x^2 - 4$

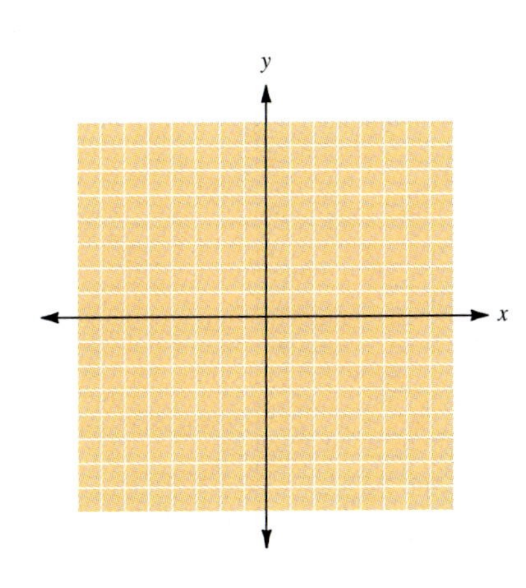

31. $y = -x^2 + 1$

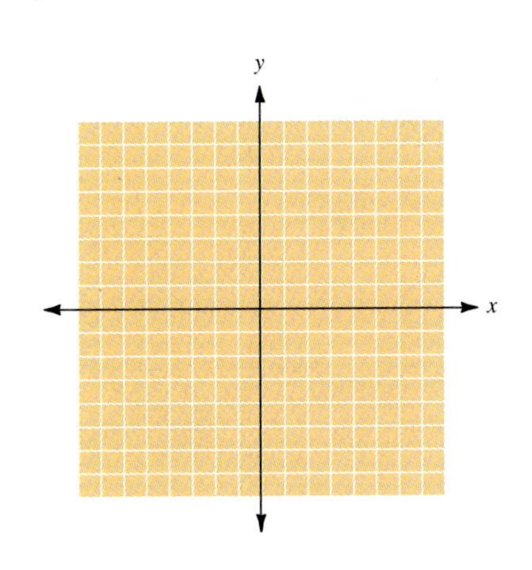

32. $y = x^2 + 4x$

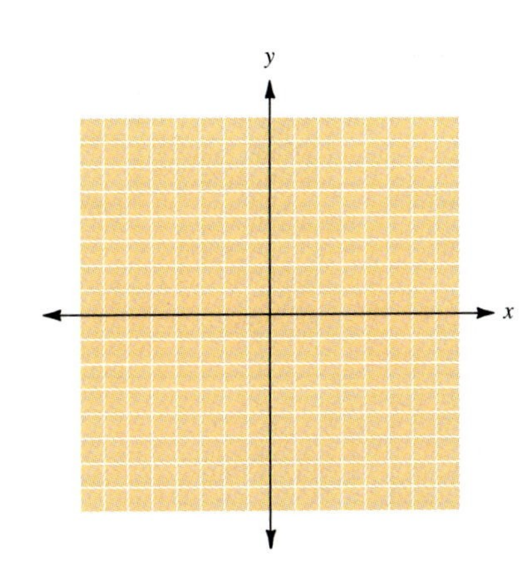

ANSWERS

27. ____________________

28. ____________________

29. ____________________

30. ____________________

31. ____________________

32. ____________________

ANSWERS

33. ______

34. ______

35. ______

36. ______

37. ______

38. ______

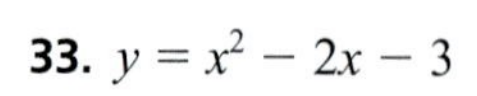

33. $y = x^2 - 2x - 3$

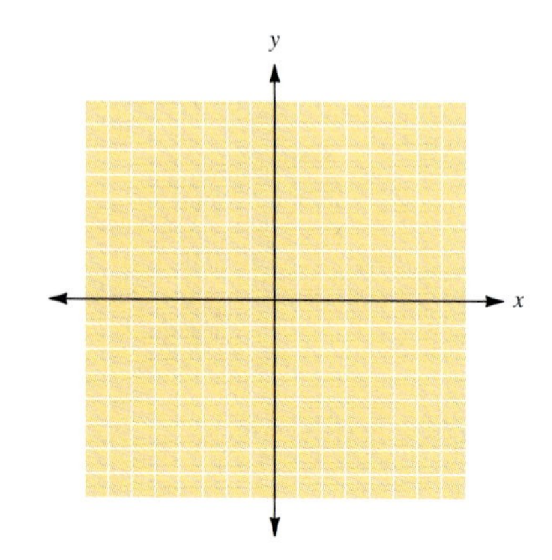

34. $y = x^2 - x - 6$

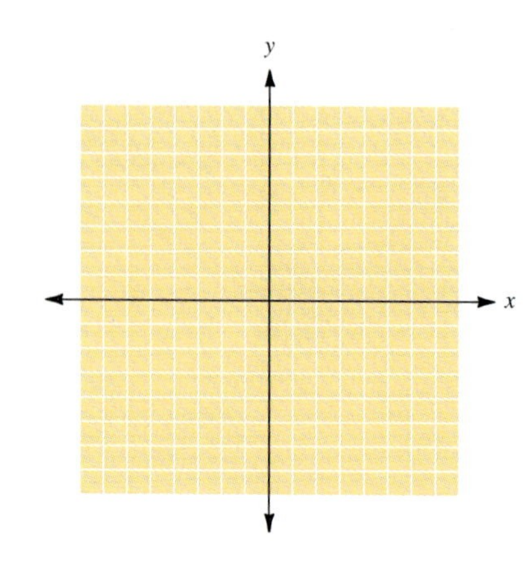

35. $y = 2x^2 - 4x + 1$

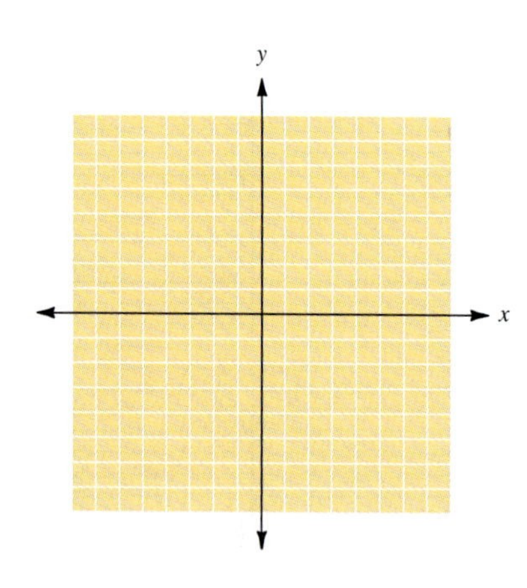

36. $y = \frac{1}{2}x^2 + x - 2$

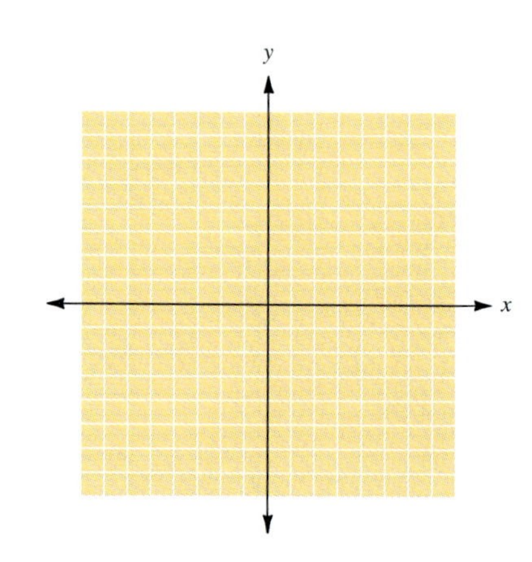

37. $y = -\frac{1}{3}x^2 + x - 2$

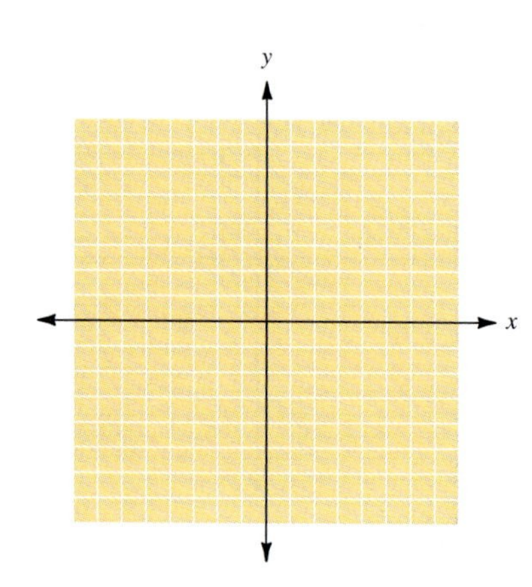

38. $y = -2x^2 + 4x - 3$

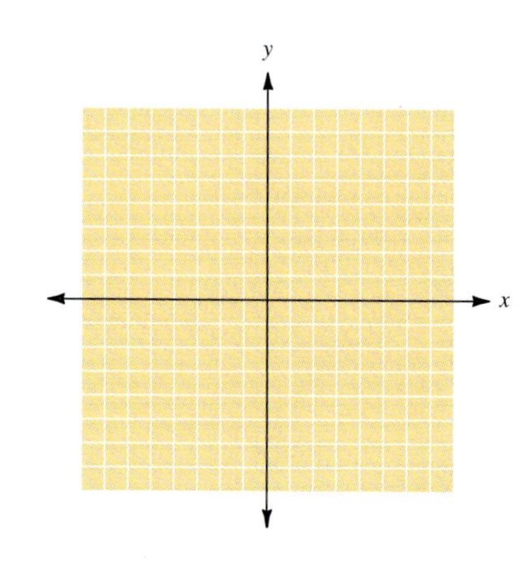

39. $y = 3x^2 + 6x - 1$

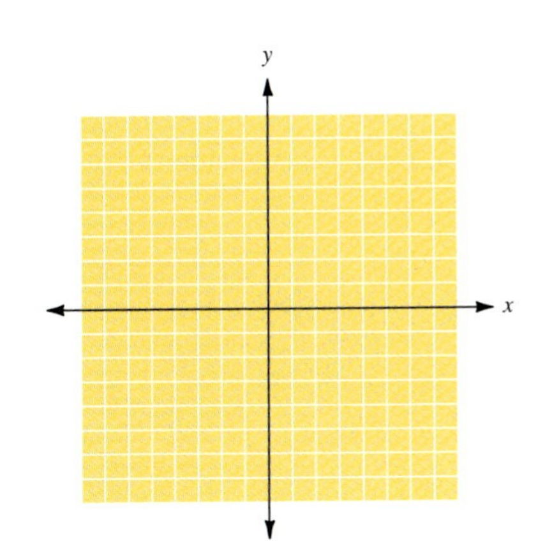

40. $y = -3x^2 + 12x - 5$

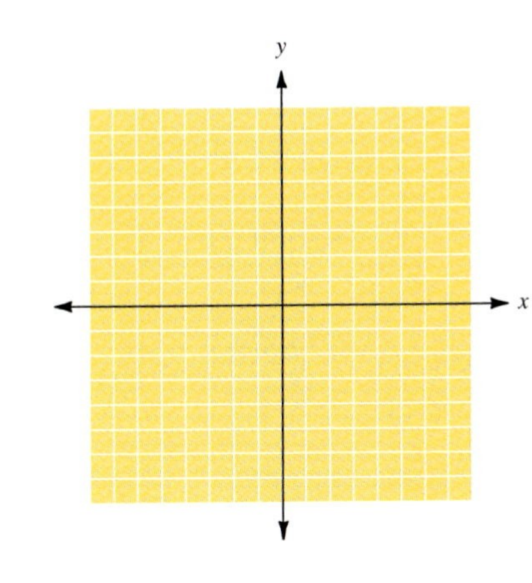

41. $x = y^2 - 4y$

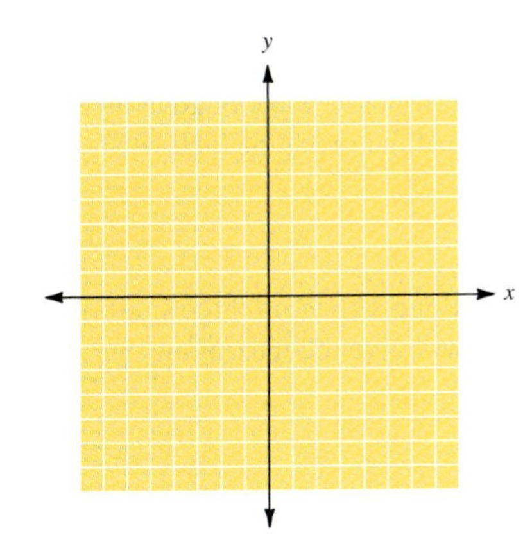

42. $x = y^2 + 3y$

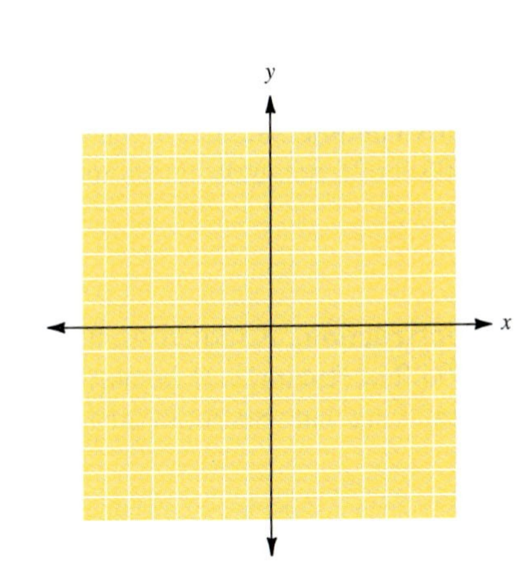

43. $x = y^2 - 3y - 4$

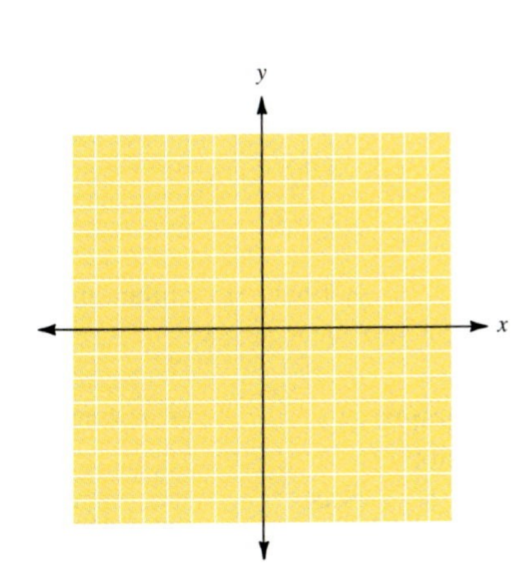

44. $x = -y^2 - y + 6$

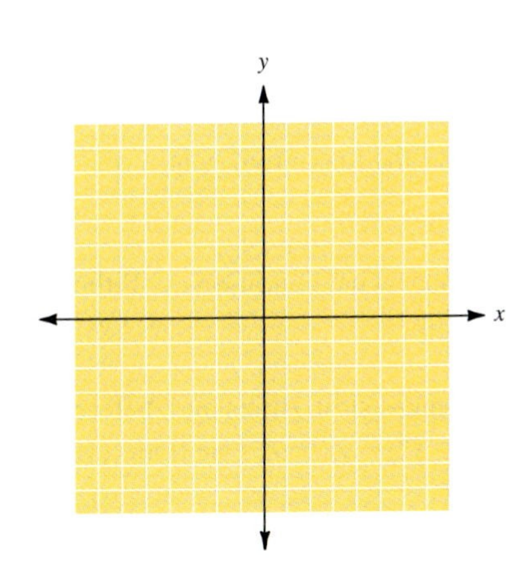

ANSWERS

39. ______

40. ______

41. ______

42. ______

43. ______

44. ______

ANSWERS

45. ____

46. ____

47. ____

48. ____

49. ____

50. ____

51. ____

52. ____

If you wish to check your work on some of the earlier exercises, you can graph $x = y^2 + 3y + 4$ using a graphing utility, rewrite the equation as a quadratic equation in y

$$y^2 + 3y + (-x + 4) = 0$$

Then use the quadratic formula to solve for y and enter the resulting equations.

$$Y_1 = \frac{-3 + \sqrt{9 - 4(-x + 4)}}{4}$$

$$Y_2 = \frac{-3 - \sqrt{9 - 4(-x + 4)}}{4}$$

45. **Profit.** A company's weekly profit P is related to the number of items sold by $P(x) = -0.2x^2 + 40x - 500$. Find the number of items that should be sold each week to maximize the profit. Then find the amount of that weekly profit.

46. **Profit.** A company's monthly profit P is related to the number of items sold by $P(x) = -0.1x^2 + 30x - 1000$. How many items should be sold each month to obtain the largest possible profit? What is that profit?

47. **Construction.** A builder wants to enclose the largest possible rectangular area with 1600 ft of fencing. What should be the dimensions of the rectangle, and what will its area be?

48. **Construction.** A farmer wants to enclose a field along a river on three sides. If 1200 ft of fencing is to be used, what dimensions will give the maximum enclosed area? Find that maximum area.

49. **Motion.** A ball is thrown upward into the air with an initial velocity of 64 ft/s. If h gives the height of the ball at time t, then the equation relating h and t is

    $h(t) = -16t^2 + 64t$

    Find the maximum height that the ball will attain.

50. **Motion.** A ball is thrown upward into the air with an initial velocity of 32 ft/s. If h gives the height of the ball at time t, then the equation relating h and t is

    $h(t) = -16t^2 + 32t$

    Find the maximum height that the ball will attain.

51. Under what conditions will the graph of $x = a(y - k)^2 + h$ have no y intercepts?

52. Discuss similarities and differences between the graphs of $y = x^2 + 3x + 4$ and $x = y^2 + 3y + 4$. Use both graphs in your discussion.

Each equation below defines a relation. Write the domain of each relation.
(*Hint*: Determine the vertex, and whether the parabola opens to the left or to the right.)

53. $x = y^2 + 6y$

54. $x = y^2 - 8y$

55. $x = -y^2 + 6y - 7$

56. $x = -y^2 - 4y + 3$

Write the domain and range of each relation.

57. $y = 3(x - 2)^2 + 1$

58. $y = (x + 1)^2 - 2$

59. $x = (y - 1)^2 - 1$

60. $x = (y + 2)^2 + 4$

Answers

1. d **3.** b **5.** a **7.** e **9.** i **11.** Left **13.** Right
15. Left **17.** $x = 0$; $(0, 0)$ **19.** $x = 0$; $(0, -5)$ **21.** $x = -3$; $(-3, 1)$
23. $x = 1$; $(1, 1)$

25.

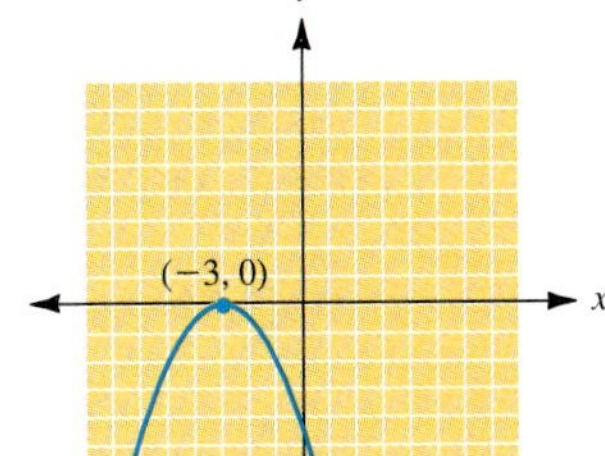

27.

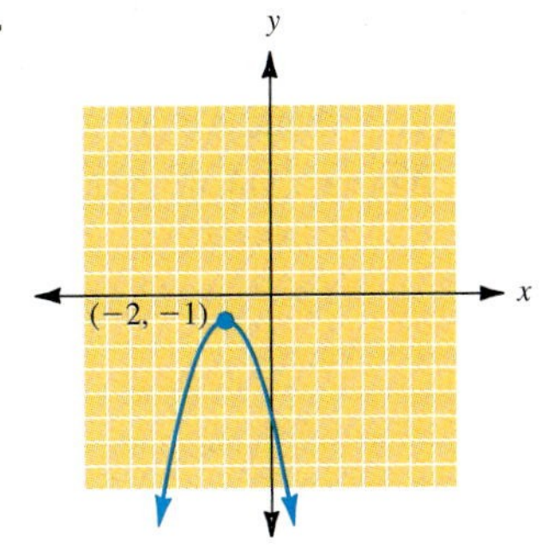

29.

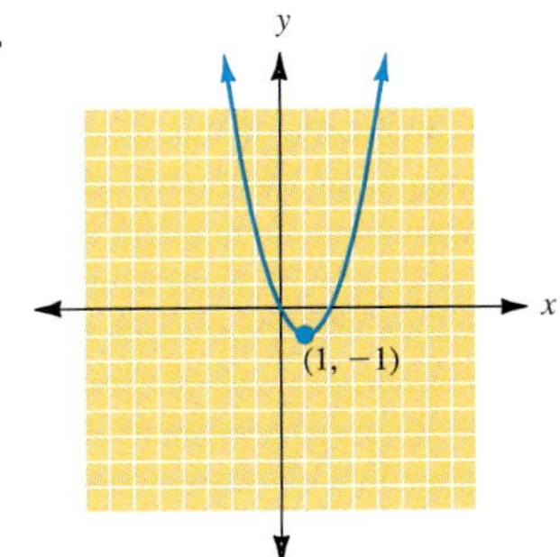

31.

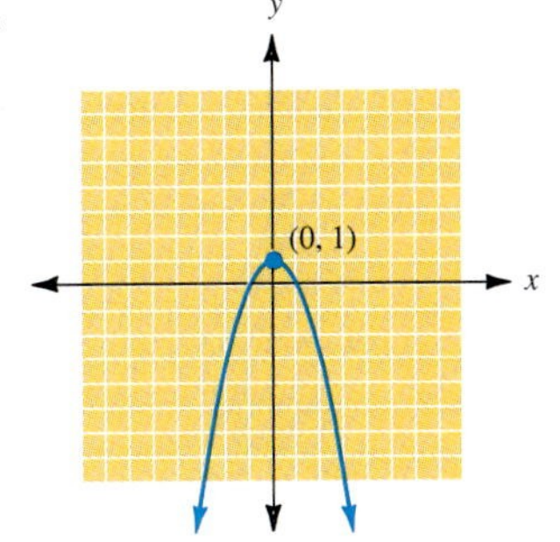

33.

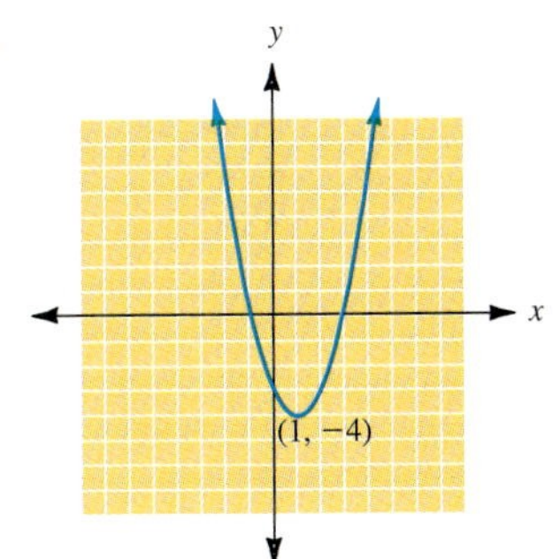

35.

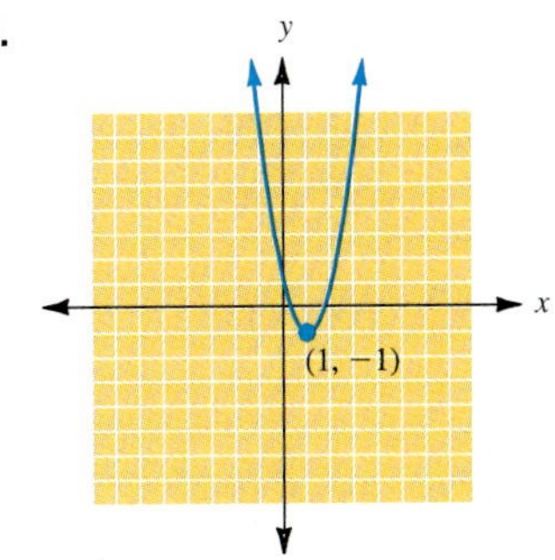

ANSWERS

53. ____________

54. ____________

55. ____________

56. ____________

57. ____________

58. ____________

59. ____________

60. ____________

37.

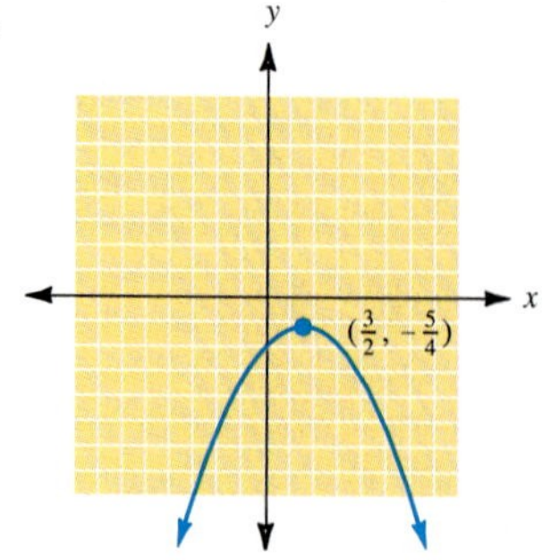

39.

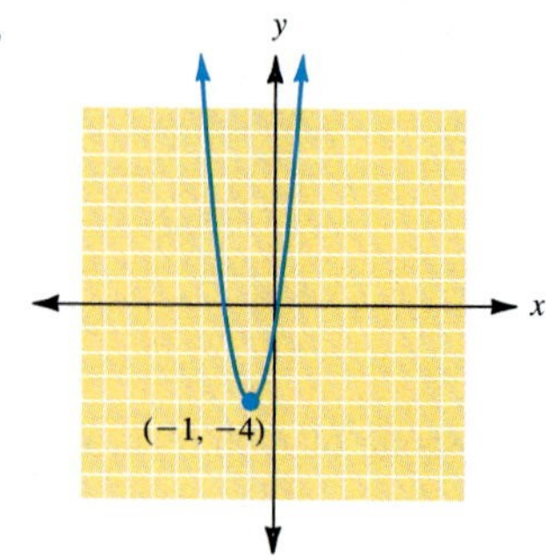

41.

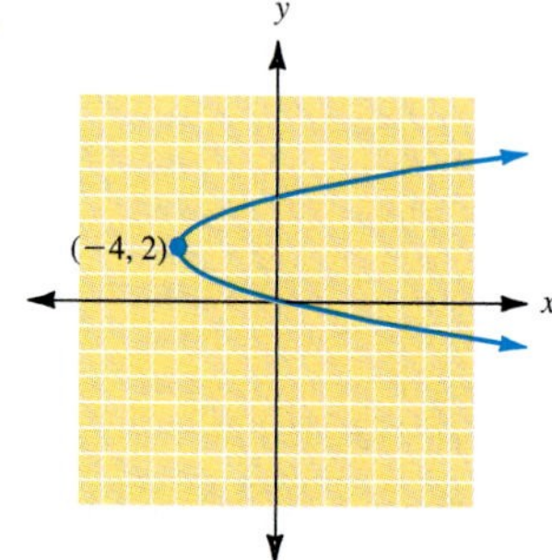

43.

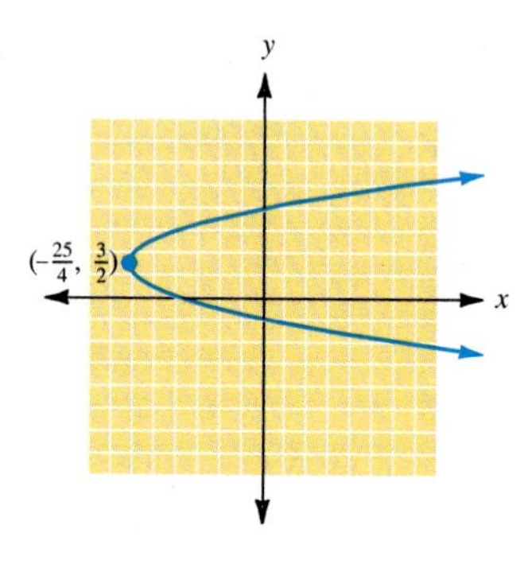

45. 100 items, \$1500 **47.** 400 by 400 ft; 160,000 ft^2 **49.** 64 ft
51. $a > 0$ and $h > 0$ or $a < 0$ and $h < 0$ **53.** $\{x \mid x \geq -9\}$ **55.** $\{x \mid x \leq 2\}$
57. D: Reals
R: $\{y \mid y \geq 1\}$
59. D: $\{x \mid x \geq -1\}$
R: Reals

10.2 The Distance Formula and the Circle

10.2 OBJECTIVES

1. Given a center and radius, find the equation of a circle
2. Given an equation for a circle, find the center and radius
3. Given an equation, sketch the graph of a circle

In Section 10.1, we examined the parabola. In this section, we turn our attention to another conic section, the circle.

Definitions: Circle

A **circle** is the set of all points in the plane equidistant from a fixed point, called the **center** of the circle. The distance between the center of the circle and any point on the circle is called the **radius** of the circle.

The distance formula is central to any discussion of conic sections.

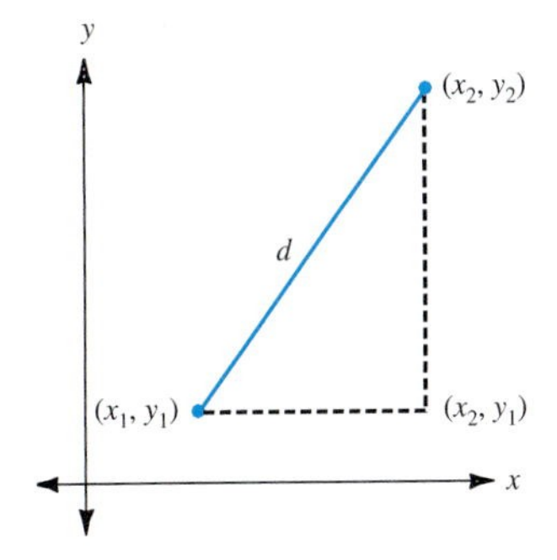

Definitions: The Distance Formula

The distance, d, between two points (x_1, y_1) and (x_2, y_2) is given by

$$d = \sqrt{(x_2 - x_1)^2 + (y_2 - y_1)^2}$$

We can use the distance formula to derive the algebraic equation of a circle, given its center and its radius.

Suppose a circle has its center at a point with coordinates (h, k) and radius r. If (x, y) represents any point on the circle, then, by its definition, the distance from (h, k) to (x, y) is r. Applying the distance formula, we have

$$r = \sqrt{(x - h)^2 + (y - k)^2}$$

Squaring both sides of the equation gives the equation of the circle

$$r^2 = (x - h)^2 + (y - k)^2$$

In general, we can write the following equation of a circle.

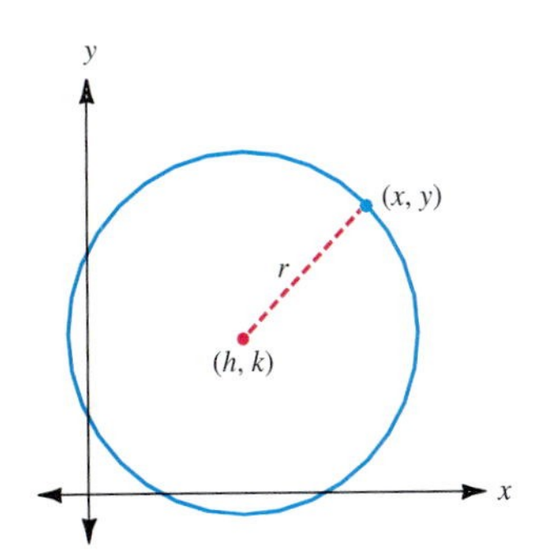

NOTE A special case is the circle centered at the origin with radius r. Then $(h, k) = (0, 0)$, and its equation is

$x^2 + y^2 = r^2$

Definitions: Equation of a Circle

The equation of a circle with center (h, k) and radius r is

$$(x - h)^2 + (y - k)^2 = r^2 \quad (1)$$

Equation (1) can be used in two ways. Given the center and radius of the circle, we can write its equation; or given its equation, we can find the center and radius of a circle.

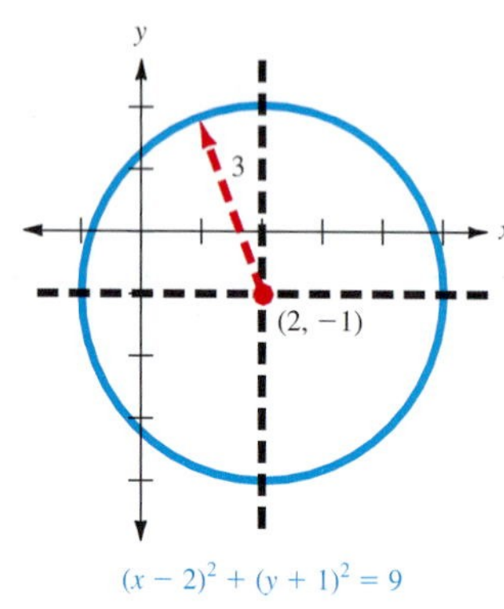

$(x-2)^2+(y+1)^2=9$

Example 1

Finding the Equation of a Circle

Find the equation of a circle with center at $(2, -1)$ and radius 3. Sketch the circle.

Let $(h, k) = (2, -1)$ and $r = 3$. Applying equation (1) yields

$$(x-2)^2 + [y-(-1)]^2 = 3^2$$
$$(x-2)^2 + (y+1)^2 = 9$$

To sketch the circle, we locate the center of the circle. Then we determine four points 3 units to the right and left and up and down from the center of the circle. Drawing a smooth curve through those four points completes the graph.

CHECK YOURSELF 1

Find the equation of the circle with center at $(-2, 1)$ and radius 5. Sketch the circle.

Now, given an equation for a circle, we can also find the radius and center and then sketch the circle. We start with an equation in the special form of equation (1).

NOTE The circle can be graphed on the calculator by solving for y, then graphing both the upper half and lower half of the circle. In this case,

$$(x-1)^2+(y+2)^2=9$$
$$(y+2)^2 = 9-(x-1)^2$$
$$(y+2) = \pm\sqrt{9-(x-1)^2}$$
$$y = -2 \pm \sqrt{9-(x-1)^2}$$

Now graph the two functions

$$y = -2 + \sqrt{9-(x-1)^2}$$

and

$$y = -2 - \sqrt{9-(x-1)^2}$$

on your calculator. (The display screen may need to be squared to obtain the shape of a circle.)

Example 2

Finding the Center and Radius of a Circle

Find the center and radius of the circle with equation

$$(x-1)^2 + (y+2)^2 = 9$$

Remember, the general form is

$$(x-h)^2 + (y-k)^2 = r^2$$

Our equation "fits" this form when it is written as

Note: $y + 2 = y - (-2)$

$$(x-1)^2 + \underbrace{[y-(-2)]^2}= 3^2$$

So the center is at $(1, -2)$, and the radius is 3. The graph is shown.

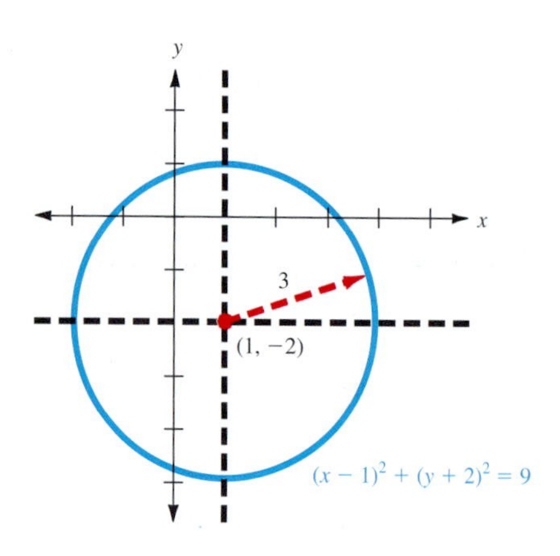

$(x-1)^2+(y+2)^2=9$

CHECK YOURSELF 2

Find the center and radius of the circle with equation

$$(x + 3)^2 + (y - 2)^2 = 16$$

Sketch the circle.

To graph the equation of a circle that is not in standard form, we *complete the square.* Let's see how completing the square can be used in graphing the equation of a circle.

Example 3

Finding the Center and Radius of a Circle

NOTE To recognize the equation as having the form of a circle, note that the coefficients of x^2 and y^2 are equal.

Find the center and radius of the circle with equation

$$x^2 + 2x + y^2 - 6y = -1$$

Then sketch the circle.

NOTE The linear terms in x and y show a translation of the center away from the origin.

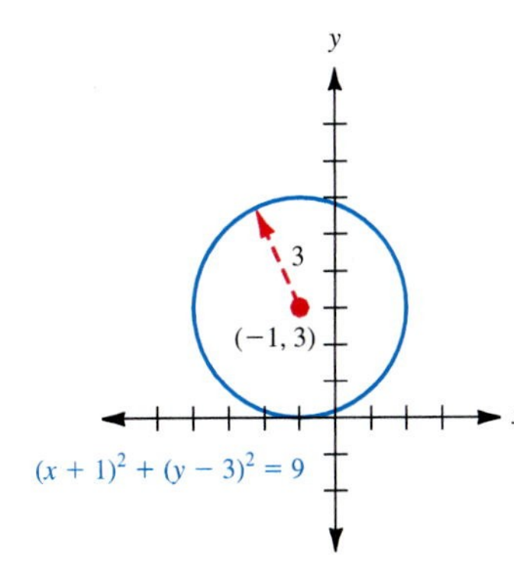

We could, of course, simply substitute values of x and try to find the corresponding values for y. A much better approach is to rewrite the original equation so that it matches the standard form.

First, add 1 to both sides to complete the square in x.

$$x^2 + 2x + 1 + y^2 - 6y = -1 + 1$$

Then add 9 to both sides to complete the square in y.

$$x^2 + 2x + 1 + y^2 - 6y + 9 = -1 + 1 + 9$$

We can factor the two trinomials on the left (they are both perfect squares) and simplify on the right.

$$(x + 1)^2 + (y - 3)^2 = 9$$

The equation is now in standard form, and we can see that the center is at $(-1, 3)$ and the radius is 3. The sketch of the circle is shown. Note the "translation" of the center to $(-1, 3)$.

CHECK YOURSELF 3

Find the center and radius of the circle with equation

$$x^2 - 4x + y^2 + 2y = -1$$

Sketch the circle.

CHECK YOURSELF ANSWERS

1. $(x + 2)^2 + (y - 1)^2 = 25$ **2.** center: $(-3, 2)$; radius $= 4$

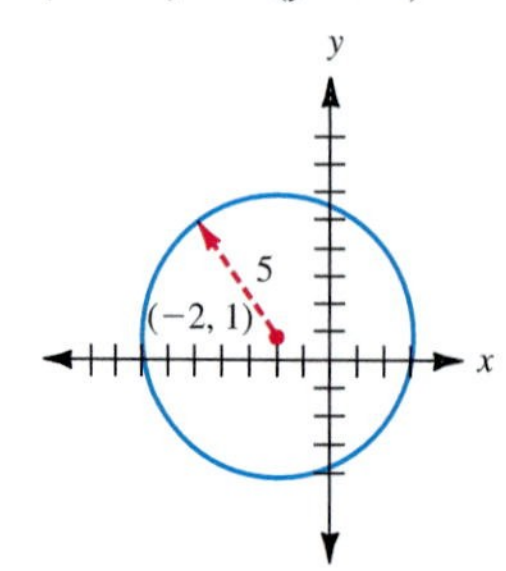

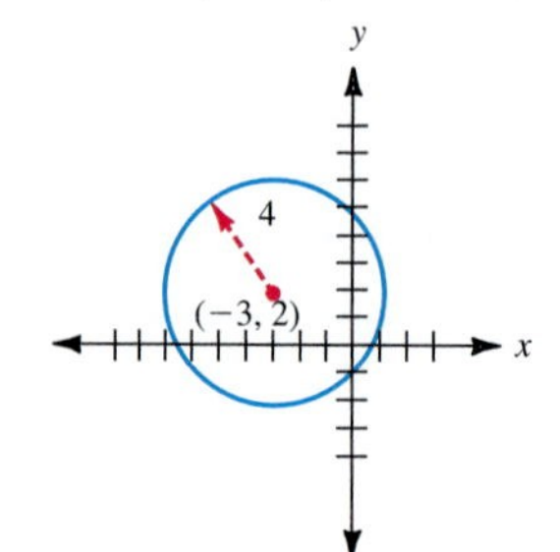

3. $(x - 2)^2 + (y + 1)^2 = 4$; center: $(2, -1)$; radius $= 2$

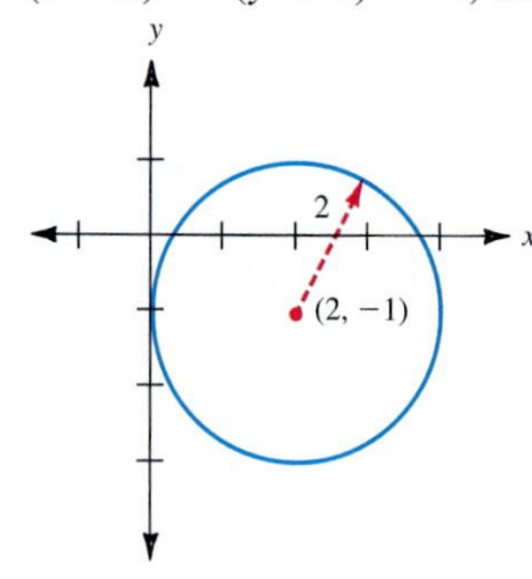

10.2 Exercises

In exercises 1 to 12, decide whether each equation has as its graph a line, a parabola, a circle, or none of these.

1. $y = x^2 - 2x + 5$

2. $y^2 + x^2 = 64$

3. $y = 3x - 2$

4. $2y - 3x = 12$

5. $(x - 3)^2 + (y + 2)^2 = 10$

6. $y + 2(x - 3)^2 = 5$

7. $x^2 + 4x + y^2 - 6y = 3$

8. $4x = 3$

9. $y^2 - 4x^2 = 36$

10. $x^2 + (y - 3)^2 = 9$

11. $y = -2x^2 + 8x - 3$

12. $2x^2 - 3y^2 + 6y = 13$

In exercises 13 to 20, find the center and the radius for each circle.

13. $x^2 + y^2 = 25$

14. $x^2 + y^2 = 72$

15. $(x - 3)^2 + (y + 1)^2 = 16$

16. $(x + 3)^2 + y^2 = 81$

17. $x^2 + 2x + y^2 = 15$

18. $x^2 + y^2 - 6y = 72$

19. $x^2 - 6x + y^2 + 8y = 16$

20. $x^2 - 5x + y^2 - 3y = 8$

In exercises 21 to 32, graph each circle by finding the center and the radius.

21. $x^2 + y^2 = 4$

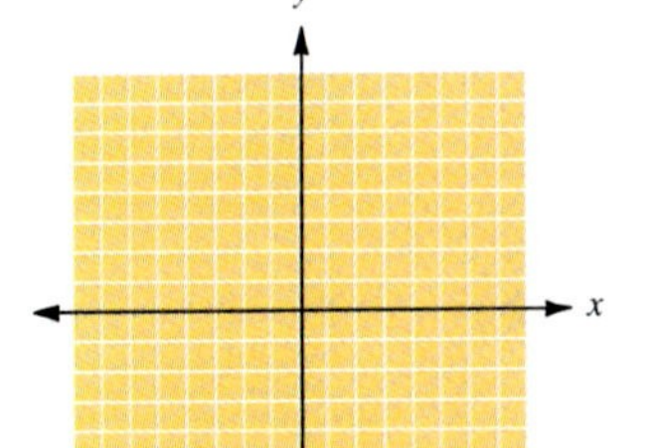

22. $x^2 + y^2 = 25$

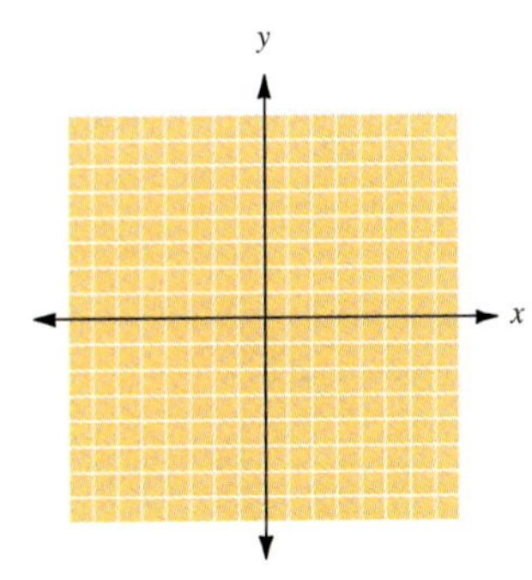

Name ______________________

Section ________ Date ________

ANSWERS

1. ______________________
2. ______________________
3. ______________________
4. ______________________
5. ______________________
6. ______________________
7. ______________________
8. ______________________
9. ______________________
10. ______________________
11. ______________________
12. ______________________
13. ______________________
14. ______________________
15. ______________________
16. ______________________
17. ______________________
18. ______________________
19. ______________________
20. ______________________
21. ______________________
22. ______________________

ANSWERS

23. ______

24. ______

25. ______

26. ______

27. ______

28. ______

23. $4x^2 + 4y^2 = 36$

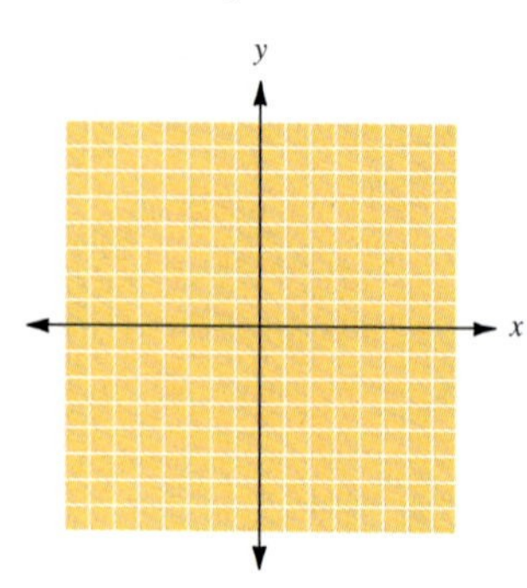

24. $9x^2 + 9y^2 = 144$

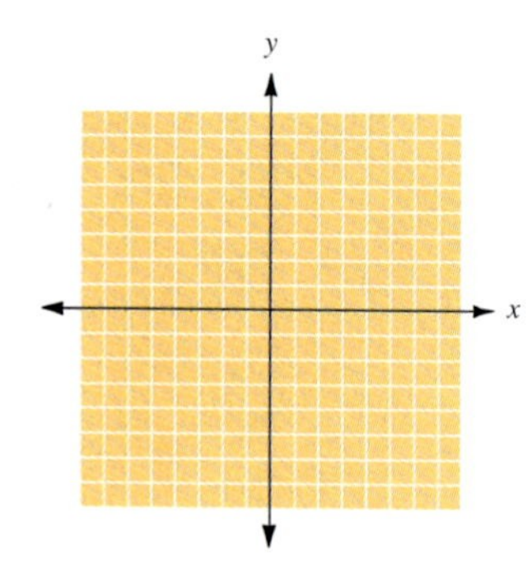

25. $(x - 1)^2 + y^2 = 9$

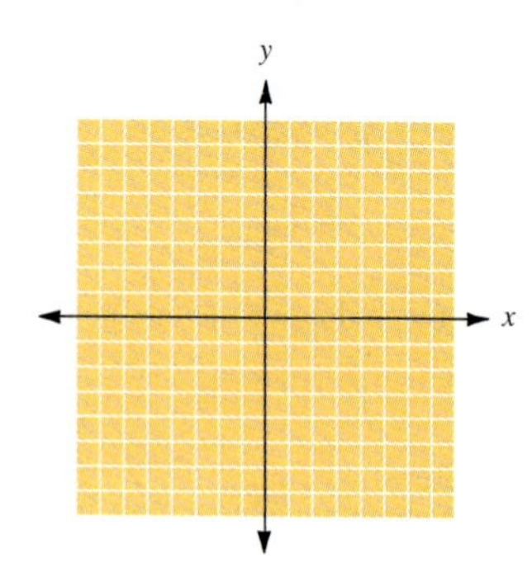

26. $x^2 + (y + 2)^2 = 16$

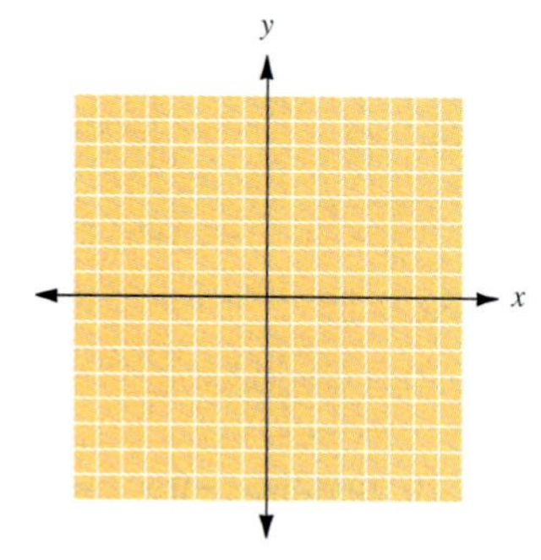

27. $(x - 4)^2 + (y + 1)^2 = 16$

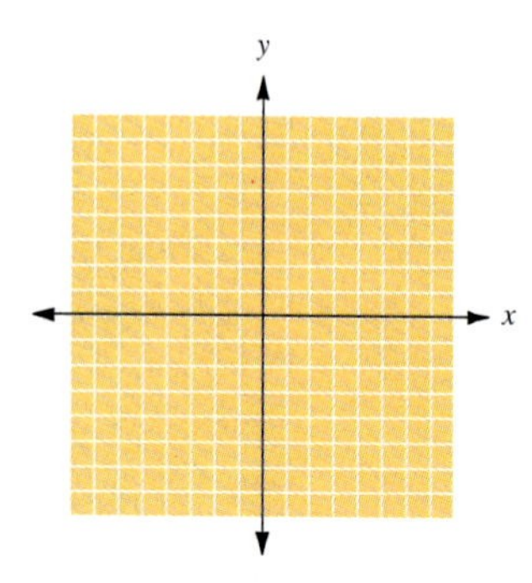

28. $(x + 3)^2 + (y + 2)^2 = 25$

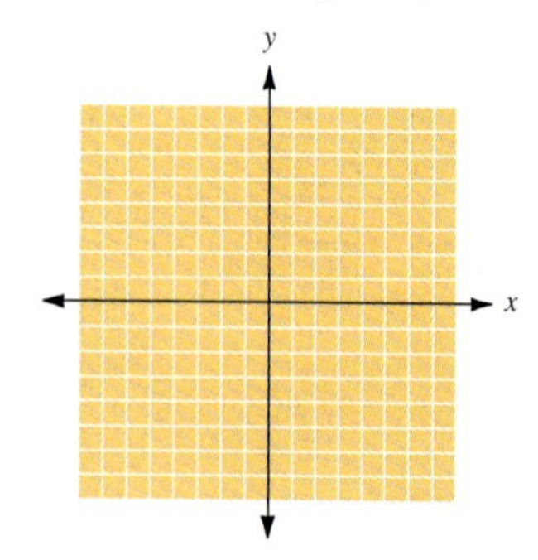

29. $x^2 + y^2 - 4y = 12$

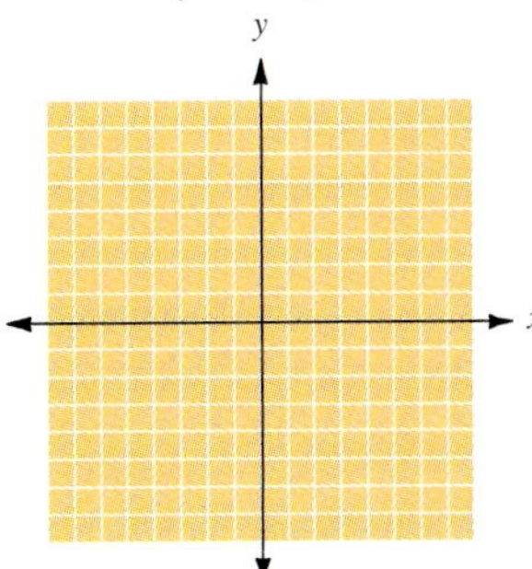

30. $x^2 - 6x + y^2 = 0$

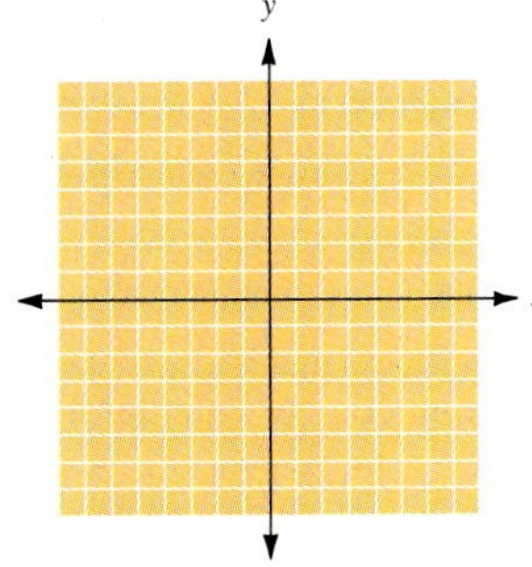

31. $x^2 - 4x + y^2 + 2y = -1$

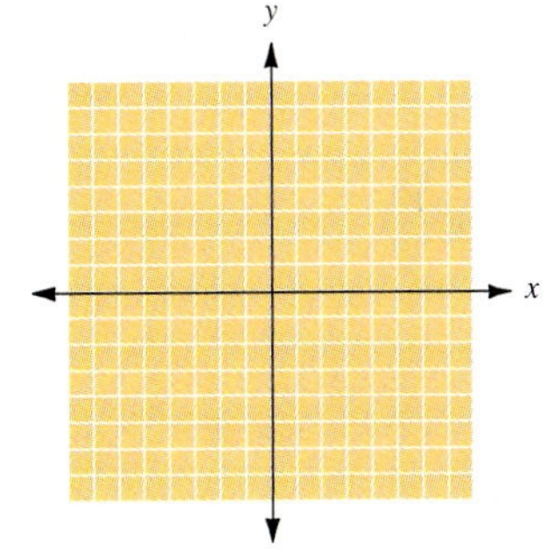

32. $x^2 - 2x + y^2 - 6y = 6$

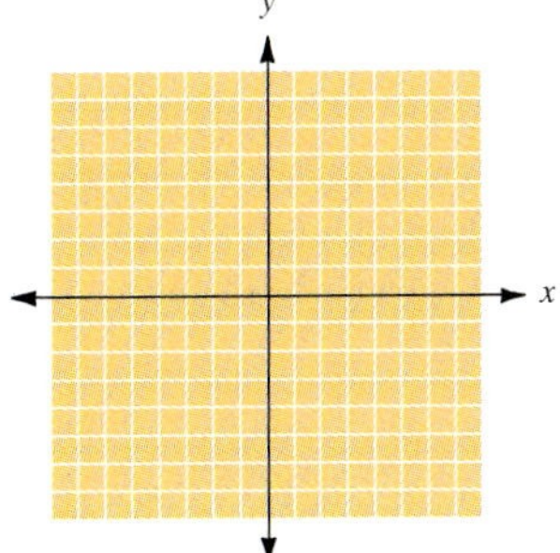

33. Describe the graph of $x^2 + y^2 - 2x - 4y + 5 = 0$.

34. Describe how completing the square is used in graphing circles.

35. A solar oven is constructed in the shape of a hemisphere. If the equation

$$x^2 + y^2 + 500 = 1000$$

describes the outer edge of the oven in centimeters, what is its radius?

36. A solar oven in the shape of a hemisphere is to have a diameter of 80 cm. Write the equation that describes the outer edge of this oven.

37. A solar water heater is constructed in the shape of a half cylinder, with the water supply pipe at its center. If the water heater has a diameter of $\frac{4}{3}$ m, what is the equation that describes its outer edge?

ANSWERS

29. ____________

30. ____________

31. ____________

32. ____________

33. ____________

34. ____________

35. ____________

36. ____________

37. ____________

38. ______

39. ______

40. ______

41. ______

38. A solar water heater is constructed in the shape of a half cylinder with circumference described by the equation

$$9x^2 + 9y^2 - 16 = 0$$

What is its diameter if the units for the equation are meters?

A circle can be graphed on a calculator by plotting the upper and lower semicircles on the same axes. For example, to graph $x^2 + y^2 = 16$, we solve for y:

$$y = \pm\sqrt{16 - x^2}$$

This is then graphed as two separate functions,

$$Y_1 = \sqrt{16 - x^2} \qquad \text{and} \qquad Y_2 = -\sqrt{16 - x^2}$$

In exercises 39 to 42, use that technique to graph each circle.

39. $x^2 + y^2 = 36$

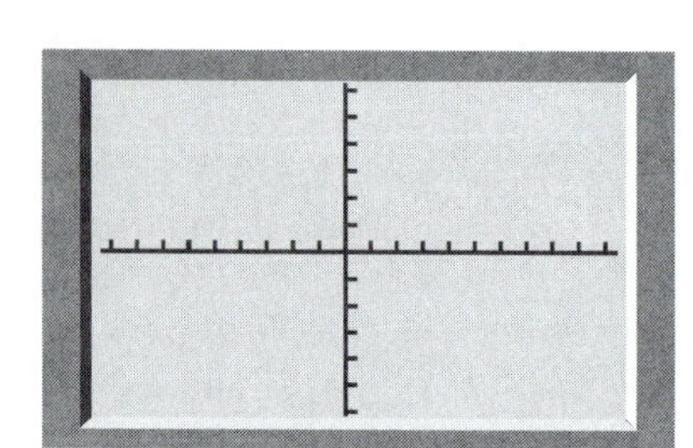

40. $(x - 3)^2 + y^2 = 9$

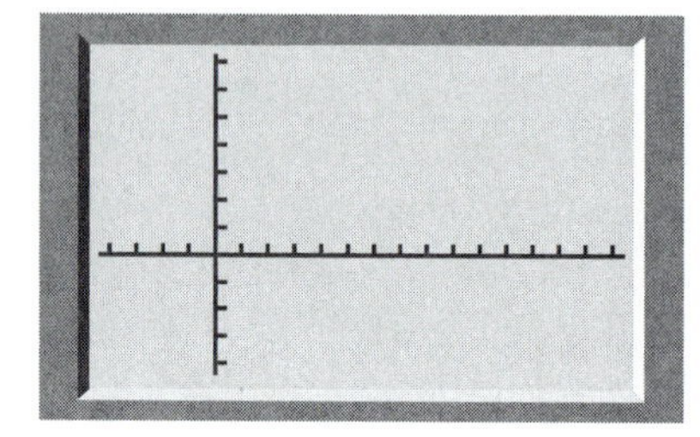

41. $(x + 5)^2 + y^2 = 36$

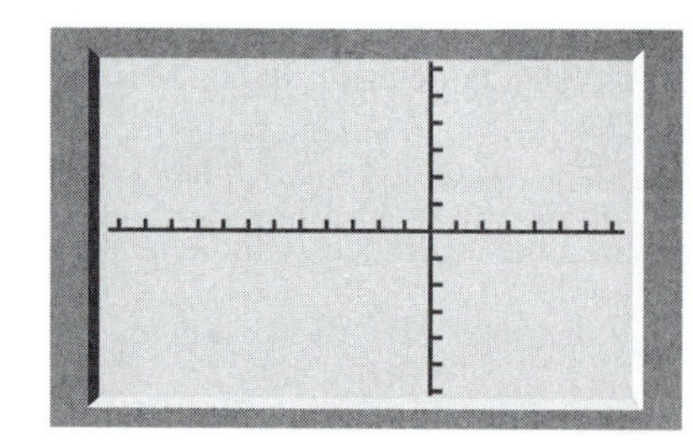

42. $(x - 2)^2 + (y + 1)^2 = 25$

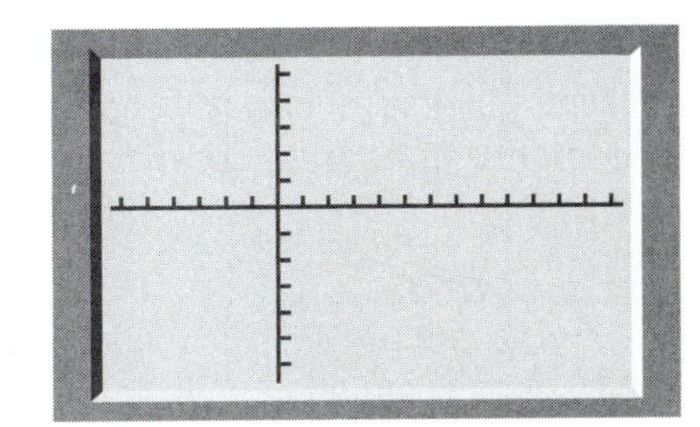

Each of the following equations defines a relation. Write the domain and the range of each relation.

43. $(x + 3)^2 + (y - 2)^2 = 16$

44. $(x - 1)^2 + (y - 5)^2 = 9$

45. $x^2 + (y - 3)^2 = 25$

46. $(x + 2)^2 + y^2 = 36$

Answers

1. Parabola **3.** Line **5.** Circle **7.** Circle **9.** None of these
11. Parabola **13.** Center: (0, 0); radius: 5 **15.** Center: (3, −1); radius: 4
17. Center: (−1, 0); radius: 4 **19.** Center: (3, −4); radius: $\sqrt{41}$

21. $x^2 + y^2 = 4$

Center: (0, 0); radius: 2

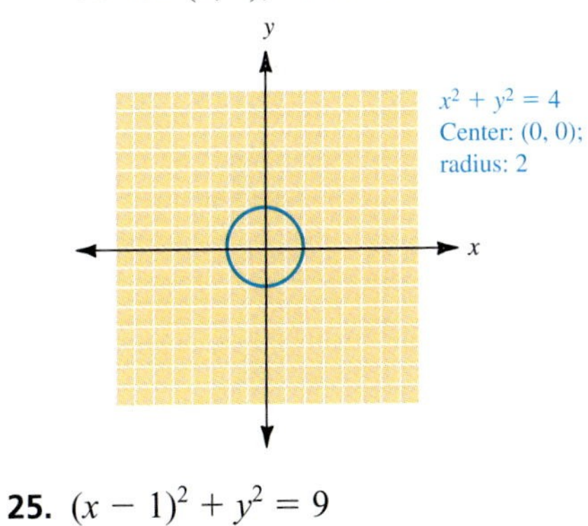

23. $4x^2 + 4y^2 = 36$

$x^2 + y^2 = 9$

Center: (0, 0); radius: 3

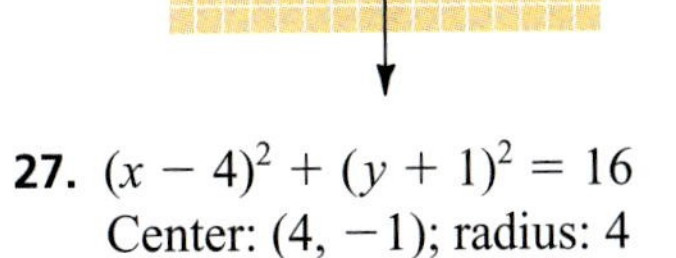

25. $(x - 1)^2 + y^2 = 9$

Center: (1, 0); radius: 3

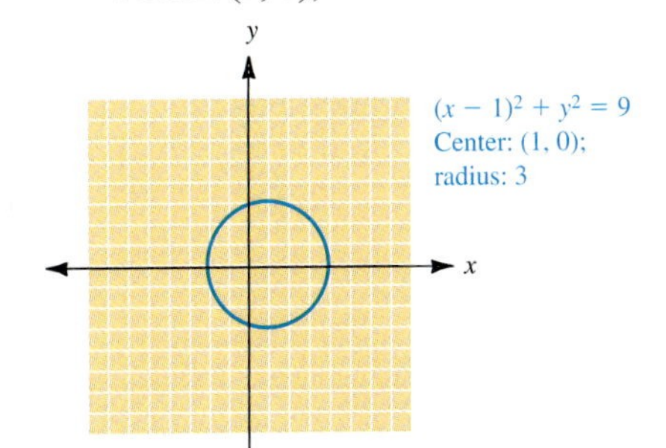

27. $(x - 4)^2 + (y + 1)^2 = 16$

Center: (4, −1); radius: 4

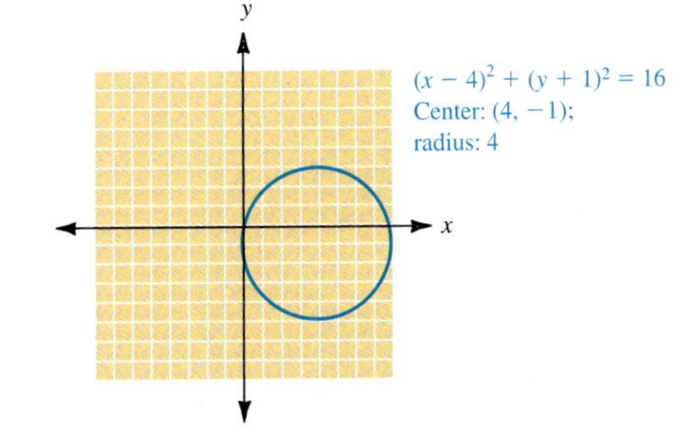

ANSWERS

42. ____________

43. ____________

44. ____________

45. ____________

46. ____________

29. $x^2 + y^2 - 4y = 12$
Center: (0, 2); radius: 4

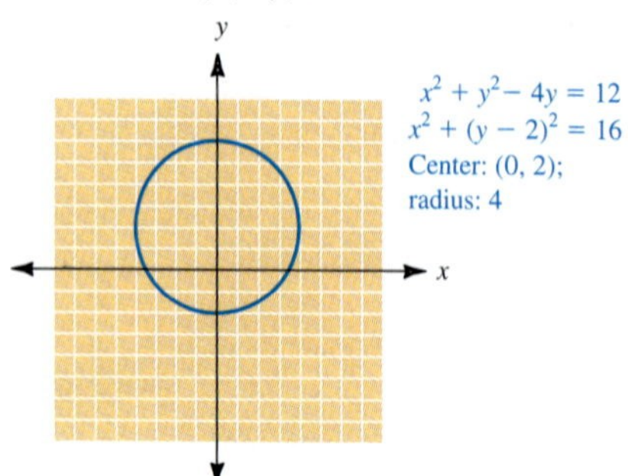

31. $x^2 - 4x + y^2 + 2y = -1$
Center: (2, −1); radius: 2

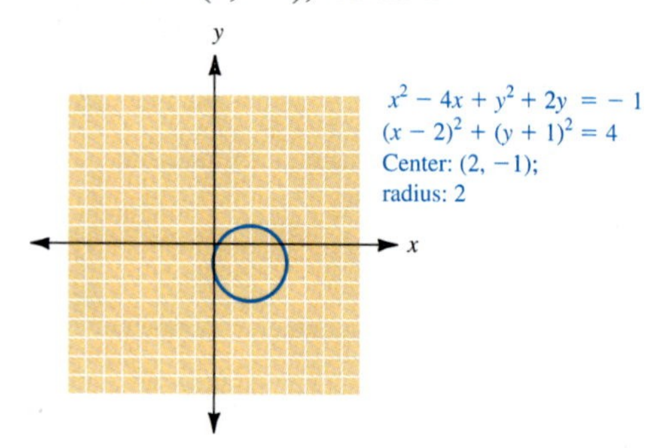

33. Circle with radius = 0; center at (1, 2)

35. $\sqrt{500} = 10\sqrt{5}$ cm $\cong 24.4$ cm

37. $x^2 + y^2 = \dfrac{4}{9}$

39. $x^2 + y^2 = 36$

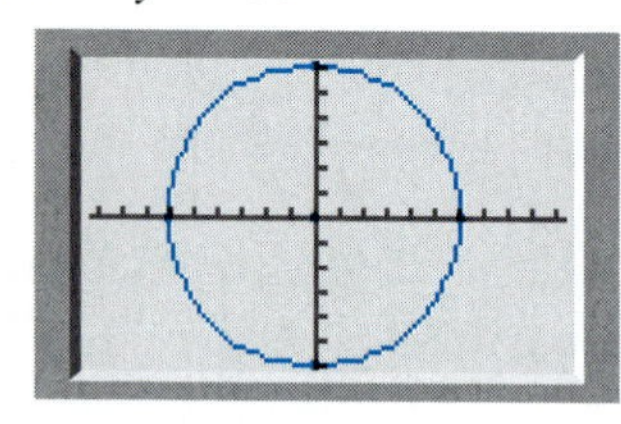

$y = \sqrt{36 - x^2}$

$y = -\sqrt{36 - x^2}$

41. $(x + 5)^2 + y^2 = 36$

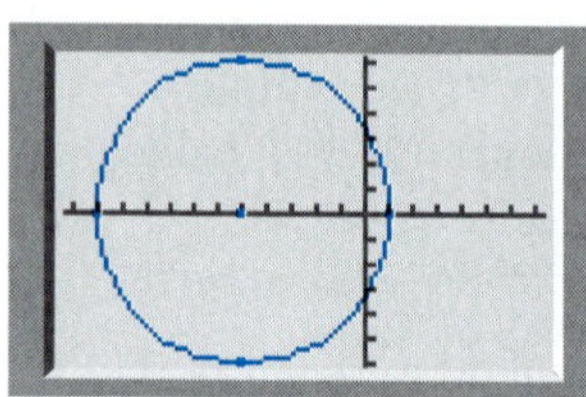

$y = \sqrt{36 - (x + 5)^2}$

$y = -\sqrt{36 - (x + 5)^2}$

43. Domain: $\{x | -7 \le x \le 1\}$
Range: $\{y | -2 \le y \le 6\}$

45. Domain: $\{x | -5 \le x \le 5\}$
Range: $\{y | -2 \le y \le 8\}$

10.3 The Ellipse and the Hyperbola

10.3 OBJECTIVES

1. Given an equation, graph an ellipse
2. Given an equation, graph a hyperbola

Let's turn now to the third conic section, the ellipse. It can be described as an "oval-shaped" curve and has the following geometric description.

NOTE Ellipses occur frequently in nature. The planets have elliptical orbits with the sun at one focus.

Definitions: Ellipse

An **ellipse** is the set of all points (x, y) such that the sum of the distances from (x, y) to two fixed points, called the *foci* of the ellipse, is constant.

NOTE The reflecting properties of the ellipse are also interesting. Rays from one focus are reflected by the ellipse in such a way that they always pass through the other focus.

The following sketch illustrates the definition in two particular cases:

1. When the foci are located on the x axis and are symmetric about the origin.
2. When the foci are located on the y axis and are symmetric about the origin.

NOTE In either case, $d_1 + d_2$ is constant.

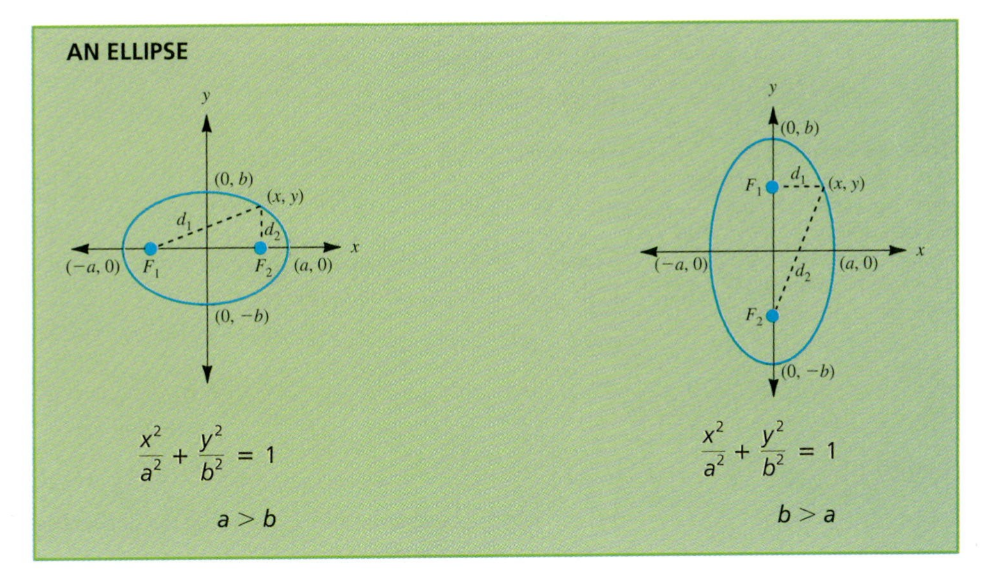

To sketch ellipses of these forms quickly, we need to determine only *four points,* the points where the ellipse intercepts the coordinate axes.

Fortunately those points are easily found when the ellipse is written in standard form:

$$\frac{x^2}{a^2} + \frac{y^2}{b^2} = 1 \tag{1}$$

NOTE To find the x intercepts, let $y = 0$ and solve for x. To find the y intercepts, let $x = 0$ and solve for y.

The x intercepts are $(a, 0)$ and $(-a, 0)$. The y intercepts are $(0, b)$ and $(0, -b)$. Let's use this information to sketch an ellipse.

Example 1

Graphing an Ellipse

Sketch the ellipse.

$$\frac{x^2}{9} + \frac{y^2}{4} = 1$$

Step 1 The equation is in standard form.

Step 2 Find the x intercepts. From equation 1, $a^2 = 9$, and the x intercepts are $(3, 0)$ and $(-3, 0)$.

Step 3 Find the y intercepts. From equation 1, $b^2 = 4$, and the y intercepts are $(0, 2)$ and $(0, -2)$.

Step 4 Plot the intercepts found above, and draw a smooth curve to form the desired ellipse.

CHECK YOURSELF 1

Sketch the ellipse.

$$\frac{x^2}{16} + \frac{y^2}{9} = 1$$

Example 2

Graphing an Ellipse

NOTE To recognize the equation as an ellipse in this form, note that the equation has both x^2 and y^2 terms. The coefficients of those terms have the *same* algebraic signs but *different* coefficients.

Sketch the ellipse with equation

$$9x^2 + 4y^2 = 36$$

Step 1 Because this equation is *not* in standard form (the right side is *not* 1), we divide both sides of the equation by the constant 36:

$$\frac{9x^2}{36} + \frac{4y^2}{36} = \frac{36}{36}$$

$$\frac{x^2}{4} + \frac{y^2}{9} = 1$$

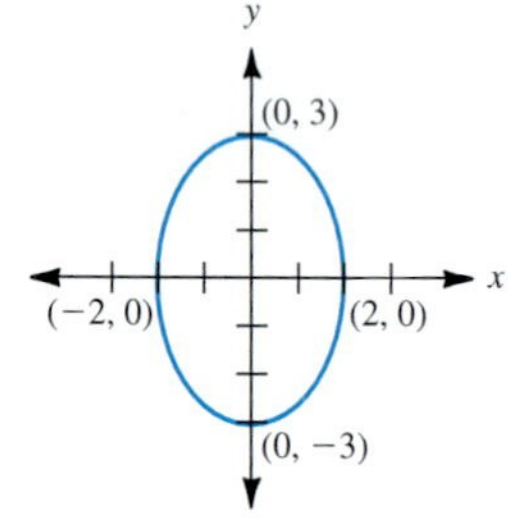

$9x^2 + 4y^2 = 36$

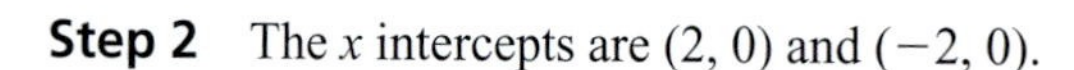

We can now proceed as before. Comparing the derived equation with that in standard form, we deduce steps 2 and 3.

Step 2 The x intercepts are $(2, 0)$ and $(-2, 0)$.

Step 3 The y intercepts are $(0, 3)$ and $(0, -3)$.

Step 4 We connect the intercepts with a smooth curve to complete the sketch of the ellipse.

CHECK YOURSELF 2

Sketch the ellipse.

$$25x^2 + 4y^2 = 100$$

Hint: First write the equation in standard form by dividing both sides of the equation by 100.

The following algorithm summarizes our work with graphing ellipses.

Step by Step: Graphing the Ellipse

Step 1 Write the given equation in standard form.
Step 2 From that standard form, determine the x intercepts.
Step 3 Also determine the y intercepts.
Step 4 Plot the four intercepts and connect the points with a smooth curve, to complete the sketch.

Our discussion will now turn to the last of the conic sections, the hyperbola. As you will see, the geometric description of the hyperbola (and hence the corresponding standard form) is quite similar to that of the ellipse.

Definitions: Hyperbola

A **hyperbola** is the set of all points (x, y) such that the absolute value of the differences from (x, y) to each of two fixed points, called the *foci* of the hyperbola, is constant.

The following sketch illustrates the definition in the case in which the foci are located on the x axis and are symmetric about the origin.

NOTE This is the first of two special cases we will investigate in this section.

NOTE The difference $|d_1 - d_2|$ remains constant for any point on the hyperbola.

THE HYPERBOLA

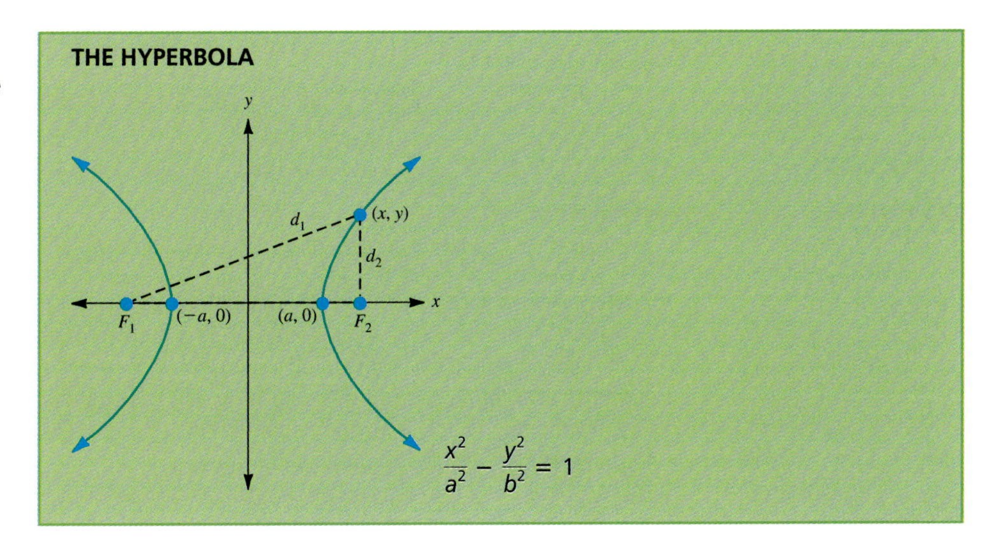

$$\frac{x^2}{a^2} - \frac{y^2}{b^2} = 1$$

Before we try to sketch a hyperbola from its equation, let's examine the standard form more carefully. For

$$\frac{x^2}{a^2} - \frac{y^2}{b^2} = 1 \qquad (2)$$

the graph is a hyperbola that opens to the right and left and is symmetric about the x axis. The points where this hyperbola intercepts the x axis are called the **vertices** of the hyperbola. The vertices of the hyperbola are located at $(a, 0)$ and at $(-a, 0)$.

As we move away from the center of the hyperbola, the **branches** of the hyperbola will approach two straight lines called the **asymptotes** of the hyperbola. The equations of the two asymptotes of the hyperbola are given by

NOTE Although we show these equations, you will see an easier method for finding the asymptotes in our first example.

$$y = \frac{b}{a}x \qquad \text{and} \qquad y = -\frac{b}{a}x$$

These asymptotes will prove to be extremely useful aids in sketching the hyperbola. In fact, for most purposes, the vertices and the asymptotes will be the only tools that we will need. The following example illustrates.

NOTE The equation of the hyperbola also has both x^2 and y^2 terms. Here the coefficients of those terms have *opposite* signs. If the x^2 coefficient is *positive*, the hyperbola will open *horizontally*.

Example 3

Graphing a Hyperbola

Sketch the hyperbola.

$$\frac{x^2}{9} - \frac{y^2}{4} = 1$$

Step 1 The equation is in standard form.

Step 2 Find and plot the vertices.

From the standard form we can see that $a^2 = 9$ and $a = 3$ or $a = -3$. The vertices of the hyperbola then occur at $(3, 0)$ and $(-3, 0)$.

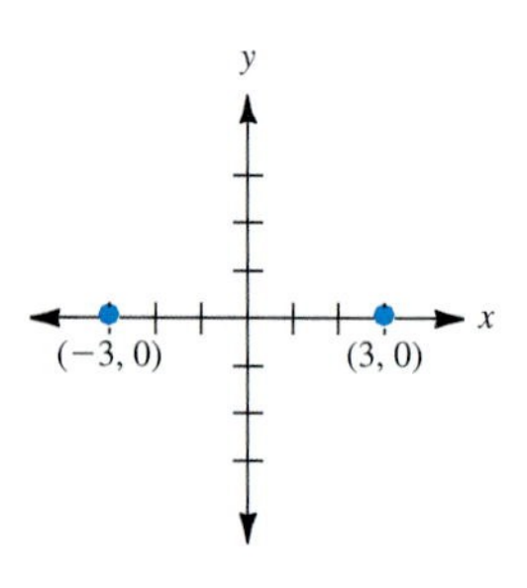

Step 3 Sketch the asymptotes.

Here is an easy way to sketch the asymptotes. Note from the standard form (2) that $b^2 = 4$, so $b = 2$ or $b = -2$. Plot the points $(0, 2)$ and $(0, -2)$ on the y axis.

Draw (using dashed lines) the rectangle whose sides are parallel to the x and y axes and that passes through the points determined in step 2 and step 3.

Draw the diagonals of the rectangle (again using dashed lines), and then extend those diagonals to form the desired asymptotes.

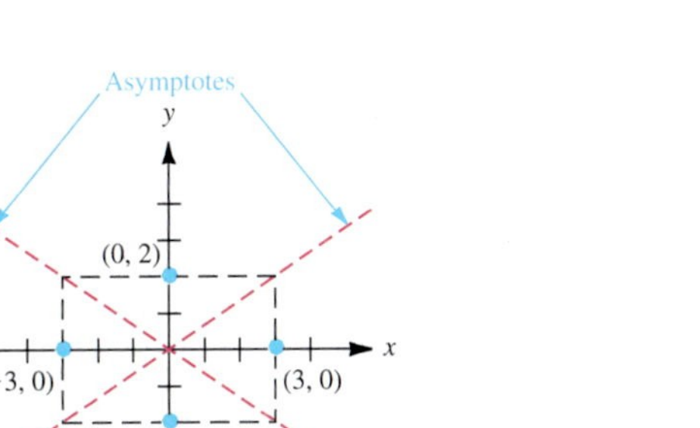

Step 4 Sketch the hyperbola.

We now complete our task by sketching the hyperbola as two smooth curves, passing through the vertices and approaching the asymptotes.

It is important to remember that the asymptotes are *not* a part of the graph. They are simply used as aids in sketching the graph as the branches get "closer and closer" to the lines.

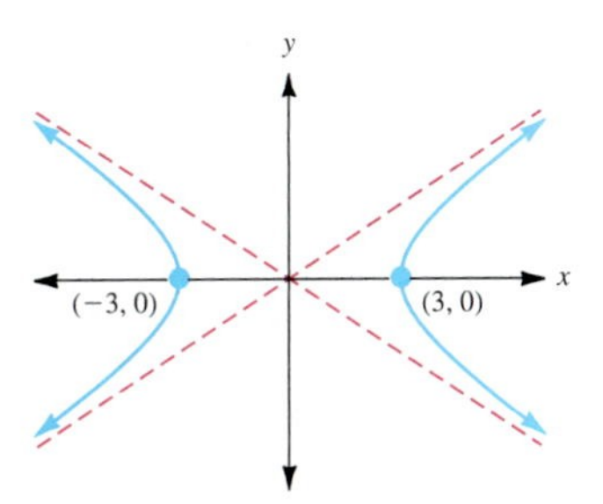

CHECK YOURSELF 3

Sketch the hyperbola.

$$\frac{x^2}{16} - \frac{y^2}{9} = 1$$

We now want to consider a second case of the hyperbola and its standard form. Suppose that the foci of the hyperbola are now on the y axis and symmetric about the origin. A sketch of such a hyperbola, and the equation that corresponds, follows.

THE HYPERBOLA

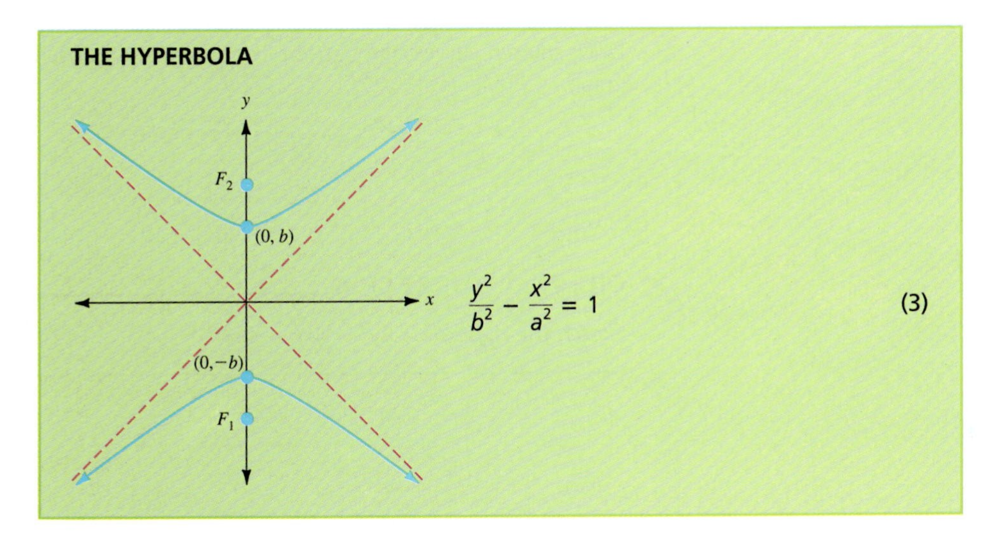

$$\frac{y^2}{b^2} - \frac{x^2}{a^2} = 1 \qquad (3)$$

Some observations about this case are in order:

NOTE Here the vertices are on the y axis.

1. The vertices of the hyperbola are now at $(0, b)$ and $(0, -b)$.
2. The asymptotes of the hyperbola have the equations

NOTE The asymptotes are the same as before.

$$y = \frac{b}{a}x \quad \text{and} \quad y = -\frac{b}{a}x$$

The following example illustrates sketching a hyperbola in this case.

Example 4

Graphing a Hyperbola

NOTE You can recognize this equation as corresponding to a hyperbola because the coefficients of the squared terms are *opposite* in sign. Because the y^2 coefficient is *positive*, the hyperbola will open *vertically*.

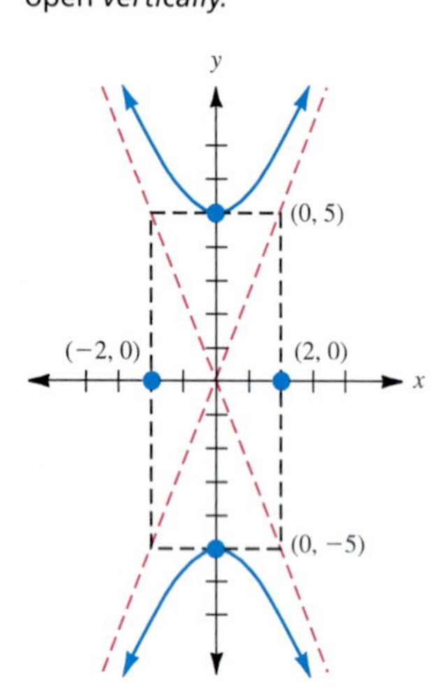

Sketch the hyperbola.

$4y^2 - 25x^2 = 100$

Step 1 Write the equation in the standard form (3) by dividing both sides by 100:

$$\frac{4y^2}{100} - \frac{25x^2}{100} = \frac{100}{100}$$

$$\frac{y^2}{25} - \frac{x^2}{4} = 1$$

Step 2 Find the vertices.

From the standard form (3), we see that because $b^2 = 25$, $b = 5$ or $b = -5$, so the vertices are at $(0, 5)$ and $(0, -5)$.

Step 3 Sketch the asymptotes.

Also from standard form (3), we see that because $a^2 = 4$, $a = 2$ or $a = -2$.

Plot $(2, 0)$ and $(-2, 0)$ on the x axis, and complete the dashed rectangle as before. The diagonals once again extend to form the asymptotes.

Step 4 Sketch the hyperbola.

Draw smooth curves, through the vertices, that approach the asymptotes to complete the graph.

CHECK YOURSELF 4

Sketch the hyperbola $9y^2 - 4x^2 = 36$.

The following algorithm summarizes our work with sketching hyperbolas.

Step by Step: Graphing the Hyperbola

Step 1 Write the given equation in standard form.

Step 2 Determine the vertices of the hyperbola.

If the x^2 coefficient is positive, the vertices are at $(a, 0)$ and $(-a, 0)$ on the x axis.

If the y^2 coefficient is positive, the vertices are at $(0, b)$ and $(0, -b)$ on the y axis.

Step 3 Sketch the asymptotes of the hyperbola.

Plot points $(a, 0)$, $(-a, 0)$, $(0, b)$, and $(0, -b)$. Form a rectangle from these points. The diagonals (extended) are the asymptotes of the hyperbola.

Step 4 Sketch the hyperbola.

Draw smooth curves through the vertices and approaching the asymptotes.

The following chart shows all the conic section equation forms.

Curve	Example	Recognizing the Curve
Line	$4x - 3y = 12$	The equation involves x and/or y to the first power only.
Parabola	$y = x^2 - 3x$ or $x = y^2 - 2y + 3$	Only one term, in x or in y, may be squared. The other variable appears to the first power.
Circle	$x^2 + 4x + y^2 = 5$	The equation has both x^2 and y^2 terms. The coefficients of those terms are equal.
Ellipse	$4x^2 + 9y^2 = 36$	The equation has both x^2 and y^2 terms. The coefficients of those terms have the same algebraic sign but different values.
Hyperbola	$4x^2 - 9y^2 = 36$ or $9y^2 - 16x^2 = 144$	The equation has both x^2 and y^2 terms. The coefficients of those terms have different algebraic signs.

CHECK YOURSELF ANSWERS

1. $\frac{x^2}{16} + \frac{y^2}{9} = 1$

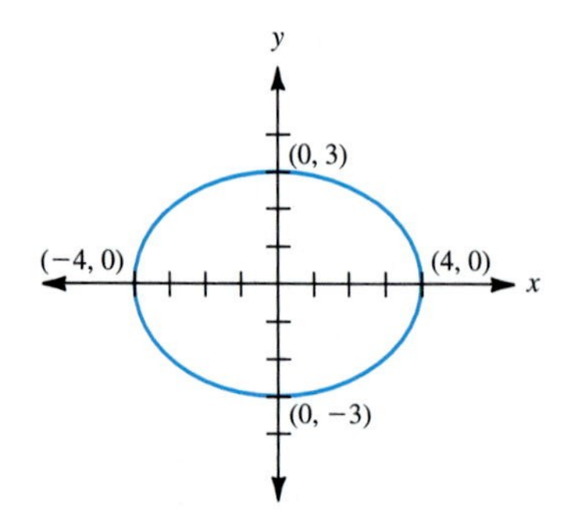

2. $\frac{x^2}{4} + \frac{y^2}{25} = 1$

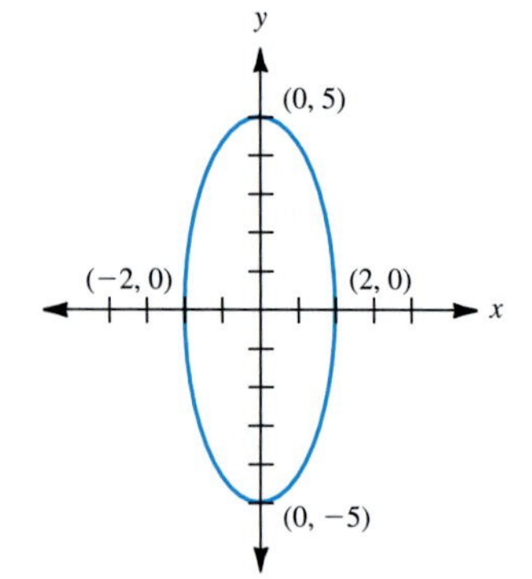

3. $\frac{x^2}{16} - \frac{y^2}{9} = 1$

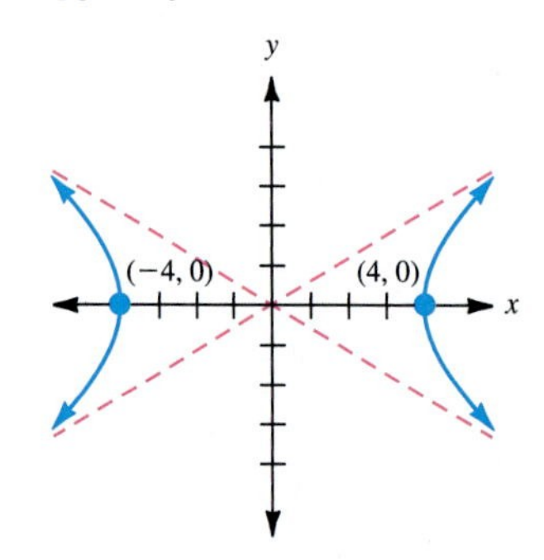

4. $\frac{y^2}{4} - \frac{x^2}{9} = 1$

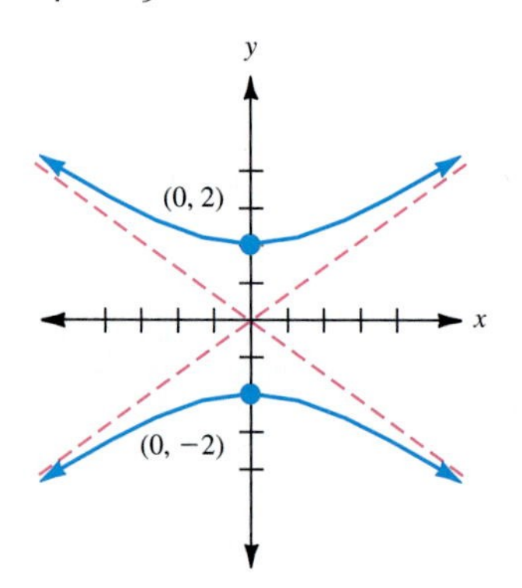

10.3 Exercises

Match each of the curves shown with the appropriate equation below.

1.

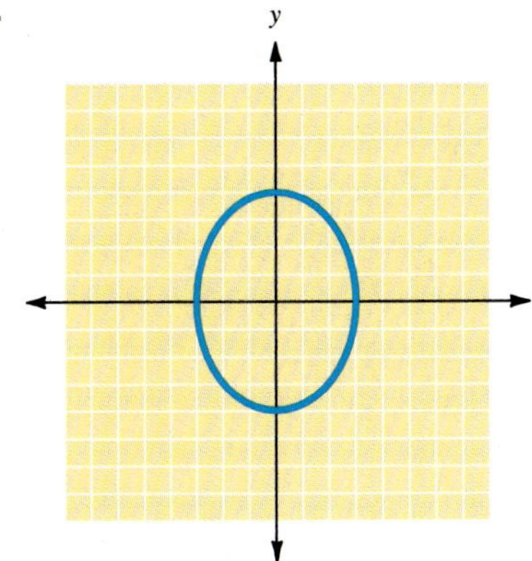

2.

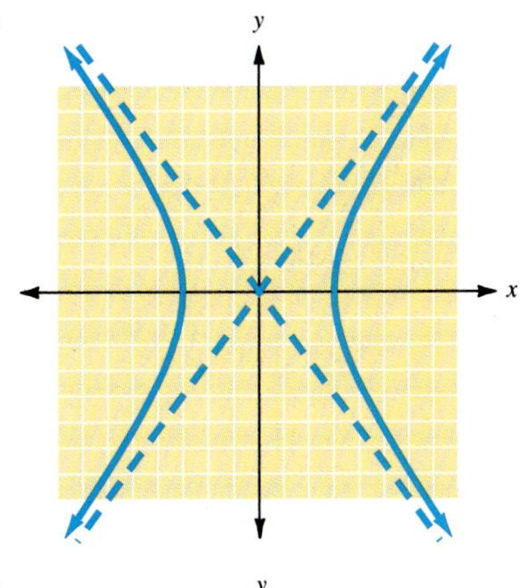

3.

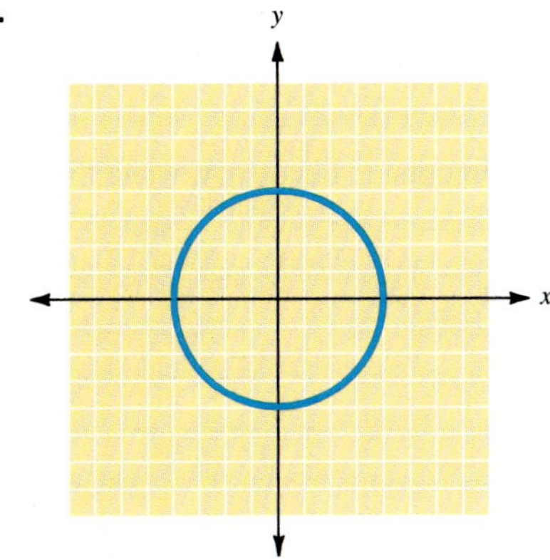

4.

5.

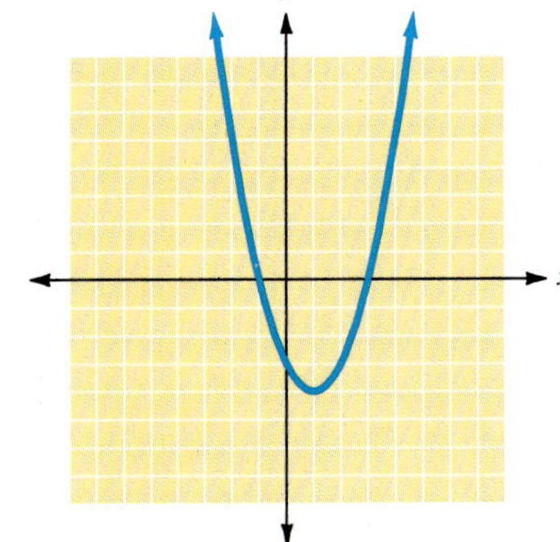

6.

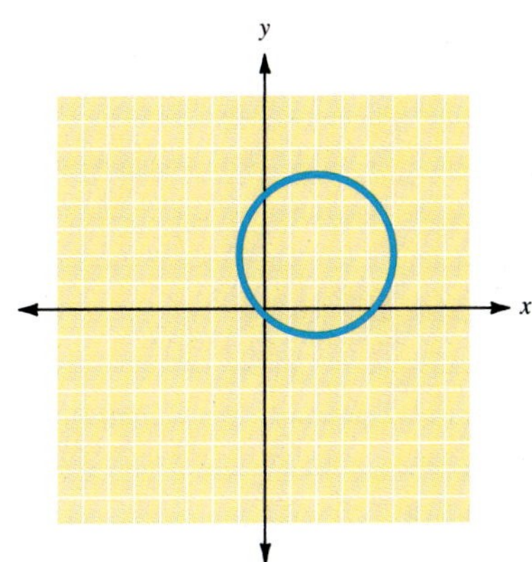

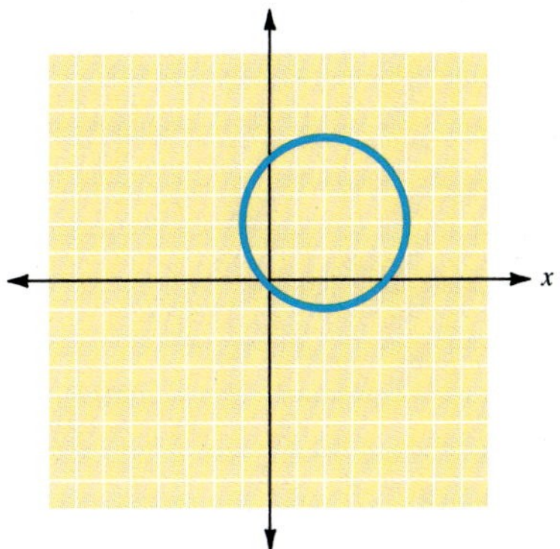

7.

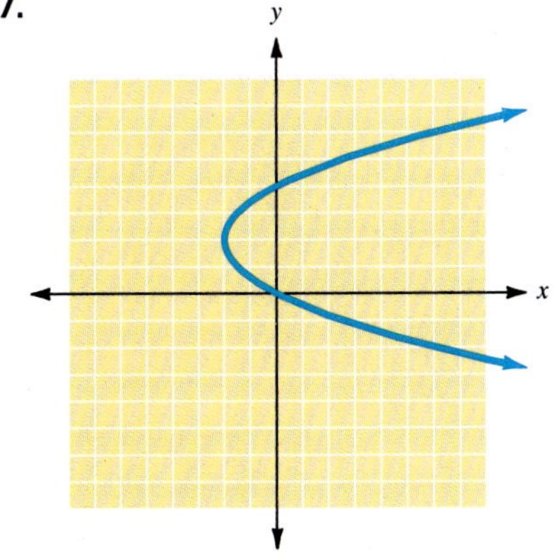

8.

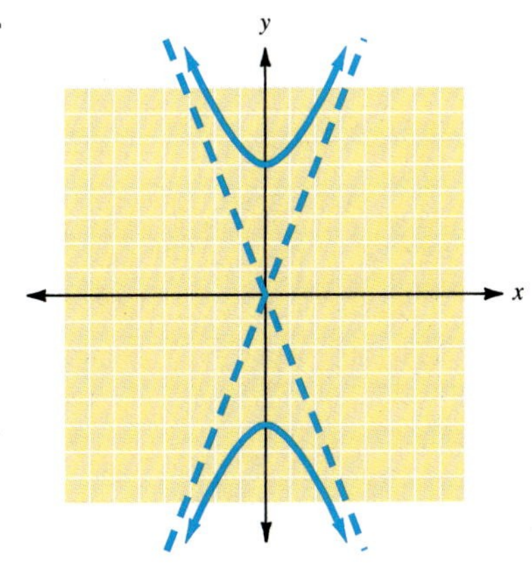

ANSWERS

1. ____________
2. ____________
3. ____________
4. ____________
5. ____________
6. ____________
7. ____________
8. ____________

Equations

(a) $4x^2 + 25y^2 = 100$

(b) $y = x^2 - 2x - 3$

(c) $x = \frac{1}{2}y^2 - 2y$

(d) $\frac{x^2}{9} + \frac{y^2}{16} = 1$

(e) $\frac{y^2}{25} - \frac{x^2}{4} = 1$

(f) $16x^2 - 9y^2 = 144$

(g) $(x - 2)^2 + (y - 2)^2 = 9$

(h) $x^2 + y^2 = 16$

ANSWERS

9. ________
10. ________
11. ________
12. ________
13. ________
14. ________
15. ________
16. ________
17. ________
18. ________
19. ________
20. ________
21. ________
22. ________
23. ________
24. ________

Identify the graph of each of the following equations as one of the conic sections (the parabola, circle, ellipse, or hyperbola).

9. $x^2 + y^2 = 16$

10. $\frac{x^2}{4} - \frac{y^2}{16} = 1$

11. $y = x^2 - 4$

12. $\frac{x^2}{16} + \frac{y^2}{9} = 1$

13. $9x^2 - 4y^2 = 36$

14. $x^2 = 4y$

15. $y^2 - 4x^2 = 4$

16. $x = y^2 - 2y + 1$

17. $x^2 - 6x + y^2 + 2x = 2$

18. $4x^2 + 25y^2 = 100$

19. $9y^2 - 16x^2 = 144$

20. $y = x^2 - 6x + 8$

Graph the following ellipses by finding the x and y intercepts. If necessary, write the equation in standard form.

21. $\frac{x^2}{4} + \frac{y^2}{9} = 1$

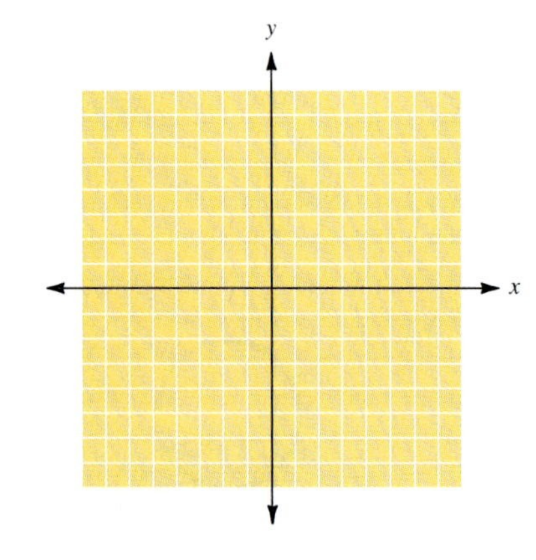

22. $\frac{x^2}{16} + \frac{y^2}{9} = 1$

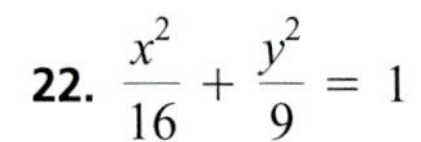

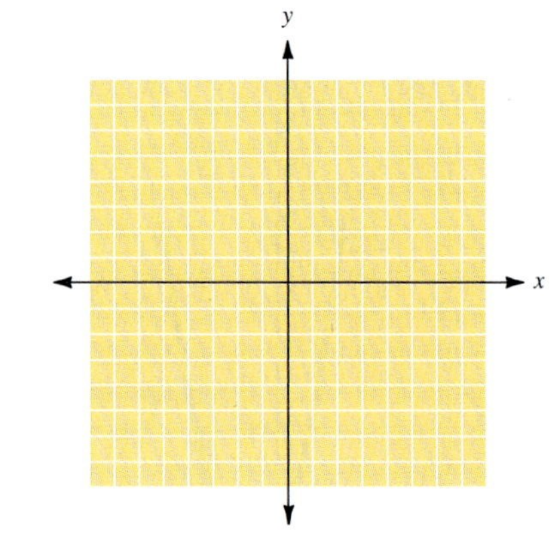

23. $\frac{x^2}{9} + \frac{y^2}{25} = 1$

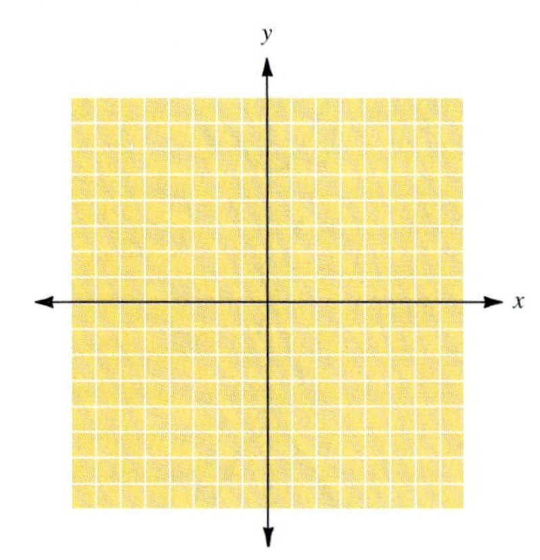

24. $\frac{x^2}{36} + \frac{y^2}{16} = 1$

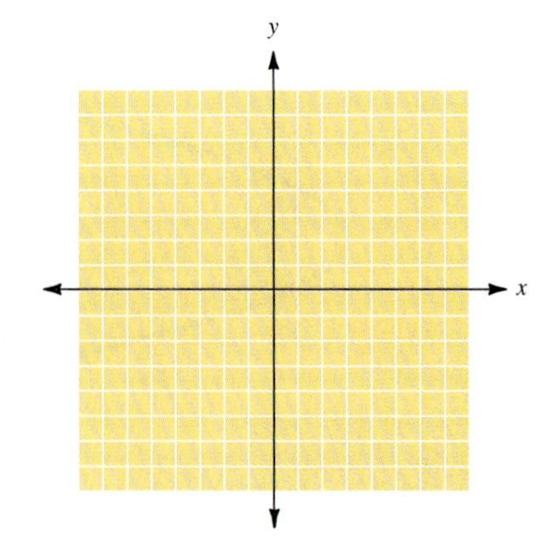

25. $x^2 + 9y^2 = 36$

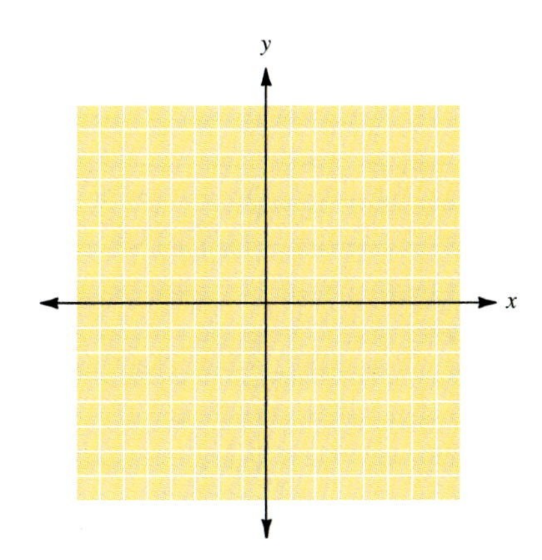

26. $4x^2 + y^2 = 16$

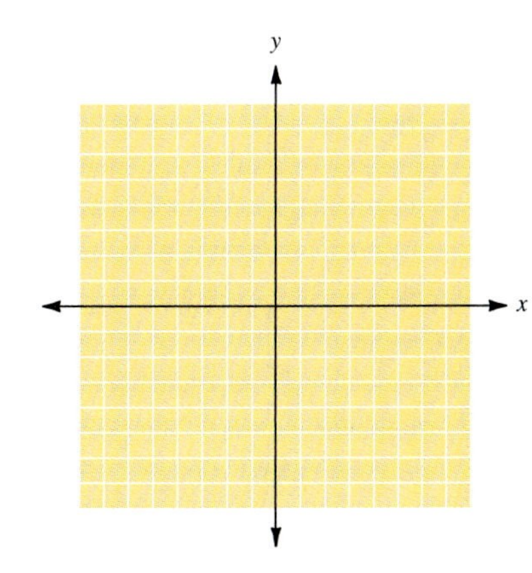

27. $4x^2 + 9y^2 = 36$

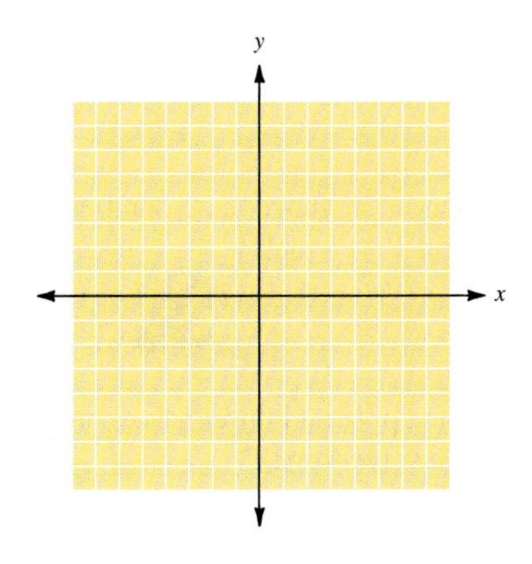

28. $25x^2 + 4y^2 = 100$

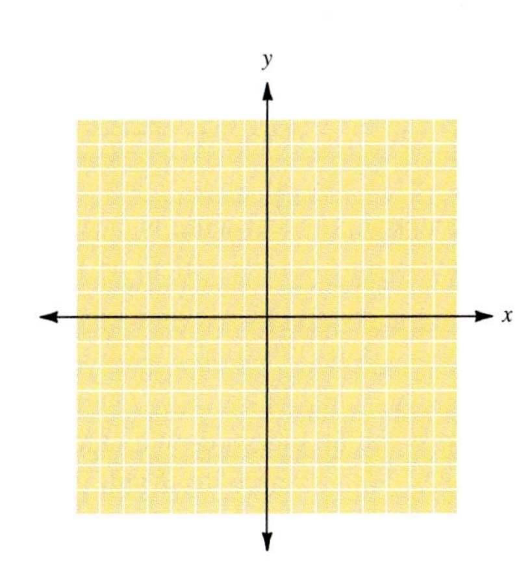

29. $4x^2 + 25y^2 = 100$

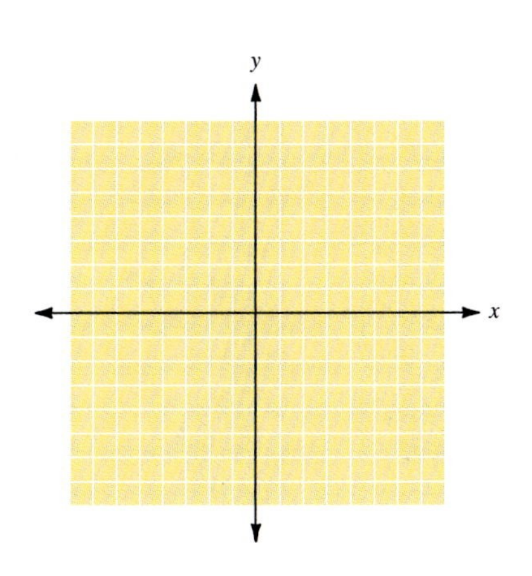

30. $9x^2 + 16y^2 = 144$

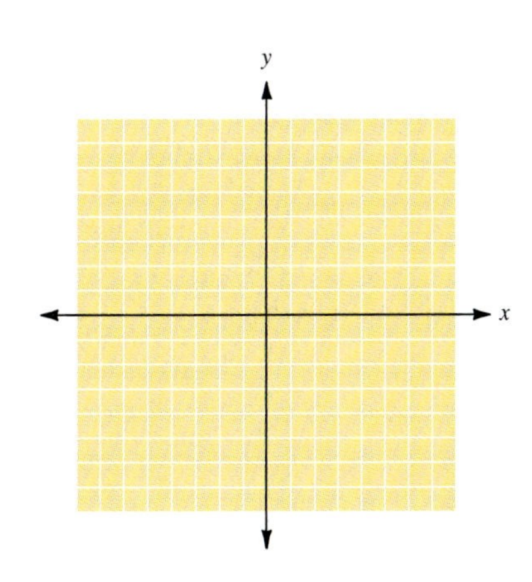

ANSWERS

25. ______

26. ______

27. ______

28. ______

29. ______

30. ______

ANSWERS

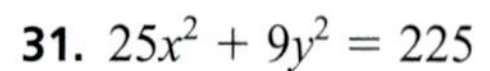

31. ______

32. ______

33. ______

34. ______

35. ______

36. ______

31. $25x^2 + 9y^2 = 225$

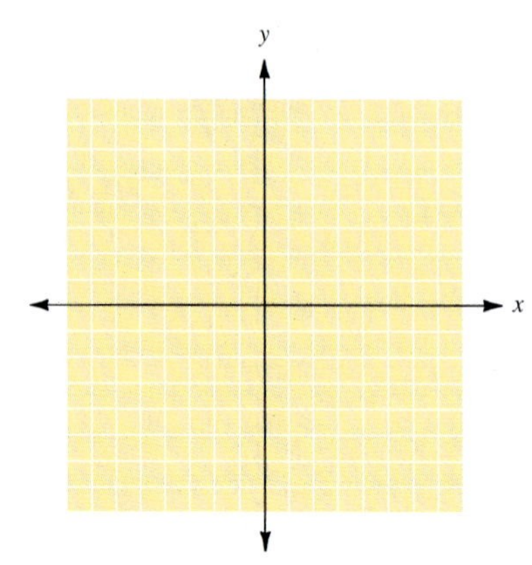

32. $16x^2 + 9y^2 = 144$

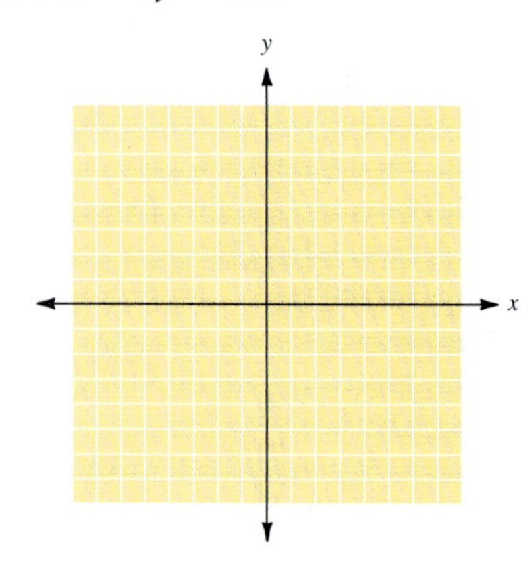

33. A semielliptical archway over a one-way road has a height of 10 ft and a width of 40 ft (see the figure). Will a truck that is 10 ft wide and 9 ft high clear the opening of the highway?

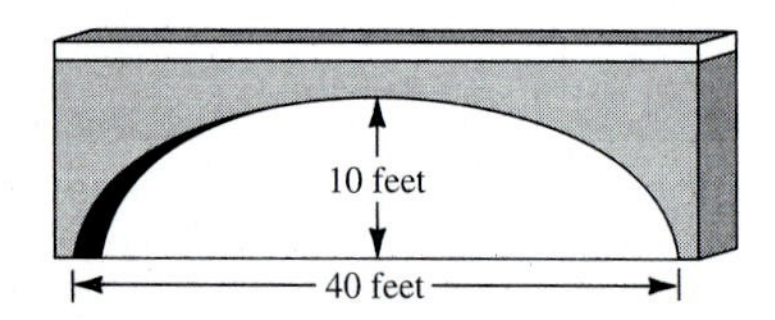

34. A truck that is 8 ft wide is carrying a load that reaches 7 ft above the ground. Will the truck clear a semielliptical arch that is 10 ft high and 30 ft wide?

An ellipse can be graphed on a calculator by plotting the upper and lower halves on the same axes. For example, to graph $9x^2 + 16y^2 = 144$, we solve for y:

$$Y_1 = \frac{\sqrt{144 - 9x^2}}{4} \quad \text{and} \quad Y_2 = -\frac{\sqrt{144 - 9x^2}}{4}$$

Then graph each equation as a separate function.

In exercises 35 to 38, use this technique to graph each ellipse.

35. $4x^2 + 16y^2 = 64$

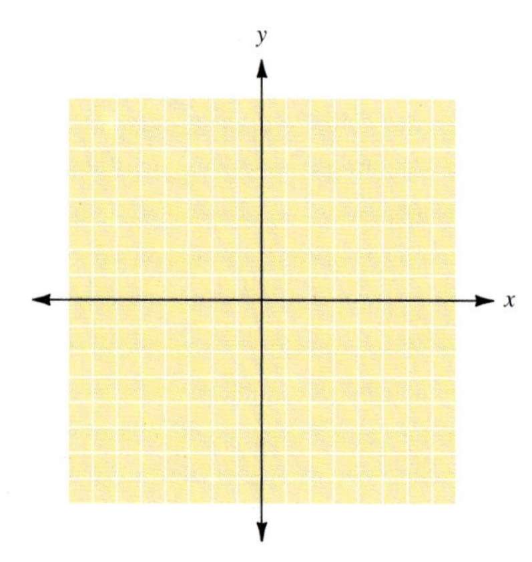

36. $9x^2 + 36y^2 = 324$

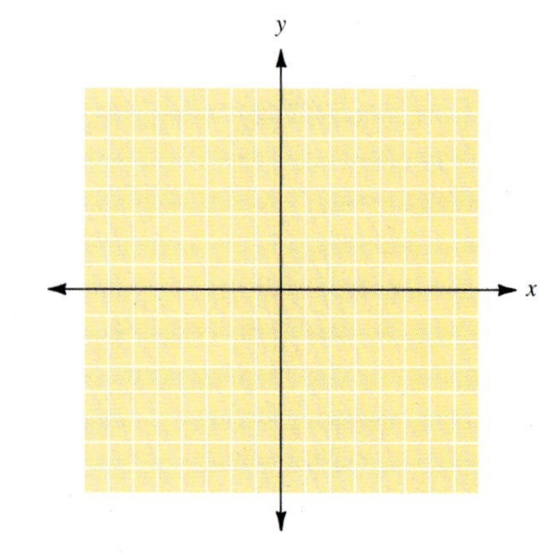

37. $25x^2 + 9y^2 = 225$

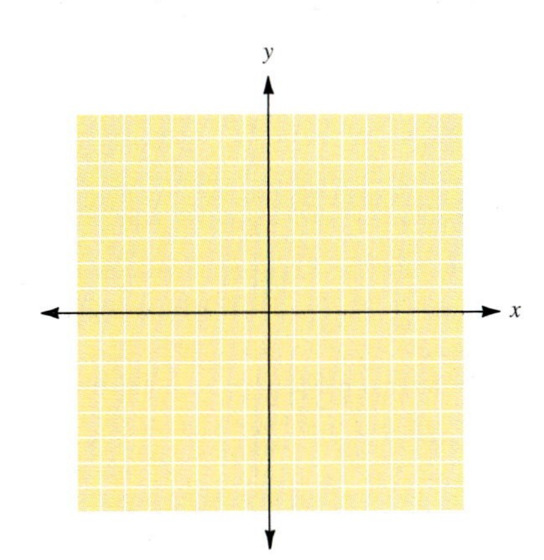

38. $4x^2 + 9y^2 = 36$

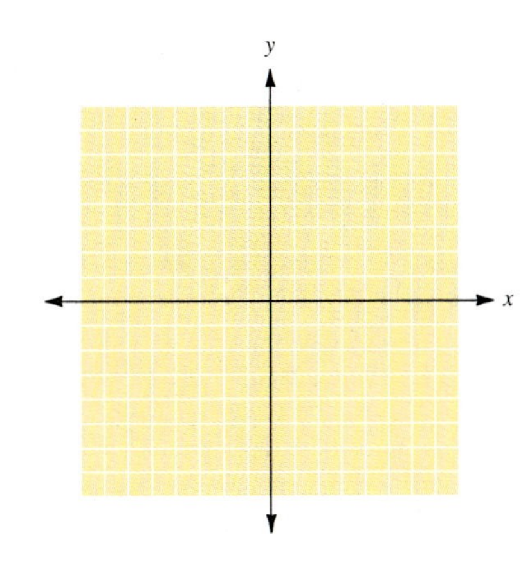

Graph the following hyperbolas by finding the vertices and asymptotes. If necessary, write the equation in standard form.

39. $\frac{x^2}{9} - \frac{y^2}{9} = 1$

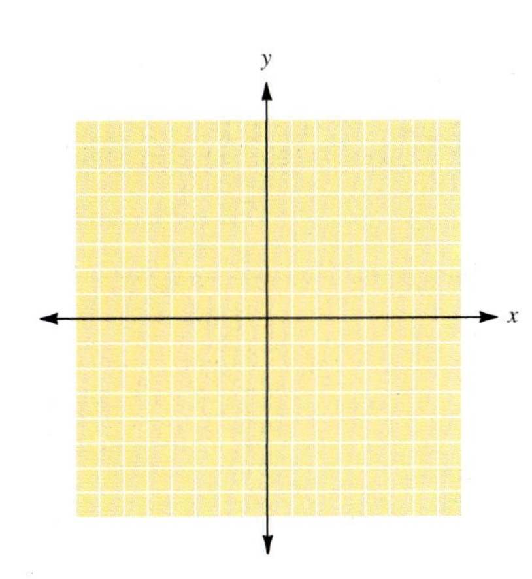

40. $\frac{y^2}{9} - \frac{x^2}{4} = 1$

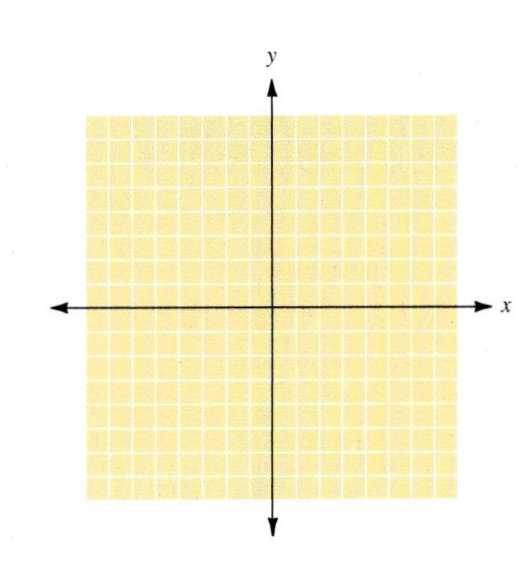

41. $\frac{y^2}{16} - \frac{x^2}{9} = 1$

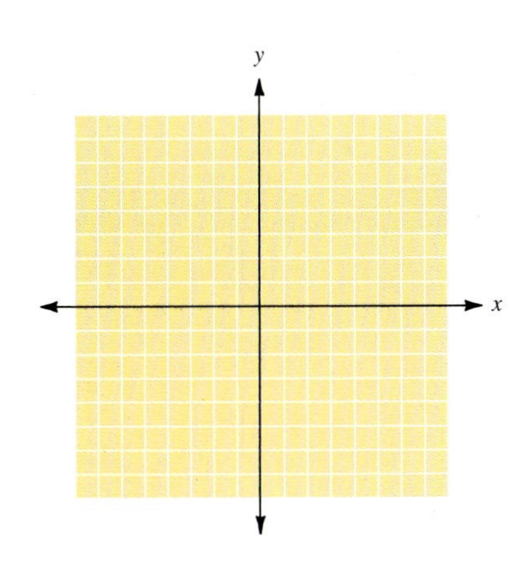

42. $\frac{x^2}{25} - \frac{y^2}{16} = 1$

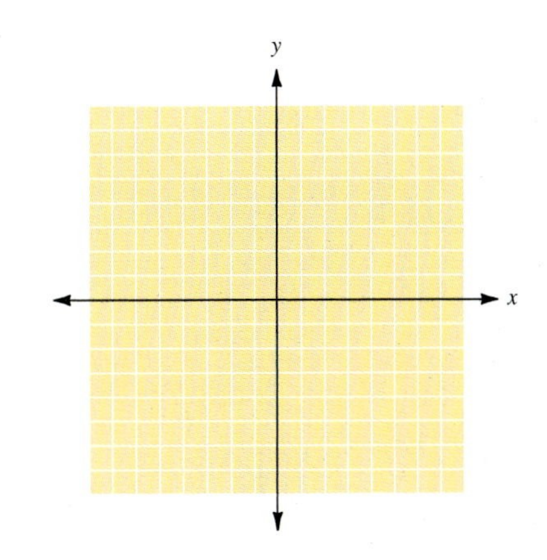

ANSWERS

37. ______

38. ______

39. ______

40. ______

41. ______

42. ______

ANSWERS

43. ____________________

44. ____________________

45. ____________________

46. ____________________

47. ____________________

48. ____________________

43. 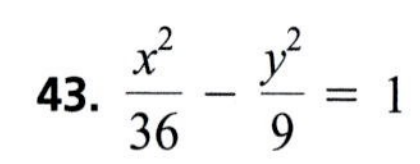 $\dfrac{x^2}{36} - \dfrac{y^2}{9} = 1$

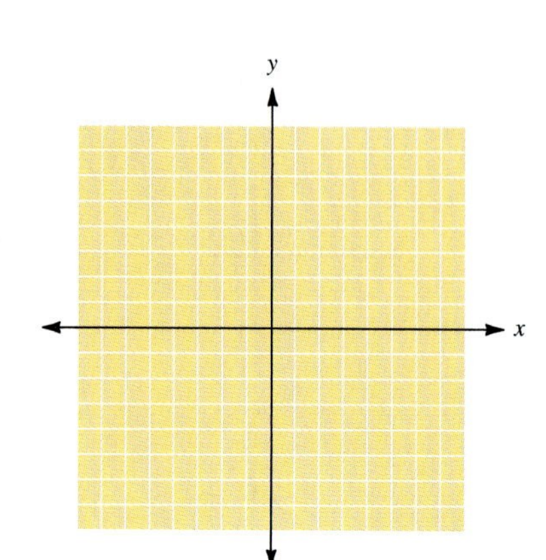

44. $\dfrac{y^2}{25} - \dfrac{x^2}{9} = 1$

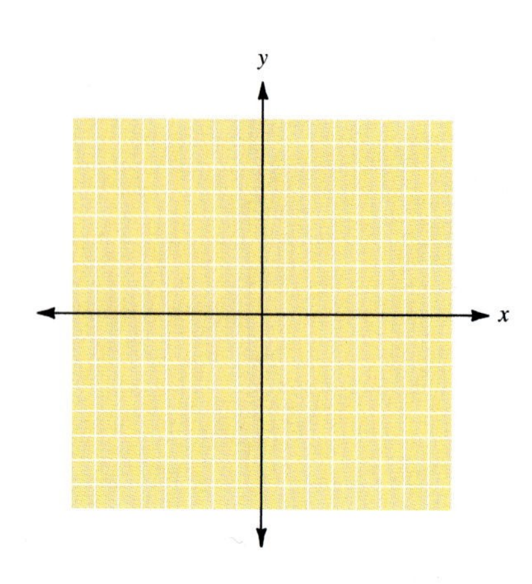

45. $x^2 - 9y^2 = 36$

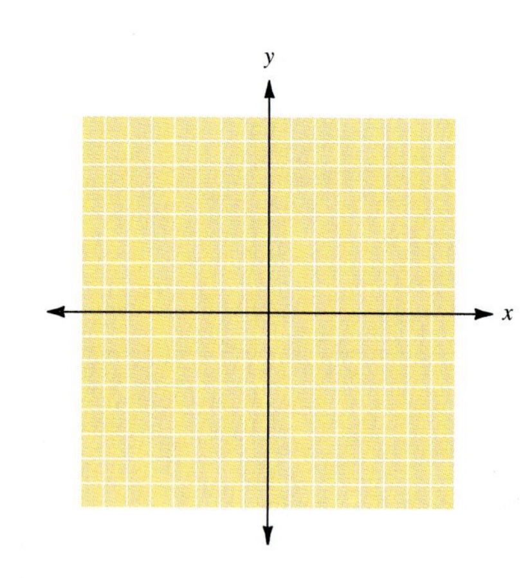

46. $y^2 - 4x^2 = 36$

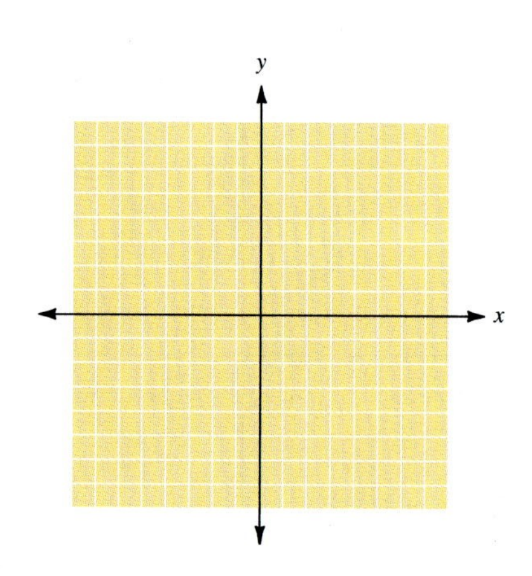

47. $9x^2 - 4y^2 = 36$

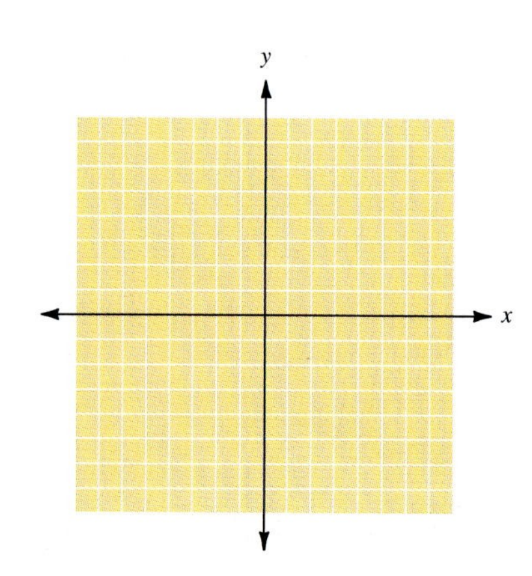

48. $9y^2 - 4x^2 = 36$

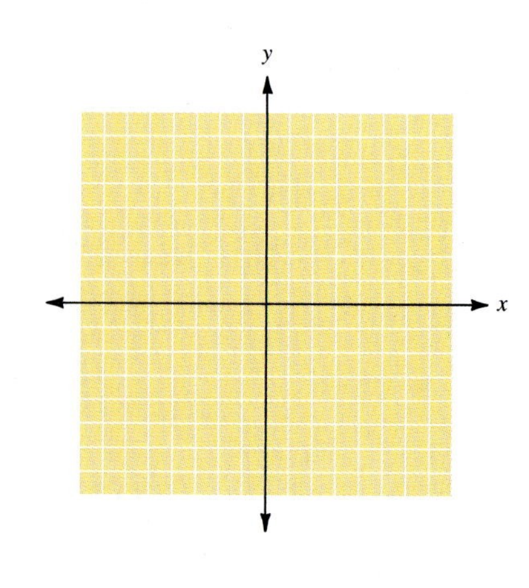

49. $25y^2 - 9x^2 = 225$

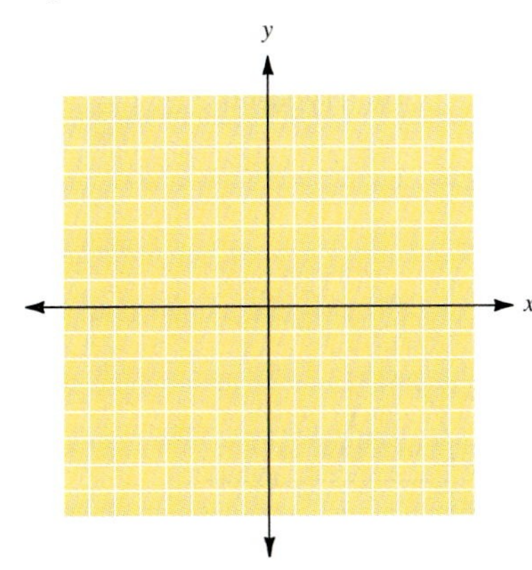

50. $4x^2 - 9y^2 = 36$

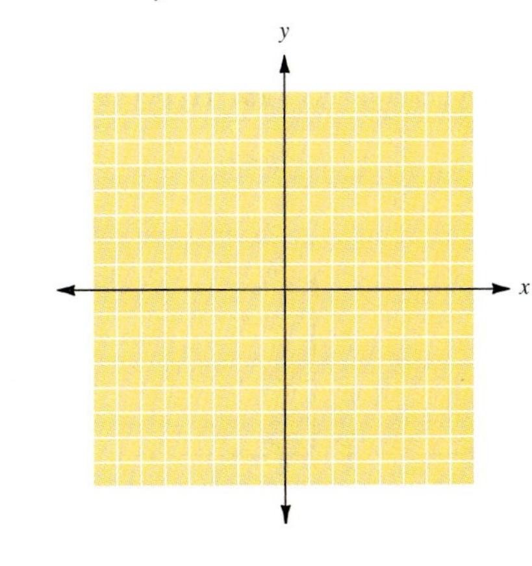

51. $25y^2 - 4x^2 = 100$

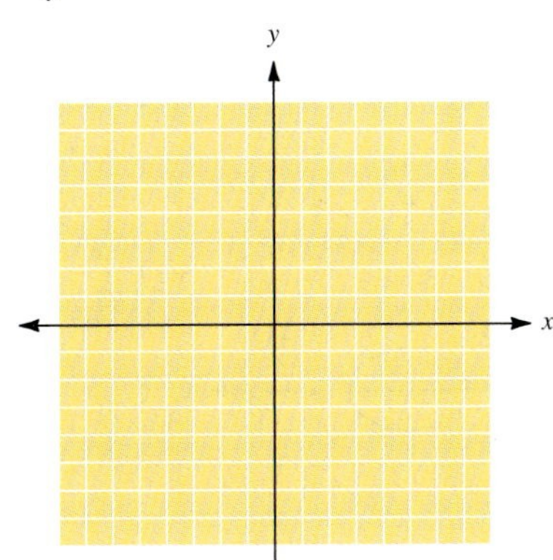

52. $9x^2 - 25y^2 = 225$

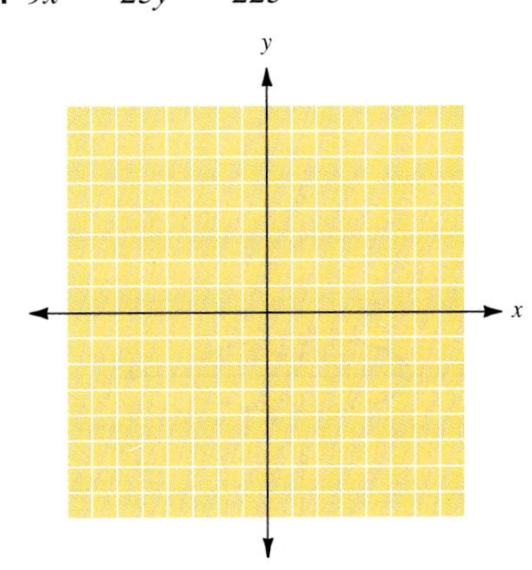

In exercises 53 to 56, use a graphing utility to graph each equation.

53. $y^2 - 16x^2 = 16$

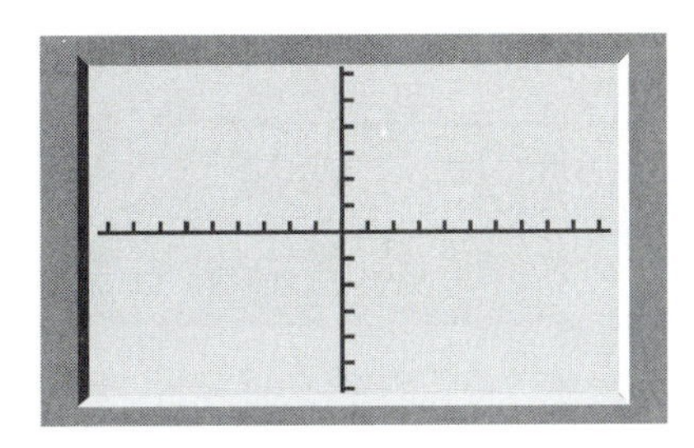

54. $y^2 - 25x^2 = 25$

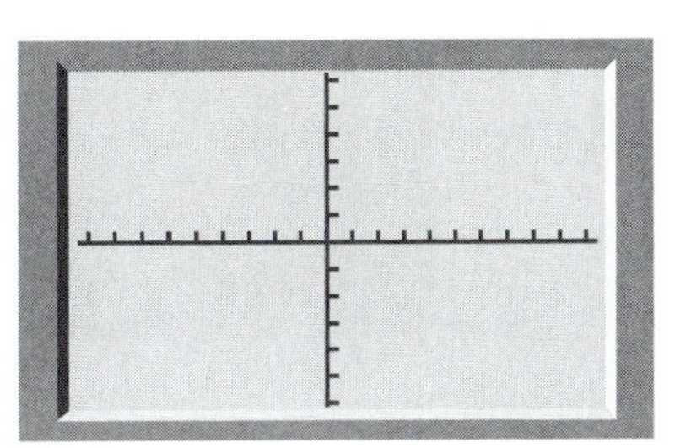

55. $16x^2 - 9y^2 = 144$

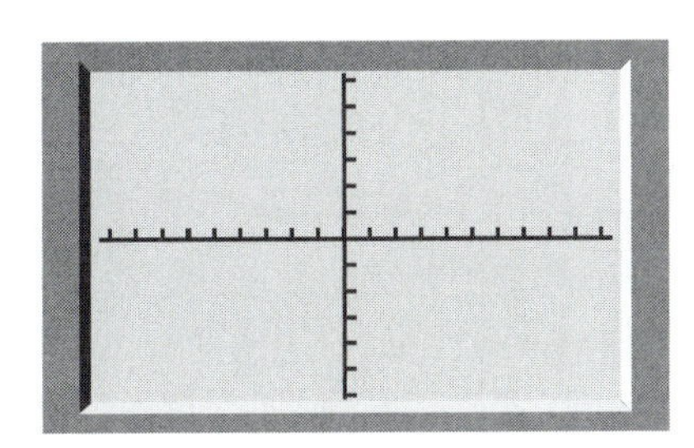

56. $25x^2 - 16y^2 = 400$

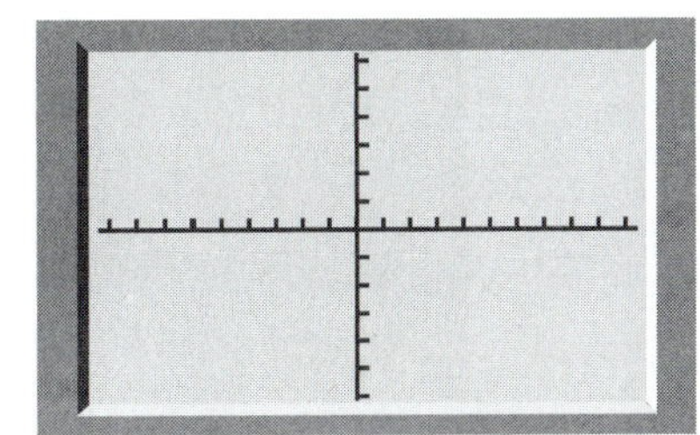

ANSWERS

49. ________

50. ________

51. ________

52. ________

53. ________

54. ________

55. ________

56. ________

Answers

1. (d) **3.** (h) **5.** (b) **7.** (c) **9.** Circle **11.** Parabola
13. Hyperbola **15.** Hyperbola **17.** Circle **19.** Hyperbola

21. $\frac{x^2}{4} + \frac{y^2}{9} = 1$

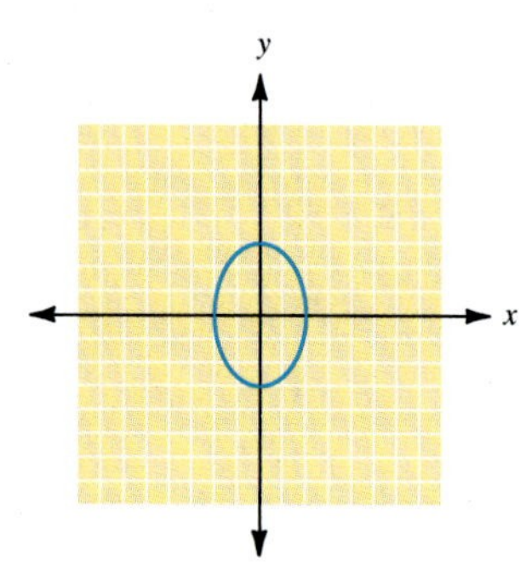

23. $\frac{x^2}{9} + \frac{y^2}{25} = 1$

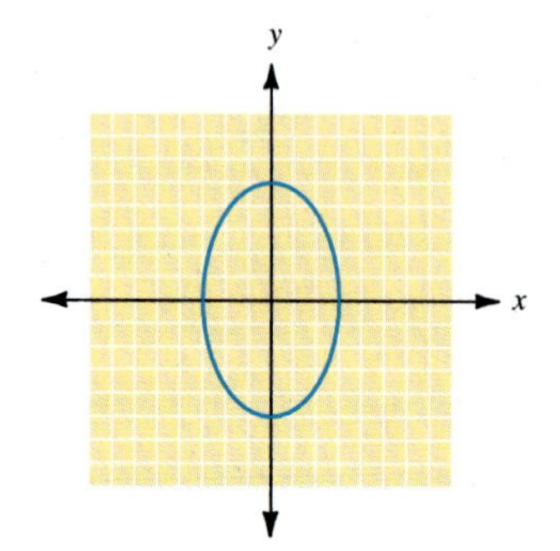

25. $x^2 + 9y^2 = 36$

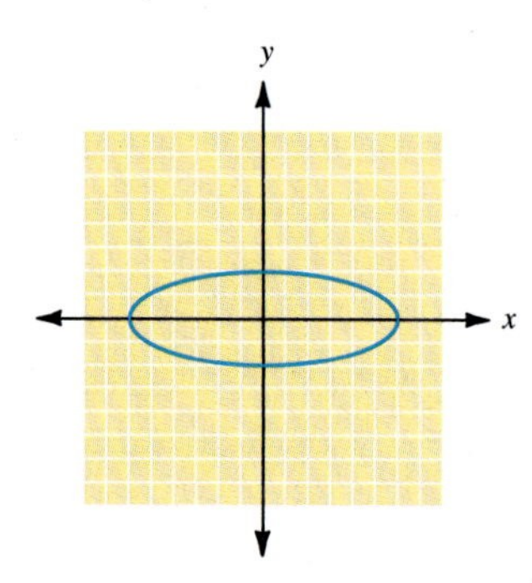

27. $4x^2 + 9y^2 = 36$

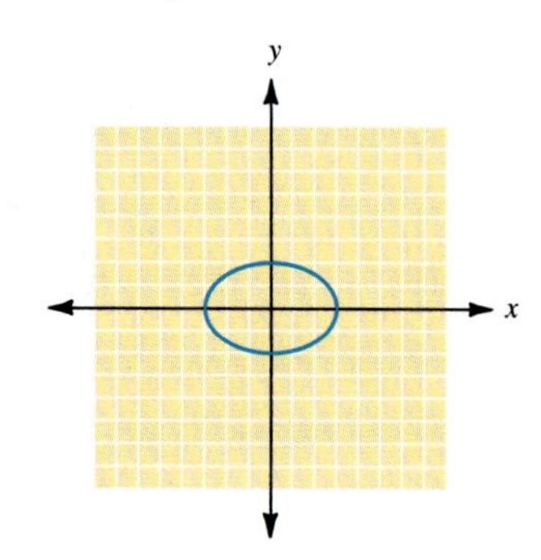

29. $4x^2 + 25y^2 = 100$

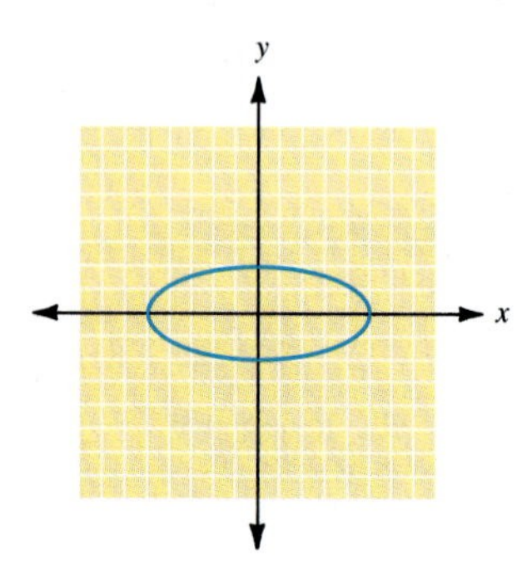

31. $25x^2 + 9y^2 = 225$

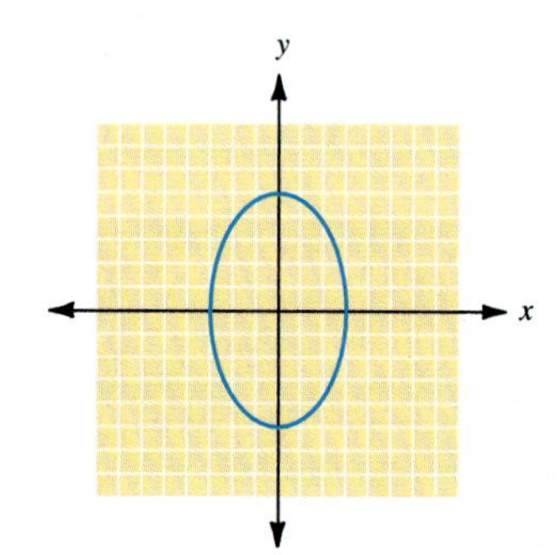

33. Yes

35. $4x^2 + 16y^2 = 64$

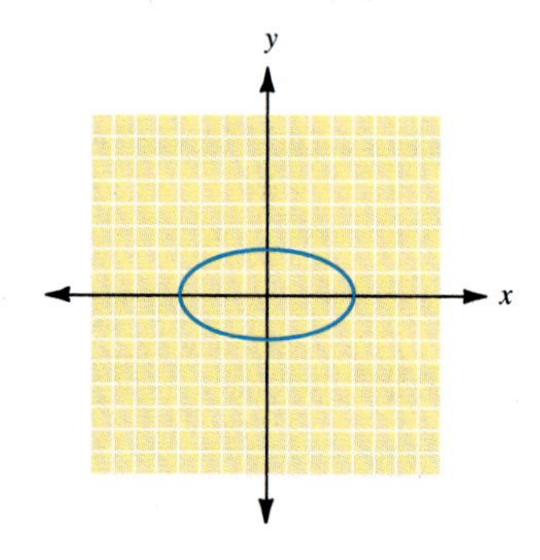

37. $25x^2 + 9y^2 = 225$

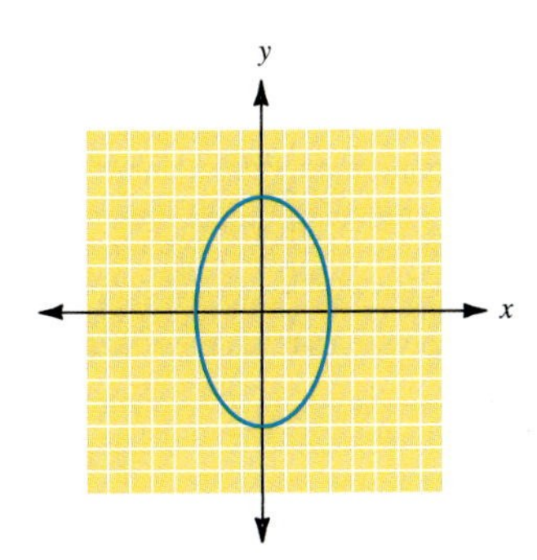

39. $\frac{x^2}{9} - \frac{y^2}{9} = 1$

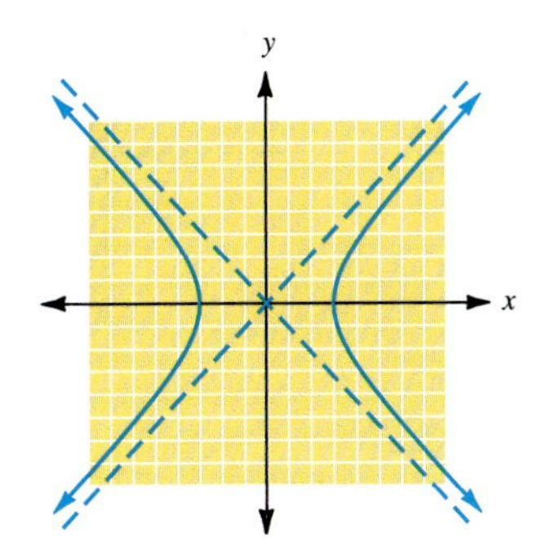

41. $\frac{y^2}{16} - \frac{x^2}{9} = 1$

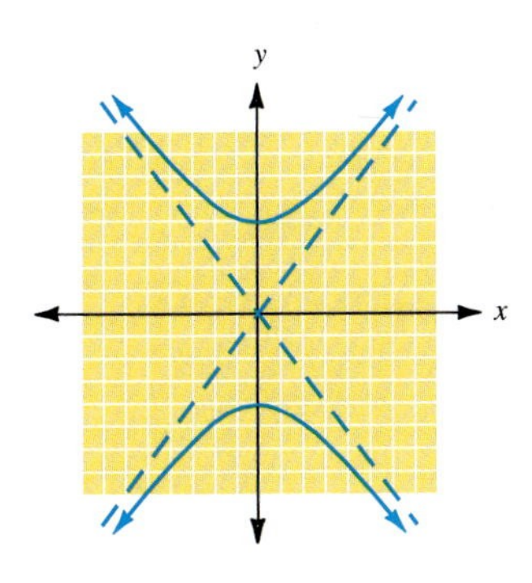

43. $\frac{x^2}{36} - \frac{y^2}{9} = 1$

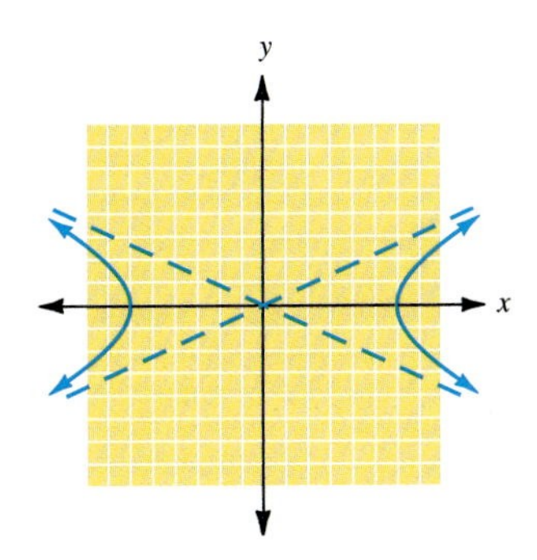

45. $x^2 - 9y^2 = 36$

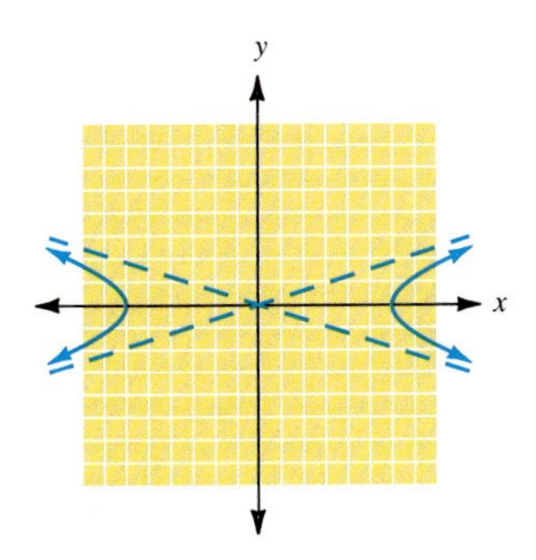

47. $9x^2 - 4y^2 = 36$

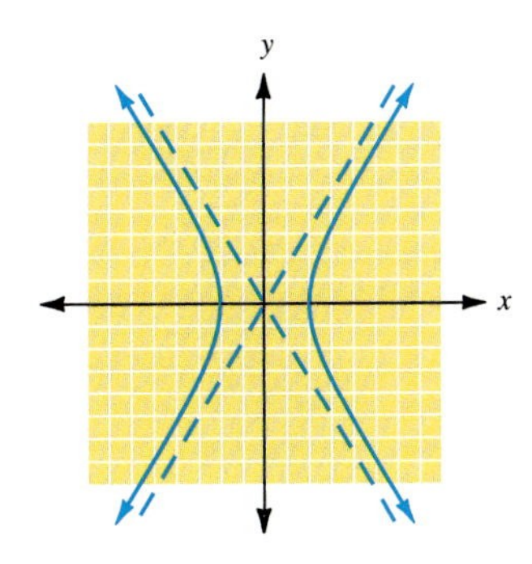

49. $16y^2 - 9x^2 = 144$

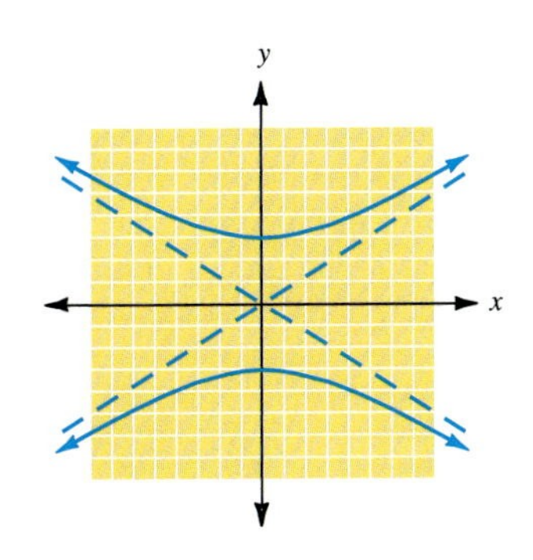

51. $25y^2 - 4x^2 = 100$

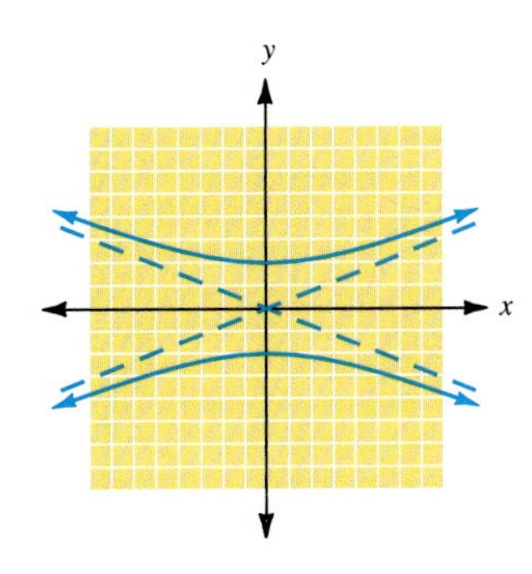

53. $y^2 - 16x^2 = 16$

55. $16x^2 - 9y^2 = 144$

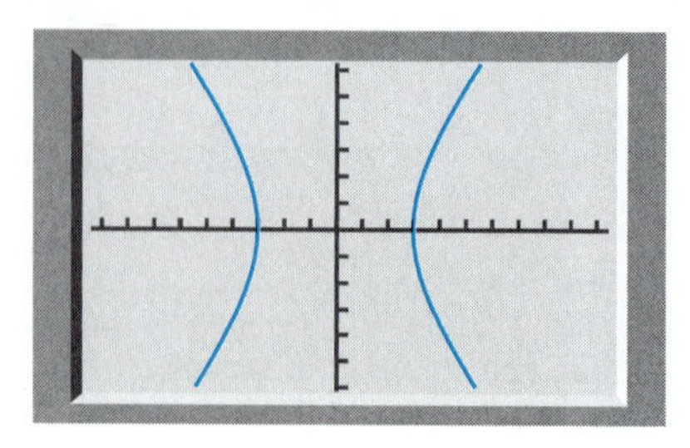

Graphing Nonlinear Systems

OBJECTIVES

1. Graph a system of nonlinear equations
2. Find ordered pairs associated with the solution set of a nonlinear system
3. Graph a system of nonlinear inequalities
4. Use substitution to find the solution set for a nonlinear system.
5. Identify the solution set of a system of nonlinear inequalities

In Section 5.1, we solved a system of linear equations by graphing the lines corresponding to those equations, and then recording the point of intersection. That point represented the solution to the system of equations. We will use a similar method to find the solution set for a nonlinear system. A system with two or more conic curves can have zero, one, two, three, or four solutions. The following graphs represent each of those possibilities.

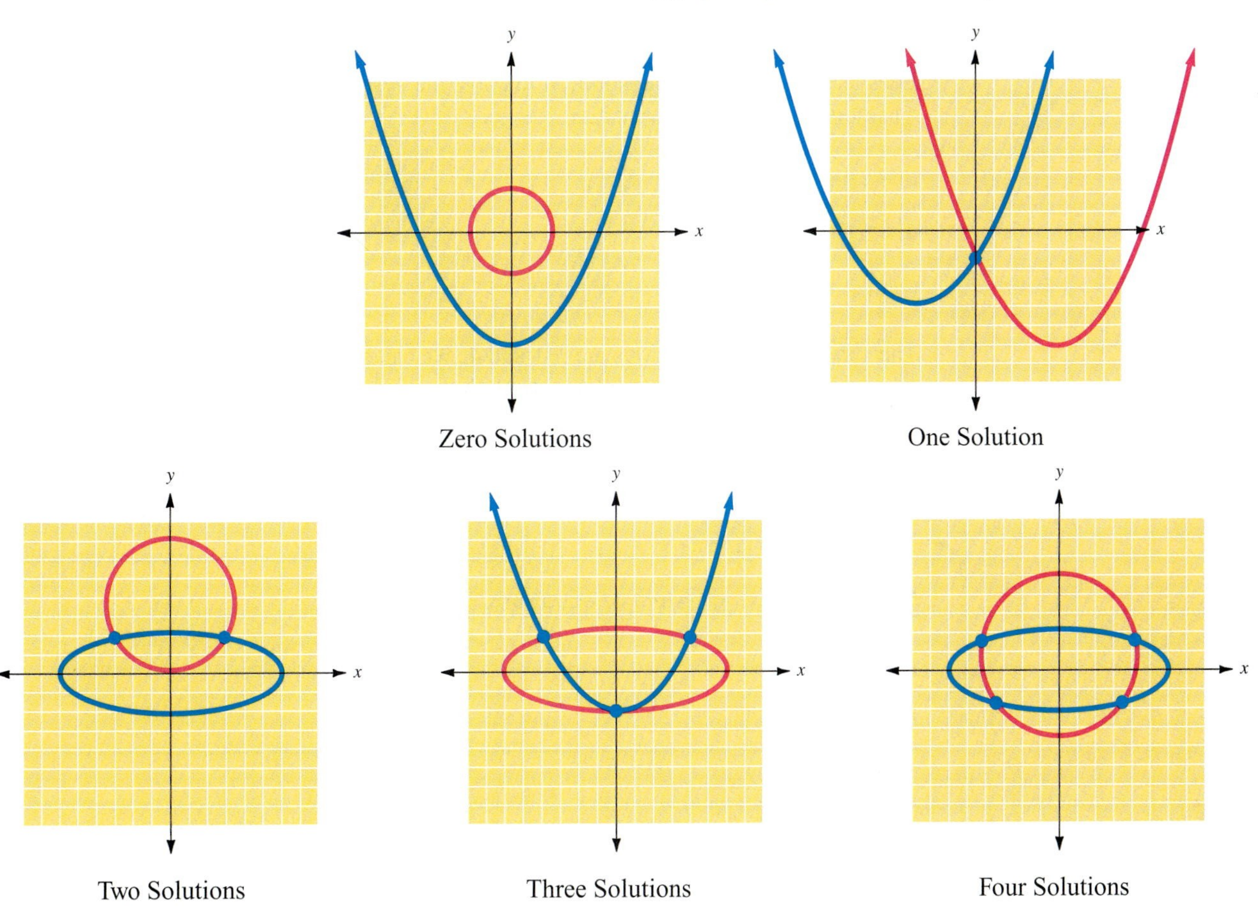

Zero Solutions

One Solution

Two Solutions

Three Solutions

Four Solutions

For the remainder of this section, we will restrict our discussion to a system that has as its graph a line and a parabola. Such a system has either zero, one, or two solutions.

Example 1

Solving a System of Nonlinear Equations

Solve the following system of equations.

$y = x^2 - 3x + 2$

$y = 6$

First, we will graph the system. From this graph we will be able to see the number of solutions. The graph will also give us a way to check the reasonableness of our algebraic results.

NOTE Use your calculator to approximate the solutions for the system.

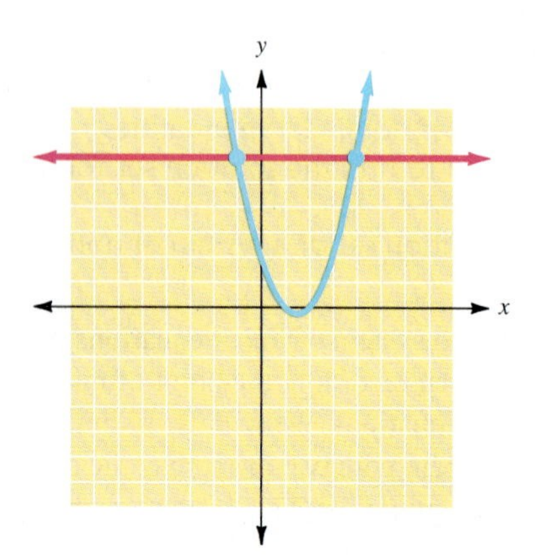

Let's use the method of substitution to solve the system. Substituting 6, from the second equation, for y in the first equation, we get

$$6 = x^2 - 3x + 2$$

$$0 = x^2 - 3x - 4$$

$$0 = (x - 4)(x + 1)$$

The x values for the solutions are -1 and 4. We can substitute these values for x in either equation to solve for y, but we know from the second equation that $y = 6$. The solution set is $\{(-1, 6), (4, 6)\}$. Looking at the graph, we see that this is a reasonable solution set for the system.

CHECK YOURSELF 1

Solve the following system of equations.

$$y = x^2 - 5x + 4$$

$$y = 10$$

Of course, not every quadratic expression is factorable. In Example 2, we must use the quadratic formula.

Example 2

Solving a Nonlinear System

Solve the following system of equations.

$$y = x^2 + x + 3$$

$$y = 7$$

Let's look at the graph of the system.

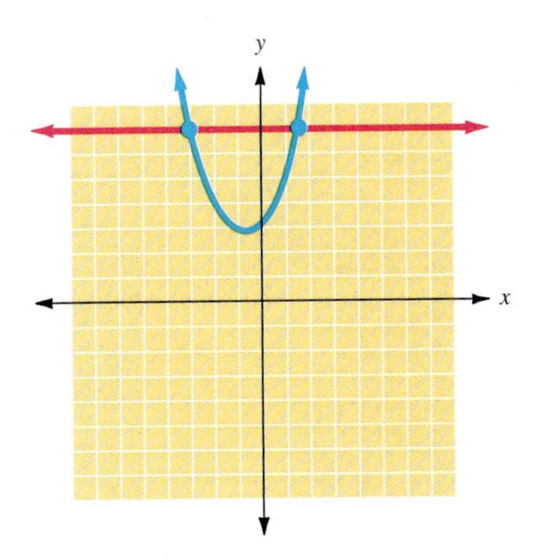

We see two points of intersection, but neither seems to be an integer value for x. Let's solve the system algebraically. Using the method of substitution, we find

$$7 = x^2 + x + 3$$

$$0 = x^2 + x - 4$$

The result is not factorable, so we use the quadratic formula to find the solutions.

$$x = \frac{-1 \pm \sqrt{1 + 16}}{2} = \frac{-1 \pm \sqrt{17}}{2}$$

The two points of intersection are $\left(\frac{-1 - \sqrt{17}}{2}, 7\right)$ and $\left(\frac{-1 + \sqrt{17}}{2}\right)$. It is difficult to check these points against the graph, so we will approximate them. The approximate solutions (to the nearest tenth) are $(-2.6, 7)$ and $(1.6, 7)$. The graph indicates that these are reasonable answers.

CHECK YOURSELF 2

Solve the following system of equations.

$$y = x^2 + x + 5$$

$$y = 8$$

As was stated earlier, not every system has two solutions. In Example 3, we will see a system with no real solution.

Example 3

Solving a System of Nonlinear Equations

Solve the following system of equations.

$$y = x^2 - 2x + 1$$

$$y = -2$$

As we did with the previous systems, we will first look at the graph of the system.

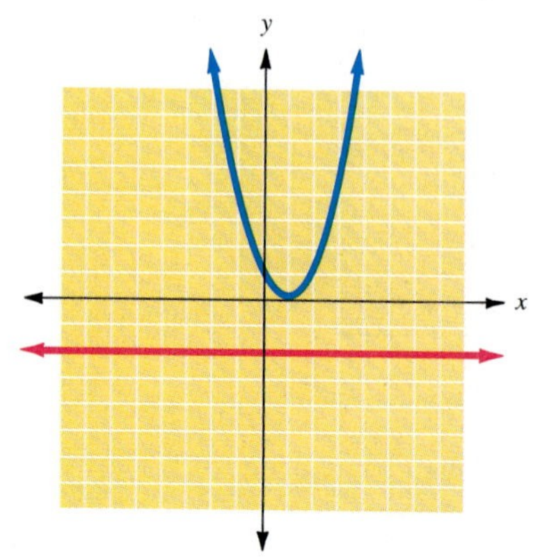

Using the method of substitution, we get

$$-2 = x^2 - 2x + 1$$

$$0 = x^2 - 2x + 3$$

Using the quadratic formula, we can confirm that there are no real solutions to this system.

$$\frac{-(-2) \pm \sqrt{(-2)^2 - 4(1)(3)}}{2(1)} = \frac{2 \pm \sqrt{-8}}{2}$$

CHECK YOURSELF 3

Solve the following system of equations.

$$y = x^2 + 3x + 5$$

$$y = 2$$

Consider the system consisting of the following two equations:

$$x^2 + y^2 = 25$$

$$3x^2 - y^2 = 11$$

The graph of the system indicates there are four solutions.

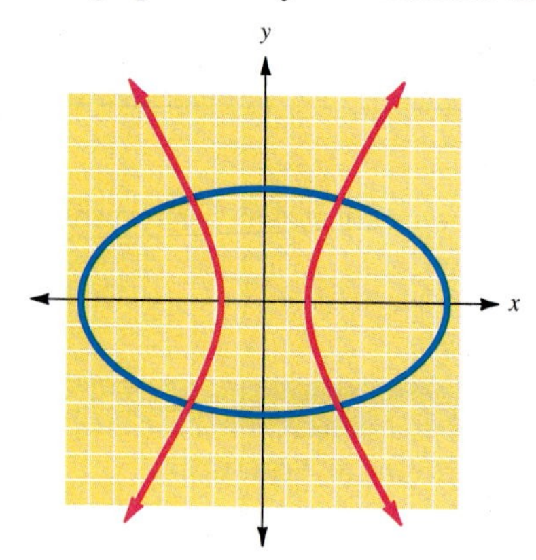

We could approximate the solutions, then check those approximations by substitution. But how could we find the solutions algebraically? Example 4 illustrates the elimination method.

Example 4

Solving a Nonlinear System by Elimination

Solve the following system algebraically.

$x^2 + y^2 = 25$

$3x^2 - y^2 = 11$

As was the case with linear systems, we can eliminate one of the variables. In this case, adding the equations eliminates the y variable.

$$\begin{array}{rcl} x^2 + y^2 &=& 25 \\ 3x^2 - y^2 &=& 11 \\ \hline 4x^2 \qquad &=& 36 \end{array}$$

Dividing by 4, we have

$x^2 = 9$, so

$x = \pm 3$

Substituting the value 3 into the first equation

$$\begin{aligned} (3)^2 + y^2 &= 25 \\ 9 + y^2 &= 25 \\ y^2 &= 16 \\ y &= \pm 4 \end{aligned}$$

Two of the ordered pairs in the solution set are $(3, -4)$ and $(3, 4)$.

Substituting the value -3 into the first equation

$$\begin{aligned} (-3)^2 + y^2 &= 25 \\ 9 + y^2 &= 25 \\ y^2 &= 16 \\ y &= \pm 4 \end{aligned}$$

The other two pairs in the solution set are $(-3, -4)$ and $(-3, 4)$.

The solutions set is $\{(-3, -4), (-3, 4), (3, -4), (3, 4)\}$

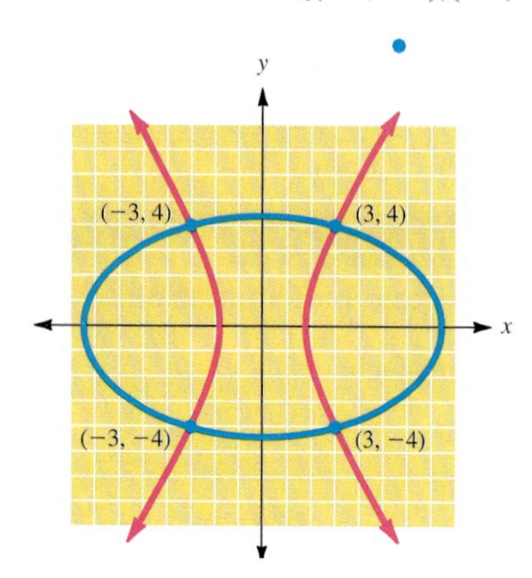

CHECK YOURSELF 4

Solve by the elimination method.

$x^2 - y^2 = 5$

$2x^2 + 3y^2 = 14$

Recall that a system of inequalities has as its solutions the set of all ordered pairs that make every inequality in the system a true statement. We almost always express the solutions to a system of inequalities graphically. We will do the same thing with nonlinear systems.

Example 5

Solving a System of Nonlinear Inequalities

Solve the following system.

$y \geq x^2 - 3x + 2$

$y \leq 6$

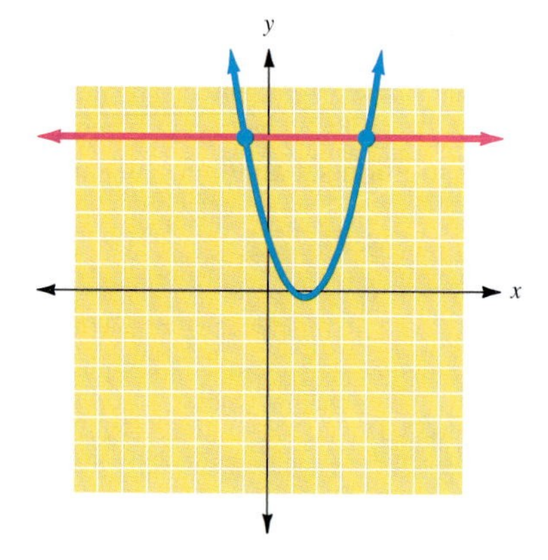

From Example 1, we have the graph of the related system of equations.

The first inequality has as its solution set every ordered pair with a y value that is greater than (above) the graph of the parabola. The second statement has as its solution set every ordered pair with a y value that is less than (below) the graph of the line. The solution set to the system is the set of ordered pairs that meet both of those criteria. Here is the graph of the solution set.

NOTE The solution set is the shaded area above the parabola and below the line.

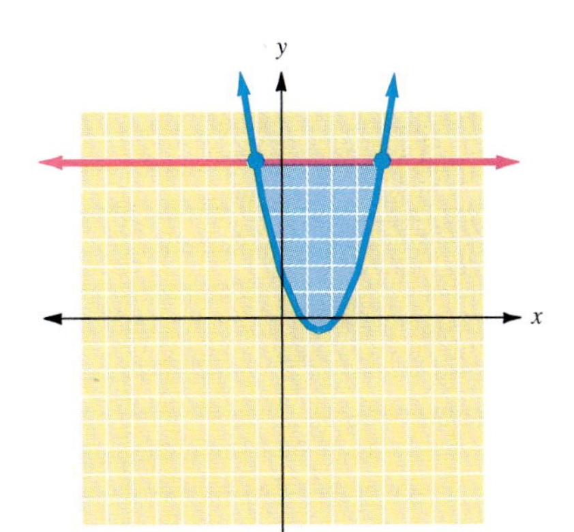

CHECK YOURSELF 5

Solve the following system.

$y \geq x^2 - 5x + 4$

$y \leq 10$

Example 6 demonstrates that, even if the related system of equations has no solution, the system of inequalities could have a solution.

Example 6

Solving a System of Nonlinear Inequalities

Solve the following system.

$y \leq x^2 - 2x + 1$

$y \geq -2$

As we did with the previous systems, we will first look at the graph of the related system of equations (from Example 3.)

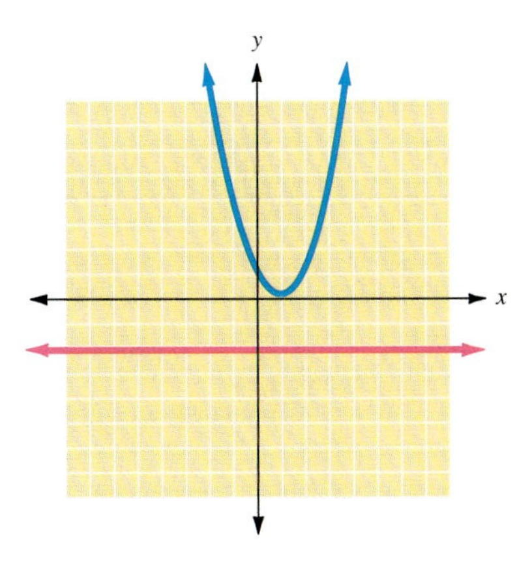

The solution set is now the set of all ordered pairs *below* the parabola ($y \le x^2 - 2x + 1$) and *above* the line ($y \ge -2$). Here is the graph of the solution set.

NOTE The solution continues beyond the borders of the grid.

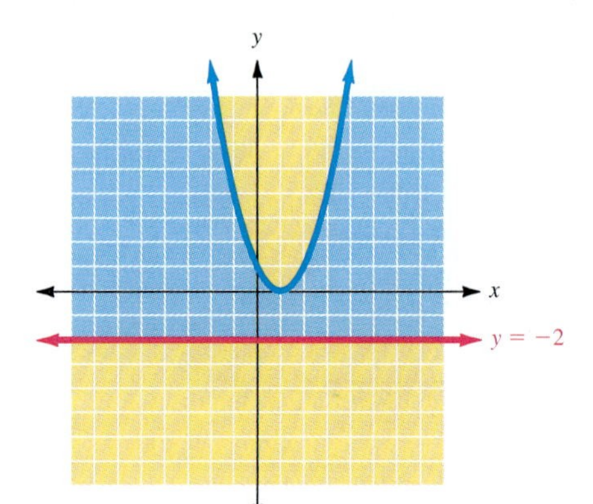

CHECK YOURSELF 6

Solve the following system.

$y \le x^2 + 3x + 5$

$y \ge 2$

CHECK YOURSELF ANSWERS

1. $\{(-1, 10), (6, 10)\}$ **2.** $\left\{\left(\frac{-1 \pm \sqrt{13}}{2}, 8\right)\right\} \approx \{(1.3, 8), (-2.3, 8)\}$

3. No real solution **4.** $\{(-1, -3), (-1, 3), (1, -3), (1, 3)\}$

5.

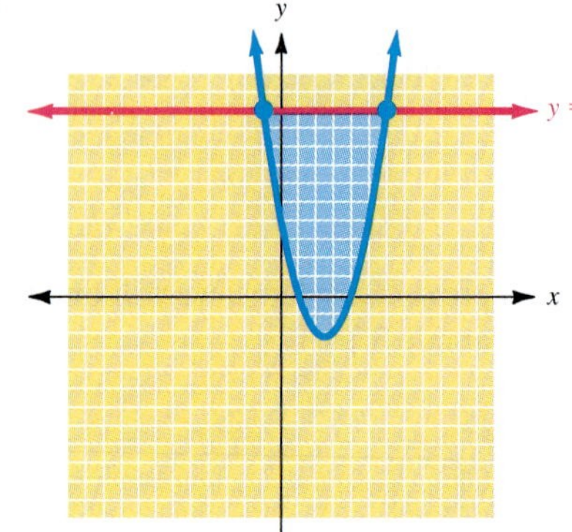

6.

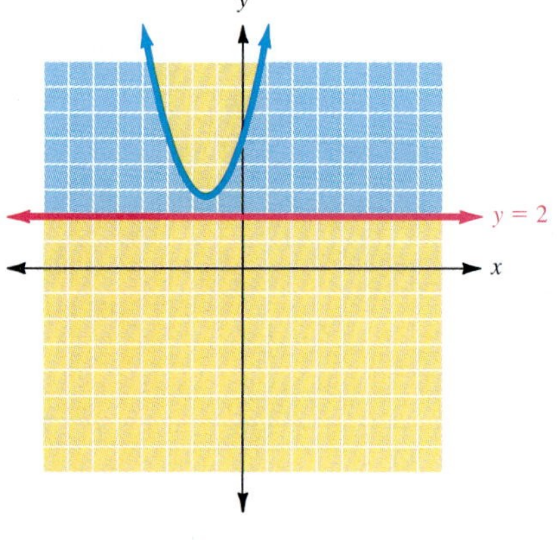

Name ______________________

Section ________ Date ________

10.4 Exercises

In exercises 1 to 8, the graph of a system of equations is given. Determine how many real solutions each system has.

1.

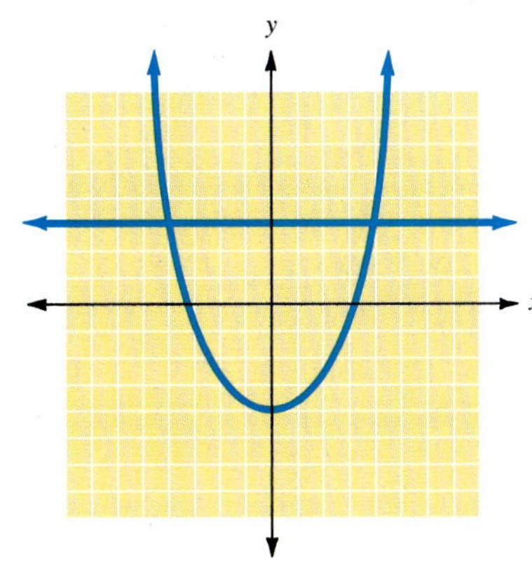

2.

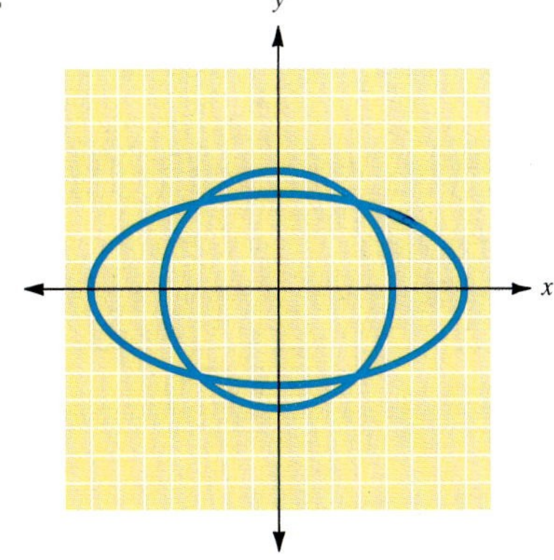

3.

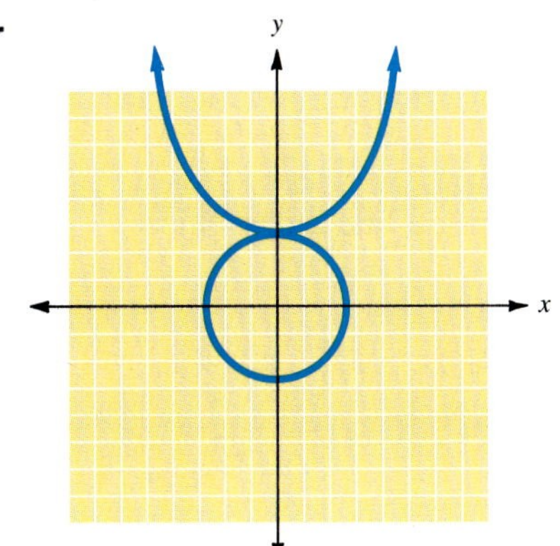

4.

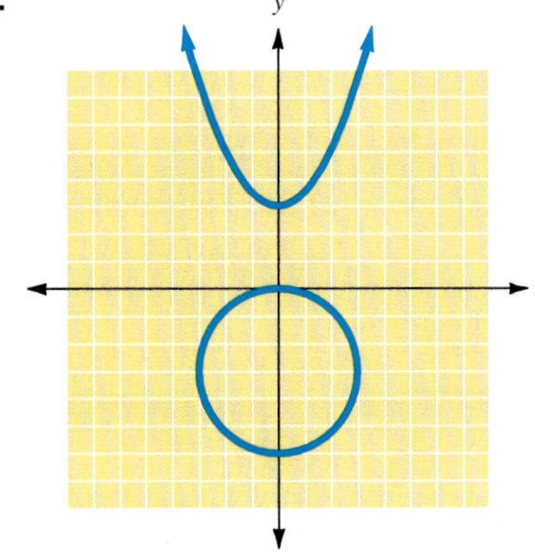

5.

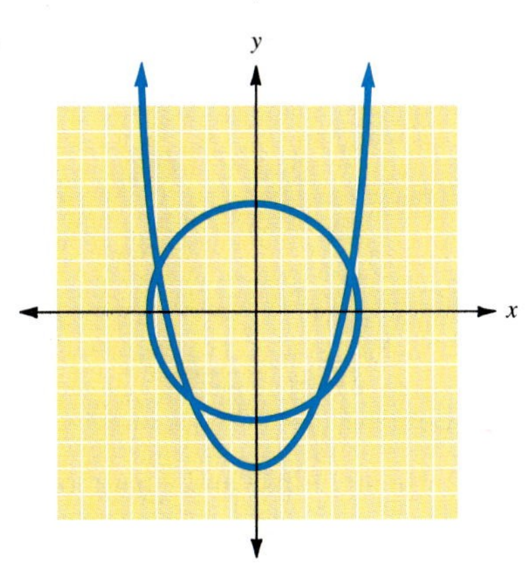

6.

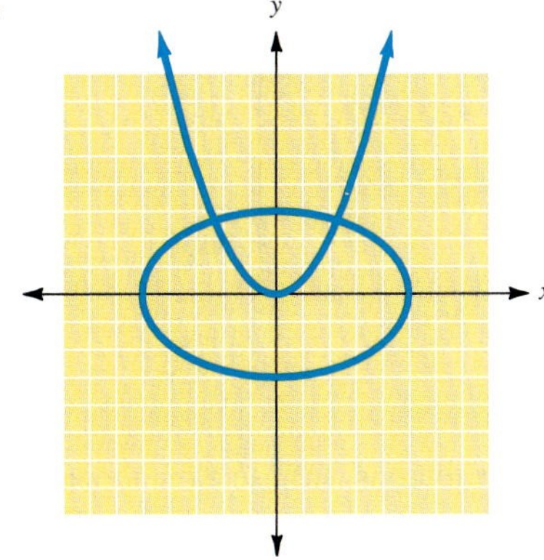

7.

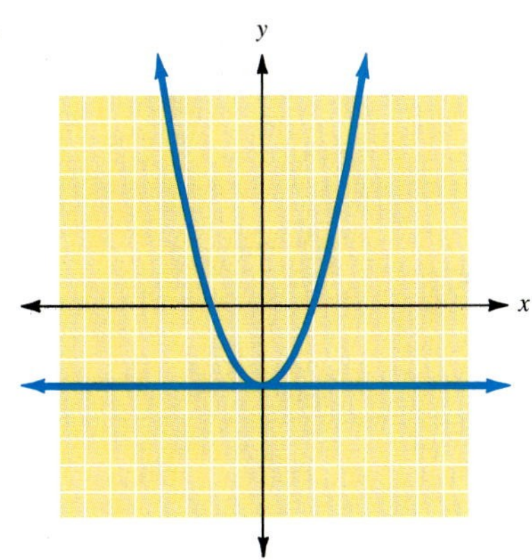

8.

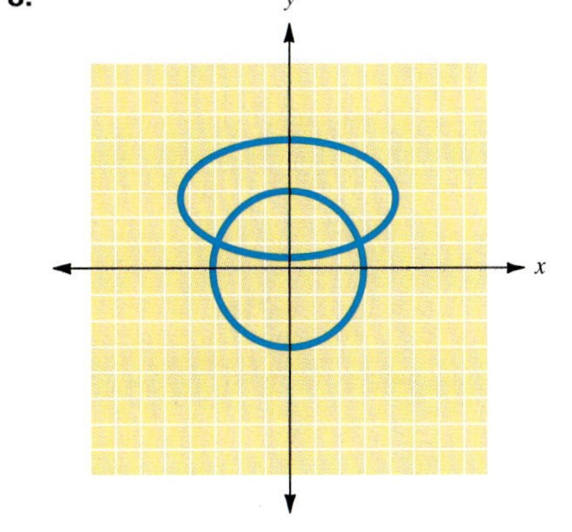

ANSWERS

1. ______________________
2. ______________________
3. ______________________
4. ______________________
5. ______________________
6. ______________________
7. ______________________
8. ______________________

ANSWERS

9. ____________

10. ____________

11. ____________

In exercises 9 to 12, draw the graph of a system that has the indicated number of solutions. Use the conic sections indicated.

9. 0 solutions: **(a)** use a circle and an ellipse, and **(b)** use a parabola and a line.

(a)

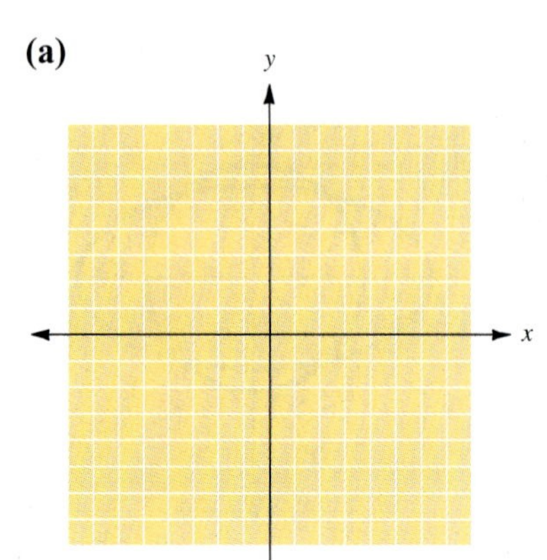

(b)

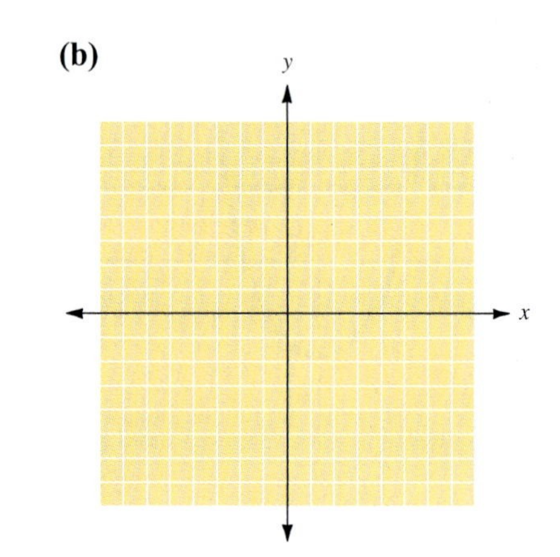

10. 1 solution: **(a)** use a parabola and a circle, and **(b)** use a line and an ellipse.

(a)

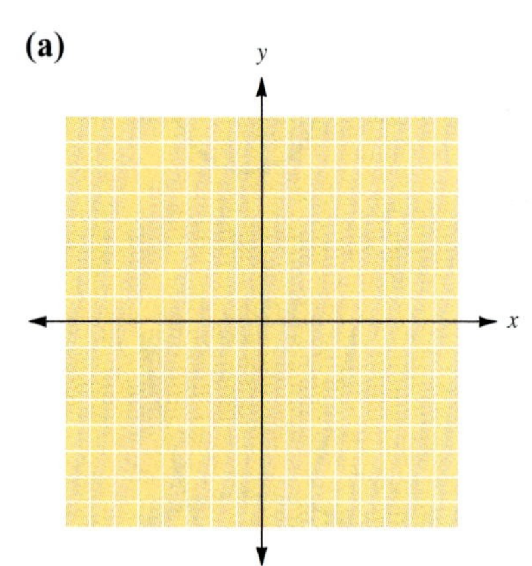

(b)

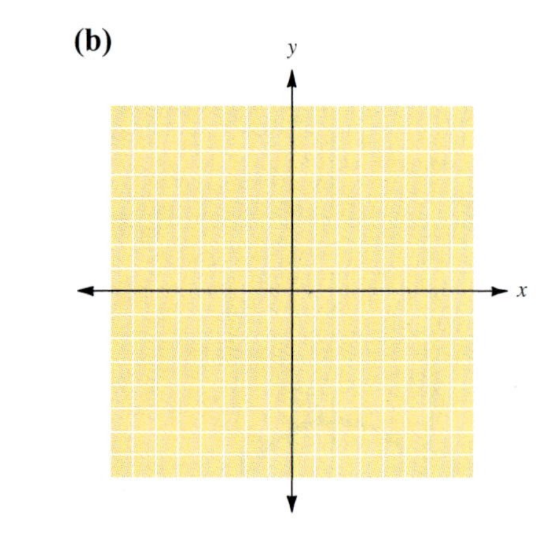

11. 2 solutions: **(a)** use a parabola and a circle, and **(b)** use an ellipse and a parabola.

(a)

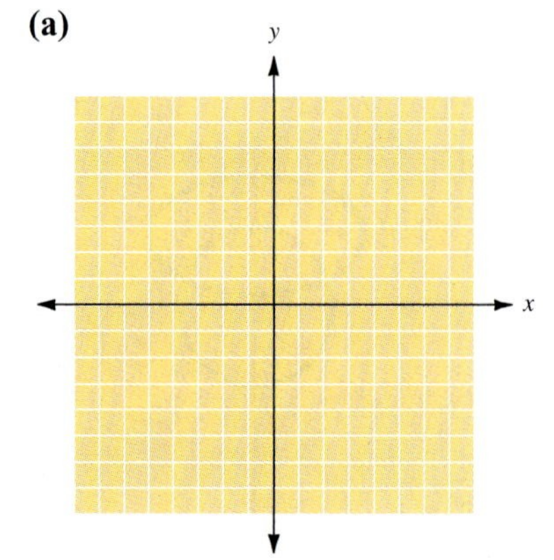

(b)

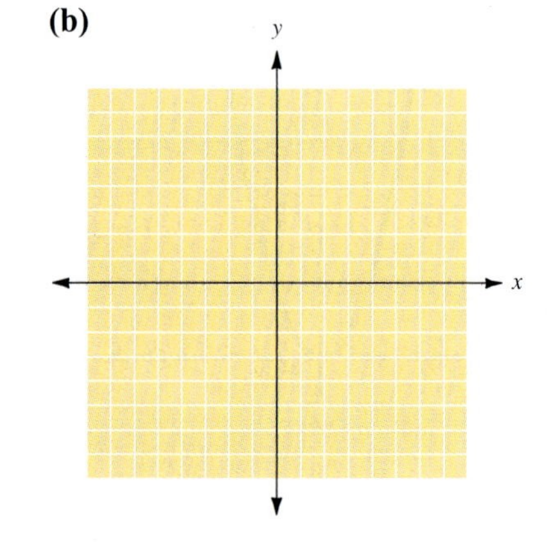

ANSWERS

12. ____

13. ____

14. ____

15. ____

16. ____

12. 4 solutions: **(a)** use a circle and an ellipse, and **(b)** use a parabola and a circle.

(a)

(b)

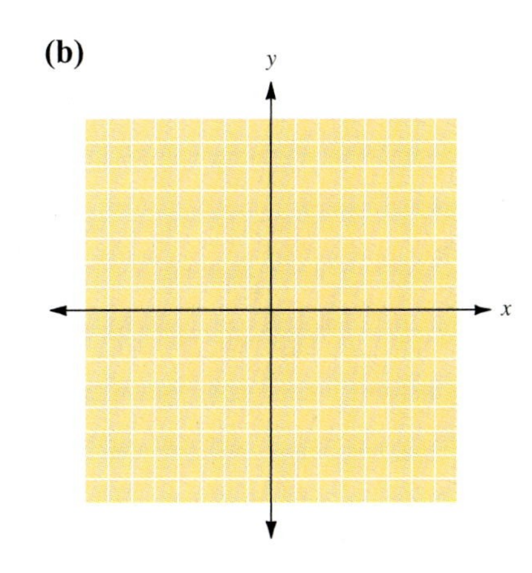

In exercises 13 to 24, graph each system and estimate the solutions.

13. $y = x^2 - x - 2$
$y = 4$

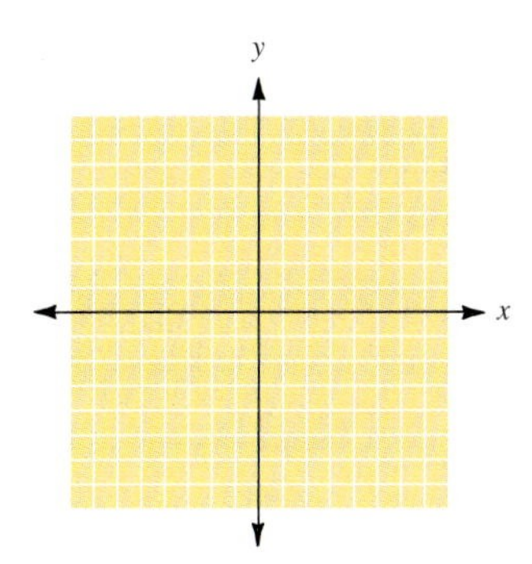

14. $y = x^2 - 3x + 2$
$y = 6$

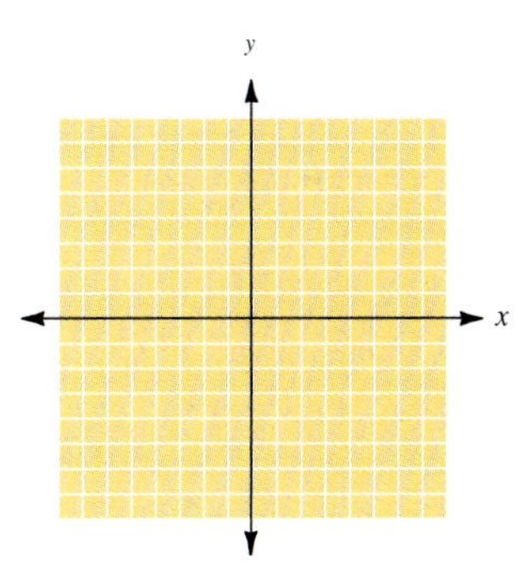

15. $y = x^2 - 5x + 7$
$y = 3$

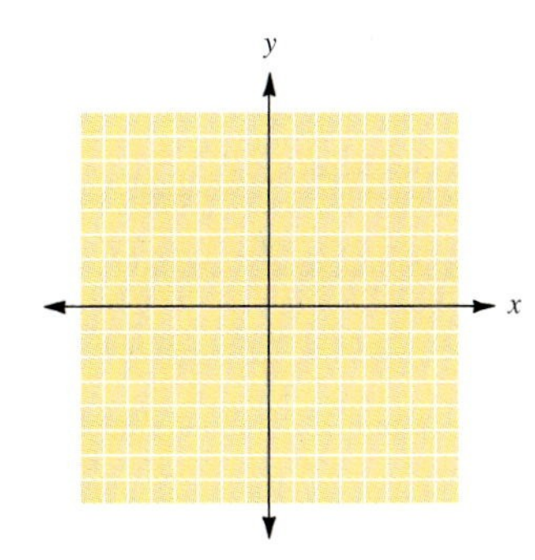

16. $y = x^2 - 8x + 18$
$y = 6$

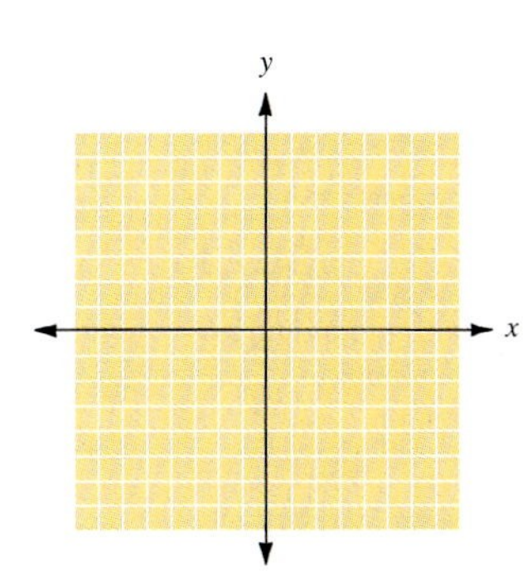

ANSWERS

17. ______

18. ______

19. ______

20. ______

21. ______

22. ______

17. $y = x^2 + 4x + 7$
$y = 4$

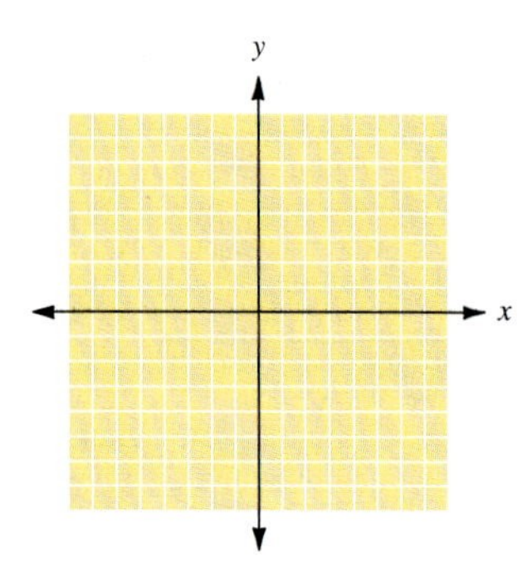

18. $y = x^2 - 6x + 7$
$y = 2$

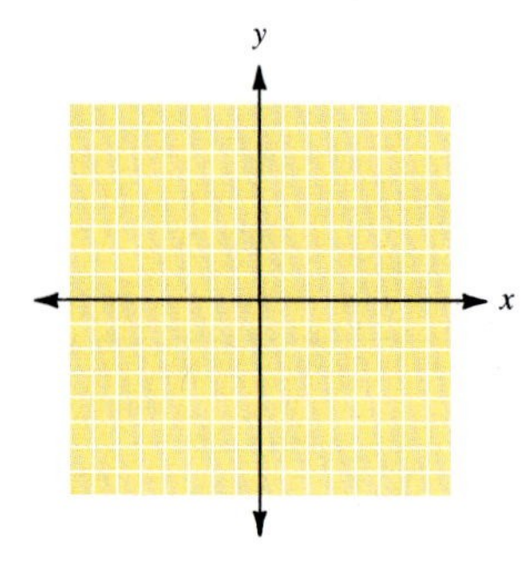

19. $y = x^2 + x + 5$
$y = 6$

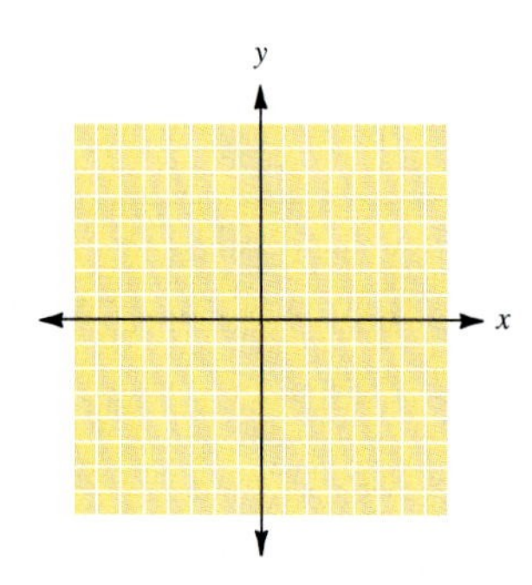

20. $y = x^2 + 8x + 17$
$y = 5$

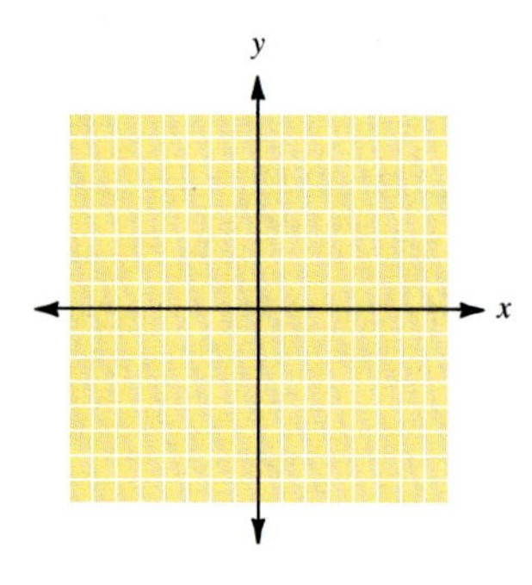

21. $y = x^2 - 7x + 11$
$y = 6$

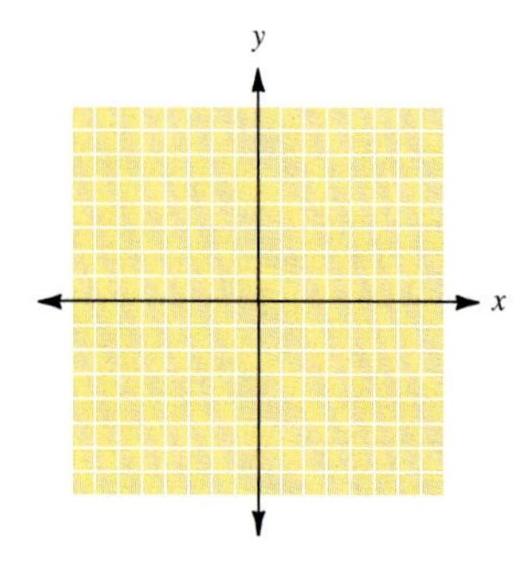

22. $y = x^2 - 2x + 2$
$y = 6$

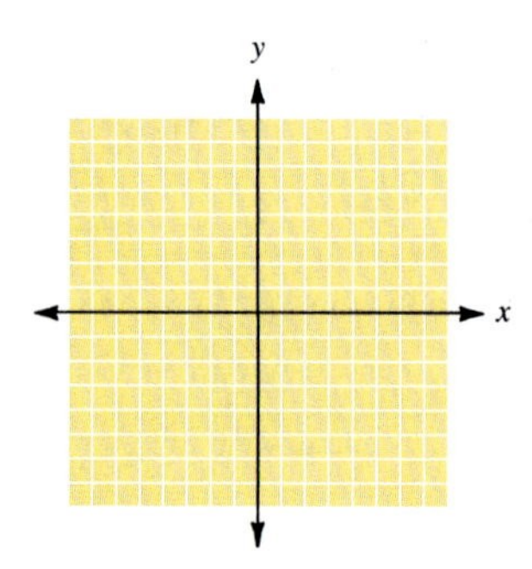

23. $y = x^2 + 5$
$y = 4$

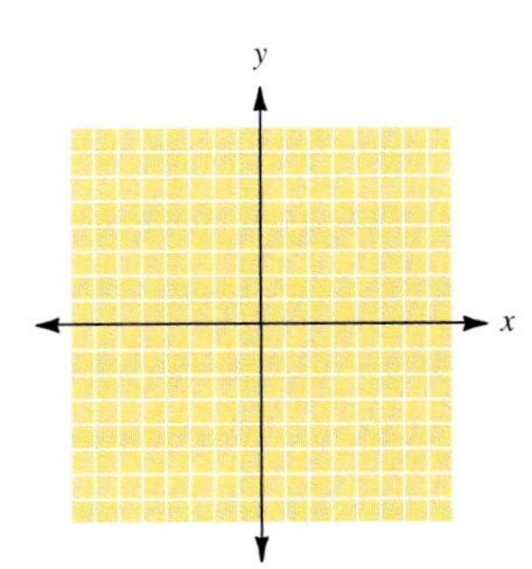

24. $y = x^2 - 4x + 9$
$y = 2$

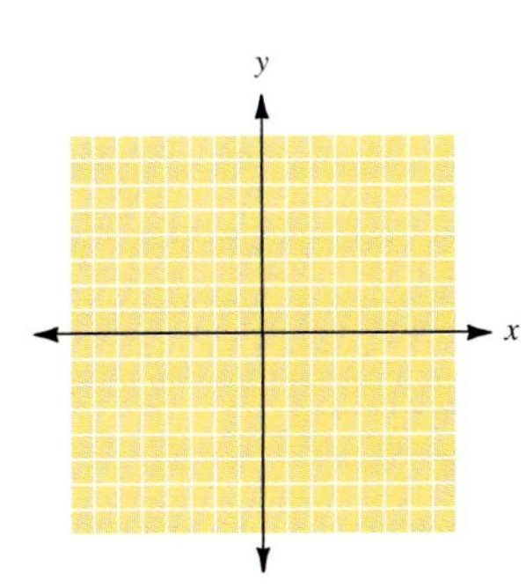

In exercises 25 to 32, solve using algebraic methods. (Note: These exercises have been solved graphically in exercises 13 to 24.)

25. $y = x^2 - x - 2$
$y = 4$
(See exercise 13.)

26. $y = x^2 - 3x + 2$
$y = 6$
(See exercise 14.)

27. $y = x^2 - 5x + 7$
$y = 3$
(See exercise 15.)

28. $y = x^2 - 8x + 18$
$y = 6$
(See exercise 16.)

29. $y = x^2 + x + 5$
$y = 6$
(See exercise 19.)

30. $y = x^2 + 8x + 17$
$y = 5$
(See exercise 20.)

31. $y = x^2 + 5$
$y = 4$
(See exercise 23.)

32. $y = x^2 - 4x + 9$
$y = 2$
(See exercise 24.)

In exercises 33 to 40, solve the systems of inequalities graphically. (*Note:* These have already been graphed as systems of equations in exercises 13 to 24.)

33. $y \geq x^2 - x - 2$
$y \leq 4$
(See exercise 13.)

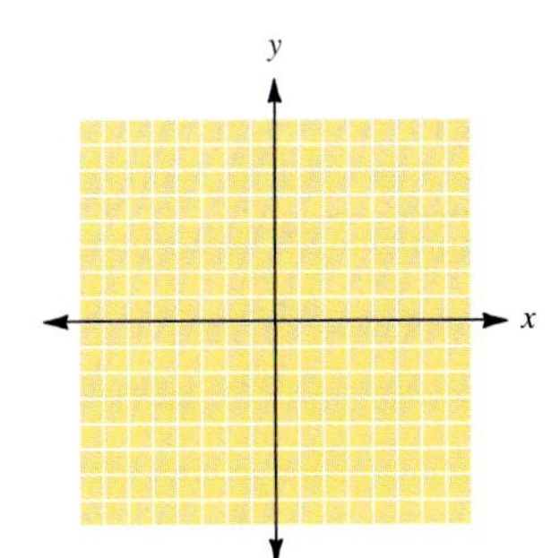

34. $y \geq x^2 - 3x + 2$
$y \leq 6$
(See exercise 14.)

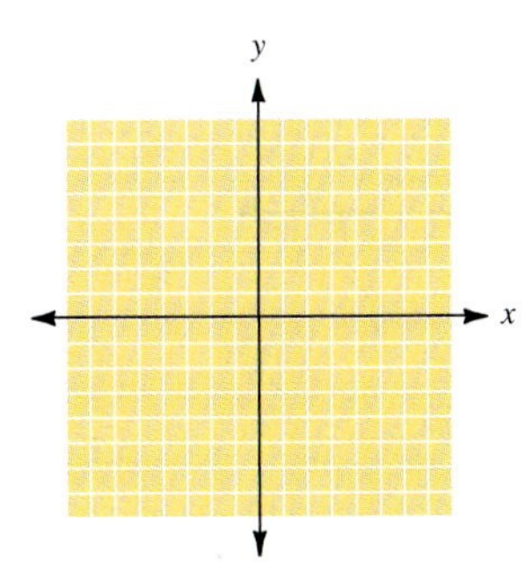

ANSWERS

23. ______
24. ______
25. ______
26. ______
27. ______
28. ______
29. ______
30. ______
31. ______
32. ______
33. ______
34. ______

ANSWERS

35. ____________

36. ____________

37. ____________

38. ____________

39. ____________

40. ____________

35. $y \geq x^2 - 5x + 7$

$y \leq 3$

(See exercise 15.)

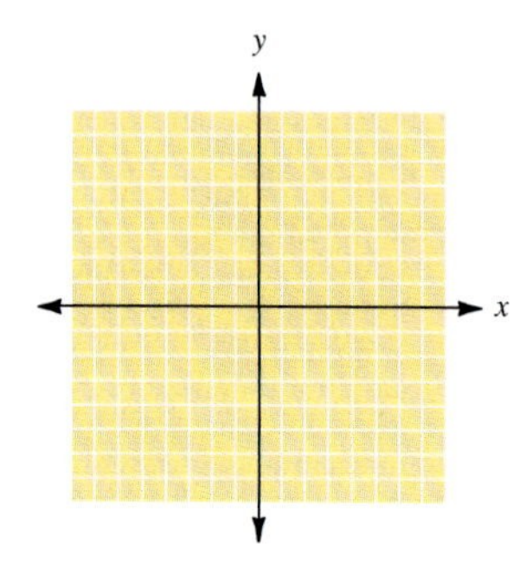

36. $y \geq x^2 - 8x + 18$

$y \leq 6$

(See exercise 16.)

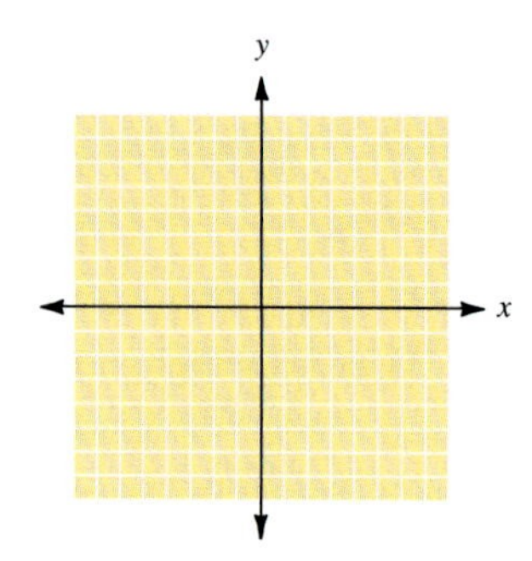

37. $y \geq x^2 + 4x + 7$

$y \geq 4$

(See exercise 17.)

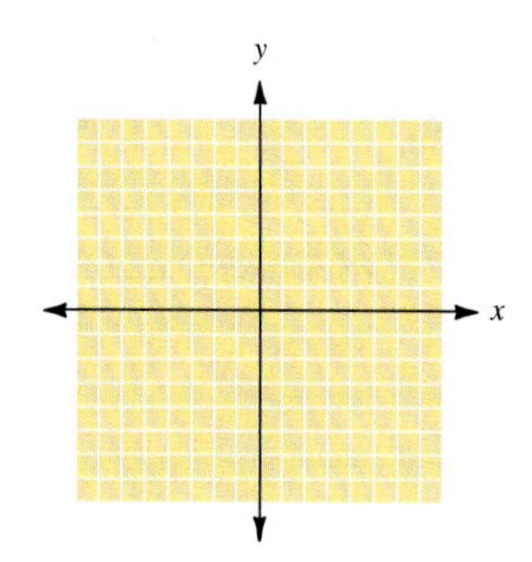

38. $y \geq x^2 - 6x + 7$

$y \geq 2$

(See exercise 18.)

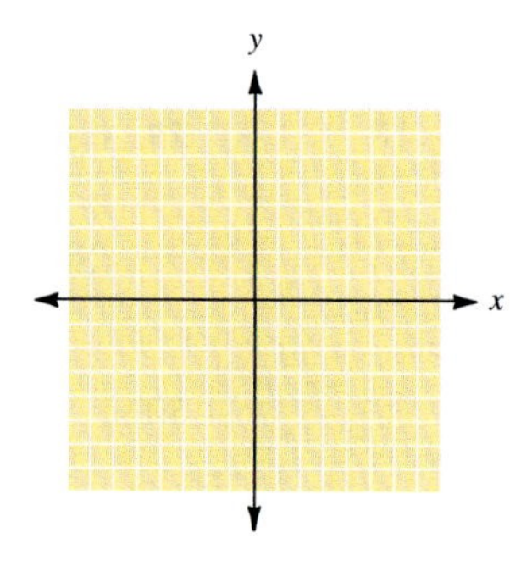

39. $y \leq x^2 + 5$

$y \geq 4$

(See exercise 23.)

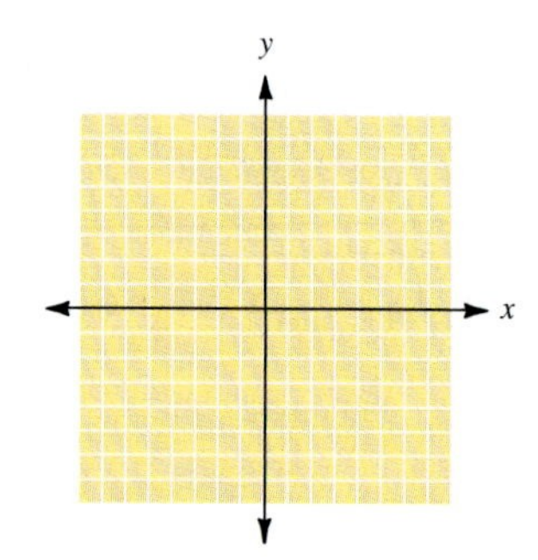

40. $y \leq x^2 - 4x + 9$

$y \geq 2$

(See exercise 24.)

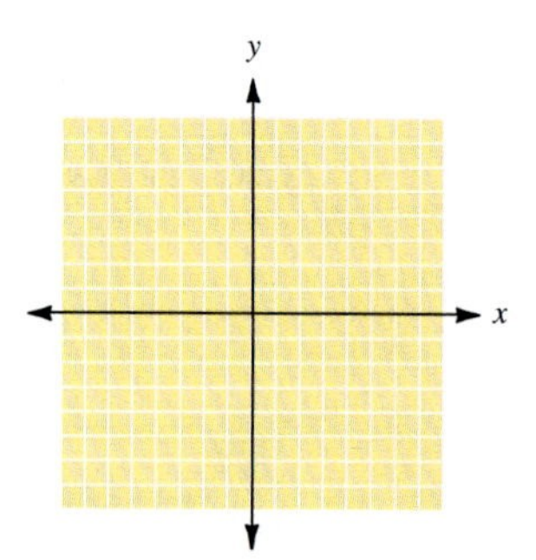

In exercises 41 to 44, **(a)** graph each system and estimate the solution, and **(b)** use algebraic methods to solve each system.

41. $y = x^2$
$x + y = 2$

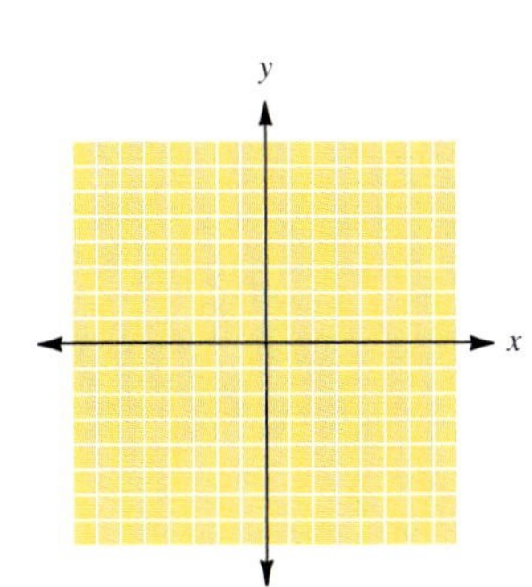

42. $y = x^2 - 6x$
$3x + y = 4$

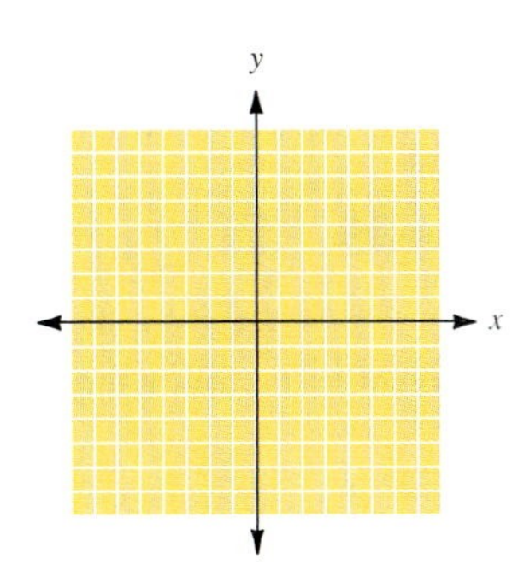

43. $x^2 + y^2 = 5$
$-3x + 4y = 2$

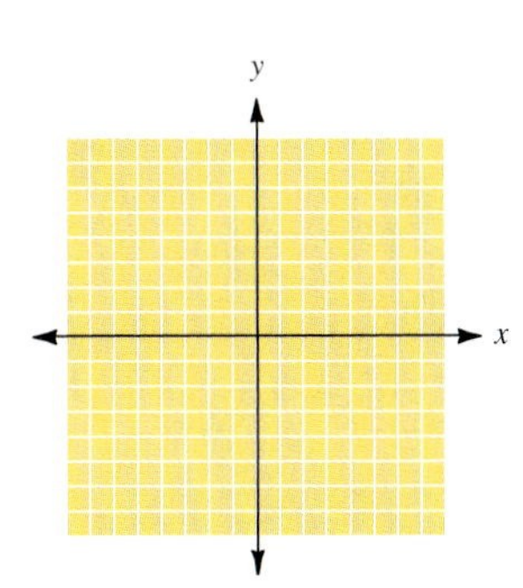

44. $x^2 + y^2 = 9$
$x + y = -3$

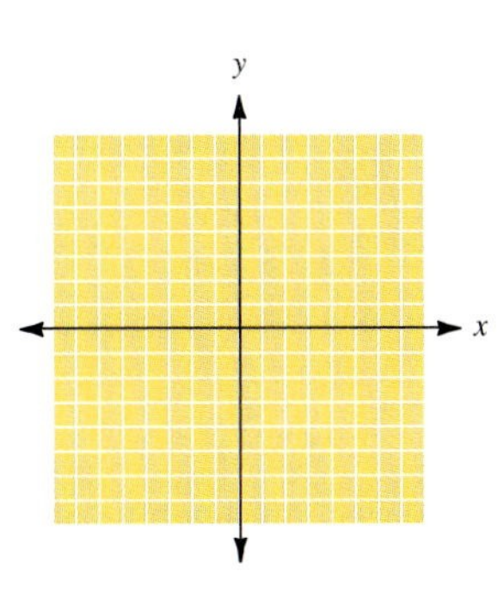

Solve the following applications.

45. The manager of a large apartment complex has found that the profit, in dollars, is given by the equation

$P = 120x - x^2$

in which x is the number of apartments rented. How many apartments must be rented to produce a profit of $3600?

46. The manager of a bicycle shop has found that the revenue (in dollars) from the sale of x bicycles is given by the following equation.

$R = x^2 - 200x$

How many bicycles must be sold to produce a revenue of $12,500?

ANSWERS

41. ______

42. ______

43. ______

44. ______

45. ______

46. ______

ANSWERS

47. ______

48. ______

49. ______

50. ______

51. ______

52. ______

53. ______

54. ______

47. Find the equation of the line passing through the points of intersection of the graphs $y = x^2$ and $x^2 + y^2 = 90$.

48. Write a system of inequalities to describe the following set of points: The points are in the interior of a circle whose center is the origin with a radius of 4, and above the line $y = 2$.

49. We are asked to solve the following system of equations.

$$x^2 - y = 5$$
$$x + y = -3.$$

Explain how we can determine, before doing any work, that this system cannot have more than two solutions.

50. Without graphing, how can you tell that the following system of inequalities has no solution?

$$x^2 + y^2 < 9$$
$$y > 4$$

Solve the following systems algebraically.

51. $x^2 + y^2 = 17$
$x^2 - y^2 = 15$

52. $x^2 + y^2 = 29$
$2x^2 - y^2 = 46$

53. $x^2 + y^2 = 8$
$2x^2 + 3y^2 = 20$

54. $2x^2 + y^2 = 3$
$3x^2 + 4y^2 = 7$

Answers

1. 2 **3.** 1 **5.** 4 **7.** 1

9. **(a)**

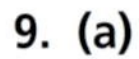

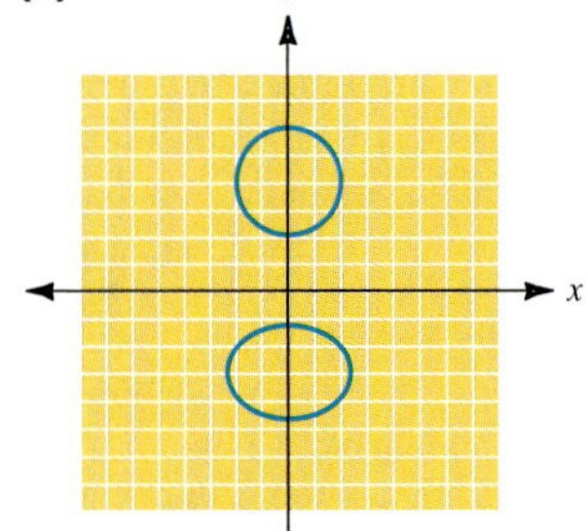

(b)

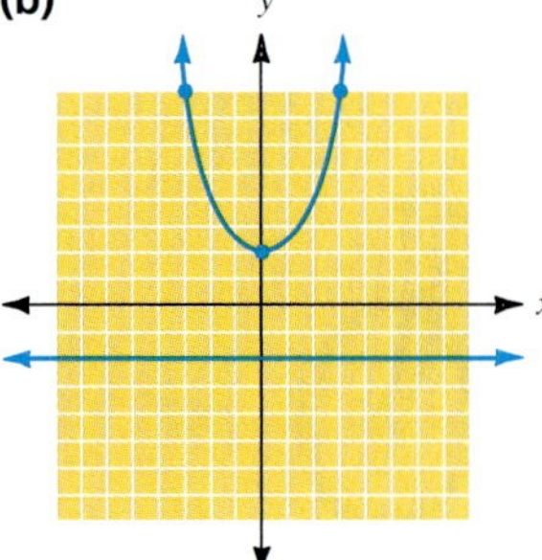

11. (a)

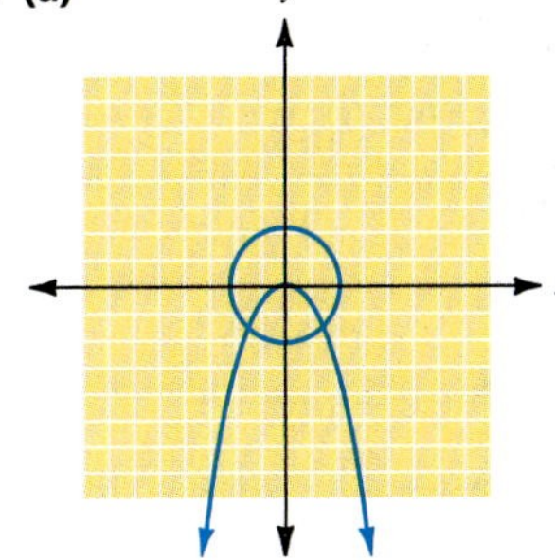

(b)

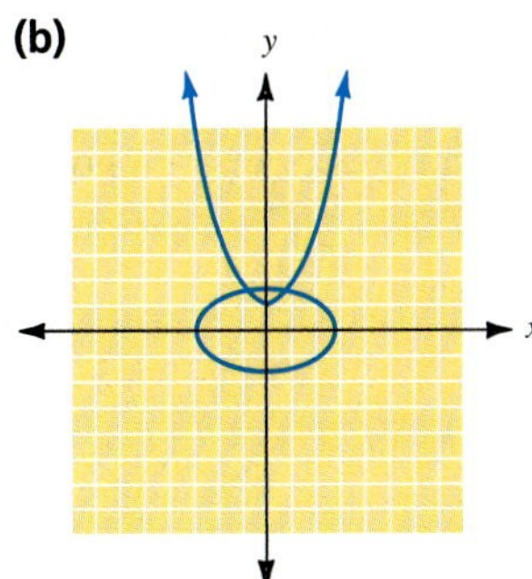

13. (3, 4) and (−2, 4)

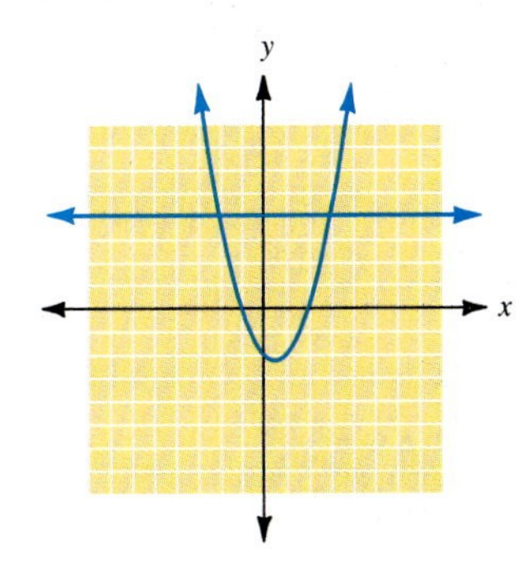

15. (4, 3) and (1, 3)

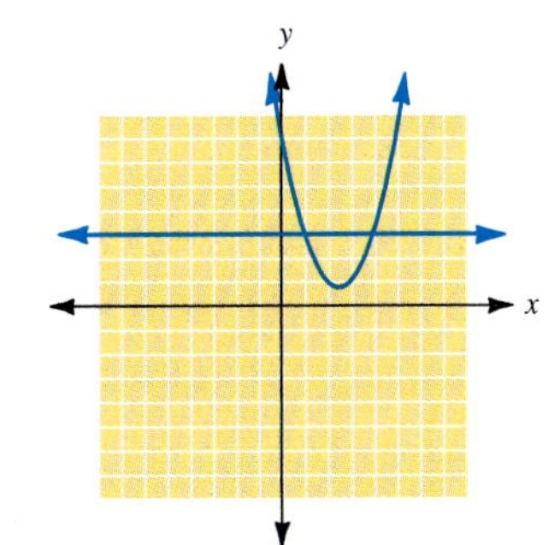

17. (−3, 4) and (−1, 4)

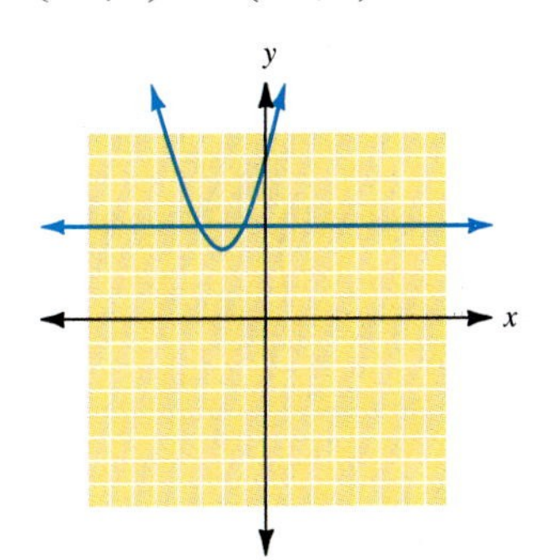

19. (0.6, 6) and (−1.6, 6)

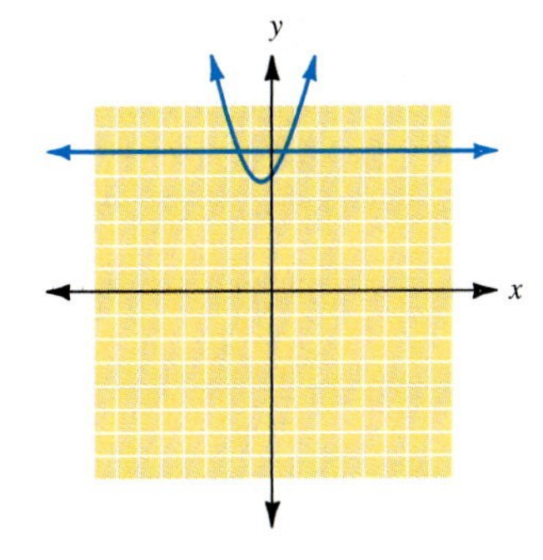

21. (6.2, 6) and (0.8, 6)

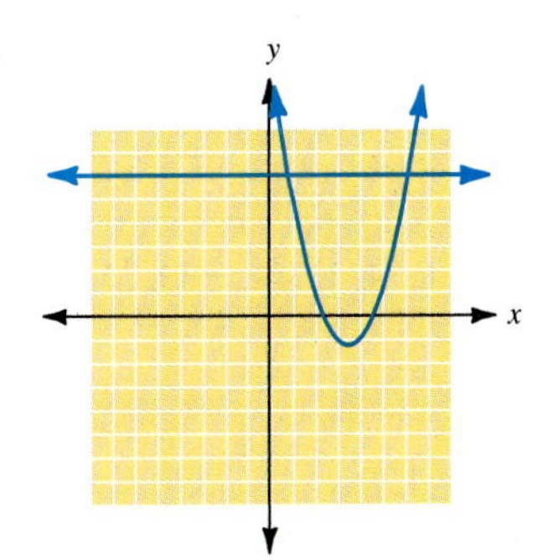

23. No solution

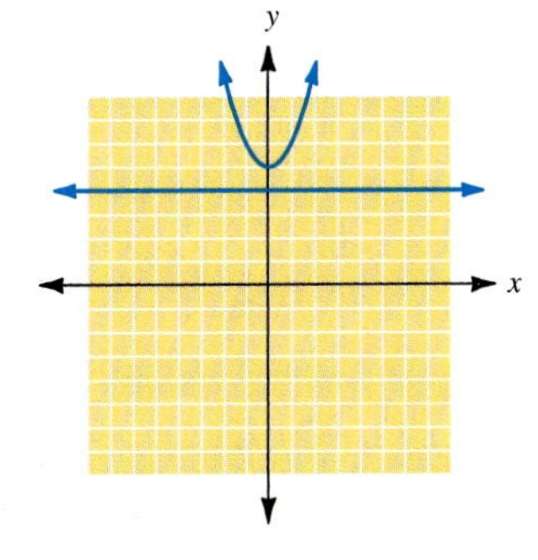

25. (3, 4) and (−2, 4) **27.** (4, 3) and (1, 3)

29. $\left(\frac{-1+\sqrt{5}}{2}, 6\right)$ and $\left(\frac{-1-\sqrt{5}}{2}, 6\right)$ or (0.618, 6) and (−1.62, 6)

31. No solution

33.

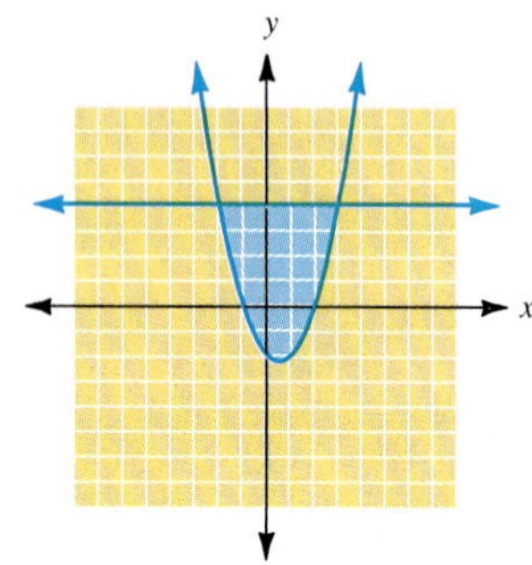

35.

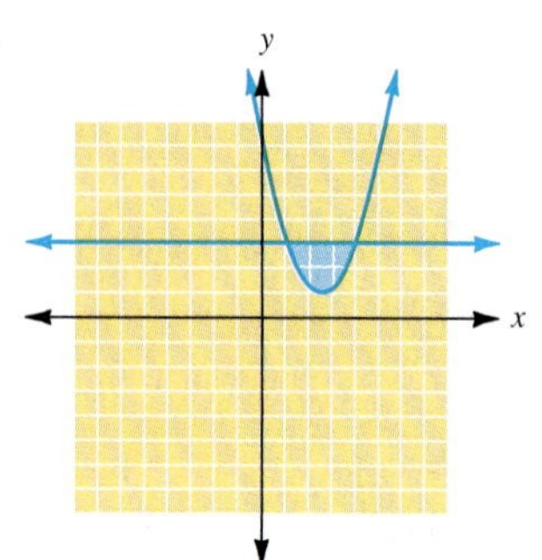

37.

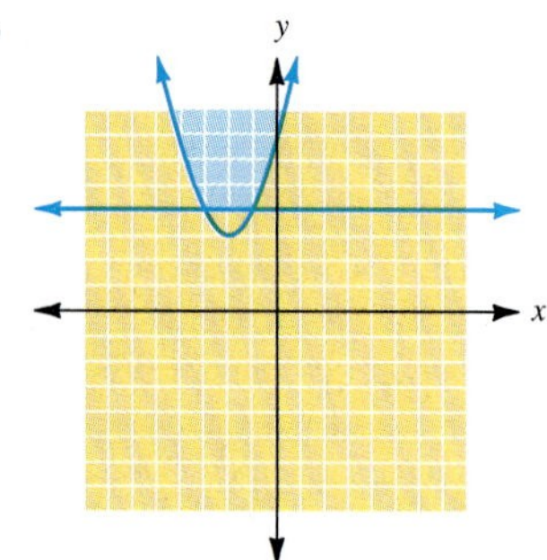

39.

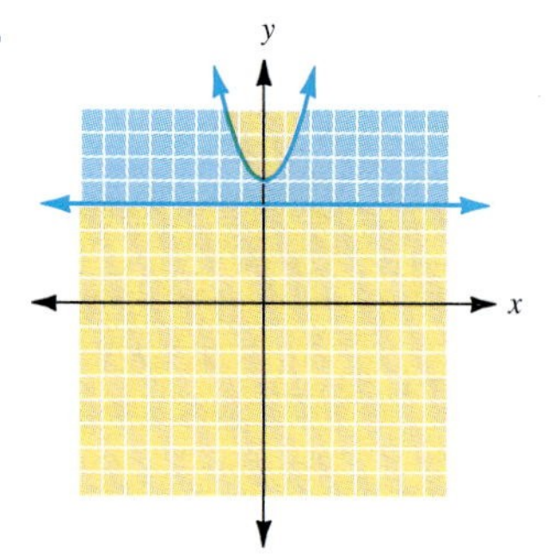

41. (1, 1) and (−2, 4)

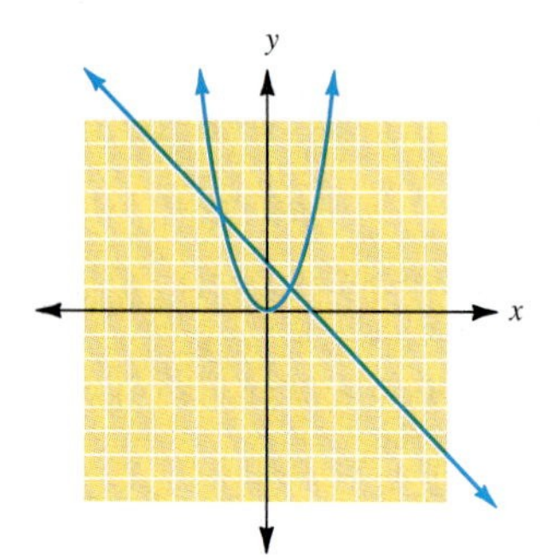

43. (−2, −1) and $\left(\frac{38}{25}, \frac{41}{25}\right)$

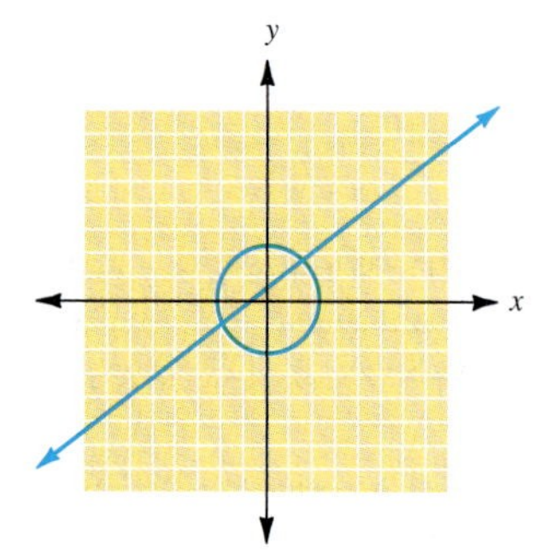

45. 60 **47.** $y = 9$ **49.** 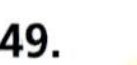  **51.** {(−4, −1), (−4, 1), (4, −1), (4, 1)}

53. {(−2, −2), (−2, 2), (2, −2), (2, 2)}

10 Summary

DEFINITION/PROCEDURE	EXAMPLE	REFERENCE
Graphing Parabolas		**Section 10.1**
The *conic sections* are the curves formed when a plane cuts through a cone. These include the *parabola, circle, ellipse,* and *hyperbola.*		p. 746
The graph of an equation quadratic in one variable and linear in the other is a parabola.	$y = x^2$ is linear in y and quadratic in x.	
If the equation is in the form $y = a(x - h)^2 + k$ the vertex of the parabola is at (h, k).	$y = (x - 2)^2 + 3$; vertex $(2, 3)$	p. 747

Continued

DEFINITION/PROCEDURE	EXAMPLE	REFERENCE
Graphing Parabolas		**Section 10.1**
The graph of $$y = ax^2 + bx + c \qquad a \neq 0$$ is a parabola. The parabola opens *upward* if $a > 0$.	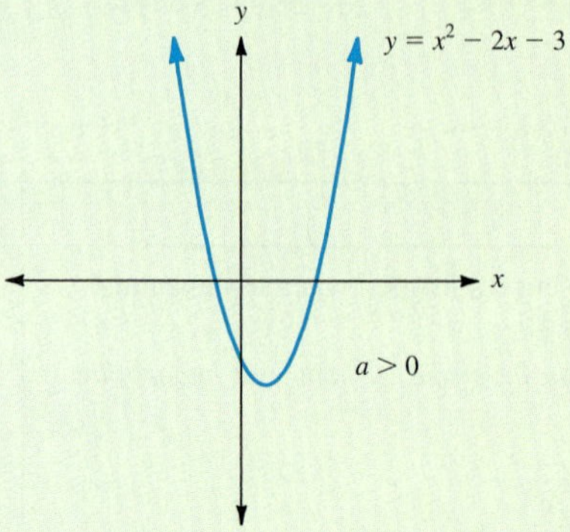 $y = x^2 - 2x - 3$; $a > 0$	
The parabola opens *downward* if $a < 0$.	$a < 0$; $y = -x^2 - 2x + 3$	
The *vertex* of the parabola (either the highest or the lowest point on the graph) is on the *axis of symmetry* with the equation $$x = -\frac{b}{2a}$$	Vertex; Axis of symmetry	
To graph a parabola: **1.** Find the axis of symmetry. **2.** Find the vertex. **3.** Determine two symmetric points. **Note:** You can use the x intercepts if the quadratic member of the given equation is factorable and the x intercepts are distinct. Otherwise, use the y intercept and its symmetric point. **4.** Draw a smooth curve connecting the points found above, to form the parabola. You may choose to find additional pairs of symmetric points at this time.	To graph $y = x^2 - 4x + 3$; (0, 3); (4, 3); (2, −1) Vertex; Axis of symmetry	p. 753

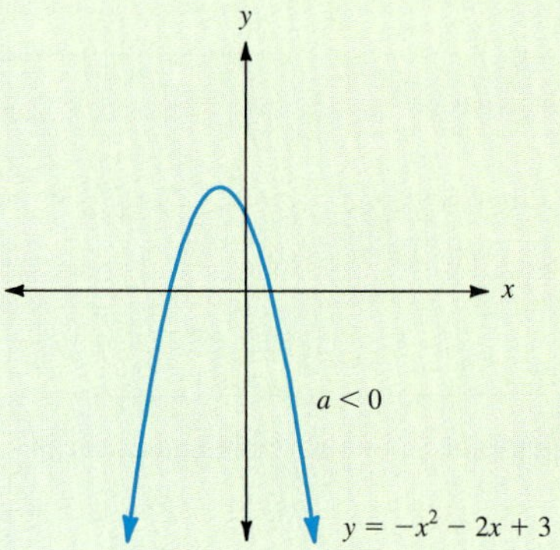

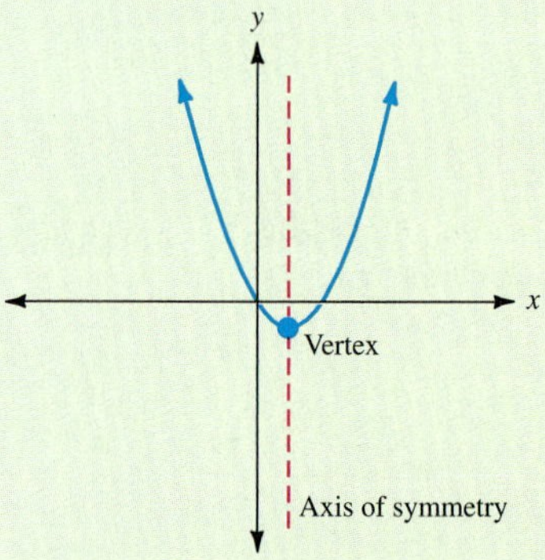

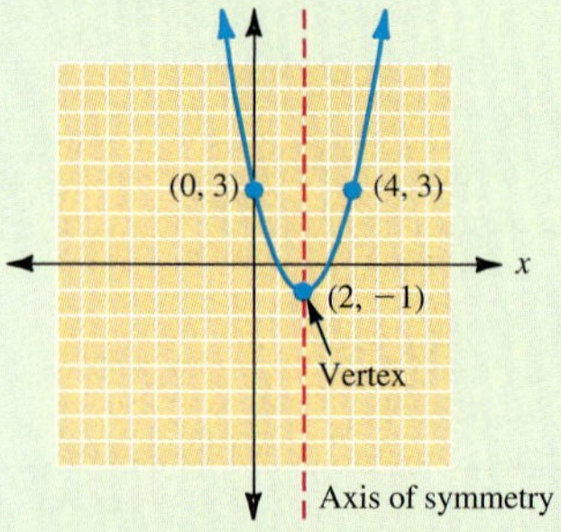

DEFINITION/PROCEDURE	EXAMPLE	REFERENCE
The Distance Formula and the Circle		**Section 10.2**
The standard form for the circle with center (h, k) and radius r is $$(x - h)^2 + (y - k)^2 = r^2$$ Determining the center and radius of the circle from its equation allows us to easily graph the circle. **Note:** Completing the square may be used to derive an equivalent equation in standard form if the original equation is not in this form.	Given the equation $$(x - 2)^2 + (y + 3)^2 = 4$$ we see that the center is at $(2, -3)$ and the radius is 2.	p. 767
The Ellipse and the Hyperbola		**Section 10.3**
Ellipse The standard form for the *ellipse,* whose foci are located on either the x or y axis and are symmetric about the origin, is $$\frac{x^2}{a^2} + \frac{y^2}{b^2} = 1$$ The x intercepts for the ellipse are $(a, 0)$ and $(-a, 0)$. The y intercepts for the ellipse are $(0, b)$ and $(0, -b)$. Determining the four intercepts of an ellipse allows us to sketch its graph. **Note:** If the given equation is not in standard form, we can divide both sides of the equation by the appropriate constant to derive the standard form.	Graph the equation $$\frac{x^2}{9} + \frac{y^2}{4} = 1$$	p. 777
Hyperbola The standard form for the *hyperbola* whose foci are located on the x axis and are symmetric about the origin is $$\frac{x^2}{a^2} - \frac{y^2}{b^2} = 1$$ The vertices for this hyperbola are on the x axis, $(a, 0)$ and $(-a, 0)$. The asymptotes of the hyperbola have the equations $$y = \pm\frac{b}{a}x$$ Determining and sketching the vertices and asymptotes of a hyperbola will allow us to sketch its graph quickly. **Note:** Again divide both sides of the given equation by the appropriate constant if the equation is not in standard form. If the foci of the hyperbola are located on the y axis and are symmetric about the origin, the standard form is $$\frac{y^2}{b^2} - \frac{x^2}{a^2} = 1$$ The vertices for this hyperbola are on the y axis $(0, b)$ and $(0, -b)$. The equations for the asymptotes of the hyperbola remain the same as before.	Sketch the hyperbola $$\frac{x^2}{4} - \frac{y^2}{9} = 1$$ The vertices are at $(2, 0)$ and $(-2, 0)$. The asymptotes are $y = \frac{3}{2}x$ and $y = -\frac{3}{2}x$.	p. 779

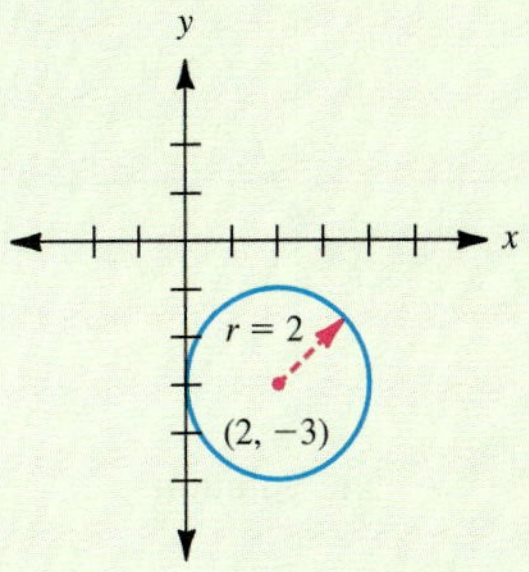

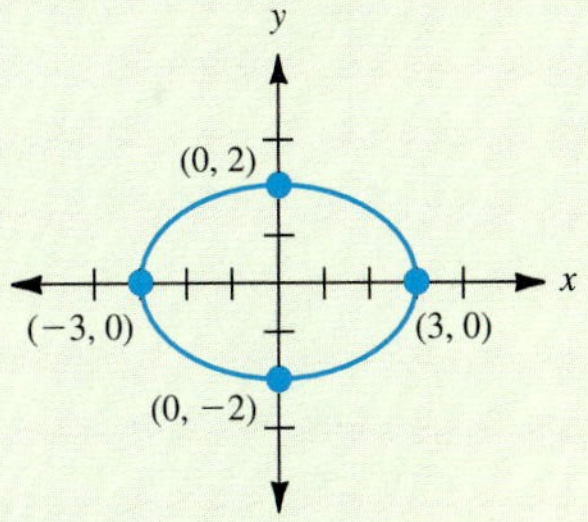

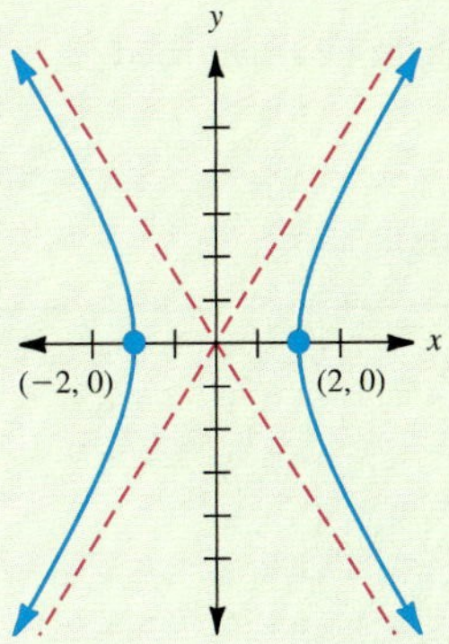

Continued

DEFINITION/PROCEDURE	EXAMPLE	REFERENCE
Graphing Nonlinear Systems		**Section 10.4**

A system with two conic curves can have zero, one, two, three, or four solutions.

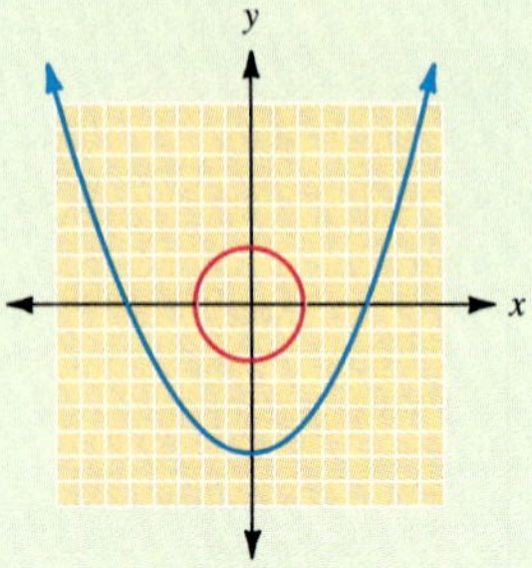

Zero Solutions

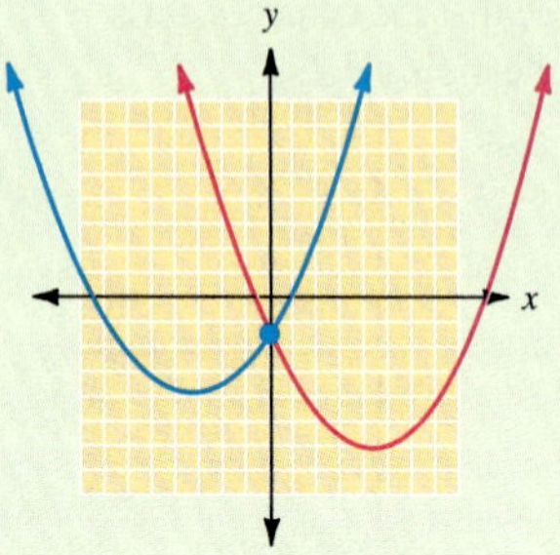

One Solution

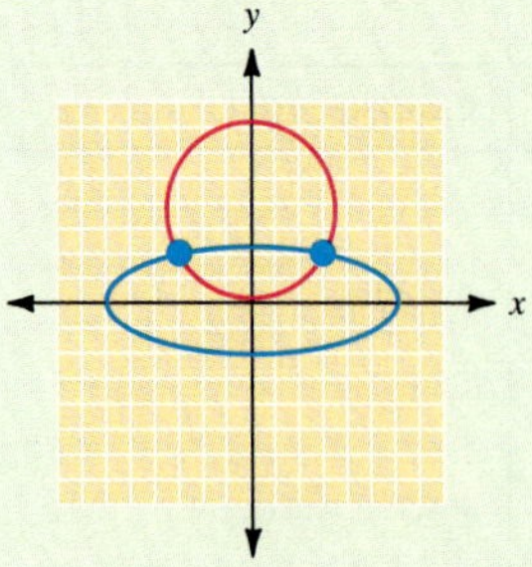

Two Solutions

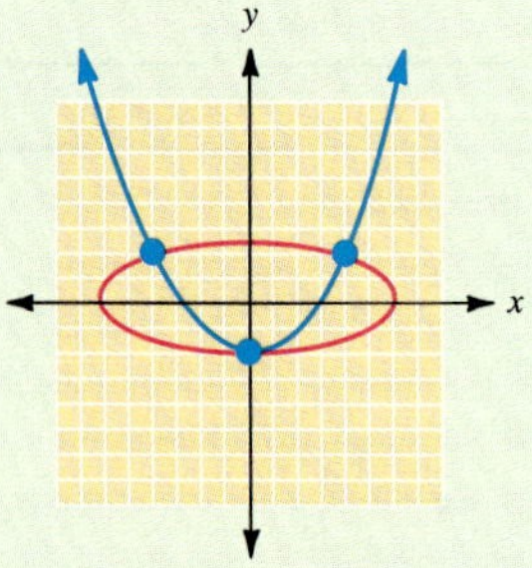

Three Solutions

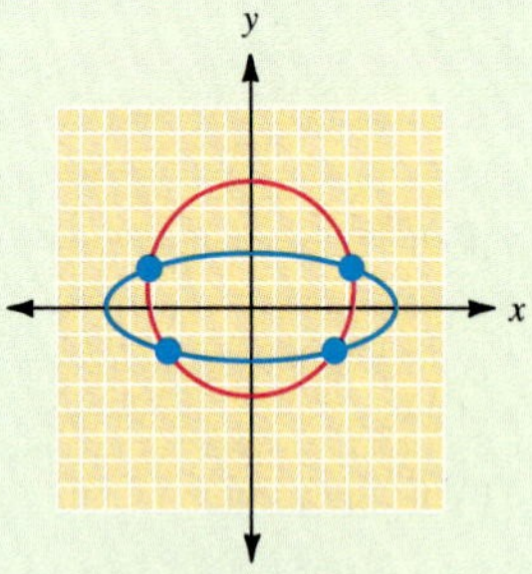

Four Solutions

A system that has as its graph a line and a parabola has, at most, two solutions. To solve such a system, use the following steps:

1. Solve both equations for y.
2. Create a new equation by setting the two right-hand expressions equal to each other.
3. Solve the equation for x.
4. Find the associated ordered pair(s).

To solve:

$$y = x^2 + x - 5$$
$$y = 7$$

let

$$7 = x^2 + x - 5$$
$$0 = x^2 + x - 12$$
$$0 = (x + 4)(x - 3)$$
$$x = 3, -4$$

$(x, y) = (3, 7)$ or $(-4, 7)$

p. 795

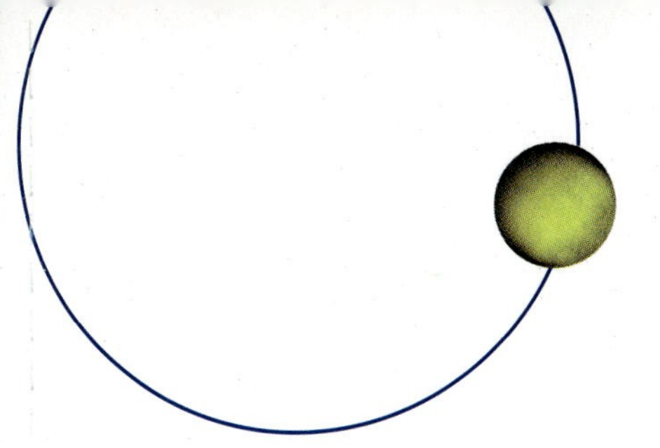

Summary Exercises

[10.1] Graph each of the following.

1. $y = x^2$

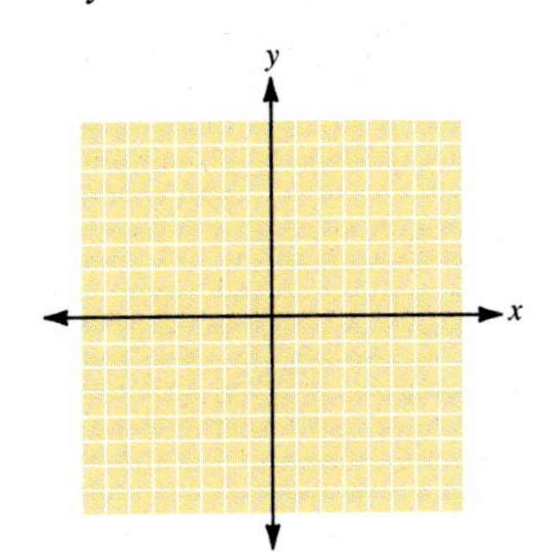

2. $y = x^2 + 2$

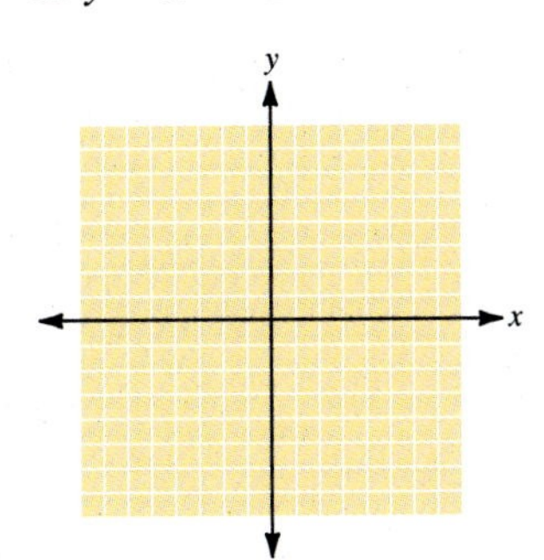

3. $y = x^2 - 5$

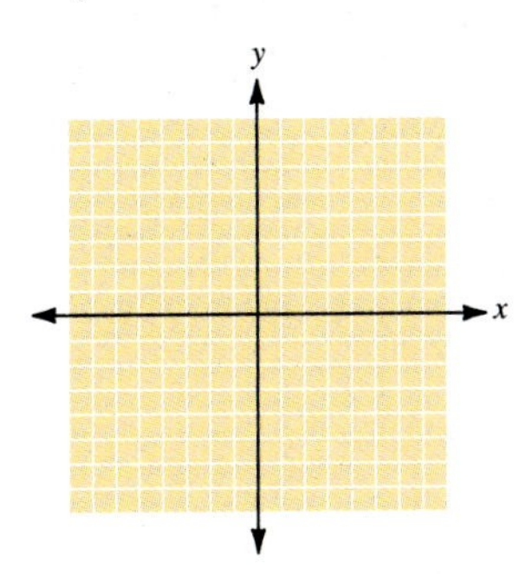

4. $y = (x - 3)^2$

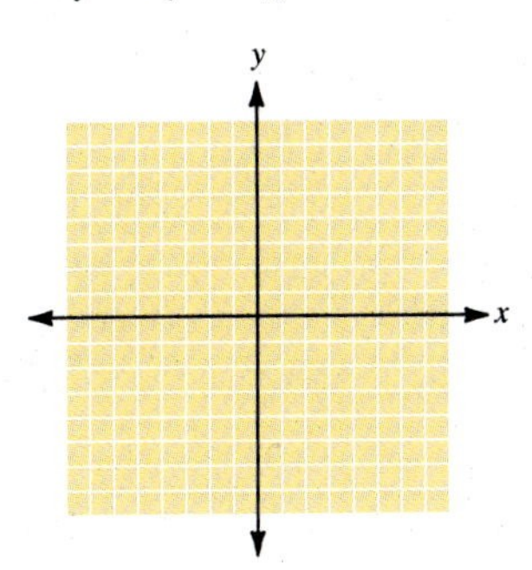

5. $y = (x + 2)^2$

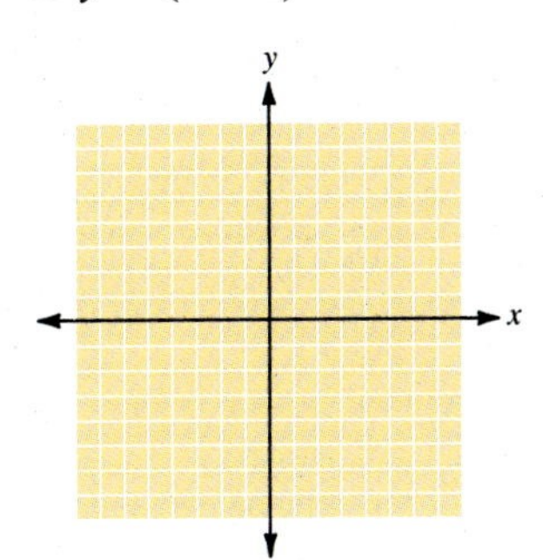

6. $y = -(x - 3)^2$

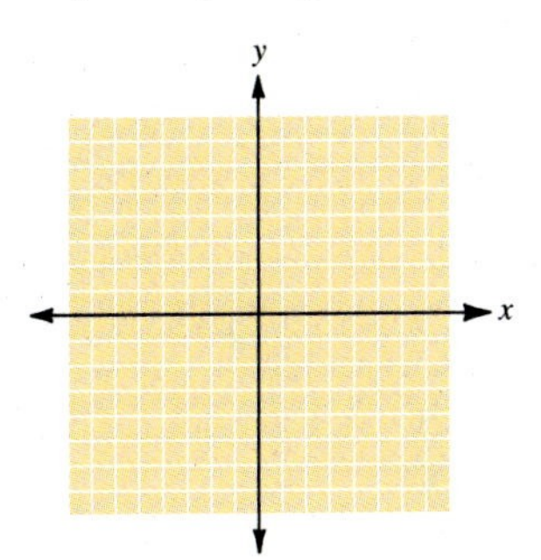

7. $y = (x + 3)^2 + 1$

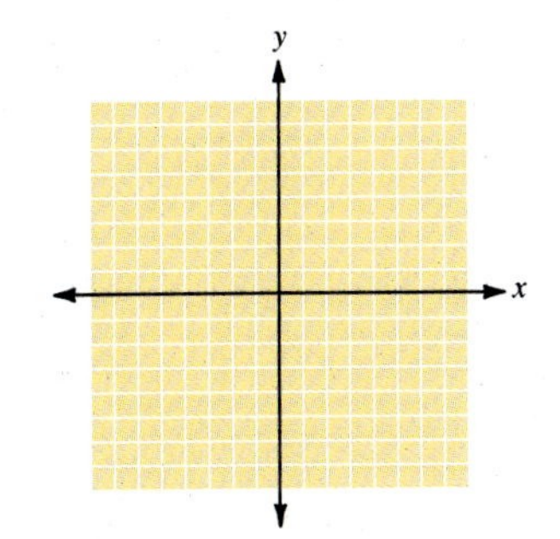

8. $y = -(x + 2)^2 - 3$

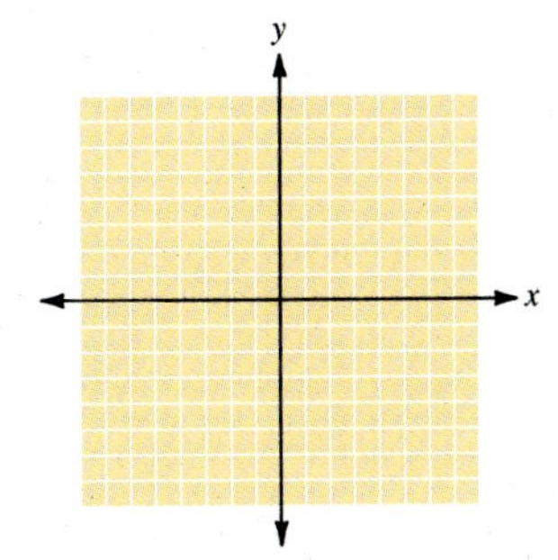

9. $y = -(x - 5)^2 - 2$

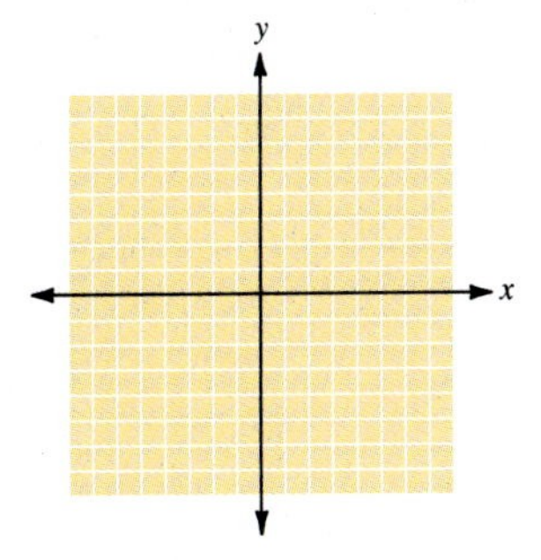

10. $y = 2(x - 2)^2 - 5$

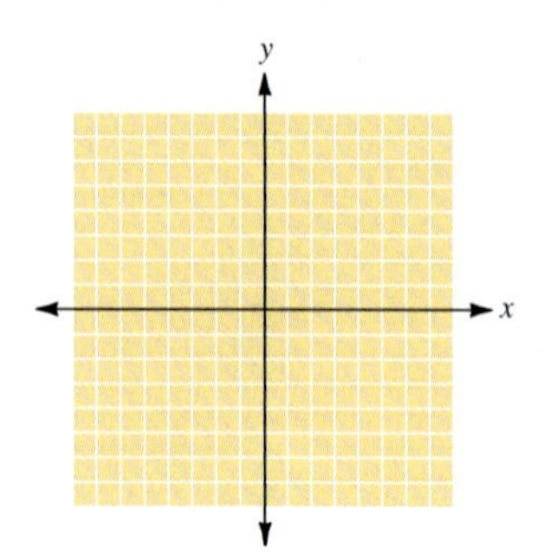

11. $y = x^2 + 2x - 3$

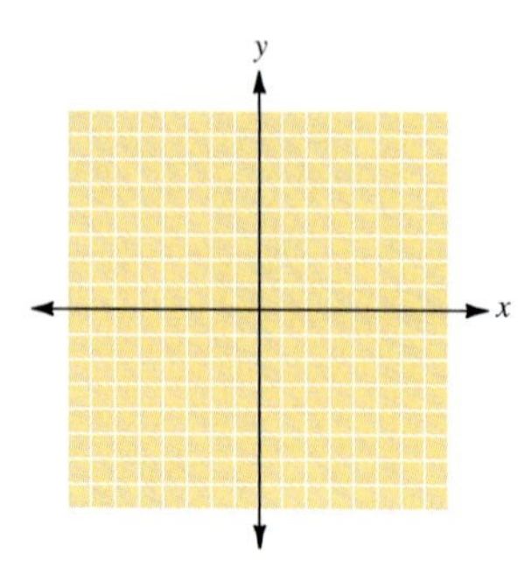

12. $y = x^2 - 4x + 3$

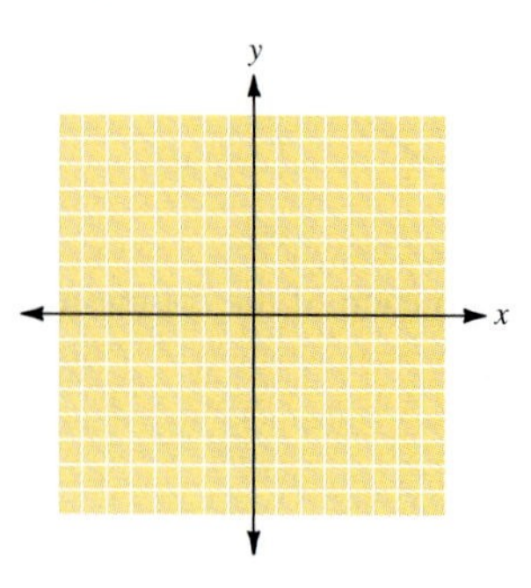

13. $y = -x^2 - x + 6$

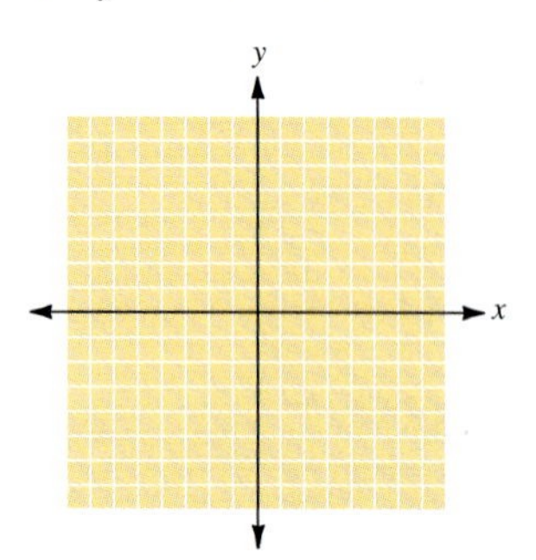

14. $y = -x^2 + 3x + 4$

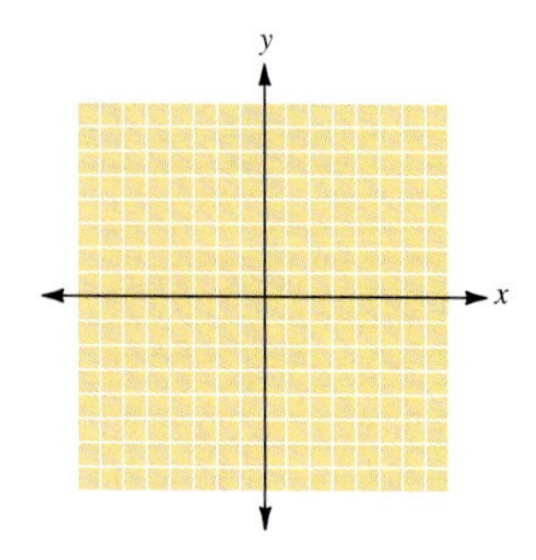

15. $y = x^2 + 4x + 5$

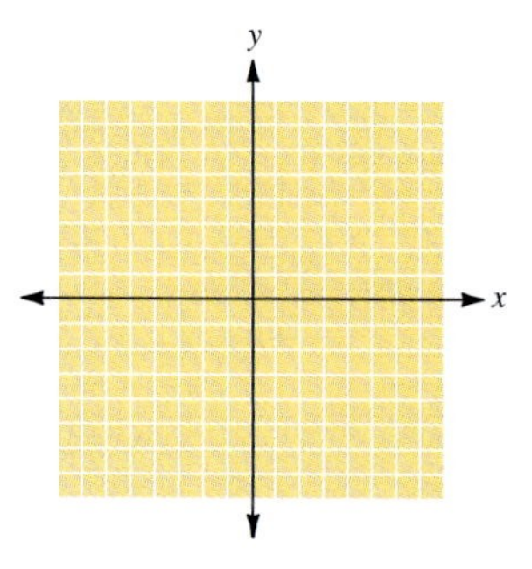

16. $y = x^2 - 6x + 4$

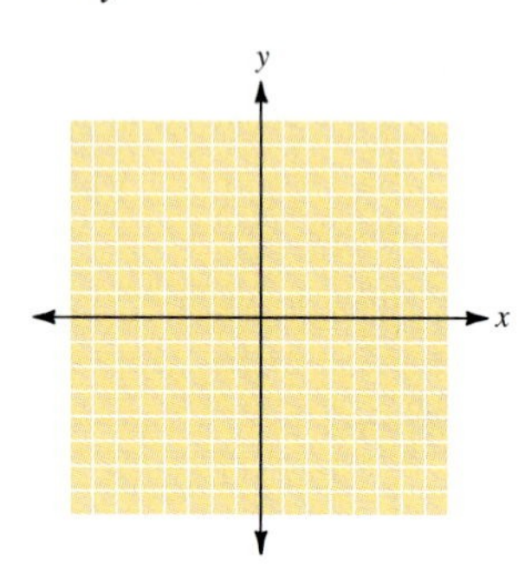

17. $y = x^2 - 2x + 4$

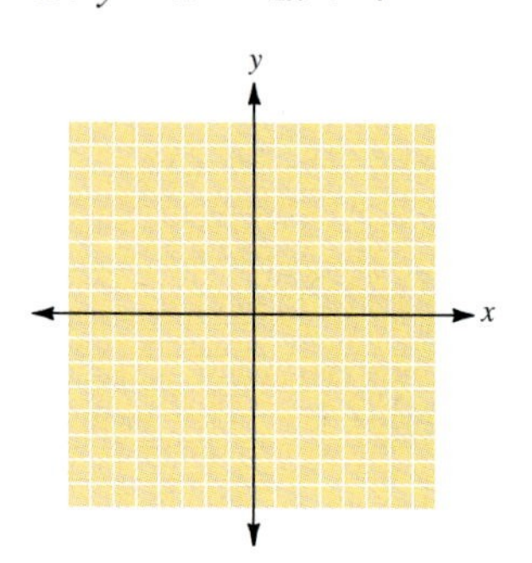

18. $y = -x^2 + 2x - 2$

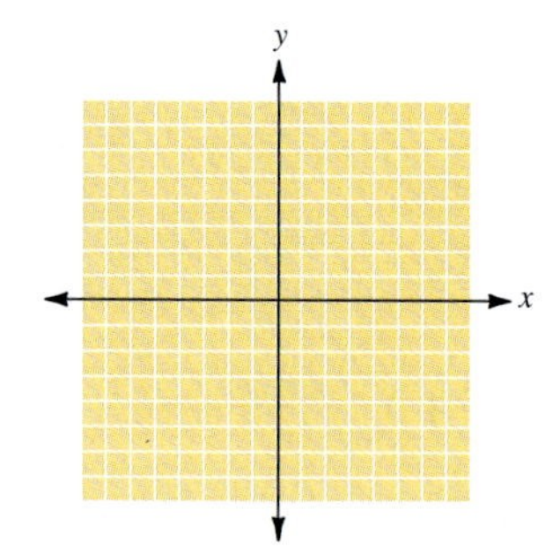

19. $y = 2x^2 - 4x + 1$

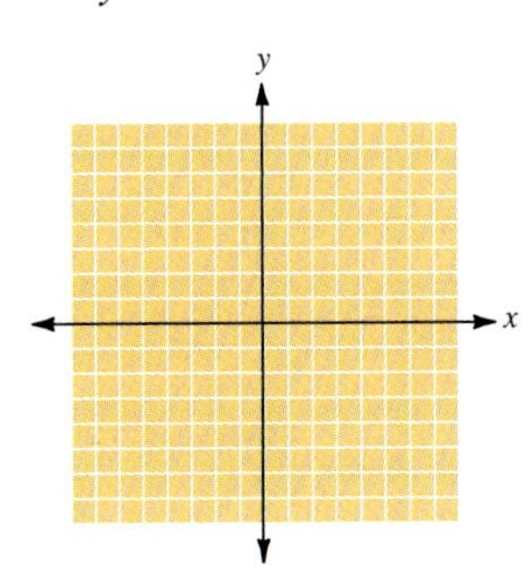

20. $y = \frac{1}{2}x^2 - 4x$

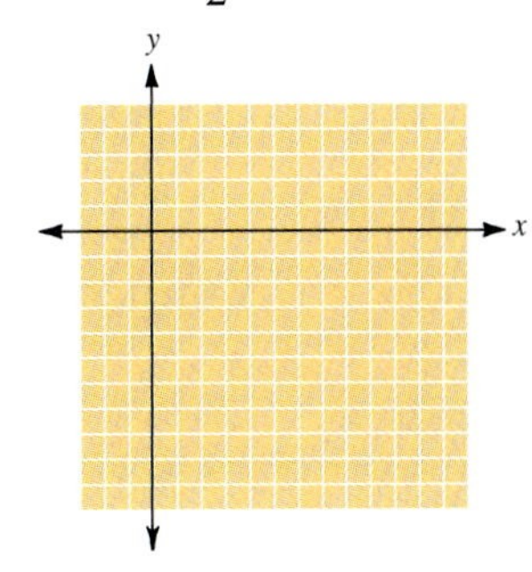

21. $x = y^2 - 4y$

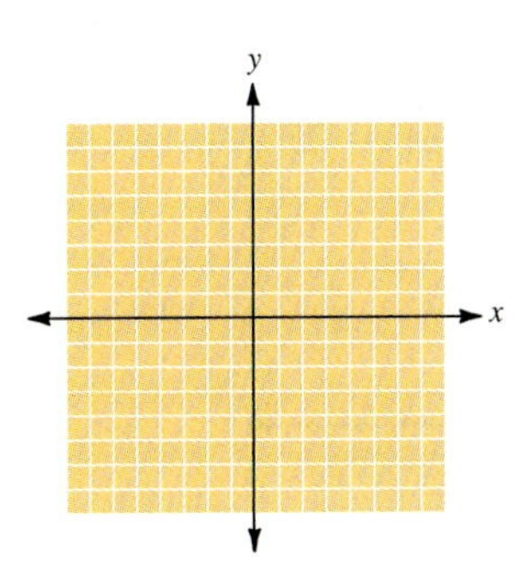

22. $x = -y^2 + 4y$

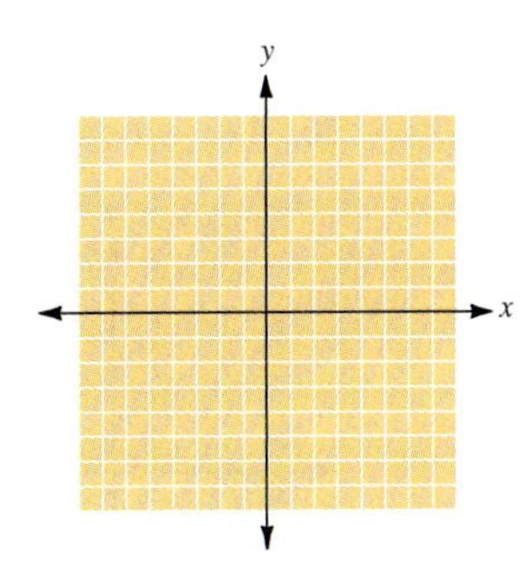

[10.2] Find the center and the radius of the graph of each equation.

23. $x^2 + y^2 = 16$

24. $x^2 + y^2 = 50$

25. $4x^2 + 4y^2 = 36$

26. $3x^2 + 3y^2 = 36$

27. $(x - 3)^2 + y^2 = 36$

28. $(x - 2)^2 + y^2 = 9$

29. $(x - 1)^2 + (y - 2)^2 = 16$

30. $x^2 + 6x + y^2 + 4y = 12$

31. $x^2 + 8x + y^2 + 10y = 23$

32. $x^2 - 6x + y^2 + 6y = 18$

33. $x^2 + y^2 - 4y - 5 = 0$

34. $x^2 - 2x + y^2 - 6y = 6$

Graph each of the following.

35. $x^2 + y^2 = 16$

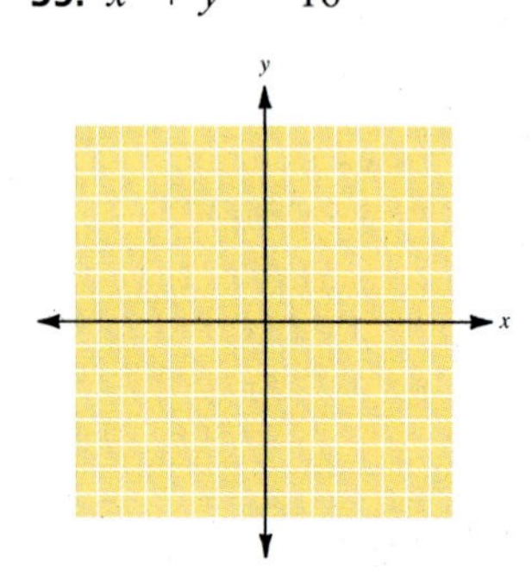

36. $4x^2 + 4y^2 = 36$

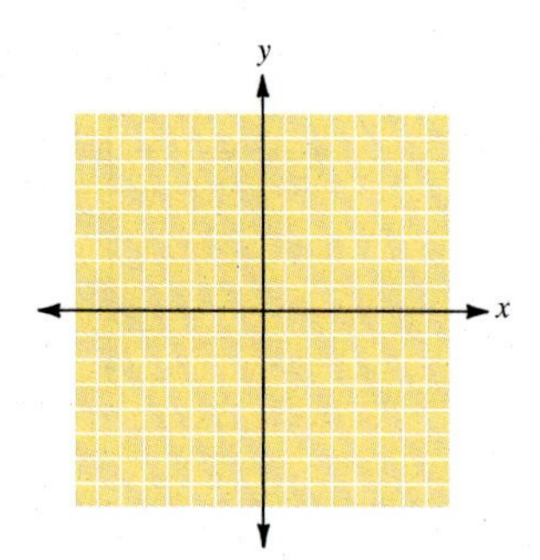

37. $x^2 + (y + 3)^2 = 25$

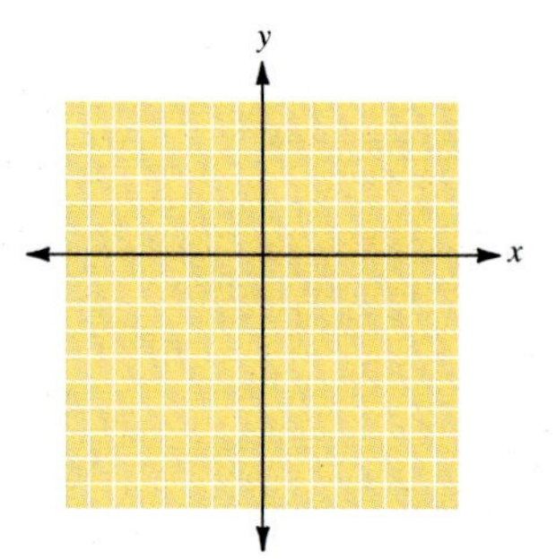

38. $(x - 2)^2 + y^2 = 9$

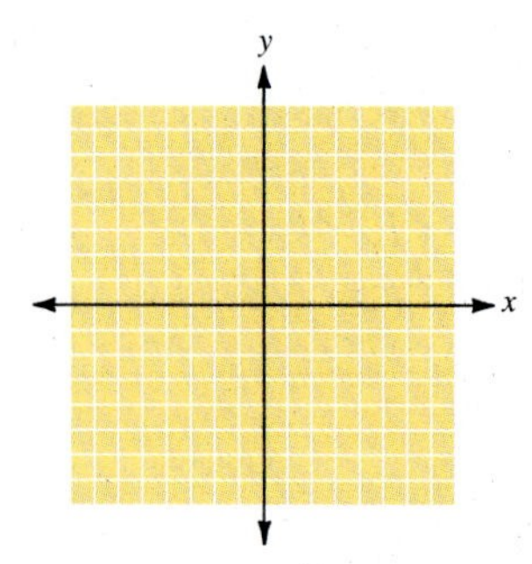

39. $(x - 1)^2 + (y - 2)^2 = 16$

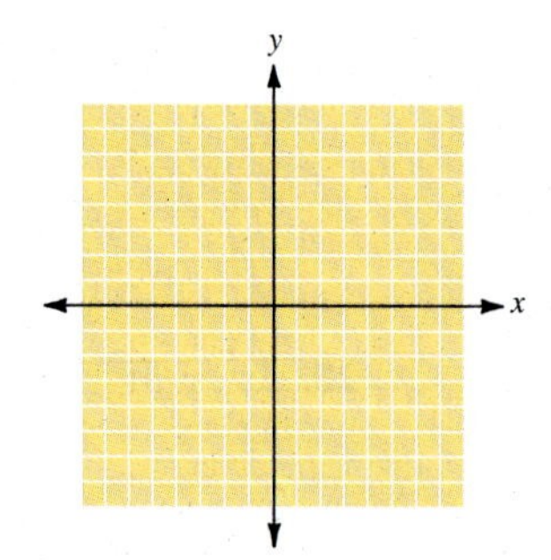

40. $(x + 3)^2 + (y + 2)^2 = 25$

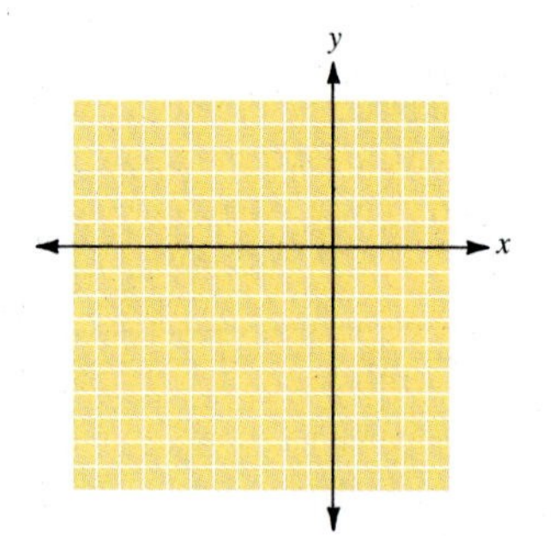

41. $x^2 + y^2 - 4y - 5 = 0$

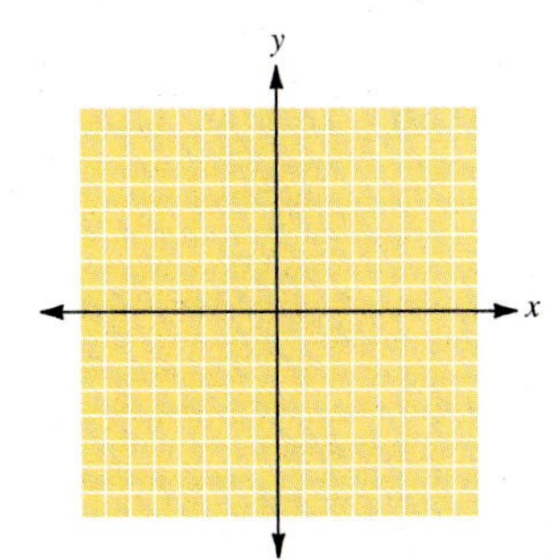

42. $x^2 - 2x + y^2 - 6y = 6$

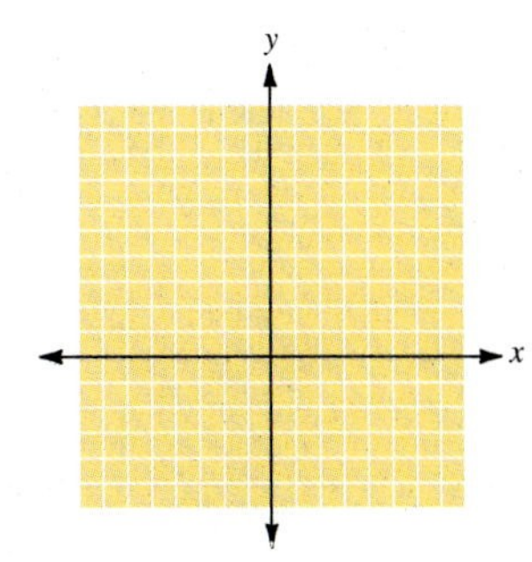

[10.1–10.3] For each of the following equations decide whether its graph is a line, a parabola, a circle, an ellipse, or a hyperbola.

43. $x + y = 16$

44. $x + y^2 = 5$

45. $4x^2 + 4y^2 = 36$

46. $3x + 3y = 36$

47. $y = (x - 3)^2$

48. $(x - 2)^2 + y^2 = 9$

49. $y = (x - 1)^2 + 1$

50. $x = y^2 + 4y + 4$

[10.3] Graph each of the following.

51. $\frac{x^2}{25} + \frac{y^2}{9} = 1$

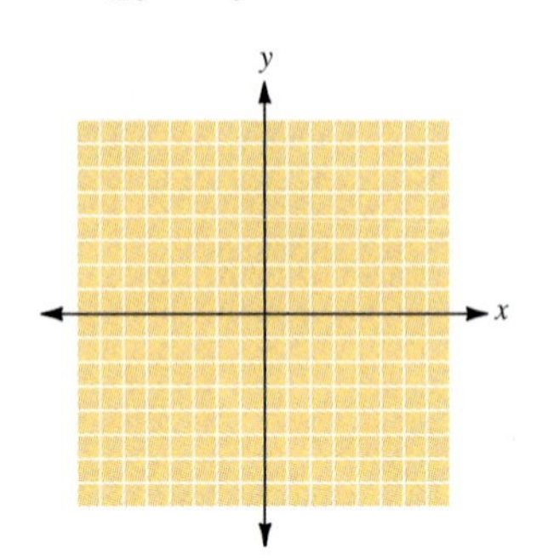

52. $\frac{x^2}{4} + \frac{y^2}{16} = 1$

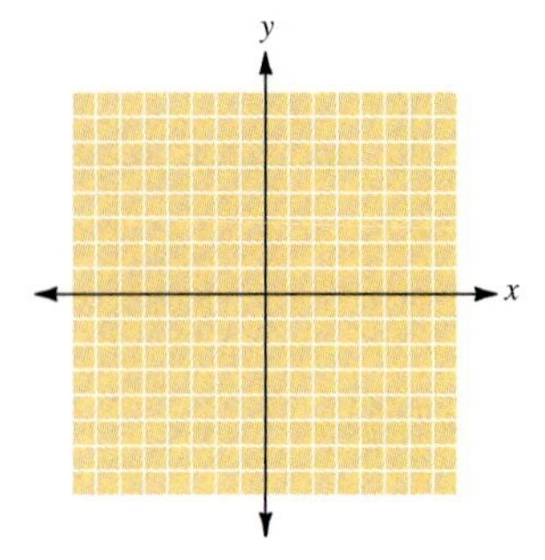

53. $9x^2 + 4y^2 = 36$

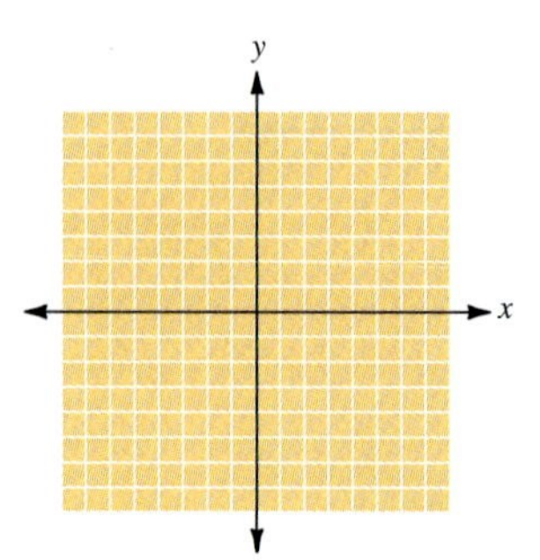

54. $16x^2 + 9y^2 = 144$

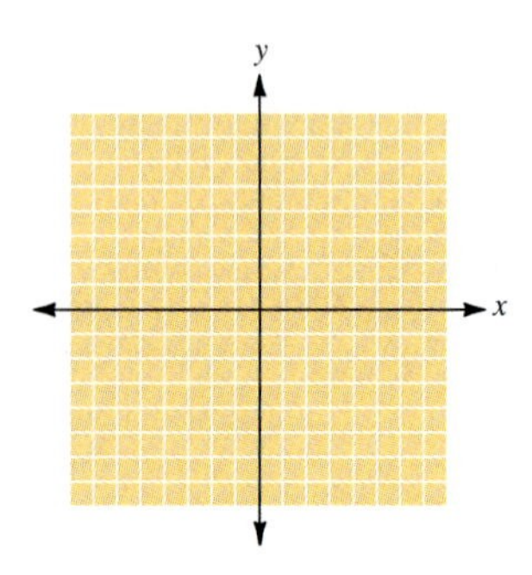

55. $\frac{x^2}{9} - \frac{y^2}{4} = 1$

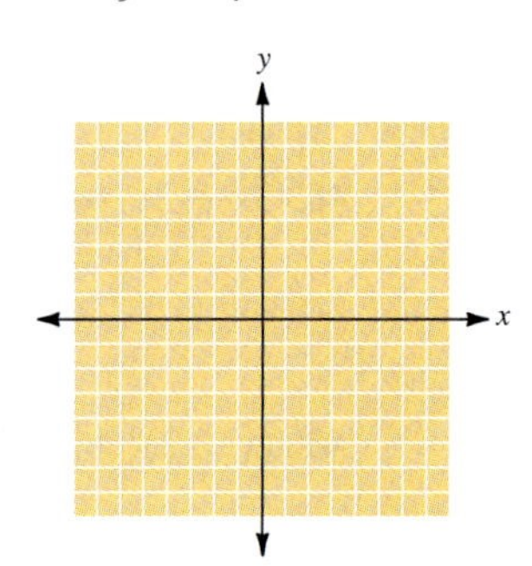

56. $\frac{y^2}{16} - \frac{x^2}{4} = 1$

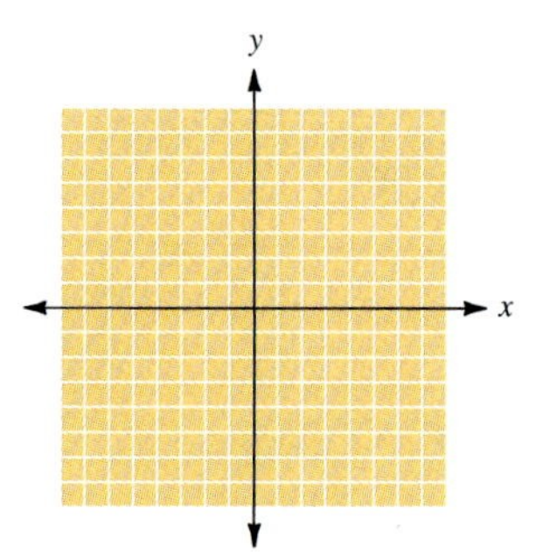

57. $4x^2 - 9y^2 = 36$

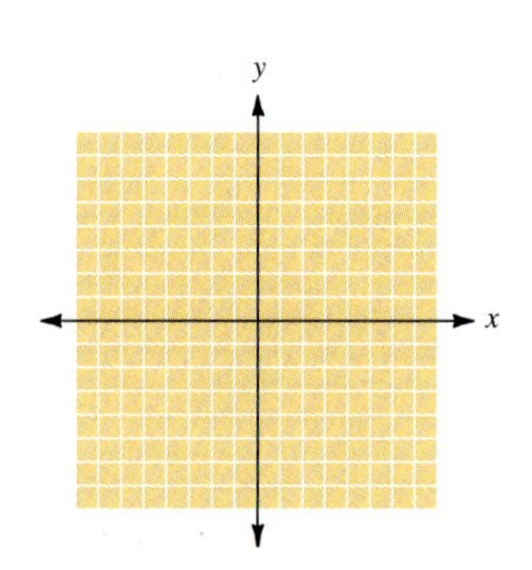

58. $16x^2 - 9y^2 = 144$

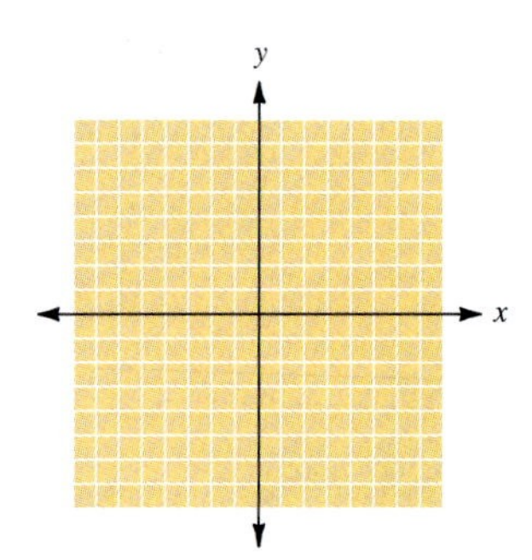

[10.4] Solve each of the following systems graphically.

59. $y = x^2 - x + 4$
$y = 4$

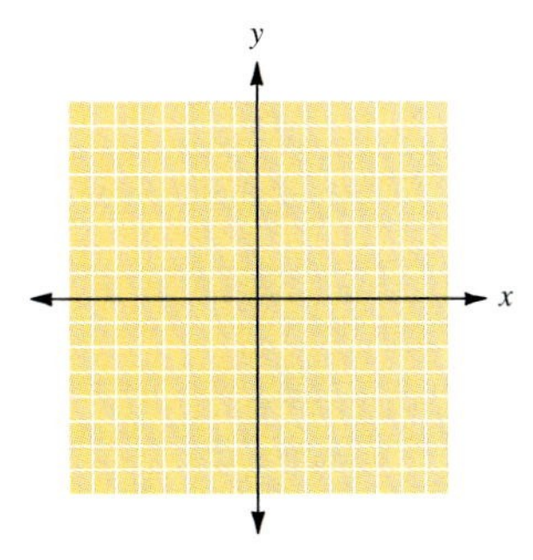

60. $y = x^2 + x - 9$
$y = 3$

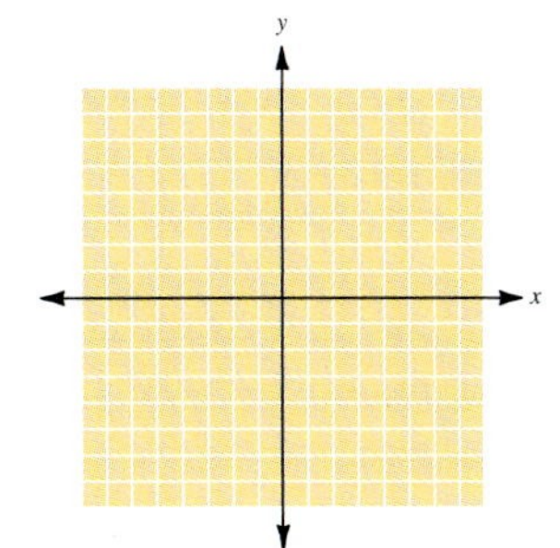

61. $y = x^2 - 3x + 1$
$y = 5$

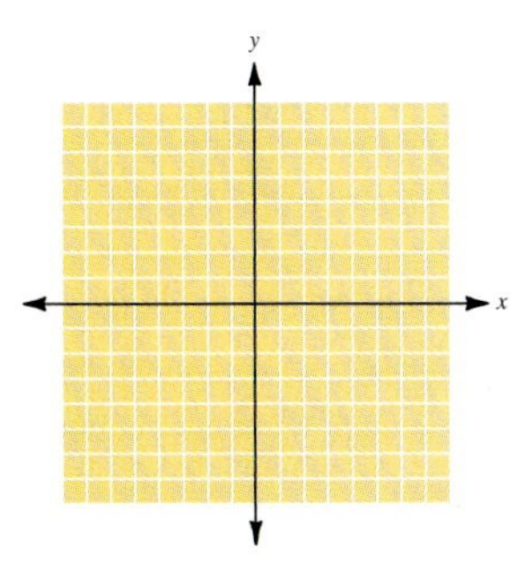

62. $y = x^2 + 12x + 40$
$y = 5$

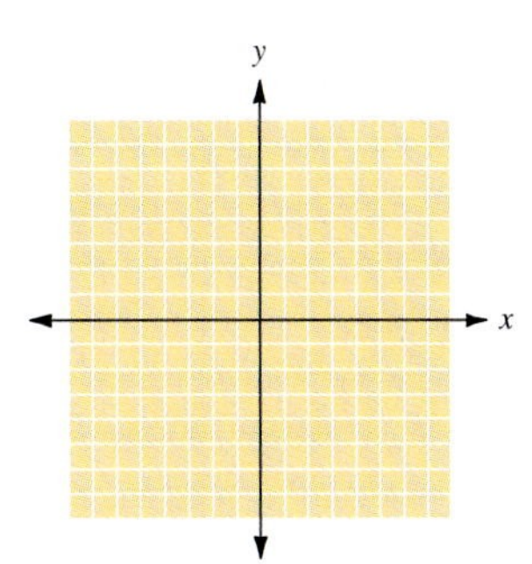

63. $y = x^2 + x - 6$
$y = 6$

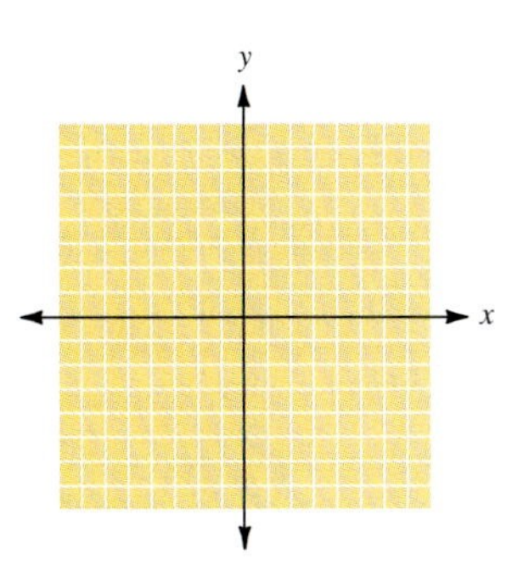

Solve the following systems algebraically.

64. $y = x^2 - x + 4$
$y = 4$
(See exercise 59.)

65. $y = x^2 + x - 9$
$y = 3$
(See exercise 60.)

66. $y = x^2 - 3x + 1$
$y = 5$
(See exercise 61.)

67. $y = x^2 + 12x + 40$
$y = 5$
(See exercise 62.)

68. $y = x^2 + x - 6$
$y = 6$
(See exercise 63.)

Solve each of the following systems of inequalities.

69. $y \geq x^2 - x + 4$
$y \geq 4$
(See exercise 59.)

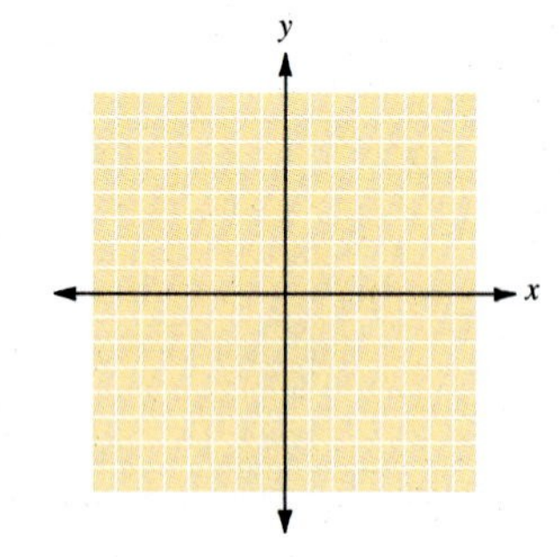

70. $y \geq x^2 + x - 9$
$y \leq 3$
(See exercise 60.)

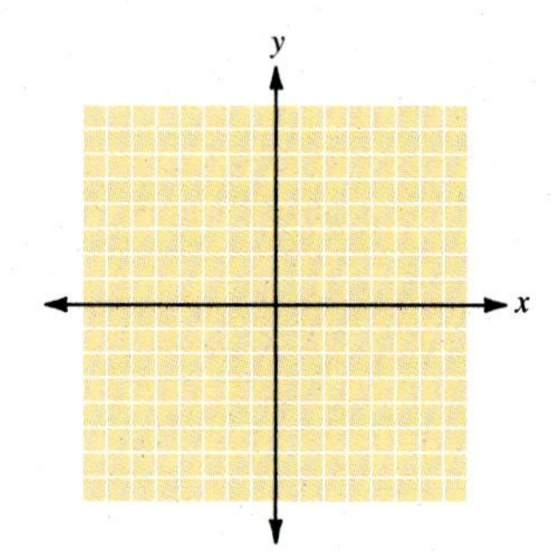

71. $y \geq x^2 - 3x + 1$
$y \leq 5$
(See exercise 61.)

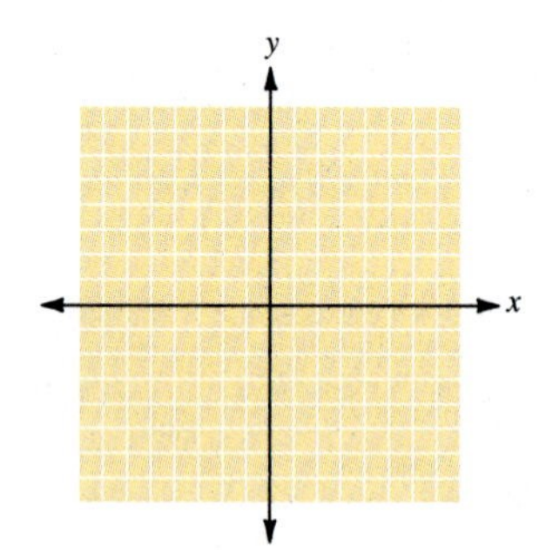

72. $y \geq x^2 + 12x + 40$
$y \leq 5$
(See exercise 62.)

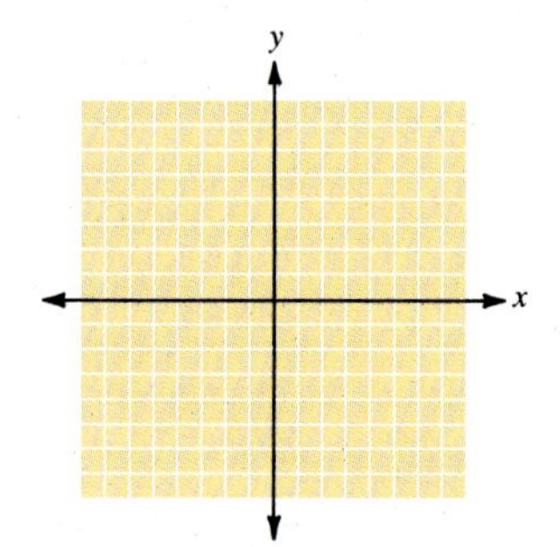

73. $y \leq x^2 + x - 6$
$y \leq 6$
(See exercise 63.)

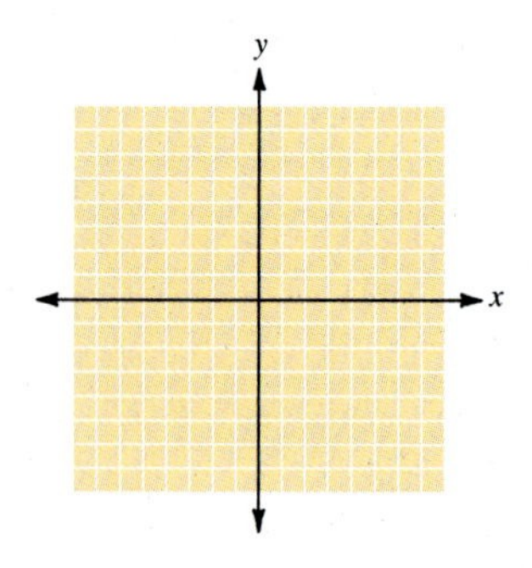

Self-Test for Chapter 10

Name ____________

Section ______ Date ______

ANSWERS

1. ____________
2. ____________
3. ____________
4. ____________

The purpose of this self-test is to help you check your progress and to review for a chapter test in class. Allow yourself about an hour to take the test. When you are done, check your answers in the back of the book. If you missed any questions, be sure to go back and review the appropriate sections in the chapter and the exercises that are provided.

Graph each of the following equations.

1. $y = (x - 5)^2$

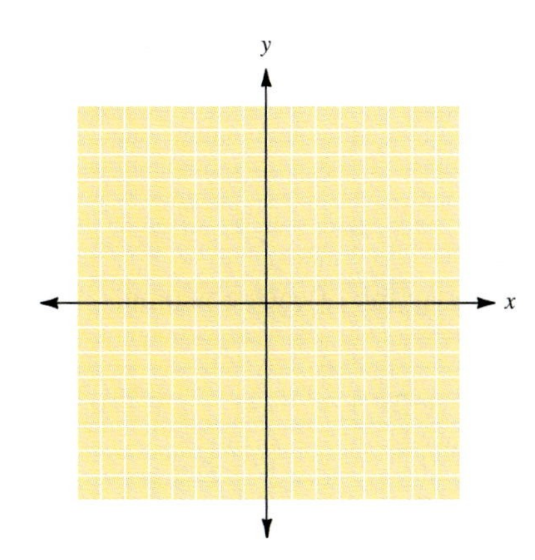

2. $y = (x + 2)^2 - 3$

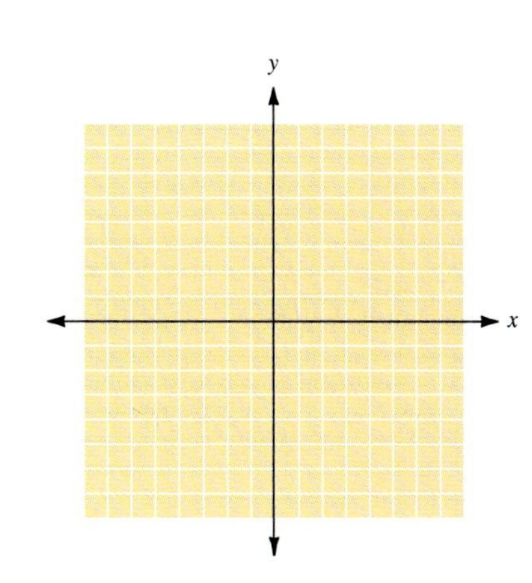

3. $y = -2(x - 3)^2 - 1$

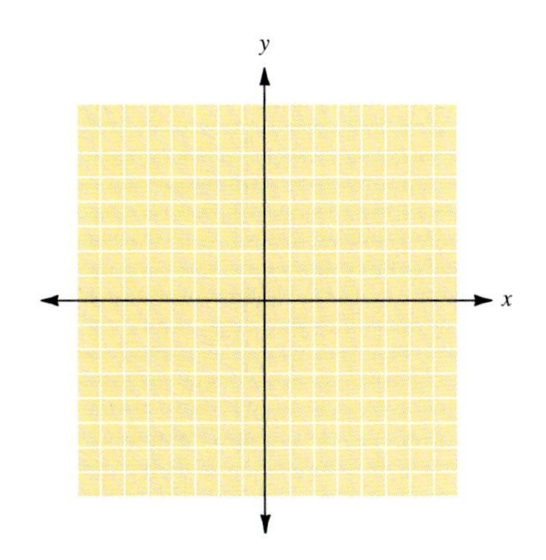

4. $y = 3x^2 + 9x + 2$

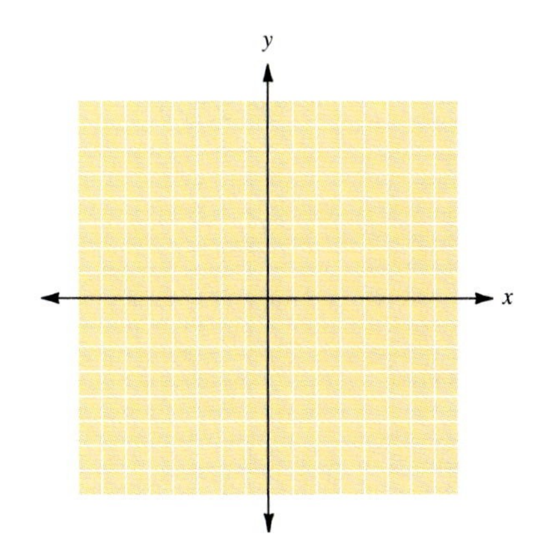

ANSWERS

5. ______

6. ______

7. ______

8. ______

9. ______

10. ______

5. $x = \frac{1}{2}(y - 4)^2 + 2$

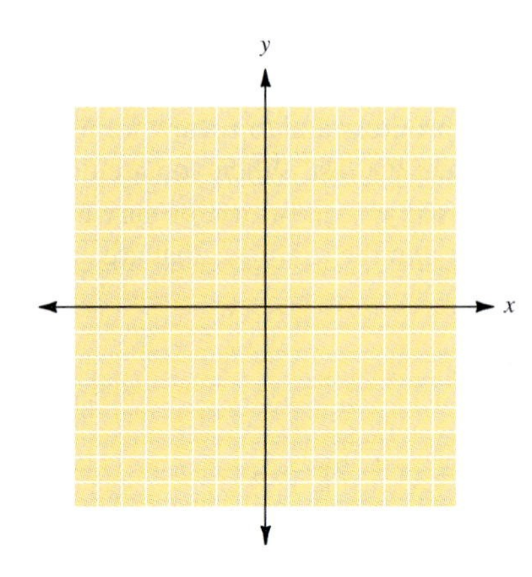

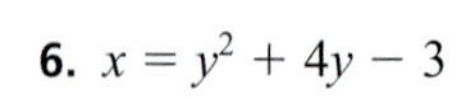

6. $x = y^2 + 4y - 3$

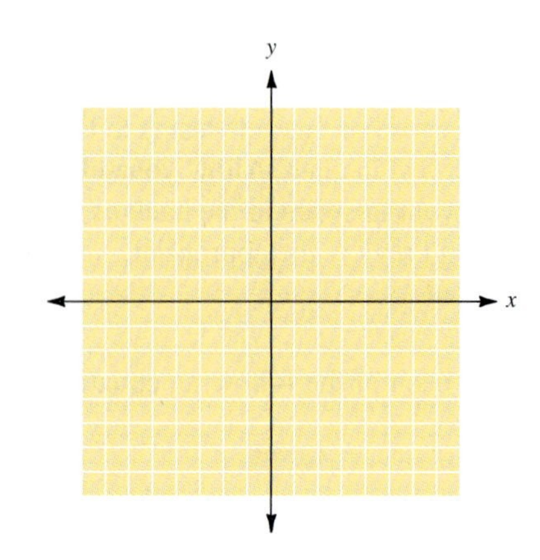

Find the coordinates for the center and the radius of the graph of each equation.

7. $(x - 3)^2 + (y + 2)^2 = 36$

8. $x^2 + 2x + y^2 - 4y - 21 = 0$

Sketch the graph of each of the following equations.

9. $(x - 2)^2 + (y + 3)^2 = 9$

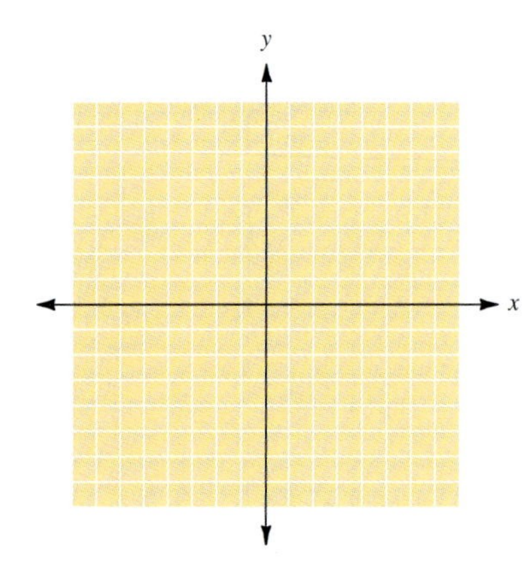

10. $\frac{x^2}{25} + \frac{y^2}{9} = 1$

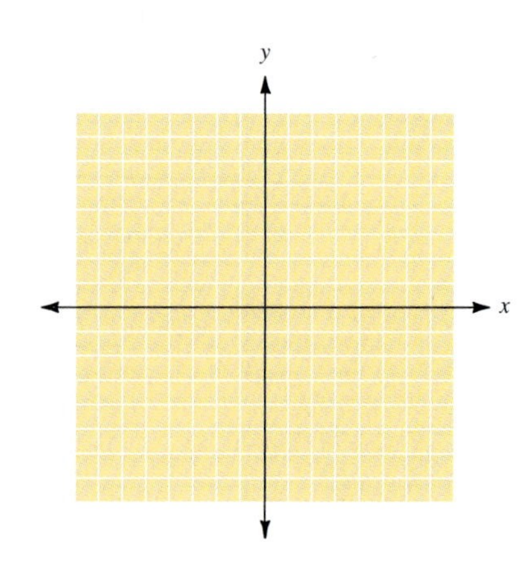

11. $\frac{x^2}{9} - \frac{y^2}{16} = 1$

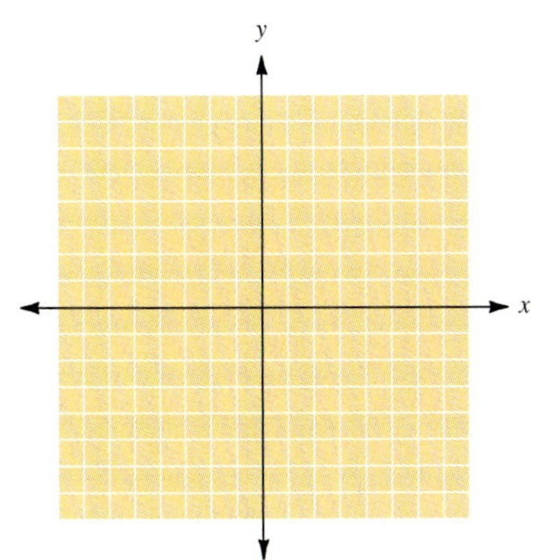

12. $4y^2 - 25x^2 = 100$

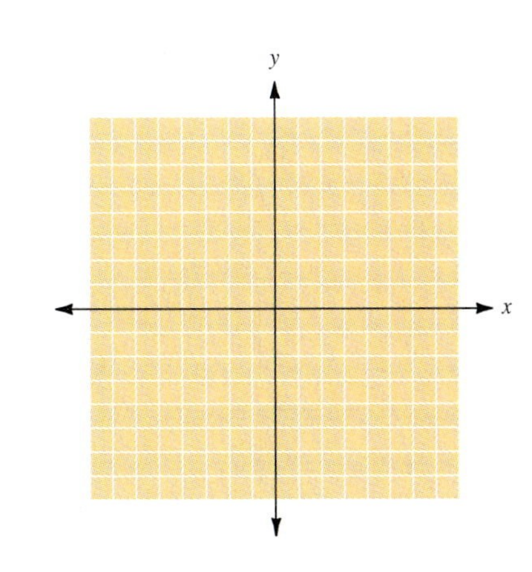

Solve each of the following systems graphically.

13. $y = x^2 + 3x - 5$
$y = 5$

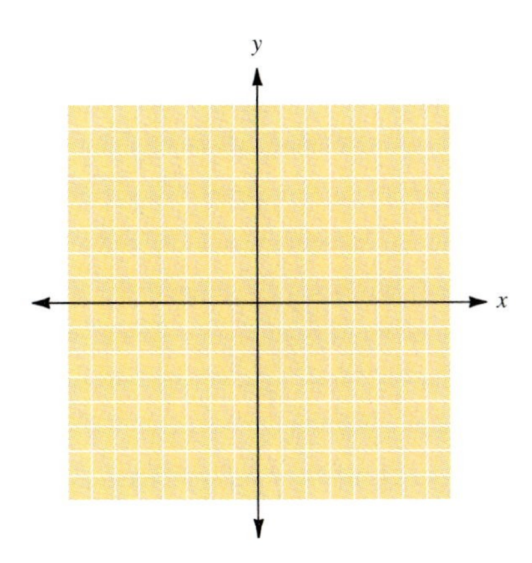

14. $y = x^2 - x - 8$
$y = 4$

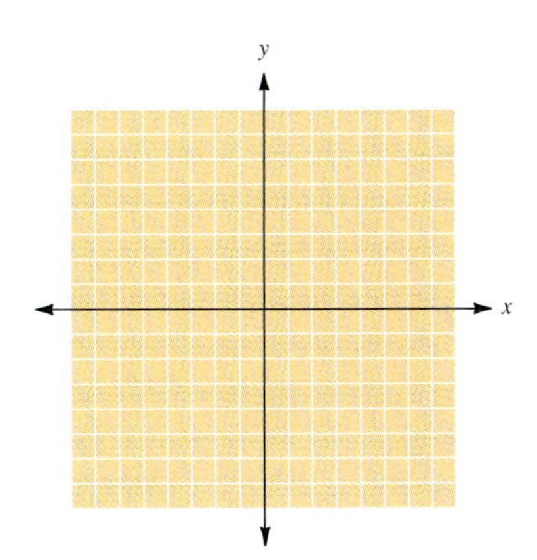

Solve each of the following systems algebraically.

15. $y = x^2 + 3x - 5$
$y = 5$
(See exercise 13.)

16. $y = x^2 - x - 8$
$y = 4$
(See exercise 14.)

ANSWERS

11. ______

12. ______

13. ______

14. ______

15. ______

16. ______

ANSWERS

17. ____________

18. ____________

19. ____________

20. ____________

Solve each of the following systems of inequalities.

17. $y \geq x^2 + 3x - 5$

$y \leq 5$

(See exercise 13.)

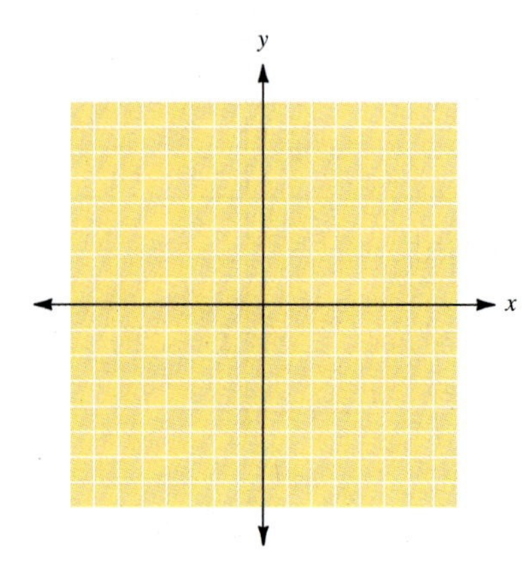

18. $y \geq x^2 - x - 8$

$y \geq 4$

(See exercise 14.)

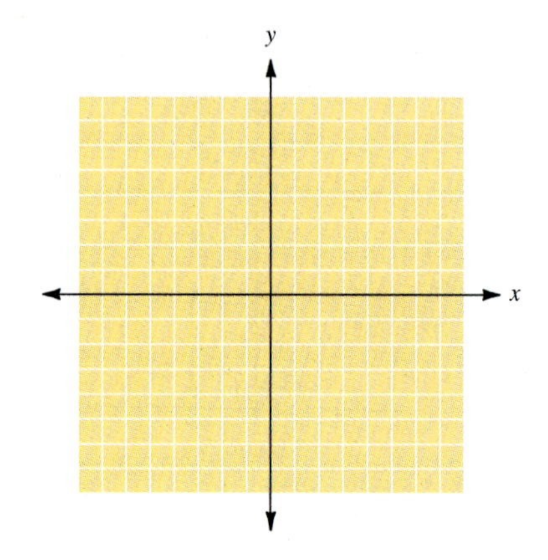

Identify the following as line, circle, parabola, ellipse or hyperbola.

19. $x^2 - 4x + y^2 + 6y - 3 = 0$

20. $9x^2 - 4y^2 = 36$

Cumulative Test for Chapters 1 to 10

Name ______

Section ______ Date ______

ANSWERS

1. ______
2. ______
3. ______
4. ______
5. ______
6. ______
7. ______
8. ______
9. ______
10. ______
11. ______
12. ______
13. ______
14. ______
15. ______
16. ______
17. ______

This test is provided to help you in the process of reviewing the previous chapters. Answers are provided in the back of the book. If you missed any answers, be sure to go back and review the appropriate chapter section.

Solve each of the following.

1. $3x - 2(x + 5) = 12 - 3x$

2. $2x - 7 < 3x - 5$

3. $|2x - 3| = 5$

4. $|3x + 5| \leq 7$

5. $|5x - 4| > 21$

6. $x^2 - 5x - 24 = 0$

7. $\sqrt{x} = \sqrt{10x - 9}$

8. $\dfrac{5y}{y + 1} - \dfrac{y}{3y + 3} = \dfrac{-56}{6y + 6}$

9. $2x^2 = 2x + 1$

Graph each of the following.

10. $5x + 7y = 35$

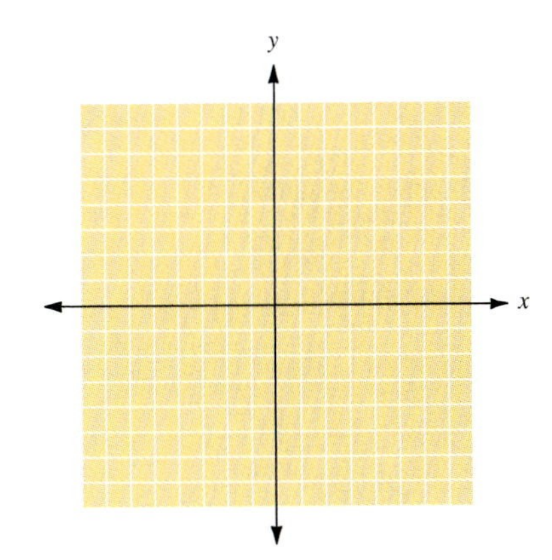

11. $2x + 3y < 6$

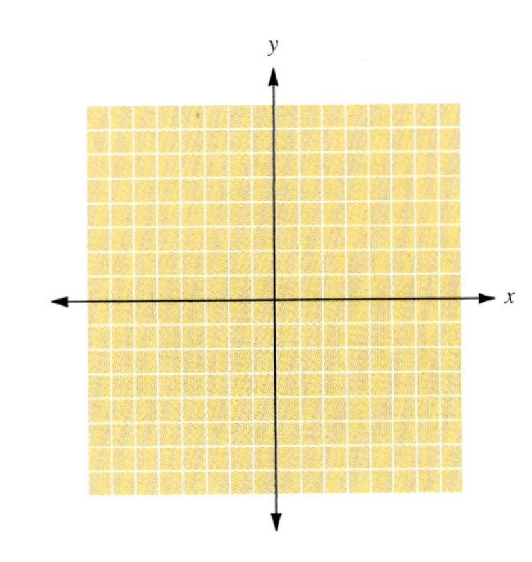

12. Find the distance between the points $(-1, 2)$ and $(4, -22)$.

13. Find the slope of the line connecting $(4, 6)$ and $(3, -1)$.

14. Write the equation of the line that passes through the points $(-1, 4)$ and $(5, -2)$.

Simplify the following polynomials.

15. $(2x + 1)(x - 3)$

16. $(3x - 2)^2$

17. Completely factor the function $f(x) = x^3 - 3x^2 - 5x + 15$.

ANSWERS

18. ______

19. ______

20. ______

21. ______

22. ______

23. ______

24. ______

25. ______

Graph the following:

18. $y = x^2 - 6x + 5$

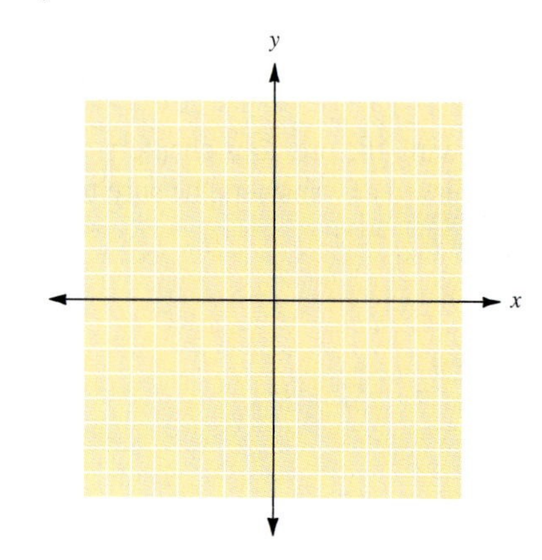

19. $(x + 1)^2 + (y - 2)^2 = 25$

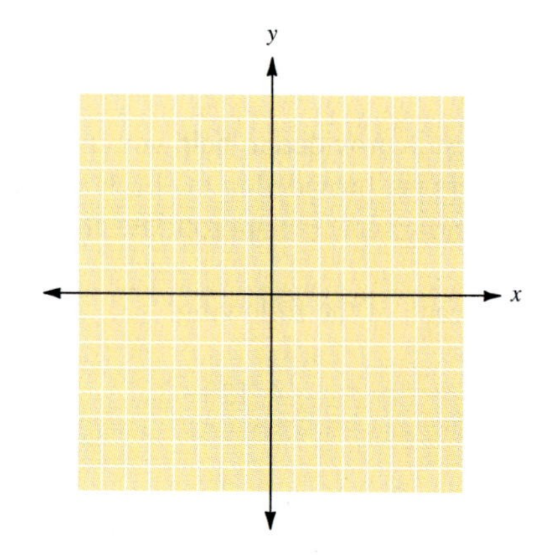

20. $\dfrac{x^2}{64} + \dfrac{y^2}{9} = 1$

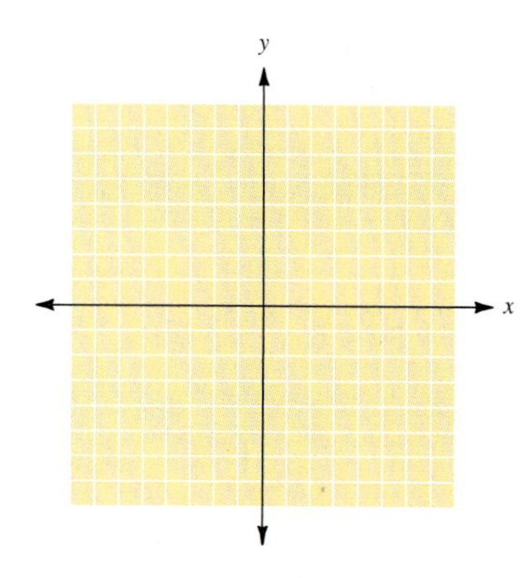

21. Simplify the expression $\dfrac{7x}{3x + 1} - \dfrac{x - 2}{3x + 1}$.

22. Simplify the expression $\sqrt{75w^3} - w\sqrt{48w}$.

Solve the following system of equations.

23. $2x + 3y = 6$
$5x + 3y = -24$

Solve each of the following applications.

24. Geometry. The length of a rectangle is 3 cm more than twice its width. If the perimeter of the rectangle is 54 cm, find the dimensions of the rectangle.

25. Number problem. The sum of the digits of a two-digit number is 10. If the digits are reversed, the new number is 36 less than the original number. What was the original number?

EXPONENTIAL AND LOGARITHMIC FUNCTIONS

11

INTRODUCTION

Pharmacologists researching the effects of drugs use exponential and logarithmic functions to model drug absorption and elimination. After a drug is taken orally, it is distributed throughout the body via the circulatory system. Once in the bloodstream, the drug is carried to the body's organs, where it is first absorbed and then eliminated again into the bloodstream. For a medicine or drug to be effective, there must be enough of the substance in the body to achieve the desired effect but not enough to cause harm. This therapeutic level is maintained by taking the proper dosage at timed intervals determined by the rate the body absorbs or eliminates the medicine.

The rate at which the body eliminates the drug is proportional to the amount of the drug present. That is, the more drug there is, the faster the drug is eliminated. The amount of a drug dosage, P, still left after a number of hours, t, is affected by the **half-life** of the drug. In this case, the half-life is how many hours it takes for the body to use up or eliminate half the drug dosage.

If P is the amount of an initial dose, and H is the time it takes the body to eliminate half a dose of a drug, then the amount of the drug still remaining in the system after t units of time is

$$A(t) = Pe^{t(-\ln 2)/H}$$

If the amount of an initial dose of a drug is 30 mg and if the half-life of the drug in the body is 4 h, the amount in mg of the drug still in the body t h after one dose is given by the following formula:

$$A(t) = 30e^{-0.173t}$$

11.1 Inverse Relations and Functions

11.1 OBJECTIVES

1. Find the inverse of a relation
2. Graph a relation and its inverse
3. Find the inverse of a function
4. Graph a function and its inverse
5. Identify a one-to-one function

Let's consider an extension of the concepts of relations and functions discussed in Chapter 3.

Suppose we are given the relation

$$\{(1, 2), (2, 4), (3, 6)\} \qquad (1)$$

If we *interchange* the first and second components (the x and y values) of each of the ordered pairs in relation (1), we have

$$\{(2, 1), (4, 2), (6, 3)\} \qquad (2)$$

which is another relation. Relations (1) and (2) are called **inverse relations,** and in general we have the following definition.

Definitions: Inverse of a Relation

The *inverse* of a relation is formed by interchanging the components of each of the ordered pairs in the given relation.

Because we know that relations are often specified by equations, it is natural for us to want to work with the concept of the inverse relation in that setting. We form the inverse relation by interchanging the roles of x and y in the defining equation. Example 1 illustrates this concept.

Example 1

Finding the Inverse of a Relation

Find the inverse of the relation.

$$f = \{(x, y) | y = 2x - 4\} \qquad (3)$$

First interchange variables x and y to obtain

NOTE Notice that x and y have been interchanged from the original equation.

$$x = 2y - 4$$

We now solve the defining equation for y.

$$2y = x + 4$$

$$y = \frac{1}{2}x + 2$$

Then, we rewrite the relation in the equivalent form.

$$f^{-1} = \left\{(x, y) | y = \frac{1}{2}x + 2\right\} \qquad (4)$$

NOTE The notation f^{-1} has a *different meaning* from the negative exponent, as in x^{-1} or $\frac{1}{x}$.

The inverse of the original relation (3) is now shown in (4) with the defining equation "solved for y." That inverse is denoted f^{-1} (this is read as "f inverse").

We use the notation f^{-1} to indicate the inverse of f when that inverse is *also a function.*

CHECK YOURSELF 1

Write the inverse relation for $g = \{(x, y) | y = 3x + 6\}$.

The graphs of relations and their inverses are related in an interesting way. First, note that the graphs of the ordered pairs (a, b) and (b, a) always have symmetry about the line $y = x$.

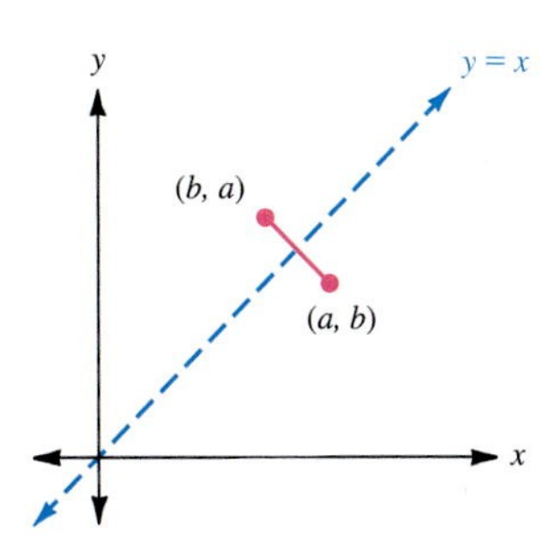

Now, with the above symmetry in mind, let's consider Example 2.

Example 2

Graphing a Relation and Its Inverse

Graph the relation f from Example 1 along with its inverse.

Recall that

$$f = \{(x, y) | y = 2x - 4\}$$

and

$$f^{-1} = \left\{(x, y) \middle| y = \frac{1}{2}x + 2\right\}$$

The graphs of f and f^{-1} are shown below.

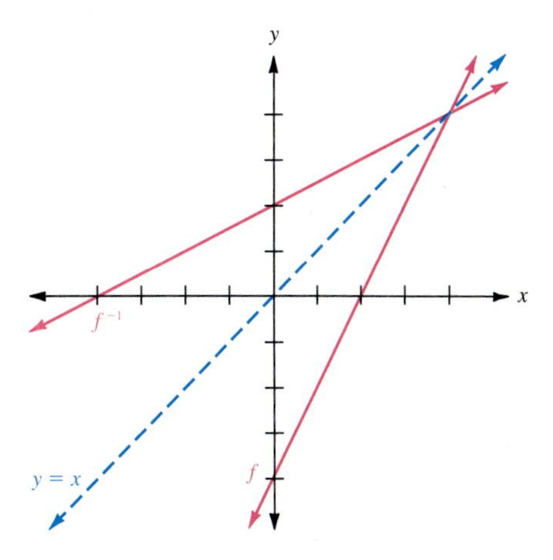

Note that the graphs of f and f^{-1} are symmetric about the line $y = x$. That symmetry follows from our earlier observation about the pairs (a, b) and (b, a) because we simply reversed the roles of x and y in forming the inverse relation.

CHECK YOURSELF 2

Graph the relation g from the Check Yourself 1 exercise along with its inverse.

From our work thus far, it should be apparent that every relation has an inverse. However, that inverse may or may not be a function.

Example 3

Finding the Inverse of a Function

Find the inverses of the following functions.

(a) $f = \{(1, 3), (2, 4), (3, 9)\}$

Its inverse is

$\{(3, 1), (4, 2), (9, 3)\}$

which is also a function.

NOTE The elements of the ordered pairs have been interchanged.

(b) $g = \{(1, 3), (2, 6), (3, 6)\}$

Its inverse is

$\{(3, 1), (6, 2), (6, 3)\}$

which is *not* a function.

NOTE It is not a function because 6 is mapped to both 2 and 3.

CHECK YOURSELF 3

Write the inverses for each of the following relations. Which of the inverses are also functions?

(a) $\{(-1, 2), (0, 3), (1, 4)\}$ **(b)** $\{(2, 5), (3, 7), (4, 5)\}$

Can we predict in advance whether the inverse of a function will also be a function? The answer is yes.

We already know that for a relation to be a function, no element in its domain can be associated with more than one element in its range.

In addition, if the inverse of a function is to be a function, no element in the range can be associated with more than one element in the domain—that is, no two distinct ordered pairs in the function can have the same second component. A function that satisfies this additional restriction is called a **one-to-one function.**

The function in Example 3(a)

$f = \{(1, 3), (2, 4), (3, 9)\}$

is a one-to-one function and its inverse is also a function. However, the function in Example 3(b)

$g = \{(1, 3), (2, 6), (3, 6)\}$

is *not* a one-to-one function, and its inverse is *not* a function.

From those observations we can state the following general result.

Rules and Properties: Inverse of a Function

A function f has an inverse f^{-1}, which is also a function, if and only if f is a one-to-one function.

Because the statement is an "if and only if" statement, it can be turned around without changing the meaning. Here we use the same statement as a definition for a one-to-one function.

NOTE This tells us that the inverse of f is also a function.

Definitions: One-To-One Function

A function f is a *one-to-one function* if and only if no two distinct domain elements are paired with the same range element.

Our result regarding a one-to-one function and its inverse also has a convenient graphical interpretation, as Example 4 illustrates.

Example 4

Graphing a Function and Its Inverse

Graph each function and its inverse. State which inverses are functions.

(a) $f = \{(x, y)|y = 4x - 8\}$

Because f is a one-to-one function (no value for y can be associated with more than one value for x), its inverse is also a function. Here,

$$f^{-1} = \left\{(x, y)\middle|y = \frac{1}{4}x + 2\right\}$$

This is a **linear function** of the form $f = \{(x, y)|y = mx + b\}$. Its graph is a straight line. A linear function, in which $m \neq 0$, is always one-to-one.

The graphs of f and f^{-1} are shown below.

NOTE The vertical-line test tells us that *both* f and f^{-1} are functions.

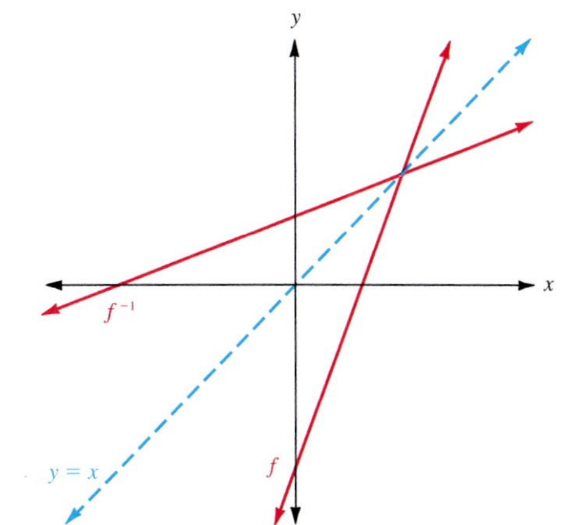

(b) $g = \{(x, y) | y = x^2\}$

This is a **quadratic function** of the form

$g = \{(x, y) | y = ax^2 + bx + c\}$ in which $a \neq 0$

Its graph is always a parabola, and a quadratic function is *not* a one-to-one function.

For instance, 4 in the range is associated with both 2 and -2 from the domain. It follows that the inverse of g

$\{(x, y) | x = y^2\}$

or

$\{(x, y) | y = \pm\sqrt{x}\}$

NOTE By the vertical-line test, we see that the inverse of g is *not* a function because g is *not* one-to-one.

is *not* a function. The graphs of g and its inverse are shown below.

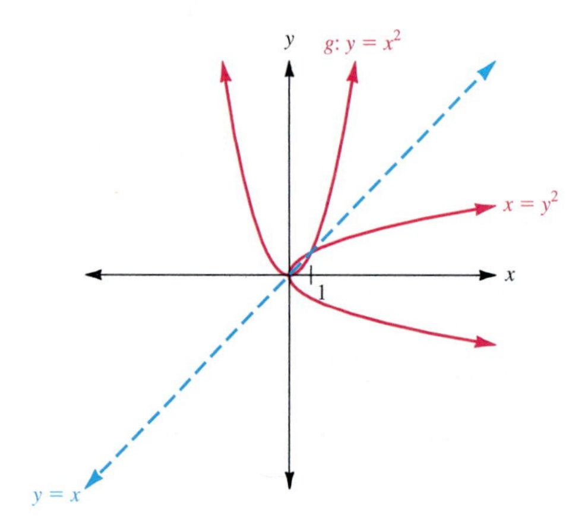

Note: When a function is not one-to-one, as in Example 4(b), we can restrict the domain of the function so that it will be one-to-one. In this case, if we redefine function g as

NOTE The domain is now restricted to nonnegative values for x.

$g = \{(x, y) | y = x^2, x \geq 0\}$

it will be one-to-one and its inverse

$g^{-1} = \{(x, y) | y = \sqrt{x}\}$

will be a function, as shown in the following graph.

NOTE The function g is now one-to-one, and its inverse g^{-1} is also a function.

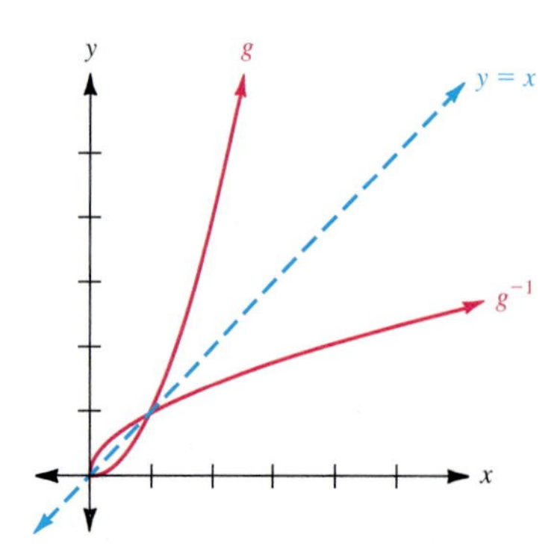

CHECK YOURSELF 4

Graph each function and its inverse. Which inverses are functions?

(a) $f = \{(x, y) | y = 2x - 2\}$ **(b)** $g = \{(x, y) | y = 2x^2\}$

NOTE In Chapter 3, we referred to the *vertical line test* to determine whether a relation was a function. The *horizontal line test* determines whether a function is one-to-one.

It is easy to tell from the graph of a function whether that function is one-to-one. If any horizontal line can meet the graph of a function in at most one point, the function is one-to-one. Example 5 illustrates this approach.

Example 5

Identifying a One-to-One Function

Which of the following graphs represent one-to-one functions?

(a)

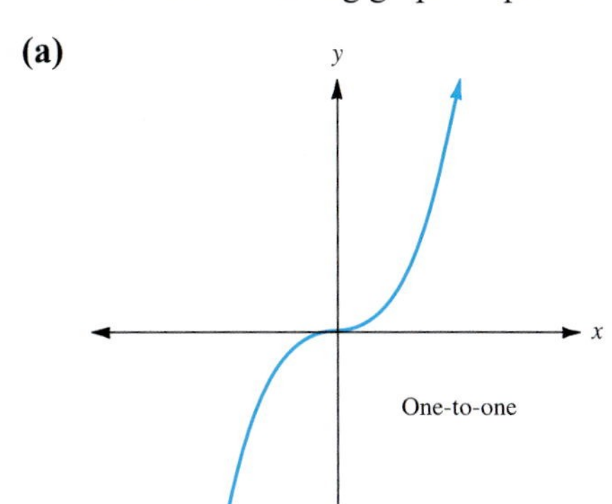

Because no horizontal line passes through any two points of the graph, *f* is one-to-one.

(b)

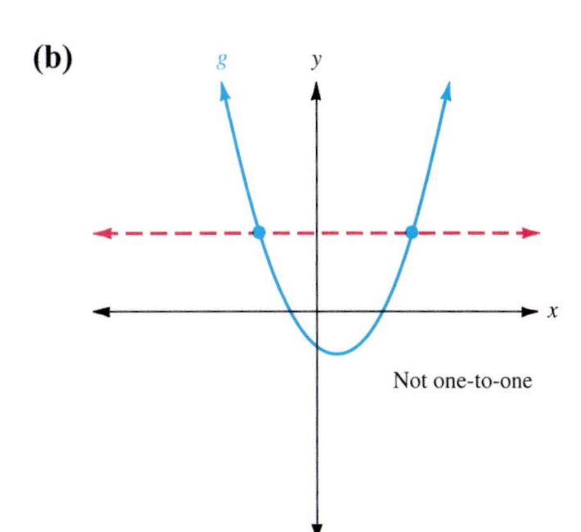

Because a horizontal line can meet the graph of function *g* at two points, *g* is *not* a one-to-one function.

CHECK YOURSELF 5

Consider the graphs of the functions of Check Yourself 4. Which functions are one-to-one?

The following algorithm summarizes our work in this section.

Step by Step: Finding Inverse Relations and Functions

Step 1 Interchange the x and y components of the ordered pairs of the given relation or the roles of x and y in the defining equation.

Step 2 If the relation was described in equation form, solve the defining equation of the inverse for y.

Step 3 If desired, graph the relation and its inverse on the same set of axes. The two graphs will be symmetric about the line $y = x$.

CHECK YOURSELF ANSWERS

1. $g^{-1} = \left\{(x, y) \middle| y = \frac{1}{3}x - 2\right\}$ **2.**

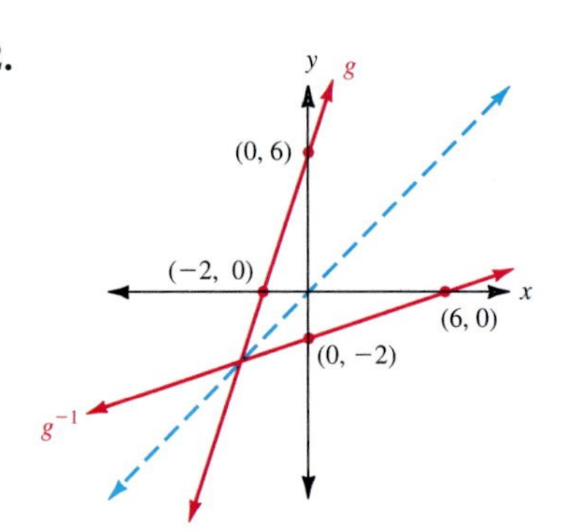

3. **(a)** $\{(2, -1), (3, 0), (4, 1)\}$—a function; **(b)** $\{(5, 2), (7, 3), (5, 4)\}$—*not* a function

4. **(a)** $f = \{(x, y) | y = 2x - 2\}$, $f^{-1} = \left\{(x, y) \middle| y = \frac{1}{2}x + 1\right\}$, the inverse is a function;

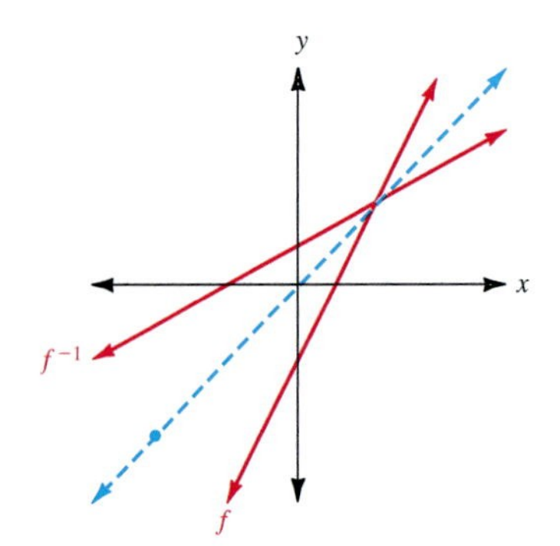

(b) $g = \{(x, y) | y = 2x^2\}$, $g^{-1} = \left\{(x, y) \middle| y = \pm\sqrt{\frac{x}{2}}\right\}$, the inverse is not a function

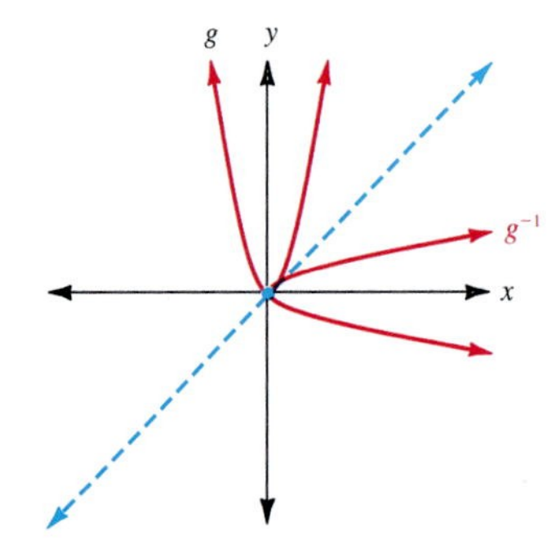

5. **(a)** Is one-to-one; **(b)** is not one-to-one

11.1 Exercises

In exercises 1 to 6, write the inverse relation for each function. In each case, decide whether the inverse relation is also a function.

1. $\{(2, 3), (3, 4), (4, 5)\}$

2. $\{(2, 3), (3, 4), (4, 3)\}$

3. $\{(1, 2), (2, 2), (3, 2)\}$

4. $\{(5, 9), (3, 7), (7, 5)\}$

5. $\{(2, 4), (3, 9), (4, 16)\}$

6. $\{(-1, 2), (0, 3), (1, 2)\}$

In exercises 7 to 16, write an equation for the inverse of the relation defined by each equation.

7. $y = 2x + 8$

8. $y = -2x - 4$

9. $y = \dfrac{x - 1}{2}$

10. $y = \dfrac{x + 1}{3}$

11. $y = x^2 - 1$

12. $y = -x^2 + 2$

13. $x^2 + 4y^2 = 36$

14. $4x^2 + y^2 = 36$

15. $x^2 - y^2 = 9$

16. $4y^2 - x^2 = 4$

In exercises 17 to 22, write an equation for the inverse of the relation defined by each of the following, and graph the relation and its inverse on the same set of axes. Determine which inverse relations are also functions.

17. $y = 3x - 6$

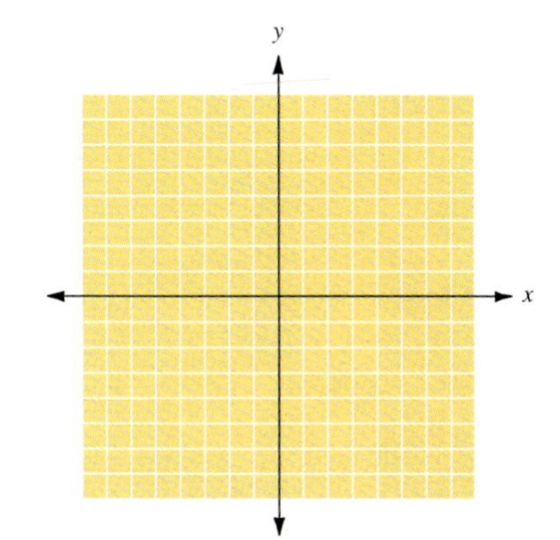

ANSWERS

1. ______
2. ______
3. ______
4. ______
5. ______
6. ______
7. ______
8. ______
9. ______
10. ______
11. ______
12. ______
13. ______
14. ______
15. ______
16. ______
17. ______

ANSWERS

18. ____________________

19. ____________________

20. ____________________

18. $y = 4x + 8$

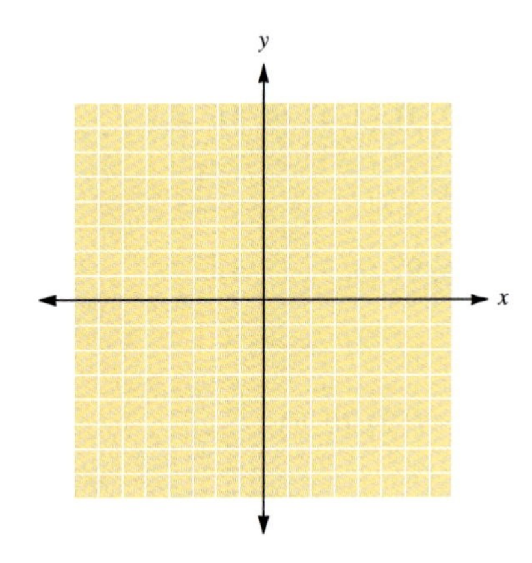

19. $2x - 3y = 6$

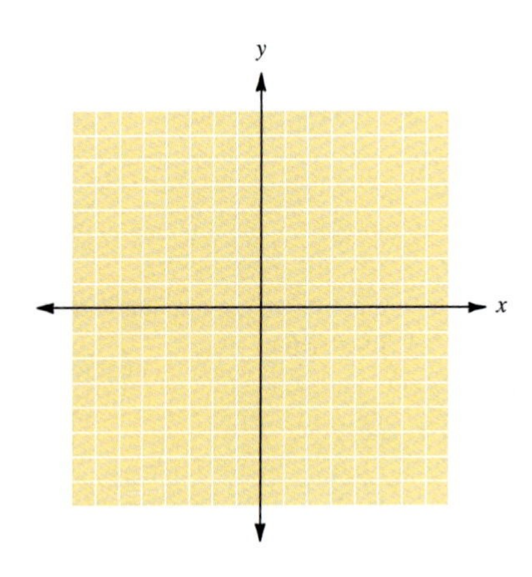

20. $y = 3$

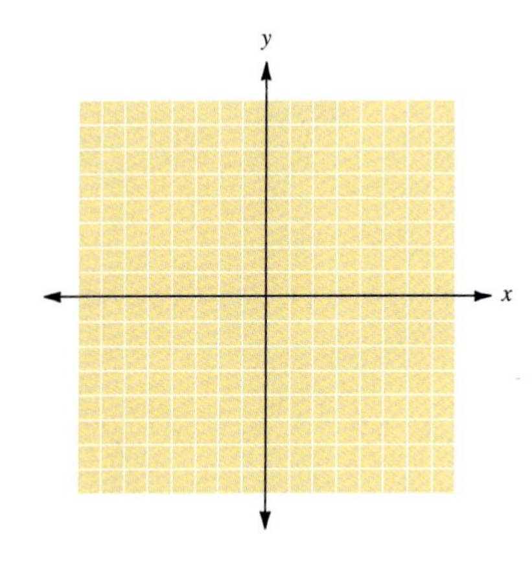

21. $y = x^2 + 1$

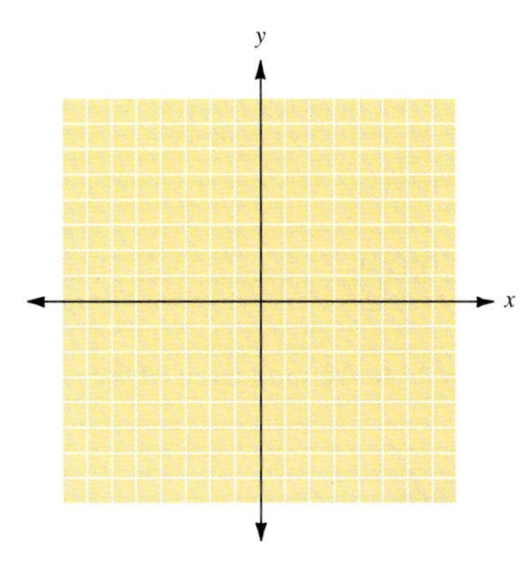

22. $y = -x^2 + 1$

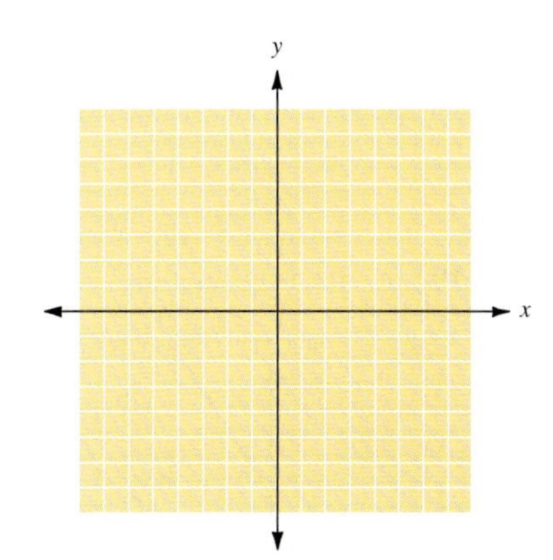

23. An inverse process is an operation that undoes a procedure. If the procedure is wrapping a present, describe in detail the inverse process.

24. If the procedure is the series of steps that take you from home to your classroom, describe the inverse process.

If $f(x) = 3x - 6$, then $f^{-1}(x) = \frac{1}{3}x + 2$. Given these two functions, in exercises 25 to 30, find each of the following.

25. $f(6)$

26. $f^{-1}(6)$

27. $f(f^{-1}(6))$

28. $f^{-1}(f(6))$

29. $f(f^{-1}(x))$

30. $f^{-1}(f(x))$

ANSWERS

21. ______

22. ______

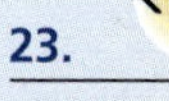

23. ______

24. ______

25. ______

26. ______

27. ______

28. ______

29. ______

30. ______

ANSWERS

31. ____
32. ____
33. ____
34. ____
35. ____
36. ____
37. ____
38. ____
39. ____
40. ____
41. ____
42. ____

If $g(x) = \frac{x + 1}{2}$, then $g^{-1}(x) = 2x - 1$. Given these two functions, in exercises 31 to 36, find each of the following.

31. $g(3)$

32. $g^{-1}(3)$

33. $g(g^{-1}(3))$

34. $g^{-1}(g(3))$

35. $g(g^{-1}(x))$

36. $g^{-1}(g(x))$

Given $h(x) = 2x + 8$, then $h^{-1}(x) = \frac{x - 8}{2}$ in exercises 37 to 42, find each of the following.

37. $h(4)$

38. $h^{-1}(4)$

39. $h(h^{-1}(4))$

40. $h^{-1}(h(4))$

41. $h(h^{-1}(x))$

42. $h^{-1}(h(x))$

Answers

1. $\{(3, 2), (4, 3), (5, 4)\}$; function **3.** $\{(2, 1), (2, 2), (2, 3)\}$; not a function
5. $\{(4, 2), (9, 3), (16, 4)\}$; function **7.** $y = \frac{1}{2}x - 4$ **9.** $y = 2x + 1$
11. $y = \pm\sqrt{x + 1}$ or $x = y^2 - 1$ **13.** $4x^2 + y^2 = 36$ **15.** $y^2 - x^2 = 9$
17. Inverse is a function **19.** Inverse is a function

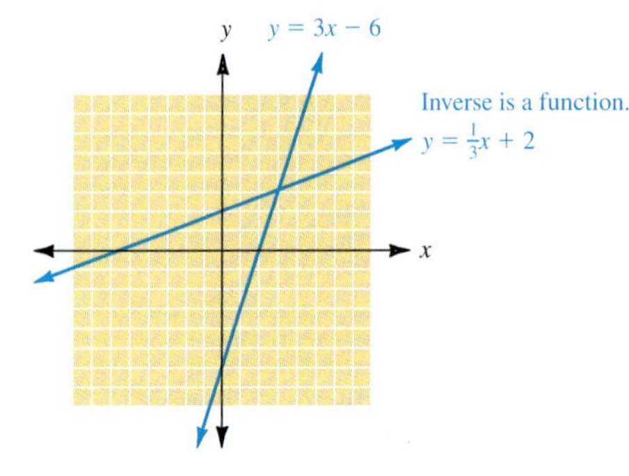

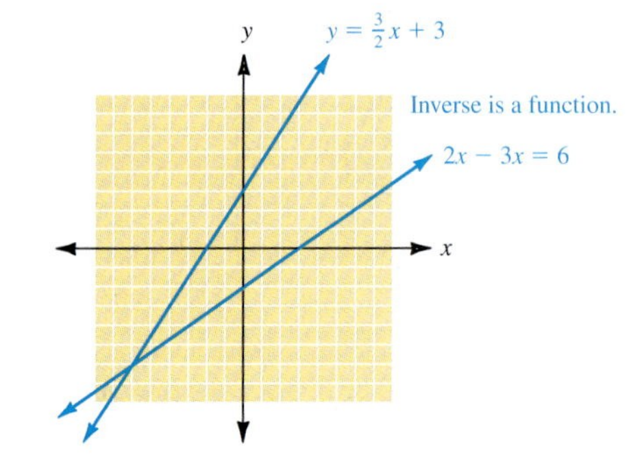

21. Inverse is not a function

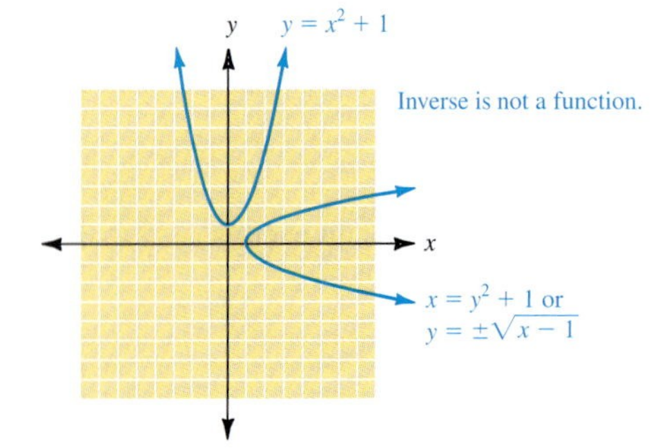

23. **25.** 12 **27.** 6 **29.** x **31.** 2 **33.** 3 **35.** x

37. 16 **39.** 4 **41.** x

11.2 Exponential Functions

11.2 OBJECTIVES

1. Graph an exponential function
2. Solve an application of exponential functions
3. Solve an elementary exponential equation

Up to this point in the text, we have worked with polynomial functions and other functions in which the variable was used as a base. We now want to turn to a new classification of functions, the **exponential function.**

Exponential functions are functions whose defining equations involve the variable as an *exponent.* The introduction of these functions will allow us to consider many further applications, including population growth and radioactive decay.

Definitions: Exponential Functions

An **exponential function** is a function that can be expressed in the form

$f(x) = b^x$

in which $b > 0$ and $b \neq 1$. We call b the **base** of the exponential function.

The following are examples of exponential functions.

$$f(x) = 2^x \qquad g(x) = 3^x \qquad h(x) = \left(\frac{1}{2}\right)^x$$

As we have done with other new functions, we begin by finding some function values. We then use that information to graph the function.

Example 1

Graphing an Exponential Function

Graph the exponential function

$$f(x) = 2^x$$

First, choose convenient values for x.

NOTE

$2^{-2} = \frac{1}{2^2} = \frac{1}{4}$

$$f(0) = 2^0 = 1 \qquad f(-1) = 2^{-1} = \frac{1}{2} \qquad f(1) = 2^1 = 2$$

$$f(-2) = 2^{-2} = \frac{1}{4} \qquad f(2) = 2^2 = 4 \qquad f(-3) = 2^{-3} = \frac{1}{8}$$

Next, form a table from these values. Then, plot the corresponding points, and connect them with a smooth curve for the desired graph.

NOTE There is no value for x such that

$2^x = 0$

so the graph never touches the x axis.

x	$f(x)$
−3	0.125
−2	0.25
−1	0.5
0	1
1	2
2	4
3	8

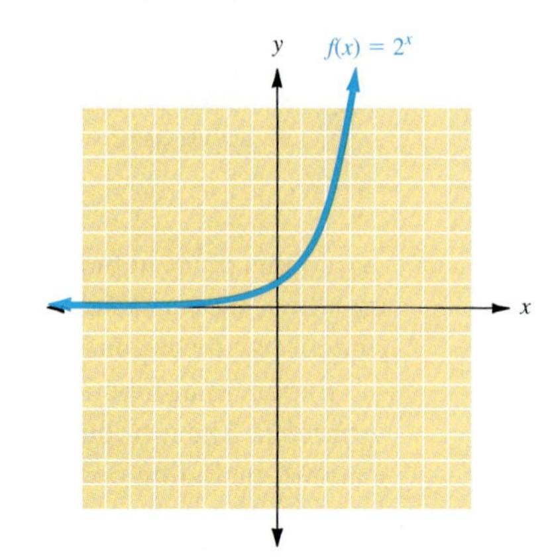

Let's examine some characteristics of the graph of the exponential function. First, the vertical-line test shows that this is indeed the graph of a function. Also note that the horizontal-line test shows that the function is one-to-one.

NOTE We call $y = 0$ (or the x axis) the **horizontal asymptote.**

The graph *approaches* the x axis on the left, but it does *not intersect* the x axis. The y intercept is (0, 1) (because $2^0 = 1$). To the right the functional values get larger. We say that the values *grow without bound.*

CHECK YOURSELF 1

Sketch the graph of the exponential function

$g(x) = 3^x$

Let's look at an example in which the base of the function is less than 1.

Example 2

Graphing an Exponential Function

Graph the exponential function

$$f(x) = \left(\frac{1}{2}\right)^x$$

NOTE Recall that

$$\left(\frac{1}{2}\right)^x = 2^{-x}$$

First, choose convenient values for x.

$$f(0) = \left(\frac{1}{2}\right)^0 = 1 \qquad f(-1) = \left(\frac{1}{2}\right)^{-1} = 2 \qquad f(1) = \left(\frac{1}{2}\right)^1 = \frac{1}{2}$$

$$f(-2) = \left(\frac{1}{2}\right)^{-2} = 4 \qquad f(2) = \left(\frac{1}{2}\right)^2 = \frac{1}{4} \qquad f(-3) = \left(\frac{1}{2}\right)^{-3} = 8$$

$$f(3) = \left(\frac{1}{2}\right)^3 = \frac{1}{8}$$

Again, form a table of values and graph the desired function.

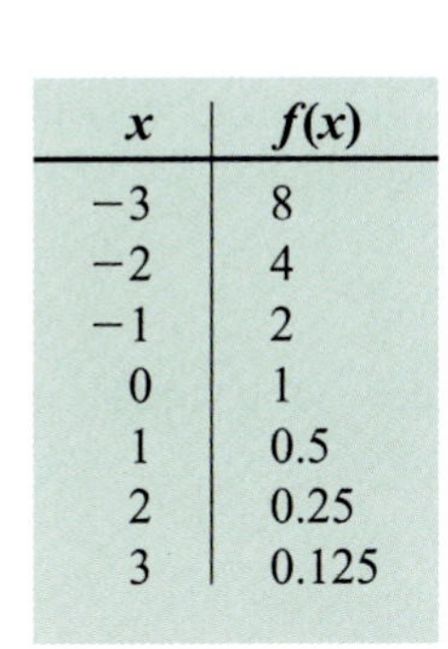

x	$f(x)$
−3	8
−2	4
−1	2
0	1
1	0.5
2	0.25
3	0.125

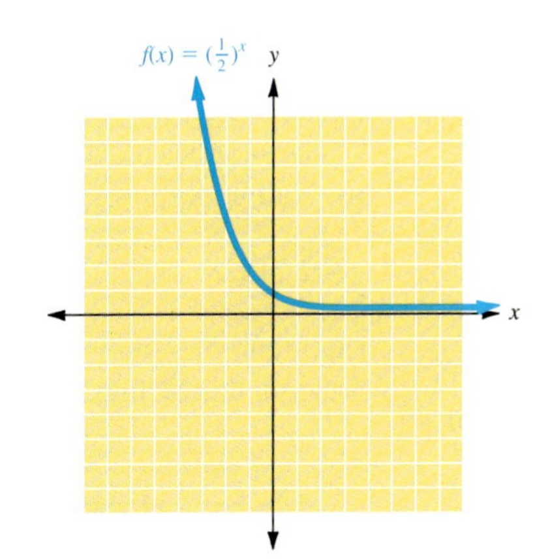

NOTE Again, by the vertical- and horizontal-line tests, this is the graph of a one-to-one function.

Let's compare this graph and that in Example 1. Clearly, the graph also represents a one-to-one function. As was true in the first example, the graph does not intersect the x axis but approaches that axis, here on the right. The values for the function again grow without bound, but this time on the left. The y intercept for both graphs occurs at 1.

Note that the graph of Example 1 was *increasing* (going up) as we moved from left to right. That function is an example of a **growth function.**

The graph of Example 2 is *decreasing* (going down) as we move from left to right. It is an example of a **decay function.**

NOTE The base of a *growth function* is *greater than* 1.

NOTE The base of a *decay function* is *less than 1* but greater than 0.

CHECK YOURSELF 2

Sketch the graph of the exponential function

$$g(x) = \left(\frac{1}{3}\right)^x$$

The following algorithm summarizes our work thus far in this section.

Step by Step: Graphing an Exponential Function

Step 1 Establish a table of values by considering the function in the form $y = b^x$.

Step 2 Plot points from that table of values and connect them with a smooth curve to form the graph.

Step 3 If $b > 1$, the graph increases from left to right. If $0 < b < 1$, the graph decreases from left to right.

Step 4 All graphs will have the following in common:
(a) The y intercept will be (0, 1).
(b) The graphs will approach, but not touch, the x axis.
(c) The graphs will represent one-to-one functions.

We used bases of 2 and $\frac{1}{2}$ for the exponential functions of our examples because they provided convenient computations. A far more important base for an exponential function

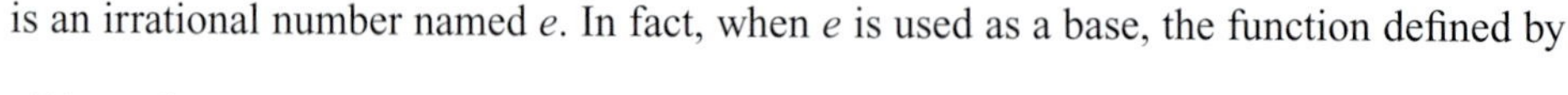

NOTE The use of the letter e as a base originated with Leonhard Euler (1707–1783) and e is sometimes called *Euler's number* for that reason.

is an irrational number named e. In fact, when e is used as a base, the function defined by

$$f(x) = e^x$$

is called *the* exponential function.

The significance of this number will be made clear in later courses, particularly calculus. For our purposes, e can be approximated as

$$e \approx 2.71828$$

NOTE Graph $y = e^x$ on your calculator. You may find the $\boxed{e^x}$ key to be the 2nd (or inverse) function to the ln x key. Note that e^1 is approximately 2.71828.

The graph of $f(x) = e^x$ is shown below. Of course, it is very similar to the graphs seen earlier in this section.

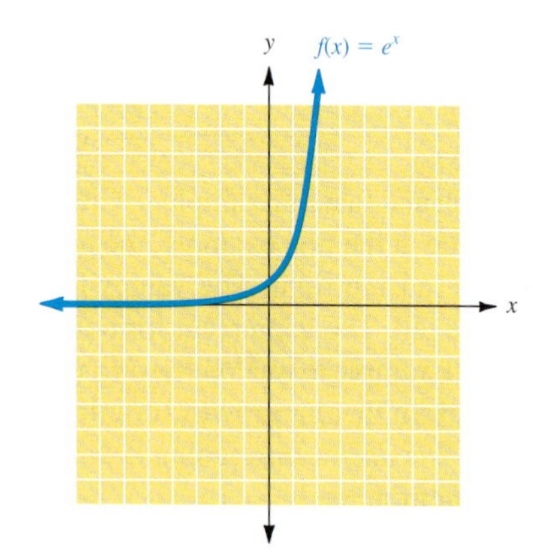

Exponential expressions involving base e occur frequently in real-world applications. Example 3 illustrates this approach.

Example 3

A Population Application

(a) Suppose that the population of a city is presently 20,000 and that the population is expected to grow at a rate of 5% per year. The equation

$$P(t) = 20{,}000e^{(0.05)t}$$

NOTE Be certain that you enclose the multiplication (0.05 × 5) in parentheses or the calculator will misinterpret your intended order of operation.

gives the town's population after t years. Find the population in 5 years.

Let $t = 5$ in the original equation to obtain

$$P(5) = 20{,}000e^{(0.05)(5)} \approx 25{,}681$$

which is the population expected 5 years from now.

NOTE Continuous compounding will give the highest accumulation of interest at any rate. However, daily compounding will result in an amount of interest that is only slightly less.

(b) Suppose $1000 is invested at an annual rate of 8%, compounded continuously. The equation

$$A(t) = 1000e^{0.08t}$$

gives the amount in the account after t years. Find the amount in the account after 9 years.

Let $t = 9$ in the original equation to obtain

NOTE Note that in 9 years the amount in the account is a little more than *double* the original principal.

$$A(9) = 1000e^{(0.08)(9)} \approx 2054$$

which is the amount in the account after 9 years.

CHECK YOURSELF 3

If $1000 is invested at an annual rate of 6%, compounded continuously, then the equation for the amount in the account after t years is

$A(t) = 1000e^{0.06t}$

Use your calculator to find the amount in the account after 12 years.

As we observed in this section, the exponential function is always one-to-one. This yields an important property that can be used to solve certain types of equations involving exponents.

Rules and Properties: Exponential Property

If $b > 0$ and $b \neq 1$, then

$$b^m = b^n \quad \text{if and only if} \quad m = n \tag{1}$$

in which m and n are any real numbers.

The usefulness of this property is illustrated in Example 4.

Example 4

Solving an Exponential Equation

(a) Solve $2^x = 8$ for x.

We recognize that 8 is a power of 2, and we can write the equation as

$2^x = 2^3$ Write with equal bases.

Applying property (1), we have

$x = 3$ Set exponents equal.

and 3 is the solution.

(b) Solve $3^{2x} = 81$ for x.

Because $81 = 3^4$, we can write

$3^{2x} = 3^4$

$2x = 4$

$x = 2$

We see that 2 is the solution for the equation.

NOTE The answer can easily be checked by substitution. Letting $x = 2$ gives $3^{2(2)} = 3^4 = 81$

(c) Solve $2^{x+1} = \frac{1}{16}$ for x.

Again, we write $\frac{1}{16}$ as a power of 2, so that

$2^{x+1} = 2^{-4}$

NOTE $\frac{1}{16} = \frac{1}{2^4} = 2^{-4}$

NOTE To verify the solution.

$2^{-5+1} \stackrel{?}{=} 2^{-4}$

$2^{-4} \stackrel{?}{=} 2^{-4}$

$\frac{1}{16} = \frac{1}{16}$

Then

$$x + 1 = -4$$

$$x = -5$$

The solution is -5.

CHECK YOURSELF 4

Solve each of the following equations for x.

(a) $2^x = 16$ **(b)** $4^{x+1} = 64$ **(c)** $3^{2x} = \frac{1}{81}$

CHECK YOURSELF ANSWERS

1. $y = g(x) = 3^x$

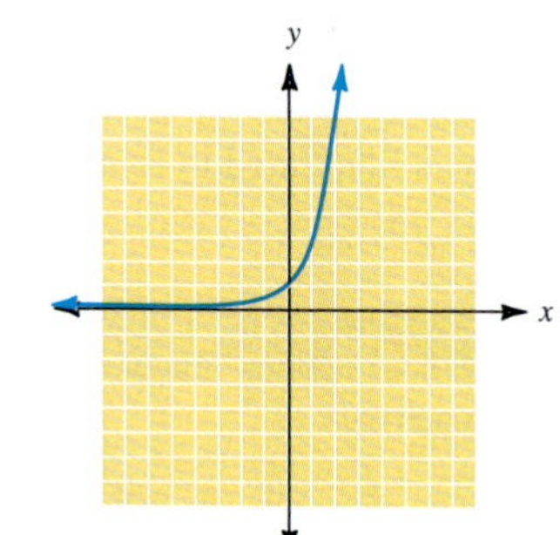

2. $y = \left(\frac{1}{3}\right)^x$

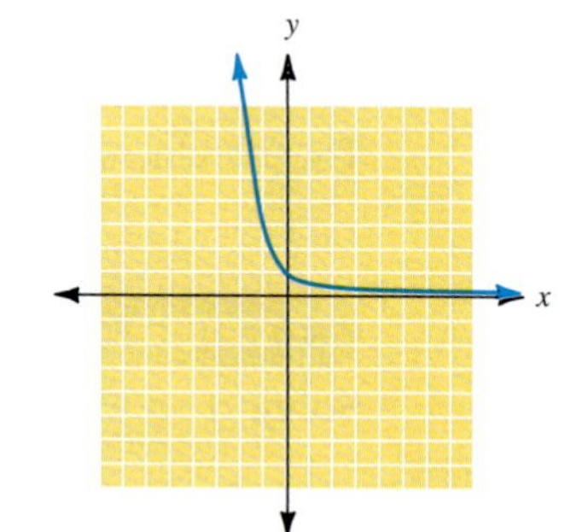

3. \$2054.43 **4.** **(a)** {4}; **(b)** {2}; **(c)** {−2}

Name ______________________

Section ________ Date ________

11.2 Exercises

Match the graphs in exercises 1 to 8 with the appropriate equation.

(a) $y = \left(\frac{1}{2}\right)^x$ **(b)** $y = 2x - 1$ **(c)** $y = 2^x$ **(d)** $y = x^2$

(e) $y = 1^x$ **(f)** $y = 5^x$ **(g)** $x = 2^y$ **(h)** $x = y^2$

ANSWERS

1. ______________________
2. ______________________
3. ______________________
4. ______________________
5. ______________________
6. ______________________
7. ______________________
8. ______________________

1.

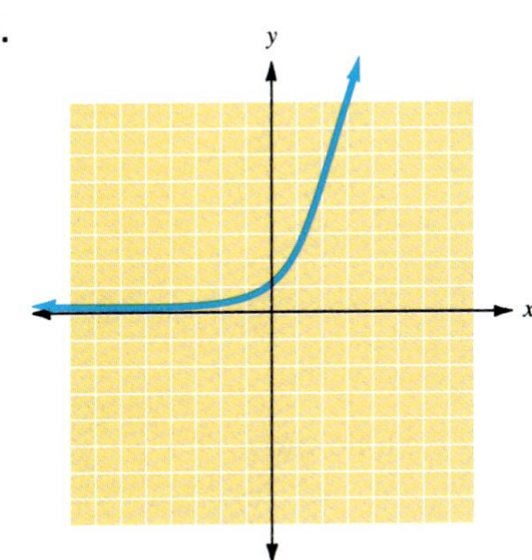

2.

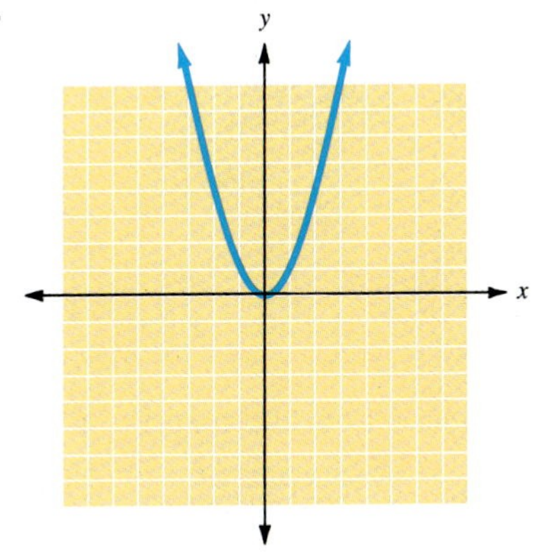

3.

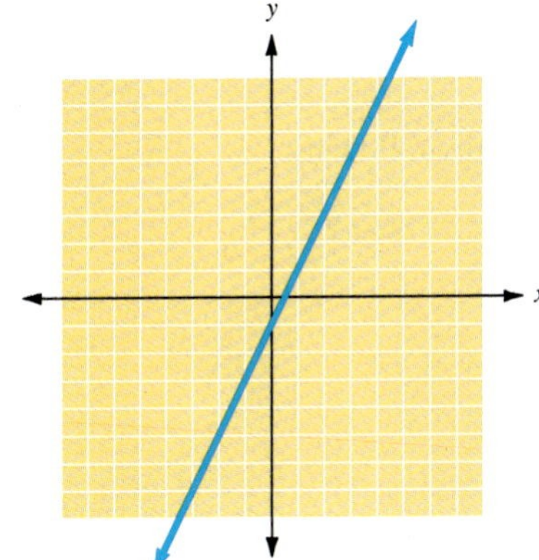

4.

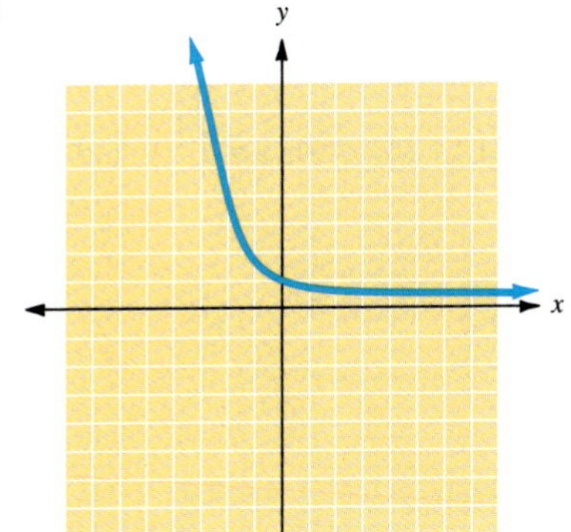

5.

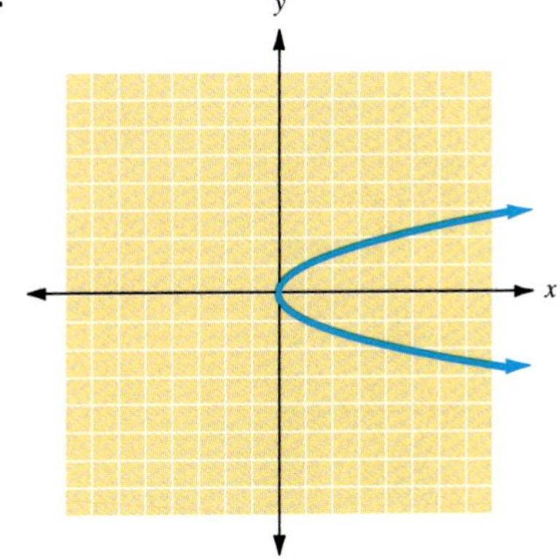

6.

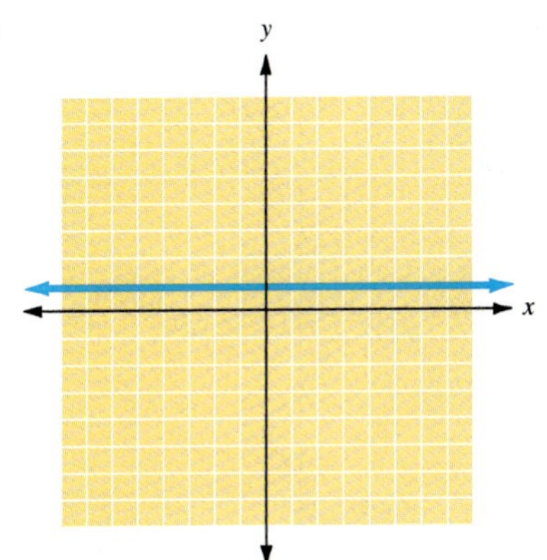

7.

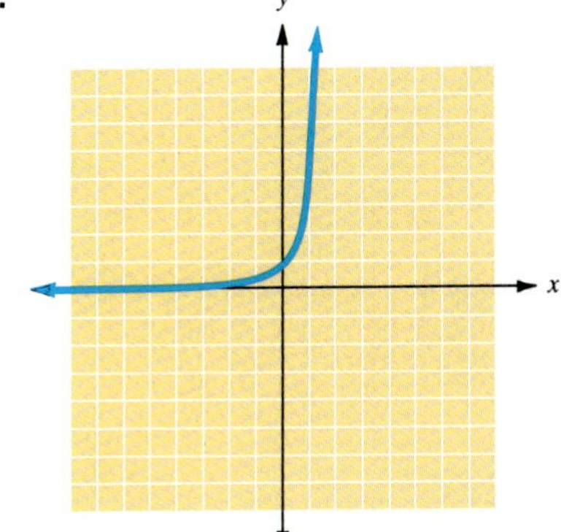

8.

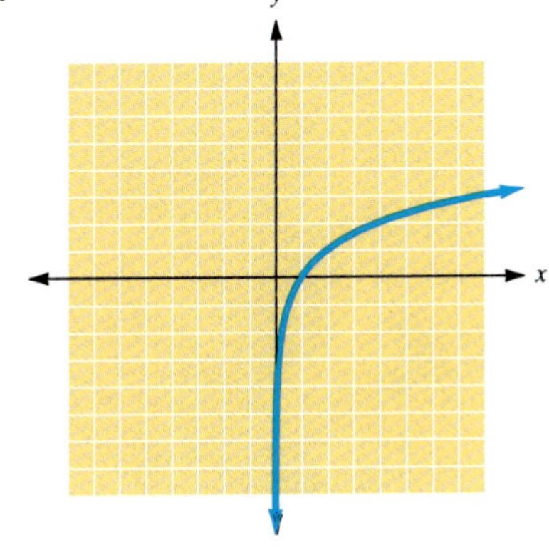

ANSWERS

9. ______

10. ______

11. ______

12. ______

13. ______

14. ______

15. ______

16. ______

17. ______

18. ______

19. ______

20. ______

21. ______

22. ______

23. ______

24. ______

25. ______

26. ______

In exercises 9 to 12, let $f(x) = 4^x$ and find each of the following.

9. $f(0)$

10. $f(1)$

11. $f(2)$

12. $f(-2)$

In exercises 13 to 16, let $g(x) = 4^{x+1}$ and find each of the following.

13. $g(0)$

14. $g(1)$

15. $g(2)$

16. $g(-2)$

In exercises 17 to 20, let $h(x) = 4^x + 1$ and find each of the following.

17. $h(0)$

18. $h(1)$

19. $h(2)$

20. $h(-2)$

In exercises 21 to 24, let $f(x) = \left(\frac{1}{4}\right)^x$ and find each of the following.

21. $f(1)$

22. $f(-1)$

23. $f(-2)$

24. $f(2)$

In exercises 25 to 36, graph each exponential function.

25. $y = 4^x$

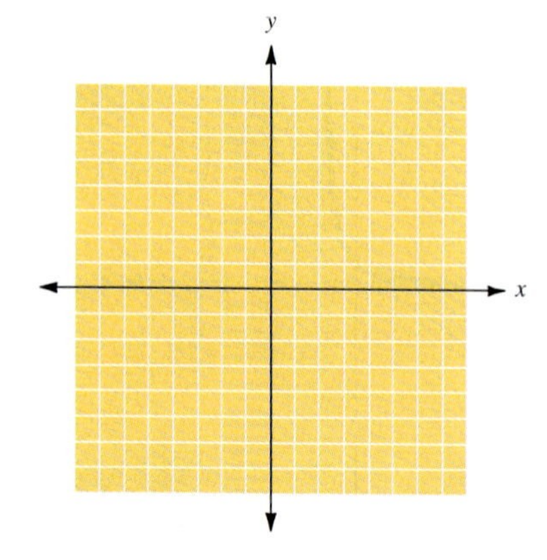

26. $y = \left(\frac{1}{4}\right)^x$

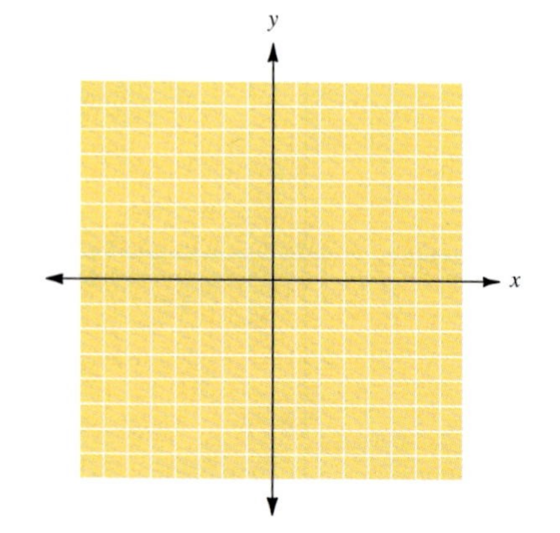

27. $y = \left(\frac{2}{3}\right)^x$

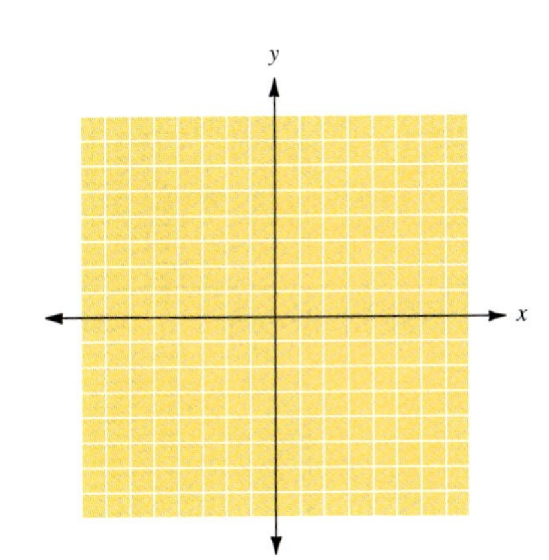

28. $y = \left(\frac{3}{2}\right)^x$

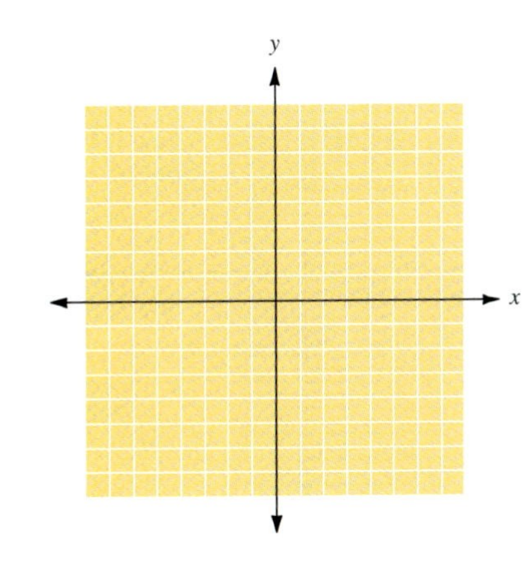

29. $y = 3 \cdot 2^x$

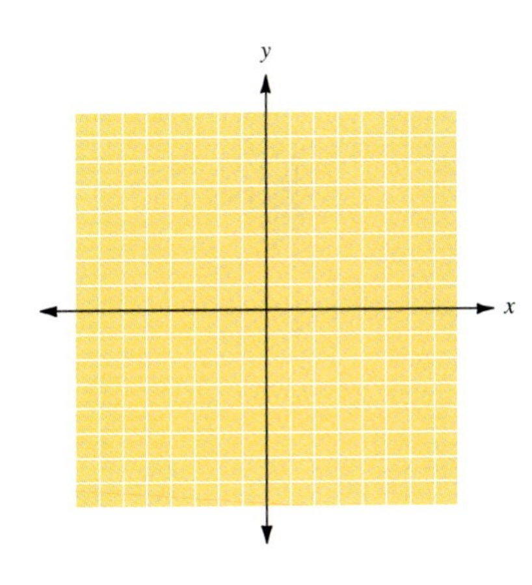

30. $y = 2 \cdot 3^x$

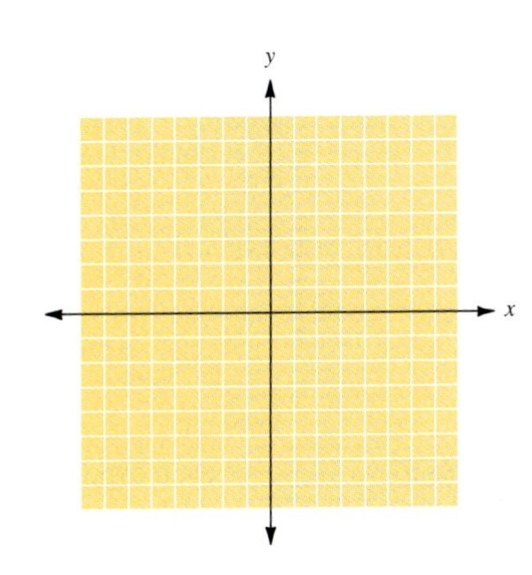

31. $y = 3^x$

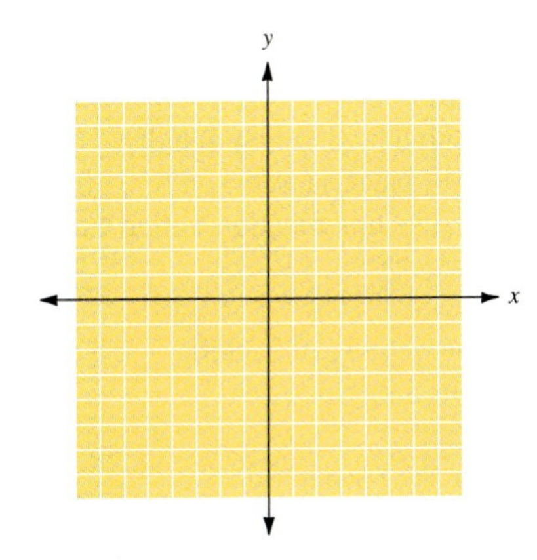

32. $y = 2^{x-1}$

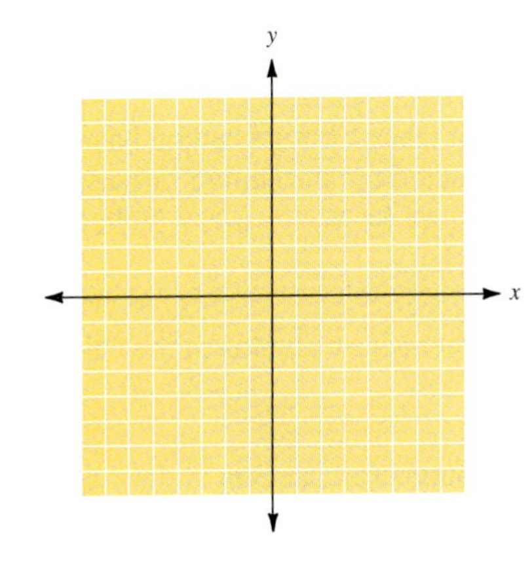

ANSWERS

27. ______

28. ______

29. ______

30. ______

31. ______

32. ______

ANSWERS

33. ______

34. ______

35. ______

36. ______

37. ______

38. ______

39. ______

40. ______

41. ______

42. ______

43. ______

44. ______

45. ______

46. ______

47. ______

48. ______

49. ______

50. ______

51. ______

33. $y = 2^{2x}$

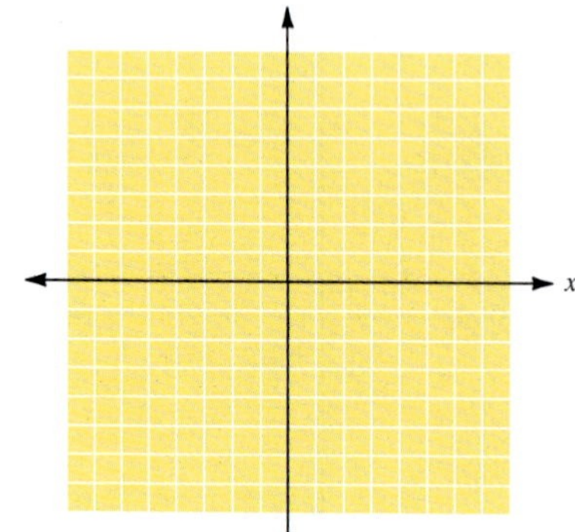

34. $y = \left(\frac{1}{2}\right)^{2x}$

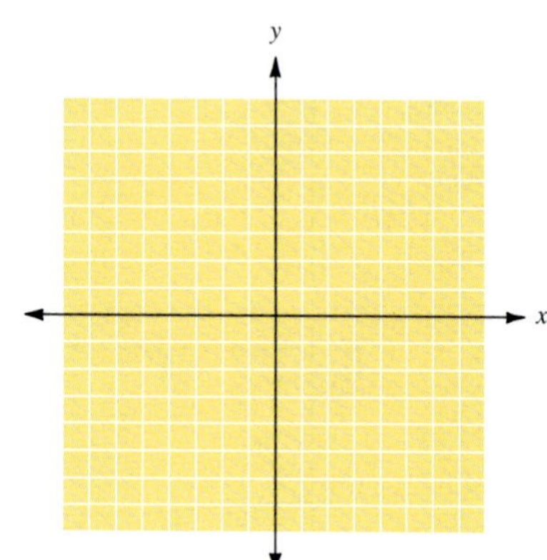

35. $y = e^{-x}$

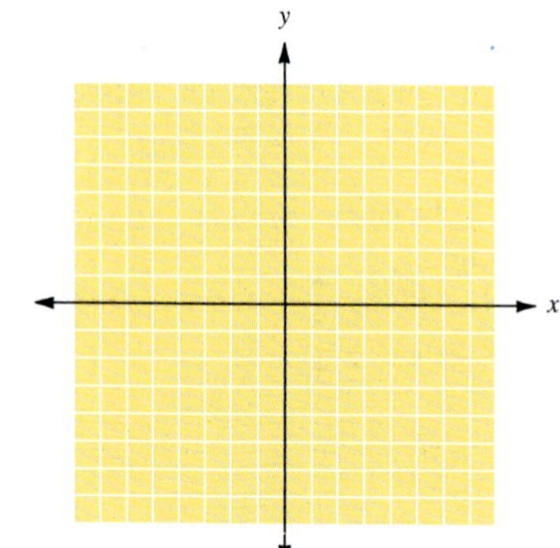

36. $y = e^{2x}$

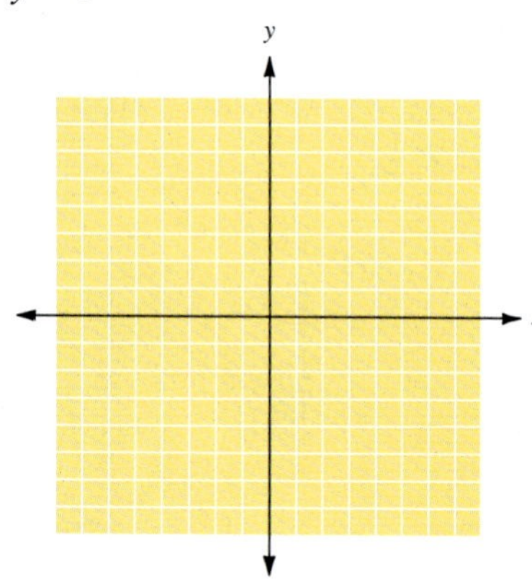

In exercises 37 to 48, solve each exponential equation for x.

37. $2^x = 32$

38. $4^x = 64$

39. $10^x = 10{,}000$

40. $5^x = 125$

41. $3^x = \frac{1}{9}$

42. $2^x = \frac{1}{16}$

43. $2^{2x} = 64$

44. $3^{2x} = 81$

45. $2^{x+1} = 64$

46. $4^{x-1} = 16$

47. $3^{x-1} = \frac{1}{27}$

48. $2^{x+2} = \frac{1}{8}$

Suppose it takes 1 h for a certain bacterial culture to double by dividing in half. If there are 100 bacteria in the culture to start, then the number of bacteria in the culture after x hours is given by $N(x) = 100 \cdot 2^x$. In exercises 49 to 51, use this function to find each of the following.

49. The number of bacteria in the culture after 2 h.

50. The number of bacteria in the culture after 3 h.

51. The number of bacteria in the culture after 5 h.

52. Graph the relationship between the number of bacteria in the culture and the number of hours. Be sure to choose an appropriate scale for the N axis.

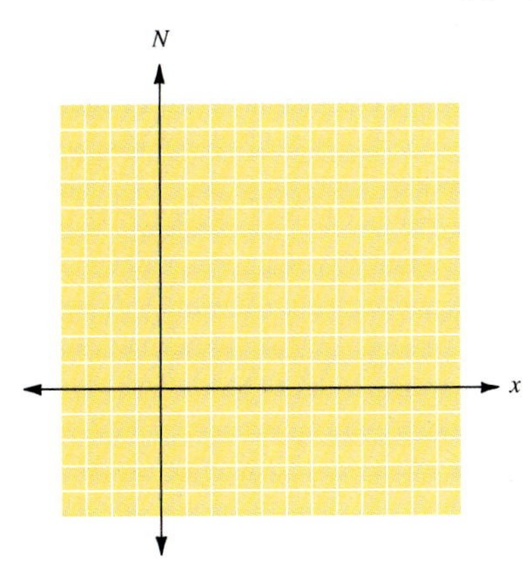

The half-life of radium is 1690 years. That is, after a 1690-year period, one-half of the original amount of radium will have decayed into another substance. If the original amount of radium is 64 grams (g), the formula relating the amount of radium left after time t is given by $R(t) = 64 \cdot 2^{-t/1690}$. In exercises 53 to 55, use that formula to find each of the following.

53. The amount of radium left after 1690 years.

54. The amount of radium left after 3380 years.

55. The amount of radium left after 5070 years.

56. Graph the relationship between the amount of radium remaining and time. Be sure to use appropriate scales for the R and t axes.

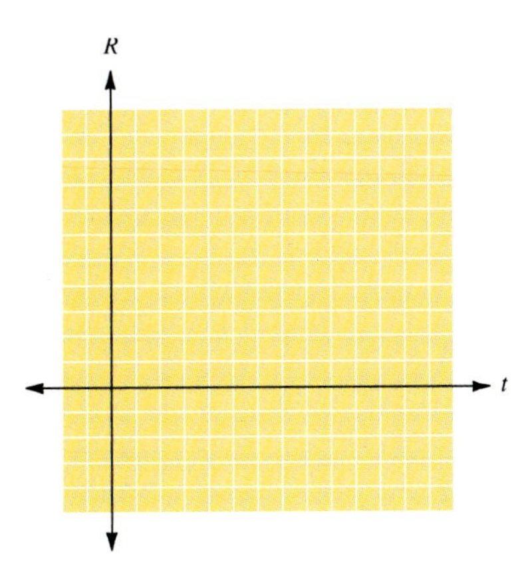

If \$1000 is invested in a savings account with an interest rate of 8%, compounded annually, the amount in the account after t years is given by $A(t) = 1000(1 + 0.08)^t$. In exercises 57 to 59, use a calculator to find each of the following.

57. The amount in the account after 2 years.

58. The amount in the account after 5 years.

ANSWERS

52. __________

53. __________

54. __________

55. __________

56. __________

57. __________

58. __________

ANSWERS

59. ______

60. ______

61. ______

62. ______

63. ______

64. ______

59. The amount in the account after 9 years.

60. Graph the relationship between the amount in the account and time. Be sure to choose appropriate scales for the A and t axes.

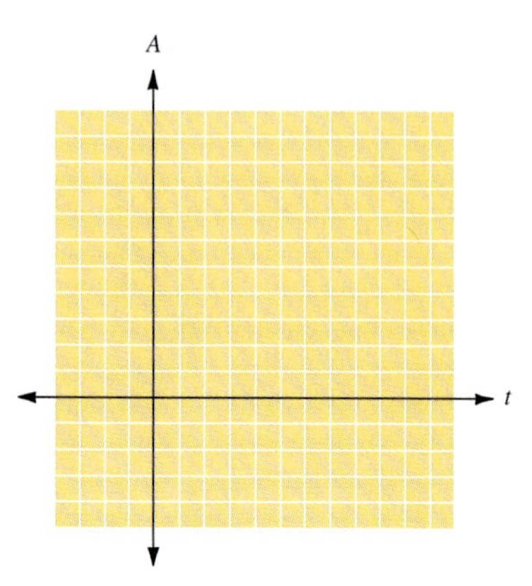

The so-called learning curve in psychology applies to learning a skill, such as typing, in which the performance level progresses rapidly at first and then levels off with time. You can approximate N, the number of words per minute that a person can type after t weeks of training, with the equation $N = 80(1 - e^{-0.06t})$. Use a calculator to find the following.

61. **(a)** N after 10 weeks, **(b)** N after 20 weeks, **(c)** N after 30 weeks.

62. Graph the relationship between the number of words per minute N and the number of weeks of training t.

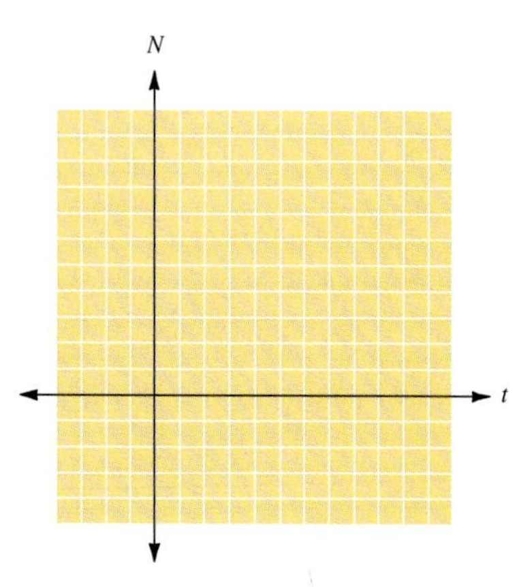

63. Find two different calculators that have $\boxed{e^x}$ keys. Describe how to use the function on each of the calculators.

64. Are there any values of x for which e^x produces an exact answer on the calculator? Why are other answers not exact?

A possible calculator sequence for evaluating the expression

$$\left(1 + \frac{1}{n}\right)^n$$

in which $n = 10$ is

[(] 1 [+] 1 [÷] 10 [)] [∧] 10 [=]

In exercises 65 to 69, use that sequence to find $\left(1 + \frac{1}{n}\right)^n$ for the following values of n.

65. $n = 100$

66. $n = 1000$

67. $n = 10{,}000$

68. $n = 100{,}000$

69. $n = 1{,}000{,}000$

70. What did you observe from the experiment above?

71. Graph the exponential function defined by $y = 2^x$.

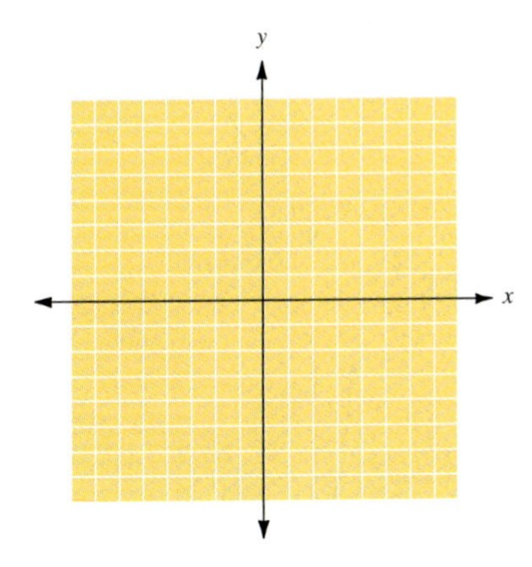

72. Graph the function defined by $x = 2^y$ on the same set of axes as the previous graph. What do you observe? *Hint:* To graph $x = 2^y$, choose convenient values for y and then find the corresponding values for x.

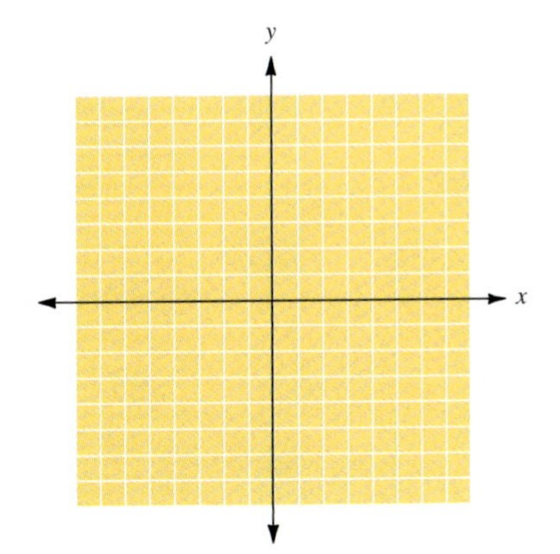

ANSWERS

65. ______

66. ______

67. ______

68. ______

69. ______

70. ______

71. ______

72. ______

Answers

1. (c) **3.** (b) **5.** (h) **7.** (f) **9.** 1 **11.** 16 **13.** 4 **15.** 64

17. 2 **19.** 17 **21.** $\frac{1}{4}$ **23.** 16

25. $y = 4^x$

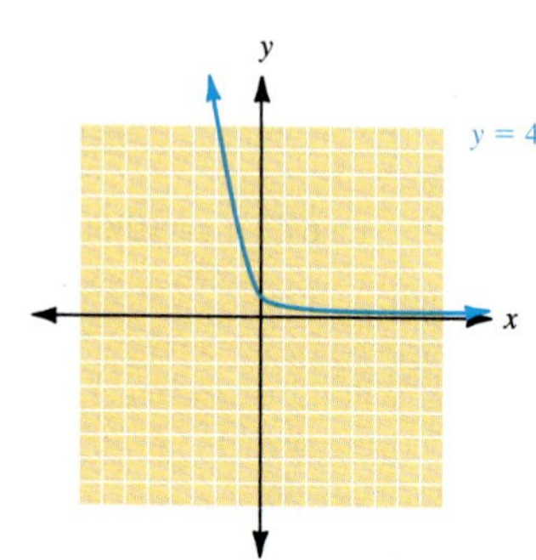

27. $y = \left(\frac{2}{3}\right)^x$

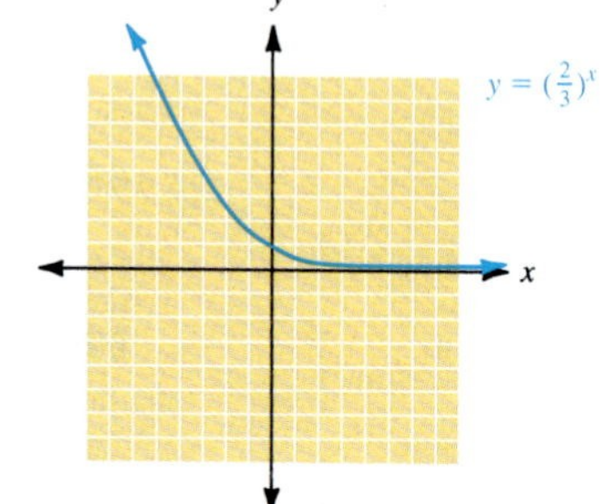

29. $y = 3 \cdot 2^x$

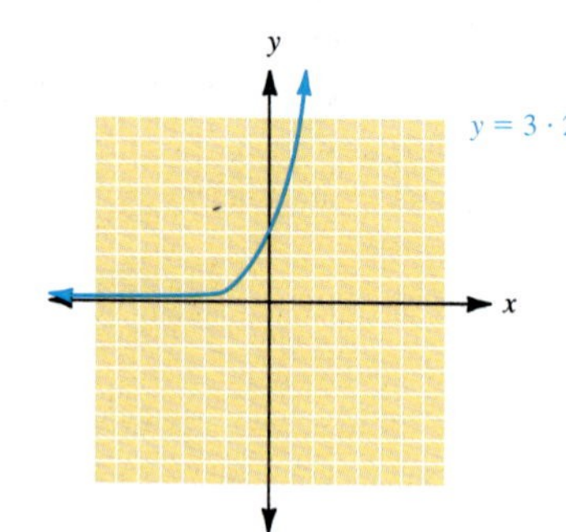

31. $y = 3^x$

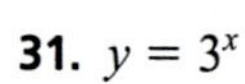

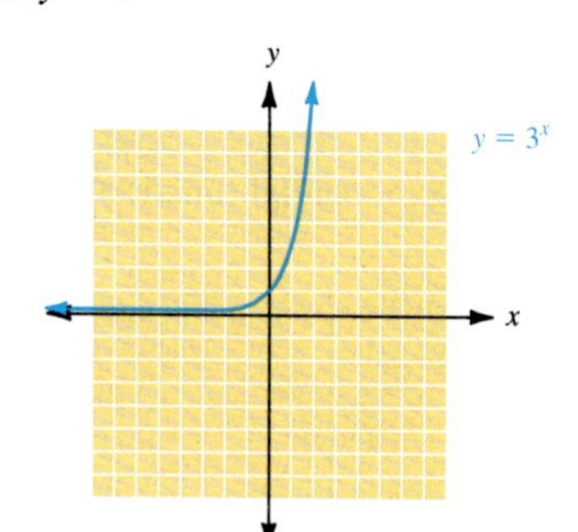

33. $y = 2^{2x}$

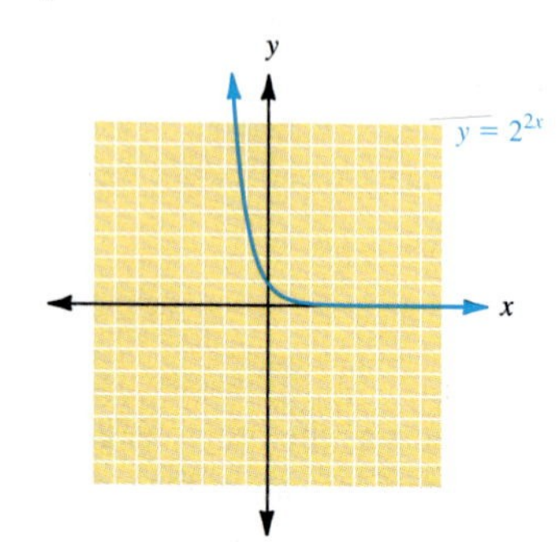

35. $y = e^{-x}$

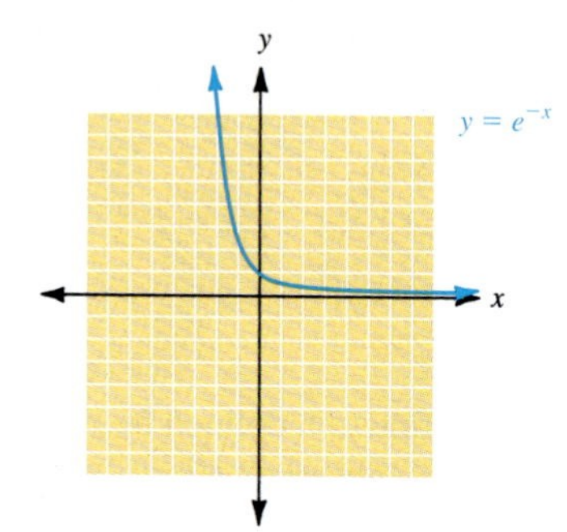

37. $\{5\}$ **39.** $\{4\}$ **41.** $\{-2\}$ **43.** $\{3\}$ **45.** $\{5\}$ **47.** $\{-2\}$

49. 400 **51.** 3200 **53.** 32 g **55.** 8 g **57.** \$1166.40 **59.** \$1999

61. **(a)** 36; **(b)** 56; **(c)** 67 **63.** **65.** 2.7048 **67.** 2.71815

69. 2.71828 **71.** See Example 1 for graph of $y = 2^x$.

Logarithmic Functions

OBJECTIVES

1. Graph a logarithmic function
2. Convert between logarithmic and exponential expressions
3. Evaluate a logarithmic expression
4. Solve an elementary logarithmic equation

Given our experience with the exponential function in Section 11.2 and our earlier work with the inverse of a function, we now can introduce the logarithmic function.

NOTE Napier also coined the word "logarithm" from the Greek words "logos"—a ratio—and "arithmos"—a number.

John Napier (1550–1617), a Scotsman, is credited with the invention of logarithms. The development of the logarithm grew out of a desire to ease the work involved in numerical computations, particularly in the field of astronomy. Today the availability of inexpensive scientific calculators has made the use of logarithms as a computational tool unnecessary.

However, the concept of the logarithm and the properties of the logarithmic function that we describe in a later section still are very important in the solutions of particular equations, in calculus, and in the applied sciences.

Again, the applications for this new function are numerous. The Richter scale for measuring the intensity of an earthquake and the decibel scale for measuring the intensity of sound both make use of logarithms.

To develop the idea of a logarithmic function, we must return to the exponential function

$$f = \{(x, y) | y = b^x, b > 0, b \neq 1\} \tag{1}$$

Interchanging the roles of x and y, we have the inverse function

NOTE Recall that f is a one-to-one function, so its inverse is also a function.

$$f^{-1} = \{(x, y) | x = b^y\} \tag{2}$$

Presently, we have no way to solve the equation $x = b^y$ for y. So, to write the inverse (2) in a more useful form, we offer the following definition.

Definitions: Logarithm

The *logarithm of x to base b* is denoted

$\log_b x$

and

$y = \log_b x$ if and only if $x = b^y$

We can now write our inverse function, using this new notation, as

NOTE Notice that the restrictions on the base are the same as those used for the exponential function.

$$f^{-1} = \{(x, y) | y = \log_b x, b > 0, b \neq 1\} \tag{3}$$

In general, any function defined in this form is called a **logarithmic function.**

At this point we should stress the meaning of this new relationship. Consider the equivalent forms illustrated here.

Definitions: Logarithm

Power or exponent

$y = \log_b x$ means the same as $x = b^y$

Base

The logarithm y is the power to which we must raise b to get x. In other words, *a logarithm is simply a power or an exponent.* We return to this thought later when using the exponential and logarithmic forms of equivalent equations.

We begin our work by graphing a typical logarithmic function.

Example 1

Graphing a Logarithmic Function

Graph the logarithmic function

$y = \log_2 x$

Because $y = \log_2 x$ is equivalent to the exponential form

NOTE The base is 2, and the logarithm or power is y.

$x = 2^y$

we can find ordered pairs satisfying this equation by choosing convenient values for y and calculating the corresponding values for x.

Letting y take on values from -3 to 3 yields the table of values shown below. As before, we plot points from the ordered pairs and connect them with a smooth curve to form the graph of the function.

NOTE What do the vertical- and horizontal-line tests tell you about this graph?

x	y
$\frac{1}{8}$	-3
$\frac{1}{4}$	-2
$\frac{1}{2}$	-1
1	0
2	1
4	2
8	3

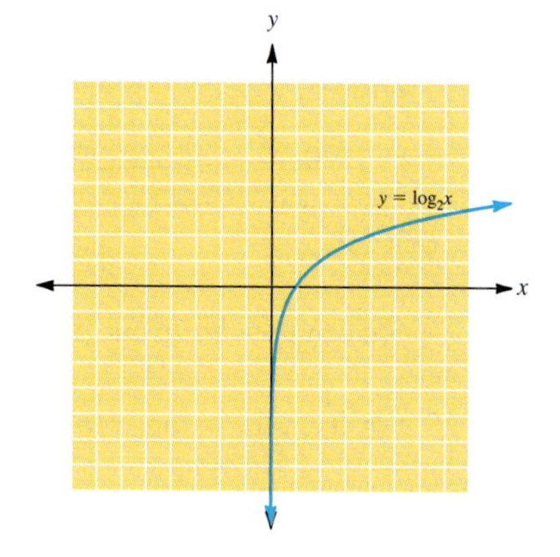

NOTE Use your calculator to compare the graphs of $y = 10^x$ and $y = \log_{10} x$. Are they inverse functions? How can you tell?

We observe that the graph represents a one-to-one function whose domain is $\{x|x > 0\}$ and whose range is the set of all real numbers.

For base 2 (or for any base greater than 1) the function will always be increasing over its domain.

Recall from Section 11.1 that the graphs of a function and its inverse are always reflections of each other using the line $y = x$. Because we have defined the logarithmic function as the inverse of an exponential function, we can anticipate the same relationship.

The graphs of

$$f(x) = 2^x \qquad \text{and} \qquad f^{-1}(x) = \log_2 x$$

are shown below.

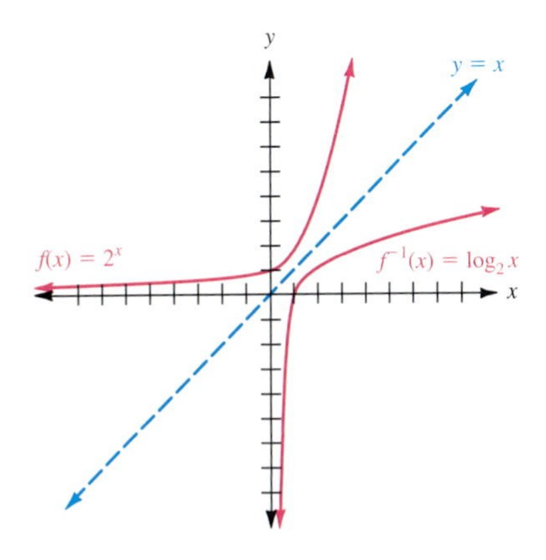

We see that the graphs of f and f^{-1} are indeed reflections of each other about the line $y = x$. In fact, this relationship provides an alternate method of sketching $y = \log_b x$. We can sketch the graph of $y = b^x$ and then reflect that graph using the line $y = x$ to form the graph of the logarithmic function.

CHECK YOURSELF 1

Graph the logarithmic function defined by

$$y = \log_3 x$$

Hint: Consider the equivalent form $x = 3^y$.

For our later work in this chapter, it will be necessary for us to be able to convert back and forth between exponential and logarithmic forms. The conversion is straightforward. You need only keep in mind the basic relationship

NOTE Again, this tells us that a logarithm is an exponent or a power.

$$y = \log_b x \qquad \text{means the same as} \qquad x = b^y$$

Look at the following example.

Example 2

Writing Equations in Logarithmic Form

Convert to logarithmic form.

NOTE The base is 3, the exponent or power is 4.

(a) $3^4 = 81$ is equivalent to $\log_3 81 = 4$.

(b) $10^3 = 1000$ is equivalent to $\log_{10} 1000 = 3$.

(c) $2^{-3} = \frac{1}{8}$ is equivalent to $\log_2 \frac{1}{8} = -3$.

(d) $9^{1/2} = 3$ is equivalent to $\log_9 3 = \frac{1}{2}$.

CHECK YOURSELF 2

Convert each statement to logarithmic form.

(a) $4^3 = 64$ **(b)** $10^{-2} = 0.01$ **(c)** $3^{-3} = \frac{1}{27}$ **(d)** $27^{1/3} = 3$

Example 3 shows how to write a logarithmic expression in exponential form.

Example 3

Writing Equations in Exponential Form

Convert to exponential form.

NOTE Here, the base is 2; the logarithm, which is the power, is 3.

(a) $\log_2 8 = 3$ is equivalent to $2^3 = 8$.

(b) $\log_{10} 100 = 2$ is equivalent to $10^2 = 100$.

(c) $\log_3 \frac{1}{9} = -2$ is equivalent to $3^{-2} = \frac{1}{9}$.

(d) $\log_{25} 5 = \frac{1}{2}$ is equivalent to $25^{1/2} = 5$.

CHECK YOURSELF 3

Convert to exponential form.

(a) $\log_2 32 = 5$ **(b)** $\log_{10} 1000 = 3$ **(c)** $\log_4 \frac{1}{16} = -2$ **(d)** $\log_{27} 3 = \frac{1}{3}$

Certain logarithms can be directly calculated by changing an expression to the equivalent exponential form, as Example 4 illustrates.

Example 4

Evaluating Logarithmic Expressions

(a) Evaluate $\log_3 27$.

If $x = \log_3 27$, in exponential form we have

NOTE Recall that $b^m = b^n$ if and only if $m = n$.

$$3^x = 27$$

$$3^x = 3^3$$

$$x = 3$$

We then have $\log_3 27 = 3$.

(b) Evaluate $\log_{10} \frac{1}{10}$.

If $x = \log_{10} \frac{1}{10}$, we can write

NOTE Rewrite each side as a power of the same base.

$$10^x = \frac{1}{10}$$

$$= 10^{-1}$$

We then have $x = -1$ and

$$\log_{10} \frac{1}{10} = -1$$

CHECK YOURSELF 4

Evaluate each logarithm.

(a) $\log_2 64$ **(b)** $\log_3 \frac{1}{27}$

The relationship between exponents and logarithms also allows us to solve certain equations involving logarithms in which two of the quantities in the equation $y = \log_b x$ are known, as Example 5 illustrates.

Example 5

Solving Logarithmic Equations

(a) Solve $\log_5 x = 3$ for x.

Because $\log_5 x = 3$, in exponential form we have

$$x = 5^3$$

$$= 125$$

(b) Solve $y = \log_4 \frac{1}{16}$ for y.

The original equation is equivalent to

$$4^y = \frac{1}{16}$$
$$= 4^{-2}$$

We then have $y = -2$ as the solution.

(c) Solve $\log_b 81 = 4$ for b.

In exponential form the equation becomes

NOTE Keep in mind that the base must be *positive*, so we do not consider the possible solution $b = -3$.

$$b^4 = 81$$
$$b = 3$$

CHECK YOURSELF 5

Solve each of the following equations for the variable cited.

(a) $\log_4 x = 4$ for x **(b)** $\log_b \frac{1}{8} = -3$ for b **(c)** $y = \log_9 3$ for y

NOTE Loudness can be measured in **bels (B)**, a unit named for Alexander Graham Bell. This unit is rather large, so a more practical unit is the **decibel (dB)**, a unit one-tenth as large.

To conclude this section, we turn to two common applications of the logarithmic function. The **decibel scale** is used in measuring the loudness of various sounds.

If I represents the intensity of a given sound and I_0 represents the intensity of a "threshold sound," then the decibel (dB) rating of the given sound is given by

$$L = 10 \log_{10} \frac{I}{I_0}$$

NOTE Variable I_0 is the intensity of the minimum sound level detectable by the human ear.

in which $I_0 = 10^{-16}$ watts per square centimeter (W/cm²). Consider Example 6.

Example 6

A Decibel Application

(a) A whisper has intensity $I = 10^{-14}$. Its decibel rating is

$$L = 10 \log_{10} \frac{10^{-14}}{10^{-16}}$$
$$= 10 \log_{10} 10^2$$
$$= 10 \cdot 2$$
$$= 20$$

(b) A rock concert has intensity $I = 10^{-4}$. Its decibel rating is

$$L = 10 \log_{10} \frac{10^{-4}}{10^{-16}}$$
$$= 10 \log_{10} 10^{12}$$
$$= 10 \cdot 12$$
$$= 120$$

CHECK YOURSELF 6

Ordinary conversation has intensity $I = 10^{-12}$. Find its rating on the decibel scale.

NOTE The scale was named after Charles Richter, a U.S. geologist.

Another commonly used logarithmic scale is the **Richter scale.** Geologists use that scale to convert seismographic readings, which give the intensity of the shock waves of an earthquake, to a measure of the magnitude of the earthquake.

The magnitude M of an earthquake is given by

NOTE A "zero-level" earthquake is the quake of least intensity that is measurable by a seismograph.

$$M = \log_{10} \frac{a}{a_0}$$

in which a is the intensity of its shock waves and a_0 is the intensity of the shock wave of a zero-level earthquake.

Example 7

A Richter Scale Application

How many times stronger is an earthquake measuring 5 on the Richter scale than one measuring 4 on the Richter scale?

Suppose a_1 is the intensity of the earthquake with magnitude 5 and a_2 is the intensity of the earthquake with magnitude 4. Then

NOTE On your calculator, the log key is actually $\log_{10} x$.

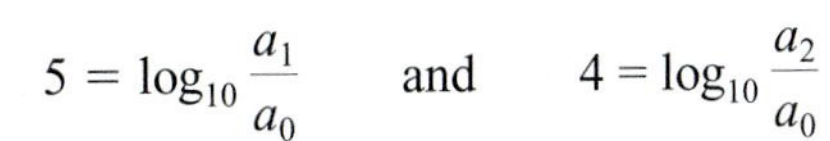

$$5 = \log_{10} \frac{a_1}{a_0} \qquad \text{and} \qquad 4 = \log_{10} \frac{a_2}{a_0}$$

We convert these logarithmic expressions to exponential form.

$$10^5 = \frac{a_1}{a_0} \qquad \text{and} \qquad 10^4 = \frac{a_2}{a_0}$$

or

$$a_1 = a_0 \cdot 10^5 \qquad \text{and} \qquad a_2 = a_0 \cdot 10^4$$

We want the ratio of the intensities of the two earthquakes, so

NOTE The ratio of a_1 to a_2 is $\frac{a_1}{a_2}$

$$\frac{a_1}{a_2} = \frac{a_0 \cdot 10^5}{a_0 \cdot 10^4} = 10^1 = 10$$

The earthquake of magnitude 5 is *10 times stronger* than the earthquake of magnitude 4.

CHECK YOURSELF 7

How many times stronger is an earthquake of magnitude 6 than one of magnitude 4?

CHECK YOURSELF ANSWERS

1. $y = \log_3 x$

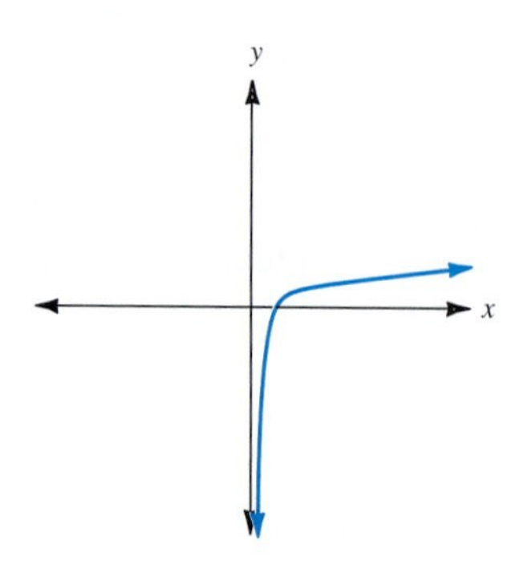

2. **(a)** $\log_4 64 = 3$; **(b)** $\log_{10} 0.01 = -2$; **(c)** $\log_3 \frac{1}{27} = -3$; **(d)** $\log_{27} 3 = \frac{1}{3}$

3. **(a)** $2^5 = 32$; **(b)** $10^3 = 1000$; **(c)** $4^{-2} = \frac{1}{16}$; **(d)** $27^{1/3} = 3$

4. **(a)** $\log_2 64 = 6$; **(b)** $\log_3 \frac{1}{27} = -3$ **5.** **(a)** $x = 256$; **(b)** $b = 2$; **(c)** $y = \frac{1}{2}$

6. 40 dB **7.** 100 times

Name ______________________

Section ________ Date ________

11.3 Exercises

In exercises 1 to 6, sketch the graph of the function defined by each equation.

1. $y = \log_4 x$

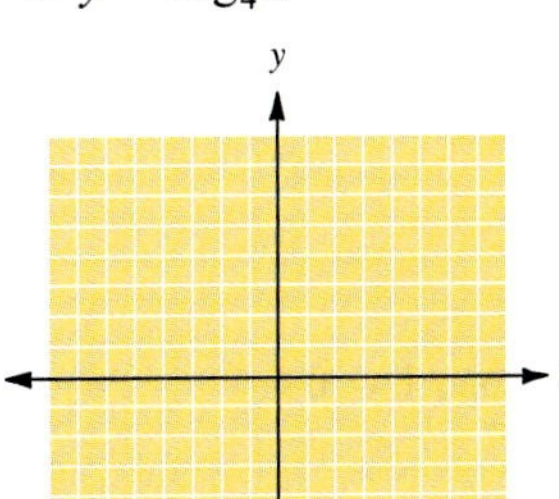

2. $y = \log_{10} x$

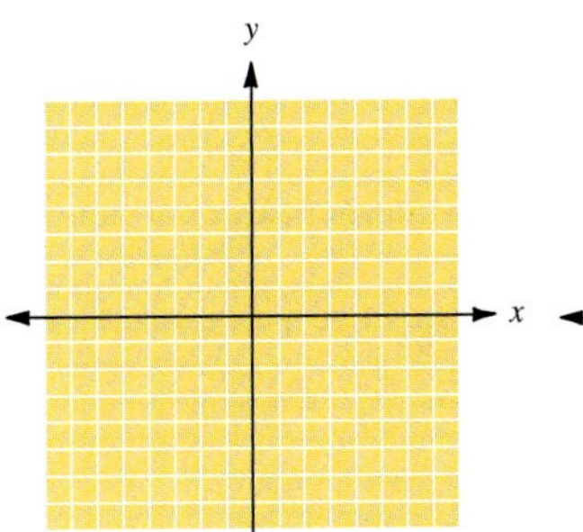

3. $y = \log_2 (x - 1)$

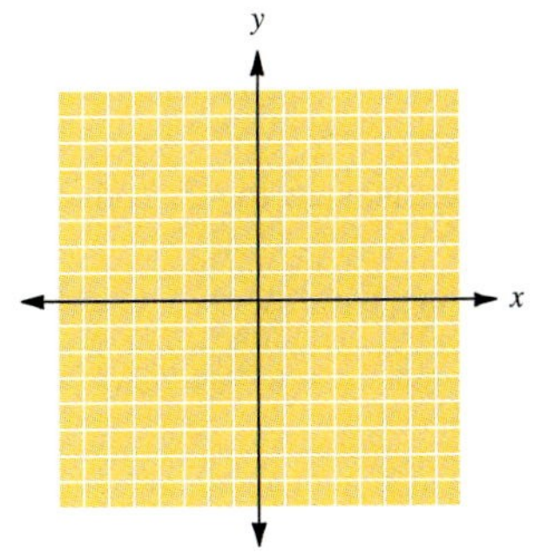

4. $y = \log_3 (x + 1)$

5. $y = \log_8 x$
(Use a graphing utility.)

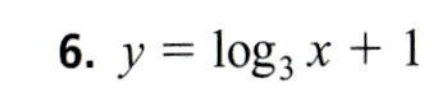

6. $y = \log_3 x + 1$

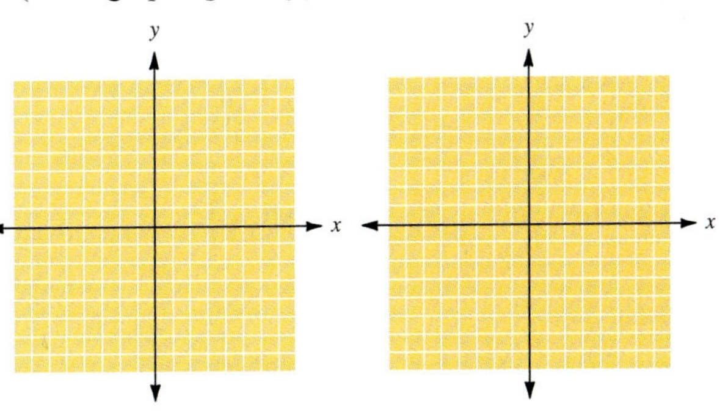

In exercises 7 to 24, convert each statement to logarithmic form.

7. $2^4 = 16$

8. $3^5 = 243$

9. $10^2 = 100$

10. $4^3 = 64$

11. $3^0 = 1$

12. $10^0 = 1$

13. $4^{-2} = \frac{1}{16}$

14. $3^{-4} = \frac{1}{81}$

15. $10^{-3} = \frac{1}{1000}$

16. $2^{-5} = \frac{1}{32}$

17. $16^{1/2} = 4$

18. $125^{1/3} = 5$

ANSWERS

1. ______________________
2. ______________________
3. ______________________
4. ______________________
5. ______________________
6. ______________________
7. ______________________
8. ______________________
9. ______________________
10. ______________________
11. ______________________
12. ______________________
13. ______________________
14. ______________________
15. ______________________
16. ______________________
17. ______________________
18. ______________________

ANSWERS

19. ______
20. ______
21. ______
22. ______
23. ______
24. ______
25. ______ 26. ______
27. ______ 28. ______
29. ______ 30. ______
31. ______ 32. ______
33. ______
34. ______
35. ______
36. ______
37. ______
38. ______
39. ______
40. ______
41. ______
42. ______
43. ______ 44. ______
45. ______ 46. ______
47. ______ 48. ______

19. $64^{-1/3} = \frac{1}{4}$ **20.** $36^{-1/2} = \frac{1}{6}$ **21.** $8^{2/3} = 4$

22. $9^{3/2} = 27$ **23.** $27^{-2/3} = \frac{1}{9}$ **24.** $16^{-3/2} = \frac{1}{64}$

In exercises 25 to 42, convert each statement to exponential form.

25. $\log_2 16 = 4$ **26.** $\log_3 3 = 1$ **27.** $\log_5 1 = 0$

28. $\log_3 27 = 3$ **29.** $\log_{10} 10 = 1$ **30.** $\log_2 32 = 5$

31. $\log_5 125 = 3$ **32.** $\log_{10} 1 = 0$ **33.** $\log_3 \frac{1}{27} = -3$

34. $\log_5 \frac{1}{25} = -2$ **35.** $\log_{10} 0.01 = -2$ **36.** $\log_{10} \frac{1}{1000} = -3$

37. $\log_{16} 4 = \frac{1}{2}$ **38.** $\log_{125} 5 = \frac{1}{3}$ **39.** $\log_8 4 = \frac{2}{3}$

40. $\log_9 27 = \frac{3}{2}$ **41.** $\log_{25} \frac{1}{5} = -\frac{1}{2}$ **42.** $\log_{64} \frac{1}{16} = -\frac{2}{3}$

In exercises 43 to 52, evaluate each logarithm.

43. $\log_2 32$ **44.** $\log_3 81$ **45.** $\log_4 64$

46. $\log_{10} 1000$ **47.** $\log_3 \frac{1}{81}$ **48.** $\log_4 \frac{1}{64}$

49. $\log_{10} \frac{1}{100}$

50. $\log_5 \frac{1}{25}$

51. $\log_{25} 5$

52. $\log_{27} 3$

In exercises 53 to 74, solve each equation for the unknown variable.

53. $y = \log_5 25$

54. $\log_2 x = 4$

55. $\log_b 64 = 3$

56. $y = \log_3 1$

57. $\log_{10} x = 2$

58. $\log_b 125 = 3$

59. $y = \log_5 5$

60. $y = \log_3 81$

61. $\log_{3/2} x = 3$

62. $\log_b \frac{4}{9} = 2$

63. $\log_b \frac{1}{25} = -2$

64. $\log_3 x = -3$

65. $\log_{10} x = -3$

66. $y = \log_2 \frac{1}{16}$

67. $y = \log_8 \frac{1}{64}$

68. $\log_b \frac{1}{100} = -2$

69. $\log_{27} x = \frac{1}{3}$

70. $y = \log_{100} 10$

71. $\log_b 5 = \frac{1}{2}$

72. $\log_{64} x = \frac{2}{3}$

73. $y = \log_{27} \frac{1}{9}$

74. $\log_b \frac{1}{8} = -\frac{3}{4}$

Use the decibel formula

$$L = 10 \log_{10} \frac{I}{I_0}$$

to solve exercises 75 to 78.

75. **Sound.** A television commercial has a volume with intensity $I = 10^{-11}$ W/cm^2. Find its rating in decibels.

76. **Sound.** The sound of a jet plane on takeoff has an intensity $I = 10^{-2}$ W/cm^2. Find its rating in decibels.

ANSWERS

49. ______ 50. ______
51. ______ 52. ______
53. ______ 54. ______
55. ______ 56. ______
57. ______ 58. ______
59. ______ 60. ______
61. ______ 62. ______
63. ______ 64. ______
65. ______ 66. ______
67. ______ 68. ______
69. ______ 70. ______
71. ______ 72. ______
73. ______ 74. ______
75. ______
76. ______

ANSWERS

77. ______

78. ______

79. ______

80. ______

81. ______

82. ______

83. ______

84. ______

77. Sound. The sound of a vacuum cleaner has an intensity of $I = 10^{-9}$ W/cm^2. Find its rating in decibels.

78. Sound. The sound of a busy street has an intensity of $I = 10^{-8}$ W/cm^2. Find its rating in decibels.

The formula for the decibel rating L can be solved for the intensity of the sound as $I = I_0 \cdot 10^{L/10}$. Use this formula in exercises 79 to 83.

79. Sound. Find the intensity of the sound in an airport waiting area if the decibel rating is 80.

80. Sound. Find the intensity of the sound of conversation in a crowded room if the decibel rating is 70.

81. Sound. What is the ratio of intensity of the sound of 80 dB to that of 70 dB?

82. Sound. What is the ratio of intensity of a sound of 60 dB to one measuring 40 dB?

83. Sound. What is the ratio of intensity of a sound of 70 dB to one measuring 40 dB?

84. Derive the formula for intensity provided above. *Hint:* First divide both sides of the decibel formula by 10. Then write the equation in exponential form.

Use the earthquake formula

$M = \log_{10} \frac{a}{a_0}$ to solve exercises 85 to 88.

ANSWERS

85. ____
86. ____
87. ____
88. ____

89. ____

90. ____
91. ____
92. ____
93. ____
94. ____
95. ____

85. Earthquakes. An earthquake has an intensity a of $10^6 \cdot a_0$, in which a_0 is the intensity of the zero-level earthquake. What is its magnitude?

86. Earthquakes. The great San Francisco earthquake of 1906 had an intensity of $10^{8.3} \cdot a_0$. What was its magnitude?

87. Earthquakes. An earthquake can begin causing damage to buildings with a magnitude of 5 on the Richter scale. Find its intensity in terms of a_0.

88. Earthquakes. An earthquake may cause moderate building damage with a magnitude of 6 on the Richter scale. Find its intensity in terms of a_0.

89. The **learning curve** describes the relationship between learning and time. Its graph is a logarithmic curve in the first quadrant. Describe that curve as it relates to learning.

90. In which scientific fields would you expect to again encounter a discussion of logarithms?

The *half-life* of a radioactive substance is the time it takes for half the original amount of the substance to decay to a nonradioactive element. The half-life of radioactive waste is very important in figuring how long the waste must be kept isolated from the environment in some sort of storage facility. Half-lives of various radioactive waste products vary from a few seconds to millions of years. It usually takes at least 10 half-lives for a radioactive waste product to be considered safe.

The half-life of a radioactive substance can be determined by the following formula.

$$\ln \frac{1}{2} = -\lambda x$$

in which λ = radioactive decay constant

x = half-life

In exercises 91 to 95, find the half-lives of the following important radioactive waste products given the radioactive decay constant (RDC).

91. Plutonium 239, RDC = 0.000029

92. Strontium 90, RDC = 0.024755

93. Thorium 230, RDC = 0.000009

94. Cesium 135, RDC = 0.00000035

95. How many years will it be before each waste product will be considered safe?

Answers

1.

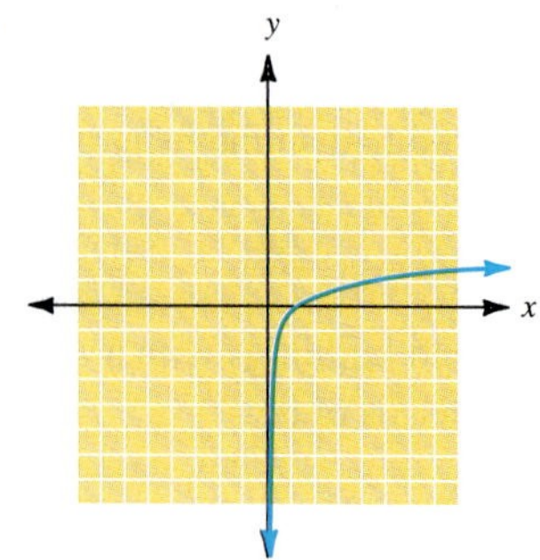

3.

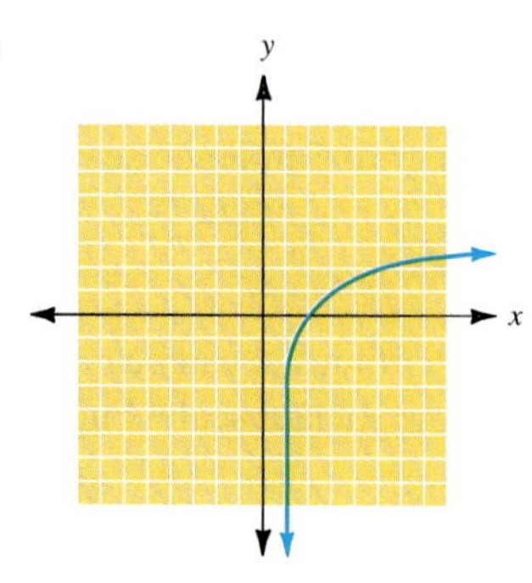

5.

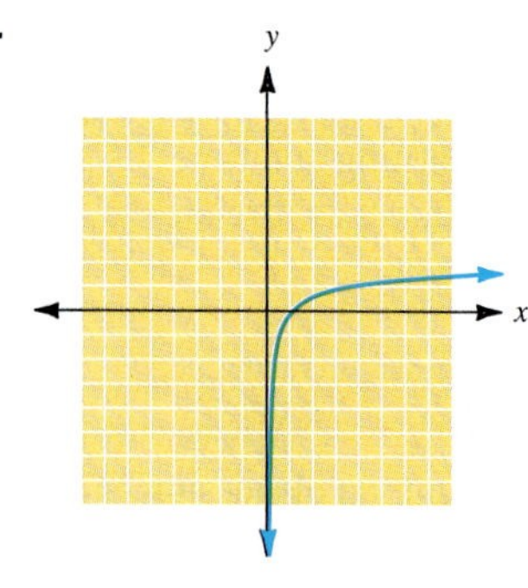

7. $\log_2 16 = 4$ **9.** $\log_{10} 100 = 2$ **11.** $\log_3 1 = 0$ **13.** $\log_4 \frac{1}{16} = -2$

15. $\log_{10} \frac{1}{1000} = -3$ **17.** $\log_{16} 4 = \frac{1}{2}$ **19.** $\log_{64} \frac{1}{4} = -\frac{1}{3}$ **21.** $\log_8 4 = \frac{2}{3}$

23. $\log_{27} \frac{1}{9} = -\frac{2}{3}$ **25.** $2^4 = 16$ **27.** $5^0 = 1$ **29.** $10^1 = 10$

31. $5^3 = 125$ **33.** $3^{-3} = \frac{1}{27}$ **35.** $10^{-2} = 0.01$ **37.** $16^{1/2} = 4$

39. $8^{2/3} = 4$ **41.** $25^{-1/2} = \frac{1}{5}$ **43.** 5 **45.** 3 **47.** -4 **49.** -2

51. $\frac{1}{2}$ **53.** $\{2\}$ **55.** $\{4\}$ **57.** $\{100\}$ **59.** $\{1\}$ **61.** $\left\{\frac{27}{8}\right\}$

63. $\{5\}$ **65.** $\left\{\frac{1}{1000}\right\}$ **67.** $\{-2\}$ **69.** $\{3\}$ **71.** $\{25\}$ **73.** $\left\{-\frac{2}{3}\right\}$

75. 50 dB **77.** 70 dB **79.** 10^{-8} **81.** 10 **83.** 1000 **85.** 6

87. $10^5 \cdot a_0$ **89.**  **91.** 24,000 y **93.** 77,000 y

95. Pu239: 240,000 y; Sr90: 280 y; Th230: 770,000 y; Cs135: 20,000,000 y

11.4 Properties of Logarithms

OBJECTIVES

1. Apply the properties of logarithms
2. Evaluate logarithmic expressions with any base
3. Solve applications involving logarithms
4. Estimate the value of an antilogarithm

As we mentioned earlier, logarithms were developed as aids to numerical computations. The early utility of the logarithm was due to the properties that we will discuss in this section. Even with the advent of the scientific calculator, that utility remains important today. We can apply these same properties to applications in a variety of areas that lead to exponential or logarithmic equations.

Because a logarithm is, by definition, an exponent, it seems reasonable that our knowledge of the properties of exponents should lead to useful properties for logarithms. That is, in fact, the case.

We start with two basic facts that follow immediately from the definition of the logarithm.

NOTE The properties follow from the facts that

$b^1 = b$ and $b^0 = 1$

Rules and Properties: Properties 1 and 2 of Logarithms

For $b > 0$ and $b \neq 1$,

Property 1. $\log_b b = 1$

Property 2. $\log_b 1 = 0$

We know that the logarithmic function $y = \log_b x$ and the exponential function $y = b^x$ are inverses of each other. So, for $f(x) = b^x$, we have $f^{-1}(x) = \log_b x$.

NOTE The inverse has "undone" whatever f did to x.

It is important to note that for any one-to-one function f,

$f^{-1}(f(x)) = x$ for any x in domain of f

and

$f(f^{-1}(x)) = x$ for any x in domain of f^{-1}

Because $f(x) = b^x$ is a one-to-one function, we can apply the above to the case in which

$f(x) = b^x$ and $f^{-1}(x) = \log_b x$

to derive the following.

NOTE For Property 3,

$f^{-1}(f(x)) = f^{-1}(b^x) = \log_b b^x$

But in general, for any one-to-one function f,

$f^{-1}(f(x)) = x$

Rules and Properties: Properties 3 and 4 of Logarithms

Property 3. $\log_b b^x = x$

Property 4. $b^{\log_b x} = x$ for $x > 0$

Because logarithms are exponents, we can again turn to the familiar exponent rules to derive some further properties of logarithms. Consider the following.

We know that

$\log_b M = x$ if and only if $M = b^x$

and

$\log_b N = y$ if and only if $N = b^y$

Then

$$M \cdot N = b^x \cdot b^y = b^{x+y} \tag{1}$$

From equation (1) we see that $x + y$ is the power to which we must raise b to get the product MN. In logarithmic form, that becomes

$$\log_b MN = x + y \tag{2}$$

Now, because $x = \log_b M$ and $y = \log_b N$, we can substitute in (2) to write

$$\log_b MN = \log_b M + \log_b N \tag{3}$$

This is the first of the basic logarithmic properties presented here. The remaining properties may all be proved by arguments similar to those presented in equations (1) to (3).

NOTE In all cases, $M, N > 0$, $b > 0$, $b \neq 1$, and p is any real number.

Rules and Properties: Properties of Logarithms

Product property

$$\log_b MN = \log_b M + \log_b N$$

Quotient property

$$\log_b \frac{M}{N} = \log_b M - \log_b N$$

Power property

$$\log_b M^p = p \log_b M$$

Many applications of logarithms require using these properties to write a single logarithmic expression as the sum or difference of simpler expressions, as Example 1 illustrates.

Example 1

Using the Properties of Logarithms

Expand, using the properties of logarithms.

(a) $\log_b xy = \log_b x + \log_b y$ — Product property

(b) $\log_b \frac{xy}{z} = \log_b xy - \log_b z$ — Quotient property

$= \log_b x + \log_b y - \log_b z$ — Product property

(c) $\log_{10} x^2y^3 = \log_{10} x^2 + \log_{10} y^3$ — Product property

$= 2 \log_{10} x + 3 \log_{10} y$ — Power property

NOTE Recall $\sqrt{a} = a^{1/2}$.

(d) $\log_b \sqrt{\frac{x}{y}} = \log_b \left(\frac{x}{y}\right)^{1/2}$ — Definition of rational exponent

$= \frac{1}{2} \log_b \frac{x}{y}$ — Power property

$= \frac{1}{2} (\log_b x - \log_b y)$ — Quotient property

CHECK YOURSELF 1

Expand each expression, using the properties of logarithms.

(a) $\log_b x^2y^3z$ **(b)** $\log_{10} \sqrt{\frac{xy}{z}}$

In some cases, we will reverse the process and use the properties to write a single logarithm, given a sum or difference of logarithmic expressions.

Example 2

Rewriting Logarithmic Expressions

Write each expression as a single logarithm with coefficient 1.

(a) $2 \log_b x + 3 \log_b y$

$= \log_b x^2 + \log_b y^3$ Power property

$= \log_b x^2y^3$ Product property

(b) $\frac{1}{2}(\log_2 x - \log_2 y)$

$= \frac{1}{2}\left(\log_2 \frac{x}{y}\right)$ Quotient property

$= \log_2 \left(\frac{x}{y}\right)^{1/2}$ Power property

$= \log_2 \sqrt{\frac{x}{y}}$

CHECK YOURSELF 2

Write each expression as a single logarithm with coefficient 1.

(a) $3 \log_b x + 2 \log_b y - 2 \log_b z$ **(b)** $\frac{1}{3}(2 \log_2 x - \log_2 y)$

Example 3 illustrates the basic concept of the use of logarithms as a computational aid.

Example 3

Evaluating Logarithmic Expressions

NOTE We have written the logarithms correct to three decimal places and will follow this practice throughout the remainder of this chapter.Keep in mind, however, that this is an approximation and that $10^{0.301}$ will only approximate 2. Verify this with your calculator.

Suppose $\log_{10} 2 = 0.301$ and $\log_{10} 3 = 0.447$. Given these values, find the following.

(a) $\log_{10} 6$ Because $6 = 2 \cdot 3$

$= \log_{10} (2 \cdot 3)$

$= \log_{10} 2 + \log_{10} 3$

$= 0.301 + 0.477$

$= 0.778$

(b) $\log_{10} 18$ Because $18 = 2 \cdot 3 \cdot 3$

$= \log_{10} (2 \cdot 3 \cdot 3)$

NOTE We have extended the product rule for logarithms.

$= \log_{10} 2 + \log_{10} 3 + \log_{10} 3$

$= 1.255$

(c) $\log_{10} \frac{1}{9}$ Because $\frac{1}{9} 5 \frac{1}{3^2}$

$= \log_{10} \frac{1}{3^2}$

NOTE Notice that $\log_b 1 = 0$ for any base b.

$= \log_{10} 1 - \log_{10} 3^2$

$= 0 - 2 \log_{10} 3$

$= -0.954$

(d) $\log_{10} 16$ Because $16 = 2^4$

$= \log_{10} 2^4 = 4 \log_{10} 2$

$= 1.204$

NOTE Verify each answer with your calculator.

(e) $\log_{10} \sqrt{3}$ Because $\sqrt{3} = 3^{1/2}$

$= \log_{10} 3^{1/2} = \frac{1}{2} \log_{10} 3$

$= 0.239$

CHECK YOURSELF 3

Given the values above for $\log_{10} 2$ *and* $\log_{10} 3$, *find each of the following.*

(a) $\log_{10} 12$ **(b)** $\log_{10} 27$ **(c)** $\log_{10} \sqrt[3]{2}$

There are two types of logarithms used most frequently in mathematics:

Logarithms to base 10

Logarithms to base e

Of course, the use of logarithms to base 10 is convenient because our number system has base 10. We call logarithms to base 10 **common logarithms,** and it is customary to omit the base in writing a common (or base-10) logarithm. So

NOTE When no base for "log" is written, it is assumed to be 10.

$\log N$ means $\log_{10} N$

The following table shows the common logarithms for various powers of 10.

Exponential Form	Logarithmic Form
$10^3 = 1000$	$\log 1000 = 3$
$10^2 = 100$	$\log 100 = 2$
$10^1 = 10$	$\log 10 = 1$
$10^0 = 1$	$\log 1 = 0$
$10^{-1} = 0.1$	$\log 0.1 = -1$
$10^{-2} = 0.01$	$\log 0.01 = -2$
$10^{-3} = 0.001$	$\log 0.001 = -3$

Example 4

Approximating Logarithms with a Calculator

Verify each of the following with a calculator.

NOTE The number 4.8 lies between 1 and 10, so log 4.8 lies between 0 and 1.

(a) $\log 4.8 = 0.681$

(b) $\log 48 = 1.681$

(c) $\log 480 = 2.681$

(d) $\log 4800 = 3.681$

(e) $\log 0.48 = -0.319$

NOTE Notice that

$480 = 4.8 \times 10^2$

and

$\log (4.8 \times 10^2)$

$= \log 4.8 + \log 10^2$

$= \log 4.8 + 2$

$= 2 + \log 4.8$

NOTE The value of log 0.48 is really $-1 + 0.681$. Your calculator will combine the signed numbers.

CHECK YOURSELF 4

Use your calculator to find each of the following logarithms, correct to three decimal places.

(a) $\log 2.3$ **(b)** $\log 23$ **(c)** $\log 230$

(d) 2300 **(e)** $\log 0.23$ **(f)** $\log 0.023$

Let's look at an application of common logarithms from chemistry. Common logarithms are used to define the pH of a solution. This is a scale that measures whether the solution is acidic or basic.

NOTE A solution is **neutral** with pH = 7, **acidic** if the pH is less than 7, and **basic** if the pH is greater than 7.

The pH of a solution is defined as

$$\text{pH} = -\log [\text{H}^+]$$

in which $[\text{H}^+]$ is the hydrogen ion concentration, in moles per liter (mol/L), in the solution.

Example 5

A pH Application

Find the pH of each of the following. Determine whether each is a base or an acid.

(a) Rainwater: $[\text{H}^+] = 1.6 \times 10^{-7}$

From the definition,

$$\begin{aligned}\text{pH} &= -\log [\text{H}^+] \\ &= -\log (1.6 \times 10^{-7}) \\ &= -(\log 1.6 + \log 10^{-7}) \\ &= -[0.204 + (-7)] \\ &= -(-6.796) = 6.796\end{aligned}$$

NOTE Notice the use of the product rule here.

NOTE Also, in general, $\log_b b^x = x$, so $\log 10^{-7} = -7$.

The rain is just slightly acidic.

(b) Household ammonia: $[H^+] = 2.3 \times 10^{-8}$

$$\begin{aligned} pH &= -\log (2.3 \times 10^{-8}) \\ &= -(\log 2.3 + \log 10^{-8}) \\ &= -[0.362 + (-8)] \\ &= 7.638 \end{aligned}$$

The ammonia is slightly basic.

(c) Vinegar: $[H^+] = 2.9 \times 10^{-3}$

$$\begin{aligned} pH &= -\log (2.9 \times 10^{-3}) \\ &= -(\log 2.9 + \log 10^{-3}) \\ &= 2.538 \end{aligned}$$

The vinegar is very acidic.

CHECK YOURSELF 5

Find the pH for the following solutions. Are they acidic or basic?

(a) Orange juice: $[H^+] = 6.8 \times 10^{-5}$

(b) Drain cleaner: $[H^+] = 5.2 \times 10^{-13}$

Many applications require reversing the process. That is, given the logarithm of a number, we must be able to find that number. The process is straightforward.

Example 6

Using a Calculator to Estimate Antilogarithms

Suppose that $\log x = 2.1567$. We want to find a number x whose logarithm is 2.1567. Using a calculator requires one of the following sequences:

NOTE Because it is a one-to-one function, the logarithmic function has an inverse.

2.1567 [10^x] or 2.1567 [INV] [log] or [2nd] [log] 2.1567

Both give the result 143.45, often called the **antilogarithm** of 2.1567.

CHECK YOURSELF 6

Find the value of the antilogarithm of x.

(a) $\log x = 0.828$

(b) $\log x = 1.828$

(c) $\log x = 2.828$

(d) $\log x = -0.172$

Let's return to the application from chemistry for an example requiring the use of the antilogarithm.

Example 7

A pH Application

Suppose that the pH for tomato juice is 6.2. Find the hydrogen ion concentration $[H^+]$.
Recall from our earlier formula that

$$\text{pH} = -\log [H^+]$$

In this case, we have

$$6.2 = -\log [H^+] \qquad \text{or} \qquad \log [H^+] = -6.2$$

The desired value for $[H^+]$ is then the antilogarithm of -6.2.

The result is 0.00000063, and we can write

$$[H^+] = 6.3 \times 10^{-7}$$

CHECK YOURSELF 7

The pH for eggs is 7.8. Find $[H^+]$ for eggs.

NOTE Natural logarithms are also called **napierian logarithms** after Napier. The importance of this system of logarithms was not fully understood until later developments in the calculus.

NOTE The restrictions on the domain of the natural logarithmic function are the same as before. The function is defined only if $x > 0$.

As we mentioned, there are two systems of logarithms in common use. The second type of logarithm uses the number e as a base, and we call logarithms to base e the **natural logarithms.** As with common logarithms, a convenient notation has developed, as the following definition shows.

Definitions: Natural Logarithm

The **natural logarithm** is a logarithm to base e, and it is denoted $\ln x$, as

$$\ln x = \log_e x$$

By the general definition of a logarithm,

$$y = \ln x \qquad \text{means the same as} \qquad x = e^y$$

and this leads us directly to the following.

NOTE In general

$\log_b b^x = x \qquad b \neq 1$

$$\ln 1 = 0 \qquad \text{because} \qquad e^0 = 1$$

$$\ln e = 1 \qquad \text{because} \qquad e^1 = e$$

$$\ln e^2 = 2 \qquad \text{and} \qquad \ln e^{-3} = -3$$

Example 8

Estimating Natural Logarithms

To find other natural logarithms, we can again turn to a calculator. To find the value of ln 2, use the sequence

[ln] 2 or [ln (] 2 [)]

The result is 0.693 (to three decimal places).

CHECK YOURSELF 8

Use a calculator to find each of the following.

(a) $\ln 3$ **(b)** $\ln 6$ **(c)** $\ln 4$ **(d)** $\ln \sqrt{3}$

Of course, the properties of logarithms are applied in an identical fashion, no matter what the base.

Example 9

Evaluating Logarithms

If $\ln 2 = 0.693$ and $\ln 3 = 1.099$, find the following.

NOTE Recall that
$\log_b MN = \log_b M + \log_b N$
$\log_b M^p = p \log_b M$

(a) $\ln 6 = \ln (2 \cdot 3) = \ln 2 + \ln 3 = 1.792$

(b) $\ln 4 = \ln 2^2 = 2 \ln 2 = 1.386$

(c) $\ln \sqrt{3} = \ln 3^{1/2} = \frac{1}{2} \ln 3 = 0.549$

Again, verify these results with your calculator.

CHECK YOURSELF 9

Use $\ln 2 = 0.693$ *and* $\ln 3 = 1.099$ *to find the following.*

(a) $\ln 12$ **(b)** $\ln 27$

The natural logarithm function plays an important role in both theoretical and applied mathematics. Example 10 illustrates just one of the many applications of this function.

Example 10

A Learning Curve Application

NOTE Recall that we read $S(t)$ as "S of t", which means that S is a function of t.

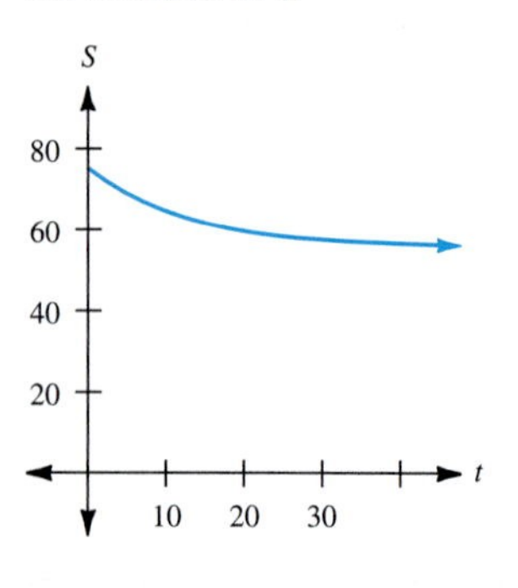

NOTE This is an example of a **forgetting curve.** Note how it drops more rapidly at first. Compare this curve to the learning curve drawn in Section 11.2, exercise 62.

A class of students took a final mathematics examination and received an average score of 76. In a psychological experiment, the students are retested at weekly intervals over the same material. If t is measured in weeks, then the new average score after t weeks is given by

$$S(t) = 76 - 5 \ln (t + 1)$$

Complete the following.

(a) Find the score after 10 weeks.

$$S(t) = 76 - 5 \ln (10 + 1)$$
$$= 76 - 5 \ln 11 \approx 64$$

(b) Find the score after 20 weeks.

$S(t) = 76 - 5 \ln (20 + 1) \approx 61$

(c) Find the score after 30 weeks.

$S(t) = 76 - 5 \ln (30 + 1) \approx 59$

CHECK YOURSELF 10

The average score for a group of biology students, retested after time t (in months), is given by

$S(t) = 83 - 9 \ln (t + 1)$

Find the average score after

(a) 3 months **(b)** 6 months

We conclude this section with one final property of logarithms. This property will allow us to quickly find the logarithm of a number to any base. Although work with logarithms with bases other than 10 or e is relatively infrequent, the relationship between logarithms of different bases is interesting in itself. Consider the following argument.

Suppose that

$x = \log_2 5$

or

$$2^x = 5 \tag{4}$$

Taking the logarithm to base 10 of both sides of equation (4) yields

$\log 2^x = \log 5$

or

$$x \log 2 = \log 5 \tag{5}$$

Use the power property of logarithms.

(Note that we omit the 10 for the base and write log 2, for example.) Now, dividing both sides of equation (5) by log 2, we have

$$x = \frac{\log 5}{\log 2}$$

We can now find a value for x with the calculator. Dividing with the calculator log 5 by log 2, we get an approximate answer of 2.3219.

Because $x = \log_2 5$ and $x = \dfrac{\log 5}{\log 2}$, then

$$\log_2 5 = \frac{\log 5}{\log 2}$$

Generalizing our result, we find the following.

Rules and Properties: Change-of-Base Formula

For the positive real numbers a and x,

$$\log_a x = \frac{\log x}{\log a}$$

Note that the logarithm on the left side has base a whereas the logarithms on the right side have base 10. This allows us to calculate the logarithm to base a of any positive number, given the corresponding logarithms to base 10 (or any other base), as Example 11 illustrates.

Example 11

Evaluating Logarithms

Find $\log_5 15$.

From the change-of-base formula with $a = 5$ and $b = 10$,

$$\log_5 15 = \frac{\log 15}{\log 5}$$

$$= 1.683$$

NOTE We have written log 15 rather than log 15 to emphasize the change-of-base formula.

NOTE $\log_5 5 = 1$ and $\log_5 25 = 2$, so the result for $\log_5 15$ must be between 1 and 2.

The calculator sequence for the above computation is

[log] 15 [÷] [log] 5 [ENTER]

CHECK YOURSELF 11

Use the change-of-base formula to find $\log_8 32$.

CAUTION

A *common error* is to write

$$\frac{\log 15}{\log 5} = \log 15 - \log 5$$

This is *not* a logarithmic property. A true statement would be

$$\log \frac{15}{5} = \log 15 - \log 5$$

but

$$\log \frac{15}{5} \quad \text{and} \quad \frac{\log 15}{\log 5}$$

are *not* the same.

Note: Recall that the $\log_e x$ is called the **natural log** of x. We use "ln x" to designate the natural log of x. A special case of the change-of-base formula allows us to find natural logarithms in terms of common logarithms:

$$\ln x = \frac{\log x}{\log e}$$

so

$$\ln x \approx \frac{\log x}{0.434} \text{ or, because } \frac{1}{0.434} \approx 2.304, \text{ then } \ln x \approx 2.304 \log x$$

Of course, because all modern calculators have both the log function key and the ln function key, this conversion formula is now rarely used.

CHECK YOURSELF ANSWERS

1. **(a)** $2 \log_b x + 3 \log_b y + \log_b z$; **(b)** $\frac{1}{2}(\log_{10} x + \log_{10} y - \log_{10} z)$

2. **(a)** $\log_b \frac{x^3y^2}{z^2}$; **(b)** $\log_2 \sqrt[3]{\frac{x^2}{y}}$ **3.** **(a)** 1.079; **(b)** 1.431; **(c)** 0.100

4. **(a)** 0.362; **(b)** 1.362; **(c)** 2.362; **(d)** 3.362; **(e)** −0.638; **(f)** −1.638

5. **(a)** 4.17, acidic; **(b)** 12.28, basic **6.** **(a)** 6.73; **(b)** 67.3; **(c)** 673; **(d)** 0.673

7. $[H^+] = 1.6 \times 10^{-8}$ **8.** **(a)** 1.099; **(b)** 1.792; **(c)** 1.386; **(d)** 0.549

9. **(a)** 2.485; **(b)** 3.297 **10.** **(a)** 70.5; **(b)** 65.5 **11.** $\log_8 32 = \frac{\log 32}{\log 8} \approx 1.667$

Name ______________________

Section ________ Date ________

11.4 Exercises

In exercises 1 to 18, use the properties of logarithms to expand each expression.

1. $\log_b 5x$

2. $\log_3 7x$

3. $\log_4 \dfrac{x}{3}$

4. $\log_b \dfrac{2}{y}$

5. $\log_3 a^2$

6. $\log_5 y^4$

7. $\log_5 \sqrt{x}$

8. $\log \sqrt[3]{z}$

9. $\log_b x^3y^2$

10. $\log_5 x^2z^4$

11. $\log_4 y^2 \sqrt{x}$

12. $\log_b x^3 \sqrt[3]{z}$

13. $\log_b \dfrac{x^2y}{z}$

14. $\log_5 \dfrac{3}{xy}$

15. $\log \dfrac{xy^2}{\sqrt{z}}$

16. $\log_4 \dfrac{x^3\sqrt{y}}{z^2}$

17. $\log_5 \sqrt[3]{\dfrac{xy}{z^2}}$

18. $\log_b \sqrt[4]{\dfrac{x^2y}{z^3}}$

ANSWERS

1. ______
2. ______
3. ______
4. ______
5. ______
6. ______
7. ______
8. ______
9. ______
10. ______
11. ______
12. ______
13. ______
14. ______
15. ______
16. ______
17. ______
18. ______

ANSWERS

19. ______
20. ______
21. ______
22. ______
23. ______
24. ______
25. ______
26. ______
27. ______
28. ______
29. ______
30. ______
31. ______
32. ______
33. ______
34. ______
35. ______
36. ______
37. ______
38. ______
39. ______
40. ______
41. ______
42. ______
43. ______
44. ______

In exercises 19 to 30, write each expression as a single logarithm.

19. $\log_b x + \log_b y$

20. $\log_5 x - \log_5 y$

21. $2 \log_2 x - \log_2 y$

22. $3 \log_b x + \log_b z$

23. $\log_b x + \frac{1}{2} \log_b y$

24. $\frac{1}{3} \log_b x - 2 \log_b z$

25. $\log_b x + 2 \log_b y - \log_b z$

26. $2 \log_5 x - (3 \log_5 y + \log_5 z)$

27. $\frac{1}{2} \log_6 y - 3 \log_6 z$

28. $\log_b x - \frac{1}{3} \log_b y - 4 \log_b z$

29. $\frac{1}{3}(2 \log_b x + \log_b y - \log_b z)$

30. $\frac{1}{5}(2 \log_4 x - \log_4 y + 3 \log_4 z)$

In exercises 31 to 38, given that $\log 2 = 0.301$ and $\log 3 = 0.477$, find each logarithm.

31. $\log 24$

32. $\log 36$

33. $\log 8$

34. $\log 81$

35. $\log \sqrt{2}$

36. $\log \sqrt[3]{3}$

37. $\log \frac{1}{4}$

38. $\log \frac{1}{27}$

In exercises 39 to 44, use your calculator to find each logarithm.

39. $\log 6.8$

40. $\log 68$

41. $\log 680$

42. $\log 6800$

43. $\log 0.68$

44. $\log 0.068$

ANSWERS

45. ______

46. ______

47. ______

48. ______

49. ______

50. ______

51. ______

52. ______

53. ______

54. ______

55. ______

56. ______

57. ______

58. ______

59. ______

60. ______

In exercises 45 and 46, find the pH, given the hydrogen ion concentration $[H^+]$ for each solution. Use the formula

$$pH = -\log [H^+]$$

Are the solutions acidic or basic?

45. Blood: $[H^+] = 3.8 \times 10^{-8}$

46. Lemon juice: $[H^+] = 6.4 \times 10^{-3}$

In exercises 47 to 50, use your calculator to find the antilogarithm for each logarithm.

47. 0.749

48. 1.749

49. 3.749

50. -0.251

In exercises 51 and 52, given the pH of the solutions, find the hydrogen ion concentration $[H^+]$.

51. Wine: pH = 4.7

52. Household ammonia: pH = 7.8

In exercises 53 to 56, use your calculator to find each logarithm.

53. $\ln 2$

54. $\ln 3$

55. $\ln 10$

56. $\ln 30$

The average score on a final examination for a group of psychology students, retested after time t (in weeks), is given by

$$S = 85 - 8 \ln (t + 1)$$

In exercises 57 and 58, find the average score on the retests:

57. After 3 weeks

58. After 12 weeks

In exercises 59 and 60, use the change-of-base formula to find each logarithm.

59. $\log_3 25$

60. $\log_5 30$

ANSWERS

61. ______

62. ______

63. ______

64. ______

65. ______

66. ______

The amount of a radioactive substance remaining after a given amount of time t is given by the following formula:

$$A = e^{\lambda t + \ln A_0}$$

in which A is the amount remaining after time t, variable A_0 is the original amount of the substance, and λ is the radioactive decay constant.

61. How much plutonium 239 will remain after 50,000 years if 24 kg was originally stored? Plutonium 239 has a radioactive decay constant of -0.000029.

62. How much plutonium 241 will remain after 100 years if 52 kg was originally stored? Plutonium 241 has a radioactive decay constant of -0.053319.

63. How much strontium 90 was originally stored if after 56 years it is discovered that 15 kg still remains? Strontium 90 has a radioactive decay constant of -0.024755.

64. How much cesium 137 was originally stored if after 90 years it is discovered that 20 kg still remains? Cesium 137 has a radioactive decay constant of -0.023105.

65. Which keys on your calculator are function keys and which are operation keys? What is the difference?

66. How is the pH factor relevant to your selection of a hair care product?

Answers

1. $\log_b 5 + \log_b x$ **3.** $\log_4 x - \log_4 3$ **5.** $2 \log_3 a$ **7.** $\frac{1}{2} \log_5 x$
9. $3 \log_b x + 2 \log_b y$ **11.** $2 \log_4 y + \frac{1}{2} \log_4 x$ **13.** $2 \log_b x + \log_b y - \log_b z$
15. $\log x + 2 \log y - \frac{1}{2} \log z$ **17.** $\frac{1}{3}(\log_5 x + \log_5 y - 2 \log_5 z)$ **19.** $\log_b xy$
21. $\log_2 \frac{x^2}{y}$ **23.** $\log_b x\sqrt{y}$ **25.** $\log_b \frac{xy^2}{z}$ **27.** $\log_6 \frac{\sqrt{y}}{z^3}$
29. $\log_b \sqrt[3]{\frac{x^2 y}{z}}$ **31.** 1.380 **33.** 0.903 **35.** 0.151 **37.** -0.602
39. 0.833 **41.** 2.833 **43.** -0.167 **45.** 7.42, basic **47.** 5.61
49. 5610 **51.** 2×10^{-5} **53.** 0.693 **55.** 2.303 **57.** 74
59. 2.930 **61.** 5.6 kg **63.** 60 kg **65.**

11.5 Logarithmic and Exponential Equations

OBJECTIVES

1. Solve a logarithmic equation
2. Solve an exponential equation
3. Solve an application involving an exponential equation

Much of the importance of the properties of logarithms developed in the previous section lies in the application of those properties to the solution of equations involving logarithms and exponentials. Our work in this section will consider solution techniques for both types of equations. Let's start with a definition.

Definitions: Logarithmic Equation

A **logarithmic equation** is an equation that contains a logarithmic expression.

We solved some simple examples in Section 11.3. Let's review for a moment. To solve $\log_3 x = 4$ for x, recall that we simply convert the logarithmic equation to exponential form. Here,

$$x = 3^4$$

so

$$x = 81$$

and 81 is the solution to the given equation.

Now, what if the logarithmic equation involves more than one logarithmic term? Example 1 illustrates how the properties of logarithms must then be applied.

Example 1

Solving a Logarithmic Equation

Solve each logarithmic equation.

(a) $\log_5 x + \log_5 3 = 2$

The original equation can be written as

$$\log_5 3x = 2$$

NOTE We apply the product rule for logarithms: $\log_b M + \log_b N = \log_b MN$

Now, because only a single logarithm is involved, we can write the equation in the equivalent exponential form:

$$3x = 5^2$$

$$3x = 25$$

$$x = \frac{25}{3}$$

(b) $\log x + \log (x - 3) = 1$

Write the equation as

NOTE Because no base is written, it is assumed to be 10.

$$\log x(x - 3) = 1$$

or

NOTE Given the base of 10, this is the equivalent exponential form.

$$x(x - 3) = 10^1$$

We now have

$$x^2 - 3x = 10$$

$$x^2 - 3x - 10 = 0$$

$$(x - 5)(x + 2) = 0$$

Possible solutions are $x = 5$ or $x = -2$.

Note that substitution of -2 into the original equation gives

NOTE Checking possible solutions is particularly important here.

$$\log (-2) + \log (-5) = 1$$

Because logarithms of negative numbers are *not* defined, -2 is an extraneous solution and we must reject it. The only solution for the original equation is 5.

CHECK YOURSELF 1

Solve $\log_2 x + \log_2 (x + 2) = 3$ *for x.*

The quotient property is used in a similar fashion for solving logarithmic equations. Consider Example 2.

Example 2

Solving a Logarithmic Equation

Solve each equation for x.

(a) $\log_5 x - \log_5 2 = 2$

Rewrite the original equation as

NOTE We apply the quotient rule for logarithms:

$$\log_b M - \log_b N = \log_b \frac{M}{N}$$

$$\log_5 \frac{x}{2} = 2$$

Now,

$$\frac{x}{2} = 5^2$$

$$\frac{x}{2} = 25$$

$$x = 50$$

(b) $\log_3 (x + 1) - \log_3 x = 3$

$$\log_3 \frac{x+1}{x} = 3$$

$$\frac{x+1}{x} = 27$$

$$x + 1 = 27x$$

$$1 = 26x$$

$$x = \frac{1}{26}$$

NOTE Again, you should verify that substituting $\frac{1}{26}$ for x leads to a positive value in each of the original logarithms.

CHECK YOURSELF 2

Solve $\log_5 (x + 3) - \log_5 x = 2$ *for* x.

The solution of certain types of logarithmic equations calls for the one-to-one property of the logarithmic function.

Rules and Properties: One-To-One Property of Logarithmic Functions

If $\log_b M = \log_b N$

then $M = N$

Example 3

Solving a Logarithmic Equation

Solve the following equation for x.

$$\log (x + 2) - \log 2 = \log x$$

Again, we rewrite the left-hand side of the equation. So

$$\log \frac{x+2}{2} = \log x$$

Because the logarithmic function is one-to-one, this is equivalent to

$$\frac{x+2}{2} = x$$

or

$$x = 2$$

CHECK YOURSELF 3

Solve for x.

$$\log (x + 3) - \log 3 = \log x$$

The following algorithm summarizes our work in solving logarithmic equations.

Step by Step: Solving Logarithmic Equations

Step 1 Use the properties of logarithms to combine terms containing logarithmic expressions into a single term.

Step 2 Write the equation formed in step 1 in exponential form.

Step 3 Solve for the indicated variable.

Step 4 Check your solutions to make sure that possible solutions do not result in the logarithms of negative numbers or zero.

Let's look now at **exponential equations,** which are equations in which the variable appears as an exponent.

We solved some particular exponential equations in Section 11.2. In solving an equation such as

$3^x = 81$

we wrote the right-hand member as a power of 3, so that

NOTE Again, we want to write both sides as a power of the same base, here 3.

$3^x = 3^4$

or

$x = 4$

The technique here will work only when both sides of the equation can be conveniently expressed as powers of the same base. If that is not the case, we must use logarithms for the solution of the equation, as illustrated in Example 4.

Example 4

Solving an Exponential Equation

Solve $3^x = 5$ for x.

We begin by taking the common logarithm of both sides of the original equation.

NOTE Again: if $M = N$, then $\log_b M = \log_b N$

$\log 3^x = \log 5$

Now we apply the power property so that the variable becomes a coefficient on the left.

$x \log 3 = \log 5$

Dividing both sides of the equation by log 3 will isolate x, and we have

CAUTION

This is *not* log 5 − log 3, a common error.

$$x = \frac{\log 5}{\log 3}$$

$= 1.465$ (to three decimal places)

Note: You can verify the approximate solution by using the $\boxed{y^x}$ key on your calculator. Raise 3 to power 1.465.

CHECK YOURSELF 4

Solve $2^x = 10$ *for* x.

Example 5 shows how to solve an equation with a more complicated exponent.

Example 5

Solving an Exponential Equation

Solve $5^{2x+1} = 8$ for x.

The solution begins as in Example 4.

$$\log 5^{2x+1} = \log 8$$

NOTE On the left, we apply $\log_b M^p = p \log_b M$

$$(2x + 1) \log 5 = \log 8$$

$$2x + 1 = \frac{\log 8}{\log 5}$$

$$2x = \frac{\log 8}{\log 5} - 1$$

NOTE On a graphing calculator, the sequence would be
[(] [log] 8 ÷ [log] 5 − 1 [)] [÷] 2

$$x = \frac{1}{2}\left(\frac{\log 8}{\log 5} - 1\right)$$

$$x \approx 0.146$$

CHECK YOURSELF 5

Solve $3^{2x-1} = 7$ for x.

The procedure is similar if the variable appears as an exponent in more than one term of the equation.

Example 6

Solving an Exponential Equation

Solve $3^x = 2^{x+1}$ for x.

$$\log 3^x = \log 2^{x+1}$$

NOTE Use the power property to write the variables as coefficients.

$$x \log 3 = (x + 1) \log 2$$

$$x \log 3 = x \log 2 + \log 2$$

NOTE We now isolate x on the left.

$$x \log 3 - x \log 2 = \log 2$$

$$x(\log 3 - \log 2) = \log 2$$

NOTE To check the reasonableness of this result, use your calculator to verify that $3^{1.710} = 2^{2.710}$

$$x = \frac{\log 2}{\log 3 - \log 2}$$

$$\approx 1.710$$

CHECK YOURSELF 6

Solve $5^{x+1} = 3^{x+2}$ for x.

The following algorithm summarizes our work with solving exponential equations.

Step by Step: Solving Exponential Equations

Step 1 Try to write each side of the equation as a power of the same base. Then equate the exponents to form an equation.

Step 2 If the above procedure is not applicable, take the common logarithm of both sides of the original equation.

Step 3 Use the power rule for logarithms to write an equivalent equation with the variables as coefficients.

Step 4 Solve the resulting equation.

There are many applications of our work with exponential equations. Consider the following.

Example 7

An Interest Application

If an investment of P dollars earns interest at an annual interest rate r and the interest is compounded n times per year, then the amount in the account after t years is given by

$$A = P\left(1 + \frac{r}{n}\right)^{nt} \qquad (1)$$

If $1000 is placed in an account with an annual interest rate of 6%, find out how long it will take the money to double when interest is compounded annually and quarterly.

(a) Compounding interest annually.

NOTE Because the interest is compounded *once* a year, $n = 1$.

Using equation (1) with $A = 2000$ (we want the original 1000 to double). $P = 1000$, $r = 0.06$, and $n = 1$, we have

$$2000 = 1000(1 + 0.06)^t$$

Dividing both sides by 1000 yields

$$2 = (1.06)^t$$

We now have an exponential equation that can be solved by our earlier techniques.

$$\log 2 = \log (1.06)^t$$

$$= t \log 1.06$$

NOTE From accounting, we have the **rule of 72,** which states that the doubling time is approximately 72 divided by the interest rate as a percentage. Here $\frac{72}{6} = 12$ years.

or

$$t = \frac{\log 2}{\log 1.06}$$

$$\approx 11.9 \text{ years}$$

It takes just a little less than 12 years for the money to double.

(b) Compounding interest quarterly.

NOTE Because the interest is compounded four times per year, $n = 4$.

Now $n = 4$ in equation (1), so

$$2000 = 1000\left(1 + \frac{0.06}{4}\right)^{4t}$$

$$2 = (1.015)^{4t}$$

$$\log 2 = \log (1.015)^{4t}$$

$$\log 2 = 4t \log 1.015$$

$$\frac{\log 2}{4 \log 1.015} = t$$

$$t \approx 11.6 \text{ years}$$

Note that the doubling time is reduced by approximately 3 months by the more frequent compounding.

CHECK YOURSELF 7

Find the doubling time in Example 7 if the interest is compounded monthly.

Problems involving rates of growth or decay can also be solved by using exponential equations.

Example 8

A Population Application

A town's population is presently 10,000. Given a projected growth rate of 7% per year, t years from now the population P will be given by

$$P = 10{,}000e^{0.07t}$$

In how many years will the town's population double?

We want the time t when P will be 20,000 (doubled in size). So

$$20{,}000 = 10{,}000e^{0.07t}$$

or

$$2 = e^{0.07t}$$

NOTE Divide both sides by 10,000.

In this case, we take the *natural logarithm* of both sides of the equation. This is because e is involved in the equation.

NOTE Apply the power property.

$\ln e = 1$

$$\ln 2 = \ln e^{0.07t}$$

$$\ln 2 = 0.07t \ln e$$

$$\ln 2 = 0.07t$$

$$\frac{\ln 2}{0.07} = t$$

$$t \approx 9.9 \text{ years}$$

The population will double in approximately 9.9 years.

CHECK YOURSELF 8

If $1000 is invested in an account with an annual interest rate of 6%, compounded continuously, the amount A in the account after t years is given by

$$A = 1000e^{0.06t}$$

Find the time t that it will take for the amount to double (A = 2000). Compare this time with the result of the Check Yourself 7 exercise. Which is shorter? Why?

CHECK YOURSELF ANSWERS

1. $\{2\}$ **2.** $\left\{\frac{1}{8}\right\}$ **3.** $\left\{\frac{3}{2}\right\}$ **4.** $\{3.322\}$ **5.** $\{1.386\}$ **6.** $\{1.151\}$ **7.** 11.58 years

8. 11.55 years; the doubling time is shorter, because interest is compounded more frequently

Name ______________________

Section ________ Date ________

11.5 Exercises

In exercises 1 to 20, solve each logarithmic equation for x.

1. $\log_4 x = 3$

2. $\log_3 x = -2$

3. $\log (x + 1) = 2$

4. $\log_5 (2x - 1) = 2$

5. $\log_2 x + \log_2 8 = 6$

6. $\log 5 + \log x = 2$

7. $\log_3 x - \log_3 6 = 3$

8. $\log_4 x - \log_4 8 = 3$

9. $\log_2 x + \log_2 (x + 2) = 3$

10. $\log_3 x + \log_3 (2x + 3) = 2$

11. $\log_7 (x + 1) + \log_7 (x - 5) = 1$

12. $\log_2 (x + 2) + \log_2 (x - 5) = 3$

13. $\log x - \log (x - 2) = 1$

14. $\log_5 (x + 5) - \log_5 x = 2$

15. $\log_3 (x + 1) - \log_3 (x - 2) = 2$

16. $\log (x + 2) - \log (2x - 1) = 1$

17. $\log (x + 5) - \log (x - 2) = \log 5$

18. $\log_3 (x + 12) - \log_3 (x - 3) = \log_3 6$

19. $\log_2 (x^2 - 1) - \log_2 (x - 2) = 3$

20. $\log (x^2 + 1) - \log (x - 2) = 1$

In exercises 21 to 38, solve each exponential equation for x. Give your solutions in decimal form, correct to three decimal places.

21. $5^x = 625$

22. $4^x = 64$

23. $2^{x+1} = \frac{1}{8}$

24. $9^x = 3$

25. $8^x = 2$

26. $3^{2x-1} = 27$

27. $3^x = 7$

28. $5^x = 30$

29. $4^{x+1} = 12$

30. $3^{2x} = 5$

31. $7^{3x} = 50$

32. $6^{x-3} = 21$

33. $5^{3x-1} = 15$

34. $8^{2x+1} = 20$

35. $4^x = 3^{x+1}$

36. $5^x = 2^{x+2}$

37. $2^{x+1} = 3^{x-1}$

38. $3^{2x+1} = 5^{x+1}$

ANSWERS

1. ______
2. ______
3. ______
4. ______
5. ______
6. ______

7.	8.
9.	10.
11.	12.
13.	14.
15.	16.
17.	18.
19.	20.
21.	22.
23.	24.
25.	26.
27.	28.
29.	30.
31.	32.
33.	34.
35.	36.
37.	38.

ANSWERS

39. ______

40. ______

41. ______

42. ______

43. ______

44. ______

45. ______

46. ______

47. ______

48. ______

49. ______

Use the formula

$$A = P\left(1 + \frac{r}{n}\right)^{nt}$$

to solve exercises 39 to 42.

39. Interest. If \$5000 is placed in an account with an annual interest rate of 9%, how long will it take the amount to double if the interest is compounded annually?

40. Repeat exercise 39 if the interest is compounded semiannually.

41. Repeat exercise 39 if the interest is compounded quarterly.

42. Repeat exercise 39 if the interest is compounded monthly.

Suppose the number of bacteria present in a culture after t hours is given by $N(t) = N_0 \cdot 2^{t/2}$, in which N_0 is the initial number of bacteria. Use the formula to solve exercises 43 to 46.

43. How long will it take the bacteria to increase from 12,000 to 20,000?

44. How long will it take the bacteria to increase from 12,000 to 50,000?

45. How long will it take the bacteria to triple? *Hint:* Let $N(t) = 3N_0$.

46. How long will it take the culture to increase to 5 times its original size? *Hint:* Let $N(t) = 5N_0$.

The radioactive element strontium 90 has a half-life of approximately 28 years. That is, in a 28-year period, one-half of the initial amount will have decayed into another substance. If A_0 is the initial amount of the element, then the amount A remaining after t years is given by

$$A(t) = A_0\left(\frac{1}{2}\right)^{t/28}$$

Use the formula to solve exercises 47 to 50.

47. If the initial amount of the element is 100 g, in how many years will 60 g remain?

48. If the initial amount of the element is 100 g, in how many years will 20 g remain?

49. In how many years will 75% of the original amount remain? *Hint:* Let $A(t) = 0.75A_0$.

50. In how many years will 10% of the original amount remain? *Hint:* Let $A(t) = 0.1A_0$.

Given projected growth, t years from now a city's population P can be approximated by $P(t) = 25{,}000e^{0.045t}$. Use the formula to solve exercises 51 and 52.

51. How long will it take the city's population to reach 35,000?

52. How long will it take the population to double?

The number of bacteria in a culture after t hours can be given by $N(t) = N_0e^{0.03t}$, in which N_0 is the initial number of bacteria in the culture. Use the formula to solve exercises 53 and 54.

53. In how many hours will the size of the culture double?

54. In how many hours will the culture grow to four times its original population?

The atmospheric pressure P, in inches of mercury (in. Hg), at an altitude h feet above sea level is approximated by $P(t) = 30e^{-0.00004h}$. Use the formula to solve exercises 55 and 56.

55. Find the altitude if the pressure at that altitude is 25 in. Hg.

56. Find the altitude if the pressure at that altitude is 20 in. Hg.

Carbon 14 dating is used to measure the age of specimens and is based on the radioactive decay of the element carbon 14. If A_0 is the initial amount of carbon 14, then the amount remaining after t years is $A(t) = A_0e^{-0.000124t}$. Use the formula to solve exercises 57 and 58.

57. Estimate the age of a specimen if 70% of the original amount of carbon 14 remains.

58. Estimate the age of a specimen if 20% of the original amount of carbon 14 remains.

59. In some of the earlier exercises, we talked about bacteria cultures that double in size every few minutes. Can this go on forever? Explain.

60. The population of the United States has been doubling every 45 years. Is it reasonable to assume that this rate will continue? What factors will start to limit that growth?

ANSWERS

50. ____
51. ____
52. ____
53. ____
54. ____
55. ____
56. ____
57. ____
58. ____
59. ____
60. ____

ANSWERS

61. ______

62. ______

63. ______

64. ______

In exercises 61 to 64, use your calculator to find the graph for each equation, then explain the result.

61. $y = \log 10^x$

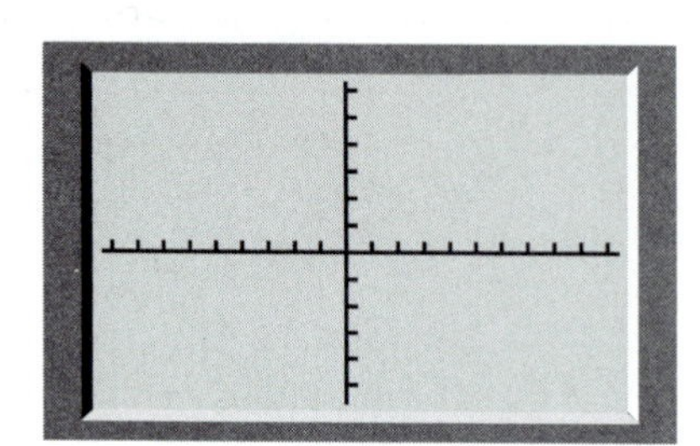

62. $y = 10^{\log x}$

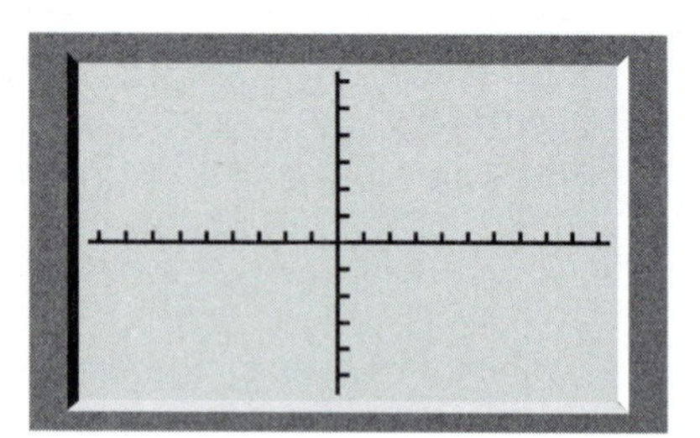

63. $y = \ln e^x$

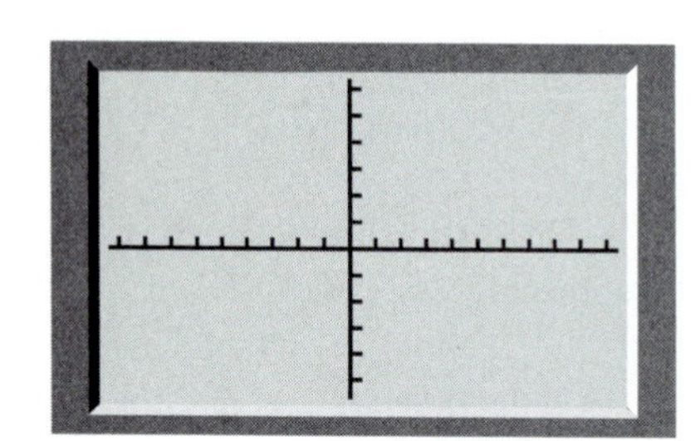

64. $y = e^{\ln x}$

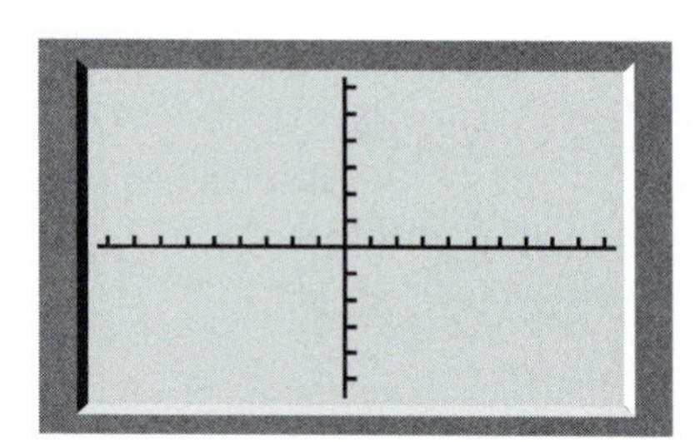

Answers

1. $\{64\}$ 3. $\{99\}$ 5. $\{8\}$ 7. $\{162\}$ 9. $\{2\}$ 11. $\{6\}$ 13. $\left\{\frac{20}{9}\right\}$ 15. $\left\{\frac{19}{8}\right\}$ 17. $\left\{\frac{15}{4}\right\}$ 19. $\{5, 3\}$ 21. $\{4\}$ 23. $\{-4\}$ 25. $\left\{\frac{1}{3}\right\}$ 27. $\{1.771\}$ 29. $\{0.792\}$ 31. $\{0.670\}$ 33. $\{0.894\}$ 35. $\{3.819\}$ 37. $\{4.419\}$ 39. 8.04 y 41. 7.79 y 43. 1.47 h 45. 3.17 h 47. 20.6 y 49. 11.6 y 51. 7.5 y 53. 23.1 h 55. 4558 ft 57. 2876 y 59. 61.

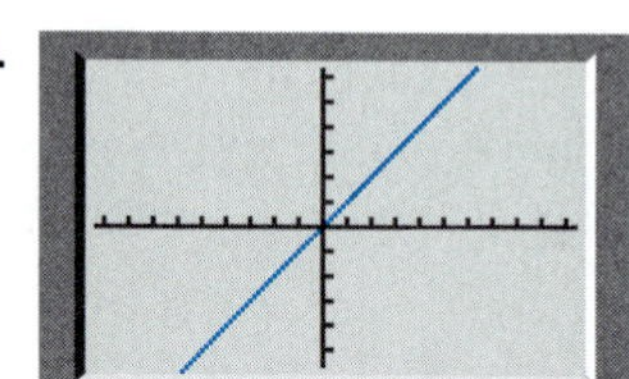

63.

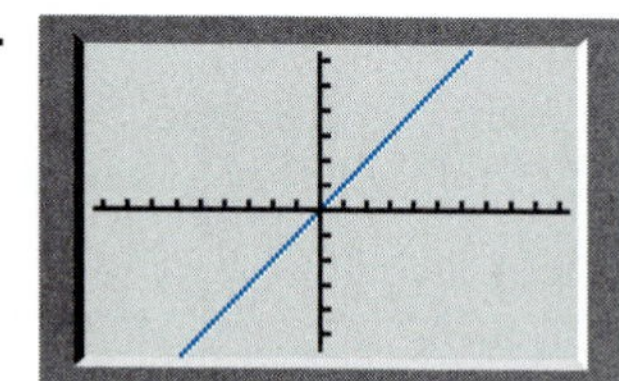

11 Summary

DEFINITION/PROCEDURE	EXAMPLE	REFERENCE
Inverse Relations and Functions		**Section 11.1**
The **inverse** of a relation is formed by interchanging the components of each ordered pair in the given relation. If a relation (or function) is specified by an equation, interchange the roles of x and y in the defining equation to form the inverse. $y = 4x - 8$ $y = \frac{1}{4}x + 2$	The inverse of the relation $\{(1, 2), (2, 3), (4, 3)\}$ is $\{(2, 1), (3, 2), (3, 4)\}$ To find the inverse of $f(x) = 4x - 8,$ $y = 4x - 8$ change y to x and x to y $x = 4y - 8$ so $4y = x + 8$ $y = \frac{1}{4}(x + 8)$ $y = \frac{1}{4}x + 2$	p. 832
The inverse of a function f may or may not be a function. If the inverse *is* also a function, we denote that inverse as f^{-1}, read "the inverse of f." A function f has an inverse f^{-1}, which is also a function, if and only if f is a **one-to-one** function. That is, no two ordered pairs in the function have the same second component. A function f is a one-to-one function if and only if it has an inverse f^{-1}, which is also a function. The **horizontal-line test** can be used to determine whether a function is one-to-one.	If $f(x) = 4x - 8$, then $f^{-1}(x) = \frac{1}{4}x + 2$ *Not* one-to-one	p. 835
Finding Inverse Relations and Functions **1.** Interchange the x and y components of the ordered pairs of the given relation or the roles of x and y in the defining equation. **2.** If the relation was described in equation form, solve the defining equation of the inverse for y. **3.** If desired, graph the relation and its inverse on the same set of axes. The two graphs will be symmetric about the line $y = x$.		p. 838

Continued

DEFINITION/PROCEDURE	EXAMPLE	REFERENCE
Exponential Functions		**Section 11.2**

An **exponential function** is any function defined by an equation of the form

$$y = f(x) = b^x \qquad b > 0, b \neq 1$$

If b is greater than 1, the function is always increasing (a **growth function**). If b is less than 1, the function is always decreasing (a **decay function**).

In both cases, the exponential function is one-to-one. The domain is the set of all real numbers, and the range is the set of positive real numbers.

The function defined by $f(x) = e^x$, in which e is an irrational number (approximately 2.71828), is called *the* exponential function.

Example: $y = 2^x$, $b > 1$

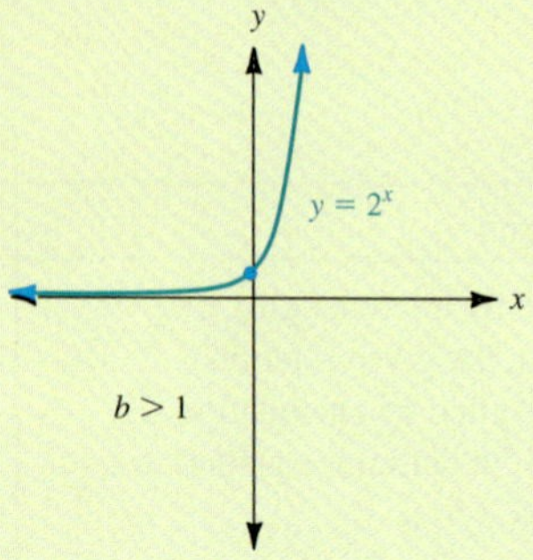

p. 843

Graphing an Exponential Function

1. Establish a table of values by considering the function in the form $y = b^x$.
2. Plot points from that table of values and connect them with a smooth curve to form the graph.
3. If $b > 1$, the graph increases from left to right. If $0 < b < 1$, the graph decreases from left to right.
4. All graphs will have the following in common:
   (a) The y intercept will be (0, 1).
   (b) The graphs will approach, but not touch, the x axis.
   (c) The graphs will represent one-to-one functions.

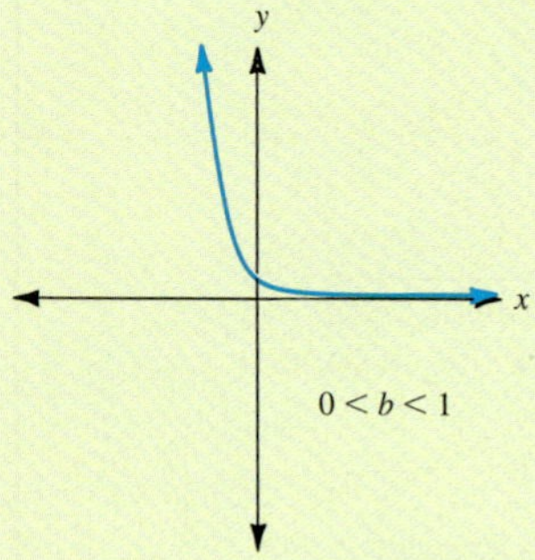

Example: $0 < b < 1$

p. 845

DEFINITION/PROCEDURE	EXAMPLE	REFERENCE
Logarithmic Functions		**Section 11.3**

In the expression

$$y = \log_b x$$

y is called the *logarithm of* x *to base* b, when $b > 0$ and $b \neq 1$.

An expression such as $y = \log_b x$ is said to be in **logarithmic form.**

An expression such as $x = b^y$ is said to be in **exponential form.**

$$y = \log_b x \qquad \text{means the same as} \qquad x = b^y$$

A logarithm is an exponent or a power. The logarithm of x to base b is the power to which we must raise b to get x.

A **logarithmic function** is any function defined by an equation of the form

$$y = f(x) = \log_b x \qquad b > 0, b \neq 1$$

The logarithm function is the inverse of the corresponding exponential function. The function is one-to-one with domain $\{x \mid x > 0\}$ and range composed of the set of all real numbers.

Example:

$\log_3 9 = 2$ is in logarithmic form.

$3^2 = 9$ is the exponential form.

$\log_3 9 = 2$ is equivalent to $3^2 = 9$.

2 is the power to which we must raise 3 to get 9.

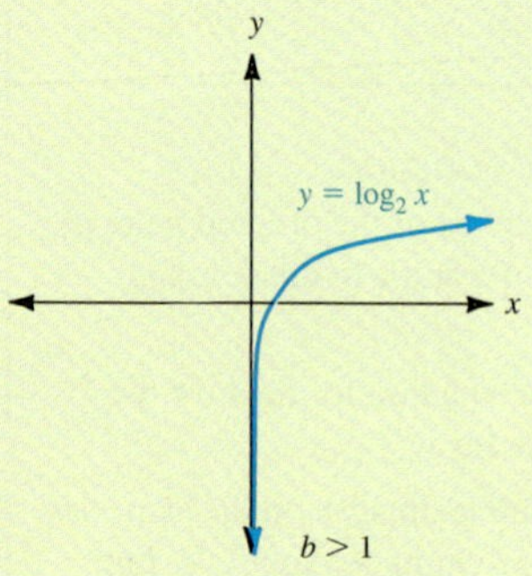

$y = \log_2 x$, $b > 1$

p. 858

DEFINITION/PROCEDURE	EXAMPLE	REFERENCE
Properties of Logarithms		**Section 11.4**
If M, N, and b are positive real numbers with $b \neq 1$ and if p is any real number, then we can state the following properties of logarithms: **1.** $\log_b b = 1$ **2.** $\log_b 1 = 0$ **3.** $b^{\log_b x} = x$ **4.** $\log_b b^x = x$	$\log 10 = 1$ $\log_2 1 = 0$ $3^{\log_3 2} = 2$ $\log_5 5^x = x$	p. 871
Product Property $\log_b MN = \log_b M + \log_b N$	$\log_3 x + \log_3 y = \log_3 xy$	p. 872
Quotient Property $\log_b \frac{M}{N} = \log_b M - \log_b N$	$\log_5 8 - \log_5 3 = \log_5 \frac{8}{3}$	p. 872
Power Property $\log_b M^p = p \log_b M$	$\log 3^2 = 2 \log 3$	p. 872
Common logarithms are logarithms to base 10. For convenience, we omit the base in writing common logarithms: $\log M = \log_{10} M$	$\log_{10} 1000 = \log 1000$ $= \log 10^3 = 3$	p. 874
Natural logarithms are logarithms to base e. By custom we also omit the base in writing natural logarithms: $\ln M = \log_e M$	$\ln 3 = \log_e 3$	p. 877
Logarithmic and Exponential Equations		**Section 11.5**
A **logarithmic equation** is an equation that contains a logarithmic expression. $\log_2 x = 5$ is a logarithmic equation.	To solve $\log_2 x = 5$: Write the equation in the equivalent exponential form to solve $x = 2^5$ or $x = 32$	p. 885
Solving Logarithmic Equations **1.** Use the properties of logarithms to combine terms containing logarithmic expressions into a single term. **2.** Write the equation formed in step 1 in exponential form. **3.** Solve for the indicated variable. **4.** Check your solutions to make sure that possible solutions do not result in the logarithms of negative numbers or zero.	To solve $\log_4 x + \log_4 (x - 6) = 2$ $\log_4 x(x - 6) = 2$ $x(x - 6) = 4^2$ $x^2 - 6x - 16 = 0$ $(x - 8)(x + 2) = 0$ $x = 8$ or $x = -2$ Because substituting -2 for x in the original equation results in the logarithm of a negative number, we reject that answer. The only solution is 8.	p. 888
An **exponential equation** is an equation in which the variable appears as an exponent. The following algorithm summarizes the steps in solving any exponential equation.	To solve $4^x = 64$: Because $64 = 4^3$, write $4^x = 4^3$ or $x = 3$	p. 888

Continued

DEFINITION/PROCEDURE	EXAMPLE	REFERENCE
Logarithmic and Exponential Equations		**Section 11.5**
Solving Exponential Equations 1. Try to write each side of the equation as a power of the same base. Then, equate the exponents to form an equation. 2. If the above procedure is not applicable, take the common logarithm of both sides of the original equation. 3. Use the power rule for logarithms to write an equivalent equation with the variables as coefficients. 4. Solve the resulting equation.	$2^{x+3} = 5^x$ $\log 2^{x+3} = \log 5^x$ $(x + 3) \log 2 = x \log 5$ $x \log 2 + 3 \log 2 = x \log 5$ $x \log 2 - x \log 5 = -3 \log 2$ $x (\log 2 - \log 5) = -3 \log 2$ $x = \dfrac{-3 \log 2}{\log 2 - \log 5} \approx 2.269$	p. 890

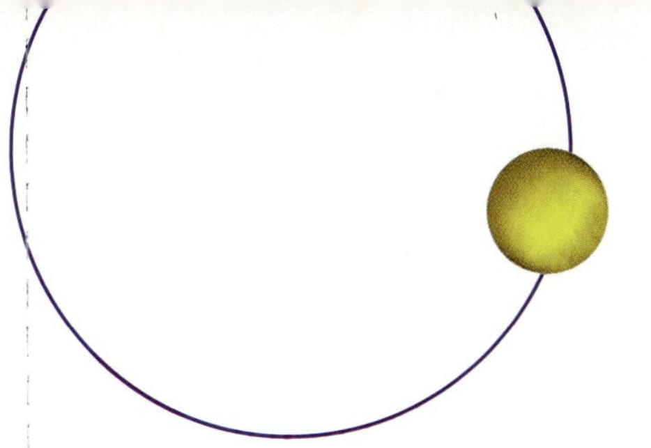

Summary Exercises

This summary exercise set is provided to give you practice with each of the objectives in the chapter. Each exercise is keyed to the appropriate chapter section. The answers are provided in the *Instructor's Manual*.

[11.1] Write the inverse relation for each of the following functions. Which inverses are also functions?

1. $\{(1, 5), (2, 7), (3, 9)\}$

2. $\{(3, 1), (5, 1), (7, 1)\}$

3. $\{(2, 4), (4, 3), (6, 4)\}$

Write an equation for the inverse of the relation defined by each of the following equations.

4. $y = 3x - 6$

5. $y = \dfrac{x + 1}{2}$

6. $y = x^2 - 2$

Write an equation for the inverse of the relation defined by each of the following equations. Which inverses are also functions?

7. $y = 3x + 6$

8. $y = -x^2 + 3$

9. $4x^2 + 9y^2 = 36$

[11.2] Graph the exponential functions defined by each of the following equations.

10. $y = 3^x$

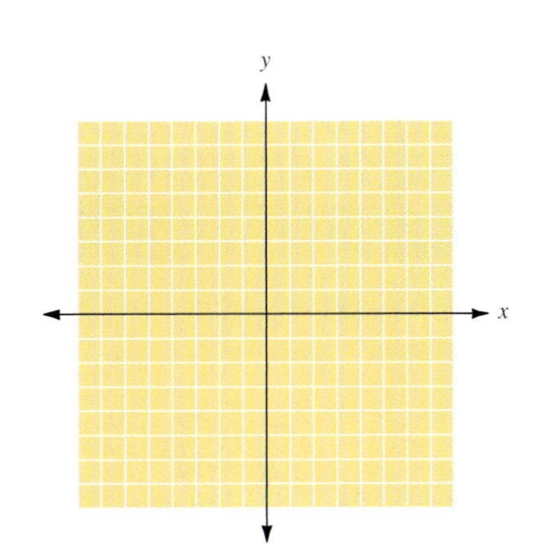

11. $y = \left(\dfrac{3}{4}\right)^x$

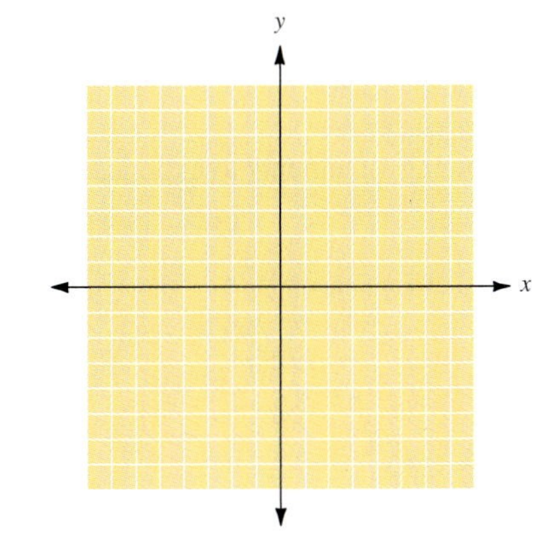

[11.3] Solve each of the following exponential equations for x.

12. $5^x = 125$

13. $2^{2x+1} = 32$

14. $3^{x-1} = \frac{1}{9}$

If it takes 2 h for the population of a certain bacteria culture to double (by dividing in half), then the number N of bacteria in the culture after t hours is given by $N = 1000 \cdot 2^{t/2}$, when the initial population of the culture was 1000. Using this formula, find the number in the culture:

15. After 4 h

16. After 12 h

17. After 15 h

Graph the logarithmic functions defined by each of the following equations.

18. $y = \log_3 x$

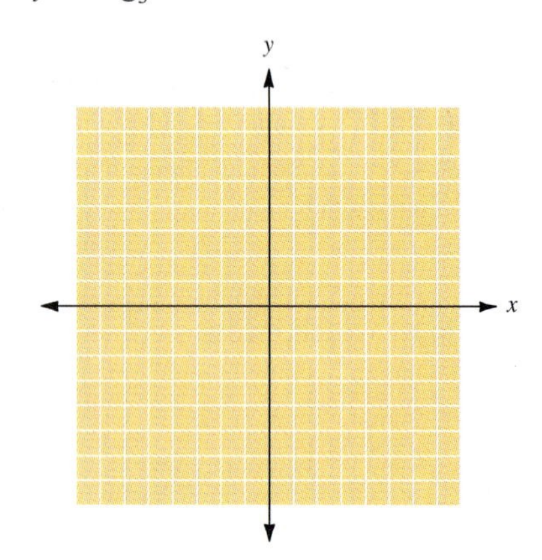

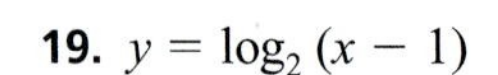

19. $y = \log_2 (x - 1)$

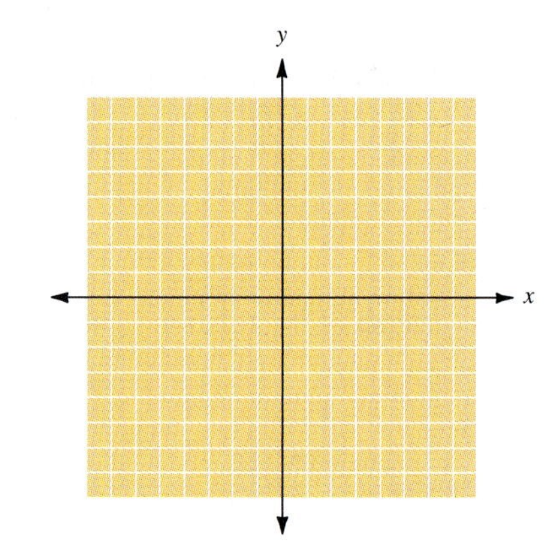

Convert each of the following statements to logarithmic form.

20. $3^4 = 81$

21. $10^3 = 1000$

22. $5^0 = 1$

23. $5^{-2} = \frac{1}{25}$

24. $25^{1/2} = 5$

25. $16^{3/4} = 8$

Convert each of the following statements to exponential form.

26. $\log_3 81 = 4$

27. $\log 1 = 0$

28. $\log_{81} 9 = \frac{1}{2}$

29. $\log_5 25 = 2$

30. $\log 0.001 = -3$

31. $\log_{32} \frac{1}{2} = -\frac{1}{5}$

Solve each of the following equations for the unknown variable.

32. $y = \log_5 125$

33. $\log_b \frac{1}{9} = -2$

34. $\log_8 x = 2$

35. $y = \log_5 1$

36. $\log_b 3 = \frac{1}{2}$

37. $y = \log_{16} 2$

38. $y = \log_8 2$

The decibel (dB) rating for the loudness of a sound is given by

$$L = 10 \log \frac{I}{I_0}$$

in which I is the intensity of that sound in watts per square centimeter and I_0 is the intensity of the "threshold" sound, $I_0 = 10^{-16}$ W/cm^2. Find the decibel rating of each of the given sounds.

39. A table saw in operation with intensity $I = 10^{-6}$ W/cm^2

40. The sound of a passing car horn with intensity $I = 10^{-8}$ W/cm^2

The formula for the decibel rating of a sound can be solved for the intensity of the sound as

$$I = I_0 \cdot 10^{L/10}$$

in which L is the decibel rating of the given sound.

41. What is the ratio of intensity of a 60-dB sound to one of 50 dB?

42. What is the ratio of intensity of a 60-dB sound to one of 40 dB?

The magnitude of an earthquake on the Richter scale is given by

$$M = \log \frac{a}{a_0}$$

in which a is the intensity of the shock wave of the given earthquake and a_0 is the intensity of the shock wave of a zero-level earthquake. Use that formula to solve the following.

43. The Alaskan earthquake of 1964 had an intensity of $10^{8.4} a_0$. What was its magnitude on the Richter scale?

44. Find the ratio of intensity of an earthquake of magnitude 7 to an earthquake of magnitude 6.

[11.4] Use the properties of logarithms to expand each of the following expressions.

45. $\log_b x^2y$

46. $\log_4 \frac{y^3}{5}$

47. $\log_3 \frac{xy^2}{z}$

48. $\log_5 x^3yz^2$

49. $\log \frac{xy}{\sqrt{z}}$

50. $\log_b \sqrt[3]{\frac{x^2y}{z}}$

Use the properties of logarithms to write each of the following expressions as a single logarithm.

51. $\log x + 2 \log y$

52. $3 \log_b x - 2 \log_b z$

53. $\log_b x + \log_b y - \log_b z$

54. $2 \log_5 x - 3 \log_5 y - \log_5 z$

55. $\log x - \frac{1}{2} \log y$

56. $\frac{1}{3} (\log_b x - 2 \log_b y)$

Given that $\log 2 = 0.301$ and $\log 3 = 0.477$, find each of the following logarithms. Verify your results with a calculator.

57. $\log 18$

58. $\log 16$

59. $\log \frac{1}{8}$

60. $\log \sqrt{3}$

Use your calculator to find the pH of each of the following solutions, given the hydrogen ion concentration $[H^+]$ for each solution, when

$\text{pH} = -\log [H^+]$

Are the solutions acidic or basic?

61. Coffee: $[H^+] = 5 \times 10^{-6}$

62. Household detergent: $[H^+] = 3.2 \times 10^{-10}$

Given the pH of the following solutions, find the hydrogen ion concentration $[H^+]$.

63. Lemonade: $\text{pH} = 3.5$

64. Ammonia: $\text{pH} = 10.2$

The average score on a final examination for a group of chemistry students, retested after time t (in weeks), is given by

$S(t) = 81 - 6 \ln (t + 1)$

Find the average score on the retests after the given times.

65. After 5 weeks

66. After 10 weeks

67. After 15 weeks

68. Graph these results.

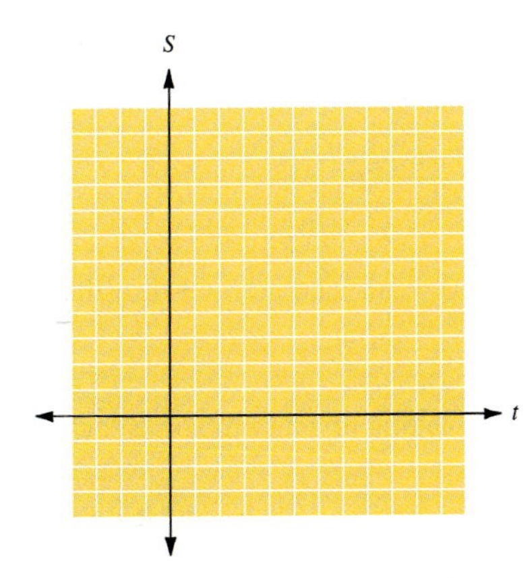

The formula for converting from a logarithm with base a to a logarithmic expression with base b is

$$\log_a x = \frac{\log_b x}{\log_b a}$$

Use that formula to find each of the following logarithms.

69. $\log_4 20$

70. $\log_8 60$

[11.5] Solve each of the following logarithmic equations for x.

71. $\log_3 x + \log_3 5 = 3$

72. $\log_5 x - \log_5 10 = 2$

73. $\log_3 x + \log_3 (x + 6) = 3$

74. $\log_5 (x + 3) + \log_5 (x - 1) = 1$

75. $\log x - \log (x - 1) = 1$

76. $\log_2 (x + 3) - \log_2 (x - 1) = \log_2 3$

Solve each of the following exponential equations for x. Give your results correct to three decimal places.

77. $3^x = 243$

78. $5^x = \frac{1}{25}$

79. $5^x = 10$

80. $4^{x-1} = 8$

81. $6^x = 2^{2x+1}$

82. $2^{x+1} = 3^{x-1}$

If an investment of P dollars earns interest at an annual rate of 12% and the interest is compounded n times per year, then the amount A in the account after t years is

$$A(t) = P\left(1 + \frac{0.12}{n}\right)^{nt}$$

Use that formula to solve each of the following.

83. If \$1000 is invested and the interest is compounded quarterly, how long will it take the amount in the account to double?

84. If \$3000 is invested and the interest is compounded monthly, how long will it take the amount in the account to reach \$8000?

A certain radioactive element has a half-life of 50 years. The amount A of the substance remaining after t years is given by

$$A(t) = A_0 \cdot 2^{-t/50}$$

when A_0 is the initial amount of the substance. Use this formula to solve each of the following.

85. If the initial amount of the substance is 100 milligrams (mg), after how long will 40 mg remain?

86. After how long will only 10% of the original amount of the substance remain?

A city's population is presently 50,000. Given the projected growth, t years from now the population P will be given by $P(t) = 50{,}000e^{0.08t}$. Use this formula to solve each of the following.

87. How long will it take the population to reach 70,000?

88. How long will it take the population to double?

The atmospheric pressure, in inches of mercury, at an altitude h miles above the surface of the earth, is approximated by $P(h) = 30e^{-0.021h}$. Use this formula to solve the following exercises.

89. Find the altitude at the top of Mt. McKinley in Alaska if the pressure is 27.7 in. Hg.

90. Find the altitude outside an airliner in flight if the pressure is 26.1 in. Hg.

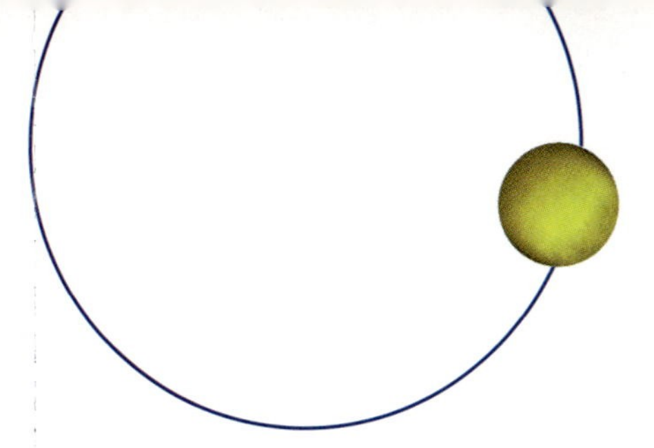

Self-Test for Chapter 11

Name ______

Section ______ Date ______

The purpose of this self-test is to help you check your progress and to review for a chapter test in class. Allow yourself about 1 hour to take the test. When you are done, check your answers in the back of the book. If you missed any answers, be sure to go back and review the appropriate sections in the chapter and the exercises that are provided.

1. Use $f(x) = 4x - 2$ and $g(x) = x^2 + 1$ in each of the following.

(a) Find the inverse of f. Is the inverse also a function?

(b) Find the inverse of g. Is the inverse also a function?

(c) Graph f and its inverse on the same set of axes.

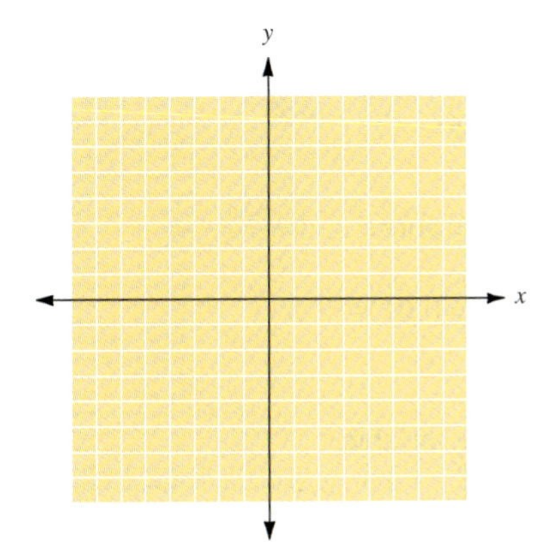

Graph the exponential functions defined by each of the following equations.

2. $y = 4^x$

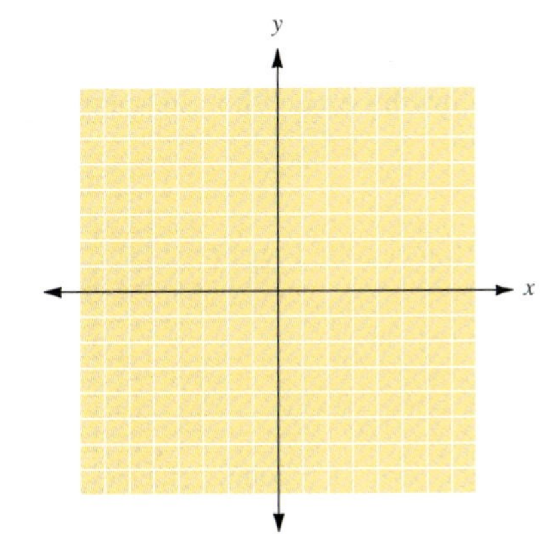

3. $y = \left(\frac{2}{3}\right)^x$

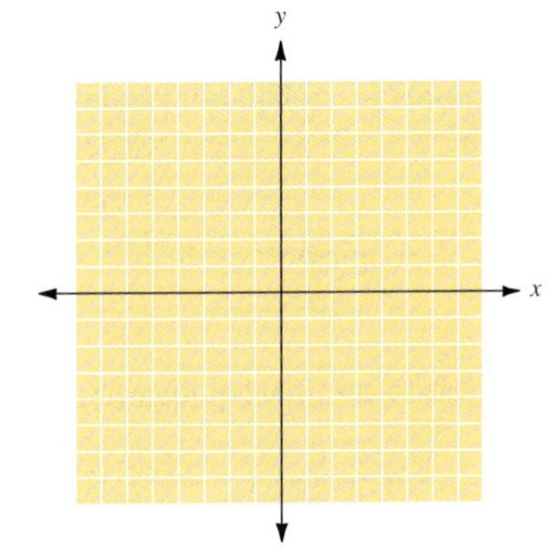

4. Solve each of the following exponential equations for x.

(a) $5^x = \frac{1}{25}$

(b) $3^{2x-1} = 81$

ANSWERS

1. (a)

(b)

(c)

2.

3.

4.

ANSWERS

5. ______
6. ______
7. ______
8. ______
9. ______
10. ______
11. ______
12. ______
13. ______
14. ______
15. ______
16. ______
17. ______
18. ______
19. ______
20. ______

5. Graph the logarithmic function defined by the following equation.

$y = \log_4 x$

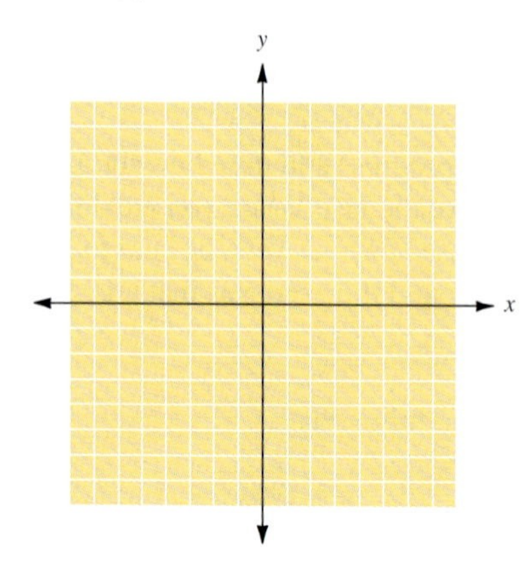

Convert each of the following statements to logarithmic form.

6. $10^4 = 10{,}000$

7. $27^{2/3} = 9$

Convert each of the following statements to exponential form.

8. $\log_5 125 = 3$

9. $\log 0.01 = -2$

Solve each of the following equations for the unknown variable.

10. $y = \log_2 64$

11. $\log_b \frac{1}{16} = -2$

12. $\log_{25} x = \frac{1}{2}$

Use the properties of logarithms to expand each of the following expressions.

13. $\log_b x^2yz^3$

14. $\log_5 \sqrt{\frac{xy^2}{z}}$

Use the properties of logarithms to write each of the following expressions as a single logarithm.

15. $\log x + 3 \log y$

16. $\frac{1}{3}(\log_b x - 2 \log_b z)$

Solve each of the following logarithmic equations for x.

17. $\log_6 (x + 1) + \log_6 (x - 4) = 2$

18. $\log (2x + 1) - \log (x - 1) = 1$

Solve each of the following exponential equations for x. Give your results correct to three decimal places.

19. $3^{x+1} = 4$

20. $5^x = 3^{x+1}$

Cumulative Test for Chapters 1 to 11

Name ______________________

Section __________ Date __________

ANSWERS

1. ______________________
2. ______________________
3. ______________________
4. ______________________
5. ______________________
6. ______________________
7. ______________________
8. ______________________
9. ______________________
10. ______________________
11. ______________________
12. ______________________
13. ______________________

This test is provided to help you in the process of reviewing the previous chapters. Answers are provided in the back of the book. If you missed any answers, be sure to go back and review the appropriate chapter section.

Solve each of the following.

1. $2x - 3(x + 2) = 4(5 - x) + 7$

2. $|3x - 7| > 5$

3. $\log x - \log (x - 1) = 1$

Graph each of the following.

4. $5x - 3y = 15$

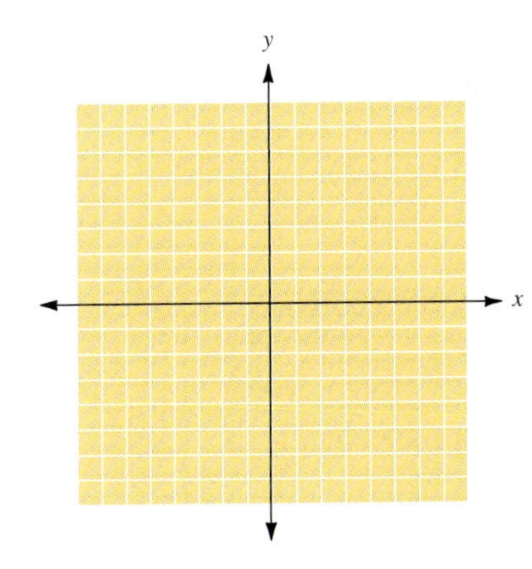

5. $-8(2 - x) \geq y$

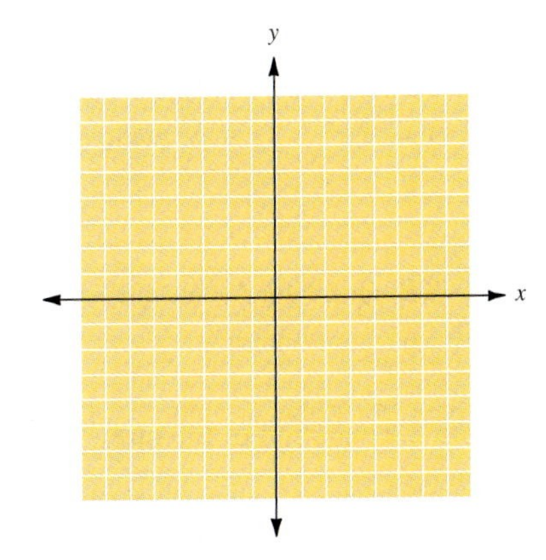

6. Find the equation of the line that passes through the points $(2, -1)$ and $(-3, 5)$.

7. Solve the linear inequality.

$3x - 2(x - 5) \geq 20$

Simplify each of the following expressions.

8. $4x^2 - 3x + 8 - 2(x^2 + 5) - 3(x - 1)$

9. $(3x + 1)(2x - 5)$

Factor each of the following completely.

10. $2x^2 - x - 10$

11. $25x^3 - 16xy^2$

Perform the indicated operations.

12. $\dfrac{2}{x - 4} - \dfrac{3}{x - 5}$

13. $\dfrac{x^2 - x - 6}{x^2 + 2x - 15} \div \dfrac{x - 2}{x + 5}$

ANSWERS

14. ______

15. ______

16. ______

17. ______

18. ______

19. ______

20. ______

21. ______

22. ______

23. ______

24. ______

25. ______

Simplify each of the following radical expressions.

14. $\sqrt{18} + \sqrt{50} - 3\sqrt{32}$

15. $(3\sqrt{2} + 2)(3\sqrt{2} + 2)$

16. $\dfrac{5}{\sqrt{5} - \sqrt{2}}$

17. Find three consecutive odd integers whose sum is 237.

Solve each of the following equations.

18. $x^2 + x - 2 = 0$

19. $2x^2 - 6x - 5 = 0$

20. Solve the following inequality:

$2x^2 + x - 3 \leq 0$

21. If $f(x) = -x^3 + 3x + 5$, evaluate $f(-1)$.

22. Solve the inequality $|2x - 5| \leq 3$.

23. Write the equation of the circle with center $(1, -2)$ and radius 3.

24. Simplify the expression $\left(\dfrac{x^{-2}w^3}{x^{-3}w^{-1}}\right)^3$.

25. Solve the equation $\dfrac{2}{x - 2} - \dfrac{3}{x + 2} = \dfrac{4}{x^2 - 4}$.

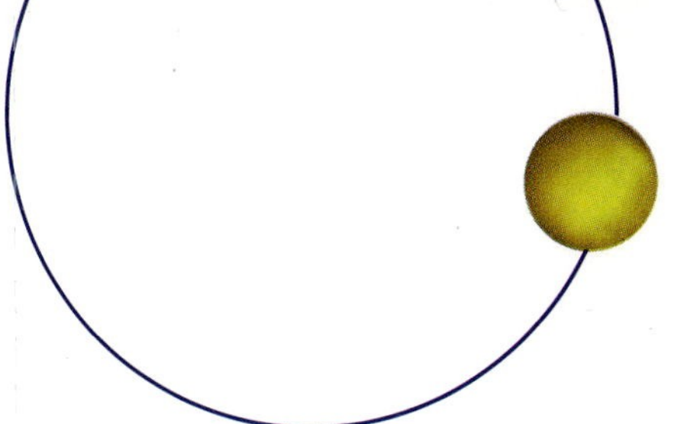

Answers to Self–Tests and Cumulative Tests

Self–Test for Chapter 1

1. **(a)** 9; **(b)** $-6, 0, 9$; **(c)** $-6, -\frac{2}{3}, 4.1, 0, 9, \frac{5}{4}, 0.\overline{78}$; **(d)** $-\sqrt{3}, \pi$; **(e)** All are real numbers **2.** $2(x + y)$ **3.** $\frac{p-8}{t}$ **4.** 47 **5.** -48
6. Associative property of addition **7.** Distributive property

8. 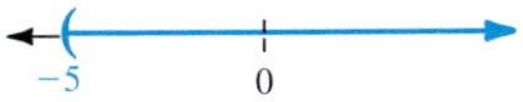

9. 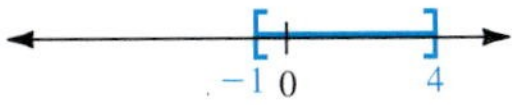

10. -3 **11.** 2 **12.** $3x^4y^7$ **13.** $4x^4y^6$ **14.** $9b^6c^8$ **15.** $\frac{x^2}{y^6}$
16. x^5y^{10} **17.** $\frac{xy^3}{8}$ **18.** $-5c - 3b$ **19.** $21a + 13$ **20.** 2.53×10^6

Self–Test for Chapter 2

1. $\left\{\frac{4}{5}\right\}$ **2.** $\{2\}$ **3.** $\{2\}$ **4.** $\left\{\frac{3}{2}\right\}$ **5.** $r = \frac{A - P}{Pt}$ **6.** $h = \frac{2A}{B + b}$
7. 3:30 P.M. **8.** 800 flashlights **9.** $\{x \mid x \le 4\}$ **10.** $\left\{x \mid x > \frac{17}{2}\right\}$
11. $\{x \mid x > 1\}$ **12.** $\{x \mid -2 \le x \le 4\}$ **13.** $\left\{-\frac{2}{3}, 4\right\}$ **14.** $\left\{\frac{2}{3}, 4\right\}$
15. $\left\{x \,\middle|\, -\frac{3}{2} < x < 3\right\}$ **16.** $\left\{x \,\middle|\, x \le -2 \text{ or } x \ge \frac{9}{2}\right\}$ **17.** 2 **18.** 75
19. 12 **20.** 500 lb/ft^2

Cumulative Test for Chapters 1 and 2

1. **(a)** 3, 5; **(b)** $-7, -5, 0, 3, 5$ **2.** $\frac{p-5}{s}$ **3.** **(a)** x^6y^{10}; **(b)** x^7y^8
4. -14 **5.** 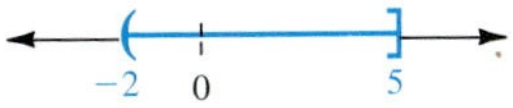

6. 4.37×10^9
7. Associative property of addition **8.** 25 **9.** $3x + 3$
10. $\left\{\frac{5}{2}\right\}$ **11.** $\left\{\frac{21}{2}\right\}$ **12.** $\left\{\frac{78}{7}\right\}$ **13.** $B = \frac{2A - hb}{h}$
14. $p = \frac{4y - 12}{7}$ **15.** $3\frac{2}{3}$ h **16.** 30 mi
17. $\{x \mid x < -7\}$ 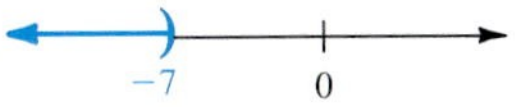

18. $\{x \mid 2 \le x \le 5\}$ 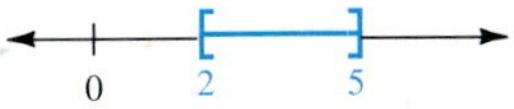

19. $\{1, 7\}$
20. All real numbers **21.** $\{x \mid x < 0 \text{ or } x > 5\}$
22. $\{x \mid -10 \le x \le -2\}$ **23.** 15 **24.** 96 **25.** 22 A

Self–Test for Chapter 3

1. b **2.** **(a)** D: $\{-3, 1, 2, 3, 4\}$, R: $\{-2, 0, 1, 5, 6\}$;
(b) D: {United States, Germany, Russia, China}, R: {101, 65, 63, 50}

3. 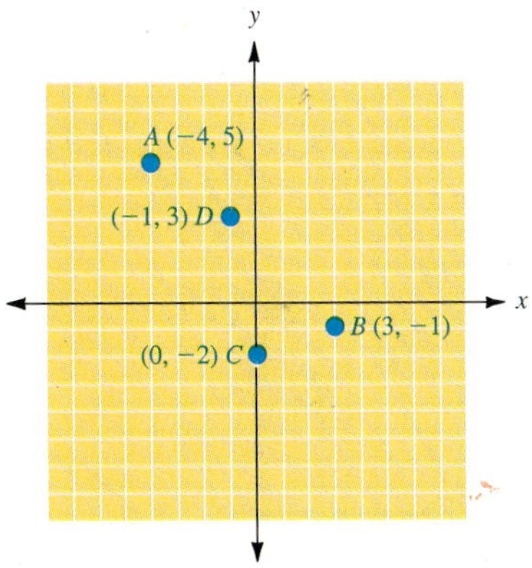

4. **(a)** Quadrant II; **(b)** x axis; **(c)** Quadrant III; **(d)** Quadrant I
5. **(a)** 6; **(b)** 12; **(c)** 2 **6.** **(a)** 2; **(b)** -5 **7.** Function
8. Not a function **9.** A: (1, 0); B: $(-3, -4)$
10. A: $(-4, -2)$; B: (1, 2) **11.** **(a)** 3; **(b)** 4; **(c)** 5
12. **(a)** -3; **(b)** -2; **(c)** 0 **13.** **(a)** 0; **(b)** 1; **(c)** 2
14. **(a)** 0; **(b)** $-2, 2$; **(c)** $-4, 4$
15. x int.: $(-1, 0)$ and (5, 0); y int.: $(0, -5)$
16. x int.: (5, 0); y int.: $(0, -5)$
17. D: $\{x \mid -7 \le x \le 7\}$; R: $\{y \mid -7 \le y \le 7\}$
18. D: $\{x \mid x \in R\}$; R: $\{y \mid y \le 9\}$

Cumulative Test for Chapters 1–3

1. **(a)** $-8, 0$; **(b)** $-8, 3.2, 0$ **2.** **(a)** x^8y^6; **(b)** x^4y^3
3. 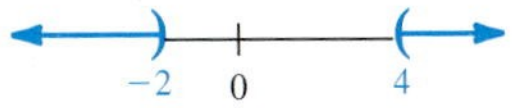

4. 6.8×10^7 **5.** $\frac{2}{7}$ **6.** $\{1\}$ **7.** $\{11\}$ **8.** $\{8\}$ **9.** $\{-1, 9\}$
10. $\{x \mid x \ge 4\}$ **11.** $\{x \mid 1 < x < 5\}$ **12.** $\{x \mid x < 2 \text{ or } x > 3\}$
13. $y = \frac{15 - 3x}{5}$ **14.** 56 mi/h going, 48 mi/h returning **15.** 6
16. D: $\{-2, -1, 2, 3, 4\}$; R: $\{0, 1, 5, 7\}$ **17.** Third
18. A: $(-2, -1)$; B: $(0, -5)$ **19.** $\{-11\}$ **20.** Function **21.** Function
22. **(a)** Not a function; **(b)** function **23.** **(a)** -3; **(b)** 0; **(c)** 3
24. x int.: (6, 0), y int.: $(0, -4)$; x int.: (7, 0), y int.: (0, 3)
25. D: $\{x \mid -6 \le x \le 6\}$; R: $\{y \mid -6 \le y \le 6\}$

Self–Test for Chapter 4

1. x intercept: (6, 0); y intercept: (0, 5)
2. x intercept: (3, 0); y intercept: none
3. D: $\{x \mid x \in R\}$; R: $\{y \mid y \in R\}$
4. D: $\{x \mid x = 4\}$; R: $\{y \mid y \in R\}$ **5.** D: $\{x \mid x \in R\}$; R: $\{y \mid y = -7\}$
6. $y = 2x - 1$ **7.** $y = \frac{2}{3}x - 4$ **8.** $y = -3x - 1$ **9.** $y = 4x - 2$
10. $y = \frac{2}{3}x + 3$ **11.** 2 **12.** Undefined **13.** 0 **14.** $-\frac{2}{5}$
15. Undefined **16.** 0 **17.** Neither **18.** Perpendicular **19.** Parallel

20.

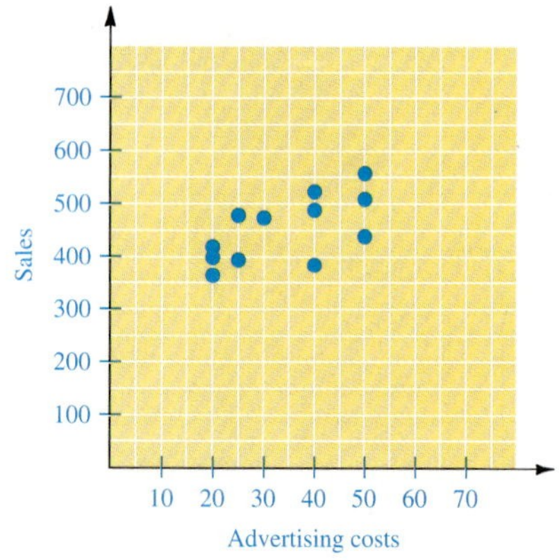

21.

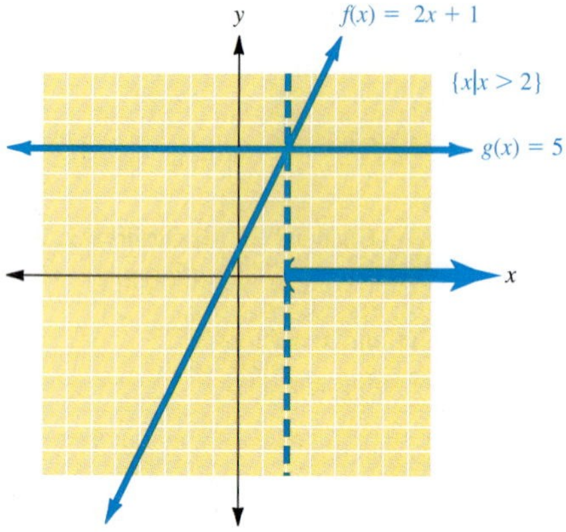

22.

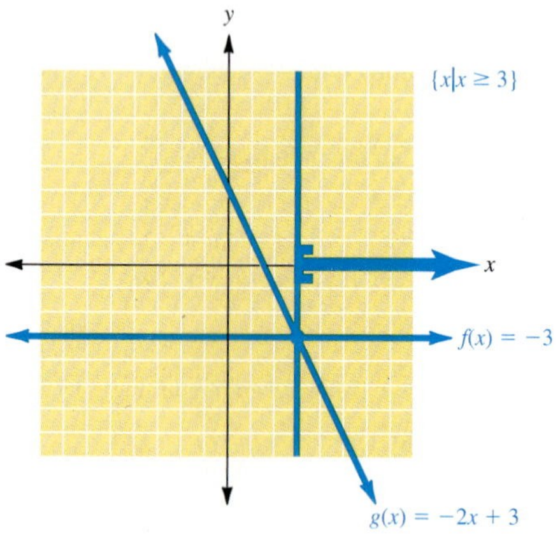

23. $\{-3, 7\}$

24.

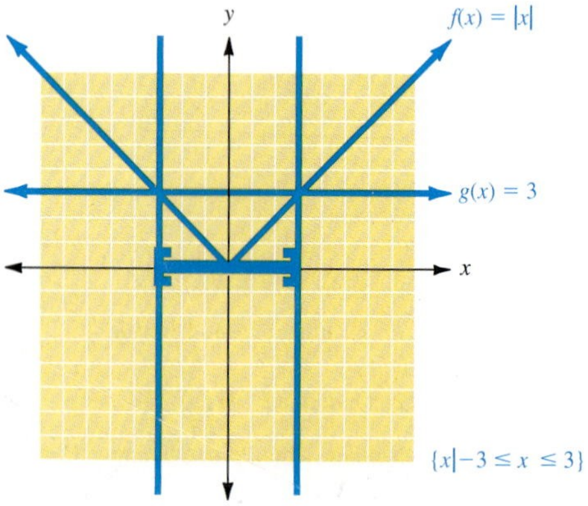

25.

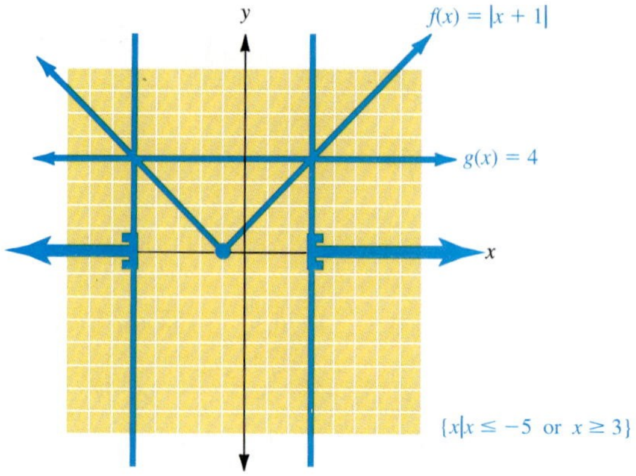

Cumulative Test for Chapters 1–4

1. Associative property of multiplication
2. Commutative property of addition **3.** Distributive property

4. 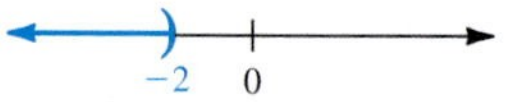

5. 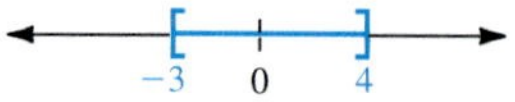

6. $\left\{\frac{5}{2}\right\}$ **7.** $\left\{\frac{21}{2}\right\}$ **8.** $\left\{\frac{78}{7}\right\}$ **9.** $B = \frac{2A - hb}{h}$ **10.** $p = \frac{4y - 12}{7}$

11. $\{x|x < -7\}$ 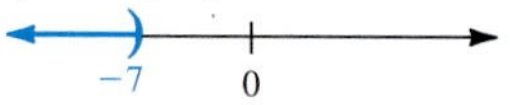

12. $\{x|2 \le x \le 5\}$ 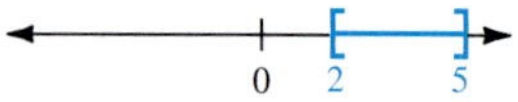

13. $\{x|x < 0 \text{ or } x > 5\}$ 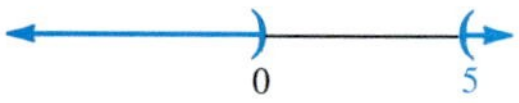

14. $-7, -9$ **15.** 94 **16.** $3\frac{2}{3}$ h **17.** 112 \$4 tickets, 60 \$5 tickets

18.

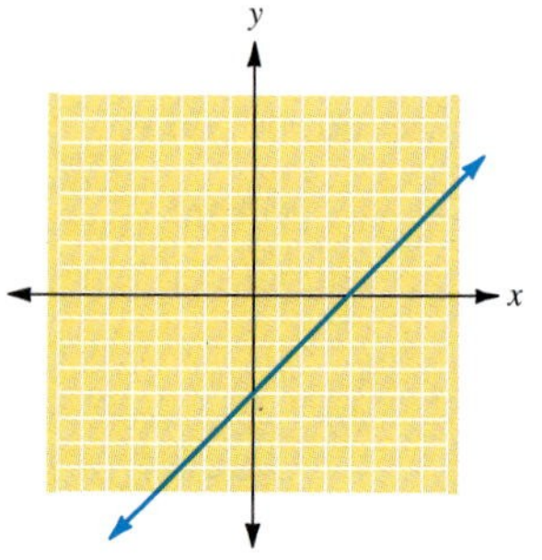

19.

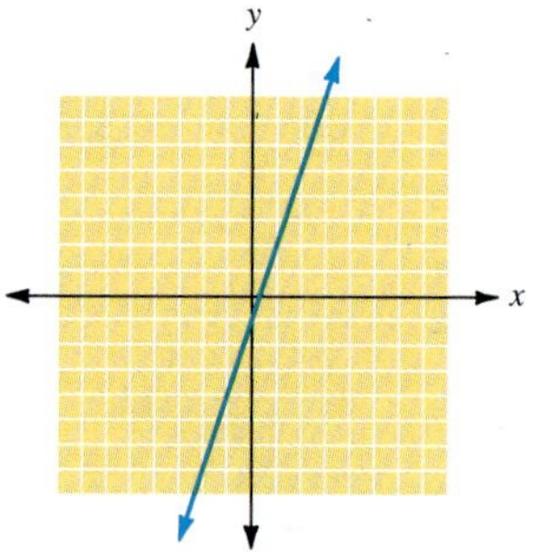

20.

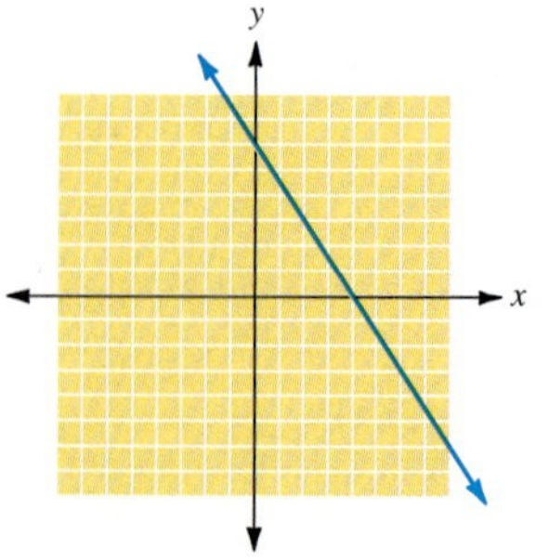

21.

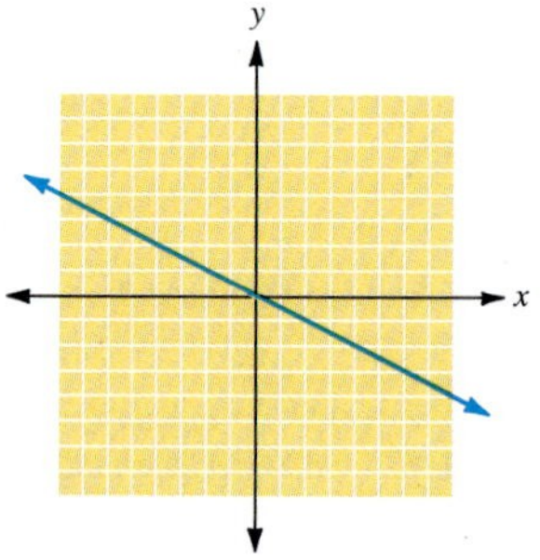

22.

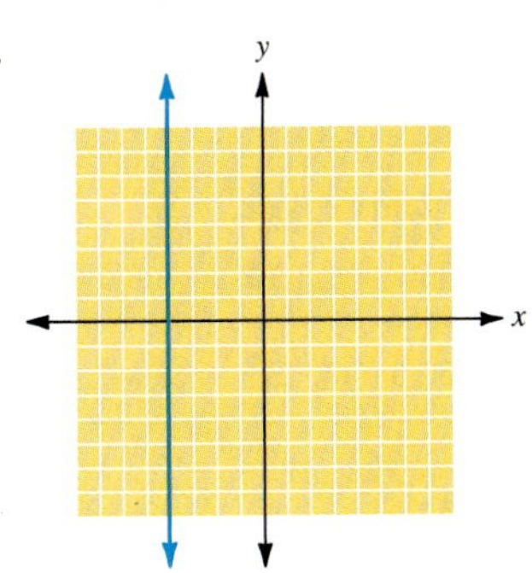

23. -2 **24.** Undefined

25. $-\frac{3}{4}$ **26.** 0 **27.** Perpendicular **28.** Neither

29. $y = -2x + 8$ **30.** $y = -2x + 7$

Self–Test for Chapter 5

1. $\{(-3, 4)\}$ **2.** Dependent system **3.** Inconsistent system

4. $\{(-2, -5)\}$ **5.** $\{(5, 0)\}$ **6.** $\left\{\left(3, \frac{-5}{3}\right)\right\}$ **7.** $\{(-1, 2, 4)\}$

8. $\left\{\left(2, -3, -\frac{1}{2}\right)\right\}$ **9.** Disks \$2.50, ribbons \$6

10. 60 lb jawbreakers, 40 lb licorice **11.** Four 5-in. sets, six 12-in. sets

12. \$8000 savings, \$4000 bond, \$2000 mutual fund **13.** 50 by 80 ft

14.

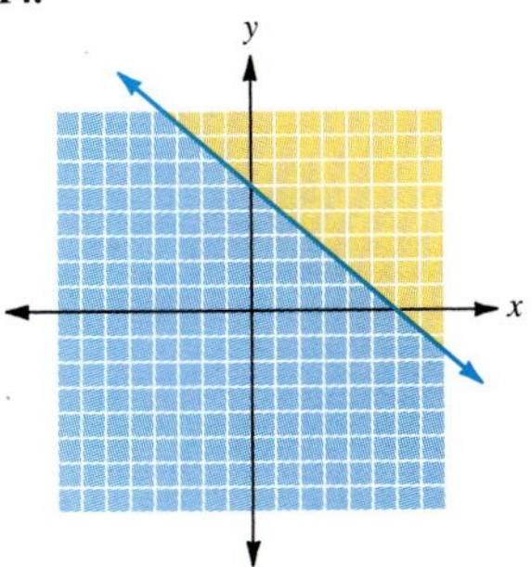

15.

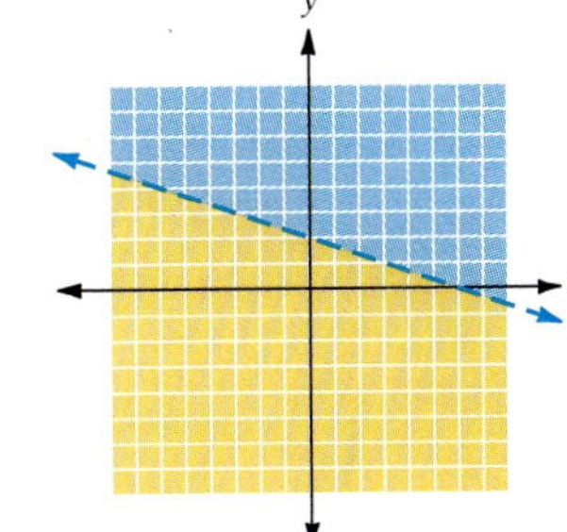

16.

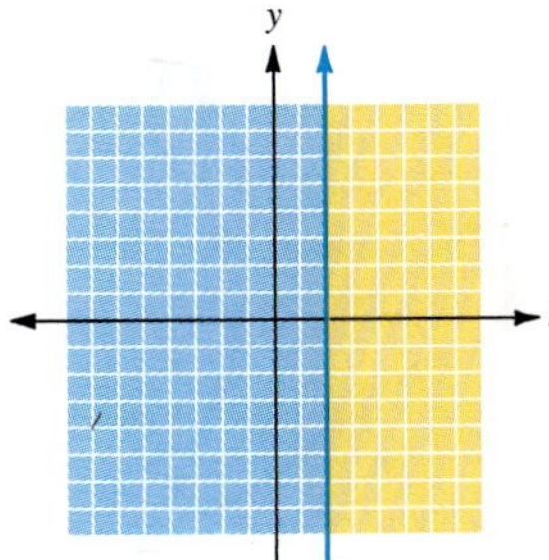

17.

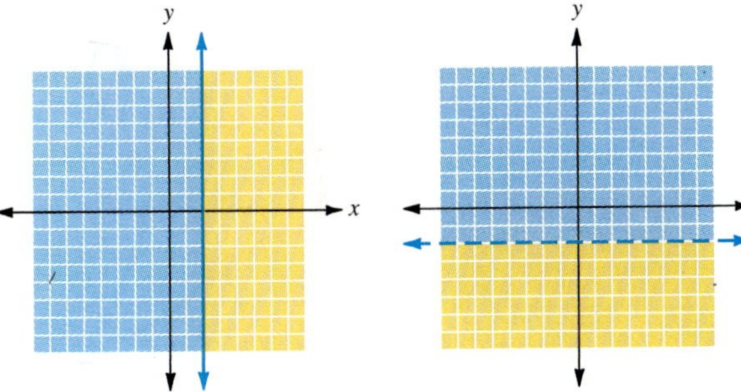

18.

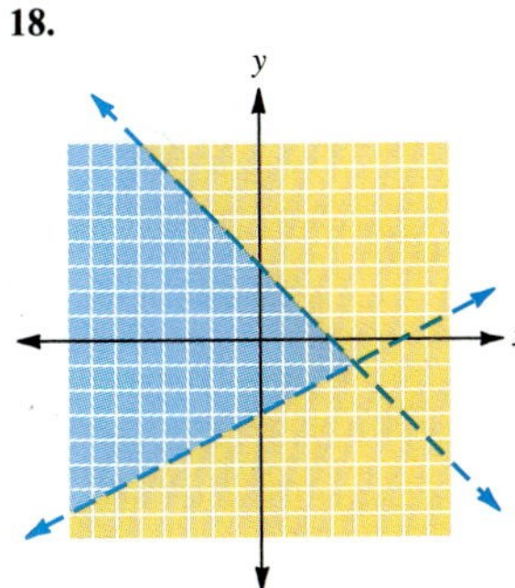

19.

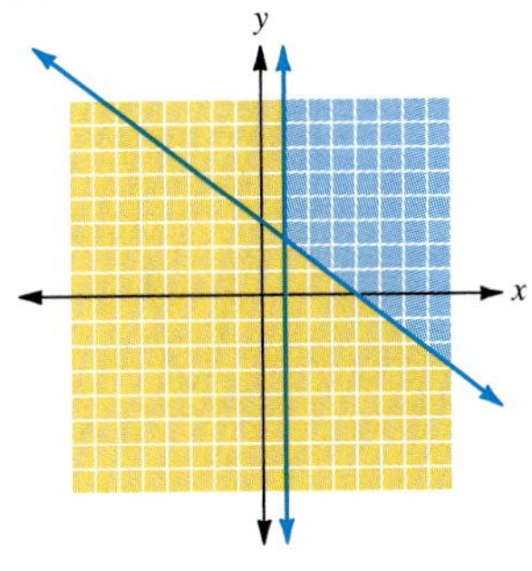

20.

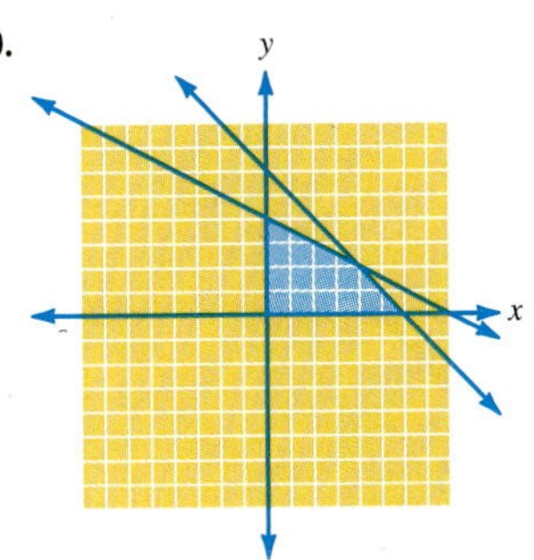

21. $\left\{\left(\frac{2}{3}, \frac{5}{3}\right)\right\}$

22. $\{(5, -2, 1)\}$

Cumulative Test for Chapters 1–5

1. $x^{18}y^6$ **2.** 62 **3.** 39 **4.** $\{15\}$ **5.** $\left\{5, -\frac{5}{3}\right\}$ **6.** $R = \frac{R_1R_2}{R_1 + R_2}$

7. $\{x|x \le 1\}$ **8.** $\{x|x < -2\}$ **9.** $\{x|-3 < x < 6\}$

10. $\{x|x > 13 \text{ or } x < -3\}$ **11.** $y = 2x - 3$ **12.** $y = \frac{2}{3}x + \frac{7}{3}$

13. $y = -\frac{5}{4}x - 2$ **14.** Function **15.** Not a function

16. Not a function **17.** Function **18.** $\{(6, 2)\}$ **19.** $\{(-6, -3)\}$

20. Inconsistent system **21.** Dependent system **22.** $\left\{\left(-2, \frac{3}{2}\right)\right\}$

23. $\{(-4, 3, 5)\}$ **24.** \$3 binder, \$2.50 paper

25. \$7000 bond, \$3000 time deposit

Self–Test for Chapter 6

1. Binomial **2.** Trinomial **3.** Not a polynomial

4. $8x^4 - 3x^2 - 7$; 8, -3, -7; 4 **5.** 25

6. **(a)** $6x^2 - 3x + 7$; **(b)** $2x^2 - 3x + 7$; **(c)** 10; **(d)** 6; **(e)** 10; **(f)** 6

7. **(a)** $-3x^3 + 3x^2 + 5x - 9$; **(b)** $-3x^3 + 7x^2 - 9x - 5$; **(c)** -4; **(d)** -10; **(e)** -4; **(f)** -10 **8.** $5x + 12$

9. $6a^2 - ab - 35b^2$ **10.** $25m^2 - 9n^2$ **11.** $4a^2 + 12ab + 9b^2$

12. $2x^3 - 13x^2 + 26x - 15$ **13.** $7ab(2ab - 3a + 5b)$

14. $(x - 3y)(x + 5)$ **15.** $(5c - 8d)(5c + 8d)$

16. $(3x - 1)(9x^2 + 3x + 1)$ **17.** $2a(2a + b)(4a^2 - 2ab + b^2)$

18. $(x - 8)(x + 6)$ **19.** $(5x - 2)(2x - 7)$ **20.** $3x(x + 3)(2x - 5)$

21. $x - 1 + \frac{-3}{3x + 1}$ **22.** $4x^2 + 3x + 13 + \frac{17}{x - 2}$

23. $3x^2 - 5$ **24.** $\left\{-3, -\frac{1}{2}\right\}$ **25.** $\left\{-\frac{5}{2}, \frac{2}{3}\right\}$

Cumulative Test for Chapters 1–6

1. $\{2\}$ **2.** -77 **3.** x int.: (6, 0); y int.: (0, 4) **4.** $y = \frac{5}{4}x - \frac{3}{4}$

5. $256x^{13}y^{12}$ **6.** 25

7. Domain: all real numbers; range: all real numbers

8. **(a)** 4; **(b)** 0; **(c)** -6 **9.** $-x^2 - 5x + 9$ **10.** $10x^2 + 7x - 12$

11. $x(3x + 2)(x - 1)$ **12.** $(4x + 5y)(4x - 5y)$ **13.** $(x - y)(3x + 1)$

14. **(a)** $3x + 7$; **(b)** $-13x - 5$; **(c)** $-40x^2 - 22x + 6$; **(d)** $\frac{-5x + 1}{8x + 6}$

15. $\{5\}$ **16.** $\{-18, -28\}$ **17.** $\{2\}$ **18.** $\{x|x \ge -1\}$

19. $\{x|-13 < x < -5\}$ **20.** $\{x|-6 < x < -4\}$ **21.** $\left\{\left(-10, \frac{26}{3}\right)\right\}$

22. $R = \frac{P - P_0}{IT}$ **23.** $\left\{3, -\frac{1}{2}\right\}$ **24.** 8 cm by 19 cm **25.** 16, 14

Self–Test for Chapter 7

1. $\frac{-3x^4}{4y^2}$ **2.** $\frac{w+1}{w-2}$ **3.** $\frac{x+3}{x-2}$

4.

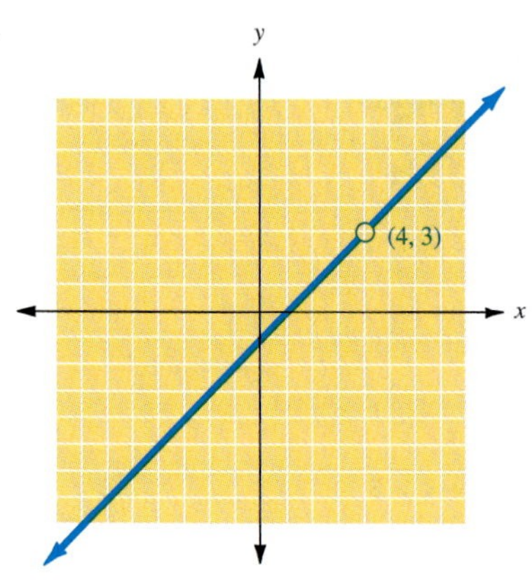

5. $\frac{4a^2}{7b}$

6. $\frac{m-4}{4}$ **7.** $\frac{2}{x-1}$ **8.** $\frac{2x+y}{x(x+3y)}$ **9.** $\frac{-5(3x+1)}{(3x-1)}$

10. $\frac{2(2x+1)}{x(x-2)}$ **11.** $\frac{2}{x-3}$ **12.** $\frac{4}{x-2}$

13. $\frac{8x+17}{(x-4)(x+1)(x+4)}$ **14.** $\frac{y}{3y+x}$ **15.** $\frac{z-1}{2(z+3)}$

16. $\frac{1}{xy(x-y)}$ **17.** $\{9\}$ **18.** $\left\{-\frac{1}{2}, 4\right\}$ **19.** 3, 6

20. Going: 50 mi/h; returning: 45 mi/h

21. Juan: 10 days; Ariel: 40 days **22.** 3 **23.** $\frac{x^8}{y^{10}}$ **24.** $\frac{c^2d^7}{2}$ **25.** $\frac{x^4}{y^{11}}$

Cumulative Test for Chapters 1–7

1. 12 **2.** -14 **3.** $6x + 7y = -21$ **4.** x intercept: $(-6, 0)$; y intercept: $(0, 7)$ **5.** $6x^2 - 5x - 2$ **6.** $2x^3 + 5x^2 - 3x$

7. Domain: $\{x \mid x = 2\}$; range: all real numbers **8.** 16

9. $x(2x+3)(3x-1)$ **10.** $(4x^8 + 3y^4)(4x^8 - 3y^4)$ **11.** $\frac{3}{x-1}$

12. $\frac{1}{(x+1)(x-1)}$ **13.** $x^2 - 3x$ **14.** **(a)** $\frac{8x^2 - 34x - 29}{x-5}$;

(b) $\frac{1}{(x-5)(8x+6)}$; **(c)** $\left\{x \mid x \neq 5 \text{ or } x \neq \frac{-3}{4}\right\}$ **15.** $\left\{\frac{7}{2}\right\}$

16. $\left\{-\frac{8}{9}, -\frac{4}{9}\right\}$ **17.** $\left\{-\frac{30}{17}\right\}$ **18.** $\{x \mid x > -2\}$

19. $\left\{x \,\middle|\, \frac{1}{5} < x < \frac{7}{5}\right\}$ **20.** $\{x \mid x \leq -4 \text{ or } x \geq -2\}$ **21.** $\{-5\}$

22. $(3, -1)$ **23.** $\frac{b^6}{a^{10}}$ **24.** Barry: 3 h; Don: 6 h

25. Width: 10 cm; Length: 18 cm

Self–Test for Chapter 8

1. $7a^2$ **2.** $-3w^2z^3$ **3.** $p^2q\sqrt[3]{9pq^2}$ **4.** $\frac{7x}{8y}$ **5.** $\frac{\sqrt{10xy}}{4y}$ **6.** $\frac{\sqrt[3]{3x^2}}{x}$

7. $3x\sqrt{3x}$ **8.** $5m\sqrt[3]{2m}$ **9.** $4x\sqrt{3}$ **10.** $3 - 2\sqrt{2}$ **11.** $\{11\}$

12. $\{4\}$ **13.** $64x^6$ **14.** $\frac{9m}{n^4}$ **15.** $\frac{8s^3}{r}$ **16.** 9.4 cm **17.** 21.9 ft

18. 3.7 cm **19.** 9.0 cm **20.** 69.3 cm **21.** $a^2b\sqrt[5]{a^4b}$ **22.** $5p^3q^2$

23. $3 + 10i$ **24.** $-14 + 22i$ **25.** $5 - 5i$

Cumulative Test for Chapters 1–8

1. $\{-15\}$ **2.** 5 **3.** $3x - 2y = 12$ **4.** $\left\{\frac{1}{3}, 3\right\}$

5. $10x^2 - 2x$ **6.** $10x^2 - 39x - 27$ **7.** $x(x-1)(2x+3)$

8. $9(x^2 + 2y^2)(x^2 - 2y^2)$ **9.** $(x+2y)(4x-5)$ **10.** $\frac{x+5}{3x-1}$

11. $\frac{x+7}{(x-5)(x-1)}$ **12.** $\frac{a+2}{2a}$ **13.** 0 **14.** $2x^4y^3\sqrt{3y}$

15. $\sqrt{6} - 5\sqrt{2} + 3\sqrt{3} - 15$ **16.** $x\sqrt{2x}$ **17.** $\frac{c^{12}}{a^6b^4}$

18.

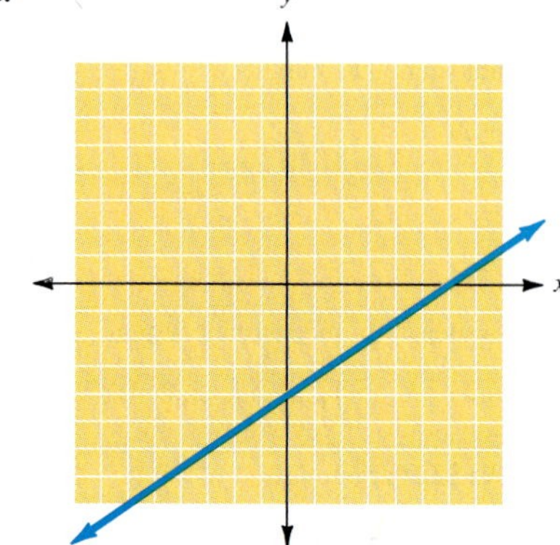

19. $15 + 18i$

20. $23 - 14i$ **21.** $\{11\}$ **22.** $(9, 1)$ **23.** $\{x \mid x \leq 17\}$

24. $\{x \mid -1 \leq x \leq 9\}$ **25.** 16

Self–Test for Chapter 9

1. $\left\{-\frac{2}{3}, 4\right\}$ **2.** $\left(-\frac{1}{2}, \frac{9}{4}\right)$ **3.** $(2, -9)$

4.

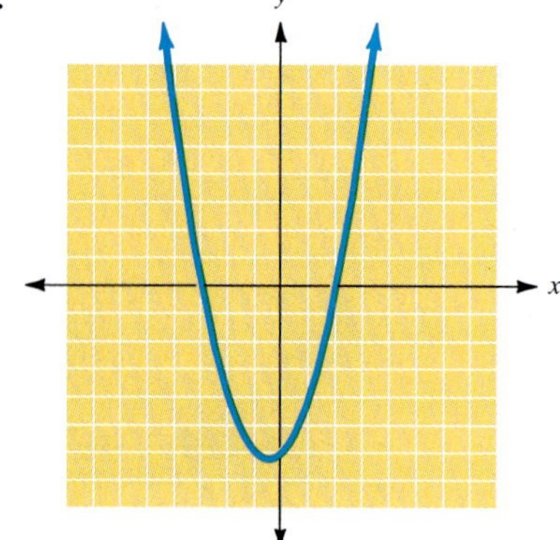

5.

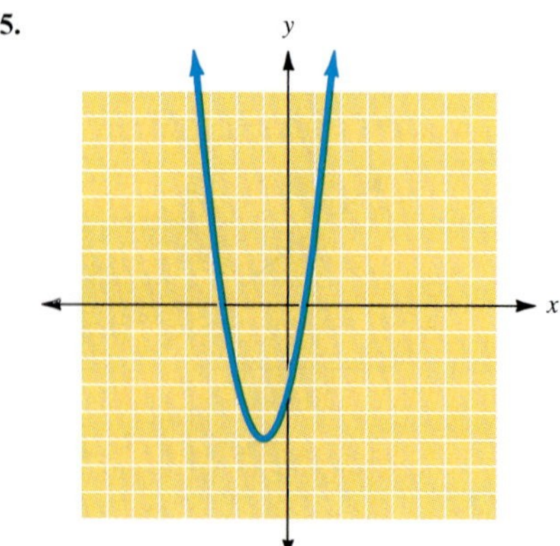

6. Width: 5 cm; length: 17 cm

7. 2.7 s **8.** $\{\pm\sqrt{5}\}$ **9.** $\{1 \pm \sqrt{10}\}$ **10.** $\left\{\frac{2 \pm \sqrt{23}}{2}\right\}$

11. $\left\{\frac{-3 \pm \sqrt{13}}{2}\right\}$ **12.** $\left\{\frac{5 \pm \sqrt{19}}{2}\right\}$ **13.** $\left\{\frac{5 \pm \sqrt{37}}{2}\right\}$

14. $\{-2 \pm \sqrt{11}\}$ **15.** $\{\pm\sqrt{3}, \pm 3\}$ **16.** $\{4, 81\}$

17. $\{-3, 5\}$ **18.** $\{3, -2\}$ **19.**

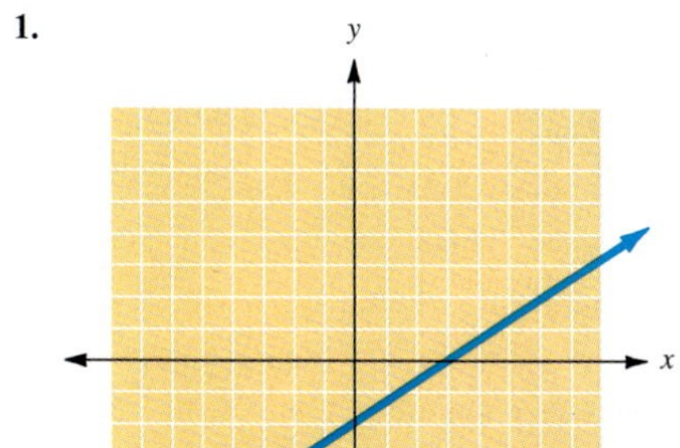

20.

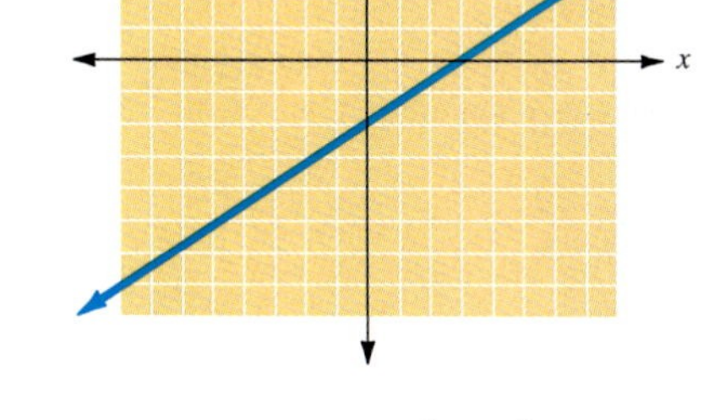

Cumulative Test for Chapters 1–9

1.

2. -3 **3.** 35 **4.** $f(x) = x^3 + 3x^2 - x - 3$ **5.** $x(x + 3)(x - 2)$

6. $\frac{-x - 4}{(x - 3)(x + 2)}$ **7.** $\sqrt{21} - \sqrt{6} + \sqrt{42} - 2\sqrt{3}$ **8.** $\left\{\frac{7}{2}\right\}$

9. $\{-4\}$ **10.** $\{-5, 3\}$ **11.** $\{1, 2\}$ **12.** $\{-10, 3\}$

13. $\left\{\frac{3 \pm \sqrt{21}}{2}\right\}$ **14.** $\{3 \pm \sqrt{5}\}$ **15.** $\{-2, 5\}$ **16.** $\{4\}$ **17.** $\{7\}$

18. $x + 2y = 6$ **19.** $\{x|x \le 9\}$ **20.** $\{x|x \le -5 \text{ or } x \ge 13\}$

21. xy^2 **22.** $3 - 4i$ **23.** -13

24. $-2 + 2\sqrt{97}, 2 + 2\sqrt{97}$ or 17.7 ft, and 21.7 ft

25. Between 21 and 59 units

Self–Test for Chapter 10

1.

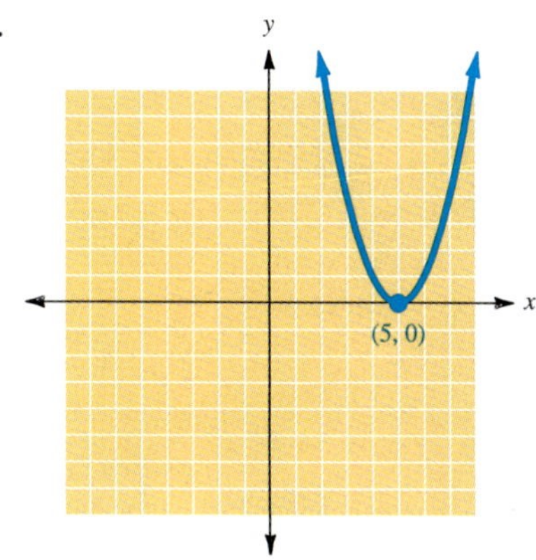

2.

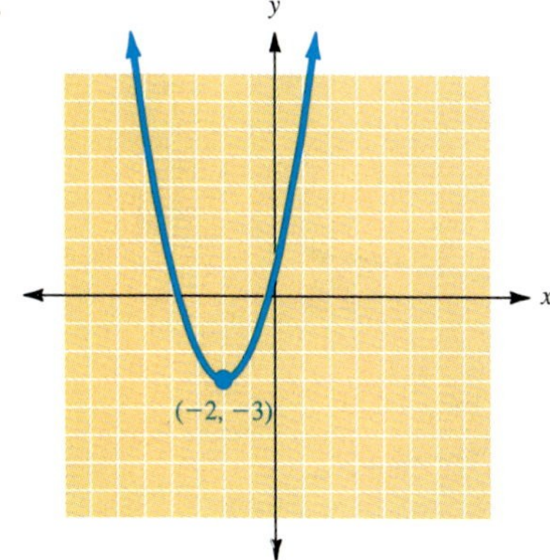

3.

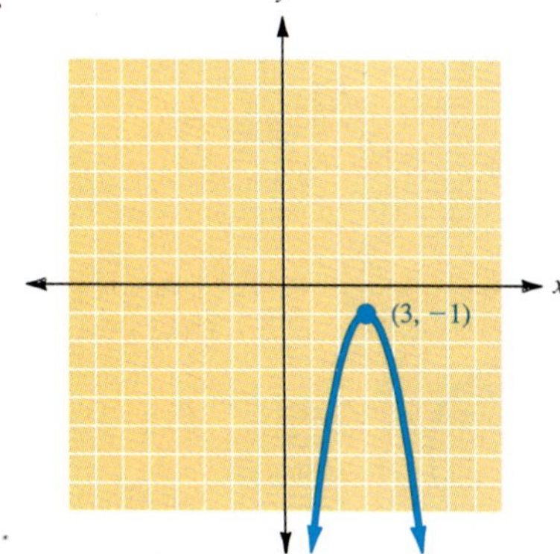

4.

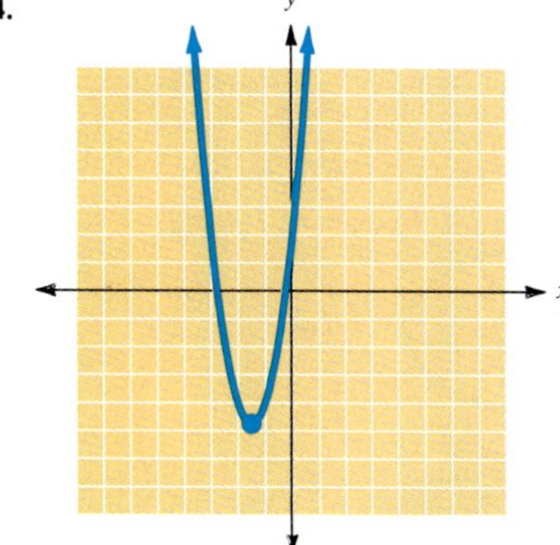

5.

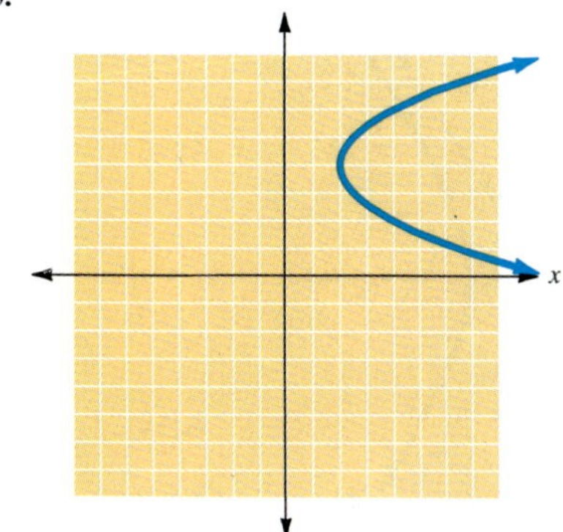

6.

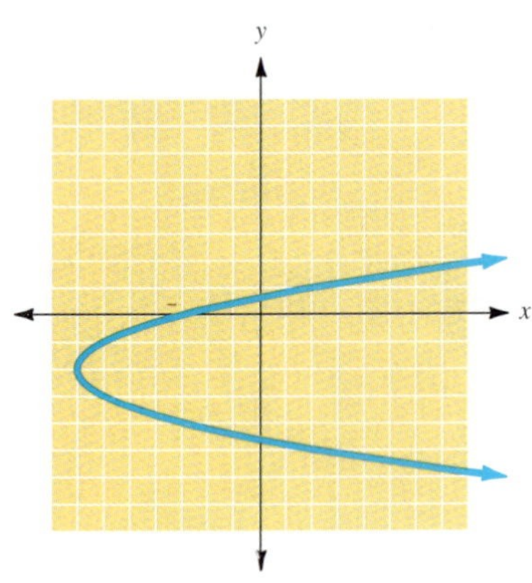

7. (3, −2); 6 **8.** (1, −2); $\sqrt{26}$

9.

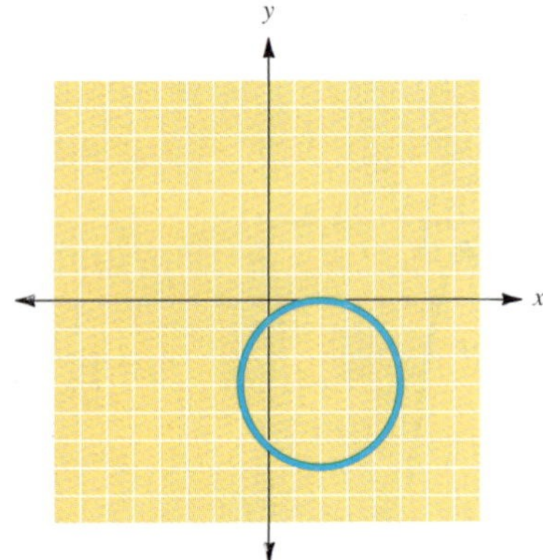

10.

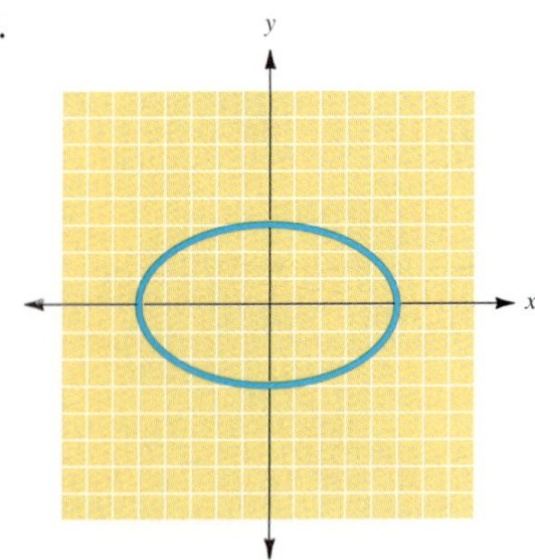

11.

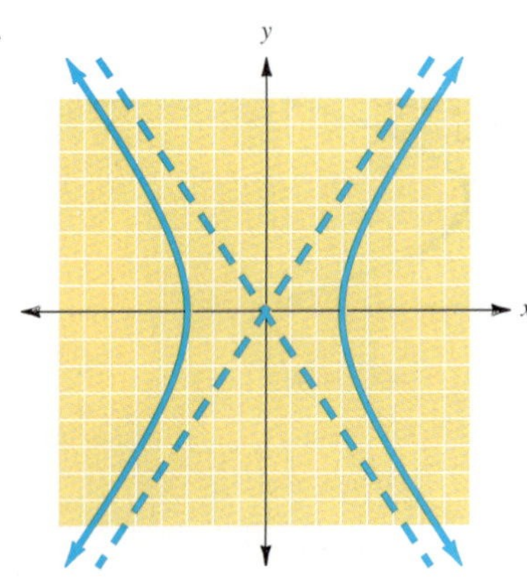

12.

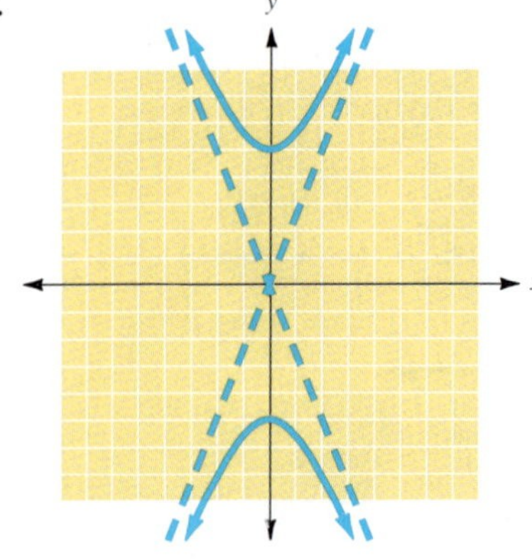

13.

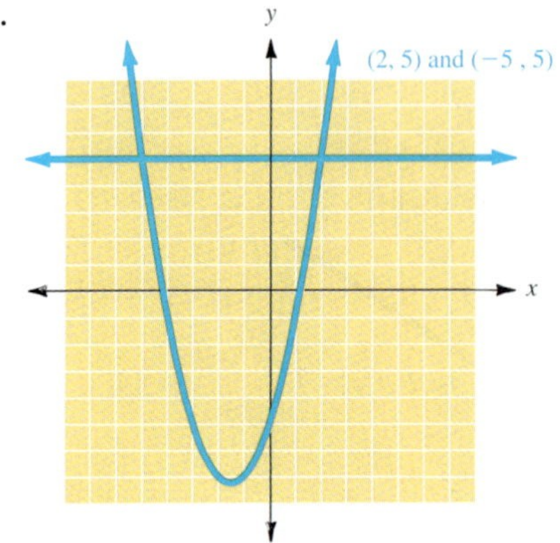

14.

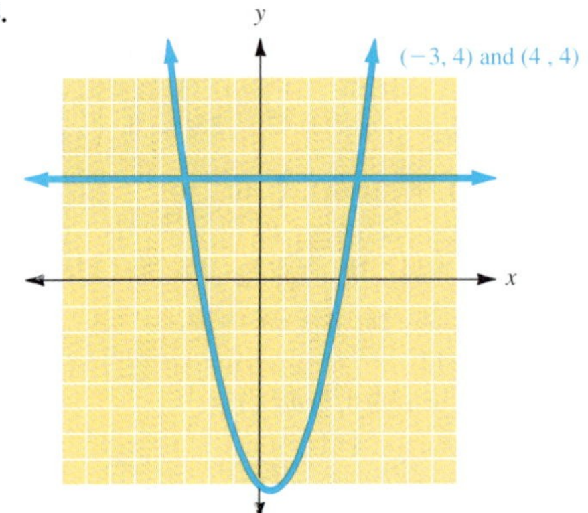

15. (2, 5) and (−5, 5) **16.** (−3, 4) and (4, 4)

17.

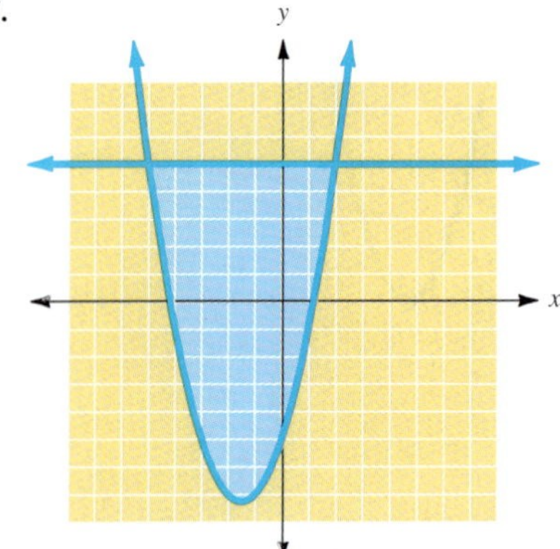

18.

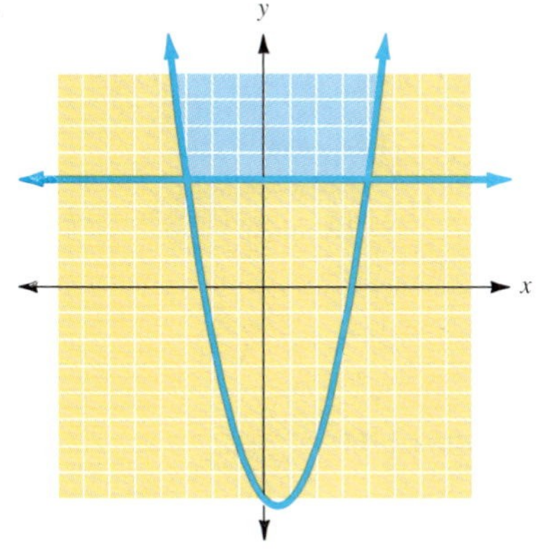

19. Circle **20.** Hyperbola

Cumulative Test for Chapters 1–10

1. $\left\{\frac{11}{2}\right\}$ **2.** $\{x|x > -2\}$ **3.** $\{-1, 4\}$ **4.** $\left\{x|-4 \le x \le \frac{2}{3}\right\}$

5. $\left\{x|x < -\frac{17}{5} \text{ or } x > 5\right\}$ **6.** $\{8, -3\}$ **7.** $\{1\}$ **8.** $\{-2\}$

9. $\left\{\frac{1 \pm \sqrt{3}}{2}\right\}$

10.

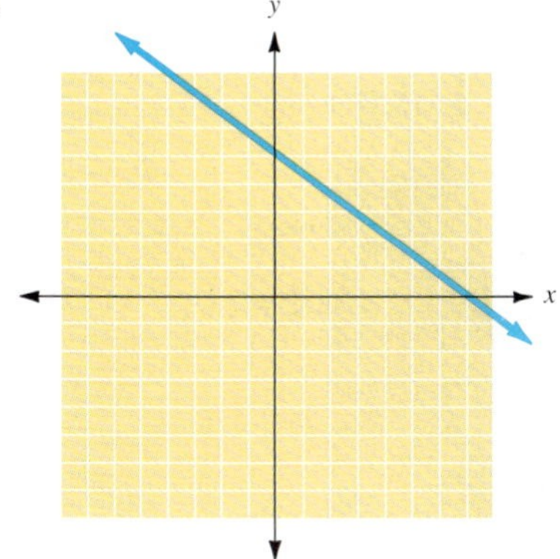

11.

12. ≈ 24.5 **13.** 7 **14.** $y = -x + 3$ **15.** $2x^2 - 5x - 3$

16. $9x^2 - 12x + 4$ **17.** $(x - 3)(x^2 - 5)$

18.

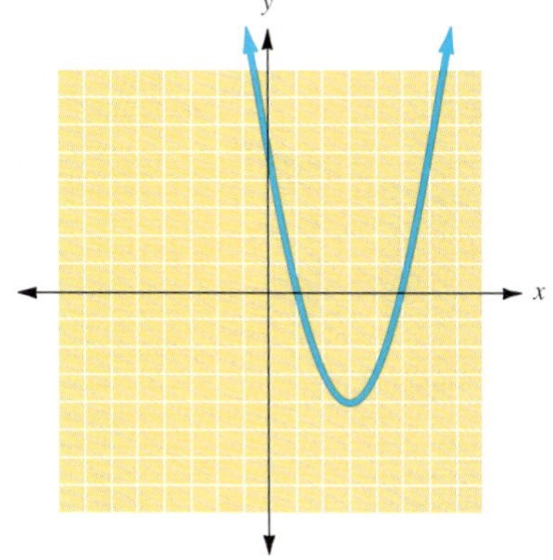

19.

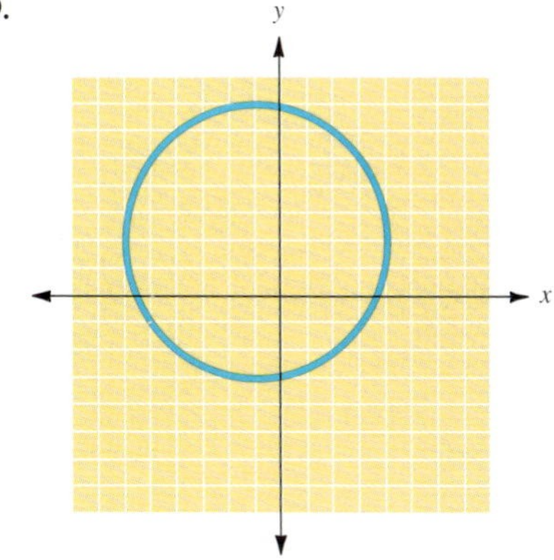

20.

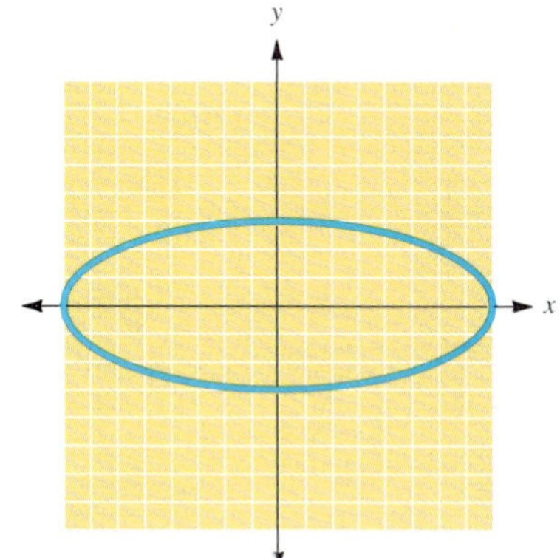

21. 2 **22.** $w\sqrt{3w}$ **23.** $\left\{-10, \frac{26}{3}\right\}$ **24.** 8 cm by 19 cm

25. 73

Self–Test for Chapter 11

1. (a) $f^{-1} = \left\{(x, y)|y = \frac{1}{4}x + \frac{1}{2}\right\}$; function

(b) $g^{-1} = \{(x, y)|y = \pm\sqrt{x - 1}\}$; not a function

(c)

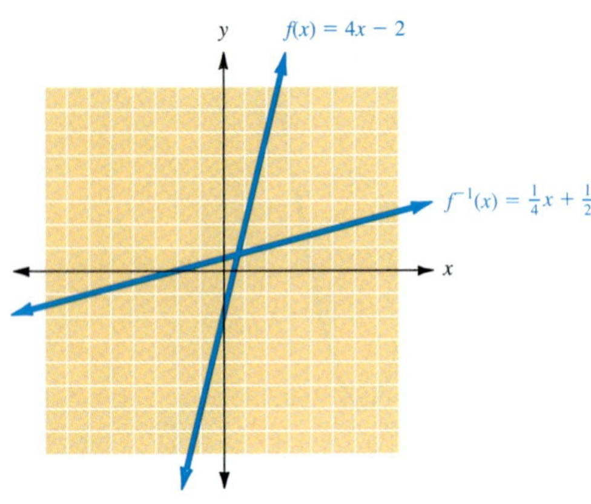

2.

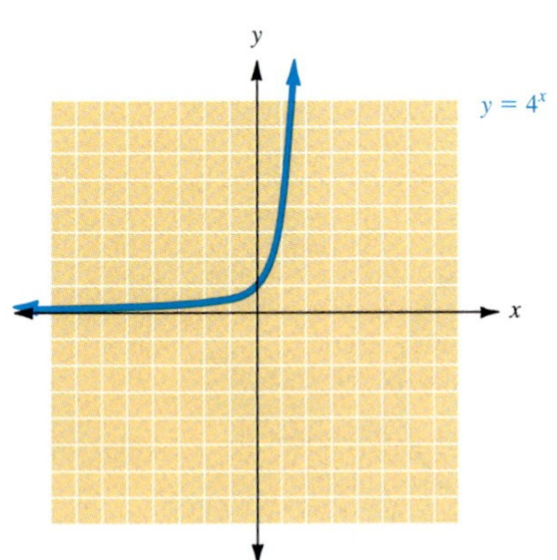

3.

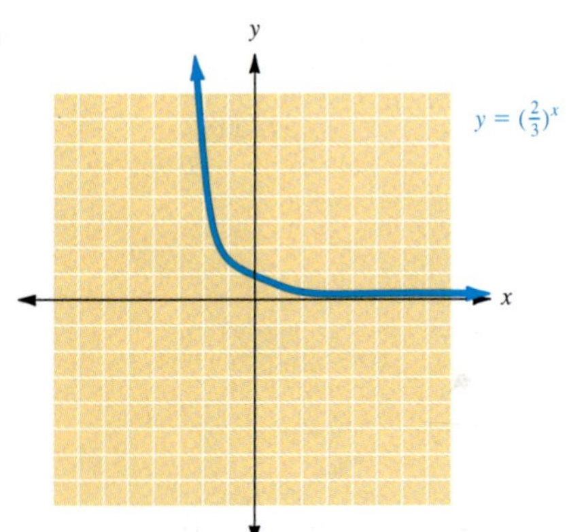

4. **(a)** $\{-2\}$; **(b)** $\left\{\frac{5}{2}\right\}$

5.

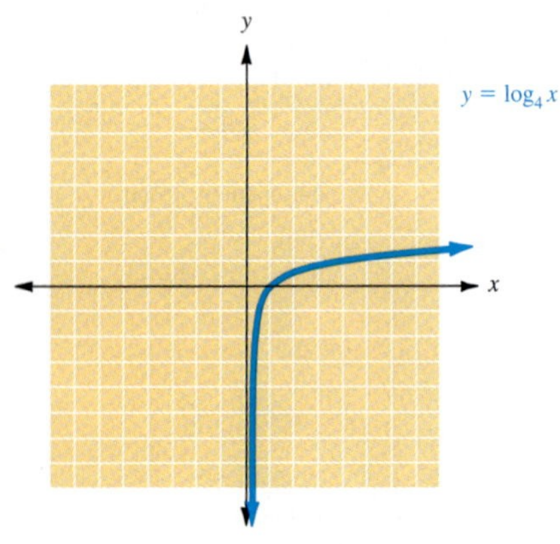

6. $\log 10{,}000 = 4$ **7.** $\log_{27} 9 = \frac{2}{3}$ **8.** $5^3 = 125$ **9.** $10^{-2} = 0.01$
10. $\{6\}$ **11.** $\{4\}$ **12.** $\{5\}$ **13.** $2 \log_b x + \log_b y + 3 \log_b z$
14. $\frac{1}{2}(\log_5 x + 2 \log_5 y - \log_5 z)$ **15.** $\log(xy^3)$
16. $\log_b \sqrt[3]{\frac{x}{z^2}}$ **17.** $\{8\}$ **18.** $\left\{\frac{11}{8}\right\}$ **19.** $\{0.262\}$
20. $\{2.151\}$

Cumulative Test for Chapters 1–11

1. $\{11\}$ **2.** $\left\{x \,\middle|\, x < \frac{2}{3} \text{ or } x > 4\right\}$ **3.** $\left\{\frac{10}{9}\right\}$

4.

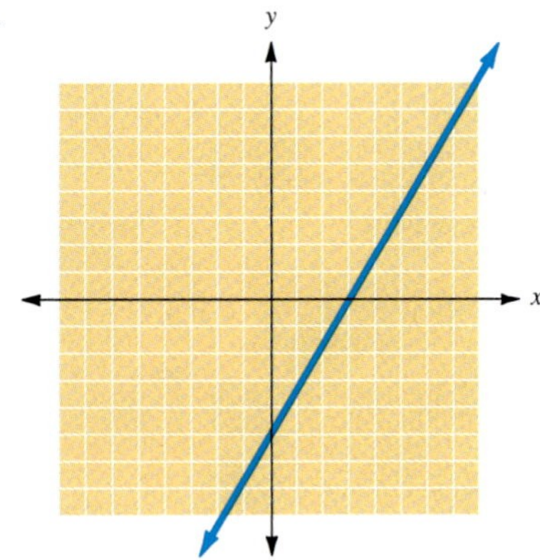

5.

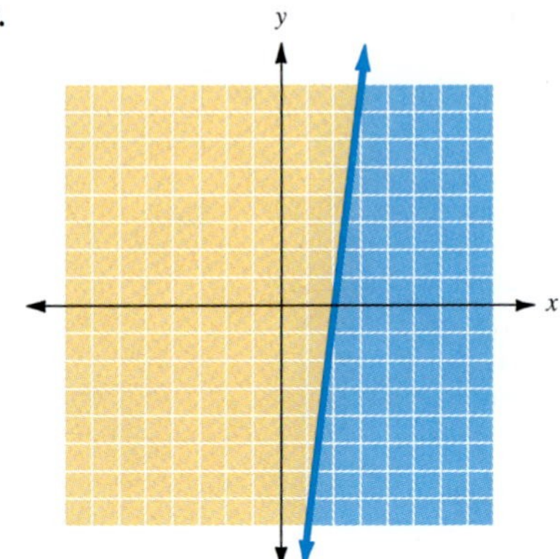

6. $6x + 5y = 7$ **7.** $\{x \mid x \geq 10\}$ **8.** $2x^2 - 6x + 1$ **9.** $6x^2 - 13x - 5$
10. $(2x - 5)(x + 2)$ **11.** $x(5x + 4y)(5x - 4y)$ **12.** $\frac{-x + 2}{(x - 4)(x - 5)}$
13. $\frac{x + 2}{x - 2}$ **14.** $-4\sqrt{2}$ **15.** $22 + 12\sqrt{2}$ **16.** $\frac{5}{3}(\sqrt{5} + \sqrt{2})$
17. 77, 79, 81 **18.** $\{-2, 1\}$ **19.** $\left\{\frac{3 \pm \sqrt{19}}{2}\right\}$
20. $\left\{x \,\middle|\, -\frac{3}{2} \leq x \leq 1\right\}$ **21.** 3 **22.** $1 \leq x \leq 4$
23. $\{x \mid (x - 1)^2 + (y + 2)^2 = 9\}$
24. x^3w^{12} **25.** $\{6\}$

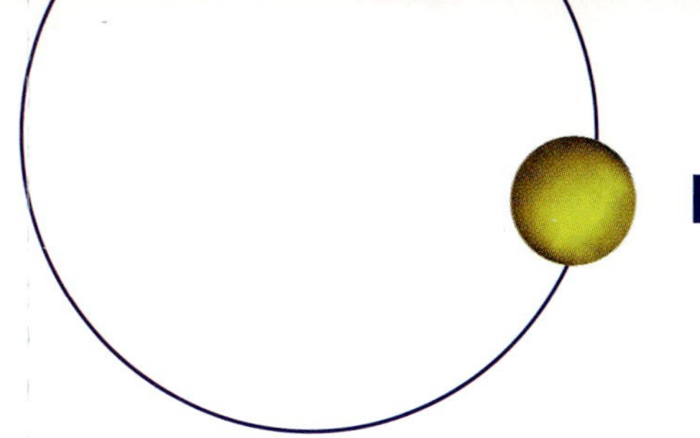

Index